The Laboratory Fish

The Handbook of Experimental Animals

Series Editors-in-Chief

Gillian Bullock
University Hospital
Department of Pathology
Ghent
Belgium

Tracie E Bunton
DuPont Pharmaceuticals Company
Department of Research and Development
Stine-Haskell Research Center
Newark, DE
USA

Series Editorial Advisory Board

Linda J Lowenstine
Pathology Microbiology & Immunology Department
School of Veterinary Medicine
University of California
Davis, CA, USA

Linda C Cork
Department of Comparative Medicine
Stanford University
Medical School Office Building
Stanford, CA, USA

Katsuhiko Arai
Department of Scleroprotein & Cell Biology
Faculty of Agriculture
Tokyo University of Agriculture & Technology
Saiwai-cho Fuchu-shi, Tokyo, Japan

Prince Masahito
c/o Director's Room
Kami Ikebukuro
Toshimo-ku, Tokyo, Japan

David Buist
Huntingdon Life Sciences
Huntingdon, Cambridgeshire, UK

Stephen W Barthold
Yale University School of Medicine
New Haven, CT, USA

Takatoshi Ishikawa
Professor of Pathology
University of Tokyo
Faculty of Medicine
Bunkyo-ku, Tokyo, Japan

Michael Sinosich
Royal North Shore Hospital
Reproductive Biochemistry and Immunology
St Leonards, NSW, Australia

Paul Herrling
Head of Corporate Research
Novartis AG
Basel, Switzerland

Sonia Wolfe-Cootes
Research Institute Medical Biophysics
Tygerberg, South Africa

Maurice Cary
Novartis AG
Toxicology, Drug Safety Department
Basel, Switzerland

The Laboratory Fish

Edited by
Gary K Ostrander

*Johns Hopkins University
Baltimore, MD
USA*

ACADEMIC PRESS

A Harcourt Science and Technology Company

San Diego   San Francisco   New York
Boston   London   Sydney   Tokyo

This book is printed on acid-free paper.

Copyright © 2000 by ACADEMIC PRESS

All Rights Reserved.
No part of this publication may be reproduced or transmitted in any form or by any means, electronic or mechanical, including photocopying, recording, or any information storage and retrieval system, without permission in writing from the publisher.

Academic Press
A Harcourt Science and Technology Company
Harcourt Place, 32 Jamestown Road, London NW1 7BY, UK
http://www.academicpress.com

Academic Press
A Harcourt Science and Technology Company
525 B Street, Suite 1900, San Diego, California 92101-4495, USA
http://www.academicpress.com

ISBN 0-12-529650-9

Library of Congress Catalog Card Number: 99-63118
A catalogue for this book is available from the British Library

On-line Version

Access for a limited period to an on-line version of The Laboratory Fish is included in the purchase price of the print edition.
This on-line version has been uniquely and persistently identified by the Digital Object Identifier (DOI)

doi:10.1006/bklf.2000

By following the link

http://dx.doi.org/10.1006/bklf.2000

from any Web Browser, buyers of The Laboratory Fish will find instructions on how to register for access

If you have any problems with accessing the on-line version, e-mail HEA@harcourt.com

Typeset by Newgen Imaging Systems (P) Ltd, Chennai, India
Printed in Great Britain by the Bath Press, Bath, Somerset

00 01 02 03 04 BP 9 8 7 6 5 4 3 2 1

Contents

List of Contributors.. xiii
Foreword.. xvii
Preface.. xix

Part 1 Introduction (Diversity of fish, Early observations and descriptions, Fish in experimentation)
David L Fabacher and Edward E Little

Part 2 Housing, Maintenance and Breeding

Chapter 1: Facilities and Husbandry (Large Fish Models) –
Jeffrey P Fisher... 13
Introduction.. 13
Overview of conceptual designs for large fish husbandry............... 14
Facilities and bioengineering... 16
Water quality requirements and maintenance............................ 22
Specific model systems for large fish................................. 35

Chapter 2: Facilities and Husbandry (Small Fish Models) –
Robin M Overstreet, Sue S Barnes, C Steve Manning, and William E Hawkins... 41
Introduction.. 41
Species... 42
Facilities.. 47
Fish culture.. 51
Fish health... 61

Chapter 3: Diet – *Douglas E Conklin*................................ 65
Introduction.. 65
History... 66
Live foods for larvae and fry... 68
Manufactured feeds.. 69
Energy – immediate nutritional needs.................................. 70
Tissue synthesis.. 71
Vitamins and minerals... 73
Summary... 75

Chapter 4: Common Diseases and Treatment — *David B Powell* 79
    Introduction 79
    Gill and skin infections 79
    Systemic bacteremial diseases 85
    Viral infections 88
    Disease prevention 90
    Disease treatment 90

Part 3 Gross Functional Anatomy

Chapter 5: Integumentary System — *Diane G Elliott* 95
    Introduction 95
    Functions 95
    General structure 97

Chapter 6: Skeletal System — *Melanie L J Stiassny* 109
    Introduction 109
    Skeletal anatomy 110

Chapter 7: Muscular System — *Melanie L J Stiassny* 119
    Introduction 119
    The cephalic musculature 119
    Main muscles of the cheek region 120
    Main muscles of the hyoid apparatus and associated structures ... 121
    Pharyngeal and associated musculature 122
    Body musculature 125

Chapter 8: Nervous System — *Ann B Butler* 129
    Introduction 130
    Brain 131
    Spinal cord and peripheral spinal nerves 142
    Meninges and the ventricular system 143
    Development of the telencephalon: evagination versus eversion ... 144
    Differences within major groups 145

Chapter 9: Respiratory System — *Kenneth R Olson* 151
    Introduction 152
    Orientation and general anatomy 152
    The gill arch 153
    Gill filament 156

Chapter 10: Circulatory System — *Kenneth R Olson* 161
    Overview 162
    The heart 162
    Arterial system 163
    Venous system 167
    Secondary circulation and lymphatics 169

Chapter 11: Digestive System — *Randal K Buddington and Victoria Kuz'mina* 173
    Regions and components of the fish digestive system 173
    Regulation of digestive system functions 178

Chapter 12: Urinary Tract – *Hartmut Hentschel, Marlies Elger,*
   Margaret Dawson and J Larry Renfro 181
   Introduction 181
   Comparative anatomy of the kidney of fishes 181
   Ontogeny 182
   Renal vascular system 184

Chapter 13: Endocrine System – *David M Janz* 189
   Introduction 189
   Hypothalamo-hypophysial axis 191
   Neurohypophysial hormones 191
   Gonadal hormones 194
   Thyroid hormones 200
   Inter-renal hormones 201
   Renin–angiotensin system 204
   Natriuretic peptides 204
   Urotensins 204
   Calcium-regulating hormones 205
   Gastro-entero-pancreatic hormones 206
   Melatonin 209
   Modulation of endocrine function by environmental chemicals 210
   Summary 212

Chapter 14: Immune System – *David B Powell* 219
   Introduction 219
   Skin, lateral line, and gills 219
   Thymus 220
   Kidney 220
   Intestinal tract 221
   Liver 222
   Spleen 222

Chapter 15: Sensory Systems 225
   15.1 Vision – *Russell D Fernald* 225
      Introduction 225
      Seeing underwater 225
      Collecting light 226
      Focusing light 227
      The retina: transforming images into neural signals 229
      Summary 233
   15.2 Mechanosensory Lateral Line: Functional Morphology and
         Neuroanatomy – *Jacqueline F Webb* 236
      Introduction 236
      Morphology of the lateral line system on the head 237
      Morphology and distribution of neuromast receptors on the head ... 238
      Morphology of lateral line canals on the trunk 239
      Neuroanatomy of the mechanosensory lateral line system 239
      Functions of the mechanosensory lateral line system 241

    15.3 Chemoreception – *Toshiaki J Hara* 245
          Peripheral olfactory organ 245
          Gustatory organ 247
          Solitary chemosensory cells....................................... 247
          Development 248
    15.4 Hearing – *Bernd Fritzsch* 250
          Overview... 250
          The semicircular canals and their function 251
          The statolithic organs and their function 252
          Hearing in water: the role of direct and indirect sound........... 254
          Sound-pressure receivers: swim bladders and
                their connection to the ear 254
          Sound production in fish 257

Chapter 16: Reproductive Systems – *J M Redding and R Patiño*................. 261
    Introduction... 261
    External anatomy 262
    Internal anatomy.. 263
    Endocrine structures and regulation 265
    Conclusion 266

Part 4 Microscopic Functional Anatomy

Chapter 17: Integumentary System – *Diane G Elliott*............................ 271
    Introduction.. 271
    Epidermis 271
    Dermis 291

Chapter 18: Skeletal System – *A Huysseune*....................................307
    Introduction.. 307
    Cartilage 308
    Bone.. 309
    Chondroid bone 314
    Teeth and dental tissues 314
    Non-osseous tissues of the dermal skeleton 315
    Conclusion 315

Chapter 19: Fish as an Experimental Model for Studying
                Muscle Function – *Lawrence C Rome*........................... 319
    Introduction.. 319
    Why fish provide a superior experimental model for exploring
           muscle function 320
    Muscle fiber types 320
    Recruitment of different fiber types................................. 322
    Muscle structure 322
    Design of the fish muscular system 323
    Summary... 329

Chapter 20: Nervous System — *Ann B Butler* 331
    Introduction 332
    Principles of sensory and motor system organization 335
    Regional anatomy 335
    Sensory and motor systems 346
    Comparative perspective 351

Chapter 21: Respiratory System — *Kenneth R Olson* 357
    Introduction 358
    Gill filament 358
    Epithelium 358
    Blood vessels 364

Chapter 22: Circulatory System — *Kenneth R Olson* 369
    Overview 370
    The heart 370
    Peripheral circulation 372
    Secondary circulation and lymphatics 375
    Retia mirabilia 377

Chapter 23: Digestive System — *Randal K Buddington*
        and Victoria Kuz'mina 379
    Introduction 379
    Mouth and pharynx 380
    Esophagus 380
    Stomach 380
    Intestine 381
    Accessory organs 383

Chapter 24: Urinary Tract — *Marlies Elger, Hartmut Hentschel,*
        Margaret Dawson and J Larry Renfro 385
    Introduction 385
    Renal vascular system 385
    Nephron and collecting duct system 386
    Glomerulus 389
    Renal tubule 397

Chapter 25: Endocrine System — *David M Janz and Lynn P Weber* 415
    Introduction 415
    Hypothalamic and pituitary hormones 417
    Gonadal hormones 425
    Thyroid hormones 427
    Gastroentero-pancreatic hormones 428
    Osmoregulatory hormones 430
    Conclusion 435

Chapter 26: Immune System — *David B Powell* 441
    Blood and lymphatic vessels 441
    Thymus 442
    Kidney 444
    Intestine 445
    Liver 446
    Spleen 446

Chapter 27: Sensory Systems 451
27.1 Vision – *Russell D Fernald* 451
Introduction 451
Phototransduction 451
Information flow in the retina 453
Retinal structure 455
Retinal cell types and connections 456
Retinal growth in teleosts 459
Summary 461
27.2 Mechanosensory Lateral Line: Microscopic Anatomy and Development – *Jacqueline F Webb* 463
Structure and function of neuromast receptor organs 463
Development of the mechanosensory lateral line system 465
27.3 Chemoreception – *Toshiaki J Hara* 471
Olfactory epithelium 471
Taste buds 473
Solitary chemosensory cells 474
Molecular basis of signal transduction 474
Neural projections and central olfactory pathways 475
Gustatory nerves and their central projections 478
27.4 Hearing – *Bernd Fritzsch* 480
Introduction 480
The hair cell as a mechanosensory transducer: correlating structure and function 481
The semicircular canals 482
The statolithic organs: opposing polarity in different ways 484
Afferent fiber connections of the fish ear 485
The efferent system of the ear and the lateral line 486

Chapter 28: Reproductive Systems – *R Patiño and J M Redding* 489
Introduction 489
Brain–Pituitary 490
Gonads 490
Conclusion 499

Part 5 Procedures

Chapter 29: Stress and Anesthesia – *Henrik Kreiberg* 503
Introduction 503
The stress response 504
Anesthesia 506
Sentience, analgesia and euthanasia 509

Chapter 30: Collection of Body Fluids – *Marsha C Black* 513
Introduction 513
Blood collection 513
Urine collection 519
Collection of fecal materials 522
Collection of gametes 525

Chapter 31: Routes of Administration for Chemical Agents –
Marsha C Black 529
- Introduction 529
- Water-borne exposures 529
- Oral administration 532
- Injection techniques 534
- Implants 538
- Topical exposure 540

Chapter 32: Fish Necropsy – *Jeffrey P Fisher and Mark S Myers* 543
- Introduction 543
- Salmoniform fish type 544
- Pleuronectiform fish type 549

Chapter 33: Surgical Techniques – *Gerald R Johnson* 557
- Introduction 557
- Operating equipment 560
- Fish surgical procedures 562
- Managing convalescents 565
- Summary 566

Chapter 34: Fixation of Fish Tissues – *John W Fournie, Rena M Krol and William E Hawkins* 569
- Introduction 569
- Purpose and principles of fixation 570
- Chemistry of fixation 570
- Types of fixatives 571
- Fixation for electron microscopy 574
- Fixation and special procedures 575
- Safety 576
- World wide web access to histological information 577
- Summary and conclusions 577

Chapter 35: Autoradiography of Fishes – *Kevin M Kleinow* 579
- Introduction 579
- Basic considerations 579
- Methodologies 581
- Quantitation 586

Part 6 Experimental Models

Chapter 36: Cancer – *Paul C Baumann and Mark S Okihiro* 591
- Introduction 591
- Fish in carcinogenicity testing 592
- Initiation and DNA adducts 594
- Medaka and trout fish tumor models 599
- Field studies of liver cancer epizootics 601
- Chromatophoromas in wild fish 604
- Damselfish neurofibromatosis 605
- Melanomas in *Xiphophorus* hybrids 605
- Viral tumors 607
- Summary: advantages of fish cancer models 609

Chapter 37: Toxicology — *Chris D Metcalfe* 617
    Introduction 617
    Factors affecting toxicity to fish 618
    Toxicity testing with fish 622
    Overview 627

Chapter 38: Cell and Tissue Culture — *Rosemarie C Ganassin,*
                      Kristin Schirmer and Niels C Bols 631
    Introduction 631
    A model of rainbow trout hemopoiesis – long-term
        hemopoietic culture 632
    Cell viability assays 638
    Induction of 7-ethoxyresorufin (EROD) activity 644
    General discussion 649

Glossary: (Terms defined in the glossary are emboldened
         in the main text) 653

Index: 663

Colour plates appear between pages 268 and 269

List of Contributors

Barnes, S S
Institute of Marine Sciences,
The University of Southern Mississippi,
Ocean Springs,
Mississippi,
USA

Baumann, P C
Field Research Station, USGS,
The Ohio State University,
Columbus,
Ohio,
USA

Black, M C
College of Agricultural and
Environmental Sciences,
The University of Georgia,
Athens,
Georgia,
USA

Bols, N C
Department of Biology,
University of Waterloo,
Waterloo,
Ontario,
Canada

Buddington, R K
Department of Biological Sciences,
Mississippi State University,
Mississippi State,
Mississippi,
USA

Butler, A B
Krasnow Institute for Advanced Study and
Department of Psychology,
George Mason University,
Fairfax,
Virginia,
USA

Conklin, D E
Department of Animal Science,
University of California,
Davis,
California,
USA

Dawson, M
Department of Physiology and Neurobiology,
University of Connecticut,
Storrs,
Connecticut,
USA

Elger, M
Institut fur Anatomie und Zellbiologie I,
Universität Heidelberg,
Heidelberg,
Germany

Elliott, D G
Western Fisheries Research Center,
Biological Resources Division,
US Geological Survey,
Seattle,
Washington,
USA

Fabacher, D L
US Geological Survey,
Biological Resources Division,
Columbia Environmental Research Center,
Columbia,
Missouri,
USA

Fernald, R D
Department of Psychology,
Stanford University,
Pal Alto,
California,
USA

Fisher, J P
Pentec Environmental, Inc.,
Edmonds,
Washington,
USA

Fournie, J W
US Environmental Protection Agency,
National Health and Environmental
Effects Research Laboratory,
Gulf Ecology Division,
Gulf Breeze,
Florida,
USA

Fritzsch, B
Department of Biomedical Sciences,
Creighton University,
Omaha,
Nebraska,
USA

Ganassin, R C
Department of Biology,
University of Waterloo,
Waterloo,
Ontario,
Canada

Hara, T J
Freshwater Institute,
Fisheries and Oceans Canada,
Winnipeg,
Manitoba,
Canada

Hawkins, W E
Institute of Marine Sciences,
The University of Southern Mississippi,
Ocean Springs,
Mississippi,
USA

Hentschel, H
Max-Planck-Institut fur molekulare Physiologie,
Dortmund,
Germany

Huysseune, A
Biology Department,
Ghent University,
Ghent,
Belgium

Janz, D M
Department of Zoology,
Oklahoma State University,
Stillwater,
Oklahoma,
USA

Johnson, G R
Department of Veterinary Pathology and
  Microbiology,
Atlantic Veterinary College,
University of Prince Edward Island,
Charlottetown,
Prince Edward Island,
Canada

Kleinow, K M
School of Veterinary Medicine,
Louisiana State University,
Baton Rouge,
Louisiana,
USA

Kreiberg, H
Canada Department of Fisheries & Oceans,
Pacific Biological Station,
Nanaimo,
British Columbia,
Canada

Krol, R M
Institute of Marine Sciences,
The University of Southern Mississippi,
Ocean Springs,
Mississippi,
USA

Kuz'mina, V
Institute for the Biology of Inland Waters,
Borok,
Yaroslavl,
Russia

Little, E E
US Geological Survey,
Biological Resources Division,
Columbia Environmental Research Center,
Columbia,
Missouri,
USA

Manning, C S
Institute of Marine Sciences,
The University of Southern Mississippi,
Ocean Springs,
Mississippi,
USA

Metcalfe, C D
Environmental and Resource Studies,
Trent University,
Peterborough,
Ontario,
Canada

Myers, M S
National Marine Fisheries Service,
Northwest Fisheries Science Center,
Seattle,
Washington,
USA

Okihiro, M S
School of Veterinary Medicine,
University of California at Davis,
Davis,
California,
USA

Olson, K R
Indiana University School of Medicine,
South Bend Center for Medical Education,
University of Notre Dame,
Notre Dame,
Indiana,
USA

Overstreet, R M
Institute of Marine Sciences,
The University of Southern Mississippi,
Ocean Springs,
Mississippi,
USA

Patiño, R
Texas Cooperative Fish & Wildlife
  Research Unit,
Texas Tech University,
Lubbock,
Texas,
USA

Powell, D B
ProFISHent, Inc.,
Redmond,
Washington,
USA

Redding, J M
Department of Biology,
Tennessee Technological University,
Cookeville,
Tennessee,
USA

Renfro, J L
Department of Physiology and Neurobiology,
University of Connecticut,
Storrs,
Connecticut,
USA

Rome, L C
Department of Biology,
University of Pennsylvania,
Philadelphia,
Pennsylvania,
USA

Schirmer, K
Department of Biology,
University of Waterloo,
Waterloo,
Ontario,
Canada

Stiassny, M L J
Department of Ichthyology,
American Museum of Natural History,
New York City,
New York,
USA

Webb, J F
Department of Biology,
Villanova University,
Villanova,
Pennsylvania,
USA

Weber, L P
Department of Zoology,
Oklahoma State University,
Stillwater,
Oklahoma,
USA

Foreword

*F*IRST came the rat and then came the fish (but not in genealogical terms!). The first title of this new series explored the whole range of structure and function of different strains of rats so that professionals and students could find a ready source of information for their everyday needs. Our series aim is to minimise the need to access many different sources for data by having it under one "roof".

Now we have moved on to the fish, taking the same series approach, but, given the different nature of the species under consideration not following exactly the same lines. It is absolutely clear from reading through the many chapters, that the Editor has done a marvellous job. Not only has he included chapters dealing with the environment and toxicological issues but he has also brought totally unexpected topics such as surgery on prize and pet fish. One of us knew very little about fish before reading the various scripts, but is now absolutely sure that this volume will find a place in many unusual corners.

Our thanks go to Gary Ostrander who brought *The Laboratory Fish* into being despite very daunting events, such as authors losing their laboratory in a hurricane or other misadventures! He has been great to work with. Thank you also to the Academic Press team who are going along the same steep learning curve as the Editors in order to bring to you the online versions of titles in the series.

Gillian Bullock
Tracie E Bunton
Editors-in-Chief
The Handbook of Experimental Animals

Preface

SCIENCE as a discipline has been marked by paradigm shifts or, as described by Kuhn in *The Structure of Scientific Revolutions*, the discovery of a common solution for a number of outstanding problems. Among the most significant in the field of Biology was Darwin's advancement of the theory of evolution in the late 1800s. Evolutionary theory provided us, over the ensuing century, with a framework for elucidating and understanding the origins and interrelationships of species. While evolution was generally accepted to occur at the level of the organism (i.e. Darwin's "survival of the fittest"), the elucidation of the structure of DNA and the subsequent recognition that an organism's DNA is comprised of genes coding for protein products, provided an insight as to how all levels of biological organization (e.g. from organelles and cells to organisms and species) evolved and subsequently function. Thus, evolutionary theory and molecular biology are inextricably linked and form the basis for all of biological science. As we move our respective disciplines forward our understanding of the function of biological systems at all levels of organization will be guided by evolutionary theory and exploited by the tools emerging from molecular biology.

While the preceding provides justification for this volume, another factor worth mentioning relates to how science is currently being conducted. In recent years investigators have come to recognize that a comparative approach is often useful for revealing new information about systems of interest. In particular, an increasing number of investigators have demonstrated that exploitation of the novel aspects of the biology of fishes (and to be fair other lower vertebrates and invertebrates as well) has yielded important new insights of form and function in all species, including higher mammals and humans. Thus, it is anticipated that this comparative approach will continue to find favor in the future.

The text that follows represents our effort to provide interested readers with a current understanding of the biology of fishes as it relates to their utility in the laboratory. No less than 42 experts in their respective fields participated in this process and for their efforts I am deeply grateful. The initial sections of this volume summarize the utility of fish as experimental animals through the ages, both in their own right and as models for understanding particular systems or phenomena among all species. Also included are sections detailing their husbandry, diet, and disease treatment and prevention. The heart of the volume includes two sections devoted to the gross and microscopic functional anatomy of 16 different systems including 5 sensory modalities, each written by an author(s) with extensive relevant research expertise in the area. A fourth section focuses on a variety of procedures for handling, manipulating, examining and sampling of fishes. A final section details some principle model systems, relevant across multiple fish species, including cell/tissue culture, toxicology, and cancer biology. Taken in total, this represents for the first time, a compendium of what is currently known about the uses of fishes as laboratory models for research. It is our hope that this will find wide value among our colleagues.

Gary K Ostrander
July 2000

PART 1

Introduction

David L Fabacher and Edward E Little
US Geological Survey, Biological Resources Division,
Columbia Environmental Research Center,
Columbia, Missouri, USA

Diversity of fish

Fish inhabit all types of aquatic environments and exhibit enormous diversity in behavior, morphology, diet, and reproduction. Approximately 25 000 species of fish are in the phylum Chordata, subphylum Vertebrata, within five distinct existing classes including: (i) Myxini; (ii) Cephalaspidomorphi; (iii) Chondrichthyes; (iv) Sarcopterygii; and (v) Actinopterygii, which includes the entire division Teleostei (teleosts) (Nelson, 1994) (Figure 1). The teleosts are the most numerous species of fish (approximately 23 640 species and 96% of all fish species) and are the most diverse group of vertebrates (38 orders, 426 families and 4064 genera).

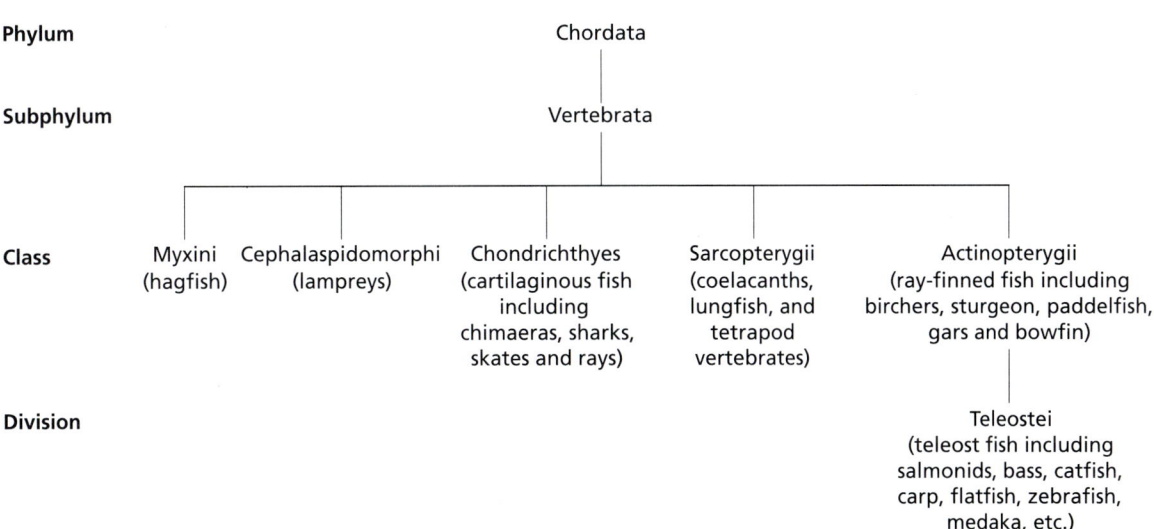

Figure 1 Higher categories of fish. (Abbreviated and modified from Nelson, 1994.)

Copyright © 2000 Academic Press

Early observations and descriptions

The following material from *History of Biology* (Gardner, 1972) provides an insight into the minds of early biologists who observed and described fish. In ancient Greece, Aristotle learned medicine from his father and attended Plato's school, the Academy of Athens. Instead of becoming master of the Academy after Plato died, Aristotle studied marine biology on the island of Lesbos. Aristotle's observations on biological organisms were surprisingly accurate. He was keenly interested in the variety of marine life and was very astute at making comparative observations on fish and other organisms. He observed that fish eggs differed in appearance from bird eggs and, like birds, fish had one umbilicus that connected the embryo with the yolk. By the seventeenth century, biologists throughout Europe refined and expanded their scientific observations. The Dutch microscopist Leewenhoek observed sperm from fish and other organisms. In England, John Ray and his student Francis Willughby traveled together observing plants and animals, and publishing scientific papers. In 1686, Willughby's manuscript of outline drawings of whole fish was published. These drawings were illustrated in sufficient detail so that a particular fish could be identified. Francesco Redi, a court physician to the Grand Duke of Tuscany, used the flesh of fish and other animals to disprove the notion that flies could develop spontaneously on putrefying flesh. G.A. Borelli was interested in the mechanics of muscular action and used mechanical models to explain how animals moved, including how fish swim. In the nineteenth century Charles Darwin found 15 new species of saltwater fish in his travels to the Galapagos Islands.

In 'The rise of fish embryology in the nineteenth century,' Wourms (1997) related how fish played a central role in the growth of developmental biology. According to the author, knowledge of fish embryology progressively developed during the nineteenth century when teleost fish became more important to biologists studying embryology. This occurred because artificial fertilization could ensure a supply of material and the transparent eggs of teleosts were readily suited for microscopic examination. Interest was then generated in comparative embryology of fish as embryological microtechniques developed. All of the above-mentioned observations and descriptions set the stage for using fish in experimentation.

Fish in experimentation

Historical use of fish

In an overview 'The fish in biological research,' Nigrelli (1953) provided a chronological synopsis of the use of fish as experimental animals and stated, 'It is safe to say that practically every important person in the history of biology and medicine had, at one time or another, used fishes in their studies', indicating how important experimentation with fish has been in the advancement of science. Early studies using fish as experimental animals are mentioned and some of the outstanding studies in experimental embryology, genetics, renal physiology, endocrinology and nerve physiology are described along with their contributions to advances in biology and medicine. Nigrelli also mentions that the literature 25 years before his overview was replete with numerous publications on the use of fish in all sorts of disciplines including evolution and genetics, anatomy and physiology, experimental morphology, morphological and chemical aspects of histology and cytology, endocrinology, nutrition, pharmacology, pathology and parasitology, aggregation and migration phenomena, reproductive and other behavioral studies, commercial aspects of fishery biology, and pollution research.

Some of the earliest uses of fish in experiments occurred in the discipline we now know as aquatic toxicology. As detailed in the book *Fundamentals of Aquatic Toxicology* by Rand and Petrocelli (1985), aquatic toxicology includes the study of toxic or otherwise adverse effects of stressors such as chemical contaminants on fish. In a review entitled 'History of acute toxicity tests with fish, 1863–1987' Hunn (1989) relates that results of the first acute toxicity tests with fish were reported in 1863 and that before the 1940s acute toxicity studies with fish developed slowly. However, after the 1940s toxicity tests with fish increased primarily as a result of the widespread use of environmentally persistent pesticides. Studies of acute toxicity tests with chemicals paved the way for using fish in bioassays to determine the effects of chemicals and other stressors on fish. In 1969, Academic Press published *Fish in Research* (edited by Neuhaus and Halver), the proceedings of a symposium on the use of fish as an experimental animal that provides further insight into the extent of experimentation that has been done with fish. This publication contains historical information on experiments with fish in cancer studies, metabolism, genetics and nutrition and shows the increasing diversity in experimentation with fish that began during the 1960s.

Atz (1986) chronicled the historical use of the mummichog (*Fundulus heteroclitus*) as an experimental animal. Experimentation with mummichog eggs began at Woods Hole as early as 1892. Mummichog eggs were subsequently used in various experiments including studies on development of the circulatory system, permeability to certain substances, and viability in different salinities and osmotic pressures. Some of the earliest citations dealt with the effects of chemical and physical factors on the mummichog's development. Other early citations dealt with the regeneration of the tail of the adult fish. Many experiments were conducted on mummichog genetics, pigmentation and endocrinology. In addition, later citations dealt with the extensive use of mummichogs in bioassays with environmental pollutants and other physiologically active substances, and even in testing for life-threatening properties in materials collected on the moon.

Beginning in the 1970s and extending especially during the 1980s and 1990s the concept of using fish as experimental models took hold. The titles of the publications in Table 1 provide an insight into the various uses (and escalation in importance) of fish as experimental models.

Advantages of using fish

Fish have characteristics that can be extremely interesting to study. As vertebrates adapted to living in water, fish possess fins and have skin that is usually covered with scales. They have two-chambered hearts that circulate venous blood, and respiration typically takes place through the gills. Some fish species can produce sound, light, electricity, or venom. In addition, they have some unique sensory abilities (e.g. response to light, production and detection of electrical impulses, olfaction) that are not found in rodents. Some fish are hermaphrodites whereas others exhibit sex reversal.

Behavioral, biochemical, physiological, or morphological adaptations can occur at different periods during the life history of a fish. Fish adaptation to extreme environments has produced remarkable morphological and physiological results, the study of which has provided a wealth of basic information about sensory transduction mechanisms, biomechanics and physiology. Such studies have demonstrated fish detecting buried prey from the electrical fields that the prey has generated and chemoreception sensitivity surpassing that of bloodhounds. As **poikilotherms**, fish are typically incapable of controlling their body temperature, which is regulated by the temperature of the surrounding water (Fry, 1971; Wolke, 1984). This offers the researcher interesting opportunities because the speed of biochemical and metabolic functions in fish should be changed by alterations in water temperature. Thus, certain biochemical pathways and physiological systems could be decreased under experimental conditions to enhance evaluation that may lead to new information on fish biochemistry and physiology. Fish are good models for studying environmental stress because they live in varied habitats and must adapt to diverse environmental factors such as temperature, dissolved oxygen, pH, water pressure, salinity and the presence of extrinsic stressors like chemical contaminants. For example, some species of fish have adapted to extremely low environmental temperatures. These fish have evolved genetically controlled antifreeze proteins to keep their blood from freezing (DeVries, 1971).

Fish have certain characteristics that make them desirable as experimental animals. Unlike mammals, fish are taxonomically and environmentally diverse and provide the researcher with a wide range of species to meet the environmental conditions the researcher is interested in studying. In general, they adapt well to captivity, can be easily held in the laboratory, can be fed commercially available diets, and some species breed readily. After being anesthetized in water, their organs, tissues and blood can be readily removed. The gene pool is easier to control because a single male can fertilize numerous eggs from a single female. Fish, especially small aquarium fish, are relatively simple to acquire, handle, and maintain and do not require as much care as mammals. They typically have multiple offspring from a single mating and many fish can be kept in the same chamber. In addition, large numbers can be used in an experiment at relatively low cost; a significant quantity of fish can therefore be periodically sampled, which strengthens statistical comparisons. Growth and metabolism of fish can be altered because they are poikilotherms. However, they lack several mammalian organ systems including lungs, certain sex glands and organs, certain

TABLE 1: Examples of fish used as experimental and biomedical models

- 'Fish as biomedical research models' (Klontz, 1971)
- 'Fish as animal models in biomedical research' (Umminger and Pang, 1979)
- 'The fish heart as a model system for the study of myoglobin' (Driedzic, 1983)
- 'The use of fish in biomedical research' (Wolke, 1984)
- 'Development of aquarium fish models for environmental carcinogenesis: tumor induction in seven species' (Hawkins et al., 1985)
- 'Fish hepatocytes: a model metabolic system' (Moon et al., 1985)
- 'The use of isolated fish opercular epithelium as a model tissue for studying intrinsic activities of loop diuretics' (Eriksson et al., 1985)
- 'Using fish as models in biomedical research' (May et al., 1987)
- 'A fish hepatocyte model for the investigation of the effects of environmental contaminants' (Baksi and Frazier, 1988)
- 'Fish as model systems' (Powers, 1989)
- 'Animal model for ultraviolet radiation-induced melanoma: platyfish–swordtail hybrid' (Setlow et al., 1989)
- 'Elasmobranchs (sharks, skates and rays) as animal models for biomedical research' (Luer, 1989)
- 'The fish pigment cell: an alternative model in biomedical research' (Karlsson et al., 1990)
- 'Fish as human disease models' (Contie, 1991)
- 'Characterization of the pufferfish (*Fugu*) genome as a compact model vertebrate genome' (Brenner et al., 1993)
- 'Transgenic fish: ideal models for basic research and biotechnological applications' (Chen et al., 1995)
- 'The goldfish as a model for studying neuroestrogen synthesis, localization, and action in the brain and visual system' (Callard et al., 1995)
- 'The zebrafish as a model system in developmental, toxicological, and transgenic research' (Lele and Krone, 1996)
- 'The zebrafish's swim to fame as an experimental model in biology' (Vascotto et al., 1997)
- 'A hemophilia model in zebrafish: analysis of hemostasis' (Jagadeeswaran and Liu, 1997)
- 'A zebrafish model for hepatoerythropoietic porphyria' (Wang et al., 1998)
- 'Positional cloning of the zebrafish *sauternes* gene: a model for congenital sideroblastic anaemia' (Brownlie et al., 1998)
- 'Zebrafish as a model system for studying neuronal circuits and behavior' (Fetcho and Liu, 1998)
- 'Fish as models for the neuroendocrine regulation of reproduction and growth' (Blazquez et al., 1998)
- 'Oxidative stress in toxicology: established mammalian and emerging piscine model systems' (Kelly et al., 1998)
- 'Mammalian immunoassays for predicting the toxicity of malathion in a laboratory fish model' (Beaman et al., 1999)
- 'Zebrafish in context: uses of a laboratory model in comparative studies' (Metscher and Ahlberg, 1999)

corresponding endocrine glands, and a mammalian-like urinary bladder. In addition, they are probably more sensitive to the stress of handling and environmental changes than mammals and are subsequently more readily prone to disease. Nevertheless, because they are lower on the evolutionary scale than mammals, fish have functions that are generally less complex and more easily studied. This can aid in understanding the mechanisms for the same functions that are more complex in mammals.

Fish are also highly effective as sentinels for environmental change, especially environmental contamination, and can provide an early warning of the release of hazardous materials in the watershed. Studies with fish also provide information about the environmental fate and persistence of chemicals, biomagnification potential of chemicals, and the inherent threat that biological concentrations of contaminants pose in the human food web. Fish are excellent animals to use while studying the environmental fate of chemical contaminants because fish, unlike mammals, can bioconcentrate chemical contaminants from the aquatic environment. In addition, all life stages of fish can be directly exposed to chemical contaminants. Many of the physiological responses of fish to contaminants can be predictive of higher vertebrate responses because such responses are phylogenetically conservative and are essentially similar among vertebrates.

Considerations for the experimental use of fish

Because fish live in water, conducting experiments requires that investigators understand all of the factors that are involved with the needs of fish in an aquatic environment. In-depth discussion of many of the following comments can be found in 'Guidelines for the care and use of fish in research' by DeTolla et al. (1995).

After thoroughly searching the literature, investigators should choose a fish species and experimental model system that will meet their specific needs and goals. In choosing a species, investigators should determine how much laboratory space is available for supporting structures such as aquaria, air pumps, etc. The level of maintenance should also be considered because some fish species are easier to maintain than others. Is the fish going to be a freshwater or marine, coldwater or warmwater species? This is an important consideration because marine and coldwater species can require more specialized care, handling and housing. Dietary needs, temperature requirements, fish density per volume of water, behavioral characteristics and resistance to disease should also be considered.

Fish bred in captivity are usually obtained from fish hatcheries or laboratory suppliers, whereas the investigators can collect fish from the wild. Researchers should be aware that the stress of collecting, handling and transporting fish from the wild can make them very susceptible to disease. Regardless of the source, investigators should use only healthy fish. Using diseased fish that are in a weakened condition can confound otherwise expected results of measured endpoints. If fish are shipped to the investigators, it is important to know the water temperature and water quality conditions during transport so that the fish can be acclimated to the water temperature and water quality conditions in the laboratory.

Fish brought into the laboratory should be quarantined and observed for disease symptoms while being acclimated to laboratory conditions (i.e. water quality, water temperature, illumination and diet). Investigators should be aware of factors that can confound the results of studies if a fish species from one source is commingled with the same fish species from another source. Some of these factors include differences in diet, growth rate, age of the fish at time of experimentation, prevalence of **biogenic** compounds in the food or water column, and genetic variation. It is important to document the species and strain, and continuity should be maintained by using the same species or strain from the same source throughout an experiment.

Fish reared and held in the laboratory will be in either some form of static or flow-through structure obtained commercially or constructed by the investigators. If the investigators construct the structure, it is important to ensure that any potentially toxic chemicals in the construction materials do not leach out into the water and affect the fish. Fish should be fed an appropriate diet to ensure optimal energy requirements and heathy condition and should be acclimated to experimental conditions before use in an experiment. If the fish are to be anesthetized before, during, or after the experiment, investigators should be familiar with the use of fish anesthetics and know whether the anesthetic has the potential of confounding the results of the experiment. Investigators should choose an acceptable means of euthanasia and a sanitary and safe method of fish disposal.

Comparative studies with fish

Fish species can respond differently to environmental stressors and researchers can explore these differences to understand the nature of the response while making new discoveries. For example, two closely related benthic fish, English sole (*Parophrys vetulus*) and starry flounder (*Platichthys stellatus*), exhibit substantial differences in the frequencies of hepatic neoplasms even though these species occur in the same chemically contaminated estuary (McCain et al., 1982; Malins et al., 1984). Further studies indicated that the differing frequencies of hepatic neoplasms may have resulted, at least in part, from differences in hepatic enzyme activities and hepatic isoenzyme profiles (Collier et al., 1992). In another example, several species of fish exhibited differential sensitivity when exposed to ultraviolet-B (UVB) radiation in a specially constructed solar simulator (Little and Fabacher, 1996). Laboratory studies showed that rainbow trout (*Oncorhynchus mykiss*) were sensitive to a certain dose of simulated solar UVB radiation whereas razorback suckers (*Xyrauchen texanus*) were unaffected by this dose of radiation. Razorback suckers contained significantly greater amounts of an apparent photoprotective substance in the outer layers of the dorsal skin than did rainbow trout (Fabacher and Little, 1995, 1998). Thus, the difference in sensitivity to UVB radiation appeared to result from differences in the amounts of this non-melanic photoprotective substance.

Fish of the same species can appear different, causing researchers initially to draw an erroneous conclusion or assumption. For example, researchers could expect the light-colored albino strain of a fish species to be more sensitive to the detrimental effects of UVB radiation than a darkly pigmented strain because albino fish lack large quantities of the pigment melanin that is known to have photoprotective properties. However, when albino and pigmented strains of medaka (*Oryzias latipes*) were exposed to UVB radiation, the effects were the same in both strains and both strains were found to have essentially the same amounts of a colorless non-melanic photoprotective substance in the outer skin layers (Fabacher *et al.*, 1999). The photoprotective substance apparently blocked the radiation before it could penetrate melanin-containing cellular layers.

Strains of the same fish species can adaptively develop and respond differently to environmental stressors. When investigators conducted laboratory studies comparing the toxicity of insecticides to mosquitofish (*Gambusia affinis*) collected from insecticide-contaminated and uncontaminated locations, the fish from the insecticide-contaminated location were found to be resistant to concentrations of insecticides that were toxic to fish from the uncontaminated location (Finley *et al.*, 1970). Fish from the different locations appeared the same, however, subsequent studies showed that insecticide-resistant fish contained more total lipid and a marked increase in cell size and lipid inclusions in hepatocytes than did insecticide-susceptible fish (Fabacher and Chambers, 1971; Yarbrough and Coons, 1975). These fish were resistant to insecticides, at least in part, because they contained more lipid to sequester lipid-soluble insecticides and prevent them from reaching the site of action. Thus, an insecticide-resistant strain of mosquitofish had developed after years of selective pressure.

Fish as substitutes for mammals

Although fish will probably never replace mammals as experimental animals, they can substitute for mammals in certain stages of complex experimental protocols. For example, certain stages of carcinogenicity testing experiments using fish can be more sensitive, conducted more rapidly, and more economical than experiments using mammals. Hoover (1984), in a National Cancer Institute Monograph includes a section on small fish species proposed for carcinogenicity testing such as fish in the genus *Poeciliopsis*; a hermaphroditic marine fish, *Rivulus marmoratus*; medaka; the Amazon molly, *Poecilia formosa*; zebrafish, *Brachydanio rerio*; guppies, *Lebistes reticulatus*; two killifish, *Cyprinodon variegatus* and *Fundulus grandis*; platyfish and swordtail hybrids, *Xiphophorus* spp.; mudminnow, *Umbra* sp.; fathead minnows, *Pimephales promelas*; and small life stages of rainbow trout. This monograph also includes sections on alternative testing systems (e.g. embryos, cell cultures) for using fish, maintaining fish in the laboratory, special problems and topics related to using fish in cancer research, and descriptions of how fish metabolize chemical contaminants and other foreign compounds.

Another report, 'Use of small fish in biomedical research, with special reference to inbred strains of medaka' (Hyodo-Taguchi and Egami, 1989), also addressed the use of small fish species. In addition to describing the general biology, husbandry (including discussions of reproductive habits, rearing and breeding) and genetics of medaka, information also included how inbred strains of medaka were established and responded to radiation and chemical carcinogens. The report included sections on using the following fish as experimental animals: fish in the genus *Xiphophorus*; moderate-sized fish such as goldfish (*Carassius auratus*); central mudminnows (*Umbra limi*); and small life stages of rainbow trout. Hightower and Renfro's (1988) review, 'Recent applications of fish cell culture to biomedical research', stimulated interest in fish cell culture as a useful complement to mammalian cell culture. The authors discussed fish cell culture techniques, cell lines, and primary tissue cultures for use in studying epithelial ion transport, endocrinology, cellular stress response, thermotolerance, environmental toxicology and cancer biology.

Experiments with fish can lead to an understanding of biological systems in other vertebrates, including humans and their diseases. **Retinoblastoma**, which occurs in children for which there is no viable rodent model, can be induced in fish eyes by exposing fish to chemical carcinogens (Van Beneden and Ostrander, 1994). Perhaps retinoblastoma in fish eyes could be used as a model that would lead to discoveries for treating this condition in humans.

The following selected examples illustrate the importance of fish as biomedical models. The pufferfish (*Fugu rubripes*) genome was reported to be the best model genome to use in discovering human genes because the pufferfish genome is about 7.5 times smaller than the human genome and pufferfish genes are similar to human genes (Brenner *et al.*, 1993). With this model researchers can ask questions about the larger human genome by readily manipulating the smaller pufferfish genome in specific ways. McKinney and Schmale (1994) reported that damselfish (*Pomacentrus partitus*) **neurofibromatosis**, a disease in which affected fish exhibit peripheral nerve sheath tumors containing morphologically abnormal Schwann cells, was the only naturally occurring animal model for human neurofibromatosis. Fish of the genus *Xiphophorus* have been intensively studied as malignant melanoma models (Setlow *et al.*, 1989; Setlow, 1996). The gene map of these fish is one of the most extensive among non-human vertebrates, and as such offers unique opportunities for studying melanoma in humans (Morizot *et al.*, 1998).

As a result of interest in human health issues surrounding toxicant-induced oxidative damage and cellular responses, fish models (in addition to traditional mammalian models) have been proposed in order to understand the oxidative stress response in more detail (Kelly *et al.*, 1998). Defects in hemoglobin synthesis cause many human anemias. In a recent report, Brownlie *et al.* (1998) described a zebrafish mutant as representing the first animal model of human congenital **sideroblastic anemia**. Another zebrafish mutant was recently reported to represent the first genetically 'accurate' animal model of human hepatoerythropoietic porphyria (Wang *et al.*, 1998).

Summary

The historical use of fish as experimental animals has been well documented, and they exhibit enormous diversity and lend themselves to various experimental uses. However, using fish in an experiment is not as simplistic as it first appears. Species and strains of fish can respond differently to the same stressor. Nevertheless, this fact can be used advantageously because comparative studies often provide more information on the effects of a stressor than when a single species or strain is used.

As in any experiment that uses animal subjects, investigators using fish in experiments should emphasize care and culture practices which reduce stress and promote health exclusive of experimental variables. Any sort of excessive stress can degrade the fish's health and allow disease to develop, which can kill the fish or otherwise influence the results of an experiment.

In some cases fish can be used as experimental models that could lead to discoveries for treating human conditions. Researchers accustomed to working with rodents may be initially overwhelmed by the perceived complexities of conducting experiments with fish in an aquatic environment. However, once the variables are examined and understood, the perceived complexity can diminish to a perception of moderate simplicity. Fish will probably never replace rodents as experimental animals in biomedical research; however, using them as substitute animals in tiered research protocols can be economical and enlightening.

Fish are interesting to study and can provide investigators with myriad characteristics to marvel at. We are not even close to understanding the tremendous potential that exists in using fish as experimental animals. As the knowledge of fish biology increases, new information will produce more and new uses of fish as experimental models. For researchers interested in fish as experimental animals, this book should be a valuable source of information well into the twenty-first century.

References

Atz, J.W. (1986). *Amer. Zool.* **26**, 111–120.
Baksi, S.M. and Frazier, J.M. (1988). *Marine Environ. Res.* **24**, 141–145.
Beaman, J.R., Finch, R., Gardner, H., Hoffmann, F., Rosencrance, A. and Zelikoff, J.T. (1999). *J. Toxicol. Environ. Health* **56**, 523–542.
Blazquez, M., Bosma, P.T., Fraser, E.J., Van Look, K.J. and Trudeau, V.L. (1998). *Comp. Biochem. Physiol. C* **119**, 345–364.

Brenner, S., Elgar, G., Sandford, R., Macrae, A., Venkatesh, B. and Aparicio, S. (1993). *Nature* **366**, 265–268.

Brownlie, A., Donovan, A., Pratt, S.J., Paw, B.H., Oates, A.C., Brugnara, C., Witkowska, H.E., Sassa, S. and Zon, L.I. (1998). *Nat. Genet.* **20**, 244–250.

Callard, G.V., Kruger, A. and Betka, M. (1995). *Environ. Health Perspect.* **103** (Suppl. 7), 51–57.

Chen, T.T., Lu, J-K, Shamblott, M.J., Cheng, C.M., Lin, C-M, Burns, J.C., Reimschuessel, R., Chatakondi, N. and Dunham, R.A. (1995). *Zoolog. Studies* **34**, 215–234.

Collier, T.K., Singh, S.V., Awasthi, Y.C. and Varanasi, U. (1992). *Toxicol. Appl. Pharmacol.* **113**, 319–324.

Contie, V.L. (1991). *Nat. Cent. Res. Resourc. Rep.* **15**, 1–4.

DeTolla, L.J., Srinivas, S., Whitaker, B.R., Andrews, C., Hecker, B., Kane, A.S. and Reimschuessel, R. (1995). *ILAR Journal*, **37**, 159–173.

DeVries, A.L. (1971). *Science* **172**, 1152–1155.

Driedzic, W.R. (1983). *Comp. Biochem. Physiol. A* **76**, 487–493.

Eriksson, O., Mayer-Gostan, N. and Wistrand, P.J. (1985). *Acta Physiol. Scand.* **125**, 55–66.

Fabacher, D.L. and Chambers, H. (1971). *Bull. Env. Contam. Toxicol.* **6**, 372–376.

Fabacher, D.L, Little, E.E. (1995). *Environ. Sci. Pollut. Res.* **2**, 30–32.

Fabacher, D.L. and Little, E.E. (1998). *Environ. Sci. Pollut. Res.* **5**, 4–6.

Fabacher, D.L., Little, E.E. and Ostrander, G.K. (1999). *Environ. Sci. Pollut. Res.* **6**, 69–71.

Fetcho, J.R. and Liu, K.S. (1998). *Ann. NY Acad. Sci.* **860**, 333–345.

Finley, M.T., Ferguson, D.E. and Ludke, J.L. (1970). *Pestic. Monit. J.* **3**, 212–218.

Fry, F.E.J. (1971). In *Fish Physiology*, vol. VI (eds W.S. Hoar and D.J. Randall), pp. 1–98. Academic Press, New York.

Gardner, E.J. (1972). *History of Biology*. 3rd edn. Burgess Publishing Company, Minneapolis, Minn. 464pp.

Hawkins, W.E., Overstreet, R.M., Fournie, J.W. and Walker, W.W. (1985). *J. Appl. Toxicol.* **5**, 261–264.

Hightower, L.E. and Renfro, J.L. (1988). *J. Exp. Zool.* **248**, 290–302.

Hoover, K.L. (ed.) (1984). *Use of Small Fish Species in Carcinogenicity Testing*. National Cancer Institute Monograph 65. National Institutes of Health (publication No. 84-2653), National Cancer Institute, Bethesda, Maryland. 409pp.

Hunn, J.B. (1989). *Investigations in Fish Control: 98*. US Fish and Wildlife Service, National Fisheries Research Center-LaCrosse, LaCrosse, Wisconsin, USA. 10pp.

Hyodo-Taguchi, Y. and Egami, N. (1989). In *Nonmammalian Animal Models for Biomedical Research* (eds A.D. Woodhead and K. Vivirito), pp. 185–214. CRC Press, Boca Raton, Florida.

Jagadeeswaran, P. and Liu, Y.C. (1997). *Blood Cells Mol. Dis.* **23**, 52–57.

Karlsson, J.O.G., Grundstrom, N., Elwing, H. and Andersson, R.G.G (1990). *ATLA* **18**, 201–224.

Kelly, K.A., Havrilla, C.M., Brady, T.C., Abramo, K.H. and Levin, E.D. (1998). *Environ. Health Perspect.* **106**, 375–384.

Klontz, G.W. (1971). In *Animal Models for Biomedical Research*, vol. 4. Symposium, pp. 27–30. National Academy of Sciences, Washington, DC.

Lele, Z. and Krone, P.H. (1996). *Biotechnol. Adv.* **14**, 57–72.

Little, E.E. and Fabacher, D.L. (1996). In *Techniques in Aquatic Toxicology* (ed. G.K. Ostrander), pp. 141–158. Lewis Publishers, Boca Raton, Florida.

Luer, C.A. (1989). In *Nonmammalian Animal Models for Biomedical Research* (eds A.D. Woodhead and K. Vivirito), pp. 121–148. CRC Press, Boca Raton, Florida.

Malins, D.C., McCain, B.B., Brown, D.W., Chan, S.-L., Myers, M.S., Landahl, J.T., Prohaska, P.G., Friedman, A.J., Rhodes, L.D., Burrows, D.G., Gronlund, W.D. and Hodgins, H.O. (1984) *Environ. Sci. Technol.* **18**, 705–713.

May, E.B., Bennett, R.D., Lipsky, M.M. and Reimschuessel, R. (1987). *Lab Animal* **16**(4), 23–28.

McCain, B.B., Myers, M.S., Varanasi, U., Brown, D.W., Rhodes, L.D., Gronlund, W.D., Elliott, D.G., Palsson, W.A., Hodgins, H.O. and Malins, D.C. (1982). *Federal Interagency Energy/Environment Research and Development Report, EPA-600/7-82-001*. Environmental Protection Agency, Washington, DC. 100pp.

McKinney, E.C. and Schmale, M.C. (1994). *Dev. Comp. Immunol.* **18**, 305–313.

Metscher, B.D. and Ahlberg, P.E. (1999). *Dev. Biol.* **210**, 1–14.

Moon, T.W., Walsh, P.J. and Mommsen, T.P. (1985). *Can. J. Fish Aquat. Sci.* **42**, 1772–1782.

Morizot, D.C., McEntire, B.B., Della Coletta, L., Kazianis, S., Schartl, M. and Nairn, R.S. (1998). *Mol. Carcinog.* **22**, 150–157.

Nelson, J.S. (1994). *Fishes of the World*, 3rd edn. John Wiley, New York. 600pp.

Neuhaus, O.W. and Halver, J.E. (eds) (1969). *Fish in Research*. Academic Press, New York. 311pp.

Nigrelli, R.F. (1953). *Trans. NY Acad. Sci.* **15**, 183–186.

Powers, D.A. (1989). *Science* **246**, 352–358.

Rand, G.M. and Petrocelli, S.R. (1985). *Fundamentals of Aquatic Toxicology*. Hemisphere Publishing Corporation, Washington, DC. 666pp.

Setlow, R.B. (1996). *Photochem. Photobiol.* **63**, 410–412.

Setlow, R.B., Woodhead, A.D. and Grist, E. (1989). *Proc. Natl. Acad. Sci.* **86**, 8922–8926.

Umminger, B.L. and Pang, P.K.T. (1979). *ILAR News* **22**, 12–18.

Van Beneden, R.J. and Ostrander, G.K. (1994). In *Aquatic Toxicology: Molecular, Biochemical and Cellular Perspectives* (eds D.C. Malins and G.K. Ostrander), pp. 295–326. Lewis Publishers, Boca Raton, Florida.

Vascotto, S.G., Beckham, Y. and Kelly, G.M. (1997). *Biochem. Cell Biol.* **75**, 479–485.

Wang, H., Long, Q., Marty, S.D., Sassa, S. and Lin, S. (1998). *Nat. Genet.* **20**, 239–243.

Wolke, R. (1984). *Comp. Path. Bull.* **16**, 1 and 5–6.

Wourms, J.P. (1997). *Amer. Zool.* **37**, 269–310.

Yarbrough, J.D. and Coons, L.B. (1975). *Chem.–Biol. Interact.* **10**, 247–254.

PART 2

Housing, Maintenance and Breeding

Contents

CHAPTER 1	**Facilities and husbandary (large fish models)**	13
CHAPTER 2	**Facilities and husbandary (small fish models)**	41
CHAPTER 3	**Diet**	65
CHAPTER 4	**Common diseases and treatment** ...	79

CHAPTER 1

Facilities and Husbandry (Large Fish Models)

Jeffrey P Fisher
Pentec Environmental, Inc., Edmonds, Washington, USA

Introduction

This chapter summarizes the biological and engineering fundamentals of facilities designed to culture large fish for experimental purposes. The term 'large fish' here refers to species that generally cannot be cultured in small aquaria without stress, where a stressor is defined as 'an environmental or biological challenge that is severe enough to require a physiological or behavioral response by a fish' (Wedemeyer, 1980). Most large fish models have not addressed the difference between the range in a species' *tolerance* and *preference*. Herein lies an essential difference between intensive aquaculture and those husbandry practices intended to support fish for experimental purposes. In commercial aquaculture high density production is critical for profitability. Such conditions are tolerated by fish but not without high engineering cost and biological risk. The husbandry of large fish for experimental purposes should therefore reflect an awareness of both the bioengineering fundamentals developed for aquaculture, and the ecological parameters natural to the fish under study.

Large fish models offer several advantages for research that are not possible or practicable with small fish models. Large fish are generally more robust and easier to handle for most non-lethal procedures such as blood sampling and surgery. Because of the studies conducted with large fish for aquaculture and fisheries research, the body of knowledge on many of the large fish species cultured exceeds that of the small fish models. For example, the extreme carcinogenicity of aflatoxin was fortuitously discovered in commercially cultured rainbow trout being fed a diet contaminated with *Aspergillus flavus* (Wolf and Jackson, 1963). Of course, large fish generally require more space and water than small fish, and these requirements can place great demands on facilities and the institutions that support them. Ultimately, studies using both large fish and small fish models will continue to be necessary to address the many questions posed in the aquatic sciences.

This chapter focuses on the husbandry requirements of the more common varieties of coldwater,

Copyright © 2000 Academic Press

coolwater, warmwater, and tropical large fish used in research, such as the salmonids (Salmonidae), catfish (Ictaluridae), flatfish (Pleuronectidae), and panfish (Centrarchidae). Coldwater fishes are commonly represented by the Salmonidae, and include those species with minimal tolerance for warm water that are naturally found in water primarily less than 18–20°C. Coolwater fish are represented by the pike family (Esocidae); they prefer slightly warmer temperatures than coldwater fishes, but tolerate cold water and warm water. Warmwater fishes are represented by the catfish (Ictaluridae) and sunfish (Centrarchidae), and exhibit the widest temperature tolerance, with a preference for slightly warmer waters than coolwater fishes. Tropical fishes such as the pompano (*Trachinotus carolinus*) and tilapia (e.g. *Oreochromis* sp.) tolerate only minimal variation in their preferred warmwater temperatures. The following discussion emphasizes temperate species, reflecting the author's predominant experience. The general principles presented here are applicable to other species as well. Of course, all 'large' fish are, for some time in their life, 'small' fish. So for certain periods at least, culture conditions that would be unacceptable for adult and subadult large fish may be perfectly adequate for their earlier life stages. An attempt is made, therefore, to summarize the appropriate conditions necessary to culture the early life stages of these species, to the extent that they are not discussed within the subsequent chapter.

Overview of conceptual designs for large fish husbandry

Culture systems for the husbandry of large fish are of two principal designs, either 'flow-through' or 'recirculating'. The engineering required for these two types of systems differs substantially because of the variable influent and effluent water quality requirements. The design of recirculating systems is driven by the necessity to maintain high quality influent waters, whereas effluent water quality is more often the driver of bioengineering for flow-through systems.

Flow-through systems

Flow-through systems are supplied with either treated city water, well water, or natural surface waters (spring, river, or lake water). Such waters may require treatment to remove chlorine, reduce suspended sediment loads, release **supersaturated gases**, and/or eliminate pathogens before delivery into tanks or ponds holding large fish. Chlorine is typically found in municipal water supplies; supersaturated gases are often problematic from well-water sources or surface-water sources taken from depth; and suspended sediments are commonly contributed by surface waters from **riverine** sources. Effluent from flow-through systems may also require biological and/or chemical treatment before release to ensure that biological and chemical waste products from the culture and/or experimental operations do not exceed relevant water quality standards of surface waters, and that potential disease organisms are not liberated. Such standards should be reviewed before any conceptual design for a culture system is put into full operation. Dependent on the experimental design, if the effluent from such facilities is directed into municipal water-treatment systems, then such effluent treatment is usually unnecessary. A typical flow-through system as used for whole effluent toxicity testing, per the design of Garton (1980) is pictured in Figure 1.1.

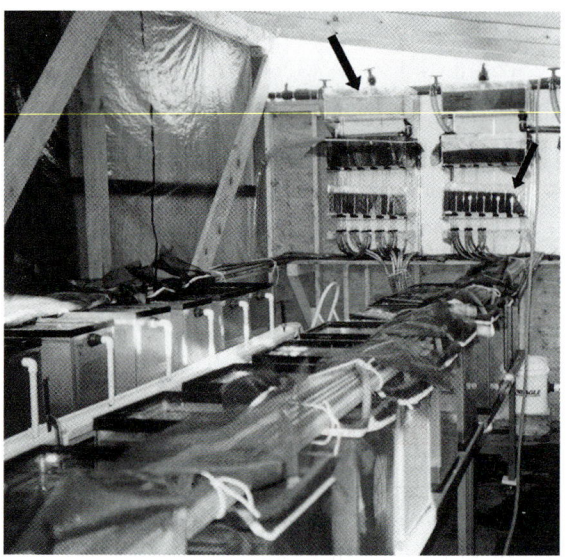

Figure 1.1 Proportional dilution design of Garton (1980) as implemented for *in-situ* whole effluent toxicity testing of river water on rainbow trout. Water is pumped from the river into a central head tank (large arrow) and diluted in sequence with filtered and dechlorinated municipal water of the same temperature (small arrow). Waters of various dilutions are then delivered to aquaria housing trout for chronic exposure testing.

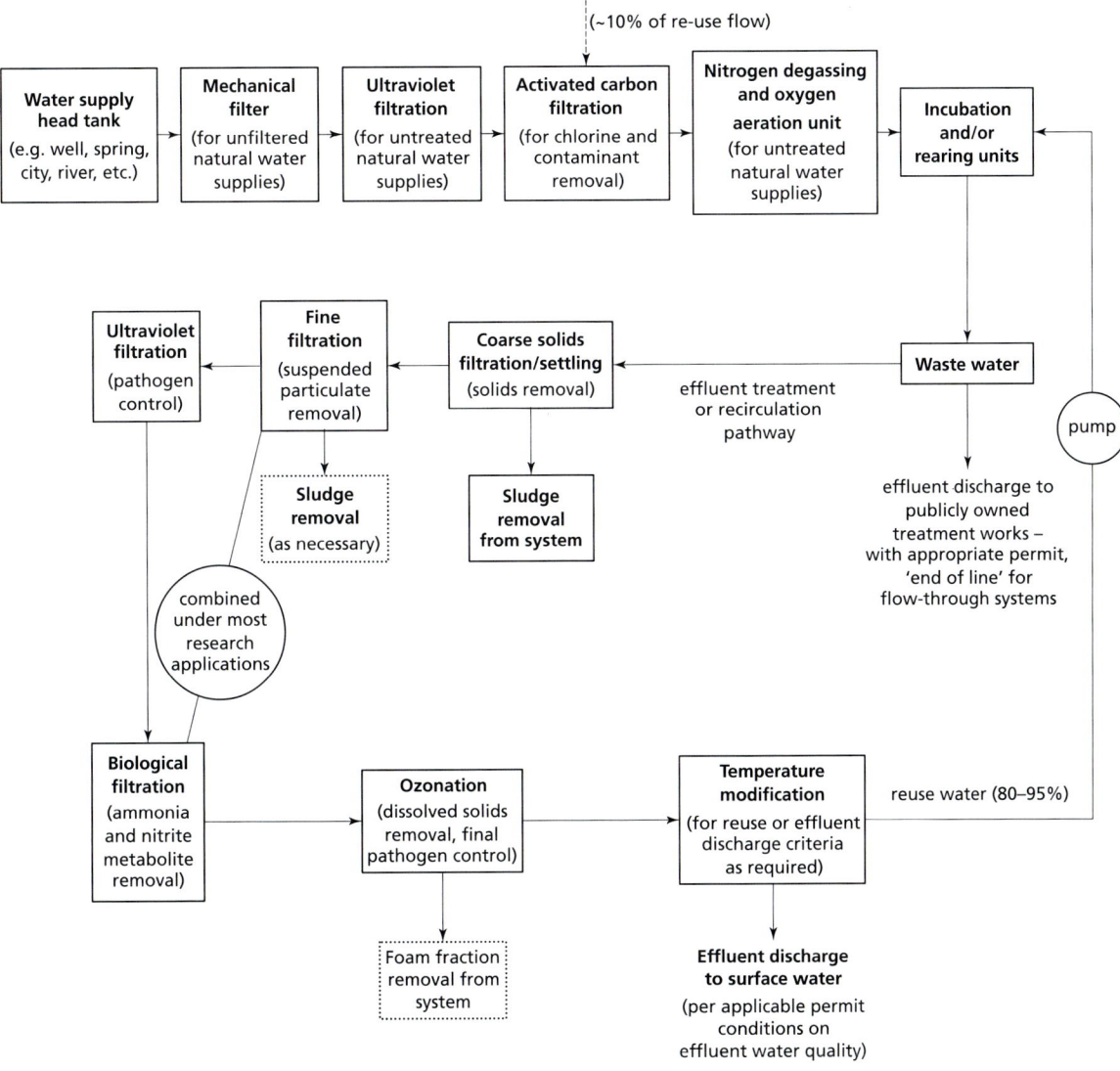

Figure 1.2 Flow chart of elements of flow-through and recirculating culture systems for the husbandry of large fish species.

A schematic flow chart representing both flow-through and recirculating system components is depicted in Figure 1.2.

Recirculating systems

Recirculating systems involve the complete or partial reuse of water by circulating it through 'closed-loop' treatment facilities that reduce biological and chemical waste loads to levels comparable to the influent waters. Blancheton and Coves (1993) promote a design for marine fish culture that sequentially provides mechanical filtration, UV irradiation, biological filtration and reoxygenation. More specifically, successful recirculating systems require staged processing that provides: (i) separation of large (settleable) suspended solids (uneaten food and feces); (ii) 'fines' filtration (removal of suspended fine particles); (iii) removal of dissolved organics (**foam fractionation**/protein skimming); (iv) ultraviolet filtration for pathogen control; (v) biological filtration/**nitrification**; (vi) toxicant removal with activated carbon (charcoal) filtration; and (vii) **ozonation** control (Figure 1.2).

Ozonation is a recommended finishing step for reuse because of the additional improvements it provides for water quality and pathogen control. For example, ozonation has been shown to reduce total suspended solids (TSS) by 35%, chemical oxygen demand (**COD**) by 36%, dissolved organic carbon (**DOC**) by 17%, and color by 82% in the culture tanks (Summerfelt et al., 1997). Ozonation also reoxygenates the water passing through the filter (Williams *et al.*, 1982), and improving the efficacy of other filtration steps (Summerfelt *et al.*, 1997). Ozonation can also be combined with foam fractionation. Because of the complexity of the treatment required, recirculating systems are more prone to failure and more sensitive to changes in environmental conditions; therefore, reliable alarm systems should be incorporated into the design for safety.

Recirculating systems offer tangible benefits in that less water is used and growth rates are generally higher because the fish are cultured in warmer waters than usually occur in flow-through systems; however, physiological stress indicators associated with acid–base imbalance from poor water quality (e.g. elevated ammonia) may occur more commonly under recirculating culture systems. These indicators could include a decrease in plasma sodium and chloride, and liver glycogen (Wedemeyer, 1980), among other typical indicators such as an increase in plasma cortisol and lactate (Pickering and Pottinger, 1987; Barnett and Pankhurst, 1998) and a variety of hematological changes (Caldwell and Hinshaw, 1994). For this reason, Timmons *et al.* (1991) recommend that 43.4 L min^{-1} kg^{-1} feed per day be used as an operating parameter in reuse systems to ensure that biological filtration maintains total ammonia levels at or below 0.5 mg L^{-1}.

Facilities and bioengineering

Incubation systems

Hatching jars, vertically stacked trays, Netart systems, Clark-Williamson troughs, perforated baskets, and hatching cups are routinely used for the incubation of large fish species eggs and their early life history stages (Piper *et al.*, 1982). Most of these incubation systems attempt to provide for upwelling through the eggs and larvae. Hatching jars, such as the McDonald type (Figure 1.3), are typically used for species with eggs that require continual suspension or agitation such as striped bass (*Morone saxatilis*) or catfish (*Ictalurus* sp.). Newly hatched larvae can be maintained within the hatching jar by screening the outlet, or they can be permitted to enter a rearing tank by volition. A modification of this design, using upwelling through gravel filled buckets, can be used for *in-situ*

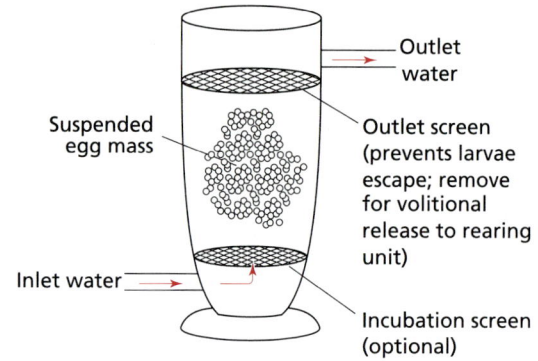

Typical for: Esocidae, *Morane* sp.

Figure 1.3 Typical hatching jar utilizing upwelling system. These systems are effective in research applications because: (i) they utilize minimal space; (ii) they easily allow lot or treatment separation; and (iii) they maintain high water quality. They are typically used for species with egg masses that stick together, but can also be used for the incubation of salmonid eggs.

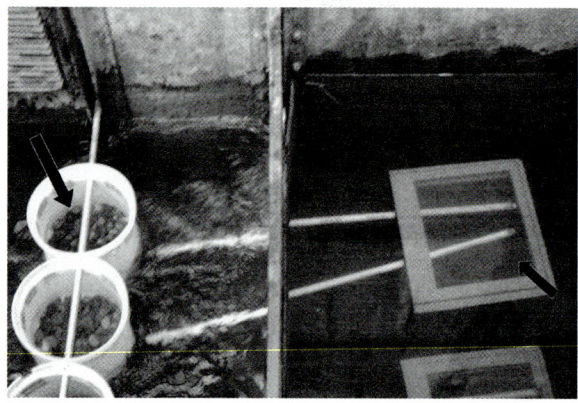

Figure 1.4 *In-situ* egg culture system for salmonids using upwelling hatching bucket design wherein inlet water is run through a coarse filter (small arrow). Water flows through the pipe into the bottom of each bucket, up through a layer of filter bed material (aquarium fiber floss), and then through a 30 cm layer of stream gravel (large arrow). Screened holes around the perimeter of the top of each bucket permit the flow of water out of each unit and the retention of fry upon emergence.

incubation to assess ecological risk to salmonids from ambient water quality (Figure 1.4).

Vertical tray systems, such as those produced by Heath^{TM}, stack salmonid eggs or larvae through a series of screens, where water upwells by gravity from the highest tray to the lowest tray and eventually exits to effluent (Figure 1.5). Vertical trays generally use the least amount of water flow, and the water passing from one tray to the next below is partially aerated in the process. These systems are typically used for salmonid culture, and egg densities should be reduced at the lower levels to account for oxygen consumption and metabolite production in eggs and larvae maintained in the upper trays. Without some limited form of substrate added to the trays, salmonid larvae will be permitted to swim around within the trays, and this added activity can result in a reduction in fry size upon the completion of yolk absorption.

The Netart system provides for salmonid egg incubation in raceways filled with gravel. Solid baffles in the raceway force water to upwell through each compartment over the gravel beds where the eggs are initially placed. One problem with the Netart system is that it prevents the monitoring of egg development because the eggs are buried within the gravel matrix. Thus, dead eggs cannot be easily removed without disturbing other eggs, and fungal overgrowth can result when dead eggs are left to decompose within the substrate.

The Clark-Williamson system employs similar upwelling compartments as used with a Netart system (Piper et al., 1982); however, eggs are held within each compartment on suspended screens stacked on top of each other (Figure 1.6). The objective of this design is to maximize the water flow upwards through the screens, and not around them, thus, the screens should be constructed to fit snugly within each compartment in the raceway. The Clark-Williamson method is effective for culturing a very high quantity of eggs with minimal water flow; however, trapped gases can form bubbles on the incubation screens, which, when released, can cause enough physical disturbance to kill young embryos – especially before eye-up. The removal of dead eggs also requires the displacement of the uppermost screens to provide access to the lower screens, a sometimes cumbersome and therefore risky task. After hatching, the photophobic

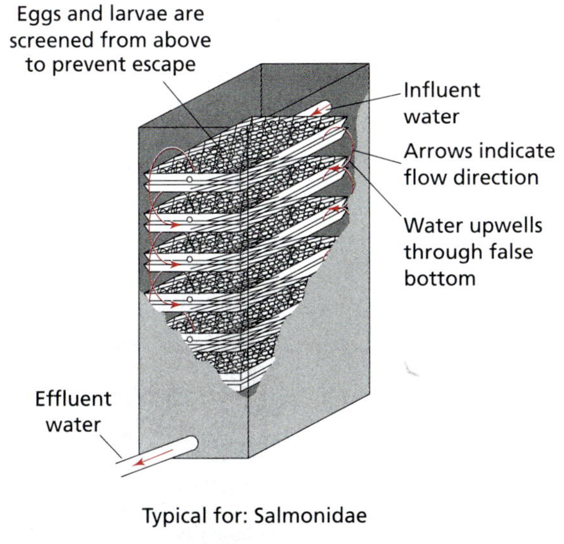

Figure 1.5 Vertical tray egg and larval incubation system as used for salmonids.

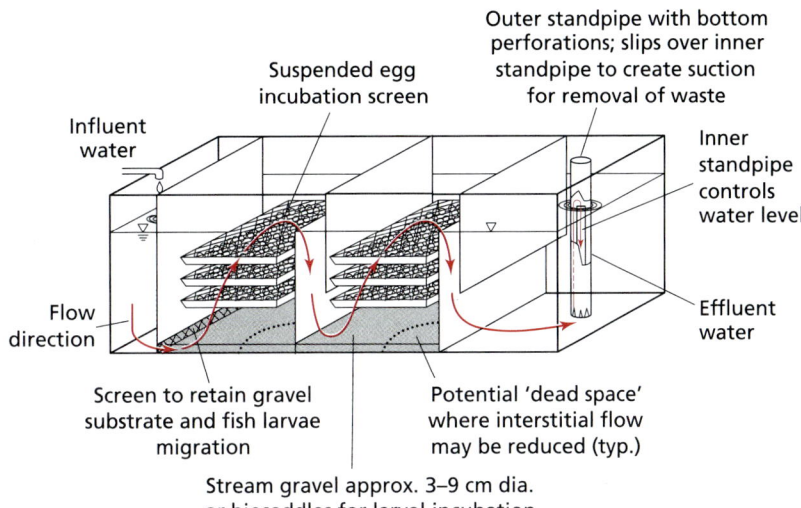

Figure 1.6 Typical Clark-Williamson/Netarts egg incubation system. The system can be compartmentalized into individual units under laboratory conditions, providing the added opportunity to implement a false bottom, thereby avoiding the potential dead space associated with the pictured design.

and geotropic salmonid **alevins** will swim down through the incubation screens (designed to a diameter to permit this movement without restriction) and seek refuge in the gravel or other substrate lining the raceway. In theory, the alevins will remain within this matrix until emergence, when yolk absorption is nearly complete and exogenous food can be provided. For both the Netart and Clark Williamson methods of incubation, uniform intra-gravel flow is rarely achieved, and hatched salmonid larvae may congregate towards those areas within the gravel where the dissolved oxygen is highest. This behavioral response can result in smothering if the larval densities are excessive.

Species such as catfish produce eggs in masses that stick together. Such egg masses require continual rotation or agitation, and can therefore be cultured effectively in upwelling hatchery jars, or in specially designed 'catfish troughs' that will provide enough aeration to continually move the egg masses (Piper *et al.*, 1982). Placement of catfish eggs into perforated baskets within troughs equipped with upwelling baffles can also be used, provided flow is adequate to gently agitate the eggs.

Perforated hatching cups or baskets suspended in the water column provide a highly effective means of maximizing experimental units with minimal water and tank space for research with many species of large fish (Figure 1.7). To ensure minimal variation in the microhabitat of the egg incubation environment, salmonid eggs should not be stacked more than two layers deep, unless a concentrated upwelling system is used.

Rearing tank designs

Tanks and other rearing units used with flow-through and recirculating culture regimes are often similar, and include raceways, rectangular tanks, circular tanks, earthen or lined ponds, artificial streams, and cages/net-pens. Although there are species-specific preferences for certain types of rearing units, these preferences are generally less specific than those required for incubation. The general operating environment and advantages and disadvantages of these systems are discussed below. The use of pipe made of copper, zinc, brass, or galvanized metal should be avoided for all types of systems. For all systems described in detail below, the flow rates used should reflect the loading density in the system, as predicated by the experimental objectives and biological

Figure 1.7 Hatching cup egg incubation system used for flow-through incubation of salmonid eggs. Eggs and larvae exhibited in cups at approximately 50% hatching (small arrow). Screened PVC cups are suspended on rocker arms in a 38 L aquarium maintained in a flow-through water bath (Fisher *et al.*, 1998). Flow is directed into the 38 L aquarium through flexible tubing from a head tank (not pictured). Aeration is provided at the inlet to ensure adequate oxygenation and nitrogen gas removal. A constant level siphon design at the outlet (large arrow) ensures water exchange at a specific height and rate (Garling and Wilson, 1976). The tops of the hatching cups are above the top of the aquarium to ensure that siphon disruption will not redistribute experimental lots separated by cup, even if the aquarium overflows. Aquaria are covered with black lids to prevent light disturbance to embryos and larvae.

limitations. Commercial aquaculture suppliers are the most affordable source of tanks, other than constructing the units oneself. The reader is referred to the Internet and other information sources (e.g. the World Aquaculture Society) for information for aquaculture equipment suppliers. Local hardware stores may also carry tanks that can be suitable for culturing large fish, with appropriate plumbing.

Raceways

The term 'raceway' is used to describe tanks of rectangular dimension that convey water from the inlet to the outlet through an open unobstructed channel (Figure 1.8). The outlet is screened to prevent the escape of fish, and effluent is either discharged (flow-through) or recirculated through the raceway after biological and chemical treatment (Figure 1.2). Raceways are usually operated at fairly shallow depths of 15 cm to 1 m. Dimensions for raceways are often developed to achieve a length/width/depth ratio of 30 : 3 : 1 (Piper *et al.*, 1982); thus, a 5-m raceway typical for the husbandry of large fish for experimental purposes would have a width of 0.5 m and a depth of 0.17 m. Side-walls vary from 90 degrees to 2H : 1V. For salmonids and other pelagic fishes (e.g. striped bass), there is actually a greater potential for losing fish from jumping out of raceways with vertical sidewalls, because these designs favor deeper pool depths. Earthen raceways with angled side-slopes, however, require extensive maintenance to keep vegetative growth in check.

Raceways can be constructed of earth, plywood, fiberglass, concrete, or aluminum; treated wood and galvanized sheet metal should be avoided to prevent potential toxicity from leached treatment chemicals (e.g. phenolic acids in the wood, zinc from the metal), although both of these materials can be 'made safe' by painting with epoxy. Large raceway systems are more typical of intensive aquaculture with salmonids and catfish, although such systems can be used for other species. Since most raceways are used under flow-through regimes, such systems are more often used in geographic areas where water supplies are abundant and cheap (Brown and Gratzek, 1980). Laboratory systems featuring raceways are generally smaller, and are better served using fiberglass or aluminum. These materials can be easily cleaned and disinfected between experiments, ensuring that disease and chemotactic cues will not bias future experiments.

The advantage of raceway systems lies in their ability to culture fish at very high density. For example, a carp farm in Japan produces roughly 2200 t (metric tonnes, MT) per hectare in raceways, which exceeds the production possible from earthen ponds by over 1000 times (Brown and Gratzek, 1980). The density of fish cultured in a raceway is primarily oxygen-controlled. Therefore, flow, the typical vehicle of oxygen delivery in flow-through systems, is usually what regulates the number and density of fish that can be cultured in a raceway. The maximal loading of fish theoretically possible in a raceway is generally not possible because the velocities created by such flows would exceed the sustained swimming capacity of the fish, resulting in their entrainment on outlet screens and ultimate mortality. The sustained swimming capacity of the fish cultured is the water velocity which the fish can tolerate for longer than 1 hour (Blake, 1983), and is highly size and species dependent. This velocity has been found to approximate 1.3 ft s^{-1} (40 cm s^{-1} or ~6.7 body lengths/second) for sockeye (*Oncorhynchus nerka*) and coho salmon (*O. kisutch*) yearlings (Brett *et al.*, 1958), but should be recognized to vary with species and exercise condition.

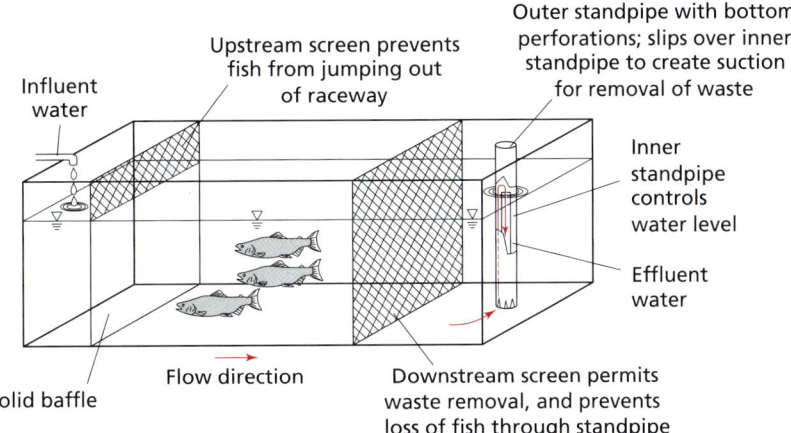

Figure 1.8 Typical raceway system for large fish husbandry.

The open-channel flow design of raceways does not reflect typical conditions that fish experience or prefer in the environment. There is little to no flow refuge as would be normally created from channel sinuosity, in-channel structure, or channel roughness (e.g. from a cobble substrate). In addition, because of the design of raceways, water quality deteriorates as it passes toward the outlet. Under intensive aquaculture conditions, gradations in oxygen and metabolic products can be measured from the front to the back, and the preference of fish for more highly oxygenated environs may be evident through their increased density near the inlet (Piper *et al.*, 1982). Such behavioral clues indicate overloading beyond the preferred range of the fish, suggesting that densities should be reduced. Concentrating fish under such conditions promotes stress and accelerates disease transmission.

Figure 1.9 Commercial on-shore aquaculture operation for salmonid rearing to smolt stage, with aerated head tank (foreground) for the gravity feed of inlet water to shaded, above-ground circular tanks (background).

Circular and square tanks

Circular tank designs take advantage of the Coriolis effect to achieve a form of 'self maintenance', whereby fecal matter and uneaten food are naturally collected toward a center standpipe that controls the depth in the tank. Tank diameter-to-depth ratios should be maintained at 5:10 to maximize the self-cleaning function of a circular tank (Timmons *et al.*, 1991). A slightly larger standpipe with perforations drilled along the bottom is slipped over the inner standpipe and debris is sucked up through the perforations like a vacuum. As a rule of thumb, the size of the perforations should not exceed roughly half the size of the smallest fish that will be cultured within the system. In the northern hemisphere inlet water should be directed counterclockwise to achieve this effect, and clockwise in the southern hemisphere. If the inlet water is directed into the tank under high pressure, such as through perforated holes in PVC pipe, the **Coriolis circulation** in the tank facilitates the repeated oxygenation of the water in the tank, thereby enabling a higher density of fish to be grown in this type of tank than would be possible with the same flow directed into a raceway system.

Circular tanks can be as small as 3.785 L (i.e. a 1-gal bucket), to as large as 30 m in diameter (Figure 1.9); keep in mind that the Coriolis effect is more pronounced with larger tanks, and young fish can be easily overwhelmed by its force, resulting in entrainment on outlet screens and associated mortality. Circular tanks are commonly constructed of concrete, metal, plastic, or fiberglass. A very affordable system can be made by fitting a commercially available 5-gal bucket with a center standpipe. Galvanized metal circular tanks produced for livestock watering can be used for large fish culture if fitted with a plastic liner. Commercial aquaculture suppliers also produce numerous sizes of circular tanks, many with depressed conical centers, which further facilitate the settling and removal of solids from the system. For experimental purposes with large fish models, the circular tank is this author's overall preference because of its low maintenance, the ability to house numerous experimental units in a small area, the ability to isolate diseased stock if necessary, the conservative use of water, and the relatively uniform water quality maintained within it.

Square tanks maximize the surface area-to-volume ratio, which is especially useful for species that do not perform well when concentrated vertically, such as the Atlantic salmon (Piper *et al.*, 1982). Perfectly square tanks are not advisable for the culture of most large fish species, however, because of the difficulty in thoroughly cleaning and disinfecting the edges and corners between experiments, and the lack of circulation around these corners – the same criticism as can be applied to the general use of rectangular aquaria. An alternative design to the square tank, however, is the 'Swedish pond'. These are square tanks with rounded corners. Each tank has a perforated screen on the bottom where solids collect and are sucked out of the system. Water level in the tank is controlled by a standpipe outside the tank that is sleeved by an outer standpipe, similar to the center standpipe design for a circular tank. Swedish ponds can vary widely in size, but most experimental systems are less than 2.44 m × 2.44 m. Inlet water can be directed to take advantage of the Coriolis effect, although turbulence created by the corners will reduce the efficiency of this process for self-cleaning.

Partitioned tanks

Another less common culture tank used to house large fish is the rectangular or semicircular tank with a center partition, called the 'Burrows pond' (Figure 1.10). This tank design operates somewhat like a circular tank in that water is flushed around the tank, permitting the Coriolis effect to self-clean the unit, albeit with less efficiency than a circular tank. Fish can distribute themselves more uniformly throughout the rearing area than with typical raceway systems, minimizing the gradations in water quality between the inlet and outlet.

Hydraulic conditions that facilitate self-cleaning and the broad distributions of fish are highly depth-controlled (Piper et al., 1982). Operations at improper depth will accumulate feces and uneaten food in dead-spaces around the unit. Thus, the flexibility of such systems is somewhat limited. These units are particularly effective for maintaining stock, such as a broodstock, at a relatively uniform density.

Horizontal partitioning of raceways can be an effective means of separating experimental treatments. Upwelling can be created along the longitudinal length of a raceway by using solid paired baffles, whereby the water is forced to travel up and over a simulated weir. This design is similar to the Netarts system, originally developed for the incubation of salmonid eggs (Figure 1.6). The design facilitates oxygenation of water as it passes along a raceway, thereby reducing the gradient in water quality observed in raceways without such additions.

Artificial streams

Artificial streams have been used for several decades to mimic more natural conditions of the species under study; thus, they have proven quite useful in behavioral studies of fish (Figure 1.11). The fundamental difference between an artificial stream and a raceway is the inclusion in the former of in-channel structure such as substrate or refuge. An artificial stream need not be sinuous, as the addition of in-channel structure can provide the hydraulic control needed to replace the role of sinuosity. For studies of the Salmonidae, artificial streams have been employed to research predator avoidance (Fisher et al., 1994), territoriality and agonism (Ostrander et al., 1990), intragravel survival, emergence, spawning behavior and learning. They can be built with any of the materials described for circular tanks, but can also be built of more transportable materials such as plywood, enabling their construction at remote sites. Commercially produced units are also available.

Figure 1.11 Author design of miniaturized artificial stream channel for the evaluation of emergence behaviors in salmonid fish in a laboratory setting. Water flow upwells through the gravel from the head end of the channel through a screened inlet (top arrow). Effluent spills over a baffle at the outlet (bottom arrow) so that emergent fry can be captured and emergence rates recorded.

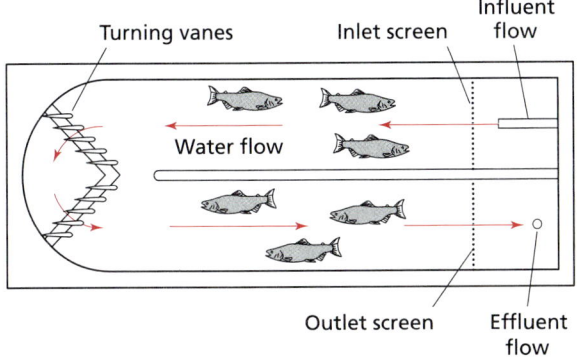

Figure 1.10 Top view of partitioned, semicircular tank for the husbandry of large fish.

Figure 1.12 Typical cage/net-pen farming system for commercial farming of salmon in Scotland. The edges of the nets are raised above the water surface to prevent the fish from jumping out (arrow). Smaller cage designs can be applied under research settings for a variety of large fish species.

Cages/net-pens

Cage designs vary from small rectangular cages or mesh bags used for catfish culture and *in-situ* exposures to support ecological risk assessments, to very large floating net-pens (Figure 1.12) or completely submersible cage designs used in commercial aquaculture. Cages are implemented as enclosures in pond, lake, fjord, or open-ocean environments. Cages and net-pens are most advisable in protected shoreline habitat areas, where stock maintenance presents few difficulties.

Water quality requirements and maintenance

Maintaining water quality parameters within a range acceptable for the species under study is essential to prevent experimental bias from stress. Conventional water quality parameters of dissolved oxygen, pH, temperature, alkalinity, suspended solids/sediment, carbon dioxide, nitrogen supersaturation, ammonia and nitrite are the most common factors that affect the stress status of fish in a culture environment. With the exception of temperature and salinity, there are few differences between species for the preferred ranges of these parameters. Tolerance ranges vary, however, with warmwater fish generally exhibiting a broader tolerance range for conventional water quality parameters in fresh water than coldwater fishes, and euryhaline fish exhibiting greater tolerance for water quality fluctuation than strictly marine species. Acceptable water quality values for the culture of large fish culture are presented in Table 1.1. These values reflect this author's personal experience with flow-through salmonid culture and closed-system culture of striped bass and a variety of marine species, as well as published information from a variety of sources (Brown and Gratzek, 1980; Piper *et al.*, 1982; Forteath, 1988; Langdon, 1988; Moe, 1992). Fish may be cultured outside the ranges specified in Table 1.1; however, acclimatization is generally required.

Water quality parameters largely determine the carrying capacity of experimental and intensive aquaculture systems for large fish. It is therefore imperative that these parameters be monitored at a frequency appropriate for the parameter, usually daily or semi-weekly. Most of these parameters can be monitored using commercially available and economical test kits or probes. More accurate spectrophotometric methods are available for many of the parameters; however, such methods are usually unnecessary for typical culture needs. The appropriate monitoring frequency for water quality parameters is also depicted in Table 1.1.

Ideally, large fish experimental systems should be bioengineered to maintain water quality parameters within the ranges preferred by the species under study, not simply the ranges tolerated by the species. When parameter values deviate from the preferred range it is most often due to exceeding the loading densities beyond the carrying capacity of the system design. Similarly, if the experimental requirements demand more fish than the system can produce without compromising water quality, then the researcher must recognize that the system has not been engineered appropriately. A brief discussion of the water quality elements of greatest influence in a fish culture environment follows.

Dissolved oxygen and oxygen saturation

Dissolved oxygen is the most critical element within culture systems for fish. With adequate oxygenation, growth will be maximized and densities may even be extended beyond advisable limits – with careful

TABLE 1.1: Preferred water quality limits for experimental culture of fish[a]

Water quality factors	Stenohaline, warm water	Stenohaline, cold water	Euryhaline species	Marine species	Monitoring frequency
Conventional parameters					
Dissolved oxygen (mg/L)	>5	>7	>5	>5.5	daily
Oxygen saturation (%)	100	100	100	100	daily
Nitrogen gas saturation (%)	<102	<100	<100	<100	weekly
Total gas saturation (%)	<110	<110	<110	<110	weekly
Total suspended solids (mg L^{-1})[b]	<2000	<55	SS[c]	SS[c]	weekly
Total dissolved solids (mg L^{-1})	<400	<400	NA[d]	NA[d]	weekly
Turbidity (in NTUs)	SS[c]	<20	<50 (SS)[c]	(SS)[c]	weekly
pH (0 to 14 log scale)	7.5–9.0	6.5–8.2	7.5–8.5	8–8.5	twice a week
Alkalinity (as mg L^{-1} CaCO$_3$)	50–400	10–400	NA[d]	NA[d]	twice a week
Hardness (as mg L^{-1} CaCO$_3$)	50–400	20–400	NA[d]	NA[d]	twice a week
Total ammonia (mg L^{-1})	<0.5	<0.5	<0.5	<0.3	twice a week
Un-ionized ammonia (mg L^{-1})	<0.02	<0.0125	<0.0125	<0.0125	twice a week
Nitrite (mg L^{-1})	<0.1	<0.2	<0.2	<0.2	twice a week
Hydrogen sulphide (µg L^{-1})	<0.002	0	0	0	weekly
Carbon dioxide (mg L^{-1})	<10	<5	<10	<10	weekly
Salinity (g L^{-1})	0.1–3.0	0.1–3.0	0.1–35	28–35	weekly
Chloride (dissolved)	10 to 1000	10 to 1000	NA[d]	NA[d]	quarterly
Chlorine (total residual, in µg L^{-1})	11	11	7.5	7.5	weekly
Temperature (see also Figure 1.13)	20–32 (SS)[c]	9 to 18 (SS)[c]	SS[c]	SS[c]	daily
Total metals[e]					
Aluminum	<87	<87	<87	<87	quarterly
Arsenic (1.0)	<190	<190	<36	<36	quarterly
Cadmium[f] (0.994)	<1	<1	<9.3	<9.3	quarterly
Chromium (III)	<180	<180	NA[d]	NA[d]	quarterly
Chromium (VI) (0.993)	<10	<10	<50	<50	quarterly
Copper[f] (0.83)	<11	<11	<3.1	<3.1	quarterly
Lead[f] (0.951)	<2.5	<2.5	<8.1	<8.1	quarterly
Mercury (0.85)	<0.012	<0.012	<0.025	<0.025	quarterly

Nickel[f] (0.990)	<160	<160	<8.2	<8.2	quarterly
Selenium (0.998)	<5	<5	<71	<71	quarterly
Silver[f] (0.85)	<0.4	<0.4	<0.4	<0.4	quarterly
Zinc[c] (0.946)	<100	<100	<81	<81	quarterly

[a]Tolerance levels may exceed these limits, especially with adaptation. See Parker and Davis (1981), Piper et al. (1982), and Moe (1992) for further information on water quality requirements of warmwater, coldwater, and marine fish, respectively.

[b]Highly variable, dependent upon particle size and angularity and species; see Newcombe and Jensen (1996) for review.

[c]Species-specific, range too broad to state definitively.

[d]Not applicable or relevant to the water type.

[e]All values from US Code of Federal Regulations for chronic waterquality criteria (EPA, 1996). Parentheses following some listings specify conversion factor to express euryhaline and marine criteria as dissolved metals fractions.

[f]Hardness-dependent criteria; values represented are based on 100 mg L^{-1} CaCO$_3$.

monitoring. Dissolved oxygen saturation decreases with increasing temperature, altitude and salinity (Table 1.2). The absolute requirement for oxygen in fish is driven ultimately by the partial pressure differences between fish blood and the dissolved oxygen concentration in the water – equilibrated to the temperature and atmospheric pressure of the culture environment. Fish use of dissolved oxygen is maximal and independent of environmental oxygen concentration when the partial pressure of oxygen (pO_2) is sufficiently high (Downey and Klontz, 1980). Not surprisingly, the blood of warmwater fish has a higher affinity for oxygen than that of coldwater fish, reflecting the generally lower dissolved oxygen concentrations of their natural environments. Venous pressure of oxygen (pO_2) in fish blood is lower than that in the water (e.g. pO_2 fish = 50–110 mmHg, vs. water pO_2 = 154–158 mmHg at sea level). Thus, a concentration gradient is established that facilitates the transfer of oxygen into fish blood. For trout, and presumably other related salmonids, a minimum pO_2 of 118 mmHg in water is required to prevent hypoxia (Forteath, 1988).

Diurnal fluctuations of dissolved oxygen will be observed in static pond systems used to culture large fish, and may also be observed in recirculating systems (Boyd, 1979). Such diurnal fluctuations require forethought. Feeding increases oxygen demand by 200% (Timmons et al., 1991). The amount of food fed will also create both biological oxygen demand (BOD) and chemical oxygen demand (COD). During the daytime there will be a gradual consumption of CO_2 as a result of plant photosynthesis in such pond systems. The result of this effect is to drive the carbonic acid/bicarbonate/carbonate cycle toward the production of carbonate, thus raising the pH, and increasing dissolved oxygen. At night, photosynthesis ceases, oxygen is consumed by both the fish and the plants, and CO_2 is not consumed, thus driving the carbonic acid cycle toward the production of carbonic acid, thereby lowering the pH. Oxygen consumption in the absence of oxygen production will obviously reduce the amount available for respiration. Therefore, an adequate supply must be maintained to support the fish, plants and COD/BOD requirements. Biological oxygen demand will vary with feed type. For example, pelleted feed with 25% protein and 10% water exhibits a BOD of approximately 140 g O_2 kg^{-1} every 24 h at 30°C (Little and Muir, 1987); these researchers provide BOD measurements for a variety of potential feedstuffs.

Temperature

The preferred temperature of most species is generally 1 to 3°C within the temperature at which maximum growth can be achieved (Timmons et al., 1991) (Figure 1.13). Similarly, the optimum temperature for growth generally far exceeds the lower limit of tolerance for nearly all large fish species commonly cultured. Spawning temperatures are usually slightly lower than the optimum temperatures for growth, and incubation temperatures in temperate species that spawn on a decreasing temperature cycle are generally lower still, reflecting the natural conditions under which the fish spawn and their resultant eggs incubate. As with many other water quality parameters, fish may be acclimated to temperatures outside their normal preference and/or maximum growth range without undue stress. However, should the experimental set-up require this, acclimation must be gradual, with no more than a maximum change of 1°C h^{-1}, and a maximum of 10°C in a 24-h period. Temperature limits that will cause mortality

TABLE 1.2: Effect of temperature on saturated oxygen solubility (mg L^{-1}) in fresh water and sea water[1]					
Temp (°C)	Fresh water @600 m	Fresh water @sea level	Sea water @20 ppth[2]	Sea water @30 ppth[2]	Sea water @35 ppth[2]
10	10.5	11.3	9.9	9.3	9.0
15	9.4	10.1	8.9	8.4	8.1
20	8.4	9.1	8.1	7.6	7.3
25	7.6	8.2	7.4	6.9	6.7
30	6.9	7.5	6.5	6.4	6.2

[1]Freshwater data obtained from Piper et al. (1982) and Redacliff (1988). Saltwater data obtained from Moe (1992).
[2]Parts per thousand (ic, g L^{-1}).

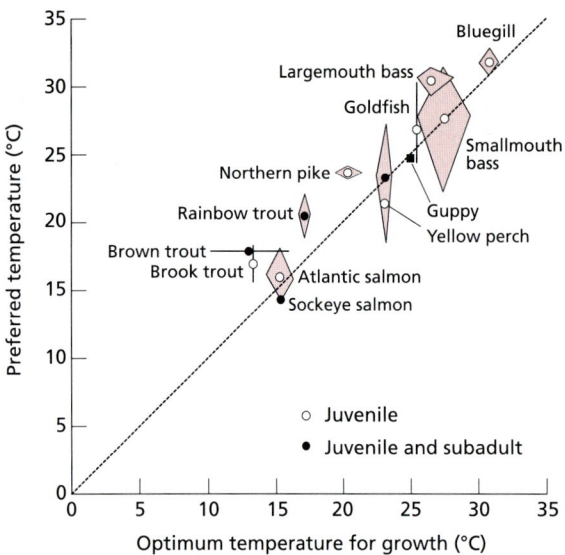

Figure 1.13 Temperature preferences for a variety of fish species cultured for research. The guppy datum represents a small fish species included for comparison. Temperature ranges and mean averages are illustrated. (Figure reproduced by permission from M. Timmons, Cornell University.)

should be recognized for each species under culture. In general, fish tolerate more abrupt increases in temperature than they do sudden decreases. Tropical fish will die at temperatures around 10°C, and have a relatively narrow temperature tolerance range. Warmwater fish, such as members of the Centrarchidae and Ictaluridae, tolerate the widest temperature variation, surviving at temperatures slightly above freezing to nearly 40°C. Coldwater species such as the salmonids will tolerate temperatures just above freezing for a longer period than warmwater fish; however, like tropical fish, they have a relatively narrow range of tolerated temperatures.

It may be necessary to heat or chill the water source to achieve a temperature preferred by the species under study. In general, water heating or cooling is not practicable at exchange rates that exceed 2% per day (Timmons et al., 1991). Cooling is generally more expensive than heating water. The percent of hot water (%h) from a heated water source required for mixing with a coldwater source to achieve an appropriate intermediate temperature (T_{mixed}) can be calculated as follows:

$$\%h = [(T_{mixed} - T_{cold})/(T_{hot} - T_{cold})] \times 100 \quad [1]$$

Thus, if one sought 15°C mixed water, and had a hotwater source of 50°C, a coldwater source of 5°C, and a total system demand of 38 L min^{-1}, the %h would be:

$$\%h = 15 - 5/50 - 5 = 22\%$$

To determine the heating requirement in British Thermal Units (**BTUs**), one may use eqn [2] after first determining the system demand based upon carrying capacity and loading density:

$$BTU/h = (\%hot \times system\ demand) \\ \times 8.33 \times 60 \times \Delta T \quad [2]$$

For the above example, the size of heater required would be calculated as follows:

$$BTU/h = (0.22 \times 38) \times 8.33 \times 60 \times (50 - 5)$$
$$= 189\,924\ BTU$$

Thus, a heater generating at least 190 000 BTUs would be adequate for a large fish system with such a flow requirement.

Alkalinity and hardness

Alkalinity refers to the buffering capacity of the water in which the fish are being held. It is determined by the proportionate concentrations of bicarbonate (HCO_3^-), carbonate (CO_3^{2-}) and hydroxide (OH^-) ions. As alkalinity increases, so too does the ability to absorb hydrogen ions without a measurable suppression in pH. Natural waters vary widely in alkalinity, from below 50 mg L^{-1} as calcium carbonate ($CaCO_3$), to over 1000 mg L^{-1}. Species tolerances and preferences have not been formally addressed *per se*; however, optimal alkalinity for fish culture is between approximately 100 and 500 mg L^{-1} as $CaCO_3$.

Alkalinity is consumed in nitrification; thus, in recirculating systems it is especially important to monitor and replace alkalinity, as necessary. Bisogni and Timmons (1991) report that 1.98 mol of HCO_3 are consumed per mole of NH_4 oxidized. Without replacement, pH may drop below levels where effective biological filtration occurs (pH 6.5 to 7.5). Bisogni and Timmons supplement alkalinity with sodium bicarbonate (Na_2CO_3) at 5% of the feeding rate, incorporating the sodium bicarbonate into the feed. Other sources of alkalinity that can be used include sodium hydroxide (NaOH), sodium carbonate (Na_2CO_3), calcium carbonate ($CaCO_3$), lime (CaO), dolomite ($CaMg(CO_3)_2$), magnesium carbonate ($MgCO_3$), and magnesium hydroxide ($MgOH_2$).

Hardness is a direct measure of the calcium and magnesium salts in a water body, and is also expressed as $mg\,L^{-1}$ $CaCO_3$. Since the carbonate that contributes to alkalinity is usually from the same source as that which creates hardness, hardness and alkalinity values are often equivalent. High hardness values are often associated with spring waters that arise from limestone sources. Low hardness values are often associated with granitic rock. Excessively hard waters could theoretically result in nephrocalcinosis, the formation of calcium crystals in the nephrons of the fish kidney when the calcium-to-magnesium ratio is too high. However, this condition is more often caused by excessive dietary calcium, or an enriched dietary Ca : Mg ratio. For example, 0.1% Mg prevented dietary nephrocalcinosis in rainbow trout fed 4% Ca (Ferguson, 1988). Soft waters of $11\,mg\,L^{-1}$ $CaCO_3$ caused white-spot disease (i.e. coagulated yolk disease) in brook trout larvae (MacKinnon, 1969). Hardness values above $500\,mg\,L^{-1}$ $CaCO_3$ caused excessive egg mortality in a New York hatchery, forcing the abandonment of that system for further salmonid culture (H. Simonin, NY Department of Environmental Conservation (DEC), personal communication); however, once hatched, waters of high hardness generally result in improved growth performance.

Carbon dioxide

Carbon dioxide (CO_2) is respired by fish and plants, and is also a gas by-product of nitrification. In water, CO_2 becomes readily hydrated to yield carbonic acid (H_2CO_3). Carbonic acid is in equilibrium with bicarbonate (HCO_3^-) and carbonate (CO_3^{2-}) ions, with the following reactions:

$$CO_2 + H_2O \longleftrightarrow H_2CO_3 + OH^-$$

$$H_2CO_2 + OH^- \longleftrightarrow HCO_3^- + H_2O$$

$$HCO_3^- + OH_- \longleftrightarrow CO_3^{-2} + H_2O$$

If a system has inadequate buffering or aeration, the production of CO_2 will drive the bicarbonate/carbonate/carbonic acid cycle toward the production of carbonic acid, and ultimately depress the pH. Fish also have specific tolerances for CO_2. Piper et al. (1982) report that CO_2 in excess of $20\,mg\,L^{-1}$ is harmful to fish. Chronically high CO_2 (~10 to 20 ppm) can result in nephrocalcinosis, whereas extremely high levels (>100 ppm) lead to anesthesia (Ferguson, 1988).

pH

The pH defines the acidity of the water. It is controlled by the carbonic acid cycle, the buffering capacity of the water, and the amount of CO_2 liberated into and out of the water via respiration and aeration, respectively. Fish tolerate an exceptionally broad range of pH (4.5 to 10.5), but culture conditions should be maintained between approximately 6.5 and 7.5 in fresh water, and between 8 and 8.5 in salt water. Maintenance of pH within acceptable ranges is especially problematic in saltwater systems, where routine buffering with alkalinity agents is often required. The use of calcareous filtrants such as dolomite, oyster shell and limestone will not prevent the decline in pH below preferred pH ranges for saltwater culture, but will prevent pH values from becoming acidic (Bower et al., 1981).

Total suspended solids

Solids in fish culture environments are represented in both dissolved and particulate form. Dissolved solids are principally organic compounds, usually proteins from undigested fecal matter or solubilized uneaten food. Particulate solids are settleable-sized particles of organic or inorganic nature that arise from fecal matter or uneaten food, or that occur naturally in the culture water (e.g. colloids). Both dissolved and particulate suspended solids contribute to turbidity, which is usually reported as an index of light absorption (nephelometric turbidity units – NTUs) or light scatter (Jackson turbidity units – JTUs). Neither of these optical techniques for expressing suspended solid loads provides information on the physical characteristics of the solids, such as distinguishing between angular or colloidal particles. Physical characteristics are usually described as the size, shape and weight of the suspended particles. An understanding of the physical, chemical and optical characteristics of suspended solids in the culture environment is important, as these factors can reflect the proportionate contribution of natural geologic components to turbidity relative to organic contributions.

Tolerance levels for suspended solids vary within and between species, dependent on the components that create the turbidity, and on species preferences (Table 1.3). For example, carp, catfish and tilapia prefer culture environments with moderate turbidity, whereas salmonids prefer much cleaner water. Suspended solids principally affect fish behavior, but

TABLE 1.3: Percentage of free ammonia with pH and temperature in fresh water and sea water[1]									
	pH								
Temp (°C)	6 Fresh	7 Fresh	7.5 Fresh	7.5 Salt	8 Fresh	8 Salt	8.5 Fresh	8.5 Salt	9 Fresh
5	0.0125	0.125	0.394	0.332	1.23	1.04	3.8	3.22	11.1
10	0.0186	0.19	0.59	0.481	1.83	1.51	5.55	4.61	15.7
15	0.0274	0.27	0.85	0.697	2.65	2.17	7.98	6.56	21.5
20	0.0397	0.4	1.24	1.01	3.83	3.12	11.18	9.24	28.4
25	0.057	0.55	1.73	1.46	5.28	4.47	14.97	12.9	36.3
30	0.081	0.8	2.48	nd	7.36	nd	20.3	nd	46.6

[1]Freshwater data obtained from Piper *et al.* (1982) and Redacliff (1988). Saltwater data obtained from Bower and Bidwell (1978) for salinities between 28 and 31 g L^{-1}.
nd = not determined.

also can be acutely lethal at extreme levels or cause sublethal stress. In salmonids, sublethal effects of reduced growth, gill flaring, coughing, reduced prey capture efficiency and habitat displacement have been documented at turbidities ranging from 20 to 265 NTUs, corresponding to TSS concentrations of 190 to 3000 mg L^{-1} (Berg and Northcote, 1985; Servizi, 1988; Sigler, 1988). Species with preferences for higher turbidity may be more susceptible to sunburn. Following the eruption of Mt St Helens, Newcombe and Flagg (1983) showed that juvenile salmonids can tolerate ash loads up to 6100 mg L^{-1}, although Emmett *et al.* (1988) demonstrated that juvenile salmonids shifted their prey preference from benthic amphipods to pelagic copepods. In the natural environment, there appears to be a window of turbidity preference wherein some turbidity facilitates feeding activity by reducing predation influence and increasing prey contrast (Hester, 1968; Gregory and Levings, 1998), but at excessive levels, visibility is reduced to a point where foraging is compromised (Hester, 1968).

In the culture environment, reduced growth is the most commonly observed manifestation of excessive suspended solids, occurring primarily from reduced feeding efficiency; however, reduced growth may be a manifestation of pathological stress caused by the turbidity. Turbidity, especially that caused by inorganic particles of angular form, can abraid gill lamellae, causing osmoregulatory stress and promoting bacterial gill disease. When the suspended solid loads are dominated by uneaten food and fecal matter, bacterial and fungal populations increase, and some normally non-pathogenic organisms can cause disease. For example, Redding and Schreck (1987) found that infection occurred among yearling steelhead exposed to 2500 mg L^{-1} of topsoil in a 48-h exposure.

Solids in culture water are especially problematic during egg incubation. High percentages of fine particles less than 0.85 mm are reported to reduce oxygen transport through the intragravel incubation environment, and to delay development or cause mortality by suffocation (Ziebell, 1960). Bjornn and Reiser (1991) report that intragravel salmonid embryo survival declines significantly when the percentage of fines less than 6.35 mm increases above approximately 20%, with species that have larger eggs, such as chinook salmon, tolerating the highest concentrations of fines (~28%).

In salmonids, approximately 0.4 lb of solids are produced per pound of food fed (400 g kg^{-1}) (Timmons *et al.*, 1991). Thus, it is critical that appropriate technology is incorporated into the culture environment to remove these solids to a level that does not cause stress or result in excessive bacterial concentrations. Bacterial gill disease has been shown to be directly related to the feeding activity, thus, overfeeding can be especially problematic (MacPhee *et al.*, 1995). Solids removal should reflect the size and settling rate of the solids to be encountered. Large solids (>200 microns (μm)) can be settled out through a settling basin. **Foam fractionators** are effective at removing solids smaller than 30 μm, and they are also useful at removing high molecular weight dissolved solids. Solids can also be removed by screening. Ultimately, the prevention of excessive suspended solids in a system is the best method to control these potential stressors, and this is done primarily by not overfeeding the fish, and by using appropriate tank designs (e.g. conical and circular tanks that take advantage of the Coriolis effect).

Ammonia metabolites, nitrification and biological filtration

Ammonia is the primary metabolite of fish, and, in its un-ionized state, can be highly toxic to fish if permitted to accumulate. Elevated ammonia is the primary cause of environmental gill disease, manifest in fish as a hypertrophy and subsequent hyperplasia of the gill lamellae, which reduces respiratory exchange capacity and osmoregulatory function (Klontz et al., 1980). Ammonia toxicity works in concert with high suspended-solid loads to promote bacterial gill disease.

The ammonia ionization state in fresh water is regulated by temperature and pH, where the percentage of un-ionized ammonia (the toxic form) increases as a multivariate function of increasing pH and temperature (Table 1.3). In sea water, pH and temperature have similar influences on the ionization state of ammonia, but there is a dampening effect asserted by increasing salinity. Tables depicting these relationships, as provided in Piper et al. (1982), and Bower and Bidwell (1978) should be consulted when evaluating water quality monitoring results from fresh water and salt water, respectively.

Biological filtration is the principal means through which ammonia is removed from fish culture waters, although ammonia can also be stripped from waters via ion exchange. The latter technique involves passing the waters through a bed of zeolite crystals (e.g. clinoptilolite) where the ammonia is trapped and subsequently released as gas (Piper et al., 1982). Biological filters provide the substrate necessary for colonization by bacteria that oxidize ammonia. Biological filtration is an essential element of recirculating systems that support large or small fish, as both ammonia and nitrite can be highly toxic to fish. Flow-through systems may also require biological filtration of the effluent to ensure compliance with water quality criteria of regulated discharge (Figure 1.1).

The nitrification process is driven by two genera of bacteria, *Nitrosomonas* sp. and *Nitrobacter* sp. Nitrate, the end-product of this process, is ultimately utilized by plant life and is essentially non-toxic to fish. The oxidation (nitrification) of ammonia to nitrate is a two-step process, as follows:

$$2NH_4^+ + 3O_2 \rightarrow 2NO_2 + 2H_2O + 4H^+$$
$$(\textit{Nitrosomonas sp.})$$

$$2NO_2 + O_2 \rightarrow 2NO_3 \ (\textit{Nitrobacter sp.})$$

Nitrosomonas and *Nitrobacter* are aerobic **autotrophs** (i.e. they require oxygen and obtain their cellular carbon from carbon dioxide), and are substrate dependent. Thus, the primary design element of a biological filter provides substrate for attachment of these '**nitrifiers**'. A variety of biological filter media may serve as substrate, including the following: zeolite, crushed oyster shell, sand, dolomite, pea gravel, diatomaceous earth, and synthetic substances such as styrofoam beads ('bio-balls'), plastic rings ('bio-rings'), or fiber filter. The efficiency of these substances to support nitrifiers is proportional to their surface area-to-volume ratios. Thus, smaller particles provide greater surface for bacterial attachment; however, very fine particles, such as sand, clog easily, and have high maintenance and risk associated with them as a result. Surface area-to-volume ratios for a variety of typical media are provided in Moe (1992). The area requirements for the media are calculated as a function of feeding rate.

Biological filters function best with adequate oxygenation and food supply (i.e. ammonia and nitrite), although nitrifiers deprived of food for several weeks can regain the majority of their oxidizing potential within 48 h (Bower and Turner, 1983). Biological filtration is limited below dissolved oxygen levels of 2 mg L^{-1} (Timmons et al., 1991), but above this range there is little alteration in the oxidation rate. Since oxygen levels of 2 mg L^{-1} would be lethal for nearly all fish species cultured, the practitioner employing a recirculating system will have greater worries than the biological filter should the dissolved oxygen level decrease to a level that could impact nitrification! Nevertheless, the design of the biofilter should reflect the following design parameters, with respect to the anticipated loading density of the fish in the system:

- The oxidation of 1 mg ammonia consumes approximately 4.6 mg of oxygen (i.e. nitrogenous oxygen demand = 4.6).
- The oxidation of ammonia produces acid that consumes approximately 7.1 mg L^{-1} alkalinity (as CaCO$_3$).
- Approximately 0.4 lb of oxygen is consumed per pound of food fed (i.e. 400 mg O$_2$ per gram of food).

Nitrifying bacteria are sensitive to sudden pH changes, and do not generally grow well outside the range of 7.2 to 8.5. Their growth can also be affected by chemicals used for disease treatment. In recirculating seawater systems for example, nitrifying bacteria were inhibited by 8 mg L^{-1} methylene blue, 66.7 mg L^{-1} neomycin sulfate, 13.3 mg L^{-1} chloramphenicol and 1.2 mg L^{-1} copper sulfate; nitrification was not significantly inhibited by gentamycin sulfate at 5.3 mg L^{-1}, nifurpirinol at 0.1 mg L^{-1} and quinacrine hydrochloride at 12 mg L^{-1} (Bower and Turner, 1982). In freshwater recirculating systems, erythromycin and methylene blue stopped nitrification at therapeutic concentrations, but formalin, malachite green, copper sulfate, potassium permanganate, sodium chloride, chloramphenical oxytetracycline, and sulfameraxine and nifurpirinol had no effect (Collins *et al.*, 1975, 1976).

Nitrogen and oxygen gas supersaturation

Supersaturated nitrogen and oxygen gas dissolved in water can result in gas bubble disease (Rucker, 1972). Smaller fish are more susceptible than large fish, although all life stages can be affected. Marine fish larvae exhibit greater sensitivity to gas bubble disease than salmonids (Colt *et al.*, 1987). This sensitivity may reflect conditions in their natural environment, where nitrogen gas supersaturation would be less likely to occur in the habitats they would generally utilize. In the culture environment, nitrogen supersaturation is a condition most often associated with well water taken from depth, but can also be a problem if using municipal water taken from deep within a lake system, or from waters derived from a water tower. Gas bubble disease can be lethal to all species of fish; thus, it is advisable to engineer your system so that the water is aerated before entering units holding fish. Degassing towers are commonly employed for this purpose, in which water is cascaded through a closed column packed with a porous substrate that will disrupt the water, permitting the escape of supersaturated nitrogen gas, and also facilitating oxygenation of the water, if oxygen levels are not at saturation (Figure 1.1) (Marking *et al.*, 1983). Packed columns are capable of reducing nitrogen saturation from approximately 130% down to near 100% saturation (Owsley, 1981). Alternatively, incoming water can be sheeted over a flat surface to create a thin microlayer.

Both methods facilitate the release of trapped gases, although sheeting is not as efficient.

Other toxicants

Other noteworthy toxicants potentially associated with waters used for fish culture include naturally occurring hydrogen sulfide, chlorine from municipal water treatment, heavy metals, synthetic organic pollutants derived from agricultural and industrial practices, and chemotherapeutics sometimes needed to treat disease. Values of metals generally regarded as safe for fish culture are provided in Table 1.1. Organic pollutants should not persist at detectable levels in fish culture waters used for experimental purposes. Chemotherapeutics are discussed in Chapter 4 and should be used only as necessary. Appropriate elimination time should be factored into experimental design if such chemicals must be used.

In most cases, the installation of an activated carbon unit will be sufficient to eliminate such potential contaminants from culture waters before entering the system. Activated carbon filtration units can be employed either before or after incoming waters have gone through a degassing unit in flow-through systems (Figure 1.2). In recirculating systems, it is imperative to prefilter solids before waters are passed through an activated carbon filter, to prevent rapid clogging.

Carrying capacity and loading density

Without question, the single most important factor influencing the overall health of cultured fish stocks is the total **biomass** housed within the husbandry unit – the loading density. As previously discussed, the biomass within a system often regulates the dissolved oxygen in a system, as well as the pH (via the carbonic acid/carbonate/bicarbonate equilibrium) and ammoniacal metabolites. The biomass that can be contained within a system without compromising water quality, fish nutrition, or fish health is referred to as the carrying capacity. Under most intensive culture situations, dissolved oxygen or ammonia waste regulate carrying capacity. Under experimental systems for large fish, such densities should be seldom realized, and carrying capacity will most often be determined by the density of fish that does not result in behavioral and/or physiological stress. Whichever method is used, a properly managed system will not load fish to a density that exceeds its carrying capacity.

Carrying capacity is a function of water flow, tank volume, exchange rate, rearing temperature, dissolved oxygen, species of fish under culture, and the production and processing of metabolic wastes. The bioenergetic assumptions used to estimate carrying capacity and loading density in salmonids were recently found to be applicable to other species as well (Kaushik, 1998). Thus, the following discussion largely reflects salmonid experience. Ultimately, the researcher must determine whether the facility will be adequate to accommodate the species under study at the time when the facilities will be most restrictive – usually when the biomass is the greatest. For example, a system strictly designed to culture 1000 fingerling striped bass would be inadequate to maintain those same fish one year later, after their size will have increased several-fold. Similarly, a system designed to hold 1000 trout yearlings at a water temperature of 8°C would likely prove stressful at that same density if the ambient water temperature was 18°C. Thus, system design must reflect the maximum biomass and water quality expected over the course of the planned experiments.

Given the previous discussion on water quality parameters important for fish culture, several important rules with respect to carrying capacity deserve emphasis:

- Carrying capacity is reduced with increasing water temperature, due primarily to the reduction in pO_2 with increasing temperature (Downey and Klontz, 1980). For example, oxygen consumption in yearling rainbow trout increases from 3 mL O_2 h^{-1} to 12 mL h^{-1}, when the water temperature is raised from 7.2°C to 20°C (Leitritz and Lewis, 1980).
- A system will generally accommodate a greater total weight of larger fish than smaller fish (i.e. fish biomass can be increased with fish size).
- The amount of oxygen consumed is proportional to the amount of food fed. Uneaten food reduces the carrying capacity through BOD. Effects of BOD are more problematic in recirculating systems.
- When water quality (i.e. dissolved oxygen or ammonia waste) does not limit carrying capacity, it is not uncommon that loading density will – due to behavioral interactions that may lead to stress at high density. In general, gregarious species (e.g. flatfish) generally tolerate higher densities than territorial species.

Although carrying capacity is a multifunction property of the water quality and spatial factors in a rearing unit, it is generally expressed as either a density measure (e.g. lb/ft³, kg/ha) – the *density index*, or in units of weight per unit flow – the *flow index*. Generally, the density index is most useful for static or recirculating systems, whereas the flow index is more often applied to flow-through systems. Carrying capacity may also be calculated on the basis of feeding rates – ultimately a factor driven by metabolic waste production (Haskell, 1955; Westers *et al.*, 1978). All means of estimating carrying capacity should be calculated and the most conservative estimate should not be exceeded under research applications. As a rule-of-thumb, the density of fish to be loaded into a system should consider the final weight of the fish anticipated at the conclusion of the experiments.

Flow index (F)

The flow index considers oxygen consumption the primary driver of carrying capacity. It is therefore advisable to first establish empirically the weight of fish that could be cultured within a unit without depressing oxygen concentrations below the preferred level for the species under culture. For salmonids consuming diets with approximately 40% protein, approximately 0.22 lb of oxygen are consumed per pound of food fed (i.e. 220 mg O_2 per gram of food). Non-feeding salmonids will consume approximately 7 mg O_2 per gram of fish per day (Speece, 1973). Thus, the oxygen consumption can be estimated without initial empirical determination. In a recirculating system, the additional oxygen consumption of the biological filter (i.e. BOD = 400 mg O_2 per gram of food) must also be considered if the flow index is to be used to estimate carrying capacity.

Once the flow index is determined (eqn [3]), the capacity of the same unit to house the same species at a different size class can be derived by solving the flow index eqn [3] for W (Piper *et al.*, 1982).

$$F = W/(L \times I) \quad [3]$$

where F = flow index in pounds fish per gallon per minute flow (or kg L^{-1} min^{-1}); W = known weight in pounds (or kg); L = length in inches (or cm); I = water inflow, in gallons per minute (or L min^{-1}).

A general rule-of-thumb for salmonids loadings, measured in kg L^{-1} min^{-1} = mg dissolved oxygen available/(2× % body weight) (Meade, 1988). This should result in an empirical reduction of dissolved oxygen in the rearing unit of 25 to 30% between the influent and effluent water. Meade (1988) specifies that under such a system (usually recirculating), aeration between tanks is imperative, so that oxygen saturation is at or above 90% before waters are passed into the next tank in the series. The maximum loading determined under this approach is that where the specific growth rate is reduced to a preselected minimum, which should correspond with the limits for dissolved oxygen use without stress to the fish. Meade (1988) also recommends full water exchange rates of three times per hour; if such an exchange rate is not feasible because of water or power limitations, supplemental infusions of pure oxygen may be used to increase loadings. To this end, Dwyer and Peterson (1993) describe the implementation of a low head oxygenator that increased carrying capacities of outdoor raceways by a factor of 2.2. Design procedures for pure oxygen infusion systems are provided by Colt and Watten (1988).

Density index (D)

The density index considers behavioral and pathological stress from overcrowding as the limiting factor in the carrying capacity of a husbandry unit for fish (Wedemeyer and Wood, 1974). When oxygen is not limiting, overcrowding stress is usually the limiting factor. The density index can be calculated from eqn [4]:

$$D = W/(V \times L) \qquad [4]$$

where D = density index in lb ft^{-3} in^{-1}; W = weight of fish (in pounds); V = volume of unit (in ft^3); L = length (in inches).

Although the density index assumes constancy with increasing weight, empirical evidence from aquaculture indicates that a greater biomass can be loaded into a system as fish size increases (Piper et al., 1982). For experimental systems, the acceptance of a constant density index will ensure that overcrowding does not occur.

Ammonia metobolite production

The preceding discussion of flow and density indices highlights how carrying capacity estimates should still be confirmed with empirical testing. The third method for determining carrying capacity reflects such an empirical model wherein ammonia concentrations are monitored, and the total biomass cultured is kept below a level that will exceed ammonia toxicity thresholds for the system, as detailed in Table 1.1. Meade (1988) modified this approach to determine production limits according to when ammonia metabolite build-up resulted in reduced growth.

Ammonia production (and hence carrying capacity) can be predicted on the basis of a calculated ammonia factor (Piper et al., 1982). The ammonia factor references the flow incoming and the amount of food fed per day (eqn [5]):

$$\text{Ammonia factor} = (\text{mg L}^{-1} \text{ total ammonia}) \times (\text{flow rate})/(\text{weight of food fed per day}) \qquad [5]$$

Ammonia factors tend to be very system and species specific. Thus, to establish the ammonia factor, one must first measure the total ammonia concentration in the tanks under existing operation several times during a 1–2 day period. Once the factor is calculated, the amount of total ammonia produced from a new stock that will be cultured within the same type of system can be estimated by rearranging eqn [5] above as eqn [6]:

$$\text{ppm total ammonia} = (\text{weight of food fed/day}) \times (\text{ammonia factor})/(\text{flow rate}) \qquad [6]$$

In recirculating systems, the estimation of carrying capacity on the basis of ammonia production carries significantly more risk than with flow-through systems simply because nitrifiers are sensitive to changes in environmental conditions, thus, the collapse of the biological filter can result in sudden, often toxic elevations of ammonia concentrations. Without some means to reduce loading density under such conditions, mortality can quickly ensue.

The health status of the fish under culture at the practised densities will ultimately provide the criteria as to whether the densities used are appropriate. McDonald and Milligan's research (1997) suggests that 'simple' responses to stress such as elevations in serum adrenaline and cortisol are inappropriate measures of ammonia-induced stress, and that 'compound' stress responses involving multiple physiological and hormonal indicators are what should be monitored to determine if fish health is compromised.

Empirically derived estimates of appropriate loading densities for different types of systems and species are provided in Table 1.4. Several rules-of-thumb regarding loading densities are commonly used in aquaculture and aquaria science; however,

TABLE 1.4: Empirical loading densities used for large fish culture

Species	Scientific name	Age/size	Rearing unit	Rearing method	Density	Reference
Trout	Oncorhynchus mykiss	fry/5 cm	raceway	recirculating	1 lb ft^{-3}	Timmons et al., 1991
		juv./20 cm	raceway	recirculating	4 lb ft^{-3}	Timmons et al., 1991
		adult/30.5 cm	raceway	recirculating	6 lb ft^{-3}	Timmons et al., 1991
		1 yr old	0.6 × 0.3 m net pens	flow-through; net pens in 3040 L circular tanks	11.1 g-fish L^{-1} cm^{-1} fish length	Kebus et al., 1992
Coho salmon	O. kisutch		raceway @17°C	flow-through	2 to 15 lb/gpm*	Piper et al., 1982
Chinook salmon	O. tsawytcha		raceway @17°C	flow-through	1.2 to 5.5 lb/gpm*	Lasordo and Timmons, 1994
Catfish	Ictalurus punctatus		ponds	static	up to 20 lb ft^{-1} (1.6 kg L^{-1})	
			ponds	static	5 to 10 fish ft^{-3} 160 000 to 240 000 kg m^{-3} s^{-1} flow	
		fingerling	0.04 ha earthen pond	static	12 500 to 50 000 fish ha^{-1}	Terhune et al., 1997
			ponds	static	100 000 to 200 000 fry acre^{-1}	Ghate and Burtle, 1993
		250 g	ponds	static	22 000 to 66 000 fish ha^{-1}	Lin and Diana, 1995
Catfish hybrid	Clarias macrocephalus × C. gariepinus	fingerling/ 13–17 g	cages	cages suspended in earthen ponds, w/ 6–7 g tilapia @ 26 g m^{-3}	275 fish m^{-3} cage (4125 g m^{-3}); 880 & 1760 fish/pond (120–240 g m^{-3})	
Largemouth bass	Micropterus salmoides	juveniles	ponds	static	30 to 150 lb acre^{-1}	Piper et al., 1982
		fry	ponds	static	50 000 to 75 000 fry acre^{-1}	Piper et al., 1982
Striped bass	Morone saxatilis	fingerling	ponds	fertilized ponds	50 000 ha^{-1} (median)	Anderson, 1993
		fingerling	ponds	fertilized ponds	75 000–125 000 fry acre^{-1}	Piper et al., 1982
		fingerling to juv.	12 500 L tanks	recirculating	36 to 144 fish m^{-3}	Nunley and Libey, 1992
Carp	Cirrhinus mrigala & rohu	fry/2–3.5 g	9580 L tanks	static or recirculating	1312 or 1642 fry/tank	Chakrabarti and Jana, 1998

TABLE 1.4: (Continued) Housing, Maintenance and Breeding — Facilities and Husbandry (Large Fish Models)

Species	Scientific name	Age/size	Rearing unit	Rearing method	Rearing method	Density	Reference
Northern pike/walleye	Esox lucius	fry	ponds	static		50 000 to 70 000 fish/acre	Piper et al., 1982
		juv./10 to 25 cm	ponds	static		5000 to 20 000 fish/acre	Piper et al., 1982
Red porgie	Pagrus pagrus	70 g	partitioned raceway	flow-through		1 kg fish/400 m^3	Kentouri et al., 1994
Gilthead sea bream	Sparus aurata	juveniles	sea cages			120 fish m^{-3}	Porter et al., 1986
Tilapia	Oreochromis sp.	fingerlings (8 g)	50 L aquaria	recirculating aquaria		130 fish/50 L (20 g L^{-1})	Cordova et al., 1996
Nile tilapia	Oreochromis niloticus	adult/141–152 g	4 m^3 cages	cages in earth ponds		30 to 70 fish m^{-3}	Yi-Yang et al., 1996
		juv./54–57 g	earthen ponds	free in ponds w/adults in cages in same pond		2 fish m^{-3}	Yi-Yang et al., 1996
		fingerling	tanks	self-contained, 4-zone, recirculating tanks		2440 fish 20 m^{-3}	Twarowska et al., 1996
Red tilapia	O. mossambicus × O. urolepis hornorum	juveniles	irrigation ditches	pulsed flow-through		10 to 70 fish m^{-3}, 20 fish m^{-3} was best performance	D'Silva and Maughan, 1995
		juveniles/75 g	concrete circular tanks	recirculating system		50, 100 and 200 fish m^{-3}, (1 kg L^{-1} min^{-1})	Suresh and Lin, 1995
Tilapia	Sarotherodon galilaeus	juveniles	42 L aquaria	recirculating system		35 to 70 fish/42 L aquaria	Hoener et al., 1987
Striped grey mullet	Mugil cephalus	juveniles	earthen ponds	static		188 kg m^{-3}	Cardona et al., 1996
Red drum	Scianops ocellatus		230 L tank	recirculating system		15.4 kg m^{-3} w/ozone, 12.9 kg m^{-3} w/ozone	

*gpm = gallons per minute.

these do not necessarily reflect eqn [3] or [4], and should therefore be used judiciously. These include:

- For low density, closed-system aquaria and similar recirculating units: 1 inch fish per gallon ($0.67\,\text{cm}\,\text{L}^{-1}$), or 1 inch/10 inch^2 surface area ($2.54\,\text{cm}/64.5\,\text{cm}^2$). Example: a 100-gal (378.5-L) unit could hold 50 2-inch (5-cm) perch.
- For high density flow-through systems, a density index of $0.5 \times$ fish length will avoid overcrowding (for trout). Example: 2-inch trout could be held at 1 lb/ft^3 (i.e. 2.2 kg/28 L); thus, a 10-ft^3 unit could hold 10 lb (i.e. 22 kg) of trout.

Specific model systems for large fish

In this section recent research findings specific to the culture of flatfish and *Morone* sp. are summarized. The preceding text represented numerous aspects of salmonid culture and other coldwater fish species routinely used as large fish models, and the tabular and graphical representations should also be examined for further information to species not specifically addressed. The flatfish and striped bass are discussed in detail here because they highlight different requirements and techniques for culture that are not necessarily practised for the culture of other large fish such as the Salmonidae, or typically pond cultured fish such as the Ictaluridae. These summaries are not meant to be exhaustive, but rather to provide representation of the variation in research findings between fish families typically cultured as large fish models.

Flatfish

Numerous flatfish species are grown in captivity. In the US, UK, Scandinavia, and southern Europe especially, the culture of Atlantic halibut (*Hippoglossus hippoglossus*) and turbot (*Scophthalmus maximus* L.) have been commercialized. In the US, the sole (*Solea solea*), winter flounder (*Pleuronectes americanus*), the witch flounder (*Glyptocephalus cynoglossus*), and the yellowtail flounder are also cultured (*Pleuronectes ferruginea*) (Nardi, 1998). Turbot and Japanese flounder (*Paralichthys olivaceus*) are the most intensively researched of the commercial species as summarized in a recent bibliography on flatfish culture (Stickney, 1997).

Recent research with greenback flounder (*Rhombosolea tapirina*) suggests that flatfish respond to stress similarly to other teleosts in captivity. For example, a rapid increase in plasma cortisol is sustained for at least 48 h if the fish are maintained under highly crowded conditions (Barnett and Pankhurst, 1998). Many species exhibit great tolerance for variable water quality conditions. For example, the southern flounder (*Paralichthys lethostigma*) and *Paralichthyes orbignyanus* can be cultured after metamorphosis at fresh and brackish water salinities (i.e. 0 to 20 parts/10^3) without compromising growth or survival (Daniels and Borski, 1998).

Tank culture and cage culture systems are primarily used for flatfish species (Nardi, 1998). For example, 30 L tanks were used to study flatfish sensitivity to acidity (Wasielesky *et al.*, 1997). Large diameter, shallow (~0.5 m), tanks with multiple inlets and slightly conical bottoms are reported to provide the best results for flatfish species (Cripps and Poxton, 1992).

Spawning is generally controlled by photoperiod and temperature manipulation (Nardi, 1996). Stripping must be matched very closely to ovulation timing otherwise hatching rates will be greatly reduced (Holmefjord, 1991).

Providing proper larval nutrition prior to metamorphosis is critical for flatfish culture. Flatfish mortality prior to metamorphosis is generally extreme, often exceeding 90%. Larval nutrition for flatfish requires the separate culture of *Artemia* sp., rotifers and copepods, along with the algal species that these zooplankton will consume; thus, in practice, a separate room is used for the culture of these food items (Nardi, 1996). Minced clam meat can also be used to support larval flatfish nutrition; however, should such natural foods be used in a research setting, the tissues should be analyzed for bioaccumulative contaminants. Sekai (1989) demonstrated that albinism was associated with the feeding of Brazilian *Artemia* nauplii and rotifers. Higher levels of albinism have also been reported from low water exchange rates (Sugimoto *et al.*, 1985). Following metamorphosis, the larvae may be weaned on to artificial, nutritionally balanced diets, at lengths of approximately 20 mm (Jones *et al.*, 1981). The continued observance of abnormal pigmentation of cultured flatfish stocks, along with a high frequency of early life stage deformities, suggests that the nutritional and water quality requirements for flatfish species are far from fully understood (Sekai and Matsumoto, 1994).

Striped bass and other Morone species

Striped bass are native to coastal areas of the eastern US, from the Canadian maritime provinces to Florida. They have also been introduced to the west coast of the US in the San Francisco Bay area. The northernmost spawning population along the east coast occurs in the Connecticut River, which drains into Long Island Sound. They are an **anadromous** species that exhibits behavioral divergence with latitude. Populations along the southern Atlantic coast reside primarily in rivers throughout their life, spending little time at sea. Populations further north spend a much greater amount of their life history in marine waters.

Culture techniques for striped bass and other Morone species have been recently compiled (Hochheimer and Wheaton, 1997). Spawning occurs in nature at temperatures of at least 14°C (Fay et al., 1983). Similar to flatfish, the production of young striped bass has associated high mortality. For example, natural egg mortality has been recorded at 80% and 94% per day on rivers in the southeastern US (Bulak et al., 1993), and, while artificial culture conditions improve upon nature's results, total egg mortality in excess of 50% is common (Geiger and Parker, 1985). Under experimental and aquaculture conditions, spawning is generally achieved by chorionic gonadotropin injection (275–200 IU/kg body weight for females; 110–165 IU/kg body weight for males) (Kerby et al., 1983). Within 2 to 4 days of injection eggs can be manually stripped, or adults can be permitted to spawn volitionally in a circular tank, with typical sex ratios of two females to four males. If eggs are manually stripped, they are mixed with semen and subsequently mixed with water. After this procedure they are suspended in McDonald hatching jars at a flow rate that will keep them in motion throughout incubation, at a density of roughly 100,000 per jar. If eggs are produced volitionally, the center standpipe must be screened to prevent egg loss, and velocity should be maintained in the tanks between 10 and 15 cm s^{-1}, with flow rates of approximately 30 to 38 L min^{-1} (Kerby et al., 1983). Hatching of eggs occurs within 1 to 3 days of spawning, dependent mostly on temperature.

Producing fry and fingerlings from eggs is the most challenging aspect of Morone culture. Typically, production of fingerlings is achieved by transferring 5- to 10-day-old prolarvae into ponds fertilized with no more than 300 μg L^{-1} nitrogen 2 to 3 days before transfer. Failure to inflate the swimbladder (usually occurring between days 4 and 6 post-hatch) is the most common cause of larval mortality during this period, and this problem is exacerbated when transfer to the ponds occurs on or later than day 7 post-hatch (Anderson, 1993). Prestocking pond fertilization stimulates planktonic growth to provide a natural food source. Optimal food levels for larval striped bass growth are between approximately 15 and 25 mesozooplankton per liter (Buchanan et al., 1994). Hatchery survival to the fingerling stage averages around 40% (Kerby et al., 1983; Geiger and Parker, 1985), and growth in ponds is generally density dependent (Nunley and Libey, 1992). Although fry and fingerling production as described above is most often practised for intensive aquaculture purposes, the principles can still be applied to small ponds set up for research.

The effects of water quality conditions on striped bass early life stage survival have been reviewed (Cheek et al., 1982; Geiger and Parker, 1985; Hall, 1991; Hall et al., 1993). Preferred values for pH range from 7 to 8.5, for temperature from 12 to 24°C, for salinity from 0.5 to 15 g L^{-1}, for hardness greater than 150 mg L^{-1} as $CaCO_3$, for alkalinity at least 20 mg L^{-1} as $CaCO_3$, and for dissolved oxygen at least 5.0 mg L^{-1}. Hall et al. (1988) also report, from *in-situ* cage culture experiments, that **prolarvae** are potentially stressed by monomeric aluminum concentrations of 0.150 mg L^{-1}, cadmium at 0.003 mg L^{-1} and copper at 0.04 mg L^{-1}. Larvae younger that 14 days post-hatch may die at pH values above 8.5 (Anderson, 1993). Their sensitivity to aquaculture chemicals has also been tabulated (Bills et al., 1993), with 96 h LC$_{50}$'s ranging from 0.129 mg L^{-1} for malachite green to 340 mg L^{-1} for erythromycin only the toxicity of chloramine T increased with increasing acidity (decreasing pH). Acclimation to high ammonia concentrations (0.15 mg L^{-1}) was not found to cause changes in hematologic or serum biochemistry beyond the normal range (Hrubec et al., 1997), although pathologically high serum creatinine and low serum chloride were observed in fish held at 200 mg nitrate per liter. Temperatures of 18°C and below will suppress the immune response in a manner similar to that found from exposure to relatively high nitrate levels of 200 mg L^{-1} (Hrubec et al., 1996).

Culture in intensive recirculating systems has been shown to induce gill lesions such a lamellar hyperplasia and fusion that may involve up to 75% of the gill filament (Smith et al., 1994). These lesions were reversible when the striped bass were transferred to clean water, but demonstrate how a lower density of fish would be advisable for experimental fish husbandry – especially under recirculating conditions.

Acknowledgments

The production assistance of Diane Shannon, Kathy Sitchin and Christopher Scott is greatly appreciated.

References

Anderson, R.O. (1993). *J. Appl. Aquacult.* **2**(3–4), 101–118.

Barnett, C.W. and Pankhurst, N.W. (1998). *Aquaculture* **162**(3–4), 313–329.

Berg, L. and Northcote, T.G. (1985). *Can. J. Fish. and Aquat. Sci.* **42**(8), 1410–1417.

Bills, T.D., Marking, L.L. and Howe, G.E. (1993). US Fish and Wildlife Service, Natl. Fish. Res. Cent, La Cross, Wisconsin.

Bisogni, J.J. and Timmons, M.B. (1991). Control of pH in closed cycle aquaculture systems. In *Proceedings of the Engineering Aspects of Intensive Aquaculture*, pp. 333–348. NRAES-49, Riley Robb Hall, Cornell University, Ithaca, New York.

Bjornn, T.C. and Reiser, D.W. (1991). American Fisheries Society Special Publication, **19**, 83–138.

Blake, R.W. (1983). *Fish Locomotion*. Cambridge University Press, Cambridge.

Blancheton, J.P. and Coves, D. (1993). *Production*. Ghent Belgium European Aquacult. Soc., vol. 18, pp. 87–94.

Bower, C.E. and Bidwell, J.P. (1978). *J. Fish. Res. Bd. Can.* **35**, 1012–1016.

Bower, C.E. and Turner, D.T. (1982). *Aquaculture* **29**, 331–345.

Bower, C.E. and Turner, D.T. (1983). *Aquaculture* **34**, 85–92.

Bower, C.E., Turner, D.T. and Spotte, S. (1981). *Aquaculture* **23**, 211–217.

Boyd, C.E. (1979). *Water Quality in Warmwater Fish Ponds*. Auburn University, Agricultural Experiment Station, Auburn, Alabama.

Brett, J.R., Hollands, M. and Alderdice, D.F. (1958). *J. Fish. Res. Bd. Can.* **15**, 587–605.

Brown, E.E. and Gratzek, J.B. (1980). *Fish Farming Handbook*. AVI Publishing Co., Westport, Connecticut.

Buchanan, C., Alden, C., Birdsong, R.W., Sellner, K.G. and Jacobs, F. (1994). Zooplankton indicators of estuarine ecosystem health. 37th Conference of the International Association for Great Lakes Research and Estuarine Research Federation Program, 166. International Assoc. for Great Lakes Research, Buffalo, New York.

Bulak, J.S., Hurley, N.M., Jr and Crane, J.S. (1993). *Water Quality and the Early Life Stages of Fishes* vol. 14, pp. 29–37.

Caldwell, C.A. and Hinshaw, J. (1994). *Aquaculture* **126**, 183–193.

Cardona, L., Torras, X., Gisbert, E. and Castello, F. (1996). *Isr. J. Aquacult. Bamidgeh* **48**(4), 179–185.

Chakrabarti, R. and Jana, B.B. (1998). *J. Appl. Aquacult.* **8**(2), 87–95.

Cheek, T.E., Van Den Avyle, M.J. and Coutant, C.C. (1982). Striped Bass Env. Risks in Fresh and Salt Water **114**(1), 67–76.

Collins, M.T., Gratzek, J.B., Dawe, D.L. and Nemetz, T.G. (1975). *J. Fish. Res. Bd. Can.* **32**, 2033–2037.

Collins, M.T., Gratzek, J.B., Dawe, D.L. and Nemetz, T.G. (1976). *J. Fish. Res. Bd. Can.* **33**, 215–218.

Colt, J. and Watten, B. (1988). *Aquacult. Eng.* **7**(6), 397–441.

Colt, J., Bouck, G. and Fidler, L. (1987). *Review of Current Literature and Research on Gas Supersaturation and Gas Bubble Trauma*. Special Publication No. 1. NTIS Order No. DE87009906/GAR, Washington, DC.

Cordova, S.M., Auro-de-O, A. and De-Buen-de-A, N. (1996). *Vet.-Mex.* **27**(2), 143–148.

Cripps, S.J. and Poxton, M.G. (1992). A review of the design and performance of tanks relevant to flatfish culture. *Aquacult. Eng.* **11**(2), 71–91.

D'Silva, A.M. and Maughan, O.E. (1995). *J. Appl. Aquacult.* **5**(1), 69–76.

Daniels, H.V. and Borski, R.J. (1998). In *Nutrition and Technical Development of Aquaculture* (eds W.H. Howell, B.J. Keller, P.K. Park, J.P. McVey, K. Takayanagi and Y. Uekita). Sea Grant Program, New Hampshire University, Durham.

Downey, P.C. and Klontz, G.W. (1980). In *Proceedings of the North Pacific Aquaculture Symposium* (eds B.R. Melteff and R.A. Neve), pp. 199–201. Alaska Sea Grant, Anchorage, Alaska.

Dwyer, W.P. and Peterson, J.E. (1993). *Prog. Fish. Cult.* **55**(2), 121–124.

Emmett, R.L., McCabe, G.T., Jr and Muir, W.D. (1988). In *Effects of Dredging on Anadromous Pacific Coast Fishes* (ed. C.A. Simenstad), pp. 75–91. University of Washington, Seattle.

EPA (US Environmental Protection Agency) (1996). *Toxics Criteria for those States not Complying with*

Clean Water Act Section 303(c)(2)(B). EPA, 40 CFR, Ch. I (7-1-96 edition), part 131. Washington, DC.

Fay, C.W., Neves, R.J. and Pardue, G.B. (1983). *Species Profiles: Life Histories and Environmental Requirements of Coastal Fishes and Invertebrates (mid-Atlantic, Striped Bass)*. US Fish and Wildlife Service, FWS/OBS-82/11.8, Washington, DC.

Ferguson, H. (1988). In *Fish Diseases*, pp. 49–54. Post Graduate Committee in Veterinary Science, University of Sydney, Australia.

Fisher, J.P., Spitsbergen, J.M., Bush, B. and Jahan-Parwar, B. (1994). In *Environmental Toxicology and Risk Assessment*, vol. 2. (eds J.W. Gorsuch, F.J. Dwyer, C.G. Ingersoll and T.W. LaPoint). ASTM STP 1216. Amer. Soc. Test. Mat., Philadelphia, Pennsylvania.

Fisher, J.P., Brown, S.B., Wooster, G.W. and Bowser, P.R. (1998). *J. Nutr.* **128**, 2456–2466.

Forteath, N. (1988). In *Fish Diseases*, pp. 145–163. Post Graduate Committee in Veterinary Science, University of Sydney, Australia.

Garling, D.L., Jr and Wilson, R.P. (1976). *Prog. Fish. Cult.* **38**(1), 52–53.

Garton, R.R. (1980). A simple continuous-flow toxicant delivery system. *Water Res.* **14**, 227–230.

Geiger, J.G. and Parker, N.C. (1985). *Prog. Fish. Cult.* **47**(1), 1–13.

Ghate, S.R. and Burtle, G.J. (1993). In *Techniques for Modern Aquaculture* (ed. J.K. Wang), pp. 177–186. American Soc. of Agricult. Eng.

Gregory, R.S. and Levings, C.D. (1998). *Trans. Amer. Fish. Soc.* **127**(2), 275–285.

Hall, L.W., Jr (1991). *Revs. Aquat. Sci.* **4**, 261–288.

Hall, L.W., Jr, Hall, W.S., Bushong, S.J. and Herman, R.L. (1987). *Aquat. Toxicol.* **10**(2–3), 73–79.

Hall, L.W., Jr, Bushong, S.J., Ziegenfull, M.C., Hall, W.S. and Herman, R.L. (1988). *Environ. Toxicol. Chem.* **7**(10), 815–830.

Hall, L.W., Jr, Finger, S.E. and Ziengenfuss, M.C. (1993). *Water Quality and the Early Life Stages of Fishes*, vol. 14, pp. 3–15.

Haskell, D.C. (1955). *Prog. Fish. Cult.* **17**(3), 117–118.

Hester, F.J. (1968). *Vision Res.* **8**, 1315–1335.

Hochheimer, H.N. and Wheaton, F.W. (1997). In *Striped Bass and Other Morone Culture* (ed. R.M. Harrell). Elsevier, New York.

Hoener, G., Rosenthal, H. and Kruener, G. (1987). *J. Aquacult. Tropics* **2**(1), 59–71.

Holmefjord, I. (1991). *Larvi-'91* **15**, 3–204.

Hrubec, T.C., Robertson, J.L., Smith, S.A. and Tinker, M.K. (1996). *Vet. Immunol. Immunopathol.* **50**(1–2), 157–166.

Hrubec, T.C., Robertson, J.L. and Smith, S.A. (1997). *Am. J. Vet. Res.* **58**(2), 131–135.

Jones, A., Prickett, R.A. and Douglas, M.T. (1981). In *Early Life History of Fish: Recent Studies*, vol. 178 (eds R. Lasker and K. Sherman), pp. 522–526.

Kaushik, S.J. (1998). *Aquat. Living Resour.* **11**(4), 211–217.

Kentouri, M., O'Neill, D., Divanach, P. and Charalambakis, G. (1994). *Aquacult. Fish. Manag.* **25**(7), 741–752.

Kerby, J.H., Woods, L.C. III and Huish, M.T. (1983). Culture of the striped bass and its hybrids: a review of methods, advances and problems. In *Proceedings of the Warmwater Fish Culture Workshop* (eds R.R. Stickney and S.P. Meyers), pp. 23–53. Louisiana State Univ., Div. Cont. Edu., Baton Rouge.

Klontz, G.W., Chacko, A.J. and Beleau, M.H. (1980). In *Proceedings of the North Pacific Aquaculture Symposium* (eds B.R. Melteff and R.A. Neve), pp. 337–339. Alaska Sea Grant, Anchorage, Alaska.

Langdon, J.S. (1988). In *Fish Diseases*, pp. 167–223. Post Graduate Committee in Veterinary Science, University of Sydney, Australia.

Lasordo, M. and Timmons, M.B. (1994). In *Aquaculture Reuse Systems: Engineering Design and Management*, vol. 27 (eds M.B. Timmons and T.M. Losordo), pp. 1–7, Elsevier, New York.

Leitritz, E. and Lewis, R.C. (1980). *Calif. Fish Bull.* No. 164, University of California, Berkeley.

Lin, C.K. and Diana, J.S. (1995). (Aquat. Living-Resour.) *Resour. Vivantes-Aquat.* **8**(4), 449–454.

Little, D. and Muir, J. (1987). *A Guide to Integrated Warm Water Aquaculture*. Institute of Aquaculture, University of Stirling, Scotland.

MacKinnon, D.F. (1969). *Prog. Fish Cult.* **32**, 74–78.

MacPhee, D.D., Ostland, V.E., Lumsden, J.S., Derksen, J. and Ferguson, H.W. (1995). *Dis. Aquat. Org.* **21**, 163–170.

Marking, L.L., Dawson, V.K. and Crowther, J.R. (1983). *Prog. Fish Cult.* **45**(2), 81, 83.

McDonald, and Milligan (1997). In *Fish Stress and Health in Aquaculture*, vol. 62 (eds G.K. Iwama, A.D. Pickering, J.P. Sumpter and C.B. Schreck), pp. 119–144. Cambridge University Press, Cambridge.

Meade, J.W. (1988). *Aquacult. Eng.* **7**, 139–146.

Moe, M.A., Jr (1992). *The Marine Aquarium Reference: Systems and Invertebrates*. Green Turtle Publications, Plantation, Florida.

Nardi, G.C. (1996). *J. Shellfish Res.* **15**, 458.

Nardi, G.C. (1998). In *Nutrition and Technical Development of Aquaculture* (eds W.H. Howell, B.J. Keller, P.K. Park, J.P. McVey, K. Takayanagi and Y. Uekita). Sea Grant Program, New Hampshire University, Durham.

Newcombe, T.W. and Flagg, T.A. (1983). *Mar. Fish. Rev.* **45**(2), 8–12.

Nunley, C.E. and Libey, G.S. (1992). In *Aquaculture 92: Growing Toward the 21st Century*, p. 192. Virginia Polytech. Inst. and State Univ., Blacksburg.

Ostrander, G.K., Anderson, J.J., Fisher, J.P., Landolt, M.L. and Kocan, R.M. (1990). *Fishery Bull.* **88**(3), 551–555.

Owsley, D.E. (1981). In *Proceedings of the Bio-Engineering Symposium for Fish Culture* (eds L.J. Allen and E.C. Kinney), pp. 71–82. American Fisheries Society, Bethesda, Maryland.

Parker, N.C. and Davis, K.B. (1981). In *Proceedings of the Bio-Engineering Symposium for Fish Culture* (eds L.J. Allen and E.C. Kinney), pp. 21–28. American Fisheries Society, Bethesda, Maryland.

Pickering, A.D. and Pottinger, T.G. (1987). *J. Fish Biol.* 30(3), 363–374.

Piper, R.G., McElwain, I.B., Orme, L.E., McCraren, J.P., Flower, L.G. and Leonard, J.R. (1982). *Fish Hatchery Management*. United States Fish and Wildlife Service, Washington, DC.

Porter, C.B., Krom, M.D. and Gordin, H. (1986). *Aquaculture* 59(3–4), 299–315.

Redacliff, G.L. (1988). In *Fish Diseases*, pp. 279–297. Post Graduate Committee in Veterinary Science, University of Sydney, Australia.

Redding, J.M. and Schreck, C.B. (1987). *Trans. Am. Fish. Soc.* 116, 737–744.

Rucker, R.R. (1972). *Gas Bubble Disease of Salmonids: A Critical Review*. US Fish and Wildlife Service, Technical Paper No. 58.

Sekai, T. (1989). In *The Early Life History of Fish; Third ICES Symposium* (eds J.H.S. Blaxter, J.C. Gamble and H. von Westernhagen), pp. 191–489. International Council for the Exploration of the Sea, Copenhagen.

Sekai, T. and Matsumoto, J. (1994). *J. World Aquacult. Soc.* 25, 78–85.

Servizi, J.A. (1988). In *Effects of Dredging on Anadromous Pacific Coast Fishes* (ed. C.A. Simenstad), pp. 57–63. University of Washington, Seattle.

Sigler, J.W. (1988). In *Effects of Dredging on Anadromous Pacific Coast Fishes* (ed. C.A. Simenstad), pp. 27–37. University of Washington, Seattle.

Smith, B.J., Pfeiffer, C.J. and Smith, C.J. (1994). In *International Symposium on Aquatic Animal Health: Program and Abstracts*, p. 91. School of Veterinary Medicine, Davis, California.

Speece, R.E. (1973). *Trans. Amer. Fish. Soc.* 2, 323–334.

Stickney, R.R. (1997). *Annotated Bibliography of Flatfish Aquaculture*. Texas Sea Grant, TAMU-SG-97-603, Bryan, Texas.

Sugimoto, M., Nakano, H., Yano, Y., Fukuda, M. and Murakami, N. (1985). *Bull. Hokkaido Reg. Fish. Res. Lab. Hokusuiken-Hokoku* 50, 63–69.

Summerfelt, S., Hankins, J.S., Weber, A.L. and Durant, M.D. (1997). Ozonation of a recirculating rainbow trout culture system II. Effects on microscreen filtration and water quality. *Aquaculture* 158, 57–67.

Suresh, A.V. and Lin, C.K. (1992). *Aquacult. Eng.* 11(1), 1–22.

Terhune, J.S., Schwedler, T.E. and Collier, J.A. (1997). *J. World Aquacult. Soc.* 28, 20–26.

Timmons, M.B., Youngs, W.D., Bowser, P.R. and Rumsey, G. (1991). *Design Principles of Water Reuse Systems for Salmonids*. New York State College of Agricultural and Life Sciences, Agricultural and Biological Engineering, Extension Bulletin 462, Cornell University, Ithaca, New York.

Twarowska, J.G., Westerman, P.W. and Losordo, T.M. (1996). In *Aquacultural Engineeering Society Proceedings II: Successes and Failure in Commercial Recirculating Aquaculture Conference*, vol. 2. Northeast Regional Agricultural Engineering Service, Ithaca, New York.

Wasielesky, W., Jr, Bianchini, A., Santos, M.H.S. and Poersch, L.H. (1997). *J. World Aquacult. Soc.* 28, 202–204.

Wedemeyer, G.A. (1980). In *Proceedings of the North Pacific Aquaculture Symposium* (eds B.R. Melteff and R.A. Neve), pp. 155–169. Alaska Sea Grant, Anchorage, Alaska.

Wedemeyer, G.A. and Wood, J.W. (1974). *Stress as a Predisposing Factor in Fish Diseases*. US Fish and Wildlife Service, Washington DC.

Westers, H. (1978). *Fish. Manag.* 9(2), 45–65.

Williams, R.C., Hughes, S.G. and Rumsey, G.L. (1982). *Prog. Fish Cult.* 44(2), 102–105.

Wolf, H. and Jackson, E.W. (1963). *Science* 142, 676–678.

Yi-Yang, Kwei-Lin, C. and Diana, J.S. (1996). *Aquaculture* 146(3–4), 205–215.

Ziebell, C.D. (1960). In *Proceedings of the Seventh Symposium on Water Pollution Research*, US Department of Health, Education and Welfare, Portland, Oregon.

CHAPTER 2

Facilities and Husbandry (Small Fish Models)

**Robin M Overstreet, Sue S Barnes,
C Steve Manning, and
William E Hawkins**
Institute of Marine Sciences, The University of Southern Mississippi, Ocean Springs, Mississippi, USA

Introduction

Small fishes offer numerous advantages as experimental animals, many of which are not offered by relatively larger fishes like trout, channel catfish, striped bass and its hybrids, centrarchids, and others. They can be bred, grown, and housed for a relatively small cost. Enough specimens of either a homogeneous or heterogeneous stock can be produced to obtain statistically significant results as well as to repeat experiments at the researcher's convenience after an initial experiment. The small fishes can be examined readily for gross behavioral or pathological alterations resulting from experimental factors. Also, they can serve as histological, biochemical, or genetic models. Small fish models often have economic advantages over relatively large fishes or homeothermic vertebrates. These advantages arise because large sample-sizes of fish can be reared inexpensively in small aquaria, tanks, or large raceways, using few supplies and minimal labor. These small test fish can then be analyzed relatively inexpensively. For example, two or few **parasagittal** sections through each fish can provide necessary histopathological information on most relevant vital organs compared with separate sections, with all the associate time and expense, for each organ in a large fish or a homeothermic test animal. Also, many recent advancements in methodology allow most analyses to be conducted on a small amount of tissue.

Small fish species are used for a variety of types of study involving (i) carcinogenesis bioassay or investigation; (ii) toxicological bioassay or investigation; (iii) field verification; (iv) laboratory demonstration; (v) physiological and behavioral activities; and (vi)

Copyright © 2000 Academic Press

parasitology and disease. Details of many of these are presented in Chapters 4, 31, 36, and 37. Unless large quantities of tissue are necessary or a highly specific enzyme or other substance offered by a large fish species only, the small fish candidate would be preferred over larger ones. In some cases, cells, tissues, or fluids from small fish species have to be pooled to obtain enough material to analyze. In other cases, recent molecular and immunological procedures can be used with material from single small fishes, and some can be performed on fixed histologiocal sections. Useful studies involving either small or large species depend on the endpoints to be measured. Examples of such endpoints include mortality, behavior, **neoplasia**, other gross and histopathological alterations, blood characteristics, genetic features, biochemical levels, and parasitic infections.

Small fishes constitute good candidates or models for toxicological studies using a variety of exposures. These exposures include those conducted with (i) direct administration through water in static, static-renewal, or flow-through systems; (ii) dietary exposure; (iii) fish injection; (iv) egg injection; (v) embryo exposure; (vi) cages in the field exposed to potentially contaminated effluents and to reference site effluents; and (vii) mobile units fit to accept passage of water pumped from sites both upstream and downstream relative to a point-source of contamination.

Cultured fishes as considered from a parasitological viewpoint can be an immensely valuable tool when assessing fish as bioassay models, models of environmental health, hosts for parasites, and parasitic infections as well as for other uses. Fishes can be valuable as either assumed normal hosts or as abnormal hosts. Specific pathogen-free fishes are bred and reared to determine features such as life cycles and life histories of parasites, host–parasite relationships, and interrelationships with *in situ* parasites and chemical compounds. For example, Fournie and Overstreet (1993) demonstrated, using a series of **atheriniform** fishes and nonrelated species including both laboratory-reared and wild individuals, a variety of different host responses to and parasite development in the hepatic coccidian *Calyptospora funduli*. In other kinds of cases where infected wild fish are used as biological indicators of host biology, general or specific environmental health, or other parameters, questions may be asked about prior infections, altered immune responses, unknown ages of host or parasite, and non-repeatable results. Experimental infections in laboratory-reared counterparts can answer some of those questions.

Large numbers of a fish can be maintained in test containers. We have maintained as many as 1500 test fish for up to 8 months in individual 300-L raceways and 1500 in an all-glass aquarium for up to 28 days beginning with fish 14 to 16 days post-hatch. Our test systems for carcinogenicity bioassays vary according to experimental protocol, but we often test as many as 1200 to 1500 medaka and 800 to 900 guppies concurrently, divided among specially constructed aquaria held within a single enclosed system under negative pressure with appropriate delivery of toxicant for periods of up to 16 months. The number of test fish in each evaluation depends on a number of factors including age of fish (embryo, fry, juvenile, or adult), number of concentrations, duration of exposure, and study endpoints (e.g. 4 concentrations $\times$ 2 replicate aquaria $\times$ 2 fish species $\times$ 100 or 200 fish per tank).

The primary purpose of this chapter is to provide information useful in housing, breeding, and rearing some frequently used small fish species. These activities can be conducted under Good Laboratory Practice (GLP) guidelines as are being met in some studies in our laboratory. We emphasize the freshwater medaka (*Oryzias latipes*) and guppy (*Poecilia reticulata*), which can also be reared in saltwater conditions, as well as the sheepshead minnow (*Cyprinodon variegatus*), a marine species. Other species are mentioned, as well as aspects of exposure facilities.

Species

A number of small fish species are available for experimental studies. The experience of the authors has centered around the use of the medaka, also known as the Japanese medaka (*Oryzias latipes*) and king cobra strain of guppy (*Poecilia reticulata*) for carcinogenesis experiments and the use of a few different marine species for a variety of purposes involving toxicology, parasites, and diseases. For the marine species, we will use the sheepshead minnow (*Cyprinodon variegatus*) as our primary example. Nevertheless, we mention several species commonly used for scientific research.

Egg-laying fishes

The medaka, an egg-layer (oviparous), has long been a common model fish used for embryological fish development, physiological and toxicological studies, carcinogenicity studies, and demonstrations (e.g. Hawkins *et al.*, 1995; Masahito *et al.*, 1996) because it is an easily reared, non-aggressive species that can be grown under a variety of water conditions. Medaka is native to Southeast Asia, and there is some confusion regarding specimens from some of the countries. The Nagoya University Medakafish Group has created an extensive research website (http://biol1.bio.nagoya-u.ac.jp:8000), which treats this and other aspects of strains, breeding, genome, genetics, embryology, neuroscience, physiology, pathology, and ecology. The website lists 13 separate species, some of which most certainly represent junior synonyms. This chapter treats *Oryzias latipes* (Japanese medaka) from Japan and more northern countries, even though we have reared *Oryzias javanicus* with similar results. Considering *Oryzias latipes*, the website lists numerous wild-type strains, mutant strains, and inbred strains. Compared to the haploid genome of zebrafish, the genome of medaka is about half the size (680 to 850 Mb); this corresponds to one-fourth to one-fifth of the entire human genome (Tanaka, 1995). Along with the zebrafish, a species discussed briefly below, it is an important model for developmental genetics; embryonic development has been described (Ishikawa, 1997), and there exists for the medaka one of the few atlases of a fish brain (Anken and Bourrat, 1998). The medaka was the first vertebrate to mate, produce eggs, and have those eggs produce normal offspring in space (Ijiri, 1995).

Medaka is an especially good fish for carcinogenicity studies because of its sensitivity to a range of carcinogens, because of its low spontaneous rate of neoplasm incidence, and because a large number of neoplasms in addition to the various hepatic types can be induced. Neoplasms and toxic responses have been induced with numerous compounds (e.g. Hatanaka *et al.*, 1982; Hawkins *et al.*, 1988, 1995). Embryos are useful to assess low levels of toxicants such as dioxin (Wisk and Cooper, 1992; Cantrell *et al.*, 1998). In our laboratory tests with carcinogens and other toxicants as well as those by others, medaka have been used successfully in dose–response studies involving mortality, neoplasia, fecundity, ability to hatch, fry survival, and fry abnormalities (Figures 2.1 and 2.2) (e.g. Hawkins *et al.*, 1998). Importantly, a consensus has been developed for criteria for diagnosing hepatic neoplasms in medaka (Boorman *et al.*, 1997). Additionally, the fish has demonstrated good reproducibility in egg production, hatching, and fry survival that can be affected by administration of a heavy metal (Walker *et al.*, 1989). The medaka serves as a good model for proliferating cell nuclear antigen immunohistochemical assays (Ortego *et al.*, 1996), for studying oncogene activation (Van Beneden *et al.*, 1990), and for assessing DNA repair (Ishikawa *et al.*, 1984).

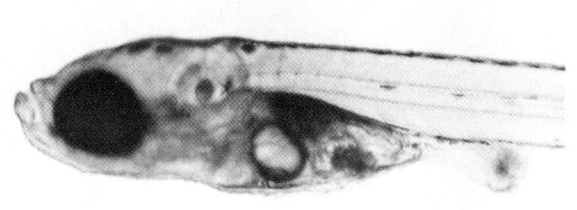

Figure 2.1 Medaka fry, normal appearance.

The medaka is typically 2 to 4 cm long, and some individuals can live 5 years (Masahito *et al.*, 1996). The male can be distinguished from the female by its more pronounced dorsal and anal fins, a deep notch between the last two rays in the dorsal fin, and numerous small papillar processes on the posterior region of the anal fin. The female has a much more pronounced urogenital papilla. Eggs measure 1.0 to 1.5 mm in diameter, are filamented, and have a transparent chorion (Figure 2.2). Large numbers can be produced with relatively little effort or expense.

Additional egg-laying species useful as laboratory fishes include the zebrafish, *Danio rerio* (often referred to as *Brachydanio rerio*), a cyprinid fish originally from India for which the genetics and nervous system are also well understood. The species is commonly used to analyze how the vertebrate nervous system is regulated at the cellular, genetic, and molecular levels. Methods have been standardized to produce haploid embryos with only the female genetic complement by fertilizing eggs with sperm that have been irradiated with ultraviolet light, gamma rays, or carcinogens to destroy the sperm DNA. These embryos can then be made diploid by applying pressure or heat shock during early developmental stages. Phenotypic mutants of this artificial **parthenogenesis** process can be readily observed. A manual, *The Zebrafish Book* (Westerfield, 1995), provides considerable useful information for rearing the fish and conducting various experiments. Additional communication can be achieved through an electronic bulletin board on line since 1990 (biosci-server@net.bio.net; subscribe zbrafish). We consider the medaka to be much easier to breed, hatch, and maintain than the zebrafish. The medaka requires less space and technical attention; the young zebrafish is smaller

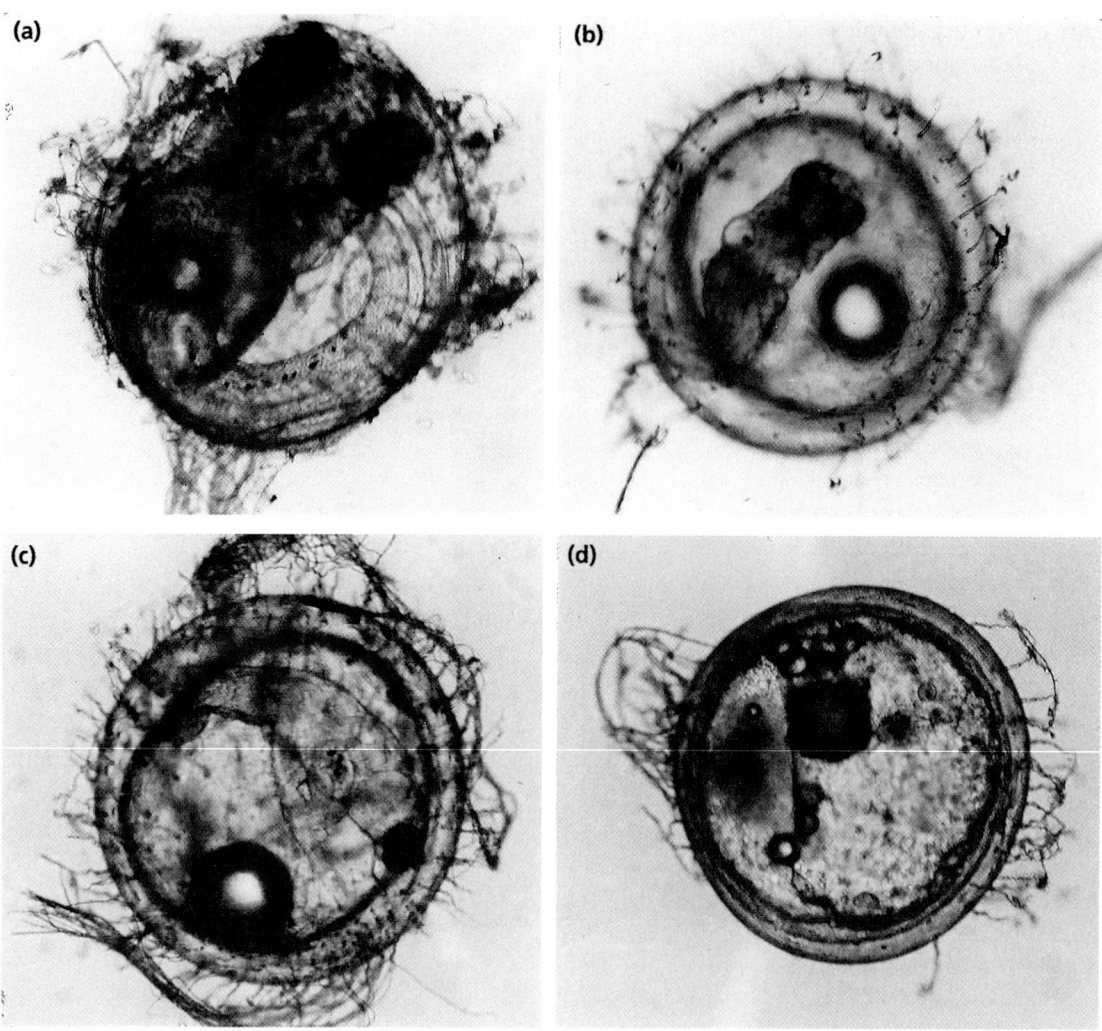

Figures 2.2a-d Medaka eggs showing filaments and a transparent chorion as well as abnormal embryonic development resulting from exposure to bis (tri-n-butyltin) oxide. 2.2a Abnormal fins and abnormal circulation. 2.2b Optic cup deformation and bilateral microphthalmia. 2.2c Cyclophthalmic embryo (single eye in center of optic area). 2.2d Lack of cell differentiation; embryo did not develop further during next 4 days before it died. Figures 2.2b–d from Walker et al. (1989).

and requires special food for a much longer period, and zebrafish eggs are highly susceptible to fungal infections. Also, because sex of zebrafish is apparently determined by a combination of genetic and environmental factors, there exists a tendency, at least in some strains, for male offspring to well outnumber females unless inbredness is eliminated by continually introducing a few individuals from other sources of fish into the brood stock. Medaka is also preferred for carcinogen studies, since it is much more susceptible to induction, at least to diethylnitrosamine (DEN). In fact, we have conducted a 'megamedaka' study involving more than 36 000 fish for histological evaluation. The study analyzed the dose–response curve for neoplasia induced from DEN. Attention was directed to the portion of the curve that had a low prevalence of hepatic neoplasms and consequently necessitated relatively large sample sizes. Such a study had been shown to be impractical and non-conclusive with rodent models (Staffa and Mehlman, 1979). Neither the medaka nor zebrafish is acceptable for use in situations where the potential for establishment within natural waters of the Americas exists because the species are not native US fishes.

The fathead minnow (*Pimephales promelas*) is typically used for freshwater bioassay studies, even though some problems exist for certain types of studies. A nest is necessary for deposition of eggs, and cut PVC pipe is adequate. Males vigorously protect the nest, and they often die after spawning, so long-term experiments are hampered. EPA bioassay methods using the fathead minnow have been established for many years (e.g. Benoit, 1982; Denny, 1987). The fish provides a good control for comparative neoplasm studies because neoplasia is difficult to induce in that species, even with genotoxic compounds like methylazoxymethanol acetate (MAM-ac) (Hawkins et al., 1988). Also, its liver constitutes a small proportion of its body weight, liver tissue is difficult to collect, and the mixed function oxidase system as determined from cytochrome P-450 levels is not as readily induced as in many other fishes (James et al., 1988).

When a marine species, especially an egg-laying species, is required for bioassay or appropriate parasite studies, we recommend the sheepshead minnow (*Cyprinodon variegatus*) as a leading candidate. This is a widespread species occurring from Cape Cod to Venezuela and the Gulf of Mexico. Successful culture methods for it have been established (Hansen, 1978). Disease-free brood-stock have been maintained in our laboratory for about 15 years, and offspring have been used in successful carcinogenicity, toxicological, and parasitological studies involving acute, chronic, and 'whole life' studies. We have maintained it successfully by selecting subordinate rather than dominant members for our brood stock. We believe this practice has reduced cannibalism and the continual need to grade fish by size. The sheepshead minnow, like some other marine fishes, had higher mixed function oxidase activities than did medaka and guppy (James et al., 1988). Couch et al. (1981) considered this species analogous to the mouse in the mouse–rat model when comparing marine fishes with mammalian toxicological bioassay animals. The Gulf killifish (*Fundulus grandis*) was considered the 'rat'. The related diamond killifish (*Adinia xenica*) can tolerate relatively low oxygen conditions, and immediately hatched progeny are in a more advanced condition than those of the sheepshead minnow. The diamond killifish demonstrates a considerable heterogeneity in hatching periods (unless incubation water is continuously exchanged) and developmental events, both features allowing for success of the fish in its intertidal salt marsh habitat (Koenig and Livingston, 1976; Cunningham and Balon, 1985). Those plus other adaptations promote survival in nature, but do not necessarily provide a homogeneous stock of test animals for many types of experiments. On the other hand, those features and the relatively large transparent eggs available year-round in the laboratory make this hardy species a potentially valuable test animal. The methods presented below for handling the sheepshead minnow will be helpful for a variety of other marine and estuarine species.

Another good marine candidate is the Gulf killifish from Florida and the Gulf of Mexico. The Gulf killifish (*Fundulus grandis*, or its counterpart, the 12-cm long mummichog, *Fundulus heteroclitus*, along the western Atlantic coast from the Gulf of St Lawrence to northeast Florida) is useful for laboratory comparisons or verifications of field observations of tolerance to or responses resulting from toxic compounds (e.g. Weis and Weis, 1989). Actually, a series of killifish species could be used, but less attention has been directed to establishing any of them as models. In natural waters, egg production for the Gulf killifish is seasonal, but reproduction occurs during most months. Production can be achieved throughout most or all of the year for it and its counterpart in the laboratory, even though spawning activity is still related to tidal and tidal/moonlight cycles (Hsiao and Meier, 1989). The Gulf killifish is larger than the other 'small species' treated in this chapter, reaching 18 cm as opposed to 3 to 8 cm for most of the other small fish species discussed. Even this larger size can result in additional work for histology and for the number of fish that can be held per liter, but it can provide more tissue for biochemical or other purposes. Moreover, the species or some related species can be used for cage studies in the wild without fear of releasing an exotic species into local waters. We have conducted studies that show that fertilized eggs of the Gulf killifish can be collected and maintained in damp sand for long periods without hatching so that concurrent, synchronous hatching can be achieved on demand (Cherie Heard and Overstreet, unpublished data). Like many of the species indicated here, the fish is a hardy species. This is a real advantage for most bioassay work. One might argue that the use of an extremely sensitive species that dies after minimal contact with an administered stress is a more acceptable fish. On the other hand, if the fish (e.g. Gulf killifish) is hardy and mortality occurs in the test group and not the controls, there is less question about a death endpoint, and it is easier to evaluate sublethal responses (e.g. Weis and Weis, 1989; Overstreet, 1993). Spawning and stages in development of *Fundulus grandis* and *Fundulus heteroclitus* were described by Armstrong and Child (1965), Wallace and Selman (1978, 1980), Hsiao and Meier (1989), and others, and related species are similar. Costello and Henley (1971) described general methods.

Silversides, with an emphasis on *Menidia peninsulae* from the Gulf of Mexico (Middaugh and Hemmer, 1984, 1987), *Menidia menidia* from the Atlantic coast (Beck and Poston, 1980; Conover and

Kynard, 1984; Conover, 1985), and *Menidia beryllina* from both Gulf and Atlantic coasts plus California (Middaugh and Hemmer, 1992), where it has been introduced, also provide good marine fishes for laboratory studies. They can be produced in large numbers on demand (Middaugh *et al.*, 1986, 1987). Unlike the sheepshead minnow but similar to medaka, silversides are too small when hatched to feed on brine shrimp. Moreover, atherinids from throughout the world offer potential as bioassay species (e.g. Overstreet, 1995). There exists a complex of silversides in the genus *Menidia* that offers a good system for biological analysis. In the Gulf of Mexico, the assemblage includes two bisexual species (*Menidia beryllina* and *Menidia peninsulae*), non-recombinant hybrids of the two (some of which are triploids), and unisexual (female) forms referred to as the *Menidia clarkshubbsi* complex, which appeared to arise from *Menidia beryllina* and a missing ancestor (Echelle *et al.*, 1988). Embryos (7- to 10-day-old) of *Menidia beryllina*, presumably like those of many other atheriniform fishes, provide a sensitive indicator for toxicity/teratogenicity; a test with the embryo was developed by USEPA to complement the freshwater test using the fathead minnow (Middaugh *et al.*, 1997).

The mangrove rivulus (*Rivulus marmoratus*) from the eastern coast of Florida to Central America and the Bahamas offers an unusual animal model because its reproduction can be by natural synchronous internal self-fertilization and consequent production of homozygous clonal offspring. Even though the male was considered to be rare (< 1% in South Florida), Davis *et al.* (1990) found as high as 25% of some populations in Belize mangrove islands to consist of males. Nevertheless, the hermaphrodites serve as valuable natural test animals to conduct a variety of experiments. For example, laboratory-reared fish have been used for toxicological bioassays (Davis, 1988); carcinogenicity studies, but with less response than by medaka and most other fishes indicated here (Koenig and Chasar, 1984; Hawkins *et al.*, 1988); and inflammatory responses (Vogelbein *et al.*, 1987). The latter article used the fish as an abnormal host to a coccidian parasite to provide two different, dependably consistent inflammatory responses, one of which (granulomatous) was not typically produced in natural hosts.

Live-bearers

The guppy, or *Poecilia reticulata*, commonly mentioned under its synonym *Lebistes reticulatus*, is native to Venezuela and the Guianas and possibly many Caribbean islands, but it has been introduced into the US and other tropical and subtropical countries. The species offers several of the same advantages provided by the medaka, including its hardiness and ability to survive in moderate salinity conditions. Additionally, it is a live-bearer. Free-swimming fry of live-bearers depend more on yolk and are considerably larger at birth than fry that hatch externally after several days, and they can be used as hardy test animals at birth. We have used the 'medaka–guppy pair' as comparative freshwater species to complement each other in carcinogenicity studies, such as the comparative laboratory rat and mouse when used simultaneously to assess carcinogens (Hawkins *et al.*, 1988). For example, the guppy develops pancreatic neoplasms, but from a low dose of MAM-ac only (Fournie *et al.*, 1987). In comparison, the medaka develops this type of neoplasm rarely, and only as a spontaneous neoplasm (Hawkins *et al.*, 1991). There is no 'most sensitive species' that can be used for all cases or for setting standards for environmental safety testing (Cairns, 1986). Details of breeding and rearing of the guppy are provided below. Numerous books, pamphlets, and magazines for tropical fish enthusiasts giving general information on the guppy (e.g. Whitney and Hahnel, 1964) and producing fancy strains (Iwasaki, 1989) are available.

Zimmerer (1984) suggested that some contradictory reports about the guppy's sensitivity to environmental variables relate to strain differences. He also suggested that establishment of carefully maintained inbred strains for cancer research was the best likely solution to understanding this problem. In addition to our laboratory's stock mentioned later in this chapter, we have inbred two phenotypes from this stock through 12 generations of breeding with selected pairs of individuals. The resulting stocks can be made available for comparative studies.

The western mosquitofish (*Gambusia affinis*) and some related mosquitofishes are similar to the guppy in that they are live-bearers, but they thrive better than the guppy in salt water. Neoplasms in the western mosquitofish are also easily induced with carcinogens (Law *et al.*, 1994) and, by assessing lesions and parasites, one can use it as an indicator of environmental health (Overstreet *et al.*, 1995; Overstreet, 1997). At least mosquitofish offspring from local stocks can be released into many areas, if necessary, without fear of exotic introductions since strains are present throughout much of the world already. The mosquitofish serves as a good live-bearer for comparisons with the guppy or an egg-layer when one needs a wild species model that can live and breed in fresh water to moderately high salinity water. The species should also be good for assessing androgenic endocrine disrupters since **masculinization** of females is common in wild fish exposed to some paper mill effluents (Bortone *et al.*, 1989; personal observations). Also, tolerance to some toxicants by the fish has been attributed to genetic variation (e.g. Hughes *et al.*, 1991). Mosquitofish livers are easy to work with for biochemical studies. Some problems can occur when rearing the mosquitofish, but these can be reduced by separating juveniles of the two sexes until at least after the period of initial breeding would have occurred. After a young, recently matured laboratory-reared female specimen mates for the first time, the 'oviduct/birth canal' often becomes occluded, causing blockage and preventing passage of eggs or embryos (a condition sometimes termed 'egg-bound') and consequently unhealthy and reproductively unsuitable specimens. Normally, the gestation period is about 3 to 4 weeks, but depends on temperature and other factors. We have counted about 200 fry in a female, but our year-old brood stock female usually holds a brood of about 30 to 50 ready to be deposited over a few days. A dated but important paper by Krumholz (1948) provides helpful information on reproduction in Illinois and introduction of fish into Michigan.

Several other poeciliid fishes (viviparous topminnows; actually ovoviviparous in that eggs are produced, hatched, and fertilized internally, ultimately producing embryos that are gaining nutrients from both mother and yolk and then being deposited as several broods from the single fertilization) can be used in laboratory studies, but most are more difficult than the guppy and mosquitofish to obtain large numbers of specimens born on the same day on demand. For example, the sailfin molly (*Poecilia latipinna*) native to the southeast US, Gulf of Mexico, and Yucatan serves as a good marine species for parasitology studies (personal observations) and genetic analyses (e.g. Angus, 1983). In the laboratory, we find that most young from a group of females are produced over a short period but not necessarily on demand.

Of course, there are many other species of livebearers and egg-layers that can be used. These will depend on the locale in which the experiments are conducted as well as the type of experiments to be conducted. Pet stores and local waters provide good sources for some needs, and recognized repositories occur for others. For example, the University of Oregon has both a Stock Center that provides a few different wild-type and mutant strains to researchers and a group that provides *The Zebrafish Science Monitor* through The Fish Net, a research website (http://zfish.uoregon.edu/). The Gulf Coast Research Laboratory (GCRL) has provided medaka to numerous research and educational facilities along with instructions for rearing and maintenance. The use of such fish assures that some features of an experiment are constant among facilities.

Facilities

Because of the small size of the fish, statistically significant numbers can be maintained in any variety of containers, depending on whether the fish are being housed for breeding, nursery, exposure, growout, or other purposes. Some of these facilities will be mentioned. Facilities to accommodate most species are similar; of course exceptions exist such as the self-fertilizing rivulus which is kept in small culture dishes stacked one above the other most of the time. Since breeding and associated facilities differ among the three primary examples, the facilities for those examples will be provided below with the fish culture. This section treats the water and general systems common to several species.

Water

Culture water used in maintenance, production, and exposures of freshwater fishes at GCRL is provided from a 177-m well and is non-chlorinated, particle-filtered, carbon-filtered, temperature-adjusted, and aerated prior to introduction into tanks and raceways. The same water is used for all stages of culture except hatching. It is pumped into a 3000-L polyethylene tank, then pumped through to a series of three carbon filters and two particle filters in a continuous loop before being directed to a central head box. Water destined for the test systems enters two carbon filters, a 14-kg and a small 7-kg one, a 16-μm cartridge, and two UV sterilizers. A manifold trunk line directs water to the brood stock and fry, the test systems, and growout facilities. We estimate that more than 19 000 L of fresh water enter the facilities daily.

Once water is in culture and growout containers with fish, individual tank filtration and aeration are accomplished with biological sponge filters (e.g. TetraMin Twin Brilliant sponge filters/aerators, JFK

Enterprises, New Orleans, LA). One of the pair of sponge filters is alternately cleaned once a week by gently swirling it in a bucket of clean water to remove the external debris. If an abundance of material occurs inside, the filter is removed; otherwise it, with its established bacterial culture, is replaced. On the other hand, in toxicity tests, we recommend to exchange one of the pair each week. When a filter is removed or after an experiment, filters are cleaned thoroughly and then autoclaved or placed in an automatic washing machine on the gentle cycle without detergent.

Well-water in southern Mississippi is typically pH 8.5 to 9.0, and that has caused no problems for rearing freshwater fishes or making artificial sea-water. It, however, may offer either an advantage or a disadvantage when testing specific compounds in toxicity tests. For example, the effect of phenol on fish is intensified in acidic conditions. Consequently, the effect of phenol in our basic water would not be as toxic. Even though phenol is not carcinogenic, environmental factors allow one to dissect the toxic effect from the carcinogenic effect of various compounds. We find DEN to be highly carcinogenic but not toxic to the cells. Many carcinogenic compounds produce toxic cellular responses.

Systems for salt water are different from those used with well water. Depending on the dilution-water needs for an experiment or fish culture, we have several options. Filtered sea water can be trucked 215 km in a 9000-L stainless steel tank directly from a US Environmental Protection Agency facility in Gulf Breeze, Florida; Mississippi Sound water can be pumped into a settling tank from a bayou adjoining our facilities; or reconstituted commercial synthetic sea salts can be prepared with well water. In any case, the sea water is stored in a variety of large fiberglass holding reservoirs and is ultimately pumped through a multiple-raceway system and through a 10-μm filter. The reservoir holding system is temperature controlled within 1°C and regulated to within 0.5 ppt salinity. Addition of well water or distilled water compensates for evaporation.

Water quality is monitored regularly in all systems where fish are maintained. In flow-through systems, we have a **kymograph** recording continuous temperature measurements. In static systems or growout aquaria, usually in temperature-controlled water baths, we use digital thermometers in representative aquaria. We maintain a constant temperature ($\pm 1°C$) in the water baths with thermostat-controlled titanium heaters (Glo-Quartz Electric Heater Co., Inc.,

Mentor, OH). At a minimum, ammonia, nitrate, nitrite, hardness, and alkalinity of the diluent water are measured on a monthly basis. Measurements of dissolved oxygen, pH, temperature, and salinity, if appropriate, are taken twice a week in maintenance tanks, brood stock raceways, and growout aquaria. Chemical values are routinely determined with commercial test kits. For example, ammonia, nitrate, and nitrite can be easily but not precisely determined using commercial test products (e.g., LaMotte testing kits, LaMotte Chemical Products Co., Chestertown, MD; Hach reagents in conjunction with laboratory spectrophotometric analysis). Precise values are obtained from our institution's analytical chemistry laboratory for water in the exposure chambers. The expense of obtaining analytical data from the test systems is not necessary for systems holding fish not being tested. If unacceptably high levels occur in maintenance and brood raceways, whatever measurements considered necessary are monitored continually to resolve the problem and re-establish good general water quality for animal health. Spotte (1991) provides a good review of water quality parameters and what they mean. From another point of view, some values for water conditions recommended for breeding such as pH and hardness that are provided in old, standard aquarist books (e.g. McInerny and Gerard, 1958) are not necessary. Methods we provide using different water conditions have proved to produce large numbers of high quality offspring that result in significant, repeatable studies.

In aquaria receiving toxicants, we monitor the toxicant concentrations using the detection limits for the most sensitive or appropriate method. Initially, to determine the appropriate periodicity for obtaining the measurements, we measure values frequently and then assess our methodology. During studies, we typically measure toxicant concentrations twice a week. We suggest that others use a similar approach. Historically, we have maintained chemical concentrations within 10% for periods longer than a year.

Water is delivered to fish-holding systems in conditioned ('water-cured') PVC pipe; diluent water directed to test systems is typically through glass or microbore polyvinyl tubing. Our laboratory is arranged so that delivery systems, exposure chambers, holding facilities, and isolated bioassay laboratories can be and are modified for each different test to accommodate the needs of the compounds involved. For example, if compounds are volatile, an assumption we make for chronic tests, systems are under negative pressure so that all air in the system exits through ducts to carbon filters and treatment units and does not come into contact with control aquaria or with researchers, who as a precaution also wear masks, gloves, and other appropriate gear. If compounds are light sensitive, we either use dark colored microbore tubing within a conduit with foil covering any exposed areas or we cover all ducts and lines with foil so that the compounds reach the fish in an unaltered state.

Testing systems

Exposure capabilities at our laboratory include all variations of static and flow-through approaches lasting for hours or months. Many toxicological processes are chronic and may be associated with prolonged latency periods requiring the use of testing systems which must accommodate lengthy test periods in which compound concentrations and environmental conditions must be carefully maintained. Exposure systems have to be designed individually to accommodate specific tests, but the primary focus of our toxicological program has been in development and use of exposure systems and methods designed to allow for continuous long-term exposure to constant concentrations of a test material. These systems and methods incorporate safeguards that limit the effect of events that may compromise the continuous operation of the system (Figure 2.3). For example, to evaluate compounds that may exert a carcinogenic effect only after a prolonged exposure, flow-through exposures with relatively unvariable test substance concentrations must be adapted for tests that span most of an organism's life. We have used commercial carcinogen glove boxes and regular hoods for short static tests, but we have also placed aquaria used for static and static renewal tests in our specially designed and constructed elongated wooden glove boxes used for flow-through experiments. These are modified from that described by Walker et al. (1985) and altered some (Hawkins et al., 1995).

The facility currently contains four isolated exposure laboratories, each with single-pass climate control, independent lighting, and capability of bench-level effluent treatment and multiple discharge options (i.e. different effluents can be directed to varied disposal mechanisms: treat, followed by release into city sewerage; hold for more thorough treatment; or hold for disposal as hazardous waste). Exposures are conducted within these isolated laboratories which each house an exposure chamber and all associated materials necessary for conducting the exposure of the test organisms to the test material. These laboratories independently receive climate-controlled air in single-pass fashion which is exhausted exterior to the building through roof vent fans.

Exposure chambers within the isolation laboratories are maintained at a slightly negative pressure by an exhaust blower that draws in air from within the laboratory in single-pass fashion. Air exhausts through the roof of the chamber and exits through an activated carbon filter to the exterior of the building. Treatment aquaria and other materials within the chamber are accessed and manipulated through sliding doors or gloves along the front or on each side of the chamber. A water bath is used to maintain temperature within the treatment aquaria. The bath temperature is maintained by both thermoregulated heaters and chilling units to within $1°C$ of the target test temperature. Waste exiting the exposure chamber is transferred to a waste treatment area through plumbing dedicated to each exposure system so that waste test materials can be handled individually.

In the evaluation of test materials a wide variety of exposure capabilities are required for testing single compounds, complex chemical mixtures, sediments, drilling fluids, and mixed effluents. Special techniques, including using high shear chemical mixing pumps, saturation columns, and stirred carboys, are needed for working with relatively insoluble materials (polyaromatic hydrocarbons, halogenated dibutyltin, chlorinated hydrocarbons) as a preferable alternative to the use of solvents or carriers. In our experience, organic solvents often promote growth of microfauna that can affect water quality and test substance stability and behavior. Water-based stocks are generally prepared in carboys of 20- to 45-L capacity to attain a dissolved test substance concentration great enough to permit the injection of small volumes of stock to

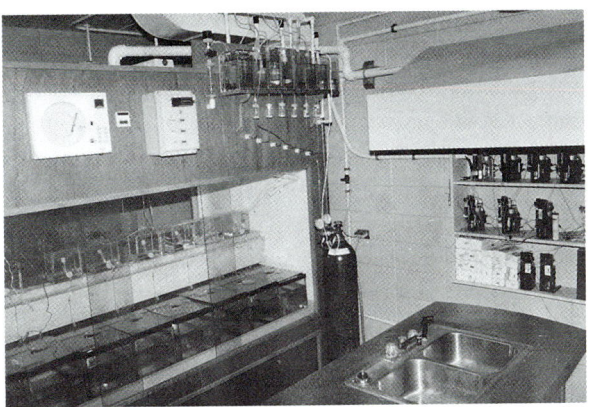

Figure 2.3 Portion of a flow-through system for exposing fresh water or salt water to test aquaria containing bioassay animals.

achieve the desired treatment concentrations. Stock carboys prepared in water are maintained with a small volume of undissolved test substance to insure a more stable maximum concentration at the water saturation level of the test material. Usually, three stock carboys are serially connected with the injection of stock from the final carboy in the series. Serially connecting the carboys permits the maintenance of a stable concentration in the last carboy of the series. Injectors and stock carboys are housed under a hood vented to the exterior of the laboratory through activated carbon.

The control of external factors within a limited degree of variability coupled with greater consistency in administering test substances to test organisms insures the accuracy and reproducibility of the test organism response. The goal of our laboratory is to produce accurate, reliable, reproducible evaluations of test substances in both acute and chronic fish studies. Test-organism exposure to the test substance is a critical phase of each evaluation. Describing these exposure systems is not the purpose of this chapter, and some will be described in detail by Paul Bauman in Chapter 36 and by Chris Metcalfe in Chapter 37. However, we provided brief mention of our facilities to complement those sections and to allow one to appreciate the potential of using fish models. Also, such containers can constitute the space where fish live most of their lives.

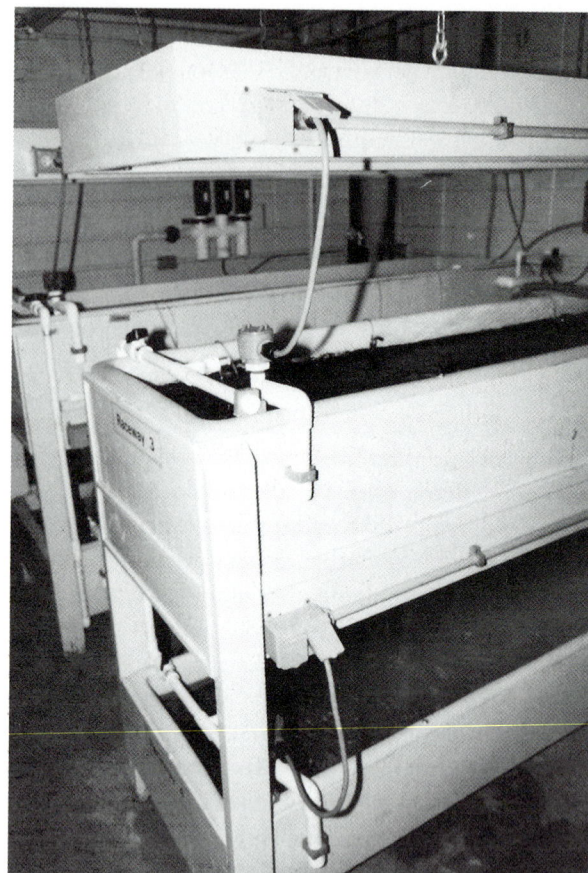

Figure 2.4 Two tiers of general raceways for maintenance and brood stock within wooden frame.

Holding and growout systems

In preparation for tests or spawning, we use several general raceways for maintenance and brood stock. Most of these are commercially-produced fiberglass ones and are of different sizes to accommodate the available space. They mostly consist of an elongated type, 242 long by 57 wide by 30 cm high and arranged in two tiers (Figure 2.4). Measurements are inside ones, taken near the top, which is slightly larger than at the bottom. A 4-cm diameter PVC outlet pipe is located at different heights to accommodate different needs, but in many it is located so that water fills 21 cm, or 300 L (80 gal), of the container. A shelf holding fluorescent lightbulbs sits 55 cm above the top of each of the two tiers of raceways. The raceways, separated by 40 cm, are set within a wooden frame constructed of plywood and studs painted with epoxy. Raceways used for growout purposes are similar but are 212 by 56 by 30 cm, with a water depth of 20 cm. As with all our facilities, we modify them to accommodate the test. For example, we can place a support within the raceway at 10 cm which will hold retention chambers so that specific groups or subsamples of fish can be assessed. These chambers consist of an appropriate size of Petri dish with a Nitex mesh screen (sleeve or collar) glued to it with silicone to make a columnar vessel extending up and out of the water. This chamber also can be removed temporarily so that fish can be condensed into the chamber's dish, photographed as an entire group, counted, and examined for mortalities and abnormalities. Photos taken periodically during a time sequence over a 28-day period can then be assessed with image analysis to document growth of individuals in response to an exposure or condition. Such chambers can be maintained in static, static renewal, or flow-through systems.

Most growout occurs in 37-L aquaria set within a wooden water-bath, supported from the bottom with split PVC pipe so water can flow, encircling the lower portion of each aquarium in the system. The bath is 240 by 121 by 17 cm, with a water depth of 10 cm.

It is plywood covered by fiberglass and a gel-coat (Figure 2.5). The typical 37-L aquarium is 490 mm long by 245 mm wide by 310 mm high (inside measurements), with a standpipe PVC drain covered by Nitex at its bottom and reaching 7 cm from the top of the aquarium (Figure 2.6). Each aquarium is covered by a glass top, primarily to assure the fish remain segregated. Each top has a handle and with two corners cut at angles to allow passage of food and air line. In some cases, each aquarium has its own dip net to prevent contamination. When large sample sizes are required, 300-L raceways are used rather than glass aquaria.

When eggs or fry need to be collected for specific purposes or in experimental situations, a plexiglass (polycarbonate) spawning chamber insert with a Nitex screen bottom with mesh large enough for eggs or fry to drop through is often the best or easiest method (Figure 2.7). Below this screen can be placed a tray with a finer mesh to collect eggs. Such a system can be introduced into static or flow-through systems so that accurate counts can be documented for number of eggs laid, eggs hatched, fry viable, and developmental abnormalities. We have used this successfully with the sheepshead minnow.

Fish culture

Medaka husbandry

The medaka has been used at the GCRL since 1982 and a fresh supply was purchased during 1984 (from Carolina Biological Supply Company, Burlington, NC) at which time all previous stocks were destroyed. There is some debate as to whether a stock should be maintained with or without periodic outbreeding

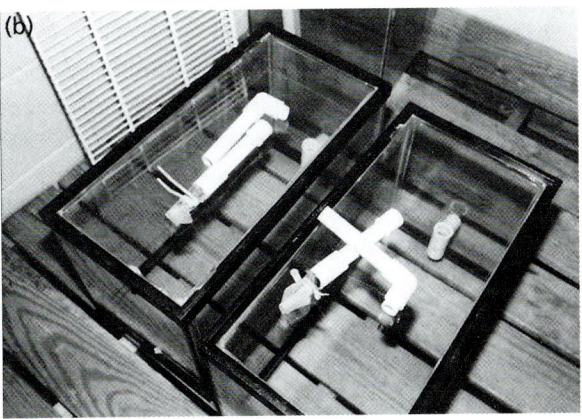

Figures 2.6a-b Aquaria (37-L) used for typical growout. 2.6a Series in water bath. 2.6b Unfilled, with disassembled PVC pipes.

Figure 2.5 Typical waterbath racks for growout, showing large raceway on right and 37-L aquaria on left.

Figure 2.7 Polycarbonate spawning chamber insert with a Nitex screen bottom with mesh large enough for eggs or fry to drop through. Note open middle portion at end for water flow, gravel in chamber to encourage spawning, and tray with fine mesh to collect eggs under chamber.

introducing new genetic material. We chose to maintain a large stock, interbreeding specimens continually without adding any new individuals. We have no empirical evidence in the form of reduced vigor, reproductive activity, and disease resistance or increase in spontaneous tumor rates that would indicate any acquired genetic weakness. Nevertheless, we pool fry from 4 to 12 raceways to initiate new brood stock and maintain some diversity. Offspring of this stock has been provided frequently to other facilities, helping them assure disease-free specimens for experimental purposes as well as providing identical material producing results that can be compared across laboratories. Sex ratio of our fish has remained about 50:50, and in a controlled longevity study at $26 \pm 1°C$, approximately 3% of the fish started dying at 9 months, with 90% mortality at 19 months when the study was terminated. We have adapted husbandry methods from those available elsewhere (e.g. Yamamoto, 1975; Kirchen and West, 1976), with special attention directed to producing large numbers of eggs and fry with a minimum of personnel labor.

Fry can be cultured with a stocking density of 50 per 19-cm inside diameter culture bowl (>50 may result in size variation and mortality) without aeration or with up to 1000 fish in flow-through 60 cm long by 40 cm wide by 14 cm high trays (24 L). Smaller commercial trays 33 cm by 28 cm by 14 cm can be used, if fewer fish are needed. Water for these trays enters through Tygon tubing, exits through a sponge filter apparatus, which serves to allow water to overflow into the water bath. Any plastic that comes in contact with fish is polycarbonate, conditioned PVC, or Tygon. In the larger containers, aeration is recommended. Aeration in the culture bowls is not necessary for medaka; however, about 80% of the water is exchanged every day, after the first feeding. On the other hand, aeration is necessary for fry of the marine fishes indicated above. In any case, the holding vessels are maintained in a fiberglass 235-cm by 76-cm by 15-cm water bath. Water in the bath is about 8 to 10 cm high, maintaining a constant temperature, typically $27 \pm 1°C$. The vessels are supported over the bottom by longitudinally split PVC pipe to allow water to flow within the water bath (Figure 2.8). Fry are exposed to a photoperiod initially consisting of 24 hours of continuous light for about 14 days to encourage feeding and growth. For fry earmarked as test organisms, the photoperiod is adjusted at least 48 hours prior to initiation of the test to simulate test conditions. Carcinogenicity and some toxicity assays are usually conducted beginning with 6- to 14-day-old

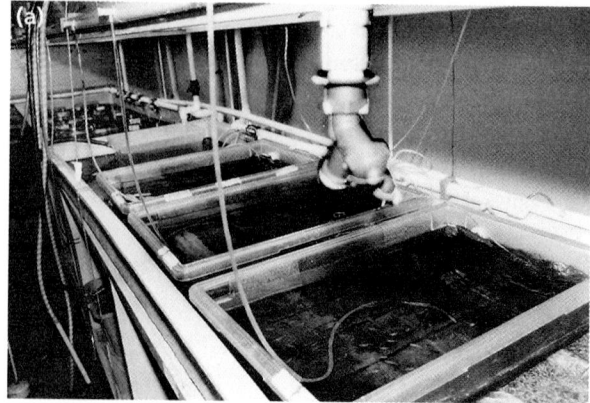

Figures 2.8a-b Containers in water bath used for first week or two of fry development. 2.8a Aerated trays in foreground are used for large numbers of fish. 2.8b Bowls are used for 50 or so fish. Aeration in bowls is necessary for marine fish but not for medaka.

fish, individuals which have rapidly developing livers and other organs. If fish are to be tested at an older age or to be used as brood stock, they are maintained for 14 days before transferring them to raceways or aquaria at a loading density of about 2.5 fish per liter.

Diet for these fry consists of a combination of a culture of a ciliate and another of a nematode from hatch to day 3. The nematode culture consists of both juvenile and adult worms and can be administered as food through day 6 or whenever the fish are transferred from the fry-development facility. Beginning with day 3 and continuing through day 14, we additionally feed nauplii of $\leq$ 24-hours old post-hydration brine shrimp two times daily, with a total of about 0.125 ml of nauplii per fish per day. After day 14, we feed a commercial flake diet from one to three times per day, depending on the final deposition of the fish, supplemented with a single daily allotment of brine shrimp late in the day to assure that the flakes are eaten.

Protocols for fish in an experiment or in an exposure system usually require a single daily feeding of commercial flake food, which is initiated at least 48 hours prior to the fish being placed into the system, and a single feeding of brine shrimp. In some cases for exposures lasting less than a month, we have fed brine shrimp only. We have conducted several 16-month continuous exposure tests, and by the time fish become adults, they receive 0.50 ml of brine shrimp per day. We use a commercial flake food with a known composition to assure adequate nutrition, including vitamins and minerals. At least one product, TetraMin, produces healthy fish, but appears to us to contain some additional special properties that act against some parasitic agents (Solangi and Overstreet, 1980), the resulting fish stock may be inadequate for use in some types of bioassays.

In regard to brine shrimp—there are several identifiable strains of *Artemia salina* and *Artemia franciscana* (one of the latter from Great Salt Lake) – one should use desiccated cysts from the same lot for each set of comparative experiments. If toxicological studies are being conducted, the lot should be analyzed for contaminants. We analyze each lot for the test substance, total organochlorine and total organophosphate pesticides, and polychlorinated phenols. Certification may be provided by some suppliers. Stocks of brine shrimp from different areas develop at different rates and provide different nutritional value. Those young nauplii, for example 24-hours old, are still nonfeeding and have a high level of fatty yolk. One may also desire to use as food older feeding brine shrimp that have been allowed to feed on and thereby concentrate specific nutritional supplements. Young brine shrimp are produced in separation funnels or special devices, so that nauplii can be separated from cyst remains before using them as a food source. We hatch cysts in multiple 44-L fiberglass hatching funnels enclosed within an opaque plexiglass housing unit. These cysts, usually 10 ml per liter, are incubated with vigorous aeration and illuminated with continuous flourescent light in either natural or artificial salt water (15 ppt, 15 g per liter, but 15 to 35 ppt all work well) at 25 to 28°C for 24 or 48 hours (depending on time necessary to hatch a specific brand of cysts). After the aeration is discontinued, the brine shrimp and associated cyst remains are poured into a dip net, allowing the water to pass through; they are then transferred to a light/dark separation chamber and allowed to migrate to the lighted end away from the cyst casings and are again captured in a dip net from a petcock or siphon (Figure 2.9). If destined for a freshwater

Figure 2.9 Brine shrimp hatching container. A variety of production systems can be used. This system has a boxed enclosure and dual containers so that brine shrimp can be collected daily from one of the two containers. After aeration is stopped and shrimp collected from spigot at bottom, additional spores and casings are removed using a separate light–dark sorting device.

static exposure aquarium where there would be a slow build-up in salt concentration, the nauplii are rinsed with aquarium diluent water; otherwise, they are rinsed with clean hatching solution and, in any case, diluted to a concentration appropriate for feeding. If necessary to purify the nauplii, they could be resuspended in the separation chamber. In any event, about 50 ml of the clean concentrated brine shrimp are added to the aquarium water to make a dispensing volume of 2.5 L, and we have calibrated this volume to contain about 820 nauplii per milliliter. The amount we feed depends on the age and species of fish. Young or older fish that eat unhatched cysts or cyst casings can develop pathological alterations in the body cavity or visceral organs, so separation of these items is important. The small amount of salt added to a flow-through system is rapidly flushed. To avoid microbial contamination to the fish from the brine shrimp, we recommend rinsing the cysts in an iodine solution (e.g. Wescodyne; 100 mg iodine for 15 minutes at pH 6.0 to 8.0 for brine shrimp cysts, and the same treatment can be used for fish eggs after the water hardening process) before introducing them into culture. Many commercial products have already gone through this process, and, for them, such a treatment is not necessary.

Nematodes, also called microworms, cultured in our laboratory consist of a monoculture of a free-living soil nematode (*Panagrellus redivivus* based on the study by Hechler [1971]). The initial culture was originally purchased from a tropical fish supplier, and subcultures have been prepared since then with instant or quick-cooking oatmeal, preferably finely ground, and granulated baking yeast (Fleischmann's RapidRise Yeast) in closed plastic containers with holes (filled with filter floss) bored in the top to provide for escape of gas and for prevention of contamination with fruit flies (Figure 2.10). The culture consists of slightly moist oatmeal made with hot water. Two pinches of yeast, 30 g of oatmeal, and a thin surface layer of a subculture of the worms at room temperature will produce a culture that lasts 2 weeks, with maximal production at about week 1. Clean masses of worms can be collected by temporarily placing clean glass plates, bottoms of Petri dishes, or plastic pipettes on top of the culture or by removing worms climbing up the side of the container with a clean cotton swab. Worms can be placed in a beaker of well water, rinsed as necessary, and concentrated on the bottom in an approximate 1:10 worm:water dilution before feeding with a pipette. We have calibrated our feeding solution as about 4300 worms per milliliter. The ciliate in our facilities is a monospecific culture of *Paramecium caudatum*. It is propagated in 11-cm diameter glass stacking culture bowls with about eight wheat seeds per 250 ml of pasteurized well water or spring water along with the subculture of the ciliate, which also contains a flagellate monoculture culture, which in turn serves as food for the ciliate. We periodically purchase and test a new starter culture from a scientific supplier (e.g. Carolina Biological Supply Company, Burlington, NC).

As suggested above, the duration of the fry care depends upon the final disposition of the fry, and, in the case of fish reared as test organisms, protocol specifications depend on their age at test initiation. Fry care may be conducted under static renewal or flow-through conditions. We prefer flow-through conditions as the general rule for medaka that will be used in testing. We maintain water temperature at $27\pm1°C$ for fry and all other stages except hatching eggs unless the protocol for a test specifies otherwise. On the other hand, the fish is a hardy species and can survive well in unheated static aquaria at room temperature, if exact conditions are not necessary.

Medaka breeding

Brood stock medaka usually produce eggs at 2 months of age, if reared optimally as indicated. If fish are not fed adequately in the first few days of life, sexual maturity does not occur until about 5 months. We maintain actively producing brood stock for about 4 months before we rotate them out of service. We recommend using only adequately fed brood stock that mature early. There occurs a tendency for the initially inadequately fed fish to ultimately develop skeletal abnormalities. Brood stock used to produce large numbers of embryos and fish for testing purposes are typically maintained in 300 L of water in the flow-through, fiberglass raceways described above, but other containers such as a 114-L (30-gal) glass aquaria could be used. A photoperiod of 16 hours light to 8 hours dark is recommended, and it can be achieved with fluorescent bulbs and incandescent bulbs with dimming ballasts that provide 15-minute periods simulating dusk and dawn. That photoperiod and a constant temperature of $27\pm1°C$ gives good production of eggs without altering the pH or hardness of our local water. These brood stock are fed four times each day, consisting of one with brine shrimp and three with ground flake food (e.g. Prime Tropical Flakes-Yellow, Zeigler Brothers, Inc., Gardners, PA). Raceways are siphoned to approximately 75% of the maximum volume at least once a week to remove

Figure 2.10 Microworm culture near end of its useful life. Pipettes are used to congregate clean masses of nematodes; additional masses can be swabbed from sides.

uneaten food and waste. If static renewal tanks are used, feces and debris are removed from each tank in conjunction with a minimum 25% water change at least twice weekly. Also, brine shrimp should be fed late in the day to assure the fish eat all the flakes, obtaining the necessary dietary requirements not present in the more preferred brine shrimp.

Female medaka expel their eggs almost immediately after ovulation, either just prior to or during the courtship and fertilization process. Usually clutches of 10 to 30 eggs remain attached to their vents. When light comes on at, for example, 07.00, eggs are usually deposited before 08.30, allowing for favorable harvest of those eggs with a hardened chorion (water-hardened) at 09.30. The clusters of eggs remain attached to the ventral surface of the females until they become dislodged through swimming actions or by brushing against a surface. In the natural environment, the eggs with their filaments become attached to vegetation. A variety of methods can be used to obtain eggs. These include vacuuming or siphoning the tank bottom, placing nets in the tank, and manually stripping the females. We, however, prefer to place several (30 to 50) cylindrical sponge filters on the bottom so that female medaka can brush their eggs onto them. These are in addition to the three to five twin filters (depending on the stocking density) already used to augment the aeration provided by one or two airstones. One can remove the sponges and squeeze water from them an hour after the eggs are deposited, without any damage to the eggs. Manual removal of the eggs is easy, and we typically obtain 1000 to 1500 eggs daily over a 3- or 4-month period from a 300-L raceway containing 500 females and 500 males. On the day prior to any egg collections, one should siphon raceways or containers thoroughly to remove any previously deposited eggs. Sponge filters should be replaced to assure that eggs collected the next day are all from that day's spawn.

Following removal of eggs from the sponges or obtaining them by other methods, we evaluate them microscopically for viability and then transfer them in groups of about 1000 or fewer into 3.8-L (1-gal) glass hatching jars containing embryo-rearing solution (Yamamoto's Isotonic Ringer Saline, prepared with 10 ml of each of the following stock solutions: NaCl, 100 g/L; KCl, 3 g/L; $CaCl_2 \cdot 2H_2O$, 4 g/L; $MgSO_4 \cdot 7H_2O$, 16.3 g/L and add those 40 ml of stock solution to 960 ml of distilled water). To the autoclaved solution can be added 1 to 5 mg/L of methylene blue to inhibit fungi; however, this or other such antifungal chemicals should not be used if the fish are to be used in carcinogenicity or toxicology tests. Yamamoto's solution permits better development than fresh water, plus it can be used to keep an unfertilized egg in fertilizable condition for artificial insemination. Cryopreserved spermatozoa will successfully fertilize nearly all such eggs, producing hatched fry (Aoki *et al.*, 1997). The contents of the hatching jars in our laboratory are vigorously aerated to keep the eggs suspended in the water column. Hatching jars are maintained in a water bath at a temperature of $24 \pm 1°C$ under continuous fluorescent lighting (Figure 2.11). This temperature is less than the $27°C$ occurring elsewhere throughout the system to discourage fungal development. Unfertilized eggs, dead embryos, and fungus-covered embryos are removed, tallied, and discarded daily, followed by replacement of a fresh container of hatching solution. To accomplish this, one removes the air line from the jar, eggs drop to the bottom, the fluid is decanted off, and a small volume of fluid with the eggs is swirled and

Figure 2.11 Aerated hatching jar in water bath. Hatching solution in the incubation chamber is changed daily, eggs are examined for contamination or other conditions, and those eggs, usually ones fouled with fungus, are removed to insure optimal production.

dumped into a clean dish for examination. Healthy eggs are then put back in the jar, which has received an exchange of rearing solution. We typically achieve about 95% survival of eggs unless they are in a static test without aeration.

Chorionic filaments on the eggs cause the eggs to clump in the hatching incubator, which in turn promotes the development of fungus and affects hatching rate. These filaments can be cut with iris scissors or teased with glass pipettes, but that is labor intensive. As a more practical solution, a large group can be gently massaged against a Nitex mesh (0.75 mm) screen (Figure 2.12) until the filaments 'ball up', leaving relatively filament-free eggs with minimal mechanical damage. No loss in sensitivity of embryos result from the massage technique, and filaments are cut only when a study must account for each egg.

Hatching begins after approximately 9 days. Newly-hatched fry are removed from the hatching vessels and distributed into approximately 24-L polycarbonate flow-through culture trays and treated as indicated above. In our system, there is a high majority of uniformly sized fry. The few large and small individuals are not used for breeding or testing. If not enough fry can be obtained from all uniform-sized fish in the group born on the tenth day, one can use fish born over a 2- to 4-day period. In regard to comparison of systems in the zebrafish model, the medaka can also produce haploid embryos. These embryos have been produced by activating ova with sperm attenuated by toluidine blue (Uwa, 1965).

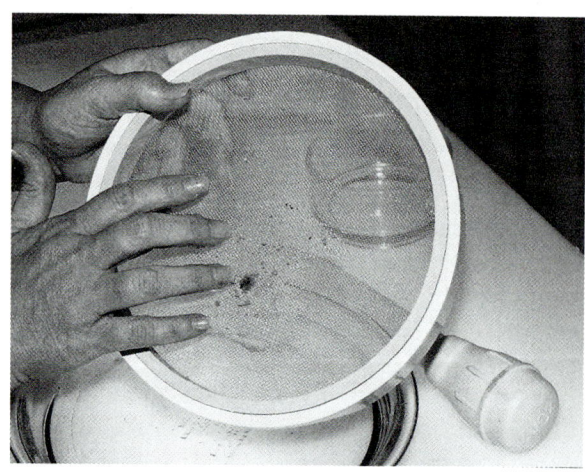

Figure 2.12 Nitex screened ring used to remove filaments from eggs. This procedure eliminates clumping of eggs in hatching incubators (such as that shown in Figure 2.11).

Guppy husbandry

Separate methods are also provided for the guppy (*Poecilia reticulata*) because it differs from the medaka by being a live-bearer, complementing the medaka in concurrent experiments, able, similar to the medaka, to grow in either fresh water or in salt water up to 12 ppt for long-term exposures. Our GCRL guppy stock was initially purchased as a couple of hundred individuals (non-pure, red king cobra strain, Aqua World, Inc., St Louis, MO) and has been maintained as an established culture without outbreeding since 1984. The fish is easy to culture and serves as a good bioassay model. In a controlled longevity study at $26\pm1°C$, it lived longer than medaka; only 15% died by 16 months and 34% by 27 calendar months when the study was terminated. Methods described below consist primarily of those that differ from those used for medaka.

Guppies destined to be brood stock are fed flake food three times daily and 0.5 ml of 24 ± 4 hours post-hydration brine shrimp nauplii per fish once per day whether young or adult. Residual food and feces are removed by siphoning at least once weekly. Prior to initiation of a study, existing fry progeny can be removed from all relevant brood tanks so parents do not get used to feeding on their young. Depending on the needs of the institution, one may remove or select young adults of specific phenotypes. Fish not required for brood stock or experiments make a good food source for other fish species being maintained. Brood stock and prospective brood stock should be kept in separate raceways.

Fry to be used in experiments are removed and distributed into 24-L polycarbonate flow-through culture aquaria containing aged and carbon-filtered well water. The aquaria, in turn, are maintained in a water bath at $\pm1°C$ of the desired test temperature under continuous fluorescent lighting. Fry are fed flake food three times daily and 24 ± 4 hours post-hydration brine shrimp nauplii once daily. Debris is siphoned from each aquarium, and 20 to 30% of the water removed at least once weekly if in a flow-through system and twice weekly, if in a static one. Temperature is measured daily in each aquarium and pH and dissolved oxygen twice weekly. Ammonia and nitrite are measured once during the culture period.

Guppy breeding

Guppy brood stock used to produce fry for study generally ranges in age from 3 to 12 months old and is maintained under flow-through conditions at an approximate two male to three female ratio and at a stocking density of not more than 1200 fish per 300 L of water, with a maximum density of 3.0 g fish per liter of water. Groups are maintained in multiple fiberglass raceways. Routinely, these groups of brood stock are rotated out of service at 12 months rather than at 6 months as done with medaka. Aged, carbon-filtered well water is provided to each raceway at a sufficient flow rate to provide a minimum of one volume addition per day. Water in each raceway is maintained at $27\pm1°C$ with heaters in various water pretreatment locations and in each raceway. A daily photoperiod of 16 hours light to 8 hours dark is provided by timer-controlled fluorescent lights. Each brood stock raceway is equipped with biological sponge filter aerators as in medaka brood stock raceways.

Sexual dimorphism is conspicuous in the guppy, with the smaller male evident from his gonopodium, an extension of the anal fin, along with modified pelvic fins used to transmit sperm in spermatophores to the female. Sperm, which can survive within the female for up to 8 months, can fertilize successive batches of eggs after prior ones develop to active fry. Shortly after being fertilized, the female develops a dark area near her vent, and it increases in size as the embryos develop. Under specific conditions, such as the lack of a male, a female can produce functional sperm.

Brood stock producing fry for experiments are fed flake food three times daily and 24 ± 4 hours post-hydration brine shrimp nauplii once daily. Production can be increased by increasing the feeding regimen and maintaining good water quality. Residual food and feces are removed by siphoning at least once weekly. Prior to initiation of a study, existing fry progeny are removed from brood tanks. Artificial refuges such as plastic vegetation, plastic filaments (plastic breeding grass), or nylon screen are provided as cover for escape of young from predation by the adults. If fish are not to be used for bioassay animals or if introduction of possible unknown or inconsistent factors makes no difference, bargain plastic Christmas trees (usually containing wire or other potential sources of contamination) or natural vegetation are practical. If live vegetation is used rather than artificial plastic grass, it can be purchased from a pet store or obtained from the wild and treated with Dylox or some other chemical product to eliminate insects and other invertebrates. If Dylox (sold as 'Lifebearer' in the aquarium trade) is used, two treatments of 0.5 to 1.0 ppm O,O-dimethyl-1-hydroxy-2-trichloro methyl phosphonate separated by 7 to 10 days is sufficient. Natural vegetation has an advantage over artificial material because it can help reduce nitrogenous wastes by using waste products, it can lower carbon dioxide levels and therefore promote fry production, and it can provide nutrition.

Collection of fry the morning after preparation of brood stock aquaria may provide the desired number necessary for an evaluation. The number of brood fish must be estimated and produced or obtained before the specific number of fry are needed. If consistency of temperature is not important, temperature can be reduced when fry are not needed, and then an increase in temperature by several degrees will promote their production. Under normal constant temperatures as indicated above, we can expect up to 1500 fry per day from a 300-L raceway at $27\pm1°C$ with about 800, well-fed, 1-year-old fish with an equal ratio of males and females. Using younger fish, we usually obtain 300 to 500 fry per day. With natural vegetation, heavy feeding, and removal of wastes every day or two, a short period of high fry production can be achieved. If necessary, fry collected and pooled over 2 to 3 days may be appropriate (depending on the type of experiment), even if 6- or 10-day-old individuals are required for the experiment. The guppy fry are well developed at birth and could be used immediately or whenever needed in tests without fear of mortality from stress. If older fish are required, fry from longer collections can be combined and randomized or indiscriminately distributed.

Because there is a specific gestation period for development of the embryos, usually about 3 weeks, depending on temperature and other factors, broods are produced optimally every 4 weeks. The new-born, free-swimming fry are released over a short period of time, rather than as a continual daily release like the eggs of medaka and many other egg-layers. Consequently, fewer guppy fry may be produced per female, and those fry are not necessarily produced in a reliable daily output.

Sheepshead minnow husbandry

A marine small fish species is often necessary for a variety of reasons. One might be specified for a test, one might be necessary for field comparisons, intestinal as opposed to gill entrance of a toxicant might be important, a fish or fish group might be necessary for investing a parasite or disease, or some other reason. One fish cannot represent all marine or estuarine species, but the sheepshead minnow serves as a good example because it is a USEPA-approved bioassay model, has been used for toxicity testing for over 25 years, and is easy to produce. We have had the species in culture without disease continuously since the early 1980s, a specific stock in culture since 1988, and separate stocks used for interbreeding on hand since 1991 and 1998.

Sheepshead minnow stocks can be easily maintained in 150- to 950-L raceways that serve as part of a four-raceway, recirculating system (Figure 2.13). In our laboratory, the largest raceway, 950 L and circular, serves as a reservoir and filter and does not hold fish. That raceway has a bed of crushed coral for a buffer as well as a heater, chiller, thermostat, float switch with solenoid to regulate addition of water, and a regulator for dissolved oxygen. To it is attached a biofilter in continuous use and a carbon filter used on occasion. The system is a reverse-flow type, with water entering from the bottom. Water from this reservoir can be directed to growout tanks if a closed system is desired or tanks can receive a separate water supply that is continually added and discharged. Raceways for holding fish also have a separate manifold, permitting a jar that holds eggs to be set in the raceway and that can be provided a continuous flow of water that overflows into the system. Our system can accommodate automatic feeders so fish can receive small portions of commercial flake food on numerous occasions throughout the day.

If there is no need to obtain large numbers of identically treated fish, the sheepshead minnow can be easily maintained in 37- to 114-L aquaria with sponge filters and a 30 to 40% water exchange once or twice a week. One can maintain a few fish with no water exchange unless water quality deteriorates. Organic material should be siphoned from the bottom once a week. Cloudy water serves as a good sign that the quality has dropped, organic levels have increased, and bacterial levels have increased. An 80% water exchange or use of a vortex filter will allow time for the culturist to correct the fundamental problem.

Sheepshead minnow culture offers different challenges than culturing medaka and guppy. Fry are carnivorous compared with medaka, and runts are usually eaten. Maintaining stocks for several years seems to have selected against the few large dominant and aggressive fish. Consequently, most fish are roughly the same size. Early experience produced a great range in size of fish, and fish had to be sized by length to decrease cannibalism, fin biting, and development of poorly feeding runts.

Sheepshead minnow breeding

The sheepshead minnow has not been inbred like the guppy and to a lesser extent the medaka, and consequently it and the other marine species indicated above have to be bred using methods that simulate natural spawning behavior. Methods that provide a good supply of healthy fish have been perfected in our laboratory. Initially we maintain our prospective brood stock, about 400, in 300-L raceways with an ambient temperature of 21 to 25°C and a photoperiod of 16 hours light and 8 hours dark. Embryo production can be enhanced by increasing the temperature to 28 to 30°C and feeding a supplement of frozen adult brine shrimp. Fish normally are fed brine shrimp once a day and flake food three times. Most males remain in spawning condition throughout the year and are easy to distinguish by the iridescent blue dorsum. Spawning is most easily achieved by using 'Schesny rings' (Figures 2.14 and 2.15). These are PVC pipes cut into rings used as chambers; rings can be 210 mm internal diameter and 50 mm high. The top is lined with nylon tulle (wedding mesh) fabric and a removable bottom insert is lined with a fine Nitex screen (mesh not to exceed 0.75 mm, but large enough to allow water flow). Smaller ones, identical except 111 mm in diameter by 40 mm high, also provide good recovery of eggs. These chambers, usually five, can be placed in a shallow water raceway with about 18 cm of water above the chambers and serve as a territory for the male fish. Six ripe males and eight ripe females are placed into the raceway. On the other hand, to

Figure 2.13 Three-tiered recirculating saltwater system for holding stocks and for spawning broodstock of marine fishes. The cylindrical raceway under the middle raceway is a sand filter.

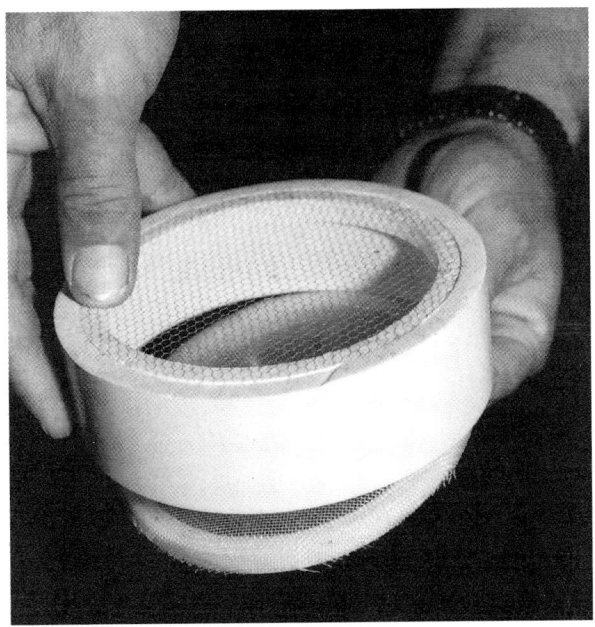

Figure 2.14 'Schesny rings' used as artificial spawning territories used to collect eggs from sheepshead minnow.

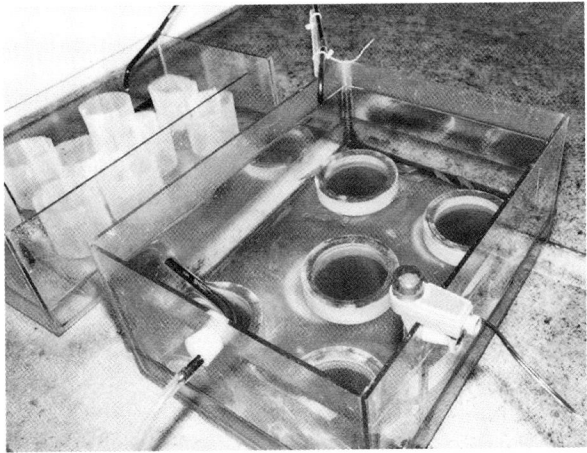

Figure 2.15 'Schesny rings' arranged in aquarium within water bath to evaluate spawns from sheepshead minnow. Note Nitex retention chambers in adjacent aquarium used to isolate individuals for a variety of purposes.

obtain large numbers of eggs, we place several rings in a large brood stock holding raceway containing an equal male to female ratio. Each male selects a territory (chamber) and attracts females to spawn over them. The fertilized eggs become trapped in the chamber and can be removed by removing the internal ring and flushing them out with a squirt bottle. They can then be accurately counted and placed into an aerated incubation jar or separatory funnel. The best method seems to be with the 21-cm ring because that size simulates a natural territory and because that ring fits well in a bucket, where they can be collected and disinfected. On the other hand, small ones can be spaced further apart in an aquarium or raceway, reducing competition among males. If competition is reduced further by placing one male and three females into a 21-cm diameter by 22-cm high, closed chamber, results are often inconsistent and an aggressive male can harm or kill the trapped females. Additionally, eggs can be collected from sponge filters added to the brood stock tank or from aquarium inserts with Nitex mesh bottoms. Gravel distributed into territories helps spawning activity. The method using the Schesny rings, however, allows better production and more accurate means of measuring egg production and analysis of abnormalities in embryos.

A special apparatus constructed from PVC pipe ('gravel station', Figure 2.16) is more difficult to build compared to the rings and has other disadvantages. However, if one requires a continuous supply of eggs or fry, preferably not for a homogeneous supply for bioassays, the station can provide a continual supply for other purposes without much labor input.

For studies in separate facilities conducting research on parasites and diseases, we routinely use laboratory-reared sheepshead minnows to cycle nitrogen in our aquaria for later introduction into aquaria of various marine animals. Basically, we use a commercial biological conditioner (e.g. Fritz-Zyme #9) as a starter culture of a mixture of nitrifying bacteria to reproduce in the sponge aerating filters. Appropriate levels of bacteria are reached after a week of daily drops of ammonia are introduced, either from a reagent bottle or as waste from a sheepshead minnow.

Eggs hatch in about 3 days at 30°C and can be treated like medaka eggs, except that fry in bowls or trays require salt water, aeration, and microworms for an entire week along with brine shrimp. Flake food and brine shrimp should constitute the diet during the second week. Egg production is high at 15 or 20 ppt. When salinity is lowered from 15 to 10 ppt, daily egg production of about 100 eggs per 4-month-old female is reduced. Production is significantly inhibited at 5 ppt and ceases or nearly ceases at 2 ppt or less.

Eggs can be obtained by injecting females that have been acclimated for a week at about 30°C with two intraperitoneal injections of 50 units of human chorionic gonadotropin (HCG) or smaller amounts of a luteinizing hormone-releasing hormone analog 24 hours apart. Two days after the last injection, eggs can be stripped by squeezing, massaging, or dissecting the female into 50 ml of water from the acclimation

Figure 2.16a-b A special apparatus termed a 'gravel station' that is useful for a continual supply of eggs or fry of sheepshead minnow. 2.16a Apparatus out of water. 2.16b Apparatus *in situ*, showing fish attracted to station.

tank. Sperm from macerated testes of a few males is added to the eggs, mixed, and incubated at 30°C for 1 hour. Eggs are then transferred to the egg incubation units. This method does not provide high rates of hatching, normal development, and survival of fry. Such a method without a homogeneous supply of fry can be used on a fish species for which the methods have not been worked out and eggs are not needed on a continual basis.

General growout methodology

In some experiments, medaka, guppy, or other fishes are exposed to a selected toxicant in a static, static-renewal, or flow-through system for a period of time and then prepared to be transferred to growout facilities. We consider a growout facility to be a system that holds fish for a period of time in noncontaminated water after prior exposure to a test substance to determine if neoplasia or other conditions develop. Following termination of test substance exposure, aquaria receive clean diluent water from the exposure system for 1 to 3 days. This 1- to 3-day interval affords the test fish a recovery period, thereby minimizing mortality due to sampling stress and eliminating test-substance exposure to technical personnel during the removal of fish from aquaria. The fish are then removed from the exposure-aquaria, counted, and transferred to growout tanks. Any dead or moribund fish discovered during this process are fixed for histopathology.

Treated and control fish removed from the exposure system are randomly assigned to growout tanks for a length of time determined by the endpoint of the study. We typically use growout tanks in duplicate, but occasionally in triplicate, that are either 37-L, all-glass aquaria in water baths or 300-L fiberglass raceways. Whether aquaria or fiberglass raceways are used depends on the number of exposed fish and the length of the holding period. A drain siphon designed to remove water from near the bottom of each glass aquarium or raceway maintains a constant depth and volume. Growout tanks are maintained at ±1°C of the desired temperature, typically 27°C using temperature-regulated, two-tiered, water baths, with the same source of filtered water delivered as diluent water from the head box to the experimental aquaria. Flow rate in aquaria is adjusted to maintain approximately a minimum of two volume additions per day and to optimize water quality. The rate is monitored at least once weekly and corrected, if necessary. Growout tanks are covered to prevent the escape of fish and are raised as necessary for feeding, tank maintenance, and observations. Covers are not raised from adjacent tanks simultaneously.

Following growout for the prescribed period, fish are removed and prepared for histopathological examination. For example, with a study beginning with 6-day-old fish, we may use an initial 100 fish per replicate 37-L aquarium and then remove subsamples of 35 for histological examination (a total of 70) at two periods sometimes between 4 and 12 months. Removing the first subsamples leaves enough room for the remaining fish to grow uninhibited. During the growout period, fish are observed for mortality and physical or behavioral abnormalities at least twice daily. Each fish observed with defined abnormalities during the sampling process is isolated for unique identification (e.g. Figures 2.17 and 2.18). Most neoplasms and other

Figure 2.17 Medaka on top exposed to chlorodibromomethane plus carbon tetrachloride, showing enlarged liver, compared with non-exposed control fish below it.

Figure 2.18 Medaka that had been exposed to methylazoxymethanol acetate, exhibiting an obvious neuroblastoma from the lateral line.

pathological alterations are not visible grossly, and fish must be sectioned to detect those lesions. Unusual observations, or absence thereof, are noted on the documentation accompanying the custody transfer of fish. Dead fish and those fish considered moribund are removed, externally examined, and fixed for histopathological analysis. Routinely, fish are fed brine shrimp nauplii and flake food once daily during growout. Uneaten food and other debris are siphoned from each growout tank at least once weekly to minimize deterioration of water quality.

Fish health

In this chapter, we have assumed that most users are interested in producing large numbers of fish for study. Our methods center around keeping a stock in the laboratory for long periods of time without introducing fish from the wild or from other sources. This procedure eliminates or reduces the opportunities for introduction of most infectious agents. We do not allow wild fish or fish from other sources in the same buildings without quarantine. We routinely monitor representative individuals for disease-causing agents. Examination for bacterial agents on culture plates and histological sections also reveals overall health of the animals. For example, the presence of granulomas caused by *Mycobacterium marina* or related bacteria could indicate a chronic stress condition (Abner *et al.*, 1994). On the other hand, the bacteria could cause a disease in both freshwater and marine fishes as well as cause granulomatous lesions on the hands and arms of aquarists (e.g. Spotte, 1991; Stoskopf, 1992).

Most researchers would probably not accept fish for use as bioassay animals if they were known to be infected with any agent or had been treated with antibiotics or other compounds. Nevertheless, there are occasions when infections develop during a test or in fish being held. David Powell (Chapter 4) treats some of those. There are volumes written in the scientific and lay literature on the subject, but we might reference at least the books by Stoskopf (1992) and Noga (1995). Herwig *et al.* (1979) list numerous compounds useful in treating diseases. Prevention is also important, and some researchers not using fish for toxicological purposes have a plethora of commercial products useful in treating or preventing disease. In the case of fungus on fish eggs, which we have mentioned and which can become serious unless the aquarist performs the labor-intensive task of removing infected (internal) or infested (external) eggs, we have already mentioned treatment with methylene blue, which is toxic to nitrifying bacteria, and an iodophor such as Wescodyne (50 to 100 ppm for 10 minutes should kill surface viruses, but not all species of bacteria). Proflavine should kill those troublesome bacteria. Salt is a safe treatment to kill saprolegnia (fungus), and 30 ppt is recommended by some, but we find that 5 ppt eliminates a common strain that infects local freshwater fish and eggs without promoting contamination by other agents. Malachite green (0.15 ppm indefinitely or 1 to 5 ppm for 1 hour) is effective but harms vegetation and might be carcinogenic; phenoxyethanol (100 ppm indefinitely) does not affect plants.

Acknowledgments

Numerous people at or from the Gulf Coast Research Laboratory Institute of Marine Sciences have taken part in developing the procedures indicated above or otherwise assisted in the presentation; these include William W. Walker, Alex 'Buck' Schesny, Nate Jordan, Ronnie Palmer, Rena Krol, Marie Wright, John Fournie, Wolfgang Vogelbein, Cherie Heard, Pam Monson, Helen Gill, and others. We thank W. Duane Hope of the National Museum of Natural History, Smithsonian Institution, for identifying the nematode used for an early diet. These studies have been supported by the National Institutes of Health (National Cancer Institute) contract N01-CP-61070; NIEHS (National Toxicology Program) contract no. NO1-ES-35371; US Environmental Protection Agency contract nos. CR-816007-01 and CR-820641-01; International Paper, and US Department of Commerce, NMFS, award no. NA86FL0476 and NA96FL0358. Our use of brand names and sources of supplies is not meant to be an endorsement by the Institute of Marine Sciences or Academic Press.

References

Abner, S.R., Frazier, C.L., Scheibe, J.S., Krol, R.A., Overstreet, R.M., Walker, W.W. and Hawkins, W.E. (1994). In *Modulators of Fish Immune Responses*, Vol. 1 (eds J.S. Stolen and T.C. Fletcher), pp. 219–234. SOS Publications, Fairhaven, NJ.

Angus, R.A. (1983). *J. Hered.* **74**, 81–84.

Anken, R. and Bourrat, F. (1998). *Brain Atlas of the Medakafish* Oryzias latipes. INRA Editions, Versailles, France. 92 pp.

Aoki, K., Okamoto, M., Tatsumi, K. and Ishikawa, J. (1997). *Zool. Sci.* **14**, 641–644.

Armstrong, P.B. and Child, J.S. (1965). *Biol. Bull.* **128**, 143–168.

Beck, A.D. and Poston, H.A. (1980). *Prog. Fish-Cult.* **42**, 138–143.

Benoit, D. (1982). User's guide for conducting life-cycle chronic toxicity tests with fathead minnows (*Pimephales promelas*). US Environmental Protection Agency, Report EPA-600/8-81-011, Environmental Research Laboratory, Duluth, MN.

Boorman, G.A., Botts, S., Bunton, T.S., Fournie, J.W., Harshbarger, J.C., Hawkins, W.E., Hinton, D.E., Jokinen, M.P., Okihiro, M.S. and Wolfe, M.J. (1997). *Toxicol. Pathol.* **25**, 202–210.

Bortone, S.A., Davis, W.P. and Bundrick, C.M. (1989). *Bull. Environ. Contam. Toxicol.* **43**, 370–377.

Cairns, J. Jr. (1986). *BioScience* **36**, 670–672.

Cantrell, S.M., Joy-Schlezinger, J., Stegeman, J.J., Tillitt, D.E. and Hannink, M. (1998). *Toxicol. Appl. Pharmacol.* **148**, 24–34.

Conover, D.O. (1985). *Fish. Bull.* **83**, 331–341.

Conover, D.O. and Kynard, B.E. (1984). *Environ. Biol. Fishes* **11**, 161–171.

Costello, D.P. and Henley, C. (1971). *Methods for Obtaining and Handling Marine Eggs and Embryos*, 2nd edn. Marine Biological Laboratory, Woods Hole, MA. 247 pp.

Couch, J.A., Courtney, L.A. and Foss, S.S. (1981). In *Phyletic Approaches to Cancer* (eds C.J. Dawe et al.), pp. 125–131. Japan Scientific Society Press, Tokyo.

Cunningham, J.E.R. and Balon, E.K. (1985). *Environ. Biol. Fishes* **14**, 115–166.

Davis, W.P. (1988). *Environ. Biol. Fishes* **21**, 81–90.

Davis, W.P., Taylor, D.S. and Turner, B.J. (1990). *Ichthyol. Explor. Freshwaters* **1**, 123–134.

Denny, J.S. (1987). Guidelines for the culture of fathead minnows (*Pimephales promelas*) for use in toxicity tests. EPA-600/3-87-001. US Environmental Protection Agency, Environmental Research Laboratory, Duluth, MN.

Echelle, A.A., Echelle, A.F., DeBault, L.E. and Durham, D.W. (1988). *J. Fish Biol.* **32**, 835–844.

Fournie, J.W. and Overstreet, R.M. (1993). *J. Parasitol.* **79**, 720–727.

Fournie, J.W., Hawkins, W.E., Overstreet, R.M. and Walker, W.W. (1987). *J. Natl. Cancer Inst.* **78**, 715–725.

Hansen, D.J. (1978). In *Bioassay Procedures for the Ocean Disposal Permit Program*, pp. 107–108. USEPA-600/9-78-010.

Hatanaka, J., Doke, N., Harada, T., Aikawa, T. and Enomoto, M. (1982). *Jpn. J. Exp. Med.* **52**, 243–253.

Hawkins, W.E., Overstreet, R.M. and Walker, W.W. (1988). *Aquat. Toxicol.* **11**, 113–128.

Hawkins, W.E., Fournie, J.W., Battalora, M.St.J. and Walker, W.W. (1991). *J. Aquat. Anim. Health* **3**, 213–220.

Hawkins, W.E., Walker, W.W. and Overstreet, R.M. (1995). In *Fundamentals of Aquatic Toxicology, Effects, Environmental Fate, and Risk Assessment*, 2nd edn (ed. G.M. Rand), pp. 421–446. Taylor & Francis, Washington, DC.

Hawkins, W.E., Walker, W.W., James, M.O., Manning, C.S., Barnes, D.H., Heard, C.S. and Overstreet, R.M. (1998). *Mutat. Res.* **399**, 221–232.

Hechler, H.C. (1971). Taxonomic notes of four species of *Panagrellus* Thorne (Nematoda: Cephalobidae). *J. Nematol.* **3**, 227–237.

Herwig, N., Garibaldi, L. and Wolke, R.E. (1979). *Handbook of Drugs and Chemicals Used in the Treatment of Fish Diseases*. Charles C. Thomas Pub., Springfield, IL. 272pp.

Hsiao, S.-M. and Meier, A.H. (1989). *J. Exp. Zool.* **252**, 213–218.

Hughes, J.M., Harrison, D.A. and Arthur, J.M. (1991). *Biol. J. Linn. Soc.* **44**, 153–167.

Ijiri, K. (1995). *The First Vertebrate Mating in Space – A Fish Story.* RICUT, Tokyo, 57pp.

Ishikawa, T., Masahito, P. and Takayama, S. (1984). *Natl. Cancer Inst. Monogr.* **65**, 35–43.

Ishikawa, Y. (1997). *Fish Biol. J. Medaka* **9**, 17–31.

Iwasaki, N. (1989). *Guppies Fancy Strains and How to Produce Them.* T.F.H. Publications, Neptune City, NJ. 144pp.

James, M.O., Heard, C.S. and Hawkins, W.E. (1988). *Aquat. Toxicol.* **12**, 1–15.

Kirchen, R.V. and West, W.R. (1976). *The Japanese Medaka: Its Care and Development.* Carolina Biological Supply Company, Burlington, NC. 36pp.

Koenig, C.C. and Chasar, M.P. (1984). *Natl. Cancer Inst. Monogr.* **65**, 15–33.

Koenig, C.C. and Livingston, R.J. (1976). *Copeia 1976*, 435–445.

Krumholz, L.A. (1948). *Ecol. Monogr.* **18**, 3–43.

Law, J.M., Hawkins, W.E., Overstreet, R.M. and Walker, W.W. (1994). *J. Comp. Pathol.* **110**, 117–127.

Masahito, P., Aoki, K., and Ishikawa, T. (1996). *Fish Biol. J. Medaka* **8**, 15–19.

McInerny, D. and Gerard, G. (1958). *All About Tropical Fish.* Macmillan, New York. 480pp.

Middaugh, D.P. and Hemmer, M.J. (1984). *Estuaries* **7**, 139–148.

Middaugh, D.P. and Hemmer, M.J. (1987). *Copeia 1987*, 727–732.

Middaugh, D.P. and Hemmer, M.J. (1992). *Copeia 1992*, 53–61.

Middaugh, D.P., Hemmer, M.J. and Rose, Y.L. (1986). *Environ. Biol. Fishes* **15**, 107–117.

Middaugh, D.P., Hemmer, M.J. and Goodman, L.R. (1987). Methods for spawning, culturing and conducting toxicity-tests with early life stages of four atherinid fishes: the inland silverside, *Menidia beryllina*, Atlantic silverside, *M. menidia*, tidewater silverside, *M. peninsulae*, and the California grunion, *Leuresthes tenuis.* EPA/600/8-87/004, Office of Research and Development, US Environmental Protection Agency, Washington, DC.

Middaugh, D.P., Beckham, N., Fournie, J.W. and Deardorff, T.L. (1997). *Arch. Environ. Contam. Toxicol.* **32**, 367–375.

Noga, E.J. (1995). *Fish Disease: Diagnosis and Treatment.* Mosby, St Louis, MO. 367pp.

Ortego, L.S., Hawkins, W.E., Krol, R.M. and Walker, W.W. (1996). In *Techniques in Aquatic Toxicology* (ed G.K. Ostrander), pp. 327–339. Lewis Publishers, Boca Raton, FL.

Overstreet, R.M. (1993). In *Pathobiology of Marine and Estuarine Organisms* (eds J.A. Couch and J.W. Fournie), pp. 111–156. CRC Press, Boca Raton, FL.

Overstreet, R.M. (1995). In *Proceedings of the Ecotoxicology Workshop.* Curtin University of Technology, Perth, Western Australia, pp. 48–52.

Overstreet, R.M. (1997). *Parassitologia* **39**, 169–175.

Overstreet, R.M., Hawkins, W.E. and Deardorff, T.L. (1995). In *Environmental Fate and Effects of Pulp and Paper Mill Effluents* (eds M.R. Servos, K.R. Munkittrick, J.H. Carey and G.J. Van Der Kraak), pp. 495–509. St Lucie Press, Delray Beach, FL.

Solangi, M.A. and Overstreet, R.M. (1980). *J. Parasitol.* **66**, 513–526.

Spotte, S. (1991). *Captive Seawater Fishes: Science and Technology.* John Wiley, New York, NY, 942pp.

Staffa, J.A. and Mehlman, M.A. (eds) (1979). *Innovations in Cancer Risk Assessment (ED_{01} Study).* National Center for Toxicological Research, US Food and Drug Administration, and The American College of Toxicology, Pathotox, Park Forest South, IL. 246pp.

Stoskopf, M.K. (1992). *Fish Medicine.* W.B. Saunders, Philadelphia, PA. 882pp.

Tanaka, M. (1995). *Fish Biol. J. Medaka* **7**, 11–14.

Uwa, H. (1965). *Embryologia* **9**, 40–48.

Van Beneden, R.J., Henderson, K.W., Blair, D.G., Papas, T.S. and Gardner, H.S. (1990). *Cancer Res. [Suppl]* **50**, 5671s–5674s.

Vogelbein, W.K., Fournie, J.W. and Overstreet, R.M. (1987). *J. Fish Biol.* **31**(Suppl. A), 145–153.

Walker, W.W., Manning, C.S., Overstreet, R.M. and Hawkins, W.E. (1985). *J. Appl. Toxicol.* **5**, 255–260.

Walker, W.W., Heard, C.S., Lotz, K., Lytle, T.F., Hawkins, W.E., Barnes, C.S., Barnes, D.W. and Overstreet, R.M. (1989). In *Oceans '89*, Vol. 2, *Ocean Pollution*, pp. 516–524. IEEE Service Center, Piscataway, NJ.

Wallace, R.A. and Selman, K. (1978). *Dev. Biol.* **62**, 354–369.

Wallace, R.A. and Selman, K. (1980). *Gen. Comp. Endocrinol.* **42**, 345–354.

Weis, J.S. and Weis, P. (1989). *BioScience* **39**, 89–95.

Westerfield, M. (1995). *The Zebrafish Book: A Guide for the Laboratory Use of Zebrafish* (Danio rerio). 3rd edn. Institute of Neuroscience, Univ. Oregon, Univ. Oregon Press, Eugene, OR. 385pp.

Whitney, L.F. and Hahnel, P. (1964). *All About Guppies.* T.F.H. Publications, Neptune City, NJ. 128pp.

Wisk, J.D. and Cooper, K.R. (1992). *Arch. Toxocol.* **66**, 245–259.

Yamamoto, T-O. (1975). *Medaka (Killifish) Biology and Strains.* Keigaku Publishing, Tokyo, Japan. 213 pp.

Zimmerer, E. J. (1984). *Natl. Cancer Inst. Monogr.* **65**, 59–64.

CHAPTER 3

Diet

Douglas E Conklin
Department of Animal Science,
University of California, Davis, USA

Introduction

Cultured fish are increasingly popular experimental animals (see Table 3.1). Part of the impetus behind this trend is the availability of a wide range of technological advances in fish husbandry arising from commercial aquaculture. Over the last decade, it has become more and more apparent that capture fisheries are nearing their productive limits. This has resulted in a growing shortfall of fisheries products to meet the demand of the burgeoning global population. The emerging field of aquaculture is seen as a means to increase supplies of fish and shellfish even as natural harvests level off (New, 1997). In moving to produce ever-higher yields, aquaculturists have forged numerous improvements in fish culture technology including the development of effective formulated diets for several important food species.

This chapter attempts to review what is known about fish nutrition as it applies to the development of formulated diets for the culture of fish specifically for laboratory research. Optimal nutrition is as important for sound research studies relating to the normal physiological and biochemical responses of fish as it would be for any other animal (Mehrle et al., 1977). While available information on nutritional needs for most species would not support the criterion of optimal nutrition, sufficient details are available to describe the general nutritional requirements of fish. Based on such information, diets can be formulated that support the culture of most species. Such diets may contain a number of nutrient excesses or produce less than optimal results but they support reasonable growth and survival and thus provide a starting point for refinements.

As with all animals, the nutritional component of fish husbandry requires meeting at least two fundamental needs. Foremost is the requirement for a constant supply of energy. Energy liberated from the metabolism of organic compounds is essential to fuel the suite of biochemical processes maintaining biological organization and function. In addition, arrays of specific organic compounds as well as some inorganic elements are needed in the synthesis of the complex organic compounds. These compounds are used either for body maintenance or production of new tissue during growth and reproduction. A balance is required both in the array of required nutrients as well as the proportion of the total organic input devoted to energy use. Ideally this assortment of nutrients which must be moved through the aqueous environment should arrive to the fish relatively unchanged in composition and in a form that induces consumption. However, meeting this suite of objectives with formulated diets is quite difficult at present in that feed

TABLE 3.1: Fish species commonly used in research		
Common name	Genus and species	Common uses
Coldwater species		
Rainbow trout and Pacific salmon	*Oncorhynchus* spp.	Fisheries enhancement
		Food production
Atlantic salmon	*Salmo salar*	Food production
Warmwater species		
Channel catfish	*Ictalurus punctatus*	Food production
Tilapia	*Tilapia* spp. and *Oreochromis* spp.	Food production
Common carp	*Cyprinus carpio*	Food production
		Comparative biology
Gold fish	*Carassius auratus*	Neuroscience
		Ornamental culture
Killifish	*Fundulus heteroclitus*	Developmental biology
Fathead minnow	*Pimephales promelas*	Aquatic toxicology
Stickleback	*Gasterosteus aculea*	Aquatic toxicology
Guppy	*Poecilia reticulata*	Ornamental culture
Medaka	*Oryzias latipes*	Developmental biology
		Carcinogenicity assays
Platyfish and swordtail	*Xiphophorus* spp.	Carcinogenesis
		Ornamental culture
Zebrafish	*Danio rerio*	Developmental biology
Electric eel and ray	*Electrophorus electricus* and *Torpedo* spp.	Neuroscience

Source: Adapted from Casebolt *et al.*, 1998.

formulation technology is still being developed and data on nutrient needs of fishes are limited.

In that sophisticated methods for accurately delivering nutrients to individual fish have yet to be developed, quantification of nutrient needs is quite coarse. The optimal dietary level of a nutrient is typically defined by the average growth of a group of fish on an 'as fed' basis. Under these conditions, nutrient losses during processing, storage, feeding and utilization are obscured. Another quandary is how to prevent intake of nutrients through routes other than the proffered ration. Feeding on available organisms from the aquatic environment or even direct uptake of nutrients, in the case of minerals, can also obscure actual requirements. Finally, in that the field of fish nutrition is relatively new, current information is not very robust. Often accepted insights into fish nutrition result from research with a single fish species or may sometimes be derived from nutritional studies with terrestrial animals and have yet to be actually tested in fish.

History

Historically, fish were cultured using live or fresh food items. It was not until after the Second World War, that serious consideration was given to the production of formulated diets for use in various culture facilities, primarily trout hatcheries (Hardy, 1989). Trout were a fortunate species to start with because they hatched from large eggs and were ready to feed on prepared food immediately after consuming their supply of yolk. Ground meats were used for early trout diets with various nutritional supplements being added over time, as they were found necessary. Eventually, the ground meat portion was replaced with isolated protein sources and the supplemental meals were replaced with more refined feedstuffs. These improvements led to the advent of effective test diets to determine specific nutritional requirements of fish in the 1950s. A vitamin test diet used to quantify vitamin requirements for rainbow trout was first developed in 1951 (Wolf, 1951). An improved vitamin test diet containing purified ingredients was published in 1953 (Halver, 1953) followed by an amino acid test diet towards the end of the decade (Halver, 1957). Trout not only readily accepted such test diets but as their husbandry requirements for cold well-oxygenated water could most easily be met by flow-through raceway culture systems, nutrient contamination from natural food sources was negligible.

Following on the heels of the trout industry was the development of salmon culture (Pennell and Barton, 1996). Again, shaped by extensive research efforts to improve hatchery production, the farming of salmon in marine cages is now a highly developed aquaculture industry. Nutritional insights developed with various species of salmon have both extended and refined the earlier information gained with the related rainbow trout. In that trout and salmon have optimum temperature requirements in the range of 10–15°C this group is often referred to as 'coldwater' species.

In contrast to coldwater species such as salmon, 'warmwater' fish are from the more temperate and tropical areas of the world. They are traditionally reared in freshwater ponds where temperatures in the range of 20–30°C promote not only rapid fish growth but also striking amounts of natural prey items. As a consequence, the need for complete nutritional information was not as critical for the aquaculture of warmwater species in that natural biota in the pond provided most or all of the food needed by the fish. Historically, methods of increasing production of warmwater pond species focused on enhancing the production of the total pond biota by using fertilizer to stimulate algal production at the base of the food chain. It was not until recently, as production density was increased beyond what could be supported by pond organisms, that the addition of supplementary feeds containing a broader and broader array of nutrients became obligatory (Hepher, 1988). As this trend continues and information on the nutritional needs of warmwater fish, such as catfish, carp, tilapia and others, continues to accumulate, formulation of pond feed is becoming less of an art and more of a science (see Wilson, 1991). For some species, such as the channel catfish grown in the southern United States, commercial formulated feeds come close to meeting all the nutritional needs of the species (Robinson, 1989).

One other group of warmwater species receiving increasing industry attention is ornamental fish (Chapman et al., 1997). Currently the majority of freshwater ornamental fish are produced in fertilized ponds. However, most marine ornamentals are collected from the wild. As with marine food fish, a combination of habitat destruction, pollution and overharvesting has led to declining natural populations of marine ornamentals. Development of appropriate culture techniques for marine ornamentals is of interest both to commercial aquaculturists and to conservation biologists (Andrews, 1990). As technologies are refined for culturing ornamental species, both freshwater and marine, development of suitable formulated diets will be a priority (Fernando et al., 1991; Earle, 1995; Kaiser et al., 1997).

The fact that fishes are the most numerous (~21 500 species) of all vertebrate groups (almost half of all vertebrate species), and that detailed nutritional information is only available for a handful is not as daunting as it might seem at first sight (Figure 3.1). Unique feeding specializations are found among fishes, such as the species of the South American catfishes that feed on the gill filaments and blood of larger fish (Baskin et al., 1980). Fortunately, such peculiar adaptations are the exception so that the vast majority of fish can be adapted to feed on pellets or granules of an appropriate size. There also are fish species feeding at every trophic level – **detritivore**, herbivore, carnivore and omnivore – although most are either carnivorous or omnivorous. Fortunately for those wanting to culture the omnivores, they do not appear to have an absolute requirement of a varied input, in spite of being adapted to feed on an array of organisms in nature. Indeed they tend to thrive on similar formulations to that originally developed for the coldwater carnivorous species such as the rainbow trout even though it might not be theoretically optimum.

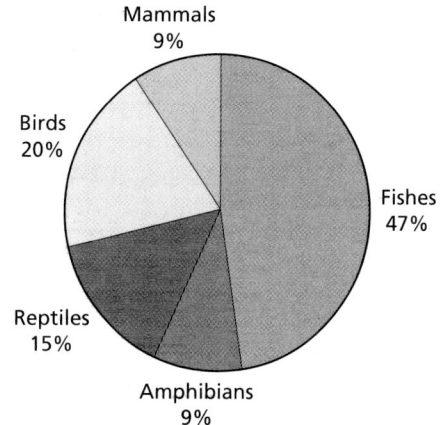

Figure 3.1 Fish as a percentage of total (~45 600) vertebrate species.

Modified salmon diets have also been used for the culture of marine food fish, one of the newest areas of interest for commercial aquaculture. These species, such as yellowtail, red sea bream, turbot, etc., are typically reared at water temperatures above what is suitable for salmon and other coldwater species. Nevertheless, with some modifications primarily with respect to fatty acids, the older juveniles of these carnivorous species adapt quite well to these types of protein-rich diets. The biggest challenge has been in providing suitable feeds for the early stages of the marine species.

The larval stages of many marine food species require restrictively fine particles at first feeding and thus these culture industries have been built around live food for the early stages of culture (Yoshimura et al., 1996). This is also true of many of the smaller warmwater species used as aquarium fish and more and more as laboratory animals (DeTolla et al., 1995; Westerfield, 1995; Kane et al., 1996; Rao et al., 1997). Early stages of these species do better on live feed organisms although once they have grown large enough to consume crumbles they can be switched to flake feeds. These commercial flake feeds intended for the home aquarist are formulated to be inclusive of any and all the nutrient needs identified for fish (Earle, 1995).

Live foods for larvae and fry

As indicated above, many small aquarium species as well as the larval stage of many larger marine fish species are either unable or reluctant to feed on prepared feeds. This may be from a combination of factors relating either directly to size of the particle, subtle organoleptic factors or the lack of characteristic prey movement (Kanazawa, 1995). For these species, a wide variety of live feeds have been examined for their usefulness. While live feeds generally give excellent results, in most cases, provision of live foods means the investigator has to become involved in the culture of the prey item in addition to the fish. This added effort and cost is the primary disadvantage of live feeds. There are also other potential disadvantages such as the possibility of variable nutritional planes and the possibility of introducing disease. Because their unique characteristics tend to ameliorate these disadvantages, two particular live food items, brine shrimp nauplii and rotifers, tend to be preferred by culturists.

Brine shrimp

One live food organism requiring minimal culture effort is the freshly hatched **nauplii** of brine shrimp *Artemia* spp. In nature, brine shrimp produce a cyst-encapsulated egg that is resistant to complete desiccation and death of the organism. These cysts are harvested, primarily from the Great Salt Lake in Utah, externally dried and packaged. The cysts will remain viable for a period of years but hatch readily within a day upon water immersion (Sorgeloos et al., 1977). Disease concerns can be minimized by first removing the bacterial contaminated cyst wall with chlorine bleach and then inducing hatching (Campton and Busack, 1989).

Brine shrimp nauplii are less than a half a millimeter in size, strongly pigmented, and have a swimming movement that attracts first-feeding fish. They are used widely both in aquaculture and in the rearing of laboratory species (Sorgeloos, 1980). Often they are used in combination with a prepared feed of unknown efficacy to enhance the overall nutrition of the fish (Abi-Ayad and Kestemont, 1994). While attractive because of their 'off the shelf' attributes, nutritionally they can be lacking and, more importantly, for some fish larvae they are too large.

Rotifers

Both a freshwater species (Lim and Wong, 1997) and several brackish-water species of rotifer are commonly available within the aquaculture industry (Hagiwara et al., 1997). Many fish and fish fry with a small mouth gape and which were previously fed on various protozoans do particularly well on rotifers, roughly half the size of brine shrimp nauplii. The **euryhaline** rotifer *Brachionus plicatilis* has proven to be essential for the effective commercial culture of marine fish larvae and is now cultured in large-scale intensive systems under carefully controlled conditions (Yoshimura et al., 1996; Lubzens et al., 1997). Related freshwater species also appears to be useful for the smaller freshwater aquarium species (Lim and Wong 1997; Chapman et al., 1998) and can be used for first feeding of zebrafish in lieu of protozoan prey (Westerfield, 1995). A disadvantage is that cysts are not readily available and rotifer use entails the additional burden of culturing either algae or yeast in order to maintain the rotifer cultures (Fulks and Main, 1991).

Enrichment of zooplankton

Experience has shown that freshly hatched brine shrimp nauplii and rotifers can be deficient in essential fatty acids particularly required by marine fish larvae (Rainuzzo et al., 1997). As a consequence, a great deal of effort has been made to develop methods to enhance the fatty acid profile of live foods for some of the long-chain unsaturated fatty acids. Feeding lipid-enriched diets (Nanton and Castell, 1998) can significantly enhance fatty acid composition of zooplankton. For example, since *Euglena gracilis* is high in these essential fatty acids, it has been found to be a good supplement to feed to rotifers and *Artemia* (Hayashi et al., 1993). The most widely used enrichment technique has been to first grow yeast in a medium containing fish oil in order to increase the percentage of desirable fatty acids in the yeast. Subsequent feeding of rotifers and brine shrimp nauplii on the enriched yeast results in a more nutritionally desirable prey item for larval fish (Furuita et al., 1996). Yeast is relatively easy to culture and the enrichment technique is now well established. Direct enrichment is also possible by manipulating the system so that the zooplankton organisms have lipid-filled guts when they are consumed by the fish larvae. Typically, emulsified fish oil droplets are added to the rotifer or brine shrimp culture medium and the zooplankton allowed to gorge on the droplets. The various techniques and strategies for presenting dietary polyunsaturated fatty acids to larval fish either through single cell organisms or through purified oils have recently been reviewed by several authors (Sargent et al., 1997; Coutteau and Sorgeloos, 1998). In addition to fatty acids, live feeds can be used to deliver other nutrients such as vitamins (Coutteau and Sorgeloos, 1998) or mendicants to larval fish (Robles et al., 1998).

Another method of introducing dietary elements which are either limited or lacking all together in live prey is to practice co-feeding of larvae with a mixture of live and inert diets. This approach appears to be a promising mechanism to reduce the amount of the live diet needed (Canavate and Fernandez-Diaz 1999) as long as the fish larvae have not been habituated to only live feed (Dutton, 1992; Fernandez-Diaz et al., 1994).

Other live feeds

While widely used, brine shrimp nauplii and rotifers are not a universal solution indicating there is still much to learn regarding live feeds. For example, for some marine fish species investigators have had better success with copepods than with brine shrimp (Shansudin et al., 1997). While it is still not completely clear why some species of fish larvae appear to do better on copepods than brine shrimp, it appears again to be associated with lipid requirements rather than size (Ronnestad et al., 1998). Copepod culture techniques are somewhat cumbersome (Uhlig, 1984; Stottrup and Norsker, 1997). In addition, many copepods have the further behavioral disadvantage of tending to stay on the culture vessel walls and thus not be available to the fish larvae. There are also fish larvae that strongly benefit from having algae in the culture media. Some species require algae as a direct food source for the first few days of endogenous feeding (Rodrigues and Hirayama, 1997) but for others the advantage conferred by the algal cells is not clear (Reitan et al., 1997). While such alternatives as copepods and necessary supplements as algae are of interest, it is likely that as the specific nutrient advantage of each is elucidated there will be an attempt to incorporate the factor into either brine shrimp nauplii or rotifers.

Manufactured feeds

While live feeds have proved necessary, ideally fish culturists would prefer manufactured feeds because of the greater control over nutrient quality and reduced labor costs. Leaching and consequent dissolution of vital nutrients into the aqueous environment has made this difficult. This not only results in nutrient deficiencies but also pollutes the enveloping aqueous environment.

In many respects, the field is just beginning to deal with the problem of delivering quantifiable amounts of nutrients to animals. Typically, significant but unknown amounts of nutrients are lost to the aqueous environment before the targeted fish consumes the feed. Even under conditions where uneaten feed is negligible, leaching of soluble ingredients from pellets can be consequential (Goldblatt et al., 1979). Loss of water-soluble vitamins can be over 50% before trout, a fish that feeds relatively rapidly, snaps up a pellet (Slinger et al., 1979). Loss of soluble amino acids from diets makes it difficult to meet essential amino acid requirements of larval fish (Lopez-Alvarado and Kanazawa, 1994; Lopez-Alvarado et al., 1994). As the accurate assessment of the amount of food actually consumed by fish is a daunting task, most fish are fed 'to excess'. Uneaten residue then serves as substrate for a variety of microbial organisms. Small fish can gain appreciable nutrition from organic detritus (Lemke and Bowen, 1998).

Microencapsulated diets

As feed particle size is reduced, the problem of leaching is greatly exacerbated. Potentially one could avoid the problem of leaching if the fine food particles could be encased with an impenetrable film. Unfortunately, conventional **microencapsulation** technology tends to use protein films, which do not prevent the movement of small water-soluble molecules. Consequently, most researchers feel that manufactured diets replacing live feeds should have a complex wall structure made up of several layers. A layer of lipid is probably a necessary component of the microcapsule wall in order to prevent leaching of water-soluble components (Lopez-Alvarado et al., 1994; Langdon and Buchal, 1998). This is a complicated research task involving defining nutrient requirements while trying at the same time to determine appropriate delivery systems. It is likely that the goal of developing effective artificial micro-feeds meeting the needs of fish larvae presently dependent on live food diets will remain elusive (Watanabe and Kiron 1994; Southgate and Partridge 1998).

Purified and semipurifed feeds

Ideally, in order to determine the exact nutrient requirements of animals, a purified diet would offer a researcher maximal control over the nutritional aspect of husbandry. Purified diets are made up of individual amino acids, simple sugars, individual fatty acids, and mixtures of required vitamins and minerals along with some sort of filler such as pure cellulose. In general, however, purified diets are not effective with fish and performance has been problematic. Both rapid leaching and poor palatability are probably to blame. Somewhat more effective for fish are semipurified diets. A semipurified diet is the most common type of diet used in nutritional studies with fish. While the ingredients are refined, complex organics rather than individual nutrients are used. For example, the amino acid in a semipurifed diet would be supplied through isolated proteins, such as casein and gelatin. Other typical ingredients would be corn starch as a carbohydrate source, corn or fish oil as a source of fatty acids along with some sort of binder such as gluten or agar. Vitamin and mineral mixtures are used, as is the case with purified diets. As indicated earlier, there are a number of such diets in the literature, which have been developed to study the nutritional requirements of various aquaculture species such as trout and catfish (National Research Council, 1993). Similar semipurifed diets can be made for the smaller laboratory fish species such as medaka (DeKoven et al., 1992) using the general insights gained with the well-known aquaculture species.

Practical and commercial feeds

Practical feeds are manufactured from readily available ingredients that have a minimal amount of processing apart from drying and grinding. Commercial feeds for aquaculture which are also sometimes used in laboratory settings are of this type. Common feedstuffs making up practical diets might be fish and soybean meals, fish oils, ground wheat and corn. As with the other types of diets, mixtures of known vitamins and minerals are also added. Due to the limited definition of their contents, nutrient input when using these feeds is somewhat uncertain. On the other hand, commercial feeds typically are designed to have a margin of safety above what is minimally needed by the target fish species and are thus suitable for a number of uses in the laboratory (Ako, 1999).

Energy – immediate nutritional needs

As **poikilothermic** (cold-blooded) animals, the temperature of the culture water primarily determines dietary energy requirements for fish. Under laboratory conditions of constant temperature and only nominal activity, energy requirements of fish for maintenance are much less that for terrestrial vertebrates because of the buoyancy of water. Key factors in determining the energy needs of fish are size and physiological state. These differences may not be noticed in laboratory culture where fish are typically fed to excess. Metabolism is higher in smaller fish than in larger ones. This is true both within a species, where the size difference is typically the result of age difference, and between different species with differing sizes at maturity. Energy is also required in processing food as well as in the synthesis of tissues. These are important elements for aquaculture where the goal is to achieve the maximum amount of growth with as little feed as possible. The details of energy utilization with regard to laboratory husbandry, however, are less important because while rapid growth may be desirable, efficient food conversion is less critical.

TABLE 3.2: Qualitative and quantitative amino acid requirements of fish

Essential amino acid requirement (g/100 g protein)	Non-essential amino acids
Arginine 3.3–5.9	Alanine
Histidine 1.3–2.1	Aspartic acid
Isoleucine 2.0–4.0	Cystine
Leucine 2.8–5.3	Glutamic acid
Lysine 4.1–6.1	Glycine
Methionine 2.2–6.5	Proline
Phenylalanine 5.0–6.5	Serine
Threonine 2.0–4.0	Tyrosine
Tryptophan 0.3–1.4	
Valine 2.3–4.0	

Tissue synthesis

In nature, much of the energy used by fish comes from protein. As protein-rich feedstuffs tend to be more expensive, aquaculturists attempt to formulate diets so that energy for metabolism comes from lipid or carbohydrate. Conserving protein for use in growth is referred to as protein sparing and is important to the economic success of commercial aquaculture. As protein is economically important to commercial aquaculturists, a great deal of effort has been devoted to defining the protein requirements of fish, such information being of interest to researchers involved with laboratory culture of fish. Efficient utilization of dietary protein will reduce ammonia production. This being the breakdown product of protein catabolism in fish. As ammonia and other nitrogenous metabolites arising from it are toxic to fish, these products have to be removed by dilution or biological filtration. Under these conditions, catabolism of excess protein will result in increased system maintenance requirements.

Protein

Fish require the same array of essential amino acids common to other vertebrate species (Table 3.2). Quantitative requirements have been successfully based on mimicking the amino acid profile of fish muscle. As muscle makes up around 80% of the tissue mass, the amino acid profile of fish muscle tissue provides a reasonable estimate of the ratio of essential amino acids that will be required for growth (van der Meer and Verdegem, 1996). There are however apparently some differences in other uses of essential amino acids. Thus, while these muscle amino acid patterns are quite similar among fish, comparisons of amino acid requirements based on growth tests suggest there are differences between salmon and carp (Akiyama et al., 1997). It is not yet known if these differences can be expanded to embrace the warmwater versus coldwater groupings.

The protein requirement of fish is a reflection of the quantitative requirement for the 10 essential amino acids. The closer the balance of essential amino acids in the protein is to the balance required by the animal, the more efficient protein utilization, thus lowering the apparent requirement for protein. The apparent requirement for protein can also be influenced by other factors such as digestibility and palatability. Rapidly growing early juvenile fish do better on high protein ($\sim$50% of the diet) rations with the optimum level of protein declining as they grow. In general, coldwater species like salmon have higher requirements for protein ($\sim$40% of the diet) than warmwater species ($\sim$30% of the diet) (Lochmann and Phillips, 1994; Wilson, 1994a).

Matching the necessary balance of essential amino acids for fish, in particular arginine, lysine and methionine, generally requires the use of animal protein. Typically this is fishmeal in practical diets and casein for purified diets, including those for the smaller ornamental species (Shim and Ng, 1988; DeKoven et al., 1992). While neutralized free amino acids can be used to examine requirements on an experimental basis (Rodehutscord et al., 1997; Twibell and Brown, 1997) supplementary amino acids are not used in practical diets. This is because protein-bound amino acids are utilized more efficiently than free amino acids (Ng et al., 1996). The better utilization of protein-bound amino acids is true even if leaching is taken into account (Zarate and Lovell, 1997).

Lipid and carbohydrate requirements

Ideally carbohydrates and lipids serve as the primary energy sources in the diet of fish in order to spare protein for growth. In contrast to mammals and birds, carbohydrates are not particularly useful as an energy source in fish. Lipids are ideal for fish in that stores can serve both as an energy reserve and as a buoyancy factor. The buoyant eggs of numerous marine fish species contain triglyceride-rich oil droplets in their eggs as the primary energy source (Heming and Buddington, 1988). Species of marine fish utilizing oil droplets are characterized by rapid development fueled by these triglycerides of the oil (Mourente and Vazquez, 1996).

While carbohydrates are not essential for fish, they are a common constituent of the diet. Cheaper than proteins and fats, carbohydrates such as cornstarch are used in fish diets as a source of energy. Fish generally utilize complex carbohydrates such as starch more effectively than simple sugars but starches must be cooked to promote digestion. Coldwater and marine fish species are able to utilize up to around 20% of the diet as carbohydrates while warmwater pond species tend to be able to use 30–40% in the diet (Wilson, 1994b). Excessive amounts of digestible carbohydrates in the diet lead to an increase in mesentery fat deposits.

Fish can effectively convert excess energy into storage lipids. Lipids, fatty acids and phospholipids are also vital elements of biological membranes. Fish are unable to synthesize the two specific fatty acids linoleic acid (18:2 n-6)[1] and linolenic acid (18:3 n-3) and thus require dietary sources (Kanazawa et al., 1979). Each of these parent fatty acids, through elongation and desaturation, gives rise to some important fatty acids of each series. Linoleic acid gives rise to arachidonic acid (20:4 n-6) and linolenic acid give rise to eicosapentaenoic acid (20:5 n-3) and docosahexaenoic acid (22:6 n-3). Physiologically more significant than the respective parent fatty acid, these long-chain polyunsaturated fatty acids (**PUFA**) play a key role in maintaining the fluidity, flexibility and permeability of cellular membranes. They are also the precursors of the hormone-like compounds, eicosanoids.[3]

Specific requirements for individual fatty acids is related to the ability of the species to desaturate and elongate them as well as the need to meet the challenge of such environmental variables as temperature and salinity (Sargent et al., 1989). All fish appear to require n-3 PUFAs of C_{18} chain length or greater. These fatty acids need to be provided in the diet at around 1–2%. Marine fish appear to lack the ability to elongate C_{18} fatty acids efficiently and thus growth and survival is enhanced with the addition of eicosapentaenoic acid (20:5 n-3) and/or docosahexaenoic acid (22:6 n-3). The requirement for dietary sources of these longer chain fatty acids is also seen in marine fish larvae (Sargent et al., 1997). Freshwater species do not seem to have this requirement for the C_{20} and C_{22} PUFA but may have additional requirements for n-6 PUFAs. In analyzing commonly used live or fresh foods used for maturation in freshwater ornamental fish Tamaru et al. (1997) suggested arachidonic acid (20:4 n-6) might be important in reproduction.

The ability to modify the carcass composition of fish to enhance the content of PUFAs has been appealing because of their purported health effects in humans (Steffens, 1997). Bell and co-workers (1995) found that the carcass composition of PUFA could easily be enriched in turbot by feeding high levels of these essential fatty acids. It should be noted however that some peroxidation in these oils is difficult to prevent even with added antioxidants (Gonzalez et al., 1992) and thus some caution in use of high levels of PUFA is appropriate. Increased lipid levels in diets with the increased oxidative challenge often requires the addition of more vitamin E (Baker and Dvies, 1996).

One effective way of adding fatty acids to the diet is through the addition of phospholipids. Dabrowski (1986) points out that fish eggs contain large amounts of phospholipids and thus it is possible that larval fish may have a requirement for phospholipids. A number of experiments have shown enhanced growth and survival when lecithin, a convenient phospholipid

---

[1] Fatty acid nomenclature – shorthand convention, in the example 18:2 n-6; the number before the colon indicates the number of carbon atoms in the chain (18), the number following the colon indicates the number of double bonds (2), and the number following the 'n' indicates the position of the first double bond from the methyl end (between the sixth and seventh carbon atoms).

source, is added to the diet of numerous fish particularly marine species (Kanazawa, 1993). Dietary phospholipids are beneficial in the diet of fish larvae in terms of survival, growth and resistance to stress (Coutteau et al., 1997; Kanazawa, 1997). The addition of lecithin was even found beneficial in a practical diet for juvenile goldfish (Lochmann and Brown, 1997).

Vitamins and minerals

Most people are familiar with the dramatic impact that a lack of vitamins, and to some extent minerals, can have on animals. Determining the impact on fish has been arduous because of the aqueous environment and the permeability of the fish gill to ions. On the one hand, water-soluble vitamins in the feed can be lost before the fish engulfs it. On the other hand, fish can gain vitamins by feeding on organisms in the water or through intestinal synthesis. Minerals can also be gained from the water by uptake of ions across the gill membranes.

Vitamins

Qualitative vitamin requirements for several fish species are known and mirror those seen in various warm-blooded vertebrates. Undoubtedly, this reflects the similarity of biochemical processes used by all vertebrates for intermediary metabolism and for the synthesis of specific compounds in the maintenance and production of tissue. While there are some differences, these typically relate to the ability or lack thereof to synthesize particular compounds rather than differences in the synthetic pathways themselves.

Quantitative vitamin requirements are less well known and estimates can vary by an order of magnitude or more depending on the source of information. For example, contrast the two left-hand columns of Table 3.3, both of which relate to estimates of the vitamin needs of trout and the related species of salmon. Some of the reasons for these large differences most certainly relate to the relative newness of the field compounded by the inherent difficulty of measuring actual intake of water-soluble nutrients presented in an aqueous environment. But there are also a number of other factors involved, an understanding of which is useful to those considering developing a diet or selecting an existing formulation.

TABLE 3.3: A comparison of vitamin needs for selected fishes

Vitamin	Salmonids[a]	Salmonids[b]	Catfish[c]	Carp[a]
Thiamin (mg kg^{-1})	10	1	1	0.5
Riboflavin (mg kg^{-1})	20	4	9	4
Niacin (mg kg^{-1})	150	10	14	28
Vitamin B_6 (mg kg^{-1})	10	2	3	5–6
Pantothenate (mg kg^{-1})	40	20	10–15	30–50
Biotin (mg kg^{-1})	1	0.1	R	1
Folacin (mg kg^{-1})	5	1	1.5	NR
Vitamin B_{12} (mg kg^{-1})	0.02	0.01	R	NR
Ascorbic acid (mg kg^{-1})	100	10	3	5–6
Choline (mg kg^{-1})	3000	500–4000	400	1500
Inositol (mg kg^{-1})	400	250	NR	440
Vitamin A (IU kg^{-1})	2500 IU	2500–5000 IU	1000–2000 IU	4000–20 000 IU
Vitamin E (IU or mg kg^{-1})	30 IU	28 IU	25 mg	100 mg
Vitamin D_3 (IU kg^{-1})	2400 IU	2400 IU	500–1000 IU	not determined
Vitamin K (mg kg^{-1})	10	0.5	R	not determined

[a] National Research Council, 1981.
[b] Woodward, 1994.
[c] National Reserch Council, 1983.
R = required; NR = not required.

As indicated above, there is clearly an element of refinement with time that is leading to better and better estimates of the actual requirements. Often early research studies covered relatively wide increments in order to make certain the experimental treatments adequately covered the potential range of deficiency to surplus. As noted, the species most often used in the United States during the early phase of research on fish nutrition were salmonids, either Pacific salmon or rainbow trout. Hatching from a relatively large egg, these species would accept formulated diets from first feeding onwards. Also, these fish quickly engulfed proffered feed enabling progress to be made in defining nutrient requirements even with diets that were not particularly water stable. A summary of this early work taken from the National Research Council's (1981) report on the nutrient requirements of salmonids is listed in the left-hand column of Table 3.3. In that these results were to provide general guidance with respect to fish nutrition, requirements were defined with caution so as to err on the side of excess. Such conservative estimates of need would provide a large margin of safety regardless of application. On the other hand, based on what was known from terrestrial vertebrates with regard to deleterious impacts with an excess of the fat-soluble vitamins A, D and E, estimates of the requirements for these vitamins were kept at a minimal level.

In contrast to the first report (1981), a subsequent report (National Research Council, 1983) on the nutrition of warmwater species such as catfish and carp enumerated minimal requirements. These differences in approach have given rise to the belief that trout and salmon and by inference other carnivorous species have higher vitamin requirements than warmwater omnivorous fish species. Woodward (1994), however, has argued that there is little difference if a range of experimental evidence is examined and consideration is given to the poor water satiability of the classic purified experimental diet. The conclusions drawn by Woodward, based on his examination of the literature as to minimal dietary requirements, are shown in the second column of Table 3.3 and are quite close to those found for catfish and carp. This suggests that the dietary requirement for vitamins such as thiamin, riboflavin, niacin, vitamin B_6, pantothenate, biotin, folate, and vitamin B_{12} are likely to be similar for most fish species. It is thought that the lack of response to removal of folic acid and vitamin B_{12} from the diet of carp reflects microbial synthesis in the gut rather than an actual species difference (Kashiwada et al., 1970, 1971).

It should be noted, however, that many of the vitamin requirements were established using practical diets in which the feedstuffs contain additional sources of vitamins. Consequently, when making up semipurified diets Kaushik and co-authors (1998) found a modest increase (20–30%) was necessary for salmon and sea bass. Researchers wanting to incorporate vitamin mixtures into research diets to rear fish for laboratory experimentation should also use a similar safety margin.

Vitamin C

While the majority of vitamins can be added in the form of a premix in order to meet established requirements, the case for vitamin C is a little more complicated. It appears that not all fish require vitamin C. A number of fish have the enzymatic capability to synthesize ascorbic acid (Yamamoto et al., 1978; Touhata et al., 1995). Both sturgeon (Moreau et al., 1996) and carp (Sato et al., 1978) apparently synthesize ascorbic acid at a rate sufficient to meet their nutritional needs. Thus it would appear that each species must be investigated as to its need for dietary ascorbic acid. There also may be a change with age. Contrary to the earlier findings for juvenile carp, Gouillou-Coustans and colleagues (1998) found dietary ascorbate was beneficial for carp larvae. As ascorbic acid is prone to rapid oxidation, most feed manufacturers now use one of the stabilized forms, either a phosphate or a sulfate derivative. Again there appear to be some important differences in the ability of various species to use both of these derivatives. While some species like tilapia (Abdelghany, 1996) can use either, others like the sea bass (Amerio et al., 1998) do best on the phosphate derivatives. Species such as trout can readily utilize the phosphate derivatives, converting ascorbyl-2-phosphate to ascorbic acid in rainbow trout (Miyasaki et al., 1992). Such species may also synthesize the phosphate form for tissue storage (Miyasaki et al., 1991).

Other vitamin-like factors

L-Carnitine promotes the utilization of long chain fatty acids (Chatzifotis et al., 1995). Dietary L-carnitine heightened membrane impermeability to fluorescein in guppies (Schreiber et al., 1997). The possible dietary requirement for carnitine may be a result of other nutritional disorders. In that vitamin C is involved in its synthesis, one might look first at defining the requirement for this known factor.

A number of investigators have suggested a possible vitamin role for carotenoids in fish (Tacon, 1981; Torrissen and Christiansen, 1995), however, demonstrating an unambiguous effect has been more difficult. While some experiments have shown improved growth and survival in groups fed diets supplemented with carotenoids such as astaxanthin no convincing metabolic explanation has been established. Particularly when purified diets are being used, even the possibility that astaxanthin could act as a feeding attractant cannot be ruled out (Christiansen *et al.*, 1994). While large amounts of pigments are deposited in the eggs of some fish such as salmon, Choubert *et al.* (1998) were unable to show any evidence of an effect of feeding carotenoid-supplemented diets to the female parent.

Minerals

As fish can take up minerals from the water via the gills, it is difficult to show dietary requirements for fish – particularly with practical diets. An exception is phosphorus in that levels in waters tend to be low. Asgard and Shearer (1997) found that ~10 g phosphorus per kilogram of diet was required by juvenile Atlantic salmon *Salmo salar*. Dietary deficiency of phosphorus initially resulted in a reduction of calcium and phosphorus tissue levels particularly in the mineral-rich bones and scales (Baeverfjord *et al.*, 1998). Development of abnormally soft bones resulted in **scoliosis** of the spine and a wrinkly appearance of the ribs. Eventually, growth was severely impaired and mortality rates elevated. Interestingly, as opposed to terrestrial animals, it appears that a high calcium to phosphorus ratio in the diet does not interfere with dietary phosphorus utilization (Vielma and Lall, 1998). It also appears that vitamin D has more minor effects on mineralization in fish than it does in terrestrial animals (O'Connell and Gatlin, 1994).

Trace mineral requirements have been identified for iron, copper, manganese, zinc and selenium (Watanabe *et al.*, 1997). Marine species can absorb most of their requirement for these trace minerals from ingested sea water. Even for freshwater fish, it is likely that under most conditions and certainly with practical diets supplementation with these elements is unnecessary although they are usually added for assurance. In that in terrestrial animals a number of additional mineral requirements have been identified, it is likely that similar requirements for minerals exist for fish. These unidentified requirements are presently being supplied either through uptake from the culture water or by accidental dietary inclusion as a component of other feedstuffs. Mineral requirements cannot be discounted until studies have been carried out where both of these possible sources are controlled. The recent studies showing that boron in the water plays a role in stimulating embryonic growth of trout (Eckhert, 1998) and zebrafish (Rowe *et al.*, 1998) are telling examples.

Summary

Information on the nutrition of a few well-studied aquaculture species provides appropriate information on which to base diet formulation for other species of fish that are being used in the laboratory but there are several caveats. While studies are still quite limited, a number of differences as to requirements for specific nutrients in various species of fish have been established. This would suggest that further differences will be discovered as the informational base on fish nutrition is expanded. Also it should be remembered that the culture of food fish is typically carried under strikingly different husbandry conditions from those fish being reared in the laboratory. The impact of husbandry conditions on the nutritional plane needed for optimal husbandry has yet to be seriously investigated.

References

Abdelghany, A.E. (1996). *J. World Aquacult. Soc.* **27**, 449–455.
Abi-Ayad, A. and Kestemont, P. (1994). Aquaculture **128**, 163–176.
Akiyama, T., Oohara, I. and Yamamoto, T. (1997). *Fish. Sci.* **63**, 963–970.
Ako, H. (1999). *Internat. Aquafeed* **8**, 30–36.
Amerio, M., Ruggi, C., Rovelli, R.M. and Volker, L. (1998). *Aquaculture* **159**, 233–237.
Andrews, C. (1990). *J. Fish Biol.* **37** (Supplement A), 53–59.
Asgard, T. and Shearer, K.D. (1997). *Aquacult. Nutr.* **3**, 17–23.
Baeverfjord, G., Asgard, T. and Shearer, K.D. (1998). *Aquacult. Nutr.* **4**, 1–11.

Baker, R.T.M. and Dvies, S.J. (1996). *Aquacult. Res.* **27**, 795–803.

Baskin, J.N., Zaret, T.M. and Mago-Leccia, F. (1980). *Biotropica* **12**, 182–186.

Bell, J.G., Tocher, D.R., MacDonald, F.M. and Sargent, J.R. (1995). *Fish Physiol. Biochem.* **14**, 373–383.

Campton, D.E. and Busack, C.A. (1989). *Prog. Fish-Cult.* **51**, 176–179.

Canavate, J.P. and Fernandez-Diaz, C. (1999). *Aquaculture* **174**, 255–263.

Casebolt, D.B., Speare, D.J. and Horney, B.S. (1998). *Lab. An. Sci.* **48**, 124–136.

Chapman, F.A., Fitz-Coy, S.A., Thunberg, E.M. and Adams, C.M. (1997). *J. World Aquacult. Soc.* **28**, 1–10.

Chapman, F.A., Colle, D.E., Rottmann, R.W. and Shireman, J.V. (1998). *Prog. Fish-Cult.* **60**, 32–37.

Chatzifotis, S., Takeuchi, T. and Seikai, T. (1995). *Fish. Sci.* **61**, 1004–1008.

Choubert, G., Blanc, J-M. and Poisson, H. (1998). *Aquacult. Nutr.* **4**, 249–254.

Christiansen, R., Lie, O. and Torrissen, O.J. (1994). *Aquacult. Fish. Manag.* **25**, 903–914.

Coutteau, P. and Sorgeloos, P. (1998). *Freshwater Biol.* **38**, 501–512.

Coutteau, P., Geurden, I., Camara, M.R., Bergot, P. and Sorgeloos, P. (1997). *Aquaculture* **155**, 149–164.

Dabrowski, K.R. (1986). *Comp. Biochem. Physiol.* **85A**, 639–655.

DeKoven, D.L., Nunez, J.M., Lester, S.M., Conklin, D.E., Marty, G.D., Parker, L.M. and Hinton, D.E. (1992). *Lab. An. Sci.* **42**, 180–189.

DeTolla, L.J., Srinivas, S., Whitaker, B.R., Andrews, C., Hecker, B., Kane, A.S. and Reimschuessel, R. (1995). *ILAR* **37**, 159.

Dutton, P. (1992). *J. Fish Biol.* **41**, 765–773.

Earle, K.E. (1995). *Vet. Quart.* **17**, S53–S55.

Eckhert, C.D. (1998). *J. Nutr.* **128**, 2488–2493.

Fernandez-Diaz, C., Pascual, E. and Yufera, M. (1994). *Mar. Biol.* **118**, 323–328.

Fernando, A.A., Phang, V.P.E. and Chan, S.Y. (1991). *Asian Fish. Sci.* **4**, 99–107.

Fulks, W. and Main, K.L. (eds), (1991). *Rotifer and Microalgae Culture Systems*. The Oceanic Institute, Honolulu, Hawaii.

Furuita, H., Takeuchi, T., Toyota, M. and Watanabe, T. (1996). *Fish. Sci.* **62**, 246–251.

Goldblatt, M.J., Conklin, D.E., Brown, D.W. (1979). In *Finfish Nutrition and Fishfeed Technology* (eds J.E. Halver and K. Tiews), pp. 117–129. Heenemann Verlagsgesellschaft mbH. Hamburg, West Germany.

Gonzalez, M.J., Gray, J.I., Schemmel, R.A., Dugan, L. Jr and Welsch, C.W. (1992). *J. Nutr.* **122**, 2190–2195.

Gouillou-Coustans, M.-F., Bergot, P. and Kaushik, S.J. (1998). *Aquaculture* **161**, 453–461.

Hagiwara, A., Balompapueng, M.D., Munuswamy, N. and Hirayama, K. (1997). *Aquaculture* **155**, 223–230.

Halver, J.E. (1953). *Transactions of the American Fisheries Society* **83**, 254–261.

Halver, J.E. (1957). *J. Nutr.* **62**, 245–254.

Hardy, R.W. (1989). In *Fish Nutrition* (ed J.E. Halver) pp. 475–548. Academic Press, San Diego, California.

Hayashi, M., Toda, K., Yoneji, T., Sato, O. and Kitaoka, S. (1993). *Nippon Suisan Gakkaishi* **59**, 1051–1058.

Heming, T.A. and Buddington, R.K. (1988). In *Fish Physiology* (eds W.S. Hoar, D.J. Randall and J.R. Brett) pp. 407–446. Academic Press, New York.

Hepher, B. (1988). *Nutrition of Pond Fishes*. Cambridge University Press, Cambridge, UK.

Kaiser, H., Britz, P., Endemann, F., Haschick, R., Jones, C.L.W., Koranteng, B., Kruger, D.P., Lockyear, J.F., Oellermann, L.K., Olivier, A.P., Rouhani, Q. and Hecht, T. (1997). *S. African J. Sci.* **93**, 351–354.

Kanazawa, A. (1993). *Essential Phospholipids of Fish and Crustaceans*. INRA Editions, Versailles, France.

Kanazawa, A. (1995). *Nutrition of Larval Fish*. AOCS Press, Champaign, Illinois.

Kanazawa, A. (1997). *Aquaculture* **155**, 129–134.

Kanazawa, A., Teshima, S. and Ono, K. (1979). *Comp. Biochem. Physiol.* **63B**, 295–298.

Kane, A.S., Gozalez, J.F. and Reimschuessel, R. (1996). *Lab An.* **25**, 33–38.

Kashiwada, K., Teshima, S. and Kanazawa, A. (1970). *Bull. Jpn. Soc. Sci. Fish.* **36**, 421–424.

Kashiwada, K., Kanazawa, A. and Teshima, S. (1971). *Mem. Fac. Fish., Kagoshima Un.* **20**, 185–189.

Kaushik, S.J., Gouillou-Coustans, M.F. and Cho, C.Y. (1998). *Aquaculture* **161**, 463–474.

Langdon, C.J. and Buchal, M.A. (1998). *Aquacult. Nutr.* **4**, 275–284.

Lemke, M.J. and Bowen, S.H. (1998). *Freshwater Biol.* **39**, 447–453.

Lim, L.C. and Wong, C.C. (1997). *Hydrobiologia* **358**, 269–273.

Lochmann, R.T. and Brown, R. (1997). *J. Am. Oil Chem. Soc.* **74**, 149–152.

Lochmann, R.T. and Phillips, H. (1994). *Aquaculture* **124**, 277–285.

Lopez-Alvarado, J. and Kanazawa, A. (1994). *Fish. Sci.* **60**, 435–439.

Lopez-Alvarado, J., Langdon, C.J., Teshima, S. and Kanazawa, A. (1994). *Aquaculture* **122**, 335–346.

Lubzens, E., Minkoff, G., Barr, Y. and Zmora, O. (1997). *Hydrobiologia* **358**, 13–20.

Mehrle, P.M., Mayer, F.L. and Johnson, W.W. (1977). In *Aquatic Toxicology and Hazard Evaluation* (eds F.L. Mayer and J.L. Hamelink), pp. 269–280. American Society for Testing and Materials, Memphis, Tennessee.

Miyasaki, T., Sato, M., Yoshinaka, R. and Sakaguchi, M. (1991). *Comp. Biochem. Physiol.* **100B**, 711–716.

Miyasaki, T., Sato, M., Yoshinaka, R. and Sakaguchi, M. (1992). *Nippon Suisan Gakkaishi* **58**, 2101–2104.

Moreau, R., Kaushik, S.J. and Dabrowski, K. (1996). *Fish Physiol. Biochem.* **15**, 431–438.

Mourente, G. and Vazquez, R. (1996). *Fish Physiol. Biochem.* **15**, 221–235.

Nanton, D.A. and Castell, J.D. (1998). *Aquaculture* **163**, 251–261.

National Research Council (1981). *Nutrient Requirements of Cold-water Fishes*. National Academy Press, Washington, DC.

National Research Council, N. R. (1983). *Nutrient Requirements of Warm-water Fishes and Shellfishes*. National Academy Press, Washington, DC.

National Research Council (1993). *Nutrient Requirements of Fish*. National Academy Press, Washington, DC.

New, M.B. (1997). *World Aquacult.* **28**, 11–19.

Ng, W.K., Hung, S.S.O. and Herold, M.A. (1996). *Fish Physiol. Biochem.* **15**, 131–142.

O'Connell, J.P. and Gatlin, D.M. III, (1994). *Aquaculture* **125**, 107–117.

Pennell, W. and Barton, B.A. (eds) (1996). *Principles of Salmonid Culture*. Elsevier, Amsterdam, The Netherlands

Rainuzzo, J.R., Reitan, K.I. and Olsen, Y. (1997). *Aquaculture* **155**, 103–115.

Rao, S.S., Metcalfe, C.D., Neheli, T.A. and Schmidt, B. (1997). *Envir. Tox. Water Qual.* **12**, 349–352.

Reitan, K.I., Rainuzzo, J.R., Oie, G. and Olsen, Y. (1997). *Aquaculture* **155**, 207–221.

Robinson, E.H. (1989). *Rev. Aqua. Sci.* **1**, 365–390.

Robles, R., Sorgeloos, P., Van Duffel, H. and Helis, H. (1998). *J. Appl. Ichthyol.* **14**, 207–212.

Rodehutscord, M., Becker, A., Pack, M. and Pfeffer, E. (1997). *J. Nutr.* **126**, 1166–1175.

Rodrigues, E.M. and Hirayama, K. (1997). *Hydrobiologia* **358**, 231–235.

Ronnestad, I., Helland, S. and Lie, O. (1998). *Aquaculture* **165**, 159–164.

Rowe, R.I., Bouzan, C., Nabili, S. and Eckhert, C.D. (1998). *Biol. Trace Elem. Res.* **66**, 261–318.

Sargent, J., Henderson, R.J. and Tocher, D.R. (1989). In *Fish Nutrition* (ed. J.E. Halver), pp. 154–219. Academic Press, London.

Sargent, J.R., McEvoy, L.A. and Bell, J.G. (1997). *Aquaculture* **155**, 117–127.

Sato, M., Yoshinaka, R., Yamamoto, Y. and Ikeda, S. (1978). *Bull. Jpn. Soc. Sci. Fish.* **44**, 1151–1156.

Schreiber, S., Becker, K., Bresler, V. and Fishelson, L. (1997). *Comp. Biochem. Physiol.* **117C**, 99–102.

Shansudin, L., Yusof, M., Azis, A. and Shuki, Y. (1997). *Aquaculture* **151**, 351–365.

Shim, K.F. and Ng, S.H. (1988). *Aquaculture* **73**, 131–141.

Slinger, S.J., Razzaque, A. and Cho, C.Y. (1979). In *Finfish Nutrition and Fishfeed Technology* vol. II (eds J.E. Halver and K. Tiews), pp. 425–434. Heenemann Verlagsgesellschaft mbH, Berlin.

Sorgeloos, P. (1980). *The Brine Shrimp* Artemia vol. 3 (eds G. Persoone, P. Sorgeloos, O. Roels and E. Jaspers), pp. 25–46. Universa Press, Wetteren, Belgium.

Sorgeloos, P., Bossuyt, E., Lavina, E., Baeza-Mesa, M. and Persoone, G. (1977). *Aquaculture* **12**, 311–315.

Southgate, P.C. and Partridge, G.J. (1998). In *Tropical Mariculture* (ed S.S. De Silva), pp. 151–169. Academic Press, San Diego, California.

Steffens, W. (1997). *Aquaculture* **151**, 79–96.

Stottrup, J.G. and Norsker, N.H. (1997). *Aquaculture* **155**, 231–247.

Tacon, A.G.J. (1981). *Prog. Fish-Cult.* **43**, 205–208.

Tamaru, C.S., Ako, H. and Paguirigan, R. (1997). *Hydrobiologia* **358**, 265–268.

Torrissen, O.J. and Christiansen, R. (1995). *J. Appl. Ichthyol.* **11**, 225–230.

Touhata, K., Toyohara, H., Mitani, T., Kinoshita, M., Satou, M. and Sakaguchi, M. (1995). *Fish. Sci.* **61**, 729–730.

Twibell, R.G. and Brown, P.B. (1997). *J. Nutr.* **127**, 1838–1841.

Uhlig, G. (1984). Special Publication of the European Mariculture Society, vol. 8, pp. 261–273.

van der Meer, M.B. and Verdegem, M.C.J. (1996). *Aquacult. Res.* **27**, 487–495.

Vielma, J. and Lall, S.P. (1998). *Aquaculture* **160**, 117–128.

Watanabe, T. and Kiron, V. (1994). *Aquaculture* **124**, 223–251.

Watanabe, T., Kiron, V. and Satoh, S. (1997). *Aquaculture* **151**, 185–207.

Westerfield, M. (1995). *The Zebrafish Book, Guide for the Laboratory Use of Zebrafish* (Danio rerio). University of Oregon Press, Eugene, Oregon.

Wilson, R.P. (ed.) (1991). *Handbook of Nutrient Requirements of Finfish*. CRC Press, Boca Raton, Florida.

Wilson, R.P. (1994a). In *Amino Acids in Farm Animal Nutrition* (ed. J.P.F. D'Mello), pp. 377–399. CAB International, Edinburgh, UK.

Wilson, R.P. (1994b). *Aquaculture* **124**, 67–80.

Wolf, L.E. (1951). *Prog. Fish-Cult.* **13**, 17–23.

Woodward, B. (1994). *Aquaculture* **124**, 133–168.

Yamamoto, Y., Sato, M. and Ikeda, S. (1978). *Bull. Jpn. Soc. Sci. Fish.* **44**, 775–779.

Yoshimura, K., Hagiwara, A., Yoshimatsu, T. and Kitajima, C. (1996). *Mar. Freshwater Res.* **47**, 217–222.

Zarate, D.D. and Lovell, R.T. (1997). *Aquaculture* **159**, 87–100.

CHAPTER 4

Common Diseases and Treatment

David B Powell
ProFISHent, Inc., Redmond, Washington State, USA

Introduction

This chapter will serve as an introductory survey of the many possible diseases which can infect fish populations. The examples described were selected on the basis of their prevalence or as well studied representative pathogens within their classification. However, because many infectious and chemical agents were not included in this review, extreme caution should be exercised to confirm the true cause of the disease signs observed. The use of therapeutic treatments should also be applied with discretion. When in doubt, an experienced professional should always be consulted. Most diseases can be prevented by starting with pathogen-free fish, clean water and uncrowded holding conditions. Table 4.1 serves as a quick reference guide to the diseases discussed in this chapter.

Gill and skin infections

Gill and skin diseases are often initially diagnosed by the observation of lesions visible to the naked eye. Unusual swimming or feeding behavior should be carefully recorded to provide clues as to the cause of the disease condition. Fish exhibiting lesions (ulcers, erosion, hemorrhaging, etc.) or abnormal behavior (such as gasping, lethargy or dark coloration) should be selected and **euthanized** by an overdose of anesthetic. Gill erosion or large parasites can sometimes be readily diagnosed without the aid of a microscope (Figure 4.1). The color of the gills should be recorded immediately in order to avoid post-mortem influences on the determination of anemia. If the gills are unusually pale, it is advisable to take a blood sample

Copyright © 2000 Academic Press

Common Diseases and Treatment

TABLE 4.1: Summary of diseases in fish: identification and possible treatments

Disease agent	Morphology	Tests to help identify	Treatments (consult with a certified specialist)
Gill and skin infections			
Columnaris *Flavobacterium columnare*	Long filamentous bacteria (2–7 µm) in 'haystack aggregates'	Characteristic 'saddleback' lesion, microscopic examination, bacterial growth at > 15°C	Bath immersion in one of the following: chloramine T, quarternary ammonium, diquat. If internal infection use oxytetracycline-HCl
Cold water disease *Flexibacter psychrophilus*	Bacterium is also a long rod (2–7 µm)	Moist, raised, yellow, convex colonies, serological indirect fluorescent antibody tests, < 17°C	Bath immersion in one of the following: chloramine T, quarternary ammonium, diquat. If internal infection use oxytetracycline-HCl
Parasitic infestations			
'Ich' or white spot disease *Ichthyophthirius multifiliis*	Up to 1 mm diameter, uniform layer of external cilia and horseshoe shaped macronucleus	Raised white spots (trophont stage), microscopic examination of skin, fin and gill lesions	Formalin treatment (~167 ppm) on alternate days during a 10-day period
Costiasis *Ichthyobodo necator*	Bean or 'tear drop' shaped flagellated cells that swim with a jerky motion	Microscopic examination, the narrow end of the organism can attach to the surface of the skin or gills	Salt (1%) or formalin treatment (~167 ppm)
Gyrodactyloidea and Dactylgyroidea	Pair of large hooks surrounded by an array of smaller hooks at the margin of its suction organ	Developing embryo can be observed inside the adult gyrodactylid. Eyespots and egg-bearing adults distinguish the dactylgyrids	Salt (1%) or formalin treatment (~167 ppm)
Trichodina	'Saucer-shaped' with a characteristic denticulate ring and circular margin of cilia	Microscopic examination of skin, fin and gills	Salt (1%) or formalin treatment (~167 ppm)
Fungal infections			
Saprolegnia	White fluffy fungal mycelia associated with eggs, skin or gills	Microscopic examination will reveal fungal hyphae and characteristic asexual zoosporangia	Formalin treatment, malachite green, or methylene blue
Systemic bacterial diseases			
Enteric Red Mouth *Yersinia ruckeri*	Short, gram-negative rods	Oxidase negative, antibody slide agglutination test	Prevention with immersion vaccines, treatment with antibiotics such as: oxytetracycline-HCl, sulfadimethoxine-ormetoprim

COMMON DISEASES AND TREATMENT

Enteric Septicemia of Catfish *Edwardsiella ictaluri*	Short, gram-negative rods	Oxidase negative, antibody slide agglutination test	Prevention with immersion vaccines, treatment with antibiotics such as: oxytetracycline-HCl, sulfadimethoxine-ormetoprim
Furunculosis *Aeromonas salmonicida*	Aggregates of short, gram-negative rods	Colonies produce diffusible brown pigment, ELISA	Prevention with injectable vaccines, treatment with antibiotics such as: oxytetracycline-HCl, sulfadimethoxine-ormetoprim
Bacterial kidney disease *Renibacterium salmoninarum*	Short, gram-positive rods	ELISA, PCR, fluorescent antibody test	Antibiotics: erythromycin, etc.
Streptococcal bacteria *Streptococcus sp.*	Short gram-positive cocci in chains	Microscopic observation of chains, antibody slide agglutination test, biochemical analysis	Prevention with injectable vaccines, antibiotics: erythromycin
Vibriosis *Vibrio anguillarum*, etc.	Short, curved gram-negative rods	Yellow colonies on Marine bile salts agar. Antibody slide agglutination test	Prevention with immersion vaccines, treatment with antibiotics such as: oxytetracycline-HCl, sulfadimethoxine-ormetoprim
Rickettsia-like organisms			
Piscirickettsia salmonis	Short, gram-negative intracellular round or 'comma shaped' rods	Acridine orange positive staining. Fluorescent antibody staining	Antibiotic treatments such as oxolinic acid have only had marginal effectiveness. Vaccines are in development
Viral infections			
Rhabdoviruses (IHNV, VHSV)	'Bullet' shaped viruses observed by electron microscopy	Cell culture serum neutralization. Polymerase chain reaction and fluorescent antibody tests	No direct treatment available. Secondary bacterial infections and reduced stress can reduce mortality
Infectious pancreatic necrosis virus (IPNV)	Birnaviruses observed by electron microscopy	Cell culture thin cytopathic cell morphology, serum neutralization, polymerase chain reaction and fluorescent antibody tests	No direct treatment available. Secondary bacterial infections and reduced stress can reduce mortality

Figure 4.1 Gill parasite lernaeid anchor worms in pond reared rainbow trout.

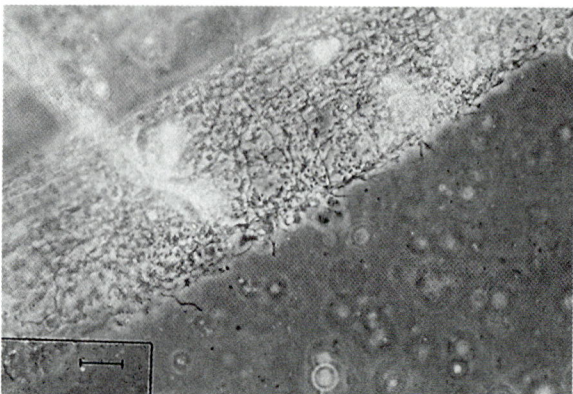

Figure 4.3 Unstained columnaris bacteria on the pectoral fin of catfish. The bar length = 25 μm.

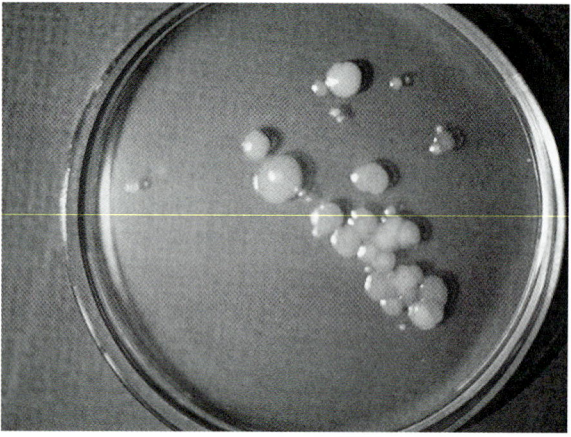

Figure 4.2 Bacterial colonies isolated on tryptic soy nutrient agar.

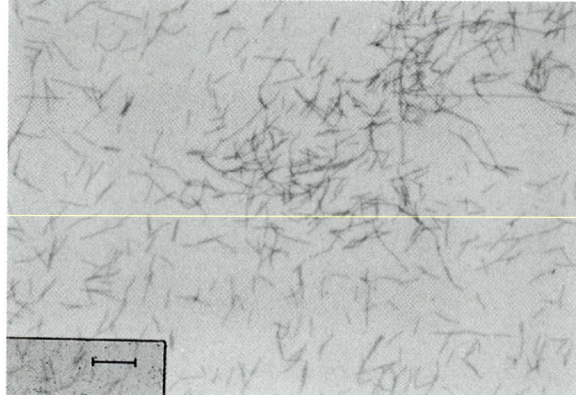

Figure 4.4 Gram-negative columnaris bacteria. The bar length = 100 μm.

from the caudal vein for further analysis (see systemic bacteremias and viruses).

Samples for bacterial culture should be taken at the start of each necropsy to avoid the introduction of contaminants. Sterile loops can be swabbed into the gills or into alcohol swabbed skin lesions and streaked onto **cytophaga** or selective media agar plates for incubation at approximately the fish rearing temperature (Anaker and Ordal, 1959; Hawke and Thune, 1992). Individual colonies should be used for identification tests to avoid the confusion of mixed strains or species of bacteria (Figure 4.2). Small tissue samples for microscopic analysis should be removed from the lesion margin, **gill rakers**, fin tips using flame sterilized scalpel and forceps. Each piece of tissue should be placed on a glass microscope slide in a drop of water (skin, gills) or physiological saline (internal organs) beneath a cover slip.

A microscope equipped with phase contrast or Nomarski optics will provide the clearest view of unstained specimens. If this equipment is not available, additional contrast for a more three-dimensional view can be gained by lowering the position of the condenser lens of a standard microscope. Restrict your search for infectious organisms along the edges of the gill lamellae and margins of fin and skin samples for a rapid diagnosis. Additional samples can be well preserved in 10% neutral buffered formalin for later histologic processing and examination.

Bacterial gill infections

Columnaris

Columnaris bacteria (*Flavobacterium columnarae*, previously *Cytophaga* or *Flexibacter columnaris*) can be a deadly infective pathogen growing rapidly on skin and the surface of the gills (Figures 4.3 and 4.4). Although the taxonomic nomenclature of this group is in transition, polymerase chain reaction polymers have been developed to identify many of the pathogenic but morphologically similar forms (Bader and Shotts, 1998).

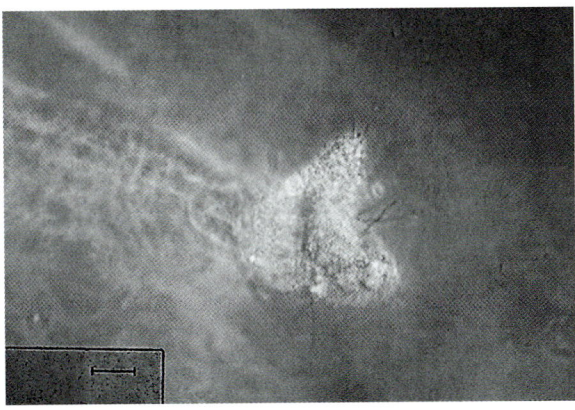

Figure 4.5 Unstained columnaris bacteria on the skin of catfish. The bar length = 100 μm.

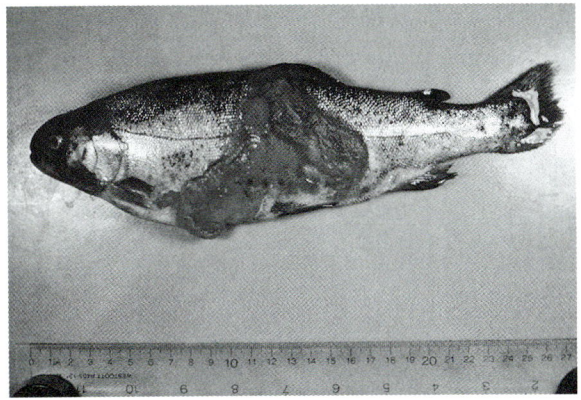

Figure 4.6 An extreme case of erosion of skin exposing underlying muscle caused by bacteria and secondary infection with fungus.

Flexibacter-like bacterial species have been isolated directly from a wide range of surface water sources as ubiquitous residents of fresh water. A related marine form known as *F. maritimus* has also been isolated from diseased fish around the world. A similar long rod bacterium called *Flavobacterium branchiophilum* specifically infects gill tissue causing 'bacterial gill disease'. Interestingly, a gill-associated antibody response against this bacteria was detected in convalescent brook trout following a bath challenge by an indirect enzyme immune assay (Lumsden *et al.*, 1993).

Mortality is generally associated with temperatures above 15°C and low to moderate levels of organic material in the water. The long filamentous bacteria agglutinate to one another forming typical round columns or 'haystacks' extending from the skin surface (Figure 4.5) or gill lamellae. The concentration of bacteria often becomes high enough to be grossly visible as white or yellow patches. Caudal fin and dorsal skin infections often lead to a characteristic 'saddleback' lesion leaving fin ray structures devoid of skin and covered with bacteria. Necrosis of skin lesions may be initiated by the secretion of chondroitin lyase, an enzyme that degrades mucopolysaccharides in connective tissues (Griffin, 1991). Of greater importance may be the production of various proteases that are capable of dissolving muscle tissue (Bertolini and Rohovec, 1992). The cause of death may be the result of toxins, respiratory failure or severe osmotic imbalances. Virulent strains of *F. columnarae* were found to adhere more readily than less pathogenic isolates (Decostere *et al.*, 1999). Adhesion was found to be increased in water with higher ionic strength, organic matter, nitrite and at higher temperatures (Decostere *et al.*, 1999).

Cold water disease

Flexibacter psychrophilus causes disease in cold water conditions below 12–14°C. However, some outbreaks have been reported to occur at temperatures as great as 16°C. It has been commonly isolated on tryptic yeast extract soy agar from skin, spleen and brain tissue of trout and salmon (Holt *et al.*, 1989). In extreme cases, the underlying muscle and caudal peduncle can become severely eroded by bacteria then secondarily infected with fungus (Figure 4.6).

Like *F. columnarae*, this bacterium is a long rod (2–7 μm) and can be cultured directly from fresh water. *Flexibacter psychrophilus* can be distinguished from columnaris by colony morphology (moist, raised, yellow, convex colonies), serological indirect fluorescent antibody tests and polymerase chain reaction tests (Holt, 1994; Bader and Shotts, 1998). The disease is particularly serious in juvenile fish and severe outbreaks have been reported in the United States, Europe, Japan, Tasmania and Chile (Schneider and Nicholson, 1980; Bernardet, 1997).

Parasitic infestations of skin and gills

Ichthyophthiriasis

'Ich' as it is commonly known or ichthyophthiriasis is caused by *Ichthyophthirius multifiliis*, a ciliate protozoan. Nearly all freshwater fish are susceptible to skin and gill infections by this parasite. Raised white spots (trophont stage) up to 1 mm in diameter are normally visible on the surface of the fish. Gill or skin preparations magnified to 40× or more will reveal the large ciliated **trophonts** characterized by a uniform layer of external cilia and a unique horseshoe-shaped macronucleus. The smaller (about 20× 50 mm) infective **theront** or **tomite** stage is free swimming and more susceptible to treatment. If the infective stages are treated on alternate days over a 10- to 14-day period, the trophonts will eventually be eliminated by attrition.

Costiasis

Ichthyobodo necator is a flagellated parasite (commonly called costia) that can cause mortality if the infection is severe (Figure 4.7). Clinical signs include a blue to white cloudy mucus layer on the skin and gasping activity of oxygen-starved fish. The organisms are very small (skin cell size) bean or 'tear drop' shaped flagellated cells that swim with a jerky motion. The narrow end of the organism can attach to the surface of the skin or gills causing gill irritation, hyperplasia and excessive mucus production. Microscopic observations at 100× or greater are needed to diagnose these infections. Very light infestations are usually no cause for concern if proper water quality is maintained.

Gyrodactyloidea and dactylgyroidea

These **monongean** parasites tenaciously attach to skin and gills by hooks on their tails. Gyrodactylids are live-bearing parasites characterized by a pair of large hooks surrounded by an array of smaller hooks attached to the margin of their suction organ. Although sometimes difficult to see, the hooks of the developing embryo can be observed inside the adult gyrodactylid (Figure 4.8).

Dactylogyrids are also monogenetic trematodes and can be distinguished by their small, paired eyespots and the fact that they are oviparous. Small numbers of these parasites on the skin are rarely cause for concern. However, large numbers in the branchial region can cause mortality because of inflammatory hyperplasia and associated reductions in oxygen transfer efficiency.

Trichodina

Trichodina are round 'saucer-shaped' external parasites with a characteristic denticulate ring and circular margin of long cilia (Figure 4.9). These complex

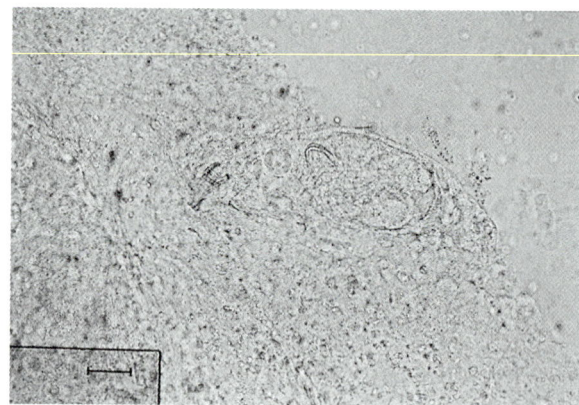

Figure 4.8 Curved hooks of the adult gyrodactylid parasite and internal developing embryo can be seen in the center of this phase contrast photo of fish skin. Bar length = 50 μm.

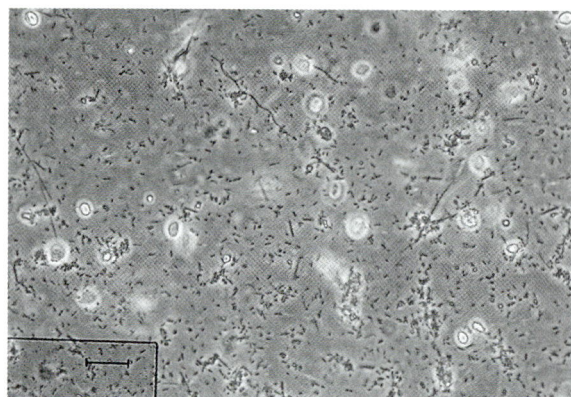

Figure 4.7 'Tear drop' shaped *Ichthyobodo necator* magnified using phase contrast microscopy. Bar length = 25 μm.

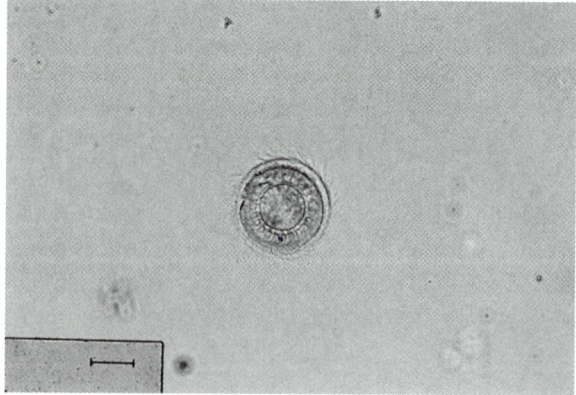

Figure 4.9 Trichodina are round 'saucer-shaped' external parasites with a characteristic denticulate ring and circular margin of long cilia. Bar length = 25 μm.

protozoa attach to the gills and skin of freshwater fish by means of a convex sucking disk composed of three concentric rings below a cytoplasmic marginal fold (Davis, 1947). Small numbers of these parasites are normally of no clinical significance. However, when fish are held in crowded conditions or are malnourished, the density of trichodina gill infestations can cause irritation, gill hyperplasia and respiratory distress (Post, 1983).

Fungal infections

Saprolegnia

Saprolegnia parasitica is the most common freshwater fungal pathogen in fish. *Saprolegnia* is usually a secondary invader because infections start in skin injuries or immunocompromised spawning fish (Figures 4.6 and 4.10). The fungal spores will quickly colonize infertile eggs or open wounds of fish to spread to healthy eggs or tissue. Infected sites are characterized by obvious white fluffy fungal mycelia. The fungus can be isolated by plating on to nutrient agar medium containing a broad spectrum antibiotic. Microscopic examination will reveal fungal hyphae and characteristic asexual zoosporangia (Neish, 1977; Neish and Hughes, 1980). The cause of death should not be attributed to fungi solely on the basis of an examination of fish collected after death. This is because *Saprolegnia* is both a **sacrotroph** and **necrotroph** so will grow quickly on the surface of moribund fish.

Systemic Bacteremial Diseases

Enteric red mouth (Yersiniosis)

Yersinia ruckeri is a gram-negative, oxidase-negative bacterium that is a serious pathogen of trout, salmon, seabass, and fathead minnows. Infected fish exhibit dark coloration, red fins, eyes and mouth (Figure 4.11). Bacteria can be readily isolated from the interior of lesions, spleen and kidney. An internal examination typically reveals scattered hemorrhaging, an enlarged spleen and necrotic intestinal tissue.

Surviving diseased fish often exhibit cycles of bacterial shedding from the lower intestine over long periods (Rucker, 1966; Busch and Lingg, 1975; Bruno and Munro, 1989). Serological and antibiotic resistance tests are necessary to confirm species identity and correctly treat antibiotic-resistant strains. Stressed fish can begin to shed bacteria and infect other fish present in the same water supply (Hunter *et al.*, 1980). The most efficient method of managing this disease is through the use of preventive immersion vaccines. These products provide very high levels of protection for relatively long periods of time (Stevenson, 1997).

Enteric septicemia of catfish

Edwardsiella ictaluri is the causative agent for enteric septicemia of catfish (Figure 4.12). As the name

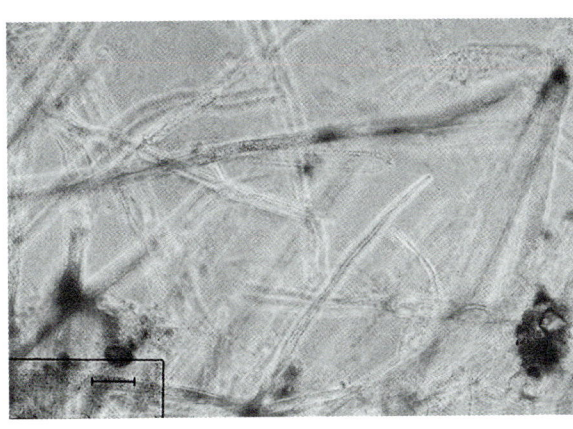

Figure 4.10 *Saprolegnia parasitica* fungal mycelium. Bar length = 100 μm.

Figure 4.11 *Yersinia ruckeri* infected fish exhibit dark coloration, red fins, eyes and mouth. **(See also Colour Plate 1.)**

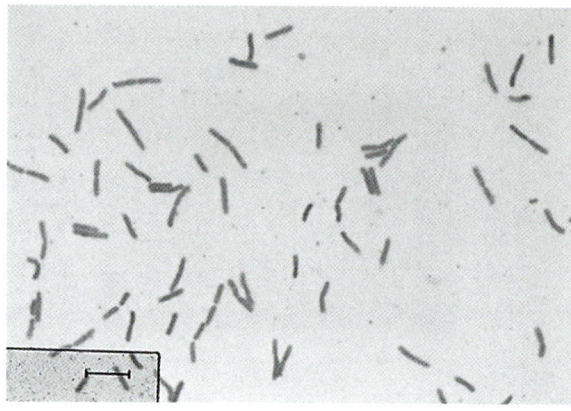

Figure 4.12 *Edwardsiella ictaluri* is the causative agent for enteric septicemia of catfish. The exact size of the bacteria may change under different fermentation conditions. Bar length = 5 μm.

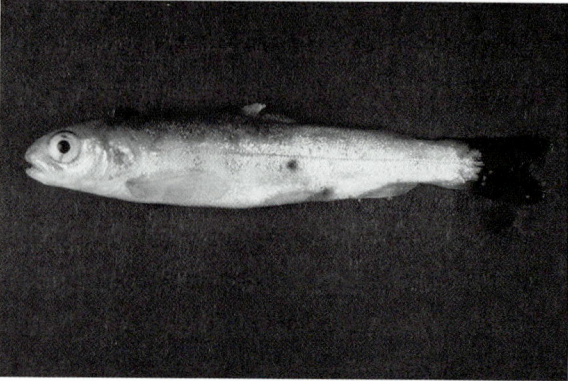

Figure 4.13 *Aeromonas salmonicida* often produces chronic muscle lesions that can form ulcers in the surface of the skin in Atlantic salmon.

implies, this organism primarily infects catfish, but has been isolated from danios and a limited number of warm, freshwater species. Although gram negative, oxidase negative and a member of the *Enterobacteriaceae* family (like *Yersinia ruckeri*), this pathogen can enter the olfactory organ and infect the brain to produce the classic 'hole-in-the-head' disease syndrome (Miyazaki and Plumb, 1985). Alternatively, *Edwardsiella ictaluri* can cause surface and intestinal hemorrhaging leading to a septicemia of the kidney, liver and spleen (Shotts *et al.*, 1986). It has been more difficult to protect against *Edwardsiella ictaluri* using conventional formalin-killed vaccines compared to *Yersinia ruckeri*. New improved vaccines in development are based on attenuated live strains with a reduced ability to infect or survive in fish tissues compared to naturally occurring pathogenic strains.

Furunculosis

Aeromonas salmonicida bacteria cause chronic muscle lesions that can form ulcers in the surface of the skin (Figure 4.13). This short rod-shaped, non-motile, auto-agglutinating, gram-negative bacterium (Figure 4.14) produces powerful proteolytic enzymes that break down connective tissue and are capable of surviving for long periods after ingestion by host macrophage cells. Surviving infected fish commonly become carriers that shed bacteria after periods of stress (handling, **smoltification**, etc.). Sublethal infections in carrier fish populations can be detected by injecting fish with cortisone to stimulate the outbreak of disease. In fish populations with a recurrent history

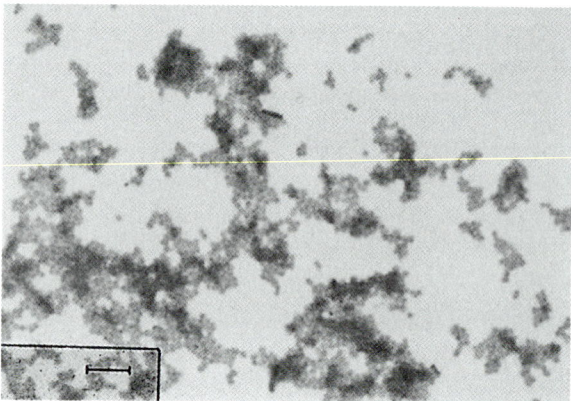

Figure 4.14 *Aeromonas salmonicida* are short non-motile, auto-agglutinating, gram-negative bacteria. Bar length = 5 μm.

of *A. salmonicida* infections, the mucus may be sampled noninvasively to detect the onset of infection (Cipriano *et al.*, 1992).

A presumptive diagnosis of typical *Aeromonas salmonicida* bacteria isolated from kidney tissue can be made by the observation of brown pigment secreted into the agar after 2–4 days. Additional evidence can be obtained by a positive staining of the colonies by Coommassie Blue (Cipriano and Bertolini, 1988). Confirmatory identification can be made with fluorescent antibody, ELISA or commercially available latex bead agglutination diagnostic kits (Austin and Austin, 1987). Slide agglutination tests should be run with adequate controls since this species has a protein outer 'A' layer that readily agglutinates to other bacteria in the preparation. Atypical isolates with unusual biochemical profiles (e.g. achromogenic subspecies, etc.) can also be isolated from similar ulcerated skin lesions in koi carp and salmonids.

Other opportunistic bacteria may overgrow the plate so selective media may be needed to separate the pathogenic invader from ubiquitous *Aeromonas hydrophila* and other less pathogenic normal flora (Noga, 1996).

Bacterial kidney disease

Bacterial kidney disease in trout and salmon is caused by *Renibacterium salmoninarum*. Young chinook salmon are particularly susceptible to this disease and it is almost never found in non-salmonid fishes. The disease derives its name from the enlargement of the kidney and high numbers of bacteria that can be observed in chronic kidney tissue infections (Figure 4.15). Renal failure often results in osmotic imbalances producing a swollen abdomen and **exophthalmic** eyes. The etiological agent is unique because it is one of the few cold-water gram-positive bacterial diseases in salmon that is vertically transmitted from maternal ovarian fluid to the developing sac fry.

Amateur observers should be cautious not to confuse the short rods of *Renibacterium salmoninarum* 0.5×1.0 μm with the dark brown refractive melanin granules common to the kidney and spleen. Other features of this bacterium include: nonmotility, PAS positive and acid fast negative staining. Long incubation periods with complex serum rich media are required for the isolation of this slow growing bacteria. *Renibacterium salmoninarum* will not grow above 25°C or on standard plates of tryptic soy agar or brain heart infusion agar (Evelyn, 1993). Confirmation of infection can be most quickly obtained by a fluorescent antibody or ELISA test on kidney tissue, ovarian fluid or spleen (Elliott and Barila, 1987; Pascho and Mulcahy, 1987).

Streptococcal bacteria

In recent years, streptococcal infections caused by *Streptococcus* spp. (*Streptococcus iniae*, *Enterococcus* spp. and/or *Lactococcus garviae*) bacteria have become increasingly common in commercial warm-water aquaculture systems. These bacteria are gram positive, 0.7–1.4 μm diameter, round or oval bacteria commonly in pairs or linked in chains (Figure 4.16). Serum cross-reactivity, biochemical tests and the degree of hemolytic activity on blood agar are useful methods to differentiate morphologically similar bacteria. Septicemias caused by streptococcal bacteria have been reported in hybrid striped bass, tilapia *Oreochromis* spp., rainbow trout *Oncorhynchus mykiss* and yellowtail *Seriola quinqueradiata* (Kitao, 1993; Bercovier et al., 1997). New outbreaks may be the result of horizontal transmission via the transportation of infected fish or the use of uncooked fish as feed.

Exophthalmia and corneal clouding and either septicemia or meningitis are common clinical findings for this disease (Bercovier et al., 1997). The warmer temperature and host range of streptococcal bacterial infections may assist the preliminary diagnosis of this bacterium from *Renibacterium salmoninarum* which is also gram positive but infects salmon and cold-water fishes below 15°C. Caution should be exercised during necropsy if fish are suspected of carrying *Streptococcus* spp. Zoonotic transmission of this infection to humans is possible in rare instances if the bacteria are allowed to enter skin cuts or abrasions. Localized swollen or inflammatory lesions should be examined by a physician.

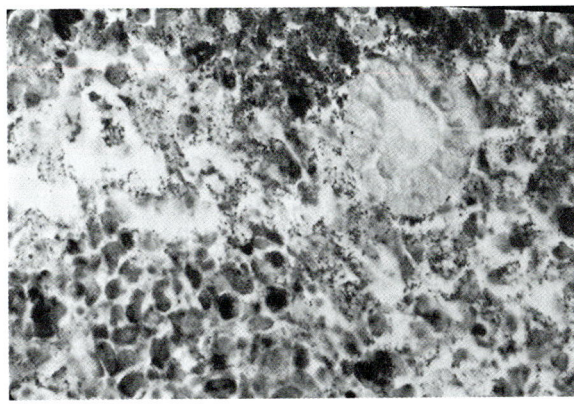

Figure 4.15 Short, dark gram-positive rods of *Renibacterium salmoninarum* 0.5×1.0 μm bacteria present in chinook salmon kidney tissue (gram stain).

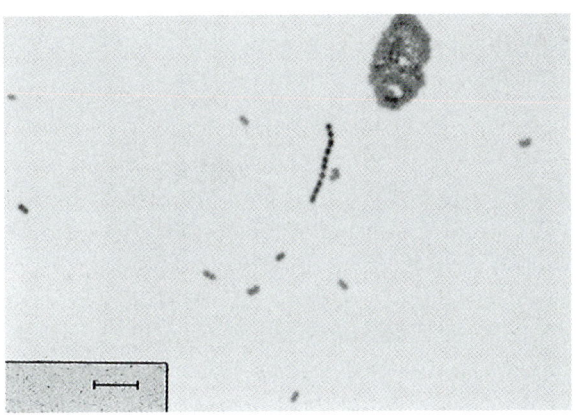

Figure 4.16 *Lactococcus garviae* are gram-positive, 0.7–1.4 μm diameter, round or oval bacteria commonly in pairs or linked in chains. Bar length = 5 μm.

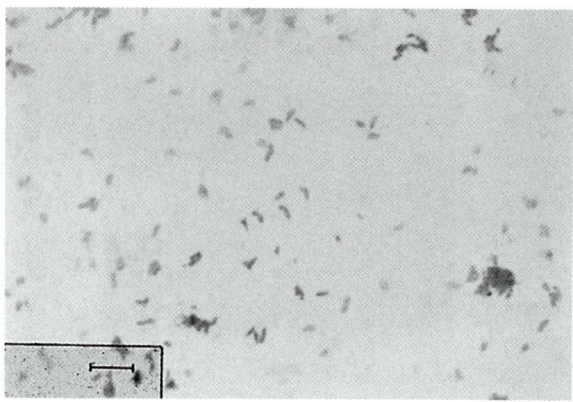

Figure 4.17 *Vibrio anguillarum* is a gram-negative, motile rod with a distinctive curved shape. Bar length = 5 μm.

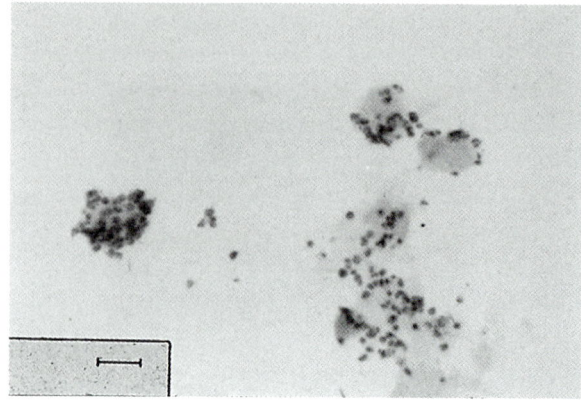

Figure 4.18 Rickettsial bacteria are pleomorphic (approximately 1 μm length), gram-negative, obligate intracellular parasites that divide in the cytoplasm of cells. This photo depicts the infection of chinook salmon embryo cells with *Piscirickettsia salmonis*. Bar length = 5 μm.

Vibriosis

Vibriosis is most commonly caused by *Vibrio anguillarum*, a gram-negative, curved motile rod (Figure 4.17) that rapidly infects stressed marine fish or fish living in brackish water. High temperatures, handling and changing salinity can result in disease outbreaks. Nearly all internal organs can become infected resulting in hemorrhaging and necrosis. In artificial saltwater holding conditions, abrasions and chronic skin lesions are commonly infected with these bacteria. *Vibrio* sp. bacteria can be readily isolated on nutrient agars supplemented with 1.5–2.5% salt.

Coldwater vibriosis or 'Hitra' disease is caused by *Vibrio salmonicida* in salmonids reared for extended periods at very cool temperatures. *Vibrio ordalii* and *Vibrio viscosus*, *Vibrio vulnificus* and *Vibrio alginolyticus* are related vibrio bacteria that can also infect and kill a variety of marine species held in suboptimal conditions.

Rickettsia-like organisms

Rickettsial bacteria are pleomorphic (approximately 1 μm length), gram-negative, obligate intracellular parasites that divide in the cytoplasm of cells (Figure 4.18). *Piscirickettsia salmonis* has been isolated from Chile, Canada, Norway and Scotland. *P. salmonis* infects coho salmon, rainbow trout and Atlantic salmon. Signs of this disease include: darkened color, lethargy and upraised scales. Internal lesions often include petechial hemorrhaging, round liver spots and enlargement of hematopoietic organs. Diagnosis is done by microscopic observations using a fluorescent antibody specific to *P. salmonis* or by examining EPC or CHSE-214 fish cell cultures exposed to diluted **axenic** fish tissues applied without antibiotics. Severe infections can be detected in blood smears or kidney tissue imprints fixed in methanol and stained with giemsa.

Viral Infections

Diagnosis

Viral diagnoses must be made using suitable fish cell lines (CHSE-214, EPC, RTG-2, etc.) at the appropriate temperature (host range) and pH (Figure 4.19). Tissue samples must be homogenized and centrifuged in media containing antibiotics and antimycotics or filter sterilized to remove particulate material and normal bacterial flora contaminants. Viruses may often be presumptively identified by the cytopathic effect on specific cell lines after a suitable time period (usually about 2 weeks). A negative finding must be confirmed by the 'blind passage' of the cell culture medium to a second set of cells to be sure there was no interference (caused by interferon, etc.) during the original test.

Confirmatory diagnosis is commonly done by demonstrating serum neutralization of the supernatant with polyclonal rabbit or mouse monoclonal

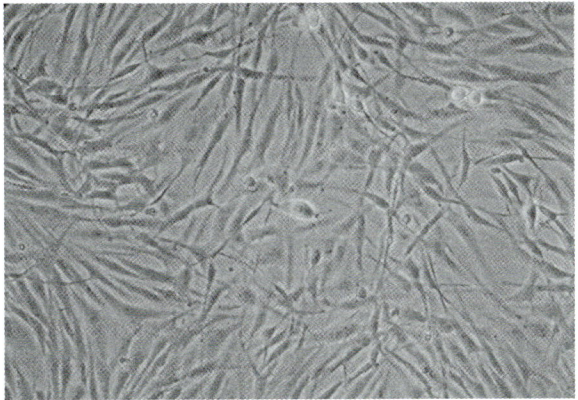

Figure 4.19 Unstained fish cell culture useful for growing viruses and rickettsia organisms. The mitotic figures indicate a healthy dividing, preconfluent population of cells.

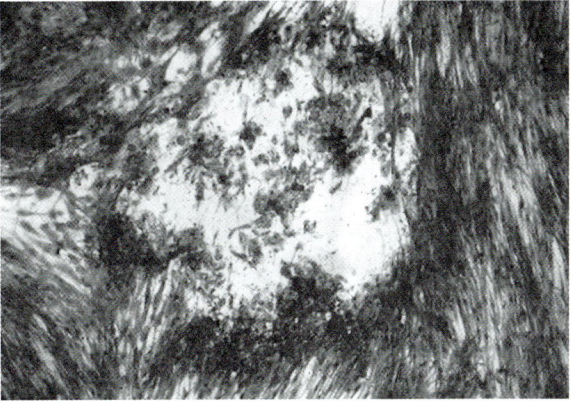

Figure 4.20 The cytopathic effect of a cell culture monolayer infected with infectious hematopoietic necrosis virus (IHNV) forming grape-like clusters of rounded cells with enlarged nuclei. Formalin fixed crystal violet stained cells.

serum containing antibodies specific for the virus in question. Morphological measurements can also be taken at extremely high magnification by electron microscopy. Due to the complexity of identifying viral diseases, it is often prudent to first eliminate the possibility that the observed disease signs may be caused by bacterial, parasitic or water quality problems.

Rhabdoviruses (IHNV, VHSV)

Infectious hematopoietic necrosis virus (IHNV), viral hemorrhagic septicemia virus (VHSV) and canine rabies virus are all members of the family *Rhabdoviridae*. Both IHNV and VHSV cause mortality in rainbow trout and in some species of salmon (but not in canine or mammalian hosts). Rhabdoviruses are 'bullet shaped' particles with an apical pore containing a single unsegmented strand of RNA. To replicate, the genome directs the formation of five viral proteins: nucleoprotein, two matrix proteins, glycoprotein, polymerase and a nonvirion-associated protein (McAllister and Wagner, 1975; Kurath and Leong, 1985). As indicated by the names of these viruses, they infect hematopoietic tissue and cause internal bleeding. Severe anemia can be seen as pale gill and kidney tissue. Hematocrit values tend to be very low and kidney failure may result in osmotic imbalances producing swollen eyes. Fish infected with either IHNV or VHSV are often observed to be dark in color and exophthalmic.

Kidney, spleen, brain and skin mucus samples can all be used to test for the presence of rhabdoviruses on cell culture (Batts and Winton, 1989; LaPatra *et al.*, 1989). With experience, giemsa-stained kidney tissue imprints can be used to help diagnose IHNV. These imprints often show large numbers of slightly enlarged 'foamy' macrophages, indicative of a viral infection. The observation of inoculated cell culture monolayers forming grape-like clusters of rounded cells is another indication that a rhabdovirus may be present (Figure 4.20). Confirmation of the viral etiology can be determined with a serum neutralization test specific for suspect viruses.

Infectious pancreatic necrosis virus (IPNV)

Infectious pancreatic necrosis virus (IPNV) is an icosahedral particle containing a bisegmented RNA genome. Unlike many viruses, IPNV does not protect itself with a lipid bilayer envelope. The viral capsid is composed of 92 capsomeres formed by three major structural components (Dobos *et al.*, 1979). IPNV has been associated with losses in juvenile rainbow trout, Pacific (sockeye, chinook, etc.) and Atlantic salmon. The carrier status and shedding of viral particles from fish infected with IPNV is well established as a means of transfer of this disease to naïve populations. In culture, IPNV-infected cell monolayers often form plaques with cells at the plaque margin exhibiting long, thin, drawn-out cytoplasmic processes (Figure 4.21).

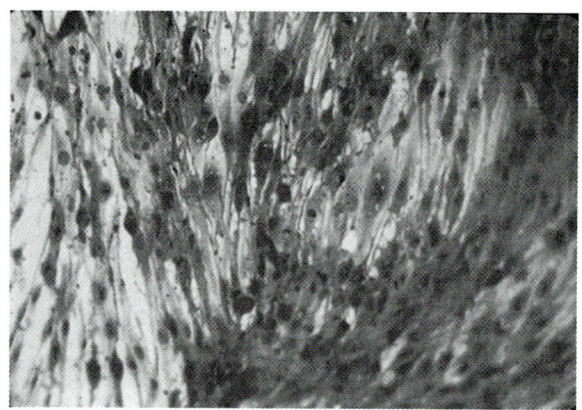

Figure 4.21 A fish cell culture monolayer infected with infectious pancreatic necrosis virus producing the characteristic cytopathic effect of elongated, thin, drawn-out cells. Formalin fixed crystal violet stained cells.

Disease prevention

Most diseases can be prevented through good husbandry practices and proper screening of incoming animals to the facility. When possible, bring in only eggs from a reputable supplier that can provide disease-free animals. Commercial blood test kits are available to screen fish for antibodies of several important fish pathogens. These kits normally do not determine active infections but can provide evidence of previous exposure by vaccination or live-disease-causing agents. More expensive laboratory DNA tests of fish tissue using the polymerase chain reaction (PCR) can provide an even higher degree of sensitivity than antibody-based tests to identify subclinical infections or previous use of vaccinations.

Isolation, rapid removal and necropsy of dead animals will reduce the spread of disease and help to provide early diagnosis and treatment of the problem. Prophylactic external (e.g. chloramine T, etc.) treatments during and after handling procedures will prevent the start of many infections. Separate nets for each tank and iodophore disinfection baths for equipment will reduce cross-contamination problems.

Vaccines are commercially available to protect against vibriosis, furunculosis, enteric red mouth and enteric septicemia bacteria. New vaccines are in development for several commercially important viral and rickettsial fish pathogens. Specialty orders of 'autogenous' vaccines can also be manufactured to protect against unique or emerging bacterial pathogens. All vaccines require that the fish be held for a period of disease-free conditions (usually 3–5 weeks) after vaccination to build up immunity before any significant exposure to infectious diseases. Vaccinations against vibriosis, enteric red mouth and enteric septicemia bacteria can be delivered to the fish by immersion. For other diseases, intraperitoneal injection is the preferred route for maximum protection and duration of immunity.

Selective breeding using quantitative genetics can be used to produce strains of fish with enhanced resistance to specific diseases. The traits for selection must be based on genes with sufficient heritability for this process to be successful. Furunculosis-resistant strains of brook trout and brown trout in the eastern United States are examples of the great potential of these efforts. Studies on skin mucus sampling for *A. salmonicida* have demonstrated the effectiveness of breeding programs to reduce the shedding of pathogenic organisms within a population (Cipriano *et al.*, 1994). Similar intensive genetic selection programs for disease resistance are underway in Norway for Atlantic salmon.

Disease treatment

Bacterial, parasitic and fungal diseases can all be controlled with chemo-therapeutants. Viral diseases are best prevented or eliminated by isolation and quarantine procedures. The key to successful treatment is the proper identification of the primary cause of the observed losses. For example, the observation of a single external parasite is not a reason for immediately beginning treatment without further investigation of other underlying infections or water quality problems. Consultation with an American Fisheries Certified Fish Health Pathologist or experienced veterinarian is strongly recommended before starting any treatment.

All compounds can have side effects and it is essential that caution be used in handling and use of any chemical. Use gloves, goggles and respirators as needed in order to reduce or eliminate human exposure. It is advisable to ask local authorities about legal restrictions on the use and disposal of chemo-therapeutants. A 96-hour pretreatment test with a small subset of animals held at comparable density to the main group will help identify any toxic reactions of new treatments before jeopardizing larger, valuable populations.

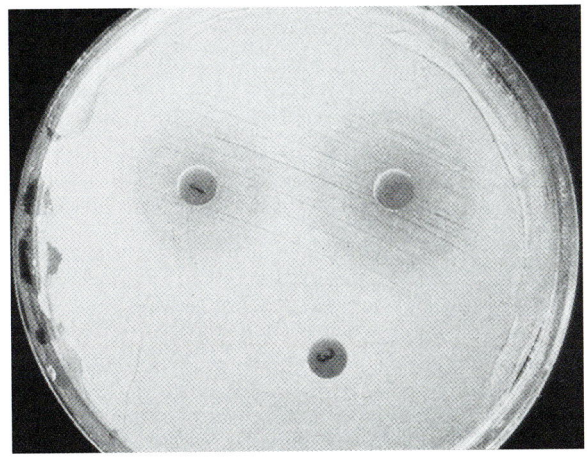

Figure 4.22 Antibiotic sensitivity testing of pathogenic bacteria on Mueller Hinton agar showing two wide diameter zones of growth inhibition and one antibiotic to which the bacteria are completely resistant.

If a pathogenic bacterium is isolated, tests for antibiotic resistance should be done to select the best drug and treatment regime (Figure 4.22). Gram-negative organisms are commonly treated with oxytetracycline or sulfamethoxazole plus trimethoprim. Gram-positive bacteria are more responsive to erythromycin or doxycycline. Caution should be exercised if the biological filter media will be exposed to these compounds because water quality may quickly deteriorate if the nitrifying bacteria are lost.

Antibiotics added to the feed may not be appropriate if the affected population of fish is refusing to eat. Fish oil additives to enhance palatability or direct injection of antibiotics are alternatives to consider before applying oral drug treatments. Particular attention must be paid to local and federal regulations regarding restrictions on the use and withdrawal periods of antibiotics in food fish.

Fungal infections of eggs can be treated with a 1–5 mg/L concentration of methylene blue or 1 ml/L formalin for 15 minutes (Kirchen and West, 1976; Schnick *et al.*, 1989). However, frequent removal of dead eggs is critical to the success of hatching survival and may limit the need for such treatments. Formalin is also useful for the treatment of external parasitic infections in juvenile or adult fish. Typical bath concentrations for up to 1 hour range from 0.125 to 0.250 mL/L with 0.167 mL/L being well documented as generally effective (Warren, 1981; Post, 1983; Jensen and Durborow, 1984; Noga, 1996). Warm water temperatures or combination with other compounds (such as chloramine T) can significantly increase the toxic side effects of formalin.

Damaged gills or skin can often be treated with 0.5–1.0% sodium chloride to improve the osmotic balance between the water and the fish tissues. Lower salt concentrations should be used on fish smaller than 10 grams. Warren (1981) suggests that dissolved salt treatments improve the effect of other surface compounds by removing debris and mucus from the gills and skin.

Judiciously applied, potassium permanganate (at about 2 mg/L), can remove surface parasites and bacteria from fish in freshwater systems (Boyd, 1979; Tucker, 1989). Care should be taken if the holding water is low in organic material or high in pH because potassium permanganate can result in toxicity or precipitation of manganese dioxide on the gills (Noga, 1996). More permanganate may be needed in tanks or ponds with high organic loads because much of the active ingredient will complex with this material.

References

Anaker, R.L. and Ordal, E.L. (1959). *J. Bacteriology* **78**, 25–32.

Austin, B. and Austin, D.A. (1987). *Bacterial Fish Pathogens: Disease in Farmed and Wild Fish*. Ellis Horwood, Chichester, UK, 384pp.

Bader, J.A. and Shotts, E.B. (1998). *J. Aquat. Anim. Health* **10**, 311–319.

Batts, W.N. and Winton, J.R. (1989). *J. Aquat. Anim. Health* **1**, 284–290.

Bercovier, H., Ghittino, C. and Elder, A. (1997). In *Fish Vaccinology*, pp. 153–160. *Developments in Biological Standardization*, Vol. 90. Karger, Basel, Switzerland.

Bernardet, J.F. (1997). In *Fish Vaccinology*, pp. 179–188. *Developments in Biological Standardization*, Vol. 90. Karger, Basel, Switzerland.

Bertolini, J.M. and Rohovec, J.S. (1992). *Dis. Aquat. Org.* **12**, 121–128.

Boyd, C.E. (1979). *Water Quality in Warmwater Fish Ponds*. Auburn University, AL. 359pp.

Bruno, D.W. and Munro, A.L.S. (1989). *Aquaculture* **81**, 205–211.

Busch, R.A. and Lingg, A.J. (1975). *J. Fish. Res. Board Can.* **32**, 2429–2432.

Cipriano, R.C. and Bertolini, J. (1988). *J. Wild. Dis.* **24**(4), 672–678.

Cipriano, R.C., Ford, L.A., Teska, J.D. and Hale, L.E. (1992). *J. Aquat. Anim. Health.* **4**, 114–118.

Davis, H.S. (1947). *US Fish Wild. Surv. Fish. Bull. US Bur. Fish.* **42**, 1–29.

Decostere, A., Haesebrouck, F., Turnbull, J.F. and Charlier, G. (1999). *J. Fish Dis.* **22**, 1–11.

Dobos, P., Hill, B.J., Kells, D.T., Becht, and Teninges, D. (1979). *J. Virol.* **32**(2), 593–605.

Elliott, D.G. and Barila, T.Y. (1987). *Can. J. Fish Aquat. Sci.* **44**, 201–210.

Evelyn, T.P.T. (1993) In *Bacterial Diseases of Fish* (eds V. Inglis, R.J. Roberts and N.R. Bromage), pp. 177–195. Inst. of Agriculture, Halsted Press, NY.

Griffin, B.R. (1991). *Trans. Am. Fish. Soc.* **120**, 391–395.

Hawke, J.P. and Thune, R.L. (1992). *J. Aquat. Anim. Health* **4**, 109–113.

Holt, R. (1994). In Blue Book Version 1. *Suggested Procedures for the Detection and Identification of Certain Finfish and Shellfish Pathogens* (ed. John C. Thoesen). AFS Fish Health Section, SOS Publications, Fair Haven, NJ.

Holt, R.A., Amandi, A., Rohovec, J.S. and Fryer, J. (1989). *J. Aquat. Anim. Health* **1**, 94–101.

Hunter, V.A., Knittel, M.D. and Fryer, J.L. (1980). *J. Fish. Dis.* **3**, 467–472.

Jensen, J. and Durborow, R. (1984). Circular ANR-414, Alabama Cooperation Extension Service, Auburn University, AL.

Kirchen, R.V. and West, W.R. (1976). *The Japanese Medaka, its Care and Development*. Carolina Biological Supply Co., Burlington, North Carolina, 36pp.

Kitao, T. (1993). In *Bacterial Diseases of Fish* (eds. V. Inglis, R.J. Roberts and N.R. Bromage). pp. 196–210. Inst. of Agriculture, Halsted Press, NY.

Kurath, G. and Leong, J.C. (1985). *J. Virol.*, **53**, 462–468.

LaPatra, S.E., Roberti, K.A., Hohovec, J.S. and Fryer, J.L. (1989). *J. Aquat. Anim. Health* **1**, 29–36.

Lumsden, J.S., Ostland, V.E., Byrne, P.J. and Ferguson, H.W. (1993). *Dis. Aquat. Org.* **16**, 21–27.

McAllister, P.E. and Wagner, R.R. (1975). *J. Virol.* **22**, 830–843.

Miyazaki, T. and Plumb, J.A. (1985). *J. Fish Dis.* **8**, 389–392.

Neish, G.A. (1977). *J. Fish. Biol.* **10**, 513–522.

Neish, G.A. and Hughes, G.C. (1980). *Fungal Diseases of Fishes (Book 6)*. In *Diseases of Fishes* (eds S.F. Snieszko and H.R. Axelrod), 159pp. TFH Publications, Neptune City, NJ.

Noga, E.J. (1996). *Fish Disease Diagnosis and Treatment*. Mosby-Year Book, St Louis, Missouri. 367pp.

Pascho, R.J. and Mulcahy, D. (1987). *Can. J. Fish. Aquat. Sci.* **44**, 183–191.

Post, G.W. (1983). *Textbook of Fish Health*. TFH Publications, Neptune City, NJ. 256pp.

Rucker, R.R. (1966). *Bull. L'Office Int. Epizoot.* **65**, 825–830.

Schneider, R. and Nicholson, B.L. (1980). *Can. J. Fish Aquat. Sci.* **37**, 1505–1513.

Schnick, R., Meyer, F. and Grey, L.D. (1989). A guide to approved chemicals in fish production and fisheries resource management. Univ. of Arkansas Cooperative Extension Service and US Fish and Wildlife Service. 27pp.

Shotts, E., Blazer, V.S. and Waltman, W.D. (1986) *Can. J. Fish. Aquat. Sci.* **43**, 36–42.

Stevenson, R.M.W. (1997). In *Fish Vaccinology*, pp. 117–124. Developments in Biological Standardizaiton, Vol. 90. Karger, Basel, Switzerland.

Tucker, C.S. (1989). *Prog. Fish Cult.* **51**, 24–26.

Warren, J.W. (1981). *Diseases of Hatchery Fish*. Ft. Snelling, Twin Cities, MN. US Fish and Wildlife Service. 91pp.

PART 3

Gross Functional Anatomy

Contents

CHAPTER 5	**Integumentary system**	95
CHAPTER 6	**Skeletal system**	109
CHAPTER 7	**Muscular system**	119
CHAPTER 8	**Nervous system**	129
CHAPTER 9	**Respiratory system**	151
CHAPTER 10	**Circulatory system**	161
CHAPTER 11	**Digestive system**	173
CHAPTER 12	**Urinary tract**	181
CHAPTER 13	**Endocrine system**	189
CHAPTER 14	**Immune system**	219
CHAPTER 15	**Sensory systems**	225
CHAPTER 16	**Reproductive systems**	261

CHAPTER 5

Integumentary System

Diane G Elliott
Western Fisheries Research Center, Biological Resources Division, US Geological Survey, Seattle, Washington, USA

Introduction

As in other vertebrates, the integument or skin of a fish is the envelope for the body that separates and protects the animal from its environment, but it also provides the means through which most of the contacts with the outer world are made. The integument is continuous with the lining of all the body openings, and also covers the fins. The skin of a fish is a multifunctional organ, and may serve important roles in protection, communication, sensory perception, locomotion, respiration, ion regulation, excretion, and thermal regulation. These functions are introduced in the following section, and the integumentary elements responsible for specific functions are described further in this chapter and in Chapter 17.

Functions

The most obvious functions of fish integument are protective. For example, mucous secretions help to keep the skin surface free of pathogens by means of constant sloughing and renewal and the presence of antimicrobial substances (Chapter 17; Pickering, 1974; Hattingh and van Warmelo, 1975; Ellis, 1981; St Louis-Cormier et al., 1984; Rambout et al., 1993; Cole et al., 1997). The precipitative action of fish mucus on suspended material in water may help to clear this material from the gills and skin of fish in highly turbid or polluted waters (Van Oosten, 1957). In addition, mucous secretions and thickened outer skin layers help to protect against abrasion in burrowing or semi-terrestrial fishes (Mittal and Banjeree, 1980; Fishelson, 1996).

Scales provide mechanical protection for deeper tissues (page 97, Burdak, 1980), and bony encasements and spines ward off attacks by predaceous enemies (page 105). Noxious substances that may be elaborated by sacciform cells in the skin and venoms produced by glands associated with spines on the fins or opercula can also help to deter such attacks (Chapter 17).

Protection can also be afforded by patterns of skin pigmentation such as cryptic coloration, countershading, or disruptive coloration, which serve to conceal a fish (Endler, 1978; Fujii, 1993). However, bright

Copyright © 2000 Academic Press

and conspicuous skin coloration may serve to advertise rather than conceal (Fujii, 1993). In fact, most color patterns are a compromise between the necessity to communicate with other members of a species and the need to avoid being eaten. Interpretation of the significance of color patterns is complicated by the ability of many fishes to change colors, either slowly, as during the maturation process, or rapidly, as during displays of aggression or courtship (Sterba, 1973; Lanzing and Bower, 1974; Kohda and Watanabe, 1982, 1983; Fujii, 1993), or for adaptive camouflage (Ramachandran et al., 1996). Additionally, certain color patterns, such as an abnormal overall darkening of the skin, can be a sign of poor health in a fish (Chapter 4). Pigment cells or chromatophores in the skin (Chapter 17) are responsible for much of the remarkable variation in coloration seen in fishes.

Among many midwater and bottom-dwelling deep-sea fishes, bioluminescent organs or **photophores** in the skin (Chapter 17) may take the place of color patterns for species and mate recognition (Lagler et al., 1977). Conversely, multiple photophores spread over the body surface may provide camouflage through countershading or by blurring the silhouette of the fish in the glow of spontaneous bioluminescence (Herring, 1982; Kashkin, 1993). Photophores may also serve to illuminate dark waters, and large light organs on the heads of some fishes are thought to be lures to attract prey (Sazonov, 1996).

Some forms of communication among fishes are facilitated by substances produced in the skin. For example, fright pheromones released into the water from damaged integumentary club cells of certain fishes alert other fish of the same species to the presence of predators (Chapter 17; Smith, 1982, 1992).

Several types of specialized sensory structures are located in the integument. Included among these are the lateral line system (Chapters 15.2 and 27.2), chemoreceptory structures or taste buds (Chapters 15.3 and 27.3), and electrosensory receptors.

Integumentary features can assist a fish during locomotion. The slippery mucus of some fishes has marked friction-reducing properties that enable a fish to move at greater speed with less expenditure of energy (Rosen and Cornford, 1971; Bernadsky et al., 1993). The passive flexural stiffness that the skin imparts to the body also aids in the generation of the undulatory waves that propel most fishes during swimming (Long et al., 1996).

The integument is an important adjunct to the breathing equipment of some fish species. Gas exchange across the skin is known to play a significant role in the respiration of larval fish (Liem, 1981; Rombough and Moroz, 1990). Certain adult fishes that are adapted to cope with temporary dewatering or short terrestrial migrations can use atmospheric oxygen by diffusion through a well-vascularized skin (Berg and Steen, 1965; Tamura et al., 1976). For most adult freshwater and marine fish species that have been studied in water, cutaneous respiration has matched the oxygen demand of the skin itself, indicating that the skin is an oxygen exchanger primarily for its own benefit (Kirsch and Nonnotte, 1977; Nonnotte, 1981; Steffensen and Lomholt, 1985).

The role of the integument in the maintenance of the balance of water and minerals inside a fish's body is not well understood. Although the skin of the primitive hagfishes and perhaps lampreys is highly permeable to water (Evans, 1993), the skin of higher fishes is generally quoted to be poorly permeable for both water and ions (Heisler, 1993). Nevertheless, a few studies in higher bony fishes (teleosts) have shown significant transfer of certain ions across the skin surface (Chapter 17; Nonnotte et al., 1979; Perry and Wood, 1985). The specialized cells (ionocytes) believed to be at least partially responsible for transepithelial ionic transfer are known to occur in the skin of several fish species (Chapter 17; reviewed by Zadunaisky, 1984 and Heisler, 1993). The superficial epithelial cells also may be involved (Chapter 17).

The function of the integument in excretion is also largely unknown, although it may have a significant role in the excretion of nitrogenous wastes, principally ammonia and urea (Wood, 1993). Limited experiments have suggested that the skin is a more important site of ammonia excretion in marine fishes than in freshwater fishes (Read, 1968; Morii et al., 1978; Sayer and Davenport, 1987). Indirect evidence also suggests that the skin of marine teleost fishes may be an important site of urea excretion (Morii et al., 1978; Sayer and Davenport, 1987). The integumentary elements involved in the excretion of nitrogenous wastes have not been identified.

Despite the fact that most fishes are obligate **poikilotherms**, they display a variety of adaptations — some of which involve the integument — to reduce the impact of temperature on physiological function. Certain tunas can maintain a body temperature several degrees above the ambient water temperature through vascular heat exchangers, which include networks of blood vessels in the skin (Hazel, 1993). Antifreeze proteins are produced by fishes living in the polar oceans (Hazel, 1993); some of these are abundant in the skin and appear to be important for protecting exterior tissues from freezing (Valerio et al., 1992; Gong et al., 1996).

General structure

The structure of the fish integument is highly adapted to carry out its various functions. As in other vertebrates, the skin of all fishes consists of two layers: an outer epidermis and an inner dermis or corium (Figures 5.1–5.5). The two layers differ in origin and structure as well as function. The epidermis is essentially a cellular material, a multilayered epithelium (Figures 5.1–5.5) derived from the ectoderm of the embryo. The dermis is primarily a fibrous structure with relatively few cells (Figures 5.1–5.5), and is derived from embryonic mesenchyme of mesodermal origin.

Epidermis

The thickness of the epidermis varies with fish species (Figures 5.1–5.5), age, different regions of the body, and environmental conditions (Chapter 17). In most species, the epidermis is thinner than the dermis. Keratinization of the epidermis to produce the horny surface layers characteristic of land vertebrates is rare among fishes (Chapter 17); fish epidermis is generally metabolically active throughout all its layers. Because the epidermis of most fishes contains little or no pigment and therefore appears largely transparent by gross observation, the structural aspects of this tissue will be described in the chapter on microscopic anatomy (Chapter 17).

Dermis

The dermal layer of skin contains blood vessels (Chapter 10), nerves (Chapter 8), scales, and adipose tissue, but the bulk of the typical dermis consists of fibrous connective tissue (Figures 5.1–5.5). When a fish is being skinned, the collagenous fibers of connective tissue that bind the skin to the underlying muscle and bone are very evident. The morphological characteristics of most dermal features are best observed microscopically, and are described in Chapter 17.

Scales

The dermal structures that are usually most prominent by gross observation are the scales, although these can vary in size from microscopic structures to huge bony plates. Scales are an important component of the dermal skeleton (Chapters 6 and 18) and usually are covered completely by epidermal tissue, though portions of scales may protrude from the epidermal surface in some species such as sharks.

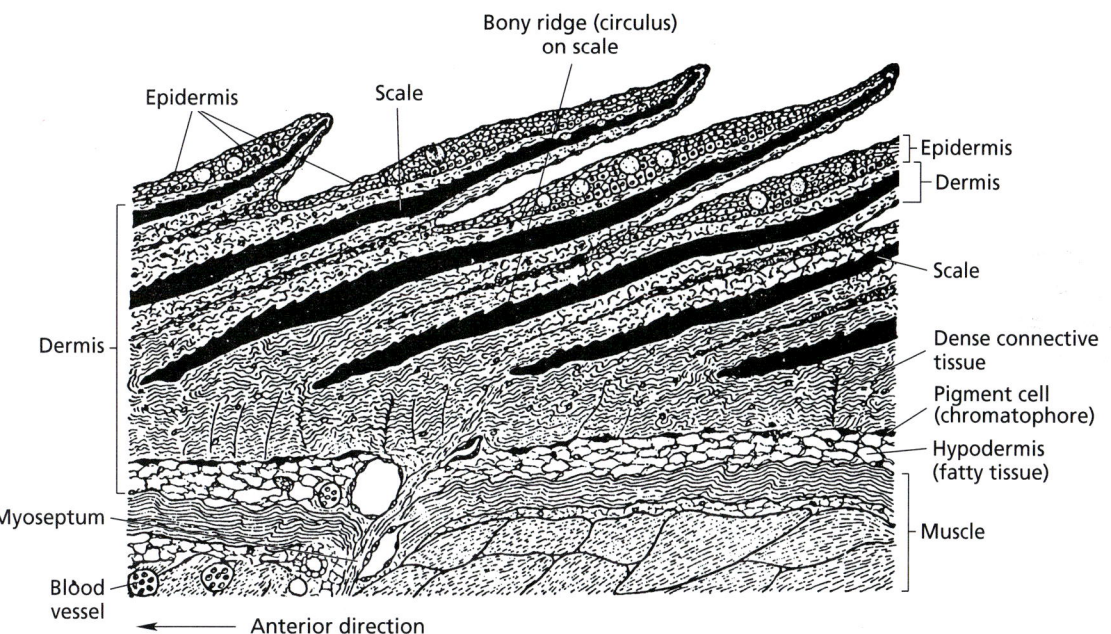

Figure 5.1 Vertical section (longitudinal to the body) of the skin of a typical teleost (bony) fish, the yellow perch, *Perca flavescens*. (Source: Ward's Natural Science Establishment, Inc.)

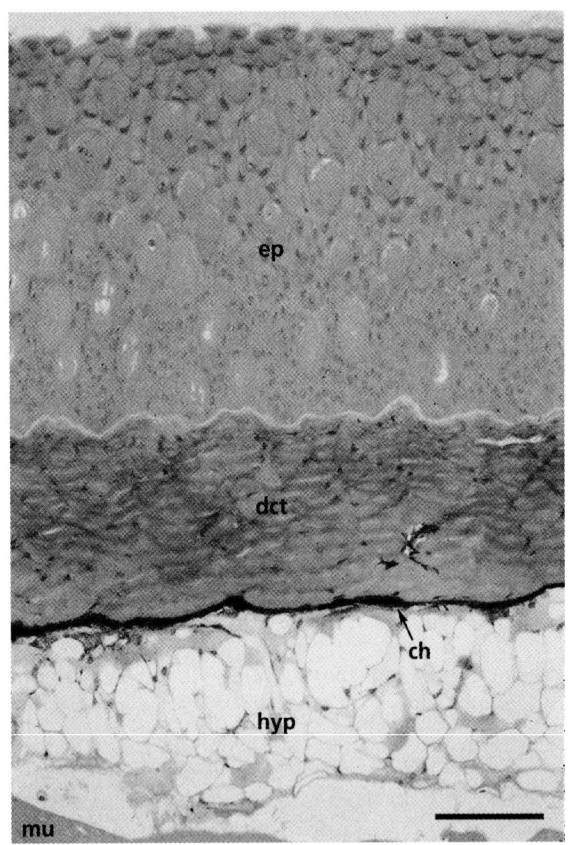

Figure 5.2 Skin layers of a jawless fish, a spawning adult Pacific lamprey, *Lampetra tridentata*. The components include a thick epidermis (*ep*) and two layers of dermis, an upper (outer) layer of dense collagenous tissue (*dct*) and a lower hypodermis (*hyp*) consisting largely of adipose tissue. Chromatophores or pigment cells (*ch*) are visible in the upper hypodermis, and a small section of subcutaneous muscle (*mu*) is visible beneath the hypodermis. Lampreys are scaleless. Scale bar = 100 μm. Hematoxylin and eosin stain. **(See also Colour Plate 2.)**

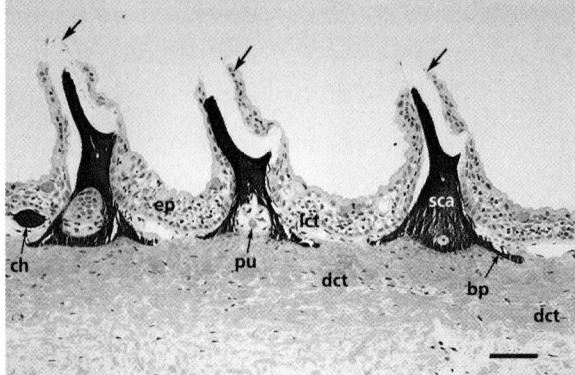

Figure 5.3 Fin tissue from a chondrichthyan, an adult female smooth dogfish shark, *Mustelus canis*. Tooth-like placoid scales (*sca*) are visible beneath the epidermis (*ep*). The basal plate (*bp*) of each scale is anchored in the dermis, and portions of pulp cavities (*pu*) are visible within the scales. The remnants of epidermis (arrowed), possibly damaged during sampling, suggest that some of the placoid scales of this species may be completely covered by epidermis. (Compare with the partially exposed scales in Figure 5.6b.) The visible dermis of the fin includes an outer layer of loose connective tissue (*lct*) and a layer of dense connective tissue (*dct*). A chromatophore or pigment cell (*ch*) is visible in the dermis. Scale bar = 100 μm. Hematoxylin and eosin stain. **(See also Colour Plate 3.)**

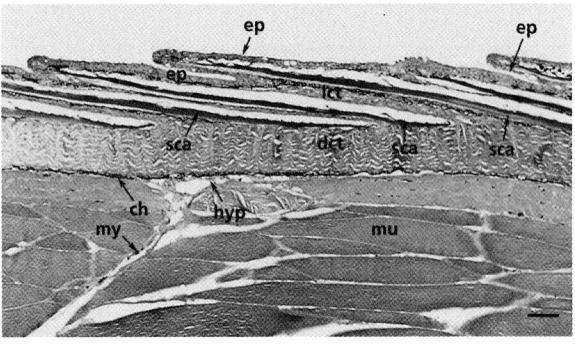

Figure 5.4 Skin layers of a juvenile steelhead trout, *Oncorhynchus mykiss*. The thin epidermis (*ep*) covers and wraps around the posterior (non-overlapped) portion of each scale (*sca*). The imbricated scales lie in pockets of dermal tissue. The layers of the dermis include an upper layer of loose connective tissue or stratum spongiosum (*lct*), a thick middle layer of dense connective tissue or stratum compactum (*dct*), and a thin lower hypodermis (*hyp*) containing chromatophores or pigment cells (*ch*) and adipose tissue. Subcutaneous muscle (*mu*), including a myoseptum (*my*), is also visible. Scale bar = 100 μm. Hematoxylin and eosin stain. **(See also Colour Plate 4.)**

Despite a resemblance to the superficial horny scales occurring on reptiles, birds, and some mammals, fish scales differ in structure and origin from these other scale types. The scales of these other vertebrates (as well as appendages such as feathers, hair, hooves, nails, claws, and quills) consist largely of keratinized epidermal tissue of ectodermal origin (Romer, 1968; Matoltsy and Bereiter-Hahn, 1986). In contrast, the scales of fishes are mineralized dermal structures that are primarily of mesenchymal origin (see also Chapters 17 and 18).

The type, number, and size of scales can divulge much information about how a fish makes its living (Moyle and Cech, 1996). On a bony fish, scalation can range from a heavy coating of mail-like armor, to a few large bony plates on the back, to a dense covering of thin, flexible scales, to a few localized prickles, to no scales at all. Bony plates or **scutes** are large modified scales that serve as armor on a number of slow-moving, bottom-oriented fishes such as sturgeons (Acipenseridae), many South American catfishes (Loricariidae, Callichthyidae and Doradidae),

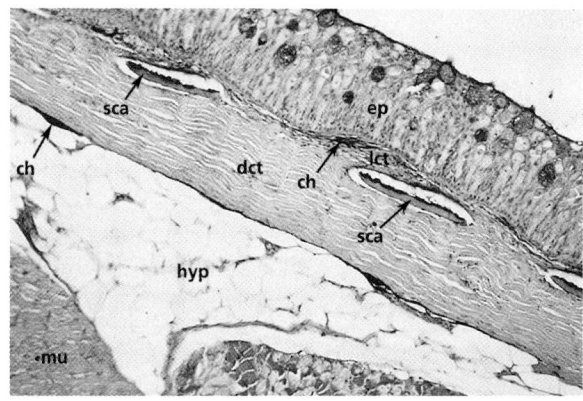

Figure 5.5 Skin layers of a short-fin eel, *Anguilla australis*. The layers include a thick epidermis (*ep*) and a dermis consisting of an upper layer of loose connective tissue or stratum spongiosum (*lct*), a middle layer of dense connective tissue or stratum compactum (*dct*), and a hypodermis (*hyp*) comprised largely of adipose tissue. Chromatophores or pigment cells (*ch*) are visible in the stratum spongiosum and hypodermis. The deeply embedded scales (*sca*) do not overlap. Muscle tissue (*mu*) is visible beneath the skin. Scale bar = 100 μm. Hematoxylin and eosin stain. (Source: Photomicrograph courtesy of Dr Barbara Nowak.) **(See also Colour Plate 5.)**

poachers (Agonidae), and pipefishes and seahorses (Syngnathidae). In contrast, the bodies of free-swimming fishes are usually covered with typical scales, which provide some protection against mechanical damage without adding excessive weight. Fast-swimming fishes and those that regularly move through fast water tend to have numerous fine scales (e.g. trout), whereas fishes that live in quiet water and do not continuously swim at high speeds often have rather coarse scales (e.g. perch and carp). Scaleless fishes are often bottom-dwellers in moving water (e.g. sculpins), fish that frequently hide in tight places such as caves and crevices (e.g. many catfish and eels), or fast-moving **pelagic** fish (e.g. swordfish and some mackerel). The primitive agnathans or jawless fishes (hagfish and lampreys) are also scaleless (Figure 5.2). Some fish that appear scaleless in fact have numerous deeply embedded scales (e.g. most tunas and anguillid eels; Figure 5.5).

As indicated by their fundamental differences in structure, scales evolved independently in cartilaginous and bony fishes (Moyle and Cech, 1996; Chapter 18). Placoid scales (also called dermal denticles or isolated odontodes) are almost exclusively restricted to the cartilaginous sharks, rays, and chimaeras (Chondrichthyes). A placoid scale typically consists of a flattened, rectangular basal plate embedded in the upper layer of dermis, and a cusp or spine that may project backward through the surface of the epidermis (Van Oosten, 1957; Raschi and Tabit, 1992; Figures 5.3 and 5.6). The protruding spines of placoid scales give shark skin its characteristic rough texture. The structure of the placoid scale resembles a tooth (Figure 5.6b; see also Chapter 18). The spine consists of a cap or cone of dentine covered with a layer of hard, transparent enameloid (Whitear, 1986; Chapter 18). A pulp cavity is enclosed within the spine. The basal plate is perforated by one or more openings through which blood vessels, nerves, and lymph channels enter the pulp cavity. Placoid scales vary in shape with modifications for a variety of functions, including protection from predators and ectoparasites, reduction of mechanical abrasion, accommodation of bioluminescent and sensory organs, and reduction of frictional drag (Raschi and Tabit, 1992). These scales may fit close together or be set apart, but they usually do not overlap except where they protect the lateral line canal (Chapters 15.2 and 27.2). **Placoid scales** do not continue to grow indefinitely; they may be replaced when old, worn out, or lost.

The scales of bony fishes (Osteichthyes) are layered plates, with bone as one of the layers (Chapter 18). One structural type of scale found in early bony fishes, the **cosmoid scale**, is known only from fossil lobe-finned fishes (Sarcopterygii) such as coelacanths and lungfishes (Van Oosten, 1957; Chapter 18) and is beyond the scope of this chapter on living laboratory fishes. Living coelacanths possess scales that resemble cycloid or **ctenoid** scales of teleost fishes (see below) except that the upper surface is studded with fixed odontodes or placoid denticles (Van Oosten, 1957; Whitear, 1986). The three genera of living lungfishes have **elasmoid scales** of the cycloid type (Van Oosten, 1957; Whitear, 1986).

One ancestral scale type of bony fishes, the **ganoid scale**, is still present in some primitive actinopterygian (ray-finned) fishes (Kerr, 1952; Van Oosten, 1957; Whitear, 1986). Ganoid scales are often rhomboidal or diamond-shaped rigid plates that slightly overlap and articulate with one another by peg-and-socket joints at the margins (Figure 5.7); they are also frequently bound together by connective tissue fibers. Ganoid scales are least modified from their ancestral form among the bichirs (Polypteridae). In these fishes, the osseous basal plate is covered by a layer of dentine, which is topped by a layer of enamel-like ganoine (Chapters 17 and 18). The scale is penetrated by canals carrying capillary blood vessels to supply the tissues exterior to the scale. The scales of gars (Lepisosteidae) still have ganoine but no dentine

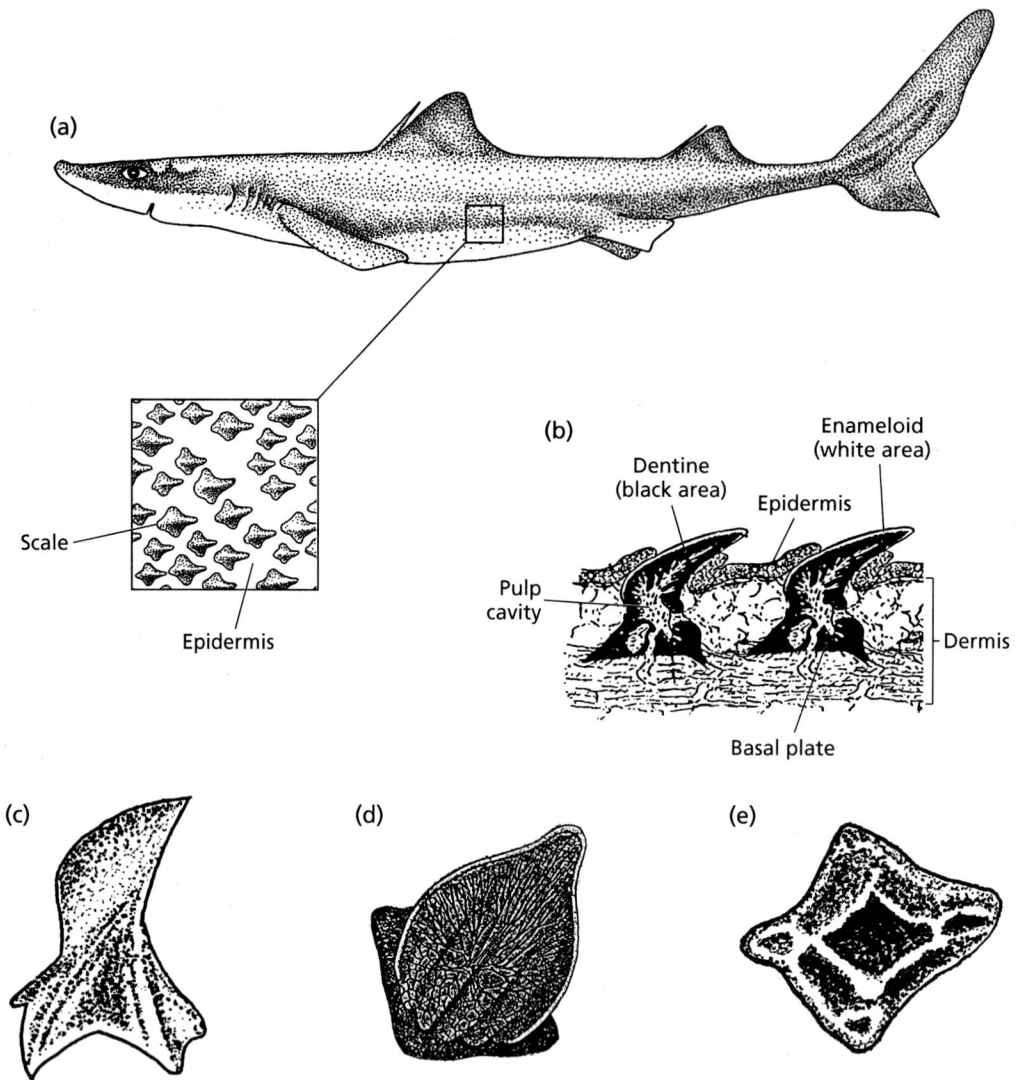

Figure 5.6 Placoid scales (dermal denticles or isolated odontodes). (a) Shark (*Squalus* sp.) and enlargement of surface view of skin showing exposed portion of placoid scales. (b) Vertical section (longitudinal to the body) of the skin and typical placoid scales of the shark *Scylorhinus canicula*. (c–e) Individual placoid scales of sharks, showing variation in form: (c) ventral body scale of *Squalus acanthias*, (d) scale of *Mustelus laevis*, (e) body scale of a type that does not erupt through the epidermis, from the luminescent shark *Isistius brasiliensis*. (Sources: (a) Drawing of fish modified from Figure 5-3A sharks: spiny dogfish, p. 71 in Moyle, 1993, after photos by Chris Mari van Dyck, courtesy of Monterey Bay Aquarium; detail of placoid scales from Figure 2-6C placoid, p. 22 in Moyle, 1993; copyright © 1993 and 2000 by Chris Mari van Dyck. (b) From Goodrich, 1909, reprinted by A. Asher & Co., 1964. (c) After Sayles and Herskowitz, 1937. (d) From Hertwig, 1874. (e) Modified from Reif, 1985; reproduced with permission from Blackwell Science Ltd.)

layer (Figure 5.7b). Odontodes or denticles may project from the surface of the scales of some bichirs and gars (Figure 5.7a and b); their presence may be transitory in gars, occurring only in juvenile or regenerating scales (Sire, 1994). The sturgeons (Acipenseridae) and paddlefish (Polydontidae) have reduced the dermal ossification to bony **scutes** without ganoine or dentine; scutes are present only as isolated vestigial denticles in paddlefish but exist as rows of large articulated bucklers and small denticles in sturgeon. In one primitive actinopterygian, the bowfin (*Amia*), the scales are not ganoid, but rather have become elasmoid scales of the cycloid type, very similar to those of teleosts (Van Oosten, 1957; Whitear, 1986; see below).

The vast majority of living bony fishes are teleosteans, and most of these possess elasmoid (bony-ridge) scales (Lagler *et al.*, 1977; Whitear, 1986). Elasmoid scales lack dentine and ganoine and are usually reduced to thin, flexible, transparent structures consisting of a basal plate of collagenous tissue, with superficial mineralization (Chapters 17 and 18). In fact, the scales of teleosts can represent an

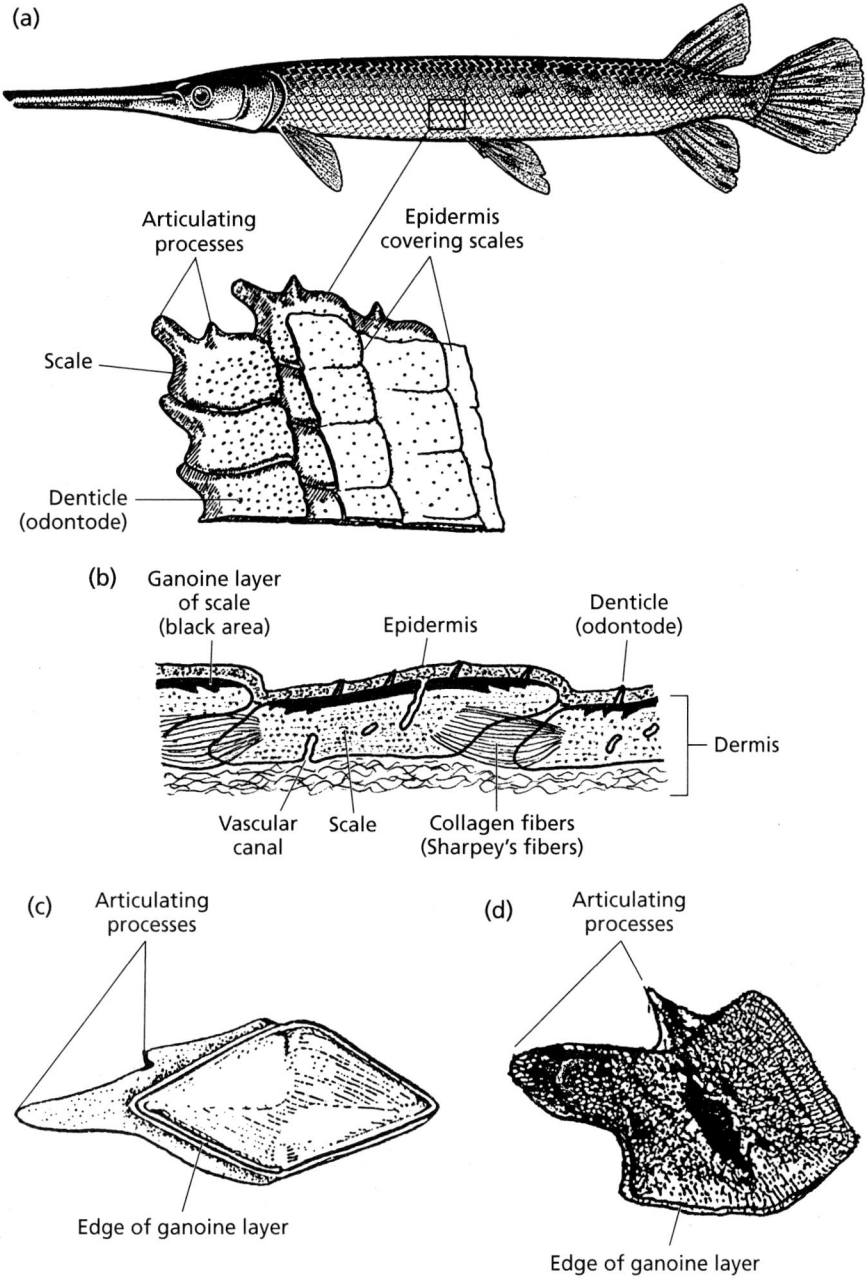

Figure 5.7 Ganoid scales. (a) Gar (*Lepisosteus osseus*) showing pattern of scalation, and enlargement of ganoid scales showing articulating processes (peg-and-socket joints) and slight overlapping of scales beneath the epidermis. (b) Vertical section (longitudinal to the scale row) of the skin and ganoid scales of a gar, *Lepisosteus*. sp. (c) Individual ganoid scale of a gar, *Lepisosteus* sp. (d) Individual ganoid scale of a bicher, *Polypterus* sp. (Sources: (a) Drawing of fish courtesy of Stewart Alcorn, copyright © by Stewart Alcorn 2000; detail of ganoid scales after Goodrich, 1909, reprinted by A. Asher & Co., 1964. (b) Modified from Goodrich, 1909, reprinted by A. Asher & Co., 1964. (c) After Lagler *et al.*, 1977, copyright ©1962, 1977 by John Wiley & Sons, Inc. Reprinted by permission of John Wiley & Sons, Inc. (d) After Kerr, 1952.)

important internal reservoir of calcium (Simkiss, 1973; Bereiter-Hahn and Zylberberg, 1993; Persson *et al.*, 1998; see also Chapter 18). Elasmoid scales (Figure 5.8) are usually marked by calcified concentric ridges (bony ridges) called circuli or striae (Figures 5.1 and 5.8a and e). The **circuli** and the backward-facing denticles that are often present on their edges (Figure 5.8e) may be involved in the mechanical anchoring of the scale into the covering dermis (Sire, 1986). Uncalcified grooves called radii or **sulci** converge on the scale focus and allow flexibility (Figure 5.8a and e).

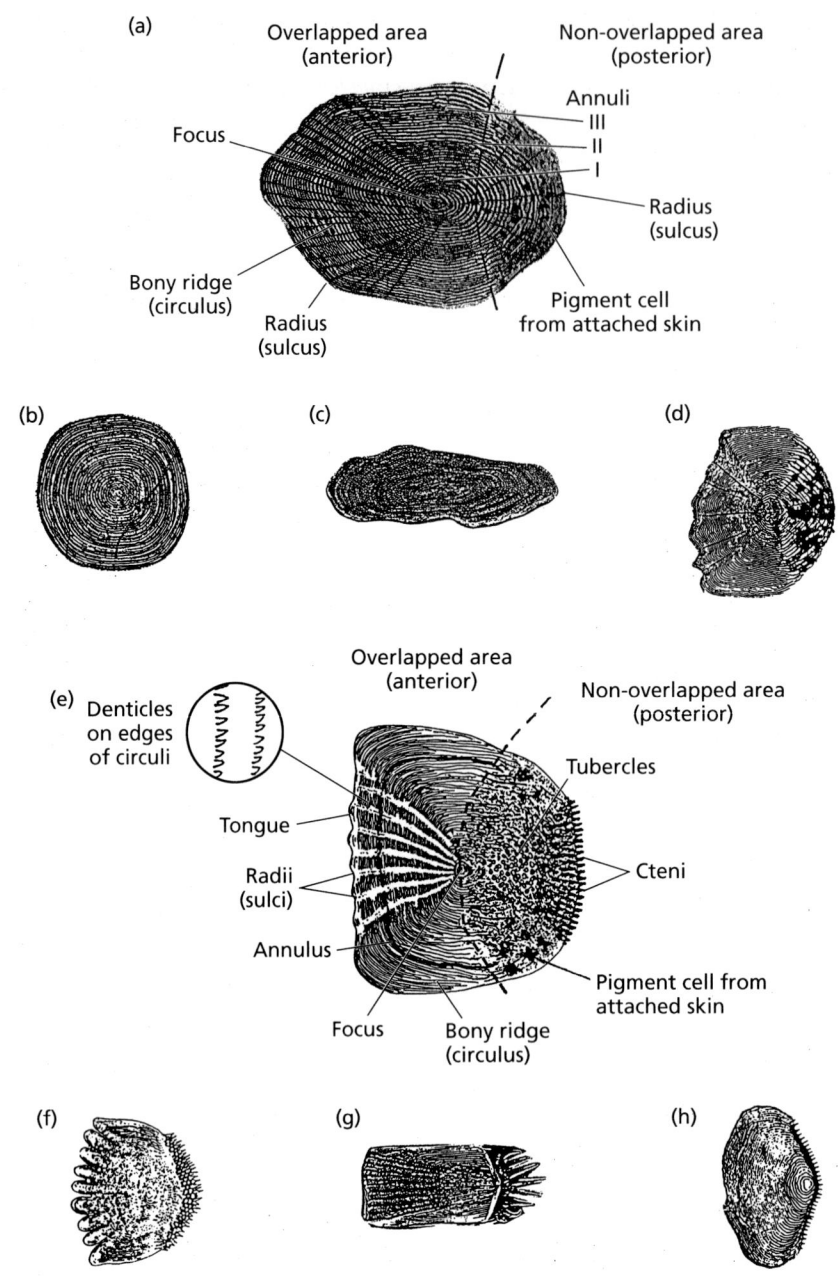

Figure 5.8 Elasmoid scales of teleost fish. (a) Cycloid scale showing morphological features. (b–d) Examples of cycloid scales. (b) Burbot, *Lota lota*. (c) eel, *Anguilla rostrata*. (d) goldfish, *Carassius auratus*. (e) Ctenoid scale showing morphological features. (f–h) Examples of ctenoid scales. (f) Perch, *Perca* sp. (g) beaked sandfish, *Gonorhynchus gonorhynchus*. (h) trout-perch, *Percopsis* sp. (Sources: (a) Modifed from a drawing courtesy of the Michigan Institute for Fisheries Research. (b) and (c) From Lagler, 1947. (d) From Yamada, 1961. (e) Modified from Sire, 1986 and Kuusipalo, 1998. (f) and (h) from Lagler et al., 1977, copyright ©1962, 1977 by John Wiley & Sons, Inc. Reprinted by permission of John Wiley & Sons, Inc. (g) After Roberts, 1993.)

Elasmoid scales are most commonly grouped into two general types, cycloid and **ctenoid**, which are characterized on the basis of their surface sculpture. Variations in scale morphology have been used in phylogenetic studies and classification of various fish groups (Lagler *et al.*, 1977; Lippitsch, 1990, 1992, 1993, 1998; Roberts, 1993; Moyle and Cech, 1996; Kuusipalo, 1998). The cycloid scale is generally a round, thin, flat scale (Figure 5.8 a–d). Ctenoid scales are similar to cycloid scales but have stiff comb-like projections (cteni or ctenii), soft projections (ciliated **cteni**), or short tubercles or grains in the posterior area (Figure 5.8e–h). The function of the cteni or other posterior projections (collectively called granulation) is not well understood, but they may improve the hydrodynamic efficiency of swimming by affecting the profile of the overlying epidermis, thus assisting in breaking the vortices caused by the swimming fish and thereby reducing drag (Burdak, 1969; Sire, 1986).

Cycloid scales are characteristic of the more primitive teleosts with soft-rayed fins – fishes such as trout (Salmonidae), minnows (Cyprinidae), and herring (Clupeidae). Ctenoid scales are found on many spiny-rayed fishes (Acanthopterygii) such as perch (Percidae), most mullets (Mugilidae), and scorpionfishes (Scorpaenidae). However, certain soft-rayed fishes such as a few of the characins (Characidae) have ctenoid-like contact organs on their scales (Lagler et al., 1977). Conversely, some spiny-rayed fishes such as the brook silverside (*Labidesthes*) and wrasses (Labridae) have cycloid scales exclusively (Lagler et al., 1977; Moyle and Cech, 1996). Both cycloid and ctenoid scales may be present on the same individual of certain fish species; the mud dab (*Limanda*) and the freshwater basses (*Micropterus*) are among the fishes in which this occurs (Van Oosten, 1957; Lagler et al., 1977). Additionally, the scale types observed may differ depending on the age (Sire, 1986; Kuusipalo, 1998) or sex (Moyle and Cech, 1996) of a fish.

Lippitsch (1990) suggested that the distinction between cycloid and ctenoid scales is superficial because numerous variations of both scale types and transitions between the two types are found, sometimes on the same fish. Furthermore, the considerable variation that occurs among scales with spine-like projections in the posterior field – ranging from outgrowths of the posterior margin (crenae) to discrete spines ('true' cteni) formed as ossifications separate from the scale plate – prompted Roberts (1993) to suggest the term 'spined' scales for all scales possessing posterior projections, to distinguish them from cycloid scales without projections. Roberts (1993) further characterized the spined scales as crenate, spinoid, or ctenoid, depending on the morphological features of the projections.

Patterns of scalation are fundamentally associated with body segmentation (Lagler et al., 1977) as manifested initially during embryonic development by the vertebrae (Chapter 6) and myomeres (Chapter 7). In arrangement, elasmoid scales are most often overlapping (imbricated) like shingles on a roof, with the free margin directed toward the posterior of the fish so as to minimize friction with the water (Figure 5.9a–c). The cteni of ctenoid scales are located on the posterior area that is not overlapped by other scales (Figure 5.8e), whereas the non-overlapped area of cycloid scales often has less distinct circuli than the anterior area (Lagler et al., 1977; Figure 5.8a). In some fishes, the posterior edge of the scale approaches the epidermis, which forms only a slight indentation or tuck beneath the scale (Figure 5.9b), whereas in other (usually pelagic) species both the scale and the covering epidermis protrude and overlap the epidermis that covers the scale behind (Figures 5.4 and 5.9c). Certain fish such as burbots (*Lota*) and immature adult freshwater eels (*Anguilla*) have mosaic patterns of scalation (Lagler et al., 1977; Pankhurst, 1982); rather than overlapping one another, the scales are minutely separated or meet adjacent scales only at the margins (Figures 5.5 and 5.9d and e).

During development of a teleost fish, the first scales usually form on the caudal peduncle and the pattern of scalation spreads from that point (Van Oosten, 1957), although scale development may begin simultaneously in more than one region of the body in some species (Sire and Arnulf, 1990). Scale counts, such as scale numbers in the lateral line or around the body of adult fish, are used in taxonomic studies (Lagler et al., 1977).

Growth of the formed elasmoid scale generally continues throughout life. Circuli mark successive stages in the growth of the scale (Figures 5.1 and 5.8). Periods of slow growth are discernible on magnified scales as closely spaced circuli; when these occur on an annual basis they are termed annuli (Figure 5.8a and e). The formation of annuli or **accessory annuli** (discontinuity of circuli that do not represent an annual cycle) may occur with decreased metabolism and appetite in cold seasons, fasting periods associated with spawning or unavailability of food, partial scale decalcification and resorption in females with developing eggs and young, or mechanical injury (Van Oosten, 1957; Moyle and Cech, 1996). Specific patterns of scale growth and morphology are used in fish life history and growth studies (Bugaev, 1992; Welch et al., 1993; Kingsford and Atkinson, 1994; Ogle et al., 1994; Moyle and Cech, 1996; Machias et al., 1998) and for stock identification (Unwin and Lucas, 1993; Okhuma, 1998). Manipulations of water temperature and feeding regimes also are used to induce distinctive marks on fish scales for batch-marking of fish for later identification of specific groups (Willett, 1994; Bigelow and White, 1996). An advantage of using scales rather than certain other hard structures such as vertebrae or otoliths (ear bones) for these studies is that scales can be sampled without killing the fish.

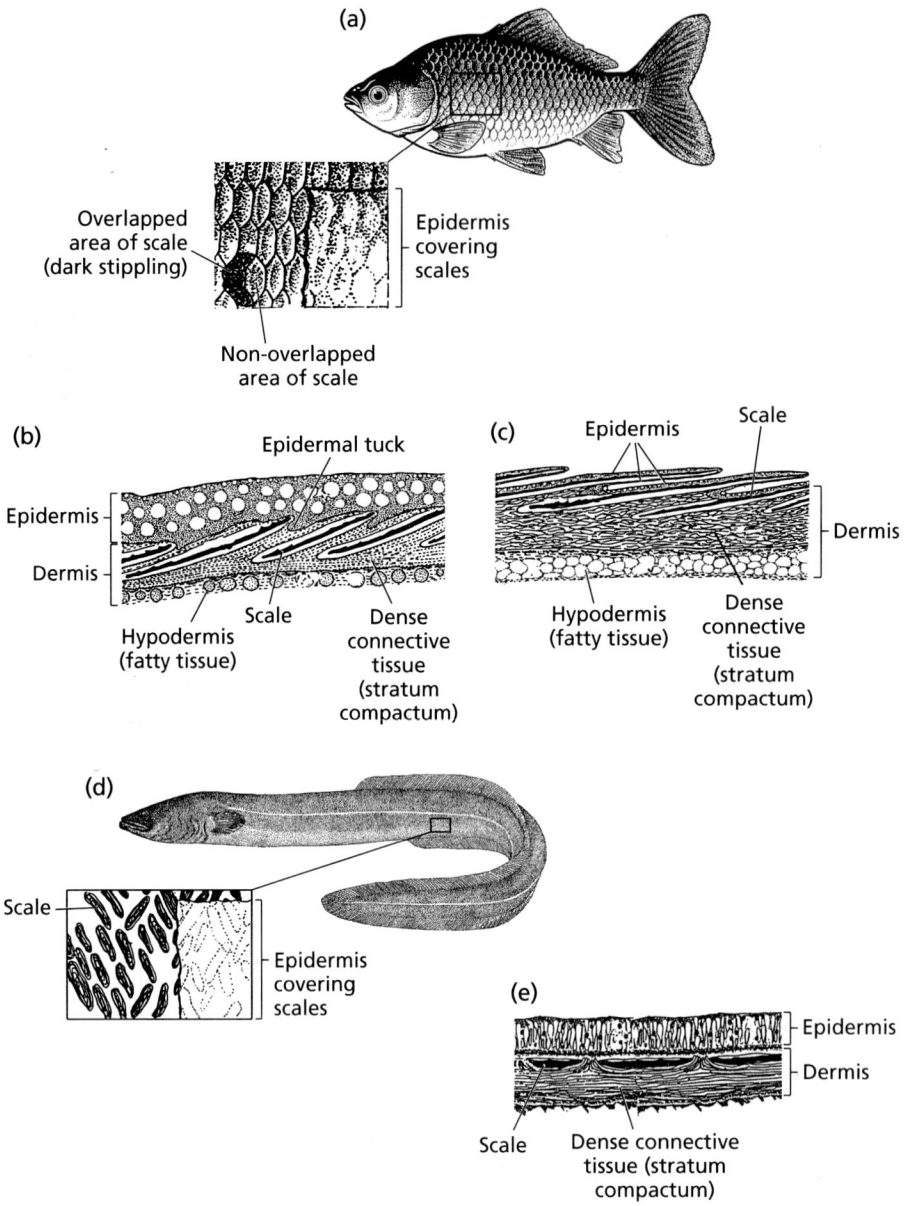

Figure 5.9 Scalation patterns of some teleost fishes. (a) Goldfish, *Carassius auratus* showing typical pattern of overlapping (imbricating) scales beneath the epidermis. (b) Vertical section (longitudinal to the body) of the skin of a minnow, *Phoxinus phoxinus*, showing a thick epidermis with slightly imbricated scales beneath. (c) Vertical section (longitudinal to the body) of the skin of an Atlantic mackerel, *Scomber scombrus*, showing protrusion of the posterior portion of each scale and its epidermal covering, which overlap the scale behind. (d) Skin of an eel, *Anguilla* sp., showing non-overlapping pattern of scales. (e) Vertical section (longitudinal to the body) of the skin of an eel, *Anguilla anguilla*, showing non-overlapping scales beneath the epidermis. (Sources: (a) Drawing of fish courtesy of Stewart Alcorn, copyright © by Stewart Alcorn 2000. (b) After Whitear, 1986, copyright ©1986 by Springer-Verlag Berlin Heidelberg. (c) After Roberts and Bullock, 1980 and Whitear, 1986. (d) Drawing of fish from Leim and Scott, 1966, courtesy of Fisheries and Oceans Canada; reproduced with the permission of Her Majesty the Queen in Right of Canada 2000. Detail of scales modified from Pankhurst, 1982. (e) Modified from Pankhurst and Lythgoe, 1982.)

Some scales (usually cycloid) lie in shallow pockets and are easily rubbed off. Deciduous scales are characteristic of fishes such as shiners (*Notropis*) and smelt (*Osmerus*). Other scales (especially ctenoid scales) are more deeply embedded and are difficult to remove, as in the pikeperch (*Stizosteodon*). In certain species scale shedding is related to specific habits or life stages. For example, considerable scale loss occurs in common gobies (*Pomatoschistus microps*) when they reach sexual maturity and engage in nest-building (burrowing) and spawning activities (Fouda and Miller, 1979), and the scales of juvenile Pacific salmon (*Oncorhynchus* spp.) become more deciduous during smoltification and migration into sea water (Bouck and Smith, 1979). Most teleost fish lose their scales at least once during their life (Bereiter-Hahn and Zylberberg, 1993). However, both the lost scales and the overlying epidermis that is torn when the scales are shed usually regenerate quickly (Bereiter-Hahn and Zylberberg, 1993; Chapter 17).

Modified scales

Placoid scales of sharks and their relatives may be enlarged or fused to form defensive weapons, which are particularly effective when associated with venom glands (Chapter 17; Van Oosten, 1957; Lagler *et al.*, 1977). Examples of denticles modified as spines include the dorsal fin spines of the spiny dogfish (*Squalus*) and chimaeroids (*Chimaera*, *Hydrolagus*), and the tail-spine or 'stinger' of sting rays (Dasyatidae).

Not all teleost scales are classically cycloid or ctenoid (or spinoid); many modifications exist (Van Oosten, 1957; Lagler *et al.*, 1977; Whitear, 1986). Specially perforated and sometimes **tubulated scales** provide surface outlets for the lateral line sensory canal (Chapter 15.2). Scales at the base of the tail of surgeonfish (*Acanthurus*) are modified into two knife-like spines that are carried in sheaths of skin but swing out in an attack. Spinous scales are also found in teleosts such as the batfish (*Ogocephalus*), frogfish (*Antennarius*), triggerfishes (*Balistes*) and filefishes (*Monacanthus*; Figure 5.10). The porcupine fishes (Diodontidae) and puffers (Tetraodontidae) bear numerous sharp bony spines that stand erect when the fish inflates its body. Other fish such as the pipefishes and seahorses (Syngnathidae) possess a dermal skeleton of bony plates arranged segmentally to form a semi-rigid case. South American armored catfishes (e.g. Callichthyidae) are protected with large, bony plates or scutes bearing tooth-like structures (odontodes) similar to those of ancient craniates (Figure 5.11; Sire and Huysseune, 1996). Perhaps the most extreme examples of protective armor are the rigid encasements of trunkfishes (Ostraciidae) which rival those of the most completely boxed turtles.

Evidence exists that certain elements of the dermal skeleton such as odontodes, some scale types, and some fin rays may have derived from dental tissues. The evolutionary relationships among these various structures are discussed in Chapter 18.

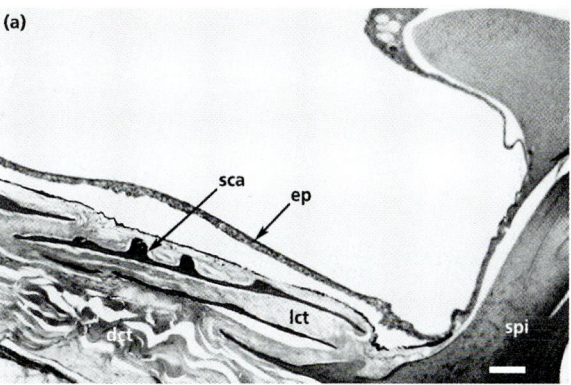

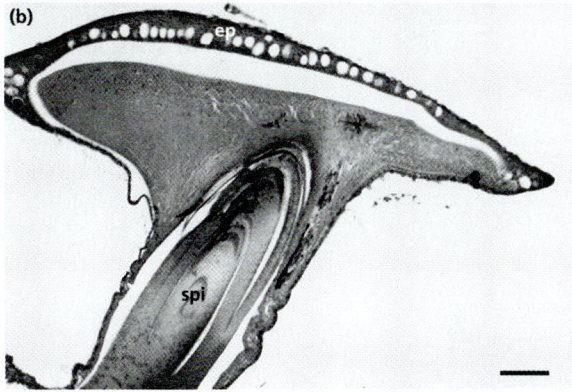

Figure 5.10 Spinous scale of a toothbrush leatherjacket, *Acanthaluteres vittiger*. This type of scale gives members of the family Monacanthidae their common name of 'filefish.' (a) Spinous scale (*spi*) embedded in the dermis. Adjacent non-spinous scales (*sca*) are also visible. The outer epidermis (*ep*) and two layers of dermis – a layer of loose connective tissue or stratum spongiosum (*lct*) and a layer dense connective tissue or stratum compactum (*dct*) – can be seen. (b) Portion of spinous scale (*spi*) that protrudes from the surface of the skin. Even the protruding portion (spinule) is covered by epidermis (*ep*). Scale bars = 100 μm. Hematoxylin and eosin stain. (Source: Photomicrographs courtesy of Dr Barbara Nowak.) **(See also Colour Plate 6.)**

Integumentary extensions

Fishes have evolved a wide variety of integumentation, including extensions of the skin. **Barbels**, for example, are extensions which have developed independently in many taxa as accessory feeding structures that carry sensory organs (Chapter 15.3). Barbels of different structure and location are present on sturgeons (Acipenseridae), marine and freshwater catfishes (Siluriformes), goatfishes (Mullidae), and some Cypriniformes such as carp (*Cyprinus carpio*) and loaches (Cobitidae).

The sargassum fish (*Histrio*) and seadragon (e.g. *Phyllopteryx*) are the most frequently cited examples of the extension of skin into flaps. Protective

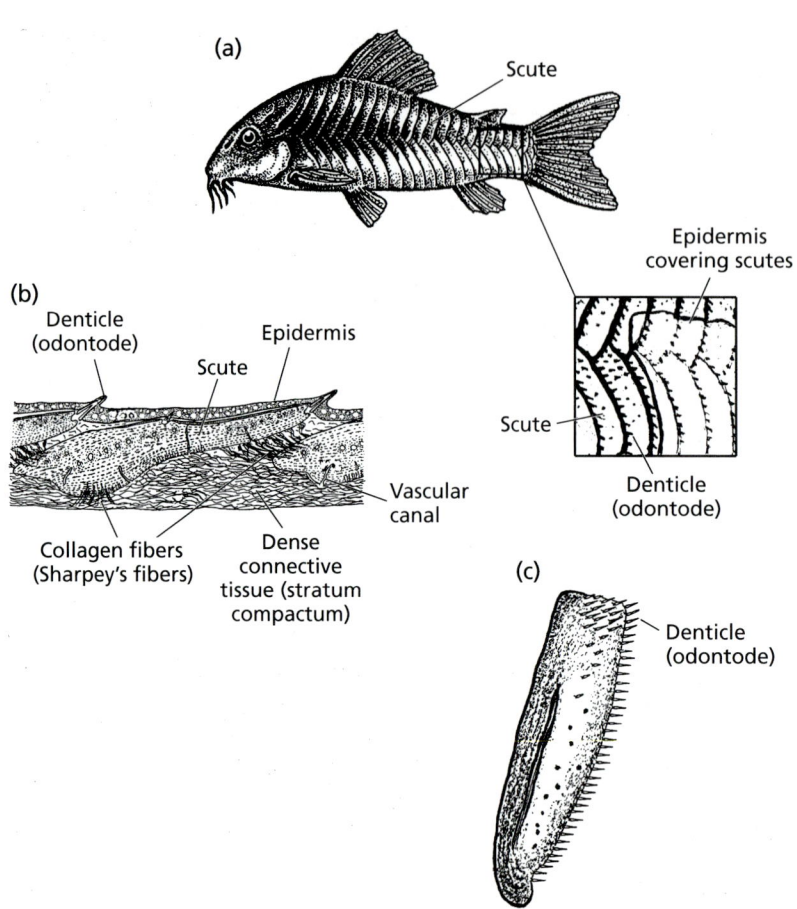

Figure 5.11 Scutes (bony plates) of a teleost, the armored catfish *Corydoras aeneus*. (a) *C. aeneus* showing location of the two rows of scutes beneath the epidermis. (b) Vertical section (longitudinal to the body) of the skin of *C. aeneus* showing orientation of scutes and denticles (odontodes) beneath the epidermis. The tips of some odontodes may protrude through the epidermis. (c) Dorsal scute from the caudal peduncle of *C. aeneus*. (Sources: (a) Drawing of fish from Sterba, 1959. (b) After Bhatti, 1938 and Whitear, 1986. (c) Modified from Sire and Huysseune, 1996; reproduced with permission from Balackwell Science Ltd.)

resemblance to the seaweed in which they hide is the function ascribed to the flaps in these fishes (Lagler *et al.*, 1977). Less prominent skin extensions in other fishes such as scorpionfishes (*Scorpaena*), combined with mottled coloration, probably aid camouflage as well.

Certain skin extensions assist in reproduction, and may be shed or reduced in size after the breeding season. Keratinized pearl organs or nuptial tubercles of Cypriniformes and certain other fishes (Chapter 17) function primarily in facilitating contact between individuals during spawning (Wiley and Collette, 1970). However, the antler-like nuptial tubercles of breeding male stonerollers (*Campostoma anomala*) are used to defend nests against intruding males (Moyle and Cech, 1996). Tissue outgrowths on the abdominal skin or fins of skin-brooding fishes enable the attachment of eggs to the skin (Wetzel *et al.*, 1997).

Acknowledgments

Among my colleagues at the Western Fisheries Research Center, I thank Carla Aiwohi for extensive technical assistance and library research, and Ronald Pascho for his helpful review of the manuscript. I am indebted to Dr Barbara Nowak, School of Aquaculture, University of Tasmania, Launceston, Tasmania, Australia, for several of the photographs used in this chapter. I thank Dr Susan Gutenberger of the Lower Columbia River Fish Health Center (U.S. Fish and Wildlife Service, Underwood, Washington, USA) for providing lamprey skin tissues, and Dr George Sanders, Department of Comparative Medicine, University of Washington, Seattle, Washington, USA, for providing shark fin tissues. I am grateful to Stewart Alcorn, School of Fisheries, University of Washington, for the drawing of a gar and goldfish.

References

Bereiter-Hahn, J. and Zylberberg, L. (1993). *Comp. Biochem. Physiol.* **105A**, 625–641.

Berg, T. and Steen, J.B. (1965). *Comp. Biochem. Physiol.* **15**, 469–484.

Bernadsky, G., Sar, N. and Rosenberg, E. (1993). *J. Fish Biol.* **42**, 797–800.

Bhatti, H.K. (1938). *Trans. Zool. Soc. Lond.* **24**, 1–102.

Bigelow, P.E. and White, R.G. (1996). *N. Am. J. Fish. Manage.* **16**, 142–153.

Bouck, G.R. and Smith, S.D. (1979). *Trans. Am. Fish. Soc.* **108**, 67–69.

Bugaev, V.F. (1992). *J. Ichthyol.* **32**(8), 1–19.

Burdak, V.D. (1969). *Zool. Zh.* (English translation) **48**, 1053–1055.

Burdak, V.D. (1980). *J. Ichthyol.* **19**(4), 101–106.

Cole, A.M., Weis, P. and Diamond, G. (1997). *J. Biol. Chem.* **272**, 12 008–12 013.

Ellis, A.E. (1981). In *Fish Biologics: Serodiagnostics and Vaccines* (eds D.P. Anderson and W. Hennessen), pp. 337–352. *Developments in Biological Standardization*, vol. 49. Karger, Basel, Switzerland.

Endler, J.A. (1978). In *Evolutionary Biology*, vol. 11 (eds M.K. Hecht, W.C. Steere, and B. Wallace), pp. 319–364. Plenum Press, New York.

Evans, D.H. (1993). In *The Physiology of Fishes* (ed. D.H. Evans), pp. 315–341. CRC Press, Boca Raton, Florida.

Fishelson, L. (1996). *Anat. Rec.* **246**, 15–29.

Fouda, M.M. and Miller, P.J. (1979). *J. Fish Biol.* **15**, 263–273.

Fujii, R. (1993). In *The Physiology of Fishes* (ed. D.H. Evans), pp. 535–562. CRC Press, Boca Raton, Florida.

Goodrich, E.S. (1909). In *A Treatise on Zoology*, Part 9, Section 1 (ed. R. Lankester), pp. 1–518. Reprint A. Asher & Co., Amsterdam, 1964.

Gong, Z., Ewart, K.V., Hu, Z., Fletcher, G.L., Hew, C.L. (1996). *J. Biol. Chem.* **271**, 4106–4112.

Hattingh, J. and van Warmelo, K.T. (1975). *Zool. Afr.* **10**, 102–103.

Hazel, J.R. (1993). In *The Physiology of Fishes* (ed. D.H. Evans), pp. 427–467. CRC Press, Boca Raton, Florida.

Heisler, N. (1993). In *The Physiology of Fishes* (ed. D.H. Evans), pp. 343–378. CRC Press, Boca Raton, Florida.

Herring, P.J. (1982). *Oceanogr. Mar. Biol. Ann. Rev.* **20**, 415–470.

Hertwig, O. (1874). *Jenaische Z. Naturwiss.* **8**, 331–404.

Kashkin, N.I. (1993). *J. Ichthyol.* **33**(8), 1–17.

Kerr, T. (1952). *Proc. Zool. Soc. Lond.* **122**, 55–78.

Kingsford, M.J. and Atkinson, M.H. (1994). *Aust. J. Mar. Freshwat. Res.* **45**, 1007–1021.

Kirsch, R. and Nonnotte, G. (1977). *Respir. Physiol.* **29**, 339–354.

Kohda, Y. and Watanabe, M. (1982). *Zool. Mag. (Tokyo)* **91**, 61–69.

Kohda, Y. and Watanabe, M. (1983). *Zool. Mag. (Tokyo)* **92**, 207–215.

Kuusipalo, L. (1998). *J. Fish Biol.* **52**, 771–781.

Lagler, K.F. (1947). *Trans. Am. Microsc. Soc.* **66**, 149–171.

Lagler, K.F., Bardach, J.E., Miller, R.R. and Passino, D.M. (1977). *Ichthyology*, 2nd edn. John Wiley, New York.

Lanzing, W.J.R. and Bower, C.C. (1974). *J. Fish Biol.* **6**, 29–41.

Liem, A.H. and Scott, W.B. (1966). *Fishes of the Atlantic Coast of Canada.* Bulletin No. 155, Fisheries Research Board of Canada, Ottawa.

Liem, K.F. (1981). *Science* **211**, 1177–1179.

Lippitsch, E. (1990). *J. Fish Biol.* **37**, 265–291.

Lippitsch, E. (1992). *J. Fish Biol.* **41**, 355–362.

Lippitsch, E. (1993). *J. Fish Biol.* **42**, 903–946.

Lippitsch, E. (1998). *J. Fish Biol.* **53**, 752–766.

Long, J., Hale, M., McHenry, M. and Westneat, M. (1996). *J. Exp. Biol.* **199**, 2139–2151.

Machias, A., Tsimenides, N., Kokokiris, L. and Divanach, P. (1998). *J. Fish Biol.* **52**, 350–361.

Matoltsy, A.G. and Bereiter-Hahn, J. (1986). In *Biology of the Integument – 2. Vertebrates* (eds J. Bereiter-Hahn, A.G. Matoltsy and K.S. Richards), pp. 1–67. Springer-Verlag, Berlin.

Mittal, A.K. and Banjeree, T.K. (1980). In *The Skin of Vertebrates*, Linn. Soc. Symp. No. 9 (eds R.I.C. Spearman and P.A. Riley), pp. 1–12. Academic Press, New York.

Morii, H., Nishikata, K. and Tamura, O. (1978). *Comp. Biochem. Physiol.* **60A**, 189–193.

Moyle, P.B. (1993). *Fish: An Enthusiast's Guide.* University of California Press, Berkeley and Los Angeles.

Moyle, P.B. and Cech, J.J., Jr (1996). *Fishes: An Introduction to Ichthyology*, 3rd edn. Prentice-Hall, Upper Saddle River, New Jersey.

Nonnotte, G. (1981). *Comp. Biochem. Physiol.* **70A**, 541–543.

Nonnotte, G., Nonnotte, L. and Kirsch, R. (1979). *Cell Tissue Res.* **17**, 387–396.

Ogle, D.H., Spangler, G.R. and Shroyer, S.M. (1994). *Can. J. Fish. Aquat. Sci.* **51**, 1721–1727.

Okhuma, K. (1998) In *NPAFC Bulletin Number 1: Assessment and Status of Pacific Rim Salmonid Stocks*, pp. 319–326. North Pacific Anadromous Fish Commission, Vancouver, BC, Canada.

Pankhurst, N.W. (1982). *J. Fish Biol.* **21**, 549–561.

Pankhurst, N.W. and Lythgoe, J.N. (1982). *J. Fish Biol.* **21**, 279–296.

Perry, S.F. and Wood, C.M. (1985). *J. Exp. Biol.* **116**, 411–433.

Persson, P., Sundell, K., Björnsson, B. Th. and Lundqvist, H. (1998). *J. Fish Biol.* **52**, 334–349.

Pickering, A.D. (1974). *J. Fish Biol.* **6**, 111–118.

Ramachandran, V.S., Tyler, C.W., Gregory, R.L., Rogers-Ramachandran, D., Duensing, S., Pilsbury, C. and Ramachandran, C. (1996). *Nature* **379**, 815–818.

Rambout, J.H., Taverne, N., van de Kamp, M. and Taverne-Thiele, A.J. (1993). *Dev. Comp. Immunol.* **17**, 309–317.

Raschi, W. and Tabit, C. (1992). *Aust. J. Mar. Freshwat. Res.* **43**, 123–147.

Read, L.J. (1968). *Comp. Biochem. Physiol.* **26**, 455–466.

Reif, W.-E. (1985). *Acta Zool. (Stockh.)* **66**, 111–118.

Roberts, C.D. (1993). *Bull. Mar. Sci.* **52**, 60–113.

Roberts, R.J. and Bullock, A.M. (1980). In *The Skin of Vertebrates*, Linn. Soc. Symp. Ser. No. 9 (eds R.I.C. Spearman and P.A. Riley), pp. 13–21. Academic Press, London.

Rombough, P.J. and Moroz, B.M. (1990). *J. Exp. Biol.* **154**, 1–12.

Romer, A.S. (1968). *The Vertebrate Body*, 3rd edn. W.B. Saunders, Philadelphia.

Rosen, M.W. and Cornford, N.E. (1971). *Nature* **234**, 49–51.

Sayer, M.D.J. and Davenport, J. (1987). *J. Fish Biol.* **31**, 561–570.

Sayles, L.P. and Hershkowitz, S.G. (1937). *Biol. Bull. (Woods Hole)* **73**, 51–66.

Sazonov, Y.I. (1996). In *Deep Sea and Extreme Shallow-water Habitats: Affinities and Adaptations*, Biosyst. Ecol. Ser. No. 11 (eds F. Ubelin, J. Ott and M. Stachowitsch), pp. 151–163. Oesterreichische Akademie der Wissenschaften, Wein, Austria.

Simkiss, K. (1973). In *Aging of Fish* (ed. T.B. Bagenal), pp. 1–12. Gresham Press, Surrey, UK.

Sire, J.-Y. (1986). *J. Fish Biol.* **28**, 713–724.

Sire, J.-Y. (1994). *Anat. Rec.* **240**, 189–207.

Sire, J.-Y. and Arnulf, I. (1990). *Jap. J. Ichthyol.* **37**, 133–143.

Sire, J.-Y. and Huysseune, A. (1996). *Acta Zool. (Stockh.)* **77**, 51–72.

Smith, R.J.F. (1982). In *Chemoreception in Fishes* (ed. T.J. Hara), pp. 327–342. Elsevier, Amsterdam.

Smith, R.J.F. (1992). *Rev. Fish Biol. Fish.* **2**, 33–63.

Steffensen, J.F. and Lomholt, J.P. (1985). *Comp. Biochem. Physiol.* **81A**, 373–375.

Sterba, G. (1959). *Süsswasserfische aus aller Welt*, Urania-Verlag, Leipzig/Jena.

St Louis-Cormier, E.A., Osterland, C.K. and Anderson, P.D. (1984). *Dev. Comp. Immunol.* **8**, 71–80.

Tamura, S.O., Morii, H. and Yuzuriha, M. (1976). *J. Exp. Biol.* **65**, 97–107.

Unwin, M.J. and Lucas, D.H. (1993). *Can. J. Fish. Aquat. Sci.* **50**, 2475–2484.

Valerio, P.F., Kao, M.H. and Fletcher, G.L. (1992). *J. Exp. Biol.* **164**, 135–151.

Van Oosten, J. (1957). In *The Physiology of Fishes*, vol. 1 (ed. M.E. Brown), pp. 207–244. Academic Press, New York.

Welch, T.J., van den Avyle, M.J., Betsill, R.K. and Driebe, E.M. (1993). *N. Am. J. Fish. Manage.* **13**, 616–620.

Wetzel, J., Wourms, J.P. and Friel, J. (1997). *Environ. Biol. Fish.* **50**, 13–25.

Whitear, M. (1986). In *Biology of the Integument – 2. Vertebrates* (eds J. Bereiter-Hahn, A.G. Matoltsy and K.S. Richards), pp. 39–64. Springer-Verlag, Berlin.

Wiley, M.L. and Collette, B.B. (1970). *Bull. Am. Mus. Nat. Hist.* **143**, 143–216.

Willett, D.J. (1994). *Fish. Manage. Ecol.* **1**, 157–163.

Wood, C.M. (1993). In *The Physiology of Fishes* (ed. D.H. Evans), pp. 379–425. CRC Press, Boca Raton, Florida.

Yamada, J. (1961). *Mem. Fac. Fish. Hokkaido Univ.* **9**, 181–226 (plus 21 plates).

Zadunaisky, J.A. (1984). In *Fish Physiology*, vol. XB (eds W.S. Hoar and D.J. Randall), pp. 129–176. Academic Press, New York.

CHAPTER 6

Skeletal System

Melanie L J Stiassny
Department of Ichthyology, American Museum
of Natural History, New York City, New York, USA

'The skeleton is what is left after the insides have been taken out and the outsides have been taken off.' Answer to junior high school science quiz, South African Airlines in-flight magazine, September 1998.

Introduction

In the following pages I will describe, in basic outline, the skeletal anatomy of a generalized teleostean fish. I should stress at the outset that what follows is a true schematic – nothing *exactly* like this fish exists, but as a guide it should provide an introduction to basic teleostean skeletal anatomy. I do not have space here to detail specific differences in the anatomy of the more common research fishes (such as trout, zebrafishes, or medaka), but reference to the following papers provides additional anatomical and developmental data for those taxa (Jollie, 1984; Stearley and Smith, 1993; Kulkarni, 1948; Langille and Hall, 1987; Cubbage and Mabee, 1996). The classification and taxonomic nomenclature used throughout this and the following chapter follows that of Nelson (1994).

Fish live in water, a dense and viscous medium that, in comparison with air, places a premium on efficient fluid propulsion mechanisms for locomotion and effective suction generation for food acquisition. Any of us who has tried suction feeding on land, or has watched a landed fish thrash its tail in attempted escape, will realize the fundamental importance of the dense aquatic medium in determining both the form and function of a fish's anatomy. There are, of course, many ways of looking at that anatomy but two predominate and these can be paraphrased as the historical/phylogenetic ('what it is') and the contemporary/functional ('what it does') approaches. Although in this essay I adopt a functional emphasis it should be remembered that most of the names applied to bones denote their historical and/or embryonic origins and, as such, they also reflect a rich phylogenetic heritage (see Stiassny, in prep.).

This contribution is based on a pre-publication version of Stiassny/ *Guide to the Anatomy of Fishes: The Musculoskeletal System*, a work to be published by Wiley-Liss, Inc., a subsidiary of John Wiley & Sons, Inc., in 2001. Reprinted by permission of Wiley-Liss, Inc. Copyright © 2001 Wiley-Liss, Inc. All rights reserved. Reproduction, adaptation or any further distribution of this material is expressly prohibited. For further information or to require permission for other uses, please contact the Permission Department, John Wiley & Sons, Inc., 605 Third Ave., New York, NY 10158-0012. Telephone (212)850-6011. Facsimile: (212) 850-6008.

Quite unlike the majority of terrestrial vertebrates, most fishes possess highly kinetic heads (capable of impressive expansion for suction generation) and their vertebral columns have a lateral flexibility and compressional rigidity capable of powering a caudal propulsion mechanism of unrivalled beauty and efficiency. It is in the context of these two basic functions, aquatic feeding by suction generation and caudal propulsion locomotion, that the basic anatomy of the teleostean fishes can perhaps be best understood.

The kinematic head

The head of most teleosts is capable of quite remarkable kinesis with over 30 moveable bony parts controlled by more than 50 muscles. As an heuristic device, the fish head can be viewed as an 'expanding cone' in which considerable suction (negative pressure) is generated when the cone (the orobranchial cavity) expands. If small enough to enter the mouth cavity, any kind of food floating or swimming in the water column can enter the mouth with the inflowing water. Additionally, pressure differences within the cone generated by differential muscle contraction, can change the shape of the cone and thereby manipulate the enclosed prey in a manner analogous to a hydrodynamic tongue. As has been elegantly shown by the studies of Muller *et al.*, 1982; Liem 1980, 1984, 1990; and Lauder, 1983, 1985, the dense aquatic medium combined with the suction generation capability of the expanding cone and fine tuning provided by the hydrodynamic tongue offer an unparalleled array of prey capture opportunities for teleosts that are not available to their terrestrial counterparts.

Caudal propulsion

The perfection of caudal locomotion has been cited as the single greatest achievement of the teleostean radiation, and indeed much of the evolutionary transformation of the group can be seen in a series of modifications of their locomotor systems. These include the strengthening of the axial skeleton, modification of the tail, and the development of flexible median and paired fins. The replacement of the notochord by a chain of short vertebrae (with vertically rigid neural and haemal spines and powerful body musculature facilitates fine control over the frequency and amplitude of lateral undulation, which is now concentrated into the caudal region. The caudal fin support is modified into a flattened, almost symmetrical plate against which the flexible fin rays of the caudal fin articulate. The median and paired fins are variously positioned, and their fin rays reduced in number and supported by a musculoskeletal system capable of providing considerable control over fin shape and posture during swimming and maneuvering.

Skeletal anatomy

The neurocranium

The teleostean neurocranium (or skull) is a complex structure formed of a large number of individual bony elements united together into a single structural unit. For ease of description the neurocranium is here divided into four regions (Figure 6.1). There are no discrete boundaries between these regions and their use here is a strictly pragmatic device to facilitate ease of description while conveying a broad notion of the functional division of the teleostean skull.

The ethmoid region (composed of lateral ethmoids, mesethmoid, vomer, and the canal bearing nasal bones) forms the anterior limit of the neurocranium. Developmentally the ethmoid derives from an anterior expansion of the trabecular cartilages – the ethmoid plate – and forms a footing for the olfactory organs. Later in development the ethmoid cartilage subdivides and is invested with an often complex array of dermal ossifications. Functionally the ethmoid serves to buttress the upper jaws and the anterior suspensorium, while also playing an important role in transmitting compressional forces along the supraorbital bridge dorsally, and the infraorbital bridge ventrally, thereby diverting pressure from the delicate orbital region.

The supraorbital bridge (composed of frontals, orbitosphenoid, and pterosphenoids) forms the roof of the orbit and the interorbital septum. Together with the ventral infraorbital bridge, the supraorbital bridge plays an important role in transmitting pressure from the ethmoid region to the cranial vault, and provides a protective housing for the eyes and associated musculature.

The infraorbital bridge (composed of parasphenoid and basisphenoid) like the supraorbital bridge, transmits pressure away from the ethmoid region to the reinforced cranial vault. Together with the basioccipital of the cranial vault, the parasphenoid forms the floor of the posterior myodome (a housing for

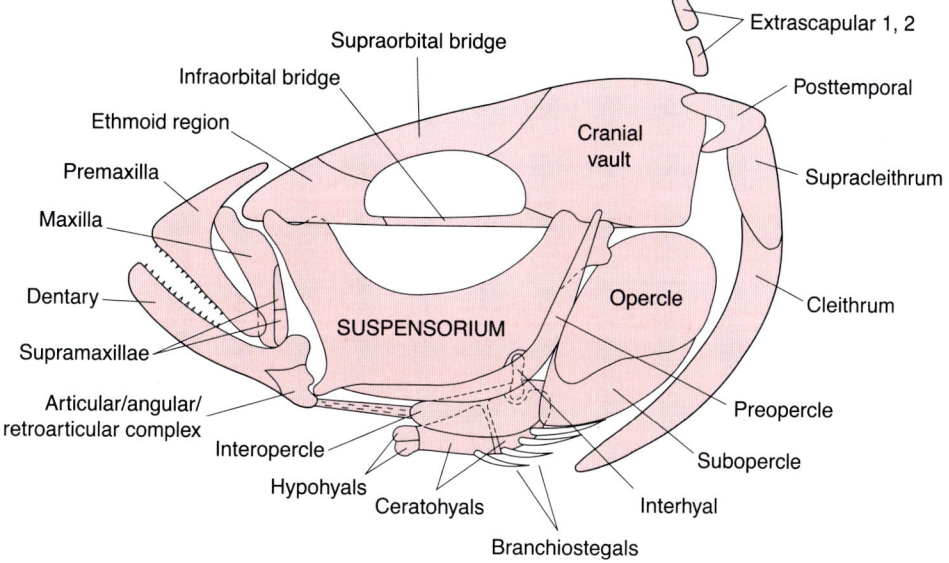

Figure 6.1 The teleostean neurocranium.

extrinsic eye musculature) and also serves as the main attachment site for the main adductor musculature of the suspensorium.

The cranial vault (composed of sphenotics, prootics, pterotics, parietals, epioccipitals, supraoccipital, exoccipitals, intercalars, and basioccipital) occupies the remainder of the postorbital neurocranium. Developmentally most of the cranial vault is derived from the fusion and subsequent ossification of the embryonic parachordal basilar plate, the occipital arch, and the auditory capsules. It constitutes the main body of the embryonic chondrocranium. Functionally the vault forms a protective housing for the brain, cranial nerves, and inner ear, while also serving as a primary attachment site for the pharyngeal apparatus, the posterior suspensorium, the pectoral girdle, the vertebral column and associated musculature (see Chapter 7 for a brief overview of teleostean cephalic myology).

Much has been written on the structure, composition, and development of the neurocranium of teleosts and the classic reviews of Goodrich (1930), Gregory (1933) and De Beer (1937) remain important references. However, for anatomical detail and an assessment of bone homologies, the seminal work by Patterson (1975) on the actinopterygian braincase is the ultimate reference in this area. Useful developmental data, particularly for commonly encountered research fishes can also be found in Jollie (1984), Langille and Hall (1987), Cubbage and Mabee (1996) and Mabee and Trendler (1996).

The suspensoria

Suspended from the neurocranium on either side of the head are two bony plates which, together with the oral jaws and associated membranes, form the walls and roof of the orobranchial cavity (Figure 6.1). Each lateral plate is termed a suspensorium, and developmentally the suspensoria (pl.) are derived from the fusion and subsequent ossification of the embryonic palatoquadrate cartilage of the mandibular arches (which form the autopalatines, metapterygoids, and quadrates) and the upper portion of the hyoid arch cartilages (which form the hyomandibulae and symplectics) of either side of the head. Additional dermal elements investing each suspensorium are the dermopalatines, ectopterygoids, endopterygoids and the canal bearing preopercles.

Functionally the suspensoria serve to suspend the oral jaws from the neurocranium, provide an attachment site for the hyoid bar and the opercular apparatus, and serve as a site of origin and insertion for the associated muscles and ligaments of the region. Anteriorly the suspensoria are suspended from the ethmoid region of the neurocranium by the palatine bones and posteriorly from the cranial vault by the hyomandibula bones. Abduction and adduction of the suspensoria can result in significant volume changes in the orobranchial chamber and thereby play a central role in suction feeding in teleosts.

Information on the structure of teleostean suspensoria is to be found scattered throughout the literature, and Arratia and Schultze (1991) provide additional information particularly on developmental aspects of the teleostean palatoquadrate arch.

The opercular–branchiostegal series

Posterior to the suspensoria and covering the gill arches on either side of the head are the opercular–branchiostegal series; a series of plate-like dermal bones (Figure 6.1). The largest and dorsalmost components are the three opercular bones: the opercles, subopercles, and interopercles (often collectively termed the 'gill cover' or 'bony operculum'). The bones of the opercular series overlie an opercular membrane and together form a protective wall for the orobranchial chamber. Abduction of the opercular series functions to permit the exit of water from the orobranchial chamber following orobranchial expansion during feeding and respiration. In teleosts the interopercle is ligamentously attached to the lower jaw such that elevation of the opercular series is transmitted via an opercular–mandibular ligament to the lower jaw thereby providing a novel linkage system mediating jaw depression.

Below the opercular series, and embedded in a ventral branchiostegal membrane, are the plate-like branchiostegal rays (Figure 6.1). The branchiostegals are arranged in series with the sub- and interopercles and are attached proximally to the ventral face of the hyoid bar. The branchiostegal rays and membranes play only a passive role in the abduction and adduction of the branchial cavity and appear to serve primarily as a ventral sealing valve.

The work of McAllister (1968) remains the most comprehensive review of branchiostegal and opercular anatomy available. Additional information is available in Arratia and Schultze (1990).

The hyoid bars

Attached to the ventromedial face of the interopercles of the opercular series, and connecting them with the medial basibranchial series of the branchial apparatus and the urohyal, are the hyoid bars. Each hyoid bar is composed of an interhyal, anterior and posterior ceratohyals, and dorsal and ventral hypohyals (Figure 6.1). Developmentally the hyoid bars are derived from the ossification of the lower portion of the embryonic hyoid arch cartilage of either side of the head. Depression and elevation of the hyoids mediate the required changes in volume of the orobranchial chamber during feeding and respiration, and as such play a central role in suction generation. The presence of an interhyal connecting the proximal tip of the hyoid bar with the suspensorium (at the symplectic–hyomandibular junction) greatly increases the efficiency of hyoid depression in teleosts.

Each ventral hypohyal is ligamentously attached to the anterior end of the median urohyal. Unlike the hyoid bar the teleostean urohyal is a plate-like membrane bone (Figure 6.2), actually an ossified tendon, embedded in the sternohyoideus musculature.

The work of McAllister (1968) remains the most comprehensive review of hyoid anatomy available.

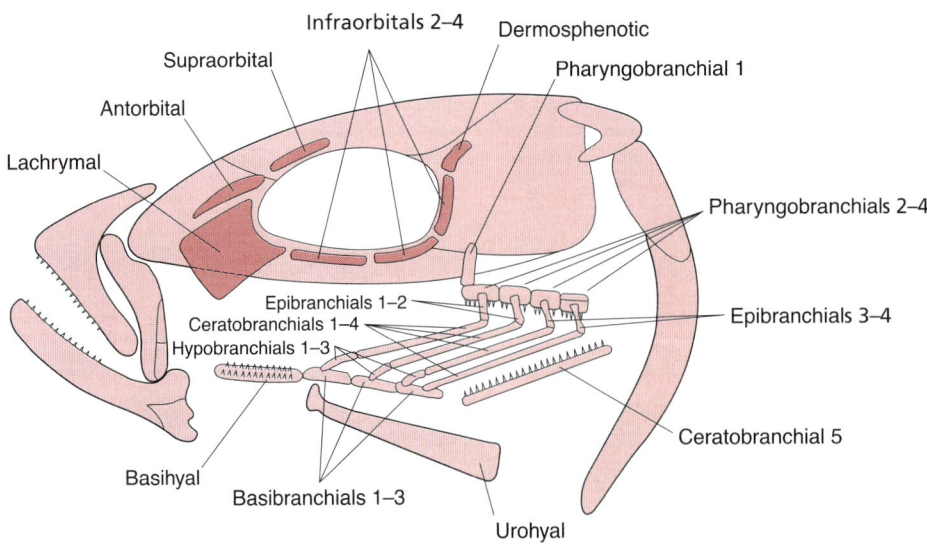

Figure 6.2 Teleostean infraorbital and supraorbital series together with branchial arches.

Additional information is available in Arratia and Schultze (1990), where a comprehensive study of urohyal development and homology is also to be found.

Circumorbital series

Situated superficially on the cheek, forming a chain of dermal bones around the eyeball is the circumorbital series. The ventral elements comprise the canal-bearing infraorbital series (Figure 6.2), which is typically composed of a broad, plate-like first infraorbital (IO1, also termed the lachrymal), followed by a variable number of additional infraorbital elements which may be flat and plate-like or narrow and tubular. The ultimate infraorbital (usually IO6) is usually firmly affixed to the postorbital process of the neurocranium and is termed the dermosphenotic. The infraorbital series carries the main trunk of the infraorbital canal of the acoustico-lateralis system that is connected to the otic and supraorbital canals by the dermosphenotic. In many percoid fishes the third infraorbital (IO3) bears a medial expansion that forms a plate of bone (the subocular shelf) which provides lateromedial support for the eyeball. Additionally, in many 'scorpaeniform' fishes the second infraorbital (IO2) bears a posteriorly directed flange of bone (the infraorbital stay).

In many teleost lineages a supraorbital series (Figure 6.2) of plate-like dermal bones is present around the upper surface of the eyeball and nasal region. These can include one or two supraorbitals and an anterior antorbital.

Information on the structure and composition of teleostean circumorbitals is scattered throughout the literature. Nelson (1969a) provides some useful discussion, as does Jollie (1975).

The oral jaws

In teleosts the upper jaw is primitively composed of three or four pairs of dermal bones: the premaxillae, maxillae, and one or two supramaxillae (Figure 6.1). The anteriormost pair are the premaxillae. Each premaxilla is composed of a median section (ascending process) which is moveably articulated with the ethmoid and a toothed-lateral arm. Posterolateral to the premaxillae are the maxillae which in most euteleosts are edentulate and are entirely excluded from the gape by the premaxillae. Above the maxillae, in many teleosts, are found one or two pairs of flat plate-like supramaxillae. The supramaxillae are always edentulate and excluded from the gape by the maxillae. Supramaxillae have been lost numerous times among teleost lineages.

Several groups of teleosts have developed protrusible premaxillae which swing forward and slide over the ethmoid block of the neurocranium as the mouth opens. The anatomy and mechanism of protrusion varies from group to group but all involve the development of elongate median processes on the premaxillae (the premaxillary ascending processes) and the presence of a median cartilage or bony nodule (a rostral cartilage in the percomorph and atherinomorph models, and a kinethmoid in the cypriniform one) that mediates the glide of the premaxillary processes over the ethmoid region.

The lower jaw, or mandible while functionally a single unit, is a complex element in teleosts. Three bones ossify in Meckel's cartilage: the dentary ossifies around the anterior end of the cartilage and is usually a tooth-bearing element; a dorsoposterior ossification of the cartilage forms an articular; and a ventroposterior ossification forms a retroarticular. Additionally a single dermal ossification, the angular is present in the lower jaw. In teleosts the dermal angular fuses with one or both of the articulars. Posteriorly on the mandible is a saddle-shaped facet for articulation with the quadrate of the suspensorium.

The pattern and sequence of fusions in the posterior mandible in teleosts is of considerable phylogenetic, but apparently of little functional, significance.

Eaton's (1935) classic study, as well as that of Schaeffer and Rosen (1961), contains much useful information on teleostean jaw structure. Alexander (1967), Motta (1984) and Westneat (1990) provide good overviews of the range of upper jaw protrusion mechanisms found in teleosts, and Nelson (1973) provides a helpful review of lower jaw structure in teleostean lineages.

The branchial arches and pharyngeal jaw apparatus

A series of chondral branchial arches are situated beneath the cranial vault from which they are suspended by a series of levatores muscles (see Chapter 7). Laterally the branchial arches are encased by the opercular series. Teleostean branchial arches are complex structures, but are perhaps best viewed as being a series of paired elements arranged into backwardly directed chevrons (Figure 6.2). The dorsal (or epal) elements are the epibranchials, of which there are usually four pairs, and the ventral (or ceratal) elements are the ceratobranchials, of which there are usually five pairs. In most teleosts the fifth ceratobranchials are variously modified into a pair of so-called lower pharyngeal jaws. Additionally, there are a series of three or four dorsalmost elements (the pharyngobranchials). The first pair are edentate and serve to suspend the epibranchials from the cranial vault while the remainder are associated with toothplates and are variously modified to form the upper pharyngeal jaws. In a few basal teleosts one or two pairs of small edentate suprapharyngobranchials, situated dorsal to the pharyngobranchials, are retained. Ventrally three pairs of hypobranchials connect the anterior three ceratobranchials to a median basibranchial series. Three basibranchial elements line the floor of the oral cavity, and anterolaterally they articulate with the ventral hypohyals of the hyoid bars, and anteromedially with a median (and often toothed) basihyal.

The ceratobranchials and epibranchials form the main skeletal supports for ventrally arrayed gill filaments and dorsally situated (dermal) gill rakers. In most teleosts there are varying numbers of toothplates arrayed over the surface of the branchial cavity, and in clupeocephalan lineages many of these toothplates are consolidated and fused to the underlying chondral elements (only the suprapharyngobranchials are never associated with toothplates). Consolidation of toothplates is variable but is most evident over the basibranchials, the fifth ceratobranchials (forming the so-called 'lower pharyngeal jaws'), and the pharyngobranchials (forming the so-called 'upper pharyngeal jaws').

There is considerable phylogenetically informative variation in branchial anatomy among teleosts but the basic configuration and the associated musculoskeletal couplings involved in mediating respiration, mastication and pharyngeal transport are remarkably stable throughout the radiation. During pharyngeal transport of food, the upper and lower pharyngeal jaws are simultaneously protracted and retracted by the action of the dorsal and ventral musculoskeletal branchial arch couplings (see Chapter 7). Lauder (1985) makes the important observation that in addition to pharyngeal transport the branchial apparatus also plays a crucial role in suction feeding by partitioning the buccal cavity from the opercular cavity and thereby decoupling patterns of volume, flow and pressure change in the two regions.

Nelson (1968, 1969b) provides an important review of teleostean pharyngeal anatomy. See also Lauder (1983) for an excellent functional analysis.

The vertebral column

The teleostean vertebral column consists of a series of rigid blocks (vertebrae) connected to one another by relatively flexible intervertebral joints (Figure 6.3). Teleostean vertebrae are usually amphicoelous (biconcave) and the intervertebral joints are formed of plugs of tissue (remnants of the embryonic notochord) filling the intervertebral spaces and serving to attach adjacent centra while also acting as shock absorbing cushions. The resultant structure provides for considerable lateral flexibility (particularly in the posterior, caudal region) while resisting compression. Replacing the notochord in ontogeny, the vertebral column forms a housing for the spinal cord within a neural canal formed dorsally above the centra by the neural arches of each vertebra. Primitively the neural (and haemal) arches are autogenous but in the majority of adult teleosts they become fully fused to the vertebral centra providing a rigid framework for muscle attachment.

The column is structurally differentiated over its length to accommodate varying functions. In many teleosts, the anterior vertebrae are modified to facilitate a firm jointed attachment to the back of the cranial vault, while posteriorly the caudal vertebrae are greatly modified to act as a basal support and framework for the caudal fin. Dorsally and ventrally the column gives support and a site for muscle attachment for the median fins and body musculature. In the abdominal region the column provides suspension and protection for the viscera, and caudally it conveys the main blood vessels ventrally beneath the centra within a protective haemal canal.

Depending on position along the vertebral column each vertebra is variously modified and adorned with processes (pre- and postzygapophyses) but two main groups of vertebrae can usefully be recognized: the precaudal and caudal vertebrae. Precaudal vertebrae form the framework for the abdominal cavity, bear laterally displaced parapophyses (to which the pleural ribs attach), and lack closed haemal arches and spines. The first caudal vertebra is that which lacks parapophyses and pleural ribs, and through which the caudal artery enters a closed haemal canal. Caudal vertebrae also usually bear well-developed haemal as well as neural spines, thus providing an

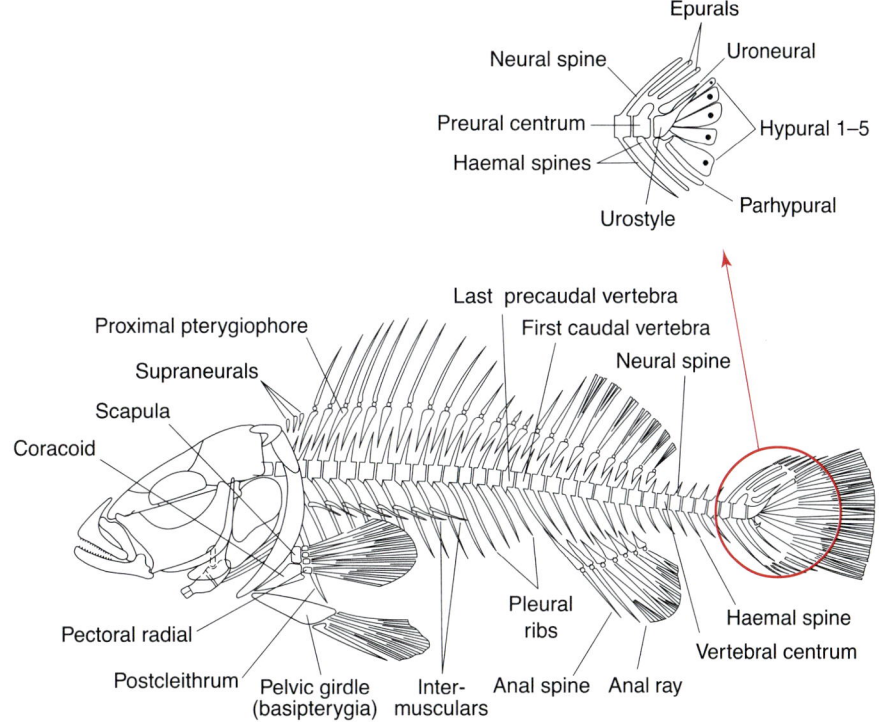

Figure 6.3 The vertebral column of the teleost.

enhanced surface area for the attachment of the caudal body musculature that powers swimming. By convention the last caudal vertebra is usually counted as the compound urostylar complex of the caudal fin skeleton (see below).

Arrayed along the vertebral column and pleural ribs are various series of so-called 'intermuscular bones'. These are ossifications (or chondrifications) of the myosepta and primitively three series are present: epineurals (usually fused to neural arches); epicentrals (usually situated in the horizontal septum); and epipleurals (situated below the horizontal septum). However, the intermuscular series have been variously modified, displaced, and often reduced in teleostean lineages, and provide a rich source of phylogenetically informative variation.

The classic work of Ford (1937) provides a good introduction to teleostean vertebral anatomy. Additional information is to be found in Lauder (1980) and Schultze and Arratia (1988, 1989). Comprehensive reviews of intermuscular bones are provided by Patterson and Johnson (1995) and Gemballa and Britz (1998).

The median fin supports

In the midline along the dorsal and posterocaudal fields are the unpaired dorsal and anal rayed fins (Figure 6.3). These are primitively supported by three series of chondral bony pterygiophores: the proximal, medial and distal series. The proximal pterygiophores are invariably the largest of the series and these elongate, blade-like elements are deeply embedded into the vertical midline, usually extending between the neural or haemal spines of alternating vertebrae. The medial pterygiophores are short cylindrical structures which articulate with both the proximal and distal series. Finally the distal pterygiophores, which often remain cartilaginous and are sometimes paired, are situated between the two halves of the branched fin rays with which they are articulated. The three series of pterygiophores supporting the dorsal and anal fins are variously modified, fused, or reduced within teleostean lineages. Similarly there is frequently a differentiation in pterygiophore structure depending on the anterior or posterior field of the fin, and on whether the pterygiophore supports a spine or a branched ray.

Anterior to the dorsal fin and embedded into the midline are a series of rod-like supraneurals ('predorsals'). Although these can be numerous in basal teleosts, they are usually reduced to two or three in number, particularly in those taxa in which the dorsal fin originates near the back of the skull.

In some teleost lineages there is a small, usually rayless, 'fin' situated along the dorsal midline behind the dorsal fin. This structure is termed the adipose fin (something of a misnomer as the fin is rarely invested with fat), and while its function is unclear, its presence is a ready marker for the identification of fishes belonging to the salmoniform, characiform, siluriform and myctophiform radiations.

In addition to Eaton's (1945) classic work, a review of median fin supports is provided by Smith and Bailey (1961), and much additional information, primarily based on developmental studies, is provided by Potthoff (1980) and Mabee (1988).

The paired girdles

The pectoral girdle

The teleostean pectoral girdle is a robust structure with a broad-based attachment site for the pectoral fin (Figure 6.3). The main body of the girdle (i.e. the dermal 'secondary girdle', so-called as it is assumed to have evolved after the 'primary' fin-bearing chondral elements) is composed of a varying number of dermal elements. The main components of which are, from dorsal to ventral, the post-temporal, supracleithrum, cleithrum and the postcleithra. The post-temporal is firmly attached to the cranial vault via two prongs of bone, one attached to the epiotic and the other to the intercalar. The body of the post-temporal carries part of the temporal branch of the acoustico-lateralis canal and connects it with the supratemporal canal (borne in a series of small, usually tubular bones, the extrascapular or tabular bones). The posttemporal overlies the dorsolateral face of a usually blade-like supracleithrum. The supracleithrum usually also bears a short section of the temporal branch of the lateralis canal which is connected posteriorly with the pored lateral line series. The supracleithrum overlies the dorsal extent of the largest element of the secondary girdle, the cleithrum. Dorsomedially the cleithrum is attached to the first vertebral centrum or to the basioccipital by a strong cord-like ligament (Baudelot's ligament). The cleithra of either side of the body are connected with the pharyngeal jaw apparatus and with the urohyal by the bilateral pharyngocleithralis musculature and the median sternohyoideus muscle. The cleithra are joined ventromedially in a strong symphysis at the junction of the sternohyoideus and hypaxialis musculature. On the medioventral face of the cleithrum on either side of the head a variable number (2–4) of elongate, scale-like postcleithra are usually present embedded in the hypaxial musculature of the body wall.

The so-called 'primary girdle' connects the pectoral fin with the cleithrum, and it is composed of a varied number of chondral bones. The main components of the primary girdle are the scapula, coracoid, and mesocoracoid and these are attached to the pectoral fin via four pectoral radials (or actinosts). The scapula is a small, square-shaped element usually pierced by a round fenestra, and it abuts the elongate coracoid situated ventral to it. Abutting the posterolateral face of the scapula are a series of four hourglass-shaped pectoral radials. These radials in turn serve as the attachment and articulation site for most of the pectoral fin rays and their associated ligaments. However, the first pectoral fin ray articulates directly with the scapula with no intervening free radial. Among teleosts there is considerable variation in the relative position and orientation of the pectoral fins (particularly in relation to the placement of the pelvic fins), and in most eurypterygian fishes the mesocoracoid is lost and the pectoral fin is reoriented and rotated up onto the lateral aspect of the flank.

In addition to the classic study of Starks (1930), the works of Jessen (1972), Brousseau (1976, 1977), and Geerlink (1979) provide much additional information on pectoral anatomy, and Jollie (1975, 1984) gives useful developmental data.

The pelvic girdle

The teleostean pelvic girdle is composed of a pair of chondral plate-like bones (the basipterygia) which may be separate, loosely bound together, or fused into a single medial element (Figure 6.3). A varying number (3–0) of pelvic radials may be associated with the bases of the pelvic fin rays, and where present these mediate attachment of the pelvic fin to the basipterygia. Primitively the girdle and fins are situated in an abdominal position in the ventral body wall where they are flanked by the ventral tips of the pleural ribs. The girdle itself is more-or-less free-floating in the hypaxial and carinal musculature, and only loosely attached to the pleural rib tips by myoseptal connective tissue strands. In many acanthomorph lineages the girdle is anteriorly displaced and occupies a thoracic or subthoracic position below the pectoral girdle. In percomorphs the basipterygia are dorsally inclined away from the ventral body wall and directly abut the cleithra or coracoids of the pectoral girdle to which they are firmly bound by connective tissue. The pelvic fin is composed of a varying number of rays, and in many percomorph lineages the first ray has been replaced by a strong pelvic spine and pelvic ray number is reduced to five.

Information on the structure and composition of teleostean pelvic anatomy is scattered throughout the literature. Stiassny and Moore (1992) provide a review of pelvic anatomy among advanced teleostean lineages.

The caudal fin support

As noted above, the posteriormost caudal vertebrae are greatly modified to form a basal supporting framework for the caudal fin (Figure 6.3). The resulting teleostean caudal fin skeleton is a complex structure composed of a varying number of modified caudal centra and a series of vertebral accessories (modified neural and haemal arches and spines). The resultant structure is flat and plate-like, and internally more-or-less symmetrical (although the terminal urostyle is slightly upturned) with the hind edges of the hypurals forming an almost vertical line. The hypural plate (a series of modified ural haemal spines) forms the articulation surface for the principal rays of the caudal fin.

In the caudal skeleton a distinction is made between the ural and preural centra. The last preural centrum (PU1) is defined as that centrum through which the caudal artery and vein issue and bifurcate to pass lateral to the modified haemal spines (hypurals) of the ural centra. The preural centra usually bear unmodified haemal arches and spines, although the haemal spine of PU1 is somewhat modified and termed the parhypural. Two ural ossifications (these are composed of a varying number of ural centra fused together) are usually present, however in many advanced lineages a compound terminal centrum (the urostyle) is formed from a fusion of PU1 plus the first, or the first and second, ural centrum. In many acanthomorph teleosts a uroneural also becomes incorporated (often fused) into the compound terminal urostyle. The hypurals (haemal arches of the ural centra) lack haemal canals and are broad flattened blades of bone against which the principal caudal fin rays articulate. Primitively in living teleosts there are seven hypurals but this number is variously reduced via loss or fusion in most lineages.

Dorsally, there are a varying number (3–0) of epurals (modified ural neural spines) and (4–0) uroneurals (modified ural neural arches). Together with the epurals the uroneurals stiffen the epaxial portion of the caudal skeleton, and dorsally they form an articulation surface for the procurrent caudal fin rays.

Perhaps in reflection of its structural and functional complexity, the teleostean caudal fin skeleton is a locus for considerable variation in structure and composition of its component elements. The resultant variation is a rich source of data for both systematic and functional studies.

Much has been written about the structure of the caudal fin support in teleostean fishes. Important references in this area are Nybelin (1963), Monod (1967), and Patterson (1968). More recently a series of comprehensive reviews, based mainly on developmental data, have been provided by Schultze and Arratia (1986, 1988, 1989). An atlas of teleostean caudal structure is provided by Fujita (1990) and useful functional studies include those of Videler (1975) and Lauder (1982, 1989).

References

Alexander, R.McN. (1967). *J. Zool., Lond.* **151**, 233–255.
Arratia, G. and Schultze, H.P. (1990). *J. Morphol.* **203**, 247–282.
Arratia, G. and Schultze, H.P. (1991). *J. Morphol.* **208**, 1–81.
Brousseau, R.A. (1976). *J. Morphol.* **148**, 89–136.
Brousseau, R.A. (1977). *J. Morphol.* **150**, 79–115.
Cubbage, C.C. and Mabee, P.M. (1996). *J. Morphol.* **229**, 121–160.
De Beer, G.R. (1937). *The Development of the Vertebrate Skull*. Oxford University Press, Oxford.
Eaton, T.H. (1935). *J. Morphol.* **58**, 157–172.
Eaton, T.H. (1945). *J. Morphol.* **76**, 193–212.
Ford, E. (1937). *J. Mar. Biol. Assoc.* **22**, 1–60.
Fujita, K. (1990). *The Caudal Skeleton of Teleostean Fishes*. Tokai University Press, pp. 1–197.
Geerlink, P.J. (1979). *Neth. J. Zool.* **29**, 9–32.
Gemballa, S. and Britz, R. (1998). *Amer. Mus. Novit.* **3241**, 1–25.
Goodrich, E.S. (1930). *Studies on the Structure and Development of Vertebrates*. Macmillan, London.
Gregory, W.K. (1933). *Trans. Am. Phil. Soc.* **23**, 75–481.
Jessen, H.L. (1972). *Fossils and Strata* **1**, 1–101.
Jollie, M. (1975). *J. Morphol.* **147**, 61–88.
Jollie, M. (1984). *Can. J. Zool.* **62**, 1757–1778.
Kulkarni, C.V. (1948). *Rec. Ind. Mus.* **48**, 65–119.
Langille, R.M. and Hall, B.K. (1987). *J. Morphol.* **193**, 135–158.
Lauder, G.V. (1980). *Paleobiology* **6**, 51–56.
Lauder, G.V. (1982). *J. Zool., Lond.* **197**, 483–495.
Lauder, G.V. (1983). *J. Exp. Biol.* **104**, 1–13.

Lauder, G.V. (1985). In *Functional Vertebrate Morphology* (eds M. Hildebrand, D.M. Bramble and K.F. Liem), pp. 210–229. Belknap Press.

Lauder, G.V. (1989). *Amer. Zool.* **29**, 85–102.

Liem, K.F. (1980). *Amer. Zool.* **20**, 295–314.

Liem, K.F. (1984). In *Trophic Interactions within Aquatic Ecosystems* (eds G.D Meyers and J.R Strickler), pp. 269–305. AAAS Selected Symposia.

Liem, K.F. (1990). *Amer. Zool.* **30**, 209–221.

Mabee, P.M. (1988). *Copeia* **4**, 827–838.

Mabee, P.M. and Trendier, T.A. (1996). *J. Morphol.* **227**, 249–287.

McAllister, D.E. (1968). *Bull. Natl. Mus. Canada* **221**, 1–239.

Monod, T. (1967). *Coil. Int. Cent. Natn, Rech. Sci.* **163**, 111–131.

Motta, P.J. (1984). *Copeia* 1984, 1–18.

Muller, M., Osse, J.W.M. and Verhagen, J.H.G. (1982). *J. Theor. Biol.* **95**, 49–79.

Nelson, G.J. (1968). In *Current Problems of Vertebrate Phylogeny* (ed. T. Orvig), pp. 129–143. Nobel Symposium 4.

Nelson, G.J. (1969a). *Am. Mus. Novit.* **2394**, 1–37.

Nelson, G.J. (1969b). *Bull. Am. Mus. Nat. Hist.* **141**, 477–552.

Nelson, G.J. (1973). In *Interrelationships of Fishes* (eds P.H. Greenwood, R.S. Miles and C. Patterson), pp. 333–349. Academic Press, London.

Nelson, J.S. (1994). *Fishes of the World*. John Wiley, New York.

Nybelin, O. (1963). *Ark. Zool.* **15**, 485–516.

Patterson, C. (1968). *Bull. Br. Mus. Nat. Hist. (Geol.)* **16**, 201–239.

Patterson, C. (1975). *Phil. Trans. Roy. Soc., London, B* **269**, 275–579.

Patterson, C. and Johnson, G.D. (1995). *Smith. Contrib. Zool.* **559**, 1–83.

Potthoff, T. (1980). *Fisheries Bull.* **78**, 277–312.

Schaeffer, B. and Rosen, D.E. (1961). *Am. Zool.* **1**, 187–204.

Schultze, H.P. and Arratia, G. (1986). *J. Morphol.* **190**, 215–241.

Schultze, H.P. and Arratia, G. (1988). *J. Morphol.*, **195**, 257–303.

Schultze, H.P. and Arratia, G. (1989). *Zool. J. Linn. Soc.* **97**, 189–231.

Smith, C.L. and Bailey, R.M. (1961). *Pap. Mich. Acad. Sci.* **XLVI**, 345–363.

Starks, E.C. (1930). *Stanf. Univ. Publ. Biol. Sci.* **6**, 147–239.

Stearley, R.F. and Smith, G.R. (1993). *Trans. Amer. Fish. Soc.* **122**, 1–33.

Stiassny, M.L.J. and Moore, J.A. (1992). *Zool. J. Linn. Soc.* **104**, 209–242.

Videler, J.J. (1975). *Neth. J. Zool.* **25**, 143–194.

Westneat, M.W. (1990). *J. Morphol.* **205**, 269–295.

CHAPTER 7

Muscular System

Melanie L J Stiassny
Department of Ichthyology, American Museum of Natural History, New York City, New York, USA

'The purpose of the skeleton is something to hitch the meat to.' Answer to school science quiz, South African Airlines in-flight magazine, September 1998.

Introduction

In Chapter 6 I provided a basic overview of the skeletal anatomy of a generalized teleostean fish. Here I will attempt to summarize the major muscle groups associated with that skeletal framework. Because of the complexity of teleostean **myology** (kinesis of the head alone is controlled by some 50 individual muscles), I have abbreviated this review and concentrate only on the main muscles associated with aquatic feeding by suction generation and pharyngeal transport, and locomotion by caudal propulsion – functions that appear to be central to the success and diversification of teleostean fishes.

With over 60 major groups subdivided into numerous individual muscles a summary of teleostean myology presents a significant challenge. Happily however, the task has been greatly facilitated by the work of Winterbottom (1974) who has compiled a comprehensive descriptive synonymy of teleostean muscles. Reference to Winterbottom's synonymy is therefore highly recommended for anyone requiring a more detailed picture of teleostean myology.

The cephalic musculature

With the exception of the extrinsic eyes muscles and the Sternohyoideus, StH, which originate from somatic muscle primordia, the muscles of the teleostean head differentiate early in ontogeny from a

This contribution is based on a pre-publication version of Stiassny/ *Guide to the Anatomy of Fishes: The Musculoskeletal System*, a work to be published by Wiley-Liss, Inc., a subsidiary of John Wiley & Sons, Inc., in 2001. Reprinted by permission of Wiley-Liss, Inc. Copyright © 2001 Wiley-Liss, Inc. All rights reserved. Reproduction, adaptation or any further distribution of this material is expressly prohibited. For further information or to require permission for other uses, please contact the Permission Department, John Wiley & Sons, Inc., 605 Third Ave., New York, N.Y. 10158-0012. Telephone (212) 850-6011. Facsimile: (212) 850-6008.

Copyright © 2000 Academic Press

visceral muscle primordium that can be divided into three main components:

1. The mandibular muscle plate (giving rise to the Levator Arcus Palatini, LAP; the Dilatator Operculi, DO; the Adductor Mandibulae, AM; and the Intermandibularis, IM). All of the muscles derived from the mandibular muscle plate are innervated by the mandibular branch of the trigeminal (V) nerve.
2. The hyoid muscle plate (giving rise to the Adductor Arcus Palatini, AAP; the Adductor Operculi, AO; the Levator Operculi, LO; the Protractor Hyoidei, PrH; the Hyohyoides Inferioris, HhI; Hyohyoides Abductores, HhAb; and the Hyohyoides Adductores, HhAd). All of the muscles derived from the hyoid muscle plate are innervated by the hyomandibular branch of the facial (VII) nerve.
3. The five branchial muscle plates (giving rise to the dorsal branchial muscles; the Levatores Externi, Le_{1-4}; Levator Posterior, LP; Levatores Interni, Li_{1-3}; the Obliqui Dorsales, ObD; the Obliquus Posterior, ObP; the Transversi Dorsales, TrD; the Retractor Dorsalis, RD; and the Adductores, Ad_{1-5}. And the ventral branchial muscles; the Sphincter Oesophagi, SpO; the Obliqui Ventrales, ObV; the Transversi Ventrales, TrV; the Recti Ventrales, RcV; the Pharyngohyoideus, PhH; and the Pharyngoclavicularis Internus and Externus, PcI, PcE). All muscles derived from the branchial muscle plates are innervated by the posttrematic branch of the glossopharyngeal (IX) and branches of the vagus (X) nerves.

Coordinated contraction of these various **cephalic** muscles mediates expansion and compression of the **orobranchial** and **opercular** cavities during respiration and suction feeding, and control pharyngeal manipulation and transport of ingested food.

While it is rarely possible to provide a simple listing of individual muscle function (many muscles perform multiple and often modulated functions, and muscle function is often contingent upon line of action in a particular position), Adriaens and Verraes (1997a) provide a useful summary of muscle activity during respiration. They recognize an initial expansive phase characterized by the synchronous contraction of the LAP, HhI, DO, and LO. These contractions result in the abduction of the suspensorium, the expansion of the branchiostegal membranes, the abduction of the gill cover, and in the depression of the lower jaw. The expansive phase is immediately followed by a compressional phase in which synchronous contractions of the AM, PrH, HhAd, and slightly later of the AAP and AO, result in mouth closure, the elevation of the hyoid bars, constriction of the branchiostegal membranes, followed by suspensorial adduction, and a final closure of the gill cover. These respiratory movements create a strong flow of water in through the oral cavity, over the gills, and out through the gill cover prior to its closure and the reinitiation of a subsequent expansive phase.

While there are some distinct differences (Lauder, personal communication), a broadly similar sequence of muscle activity is found during suction feeding although the amplitude of movement and velocity of action are considerably greater (Alexander, 1975). (For details of what is in fact an extremely complex series of couplings with modulated action, amplitude and timing, see e.g. Osse, 1969; Liem, 1978; Lauder and Liem, 1980; and Lauder, 1985.)

Main muscles of the cheek region

The adductor mandibulae complex (AM, Figure 7.1)

This is the largest muscle mass of the cheek and it occupies most of the lateral face of the suspensorium and inserts, via a number of tendons, onto the oral jaws. The number and nature of subdivision of the adductor mandibulae is highly variable among teleosts, and there may be anything from a single undifferentiated muscle mass up to ten distinct parts present.

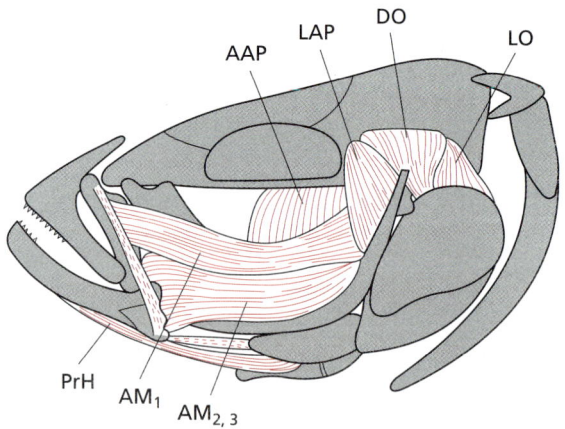

Figure 7.1 Lateral view of the teleost head musculature.

Generally in teleosts four levels of the adductor mandibulae can be recognized; AM_1, AM_2, AM_3 and A_ω which together connect the suspensorium and the oral jaws and function primarily in closing the mouth. For most teleosts the following anatomical definitions are probably appropriate:

Part A_1 is that component that originates on the lateral face of the suspensorium and inserts wholly, or in part, onto the maxilla.

Part A_2 is that component that usually occupies the ventrolateral region of the suspensorium and inserts on the medial face of the lower jaw.

Part A_3 is that component usually situated medial to both A_1 and A_2 and inserts onto a sesamoid ossification on Meckel's cartilage (the coronomeckelian) on the medial face of the lower jaw via a short tendon. Parts A_2 and A_3 are frequently consolidated into a single body, $A_{2,3}$. When this is the case it is usual for the tendon of A_3 (tA_3) to be retained and to insert onto the coronomeckelian.

Part A_ω is that component that occupies the Meckelian fossa on the medial face of the lower jaw.

While the primary function of AM is to mediate mouth closure, action of a maxillary component (A_1) can also play a central role in the jaw protrusion mechanisms of many teleosts (see Motta, 1984, for a discussion of jaw protrusion in teleosts).

The levator arcus palatini (LAP, Figure 7.1)

Typically the LAP occupies an area at the rear of the orbit passing from its origin on the sphenotic of the neurocranium to a musculous insertion on the lateral face of the hyomandibula of the suspensorium.

The primary function of the LAP is to mediate suspensorial abduction.

The dilatator operculi (DO, Figure 7.1)

Usually somewhat conical in shape, this muscle connects the postorbital region of the neurocranium with the operculum.

Usually working in concert with the LAP, the primary function of the DO is to mediate abduction of the opercle.

The adductor arcus palatini (AAP, Figure 7.1)

The AAP is variously developed, and in some forms may remain contiguous with the adductor operculi. Although usually confined to the posterior wall of the orbit, in certain advanced acanthomorphs the adductor is well developed and extends rostrad to fill the fissura infraorbitalis and form the floor of the orbit between the neurocranium and the suspensorium.

The primary function of the AAP is to mediate suspensorial adduction.

The adductor operculi (AO)

This cylindrical muscle connects the neurocranium with the medial face of the operculum at a point adjacent to its articulation with the hyomandibula.

Usually working in concert with the AAP, the primary function of the AO is to mediate adduction of the opercle.

The levator operculi (LO, Figure 7.1)

This muscle passes from the lateral neurocranial wall and (usually) inserts onto the dorsomedial face of the opercle caudal to the insertion of the dilatator operculi.

Action of the LO is an important mechanism for mediating lower jaw depression via the so-called 'four-bar linkage system' in which the force of contraction of LO is transmitted through the opercular series and the interoperculomandibular ligament to the lower jaw (see Barel *et al.*, 1977).

Main muscles of the hyoid apparatus and associated structures

The protractor hyoidei (PrH, Figures 7.1 and 7.2)

This embryologically and functionally complex muscle (often incorrectly referred to as the 'genohyoideus', see Winterbottom, 1974; Adriens and Verraes, 1997b) connects the lower jaw with the hyoid bars. The anterior site of attachment is usually to the medial face of the dentaries near their **symphysis** and its insertion is usually onto the lateral face of the hyoid bars.

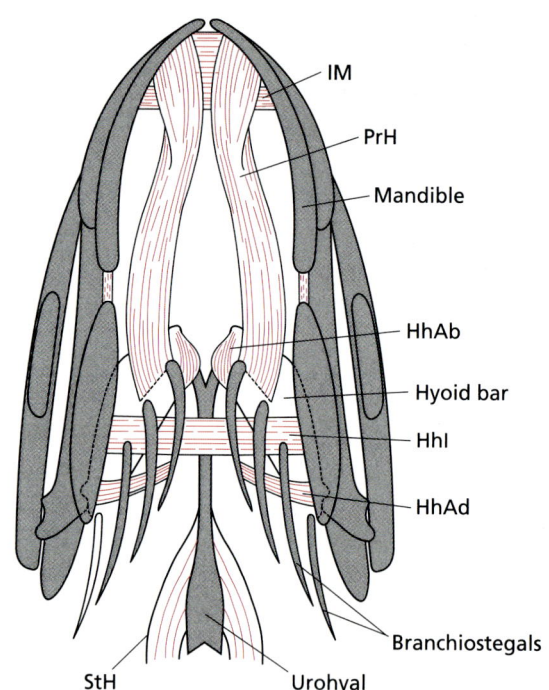

Figure 7.2 Ventral view of the teleost head musculature.

Functionally the PrH is a complex muscle, and Osse (1969) has demonstrated that its anterior and posterior sections may contract differentially during different phases of respiration. Nonetheless, as a broad generality the PrH can be said to play a primary role in the elevation (protraction) of the hyoid bars, as well as functioning to lower the mandible.

The intermandibularis (IM, Figure 7.2)

Variously developed in teleosts, the IM passes between contralateral dentaries, inserting usually in the region of their symphysis.

The hyohyoides inferioris (HhI, Figure 7.2)

Generally the HhI passes between contralateral hyoid bars, inserting on their ventrolateral faces beneath the first few branchiostegal rays. The fibers of each side either connect in the ventral midline via a median **aponeurosis** overlying the urohyal, or they may insert directly onto the urohyal bone.

There is little commentary in the literature regarding the function of HhI but adduction of the hyoid bars is suggested by its position and presumed line of action.

The hyohyoides abductores (HhAb, Figure 7.2)

The HhAb usually connects the first few branchiostegal rays with the rostral tip of the hyoid bars although this muscle exhibits considerable variation in size and the site of insertion.

Functionally the HhAb mediates expansion of the branchiostegal membranes.

The hyohyoides adductores (HhAd, Figure 7.2)

These are present as variously developed sheets of fibers connecting the distal portions of the branchiostegal rays with each other, and with the opercular bones.

Contraction of HhAd mediates constriction of the branchiostegal membranes.

The sternohyoideus (StH, Figures 7.2 and 7.3)

The StH is usually a large cone-shaped muscle that passes from the horizontal limb of the cleithrum (and often coracoid) of the pectoral girdle to insert around the urohyal. The anterodorsal portions of StH are often ligamentously attached to the hypobranchial bone. The StH usually consists of three myomeres separated by two myocommata that are often readily visible externally. In most teleosts the posterior fibers of StH are aponeurotically associated with those of the inferior obliquus of the body wall.

The StH plays a major role in mediating hyoid depression, and through a series of mechanical linkages plays an important role in mouth opening and suspensorial abduction (see e.g. discussions of StH function in Lauder and Liem, 1984).

Pharyngeal and associated musculature

The anatomical couplings and muscle activity patterns during pharyngeal maceration and transport

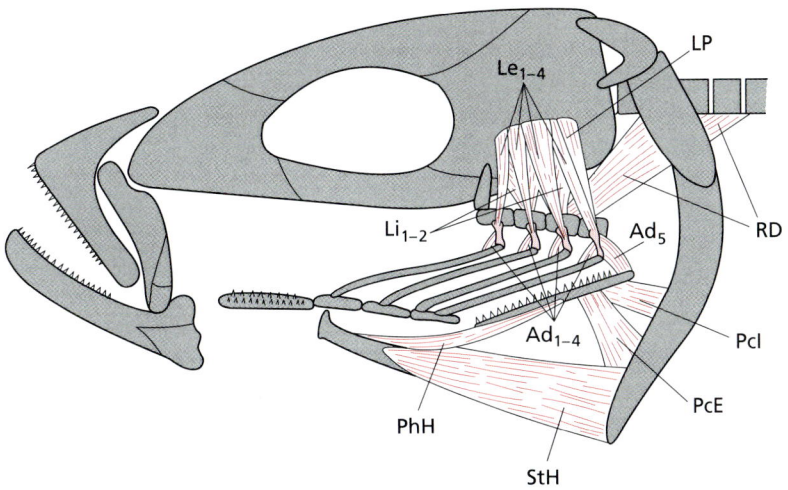

Figure 7.3 Lateral view of the teleost head musculature (suspensorium and opercular apparatus removed).

are extremely complex (see e.g. Liem, 1974; Lauder, 1983; Wainwright, 1989; Galis *et al.*, 1994; Galis and Druker, 1996), and the functioning of individual muscles is often highly variable. For example, Lauder (1983) notes that although the anterior levatores externi and interni function to retract the upper pharyngeal jaw in *Esox* the same muscles, due to their anterior excursion, actually protract the jaw in *Ambloplites*. Similarly Liem (1974) notes that in cichlids the fourth levator externus muscle is actively firing throughout all phases of pharyngeal food processing and deglutination and performs differing antagonistic functions during different phases. The assignment of a single function to the Le_4 is therefore inappropriate given these findings.

In a broad generality Liem (1974) partitions pharyngeal transport into two main phases, each with a unique electromyographic profile. During the 'retracted–adducted' phase activity of the LP, PrH, PcL, and RD retract and rotate the pharyngeal jaws. After a short transitional phase, this is followed by a 'protracted–abducted' phase resulting from contraction of LE_4, PrH and Ad_5, then StH and PcE. Additionally Wainwright (1989) provides a useful summary of pharyngeal muscle action in a more generalized **percomorph**.

Action of much of the intrinsic pharyngeal musculature (e.g. ObD, ObP, TrD, SpO, ObV, TrV, and RcV), which is often hard to locate with electrodes in living animals, has been less well analyzed electromyographically and most functional interpretation in the literature is inferential, based on the sites of origin and insertion of the various muscles.

Dorsal gill arch musculature

The levatores externi (Le_{1-4}, Figures 7.3 and 7.4)

Usually four pairs of external levators originating on the lateral wall of the cranial vault insert onto the dorsolateral faces of the epibranchials.

Function of the various levators is highly dependent upon orientation and the resultant line of action (see above). Nonetheless, for the most part the external levators function to adduct the UPJ.

The levator posterior (LP, Figures 7.3 and 7.4)

Originating from the lateral wall of the cranial vault, the LP is usually well-separated from the remaining Le series. The LP usually inserts with Le_4 onto the fourth epibranchial.

The primary function of the LP is also to mediate powerful adduction (and often also retraction) of the UPJ.

The levatores interni (Li_{1-3}, Figures 7.3 and 7.4)

Generally originating medial to the Le series on the lateral wall of the cranial vault two (or three) paris of Li insert onto the dorsal face of the second and third (and when present also the fourth) pharyngobranchials.

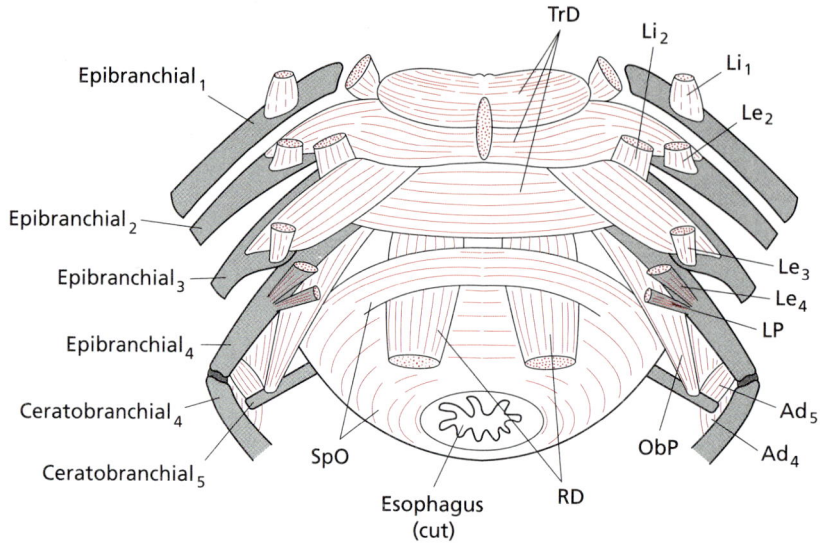

Figure 7.4 Dorsal view of pharyngeal muscles.

The Li series function, in concert with the Le series primarily to adduct the UPJ.

The transversi dorsales (TrD, Figures 7.3 and 7.4)

The TrD is frequently subdivided into a complex anterior muscle. Most fibers of the TrD originate from a connective tissue septum in the dorsal midline and insert onto the anterior epibranchials and pharyngobranchials. A posterior subdivision is often present passing between the heads of the third epibranchials and covering the posterior portion of the third pharyngobranchials. The posterior section of TrD is often overlain by the ObD.

The obliqui dorsales (ObD, Figures 7.3 and 7.4)

An ObD usually connects the second and third pharyngobranchials with the third epibranchial. Fibers pass underneath the anterior TrD.

The obliquus posterior (ObP, Figures 7.3 and 7.4)

When present the ObP connects the head of the fourth epibranchial to the posterodorsal tip of the fifth ceratobranchial (LPJ).

The retractor dorsalis (RD, Figures 7.3 and 7.4)

In neoteleosts this pair of muscles connect the posterodorsal face of the third (and sometimes fourth) pharyngobranchials with the vertebral column. The phylogenetic history and homology of these muscles is complex and they may have arisen independently a number of times within the Teleostei.

Functionally the RD muscles serve as powerful retractors of the UPJ.

The adductores (Ad_{1-5}, Figures 7.3 and 7.4)

An adductores series of five muscles connect the epibranchial and ceratobranchial elements of each arch. Functionally the anterior adductores facilitate jaw adduction. However, as Liem (1974) has noted the posteriormost of the series, Ad_5, often acts to abduct the jaws.

Ventral gill arch musculature

The sphincter oesophagi (SpO, Figures 7.4 and 7.5)

A discrete SpO is variously differentiated from the connective tissue around the anterior esophagus and in many lineages a distinctive muscle sheet surrounding the esophagus is present. A number of the dorsal

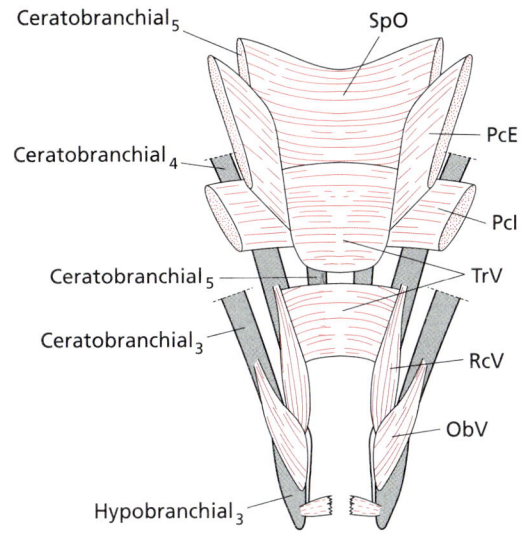

Figure 7.5 Ventral view of pharyngeal muscles.

gill arch muscles discussed here (e.g. ObD, TrD, and RD) develop from subsequent subdivision of SpO.

The obliqui ventrales (ObV, Figure 7.5)

A variable number (up to three pairs) of ObV muscles span the joints between ceratobranchial and hypobranchial elements of the anterior three arches. In many taxa these muscles are reduced in number and represented only by a single pair.

The recti ventrales (RcV, Figure 7.5)

A variable number (up to three pairs) of RcV muscles interconnect the hypobranchials with the ceratobranchial elements of the preceding arch. In many taxa these muscles are reduced in number and represented only by a single pair.

The transversi ventrales (TrV, Figure 7.5)

Usually two TrV muscles span the midline between the ventromedial faces of the fourth and fifth ceratobranchials.

The pharyngohyoideus (PhH, Figures 7.1, 7.2 and 7.3) and rectus communis

The PhH is an elongate muscle that, in euacanthomorphs, originates musculously from along the dorsolateral face of the urohyal bone and is tendinously attached to the lateral face of the LPJ. In the more generalized lineages this muscle originates on the third hypobranchial elements rather than the urohyal. Lauder (1983) suggests that the term 'rectus communis' be retained for the latter configuration.

The PhH serves primarily to protract the LPJ.

The pharyngoclavicularis internus (PcI, Figures 7.3 and 7.5)

The PcI originates (usually musculously) from the anterior face of the vertical limb of the cleithrum and fibers pass in a broad sheet to insert (often via a short tendon) onto the lateral face of the LPJ.

Acting as the functional antagonist of PhH the PcI mainly mediates retraction (and adduction) of the LPJ.

The pharyngoclavicularis externus (PcE, Figures 7.3 and 7.5)

The PcE originates musculously from the anteroventral face of the vertical limb of the cleithrum and fibers pass in a broad sheet to insert musculously along the lateral arm of the LPJ, external to the insertion of PcI.

The PcE functions mainly to mediate abduction of the LPJ.

Body musculature

Comprising over half of the total body weight, the bulk of fish musculature is found in the trunk and caudal region. Here the musculature is segmentally arranged into a series of blocks, or myomeres, separated by tendinous sheets called myosepta or myocommata. Teleosts, in common with all gnathostomes, have two main septa running much of the body length; a vertical septum (found also in lampreys) and a second, horizontal septum. As is readily evidenced by a quick viewing of any fish 'steak', these septa partition the musculature into quadrants. The myotomes connect externally to the connective tissues of the skin and internally to the vertical and horizontal septa, and to adjacent myosepta. The orientation and arrangement of muscle fibers in myotomes can be extremely complex (see e.g. Alexander, 1969, and Jayne and Lauder, 1995, for a detailed analysis of myotome function).

Individual teleostean myomeres are complex and folded and, just below the skin, resemble letter Ws displaced onto their sides. Internally each myomere bears one anteriorly directed and two posteriorly directed flanges of muscle that interdigitate with those of proximate myomeres. This complex structure and arrangement of the myomeres allows the pull of any given myomere to be exerted over several vertebral segments and thereby allows for stronger and more controlled lateral undulations to power forward locomotion. Strengthening of the axial skeleton, and the replacement of the notochord by a chain of vertebrae, facilitates fine control over both the frequency and amplitude of lateral undulation, and this is particularly true in the caudal region where the main powering of forward locomotion now resides (Gosline, 1971; Alexander, 1978; Webb and Blake, 1985; Videler, 1993).

Winterbottom (1974) recognizes three main divisions of the body musculature: a dorsal *expaxialis* (Epx), a ventral *hypaxialis* (Hpx), and a *lateralis superficialis* (LtS) division. The numerous small muscles controlling the movements of the various fin supports and rays are derived from these somatic muscles.

The epaxialis (Epx, Figure 7.6)

The Epx forms the dorsalmost component of the body musculature. The degree of subdivision and specialization of the epaxialis varies considerably among teleosts but anteriorly the muscle always attaches to the back of the cranial vault and the dorsal elements of the pectoral girdle. In many of the more advanced acanthomorph lineages epaxial muscle bundles extend forward over ridges and bony flanges on the roof of the skull. The Epx is bound to the skin by fibrous connective tissue, and internally it inserts over the dorsal portions of the vertebral centra and fin **pterygiophores**, and the neural spines. Posteriorly, the Epx overlies the modified muscles of the caudal fin, and inserts onto the dorsal ray bases of the fin.

The Epx, particularly in the caudal region, plays a central role in propulsive locomotion, but the anterior portion of the muscle also plays an important role in feeding. Contraction of the anterior Epx elevates the neurocranium and thereby plays an important role during the expansion of the orobranchial cavity (see discussions of Epx function in Liem and Osse, 1975 and Liem, 1978).

The hypaxialis (Hpx, Figure 7.6)

The Hpx forms the ventral mass of the body musculature and is situated below the horizontal septum. It too is bound to the skin by fibrous connective tissues while internally it inserts over the ventral portion of the vertebral centra and anal fin pterygiophores, pleural ribs and the haemal spines. Posteriorly, the Hpx overlies the modified muscles of the caudal fin, and inserts onto the bases of the ventral rays of the fin. In many neoteleosts the fiber bundles of the hypaxialis are subdivided, particularly in the thin-walled region surrounding the abdominal cavity, into recognizable components (named the *obliquus superioris* and the *obliquus inferioris* by Winterbottom, 1974).

Coupled with the Epx, the caudal components of the Hpx play a central role in propulsive locomotion. The anterior fibers of Hpx insert along the ventral limb of the cleithrum of the pectoral girdle and thus their contraction exerts a backward pull on the

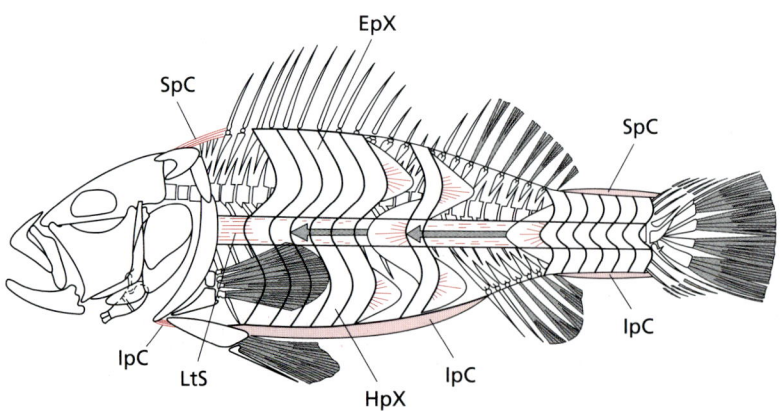

Figure 7.6 Teleost trunk and caudal musculature.

cleithrum. As the cleithra are attached to the urohyal by the StH, together these two muscles operate to facilitate hyoid depression.

The lateralis superficialis (LtS, Figure 7.6)

The variously developed LtS is a sheet of muscle overlying the horizontal septum along most of its length. Although the LtS is usually thickest in the caudal region it also covers the ventral portion of the Epx and the dorsal portion of the Hpx along most of the midline of the body. The myoglobin-rich, highly vascularized red fibers of the LtS are capable of prolonged activity to power sustained cruising and are well supplied with lipids and mitochondria.

Because of its role in powering sustained swimming the LtS is frequently particularly well developed in mesopelagic fishes such as tunas where it can comprise from anywhere up to 30% of the total body musculature. Additionally, contraction of the anterior portion of LtS will presumably rotate the pectoral girdle backwards.

Additional Epx and Hpx derivatives are the so-called *carinal* series (the dorsal *supracarinalis*, SpC and the ventral *infracarinalis*, IpC). These paired, cord-like muscles pass along the dorsal and ventral midline and interconnect the neurocranium and the dorsal median and caudal fins, and the cleithra with the pelvic, ventral median and caudal fins (Figure 7.6).

Median fin muscles

The erectors (*Erectores Dorsales* and *Anales*), depressors (*Depressores Dorsales* and *Anales*), and inclinators (*Inclinatores Dorsales* and *Anales*) of the median fins are apparently all derived from carinal anlagen. The muscles of the pectoral, pelvic and caudal fins, however, are derived more directly from the Epx and Hpx muscle plate anlagen.

Pectoral fin muscles

As summarized by Winterbottom, 1974 (see also Adriens *et al.*, 1993, for a useful summary of the functional anatomy of the pectoral girdle), the main muscles of the pectoral fin are the abductors, adductors, and arrectors. Contraction of the abductors (*Abductor superficialis* and *Profundus*) generates a forward rotation of the fin rays. Contraction of the superficial component of the adductors (*Adductor Superficialis* and *Profundus*) rotates the fin backwards.

The function of the deeper adductor component is unclear as its origin and insertion are both on more-or-less immoveable structures. The arrectors (*Arrector dorsalis* and *Ventralis*) both insert onto the base of the first pectoral ray and presumably function to raise and lower the ray. Additionally, in many neoteleosts a *Coracoradialis* muscle passes from the dorsomedial face of the coracoid and inserts onto the ventromedial face of the fourth radial. The function of the coracoradialis is unclear.

Pelvic fin muscles

There are usually six main pairs of pelvic fin muscle present: the ventrally situated *Arrector ventralis pelvicus*, and two abductors (*Abductor superficialis pelvicus* and *Profundus pelvicus*) and the medially situated *Arrector dorsalis pelvicus*, and the two adductors (*Adductor superficialis pelvicus* and *Profundus pelvicus*).

In addition to anchoring the pelvic fin firmly to the ventral body wall, the pelvic musculature, which is usually well differentiated, is capable of differential erection and retraction of the pelvic fins. Harris's (1938) classic review remains an important reference for understanding the functional role of the paired fins in teleosts.

Caudal fin muscles

As summarized by Winterbottom, 1974 (see also Lauder, 1982, 1989, for an excellent functional analysis), the main muscles of the caudal fin are the series of fin flexors attaching the neural and haemal spines to the principal caudal fin rays (i.e. the *Flexor dorsalis* and *Flexor dorsalis superior*, and the *Flexor ventralis* and *Flexor ventralis inferior* and *externus*), the *Hypochordal longitudinalis* connecting the lower hypurals to the bases of the dorsal caudal rays, and the *Interradialis* interconnecting the median principal caudal fin rays. In addition to changes in the span of the fin these muscles can also function to change the angle of attachment of differing regions resulting in a device capable of remarkably fine-tuned movement (see Lauder, 1989).

References

Adriaens, D. and Verraes, W. (1997a). *Neth. J. Zool.* 47, 61–89.

Adriaens, D. and Verraes, W. (1997b). *Zool. J. Linn. Soc.* **121**, 105–128.

Adriaens, D., Decleyre, D. and Verraes, W. (1993). *Belg. J. Zool.* **123**, 135–157.

Alexander, R. McN. (1969). *J. Mar. Biol. Ass. UK* **49**, 263–290.

Alexander, R. McN. (1975). *The Chordates*. Cambridge University Press.

Alexander, R. McN. (1978). *Functional Design in Fishes*. Hutchinson University Library.

Barel, C.D.N., van der Meulen, J.W. and Berkhoud, H. (1977). *Anat. Anz.* **142**, 21–31.

Galis, F. and Druker, E.G. (1996). *J. Evol. Biol.* **9**, 641–670.

Galis, F., Terlow, A. and Osse, J.W.M. (1994). *J. Fish Biol.* **45**, 13–26.

Gosline, W.A. (1971). *Functional Morphology and Classification of Teleostean Fishes*, University Press of Hawaii.

Harris, J.E. (1938). *J. Exp. Biol.* **15**, 32–47.

Jayne, B.C. and Lauder, G.V. (1995). *J. Exp. Biol.* **198**, 805–815.

Lauder, G.V. (1982). *J. Zool. Soc., Lond.* **97**, 483–495.

Lauder, G.V. (1983). *Zool. J. Linn. Soc.* **77**, 1–38.

Lauder, G.V. (1985). In *Functional Vertebrate Morphology* (eds M. Hildebrand, D.M. Bramble, K.F. Liem and D.B. Wake), pp. 210–229. Belknap Press.

Lauder, G.V. (1989). *Amer. Zool.* **29**, 85–102.

Lauder, G.V. and Liem, K.F. (1980). In *Charrs, salmonid fishes of the genus* Salvelinus (ed. E.K. Balon), pp. 365–390. Dr W. Junk Publ.

Liem, K.F. (1974). *Syst. Zool.* **22**, 425–441.

Liem, K.F. (1978). *J. Morph.* **158**, 323–360.

Liem, K.F. and Osse, J.W.M. (1975). *Amer. Zool.* **15**, 427–454.

Motta, P.J. (1984). *Copeia*, 1984, 1–18.

Osse, J.W.M. (1969). *Neth. J. Zool.* **19**, 289–392.

Videler, J.J. (1993) *Fish Swimming*, Chapman and Hall.

Wainwright, P.C. (1989). *J. Exp. Biol.* **141**, 359–375.

Webb, P. and Blake, M. (1985). In *Functional Vertebrate Morphology* (eds M. Hildebrand, D.M. Bramble, K.F. Liem and D.B. Wake), pp. 110–128, Belknap Press.

Winterbottom, R. (1974). *Proc. Philad. Acad. Sci.* **125**, 225–317.

CHAPTER 8

Nervous System

Ann B Butler
Krasnow Institute for Advanced Study and Department of Psychology, George Mason University Fairfax, Virginia, USA

Abbreviations

I	Olfactory nerve
II	Optic nerve
III	Oculomotor nerve
IV	Trochlear nerve
V	Trigeminal nerve
VI	Abducens nerve
VII	Facial nerve (dorsal and ventrolateral components)
VIII	Vestibulocochlear nerve
IX	Glossopharyngeal nerve (dorsal and ventrolateral components)
X	Vagus nerve (dorsal and ventrolateral components)
Cb	Cerebellum
CCb	Cerebellar corpus
D	Diencephalon
DP	Dorsal pallium
E	Epiphysis
Eg	Eminentia granularis
Epen	Ependyma
Ep	Epiphyseal nerve
ELLL	Electrosensory lateral line lobe
H	Hindbrain
Ha	Habenula
Hp	Hypophysis (pituitary gland)
Hy	Hypothalamus
LVII	Facial lobe
LX	Vagal lobe
LI	Inferior lobe of hypothalamus
LLL	Lateral line lobe
LP	Lateral pallium
LSp	Spinal lobes
LV	Lateral ventricle
MP	Medial pallium
Nlla	Anterior lateral line nerves
Nllp	Posterior lateral line nerves
OB	Olfactory bulb
OlN	Olfactory nerve
OlT	Olfactory tract

Copyright © 2000 Academic Press

ON	Optic nerve
OpT	Optic tract
OT	Optic tectum
P1	P1 pallial area
P2	P2 pallial area
P3	P3 pallial area
P	Pallium
Pr	Pretectum
Pro	Profundus nerve
Sp	Subpallium
SV	Saccus vasculosus
T	Telencephalon
Te	Terminal nerve
TLa	Torus lateralis
TLo	Torus longitudinalis
VCb	Valvula cerebelli

Introduction

As in all vertebrates, the nervous system of fishes can be divided into central and peripheral parts. The central nervous system comprises the brain, spinal cord and two of the cranial nerves, while the peripheral nervous system comprises the rest of the cranial nerves and the spinal nerves. To appreciate the gross structure of the nervous system, its cellular components need to be introduced. It is predominantly composed of various sets of neurons, which have cell bodies and processes – axons and dendrites (see Chapter 20). Neurons can be classified according to the shape of the cell body: bipolar neurons, for example, have cell bodies with only two 'poles' where processes arise, and multipolar neurons have multiple poles, i.e. more than two. In vertebrates, neurons may have multiple poles giving rise to dendrites but only one axon. Most of the bipolar neurons are located within the peripheral nervous system, while the central nervous system is characterized by multipolar neurons of a great variety of form and size.

The neuronal cell bodies frequently occur in groups called nuclei, ganglia, or other related terms, while other collections of them are arranged in multiple layers that have particular architectural relationships and are called cortex (see Chapter 20). The axons of these various sets of neuronal cell bodies form groups that are efferent (outgoing) from their cell bodies of origin and pass between other nuclei and cortices to terminate in their target structures. The term projection is used to describe this relationship, e.g. nucleus 'A' projects to nucleus 'B'. The dendrites of a given set of neuronal cell bodies are the most frequent receptive site of the afferent (incoming) axons to them. In most cases, these sets of axons (or fibers) are called tracts in the central nervous system and nerves in the peripheral nervous system.

The central and peripheral nervous systems also can be divided into somatic and autonomic parts. In general terms, the somatic nervous system supplies the various sensory receptor systems and the somatic muscles of the body wall and head, while the autonomic nervous system (which has sympathetic and parasympathetic divisions) supplies sensory and motor innervation to viscera and glands. Somatic motor innervation is via sets of single neurons with cell bodies in the central nervous system and axons that run in a cranial or spinal nerve to the muscle. Autonomic motor innervation is via a two-neuron chain with the first cell body in the central nervous system and the second within a peripherally located ganglion. On the sensory side, bipolar sensory neurons for most somatic and all autonomic nerves have cell bodies within peripherally located ganglia and project to multipolar neurons within the central nervous system.

Brain structure varies dramatically across fishes – both between and within major groups. Part of this variation derives from fundamental differences in the embryological development of the brain between ray-finned fishes (actinopterygians) and other fishes. Another part derives from marked differences in the degree to which cell proliferation and cell migration occur within the brains of fishes of each major group and in the degree to which various sensory and motor systems are elaborated. Since the preponderance of fishes used in laboratory studies are ray-finned fishes, their nervous system anatomy is emphasized in this chapter. Nevertheless, due to important differences as well as similarities across all fishes, some comments on jawless (agnathans) and cartilaginous fishes (chondrichthyans) are included as well. Only minor comments are made on the fleshy-finned (sarcopterygian) fishes, which include lungfishes and the crossopterygian coelacanth fish *Latimeria*, since they are infrequent laboratory inhabitants, and their brain anatomy speaks to brain evolution in tetrapods more than in most **extant** fishes.

Brain

The brain (Figures 8.1 and 8.2) consists of three major parts – the forebrain (prosencephalon), midbrain (mesencephalon), and hindbrain (rhombencephalon). The forebrain is further subdivided into a more rostral part, the telencephalon, and a more caudal part, the diencephalon. The midbrain is not further subdivided along the rostrocaudal dimension. The hindbrain includes the metencephalon rostrally, which comprises the pons and the cerebellum, and the myelencephalon caudally, which is continuous with the spinal cord. The term brainstem is a loose term that is variably used to refer to only the hindbrain (with or without the cerebellum) and midbrain, to the hindbrain, midbrain and diencephalon, or to the latter three regions plus a ventral portion of the telencephalon. The central nervous system develops initially as a tubular structure. Its dorsal portion is derived from tissue called the alar plate, which gives rise to most of the forebrain and to the roof structures of the midbrain and hindbrain, including the cerebellum, and to the sensory nuclei of cranial nerves. In the spinal cord, the alar plate forms the dorsal, sensory half of the cord. The ventral portion of the developing neural tube is called the basal plate; this tissue gives rise to the ventral-most portion of the diencephalon and to some of the more ventral components of the midbrain and hindbrain, particularly the motor nuclei of cranial nerves. In the spinal cord, the basal plate gives rise to the ventral, motor half of the cord.

Forebrain

The forebrain (Figures 8.1 and 8.2) consists of the diencephalon and the telencephalon, which are extensively interconnected. The diencephalon has four major divisions, which, from dorsal to ventral, are the epithalamus, dorsal thalamus, ventral thalamus and hypothalamus. Caudal to these divisions are two additional diencephalic regions, the pretectum and the posterior tuberculum. The telencephalon can be divided into a pallium dorsally and a subpallium ventrally. The pallium has medial, dorsal and lateral parts. In some fishes, these parts are readily identifiable on topological grounds, but in most fishes, particularly among the ray-finned fishes, migration of neuronal cell groups during development obscures relationships. The pallial regions receive olfactory and ascending sensory inputs, with most of the latter being relayed via diencephalic nuclei. The subpallium consists of several regions, including the striatum and the septal region. Generally speaking, the subpallial structures are the major output regions of the telencephalon for long descending projections back to brainstem regions.

The ratio of brain weight to body weight varies considerably across members of each of the three major groups of fishes. For a number of fish species with relatively large brains, the telencephalon makes a major contribution to the size differential. The brain/body ratios are usually graphed as a log function and represented for each group as a minimum convex polygon, as shown in Figure 8.3 for jawless fishes, cartilaginous fishes and ray-finned fishes in comparison with other vertebrate groups (Jerison, 1973; Northcutt, 1977, 1978). Each polygon is the outer boundary of the set of ratios for the individual species of each group. This method of making relative comparisons is called allometry.

Note that the convex polygon for the brain/body ratios of ray-finned fishes overlaps that for diapsid reptiles (which include lizards, snakes and crocodiles) and turtles. Moreover, the upper surface of the ray-finned fish polygon is considerably higher than that of the reptile polygon, which indicates that the brains

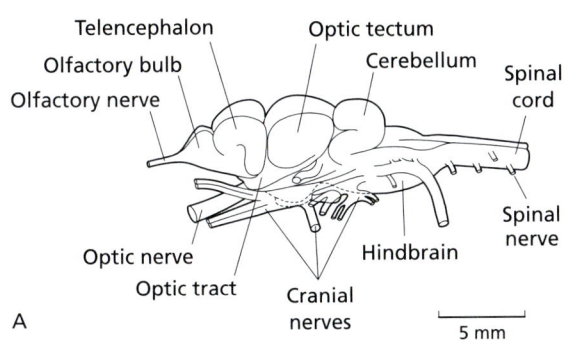

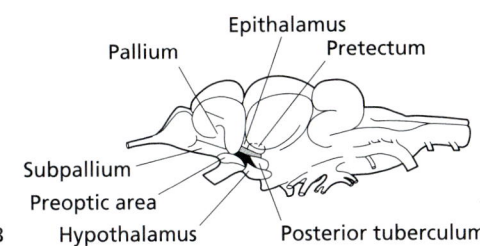

Figure 8.1 Drawing of a lateral view of the brain of a ray-finned fish, the longnose gar (*Lepisosteus osseus*). Rostral is toward the left. (A) Major parts of brain and cranial nerves. (B) Major divisions of telencephalon and diencephalon. (Adapted from Butler and Hodos (1996) after Northcutt and Butler (1976).)

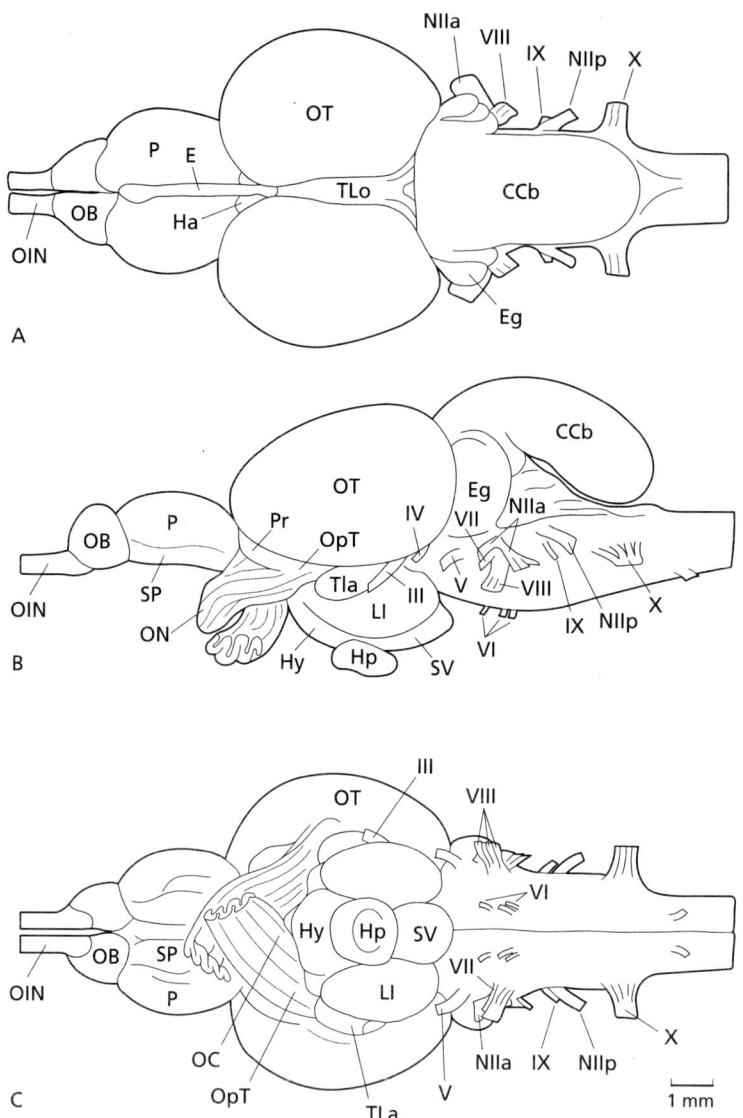

Figure 8.2 Drawings of (A) dorsal view, (B) lateral view and (C) ventral view of the brain of the rainbow trout *Salmo gairdneri*. Rostral is toward the left. (For abbreviations see list on page 129.) (Adapted from Nieuwenhuys et al., 1998.)

of a number of ray-finned fishes are actually heavier relative to a given body weight than the brains of reptiles.

The convex polygon for jawless fishes lies below that of any jawed vertebrate. Nevertheless, extant jawless fishes also exhibit a range of variation in brain development, with a major portion of this variation provided by telencephalic size (Platel and Delfini, 1981, 1986; Wicht, 1996). In contrast to jawless fishes, most cartilaginous fishes have brain/body ratios that exceed those of the ray-finned fishes. The brains in the upper range of these ratios actually overlap those of some mammals and birds (Northcutt, 1977, 1978).

This **allometric analysis** reveals an important, repeating pattern in brain evolution, as recognized by Northcutt (1981a). In each major vertebrate group, including each of the major groups of extant fishes, brain enlargement and elaboration – most often involving the telencephalon – has occurred independently for some species and been strongly selected for. Other species have been evolutionarily successful with brains that are relatively smaller and more simply organized. The wide range of variation within each group mirrors the wide range of niches that can be successfully exploited.

The telencephalon in all fishes comprises the olfactory bulbs rostrally, the telencephalic pallial regions and the subpallial regions. In teleosts, the olfactory bulbs are either sessile, i.e. they lie juxtaposed to the rostral end of the telencephalic

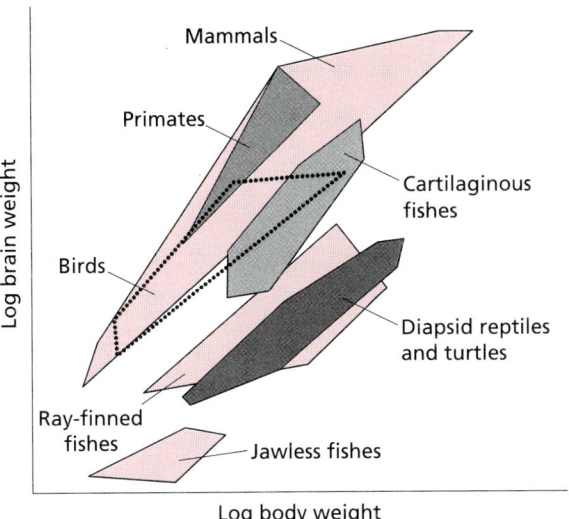

Figure 8.3 Log brain weight plotted as a function of log body weight for 198 species of vertebrates. Minimum convex polygons enclose all the data points for each major group: jawless fishes, cartilaginous fishes, ray-finned fishes, diapsid reptiles (which include lizards, snakes and crocodiles) and turtles, mammals (with that for primates indicated separately) and birds. Data points that lie along the long axis of a polygon are for species with a brain weight/body weight ratio that is representative for the group. Data points at the top or bottom of a polygon are for species with brain weights greater or less than expected for their body weights, respectively, within the group. (Adapted from Butler and Hodos (1996) after Jerison (1973) and Northcutt, 1977, 1978.)

hemispheres, as in trout (*Salmo*), eels (*Anguilla*) and sticklebacks (*Gasterosteus*), or stalked, i.e. they lie at a distance from the hemispheres and are connected by an elongated olfactory tract, as in the goldfishes (*Carassius*), catfishes (*Ictalurus*) and mormyrids (*Gnathonemus*) (Nieuwenhuys et al., 1998). Among jawless vertebrates (Figure 8.4), the olfactory system input to the pallium is extensive in both lampreys (Northcutt and Puzdrowski, 1988) and hagfishes, where it dominates the bulk of the enlarged telencephalic pallium (Wicht and Northcutt, 1993). In jawed fishes, olfactory projections are restricted to the topologic lateral pallial region (Ebbesson and Heimer, 1970; Levine and Dethier, 1985), leaving other pallial regions available for the receipt of a wide variety of other sensory inputs – a finding which contradicts the myth of fishes having only a 'smell-brain'. Among cartilaginous fishes (Figure 8.5), the telencephalon is markedly enlarged in some sharks and in skates and rays, exhibiting multiple lobes (Northcutt, 1977; Smeets, 1990). Similar extensive variation occurs in ray-finned fishes (Figure 8.6), from the smooth-surfaced, rostrocaudally elongated telencephalons of bichirs to the elaborated, multilobed telencephalons of many teleosts (Northcutt and Davis, 1983; Nieuwenhuys and Meek, 1990).

The diencephalon (Figures 8.4–8.6) lies between the telencephalon and the midbrain. It varies in size

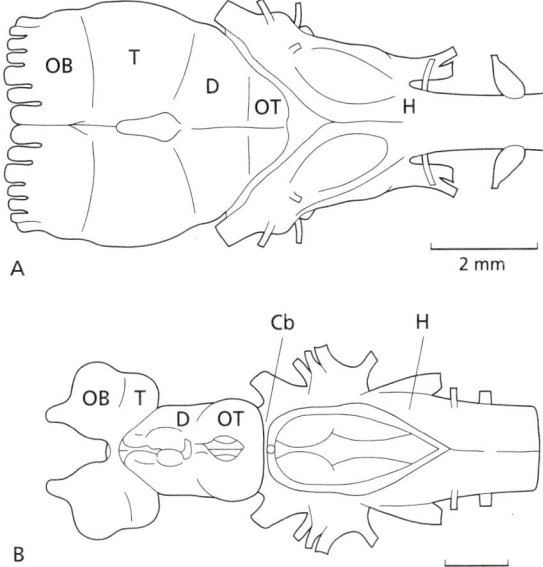

Figure 8.4 Drawings of dorsal views of the brains of (A) a Pacific hagfish (*Eptatretus stouti*) and (B) a silver lamprey (*Ichthyomyzon unicuspis*). Rostral is toward the left. (For abbreviations see list on page 129.) (Adapted from Butler and Hodos (1996) after Wicht and Northcutt (1992) and Northcutt and Puzdrowski (1988), respectively.)

partly in proportion to the telencephalon but also in regard to the degree of development of its most ventral part, the hypothalamus. In most fishes, the hypothalamus has extensive inferior lobes that form large bulges

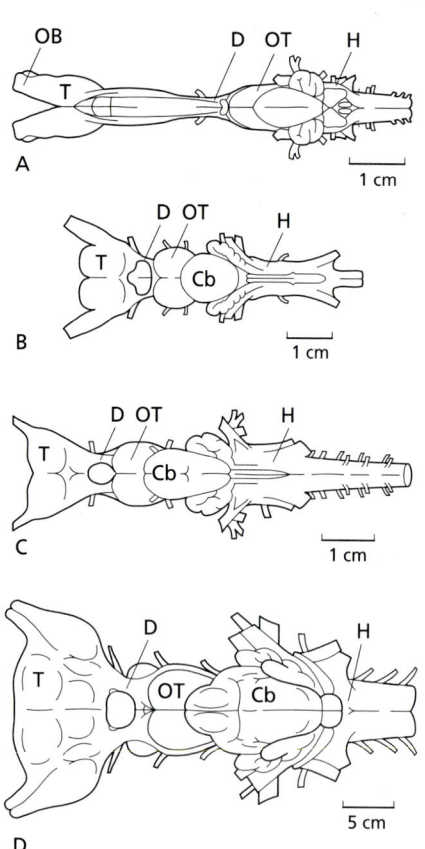

Figure 8.5 Drawings of dorsal views of the brains of (A) a chimaera (*Hydrolagus colliei*), (B) a cow shark (*Notorynchus maculatus*), (C) spiny dogfish shark (*Squalus acanthias*) and (D) a clearnose skate (*Raja eglanteria*). For the latter three species, the olfactory bulbs are at the distal ends of long olfactory tracts and are omitted from this figure. Rostral is toward the left. (For abbreviations see list on page 129.) (Adapted from Butler and Hodos (1996) with A and B after Northcutt (1978) and C and D after Northcutt (1979).)

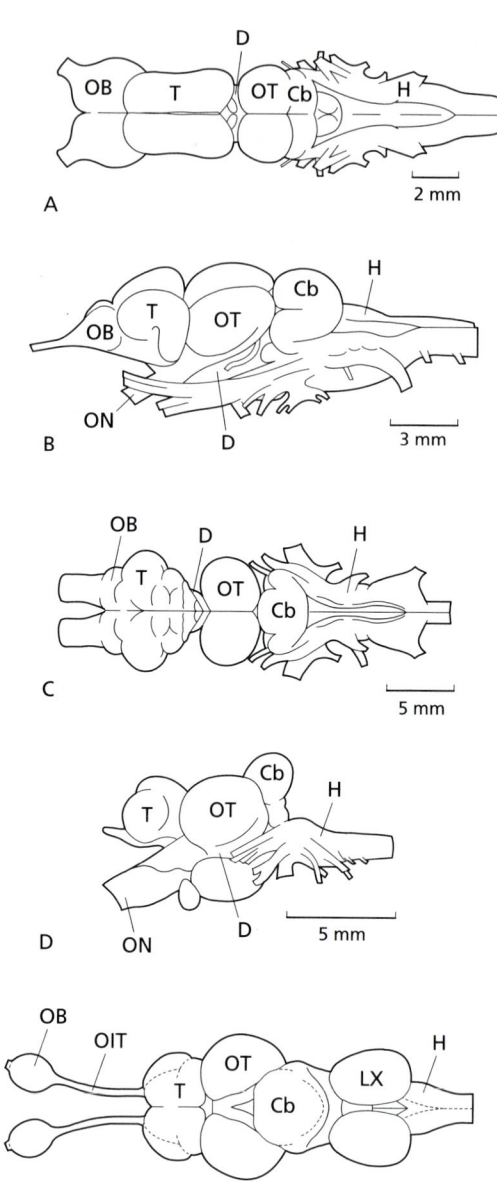

Figure 8.6 Drawings of (A) a dorsal view of the brain of a bichir (*Polypterus palmas*), (B) a lateral view of the brain of a longnose gar (*Lepisosteus osseus*), (C) a dorsal view of the brain of the bowfin (*Amia calva*), (D) a lateral view of the brain of a sunfish (*Lepomis cyanellus*) and (E) a dorsal view of the brain of a goldfish (*Carassius auratus*). Rostral is toward the left. (For abbreviations see list on page 129.) (Adapted from Butler and Hodos (1996) after Reiner and Northcutt (1992), Northcutt and Butler (1976), Butler and Northcutt (1992), Northcutt and Butler (1991) and Nieuwenhuys and Meek (1990), respectively.)

on the ventral surface of the brain (Figure 8.2). The diencephalon contains a number of nuclei situated within the dorsal thalamus, hypothalamus and posterior tuberculum that relay visual, auditory, gustatory, lateral line and other sensory information to the telencephalon (see Butler and Hodos, 1996; Nieuwenhuys *et al.*, 1998). Its most dorsal part, the epithalamus, contains the **habenular** nuclei, which form a small ridge on the surface of the diencephalon, and the epiphysis, which consists of the photoreceptive pineal organ in cartilaginous and ray-finned fishes. The **epiphysis** in lampreys additionally comprises a photoreceptive parapineal organ. The epiphysis in lampreys is the major light-sensing system during their larval period, while in contrast, hagfishes lack any epiphyseal structures (Nieuwenhuys *et al.*, 1998).

Midbrain

The midbrain consists of the tectum, which means 'roof', dorsally and a tegmentum ventrally. The tectum is divided into several parts. The optic tectum is the most rostral and dorsal part of the tectum. It is a major visual structure (Northcutt, 1983; Ebbesson, 1984; Vanegas et al., 1984) that is particularly concerned with the location of objects in space. The optic tectum forms a single large lobe in most fishes – one on each side of the midline (Figures 8.1, 8.2, 8.4–8.6) – but in some ray-finned fishes, such as some of the clupeomorphs (herrings), the tectum is a more elaborate bilobed structure (Butler, 1992). In the latter cases, distortion of the internal structure occurs, but the topologic relationships of circuitry are preserved (Saidel and Butler, 1991).

The more ventral torus semicircularis (torus means 'bulge'), which is a major auditory and lateral line structure (Boord and Northcutt, 1982; Echteler, 1984; Barry, 1987; and see Bullock and Heiligenberg, 1986), is concealed from the surface view but is present in all fishes. Unique to ray-finned fishes are two other tori, the torus longitudinalis, which receives inputs from several sources, projects to the optic tectum, and lies at the medial extent of the tectal lobe (Ito and Kishida, 1978; Northmore et al., 1983), and the torus lateralis, which lies on the lateral surface of the midbrain ventral to the optic tectum. The latter structure is known to project to the telencephalon in reedfishes (bichirs) (Northcutt, 1981b) and is quite large in these fishes and in sturgeons and gars. It is particularly large in the bowfin *Amia calva* (Nieuwenhuys and Pouwels, 1983) but is only of modest size in teleosts (e.g. Parent et al., 1978).

The midbrain tegmentum lies ventral to the tectum and its associated tori. Its ventral surface is visible on the ventral aspect of the brain caudal to the hypothalamus of the diencephalon. It contains a number of nuclei as well as numerous fiber tracts that interconnect more rostral and caudal brain regions. A more caudal isthmal region forms a transitional zone between the midbrain proper and the hindbrain.

Hindbrain

The hindbrain consists of a rostral part (metencephalon) that contains the ventrally lying pons (pontine tegmentum) with the cerebellum dorsal to it and a caudal part, the medulla (myelencephalon). Like the midbrain tegmentum, the pons and medulla contain numerous nuclei and fiber tracts. The cerebellum exhibits an extreme degree of variation in its relative development across fishes (see Butler and Hodos, 1996; Nieuwenhuys et al., 1998). Jawless fishes have only a small region in the dorsolateral part of the hindbrain that can be identified as cerebellar tissue. The cerebellum in cartilaginous fishes varies considerably in size and external form across species (Smeets, 1990). It contains rostrocaudally oriented, cylindrical cell columns called prominentiae granulares that are covered dorsally by the dome of the corpus cerebelli. The cerebellum in ray-finned fishes is also characterized by a similar pair of cell columns, the eminentiae granulares, with a dorsally lying corpus. The degree of cerebellar development varies markedly across species of ray-finned fishes (Figures 8.7 and 8.8).

Both the cerebellar corpus and a more medial structure, the valvula, are extremely enlarged in electrosensory ray-finned fishes, such as mormyrids (Meek, 1992). In these fishes (Figures 8.7 and 8.8) the cerebellum is hypertrophied to such an extent that it overflows the dorsal surfaces of the midbrain and forebrain, concealing them from view. The valvula itself is so enlarged that it protrudes caudally beyond the hemispheres of the cerebellar corpus. A structure closely related to the cerebellum, the electrosensory lateral line lobe, is also greatly enlarged in electrosensory fishes. This lobe is a hypertrophied part of the octavolateralis area, which is a set of nuclei in the brainstem that receives octaval (vestibulocochlear) and lateral line inputs. In mormyrids, for example, the electrosensory lateral line lobe is of substantial size and is interconnected with the correspondingly large valvula via the torus semicircularis (Bell et al., 1981; Finger et al., 1981; Meek et al., 1986). Electrosensory cartilaginous fishes likewise have elaborated portions of their brainstems, such as the large electric lobe present in the electric ray *Torpedo* (Smeets, 1990).

In the dorsal part of the hindbrain caudal to the corpus of the cerebellum is the set of neurons that receives gustatory input via the VIIth (facial), IXth (glossopharyngeal) and Xth (vagus) cranial nerves. This nucleus – referred to either as the nucleus of the solitary tract or the gustatory nucleus – cannot be detected grossly in most fishes. In some teleosts, however – particularly cyprinids (carps) and ictalurids (catfishes) – the gustatory region is hypertrophied into two to three lobes on each side (Finger, 1983, 1988). The rostral part of the region enlarges to form a pair of facial lobes (Figure 8.7), which in cyprinids fuse across the midline. The facial lobes are even more elaborate in ictalurids, in which they are unfused and comprise six lobules on each side. A second pair of lobes, the glossopharyngeal lobes, are present in an intermediate position in cyprinids (Morita et al., 1983). Another pair of lobes, the vagal lobes

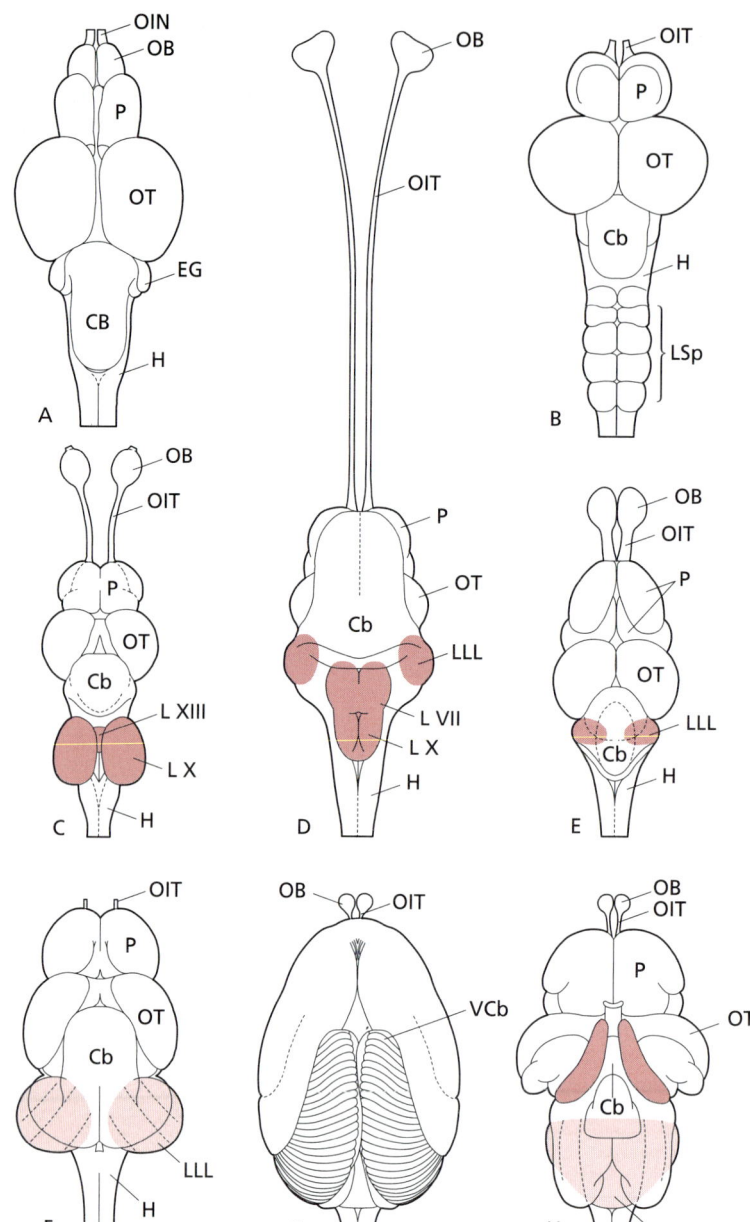

Figure 8.7 Drawings of (A) dorsal view of the brain of the rainbow trout *Salmo* in comparison with dorsal views of the brains of (B) the sea robin *Trigla*, which has specialized spinal accessory lobes for its free sensory pectoral fin rays (see Nieuwenhuys et al., 1998), (C) the goldfish *Carassius*, (D) the catfish *Clarias*, (E) the electrosensory osteoglossomorph fish *Xenomystus*, (F) the weakly electric gymnotid fish *Eigenmannia*, (G) the weakly electric mormyrid fish *Gnathonemus* and (H) *Gnathonemus* with the large valvula cerebelli removed to show underlying brain structures. Rostral is toward the top. (For abbreviations see list on page 129.) (Adapted from Nieuwenhuys et al. (1998).) Oblique hatching to right below indicates specialized viscerosensory (gustatory) regions, and oblique hatching to left below indicates specialized electrosensory regions for passive electroreception (unbroken lines) or active electroreception (dashed lines).)

(Figure 8.7), lie further caudally and are substantially enlarged in cyprinids in correlation with their receiving input from the palatal organ, a specialized chemosensory structure in the oropharyngeal cavity (see Finger, 1988). A more dramatic and apparently unique vagal lobe specialization occurs in the osteoglossid fish *Heterotis niloticus*, which has paired epibranchial organs in the pharyngeal region that aid

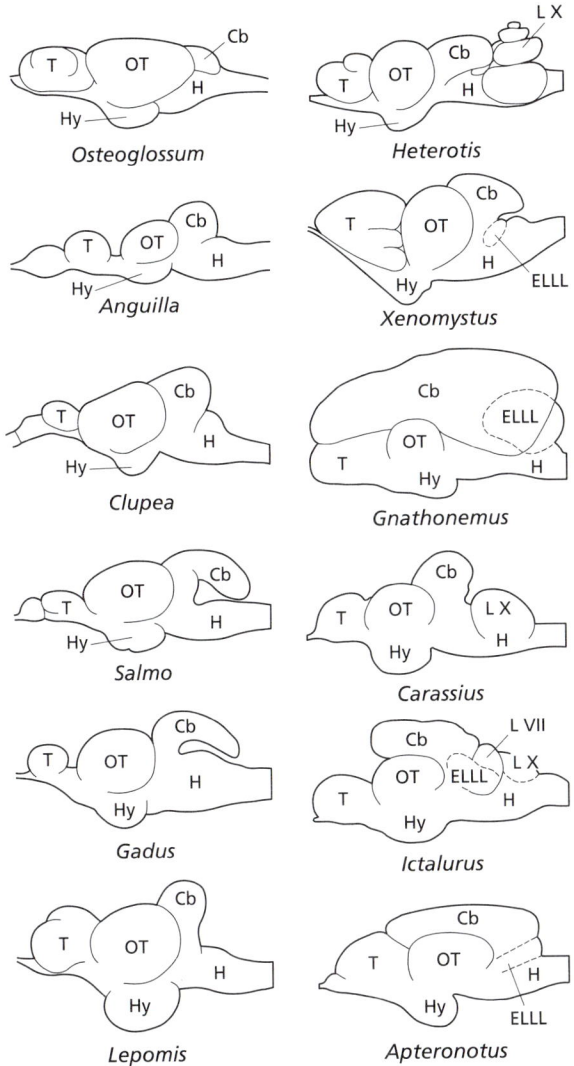

Figure 8.8 Drawings of lateral views of the brains of the osteoglossomorph fishes *Osteoglossum* (arawana), *Heterotis* (arapaimid), *Xenomystus* (African knifefish) and *Gnathonemus* (elephantnose fish); the elopomorph fish *Anguilla* (freshwater eel); the clupeomorph fish *Clupea* (herring); and the euteleost fishes *Salmo* (trout), *Gadus* (cod), *Lepomis* (sunfish), *Carassius* (goldfish), *Ictalurus* (channel catfish) and *Apteronotus* (apteronotid eel). Rostral is toward the left. (For abbreviations see list on page 129.) (Adapted from Nieuwenhuys *et al.* (1998).)

in concentrating and swallowing food and are densely studded with taste buds. These organs are coiled into a spiral, and, remarkably, the epibranchial portion of each vagal lobe that receives the gustatory input from this organ is likewise elaborated into a spiral (Braford, 1986).

Cranial nerves

As many as 22 cranial nerves can be recognized in most fishes (Butler and Hodos, 1996). Some of the cranial nerves are purely sensory, some are purely motor, and the rest are mixed nerves with both sensory and motor components (Table 8.1). The first four cranial nerves (Te, I, II and Ep) are in the forebrain and are purely sensory. The midbrain has two motor nerves (III and IV), and the remaining sensory, motor and mixed cranial nerves are associated with the hindbrain, including the gustatory components of VII, IX and X, which can arguably be regarded as individual nerves.

The 22 cranial nerves are present in lampreys, cartilaginous fishes and ray-finned fishes. Hagfishes lack a visual cranial nerve, the epiphyseal (Ep), and also lack extraocular muscles and their three motor cranial nerves (III, IV and VI). All fishes lack the distinct accessory (XI) and hypoglossal (XII) cranial nerves that are present in tetrapods in association with their innervated neck and tongue muscles. Each of the cranial nerves and/or their individual components have traditionally been classified as to whether it is somatic or visceral, afferent (sensory and with axons incoming to the brain) or efferent (motor or effector to glands and with axons outgoing from the brain), and general or special. The latter pair of terms can now be discarded in light of recent new information on the embryological origin of head musculature and the sensory bipolar neurons of most of the cranial nerves. Due to the widespread use of the traditional classification scheme, it is presented here along with the simpler new scheme that can replace it (Table 8.1). The list presented in Table 8.1 is not necessarily exhaustive, as comparison with all the components of cranial nerves in mammals might indicate (see Parent, 1996; Butler, 2000), but includes the major components as they are currently understood.

Studies of the embryological development of the head and cranial nerves have provided key information on the relationships of the various structures (Noden, 1991; Gilland and Baker, 1993; Northcutt, 1993, 1996; Webb and Noden, 1993; and see Walker and Liem, 1994 and Butler and Hodos, 1996). A population of cells known as neural crest arise at the dorsolateral aspect of the developing neural tube and contribute to a wide variety of tissues in the head and body, including the bipolar sensory neurons for all of the somatic and autonomic ganglia of spinal nerves. These neural crest cells also contribute to the sensory ganglia of most of the cranial nerves as do cells derived from neurogenic placodes, which are patches of thickened ectoderm in the head region (see Noden, 1991; Webb and Noden, 1993; Northcutt, 1996). The bipolar neurons that form the sensory ganglia of cranial

TABLE 8.1: Cranial nerve components in fishes

Nerve number or letter	Nerve name	Component (and innervation)	Traditional classification	New classification
T	Terminal	sensory: ?	—	SA
I	Olfactory	sensory: smell (nasal septum)	—	SA
II	Optic	sensory: light (retina)	—	NTA
E	Epiphyseal	sensory: light (epiphysis)	—	NTA
III	Oculomotor	motor (extraocular muscles)	GSE	SE
		parasympathetic (intraocular muscles)	GVE	VE
IV	Trochlear	motor (extraocular muscle)	GSE	SE
P	Profundus	somatosensory (upper face region)	—	SA
V	Trigeminal	somatosensory (face region)	GSA	SA
		motor (visceral arch muscles)	SVE	SE
VI	Abducens	motor (extraocular muscle)	GSE	SE
VII$_D$	Dorsal facial	motor (visceral arch muscles)	SVE	SE
VII$_{VL}$	Ventrolateral facial	sensory: gustatory (anterior oral cavity, lips and body surface)	SVA	VA + SA
VIII	Vestibulocochlear	sensory: vestibular and auditory (semicircular canals, utriculus, sacculus and lagena)	SSA	SA
LL (6)	Lateral line	sensory: electrosensory and/or mechanosensory (lateral line receptors)	—	SA
IX$_D$	Dorsal glossopharyngeal	motor (visceral arch muscles)	SVE	SE
IX$_{VL}$	Ventrolateral Glossopharyngeal	sensory: gustatory (first branchial arch region)	SVA	VA
X$_D$	Dorsal vagus	motor (visceral arch muscles)	SVE	SE
		parasympathetic (viscera of body cavities)	GVE	VE
		sensory: visceral (viscera of body cavities)	GVA	VA
X$_{VL}$	Ventrolateral vagus	sensory: gustatory (remainder of branchial arch region)	SVA	VA

nerves Te, I, Pro, V, VII, VIII, IX and X and the six lateral line nerves are thus all derived from neurogenic placodes and/or neural crest.

The traditional classification of sensory cranial nerve components has included the two basic categories also found in spinal sensory nerves – general somatic afferent (GSA) and general visceral afferent (GVA), with the term 'general' being used to distinguish these components from two additional 'special' components of head cranial nerves. The GSA category has been assigned to the trigeminal nerve (V) with its innervation of the skin and related structures of the head, while the GVA category has been assigned to the afferent fibers of cranial nerve X that supply sensation from the viscera of the body cavities. The 'special' category has been applied to the gustatory components of nerves VII, IX and X (see Finger, 1993) as special visceral afferent (SVA) and to cranial nerve VIII as special somatic afferent (SSA). In fishes, the SSA designation also would be applicable to the lateral line nerves, but these nerves are usually ignored in mammal-based classification schemes. The recent findings that the sensory ganglion cells for all of these components – GSA, GVA, SVA and SSA – are similarly derived embryologically obviates the need to differentiate between general and special categories and one can then collapse this traditional (and unwieldy) scheme to the two simple categories of somatic afferent (SA) and visceral afferent (VA). Somatic afferent fibers innervate ectodermally-derived tissues and muscle tissue derived from the somatic part of the mesoderm, while visceral afferent fibers innervate endodermally-derived tissues and muscle tissue derived from the visceral part of the mesoderm.

The traditional classification of motor cranial nerve components likewise has included the two basic categories also found in spinal motor nerves – general somatic efferent (GSE) and general visceral efferent (GVE) – as well as one 'special' category, special visceral efferent (SVE). In fishes, the only muscles of the head traditionally recognized as somatic are the extraocular muscles innervated by cranial nerves III, IV and VI. Two cranial nerves have GVE components – parasympathetic innervation of the eye via cranial nerve III and of the viscera of the body cavities via cranial nerve X. The SVE category has been assigned to nerve components that innervate the muscles of the visceral arches.

The visceral arches are arched skeletal structures formed by neural crest in the pharyngeal and gill region (see Walker and Liem, 1994). A series of pharyngeal pouches form between the arches and all except the first pouch open to the surface to form the gills (see Chapter 9). The first and second visceral arches are the mandibular arch and the hyoid arch; they are associated with the muscles of the jaw and of the hyoid apparatus, respectively. The remaining arches are called branchial arches in reference to the gills and their associated muscles.

The embryonic origin of the muscles of the visceral arches was long believed to be of 'visceral' origin. In the body during development, the mesoderm is divisible into three parts – the dorsally located epimere, the ventrally located hypomere and the mesomere in an intermediate position (Figure 8.9). The latter

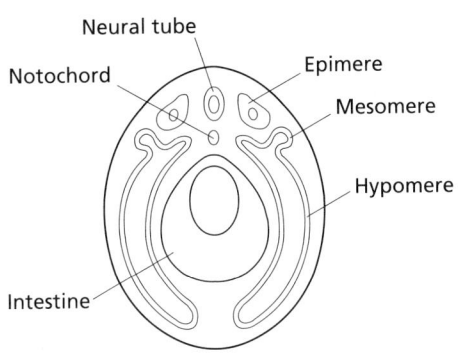

Figure 8.9 Schematic transverse section through a vertebrate embryo. The mesoderm consists of epimere dorsally, hypomere ventrally and mesomere in an intermediate position. Dorsal is toward the top. (Adapted from Butler and Hodos (1996).)

forms the kidney, while the hypomere forms the smooth muscle of the gut and cardiac muscle. The epimere forms somites, which are segmentally arrayed masses of tissue that give rise to the striated skeletal muscles of the body wall. Recent work of Noden (1991) has demonstrated that the epimere in the head gives rise to a series of partially fused somites, called somitomeres, as well as several somites in the caudal head region. These somitomeres and somites give rise to the extraocular muscles and to all the muscles of the visceral arches. Thus, all of these muscles are of somatic origin. The hypomere does not contribute to them. These recent findings allow us to discard the special visceral efferent category – and thus also the 'general' category used in contrast to it – and collapse the traditional scheme to the two simple categories of somatic efferent (SE) and visceral efferent (VE). Somatic efferent fibers innervate the somatic muscles of the head derived from somitomeres and somites, while visceral efferent fibers supply parasympathetic innervation to the eye and the viscera of the body cavities. The nuclei with the neuronal cell bodies that give rise to the nerves that innervate muscles of the visceral arches (V, VII, IX and X) are laterally displaced within the brainstem with respect to the more medially lying oculomotor nuclei (III, IV and VI), but this anatomical difference is most probably related to the evolutionarily more recent acquisition of the oculomotor system versus the acquisition of visceral arch musculature in the earliest vertebrates.

Using these new insights the cranial nerves will now be discussed with reference to both their traditional classification and the new, simpler classification. The only two exceptions to both the traditional and new classification schemes are the two visual cranial nerves, which have sensory neurons that are derived from the neural tube rather than from placodes or neural crest and are actually part of the brain itself. The motor and sensory nuclei of the cranial nerves, while not grossly visible in most cases, will be mentioned here for the sake of continuity with the material on central pathways covered in Chapter 20.

The terminal nerve (Te) was first described in cartilaginous fishes by Locy in 1905 (see Nieuwenhuys et al., 1998). Its bipolar neurons have free nerve endings located in the nasal mucosa, but their modality (possibly chemosensory or tactile) has not yet been determined. These sensory neurons are derived from the olfactory placode and project to regions in the ventral forebrain and – uniquely in ray-finned fishes – to the retina (Demski and Northcutt, 1983; Springer, 1983).

The olfactory nerve (I) is composed of the efferent processes of the olfactory bipolar cells. The nerve itself is visible grossly in fishes with sessile olfactory bulbs (Figure 8.1). The axons of the next set of neurons in the system, which lie in the olfactory bulb, form the olfactory tract, which courses caudally into the telencephalon proper. The latter is elongated and thus particularly noticeable in fishes with rostrally displaced olfactory bulbs (Figure 8.7).

The optic nerve (II) is the large cranial nerve that arises from the retina. The retinal receptor cells project to retinal bipolar cells that occupy a place in the sensory relay chain comparable to that of the sensory bipolar neurons of other cranial nerves, but they are derived from the neural tube rather than from neural crest and/or placodes, and they lie within the central nervous system rather than in a peripheral ganglion. The set of retinal axons that correspond to other cranial *nerves* are actually the axons of these bipolar neurons. The terminology differs for this visual system, however, so that the initial portions of the axons of the next set of neurons in the sensory relay chain, the retinal ganglion cells, are referred to as the optic nerve. The optic nerve courses caudally and medially from the surface of the orbit. The two optic nerves meet and intermingle at the optic chiasm (Figure 8.2), where most of the axons decussate (i.e. cross over) to the opposite side. Caudal to the optic chiasm, these same retinal ganglion cell axons are referred to as the optic tract (Figures 8.1 and 8.2). They project to multiple sites in the diencephalon – including nucleus anterior of the dorsal thalamus and several pretectal nuclei – and to the optic tectum (e.g. Repérant *et al.*, 1976; Springer and Landreth, 1977; Northcutt and Butler, 1991; and see Butler and Hodos, 1996).

A second visual nerve is present in the diencephalon in lampreys and jawed fishes – the epiphyseal (Ep) nerve. This nerve arises from epiphyseal nerve cell bodies that receive visual input from photoreceptors within the pineal organ. This system is involved with the regulation of the dark–light circadian cycle (Ekström, 1987). The pineal input is relayed to the habenular nuclei of the epithalamus and also to a number of nuclei that also receive visual input from the retina (see Nieuwenhuys *et al.*, 1998). As in the retinal visual system, the epiphyseal receptors and nerve arise embryologically from the neural tube rather than from neural crest and/or placodes and are located in the central nervous system. Both the optic and epiphyseal nerves can be classified as neural tube afferents (NTA, Table 8.1).

Three nerves (III, IV and VI) are oculomotor nerves that innervate the extraocular muscles. Some differences occur in the innervation and muscular pattern between lampreys and ray-finned fishes (Fritzsch *et al.*, 1990); the description given here is for the latter group (e.g. Luiten and Dijkstra-de Vlieger, 1978; Leonard and Willis, 1979). The oculomotor nerve (III) exits the brain near the ventral midline and innervates most of the extraocular muscles – the medial rectus, inferior rectus and inferior oblique on the ipsilateral side and the superior rectus on the contralateral side. The trochlear nerve (IV) uniquely exits the brain from its dorsal surface and, also uniquely, completely decussates to the contralateral side. It innervates the contralateral superior oblique muscle. The abducens nerve (VI), like the IIIrd cranial nerve, exits the brain near the ventral midline. It innervates the ipsilateral lateral rectus muscle (Szekley and Matesz, 1993). These nerve components are classified as (general) somatic efferent. The oculomotor nerve additionally has a (general) visceral efferent component that arises from the Edinger-Westphal nucleus (within the oculomotor nuclear complex) and provides parasympathetic innervation via the ciliary ganglion to intrinsic eye muscles for accommodation and pupillary light reflexes (Wathey, 1988). A specialized portion of the somatic efferent component of the oculomotor nerve is present in the stargazer *Astroscopus*, which has an electric organ in the roof of the skull that is derived from extrinsic eye muscles (see Nieuwenhuys *et al.*, 1998). A group of cells in the vicinity of the oculomotor nucleus proper projects to this organ (Leonard and Willis, 1979).

Sensory innervation to the head and motor innervation to mandibular muscles are provided by two cranial nerves present in fishes. Profundus and trigeminal placodes contribute neurons to the sensory ganglia of the respectively named nerves (Noden, 1991). In agnathans, some teleosts and members of the sarcopterygian radiation, these ganglia fuse, whereas they remain separate in cartilaginous and some ray-finned fishes (Northcutt and Bemis, 1993). The profundus nerve (Pro) is a purely sensory cranial nerve that may correspond to the deep ophthalmic branch of the trigeminal nerve in mammals (Northcutt, 1979). In the coelacanth *Latimeria* (Crossopterygii), the profundus nerve is known to innervate the skin of the snout and the mucosal walls of rostrally-lying tubular structures that house lateral line receptors (which are separately innervated by a ramus of the anterodorsal lateral line nerve) (Northcutt and Bemis, 1983).

The trigeminal nerve (V) emerges from the lateral aspect of the metencephalon (Figure 8.2) and runs rostrally. Along with the profundus nerve, the trigeminal nerve's somatic afferent component provides sensory innervation for the head region, including part of the oral cavities. The latter innervation could arguably be classified as visceral afferent, but it is customarily lumped with the somatic afferent trigeminal component. At least some trigeminal afferent fibers are also sensitive to magnetic intensity, as recently demonstrated in some teleost fishes (Walker et al., 1997). Trigeminal afferent neurons terminate within several sensory nuclei that form an elongated, rostrocaudally aligned cell column – a trigeminal sensory nucleus, a nucleus of the descending trigeminal tract, a spinal trigeminal nucleus and a medial funicular nucleus in the upper spinal cord (Puzdrowski, 1988).

The trigeminal nerve also contains a somatic efferent component (traditionally classified as special visceral efferent) to the muscles of the first visceral arch (Song and Boord, 1993). This component arises from neurons in the trigeminal motor nucleus and innervates the ipsilateral adductor and abductor muscles of the mandible (Luiten, 1976; Gorlick, 1989). This trigeminal motor innervation contributes to respiratory movements. In the Siamese fighting fish *Betta splendens*, the trigeminal motor component also innervates the dilator operculi muscle, which is used in aggressive opercular display behavior (Gorlick, 1989, 1990; and see Nieuwenhuys et al., 1998). Agnathans also have a trigeminal motor nucleus; a robust trigeminal motor nucleus is present in lampreys, in which it innervates the muscles used in feeding, including the sucker mouth and the rasping organ (Nieuwenhuys et al., 1998).

Gustatory cranial nerve components are present in three cranial nerves – VII, IX and X – and listed separately in Table 8.1 as VII_{VL}, IX_{VL} and X_{VL} in reference to the ventrolateral series of placodes from which their sensory bipolar neurons arise. The gustatory components emerge from the lateral aspect of the brainstem with the motor components of the three respective nerves (Figure 8.1). They have been traditionally classified as special visceral afferent (see Finger, 1993) but can now be recognized simply as visceral afferent or somatic afferent, depending on the embryonic origin and location of the taste receptor cells that they innervate. The gustatory component of VII innervates taste buds derived from endoderm in the anterior part of the oral cavity (visceral afferent), including the palatal organ in cyprinids (see Finger, 1988) as discussed above and also taste buds derived from general ectoderm (see Northcutt, 1996) on the lips and over the entire surface of the body, which are particularly extensive in some teleosts, such as cyprinids and ictalurids. In these and other teleost fishes, afferent VIIth nerve fibers project to the gustatory nucleus (nucleus of the solitary tract) (Morita et al., 1983; Puzdrowski, 1987; Finger, 1988; Díaz-Regueira and Anadón, 1992; Lázár et al., 1992), which is enlarged to form a facial lobe, as discussed above (Figure 8.7). In cyprinids, a distinct glossopharyngeal lobe is present to which the gustatory component of this nerve projects from the region of the first branchial arch (Morita et al., 1980, 1983). In other teleosts that lack a glossopharyngeal lobe, such as ictalurids, IX_{VL} projects partly to the facial lobe and partly to the lobe associated with X_{VL}, the vagal lobe (Kanwal and Caprio, 1987). The projection to the latter structure is predominantly from X_{VL}, which innervates the remainder of the branchial arch region (see Morita and Finger, 1985), including the specialized epibranchial organ in an osteoglossid fish (Braford, 1986) as discussed above. It should be noted that VII, IX and X also supply tactile and/or proprioceptive innervation to these same respective pharyngeal regions (e.g. see Mauri and Caprio, 1982; Kanwal and Caprio, 1983, 1987).

Vestibulocochlear and lateral line sensory inputs enter the brainstem via the VIIIth cranial nerve and up to six lateral line nerves, respectively. The VIIIth nerve enters the lateral aspect of the brainstem, while a set of three anterior lateral line nerves enter rostral to it (Nlla, Figure 8.2) and a second set of three posterior lateral line nerves enter caudal to it (Nllp, Figure 8.2). The sensory bipolar neurons for all of these nerves arise from a series of dorsolateral placodes in the region (Noden, 1991). The vestibulocochlear nerve traditionally has been classified as special somatic afferent, which would presumably also apply to the lateral line nerves, but the simpler designation of somatic afferent can now be applied to all of these nerves.

The vestibulocochlear nerve innervates the inner ear organs, which consist of semicircular canals that detect angular acceleration, the utriculus that detects displacements caused by gravity and the sacculus and lagena that detect displacements caused by sound waves (Platt, 1983; Popper, 1983; Platt et al., 1989; Popper and Fay, 1993). The afferent VIIIth nerve fibers project to octaval (vestibular and auditory) nuclei in the brainstem.

The lateral line nerves number up to six – anterodorsal, anteroventral, otic, middle, supratemporal and posterior (e.g. Puzdrowski, 1989 and see Butler and Hodos, 1996). The lateral line system is purely mechanosensory in most teleosts and is used in the detection of water movements. In four groups, however – ictalurids, notopterids, gymnotids and mormyrids – an electrosensory lateral line system has also evolved (Bullock *et al*., 1983). The lateral line system in these fishes can detect small electric currents or potentials in addition to water movements (see Nieuwenhuys *et al*., 1998). The lateral line nerves project to lateral line nuclei in the brainstem, which form lobes in some teleosts, as discussed above for mormyrids (Figures 8.7 and 8.8).

Lateral line electrosensory receptors are not covered separately in this volume but can be briefly considered here (Zakon, 1986, 1988). Two major types of electroreceptor organs occur – ampullary and tuberous. Ampullary organs generally consist of a jelly-filled, transepidermal canal that ends as a pouch in which the receptor cells are located. The receptor cells lie within an epithelium of support cells. The receptor cells of tuberous organs occur in a capsule that is situated beneath the epidermis, but in contrast to ampullary organs, the canal of the tuberous organ contains loosely coupled epithelial cells, and contact with the aqueous environment occurs via the extensive extracellular space. Electroreceptor morphology varies across species. Mormyrids, for example, have two types of tuberous organs – Knollenorgans in which each receptor cell is encased in its own capsule and mormyromasts, which are more complexly organized organs with two types of receptor cells. The various types of electrosensory receptors are innervated by the distal branches of the lateral line nerves.

The motor components of the facial (VII_D), glossopharyngeal (IX_D) and vagal (X_D) nerves, so designated as part of the dorsal component of each nerve, supply visceral arch muscles and emerge from the lateral aspect of the brainstem (Figure 8.2) with their other respective components. As with the motor component of the trigeminal nerve, these nerves traditionally have been classified as special visceral efferent, but due to the derivation of the visceral arch muscles from somitomeres and somites, as discussed above, these nerve components can be reclassified simply as somatic efferent (Finger, 1993). The facial motor component arises from the facial motor nucleus and supplies the adductors and abductors (levator) of the operculum (e.g. Luitin, 1976). The glossopharyngeal and vagal motor components supply the muscles of the first gill arch (third visceral arch) and of the remainder of the gill (branchial) arches, respectively, which are all involved in respiratory and related movements (Kanwal and Caprio, 1987; Morita and Finger, 1987; Díaz-Regueira and Anadón, 1992; Lázár *et al*., 1992). These components arise from neurons within the visceromotor column in the brainstem that as a group appear comparable to the nucleus ambiguus of amniotes (see Kanwal and Caprio, 1987; Nieuwenhuys *et al*., 1998). In cyprinids a distinct group of vagal motor neurons innervate the muscles of the palatal organ and the gill rakers (e.g. Morita and Finger, 1985).

The additional components of the vagus nerve are related to the viscera of the body cavities. The visceral sensory component innervates these viscera and projects to a nucleus called the general viscerosensory nucleus or the nucleus commissuralis of Cajal (Nieuwenhuys *et al*., 1998). The (general) visceral efferent component arises from neurons in the vagal motor nucleus and projects to viscera including the heart, gastrointestinal tract, swimbladder and pneumatic duct (Morita and Finger, 1987; Hornby and Demski, 1988).

Spinal cord and peripheral spinal nerves

The spinal cord lies within the vertebral canal in most teleosts with some notable exceptions. In the tetraodontiform *Mola*, for example, both the brain and spinal cord are located entirely within the skull (Nieuwenhuys, 1964) and in some other tetraodontiforms it extends only through the rostral part of the canal (Uehara and Ueshima, 1986). In contrast, the spinal cord extends the entire length of the vertebral canal and beyond in the gymnotiform knifefish *Eigenmannia* (Kirschbaum *et al*., 1978).

The spinal cord (Figure 8.10) comprises a series of segments, which can be divided into trunk and caudal regions, as described by Walker and Liem (1994). In lampreys each segment has a pair of dorsal and ventral spinal nerves. The ventral spinal nerves carry (general) somatic efferent fibers and innervate the corresponding segmental myomeres. The dorsal spinal nerves carry (general) somatic afferent fibers from the body and both (general) visceral afferent and (general)

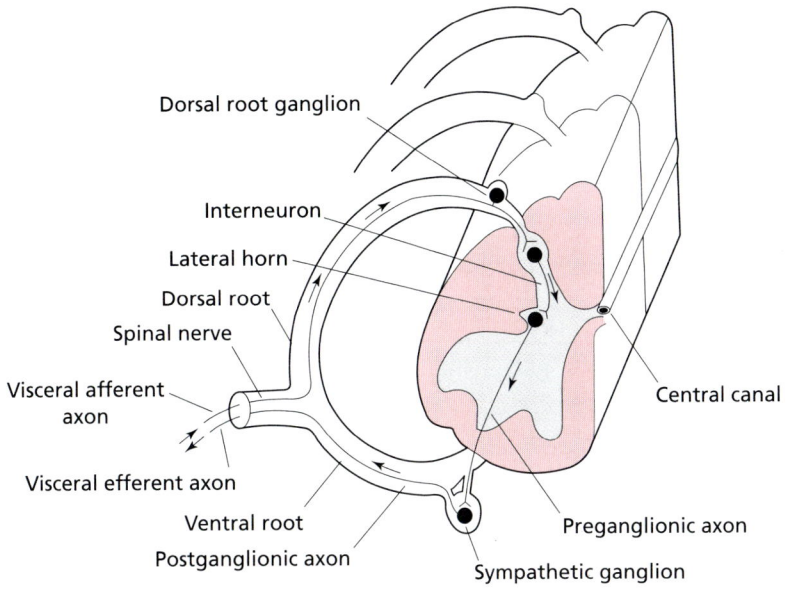

Figure 8.10 Schematic drawing of a transverse section through the left half of a spinal cord segment in a generalized vertebrate. Rostral is toward the upper right. (Adapted from Butler and Hodos (1996).)

visceral efferent fibers for innervation of the viscera of the body cavities. The spinal cord in lampreys is relatively flattened dorsoventrally and is poorly vascularized. In hagfishes and jawed fishes, the spinal cord assumes a larger, more rounded shape and is well vascularized. The dorsal and ventral spinal nerves of lampreys are represented by dorsal and ventral roots which fuse near the spinal cord to form a single pair of spinal nerves for each segment. Some visceral motor fibers exit the spinal cord via the dorsal root as in lampreys, but others exit via the ventral root, as all of them do in amniotes.

Also as discussed by Walker and Liem (1994), the bipolar neuron cell bodies of the somatic and visceral afferent fibers lie within the dorsal root ganglion that lies on the surface of the dorsal root (or the dorsal spinal nerve in lampreys) near the spinal cord. These sensory neurons are all derived from neural crest. The neuronal cell bodies for both the somatic and the visceral efferent fibers lie within the ventral portion of the spinal cord. The somatic efferent fibers directly innervate their respective body wall muscles. The visceral efferent fibers that belong to the sympathetic division of the autonomic nervous system exit the spinal cord and then synapse in the corresponding segmental ganglion, which lies near the cord and gives rise to postganglionic sympathetic fibers that innervate the viscera of the body cavities (see Walker and Liem, 1994).

Meninges and the ventricular system

In all vertebrates the brain and spinal cord are covered by one or more layers of connective tissue, which are derived from neural crest (see Northcutt, 1996). Fishes have only a single meningeal layer, called the primitive meninx (see Butler and Hodos, 1996). This condition contrasts with the two meningeal layers present in amphibians and reptiles and the three meningeal layers present in birds and mammals.

Within the developing neural tube, the ventricular system forms and consists of a rostrocaudal series of spaces (Figure 8.11). Cerebrospinal fluid circulates through the ventricular and meningeal spaces. In agnathans and cartilaginous fishes, a pair of lateral ventricles are present in the developing telencephalic hemispheres, similar to the situation in tetrapods. In hagfishes the continued expansion of the telencephalic hemispheres during embryological development secondarily obliterates the forebrain part of the ventricular system, so that the more caudal part of the ventricular system ends rostrally in a blind pocket within the diencephalon (Nieuwenhuys et al., 1998). In ray-finned fishes, equivalent spaces are

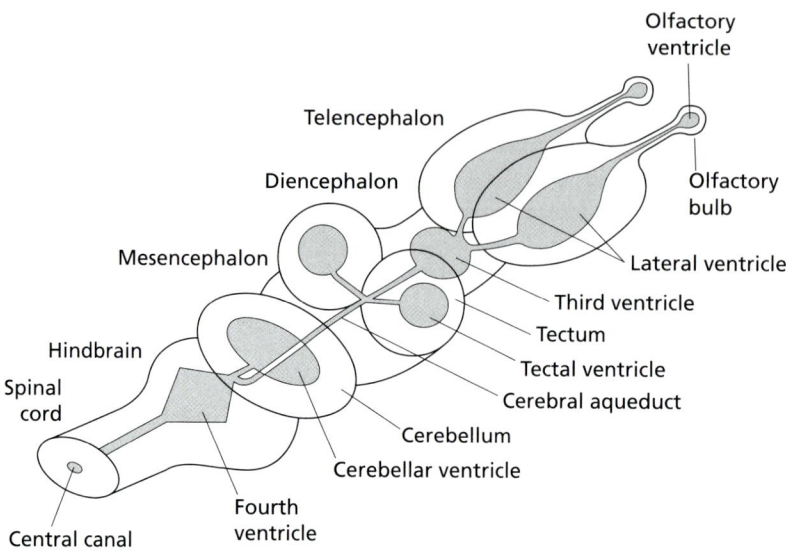

Figure 8.11 Schematic drawing of the ventricular system in a generalized fish brain, with the ventricular spaces projected on to the drawing and indicated by shading. Rostral is toward the upper right.

present in the telencephalon, but they extend over the surface of the hemispheres, covered only by a thin layer of ependyma (see Northcutt and Davis, 1983). The developmental differences that result in this unusual condition and their consequences for the anatomical relationships of cell groups within the telencephalon are discussed below and in Chapter 20.

The ventricular system caudal to the telencephalon is similar in all fishes. A thin, third ventricle is present in the diencephalon and several recesses and/or extensions of the ventricular system are present, depending upon the extent of development of various structures. For example, a sizeable extension of the ventricular system into each of the large inferior lobes of the hypothalamus is present. The ventricular system narrows at the caudal end of the diencephalon to form a narrow passageway, the cerebral aqueduct, that continues through the midbrain. Within the midbrain, a similar extension of the ventricular system into the lobes of the optic **tectum** occurs, forming the sizeable tectal ventricles. In the hindbrain, the cerebral aqueduct opens into the larger space of the fourth ventricle. Extensions and distortions of the ventricular space also occur in the hindbrain in association with the varying degrees of development of the cerebellum, lateral line lobes, vagal lobes and/or facial lobes. The fourth ventricle is continuous caudally with the small, narrow central canal of the spinal cord.

Development of the telencephalon: evagination versus eversion

In jawless and cartilaginous fishes the embryological development of the telencephalon is similar to that in lungfishes and tetrapods (see Northcutt and Davis, 1983; Butler and Hodos, 1996; Nieuwenhuys *et al.*, 1998). The ventricular space extends into the developing telencephalic hemispheres as they expand by the process of evagination (Figure 8.12A). The hemispheres enlarge and bulge out laterally on each side to form the paired lateral ventricles (Figure 8.12B). These ventricles communicate with the third ventricle of the diencephalon via smaller spaces called interventricular foramina. The result of this process of development is that the topographic relationships of the various parts of the pallium – medial, dorsal and lateral – are maintained from their first embryonic appearance to the adult condition.

In ray-finned fishes a different process of telencephalic expansion occurs during embryogenesis (Nieuwenhuys, 1962, 1963). Instead of bowing and bulging outward as in most other vertebrate groups,

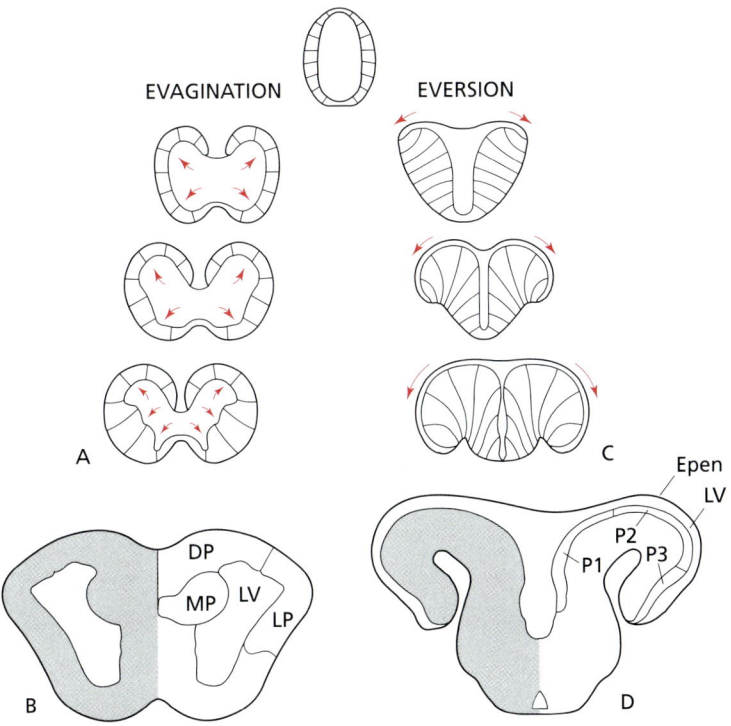

Figure 8.12 (A) Schematic drawing of the developmental process of evagination with transverse sections that are sequential in time and (B) a transverse section through the evaginated telencephalon in the adult shark *Squalus acanthias* with the originally medial part of the pallium maintained in a medial position. (C) Schematic drawing of the developmental process of eversion with transverse sections that are sequential in time and (D) a transverse section through the everted telencephalon in the adult bichir *Polypterus palmas* with the originally medial part of the pallium everted to a lateral position (P3). Dorsal is toward the top. (For abbreviations see list on page 129.) (A and C adapted from Nieuwenhuys *et al.* (1998); B and C adapted from Butler and Hodos (1996) after Northcutt *et al.* (1988) and Northcutt and Davis (1983), respectively.)

the telencephalic hemispheres expand upward and then turn outward by a process called eversion (Figure 8.12C). At the dorsal midline the connecting tissue between the two hemispheres becomes thinned out and elongated to form a simple layer of ependyma (which is the same thin layer of cells that lines the ventricular cavities throughout the brain and spinal cord). As the eversion process continues, the lateral ventricles form dorsal to the outward-turning surface of the hemispheres, enclosed dorsally by only the thin layer of ependyma. In the adult ray-finned fish (Figure 8.12D) the net result of this process is a reversal of topographic relationships even though the topologic, i.e. sequential order, is preserved. The part of the pallium that would come to lie in a lateral position in evaginated brains lies in a medial position in everted brains, and the part that would be maintained in a medial position in evaginated brains lies in a lateral position in everted brains. The reedfish *Polypterus* used for illustration in Figure 8.12A maintains a relatively simple telencephalic structure in the adult, but, as will be discussed in Chapter 20, more extensive cell proliferation and migration within the telencephalon in non-cladistian teleosts secondarily obscures and complicates the primary topologic relationships of the everted condition. An interesting footnote to this developmental aspect of telencephalic anatomy is that in the crossopterygian fish *Latimeria*, the telencephalic pallial regions develop by a complex mix of evagination and eversion (Nieuwenhuys, 1965).

Differences within major groups

The differences that occur both between and within the major groups of fishes are legion. Some examples that can be appreciated at the gross level are given here. The degree of neuronal proliferation and migration in various parts of the brain results in **allometric** differences in brain size relative to body size within each major group and differences in the elaboration of both sensory and motor systems also account for a wide range of variation.

Laminar versus elaborated brains

Within each major group of fishes, as within tetrapods, a wide range of variation exists in the degree to which neurons are generated from the periventricular germinal zone during embryological development and in the degree to which these neurons migrate superficially away from their original periventricular position. Differences in neuronal generation and migration occur throughout the length of the neuraxis in various species, depending upon individual specializations and the elaboration of various systems within the brain. Some of these differences will be examined in greater detail in Chapter 20. In terms of gross anatomical consequences, the overall size of the brain varies considerably within each major group (Northcutt, 1977, 1978, 1981a; Butler and Hodos, 1996), as discussed above concerning allometry. Those species in which relatively little cell migration occurs during embryological development, designated Group I, include lampreys among agnathans, some sharks and the chimaeras among cartilaginous fishes and cladistia, such as *Polypterus*, among ray-finned fishes. Sturgeons and paddlefishes (Chondrostei), gars (Ginglymodi) and the bowfin *Amia calva* (Halecomorphi) could arguably be placed in this group as well, as done in Table 8.2, although members of these groups have more elaborated telencephalons than the cladistia. The Group I fishes have what can be referred to as laminar brain organization and, in general, their brain/body ratios are less than those of fishes with elaborated brains, i.e. brains in which more neuronal proliferation and migration occurs during development. The latter set of fishes, designated as Group II (Table 8.2), includes hagfishes among agnathans, some sharks and the skates and rays among cartilaginous fishes, and the teleosts.

This diversity (Figures 8.2 and 8.4–8.8) that has been independently achieved over evolution (Northcutt, 1981a) reflects several key points concerning brain evolution. First, both simply and complexly organized brains can be evolutionarily successful. Each species occupies its own niche, and selective pressures vary across the range of niches. Second, evolution does not proceed according to the antiquated (but unfortunately not commonly abandoned) idea of a scale of nature leading through a series from jawless fishes to cartilaginous fishes to ray-finned fishes and then on through frogs, rats and monkeys to humans. The telencephalons of cartilaginous fishes are relatively much larger and more elaborated than those of agnathans, ray-finned fishes and reptiles. Among vertebrates, cartilaginous fishes, birds and mammals share the distinction of having the larger brain and body weights. Third, rather than concluding that evolution has direction – from simple to complex – we need to recognize that complexity as well as

TABLE 8.2: Brain organization in Group I (laminar) versus Group II (elaborated) fishes

	Group I (laminar)	Group II (elaborated)
Agnatha	Lampreys	Hagfishes
Chondrichthyes	Squalomorph sharks	Galeomorph sharks
	Squantinomorph sharks	Skates
	Holocephalians (chimeras)	Rays
Actinopterygii	Cladistians (bichirs)	Teleosts
	Chondrosteans (sturgeons and paddlefishes)	
	Ginglymodians (gars)	
	Halecomorphans (bowfin)	
Sarcopterygii	Actinistians (coelacanth)	Mammals
	Dipnoans (lungfishes)	Diapsid reptiles
	Amphibians	Birds
		Turtles

simplicity arise and can be selected for based on the availability of niches: great diversity in available niches allows for great diversity in species' characters, including brain structure.

Relative and differential development of various parts of the brain

Differences among species in the relative development of the brainstem lateral line lobes and of the cerebellar corpus and valvula, particularly in regard to the evolution of electroreception and differences in the relative development of the various lobes in the brainstem that are associated with the gustatory system are dramatic. Likewise, the electromotor system of *Astroscopus* that consists of a modified extraocular muscle innervated by cranial nerve III is a highly divergent and specialized evolutionary achievement. The olfactory bulbs can be pedunculated or sessile. The telencephalon may develop by eversion or evagination with profound topological consequences. The optic tectum may have single or double lobes. The spiral epibranchial organ and spiral vagal lobe of *Heterotis* are a marvel of peripheral–central correlation and specialization. Such variations, as discussed above, illustrate the impressive extent to which brain structure can vary across species. It should not, in fact, be surprising that such variation occurs across the almost 23 000 extant species of ray-finned fishes (Lauder and Liem, 1983), let alone the additional numbers of extant species of cartilaginous and jawless fishes, given the breadth of variation in available niches. As documented in a number of the other chapters in this volume, other systems – skeletal (Chapter 6), respiratory (Chapter 9), integumentary (Chapter 5), and so on – also exhibit marked variation across species.

Most instances of variation seem to involve the parts of the brain derived from the alar plate (i.e. forebrain and sensory structures) rather than from the basal (i.e. motor) plate. Lateral line specializations, including the cerebellum, enlargement of the gustatory lobes, telencephalic developmental variations and tectal lobe variations all involve alar plate structures. Dorsally directed expansion, lobulation and lamination of alar plate components are common. Variation involving basal plate derivatives is less frequent but also of note. The gain of the medially lying column of oculomotor nuclei subsequent to the origin of the earliest vertebrates was a seminal event. The evolution of the electric organ in *Astroscopus* with its distinct pool of oculomotor neurons for innervation represents further specialization within this basal plate column.

At the most fundamental level, the basic plan of the brain across fishes is like that of all vertebrates, with the basic components of the forebrain, midbrain, hindbrain and spinal cord. Similarly, most cranial and spinal nerves of the peripheral nervous system correspond to their counterparts in other vertebrates. The brains of extant fishes are in many ways highly derived *vis-à-vis* current models of ancestral vertebrate brains, with each group exhibiting marked and numerous specializations consistent with selective pressures for their individual niches.

Acknowledgments

The monumental work *The Central Nervous System of Vertebrates* by Nieuwenhuys *et al.* (1998) was of major help in crafting this chapter and was drawn on frequently for its broad and comprehensive perspective. This work was partially supported by NSF grant # IBN 9728155.

References

Barry, M.A. (1987). *J. Comp. Neurol.* **266**, 457–477.
Bell, C.C. (1981). In *Hearing and Sound Communication in Fishes* (eds W.N. Tavolga, A.N. Popper and R.R. Fay), pp. 383–392. Springer-Verlag, Berlin
Boord, R.L. and Northcutt, R.G. (1982). *J. Comp. Neurol.* **207**, 274–282.
Braford, M.R., Jr (1986). *Science* **232**, 489–491.
Bullock, T.H. and Heiligenberg, W. (1986). *Electroreception*. Wiley, New York.
Bullock, T.H., Bodznick, D. and Northcutt, R.G. (1983). *Brain Res. Rev.* **6**, 2–46.
Butler, A.B. (1992) *Brain, Behav. Evol.* **40**, 256–272.
Butler, A.B. (2000). In *Encyclopedia of the Human Brain*, (ed. V.S. Ramachandran). Academic Press, New York (in press).
Butler, A.B. and Hodos, W. (1996). *Comparative Vertebrate Neuroanatomy, Evolution and Adaptation*. Wiley-Liss, New York.
Butler, A.B. and Northcutt, R.G. (1992). *Brain, Behav. Evol.* **39**, 169–194.
Demski, L.S. and Northcutt, R.G. (1983). *Science* **220**, 435–437.

Díaz-Regueira, S. and Anadón, R. (1992). *Brain, Behav. Evol.* **40**, 297–310.
Ebbesson, S.O.E. (1984). In *Comparative Neurology of the Optic Tectum* (ed. H. Vanegas). pp. 33–46. Plenum Press, New York.
Ebbesson, S.O.E. and Heimer, L. (1970). *Brain Res.* **17**, 47–55.
Echteler, S.M. (1984). *J. Comp. Neurol.* **230**, 536–551.
Ekström, P. (1987). *J. Neurosci.* **7**, 987–995.
Finger, T.E. (1983) In *Fish Neurobiology, Vol. 1: Brain Stem and Sense Organs* (eds R.G. Northcutt and R.E. Davis), pp. 285–309. University of Michigan Press, Ann Arbor.
Finger, T.E. (1988) *Brain, Behav. Evol.* **31**, 17–24.
Finger, T.E. (1993). *Acta Anat.* **148**, 132–138.
Finger, T.E., Bell, C.C. and Russell, C.J. (1981). *Exp. Brain Res.* **42**, 23–33.
Fritzsch, B., Sonntag, R., Dubuc, R., Ohta, Y. and Grillner, S. (1990). *J. Comp. Neurol.* **294**, 491–506.
Gilland, E. and Baker, R. (1993). *Acta Anat.* **148**, 110–123.
Gorlick, D.L. (1989). *J. Comp. Neurol.* **290**, 412–422.
Gorlick, D.L. (1990). *Brain, Behav. Evol.* **36**, 227–236.
Hornby, P.J. and Demski, L.S. (1988). *Brain, Behav. Evol.* **31**, 181–192.
Ito, H. and Kishida, R. (1978). *J. Comp. Neurol.* **181**, 465–476.
Jerison, H.J. (1973). *The Evolution of Brain and Intelligence*. Academic Press, New York.
Kanwal, J.S. and Caprio, J. (1983). *J. Comp. Physiol.* **150**, 345–357.
Kanwal, J.S. and Caprio, J. (1987). *J. Comp. Neurol.* **264**, 216–230.
Kirschbaum, F., Meunier, F. and Tsuji, S. (1978). *Cell Tiss. Res.* **187**, 263–269.
Lauder, G.V. and Liem, K.F. (1983). *Bull. Mus. Comp. Zool.* **150**, 95–197.
Lázár, G., Szabo, T., Libouban, S., Ravaille-Veron, M., Toth, P. and Bräntle, K. (1992). *J. Comp. Neurol.* **325**, 343–358.
Leonard, R.B. and Willis, W.D. (1979). *J. Comp. Neurol.* **183**, 397–414.
Levine, R.L. and Dethier, S. (1985). *J. Comp. Neurol.* **237**, 427–444.
Luiten, P.G.M. (1976). *J. Comp. Neurol.* **166**, 191–200.
Luiten, P.G.M. and Dijkstra-de Vlieger, H.P. (1978). *J. Comp. Neurol.* **179**, 669–676.
Mauri, T. and Caprio, J. (1982). *Brain Res.* **231**, 185–190.
Meek, J. (1992). *Europ. J. Morphol.* **32**, 279–282.
Meek, J., Nieuwenhuys, R. and Elsevier, D. (1986). *J. Comp. Neurol.* **281**, 362–383.
Morita, Y. and Finger, T.E. (1987). *J. Comp. Neurol.* **264**, 231–249.
Morita, Y., Ito, H. and Masai, H. (1980). *J. Comp. Neurol.* **191**, 119–132.
Morita, Y., Murakami, T. and Ito, H. (1983). *J. Comp. Neurol.* **218**, 378–394.
Nieuwenhuys, R. (1962). *J. Morphol.* **111**, 69–88.
Nieuwenhuys, R. (1963). *J. Hirnforsch.* **6**, 171–200.
Nieuwenhuys, R. (1964). *Prog. Brain Res.* **11**, 1–57.
Nieuwenhuys, R. (1965). *J. Morphol.* **117**, 1–24.
Nieuwenhuys, R. and Meek, J. (1990). In *Cerebral Cortex, Vol. 8A: Comparative Structure and Evolution of Cerebral Cortex, Part I* (eds E.G. Jones and A. Peters), pp. 31–73. Plenum Press, New York.
Nieuwenhuys, R. and Pouwels, E. (1983). In *Fish Neurobiology, Vol. 1. Brain Stem and Sense Organs* (eds R.G. Northcutt and R.E. Davis), pp. 25–87. University of Michigan Press, Ann Arbor.
Nieuwenhuys, R., ten Donkelaar, H.J. and Nicholson, C. (1998). *The Central Nervous System of Vertebrates*. Springer-Verlag, Berlin.
Noden, D.M. (1991). *Brain, Behav. Evol.* **38**, 190–225.
Northcutt, R.G. (1977). *Amer. Zool.* **17**, 411–429.
Northcutt, R.G. (1978). In *Sensory Biology of Sharks, Skates and Rays* (eds E.S. Hodgson and R.F. Mathewson), pp. 117–193. Office of Naval Research, Arlington, Virginia.
Northcutt, R.G. (1979). In *Hyman's Comparative Vertebrate Anatomy* (ed. M.H. Wake), pp. 615–769. University of Chicago Press, Chicago.
Northcutt, R.G. (1981a). *Ann. Rev. Neurosci.* **4**, 301–350.
Northcutt, R.G. (1981b). *Neurosci. Lett.* **22**, 219–222.
Northcutt, R.G. (1983). In *Fish Neurobiology, Vol. 2: Higher Brain Areas and Functions* (eds R.E. Davis and R.G. Northcutt), pp. 1–42. University of Michigan Press, Ann Arbor.
Northcutt, R.G. (1993). *Acta Anat.* **48**, 71–80.
Northcutt, R.G. (1996). *Israel J. Morphol.* **42**, S-273–S-313.
Northcutt, R.G. and Bemis, W.E. (1993). *Brain, Behav. Evol.* **42**, S1.
Northcutt, R.G. and Butler, A.B. (1976). *J. Comp. Neurol.* **166**, 1–16.
Northcutt, R.G. and Butler, A.B. (1991). *Brain, Behav. Evol.* **37**, 333–354.
Northcutt, R.G. and Butler, A.B. (1993). *Brain, Behav. Evol.* **41**, 57–81.
Northcutt, R.G. and Davis, R.E. (1983). In *Fish Neurobiology, Vol. 2: Higher Brain Areas and Functions* (eds R.E. Davis and R.G. Northcutt), pp. 203–236. University of Michigan Press, Ann Arbor.
Northcutt, R.G. and Puzdrowski, R.L. (1998). *Brain Behav. Evol.* **32**, 96–197.
Northcutt, R.G., Reiner, A and Karten, H.J. (1988). *J. Comp. Neurol.* **277**, 250–267.
Northmore, D.P.M., Williams, B. and Vanegas, H. (1983). *J. Comp. Physiol.* [A] **150**, 39–50.
Parent, A. (1996). *Carpenter's Human Neuroanatomy*, 9th edn. Williams & Wilkins, Baltimore.
Parent, A., Dube, L., Braford, M.R. Jr and Northcutt, R.G. (1978). *J. Comp. Neurol.* **182**, 495–516.
Platel, R. and Delphini, C. (1981). *Cahiers de Biologie Marine* **22**, 407–430.

Platel, R. and Delphini, C. (1986). *J. Hirnforsch.* **27**, 279–293.

Platt, C. (1983). In *Fish Neurobiology, Vol. 1: Brain Stem and Sense Organs* (eds R.G. Northcutt and R.E. Davis), pp. 89–124. University of Michigan Press, Ann Arbor.

Platt, C., Popper, A.N. and Fay, R.R. (1989). In *The Mechanosensory Lateral Line: Neurobiology and Evolution* (eds S. Coombs, P. Görner and H. Münz), Springer-Verlag, New York. pp. 633–651.

Popper, A.N. (1983). In *Fish Neurobiology, Vol. 1: Brain Stem and Sense Organs* (eds R.G. Northcutt and R.E. Davis), pp. 125–178. University of Michigan Press, Ann Arbor.

Popper, A.N. and Fay, R.R. (1993). *Brain, Behav. Evol.* **41**, 14–38.

Puzdrowski, R.L. (1987). *J. Comp. Neurol.* **259**, 382–392.

Puzdrowski, R.L. (1988). *J. Morphol.* **198**, 1–10.

Puzdrowski, R.L. (1989). *Brain, Behav. Evol.* **34**, 110–131.

Reiner, A. and Northcutt, R.G. (1992). *J. Comp. Neurol.* **319**, 359–386.

Repérant, J., Lemire, M., Miceli, D. and Peyrichoux, J. (1976). *Brain Res.* **118**, 123–131.

Saidel, W.M. and Butler, A.B. (1991). *Brain, Behav. Evol.* **38**, 154–168.

Smeets, W.J.A.J. (1990). In *Cerebral Cortex, Vol. 8A: Comparative Structure and Evolution of Cerebral Cortex, Part I* (eds E.G. Jones and A. Peters), pp. 3–30. Plenum Press, New York.

Song, J. and Boord, R.L. (1993). *Acat Anat.* **148**, 139–149.

Springer, A.D. (1983). *J. Comp. Neurol.* **214**, 404–415.

Springer, A.D. and Landreth, G.E. (1977). *Brain Res.* **124**, 533–537.

Szekley, G. and Matesz, C. (1993). *Adv. Anat. Embryol. Cell Biol.* **128**, 1–19.

Uehara, M. and Ueshima, T. (1986). *J. Morphol.* **190**, 325–333.

Vanegas, H., Ebbesson, S.O.E. and Laufer, M. (1984). In *Comparative Neurology of the Optic Tectum* (ed. H. Vanegas), pp. 93–120. Plenum Press, New York.

Walker, M.M., Diebel, C.E., Haugh, C.V., Pankhurst, P.M., Montgomery, J.C and Green, C.R. (1997). *Nature* **390**, 371–376.

Walker, W.F., Jr and Liem, K.F. (1994). *Functional Anatomy of the Vertebrates: An Evolutionary Perspective*, 2nd edn. Saunders College Publishing, Fort Worth.

Wathey, J.C. (1988). *J. Comp. Physiol. [A]* **162**, 511–524.

Webb, J.F. and Noden, D.M. (1993). *Amer. Zool.* **33**, 434–447.

Wicht, H. (1996). *Brain, Behav. Evol.* **48**, 248–261.

Wicht, H. and Northcutt, R.G. (1992). *Brain, Behav. Evol.* **40**, 25–64.

Wicht, H. and Northcutt, R.G. (1993). *J. Comp. Neurol.* **337**, 529–542.

Zakon, H.H. (1986). In *Electroreception* (eds T.H. Bullock and W. Heiligenberg), pp. 103–156. John Wiley, New York.

Zakon, H.H. (1988). In *Sensory Biology of Aquatic Animals* (eds J. Atema, R.R. Fay, A.N. Popper and W.N. Tavolga), pp. 813–850. Springer-Verlag, New York.

CHAPTER 9

Respiratory System

Kenneth R Olson
Indiana University School of Medicine, South Bend Center for Medical Education, University of Notre Dame, Notre Dame, Indiana, USA

Abbreviations

ABA	afferent branchial artery
ABM	abductor muscle
ADM	adductor muscle
AFA	afferent filamental artery
ALA	afferent lamellar arteriole
AS	arch skeleton
AVA	postlamellar arteriovenous anastomosis
B	buccal chamber
BB	basibranchial
BF	blood flow
BV	branchial vein
CAB	concurrent afferent branchial
CB	ceratobranchial
CR	cartilage ray
EB	epibranchial
EBA	efferent branchial artery
EFA	efferent filamental artery
ELA	efferent lamellar arteriole
F	gill filament
FV	filamental vein
GR	gill raker
HB	hypobranchial
IL	interlamellar vessel
IS	interbranchial septum
LSML	longitudinal smooth muscle ligament
NA	nutrient artery
NC	nutrient capillary
O	operculum
OC	opercular cavity
OMC	outer marginal channel
PAVA	prelamellar arteriovenous anastomosis
PB	pharyngeobranchial
PC	pillar cell
RAB	recurrent afferent branchial
RL	respiratory lamella
SM	smooth muscle
TSML	transverse smooth muscle ligament
W	water flow
1–4	gill arches 1, 2, 3, 4

Copyright © 2000 Academic Press

Introduction

Fish live in an environment that is considerably different from that encountered by terrestrial vertebrates. Compared with air, water is 800 times more dense, 60 times more viscous, has 1/30 the capacity to hold oxygen and oxygen diffuses at 1/8000 the rate. Thus this respiratory medium is more abrasive and oxygen depleted than air. An aquatic environment also fosters the untoward transfer of water, salts, acid or base, as well as toxic substances. These necessitate corrective measures, many of which also occur at the gill.

The fish gill has become a multifunctional organ designed to deal with the vagaries of an aquatic medium. The gill is not only the primary site for respiration, it is also the principal, and often exclusive, site for osmoregulation, acid–base balance, and metabolism of circulating hormones and perhaps xenobiotics (Maetz, 1971; Hughes and Morgan, 1973; Heisler, 1984; Payan *et al.*, 1984; Randall and Daxboeck, 1984; Olson, 1998). The anatomy of gill tissue and its vasculature is as diverse and specialized as the functions it performs and the gill may well be the most highly differentiated of any vertebrate organ. Because all of these activities are predicated upon cellular contact with environmental water or blood, it is not surprising that the hallmark of gill tissue is an extensive elaboration of both epithelial and vascular surfaces. The general organization of the gill, gill tissues, and gill vessels are described in this chapter. The fine structure of gill tissues is described in Chapter 21.

Orientation and general anatomy

The embryonic gill originates along a series of paired pouches in the lateral walls of the pharynx. This lateral row of gill slits forms a water pathway from the pharynx to the exterior and the septal tissue between the slits becomes the gill arch and supports the gill filaments. In teleosts, the septum is reduced to a relatively simple bow-shaped structure with the ends attached to the dorsal and ventral surfaces of the oropharynx (more commonly called the buccal cavity) and the curved portion projecting posterolaterally (Figure 9.1).

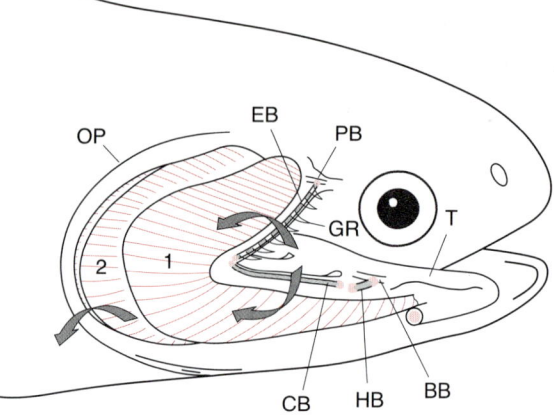

Figure 9.1 General orientation of the gill arches in the trout head. The bony arch skeleton is shown in gray. Gill rakers on the ventral portion of arch are not shown. Arrows indicate direction of water flow. (For abbreviations see list on page 151.)

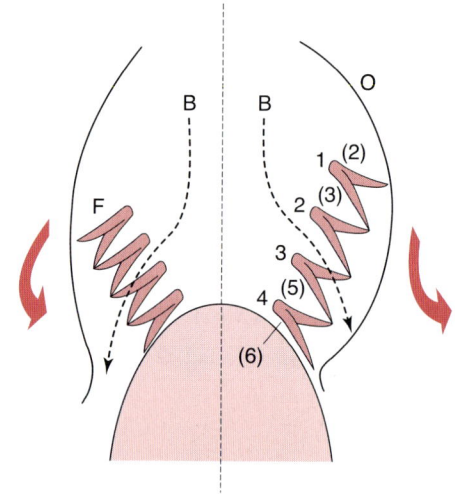

Figure 9.2 Dorsal view showing a cross-section through the four pairs of gill arches (large numbers 1–4). The operculum (O) is abducted on the right half of the figure permitting water to flow (dashed arrow) between the buccal (B) and opercular chambers. Opercular adduction (left side) forces water out of the opercular chamber. Rotation of the gill arches anterolaterally and abduction of the gill filaments (F) maintains the continuity of the gill curtain as the operculum is abducted. Gill slits are numbered in parentheses. Large arrows show movement of the operculum.

Teleost fish have eight gill arches arranged in four pairs on either side of the buccal cavity (Figure 9.2). An additional primordial gill hemiarch, the pseudobranch, is also present in most species, notable exceptions being the catfish. Structurally, the pseudobranch resembles a vestigial gill arch (Laurent and Dunel-Erb, 1984). It contains filaments similar to those on other gill arches, but they are generally covered by a relatively thick epithelium that prevents direct communication with the environment. Furthermore, the vascular supply to the pseudobranch is derived from the postbranchial (systemic) circulation (see Chapter 10). The pseudobranch is associated with the first gill slit which is generally closed in teleosts (Laurent, 1984). The second gill slit is between the operculum and the first pair of gill arches. Gill slits 3–5 are between arches 1–4, respectively, and the sixth slit is between the fourth arch and the cleithrum.

Each gill arch supports two rows of filaments. Filaments of one row constitute a hemibranch, and the two hemibranchs of a gill arch form the holobranch. Filaments from two hemibranchs on a single arch face different water channels (gill slits), whereas hemibranchs from consecutive arches (the posterior hemibranch of one arch and the anterior hemibranch of the next arch) share a common water channel. The second and sixth gill slits (anterior to the first arch, and posterior to the fourth, respectively) are the exceptions as they are only lined by a single row of hemibranchs. Hemibranchs of a holobranch radiate away from each other to the extent that the distal tips of filaments from two adjacent hemibranchs lining a common water channel nearly touch each other (Figures 9.2 and 9.3). This forms an effective gill 'curtain' or sieve which promotes water flow across the filament surface rather than between the hemibranchs.

Filaments of two hemibranchs appear to originate from the arch in pairs although they are actually staggered. The number of filaments on a gill arch varies from arch to arch, with the greatest number usually on the first or second arches and often the anterior hemibranch will have a few more filaments than the posterior one. Filament number will continue to increase while the fish is growing. For example, a 0.35 g trout has around 770 total filaments, whereas a 400 g trout has approximately 1600 (Hughes and Morgan, 1973). Filament length also varies as a function of location on the arch and hemibranch and there is considerable interspecies variation (Hughes, 1984). In trout the longest filaments are found in two locations: midway along the ceratobranchial skeleton and midway along the epibranchial skeleton.

The filament is the functional unit of the gill and structural orientation is usually defined relative to the filament. Thin plate-like respiratory lamellae originate in staggered fashion from alternative sides of each of the two lateral surfaces of the filament. The outer margins of lamellae from adjacent filaments

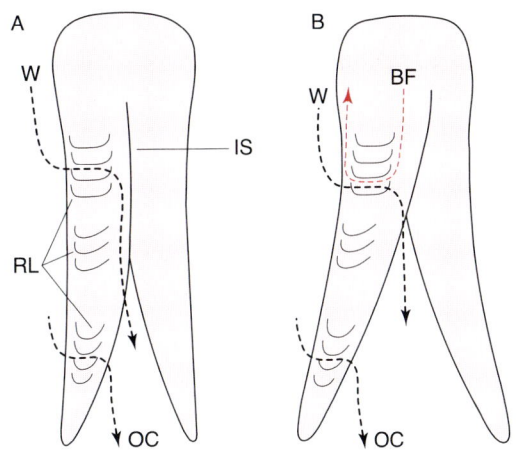

Figure 9.3 Relationship between the gill arch, interbranchial septum, and gill filaments in fish. A: In the primitive trout the filaments are attached to the interbranchial septum (IS) for half of their length. After passing over the surface of the respiratory lamellae (RL), respiratory water (W) must flow along the septum before reaching the opercular cavity (OC). B: In advanced fish, such as the perch, the septum is reduced or nearly absent and water flows directly across the lamellae to the opercular cavity. Water flow across the lamella (W, dotted line) is counter-current to blood flow through the lamella (BF, dashed line). The figure also shows some of the variations in lamellar shape found on a single filament and/or between different species.

nearly interdigitate, thus forcing respiratory water to flow between the lamellae and along the lamellar surface, thereby enhancing the effectiveness of the gill curtain. In all but a few fish, respiratory water and blood flow is antiparallel (Figures 9.3B and 9.7) and this countercurrent arrangement optimizes gas exchange efficiency.

Gill rakers are spiny projections on the anterior edge of the gill arch and opposite to the filaments (Figure 9.1). The gill rakers project into the water channel anterior to the arch and serve to protect delicate filamental structures by sieving particulates out of the respiratory current. Rakers may also help guide food posteriorly into the esophagus. In a few fish, paired gill rakers project into both anterior and posterior water channels.

The gill arch

Arch support skeleton

Three bones, the hypobranchial, ceratobranchial and epibranchial, form the gill arch skeleton. The hypobranchial bone on the ventral segment of the arch articulates with the basibranchial bone (copula) in the midline of the tongue and the dorsal, epibranchial, bone is attached to the pharyngobranchial bone in the roof of the mouth. The ceratobranchial bone is generally the longest of the three and travels most of the length of the ventral segment of the arch. The bony skeleton and its attachments to the body are hinged to permit dorsoventral extension and lateral movements necessary for both feeding and respiration. Other arch and filamental elements may also serve a structural function and/or provide anchoring sites for muscles. These vary from fish to fish and depend to a large extent on the general organization of the filaments and the extent of the interbranchial septum in the arch.

Interbranchial septum and arch musculature

Interbranchial septum

The interbranchial septum, like the arch support tissue, is a remnant of the body wall that has, through the course of evolution, been incorporated into the gill arch. The septum extends from the arch into the area between the hemibranchs. Teleost gills follow one of two or three general anatomical categories that are based on the extent of the interbranchial septum and the organization of skeletal and smooth muscles that connect the arch skeleton to the filaments (Bijtel, 1949; Hughes, 1984; Laurent, 1984; Nilsson, 1985). In the more primitive teleosts, such as trout, the interbranchial septum extends out for one-third to two-thirds of the length of the filaments and the margin of the filaments along the septum is embedded in septal tissue (Figure 9.3A). In these fish respiratory water that has passed over the more basal lamellae must flow along the septum before entering the opercular cavity. Septal tissue is considerably reduced in more advanced teleosts such as perch (Figure 9.3B) and the filaments may be free for most or all of their length. Here, water flows directly from buccal to opercular chambers.

Arch musculature

The presence or absence of an interbranchial septum not only affects water flow across the respiratory lamellae, but it can affect movement of the filaments. A number of strategies have appeared to allow the hemibranchs to maintain the gill curtain during respiration (Hughes, 1984) and these are perhaps best described in the elegant studies of Dunel-Erb and Bailly (1987).

In fish with an extended interbranchial septum, such as trout, opposing cartilage rods from filaments on adjacent hemibranchs are attached by striated adductor muscles to a smooth muscle band located in the interbranchial septum (Figure 9.4a). A second striated muscle, the abductor muscle, is attached to the gill arch skeleton and the cartilaginous rod of the filament on that side of the arch. During respiration, contraction of the abductor muscle presumably helps to pull the filaments apart and contraction of the adductor brings them together, although movement of the filaments must be constrained by their attachment to the septum. The longitudinal smooth muscle band in the septum may help adjust interfilamental spacing along a hemibranch as the arc of the gill arch opens and closes with movement of the buccal floor.

The arch musculature of advanced fish is more complex and versatile. In the perch (Figures 9.4b and 9.5), adductor muscles are attached between a basal extension of the cartilage rod of filaments from one hemibranch to a more distal portion of the rod on filaments from the opposite hemibranch. Adductor

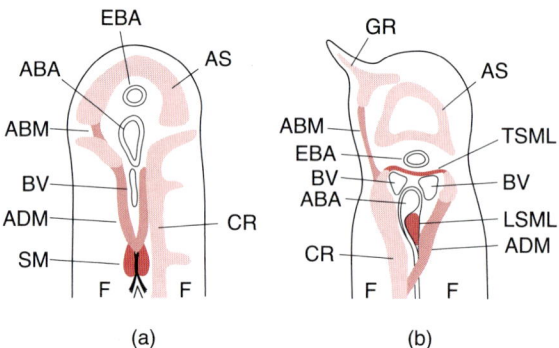

Figure 9.4 Structural and muscular organization of the gill arch of two teleosts. The basal portion of the filaments is attached to the interbranchial septum in more primitive fish such as trout (a) and filament mobility is limited. A short abductor muscle (ABM) pulls the filamental cartilage ray (CR) toward the arch support skeleton (AS) to spread the free tips of the filaments and an adductor muscle (ADM) anchored to a longitudinal smooth muscle (SM) band in the septum pulls the tips together. In advanced fish such as perch (b) the septum is reduced increasing filament mobility. Additional features, such as cross-fixation of the adductor muscle directly on opposite cartilaginous rays and a transverse smooth muscle lamina (TSML) enhance filament mobility. See list on page 151 for abbreviations. Additional details are provided in the text. (Redrawn from Dunel-Erb and Bailly, 1987, with permission.)

contraction pulls the distal filament tips together. The abductor muscles connect the gill raker and the basal end of the rod on the same hemibranch and pull the filament tips apart. There are also two bands of smooth muscle in these gills. A transverse smooth muscle lamina passes between the efferent and afferent branchial arteries and connects the basal ends of the cartilage rods of filaments on opposite hemibranchs. Contraction of this muscle by catecholamines or sympathetic nerves (Nilsson, 1985) will draw the basal ends of the cartilage rods together and, with the afferent branchial artery and surrounding tissue serving as a fulcrum, the free tips of the filaments will be pulled apart. The longitudinal smooth muscle lamina is embedded in the afferent branchial artery and runs the length of the gill arch. Contraction of this muscle presumably decreases interfilamental spacing along the hemibranch.

Arch vasculature

The vasculature of the gill arch has been reviewed by Laurent (1984) and Olson (1991) and is shown in Figures 9.5 and 9.6. The afferent branchial artery enters the anteroventral end of the arch and travels approximately one-quarter to one-third of the length of the ventral segment of the arch before bifurcating into recurrent and concurrent branches. The recurrent branch returns to the anterior end of the arch, where it ends, while the concurrent branch continues posteriorly along the arch and ends in the posterodorsal end of the arch.

Individual afferent filamental arteries arise repeatedly from the afferent branchial artery and travel to the tip of the filament, though rarely a common trunk might feed two filaments on opposing hemibranchs. The filament is drained by efferent filamental arteries and filamental veins. Individual efferent filamental arteries anastomose with the efferent branchial artery and exit the arch to become the systemic arterial circulation also known as the arterio-arterial pathway. Filamental veins drain into one of several branchial veins and blood is returned directly to the heart; thus completing the arteriovenous pathway.

A single efferent branchial artery lies dorsal to the afferent branchial artery (Figures 9.4–9.6) except in

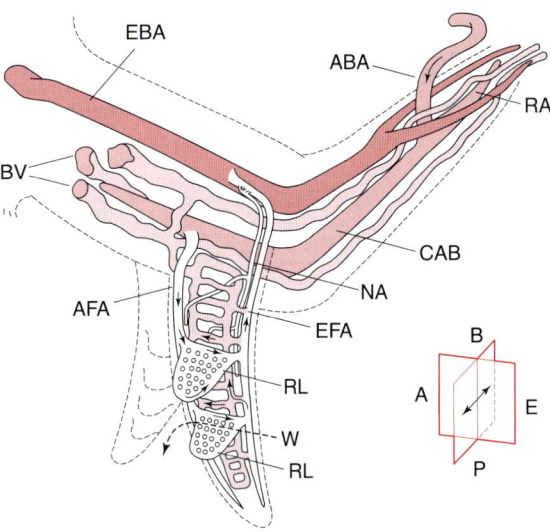

Figure 9.6 Blood vessels in the gill arch and filament of the trout. The afferent branchial artery (ABA) enters the arch and bifurcates into concurrent and recurrent branches (CAB and RAB, respectively). These branches travel along the arch and give rise to numerous afferent filamental arteries (AFA) that enter the filament and supply the respiratory lamellae (RL). Blood draining from the lamellae enters the efferent filamental artery (EFA) and then drains into either the efferent branchial artery (EBA) to become systemic arterial blood (>90% of total) or re-enters the filament as a nutrient supply and is returned to the heart via branchial veins (BV). Other nutrient arteries (NA) originate from the EFA near the base of the filament and return to the filament along the EBA. Branches from the NA travel medial to the EFA and AFA and in the middle of the filament. Water flow (W; dashed arrow) across the lamella is countercurrent to lamellar blood flow. **Inset** reference planes for filament nomenclature: B, basal filament (closest to the arch); P, peripheral, or tip of the filament; A, edge of the filament nearest the afferent filamental artery; E, filament edge nearest efferent filamental artery. Medial area of filament is in the center of the A–E plane and arrows point toward lateral borders. (Redrawn from Olson, 1991, with permission.)

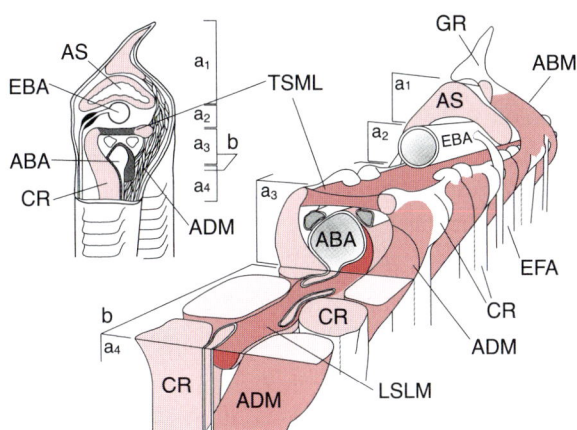

Figure 9.5 Relationships between gill arch and filament support tissues, musculature, and vasculature in the gill arch of the perch. Lower case letters refer to plane of section. See list on page 151 for abbreviations and text for details. (Redrawn from Dunel-Erb and Bailly, 1987, with permission.)

the anterior end of the arch. Here, the efferent branchial artery bifurcates and the branches pass on either side of the incoming afferent branchial artery. These branches continue independently, and they are displaced somewhat laterally, along the recurrent branch of the afferent branchial. The branches only drain filaments from their respective hemibranch. The lateral (opercular side) branch from the first gill arch continues on out of the arch and supplies the jaw and in some fish, such as trout, the pseudobranch (see Chapter 10). The medial (buccal side) branch from gill arches two and/or three (rarely one or four) also exit the arch and form the hypobranchial (systemic) arterial system. The remaining branches end within the gill arch. Anteroventral branches carry only a small fraction of the cardiac output. The bulk of the arterio-arterial flow exits the arch via the single posterodorsal efferent branchial artery.

The largest branchial veins usually travel close to the afferent branchial artery (Figures 9.5 and 9.6). They exit the arch from both anteroventral and posterodorsal ends and drain into the jugular or anterior cardinal veins, respectively.

Nutrient supply to the arch tissues, including gill rakers, comes from the oxygenated postlamellar circulation. Numerous small arterioles arise from either the efferent branchial (usually more so in the midregion of the arch) or from the basal end of individual efferent filamental arteries and they anastomose repeatedly with each other to form the systemic arterial circulation (also known as the gill secondary circulation) of the arch (see below and Chapter 21 for additional details). Nutrient vessels ultimately drain into the branchial veins.

Innervation of the arch

The pattern of gill innervation has been summarized by Nilsson (1984) and the neuroanatomy described by Laurent (1984). Teleost gills are innervated by cranial nerves IX (glossopharyngeal) and X (vagus). These nerves are classified as somatic sensory, visceral sensory and visceral motor. All are of dorsal root origin and the majority are sensory (Nilsson, 1984). The first gill arch is innervated by a post-trematic ramus of the glossopharyngeal and a pretrematic ramus of the vagus. The remaining arches are innervated by pre- and post-trematic branches of the vagus. Spinal autonomic (sympathetic) nerves enter the cranial nerves via the gray rami and travel with them to gill tissue.

The nerves to the gills are often called branchial nerves and enter the gill arch via the posterodorsal end. They either travel together as a single branch, or in two to three bundles in the arch, usually between the arch skeleton and either the efferent branchial artery or the basal extension of the efferent filamental artery. These fibers supply the arch tissues and penetrate into the filaments.

Gill filament

The filament is a long, flattened knife blade-like structure designed to provide structural support to an extensive lamellar surface yet offers minimal resistance in the respiratory water current. In cross-section, filaments appear somewhat dumb-bell in shape; rounded along the borders of the afferent and efferent filamental arteries and thinner in between. Orientation planes for filamental and lamellar nomenclature are described in Figure 9.6.

Cartilage and muscle

Mechanical support to the filament is provided by a cartilaginous rod that extends from the gill arch to the tip of the filament. The rod is located near the afferent filamental artery (Figure 9.7) and in cross-section it often has a tear-drop shape with the narrow margin pointing toward the efferent border of the filament. In some fish, such as trout, several to a dozen (depending on the length of the filament) thin spines project from the rod two-thirds of the way across the body of the filament. Presumably, these spines provide axial stabilization of the efferent edge of filament against the incoming water current. Movement of the cartilage rod is controlled by striated and smooth muscle in the arch and base of the filament (see above).

Respiratory lamellae

The thin plate-like respiratory lamellae are designed to maximize surface area, and minimize water–blood diffusion distance. There is an allometric relationship between surface area and body weight ($Y = aW^b$; where Y is total gill area (mm^2), a and b are constants,

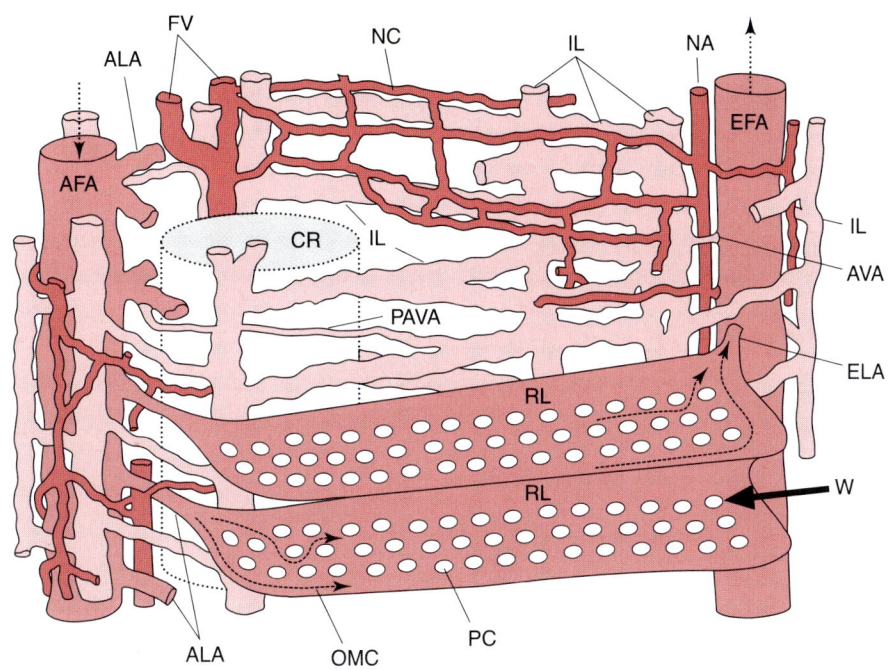

Figure 9.7 Blood vessels in the gill filament. Blood flowing through the arterio-arterial pathway passes through the afferent filamental artery (AFA), afferent lamellar arteriole (ALA), respiratory lamellae (RL), efferent lamellar arteriole (ELA), and exits the filament via the efferent filamental artery (EFA). Blood traversing the lamellar sinus may percolate between the post-like pillar cells, or follow the outer marginal channel (OMC), a wide-bore pathway around the outer edge of the lamella. Blood flow is countercurrent to the flow of respiratory water (W, thick arrow). Blood may enter the body of the filament through short arteriovenous anastomoses (AVA) from the medial border of the EFA, via prelamellar arteriovenous anastomoses (PAVA) from the ALA (very rare in most teleosts), or from nutrient arteries (NA) that arise from the basal EFA or EBA. The body of the filament often has two vessel types, small nutrient capillaries (NC) and larger, more amorphous vessels comprising the interlamellar system (IL). Filamental veins (FV) drain both interlamellar and nutrient systems. Lamellar, nutrient and interlamellar vessels are found on both sides of the filament and surround the cartilagenous rod (CR) that supports the filament.

and W is body weight in grams). Both a and b are species-specific; for most fish, a ranges from 100 to 500 (but up to 5000 in tuna) and b is generally between 0.8 and 1.0 (Muir, 1969; Hughes, 1984). This roughly equates to a total gill area of approximately 1.5–2 cm^2/g body weight for sluggish fish, 2–5 cm^2/g for moderately active fish and $> 5\ cm^2/g$ for very active fish (Hughes and Morgan, 1973).

Lamellar shape may vary between species and even along a single filament (Hughes, 1984). Lamellae may be semi-circular, rectangular or form a right triangle (Figure 9.3). Some variation of a right triangle is the most common shape with the edge of the lamella nearest the efferent filamental artery (and, therefore, the edge facing the incoming water) as the right arm. This presumably increases the efficiency of the countercurrent (water–blood) exchange and maximizes oxygen loading of the blood just prior to exiting the lamella. Additional details on lamellar structure can be found in Chapter 21.

Filamental vasculature

Arterio-arterial pathway

The major vessels of the filament are shown in Figures 9.6 and 9.7. Afferent filamental arteries travel the length of the filament and give rise to the afferent lamellar arterioles. In some fish with a prominent interbranchial septum, there is a bellows-like dilation (ampulla) in the artery near the distal margin of the septum and the arterial wall appears attached to the cartilage rod and it is devoid of smooth muscle. Also, in a few of these fish, arteries of adjacent filaments, or those of opposite hemibranchs, are interconnected in the region of the ampullae.

In the basal filament, a single afferent lamellar arteriole may branch to perfuse three to five lamellae on one side of the filament. In the mid-filament, a single lamellar arteriole usually supplies individual lamellae, whereas in the peripheral filament, a single arteriole may originate from the medial margin of the filamental artery and bifurcate to supply one lamella on each side of the filament.

The lamella consists of two layers of epithelium sandwiched over a thin, flat, vascular sinus. Post-like pillar cells separate the epithelial layers, delineating the vascular space and providing a non-thrombogenic surface. In some fish, the pillar cells may be arranged in rows thereby forming direct vascular channels across the lamella. The outer free edge of the lamellar sinus, the outer marginal channel, is typically dilated forming a preferential pathway for blood. Endothelial cells line all but the medial border of the outer marginal channel; this is the only place in the lamella where a non-pillar cell endothelium is found. The outer marginal channel is of obvious advantage in gas exchange because here convective water flow over the lamellar surface is optimal, the water–blood diffusion distance is minimal, and red cells are exposed to the exchange surface on three sides.

Some 15 to 20% of the lamellar vasculature is embedded in the body of the filament and is not available for gas exchange. Here, the pillar cells are packed closer together, apparently restricting red cell access and favoring plasma skimming. Pillar cells contain a number of enzymes capable of metabolizing circulating biomolecules (e.g. hormones) and the pattern of pillar cell distribution in the lamella suggests that red cells are diverted to the outer margins for gas exchange, whereas the plasma is diverted to the inner areas where pillar cell enzymes can exert their metabolic activity (Olson, 1991, 1998).

After percolating through the lamella, blood is collected into a short efferent lamellar arteriole that is connected to the lateral wall of the efferent filamental artery. From here, most of the blood (>90%) drains from the filament into the efferent branchial artery and continues on to the systemic circulation. A smooth muscle sphincter has been observed in the efferent filamental artery, just prior to its anastomosis with the efferent branchial artery (Laurent, 1984). The smooth muscle layer in the sphincter is twice as thick as that in the more peripheral efferent filamental artery and the sphincter muscle is highly innervated (see below), suggesting it is an important site for flow regulation.

Arteriovenous pathway

The small fraction of blood flow that does not exit the gill re-enters the body of the filament and becomes equivalent to a systemic pathway in the gill. This circulation, often described as a 'simple sac-like structure' is actually quite complex and neither its anatomy nor physiology is completely understood. Two distinct, yet interconnected, vascular networks are present in most fish. The first consists of narrow-bore capillaries and it is believed to form the nutrient circulation of the filament. Nutrient vessels are supplied either by nutrient arteries that originate from the wall of the efferent filamental artery near the anastomosis of the efferent filamental artery with the efferent branchial artery, or from the efferent branchial artery itself. Typically, numerous small, tortuous arterioles arise from the efferent arteries and they repeatedly re-anastomose with each other to form progressively larger nutrient arteries that supply arch tissues and filaments. The filamental nutrient artery usually travels singularly (or rarely branched) along and medial to the efferent filamental artery. Upon entering the filament, one or several branches from the main nutrient artery travels across the filament and these branches then proceed, in either the mid-filament or along the afferent filamental artery, to the filament tip. In some instances narrow arterioles arise directly from the medial wall of the peripheral efferent filamental artery and directly supply nutrient capillaries.

The second vascular system in the filament is the interlamellar system. These vessels have a thin wall and their diameter is two to ten times that of the nutrient capillaries. Interlamellar vessels traverse the body of the filament in register with, but between and medial to, the inner margins of the respiratory lamella. An interlamellar system is found on both sides of the filament along the afferent filamental artery and around the cartilage support rod. Toward the efferent border of the filament, interlamellar vessels from the two sides come together and form a common system. Extensions of the interlamellar vessels wrap around both afferent and efferent filamental arteries. Interlamellar vessels are supplied by narrow-bore arterioles from the medial wall of the efferent filamental artery, or by nutrient arterioles. In a few fish, such as eel and catfish, a portion of the interlamellar system is supplied by narrow-bore arterioles from the afferent filamental artery or afferent lamellar arteriole (prelamellar arteriovenous anastomoses, AVAs). However, prelamellar AVAs are rare, or non-existent in most fish.

Common filamental veins drain both nutrient and interlamellar networks in most fish. In a few species, such as catfish, they remain separate in the filament and join in the arch.

Filamental nerves

The main filamental nerve usually runs along the efferent filamental artery between the artery and the interlamellar vessels or nutrient artery. Occasional branches extend out into the filament body and additional nerves are found along the afferent filamental artery. Many of these nerves carry sensory information from proprioceptors in the filament and chemoreceptors which are plentiful in the epithelium (see Chapter 21). Afferent fibers include cholinergic motor fibers to striated muscles, and a variety of autonomic fibers to the filamental vasculature.

There is considerable variation in the type and degree of innervation of filamental vascular smooth muscle in different teleosts (Donald, 1984; Nilsson, 1984; Bailly and Dunel-Erb, 1986; Dunel-Erb and Bailly, 1986; Donald, 1987; Dunel-Erb et al., 1989). In general: (i) the basal third of the filament is the most densely innervated and innervation of the distal third is sparse; (ii) both parasympathetic cholinergic and sympathetic adrenergic fibers are present (as well as serotoninergic and a few others); and (iii) the efferent lamellar arterioles and efferent filamental arteries are more innervated than the afferent vessels in some species, whereas the converse is true in others. Nutrient and interlamellar vessels are also innervated to some degree although nerve fibers have not been found in the respiratory lamellae.

The sphincter area in the base of the efferent filamental artery has received particular attention because of its heavy smooth muscle coat and its dense innervation (Bailly and Dunel-Erb, 1986; Dunel-Erb and Bailly, 1986). Postganglionic parasympathetic (cholinergic) fibers directly innervate, and when stimulated constrict, the sphincter. Sympathetic, adrenergic fibers, however, do not directly innervate the muscle, but appear to innervate the postganglionic terminals of the parasympathetic neurons and modulate (decrease) neurotransmitter release by the latter.

Acknowledgments

The author wishes to thank Dr S. Dunel-Erb for providing several figures used in this article. The author is also grateful to the National Science Foundation for past and current support through NSF Grant Nos. PCM 79-23703, PCM 84-048897, DCB 86-16028, INT 83-00721, INT 86-02965, INT 86-18881, IBN 91-05247, and IBN 97-23306.

References

Bailly, Y. and Dunel-Erb, S. (1986). *J. Morph.* **187**, 219–237.
Bijtel, J.H. (1949). *Arch. Neerl. Zool.* **8**, 267–288.
Donald, J. (1984). *J. Morph.* **182**, 307–316.
Donald, J.A. (1987). *J. Morph.* **193**, 63–73.
Dunel-Erb, S. and Bailly, Y. (1986). *J. Morph.* **187**, 239–246.
Dunel-Erb, S. and Bailly, Y. (1987). *Cell Tiss. Res.* **247**, 339–350.
Dunel-Erb, S., Bailly, Y. and Laurent, P. (1989). *Cell Tissue Res.* **255**, 567–573.
Heisler, N. (1984). In *Fish Physiology Vol. X Gills. Part A: Anatomy, Gas Transfer and Acid–Base Regulation* (eds W.S. Hoar and D.J. Randall), pp. 315–401. Academic Press, New York.
Hughes, G.M. (1984). In *Fish Physiology Vol. X Gills. Part A: Anatomy, Gas Transfer, and Acid–Base Regulation* (eds W.S. Hoar and D.J. Randall), pp. 1–72. Academic Press, New York.
Hughes, G.M. and Morgan, M. (1973). *Biol. Rev.* **48**, 419–475.
Laurent, P. (1984). In *Fish Physiology, Vol. X Gills. Part A: Anatomy, Gas Transfer, and Acid–Base Regulation* (eds W.S. Hoar and D.J. Randall), pp. 73–183. Academic Press, New York.
Laurent, P. and Dunel-Erb, S. (1984). In *Fish Physiology, Vol. X Gills. Part B: Ion and Water Transfer* (eds W.S. Hoar and D.J. Randall), pp. 285–323. Academic Press, New York.
Maetz, J. (1971). *Phil. Trans. Roy. Soc. Lond.* **262**, 209–249.
Muir, B.S. (1969). *J. Fish. Res. Board Can.* **26**, 165–170.
Nilsson, S. (1984). In *Fish Physiology Vol. X Gills. Part A: Anatomy, Gas Transfer, and Acid–Base Regulation* (eds W.S. Hoar and D.J. Randall), pp. 185–227. Academic Press, New York.
Nilsson, S. (1985). *J. Exp. Biol.* **118**, 433–437.
Olson, K.R. (1998). *Comp. Biochem. Physiol.* **119A**, 55–65.
Olson, K.R. (1991). *J. Electron Microsc. Technique* **19**, 389–405.
Payan, P., Girard, J.P. and Mayer-Gostan, N. (1984). In *Fish Physiology Vol. X Gills. Part B: Ion and Water Transfer* (eds W.S. Hoar and D.J. Randall), pp. 39–63. Academic Press, New York.
Randall, D. and Daxboeck, C. (1984). In *Fish Physiology Vol. X Gills. Part A: Anatomy, Gas Transfer, and Acid–Base Regulation* (eds W.S. Hoar and D.J. Randall), pp. 263–314. Academic Press, New York.

CHAPTER 10

Circulatory System

Kenneth R Olson
Indiana University School of Medicine, South Bend Center for Medical Education, University of Notre Dame, Notre Dame, Indiana, USA

Abbreviations

A	atrium
AB v.	abdominal vein
AC v.	anterior cardinal vein
AP	afferent pseudobranch artery
B	bulbus arteriosus
BL v.	bladder vein
BR	brain
C	commissure vessel
CA	caudal artery
CA v.	caudal vein
CC	coracoid artery
CH	choroid artery
CHT	caudal heart
CM	celiacomesenteric artery
CO	coronary artery
DA	dorsal aorta
DC	ductus Cuvier
DI	dorsal intestinal artery
DI v.	dorsal intestinal vein
DISeg	dorsal intersegmental artery
DISeg v.	dorsal intersegmental vein
DS	duodenosplenic artery
DSeg	dorsal segmental artery
DSeg v.	dorsal segmental vein
EC	external carotid artery
EG	epigastric artery
EG v.	epigastric vein
EP	efferent pseudobranch artery
GA	gastric artery
GI	gastrointestinal artery
GO	gonadal artery
GS	gastrosplenic artery
H	hepatic artery
H v.	hepatic vein
HP v.	hepatic portal vein
HB	hypobranchial artery
HY	hyoidean artery
IA	intercostal artery
IC	internal carotid artery
ICos	intercostal artery
ICos v.	intercostal vein

Copyright © 2000 Academic Press

IN	intestinal artery	SA	swimbladder artery
J v.	jugular vein	SC	subclavian artery
K	kidney	SC v.	subclavian vein
LA	lateral aorta	SM	secondary muscular artery
LC v.	lateral cutaneous vein	SM v.	secondary muscular vein
LISeg	lateral intersegmental artery	SV	sinus venosus
LISeg v.	lateral intersegmental vein	TH	thyroidean artery
LSeg	lateral segmental artery	TM	tertiary muscular artery
LSeg v.	lateral segmental vein	TM v.	tertiary muscular vein
MA	mandibular artery	V	ventricle
N	nasal artery	Vb	vertebra
ON	orbito-nasal artery	VA	ventral aorta
OP	ophthalmic artery	VI	ventral intestinal artery
OT	optic artery	VI v.	ventral intestinal vein
P	pseudobranch	VISeg	ventral intersegmental artery
PC v.	posterior cardinal vein	VISeg v.	ventral intersegmental vein
PI	posterior intestinal artery	1–4	afferent branchial arteries to gill arches 1–4
PI v.	posterior intestinal vein	1′–4′	efferent branchial arteries from gill arches 1–4
RA	renal artery		
RP v.	renal portal vein		
S	spleen		

Overview

The teleost cardiovascular system is a simple loop with the heart, gills and systemic circulations in series. Deoxygenated blood pumped from the single ventricle is initially distributed to the gill capillaries and oxygenated blood leaving the gills is then delivered directly to systemic tissues. Blood vessels may serve one or more functions depending on their location and structure. The arterial system dampens the pressure pulse (compliance), distributes blood to different sites (**conductance**), and regulates flow through the capillaries (resistance). Capillaries permit transfer between blood and tissues (exchange) and may alter the local environment (metabolic), while veins serve as blood reservoirs (**capacitance**) and return conduits (conductance). In this chapter the general organization of the heart and circulation will be described with respect to features important in distribution of blood to and from the tissues. Emphasis will be placed on the anatomy of the trout unless otherwise specified, although the general vascular pattern is surprisingly consistent in all teleosts (Satchell, 1991). The secondary circulation, an enigmatic vasculature unique to bony fish will also be briefly described. The anatomical attributes necessary for the functional expression of the different elements of the cardiovascular system will be considered in Chapter 22.

The heart

The heart of bony fish lies in a wedge-shaped vault bounded posteriorly by the transverse septum, dorsally by the esophagus, and laterally by the pectoral girdle and posterior branchial skeleton. In most fish the pericardium is attached to adjacent tissues and may only appear as a free membrane along the ventral aspect of the heart. The heart consists of three chambers: sinus venosus, atrium, and ventricle (Figures 10.1 and 10.2).

The **sinus venosus** is a thin-walled reservoir consisting of connective tissue and some cardiac muscle enabling a weak, but probably insignificant contraction (Santer, 1985). Valves are absent from the venous inflow tract into the sinus venosus, whereas the **sinoatrial ostium** is valved to prevent retrograde flow from the atrium during **atrial systole**. The cardiac pacemaker is located near the sinoatrial junction in many fish; in others it is elsewhere in the sinus (Santer, 1985; Satchell, 1991).

The walls of the atrium are thicker than those of the sinus, although atrial mass is only 10–25% of that of the ventricle (Farrell and Jones, 1992). Strands of atrial trabeculae extend from near the sinoatrial valve across the atrium (Figure 10.1) and assist in atrial ejection, the volume of which may be equivalent to the entire ventricular stroke volume (Farrell and Jones,

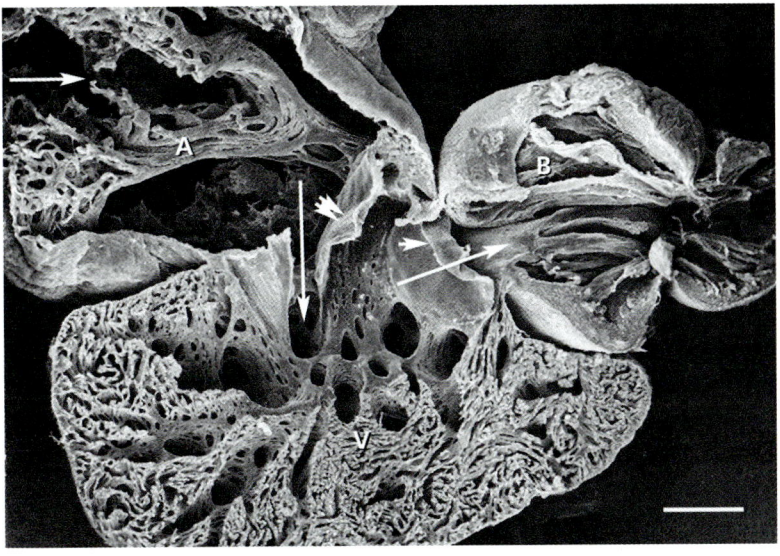

Figure 10.1 Scanning electron micrograph of a longitudinal section through the heart of the catfish, *Herteropneustes fossilis*. The atrium (A) has long trabeculae that extend from the posterior wall (left of micrograph) near the entrance from the sinus venosus (short arrow) to just dorsal to the atrioventricular valve (double arrowhead). An extensive trabecular system forms the spongy myocardium of the ventricle (V) and assists in ventricular ejection; a compact myocardium is not present in these fish. Longitudinal trabeculae are also found in the bulbus (B) for mechanical strength. Arrows indicate flow through the valves; bar equals 1 mm. (From K.R. Olson and J.S.D. Munshi, unpublished observation.)

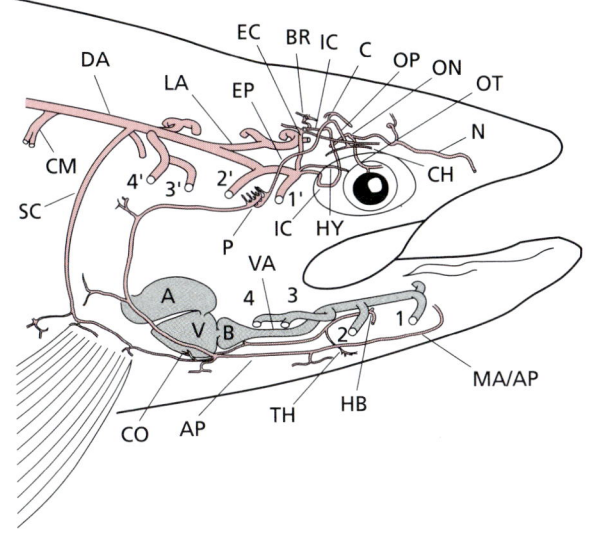

(a)

Figure 10.2 Prebranchial (gray) and postbranchial (red) arteries in the head of the trout. (For abbreviations see list on page 161.)

1992). The atrioventricular ostium is also valved and it closes during ventricular systole.

Ventricular structure varies both in its general shape and in the proportion of **trabeculated** (spongy) and compact myocardium (see Chapter 22). The ventricle is usually tubular in fish with elongated bodies, pyramidal with a triangular base in others (especially active fish, such as trout) or it may be sac-like as in some marine forms. A compact myocardium is absent in sedentary and relatively inactive fish and appears in increasing proportions as the activity level increases (trout have ~35% compact myocardium). Trabeculae are organized to form branching lumina in the ventricle and this appears to direct outflow toward the bulbus arteriosus and ventral aorta (Figure 10.1). The orientation of trabeculae may also enhance mechanical efficiency during systole (Santer, 1985; Satchell, 1991). The ventriculobulbar junction is also valved.

Arterial system

Pre-gill (branchial) arteries

Blood pumped by the ventricle passes through the bulbus arteriosus, ventral aorta and branched afferent branchial arteries before entering the gill tissue (Figure 10.2). The bulbus is a white tear-shaped vessel contained within the pericardial vault. It has thick elastic walls well endowed with vascular smooth muscle. The bulbus is over thirty times more compliant than the mammalian aorta and able to store from 25 to 100% of the stroke volume, thereby serving an important role in dampening ventricular pulse pressure (Bushnell *et al.*, 1992). (The **conus arteriosus** of elasmobranchs and other subteleostean fish occupies a similar position and function, but it consists of cardiac muscle.) In many bony fish, including trout, a dozen or more longitudinal septa line the lumen of the bulbus and extend nearly to the center (Priede, 1976). These convey structural advantage when the bulbus is distended during ventricular systole.

The ventral aorta passes through the anterior end of the pericardial vault and while doing so turns dorsally then again anteriorly in an 'S' shape. A pair of vessels arise from the lateral walls of the ventral aorta as it makes its final anterior turn and they travel posteriorly, each branching to form the third and fourth pair of afferent branchial arteries to the third and fourth pair of gill arches. The main trunk of the ventral aorta continues anteriorly, first giving rise to a pair of afferent branchial arteries that supply the second gill arches, and finally bifurcating into the afferent branchial arteries to the first pair of gill arches. These four, paired, afferent branchial arteries enter the gill arches and give rise to the gill filamental circulation (see Chapter 9). The ventral aorta and branchial arteries are elastic arteries and are structurally similar, although their walls are substantially thinner than that of the bulbus. There are no valves or intravascular septa in these vessels.

Post-gill (systemic) arteries

Systemic arterial vessels exit the gills from both the anteroventral and posterodorsal ends of the gill arches via efferent branchial arteries. The bulk of the cardiac output takes the latter route.

Anteroventral arteries

Extensions from the efferent branchial arteries of the first, second, and sometimes third and fourth gill arches exit the anteroventral portion of the arch and supply the lower jaw, pseudobranch (in trout) and heart (Figure 10.2). Efferent branchial arteries are paired in the anteroventral arch because the artery forks around the afferent branchial as the latter enters the arch. The lateral (outside fork) efferent branchial from the first gill arch becomes the mandibular artery. Side branches of this artery supply the lower jaw, whereas a main branch, now the afferent pseudobranch artery, makes an abrupt hairpin turn and proceeds ventroposteriorly and then dorsally following the arc of the operculum, to supply the pseudobranch. *En route*, a number of smaller arteries branch from the afferent pseudobranch artery and supply the operculum. (In non-salmonids, the afferent pseudobranch artery may originate from the posterodorsal arteries via the cephalic circle; Laurent and Dunel-Erb, 1984.) The pseudobranch is drained by the efferent pseudobranch artery, which becomes a singular ophthalmic artery to the **choroid gland** of the eye. A short transverse **commissural vessel**, about midway between the pseudobranch and choroid, connects the two ophthalmic arteries from the opposite sides of the head.

Ventral extensions from the medial (inside fork) efferent branchial arteries from the second pair of gill arches unite with each other to form the median hypobranchial artery. This artery travels posteriorly along the ventral surface of the ventral aorta and it bifurcates as it passes under the bulbus to form the left and right coronary arteries. One or more small branches of the coronary artery may continue posteriorly to anastomose with branches from the subclavian artery.

Posterodorsal arteries

The efferent branchial artery from the first gill arch becomes the lateral dorsal aorta as it turns posteriorly (Figure 10.2). Three other vessels, the internal and external carotids and the hyoidean artery, originate from the first gill arch efferent branchial artery as the latter makes its posterior turn. The internal carotid travels anteriorly and, after a short ventral dip, continues dorsally to the brain. Posterior branches of internal carotids from both sides of the head unite in the brain case to form the cephalic circle. The orbitonasal artery arises from the internal carotid about midway between the dip in the carotid and the braincase and travels anteriorly. One branch travels to the olfactory organ, the other becomes the optic artery and goes to the retina and lentiform body of the eye. The external carotid, or orbital artery, courses dorsally from the efferent branchial artery then branches anteriorly and posteriorly to perfuse the more superficial structures of the head. The hyoidean artery travels anteriorly to the anterior head region.

As the paired lateral dorsal aortas proceed posteriorly they are joined by efferent branchial arteries from the second gill arches (Figure 10.2). Continuing posteriorly, the two lateral dorsal aortas merge into a single common aorta that travels the longitudinal axis of the body ending in the caudal peduncle anterior to the caudal (tail) fin. Efferent branchial arteries from the third and fourth gill arches anastomose with each other and the single vessel thus formed – the epibranchial artery – joins a similar vessel from gill arches of the opposite side and together they drain into the ventral wall of the common (dorsal) aorta.

The entire common aorta is generally called the dorsal aorta, although, more precisely, it becomes the caudal artery posterior to the peritoneal cavity (Figure 10.3). The dorsal aorta travels just beneath the vertebra and above the dorsal surface of the kidney. Caudal to the peritoneum, ventral processes of the spine become fused forming the hemal arch and the caudal artery travels within this arch (the hemal canal) along with the caudal vein and several secondary veins. The dorsal aorta and caudal artery take on a characteristic undulating form due to their close association with the vertebrae. In some fish, including trout, the dorsal wall of the dorsal aorta and caudal artery tightly adhere to the vertebrae and the vessel cannot be dissected free. Priede (1975) has also described the presence of a longitudinal elastic ligament in the lumen of the trout dorsal aorta that is attached to the dorsal wall of the vessel via the basi-occipital bone and nearly completely divides the long axis of the vessel. Presumably this ligament acts as a blood propulsor when the trunk undulates during active swimming.

No systemic arteries arise directly from efferent branchial arteries from gill arches two, three and four, or from the lateral dorsal aortas. However, a pair of segmental arteries arises from the dorsal aorta shortly after the union of the lateral aortas and anterior to the anastomosis of the epibranchial arteries with the dorsal aorta. Paired subclavian arteries arise from the lateral wall of the dorsal aorta just posterior to the epibranchial anastomoses (Figure 10.3). The subclavian arteries run ventrally and supply the pectoral girdle, pectoral fins, and adjacent musculature. Branches of the subclavian arteries may connect directly to the epigastric arteries or may communicate with them via the coracoid artery (Sikorowa, 1946). Anteriorly, the subclavian arteries anastomose with the hypobranchial circulation (Figure 10.3). As the dorsal aorta travels posteriorly from the origin of the subclavian arteries it gives rise to numerous segmental arteries, a single celiacomesenteric artery to the viscera, small renal arteries and one or two small arteries to the posterior gut.

Segmental arteries

Segmental arteries (Figures 10.3 and 10.4) supply blood to the trunk musculature and possibly skin, although the secondary circulation may perfuse the latter (see below). Górkiewicz (1947) described three groups of segmental arteries: dorsal, lateral, and ventral. Dorsal and lateral segmental arteries and their corresponding veins travel in the septa between myotomes and small branches leave periodically to enter the myotomes and perfuse the muscle fibers and to perfuse the skin. Ventral segmental arteries follow a slightly different pathway but serve a similar purpose. Often segmental arteries and veins appear in alternating myosepta.

Dorsal segmental arteries leave the dorsolateral wall of the dorsal aorta and travel dorsally, briefly skirting laterally around the vertebrae *en route*. They then accompany the dorsal spines and continue to the dorsal surface. Dorsal segmental arteries of trout are paired, whereas in many other fish one of the pair is reduced. Midway along their route they give rise to perpendicular secondary muscular arteries that branch profusely to perfuse the dorsal musculature. Further dorsally, tertiary muscular arteries branch to perfuse the muscle beneath the dorsal fins.

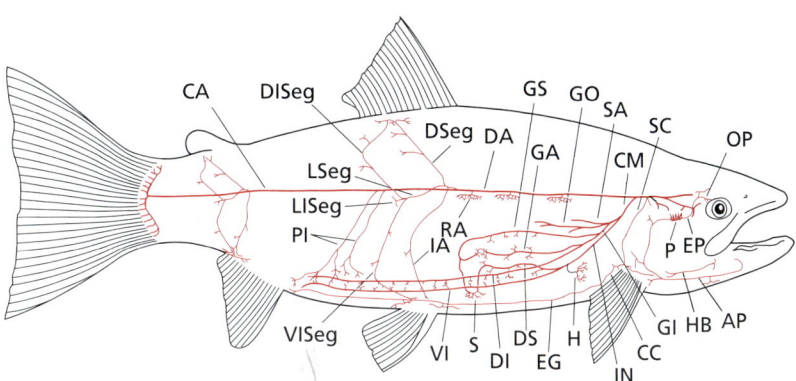

Figure 10.3 Major systemic arteries in the trout. The hypobranchial and afferent pseudobranch arteries arise from ventral extensions of the efferent branchial arteries. All other systemic arteries shown are supplied from the dorsal efferent branchials. (For abbreviations see list on page 161.)

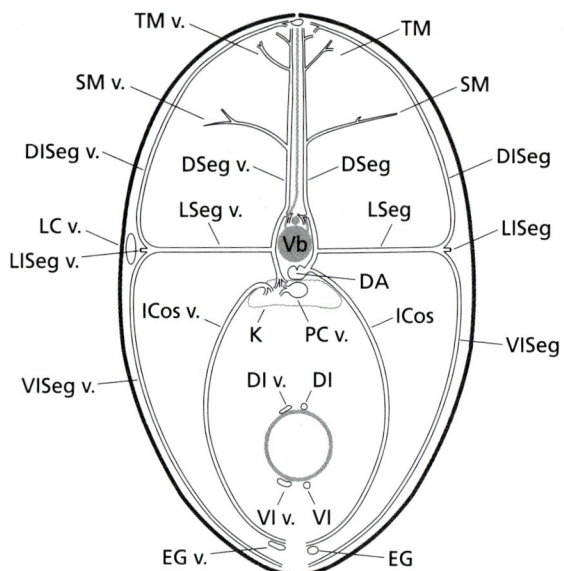

Figure 10.4 Segmental vessels to the trunk of the trout. Arterial vessels are drawn on the left and venous vessels on the right. The venous system essentially mirrors the arterial except that segmental veins drain into the renal portal sinusoids and blood re-enters the posterior cardinal vein before returning to the heart. (For abbreviations see list on page 161.) (Redrawn from Górkiewicz, 1947, and Olson, unpublished.)

Lateral segmental arteries arise from the dorsal segmental arteries at the level of the ventral border of the vertebrae and run transversely through the horizontal myosepta of the axial muscles to a point just beneath the red muscle. Here they divide into three branches, forming the dorsal, lateral, and ventral intersegmental arteries. The dorsal branch travels dorsally between the muscle and skin, roughly following the path of the myotomes and its fine terminal branches anastomose with the tertiary arteries of the dorsal segmental vessels. The ventral branch follows a similar ventral-ward course and its terminal branches ultimately anastomose with terminal vessels of the epigastric artery. Numerous branches along the length of both dorsal and ventral intersegmental arteries supply the underlying myotomes and superficial skin. The lateral intersegmental artery is considerably shorter and smaller in diameter than the other two and it continues laterally to perfuse the red muscle.

The ventral segmental arteries, more commonly known as the intercostal arteries, leave more ventrally from the aorta and travel almost perpendicularly toward the ribs. They then follow the ribs ventrally. Small branches along their length supply the ventrolateral muscles of the trunk, peritoneum, and ribs. The distal fine branches also anastomose with the epigastric vessels. Intercostal arteries are smaller than the lateral segmental arteries.

Paired epigastric arteries travel longitudinally along the ventral aspect of the fish between the peritoneum and abdominal muscles and link the pectoral and pelvic girdles. They are not true segmental vessels but have numerous capillary anastomoses with the subclavian, intercostal, and ventral intersegmental arteries.

Visceral arteries

Visceral arteries in salmonids (Figure 10.3) have been described by Grodzinski (1938), Koniar (1947), Smith and Bell (1975) and more recently by Olson and Meisheri (1989) and Thorarensen *et al.* (1991). In most trout, the celiacomesenteric artery originates as a single vessel from the ventral wall of the dorsal aorta, slightly posterior to the anastomosis of the latter with the third–fourth arch epibranchials (Figures 10.2 and 10.3). The celiacomesenteric artery travels posteroventrally, usually passing to the right of the esophagus, and soon divides into the celiac (anterior) and mesenteric (posterior) branches, although these are more commonly called the intestinal and gastrointestinal arteries, respectively (Grodzinski, 1938; Thorarensen *et al.*, 1991). Rarely in trout, the celiac and mesenteric arteries may arise from the dorsal aorta as individual vessels. In some non-salmonids, the mesenteric artery may originate quite posteriorly to the celiac. Other than a few segmental arteries that supply the very posterior gut, the celiacomesenteric artery is the only visceral artery in trout.

The intestinal artery travels ventrally and gives rise first to the hepatic artery (to the liver) and then turns posteriorly and branches to form the duodenosplenic artery (to the pyloric ceca and spleen) and ventral intestinal artery to the ventral intestine. Branches from the gastrointestinal artery include paired gonadal arteries (to the gonads), gastric (to dorsal stomach), gastrosplenic (to stomach and spleen), and dorsal intestinal (to dorsal intestine) arteries. In a number of instances, small terminal arteries from different sources anastomose to form a collateral arterial circulation. This is seen in the union of the gastrosplenic and duodenosplenic arteries in the spleen, the gastric and duodenosplenic arteries in the stomach, the dorsal and ventral intestinal arteries along the intestine, and one or two segmental arteries (posterior intestinal arteries) that travel from the distal dorsal aorta to the rectum and very posterior intestine and anastomose with both dorsal and ventral intestinal arteries (Olson and Meisheri, 1989; Thorarensen *et al.*, 1991). These segmental arteries are unlike the intercostal arteries in that they pass ventrally through the midline of the kidney and travel directly through the peritoneal cavity *en route* to the intestine.

The swimbladder in **phystostomes** such as the trout and eel is perfused by a branch from the celiacomesenteric artery (Steen, 1970). This vessel branches into a capillary network in the resorbent portion and then via retia forms a capillary bed in the secretory end. In **physoclists** the celiac and intercostal arteries supply the secretory and resorbent tissues, respectively (Steen, 1970). The kidney is supplied by numerous small renal arteries that arise either directly from the dorsal aorta, from segmental arteries, or from the venous renal portal system, described below.

Venous system

Many veins in fish, as in other vertebrates, are closely associated with the artery whose blood they return. By convention these veins are given the same name as their corresponding artery and, unless noteworthy, will not be recounted here. There are a number of exceptions to this, e.g. jugular, cardinal and abdominal veins, and these are described in the following paragraphs. In addition, there are several aspects of the piscine venous system not common to all vertebrates such as a renal portal system and the venous caudal 'heart'. These will be described in greater detail.

The salmonid venous system has been the subject of numerous studies (Grodzinski, 1938; Koniar, 1947; Olson and Meisheri, 1989; Thorarensen *et al.*, 1991). The most recent application of corrosion casting techniques (Thorarensen *et al.*, 1991) has provided the greatest detail of the complexity of the system and will be relied upon in this discussion.

Central venous system

The central venous system is comprised of the ductus Cuvier that delivers blood to the sinus venosus and the large longitudinal (axial) veins that return blood from the head and trunk (Figures 10.5 and 10.6). Fish veins do not have parietal valves. These one-way valves, found throughout the length of longitudinal veins of higher vertebrates, are presumed not necessary in fish because they live in a neutrally buoyant environment and there is little need to assist venous return against the forces of gravity (Satchell, 1991). However, ostial valves, located at the opening of tributary veins into longitudinal veins, are plentiful and prevent back-flow during contraction of skeletal muscles and these skeletal muscle 'pumps' are employed

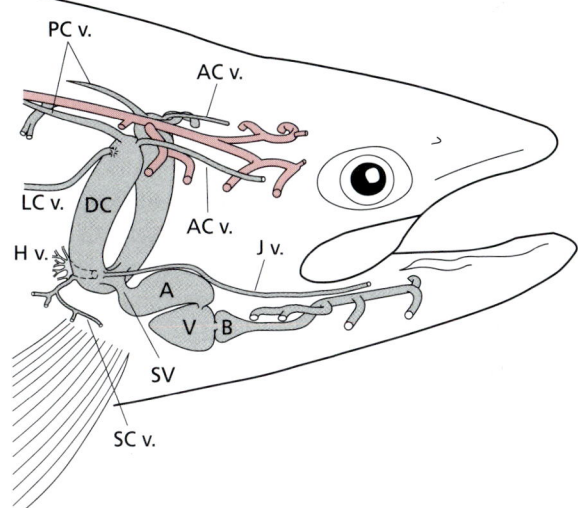

Figure 10.5 Central venous system (gray) in the trout head. (For abbreviations see list on page 161.)

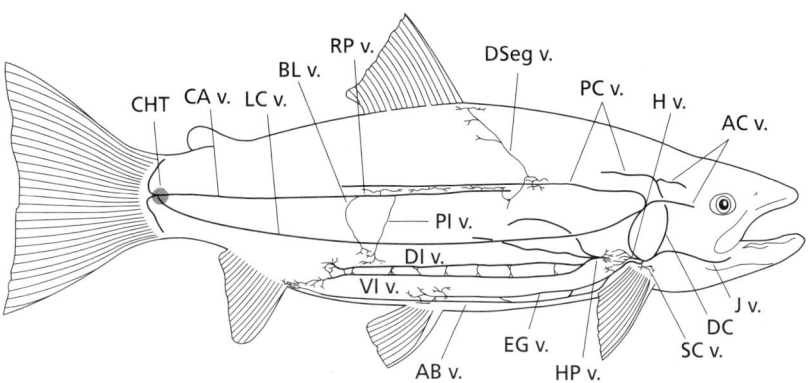

Figure 10.6 Major systemic veins in the trout. The two portal systems are evident. (For abbreviations see list on page 161.)

extensively throughout the venous system to assist venous return.

The ductus Cuvier is a 'U'-shaped vessel embedded in the transverse septum – the latter separating the pericardial and peritoneal cavities (Figure 10.5). The ostium connecting the ductus with the sinus venosus is located ventrally (near the bottom of the 'U') as is the entrance of the hepatic vein in the peritoneal side. The anterior and posterior cardinal veins drain into the dorsal horns of the ductus, and the ductus is sometimes referred to as the common cardinal vein. A prominent pair of ostial valves is located at the junction of the anterior cardinal veins with the ductus Cuvier.

Venous drainage of the head

Three principal veins drain the head: a singular jugular and paired anterior cardinals (Figure 10.5). The jugular vein lies in the midline of the lower jaw, dorsal to the ventral aorta and ventral to the esophagus. The jugular drains the thyroid gland, pharyngeal muscles and tissues of the lower jaw, the venous drainage from the anteroventral ends of the gill arches, and a portion of the pericardium. It drains directly into the sinus venosus slightly to the right of the midline (Sikorowa, 1946). The anterior cardinal veins are the central extensions of the orbito-nasal veins and drain the snout, olfactory rosettes, eyes, braincase (via anterior, medial, and posterior cerebral veins), and other tissues in this region.

Venous drainage of the trunk

Renal portal system

Much of the venous return from the trunk musculature and skin, and nearly all of it from the postabdominal region and tail fin, must pass through the renal portal system before it is returned to the heart (Satchell, 1991). Segmental veins in the postabdominal trunk and primary veins from the caudal fin drain into the central caudal vein located in the hemal arch. This vein also receives flow from secondary vessels in the caudal fin and lateral cutaneous veins via the caudal heart (described below) as well as from the urophysis. As the caudal vein approaches the kidney it is joined by the bladder vein and within the kidney a branch from the caudal vein connects with the posterior intestinal vein. The bladder may also be connected directly to the posterior cardinal vein via a second vein that passes through the gonads. It is not clear if the direction of flow in these two branches is toward or away from the caudal vein. If the latter, they may constitute distinct portal systems in parallel with the renal portal system (Satchell, 1991). Nevertheless, the majority of caudal venous blood enters the kidney and here the vein arborizes and the blood percolates through the renal parenchyma around the renal tubules.

Segmental vessels also contribute to the renal portal system. In the abdominal region of the trunk, the dorsal and lateral segmental veins anastomose and a short common vessel directly delivers blood to the renal parenchyma, as do the intercostal veins.

Renal portal blood and postglomerular renal arterial blood are then collected into the posterior cardinal vein as it passes through the kidney and the blood is delivered to the ductus Cuvier. In most fish, including the trout, the left posterior cardinal vein is either considerably reduced, or non-existent, and the right posterior cardinal vein assumes a central position and serves as the main conduit for venous return.

Hepatic portal system

Essentially all drainage from the viscera flows into a short hepatic portal vein and subsequently perfuses the liver sinusoids. Numerous anastomotic connections between splanchnic vessels form a diffuse system that provides a number of pathways for venous return. Thus the gastrosplenic vein connects the splenic vein with the lateral gastric vein, the duodenosplenic vein connects the spleen with the pyloric ceca and the duodenum, and venous anastomoses within the spleen form a patent gastrodudeno-splenic loop that has several connections with the hepatic portal vein (Thorarensen et al., 1991). The dorsal and ventral intestinal veins are similarly connected with each other through circumferential veins that encircle the intestine. The dorsal and ventral veins travel along the respective borders of the intestine and directly anastomose with the hepatic portal vein.

Hepatic arterial and hepatic portal blood mix in the liver sinusoids. This blood is collected into a very short hepatic vein that subsequently passes through the transverse septum into the medioventral wall of the ductus Cuvier and enters the sinus venosus. Usually trout have a single hepatic vein, although one and sometimes two additional smaller veins may also be present.

Cutaneous veins

There are four noticeable longitudinal vessels that run beneath the skin, one each on the dorsal and ventral margins and a pair laterally, approximately medial to the lateral line. There is substantial confusion regarding the nature of these vessels, their connections with other vessels, the direction of flow, and even the composition of fluid transported by them. The cutaneous vessels drain the skin, however, the fluid in them often appears relatively devoid of red cells and there is currently considerable debate whether these vessels are lymphatic vessels or systemic veins of the secondary circulation (see below).

The abdominal vein extends ventrally from the anus (although there are reports that a branch continues on to the tail) along the ventral midline to the region of the pectoral fins. Here it usually turns laterally and anastomoses with either the left epigastric or subclavian vein or drains directly into the ductus Cuvier (Górkiewicz, 1947; Sikorowa, 1946). The abdominal vein anastomoses repeatedly with the epigastric veins and in the pelvic region with the iliac vein. The dorsal vein follows a similar course beneath the dorsal skin. In the region of the dorsal fin it often divides into three branches, two of which travel lateral, and one medial, to the fin rays.

Lateral cutaneous veins connect at the caudal end with the caudal heart and, anteriorly, appear to drain into the ductus Cuvier, although the actual connections remain unclear. It has been suggested (Satchell, 1991) that undulating swimming movements would propel blood in the trunk away from the heart and the function of the caudal heart is to collect this blood and deliver it to the caudal vein, where the walls of the latter are insulated from muscular contraction by the bony hemal arch. These same skeletal muscle contractions also impart the pumping action of the caudal heart. Vogel (1985b) has provided anatomical evidence that the caudal heart in the trout pumps blood from the secondary veins draining the caudal fin (see below) and he proposes that this is necessary to elevate pressure in the secondary veins to that of the primary veins. Typically, caudal hearts are small ($\sim$ 1–2 mm in adult trout) and because their capacity must be limited, their overall importance remains unclear. Furthermore, in non-swimming fish most flow through the lateral cutaneous veins appears to be forward and fluid is delivered to the heart via anastomoses between the cutaneous vein and the ductus Cuvier (Figures 10.6 and 10.7; M. Russell, personal communication).

Secondary circulation and lymphatics

The anatomy of the fish lymphatic vessels has received considerable attention, especially by anatomists around the turn of the twentieth century and the field was elegantly summarized by Kampmeier in 1969. These studies provided considerable detail on numerous and extensive lymph-like vessels in a wide variety of fish.

However, recent electron microscopic analysis of tissue and of vascular corrosion replicas led Vogel (see reviews: Vogel, 1985a; Steffensen and Lomholt, 1992; Olson, 1996) to conclude that fish do not have a lymph system, but rather a second vascular system (the secondary circulation) that is derived from, and directly connected to, the primary circulation. The secondary circulation has been observed in gills, skin, fins, lining of the peritoneum and oral mucosa, and in the heat exchange muscles of the tuna. It has not been found in other skeletal muscle, brain, or splanchnic tissues, in fact neither secondary nor lymph vessels have been observed in the intestine or pyloric ceca.

The origin of secondary vessels has been well documented. Small, narrow-bore and usually tortuous arterioles arise profusely from primary arteries of the gill (efferent filamental and branchial arteries), along the dorsal aorta and from some segmental arteries (see Chapter 22 for additional anatomical details). These arterioles then anastomose with each other to form large diameter arteries of the secondary circulation that then travel to the aforementioned tissues where they branch to form capillary networks. The secondary system is presumed to drain into the cutaneous veins, the caudal vein via the caudal pump, and the branchial veins of the gill.

Lametschwandtner's group has arrived at a somewhat different conclusion regarding the secondary circulation (Lahnsteiner et al., 1990). Their studies also indicate that systemic secondary vessels arise from segmental arteries, however the tortuous vessels connect directly to segmental veins and then to the dorsal cutaneous vein without supplying a nutritive capillary network (Figure 10.7). Clearly this needs additional investigation. Lahnsteiner's study (Lahnsteiner et al., 1990) is also instructive in that it shows the relationship between the dorsal segmental vessels, the neural

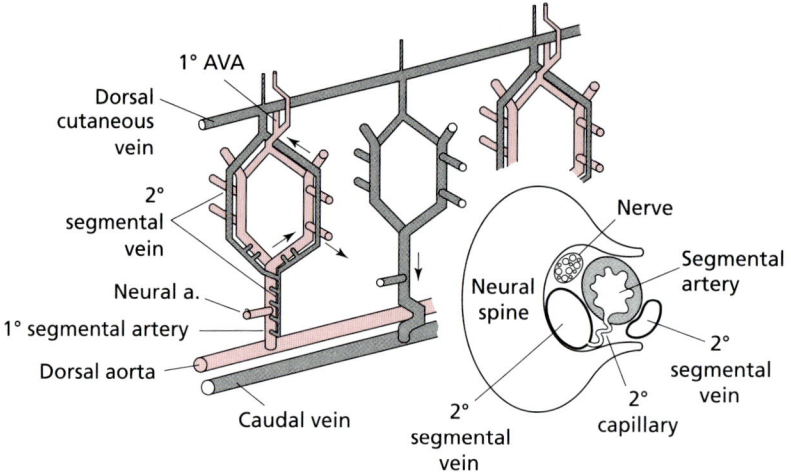

Figure 10.7 Primary and secondary dorsal segmental vessels in *Blennius pavo* and *Zosterisessor ophiocephalus*. A single dorsal segmental artery arises from the dorsal aorta at alternating metameres and bifurcates, sending tributaries to skin and muscle. A lesser branch returns medially to anastomose with the contralateral branch and the resultant vessel proceeds dorsally, eventually bifurcating into a dorsal fin arteriole and a primary arteriovenous anastomosis with the dorsal cutaneous vein. Other vessels (secondary capillaries) arise from the segmental artery and connect to parallel secondary segmental veins, the latter draining into the dorsal cutaneous vein. Neural arteries and veins arise from alternating segmental vessels on alternating metameres. Segmental veins in adjacent metameres (unlabelled) drain into the caudal vein and these segmentals also drain the neural vein. **Inset** shows the relationship (in cross-section) between segmental artery and secondary vessels. (Redrawn from Lahnsteiner *et al.*, 1990, with permission.)

vessels to the spinal cord, and the dorsal cutaneous circulation (Figure 10.7).

Although a number of ideas have been presented (see above references), the physiological function of the secondary circulation is unclear. The small tortuous vessels restrict access of red cells to this system and hematocrit of fluid from secondary veins often is less than 1, implying that transport of respiratory gases is not a major factor. The secondary system may indeed be the progenitor of the lymphatic system of other vertebrates.

References

Bushnell, P.G., Jones, D.R. and Farrell, A.P. (1992). In *Fish Physiology Vol. XII. Part A: The Cardiovascular System* (eds W.S. Hoar, D.J. Randall and A.P. Farrell), pp. 89–120. Academic Press, San Diego.

Farrell, A.P. and Jones, D.R. (1992). In *Fish Physiology Vol. XII. Part A: The Cardiovascular System* (eds W.S. Hoar, D.J. Randall and A.P. Farrell), pp. 1–73. Academic Press, San Diego.

Górkiewicz, G. (1947). *Bulletin International de l'Academie Polanaise des Sciences et des Lettres. Classe des Sciences Mathematiques et Naturelles. Serie B: Sciences Naturelles* **III**, 241–261.

Grodzinski, Z. (1938). In Anonymous *Klassen und Ordnungen des Tieffeichs, 6, 1, 2. Echte Fische Teil 22, 1. Liferung*, pp. 1–70. Akademie Verlagsgesellschaft, Leipzig.

Kampmeier, O.F. (1969). *Evolution and Comparative Morphology of the Lymphatic System*. Charles C. Thomas, Springfield, IL.

Koniar, S. (1947). *Bulletin International de l'Academie Polanaise des Sciences et des Lettres. Classe des Sciences Mathematiques et Naturelles. Serie B: Sciences Naturelles*. **III**, 261–275.

Lahnsteiner, A., Lametschwandtner, A. and Patzner, R.A. (1990). *Scan. Microsc.* **4**, 111–124.

Laurent, P. and Dunel-Erb, S. (1984). In *Fish Physiology Vol. X. Gills. Part B: Ion and Water Transfer* (eds W.S. Hoar and D.J. Randall), pp. 285–323. Academic Press, New York.

Olson, K.R. (1996). *J. Exp. Zool.* **275**, 172–185.

Olson, K.R. and Meisheri, K.D. (1989). *Am. J. Physiol. Regul. Integr. Comp. Physiol.* **256**, R10–R18.

Priede, I.G. (1975). *J. Zool., Lond.* **175**, 39–52.

Priede, I.G. (1976). *J. Fish Biol.* **9**, 209–216.

Santer, R.M. (1985). *Adv. Anat., Embryol. Cell Biol.* **89**, 1–102.

Satchell, G.H. (1991). *Physiology and Form of Fish Circulation*. Cambridge University Press, New York. 235 pp.

Sikorowa, L. (1946). *Bulletin International de l'Academie Polanaise des Sciences et des Lettres. Classe des Sciences Mathematiques et Naturelles. Serie B: Sciences Naturelles* **III**, 299–308.

Smith, L.S. and Bell, G.R. (1975). *A Practical Guide to the Anatomy and Physiology of the Pacific Salmon*, 27th edn. Canadian Fisheries and Marine Service Miscellaneous Publication.

Steen, J.B. (1970). *Fish Physiology Vol. IV* (eds W.S. Hoar and D.J. Randall), pp. 413–443. Academic Press, London.

Steffensen, J.F. and Lomholt, J.P. (1992). In *Fish Physiology Vol. XII. Part A: The Cardiovascular System* (eds W.S. Hoar, D.J. Randall and E.P. Farrell), pp. 185–213. Academic Press, San Diego.

Thorarensen, H., McLean, E., Donaldson, E.M. and Farrell, A.P. (1991). *J. Fish Biol.* **38**, 525–531.

Vogel, W.O.P. (1985a). In *Cardiovascular Shunts. Phylogenetic, Ontogenetic and Clinical Aspects* (eds K. Johansen and W. Burggren), pp. 143–159. Munksgaard, Copenhagen, Denmark.

Vogel, W.O.P. (1985b). *Acta Anat.* **121**, 41–45.

CHAPTER 11

Digestive System

Randal K Buddington
Department of Biological Sciences, Mississippi State University, Mississippi, USA

Victoria Kuz'mina
Institute for the Biology of Inland Waters, Borok, Yaroslavl, Russia

Organ systems are modified over evolutionary time so that the structure and functions are matched to the demands placed on them. This is particularly true for the digestive system, which has been highly modified during the evolution of fish allowing the numerous species to exploit a wide diversity of diet types. Arguably, the structure of the digestive system is more variable than that of any other organ system. This is evident from the variation so well described in the comprehensive study of Suyehiro (1941).

Despite the variation in structure, the functions of the digestive system are consistent among species and include: (i) digestion of feedstuffs; (ii) osmoregulation; (iii) secretion of hormones involved in regulation of digestion, metabolism, and other bodily functions; and (iv) defending the body against invasion by pathogens and other harmful components of the environment.

The objective of this chapter is to familiarize readers with the gross anatomy of the principal regions of the digestive system, including the associated organs, and how they are closely related to functions. It must be remembered that the alimentary canal is a tube that passes through the body of a fish. What is inside the alimentary canal is actually outside of the fish. Therefore, the alimentary canal represents a critical interface that regulates the exchange of energy and matter between the internal and external environments of fish.

Regions and components of the fish digestive system

The first region of the digestive system includes the mouth, oral cavity, and pharynx. After food is swallowed it enters the alimentary canal proper and proceeds via the esophagus to the stomach followed by

Copyright © 2000 Academic Press

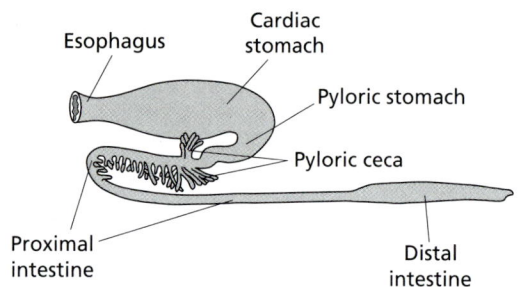

Figure 11.1 The alimentary canal of the rainbow trout.

the intestine. A generalized alimentary canal of a fish is presented in Figure 11.1. This basic plan has been highly modified to meet the wide variation in functional demands placed on the digestive system by the diversity of diets consumed by fish. Fish generally lack a region analogous to the large intestine of other vertebrates and digested food items enter a short rectum before they are voided through the vent (anus). Organs associated with the alimentary canal that provide secretions and are critical for normal functions include the liver, associated gall bladder, and the pancreas.

It needs to be noted that the alimentary canal is located in the peritoneal cavity, which also contains most of the other viscera. The peritoneal cavity is lined by a membrane, the peritoneum. The viscera, including the alimentary canal, are not actually in the cavity. Instead, they are surrounded by the peritoneum, much like sticking a finger into a balloon. The resulting double-layered membrane that attaches the viscera to the body wall is called mesentery. It can be seen as a thin sheet of tissue with embedded fat and lymph nodes. The mesentery functions to hold the viscera in position, but at the same time allows for some movement.

Mouth, oral cavity, and pharynx

The mouth, oral cavity, and pharynx are responsible for recognizing, acquiring, and the initial processing of food items. Food items consumed by fish range from large items that are consumed individually, down to small individual particles (e.g. phytoplankton and zooplankton) that are filtered from the water by the gills. Corresponding with this, both the location and size of the mouth are highly variable. For example, the mouth is generally located at the tip of the rostrum of species that consume food items present in the water column (e.g. salmonids), whereas the mouths of species that are bottom feeders tend to be ventrally situated, such as seen in the carp and sturgeons. Mouth size also spans a wide range. Filter feeders that must process large volumes of water can have expansive mouths (e.g. the paddlefish). The mouth includes skeletal features that constitute the jaw (see Chapter 6), and these are highly variable in shape corresponding with the variety of different feeding habits. However, not all fish have jaws (e.g. lamprey and hagfishes).

The mouth opens into the oral (or buccal) cavity and this region extends back to the pharynx and includes the tongue. Although fish have tongues, they are not nearly as well developed and capable of manipulating food items as the tongues of other vertebrates. Food and water pass through the buccal cavity to the pharynx, with the water directed over the gills, whereas food passes into the esophagus. The pharynx has six pairs of pharyngeal pouches (more in cyclostomes), including the gill chambers. Because of the different structures present in the pharynx, this region can be considered as being part of the digestive and respiratory systems. Some structures can participate in both functions. Exemplary are the gills of some species which have gill rakers that have been modified to form a sieve for filtering the water to obtain food items.

Most fish have teeth, but this is not universal. When teeth are present, they can be associated throughout the mouth and buccal cavity, including the lips and tongue. Furthermore, many, if not most, fish have teeth present on the upper and lower pharyngeal bones. In some species the pharyngeal teeth allow for the grinding of food items and form what is known as a 'pharyngeal mill'. Tooth shape is highly variable among fish, corresponding with the diversity of feeding habits. The teeth of carnivores tend to be conical (simple or with cusps), and the tips are pointed back to assist in seizing and holding food items, and directing them distally. In contrast, fish that consume diets dominated by plant materials (e.g. carp, tilapia) or crush food items (skates and rays) often have large, flattened, molar-like teeth that triturate the food, reduce particle size, and increase digestibility. The teeth of other fish can be shaped for tearing and scraping. When all of the teeth are of the same form (but maybe different size) this is referred to as homodont dentition, with heterodont dentition referring to the presence of different forms of teeth.

Fish are able to replace lost teeth, with the replacements located under and moving up into position when needed. This is easily seen by examining the jaws of sharks. The teeth of fish are covered by a substance known as enameloid, which differs from the enamel covering the teeth of higher vertebrates by being innervated.

Members of the shark family and certain other species differ from mammals in that the teeth are not directly connected to the jaw, but are instead held in place by collagenous fibers (acrodont dentition). They are routinely shed or lost during eating, and are rapidly replaced. The teeth of other species are more directly connected to the jaw and are set in sockets (thecodont dentition).

Esophagus

After food is swallowed, it enters the esophagus, which is considered by many to be the first region of the alimentary canal. The esophagus functions as a conduit leading to the stomach and remainder of the alimentary canal. The esophagus of fish is short and is characterized by the presence of longitudinal folds that allow for expansion to accommodate the passage of large food items. The esophagus leads directly to the intestine in primitive fish (e.g. hagfish) and some modern fishes that lack a secretory stomach (e.g. carp), but in most fish the esophagus terminates at the stomach. In some species a sphincter separates the esophagus from the stomach.

In many species, an elongate sac known as the air bladder lies dorsal to the alimentary canal. There is some speculation that the air bladder is homologous to the lungs of higher vertebrates. The air bladder is a hydrostatic organ that can be filled or emptied to regulate density, hence buoyancy, of the fish. In primitive species the air bladder remains connected to the esophagus by the pneumatic duct (physostomous). In higher teleosts the connection to the alimentary canal is lost during early development (physoclistous) and regulation of air bladder filling is dependent on a 'gas gland'.

Stomach

As mentioned above, not all fish possess a secretory (true) stomach as adults and in those species the esophagus is directly connected to the intestine. A functional stomach is also lacking in larvae of virtually all fish. Even when a true, secretory stomach is not present, an expanded region of the esophagus can serve to store food. The presence of a 'storage organ' allows fish, and other vertebrates, to eat infrequently and take advantage of abundant food resources when they are available.

The gross anatomy of the stomach is highly variable with regard to both size and structure (Figure 11.2). The capacity of the stomach relative to the body weight is related to the size of meals that are consumed, being largest in species that consume large, single prey items or sporadically eat high volumes of food, whereas the stomach is smaller in planktivores that feed much more frequently or almost continuously. The shape of the stomach has been modified and can be in the shape of simple tubes that are straight or curved, sac-like structures, or various other configurations. In some species the stomach has a diverticulum that is often called a gastric cecum. The walls of the stomach are thick, particularly in the distal (pyloric) region and are generally folded, allowing for distention. The longitudinal folding is more varied than that seen in the esophagus, with the specific patterns of folding differing among species and regions.

The stomach can be separated into three regions. The first is often called the cardiac region and is directly connected to the esophagus. The cardiac region is non-secretory, and although food is 'stored' in all regions of the stomach, the main function of the cardiac region is storage of food, particularly in species that ingest large meals. The remaining two regions of the stomach (fundic and pyloric) are secretory in fish with true stomachs. The walls of the stomach are thicker, and more muscular than other regions. In some species the muscle layers in the pyloric region are especially well developed, forming a grinding organ (gizzard) that allow it to be distinguished from the adjacent fundic region.

The well developed muscle layers of the stomach physically disrupt the food and mix it with gastric secretions resulting in smaller particles, or even a semi-liquid paste. The food leaves the stomach and

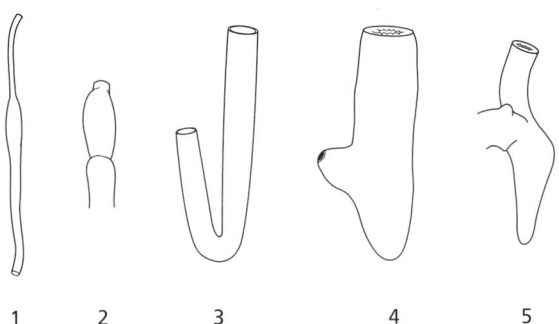

Figure 11.2 Different shapes of the stomach in fish, ranging from straight (1 and 2), to U-shaped (3) and T-shaped (4 and 5) (Adapted from Suyehiro, 1941.)

enters the intestine through the pyloric sphincter. The sphincter, when closed, physically separates the stomach from the intestine and prevents the passage of food items. The nervous system and hormones influence the state of pyloric sphincter contraction and thereby regulate the movement of food into the intestine. This is critical to insure that the hydrolytic and absorptive capacities of the intestine are not exceeded and the food is efficiently digested.

Intestine

The intestine is the principal site of digestion and its structure is highly variable corresponding with the wide diversity of feeding habits and functional demands (Figure 11.3). The most visible difference among fish is the length of the intestine (from the pyloric sphincter to the anus). In some species, notably carnivores, the length of the intestine can be as short as only about 20% of body length. In such species the intestine extends directly from the stomach to the anus. In contrast, the intestines of some herbivores can be over 20 times the length of the body. As the length increases, the intestine changes directions, and for the longest examples, the intestine can be highly coiled to fit into the limited space available in the body cavity.

Fish differ from other vertebrates by lacking a distinct colon, or large intestine, and functions considered to be characteristic of the small intestine (digestion and absorption) can be detected virtually from the pyloric sphincter to the vent (anus). The intestines of some fish can include a distal (caudal) region that is distinguished from a more proximal (cranial) region by a greater diameter. This region can also be darker and have annular rings, as seen in salmonids, or have other structural characteristics that allow it to be distinguished from more proximal regions. The proximal and distal regions of the intestine can be partly separated by a sphincter. In some sharks, a tubular gland is attached to the distal-most region of the intestine adjacent to the vent. The gland secretes salt and is important for osmoregulation.

Fish have developed four anatomical strategies to increase digestive surface area. All include modifying the basic structure of the intestine. The first is simply to lengthen the intestine. This strategy is common among fish that consume diets dominated by plants or phytoplankton. The intestines, though long in such species, are thin and the mucosa and underlying tissue layers reduced and less developed. A second strategy is the development of a thick mucosa with a very complex architecture that has extensive folding of the mucosa to increase surface area. Generally, this strategy is used by species with short intestines (e.g. carnivores).

The third adaptation is the development of diverticula that are evaginations of the proximal intestine, just distal to the pyloric sphincter. The diverticula, which are called pyloric ceca, range in number from one to thousands, as in the Thunnidae (tunas), and are highly variable in size and shape (Figure 11.4). Pyloric ceca are unique to fish, are not seen in any other vertebrates, and can exist as separate structures (salmonids, centrarchids). They can be organized by connective tissue into what appears to be a single mass (tuna), or can be fused into what appears to be a single organ (sturgeons). The ceca are restricted to the proximal small intestine and therefore differ from the ceca of other vertebrates that are found associated with the proximal colon. The ceca of some fish are found immediately adjacent to the pyloric sphincter, but in other species they can extend for several centimeters along the intestine. The ceca have the same histological features and functions as the adjacent proximal intestine and are known to be an important mechanism to increase digestive surface area.

The fourth adaptation to increase surface area is the presence of an internal epithelial fold called a

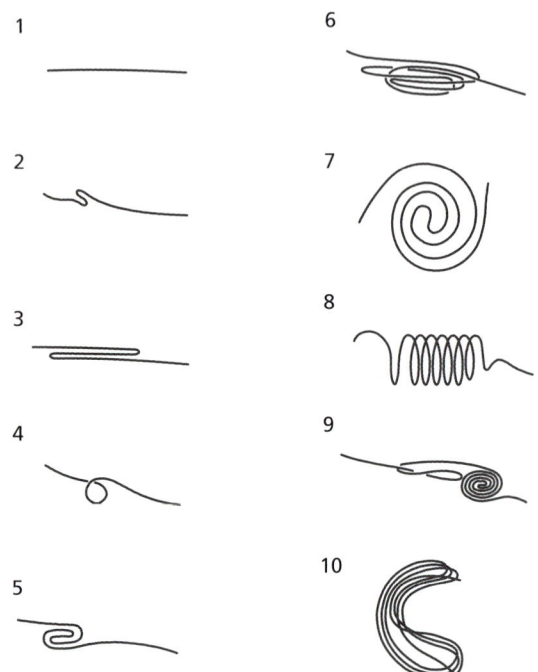

Figure 11.3 The length and configuration of the intestine vary widely among fish as evident from a short, straight intestine (1) to much longer intestines, which are made possible by an increasing complexity of twisting, spiraling, and turns (2–10). (Adapted from Suyehiro, 1941.)

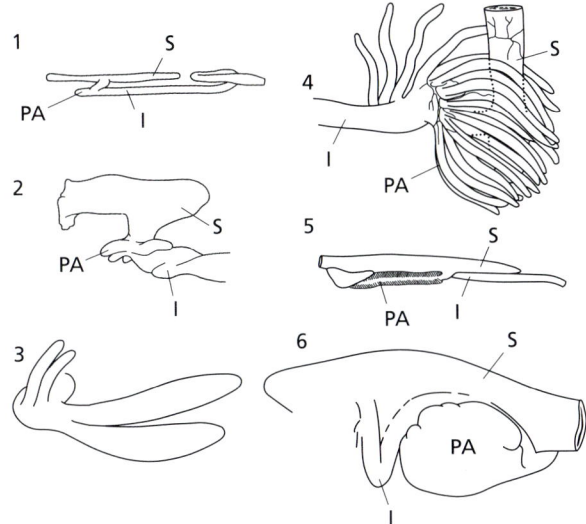

Figure 11.4 The pyloric ceca (PA) associated with the proximal intestine (I), when present range from a single, short appendage (1), to multiple appendages of varying lengths (2–4), to a single, fused mass of multiple ceca (5). S represents the stomach. (Adapted from Suyehiro, 1941.)

spiral valve in the distal region of the intestine. Spiral valves can twist several times and effectively increase the length (hence surface area) of the intestine through which the food must traverse. Spiral valves are found in sharks, acipenserids, and many other species.

Various combinations of the four strategies to increase digestive surface area can be seen in fish. For example, striped bass and salmonids have pyloric ceca and a thick mucosa with a complex architecture. Sturgeons use a combination of a thick mucosa along with fused ceca and a spiral valve.

Although a number of species consume plant materials, only a few fish have evolved expanded regions of the intestine that apparently store food and are used as chambers to enhance microbial digestion (Kyphosidae). Although many species of fish consume plant materials, transit times for most herbivorous fish are too short (< 12 h) for effective microbial digestion. Furthermore, changes in environmental temperatures and other conditions make it difficult for most fish to harbor consistent bacterial assemblages. As a result, the ability of most fish to obtain the nutrients associated with the diet is more dependent on chemical and physical digestive processes than on microbial symbioses.

The terminal region of the intestine can be expanded forming what is sometimes considered to be a rectum. Although some reports have considered the expanded, darker region of the salmonid intestine with annular rings to be the rectum, because the associated enterocytes possess digestive enzymes and nutrient transporters, this region is more appropriately considered as the posterior intestine and the rectum is the very short segment immediately adjacent to the vent.

Functional anatomy of the intestine

Although intestinal structure is highly variable among fish, the associated functions are relatively consistent across species. Food items that enter the intestine are known as chyme and are mixed with aqueous secretions from the intestine itself and from other organs (liver/gall bladder and pancreas). The secretions are essential for hydrolysis of food items, and include digestive enzymes and other components that enhance the activities of the enzymes by modifying the chemical environment. For example, bicarbonate from the pancreas neutralizes stomach acids and increases luminal pH to enhance the activities of enzymes from the pancreas and intestine. Other secretions reduce particle sizes, e.g. bile acids from the gall bladder emulsify fats and by doing so increase the surface area available for lipases.

The processes of digestion that occur in the lumen of the intestine reduce the complex polymers that are present in food items (e.g. proteins, carbohydrates, fats) into smaller fragments. The final stages of hydrolysis occur at or near the membrane of the enterocytes and release the 'building blocks' of the polymers (e.g. monosaccharides, amino acids, fatty acids and mono- and diglycerides), which are subsequently absorbed by the enterocytes.

Accessory organs

In addition to the alimentary canal proper, the digestive system includes the liver, pancreas, and gall bladder. These organs are derived from the alimentary canal during development and remain attached by ducts that carry the respective secretions to the alimentary canal where they are mixed with the chyme.

The liver is a large multilobed organ that is positioned at the level of the stomach. Because of its multitude of functions, the liver is often considered to be the largest organ and gland in the body. In most species of fish there are two distinct liver (hepatic) lobes, but there is little consistency in gross structure even within a species. The size and color of the liver are also variable, generally fluctuating between seasons and feeding states.

The majority of hepatic tissue is dedicated to metabolism. Blood draining the intestine, and therefore carrying absorbed nutrients, is directed to the liver via the hepatic portal vein. Therefore, one function of the liver is to 'filter' the blood draining the intestine before it enters the general systemic circuit. The metabolic capacities in conjunction with receiving blood from the intestine allow the liver to modify the concentrations of nutrients present in the blood and available to the rest of the body. In addition, the liver filters poisons and other toxic materials from the blood and the metabolic capacities transform many of them into other compounds. The toxic compounds or the associated metabolites can be diverted to the gall bladder followed by secretion into the intestine and elimination. Alternatively, the compounds are placed back in the blood and eliminated by the kidneys.

Another portion of the hepatic tissue is glandular and is involved with the production of bile. In most fish, after the bile is produced it is collected in small ducts called bile canniculi that coalesce and drain into a gall bladder. The gall bladder is generally spherical, though in some species it can be elongated. With appropriate stimulation, the gall bladder contracts and the bile is forced through the bile duct and enters the intestine through a papilla usually located just distal to the pyloric sphincter. If a gall bladder is not present (such as in sharks and their relatives), the bile drains directly into the proximal intestine.

The pancreas exists as diffuse tissue in most species of fish and can be found scattered throughout the peritoneal cavity where it is associated with mesenteries, blood vessels, and other organs (e.g. liver, spleen, intestine). Only in a relatively few can it be seen as a distinct organ (e.g. European eel).

The pancreas consists of both exocrine and endocrine tissues. Secretions of the exocrine pancreas include water, digestive enzymes, and bicarbonate and other ions. These collect in one or more ducts that drain into the proximal intestine and ceca, if present. The types of enzymes present in the exocrine secretions are similar to those described for mammals and other vertebrates. They include proteases (i.e. the endoproteases trypsin, chymotrypsin, and elastase and the exopeptidases, carboxypeptidases A and B), amylase, lipase, and deoxy- and ribonucleases. The relative proportions of the different enzymes are matched to the composition of the diet. For example, the exocrine pancreas of carnivorous fish synthesizes and secretes less amylase than the pancreas of herbivorous and omnivorous fish. Of critical importance to those raising fish is whether the digestive enzymes produced by the pancreas can be adapted to match changes in the composition of the diet, and there is evidence that this is possible in some species. In addition to diet, environmental conditions can also influence the activities of the enzymes.

The endocrine tissue of the pancreas secretes at least four different types of hormones, with insulin, glucagon, somatostatin, and pancreatic polypeptide the best known. The rate of pancreatic hormone secretion is known to be dependent on feeding state, other metabolic considerations, and environmental conditions.

Regulation of digestive system functions

There are three basic digestive functions that are regulated. These are motility, secretion, and absorption. Regulation of these functions is accomplished by the combined actions of nervous system inputs and chemical signals. Nervous system inputs can originate from the central nervous system via the autonomic nervous system, but are also from the enteric nervous system associated with the alimentary canal. The enteric nervous system includes the large number of neurons that extend throughout the alimentary canal. In fact, because the enteric nervous system is so well developed and includes so many neurons, it has been considered as a 'mini brain' capable of integrating sensory inputs and eliciting responses without central nervous system inputs.

There is a very wide diversity of chemicals that are released by cells and act to regulate digestive functions. In addition to a variety of neurotransmitters that affect innervated cells, there is a diversity of chemicals that are distributed in the extracellular fluids (paracine) or blood (endocrine/hormones). Secretin, the first hormone to be described, along with cholecytokinin regulate pancreatic functions and are released by the intestine in response to presence of acid and nutrients in the proximal intestine. Hormones that regulate digestive system functions and the respective actions are much better known for mammals. However, the increasing database for fish indicates that many of the signalling compounds, pathways, and responses are shared with mammals and other vertebrates, and have therefore been conserved during evolution. Not to be forgotten are the nutrients present in food. Although many chemical signals are released when nutrients are present in the alimentary canal, it is likely that some nutrients can directly influence digestive functions.

Much less is known about how the osmoregulatory and immune functions of the alimentary canal are regulated. A better understanding of the functional anatomy of the fish gastrointestinal tract and associated organs will be instrumental for efforts directed at providing this information.

Further reading

Buddington, R.K. Krogdahl, A. and Bakke-McKellep, A.M. (1997). *Acta Physiol. Scand.* **161,** Suppl 638, 67–80.

Fänge, R. and Grove, D. (1979). In *Fish Physiology*, vol. 8 (eds W.S. Hoar, D.J. Randall and J.R. Brett), pp. 161–260. Academic Press, New York.

Iwama, G. and Nakamishi, T. (eds) (1996). In *The Fish Immune System: Organism, Pathogen, and Environment*. Academic Press, New York.

Kuz'mina, V.V. and Gelman, A.G. (1997). *Rev. Fish Sci.* **5,** 99–129.

Reinecke, M., Müller, C. and Segner, H. (1997). *Anat. Embryol.* **195,** 87–102.

Romer, A.S. (1970). *The Vertebrate Body*. W.B. Saunders, Philadelphia.

Suyehiro, Y. (1941). *Jap. J. Zool.* **10,** 1–303.

Yasutake, W.T. and Wales, F.H. (1983). *Microscopic Anatomy of Salmonids: An Atlas*. United States Department of the Interior, Resource Publication 150, Washington, DC.

Zapata, A.G. and Cooper, E.L. (eds) (1990). In *The Immune System: Comparative Histology*. John Wiley, New York.

CHAPTER 12

Urinary Tract

Hartmut Hentschel
Max-Planck-Institut fur molekulare Physiologie, Dortmund, Germany

Marlies Elger
Institut fur Anatomie und Zellbiologie I, Universität Heidelberg, Heidelberg, Germany

Margaret Dawson and J Larry Renfro
Department of Physiology and Neurobiology, University of Connecticut, Connecticut, USA

Introduction

The kidneys of fishes, like those of other vertebrate classes, participate in body fluid homeostasis. A wide variety of kidney types have been found among the jawless and jawed fish (for review see Hentschel and Elger, 1987). In general, the structure of the kidney corresponds to the type of habitat in which the fish is found, namely, marine, brackish, or fresh water. Hagfish and the majority of marine teleosts have comparatively simple renal tubules that participate in ionic and osmotic regulation. Such regulation in fresh water appears to require a more elaborate organization; almost all freshwater fish have kidneys with well-developed multisegmental nephrons. Elasmobranchs have the most complex kidneys among the lower vertebrates; marine sharks and skates have zonate kidneys with a complicated countercurrent arrangement of early and late nephron segments that make up lateral bundles. This elaborate structure may be related to the retention of urea, an important factor in osmoregulation for these fishes. The kidneys of the true freshwater skates, the river rays, Potamotrygonidae, are not zonate and do not contain lateral bundles; their kidneys resemble those of other freshwater fish. This suggests that the elaborate structure of marine elasmobranch kidneys is related to the retention of urea, an important factor in osmoregulation for these fishes. Potamotrygonidae cannot elevate plasma concentrations of urea when they are transferred to water of high salinity.

Comparative anatomy of the kidney of fishes

The renal tissue is generally located retroperitoneally as discrete organs (Figures 12.1 and 12.2). An

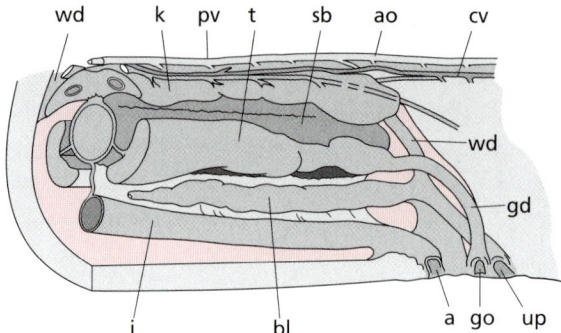

Figure 12.1 Location of kidney in caudal end of abdominal cavity of male pike, *Esox lucius*, left-side view, showing the median apertures of the rectum, genital ducts, and kidney ducts. a, anus; ao, aorta dorsalis; bl, urinary bladder; cv, caudal vein; go, gonadal duct; i, intestine; k, kidney (opisthonephros); pv, portal vein; sb, swimbladder; t, testis; up, uropore; wd, Wolffian duct. (After Goodrich; in Harder, 1975, modified.)

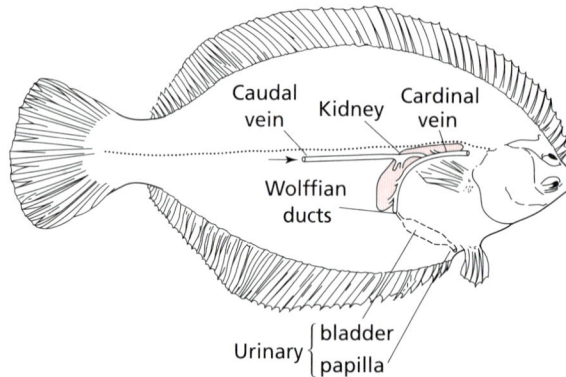

Figure 12.2 Dorsal view of a winter flounder, *Pleuronectes americanus*, showing the position of the kidney and the veins associated with the renal portal circulation. The kidney position is based on actual measurements. (After Mackenzie *et al.*, 1977, modified.)

exception is the unique organization of the kidney in hagfish (Myxinidae) that displays separate segmental glomeruli connected via vestigial tubules (necks) with the archinephric ducts. The shapes of fish kidneys are varied and correspond to the body type of their respective species. At one extreme are kidneys with great longitudinal extension, generally found in species with elongated bodies, such as lampreys, eels and eel-like fish, and dogfish. In most species, these elongated kidneys form a single unpaired organ, at least in the caudal portion. In other species, they form a pair of strap-like organs. The other extreme is found in stout fish species and in species with dorsoventrally compressed bodies, such as skates and anglerfish; kidneys in these species tend to be paired, completely separate, relatively short, compact organs (Figures 12.3 and 12.4). (For review see also Van den Broek and van Oordt, 1938.)

The excretory system of the kidney is complemented by a well-developed urinary bladder. Usually, the paired primary ureters, the Wolffian (archinephric) ducts join to form a urinary bladder, which, with a short urethral duct, opens to the body surface on a urogenital papilla behind the anus. Cytologically, the bladder may be considered as a dilation of the archinephric duct, i.e. the epithelium is similar to that of the latter. A single urinary bladder is present in most teleosts (Figure 12.5). In elasmobranchs, the large excretory ducts including the bladder differ between males and females. In the female shark, *Scyliorhinus caniculus*, the Wolffian ducts serve as ureters, and, at their end, dilate to form paired, so-called urinary bladders. These join to form a urinary sinus, which opens into the cloaca on a papilla caudally from the orifices of the oviducts. In addition, the collecting ducts originating from the most caudal portion of the kidney join to form several secondary ureters, which directly merge with the urinary sinus. In the male, the anterior portion of the opisthonephros has transformed into an epididymal organ, and the Wolffian ducts serve exclusively as sperm ducts. Segmental secondary ureters arise from the caudal, excretory portion of the (paired) opisthonephros. Several of these secondary ureters (the anterior ones) join to form a common (paired) duct, which, at the end, dilates to form a (paired) 'urinary bladder'. The two urinary bladders and the posterior secondary ureters join the unpaired urogenital sinus, which is also joined by the seminal vesicles. The urogenital sinus opens via a urogenital papilla into the cloaca. In the female, the dilated end portions of the Wolffian ducts make up the bipartite urinary bladders, whereas in the male, these are formed by the secondary ureters (Figure 12.5b).

Ontogeny

The kidney of fishes, like that of other vertebrates, develops from nephrogenic material of mesodermal origin that lies between the celomic cavity and the vertebral column. The first embryonic kidney, the pronephros, is functional in larvae of lampreys and many bony fishes including teleosts. In teleosts, such as trout and zebrafish (Tytler, 1988; Hentschel and Elger, 1996; Drummond *et al.*, 1998), the pronephric kidney consists of only one or a few pairs of glomeruli

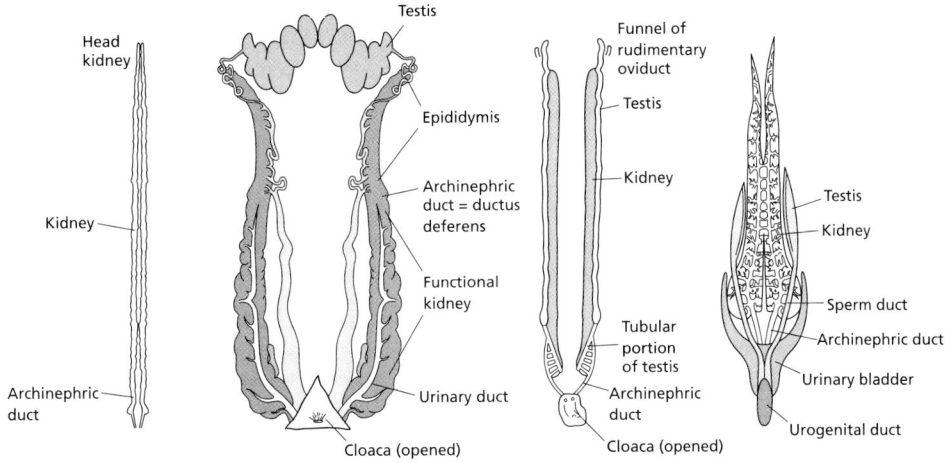
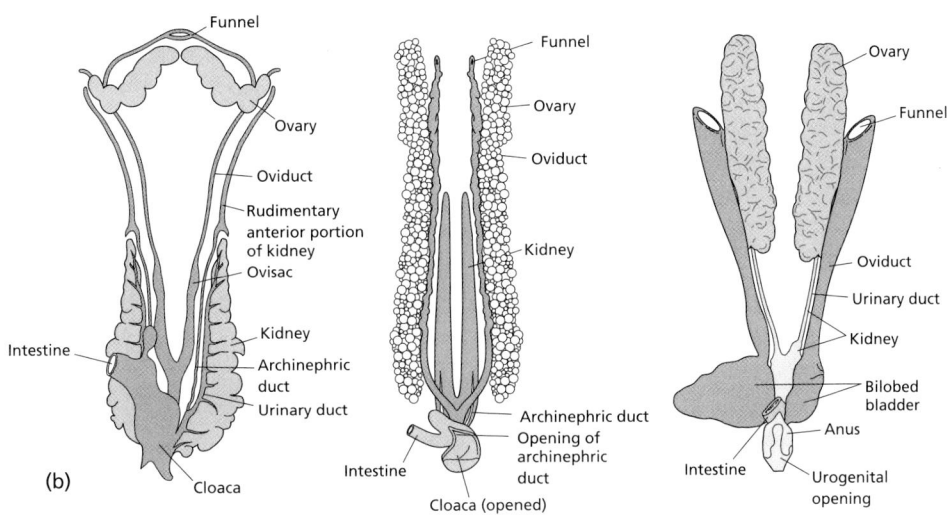
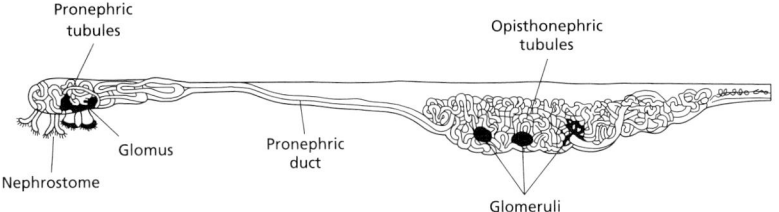

URINARY TRACT

GROSS FUNCTIONAL ANATOMY

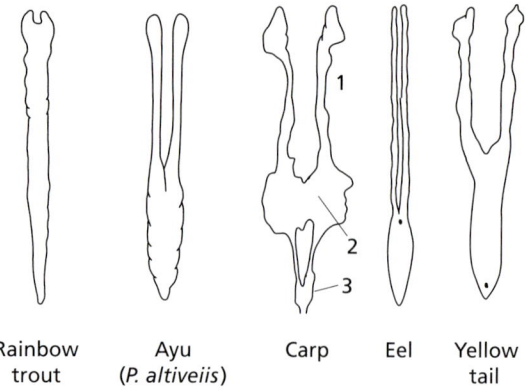

Figure 12.4 Teleostean kidney shapes. The regions of the kidney in bony fish can be assigned to (1) head kidney, (2) trunk kidney, and (3) caudal or tail kidney according to their relative position. However, it is important to note that these topographical terms of head kidney and trunk kidney are not coincident with the division of the kidney into pronephros and opisthonephros. While all portions including head kidney may contain opisthonephric nephrons, the head kidney may in addition contain remnants of pronephric nephrons (e.g. in poecilid fish). In most species the head kidney is a lymphoid organ with a minor amount of excretory kidney tissue. (After Harder, 1975, modified.)

tematic groups are aglomerular; that is, glomeruli do not develop in their opisthonephric kidneys, and their tubules perform excretory functions without glomerular ultrafiltration.

In the cranial portion of the opisthonephros, the segmental organization is generally retained during development. The more caudal portions of the opisthonephros frequently lack the segmental arrangement found in most vertebrate groups including fish, because in this region the kidney may develop from compact nephrogenic masses, rather than from segmental **anlagen**.

Several types of opisthonephric kidneys exist in teleosts (Figure 12.6). Opisthonephros with segmental glomerular nephrons and archinephric ducts serving as the only excretory ducts, can be considered a primitive type. In more advanced types of opisthonephric kidneys, the segmental organization is obscured by the subsequent growth of further nephron generations (secondary, tertiary, and higher order) that join the segmental tubules. A pars sexualis is often present in ancient, ganoid fish and in elasmobranchs, frequently in a cranial position from the pars renalis.

and pronephric tubules. It remains functional in juvenile trout and persists throughout adult life in several teleost species (e.g. in poecilids). After the formation of the pronephros, the opisthonephric kidney begins to develop. This is the adult-type kidney in most fish. It resembles the mesonephros of mammals but, unlike the mesonephros, persists as a functional kidney and, therefore, is designated as the opisthonephros (Romer and Parsons, 1977; Starck, 1982). Tubules belonging to the opisthonephric kidney differentiate caudal to the pronephros and their distal ends merge with the archinephric duct. In most species of teleosts, glomeruli develop where the proximal end of these tubules approach the intrarenal arterial system. However, certain teleost species from different sys-

Renal vascular system

The renal vasculature reflects the diversity of kidney types in the fishes. In all cartilaginous and bony fish studied, a renal portal system exists in addition to the arterial circulation. The portal system is present in all regions of the kidneys of bony fishes. In skate and dogfish, the zone of lateral countercurrent bundles is not irrigated with portal blood. (For review of literature see Hentschel and Elger, 1989; Hentschel et al., 1998.)

Figure 12.3 a–d: General shapes of kidneys from major systematic groups. (a) Urogenital systems in ventral view of males. (1) Slime hag, *Bdellostoma (Eptatretus)*; (2) elasmobranch *Torpedo*; (3) lungfish *Protopterus*; (4) teleost, the seahorse *Hippocampus*. In (1) the testis, not shown, is pendent from a mesentery lying between the two kidneys and has no connection with them. In (2) the testis has appropriated the anterior part of the kidney as an epididymis, much as in most land vertebrates, and utilizes the entire length of the archinephric duct as a sperm duct. In (3) the testis ducts drain, on the contrary, only into the posterior part of the kidney and thence to the archinephric duct. In (4) the sperm duct is entirely independent of the kidney system (1 after Conel; 2 after Borcea; 3 after Kerr, Parker; 4 after Edwards; from Romer, 1964.) (b) Urogenital systems in ventral view of females. (1) Elasmobranch *Torpedo*; (2) lungfish *Protopterus*; and (3) primitive actinopterygian *Amia*. (1 after Borcea; 2 after Parker, Kerr; 3 after Hyrtl, Goodrich; from Romer, 1964.) (c) Kidney of a lamprey larva (*Petromyzon*) of 22 mm showing relationship of pronephric and opisthonephric portions. (Based on Wheeler in Bolk et al.; from Lagler et al., 1962.) (d) Diagram of relationships of kidney, gonads, urinary ducts, urogenital sinus, and blood vessels in a female and male holostean fish, gar (*Lepisosteus*). (Based on Pfeiffer (male) and Balfour and Parker (female) in Bolk et al.; from Lagler et al., 1962.)

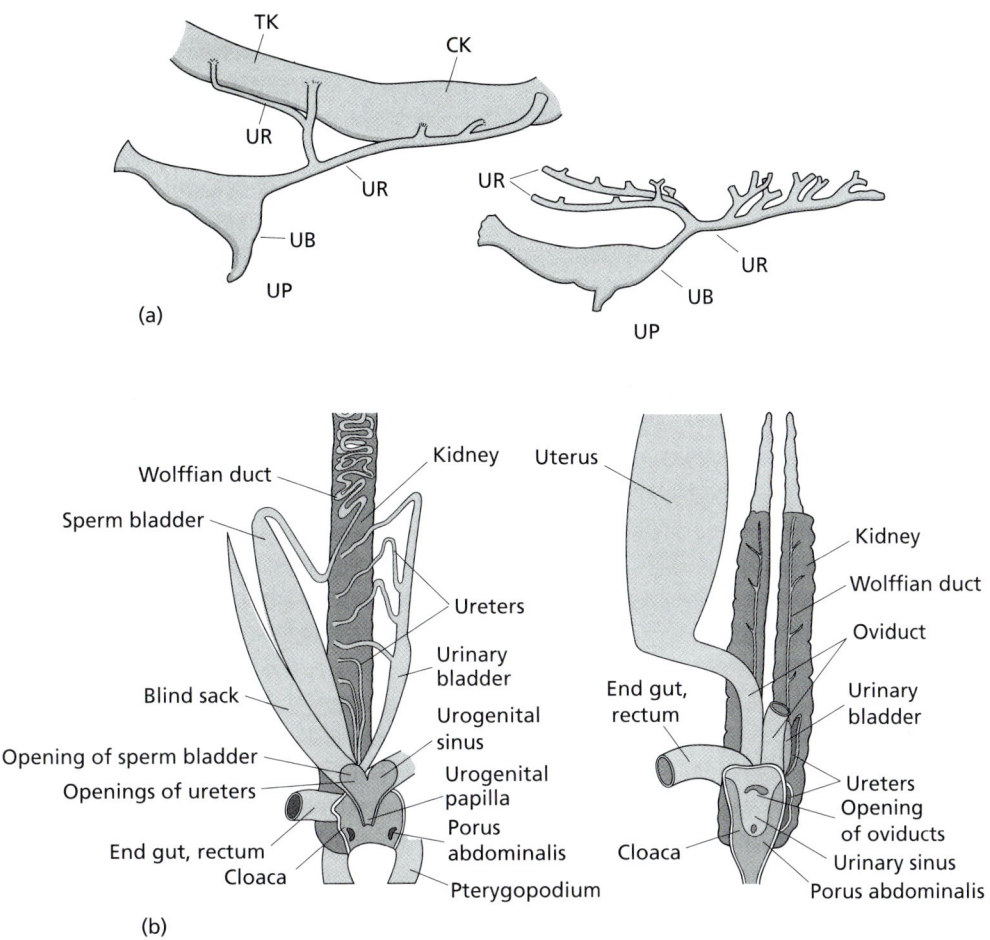

Figure 12.5 (a) Urinary system of the common eel, *Anguilla anguilla*. Lateral view of two preparations, specimen on the left with kidney tissue, on the right the kidney is removed. Three ureters are present, a pair drains the trunk kidney (TK) and a third one is formed at the caudal kidney (CK). UB, urinary bladder; UP, uropore; UR, ureter. (After Audigé, from Bolk *et al.*, in Harder, 1975.) (b) Urogenital system of the male and female shark, *Scyliorhinus caniculus*. (After Parker in Kükenthall and Matthes, 1960.)

Arterial system

The renal arteries supply arterial blood to the kidneys. These arteries originate from the dorsal aorta, either directly or from trunks in common with the segmental intercostal arteries; in a few species, the renal arteries originate instead from the **rami** of the iliac arteries. The renal arteries generally enter through the dorsal sides of the kidneys (dorsomedial side in dorsoventrally compressed species, such as skate) and divide within the renal tissue into numerous intrarenal arteries. These branch further into the afferent arterioles of the glomeruli and ultimately enter the muscular layers of the urinary ducts. In marine elasmobranchs, special bundle arteries supply blood to the lateral bundle zone (Hentschel, 1987, 1988). Numerous renal arteries, often one or more per body segment, frequently supply elongate kidneys.

Compact kidneys generally have fewer renal arteries and the metamerical organization is not evident (Figure 12.7).

Renal portal system

In general, the kidneys of fish and all other submammalian vertebrate classes have a renal portal system which conducts venous blood to renal tissue. Exceptions are the cyclostomes, whose kidneys are supplied only by the arterial system. In other fishes a major part of the peritubular circulation derives from the renal portal system; the postglomerular arterial blood contributes only a minor portion. The intercostal veins and (or) the caudal vein conduct venous blood from the musculature of the trunk and tail, respectively, via afferent renal portal veins to afferent

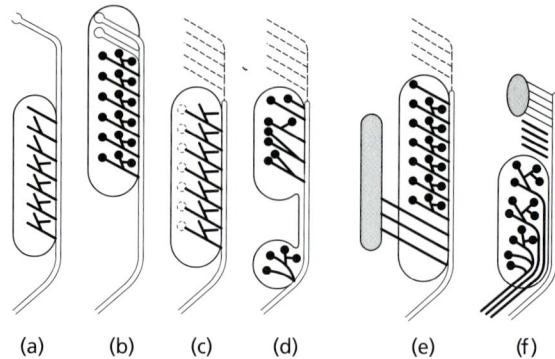

Figure 12.6 Types of definitive kidneys in fish. (a) Aglomerular kidney with persisting pronephros (examples: advanced teleosts, such as several species of gobiesocids). (b) Glomerular kidney consisting of pronephros and opisthonephros (examples: cyprinodonts, such as Poecilidae and *Fundulus*). (c) Aglomerular opisthonephros that has developed from a glomerular stage during ontogeny (example: anglerfish, such as *Lophius* sp. (d) Opisthonephros, consisting of trunk kidney and caudal kidney (examples: many lower teleost species, such as carp and catfish). (e) Opisthonephros with a caudal pars sexualis (examples: males of lungfish). (f) Opisthonephros with a pars sexualis and a pars excretoris with multiple ureters; the archinephric duct has lost the excretory function (examples: males of Elasmobranchiomorphii, such as dogfish and skates). (From Hentschel and Elger, 1989, with permission.)

intrarenal veins. The afferent intrarenal veins ramify, forming an extended system of sinusoid capillaries that surround the renal tubules (see Figures 24.1 and 24.2 in Chapter 24). Generally, the efferent vessels from the glomeruli also conduct blood to the sinuses. In the trout, efferent arterioles frequently enter the muscular sheath around the collecting ducts (Elger *et al.*, 1984). The venous blood in the sinuses is drained via small intrarenal veins to large efferent renal veins that join the large cardinal veins (Figure 12.8).

Lymphatic vessels

Lymphatic vessels similar to those found in other vertebrates have not been described in fish kidneys. Their function is served by a system of secondary blood vessels that has evolved in bony fish (see also Chapter 10). However, a lymph capillary-like vessel (central vessel) has recently been described in the renal countercurrent bundles of marine elasmobranchs (Hentschel *et al.*, 1998) (see also Figure 24.5 in Chapter 24).

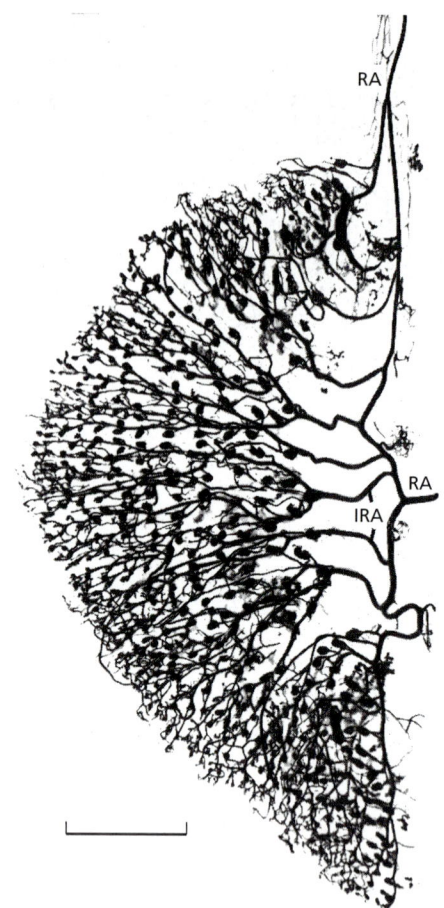

Figure 12.7 Arterial system of the left kidney of a female skate, *Raja erinacea*, age 2 to 3 years. Mercox cast. The largest (oldest) glomeruli (G) are located medially at the primary ramifications of the multiple intrarenal arteries (IRA). The lateral convex margin of the kidney displays the growth zone where nephrogenesis occurs. Here the developing glomeruli are very small. RA, renal artery. Bar = 2.5 mm. (Hentschel and Elger 1989, with permission.)

Interstitium

The blood vascular system and the renal tubules of most fish kidneys are embedded in an elaborate interstitium occupying a variable proportion of the renal tissue (see Figures 24.1 and 24.2 in Chapter 24). Non-renal tissue, comprised of hemopoietic cells and fibroblasts, as well as macrophages and melanocytes, may comprise more than 50% of kidney volume (e.g. in the trunk kidney of winter flounder (Elger *et al.*, 1998)). Melano-macrophage centers are seen even in native, unstained tissue sections. The countercurrent bundles of marine elasmobranchs are embedded in an interstitium, which consists of fibroblasts and fibrocytes. The head kidney consists principally of hemopoietic and immunocompetent cells (see also Chapter 14).

Figure 12.8 Renal portal system of the Prussian carp, *Carassius auratus gibelio*, filled with Microfil via the caudal vein. On the right two portal veins (P) originating from the caudal vein are demonstrated. Another type of portal vein, an intercostal vein, is also filled (bottom of figure). Venous blood of the kidney is drained by efferent intrarenal veins which join to form the cardinal vein (C), and the blood is conveyed to the ducts of Cuvier (DC) at the heart. Bar = 5 mm. (From Hentschel and Elger, 1989, with permission.)

References

Drummond, I.A., Majumdar, A., Hentschel, H., Elger, M., Solnica-Krezel, L., Schier, A.F., Neuhauss, S.C., Stemple, D.L., Zwartkruis, F., Rangini, Z., Driever, W. and Fishman, M.C. (1998). *Development* **125**, 4655–4667.

Elger, M., Wahlqvist, I. and Hentschel, H. (1984). *Cell Tissue Res* **237**, 451–458.

Elger, M., Werner, A., Herter, P., Kohl, B., Kinne, R.K. and Hentschel, H. (1998). *Am. J. Physiol.* **274**, F374–F383.

Harder, W. (1975). *Anatomy of Fishes*, Part II. E. Schweizerbart'sche Verlagsbuchhandlung (Nagele und Obermiller), Stuttgart.

Hentschel, H. (1987). *Zoomorphology* **107**, 115–125.

Hentschel, H. (1988). *Am. J. Anat.* **183**, 130–147.

Hentschel, H. and Elger, M. (1987). *Adv. Anat. Embryol. Cell Biol.* **108**, 1–151.

Hentschel, H. and Elger, M. (1989). *Comparative Physiology. Structure and Function of the Kidney* (ed. R.K.H. Kinne), p. 1. Karger, Basel.

Hentschel, H. and Elger, M. (1996). *J. Am. Soc. Nephrol.* **7**, 1598 (abstract).

Hentschel, H., Storb, U., Teckhaus, L. and Elger, M. (1998). *Anat. Embryol. (Berl.)* **198**, 73–89.

Kükenthal, W. and Matthes, E. (1960). *Leitfaden für das Zoologische Praktikum*. Fischer, Stuttgart.

Lagler, K.F., Bardach, J.E. and Miller, R.R. (1962). *Ichthyology*. Wiley, New York.

MacKenzie, D.D.S., Maack, T. and Kinter, W.B. (1977). *J. Exp. Zool.* **199**, 449–459.

Romer, A.S. (1964). *The Vertebrate Body*, 3rd edn. Saunders, Philadelphia.

Romer, A.S. and Parsons, T.S. (1977). *The Vertebrate Body*, 5th edn. Saunders, Philadelphia.

Starck, D. (1982). *Vergleichende Anatomie der Wirbeltiere*, vol. 3. Springer, Berlin.

Tytler, P. (1988). *J. Morphol.* **195**, 189–204.

Van den Broek, A.S.P. and van Oordt, G.S. (1938). In *Handbuch der vergleichenden Anatomie der Wirbeltiere*, vol. 5 (eds Bolk, L., Göppert, E., Kallius, E. and Lubosch, W.), p. 683. Urban & Schwarzenberg, Berlin.

CHAPTER 13

Endocrine System

David M Janz
Department of Zoology, Oklahoma State University, Stillwater, Oklahoma, USA

Introduction

The evolution of multicellular organisms required communication between cells in order to coordinate multiple physiological processes and maintain homeostasis. Historically, the nervous system and endocrine system were considered to provide complementary actions for homeostatic maintenance. However, scientific developments in the past few decades have revealed a complex integration of both systems, including the importance of neuroendocrine mechanisms of action, 'sharing' of intracellular signaling pathways, and involvement of the immune system. The endocrine system is primarily involved with chemical communication, and it serves to regulate and coordinate diverse physiological processes including development, reproduction, growth, maintenance of the internal environment, energy availability, and behavior. Therefore, the science of endocrinology encompasses all levels of biological organization and explains the bases for regulation of physiological and behavioral processes. The advent of modern biological techniques, particularly in the areas of molecular and cellular biology, has caused an explosive growth in the number of known chemical messenger molecules and has thrust endocrinology into the forefront of scientific disciplines.

Providing a comprehensive review of fish endocrinology is a daunting, if not nearly impossible, task. There are over 25 000 species of fish belonging to six major groups: cyclostomes (e.g. lampreys); chondrichthyes (e.g. sharks and rays); dipnoans (e.g. lungfishes); chondrosts (e.g. sturgeon); holosts (e.g. gar and bowfin); and teleosts. The teleosts represent more than 95% of all fish species and almost half of all vertebrate species. The tremendous adaptive radiation of teleost fishes in freshwater, and subsequently seawater, environments has resulted in a vast array of physiological adaptations involving development, growth, and reproduction. The majority of endocrinology research has focused on teleosts, particularly salmonids, and there is a bias towards species that are readily available and of economic value. Since teleosts are the predominant group utilized for laboratory research, this chapter will focus almost entirely on teleostean endocrine physiology. It must be emphasized, however, that only few of the many teleost species have been studied to date, so that extrapolation between species may not be valid. For recent articles

Copyright © 2000 Academic Press

providing insight into evolutionary aspects of endocrinology among fishes, the reader is referred to Norris and Jones (1987), Wendelaar Bonga (1993), Hazon and Balment (1997), Norris (1997), Bentley (1998) and articles within several volumes of the *Fish Physiology* series (eds W.S. Hoar and D.J. Randall, Academic Press).

In recent years exciting discoveries in fish endocrinology have paralleled the development of new techniques in molecular and cellular biology. The use of methods such as radioimmunoassay (RIA), enzyme-linked immunosorbent assay (ELISA) and immunocytochemistry have greatly increased our knowledge of fish endocrinology. Such techniques require the use of specific antibodies against hormones in order to identify and quantify their presence. Analytical techniques such as high performance liquid chromatography (HPLC) allow the sensitive determination of closely related hormonal agents with high sensitivity. In addition, the recent development of various molecular biological techniques has caused an explosive rise in our knowledge of fish endocrinology at the level of nucleic acids. The main focus of this chapter in terms of methodologies will be the application of immunoassays to fish endocrinology. Chapter 25 on molecular endocrinology will provide detailed information regarding the application of molecular biological approaches to understanding the endocrine systems of fishes.

Fortunately for the comparative endocrinologist, hormones and endocrine regulation have been highly evolutionarily conserved among vertebrates so that experimental approaches and techniques developed in mammalian endocrinology will generally be applicable to the study of fishes. For example, the catecholamines and thyroid hormones are identical across vertebrates, as are the majority of steroid hormones. Steroid hormones are synthesized by common pathways and when differences are present they are usually due to changes in the metabolic conversion of a common precursor. The highly conserved nature of vertebrate amine, thyroid and steroid hormones generally allows the use of heterologous immunoassays for their determination since antibodies created against mammalian hormones will usually recognize piscine hormones with high specificity.

Evolution has also provided a variety of molecular, biochemical, and anatomical modifications in vertebrate endocrine systems. For example, there is significant divergence in the structure of peptide hormones between taxa and indeed between fish species. Such differences in the primary structure of peptide hormones have consequences when attempting to measure biological activity using reagents developed in higher vertebrates. Differences in immunological cross-reactivity often necessitate purification of hormones and development of homologous (species-specific) immunoassays for the quantification of peptide hormones. Even when there is significant cross-reaction of antibodies produced against heterologous peptides, it is essential to fully validate immunoassays for use in a particular fish species. Researchers must also exercise caution when interpreting the reported actions of many hormones in fishes. Many earlier studies used antibodies produced in mammals, which often produced inconsistent results due to cross-reactivity with other related peptides. In addition, the recent development of more specific and sensitive endocrine assays caused realization that many traditional views of hormonal actions in fishes were due to pharmacological and not to physiological effects.

There are several examples of unique anatomical and biochemical modifications in the endocrine system of fishes, most of which are related to inhabiting an aquatic environment. Teleost fishes possess a grouping of caudal neurosecretory neurons, the **urophysis**, which secrete hormones (**urotensins**) that aid in osmoregulation (Loretz *et al.*, 1981). Interestingly, although higher vertebrates do not have a homologous structure, piscine urotensins have actions in mammals indicating the presence of receptors. Fishes lack an organized adrenal gland, and inter-renal tissues located primarily in the anterior (head) kidney region secrete corticosteroids. Teleost fishes lack the hypothalamic–hypophysial portal system present in higher vertebrates. Thus the adenohypophysis (anterior pituitary gland) receives neuropeptides directly via neurons that project from the hypothalamus. The corpuscles of Stannius produce stanniocalcin, a glycoprotein hormone unique to teleostean and holostean fishes that participates in calcium regulation. Similarly, the hormone somatolactin has recently been identified in the pituitary of teleost fishes and appears to be involved in growth, osmoregulation, and reproduction (Rand-Weaver *et al.*, 1991).

Research activity in fish endocrinology has also been stimulated due to increasing use and demand of fishes in the aquaculture industry, as sport and commercial species, and as a source of novel hormonal agents that may be of value in treating various human diseases. In addition, there has been increasing awareness in recent years that certain environmental chemicals originating from industrial, agricultural, domestic, and natural sources may adversely affect the health of fish populations by interfering with the endocrine system (Donaldson, 1990; McMaster et al., 1996; Jobling et al., 1998). This chapter will provide an overview of teleost endocrine physiology in terms of both the diverse and conserved nature of hormones in fishes. The primary sources, target cells, and actions of the major hormones in teleost fishes are summerized in Table 13.1. Whenever possible, techniques available for measuring hormones in body fluids will be mentioned and approximate circulating levels in representative species discussed.

Hypothalamo-hypophysial axis

Despite adaptations to a diverse range of aquatic habitats, pituitary endocrine physiology and morphology are remarkably similar in all fishes. As in higher vertebrates the pituitary or hypophysis of fishes consists of two distinct tissues capable of producing hormones: the neurohypophysis and adenohypophysis. The neurohypophysis (pars nervosa) originates from the diencephalon region of the floor of the brain. The adenohypophysis originates from ectodermal (pharyngeal) tissues and in most fishes is differentiated into a rostral and proximal pars distalis and a pars intermedia (Figure 13.1). Schreibman (1986) provides a detailed review of the embryology and morphology of the piscine hypophysis. The neurosecretory nuclei of the hypothalamus produce neurohormones that control pituitary hormone secretion, thus constituting the hypothalamo-hypophysial axis. Considerable progress has been made in recent years regarding hypothalamic control of hypophysial function in fishes. Relevant hypothalamic neuroendocrine factors will be mentioned where appropriate in this chapter, however a more detailed description of the current knowledge of the piscine hypothalamus will be covered in Chapter 25.

Neurohypophysial hormones

The neurohypophysis consists of many nerve axon termini that project from the hypothalamus and terminate on an organized capillary network or **neurohemal organ**. Proneurohormones are synthesized mainly by hypothalamic neurosecretory cells in the magnocellular region of the preoptic nucleus. In nerve terminals of the neurohypophysis, neurohormones are processed, stored and released into the bloodstream following appropriate stimulation.

The neurohypophysial hormones of all vertebrates studied to date are nonapeptides categorized into either vasopressin-like or oxytocin-like peptides. Except for agnathans, at least one peptide of each category is present in the neurohypophysis of vertebrates. In all fishes the vasopressin-like neurohypophysial hormone appears to be vasotocin. Although several types of oxytocin-like hormones are present in elasmobranchs and lungfishes, isotocin is the only member of the oxytocin family identified to date in teleost fishes (Hazon and Balment, 1997). However, the universality of isotocin is not definitive since neurohypophysial hormones have only been purified and characterized from very few of the 20 000 extant teleost species.

Although there has been great progress in our knowledge of the biochemistry and phylogeny of neurohypophysial hormones, their physiological functions in fishes remain unclear. There are many reports suggesting a role for vasotocin in fish osmoregulation although the precise role has not yet been established (Maetz and Lahlou, 1974; Perks, 1987; Takei, 1993; Hazon and Balment, 1997). Functional receptors have been identified in the gill (Guibollini and Lahlou, 1990) and kidney (Perrott et al., 1993) of teleost fishes. Circulating vasotocin concentrations have been determined using RIA (Holder et al., 1982) in several marine and aquatic teleosts and are normally present in body fluids in the range of 10^{-12} to 2×10^{-11} M (Warne et al., 1994). Physiological doses of vasotocin are antidiuretic, causing a decrease in the

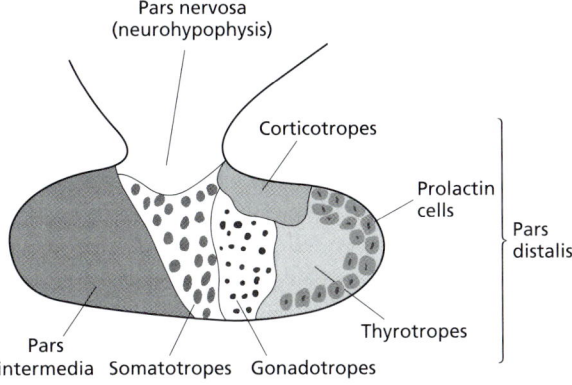

Figure 13.1 Generalized diagram of different regions and cellular zonation of trophic hormone producing cells in the adenohypophysis of a teleost fish.

glomerular filtration rate of perfused trout kidney (Amer and Brown, 1995). Thus, vasotocin appears to be an antidiuretic hormone in fishes similar to the role of vasopressin-like peptides in higher vertebrates. In support of this, vasotocin concentrations in the pituitary and plasma of freshwater-acclimated flounder (*Platichthys flesus*) and rainbow trout (*Oncorhynchus mykiss*) were greater than in seawater-acclimated fishes (Perrott *et al.*, 1991). However, comparison of plasma vasotocin levels in several marine and aquatic fishes revealed no consistent differences among species (Warne *et al.*, 1994). More knowledge is required to understand the osmoregulatory role(s) of vasotocin in fishes.

Very little attention has been given to the actions of oxytocin-like peptides, although there is evidence for the presence of isotocin receptors in gill and liver of trout (Guibollini and Lahlou, 1990). It has been suggested that isotocin may play a role in fish osmoregulation, although its effects are weak in comparison to vasotocin. Several studies have reported the involvement of isotocin and/or vasotocin in spawning (oviparous fish) and parturition (viviparous fish) of some teleosts (Maetz and Lahlou, 1974). Both isotocin and vasotocin increase testosterone production *in vitro* in rainbow trout testes (Rodriguez and Specker, 1991). However, the role of neurohypophysial hormones in fish reproduction is unclear compared to other vertebrate classes.

Adenohypophysial hormones

The location of adenohypophysial cell types has been studied in a variety of fish species. There is a distinct zonation of endocrine cell types in the adenohypophysis of all fishes studied with the exception of agnathans (Figure 13.1). Unlike all other vertebrates, teleost fish lack a hypothalamo-adenohypophysial portal system and receive direct functional innervation by the hypothalamus. Immunocytochemical identification of specific cell types has been conducted, initially with antisera raised against mammalian hormones and more recently with antisera raised against piscine hormones. Such studies have identified the presence of somatotropic, thyrotropic, corticotropic, gonadotropic, melanotropic and prolactin activity in both chondrichthyan (cartilaginous) and osteichthyan (bony) fishes. Osteichthyans possess an additional cell type located in the pars intermedia known as somatolactin or PAS-positive cells (Ono *et al.*, 1990; Rand-Weaver *et al.*, 1991). Adenohypophysial cells produce three main families of hormones: (i) thyroid-stimulating hormone and gonadotropins; (ii) the prolactin family which in fishes includes prolactin, growth hormone and somatolactin; and (iii) adrenocorticotropin and melanotropin.

Thyroid-stimulating hormone (TSH) and gonadotropins (GTHs)

Thyrotropins and gonadotropins are glycoproteins composed of α and β subunits. In a given species, the α subunit is identical for both trophic hormones while the β subunit is hormone specific. In general, development of reliable immunoassays for the glycoprotein hormones has been difficult due to the species-specificity of the β subunits of TSH and GTHs. However, a number of RIAs and ELISAs have recently been developed for both TSH and GTHs in teleost fishes, particularly salmonids. There will undoubtedly be a dramatic increase of our knowledge concerning regulation and actions of piscine TSH and GTHs in the near future.

The actions of TSH in fishes are similar to higher vertebrates, stimulating synthesis and release of thyroxine from the thyroid gland, and secondarily increasing iodide uptake by thyroid cells. Examination of the hypothalamic regulation of TSH release and subsequent actions on the thyroid gland has been limited due to the lack of a reliable assay technique for TSH. Using a RIA recently developed for coho salmon (Moriyama *et al.*, 1997), Larsen and colleagues (1998) found that TSH secretion from pituitary cells was stimulated by corticotropin-releasing hormone (CRH) family peptides (urotensin I, frog sauvagine and ovine CRH) but not by mammalian thyrotropin-releasing hormone (TRH), salmon growth hormone-releasing hormone (sGHRH) or salmon gonadotropin-releasing hormone (sGnRH). However, the lack of TSH release using mammalian TRH in these experiments may be due to dissimilarity in primary structure of TRH peptides among vertebrates (Ohide *et al.*, 1996). TRH receptors have been identified in the pituitary gland of rainbow trout and goldfish (Schwartzentruber and Omeljaniuk, 1995). Not surprisingly, a negative feedback relationship between circulating thyroid hormones and TSH release is present (Moriyama *et al.*, 1997), similar to higher vertebrates.

Although initial reports in teleosts suggested a single gonadotropin, subsequent studies have identified the presence of two distinct gonadotropins, GTH-I and GTH-II, in species representing four orders of teleosts (Idler and Ng, 1983; Kawauchi *et al.*, 1989; Swanson, 1991; Quérat, 1995; Elizur *et al.*, 1996). An exception to the duality of GTH structure and function may occur in catfish, where only a single gonadotropin (GTH-II) has been identified to date (Schulz *et al.*, 1995). The regulation, structure, and function of gonadotropins in lower fishes are not well characterized. GTH-I and GTH-II are considered to be functionally analogous to mammalian follicle-stimulating hormone (FSH) and luteinizing hormone (LH), respectively. Unlike in tetrapods, GTH-I and GTH-II are produced by distinct pituitary cell types in teleost fish (Nozaki *et al.*, 1990). The pituitary content and circulating levels of GTH-I are much greater than those of GTH-II in salmonids undergoing vitellogenesis and spermatogenesis (Swanson, 1991; Prat *et al.*, 1996). However, both GTH-I and GTH-II appear to have similar potencies based on their steroidogenic effects in teleost ovarian and testicular cells (Swanson, 1991). Studies based primarily on salmonids indicate that GTH-I is important in early gonadal development, vitellogenesis and spermatogenesis. In contrast, GTH-II levels are low and often undetectable until just prior to final oocyte maturation in females and spermiation in males.

The seasonal profiles of circulating GTH-I and GTH-II levels in maturing coho salmon and rainbow trout have been reported using homologous RIAs (Swanson, 1991; Prat *et al.*, 1996; Breton *et al.*, 1998). Basal levels of both gonadotropins were less than 5 ng/mL in males and females. Plasma GTH-I concentrations increase to 30–50 ng/mL during vitellogenesis and spermatogenesis, then decrease prior to spawning. Just prior to ovulation or spermiation, GTH-II levels increase to approximately 10–30 ng/mL. The effects of GTH-I and GTH-II on gonadal steroidogenesis will be discussed below in the section on gonadal hormones.

Prolactin family

Prolactin

Prolactin is involved in a wide variety of physiological functions in vertebrates including fishes, although its major functions remain unclear. The actions of prolactin can be grouped into seven general categories: (i) reproduction; (ii) water and electrolyte balance; (iii) growth and morphogenesis; (iv) metabolism; (v) behavior; (vi) immunoregulation; and (vii) effects on the ectoderm and epidermis (Kelly *et al.*, 1991). Although this range of actions has been demonstrated in teleosts, osmoregulation (involving mainly water and Na^+) in freshwater fishes appears to be the predominant action (Hirano, 1986) and may represent the most primitive role for prolactin (Norris, 1997). Prolactin also has hypercalcemic effects in some teleosts, although it does not appear to possess a specific function in calcium homeostasis (Wendelaar Bonga and Pang, 1991). Using a classical endocrinology approach, Pickford and Phillips (1959) showed that **hypophysectomy** of killifish (*Fundulus heteroclitus*) results in ion loss that is lethal unless exogenous prolactin is administered. The actions of prolactin appear to be most important during migration of euryhaline fishes from sea water to fresh water (Hirano, 1986; Wendelaar Bonga and Pang, 1991). Environmental stressors are also known to cause increased prolactin levels in both fresh water and sea water (Avella *et al.*, 1991). Prolactin-secreting cells have been identified immunocytochemically in the rostral pars distalis of all fishes examined with the exception of agnathans. Homologous RIAs have been developed for tilapia (*Oreochromis aureus*) (Nicoll *et al.*, 1981), Pacific salmon (Hirano *et al.*, 1985) and eel (Suzuki and Hirano, 1991). Consistent with the presumed osmoregulatory role of prolactin, plasma levels are consistently higher in freshwater (5–60 ng/mL) than sea water (0.1–1.0 ng/mL) fishes (Nicoll *et al.*, 1981; Hirano, 1986; Fargher and McKeown, 1990; Suzuki and Hirano, 1991).

Growth hormone

Growth hormone, like prolactin, is present in the rostral pars distalis of all fish except agnathans. Two distinct structural forms of growth hormone have been identified in the teleosts examined to date (Melamed *et al.*, 1998). Although this hormone is primarily involved in growth, it has been reported to have an osmoregulatory role in salmonids (Sakamoto *et al.*, 1993). Homologous RIAs have been developed for carp (*Cyprinus carpio*) (Cook *et al.*, 1983), Pacific salmon (Wagner and McKeown, 1986), cod (Rand-Weaver *et al.*, 1989), African catfish (*Clarias gariepinus*) (Lescroart *et al.*, 1996), and eel (*Anguilla anguilla*) (Marchelidon *et al.*, 1996). Detection limits range between 0.1 and 1 ng/mL. Circulating growth hormone levels in resting fish are approximately 5 ng/mL; however starvation, exercise (Barrett and McKeown, 1988), sea water transfer (Sweeting and McKeown, 1987; Sakamoto *et al.*, 1990), smoltification (Prunet *et al.*, 1989), increased temperature (Barrett and McKeown, 1989), or sampling blood at different times of the day (Bates *et al.*, 1989) have been reported to cause between two- and eightfold increases above this basal level. The actions of growth hormone are either direct via binding to growth hormone receptors, or as in higher vertebrates, indirect via mediation by insulin-like growth factors (IGF-I and IGF-II) (Bern *et al.*, 1991). Growth hormone may also play a role in the actions of thyroid hormones (Eales and MacLatchy, 1989) and cortisol (Madsen, 1990), and has been shown to stimulate ovarian steroidogenesis (Van Der Kraak *et al.*, 1990) in fishes.

Somatolactin

Somatolactin is a recently discovered teleost hormone (Ono *et al.*, 1990; Rand-Weaver *et al.*, 1991) that has been referred to as 'a hormone in search of a function' (Rand-Weaver, 1997). Released from the pars intermedia, it shares significant structural homology with prolactin and growth hormone (Ono *et al.*, 1990). Changes in water conditions such as calcium, sodium, pH and background color affect the secretion of somatolactin (Kaneko and Hirano, 1993). In salmonids, plasma somatolactin concentrations increase during stress (Rand-Weaver *et al.*, 1993), reproductive maturation and **smoltification** (Rand-Weaver *et al.*, 1992; Rand-Weaver and Swanson, 1993). Homologous RIAs have been developed for Pacific salmon (Rand-Weaver *et al.*, 1992), red drum (*Sciaenops ocellatus*) (Zhu and Thomas, 1995) and halibut (*Hippoglossus hippoglossus*) (Johnson *et al.*, 1997).

Adrenocorticotropin (ACTH) and melanotropin (MSH)

Cells synthesizing and secreting ACTH and MSH have been identified in the adenohypophysis of all groups of fishes (Schreibman, 1986). Similar to higher vertebrates, ACTH and MSH in fishes are derived from a common precursor peptide, proopiomelanocortin (POMC) (Kawauchi, 1988). Details concerning processing of the prohormone POMC are specific for each cell type, and are discussed in Chapter 25. Briefly, in corticotropes *N*-terminal peptide, ACTH and β-lipotropin are produced. In melanotropes further cleavage of ACTH, into α-MSH and corticotropin-like intermediate lobe peptide (CLIP), and of β-lipotropin, into β-endorphin and γ-lipotropin, occurs. The processing mechanisms for POMC appear similar in fishes compared to higher vertebrates (Kawauchi, 1988).

Similar to higher vertebrates, ACTH is believed to act as the primary regulator of corticosteroid synthesis and secretion in fishes, although α-MSH also appears to play an important role. Both ACTH and α-MSH were shown to exhibit corticotropic actions during larval development of carp (Stouthart *et al.*, 1998). Further discussion of the role of ACTH and α-MSH in regulating corticosteroid release in response to various environmental stressors will be discussed below in the section on inter-renal hormones. The role of the *N*-terminal peptide is unknown. Although α-MSH was named for its function in controlling dispersion of melanin in melanophores of amphibian skin, this role has only been identified in limited teleost species (Wendelaar Bonga, 1993). β-Endorphin is an endogenous opioid that is believed to act in combination with ACTH and α-MSH to modulate the stress response in fishes (Balm and Pottinger, 1995; Mosconi *et al.*, 1998).

Gonadal hormones

With their diversity of reproductive styles (oviparity, **ovoviviparity, viviparity**) fish have evolved the widest variety of reproductive strategies among vertebrates. Reproductive cycles in mature male and female fishes are dynamic processes with respect to the varying extent of cell division, differentiation, and death (apoptosis) occurring. These processes are dependent on the coordinated actions of a wide array of hormones associated with the brain–hypothalamus–pituitary–gonadal axis. Ovarian and testicular functions are controlled not only by the pituitary gonadotropins (GTH-I and GTH-II) but also by multiple hormones and growth factors that act in an endocrine, autocrine, or paracrine manner. The final response in a given cell type results from the integrated effects of these regulatory factors on the intracellular signal transduction pathways. This section will focus primarily on the physiological actions of sex steroid hormones and other major hormonal factors involved in reproduction.

Ovarian hormones
Stages of oogenesis

Oogenesis in fish refers to a complete cycle of egg development, which begins with the recruitment of oogonia (primordial germ cells) and ends at ovulation. Oogenesis is a continuous process, but can be divided into four stages based on morphological and physiological changes (Wallace and Selman, 1981):

1. A cohort of oogonia is recruited into the clutch for that breeding cycle.
2. During previtellogenesis a single layer of granulosa cells develops around the oocyte and the connective tissue surrounding the granulosa forms a second cell layer, the theca. Granulosa and thecal cells are the sites of sex steroid hormone production (steroidogenesis) as well as synthesis of several other biologically active molecules such as growth factors, prostaglandins and cytokines (Van Der Kraak et al., 1997). A noncellular layer, the zona radiata, forms between the granulosa cells and oocyte of each ovarian follicle.
3. Most of the massive oocyte growth occurs during the vitellogenic stage.
4. Following vitellogenesis the oocyte undergoes maturation just prior to ovulation. The coordinated sequence of events occurring throughout oogenesis is a dynamic process regulated predominantly by a variety of hormones and growth factors.

The follicular somatic cells play a critical role in the regulation of ovarian development during oogenesis. Much of our current knowledge concerning the hormonal control of oogenesis comes from studies in salmonids since they have relatively large ovarian follicles (3–5 mm diameter) that are easily manipulated *in vitro*. This work has resulted in the proposed 'two-cell type model' (Figure 13.2) for both the production of 17β-estradiol during oocyte growth and the production of maturation-inducing hormone prior to ovulation (Kagawa et al., 1982; Idler and Ng, 1983; Nagahama et al., 1994). Although GTH-I and GTH-II are of primary importance in controlling oocyte growth and maturation (Kawauchi et al., 1989), the actions of gonadotropins on oogenesis are not direct but are mediated via steroid hormone production in granulosa and thecal cells.

Vitellogenesis

During vitellogenesis in many teleosts, GTH-I stimulates testosterone biosynthesis in thecal cells (Figure 13.3), which then diffuses into the granulosa cell layer and is converted to 17β-estradiol by aromatase. In all teleost species studied to date, high circulating levels of 17β-estradiol have been reported during vitellogenesis. 17β-Estradiol is released into the bloodstream and stimulates the hepatic synthesis of the glycolipophosphoprotein egg yolk precursor, vitellogenin. Once secreted from hepatocytes into the circulation, vitellogenin passes through the granulosa and thecal cell layers, binds to specific receptors on the oocyte surface, and is sequestered via receptor-mediated endocytosis (Specker and Sullivan, 1994). The rate at which oocytes sequester vitellogenin changes significantly during oogenesis (Tyler et al., 1991) and may be controlled by the concentration or affinity of vitellogenin receptors that share homology with the LDL family of receptors (Tyler et al., 1995).

Oocyte maturation

Following vitellogenesis, oocytes enter into a distinct maturational phase in preparation for ovulation (Goetz, 1983; Idler and Ng, 1983; Van Der Kraak and Donaldson, 1986). In general, oocyte maturation is achieved in most teleosts by three sequential mediators (Nagahama et al., 1994). The primary mediators of oocyte growth and maturation in fishes as in higher vertebrates are gonadotropins. In salmonids, GTH-II was shown to be the predominant gonadotropin in the plasma and pituitary during final oocyte maturation (Suzuki et al., 1988; Swanson and Dickhoff, 1990; Swanson, 1991; Prat et al., 1996). Postvitellogenic follicles of teleosts can be induced to mature and ovulate following injection of female fish with a variety of gonadotropin preparations. Intact ovarian follicles can also be induced to undergo germinal vesicle breakdown *in vitro* following incubation with gonadotropin preparations. However, removal of granulosa and thecal cell layers eliminates the ability of gonadotropins to induce maturation *in vitro*. Together, these observations indicated that the actions of gonadotropins are dependent on a secondary mediator of maturation.

A number of C21 steroids have been shown to initiate germinal vesicle breakdown *in vitro* (reviewed in Nagahama et al., 1994). However, only two steroids have been identified as the endogenous maturation-inducing hormone (MIH) in fishes. 17α,20β-Dihydroxy-4-pregnen-3-one (17α,20β-DP) has been identified as the endogenous MIH in most fishes studied including salmonids (Nagahama et al., 1983), killifish (Petrino et al., 1993), medaka (*Oryzias latipes*) (Fukada et al., 1994), and a number of catfish (*Clarias* spp.) (Sundararaj et al., 1985; Suzuki et al., 1987). In the Atlantic croaker (*Micropogonias undulatus*) and spotted seatrout (*Cynoscion nebulosus*), 17α,20β,21-trihydroxy-4-pregnen-3-one (20β-P) was identified

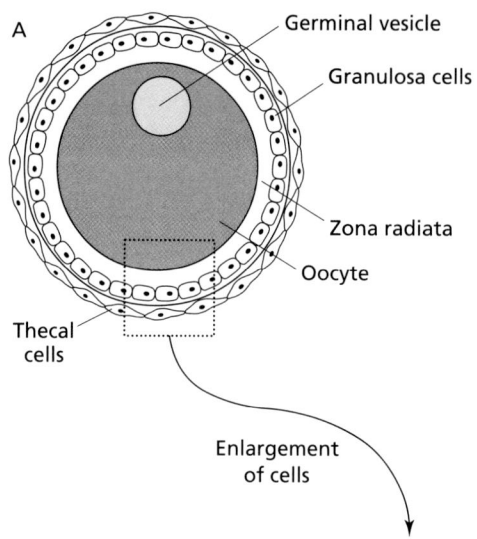

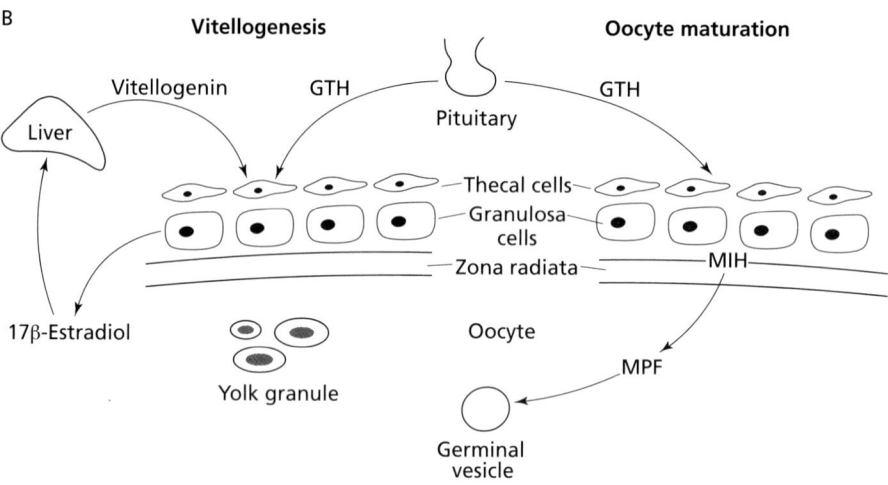

Figure 13.2 Functional diagram showing hormonal mechanisms that control vitellogenesis and oocyte maturation in salmonids and many teleosts. (A) Cross-section of an ovarian follicle. (B) Important steps in the hormonal control of vitellogenesis and oocyte maturation. See text for explanation. GTH, gonadotropin; MIH, maturation-inducing hormone; MPF, maturation-promoting factor.

as the endogenous MIH (Thomas and Trant, 1989; Trant and Thomas, 1989). A two-cell type model was proposed after studies using intact follicles and isolated granulosa and thecal cells from Pacific salmon determined that both cell layers were required for gonadotropin-stimulated MIH synthesis (Young et al., 1986; Nagahama et al., 1994).

During oocyte maturation 17α-hydroxyprogesterone (17α-P) is produced in thecal cells in response to gonadotropin stimulation. 17α-P then diffuses into the granulosa cells and provides the substrate for synthesis of 17α,20β-DP (Nagahama et al., 1994), catalyzed by 20β-hydroxysteroid dehydrogenase (20β-HSD). Thus, just prior to ovulation there is a shift in the steroidogenic pathway resulting in increased production of 17α,20β-DP and marked reductions in 17β-estradiol synthesis (Figures 13.2 and 13.3). It is unclear how this shift in steroidogenesis is produced, although it is believed that GTHs cause upregulation of 20β-HSD and downregulation of aromatase activities (Nagahama et al., 1994; Van Der Kraak et al., 1997). However, the two-cell type model may not apply to all fishes as mentioned previously. Other species such as killifish (Petrino et al., 1989) and medaka (Onitake and Iwamatsu, 1986) do not require the presence of two cell types in order to produce 17α,20β-DP since only granulosa cells are present in ovarian follicles.

Both 17α,20β-DP (Yoshikuni et al., 1993) and 20β-P (Patiño and Thomas, 1990) were subsequently

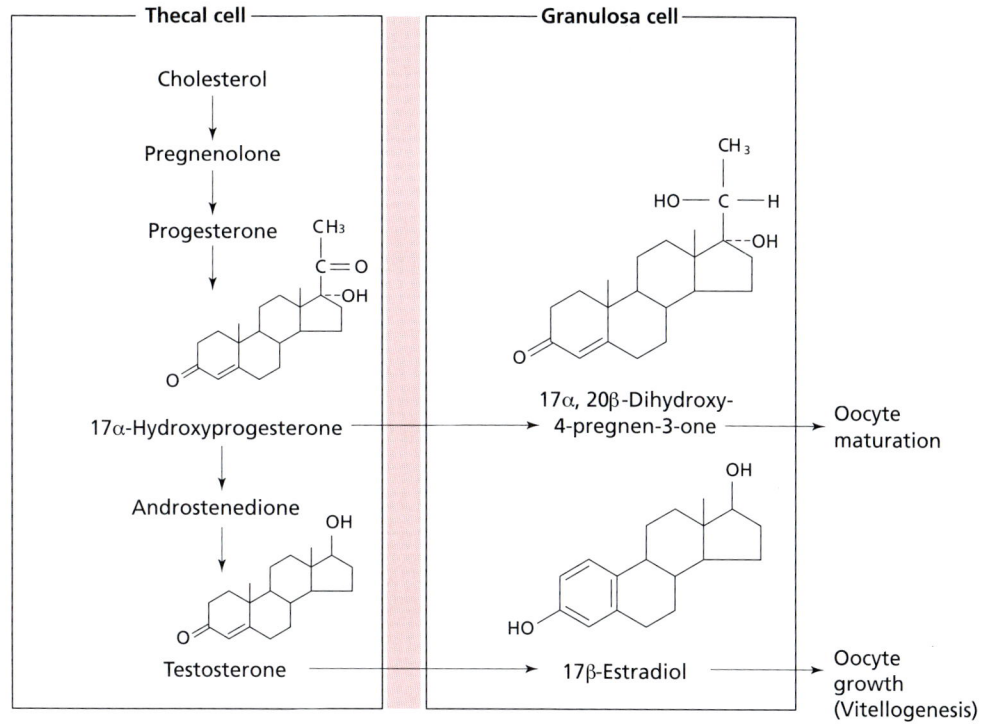

Figure 13.3 The two-cell type model for the biosynthesis of 17β-estradiol and maturation inducing hormone, 17α,20β-dihydroxy-4-pregnen-3-one (17α,20β-DP) in salmonids and many teleosts.

found to bind specifically and with high affinity to oocyte plasma membranes of rainbow trout and spotted seatrout, respectively. This led to the identification of the tertiary mediator of oocyte maturation, maturation-promoting factor (MPF). Activation of MPF acts as a cellular 'switch', stimulating the oocyte to resume meiosis and undergo ovulation. MPF has been determined in carp and goldfish (*Carassius auratus*) to consist of two components, cdc2 kinase and cyclin B (Nagahama *et al.*, 1994).

Other chemical regulators involved in oogenesis

Although the sex steroids mentioned previously are the primary hormones produced by the ovary, a variety of other agents are reported to be involved in oogenesis. Insulin-like growth factor-I (IGF-I) and insulin receptors have recently been identified in salmonid granulosa and thecal cells and may play a significant role in the regulation of ovarian growth and steroidogenesis prior to maturation (Maestro *et al.*, 1997). Epidermal growth factor (EGF) receptors have been identified in goldfish ovarian follicles and appear to be involved in modulating the production of steroids and prostaglandins (Pati *et al.*, 1996; MacDougall and Van Der Kraak, 1998). Peptide hormones such as insulin, prolactin, and in particular growth hormone are also known to influence oogenesis in fish although their precise role(s) and physiological importance are unclear.

Plasma levels of ovarian hormones

In sexually immature females, plasma sex steroid levels are very low (< 0.3 ng/mL). Sexual maturation is associated with elevated plasma levels as well as increases in the frequency and amplitude of diurnal fluctuations of sex steroids. Although the main circulating sex steroid hormone in female fishes is 17β-estradiol, significant levels of testosterone are also present since it is the immediate precursor in 17β-estradiol biosynthesis. Both 17β-estradiol and testosterone are routinely measured using commercially available RIAs. Peak plasma levels of 17β-estradiol, which occur during vitellogenesis, are approximately 30 ng/mL for salmonids and 10 ng/mL for cyprinids (reviewed in McDonald and Milligan, 1992). Plasma 17β-estradiol levels decrease substantially in parallel with increases in 17α,20β-DP during final oocyte maturation. Peak 17α,20β-DP levels, determined using HPLC, are as high as 300–600 ng/mL during the preovulatory period in salmonids (Dye *et al.*, 1986).

Testicular hormones

In comparison to ovarian endocrine physiology, surprisingly little is known regarding testicular hormone regulation in fishes. Testicular morphology is diverse among fishes and is reviewed by Lofts (1987) and Callard (1991). In general, **Sertoli cells** are directly associated with germ cells and provide physical and chemical support during spermatogenesis. **Leydig cells** are found in adjoining connective tissue where they function primarily to synthesize sex steroid hormones responsible for spermatogenesis, expression of secondary sexual characteristics, and feedback regulation of pituitary and hypothalamic secretions.

Steroidogenesis

Pituitary gonadotropins stimulate the testicular synthesis of a variety of steroids in teleosts such as testosterone, 11-ketotestosterone and androstenedione (Schulz, 1986; Fostier et al., 1987). 11-Ketotestosterone is present in many teleosts and may represent the piscine homolog of the potent circulating mammalian androgen, 5α-dihydrotestosterone. Our understanding of the relative involvement of GTH-I and GTH-II in testicular steroidogenesis is still emerging. GTH-I and GTH-II do not appear to differ qualitatively in their potency for stimulating *in vitro* testosterone production in goldfish (Van Der Kraak et al., 1992a), coho salmon (Planas and Swanson, 1995) or seabream (Tanaka et al., 1995). However, the relative potencies of the two gonadotropins appear to be dependent on the stage of spermatogenesis (Sakai et al., 1989; Schulz and Blum, 1990). It is possible that similar to oogenesis, GTH-I and GTH-II are primarily associated with spermatogenesis and **spermiation**, respectively (Swanson, 1991). In support of this, during late spermatogenesis GTH-II was more potent than GTH-I in stimulating $17\alpha,20\beta$-DP, and increased production of $17\alpha,20\beta$-DP was associated with the decrease in testosterone and 11-ketotestosterone production that occurs during the transition from spermatogenesis to spermiation (Sakai et al., 1989; Swanson, 1991). During later stages of spermatogenesis in coho salmon, GTH-II was reported to have an increased potency in terms of *in vitro* production of testosterone, 11-ketotestosterone, 17α-P and $17\alpha,20\beta$-DP (Planas and Swanson, 1995). However, the precise roles of GTH-I and GTH-II in the control of steroidogenesis will require further study.

Other chemical regulators involved in spermatogenesis

Although spermatogonial proliferation is controlled primarily by gonadotropins and androgens, a variety of other autocrine/paracrine factors also appear to be involved. IGF-I, IGF-II, and to a lesser extent insulin, were reported to stimulate testicular DNA synthesis as determined by [^{3}H]thymidine incorporation in rainbow trout spermatogonia (Loir and Le Gac, 1994). IGF-I and IGF-II are expressed in the trout testis, and the identification of type 1 IGF receptors supports the involvement of IGFs in germ cell proliferation and differentiation (Le Gac and Loir, 1995). The potential role of **activin** has also been examined in teleost fishes. In the eel, 11-ketotestosterone synthesized in Leydig cells stimulates the production of activin B within Sertoli cells (Miura et al., 1995). Activin B in turn causes proliferation of spermatogonia. The potential modulatory roles of other signaling agents such as prostaglandins, cytokines and polyunsaturated fatty acids in the teleost testis are not known.

Plasma levels of testicular hormones

Similar to females, circulating levels of sex steroids in immature male fishes are very low. The main circulating androgens in teleosts, 11-ketotestosterone and testosterone, can be determined using commercially available RIAs. Plasma concentrations of both androgens in male salmonids may reach 45 ng/mL (McDonald and Milligan, 1992). In male carp, peak 11-ketotestosterone and testosterone levels were 25 and 12 ng/mL, respectively (Barry et al., 1990).

Hormonal regulation of apoptosis

Although cell division and differentiation have traditionally been regarded as the primary influences on cellular homeostasis, recent studies on apoptosis (programmed cell death) have elucidated the importance of the balance between apoptosis-suppressing and apoptosis-initiating cellular signals in determining the fate of a cell (Steller, 1995; LeGrand, 1997). Apoptosis is an evolutionarily conserved physiological process involved in tissue remodeling, differentiation and degeneration in a variety of cell types (Kerr et al., 1972; Steller, 1995). It has been hypothesized that all cells possess a default program to undergo apoptosis and that only the continuous presence of certain cell type-specific intracellular and intercellular survival factors prevents a cell from undergoing apoptosis (Raff, 1992). It is now widely accepted that apoptosis plays a critical role in the development and homeostasis of all metazoan animals. This section will discuss recent work examining the hormonal regulation of apoptotic cell death in teleost ovarian follicles, although the role of apoptosis will probably be similar during spermatogenesis.

Atresia is a degenerative process by which ovarian follicles lose their integrity and are eliminated prior to ovulation in all vertebrate classes. For example, between 75 and 99.9% of follicles undergo atresia during all stages of ovarian development in various mammalian species (Byskov, 1978). Several studies in fish have reported the normal occurrence of atresia in ovarian follicles at all stages of oogenesis (reviewed in Saidapur, 1978). Recent studies in mammals and birds have established that the underlying molecular mechanism for ovarian follicular atresia is apoptotic cell death (Hughes and Gorospe, 1991; Tilly et al., 1991; Hsueh et al., 1994; Tilly, 1996; Kaipia and Hsueh, 1997). Many of the factors which initiate or suppress apoptosis in the mammalian ovary are hormonal (Kaipia and Hsueh, 1997). Gonadotropins (Chun et al., 1994), estrogens (Billig et al., 1993), EGF, TGFα, basic fibroblast growth factor (Tilly et al., 1992) and IGF-I (Chun et al., 1994) suppress apoptosis in mammalian ovarian follicles. In contrast, GnRH (Billig et al., 1994), androgens (Billig et al., 1993), and certain cytokines (Gorospe and Spangelo, 1993) have been shown to induce apoptosis in ovarian follicles. Therefore, a number of hormonal survival factors and **atretogenic factors** are involved in the regulation and control of apoptotic cell death in ovarian follicles of higher vertebrates. The vertebrate ovary thus represents an excellent model system to investigate apoptosis in an endocrine-responsive organ.

Recent work in teleosts has examined the extent of apoptotic DNA fragmentation during different stages of follicular development and the importance of selected hormones as follicle survival factors (Janz and Van Der Kraak, 1997). The extent of apoptosis was determined using $3'$-end labeling of ovarian cell DNA with $[^{32}P]$dideoxy-ATP and terminal transferase, fractionation using agarose gel electrophoresis, and quantification of apoptotic DNA fragments using autoradiography and liquid scintillation counting. In ovarian follicles frozen immediately following removal from fish, there was an eightfold greater extent of internucleosomal DNA fragmentation in preovulatory compared to previtellogenic follicles of rainbow trout, suggesting changes in the extent of apoptosis at different stages of ovarian development. Developmental differences were also observed in white sucker, where vitellogenic follicles had a greater extent of apoptosis compared to preovulatory follicles (Janz et al., 1997). Preliminary studies in goldfish (D.M. Janz and G. Van Der Kraak, unpublished data), fathead minnow (*Pimephales promelas*) and channel catfish (*Ictalurus punctatus*) (D.M. Janz and M. Savabiaesfahani, unpublished data) have also identified developmental differences in ovarian cell apoptosis. Thus, it appears that there is a greater extent of ovarian cell apoptosis occurring during developmental stages with high levels of cell proliferation.

The importance of various hormones and growth factors was examined in cultured ovarian follicles from rainbow trout (Janz and Van Der Kraak, 1997). Intact follicles cultured for 24–48 hours in serum-free medium exhibit a spontaneous increase in apoptosis due to the absence of survival factors, primarily gonadotropins. Treatment of cultured preovulatory follicles with either partially purified salmon gonadotropin (SG-G100) or EGF suppressed apoptotic DNA fragmentation by 31% and 41%, respectively, in comparison to follicles cultured in serum-free medium alone. Treatment of follicles with 17β-estradiol also caused a concentration-dependent suppression of apoptotic DNA fragmentation (Janz and Van Der Kraak, 1997). Collectively, these results suggest that apoptosis is involved in teleost ovarian development and that several of the hormonal factors acting as follicle survival factors in mammalian and avian ovaries may play a similar role in teleost ovarian follicles. Interestingly, ovarian follicles from fish with an asynchronous pattern of oogenesis (goldfish and fathead minnow) appear to be resistant to apoptosis compared to follicles from synchronous spawners (rainbow trout and channel catfish) (D.M. Janz et al., unpublished data).

Recent studies suggest certain environmental toxicants that affect reproductive development in fish may do so in part through stimulating apoptosis. White sucker populations downstream of kraft pulp mills in Ontario consistently display evidence of reduced gonad size, decreased fecundity, reduced plasma sex steroid hormone concentrations, and delayed sexual maturation (Van Der Kraak et al., 1992b; McMaster et al., 1996). Recent studies have revealed significantly elevated apoptotic DNA fragmentation in vitellogenic and preovulatory ovarian follicles collected from white sucker chronically exposed to pulp mill effluent (Janz *et al.*, 1997). Thus it appears that certain components of bleached pulp mill effluent act to stimulate apoptotic cell death either directly or indirectly via alterations in endocrine homeostasis.

The extent to which other stressors affect cell death processes in the fish ovary is poorly understood although factors such as temperature, hypoxia, inadequate nutrition, handling, and low pH are known to induce atresia in fish (Ball, 1960; De Montalembert *et al.*, 1978; Saidapur, 1978; Wallace and Selman, 1981; Hunter and Macewicz, 1985; Donaldson, 1990; Leino *et al.*, 1990; Tam *et al.*, 1990; Kjesbu *et al.*, 1991; McMaster *et al.*, 1996; Pankhurst *et al.*, 1996; Van Der Kraak and Pankhurst, 1996; Pankhurst and Van Der Kraak, 1997; Coward *et al.*, 1998). However, the mechanisms responsible for initiating ovarian follicle atresia in fish, as in higher vertebrates, remain unclear. Since ovarian follicular atresia may represent a significant determinant of the overall reproductive success in wild fish populations (Saidapur, 1978; N'Da and Déniel, 1993), further studies investigating these linkages are warranted.

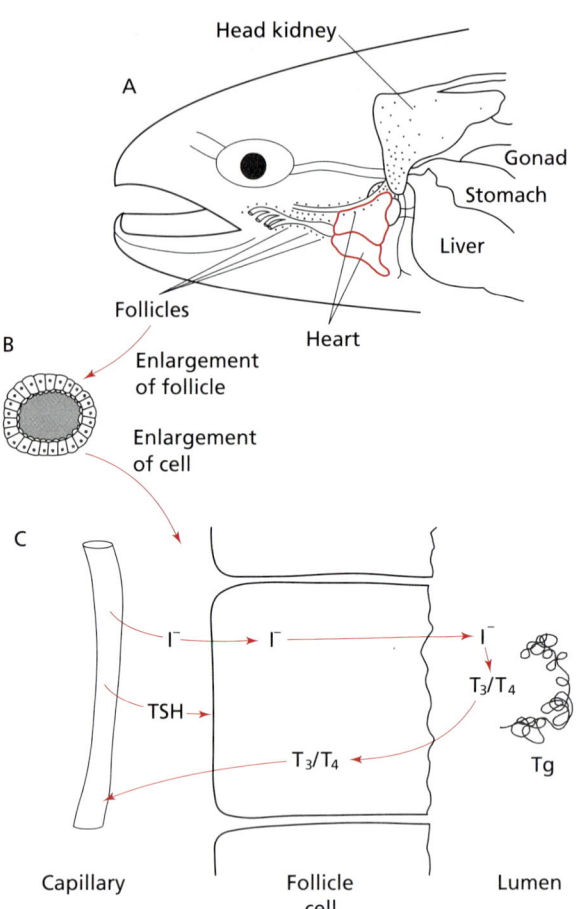

Figure 13.4 Generalized diagram of thyroid anatomy and production of thyroid hormones in a teleost fish. (A) Diffuse (heterotropic) localization of thyroid follicles typical of many teleosts. (B) A thyroid follicle, showing follicular cells surrounding a colloid-filled lumen. (C) A thyroid follicular cell. Thyroid-stimulating hormone (TSH), released from the adenohypophysis, binds to its receptor on the cell surface and stimulates several responses, including iodide (I^-) uptake, synthesis of thyroglobulin (Tg) and colloid, and synthesis and secretion of thyroid hormones (T_3 and T_4).

Thyroid hormones

Anatomy, synthesis, and transport

The functional unit of all vertebrate thyroid glands is the follicle, composed of a single layer of epithelial cells surrounding a colloid fluid-filled space (Dent, 1986). In contrast to higher vertebrates, which display compact or multilobular glands, thyroid tissue in most fishes including teleosts has a diffuse distribution in the ventral pharyngeal region. Although the majority of thyroid tissue is located between the second and fourth aortic arches, large numbers of thyroid follicles are also found embedded within the head kidney and occasionally in other locations such as the ovary and pericardium (Figure 13.4A). Such accessory or **heterotropic** thyroid tissue makes assessment of thyroid function problematic and does not allow classical endocrinological approaches such as thyroidectomy to be easily employed.

The thyroid hormones, thyroxine (T_4) and triiodothyronine (T_3), are synthesized and released by the thyroid follicles following stimulation by TSH (Figure 13.4B and C). In teleost fishes T_4 is the primary, if not only, thyroid hormone released from thyroid follicles (Eales and MacLatchy, 1989). T_3, produced by deiodination of T_4 in peripheral tissues such as the liver, is considered the physiologically active thyroid hormone since it binds with much higher affinity than T_4 to nuclear thyroid hormone receptors (Eales, 1985). In most fishes the vast majority of circulating T_3 and T_4 are transported in the bloodstream bound to plasma proteins. Both free and total (free + bound) T_3 and T_4 can be measured in plasma using commercially available RIA kits with detection limits (~ 0.1 ng/mL) suitable for quantification of even minimum total circulating plasma concentrations (approximately 0.1–1.0 ng/mL) of each hormone in fishes (McDonald and Milligan, 1992).

Biological actions of thyroid hormones

The variety of organs and metabolic processes influenced by thyroid hormones is greater than for any other hormone. In fishes thyroid hormones are important in the control of development, growth, metabolism and osmoregulation, and may often act in concert with cortisol and growth hormone (Hazon and Balment, 1997). Thyroid hormones play a key role in the posthatching metamorphosis of many teleosts. For example, a dramatic increase in plasma T_4 concentration is associated with migration of the eye and related neural structures to one side of the head, and of the mouth structures to the opposite side, during metamorphosis in flounder and other flatfishes (Grace de Jesus *et al*., 1991). Another example is the process of smoltification in salmonids which appears to be regulated by the interplay between peaks of plasma T_4, cortisol, insulin, and prolactin levels (Specker, 1988; Prunet *et al*., 1989; Young *et al*., 1989). The elevation of T_4 during smoltification in salmonids appears to result in the highest plasma levels observed, with peak levels reaching approximately 10–15 ng/mL (Prunet *et al*., 1989; Young *et al*., 1989). Average daily T_3 and T_4 concentrations in rainbow trout plasma were reported to be 2.6 and 5.5 ng/mL, respectively, with significant diurnal fluctuations observed (Gomez *et al*., 1997).

Inter-renal hormones

There is close association between tissues secreting corticosteroids and catecholamines in most fishes. In mammalian adrenal glands the inner medullary tissue is of neural crest origin and consists of chromaffin cells that produce catecholamines. The medulla is surrounded by a cortex of tissue producing glucocorticoids and mineralocorticoids which is of mesodermal origin. In contrast, such tissues are very heterogeneous in fishes and do not display a true medulla or cortex. In teleosts the steroid-producing adrenocortical tissue is usually referred to as inter-renal tissue and forms a diffuse organ at the anterior kidney (head kidney) where it is closely associated with catecholamine-producing chromaffin cells (Figure 13.5). Both cell types surround the dorsal posterior cardinal veins and their branches. In contrast, the adrenocortical and chromaffin tissues of elasmobranchs are anatomically separated (Chester-Jones *et al*., 1980). The actions of corticosteroids and catecholamines are similar in all vertebrates and generally include increased oxygen uptake, mobilization of energy stores, reallocation of energy away from growth and reproduction, and immunosuppression (Norris, 1997).

The release of corticosteroids upon stimulation of the brain–pituitary–inter-renal axis and release of catecholamines upon stimulation of the brain–sympathetic–chromaffin cell axis together constitute what is referred to as the primary stress response in fish (Figure 13.6). There is extensive literature concerning the stress response in fishes since a variety of environmental stressors such as poor water quality, acidification, confinement, handling, heavy metals and organic toxicants are known to stimulate these axes (Pickering, 1981; Adams, 1990; Wedemeyer *et al*., 1990; Barton and Iwama, 1991; Wendelaar Bonga, 1997; Hontela, 1998). This is due primarily to the intimate contact of fishes with the ambient aquatic environment via the gills. Although providing an adaptive response in the short term, sustained elevations in the release of corticosteroids and catecholamines combine to result in secondary (e.g. increased cardiac output, increased oxygen uptake, mobilization of energy substrates and altered water and ion balance) and tertiary (e.g. inhibition of growth, reproduction and immune function) stress responses. These responses involve all levels of animal organization and are collectively termed physiological stress or the integrated stress response.

Consideration of factors to minimize and control for physiological stress is an essential component of any experimental research involving fish. For extensive reviews of the physiological changes and methodologies used to assess the integrated stress response in fishes the reader is referred to reviews by Adams (1990), Donaldson (1990), Wedemeyer *et al*. (1990) and Wendelaar Bonga (1997).

Corticosteroids

Cortisol is the main ($>80\%$) circulating corticosteroid in teleosts, with corticosterone, deoxycorticosterone, and aldosterone present in lesser quantities. The inter-renal tissue of elasmobranchs releases an unusual corticosteroid, 1α-hydroxycorticosterone. Cortisol is secreted primarily in response to ACTH although α-MSH, atrial natriuretic factor, urotensins, angiotensin II, growth hormone, thyroxine, arginine vasotocin, and catecholamines all have corticotropic

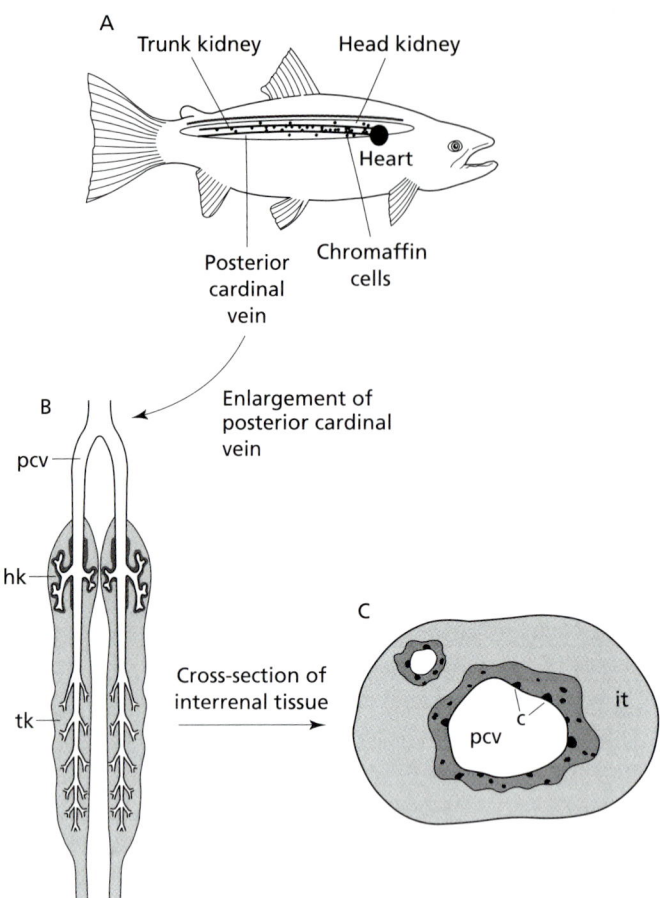

Figure 13.5 Inter-renal tissue anatomy in teleost fishes. (A) Diagram showing location of posterior cardinal vein and chromaffin cells. (B) Distribution of inter-renal tissues in the head kidney (hk) and trunk kidney (tk) and association with posterior cardinal vein (pcv). (C) Cross-section of inter-renal tissue (it) showing locations of chromaffin cells (c) associated with posterior cardinal vein (pcv).

actions (reviewed in Wendelaar Bonga, 1997). Unlike mammals, cortisol performs both glucocorticoid and mineralocorticoid functions in teleosts. The primary target tissues for cortisol are the gills, liver, and intestinal epithelia. Thus, the main functions of cortisol involve energy metabolism, ion regulation and the stress response. Cortisol activates a number of important enzymes controlling intermediary metabolism in the liver (Vijayan et al., 1991). The hyperglycemic actions of cortisol relate to stimulation of glycolysis and gluconeogenesis from proteins and lipids, although the mechanisms involved are unclear. The mineralocorticoid functions of cortisol are important in water and ion balance, acting in concert with hormones such as prolactin in freshwater fish and growth hormone in seawater fish. Cortisol levels are also elevated transiently during migration or transfer from both fresh water to sea water and sea water to fresh water (Mayer-Gostan et al., 1987; Laurent and Perry, 1990). The cellular mechanism responsible for maintaining hydromineral balance is the cortisol-stimulated proliferation of branchial chloride cells and increased activity of ion-transporting enzymes such as Na^+/K^+-ATPase in gills, intestine, and kidneys (Chester Jones et al., 1980; McCormick, 1995).

Cortisol and the stress response

An elevation of plasma cortisol levels is the most commonly used indicator of the stress response, and cortisol is probably the most frequently measured hormone in fishes (Donaldson, 1981; McDonald and Milligan, 1992). Cortisol is routinely measured using commercially available RIA kits with detection limits in the range of 3–8 ng/mL. Resting plasma concentrations range from as low as 5 ng/mL for salmonids (Pickering and Pottinger, 1985) to between 10 and 50 ng/mL in other teleosts (Davis and Parker, 1986). The variation in resting levels of plasma cortisol are

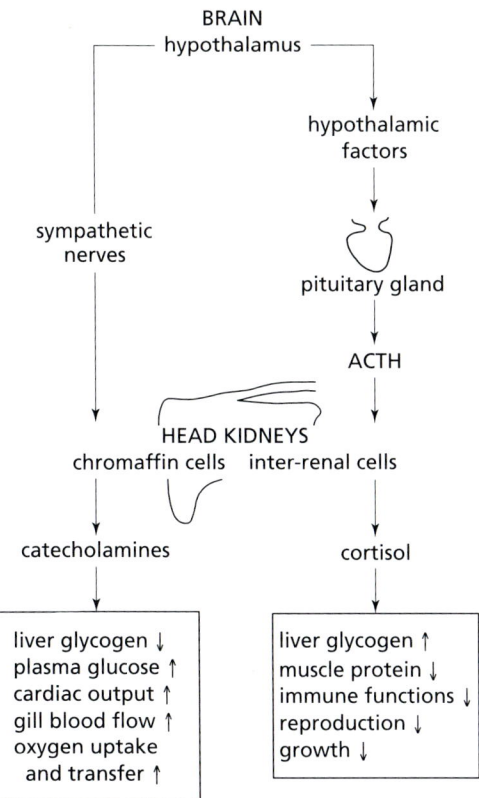

Figure 13.6 Diagrammatic representation of the stress response in fishes.

Randall and Perry, 1992). As mentioned previously, the catecholamines originate from chromaffin cells located primarily in the head kidney and postcardinal veins (Figure 13.5). Catecholamine release is mediated predominantly by preganglionic cholinergic fibers of sympathetic nerves originating in the brain (Figure 13.6). However, as with cortisol, complex control of catecholamine release is apparent and a number of noncholinergic release mechanisms exist (Randall and Perry, 1992).

Catecholamines and the stress response

Resting levels of adrenaline are usually less than 5 nmol/L, and increase rapidly to levels of greater than 1000 nmol/L within 1–3 minutes following a severe acute stressor (Mazeaud and Mazeaud, 1981). During chronic stress, catecholamine levels may remain elevated for hours or days (McDonald and Milligan, 1992; Randall and Perry, 1992; Wendelaar Bonga, 1997). The most common technique for measuring plasma catecholamine levels is HPLC, although radioenzymatic and fluorimetric procedures are also employed (McDonald and Milligan, 1992). Detection limits are less than 1 nmol/L. There are marked species differences in the circulating levels of catecholamines following acute stress, as well as seasonal and water quality related variation (Mazeaud and Mazeaud, 1981; Randall and Perry, 1992).

Actions of catecholamines on gill function

Catecholamines, in particular adrenaline, play an important role in controlling gill vasculature and function (Mayer-Gostan et al., 1987). The actions of catecholamines primarily involve β-adrenergic mechanisms and include increases in oxygen uptake, ventilation rate, branchial blood flow and oxygen transport efficiency of the blood (Randall and Perry, 1992; Wendelaar Bonga, 1997). The cardiovascular system of teleosts is under predominantly neuronal adrenergic control and only under severely stressful conditions do circulating catecholamines have an effect on cardiac output or vasoconstriction (Hazon and Balment, 1997). The hyperglycemic response associated with increased catecholamine release is due mainly to hepatic glycogenolysis and to a lesser extent to gluconeogenesis and inhibition of glycolysis (Wright et al., 1989).

likely due to differences in assay specificity and procedures, fish husbandry and/or capture procedure (Barton and Iwama, 1991). The response to stress is rapid, with elevated circulating cortisol levels appearing within 10 minutes (Sumpter et al., 1986). There is considerable variation in peak cortisol levels, which appears to be related to metabolic scope. Thus, salmonids display the greatest elevations in plasma cortisol levels, reaching 400–600 ng/mL, while less active species generally have peaks around 50–200 ng/mL (Davis and Parker, 1986; McDonald and Milligan, 1992). Cortisol levels exhibit considerable natural variation related to diurnal and seasonal cycles, final gonad maturation and spawning, nutritional status, and smoltification (Wendelaar Bonga, 1997).

Catecholamines

The main circulating catecholamines in teleosts are adrenaline and noradrenaline; dopamine is present to a much lesser extent (McDonald and Milligan, 1992;

Renin–angiotensin system

The renin–angiotensin system is involved in maintenance of extracellular ion and fluid balance in most vertebrates including fishes (Olson, 1992). A renin–angiotensin system has been found in all groups of fishes with the exception of agnathans (Sokabe and Ogawa, 1974; Takei et al., 1993). Angiotensin II is the major product and believed to be the most biologically active member of the renin–angiotensin system in fishes. It originates from the precursor angiotensinogen, which is processed by renin to angiotensin I and subsequently by angiotensin-converting enzyme (ACE) to angiotensin II in several organs. The actions of angiotensin II include increases in blood pressure, aldosterone secretion and drinking behavior, the latter being most evident in seawater teleosts. The effects on blood pressure appear to be indirect and are mediated by catecholamines (Nishimura, 1987). Adaptation to sea water necessitates antidiuresis and a reduced glomerular filtration rate, thus angiotensin II appears to be an essential hormone for sea water adaptation (Brown et al., 1990). Angiotensins are present in the plasma, brain, kidney, and rectal gland of fishes (Takei et al., 1993). It appears that the actions of angiotensin II and atrial natriuretic peptide oppose each other to provide precise control of fluid balance in fishes (Galli and Phillips, 1996).

Natriuretic peptides

A relatively recent development in vertebrate endocrinology is the discovery of a family of structurally related peptides with natriuretic and diuretic properties (de Bold et al., 1981; Takei, 1994). Four major types of natriuretic peptides have been identified thus far, three of which have been reported in fishes (Takei, 1994): A-type natriuretic peptide (ANP); C-type natriuretic peptide (CNP); and ventricular natriuretic peptide (VNP). They are present primarily in the atrium (ANP) and to a lesser extent in the ventricle (VNP), brain (CNP) and blood (ANP, CNP, and VNP). The physiological functions of the natriuretic peptides in fishes are slowly emerging (Evans, 1990, 1995). In teleosts, the main targets of natriuretic peptides are the kidneys, gills, intestine, and cardiovascular system. Earlier reports of the roles of natriuretic peptides in fishes are confused by the use of mammalian sources and heterologous assays, although Kaiya and Takei (1996) have recently developed homologous RIAs for A-type natriuretic peptide and ventricular natriuretic peptide in eel. Although our knowledge of the roles of natriuretic peptides in fishes is limited, salt regulation appears to be of greater importance than fluid volume regulation, and this function is of larger significance in seawater fishes (Evans, 1990, 1995). O'Grady (1989) reported that mammalian ANP inhibits the activity of the $Na^+/K^+/2Cl^-$ cotransporter in the intestine of winter flounder. Furthermore, eel ANP was shown to decrease water and Na^+ absorption by the intestine of seawater-adapted eel (Ando et al., 1992). These results appear inconsistent with the putative role of ANP in sea water adaptation. However, only limited work has been done with homologous hormones, and only a few species have been examined to date. Thus, we are far from understanding the roles of ANP, CNP, and VNP in teleost fishes.

Urotensins

The caudal neurosecretory system is unique to gnathostome (jawed) fishes; in teleosts it is well developed and called the **urophysis**. The urophysis is typically a ventral swelling almost at the end of the spinal cord. Dahlgren cells projecting from the spinal cord terminate in a neurohemal organ, the urophysis, which is structurally similar to the neurohypophysis. The urophysis of teleosts contains two predominant hormones, urotensin I and urotensin II. High concentrations of acetylcholine are also present in the urophysis although the physiological significance of this remains unknown. Urotensin I is structurally similar to corticotropin-releasing hormone and thus may have some effects on ACTH release, and subsequently, cortisol secretion (Woo et al., 1985). In contrast, urotensin II shares structural homology to somatostatin. Immunoreactivity to urotensins has been demonstrated in the brain, where they may act as neurotransmitters and as hypophysiotropic factors (Bern, 1990).

The main function of urotensins appears to involve osmoregulation, primarily during sea water adaptation (Arnold-Reed et al., 1991; Larson and Madani, 1991). Involvement of urotensins in the stress response has also been reported (Woo et al., 1985;

Arnold-Reed and Balment, 1989, 1994). Overall, there appears to be a complex interaction between the urophysis and the brain–pituitary–inter-renal axis in fishes. A recent study indicates that urotensins may regulate inter-renal cortisol secretion independently of the hypophysial axis, possibly in response to osmoregulatory challenge or other inducers of the stress response (Kelsall and Balment, 1998). Homologous RIAs have been developed for urotensins I and II from white sucker (*Catostomus commersoni*) (Kobayashi *et al.*, 1986; Suess *et al.*, 1986). Circulating levels of both urotensins are less than 100 pg/mL and are difficult to measure (Hontela *et al.*, 1989).

Calcium-regulating hormones

In addition to food sources, sea water and even most fresh water provides a continuous supply of calcium to fishes. Unless water calcium levels are very low, the main physiological problem involves restricting the entry of calcium through the gills. Teleosts have thus evolved hypocalcemic mechanisms to maintain extracellular calcium levels within very precise homeostatic limits. This situation is in contrast to terrestrial vertebrates in which the main hormonal factor controlling calcium levels, parathyroid hormone, is hypercalcemic. Regulation of calcium in fishes is afforded primarily by stanniocalcin and to a lesser extent by calcitonin and vitamin D_3.

Stanniocalcin

The corpuscles of Stannius are a structure unique to teleost and holostean fishes, producing the glycoprotein hormone stanniocalcin. The corpuscles of Stannius are most commonly paired structures located in or adjacent to the kidneys, although more 'primitive' teleosts exhibit considerable variety in anatomical distribution (Figure 13.7). Stanniocalcin is the predominant hypocalcemic hormone in teleosts, acting rapidly to reduce calcium uptake at the gills (Verbost *et al.*, 1993) and intestine (Sundell *et al.*, 1992) in both seawater and freshwater teleosts. Surgical removal of the corpuscles of Stannius results in pronounced hypercalcemia in freshwater and to a greater extent in seawater fishes.

Stanniocalcin appears to act as a calcium channel blocker, preventing the passive uptake of calcium by chloride cells in the gills (Flik and Verbost, 1993). A homologous ELISA has been developed for Atlantic salmon stanniocalcin with a detection limit of 0.2 ng/mL (Mayer-Gostan *et al.*, 1992). Plasma stanniocalcin concentrations in anadromous Atlantic salmon were 40 and 150 ng/mL in freshwater- and seawater-adapted individuals, respectively.

Calcitonin

Calcitonin is produced by the ultimobranchial bodies that are located in sheets of connective tissue surrounding the heart. The function, if any, of calcitonin in the calcium regulation of fishes is still unclear although it is present in significant levels in the blood

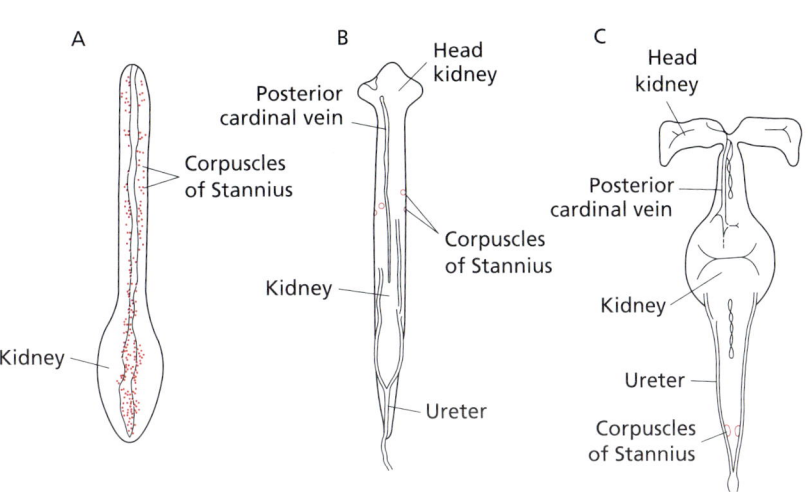

Figure 13.7 Representative diagrams of inter-renal tissue showing variation in the location and number of corpuscles of Stannius among fish groups. (A) holostean fish; (B) salmonid; (C) cyprinid.

(McDonald and Milligan, 1992). The main action of calcitonin may involve stimulation of bone formation by osteoblasts (Wendelaar Bonga and Lammers, 1982). During ovarian development in fishes, plasma calcium concentrations increase dramatically in response to hepatic synthesis and oocyte uptake of the calcium-containing yolk protein precursor, vitellogenin. Circulating calcitonin levels also increase during this period, from around 1 ng/mL in immature fish to 40–50 ng/mL (Yamauchi et al., 1978; Fouchereau-Peron et al., 1990). However, the reason for this increase in calcitonin levels may not be related to calcium regulation, since males exhibit similar increases in calcitonin during sexual maturation. Calcitonin may somehow be associated with the increase in certain plasma sex steroids and/or gonadotropins during sexual maturation in both females and males (Björnsson et al., 1989; Brown and Bern, 1989). It is also possible that calcitonin may play a role in calcium homeostasis by protecting bone during periods of oogenesis with high plasma calcium demand (Björnsson et al., 1986).

Vitamin D_3

The livers of marine teleosts have long been used by humans as a source of vitamin D_3. However, the function of vitamin D_3 (cholecalciferol) and its active metabolites in fishes remains unclear. In mammals 1,25-dihydroxycholecalciferol ($1,25(OH)_2D_3$) is the most potent metabolite involved in stimulating intestinal calcium uptake, calcium reabsorption in the kidney and bone remodeling. Teleosts are able to produce $1,25(OH)_2D_3$ and it has been shown to stimulate intestinal calcium uptake in several species.

Gastro-enteropancreatic hormones

Food digestion, absorption, and management of energy substrates represent complex physiological processes that are regulated to a large extent by an array of hormonal factors associated with the gastrointestinal (GI) tract and pancreas. Many of these hormones are produced and secreted by endocrine cells distributed diffusely throughout the epithelial cells lining the GI tract. The GI endocrine cells of fishes secrete many of the same hormones as the pancreatic islets: glucagon, glucagon-like peptide, somatostatin, pancreatic polypeptides, pancreastatin, and in agnathans and chondrichthyans, insulin (Wendelaar Bonga, 1993). Both gut and pancreas are believed to be endodermal in origin. These observations have led to the use of gastro-entero-pancreatic (GEP) endocrine system to describe the overall endocrine physiology of the GI tract and pancreas (Epple and Brinn, 1987; Hazon and Balment, 1997).

The endocrine cells of the GI tract often have microvilli that project into the lumen allowing rapid response to signals originating from gut contents in addition to humoral and neural inputs. Hormones may be secreted into the blood to act in an endocrine fashion or be released locally to produce direct paracrine effects on adjacent cells. At least 10 cell types have been identified in the GI tract, each producing a different hormone. In addition to the hormones mentioned above, a variety of neuropeptides have been identified, largely using immunocytochemical techniques with mammalian antibodies. These include gastrin, cholecystokinin, vasoactive intestinal peptide, secretin, bombesin, enkephalin, neuropeptide Y, tachykinins, gastric inhibitory peptide, neurotensin and serotonin (Bjenning and Holmgren, 1988; Burckhardt-Holm and Holmgren, 1989). Our understanding of the physiological roles of many of these neuropeptides in fishes is limited, and undoubtedly other hormonal factors await identification.

In higher vertebrates the pancreas contains islets of endocrine tissue embedded in exocrine pancreatic tissue. In fishes the structural organization of the islets is extremely variable and an association with the exocrine pancreatic tissue is not always evident (Epple and Brinn, 1987). In jawed fishes the islet cells evolved as a cell mass which in many teleosts is associated with exocrine pancreatic tissue in structures called **Brockmann bodies**. The islet cells of teleosts produce insulin, glucagon, glucagon-like peptide, somatostatin-25, somatostatin-14, and pancreatic peptides (Plisetskaya, 1989, 1990). Although the functions of the islet cell hormones are not clearly understood in fishes, most of the evidence indicates that their actions are somewhat similar to those in mammals. This section will review the actions of three predominant GEP hormones, insulin, glucagon, and somatostatin.

TABLE 13.1: Summary of the major hormones involved in endocrine physiology of teleost fishes

Source	Hormone	Nature	Primary target(s)[a]	Primary action(s)
Adenohypophysis (pars distalis and pars intermedia)	Gonadotropins (GTH-I and GTH-II)	Glycoprotein	Gonads (ovary and testis)	Folliculogenesis and spermatogenesis (GTH-I); gamete maturation (GTH-II)
	Thyroid-stimulating hormone (TSH)	Glycoprotein	Thyroid gland	Stimulates synthesis and release of thyroid hormones
	Growth hormone (GH)	Peptide	Liver, connective tissue, muscle	Predominantly somatic growth and osmoregulation
	Prolactin	Peptide	Diverse	Predominantly osmoregulation
	Somatolactin	Peptide	Unknown	Unknown
	Adrenocorticotropin (ACTH)	Peptide	Inter-renal tissue	Stimulates synthesis and release of corticosteroids
	Melanotropin (MSH)	Peptide	Melanocytes	Dispersion of melanin (some teleosts); possible role in stress response
Neurohypophysis (pars nervosa)	Vasotocin	Peptide	Kidney	Osmoregulation
	Isotocin	Peptide	Kidney and gonads?	Osmoregulation and reproduction?
Epiphysis (pineal gland)	Melatonin	Indoleamine	Diverse	Physiological and behavioral adjustments to photoperiod
Thyroid follicles	Triiodothyronine (T_3) and thyroxine (T_4)	Iodinated tyrosine derivatives	Diverse	Development, growth, metabolism, osmoregulation, reproduction
Gonads (ovary and testis)	17β-Estradiol	Steroid	Liver	Vitellogenesis (female)
	Testosterone and 11-ketotestosterone	Steroid	Testis	Spermatogenesis
	17α,20β-Dihydroxy-4-pregnen-3-one	Steroid	Ovary and testis	Gamete maturation
Inter-renal tissue	Cortisol	Steroid	Gills, liver, intestine	Energy metabolism, ion regulation, stress response
Chromaffin tissue	Adrenaline and noradrenaline	Catecholamine	Gills	Oxygen uptake and branchial blood flow
Kidney (mainly)	Angiotensin	Peptide	Diverse	Extracellular fluid homeostasis

GROSS FUNCTIONAL ANATOMY

ENDOCRINE SYSTEM

Heart (mainly)	Natriuretic peptides	Peptide	Kidney, gills, intestine, cardiovascular system	Ion regulation
Urophysis	Urotensins	Peptide	Inter-renal tissue, brain	Osmoregulation
Corpuscles of Stannius	Stanniocalcin	Glycoprotein	Gills and intestine	Calcium regulation
Ultimobranchial bodies	Calcitonin	Peptide	Bone	Bone formation?
Liver	Cholecalciferol (vitamin D_3)	Steroid	Intestine and kidney	Calcium regulation?
Pancreas	Insulin	Peptide	Liver and muscle	Glucose uptake and glycogen storage
	Glucagon	Peptide	Liver and muscle	Opposing effects to insulin
	Somatostatin	Peptide	Liver and muscle	Inhibition of insulin and glucagon release

[a]Note that an important target of many secreted hormones is the hypothalamus and/or pituitary gland, where they regulate their own synthesis and release via negative feedback control.

Insulin

Insulin is the predominant GEP hormone in most teleosts. Insulin stimulates glucose uptake by skeletal muscle and liver and promotes gluconeogenesis in these tissues. In teleosts a major function of insulin appears to be anabolic, promoting amino acid incorporation into tissues such as skeletal muscle (Epple and Brinn, 1987). Insulin also promotes fatty acid incorporation and lipogenesis in the liver (Plisetskaya and Duguay, 1993). In regularly fed teleosts, plasma levels consistently range from 5 to 20 ng/mL (McDonald and Milligan, 1992). Although similar in structure to insulin of higher vertebrates, piscine insulins show low cross-reactivity with antibodies against mammalian sources making commercial RIAs unsuitable. Homologous RIAs have been developed for insulin in Pacific salmon, tuna, and cod (McDonald and Milligan, 1992). Prolonged starvation decreases plasma insulin to very low levels. Amino acids are the most potent stimulators of insulin secretion, particularly arginine (Plisetskaya, 1989). Insulin levels display considerable seasonal variation, which may be due to food supply and/or water temperature (Gutiérrez et al., 1987). Circulating insulin levels increase during the parr–smolt transition in salmonids (Plisetskaya, 1990) and decrease during spawning (Gutiérrez et al., 1987).

Glucagon

Glucagon belongs to a family of hormones that includes vasoactive intestinal peptide, secretin, gastric inhibitory peptide and growth hormone-releasing hormone. The functions of glucagon largely oppose those of insulin and include stimulation of hepatic glycogenolysis and lipolysis resulting in hyperglycemia. As with insulin, the actions of glucagon are modified by reproductive status, season and temperature (Plisetskaya, 1990). Glucagon is a highly conserved molecule and can be easily measured using mammalian RIAs. Plasma levels are lower than insulin in Pacific salmon, ranging from 0.5 to 2.0 ng/mL (Plisetskaya, 1990). Brockmann bodies also secrete glucagon-like peptide, which causes similar physiological effects to glucagon (i.e. hepatic glycogenolysis and lipolysis) but acts via different receptors and signal transduction pathways. Although the pancreas is the main target organ for glucagon-like peptide in mammals, the liver appears to be the primary target in fishes with little or no pancreatic effects (Mommsen and Moon, 1989).

Somatostatin

Depending on the site of action, somatostatins may have endocrine, paracrine or neurotransmitter roles. Teleosts are reported to have somatostatin-14 and somatostatin-25 (Plisetskaya, 1989, 1990), although others may be present. Somatostatins contribute to the local control of insulin and glucagon, inhibiting their release. The inhibitory effect of somatostatin on pancreastatin release observed in mammals has not been demonstrated in fishes. Somatostatins also stimulate lipolysis and hyperglycemia (Eilertson et al., 1991). Pancreatic somatostatin secretion may also be modulated by pancreatic polypeptides (possibly peptide Y), which have been identified in a number of fish species (Plisetskaya and Duguay, 1993). Pancreastatin, which inhibits insulin secretion in mammals, has been identified in the islet cells of hagfish, rays and teleosts (Reinecke et al., 1991).

Melatonin

The pineal gland or epiphysis receives and processes light stimuli via hormonal and neural responses in fishes, amphibians, and reptiles. In fishes the pinealocytes are true photoreceptor cells, receiving light via the pineal stalk and window. In contrast, in birds and mammals this function disappears following early development, whereupon light stimuli are received via the eyes. Melatonin, an indole amine, is the major hormone produced by the pineal gland although other hormonal factors have been identified in mammals. Melatonin is secreted mainly during the dark phase of the day. The main function of melatonin is believed to involve adjustment of physiological and behavioral activities to changes in diurnal photoperiod and seasonal daylength (Iigo et al., 1991). Thus melatonin has a potential role in a variety of physiological processes including growth, reproduction, migration, basal metabolic rate, and skin pigmentation.

Modulation of endocrine function by environmental chemicals

A number of environmental stressors are known to compromise growth, development and reproduction of fishes, including toxicants, temperature, hypoxia, inadequate nutrition, and low pH (Ball, 1960; Wallace and Selman, 1981; Donaldson, 1990; Pankhurst et al., 1996; Van Der Kraak and Pankhurst, 1996; Pankhurst and Van Der Kraak, 1997). One of the most controversial issues in environmental science today is the potential for a variety of natural and anthropogenic environmental chemicals to interfere with endocrine homeostasis and thus impair development, growth, and reproduction of both humans and wildlife (Colborn and Clement, 1992; Tyler et al., 1998). Most of the studies to date have focused on chemicals that mimic or antagonize steroid hormones, particularly estrogens. Fish are particularly susceptible to such endocrine modulating chemicals (EMCs) due to their intimate contact with the aquatic environment. Several studies have reported endocrine-mediated toxicities in fish populations exposed to domestic and industrial discharges (e.g. Mac and Edsall, 1991; Van Der Kraak et al., 1992b; Walker and Peterson, 1994; Sandström, 1996; Jobling et al., 1998). The hypothalamus–pituitary–gonadal axis (i.e. reproduction) and hypothalamus–pituitary–inter-renal axis (i.e. stress response) appear to be the most sensitive endocrine systems affected by EMCs (Donaldson, 1990; Wendelaar Bonga, 1997; Hontela, 1998; Van Der Kraak et al., 1998). Early life stages of fish are particularly sensitive to EMCs due to critical endocrine-dependent developmental processes such as sexual differentiation.

This section will outline a variety of known and potential mechanisms of endocrine modulation by environmental chemicals in fish, including effects on hormone biosynthesis, transport, clearance, and action. It should be kept in mind that endocrine regulation, as well as dysregulation, involves complex integrated pathways of communication within an organism. Interpretation of endocrine 'disruption' must therefore take into consideration: (i) the relevance of extrapolating *in vitro* responses to *in vivo* effects; (ii) the importance of hormone feedback loops; (iii) the normal homeostatic range of hormone levels within the bloodstream or tissue; (iv) 'crosstalk' between different intracellular signaling pathways; and (v) the inherent redundancy of the endocrine system and how this may influence the interpretation of hormone alterations.

Hormone biosynthesis

The synthesis of steroid hormones occurs via a sequence of reactions catalyzed by monooxygenases associated with the cytochrome P-450 family. These reactions take place in the smooth endoplasmic reticulum and mitochondria, and occur both in the original producing cell and in target tissues following release and transport in the bloodstream. Alterations in steroid hormone biosynthesis may result from changes in availability of precursors such as cholesterol or altered activity of intermediary enzymes involved in synthesis. Since a variety of **xenobiotics** are known to induce or inhibit the activity of P-450-dependent monooxygenases (Stegeman and Kloepper-Sams, 1987), alterations in the activity of specific enzymes involved in steroid biosynthesis have the potential to significantly modulate hormone production. There are several examples of alterations in the circulating levels of sex steroid hormones in fishes exposed to toxicants (e.g. Freeman and Idler, 1975; Thomas, 1988; Van Der Kraak et al., 1992b; Sandstrom, 1996; reviewed in McMaster et al., 1996). However, in most cases the mechanism(s) responsible for such alterations in hormone biosynthesis are unclear.

Peptide hormones are synthesized in the rough endoplasmic reticulum into prehormones that are processed into active hormones at the original site or in target cells. Thus in addition to gene expression a number of post-translational modifications such as peptide cleavage, phosphorylation or glycosylation are involved in the regulation of peptide hormone synthesis and secretion. In comparison to steroid hormones, there are very few reports of toxicants altering the synthesis or secretion of peptide hormones in vertebrates including fish, although pulp mill effluents were found to decrease the circulating levels of GTH-II in white sucker (Van Der Kraak et al., 1992b).

Hormone transport

Peptide hormones are relatively water soluble and therefore circulate freely in the bloodstream. In contrast, steroid and thyroid hormones are very lipid soluble and bind to a large extent (98–99.9%) to albumin as well as globulins that contain saturable, high affinity binding sites (Van Der Kraak et al., 1998). Only the free, or unbound, hormone is available for diffusion into target cells or clearance from the body. With respect to the potential influence of xenobiotics on hormone transport, the most likely effects would involve perturbations in binding of the native hormone to high affinity binding globulins such as sex hormone binding globulin, transcortin, or transthyretin. For example, a variety of natural and anthropogenic chemicals are know to act as ligands for steroid hormone receptors and may be present in significantly higher concentrations in the bloodstream than endogenous hormones. Such chemicals may competitively bind to sex-hormone-binding globulin and displace the endogenous hormone from binding sites. The effect of this would be to increase clearance of the endogenous hormone, resulting in less availability in target tissues. Although there are no reports of such effects in fishes, the ability of xenobiotics to compete with endogenous hormones for binding sites on globulins *in vivo* represents a potentially significant response to EMCs.

Hormone clearance

Another mechanism by which xenobiotics may increase steroid and thyroid hormone clearance is by alterations in the activities of specific enzymes mediating their catabolism and excretion. Steroid hormones are initially hydroxylated, oxidated or demethylated by P-450-dependent monooxygenases prior to conjugation with glucuronide or sulfate. Thyroid hormones are deaminated or decarboxylated prior to conjugation with glucuronide or sulfate. Conjugated metabolites are primarily excreted in the urine. Similar to biosynthesis, induction or inhibition of specific P-450-dependent monooxygenases or conjugation enzymes by xenobiotics has the potential to affect the clearance of steroid and thyroid hormones. In contrast to steroid and thyroid hormones, peptide hormones and amines are degraded mainly in the liver and kidney into constituent amino acids that are easily excreted or recycled. Although there are examples of xenobiotics increasing the rates of enzymatic degradation and conjugation of steroid and thyroid hormones in mammals (e.g. Goldstein and Safe, 1989), very few studies have examined this potential mechanism of endocrine modulation in fishes.

Hormone action

The actions of hormones involve a variety of receptors and intracellular signaling pathways that are reviewed extensively in the Chapter 25. Members of the nuclear receptor superfamily (i.e. steroids, thyroid hormones, retinoids, and peroxisome proliferators) are believed to play a major role in the adverse effects elicited by EMCs. There are several ways that environmental chemicals can interact with nuclear receptor function, and molecular structure of an EMC is a critical determinant in eliciting a response. Modulation of nuclear receptor function by environmental chemicals is complex, and is dependent on: (i) the relative binding affinity of an EMC to receptors; (ii) ability of the EMC–receptor complex to recognize hormone responsive elements and their promoter sequences on DNA; (iii) the ability of the EMC–receptor complex to recruit transcription factors; and (iv) the effects on mRNA stability and translation (Van Der Kraak et al., 1998).

As in higher vertebrates, studies to date examining the effects of EMCs in fish have focused largely on responses mediated via the estrogen receptor. There is a broad range of compounds that bind to the estrogen receptor and act as agonists, partial agonists, or antagonists. Most of the environmentally relevant EMCs (e.g. phytoestrogens, alkylphenols, phthalate esters, o,p'-DDT, hydroxy-PCBs, bisphenol A) act as weak agonists or partial agonists for the estrogen receptor. One of the most common responses used as a biochemical marker of xenoestrogen exposure is the expression of vitellogenin in male fish (Folmar et al., 1996; Jobling et al., 1996; Kramer et al., 1998). Although male fish have hepatic estrogen receptors and the vitellogenin gene, circulating estrogen levels are normally too low to activate estrogenic responses. However, xenoestrogens have been shown to stimulate vitellogenin expression in male fish by activating the estrogen receptor-mediated expression of vitellogenin. Although it appears that plasma vitellogenin levels in male fish provide a sensitive biochemical marker of *in situ* exposure to xenoestrogens, a current question is whether the elevated circulating vitellogenin levels are deleterious to male fitness. It has been suggested that elevated vitellogenin levels may inhibit testicular growth (Jobling et al., 1996) and promote kidney failure (Herman and Kincaid, 1988) in male fish. A more significant effect of xenoestrogens is their potential to affect sexual differentiation in fish (Blasquez et al., 1998). Recent evidence indicates that a variety of xenoestrogens are able to cause irreversible effects on gonadal differentiation in medaka, including a high occurrence of an intersex condition known as testis–ova (Gray and Metcalfe, 1996; Metcalfe et al., 1998).

The considerable attention given to estrogenic environmental chemicals is understandable given the importance of estrogens in reproduction and growth as well as the potential relationship to a number of human diseases such as breast cancer. However, very little research has focused on the potential effects of EMCs on other hormones, including other members of the nuclear receptor family. Given the broad range of hormonal actions in maintaining physiological homeostasis, it seems likely that EMCs may affect endocrine pathways other than those mediated by the estrogen receptor. Potential effects on signal transduction components such as second messengers or kinase activities have largely been ignored. With respect to fish (and other wildlife), a recent recommendation by various workshops examining the EMC controversy (Kavlock et al., 1996; Ankley et al., 1998) was that in most species our knowledge of normal hormone levels is lacking so that we have very little baseline information available for interpreting potential adverse effects on endocrine homeostasis. Given the considerable amount of research effort currently focusing on the potential adverse effects of EMCs in fishes, important questions surrounding the physiological and ecological relevance of such chemicals may be answered in the near future.

Summary

Recent studies have shown that endocrine regulation in fishes is far more complex than previously recognized. The control of physiological processes such as growth, osmoregulation, and reproduction involve the integrated actions of multiple hormones and intracellular signaling pathways. In many cases we are still searching for the most physiologically important endocrine mediators. Recent advances in molecular and cellular biological approaches have provided new insight into the actions of all hormones. Such knowledge will enhance our understanding of the evolution of the vertebrate endocrine system as well as provide applied benefits in aquaculture, recreational fisheries, and biomedical sciences.

References

Adams, S.M. (1990). *Am. Fish. Soc. Symp.* **8**, 1–181.
Amer, S. and Brown, J.A. (1995). *Am. J. Physiol.* **269**, R775–R780.
Ando, M., Kondo, K. and Takei, Y. (1992). *J. Comp. Physiol.* **162B**, 436–439.
Ankley, G., Mihaich, E., Stahl, R., Tillitt, D., Colburn, T., McMaster, S., Miller, R., Bantle, J., Campbell, P., Denslow, N., Dickerson, R., Folmar, L., Fry, M., Giesy, J., Gray, L.E., Guiney, P., Hutchinson, T., Kennedy, S., Kramer, V., LeBlanc, G., Mayes, M., Nimrod, A., Patino, R., Peterson, R., Purdy, R., Ringer, R., Thomas, P., Touart, L., Van Der Kraak, G. and Zacharewski, T. (1998). *Environ. Toxicol. Chem.* **17**, 68–87.
Arnold-Reed, D.E. and Balment, R.J. (1989). *Gen. Comp. Endocrinol.* **76**, 267–276.
Arnold-Reed, D.E. and Balment, R.J. (1994). *Gen. Comp. Endocrinol.* **96**, 85–91.
Arnold-Reed, D.E., Balment, R.J., McCrohan, C.R. and Hackney, C.M. (1991). *Comp. Biochem. Physiol.* **99A**, 137–143.
Avella, M., Schreck, C.B. and Prunet, P. (1991). *Gen. Comp. Endocrinol.* **81**, 21–27.
Ball, J.N. (1960). *Symp. Zool. Soc. London* **1**, 105–135.
Balm, P.H. and Pottinger, T.G. (1995). *Gen. Comp. Endocrinol.* **98**, 279–288.
Barrett, B.A. and McKeown, B.A. (1988). *Can. J. Zool.* **66**, 853–855.
Barrett, B.A. and McKeown, B.A. (1989). *Comp. Biochem. Physiol.* **94A**, 791–794.
Barry, T.P., Santos, A.J.G., Furukawa, K., Aida, K. and Hanyu, I. (1990). *Gen. Comp. Endocrinol.* **80**, 223–231.
Barton, B.A. and Iwama, G.K. (1991). *Annu. Rev. Fish Dis.* **1**, 3–26.
Bates, D.J., Barrett, B.A. and McKeown, B.A. (1989). *Can. J. Zool.* **67**, 1246–1248.
Bentley, P.J. (1998). *Comparative Vertebrate Endocrinology*, 3rd edn. Cambridge University Press, Cambridge, UK. 526 pp.
Bern, H.A. (1990). In *Progress in Comparative Endocrinology* (eds A. Epple, C.G. Scanes and M.H. Stetson), pp. 242–249. Wiley-Liss, New York.
Bern, H.A., McCormick, S.D., Kelley, K.M., Gray, E.S., Nishioka, K.S., Madsen, S.S. and Tsai, P.I. (1991). In *Modern Concepts of Insulin-like Growth Factors* (ed. E.M. Spencer), pp. 85–96. Elsevier, New York.
Billig, H., Furuta, I. and Hsueh, A.J.W. (1993). *Endocrinology* **133**, 2204–2212.
Billig, H., Furuta, I. and Hsueh, A.J.W. (1994). *Endocrinology* **134**, 245–252.
Björnsson, B.T., Haux, C., Forlin, L. and Deftos, L.J. (1986). *J. Endocrinol.* **108**, 17–23.

Björnsson, B.T., Haux, C., Bern, H.A. and Deftos, L.J. (1989). *Endocrinology* **125**, 1754–1760.

Bjenning, C. and Holmgren, S. (1988). *Histochemistry* **88**, 155–163.

Blasquez, M., Bosma, P.T., Fraser, E.J., Van Look, K.J.W. and Trudeau, V.L. (1998). *Comp. Biochem. Physiol.* **119C**, 345–364.

Breton, B., Govoroun, M. and Mikolajczyk, T. (1998). *Gen. Comp. Endocrinol.* **111**, 38–50.

Brown, C.L. and Bern, H.A. (1989). In *Hormones in Maturation, Aging and Senescence of the Neuroendocrine System* (eds M.P. Schreibman and C. Scanes), pp. 289–317. Academic Press, New York.

Brown, J.A., Gray, C.J. and Taylor, S.M. (1990). *Cell Tissue Res.* **260**, 315–320.

Burckhardt-Holm, P. and Holmgren, S. (1989). *Cell Tissue Res.* **255**, 245–254.

Byskov, A.G. (1978). In *The Vertebrate Ovary* (ed. R.E. Jones), pp. 533–562. Plenum Press, New York.

Callard, G.V. (1991). In *Vertebrate Endocrinology: Fundamentals and Biomedical Implications*, vol. 4A (eds P.K.T. Pang and M.P. Schreibman), Ch. 6. Academic Press, New York.

Chester Jones, I., Mosley, W., Henderson, I.W. and Garland, H.O. (1980). In *General, Comparative and Clinical Endocrinology of the Adrenal Cortex*, vol. 3 (eds I. Chester-Jones and I.W. Henderson), pp. 396–523. Academic Press, London, UK.

Chun, S-Y., Billig, H., Tilly, J.L., Furuta, I., Tsafriri, A. and Hsueh, A.J.W. (1994). *Endocrinology* **135**, 1845–1853.

Colborn, T. and Clement, C. (1992). *Chemically-Induced Alterations in Sexual and Functional Development: The Wildlife/Human Connection.* Princeton Scientific Press, Princeton, NJ.

Cook, A.F., Wilson, S.W. and Peter, R.E. (1983). *Gen. Comp. Endocrinol.* **50**, 335–347.

Coward, K., Bromage, N.R. and Little, D.C. (1998). *J. Fish Biol.* **52**, 152–165.

Davis, K.B. and Parker, N.C. (1986). *Trans. Am. Fish. Soc.* **115**, 495–499.

De Bold, A.J., Borenstein, H.B., Veres, A.T. and Sonnenberg, H. (1981). *Life Sci.* **28**, 89–94.

De Montalembert, G., Jalabert, B. and Bry, C. (1978). *Ann. Biol. Anim. Biochim. Biophys.* **18**, 969–975.

Dent, J. (1986). In *Vertebrate Endocrinology: Fundamentals and Biomedical Implications*, vol. 1 (eds P.K.T. Pang and M.P. Schreibman), pp. 175–216. Academic Press, New York.

Donaldson, E.M. (1981). In *Stress in Fish* (ed. A.D. Pickering), pp. 11–47. Academic Press, New York.

Donaldson, E.M. (1990). *Am. Fish. Soc. Symp.* **8**, 109–122.

Dye, H.M., Sumpter, J.P., Fagerlund, U.H.M. and Donaldson, E.M. (1986). *J. Fish Biol.* **29**, 167–176.

Eales, J.G. (1985). *Can. J. Zool.* **63**, 1217–1231.

Eales, J.G. and MacLatchy, D.L. (1989). *Fish Physiol. Biochem.* **7**, 289–297.

Eilertson, C.C., O'Conner, P.K. and Sheridan, M.A. (1991). *Gen. Comp. Endocrinol.* **82**, 192–199.

Elizur, A., Zmora, N., Rosenfeld, H., Meiri, I., Hassin, S., Gordin, H. and Zohar, Y. (1996). *Gen. Comp. Endocrinol.* **102**, 39–47.

Epple, A. and Brinn, J.E. (1987). *The Comparative Endocrinology of the Pancreatic Islets.* Springer-Verlag, Berlin.

Evans, D.H. (1990). *Annu. Rev. Physiol.* **52**, 43–60.

Evans, D.H. (1995). In *Advances in Environmental and Comparative Physiology – Mechanisms of Systemic Regulation in Lower Vertebrates*, vol. 2 (ed. N. Heisler), pp. 119–152. Springer-Verlag, Heidelberg.

Fargher, R.C. and McKeown, B.A. (1990). *Gen. Comp. Endocrinol.* **75**, 129–133.

Flik, G. and Verbost, P.M. (1993). *J. Exp. Biol.* **184**, 17–30.

Folmar, L.C., Denslow, N.D., Rao, V., Chow, M., Crain, D.A., Enblom, J., Marcino, J. and Guillette, L.J. (1996). Vitellogenin induction and reduced serum testosterone concentrations in feral male carp (*Cyprinus carpio*) captured near a major metropolitan sewage treatment plant. *Environ. Health Persp.* **104**, 1096–1101.

Fostier, A., Le Gac, F. and Loir, M. (1987). In *Proceedings of the 3rd International Symposium on the Reproductive Physiology of Fish* (eds D.R. Idler, L.W. Crim and J.M. Walsh), pp. 239–241. Memorial University, St. Johns, Newfoundland.

Fouchereau-Peron, M., Arlot-Bonnemains, Y., Maubras, L., Milhaud, G. and Moukhtar, M.S. (1990). *Gen. Comp. Endocrinol.* **78**, 159–163.

Freeman, H.C. and Idler, D.R. (1975). *Can. J. Biochem.* **53**, 666–670.

Fukada, S., Sakai, N., Adachi, S. and Nagahama, Y. (1994). *Dev. Growth Differ.* **36**, 81–88.

Galli, S.M. and Phillips, M.I. (1996). *Proc. Soc. Exp. Biol. Med.* **213**, 128–137.

Goetz, F.W. (1983). In *Fish Physiology*, vol. IXA (eds W.S. Hoar, D.J. Randall and E.M. Donaldson), pp. 117–170. Academic Press, New York.

Goldstein, J.A. and Safe, S.H. (1989). In *Halogenated Biphenyls, Terphenyls, Naphthalenes, Dibenzodioxins and Related Products* (eds R.D. Kimbrough and A.A. Jensen), pp. 239–293. Elsevier, Amsterdam, Netherlands.

Gomez, J.M., Boujard, T., Boeuf, G., Solari, A. and Le Bail, P.-Y. (1997). *Gen. Comp. Endocrinol.* **107**, 74–83.

Gorospe, W.C. and Spangelo, B.L. (1993). *Endocr. J.* **1**, 3–9.

Grace de Jesus, E., Hirano, T. and Inui, Y. (1991). *Gen. Comp. Endocrinol.* **82**, 369–375.

Gray, M.A. and Metcalfe, C.D. (1996). *Environ. Toxicol. Chem.* **16**, 1082–1086.

Guibollini, M.E. and Lahlou, B. (1990). *Am. J. Physiol.* **268**, R3–R9.

Gutiérrez, J., Fernandez, J., Carillo, M., Zanuy, S. and Planas, J. (1987). *Fish Physiol. Biochem.* **4**, 137–141.

Hazon, N. and Balment, R.J. (1997). In *The Physiology of Fishes*, 2nd edn (ed. D.H. Evans), pp. 441–463. CRC Press, Boca Raton, Florida.

Herman, R.L. and Kincaid, H.L. (1988). *Aquaculture* 72, 165–172.

Hirano, T. (1986). In *Comparative Endocrinology, Developments and Directives* (ed. C.L. Ralph), pp. 53–74. Alan R. Liss, New York.

Hirano, T., Prunet, P., Kawauchi, H., Takahashi, A., Ogasawara, T., Kubota, J., Nishioka, R.S., Bern, H.A., Takada, K. and Ishii, S. (1985). *Gen. Comp. Endocrinol.* 59, 266–276.

Holder, F.C., Schroeder, M.D., Pollatz, M., Guerne, J.M., Vivien-Roels, B., Pevet, P., Buijs, R.M., Dogterom, J. and Meiniel, A. (1982). *Gen. Comp. Endocrinol.* 47, 483–491.

Hontela, A. (1998). *Environ. Toxicol. Chem.* 17, 44–48.

Hontela, A., Roy, Y., Van Coillie, R., Lederis, K. and Chevalier, G. (1989). *J. Fish Biol.* 35, 265–273.

Hsueh, A.J.W., Billig, H. and Tsafriri, A. (1994). *Endocrine Rev.* 15, 707–724.

Hughes, F.M. and Gorospe, W.C. (1991). *Endocrinology* 129, 2415–2422.

Hunter, J.R. and Macewicz, B.J. (1985). *Fish. Bull.* 83, 119–136.

Idler, D.R. and Ng, T.B. (1983). In *Fish Physiology* (eds W.S. Hoar, D.J. Randall and E.M. Donaldson), pp. 187–221. Academic Press, New York.

Iigo, M., Kezuka, H., Aida, K. and Hanyu, I. (1991). *Gen. Comp. Endocrinol.* 83, 152–159.

Janz, D.M. and Van Der Kraak, G. (1997). *Gen. Comp. Endocrinol.* 105, 186–193.

Janz, D.M., McMaster, M.E., Munkittrick, K.R. and Van Der Kraak, G. (1997). *Toxicol. Appl. Pharmacol.* 147, 391–398.

Jobling, S., Sheahan, D., Osborne, J.A., Matthiessen, P. and Sumpter, J.P. (1996). *Environ. Toxicol. Chem.* 15, 194–202.

Jobling, S., Nolan, M., Tyler, C.R., Brighty, G. and Sumpter, J.P. (1998). *Environ. Sci. Technol.* 32, 2498–2506.

Johnson, L.L., Norberg, B., Willis, M.L., Zebroski, H. and Swanson, P. (1997). *Gen. Comp. Endocrinol.* 105, 194–209.

Kagawa, H., Young, G., Adachi, S. and Nagahama, Y. (1982). *Gen. Comp. Endocrinol.* 47, 440–448.

Kaipia, A. and Hsueh, A.J.W. (1997). *Annu. Rev. Physiol.* 59, 349–363.

Kaiya, H. and Takei, Y. (1996). *J. Endocrinol.* 149, 441–447.

Kaneko, T. and Hirano, T. (1993). *J. Exp. Biol.* 184, 31–45.

Kavlock, R.J., Daston, G.P., DeRosa, C., Fenner-Crisp, P., Gray, L.E., Kaattari, S., Lucier, G., Luster, M., Mac, M.J., Maczka, C., Miller, R., Moore, J., Rolland, R., Scott, G., Sheehan, D.M., Sinks, T. and Tilson, H.A. (1996). *Environ. Health Persp.* 104 (Suppl. 4), 715–740.

Kawauchi, H. (1988). In *The Melanotropic Peptides* (ed. M.E. Hadley), pp. 39–57. CRC Press, Boca Raton, Florida.

Kawauchi, H., Suzuki, K., Itoh, H., Swanson, P., Naito, N., Nagahama, Y., Nozaki, M., Nakai, Y. and Itoh, S. (1989). *Fish Physiol. Biochem.* 7, 29–38.

Kelly, P.A., Djiane, J., Postel-Vinay, M.C. and Edery, M. (1991). *Endocrine Rev.* 12, 235–251.

Kelsall, C.J. and Balment, R.J. (1998). *Gen. Comp. Endocrinol.* 112, 210–219.

Kerr, J.F., Wyllie, A.H. and Currie, A.R. (1972). *Br. J. Canc.* 26, 239–257

Kjesbu, O.S., Klungsøyr, J., Kryvi, H., Witthames, P.R. and Greer Walker, M. (1991). *Can. J. Fish. Aquat. Sci.* 48, 2333–2343.

Kobayashi, Y., Lederis, K., Rivier, J., Ko, D., McMaster, D. and Poulin, P. (1986). *J. Pharm. Meth.* 15, 322–333.

Kramer, V.J., Miles-Richardson, S., Pierens, S.L. and Geisy, J.P. (1998). *Aquat. Toxicol.* 40, 335–360.

Larsen, D.A., Swanson, P., Dickey, J.T., Rivier, J. and Dickhoff, W.W. (1998). *Gen. Comp. Endocrinol.* 109, 276–285.

Larson, B.A. and Madani, Z. (1991). *Gen. Comp. Endocrinol.* 83, 379–387.

Laurent, P. and Perry, S.F. (1990). *Cell Tissue Res.* 259, 429–442.

Le Gac, F. and Loir, M. (1995). In *Proceedings of the 5th International Symposium on the Reproductive Physiology of Fish* (eds F. Goetz and P. Thomas), pp. 354–357. Austin, Texas.

LeGrand, E.K. (1997). *Quart. Rev. Biol.* 72, 135–147.

Leino, R.L., McCormick, J.H. and Jensen, K.M. (1990). *Can. J. Zool.* 68, 234–244.

Lescroart, O., Roelants, I., Mikolajczyk, T., Bosma, P.T., Schulz, R.W., Kuhn, E.R. and Ollevier, F. (1996). *Gen. Comp. Endocrinol.* 104, 147–155.

Lofts, B. (1987). In *Hormones and Reproduction in Fishes, Amphibians and Reptiles* (eds D.O. Norris and R.E. Jones), pp. 283–326. Plenum Press, New York.

Loir, M. and Le Gac, F. (1994). *Biol. Reprod.* 51, 1154–1159.

Loretz, C.A., Bern, H.A., Foskett, J.K. and Mainoya, J.R. (1981). In *Neurosecretion: Molecules, Cells, Systems* (eds D.S. Farner and K. Lederis), pp. 319–328. Plenum Press, New York.

Mac, M.J. and Edsall, C.C. (1991). *J. Toxicol. Environ. Health* 33, 375–394.

MacDougall, T.M. and Van Der Kraak, G. (1998). *Gen. Comp. Endocrinol.* 110, 46–57.

Madsen, S.S. (1990). *Gen. Comp. Endocrinol.* 79, 1–8.

Maestro, M.A., Planas, J.V., Moriyama, S., Gutierrez, J., Planas, J. and Swanson, P. (1997). *Gen. Comp. Endocrinol.* 106, 189–201.

Maetz, J. and Lahlou, B. (1974). In *Handbook of Physiology*, vol. IV (eds E. Knobil and W.H. Sawyer), pp. 521–544. American Physiological Society, Washington, DC.

Marchelidon, J., Schmitz, M., Houdebine, L.M., Vidal, B., Le Belle, N. and Dufour, S. (1996). *Gen. Comp. Endocrinol.* **102**, 360–369.

Mayer-Gostan, N., Wendelaar Bonga, S.E. and Balm, P.H.M. (1987). In *Vertebrate Endocrinology: Fundamentals and Biomedical Implications*, vol. 2 (eds P.K.T. Pang and M.P. Schreibman), pp. 211–238. Academic Press, Orlando, Florida.

Mayer-Gostan, N., Flik, G. and Pang, P.K.T. (1992). *Gen. Comp. Endocrinol.* **86**, 10–19.

Mazeaud, M.M. and Mazeaud, F. (1981). In *Stress in Fish* (ed. A.D. Pickering), pp. 49–75. Academic Press, New York.

McCormick, S.D. (1995). In *Cellular and Molecular Approaches to Fish Ionic Regulation* (*Fish Physiology XIV*) (eds C.M. Wood and T.J. Shuttleworth), pp. 285–315. Academic Press, San Diego, California.

McDonald, D.G. and Milligan, C.L. (1992). In *Fish Physiology*, vol. XIIB (eds W.S. Hoar, D.J. Randall and A.P. Farrell), pp. 55–133. Academic Press, New York.

McMaster, M.E., Van Der Kraak, G.J. and Munkittrick, K.R. (1996). *J. Great Lakes Res.* **22**, 153–171.

Melamed, P., Rosenfeld, H., Elizur, A. and Yaron, Z. (1998). *Comp. Biochem. Physiol.* **119C**, 325–338.

Metcalfe, C.D., Metcalfe, T.L., Gray, M.A., Kiparissis, Y. and Niimi, A.J. (1998). Soc. Environ. Toxicol. Chem. 19th Annual Meeting, p. 80. SETAC Press, Pensacola, Florida.

Miura, T., Miura, C., Yamauchi, K. and Nagahama, Y. (1995). In *Proceedings of the 5th International Symposium on the Reproductive Physiology of Fish* (eds F. Goetz and P. Thomas), pp. 284–287. Austin, Texas.

Mommsen, T.P. and Moon, T. (1989). *Fish Physiol. Biochem.* **7**, 279–288.

Mommsen, T.P. and Walsh, P.J. (1988). In *Fish Physiology*, vol XIA (eds W.S. Hoar and D.J. Randall), pp. 347–406. Academic Press, New York.

Moriyama, S., Swanson, P., Larsen, D.A., Miwa, S., Kawauchi, H. and Dickhoff, W.W. (1997). *Gen. Comp. Endocrinol.* **108**, 457–471.

Mosconi, G., Gallinelli, A., Polzonetti-Magni, A.M. and Facchinetti, F. (1998). *Neuroendocrinology.* **68**, 129–134.

N'Da, K. and Déniel, C. (1993). *J. Fish Biol.* **43**, 229–244.

Nagahama, Y., Hirose, K., Young, G., Adachi, S., Suzuki, K. and Tamaoki, B. (1983). *Gen. Comp. Endocrinol.* **51**, 15–23.

Nagahama, Y., Yoshikuni, M., Yamashita, M. and Tanaka, M. (1994). In *Fish Physiology*, vol. XIII (eds N.M. Sherwood and C.L. Hew), pp. 393–439. Academic Press, San Diego, California.

Nicoll, C.S., Wilson, S.W., Nishioka, R.S. and Bern, H.A. (1981). *Gen. Comp. Endocrinol.* **44**, 365–373.

Nishimura, H. (1987). In *Vertebrate Endocrinology: Fundamentals and Biomedical Implications*, vol. 2 (eds P.K.T. Pang and M.P. Schreibman), pp. 157–188. Academic Press, New York.

Norris, D.O. (1997). *Vertebrate Endocrinology*, 3rd edn. Academic Press, New York. 634pp.

Norris, D.O. and Jones, R.E. (1987). *Hormones and Reproduction in Fishes, Amphibians and Reptiles*. Plenum Press, New York. 613pp.

Nozaki, M., Naito, N., Swanson, P., Miyata, K., Nakai, Y., Oota, Y., Suzuki, K. and Kawauchi, H. (1990). *Gen. Comp. Endocrinol.* **77**, 348–357.

O'Grady, S.M. (1989). *Am. J. Physiol.* **256**, C142–C146.

Ohide, A., Ando, H., Yanagisawa, T. and Urano, A. (1996). *J. Neuroendocrinol.* **8**, 695–701.

Olson, K.R. (1992). In *Fish Physiology*, vol. XII (eds W.S. Hoar, D.J. Randall and A.P. Farrell), pp. 135–254. Academic Press, San Diego, California.

Onitake, K. and Iwamatsu, T. (1986). *J. Exp. Zool.* **239**, 97–103.

Ono, M., Takayama, Y., Rand-Weaver, M., Sakata, S., Yasunaga, T., Noso, T. and Kawauchi, H. (1990). *Proc. Natl. Acad. Sci. USA* **87**, 4330–4334.

Pankhurst, N.W. and Van Der Kraak, G. (1997). In *Fish Stress and Health in Aquaculture* (eds G.K. Iwama, A.D. Pickering, J.P. Sumpter and C.B. Schreck), pp. 73–93. Cambridge University Press, Cambridge, UK.

Pankhurst, N.W., Purser, G.J., Van Der Kraak, G., Thomas, P.M. and Forteath, G.N.R. (1996). *Aquaculture* **146**, 277–290.

Pati, D., Balshaw, K., Grinwich, D.L., Hollenberg, M.D. and Habibi, H.R. (1996). *Am. J. Physiol.* **270**, 1065–1072.

Patiño, R. and Thomas, P. (1990). *Gen. Comp. Endocrinol.* **78**, 204–217.

Perks, A.M. (1987). In *Vertebrate Endocrinology: Fundamentals and Biomedical Implications*, vol. 2 (eds P.K.T. Pang and M.P. Schreibman), pp. 9–44. Academic Press, San Diego, California.

Perrott, M.N., Carrick, S. and Balment, R.J. (1991). *Gen. Comp. Endocrinol.* **83**, 68–74.

Perrott, M.N., Sainsbury, R.J. and Balment, R.J. (1993). *Gen. Comp. Endocrinol.* **89**, 387–395.

Petrino, T.R., Lin, Y.-W.P. and Wallace, R.A. (1989). *Gen. Comp. Endocrinol.* **73**, 147–156.

Petrino, T.R., Lin, Y.-W.P., Netherton, J.C., Powell, D.H. and Wallace, R.A. (1993). *Gen. Comp. Endocrinol.* **92**, 1–15.

Pickering, A.D. (1981). *Stress and Fish*. Academic Press, London.

Pickering, A.D. and Pottinger, T.G. (1985). In *Current Trends in Endocrinology* (eds B. Lofts and W.N. Holms), pp. 1239–1242. Hong Kong University, Hong Kong.

Pickford, G.E. and Phillips, J.G. (1959). *Science* **130**, 454–455.

Planas, J.V. and Swanson, P. (1995). *Biol. Reprod.* **52**, 696–701.

Plisetskaya, E.M. (1989). *Fish Physiol. Biochem.* **7**, 39–48.

Plisetskaya, E.M. (1990). *Zool. Sci.* **7**, 335–353.

Plisetskaya, E.M. and Duguay, S.J. (1993). In *The Endocrinology of Growth, Development and Metabolism in Vertebrates* (eds M.P. Schreibman, C.G. Scanes and P.K.T. Pang), pp. 265–287. Academic Press, New York.

Prat, F., Sumpter, J.P. and Tyler, C.R. (1996). *Biol. Reprod.* **54**, 1375–1380.

Prunet, P., Boeuf, G., Bolton, J.P. and Young, G. (1989). *Gen. Comp. Endocrinol.* **74**, 355–364.

Quérat, B. (1995). In *Proceedings of the 5th International Symposium on the Reproductive Physiology of Fish* (eds F. Goetz and P. Thomas), pp. 7–12. Austin, Texas.

Raff, M.C. (1992). *Nature* **356**, 397–400.

Randall, D.J. and Perry, S.F. (1992). In *Fish Physiology*, vol. XIIB (eds W.S. Hoar, D.J. Randall and A.P. Farrell), pp. 255–300. Academic Press, New York.

Rand-Weaver, M. (1997). In *Advances in the Molecular Endocrinology of Fish*. 2nd IUBS Toronto Symposium, Abstract S4.

Rand-Weaver, M. and Swanson, P. (1993). *Fish Physiol. Biochem.* **11**, 175–182.

Rand-Weaver, M., Carragher, J.C. and Sumpter, J.P. (1989). In *Proceedings, 11th International Symposium of Comparative Endocrinology, Malaga*. Abstract P-284.

Rand-Weaver, M., Noso, T., Muramoto, K. and Kawauchi, H. (1991). *Biochemistry* **30**, 1509–1515.

Rand-Weaver, M., Swanson, P., Kawauchi, H. and Dickhoff, W.W. (1992). *J. Endocrinol.* **133**, 393–403.

Rand-Weaver, M., Pottinger, T.G. and Sumpter, J.P. (1993). *J. Endocrinol.* **138**, 509–515.

Reinecke, M., Hoog, A., Ostenson, C.-G., Egendic, S., Grimelius, L. and Falkmer, S. (1991). *Gen. Comp. Endocrinol.* **83**, 167–182.

Rodriguez, M. and Specker, J.L. (1991). *Gen. Comp. Endocrinol.* **83**, 167–182.

Saidapur, S.K. (1978). *Int. Rev. Cytol.* **54**, 225–244.

Sakai, N., Ueda, H., Suzuki, N. and Nagahama, Y. (1989). *Gen. Comp. Endocrinol.* **75**, 231–240.

Sakamoto, T., Ogasawara, T. and Hirano, T. (1990). *J. Comp. Physiol.* **160B**, 1–6.

Sakamoto, T., McCormick, S.D. and Hirano, T. (1993). *Fish Physiol. Biochem.* **11**, 155–164.

Sandström, O. (1996). In *Environmental Fate and Effects of Bleached Pulp Mill Effluents* (eds M.R. Servos, K.R. Munkittrick, J.H. Carey and G. Van Der Kraak), pp. 449–457. St Lucie Press, Delray Beach, Florida.

Schreibman, M.P. (1986). In *Vertebrate Endocrinology: Fundamentals and Biomedical Implications*, vol. 1 (eds P.K.T. Pang and M.P. Schreibman), pp. 11–55. Academic Press, Orlando, FL.

Schulz, R. (1986). *Fish Physiol. Biochem.* **1**, 55–61.

Schulz, R. and Blum, V. (1990). *Gen. Comp. Endocrinol.* **80**, 189–198.

Schulz, R.W., Boegerd, J., Bosma, P.T., Peute, J., Rebers, F.E.M., Zandbergen, M.A. and Goos, H.J. (1995). In *Proceedings of the Fifth International Symposium on the Reproductive Physiology of Fish* (eds F.W. Goetz and P. Thomas), pp. 2–5. Austin, Texas.

Schwartzentruber, R.S. and Omeljaniuk, R.J. (1995). *Gen. Comp. Endocrinol.* **97**, 209–219.

Sokabe, H. and Ogawa, M. (1974). *Int. Rev. Cytol.* **37**, 271–275.

Specker, J.L. (1988). *Amer. Zool.* **28**, 337–349.

Specker, J.L. and Sullivan, C.V. (1994). In *Perspectives in Comparative Endocrinology* (eds K.G. Davey, R.E. Peter and S.S. Tobe), pp. 59–63. National Research Council of Canada, Ontario, Ottawa.

Stegeman, J.J. and Kloepper-Sams, P.J. (1987). *Environ. Health Persp.* **71**, 87–95.

Steller, H. (1995). *Science* **267**, 1445–1449.

Stouthart, A.J.H.X., Lucassen, E.C.H.E.T., van Strien, F.J.C., Balm, P.H.M., Lock, R.A.C. and Wendelaar Bonga, S.E. (1998). *J. Endocrinol.* **157**, 127–137.

Suess, U., Lawrence, J., Ko, D. and Lederis, K. (1986). *J. Pharm. Meth.* **15**, 335–346.

Sumpter, J.P., Dye, H.M. and Benfey, T.G. (1986). *Gen. Comp. Endocrinol.* **62**, 377–385.

Sundararaj, B.I., Goswami, S.V. and Lamba, V. (1985). *J. Steroid Biochem.* **11**, 701–707.

Sundell, K., Björnsson, B.T., Itoh, H. and Kawauchi, H. (1992). *J. Comp. Physiol.* **162B**, 489–495.

Suzuki, K., Tan, E.S.P. and Tamaoki, B.I. (1987). *Gen. Comp. Endocrinol.* **66**, 454–456.

Suzuki, K., Kawauchi, H. and Nagahama, Y. (1988). *Gen. Comp. Endocrinol.* **71**, 459–467.

Suzuki, R. and Hirano, T. (1991). *Gen. Comp. Endocrinol.* **81**, 401–409.

Swanson, P. (1991). In *Proceedings of the Fourth International Symposium on the Reproductive Physiology of Fish* (eds A.P. Scott, J.P. Sumpter, D.E. Kime and M.S. Rolfe), pp. 2–7. Norwich, UK.

Swanson, P. and Dickhoff, W.W. (1990). In *Progress in Comparative Endocrinology* (eds A. Epple, C.S. Scanes and M. Stetson), pp. 349–356. Wiley-Liss, New York.

Sweeting, R.M. and McKeown, B.A. (1987). *Comp. Biochem. Physiol.* **88A**, 147–151.

Takei, Y. (1993). In *Fish Ecophysiology* (eds J.C. Rankin and F.B. Jensen), pp. 136–160. Chapman & Hall, London, UK.

Takei, Y. (1994). In *Perspectives in Comparative Endocrinology* (eds K.G. Davey, R.E. Peter and S.S. Tober), pp. 149–154. National Research Council Canada, Ottawa, Ontario.

Takei, Y., Hasegawa, Y., Watanabe, T.X., Nakajima, K. and Hazon, N. (1993). *J. Endocrinol.* **139**, 281–285.

Tam, W.H., Fryer, J.N., Valentine, B. and Roy, R.J.J. (1990). *Can. J. Zool.* **68**, 2468–2476.

Tanaka, H., Kagawa, H. and Hirose, K. (1995). In *Proceedings of the 5th International Symposium on the Reproductive Physiology of Fish* (eds F. Goetz and P. Thomas), pp. 10–14. Austin, Texas.

Thomas, P. (1988). *Mar. Environ. Res.* **24**, 179–183.

Thomas, P. and Trant, J.M. (1989). *Fish Physiol. Biochem.* **7**, 185–191.

Tilly, J.L. (1996). *Rev. Reprod.* **1**, 162–172.

Tilly, J.L., Kowalski, K.I., Johnson, A.L. and Hsueh, A.J.W. (1991). *Endocrinology* **129**, 2799–2801.

Tilly, J.L., Billig, H., Kowalski, K.I. and Hsueh, A.J.W. (1992). *Mol. Endocrinol.* **6**, 1942–1950.

Trant, J.M. and Thomas, P. (1989). *Gen. Comp. Endocrinol.* **75**, 397–404.

Tyler, C.R., Sumpter, J.P., Kawauchi, H. and Swanson, P. (1991). *Gen. Comp. Endocrinol.* **84**, 291–299.

Tyler, C.R., Lubberink, K., Brooks, S. and Coward, K. (1995). In *Proceedings of the 5th International Symposium on the Reproductive Physiology of Fish* (eds F. Goetz and P. Thomas), pp. 339–342. Austin, Texas.

Tyler, C.R., Jobling, S. and Sumpter, J.P. (1998). *Crit. Rev. Toxicol.* **28**, 319–361.

Van Der Kraak, G. and Donaldson, E.M. (1986). *Fish Physiol. Biochem.* **1**, 179–186.

Van Der Kraak, G. and Pankhurst, N.W. (1996). In *Global Warming: Implications for Freshwater and Marine Fish* (eds C.M. Wood and D.G. McDonald), pp. 159–176. Cambridge University Press, Cambridge, UK.

Van Der Kraak, G., Chang, J.P. and Janz, D.M. (1997). In *The Physiology of Fishes*, 2nd edn (ed. D.H. Evans), pp. 465–488. CRC Press, Boca Raton, Florida.

Van Der Kraak, G., Rosenblum, P.M. and Peter, R.E. (1990). *Gen. Comp. Endocrinol.* **79**, 233–241.

Van Der Kraak, G., Suzuki, K., Peter, R.E., Itoh, H. and Kawauchi, H. (1992a). *Gen. Comp. Endocrinol.* **85**, 217–229.

Van Der Kraak, G.J., Munkittrick, K.R., McMaster, M.E., Portt, C.B. and Chang, J.P. (1992b). *Toxicol. Appl. Pharmacol.* **115**, 224–233.

Van Der Kraak, G., Zacharewski, T., Janz, D.M., Sanders, B.M. and Gooch, J.W. (1998). In *Principles and Processes in Evaluating Endocrine Disruption in Wildlife* (eds R. Kendall, R. Dickerson, J.P. Giesy and W. Suk), pp. 97–109. SETAC Press, Pensacola, Florida.

Verbost, P.M., Flik, G., Fenwick, J.C., Greco, A.M., Pang, P.K.T. and Wendelaar Bonga, S.E. (1993). *Fish Physiol. Biochem.* **11**, 205–215.

Vijayan, M.M., Ballantyne, J.S. and Leatherland, J.F. (1991). *Gen. Comp. Endocrinol.* **82**, 476–483.

Wagner, G.F. and McKeown, B.A. (1986). *Gen. Comp. Endocrinol.* **62**, 452–458.

Walker, M.K. and Peterson, R.E. (1994). *Environ. Toxicol. Chem.* **13**, 817–820.

Wallace, R.A. and Selman, K. (1981). *Amer. Zool.* **21**, 325–343.

Warne, J.M., Hazon, N., Rankin, J.C. and Balment, R.J. (1994). *Gen. Comp. Endocrinol.* **96**, 438–444.

Wedemeyer, G.A., Barton, B.A. and McLeay, D.J. (1990). In *Methods for Fish Biology* (eds C.B. Schreck and P.B. Moyle), pp. 451–489. American Fisheries Society, Bethesda, Maryland.

Wendelaar Bonga, S.E. (1993). In *The Physiology of Fishes* (ed. D.H. Evans), pp. 469–503. CRC Press, Boca Raton, Florida.

Wendelaar Bonga, S.E. (1997). *Physiol. Rev.* **77**, 591–625.

Wendelaar Bonga, S.E. and Lammers, P.I. (1982). *Gen. Comp. Endocrinol.* **48**, 60–70.

Wendelaar Bonga, S.E. and Pang, P.K.T. (1991). *Int. Rev. Cytol.* **128**, 139–148.

Woo, N.Y.S., Hontela, A., Fryer, J.N., Kobayashi, Y. and Lederis, K. (1985). *Am. J. Physiol.* **238**, R197–R201.

Wright, P.A., Perry, S.F. and Moon, T.W. (1989). *J. Exp. Biol.* **147**, 169–178.

Yamauchi, H., Orimo, H., Yamaguchi, K., Takana, K. and Takahashi, H. (1978). *Gen. Comp. Endocrinol.* **36**, 526–529.

Yoshikuni, M., Shibata, N. and Nagahama, Y. (1993). *Fish Physiol. Biochem.* **11**, 15–24.

Young, G., Adachi, S. and Nagahama, Y. (1986). *Dev. Biol.* **118**, 1–8.

Young, G., Björnsson, B.T., Prunet, P., Lin, R.J. and Bern, H.A. (1989). *Gen. Comp. Endocrinol.* **74**, 335–345.

Zhu, Y. and Thomas, P. (1995). *Gen. Comp. Endocrinol.* **99**, 275–288.

CHAPTER 14

Immune System

David B Powell
ProFISHent, Inc., Redmond, Washington State, USA

Introduction

This discussion of immune function as it relates to gross anatomy must by its very nature be somewhat generalized. Additional details of cellular immune functions of these tissues are presented in the microscopic functional anatomy section of this book (Chapter 26). This chapter will focus on the identification and discussion of grossly visible tissues and changes in organs responsible for immunity in fish.

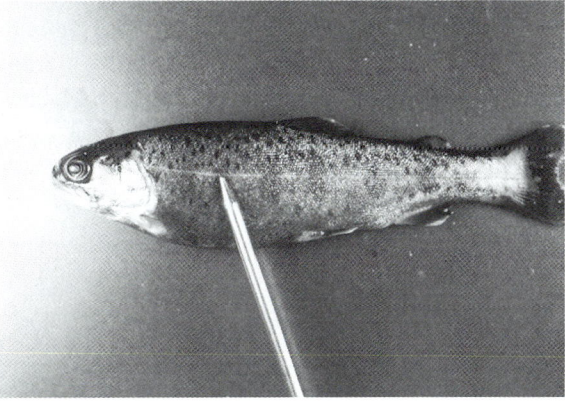

Figure 14.1 The skin and lateral line (at the probe tip) of rainbow trout (*Oncorhynchus mykiss*).

Skin, lateral line, and gills

The skin, lateral line, and gills serve as a first line of defense and barrier to the entrance of infectious agents (Figures 14.1 and 14.2). The presence of antibody-containing mucus, bacteria-inactivating lysosymes, and gill surface macrophages, appear to be important factors in preventing disease and initiating an immune

Figure 14.2 The gills have been exposed by the removal of the operculum, striped bass (*Morone saxatilis*).

Copyright © 2000 Academic Press

response. Wounds, ulcers, or parasitic infestations of the skin or gills provide a portal of entry for waterborne fungal, bacterial, and viral diseases. A variety of infectious agents and chemical toxicants will stimulate an increase in the production of skin mucus.

Sampling surface mucus is a valuable noninvasive method for detecting pathogens in valuable fish stocks just prior to an epizootic. For example, asymptomatic Atlantic salmon and brown trout have been found to have a high prevalence of *Aeromonas salmonicida* bacteria in external mucus prior to a disease outbreak (Cipriano and Ford, 1992). In the case of *Aeromonas salmonicida*, the mere presence of these bacteria on the skin surface does not confer substantial protection to the fish. Immersion **bacterins** against **furunculosis** have shown much less efficacy than preparations directed against *Vibrio anguillarum* and *Yersinia ruckeri* bacteria. Although the reasons for this discrepancy are still unclear, it has been suggested that the protective antigens and their uptake are very different in the two groups of pathogens.

Suspensions of formalin killed *Vibrio anguillarum* and *Yersinia ruckeri* bacteria trigger strong protective immunity in rainbow trout and salmon following only 30 seconds of immersion exposure (Johnson and Amend, 1983; Hastein and Refsti, 1986). The duration and concentration of antigen exposure by immersion have been linked to the level of immunity induced (Johnson *et al.*, 1982). Uptake of both particulate and soluble antigens has been demonstrated, and increased immunity has been shown following extended periods of immersion with diluted vaccines (Ototake *et al.*, 1997). *Vibrio* and *Yersinia* species produce an extracellular lipopolysaccharide (LPS) material that confers protection by immersion (Croy and Amend, 1977). On the other hand, extracts of LPS from *Aeromonas salmonicida* do not function as good vaccines unless they are injected with a potent adjuvant (Cipriano and Pyle, 1985).

Thymus

In teleost fish, the thymus is located in the upper quadrant of each branchial chamber just under the operculum (Figure 14.3). This organ is normally pale white. However, severe systemic infections that penetrate the thymus result in a red and hemorrhagic surface. The thymus is enclosed in an epithelial capsule that is in direct contact with water. Recent studies by Castillo *et al.* (1998) indicate that this pharyngeal

Figure 14.3 In teleost fish the thymus is located in the upper quadrant of each branchial chamber just under the operculum. The probe tip indicates the position of the thymus in a rainbow trout (*Oncorhynchus mykiss*).

epithelium prevents the direct entry of waterborne antigens to the inner thymic parenchyma. Scientists are only beginning to elucidate the complex immune functions of the thymus in fish.

The thymus produces T-cell lymphocytes that appear to be involved in **allograph** rejection, enhanced macrophage function, and antibody producing B-cell stimulation (Warr, 1997). Functional T cells are thought to be required for immune responses to soluble protein antigens, and for this reason, these antigens are described as T-cell dependent. The size and growth of the thymus are dependent on season, age, and stress. Commonly, this organ becomes more flattened or involuted with age or in response to long periods of stress (Chilmonczyk, 1992). Reproductive changes associated with spawning are also correlated with a decrease in thymus size in striped bass and gobiid fishes (Tamura and Honma, 1977; Groman, 1982).

Kidney

The kidney of teleost fishes is a dark red to black symmetrical organ located along the ventral surface of the vertebral column adjacent to the body cavity above the swim bladder (Figure 14.4). The size of the bilateral lobes at the anterior tip of the kidney (Figure 14.5) varies by species, with catfish, for example, having an almost 'saddle-shaped' structure compared to the long narrow shape in trout and salmon. The outer surface is surrounded by a connective tissue capsule to separate this tissue from the peritoneal cavity.

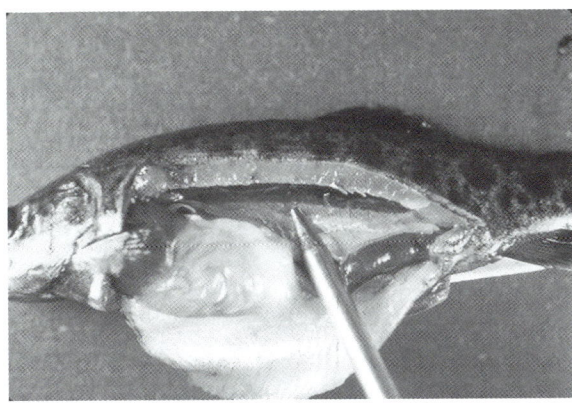

Figure 14.4 The kidney of many teleost fishes is an elongated symmetrical organ located along the ventral surface of the vertebral column inside the body cavity above the swim bladder.

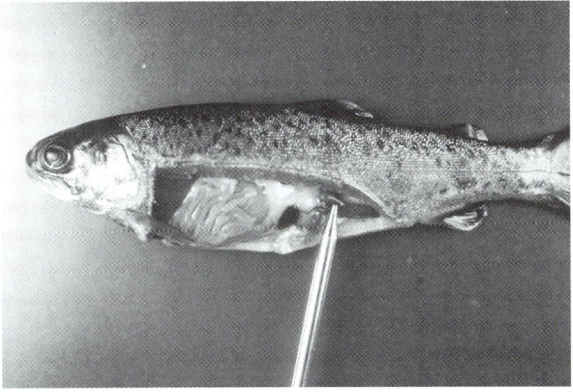

Figure 14.6 The large intestine is located along the ventral portion of the posterior peritoneal cavity in rainbow trout (*Oncorhynchus mykiss*).

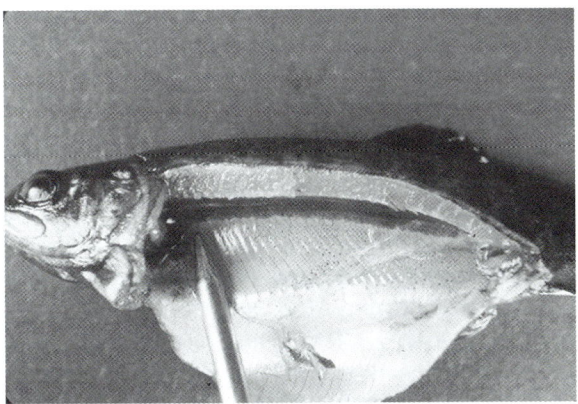

Figure 14.5 The anterior (hematopoietic) portion of the kidney of rainbow trout (*Oncorhynchus mykiss*).

The anterior kidney of fish appears to perform many of the same basic functions as the hematopoietic bone marrow in higher vertebrates. The anterior section typically functions as a major blood-forming tissue, while the posterior section filters the blood plasma for the excretion of metabolic waste products and control of osmotic balance. While white nodules may be an indication of a chronic infection or mineral deposits, the kidney often becomes distended during acute or subacute bacterial, viral, and protozoan infections. The distention may be the result of fluid accumulation or, more commonly, hyperplasia in response to pathogen-induced inflammatory reactions. Chronic infections of the kidney and spleen with *Mycobacteria* spp. produce characteristic white or pale yellow granulomatous lesions particularly common in aquarium fish species (Noga *et al.*, 1996).

The immune functions of the kidney include phagocytosis, antigen processing, IgM-like antibody formation and immunologic memory. Antigen from an intraperitoneally injected vaccine is transported, processed, and localized in areas adjacent to melano-macrophage aggregates in kidney, spleen, and liver (Lamers, 1985). Leucocytes from the kidney of immunized fish exhibit increased production of pathogen-specific macrophage activation factor when stimulated with formalin-inactivated bacteria *in vitro* (Ellis, 1997).

Intestinal tract

The lower intestinal tract has been demonstrated as an area of antigen uptake and processing. The large intestine is located along the lowest portion of the posterior peritoneal cavity in trout and salmon (Figure 14.6). Little information is available on how the intestine of fish can distinguish the myriad of feed ingredients and normal bacterial flora from pathogenic organisms. Unlike mammals, the lower intestine of fish does not contain the large aggregates of lymphocytes known as Peyer's patches. However, the granular cells of stratum granulosum have been suggested to play a role in mucosal immunity in fish (Ezeasor and Stokoe, 1980).

Vaccination preparations delivered orally may stimulate even greater immunity when delivered anally (Johnson and Amend, 1983). This suggests that the harsh acidic conditions of the stomach can denature or alter the immunogenic properties of vaccines.

Oral vaccinations work most effectively with bacteria that possess significant amounts of lipopolysaccharides that are largely unaffected by stomach acids (Lillehaug, 1989; Vigneulle and Baudin-Laurencin, 1991).

In the course of a variety of enteric infections, the intestinal tract often exhibits characteristic signs of disease. For example, bacterial infections with *Yersinia ruckeri*, *Edwardsiella ictaluri* and *Aeromonas hydrophila* result in petichial hemorrhaging of the small and large intestine. Pathologists frequently see changes in the color or composition of fecal material as a result of increased mucus production and altered feeding habits of sick fish.

Liver

The liver of fish has many features in common with the liver of other animals. This relatively large organ is positioned in the anterior portion of the peritoneal cavity and is characterized by its typically pale reddish brown color (Figure 14.7). The liver filters blood through a network of sinusoids formed by cuboidal hepatocytes. Unlike mammals, the liver in fish does not exhibit obviously discrete lobules bordered by septa, portal veins, and bile ducts. Also, in many fish species, the pancreatic tissue may be an integral part of the organ and is thus termed the hepatopancreas.

The major immune function of the liver is thought to be the phagocytosis and presentation of particulate antigens (bacteria, viruses, etc.) to other lymphoid cells (Roitt *et al.*, 1993). Infections of this organ can sometimes be grossly observed as heterogeneous coloring (mottling), hemorrhages or white granulomatous foci (*Piscirickettsia salmonis*, etc.). Exposure to carcinogenic compounds has long been known to produce liver tumors that may be grossly visible as flat or raised nodules on the surface of the liver (Sinnhuber *et al.*, 1977). In cases of anemia (nutritional or late stage viral infections), liver tissue often appears very pale. However, caution should be used in ascribing a cause to coloration changes of the liver. Normal fluctuations in glycogen storage, fat accumulation, and reproductive development can influence hepatic tissue color.

Spleen

The spleen is positioned in the lower posterior abdominal cavity, often surrounded by visceral fat (Figure 14.8). Splenomeglia, or enlargement of the spleen, can occur as a result of systemic bacterial or viral infections. The texture of the organ is normally smooth, with a dark red color due to its engorgement with red blood cells. White nodules may appear in the spleen as the result of granulomatous lesions formed following chronic bacterial (e.g. *Renibacterium salmoninarum* or *Mycobacterium* spp.) or parasitic infections. The spleen traps antigen and may facilitate the stimulation and proliferation of lymphocytes.

Anderson (1992) developed the use of *in vitro* immunization of individual spleen sections from a single fish as a model system to evaluate the relative effects of various immunostimulants, vaccines or toxicants on immunity. The spleen was selected because of its size and uniform distribution of

Figure 14.7 The liver is positioned in the anterior portion of the peritoneal cavity and is characterized by its typically pale reddish brown color.

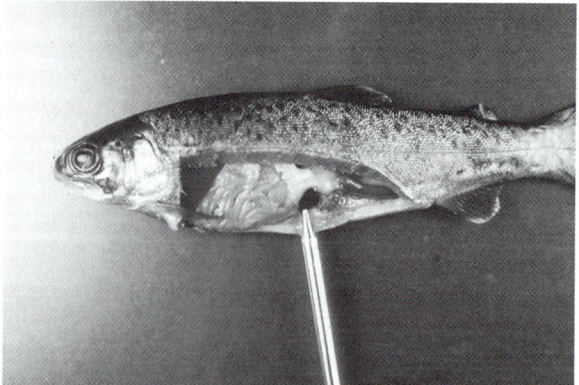

Figure 14.8 The dark red spleen in rainbow trout (*Oncorhynchus mykiss*) is positioned in the lower posterior abdominal cavity, often surrounded by visceral fat.

immunologically active cells. This procedure is one approach to control for the confounding effects of fish to fish variability commonly observed in whole animal tests. The immunomodulator levamisole was shown to stimulate phagocytosis, antibody production, and reaction with NBT (nitroblue tetrazolium) in spleen sections injected with *Yersinia ruckeri* (Siwicki *et al.*, 1990; Anderson 1992). Although the kinetics of these responses was dependent on temperature, the peak NBT reaction in salmonids occurred during the first 2–5 days. This was followed by the expansion of specific antibody-producing cells on days 12–24 (as measured by the plaque assay) and a subsequent elevated antibody titer in the cell media (Anderson, 1992).

References

Anderson, D.P. (1992). In *Techniques in Fish Immunology – 2* (eds J. Stolen, T.C. Fletcher, D.P. Anderson, S.L. Kaattari and A.F. Rowley), pp. 79–88. SOS Publications, New Jersey, USA.

Castillo, A., Razquin, B., Villena, A.J. and López-Fierro, P. (1998). *Fish Shell. Imm.* **8**, 157–170.

Chilmonczyk (1992). *Ann. Rev. Fish Dis.* 181–200.

Cipriano, R.C. and Ford, L.A. (1992). *J. Aquat. Anim. Health.* **4**, 114–118.

Cipriano, R.C. and Pyle, S.W. (1985). *Can. J. Fish Aquat. Sci.* **42**, 1290–1295.

Croy, T.R and Amend, D.F. (1977). *Aquaculture* **12**, 317–325.

Ellis, A.E. (1997). In *Fish Vaccinology*, vol. 90 (eds R. Gudding, A. Lillehaug, P. J. Midtlyng and F. Brown), pp. 107–116. *Developments in Biological Standardization*, Karger, Basel.

Ezeasor, D.N. and Stokoe, W.M. (1980). *J. Fish Biol.* **17**, 619–634.

Groman, D.B. (1982). *Histology of the Striped Bass*. American Fisheries Society Monograph No. 3, Bethesda, Maryland, pp. 1–116.

Hastein, T. and Rafsti, T. (1986). *Bull. Eur. Ass. Fish Pathol.* **6**, 45–49.

Johnson, K.A. and Amend, D.F. (1983). *J. Fish Dis.* **6**, 473–476.

Johnson, K.A., Flynn, J.K. and Amend, D.F. (1982). Duration of immunity of salmonids vaccinated by direct immersion with *Vibrio anguillarum* and *Yersinia ruckeri* bacterins. *J. Fish Dis.* **5**, 207–213.

Lamers, C.H.J. (1985). Thesis dissertation. Dept. Experimental Animal Morphology and Cell Biology, Agricultural University, Wageningen, Netherlands, pp. 1–255.

Lillehaug, A. (1989). *J. Fish Dis.* **12**, 579–584.

Noga, E.J. (1996). *Fish Disease Diagnosis and Treatment*. Mosby-Year Book Inc., St Louis, Missouri USA. 367pp.

Ototake, M., Moore, J.D. and Nakanishi, T. (1997). In *Fish Vaccinology*, vol. 90 (eds R. Gudding, A. Lillehaug, P. J. Midtlyng and F. Brown), pp. 440–441. *Developments in Biological Standardization*, Karger, Basel.

Roitt, I.M., Brostoff, J. and Male, D.K. (1993). In *Immunology*, 3rd edn. Mosby-Year Book Europe Ltd. pp. 1–12.

Sinnhuber, R.O., Hendricks, J.D, Wales, J.H. and Putnam, G.B. (1977). *Ann. NY Acad. Sci.* **298**, 389–408.

Siwicki, A.K., Anderson, D.P. and Dixon, O.W. (1990). *Dev. Comp. Immunol.* **14**, 231–237.

Tamura, E. and Honma, Y. (1977). *Bull. Jap. Soc. Sci. Fish.* **43**, 963–974.

Vigneulle, M. and Baudin-Lavrencin, F. (1991). *Dis. Aquat. Org.* **11**, 85–92.

Warr, G.W. (1997). In *Fish Vaccinology*, vol. 90 (eds R. Gudding, A. Lillehaug, P. J. Midtlyng and F. Brown) pp. 15–21. *Developments in Biological Standardization*, Karger, Basel.

CHAPTER 15

Sensory Systems

15.1 Vision

Russell D Fernald
Department of Psychology, Stanford University,
Pal Alto, California, USA

Introduction

How fish see puzzles anyone who opens their eyes underwater to a shimmering, unfocused world. The particular difficulty of seeing underwater has led to numerous adaptations that allow fish to see in virtually every living space in the hydrosphere. The remarkable diversity of habitats has shaped numerous variations in the detailed anatomy and function of fish eyes. In addition to these functional differences, fish eyes continue to grow throughout life. This makes them profoundly different from eyes of most other vertebrates. Through this growth, fish eyes are in a continuous state of flux, adding new cells throughout life. Consequently, fish eye growth has become a model for developmental neurobiologists analyzing how retinas develop as well as scientists studying ocular function. Here, the gross functional anatomy of the fish eye is reviewed, identifying putative selective pressures responsible for particular ocular structures. Fish eyes can be best understood in an evolutionary context (Land and Fernald, 1992; Fernald, 1997).

Fish eye adaptations subserve the three primary functions of any eye: (i) collecting light; (ii) focusing images on the retina; and (iii) transforming images into the neural signals necessary for behavioral decisions. Macroscopic features of fish eye are described here, and the microscopic details can be found in Chapter 27.1. Several other reviews of fish visual system are available (e.g. Nicol, 1989; Douglas and Djamgoz, 1990; Fernald, 1993).

Seeing underwater

The optical characteristics of water affect illumination intensity, spectral quality, directional distribution and polarization. Water is also the background against which stimuli are viewed (e.g. Muntz, 1990). Comparing species which live in water with distinctly different optical properties has shown how mutual adaptations occur among stimulus, environment and detector (e.g. Bowmaker, 1990).

Copyright © 2000 Academic Press

Light from the sun, moon and stars is filtered by the atmosphere before entering the water (Lowe and McFarland, 1990). Photons are reflected, refracted and to some extent polarized at the water's surface before becoming available to fish below. Most such physical transformations depend critically on the surface of the water since typically more light will enter through a still surface than a rough one. This means that the visual environment of a fish depends on surface conditions, and will be quite different at dawn and dusk, when sunlight arrives at a lower angle via a longer atmospheric path, producing much lower intensities, shifted towards the shorter wavelengths (see figure 1 in Munz and McFarland, 1977). Interestingly, ultraviolet wavelengths penetrate water sufficiently to be used for fish vision (McFarland, 1986) and recent discovery of UV detection systems in some fish species (Avery *et al.*, 1983) supports this notion. Properties of light in water will influence the evolution of spectral sensitivity and the absolute sensitivity of fish in each particular ecosystem.

The changes in lighting characteristics are due to differential absorption or reflection of photons of different wavelengths at the surface, but once light enters water, scatter or the redirection of photons due to reflection, refraction and diffraction determine how light forms images. Scatter decreases the penetration of light through water and diminishes its directionality, essentially eliminating shadows and, consequently, directionality cues. Light reflected from an image is also scattered, reducing brightness and generally reducing information about the image (Lythgoe, 1972). Fluctuations in time and space of the intensity of light underwater result from surface wave actions, and these effects can be seen at depths up to 70 m (Lowe and McFarland, 1990).

Polarization has proved harder to understand. Polarized light can be detected by some fish, though it is not clear whether or how this information is used (cf. Harwryshyn, 1992). Polarization underwater arises from three sources: direct transmission from polarized skylight; reflections at the air/water interface; and scatter by water and suspended particles underwater (Waterman, 1954), making its distribution extremely complex (e.g Lowe and McFarland, 1990; Hawryshyn, 1998).

Fish, viewing objects underwater, need to make important judgments about whether to eat, mate or flee. This requires extracting information from spectral irradiance and possibly also from polarization. Since irradiance and polarization distributions are not well defined underwater it is not yet known which cues are most salient (but see Lowe and McFarland, 1990). Direct experimentation has been the method of choice for discovering what is important for fish to see.

Seeing objects underwater requires that they reflect differently from the background as can be inferred by the numerous adaptations in fish coloration which reduce conspicuousness. For example, fish are typically dark on the dorsal surface and light on the ventral surface (countershading) because downwelling light is brighter making animals appear more uniformly colored (Cott, 1940). Silvery sides are also common among fish because symmetrical underwater light distribution makes the light entering the eye the same whether the fish is present or not (Muntz, 1990). These strategies clearly have evolved to confuse predators because predators have evolved countermeasures to overcome them. For example, fish capture prey by viewing them in silhouette, defeating both countershading and mirror strategies. In turn, some fish have evolved ventral **photophores** which generate light, obscuring their silhouette (Denton, 1970; Lawrey, 1974). Many methods of concealment have been used to illuminate how the fish visual system might work (e.g. Muntz, 1990).

Collecting light

Eye size and light collection

Both sensitivity and acuity of vision depend on the brightness of the image reaching the retina. In viewing a point of light, the brightness of the image increases in proportion to both the intensity of the source and area of the pupil, and decreases in proportion to the square of the distance from the source to the plane of the pupil (Fernald, 1988). For deep-sea fish, there are point sources of light, such as bioluminescent organisms or light organs of other fish, so light-gathering ability has probably been the primary selective force leading to large eyes in these creatures.

However, fish living in higher levels of ambient daytime light near the surface of the water view scenes illuminated by extended sources, for which the retinal illuminance is related both to pupillary area and the focal length of the eye (Fernald, 1990a). Because the focal length of the fish lens is proportional to the lens radius (Matthiessen, 1882, see below) as is the pupillary area, retinal luminance depends *only* on the intensity of illumination for surface dwelling fish (Fernald, 1990a). Consequently, proportionately larger eyes in this habitat do not collect proportionately more light, suggesting there must be other selective forces favoring larger eyes.

Focusing light

Lenses obey fundamental physical laws providing a straightforward strategy for analysis. Despite this theoretical knowledge, however, details of exactly how the optics of the fish eye work have been assessed in only a handful of species.

The optics of animals living underwater are different from those of animals living in air in one important respect: underwater there is no air–cornea interface which provides significant dioptric strength to land living vertebrates. In humans, for example, the air–cornea interface provides about 43 diopters of refractive power, with the lens responsible for an additional 13 diopters which is changed for focusing (Westheimer, 1968). Eyes underwater have no optical benefit from the cornea because its refractive index is nearly identical to that of water. This is evident to swimmers, who, deprived of 80% of their optical power, strain to focus anything underwater. Thus, in all underwater eyes, since the refractive power of the cornea is neutralized, the lens must provide all the dioptric strength. In addition, lenses of fish must have a short focal length to minimize eye size and preserve their hydrodynamic profile. Together, these constraints have been met in fish through the evolution of a spherical lens with very high refractive power. Exceptions to this rule include skates and other rays, which have proportionally larger eyes and presumably, therefore, aspherical lenses (Cohen, 1990), as well as a few species of fish that live at the air–water interface, including the 'four-eyed' minnow, *Anableps* sp., which has half a lens looking up and half looking down, and the amphibious fish *Dialommus fuscus*, which has a flattened cornea for vision in air.

Fish eyes are nearly hemispheric, with a short axial length. The actual shape of the eye depends on the amount and nature of focusing (accommodation) in the species, since the lens moves within the globe to achieve different states of focus (see below). Deep-sea fishes have eyes specially adapted to allow greater photon capture or visual field size (see Lockett, 1977). Many mesopelagic fish species have evolved asymmetrical tubular eyes, presumably because space does not allow enlargement of a spherical eye. This type of eye has been found in 30 of the 750 mesopelagic species, distributed among 11 families (Marshall, 1971), suggesting that it may have evolved several times. In these eyes, which appear like sections of a larger eye, the lens is spherical and often fills the dorsal part of the eye. The main retina lies at the end of the 'tube' and satisfies **Matthiessen's ratio**, meaning that it is at a distance appropriate for a typical fish eye lens. There is an accessory retina, which extends up the medial wall of the eyecup, from the choroid fissure to the dorsal region of the eye. In some fish, this accessory retina has light directed on to it by a bundle of transparent fibers (Lockett, 1977). These tubular eyes, similar to those which have evolved in some species of owls, are largely immobile. The second adaptation to the deep-sea environment is the appearance of retinal diverticula presumed to increase the visual field (Munk, 1966; Fredrikson, 1973). The diverticula project from the eyecup laterally through a slit-like opening. They presumably receive light via reflecting crystals located in a well-developed argenteum located laterally near the primary lens. Retinas in these fish contain rod-like photoreceptors that are significantly longer than those in surface dwelling fish.

Lens quality

Fish lenses are spherical which could be subject to **spherical aberration** or any of five other primary aberrations characteristic of optical systems (Fincham, 1959). Of these, only **chromatic aberration** is a function of the material of the lens itself, while the rest depend on the structure of the lens and the shape of its focusing surfaces. Four aberration types represent failures of the lens to focus an image on to a flat plane (Figure 15.1.1). Since the retina is curved, this is not a constraint in physiological optics. However, spherical or chromatic aberration or both could adversely affect the image quality of spherical fish lenses. Spherical aberration occurs when light rays entering at different distances from the optic axis are focused at different distances from the lens while chromatic aberration occurs when light of different wavelengths focuses at different distances from a lens along its main axis (longitudinal chromatic aberration) or different distances off the main axis of the lens (lateral chromatic aberration).

Fish lenses are very high quality (Fernald and Wright, 1983; Fernald and Wright, 1985a), which is particularly necessary because most fish eyes have no

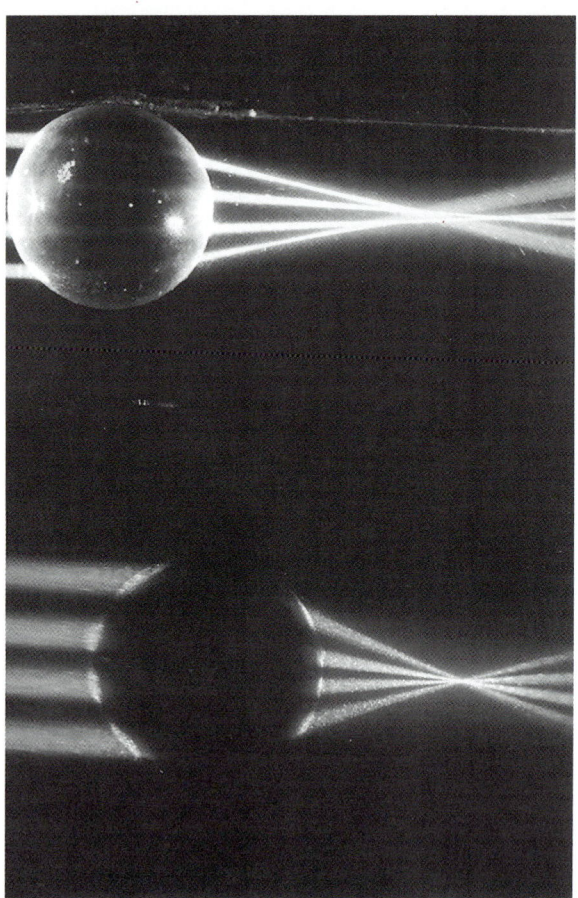

Figure 15.1.1 Top: glass sphere of uniform refractive index ($n \sim 1.53$) focusing four laser beams illustrating longitudinal spherical aberration. Note that there is no single point of focus. (Argon laser beam, $\lambda = 494$ nm, Coherent Radiation.) Bottom: freshly excised lens from an African cichlid fish, *Haplochromis burtoni*, suspended by its ligament in oxygenated fish Ringer's solution and illuminated as above. The focused cone of light is typical of fish lenses of all sizes tested. (After Fernald, 1984.)

pupil to restrict the light path through the lens center as occurs in terrestrial vertebrates. It has long been suggested that the spherical fish lens is of high quality because the lens has a refractive index gradient. The index was thought to be high in the center of the lens, decreasing continuously and symmetrically with radius in all directions. Maxwell (Maxwell, 1854) allegedly first postulated this gradient while contemplating his breakfast herring (cited in Pumphrey, 1961), though the idea arose independently in the writing of Young (1801), Maxwell (1854) and Matthiessen (1886). Brewster (1816) even described his attempts at producing a spherical lens with a refractive index gradient to test directly this hypothesis.

Matthiessen (1882) originally proposed that the refractive index (n) would vary with distance r from the lens center as $n^2 = a - br^2$. Only recently, Axelrod et al. (1988) devised a technique to measure the gradient within an intact lens. Although the function is not exactly that proposed by Matthiessen, the gradient is roughly as he predicted over 100 years ago! Recent measurements (Kroeger et al., 1994) have shown that the gradient is adjusted as fish grow to allow maintenance of the optical quality of the lens at all ages.

The optical resolution of the fish lens approaches the theoretical diffraction limit (Fernald, 1990a). Interestingly, the fish lens resolution appears to be approximately 10 times better than the retinal resolution would predict (Northmore and Dvorak, 1979; Fernald and Wright, 1985a). Only at very small lens sizes does the theoretical limit of the lens resolution match retinal resolution suggesting that, for fish which do not live in the deep sea, growth of the eyes serves primarily to increase the optical resolution and perhaps to improve contrast detection. Ultimately resolution available to the animal depends not only on lens quality but also on receptor size, receptor spacing and the convergence of receptors on higher order cells.

The chromatic aberration of teleost lenses has been measured by numerous investigators and ranges from 2 to 5% of the focal length which is below the detectable level (reviewed in Fernald, 1990a).

Accommodation

The visual image on the retina depends both on the quality of the lens and on focusing or accommodation. Focusing in fishes results from the movement of the spherically shaped lens within the globe, rather than by changing lens shape as in terrestrial vertebrates. Where it is known, the fish lens moves approximately parallel to the plane of the pupil, usually along an axis of pupillary eccentricity (e.g. Fernald, 1990a). This means that the different regions of the retina are focused at different distances for different parts of the visual world simultaneously. For example, in the relaxed accommodative state in fish, the lens lies nearer to the nasal pole of the eye (Figure 15.1.2). Thus the temporal part of the retina (nasal visual field) is focused for near vision and the nasal pole retina (temporal visual field) for far vision (Fernald and Wright, 1985b).

Beginning in the nineteenth century (Beer, 1894), scientists measured the refractive states of fish, which is difficult because the lens moves in the ocular globe

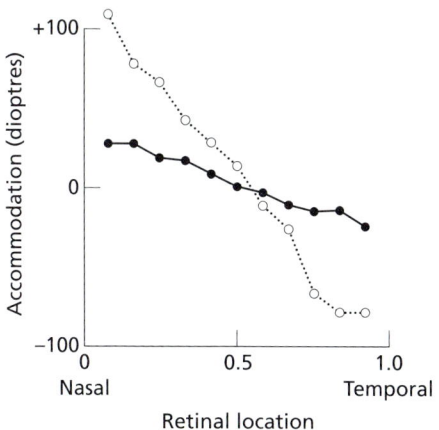

Figure 15.1.2 Maximum accommodative amplitude calculated from measurements of lens excursion for two animals of different size plotted with respect to retinal position. Retinal location units are arbitrary and scaled for the two different-sized animals. Broken curve, lens diameter 1.62 mm; solid curve, lens diameter 2.80 mm. Both lenses from an African cichlid fish, *Haplochromis burtoni*. Curves show that the maximum accommodative amplitude varies systematically as a function of retinal location and hence visual field location. This means that objects viewed along the central axis of the eye (0.5 on the retinal location scale) cannot be focused using accommodative lens movement. In contrast, objects imaged at the edges of the retina can be focused over a wide range of distances. (After Fernald and Wright, 1985b.)

with eye size, but the power of the lens decreases, so that the net effect is nearly neutralized. As accommodation is scaled to body size, with growth the distance to the nearest focal point increases from about 2 cm to 4.5 cm so that the visual near point maintains its same *relative* position in the front of the body.

Because accommodative amplitude depends strongly on retinal location, accommodative movement has substantially different effects at different retinal locations (Fernald, 1990a). The retinal poles can have large accommodative changes of different sign while the central retina does not. Thus images at a wide range of distances can be focused on either the temporal or nasal retina, in contrast to the central retina which remains focused on the intermediate distance. Moreover, if the temporal pole of the retina is focused on a near object, the nasal pole must be focused on a distant object and vice versa.

Although there has been some dispute about the state of focus of the fish eye, much may be due to variability in the measurement techniques suggesting that as Nicol (1963) stated: 'It still has to be demonstrated, convincingly, that the plane of focus in the fish eye at rest lies elsewhere than in the receptor layer; the so-called error may be one not of refraction but rather of method.'

so the state of accommodation depends on the angle of view since (Fernald and Wright, 1985b).

Retinoscopic measurement of refractive state is difficult in fish eyes because they are small. Moreover, because the measurements are made underwater, the vergence of the retinoscopic reflection must be corrected for in the air–water interface, or the degree of **ametropia** in either myopic or hyperopic eyes will be overestimated (Hueter and Gruber, 1980). An additional difficulty is where to consider the source of the retinoscopic reflection in the eye. In many fish, there are no retinal blood vessels to use as landmarks, since the **choroid reet** provides oxygen to the eye. Assuming the wrong reflective surface can lead to sizeable errors, particularly in small eyes, because the retinal thickness is nearly constant over a wide range of eye sizes.

Accommodation, refractive state and lens movement have been comprehensively analyzed in a cichlid fish, *Haplochromis burtoni* (Fernald and Wright, 1985b). This species is particularly suited for such studies because it depends critically on vision for its social interactions. In *H. burtoni*, the magnitude of the lens movement responsible for accommodation increases

The retina: transforming images into neural signals

Overview

Images focused by the lens fall on the retina, a thin, transparent laminar structure comprised of seven different cell types (six neuronal cells: photoreceptor, bipolar, horizontal, ganglion, amacrine and one glial cell: Müller) located at the back of the eye. The retina is sandwiched between the vitreous body towards the lens and the pigmented epithelium towards the sclera. The eyes initially form as an evagination from the diencephalon and hence are the only peripheral sensory organ that is part of the central nervous system. The light absorbing photoreceptors lie at the very back of the eye so visual information arrives as photons and flows from the back to the front of the eye from photoreceptor cells to bipolar cells to ganglion cells. Ganglion cell axons then send information encoded in action potentials from the eye to the brain through the optic nerve. Information transmitted via this photoreceptor–bipolar cell–ganglion cell pathway is modified by two other neuron types, horizontal cells in the outer retina and amacrine cells in the inner retina.

Macroscopic retinal structure

Fish retinas are fundamentally similar to all other vertebrate retinas but exhibit considerable structural variety due to the enormous number of fish species living in diverse habitats. There is a long tradition of studying fish retinas, beginning with Cajal (1892). The many studies comparing retinas among fish species (e.g. Wunder, 1925; Verrier, 1928; Dathe, 1969; Bathelt, 1970; Wagner, 1972), including a remarkable descriptive catalog (Ali and Anctil, 1976), show convincingly that behavioral and ecological constraints predict the details of retinal morphology far better than phylogenetic relationships. Closely related species, one nocturnal and solitary, the other highly social and diurnal, have radically different retinas. The primary differences produced by differential selective pressures are in: (i) overall retinal thickness; (ii) diversity of retinal cell subtypes, especially photoreceptors; and (iii) regional specializations of retinal cells reflecting specific behavioral requirements.

Cell phenotypes

Pigmented epithelium

The pigmented epithelium (PE) consists of cuboidal cells joined along their common boundaries by complex junctions. Towards the retina, long processes of the PE interdigitate with the retina, surrounding the light-sensitive tips of the photoreceptors (Kuwabara, 1979; Bok, 1985). Within the cells are rod-shaped pigment granules which are highly light absorbent and serve to optically isolate the cone outer segments and protect rod outer segments during bright illumination (see below). At low levels of illumination, these pigment granules migrate towards the back of the eye, into the cell bodies of the PE cells. In addition to these optical functions, PE cells phagocytize the tips of rod and cone outer segments which are continuously renewed (Young, 1970). Photoreceptor tip shedding and hence phagocytosis is highly rhythmic at both the cellular and molecular levels, having both circadian and diurnal features (cf. Korenbrot and Fernald, 1989).

Photoreceptors

Photoreceptors transduce photon energy into electrical signals to be interpreted by the nervous system. Müller (1851, 1857) first observed that there were two types of photoreceptors, cones which were common in animals with diurnal activity cycles and rods which were common in animals active at night. From this, Schultze (1866) proposed a possible functional distinction, namely that cones were responsible for vision in bright light and rods for vision in dim light. This 'duplicity theory' of photoreceptor function has proven to be a cornerstone of our understanding of visual function. The originally observed structural differences have been confirmed and extended to numerous ultrastructural features (e.g. Cohen, 1972). Moreover, the basic mechanisms responsible for transducing photons into neural signals has been elucidated, although many details remain unknown (e.g. Yau and Baylor, 1989).

Differences between rod and cone photoreceptors are clear at the level of the light microscope. Rod photoreceptors have long, cylindrical outer segments while cones have shorter, conical outer segments. Within the rod outer segment, the light absorbing **opsin** is localized on discs which are arranged in stacks contained within an outer membrane. Opsin in the cone outer segments is also located on discs, but these discs are contiguous with the outer membrane. Rod and cone inner segments also differ significantly. Rod inner segments are small and the cell body, located **vitread** of the inner limiting membrane, contains darkly staining chromatin. Synaptic connections are spherical, and hence called spherules, which contain one or two ribbon-associated synaptic complexes. Cone inner segments are nearly as large as the outer segments, containing large mitochondria, endoplasmic reticulum and the cell nucleus which is located at the outer limiting membrane. Synaptic connections to the cone pedicles are large, containing up to 100 dendrites from bipolar and horizontal cells. There is substantial evidence that the acidic amino acids L-glutamate or L-aspartate are the photoreceptor neurotransmitters (reviewed in Lasater, 1990).

As in all vertebrates, cone photoreceptor subtypes differ in their spectral sensitivity, and this difference is typically reflected in cone outer segment length in teleosts. The shortest cones are sensitive at the shortest wavelengths, with the very smallest sensitive in the ultraviolet range (Downing et al., 1986) and the longest cones are sensitive to the longest wavelengths (Levine and McNichol, 1982). Although these differences in length were once postulated to compensate for chromatic error in the lens (Eberle, 1968), the size differences are too small to provide complete correction (Fernald, 1988). In teleost retinas, cones also appear in pairs which may or may not have similar spectral sensitivities. In fish which use vision extensively, the cone photoreceptors are arranged in highly ordered mosaics as first noted by Ryder (1895). These mosaics appear organized in alternating rows of single and double cones ('row pattern') or in squares with four cone pairs set around a single, central cone ('square pattern') (e.g. Eigenmann and Shafer, 1900; Wagner, 1990). In animals with different cone types, the chromatic organization of the cone mosaic optimizes chromatic detection (Fernald, 1981). Transitions from one mosaic type to another associated with a change in life habits have been reported in salmon and trout (Lyall, 1957a,b), and in perch (Ahlbert, 1969). Changes in the cone phenotypes and mosaic occur in fish which metamorphose (Evans and Fernald, 1990). In elasmobranchs, there seems to be only one cone type which is rare so no mosaics exist (Cohen, 1990) (Figure 15.1.3).

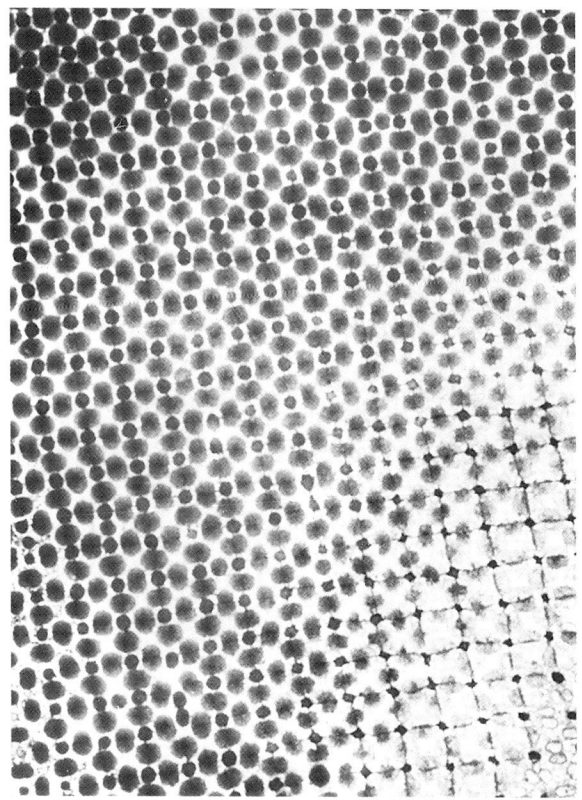

Figure 15.1.3 Tangential section through a light adapted, mature male *Haplochromis burtoni* (3 μm section, toluidine blue). Plane of section is slightly oblique through the inner segments of the cone photoreceptors. Note the orderly array of photoreceptor cross-sections.

Fish photoreceptors and pigmented epithelium exhibit some remarkable transformations in synchrony with the light/dark cycle. Every day there is a wholesale rearrangement of rod and cone photoreceptor positions, called retinomotor movements, that allow the function of each photoreceptor phenotype to be maximized. During daylight, fish have an all cone retina while at night they have an all rod retina. In the light adapted retina, cone photoreceptors move vitread, directly adjacent to the outer limiting membrane where their outer segments are in the plane of focus of the lens. Also, the pigment granules within the PE cells migrate vitread into the processes around the cone outer segments forming a dense absorbing layer **sclerad** to the cone outer segment boundary. During daytime, rod photoreceptors extend through elongation of the **myoids** beyond the pigment layer so they effectively no longer receive photic input. At night, rod myoids contract bringing the rod photoreceptors adjacent to the outer limiting membrane and the pigment granules in the PE migrate to the perikarya of the pigmented epithelial cells. In addition, the cone myoids elongate, placing the cones sclerad to the outer limiting membrane but vitread to the pigmented epithelial band. Consequently, the cone photoreceptors can receive light and may participate in vision at low light levels. In teleosts, these movements can be as much as 40–80 μm and the pigment may migrate *ca.* 100 μm. They are regulated by both the light/dark cycle and circadian rhythm (Burnside and Nagle, 1983).

Why did retinomotor movements evolve? First, retinomotor movements produce two separate functional retinas: cones for photopic vision and rods for scotopic vision (Herzog, 1905). Second, the dispersed pigment granules contribute to photopic acuity by absorbing scattered rays refracted out of the cone outer segments (Garten, 1907). Third, the pigmented epithelium shields rod outer segments from bleaching during bright light (Garten, 1907). Fourth, retinal acuity is increased significantly, especially in animals with small eyes (Fernald, 1988). For example, the retinal acuity of an all cone retina resulting from retinomotor movements is twice what it would be if rod photoreceptors were also in place (Fernald, 1988). Conversely, the uniform field of rod photoreceptors during periods of low light intensity must increase threshold detection since no photons fall on low sensitivity cones. Thus retinomotor movements provide different retinas for different conditions.

Bipolar cells

Information about photons captured by photoreceptors is transmitted to the brain via bipolar and then ganglion cells. Bipolar cells in the inner nuclear layer serve as the main conduit from photoreceptors to ganglion cells and also have synapses from horizontal cells which regulate the lateral interactions across the retina. In a pioneering work, Scholes (1975) examined teleost bipolar cell connections identifying at least 10 types. This and subsequent work revealed that the connectivity of the bipolar cells produces cell-specific responses (see Chapter 27.1 for details). He found that selective bipolar cells connect only to various spectral cone types (cone bipolar cells), that some bipolar cells connect predominantly to cones but also receive some rod input (mixed bipolar cells) and that still other bipolar cells have massive input, primarily from rods with a few red-sensitive cones included. This linking of the scotopic/rod pathway to the red cone bipolars is different from mammalian retinas which have an exclusive cone (scotopic) pathway (e.g. Sterling *et al*., 1986). The bipolar cells are organized with antagonistic receptive fields with the 'OFF' center cells located distal to the 'ON' center cells in the inner plexiform layer. Although serotonin has been indicated as a neurotransmitter in bipolar cells of elasmobranchs (Brunn *et al*., 1984), in teleosts it is unknown what bipolar cells use to communicate with their partners.

Horizontal cells

Horizontal cells comprise a large, distinct cell class located just vitread of the external limiting membrane. These cells are also organized in a mosaic pattern, particularly visible in young animals (Hagedorn and Fernald, 1992). There are three horizontal cells with long axons (*ca*. 500 μm) connected to cones (H1, H2, H3) and one connected to rods (H4) which has no axon (Stell *et al*., 1975; Weiler, 1978; Downing, 1983). Following the initial proposals of Stell *et al*. (1975), four interrelated roles for horizontal cells in transforming photic information can be stated (Wagner, 1990). First they mediate chromatic interactions between different spectral cone types; second, they generate the antagonistic surround of bipolar cell receptive fields; third, they modulate spatially summed signals in the outer plexiform layer via gap junctions; and fourth, they constitute an additional pathway between the outer plexiform and inner nuclear layers. The H1 cells use GABA as a neurotransmitter (Lam and Steinman, 1971) but the neurotransmitters in the remaining horizontal cell types are unknown (e.g. Lasater, 1990).

Amacrine cells

Amacrine cells are local circuit neurons which contact with every type of retinal cell except photoreceptors (Dowling, 1979). These cells may be the most diverse cell class in the CNS with a record number of 48 distinct subtypes identified based on structure in one teleost roach (Wagner and Wagner, 1988). Matching this diversity is that found in amacrine neurotransmitter types (Lasater, 1990). GABA, glycine and acetylcholine have been found in **amacrine cells** as have numerous neuropeptides. Given the enormous diversity, amacrine cell function is difficult to characterize. Wagner (Wagner, 1990) suggests three possible principles of operation. First, populations of amacrine cells may act as multicellular aggregates via gap junction coupling; second, individual amacrine cells might function alone under certain conditions (Miller, 1979); third, parts of the dendritic field of amacrine cells may be functional microcircuits under certain conditions (Bloomfield and Miller, 1982).

Ganglion cells

Ganglion cells integrate visual information and send it to the brain via action potential. Beginning with Cajal (1892), ganglion cells have been characterized morphologically in many teleosts (e.g. Kock and Reuter, 1978; Stell and Witkovsky, 1973), cichlids (Wagner, 1973) and classified into two major groups: large (G) and small (S). Neither the functional significance of these groupings nor the dendritic field size and shape have a bearing on the function of the ganglion cell. Ganglion cells may interact with each other (Sakai and Naka, 1986) but this also is speculative. The functional organization of their receptive fields is an antagonistic center–surround organization, reminiscent of bipolar cells. Considering the chromatic information, Daw (1968) described four types of cells: (i) non-selective units which respond with a center and surround organization to all wavelengths; (ii) P units which have color opponency only in the center of the receptive field; (iii) O units which have color opponency throughout the receptive field; and (iv) Q units which have double color opponency and are driven by the green cones.

Variability in fish retinas

Retinal thickness measured from the optic fiber layer to the outer plexiform layer is typically about 150 μm in most teleost retinas. The relative thickness of the photoreceptor layer is diagnostic for the light intensity of the habitat. For example, in diurnal species which live in relatively bright habitats, the ratio of photoreceptor thickness to neural retina is 1:1 whereas fish in low light level habitats have a ratio of 1.3:1, with the lowest values found in deep-sea fish (Wagner, 1990). Animals which live in low light levels often have strikingly different photoreceptors, including arrangements of bundles of photoreceptors (Wagner and Ali, 1978).

The variation in the density of different cell types across the retina also predicts habitat and habits of the fish. Most commonly found are increases in the photoreceptor density, often in the form of a fovea-like concentration. These regions are typically in the temporal retina which registers scenes from the front of the animal (Verrier, 1928; Kahmann, 1936; Wagner, 1972; Schwassmann, 1974; Fernald, 1983; Collin and Pettigrew, 1989). In regions of specialization, the ratio of photoreceptors to other neural cell types may remain constant so there is a general increase in cell density (Fernald, 1983). Although most of the regions of specialization are related to improved scotopic vision and correspond to increased cone densities, a few cases of rod foveas have been found (Munk, 1964; Lockett, 1977).

Another variation in teleost retinas which has been of great interest is the relationship of the cone visual pigment spectral absorbances to the wavelength distributions characteristic of the habitats. To maximize quantum catch, the visual pigment spectral characteristics should overlap with the spectral irradiance in the water. For deep-living animals, this is certainly the case (Crescitelli *et al.*, 1985; Bowmaker *et al.*, 1988; Partridge *et al.*, 1988). For many surface-living fish, this proved not to be the case, leading Lythgoe (1972) to propose that the absorbance properties of visual pigments may provide needed contrast when they are offset from the peak ambient illumination. This hypothesis has been confirmed and extended by McFarland and Munz (1975) and Munz and McFarland (1977). In animals where color vision is used in behavioral interactions (Fernald, 1990b), the cone spectral distribution is greater than expected for simple contrast detection (Fernald and Liebman, 1980). Additional data and discussion have appeared in numerous reviews (Levine and MacNichol, 1979; Lythgoe, 1984; Bowmaker, 1990).

Summary

Fish eyes are interesting because evolution has shaped their basic form in so many different ways we can understand selective pressures as they acted on elements of the eye. The quality of the lens is remarkable, particularly since making a spherical lens is so difficult. Light focused by the lens is transformed into neural signals in a similar manner to all other vertebrates, via the specialized protein opsin in combination with the chromophore. Once absorbed, the photic information is transformed by a complex retinal circuit to form receptive fields that begin to analyze the image. How fish interpret these signals remains, however, largely mysterious.

References

Ahlbert, I.-B. (1969). *Arkiv for Zoologie* **22**, 445–481.
Ali, M.A. and Anctil, M. (1976). *Retinas in Fish: An Atlas*. Springer-Verlag, Berlin.
Avery, J.A., Bowmaker, J.K., Djamgoz, M.B.A. and Downing, J.E.G. (1983). *J. Physiol., Lond.* **334**, 23–24.
Axelrod, D., Lerner, D. and Sands, P.J. (1988). *Vision Res.* **28**, 57–65.
Bathelt, D. (1970). *Zool. Jb. Abt. Anat.* **87**, 402–470.
Beer, T. (1894). *Pflugers Arch. ges. Physiol.* **58**, 523–650.
Bloomfield, S.A. and Miller, R.F. (1982). *J. Comp. Neurol.* **208**, 288–303.
Bok, D. (1985). *Invest. Ophthalmol. Vis. Sci.* **126**, 1659–1694.
Bowmaker, J.K. (1990). In *The Visual System of Fish* (eds R.H. Douglas and M.B.A. Djamgoz), pp. 81–108, Chapman & Hall, New York.
Bowmaker, J.K., Dartnall, H.J.A. and Herring, P.J. (1988). *J. Comp. Physiol.* **163**, 685–698.
Brewster, D. (1816). *Phil. Trans. R. Soc.* 311–317.
Brunn, A., Ehinger, B. and Sytsma, V.M. (1984). *Brain Research* **295**, 233–248.
Burnside, B. and Nagle, W. (1983). In *Progress in Retinal research*, vol. 2 (eds N. Osborne and G. Chader), pp. 67–109. Pergamon Press, Oxford.
Cajal, S.R. (1892). *Cellule* **9**, 121–225.

Cohen, A.I. (1972). In *Physiology of Photoreceptor Organs* (ed. M.G.F. Fuortes), pp. 63–110. Springer-Verlag, Berlin, Heidelberg, New York.

Cohen, J.L. (1990). In *The Visual System of Fish* (eds R.H. Douglas and M.B.A. Djamgoz), pp. 465–490. Springer-Verlag, New York.

Collin, S.P. and Pettigrew, I.D. (1989). *Brain. Behav. Evol.* **34**, 184–192.

Cott, H.B. (1940). *Adaptive Coloration in Animals*. Methuen, London.

Crescitelli, F., McFall-Ngai, M. and Horowitz, J. (1985). *J. Comp. Physiol.* **157**, 323–333.

Dathe, H.H. (1969). *Z. mikrosk.-anat. Forsh.* **80**, 269–319.

Daw, N.W. (1968). *J. Physiol. Lond.* **197**, 567–592.

Denton, E.J. (1970). *Phil. Trans. R. Soc. London Ser.* **258**, 285–313.

Douglas, R.H. and Djamgoz, M.B.A. (1990). *The Visual System of Fish*. Chapman & Hall, London.

Dowling, J.E. (1979). In *The Neurosciences. Fourth Study Program* (eds F.O. Schmitt and F.G. Warden), pp. 163–181. MIT Press, Cambridge, Mass.

Downing, J.E.G. (1983). Ph.D. thesis, University of London.

Downing, J.E.G., Djamgoz, M.B.A. and Bowmaker, J.K. (1986). *J. Comp. Physiol.* **159**, 859–868.

Eberle, H. (1968). *Zool. Jb., Allgemeine Zool. Physiol. Tiere* **74**, 121–154.

Eigenmann, C. and Shafer, G. (1900). *Amer. Naturalist.* **XXXIV**, 109–118.

Evans, B.I. and Fernald, R.D. (1990). *Neurobiology* **21**, 1037–1052.

Fernald, R.D. (1981). *Vision Res.* **21**, 1749–1753.

Fernald, R.D. (1983). In *Advances in Vertebrate Neuroethology* (eds J.-P. Ewert, R.R. Capranica and D.J. Ingle), pp. 569–580. Plenum Press, New York.

Fernald, R.D. (1984). *American Scientist* **72**, 58–65.

Fernald, R.D. (1988). In *Sensory Biology of Aquatic Animals* (eds J. Atema, R.R. Fay, A.N. Popper and W.N. Tavolga), pp. 185–208. Springer-Verlag, New York.

Fernald, R.D. (1990a). In *The Visual System of Fish* (eds R.H. Douglas and M.B.A. Djamgoz), pp. 45–61. Chapman & Hall, New York.

Fernald, R.D. (1990b). In *The Visual System of Fish* (eds R.H. Douglas and M.B.A. Djamgoz), pp. 443–464. Chapman & Hall, New York.

Fernald, R.D. (1993). In *The Physiology of Fishes* (ed. D.H. Evans), pp. 161–189. CRC Press, Boca Raton.

Fernald, R.D. (1997). *Brain, Behav. Evol.* **50**, 253–259.

Fernald, R.D. and Liebman, P. (1980). *Vision Res.* **20**, 857–864.

Fernald, R.D. and Wright, S. (1983). *Nature* **301**, 618–620.

Fernald, R.D. and Wright, S. (1985a). *Vision Res.* **25**, 155–161.

Fernald, R.D. and Wright, S. (1985b). *Vision Res.* **25**, 163–170.

Fincham, W.H.A. (1959). *Optics*. Hatton Press, London.

Fredrikson, R.D. (1973). *Vidensk. Meddr Dansk Naturh. Foren.* **136**, 233–244.

Garten, S. (1907). In *Graefe-Saemisch Handbuch der gesamten Augenheilkunde*. Leipzig.

Hagedorn, M. and Fernald, R.D. (1992). *J. Comp. Neurol.* **321**, 193–208.

Hawryshyn, C. (1992). *American Scientist* **80**, 164–175.

Hawryshyn, C. (1998). In *The Physiology of Fishes* (ed. D.H. Evans), pp. 345–374. CRC Press, Boca Rotan.

Herzog, H. (1905). *Arch. Anat. Physiol. (Physiol. Abstr.)* **516**, 413–464.

Hueter, R.E. and Gruber, S.H. (1980). *Vision Res.* **20**, 197–200.

Kahmann, H. (1936). *Albrecht v. Graefes Arch. Ophthalm.* **135**, 265–276.

Karenbrot, J.I. and Fernald, R.D. (1989). *Nature* **337**, 454–457.

Kock, J. and Reuter, T. (1978). *Journal of Comparative Neurology* **179**, 549–568.

Kroeger, R.H., Campbell, M.C.W., Munger, R. and Fernald, R.D. (1994). *Vision Res.* **34**, 1815–1822.

Kuwabara, T. (1979). In *The Retinal Pigment Epithelium* (eds K.M. Zinn and M.F. Marmor), pp. 58–62. Harvard University Press, London.

Lam, D.M.K. and Steinman, L. (1971). *Proc. Natl. Acad. Sci. USA.* **66**, 2777–2781.

Land, M.F. and Fernald, R.D. (1992). *Annu. Rev. Neurosci.* **15**, 1–29.

Lasater, E.M. (1990). In *The Visual System of Fish* (eds R.H. Douglas and M.B.A. Djamgoz), pp. 211–238. Chapman & Hall, New York.

Lawrey, J.V. (1974). *Nature* **247**, 155–157.

Levine, J. and MacNichol, E. (1979). *Sens. Process* **3**, 95–131.

Levine, J.S. and MacNichol, E.F. (1982). *Sci. Am.* **246**, 108–171.

Lockett, N.A. (1997). In *Handbook of Sensory Physiology*, vol. 5 (ed. R. Crescitelli), pp. 67–192. Springer-Verlag, Berlin.

Lowe, E.R. and McFarland, W.N. (1990). In *The Visual System of Fish* (eds R.H. Douglas and M.B.A. Djamgoz), pp. 1–44. Chapman and Hall, New York.

Lyall, A.H. (1957a). *Q. J. Microsc. Sci.* **98**, 101–110.

Lyall, A.H. (1957b). *Q. J. Microsc. Sci.* **98**, 198–201.

Lythgoe, J.N. (1972). In *The Handbook of Sensory Physiology* (ed. H.J.A. Dartnall), pp. 566–603. Springer-Verlag, Berlin.

Lythgoe, J.N. (1984). *Vision Res.* **24**, 1539–1550.

Marshall, N.B. (1971). *Explorations in the Life of Fishes*. Harvard University Press, Cambridge, Mass.

Matthiessen, L. (1882). *Pflugers Arch. ges. Physiol.* **27**, 510–523.

Matthiessen, L. (1886). *Pflugers Arch. ges. Physiol.* **38**, 521–528.

Maxwell, J.C. (1854). *Cambridge and Dublin Mathematical Journal* **8**, 188–195.

McFarland, W.N. (1986). *Am. Zool.* **26**, 389.

McFarland, W.N. and Munz, F.W. (1975). *Vision Research* **15**, 1063–1070.

Miller, R.F. (1979). In *The Neurosciences: Fourth Study Program* (eds F.O. Schmitt and F.G. Warden), pp. 227–245. MIT Press, Cambridge, Mass.

Muller, H. (1851). *Z. wiss. Zool.* **3**, 234–237.

Muller, H. (1857). *Z. wiss. Zool.* **8**, 1–112.

Munk, O. (1964). *Galathea Rep.* **7**, 137–149.

Munk, O. (1966). *Dana Rep. Carlsberg Found.* **70**, 1–62.

Muntz, W.R.A. (1990). In *The Visual System of Fish* (eds R.H. Douglas and M.B.A. Diagoz), pp. 491–512. Chapman & Hall, New York.

Munz, F.W. and McFarland, W.N. (1977). In *The Handbook of Sensory Physiology* (ed. F. Crescitelli), pp. 193–274. Springer-Verlag, Berlin.

Nicol, J.A.C. (1963). *Adv. Mar. Biol.* **1**, 171–201.

Nicol, J.A.C. (1989). *The Eyes of Fishes.* Clarendon Press, Oxford.

Northmore, D.P.M. and Dvorak, C.A. (1979). *Vision Res.* **19**, 225–261.

Partridge, J.C., Archer, S.N. and Lythgoe, J.N. (1988). *J. Comp. Physiol.* **162**, 543.

Pumphrey, R.J. (1961). In *The Cell and Organism* (ed. J.A. Ramsey), pp. 193–208. Cambridge University Press, Cambridge.

Ryder, J. (1895). *Proc. Acad. Nat. Sci.* 161–167.

Sakai, H.M. and Naka, K.I. (1986). *J. Comp. Neurol.* **245**, 107–115.

Scholes, J. (1975). *Phil. Trans. R. Soc. Lond. B.* **270**, 61–118.

Schultze, M. (1866). *Arch. mikr. Anat. Entw. Mech.* **2**, 175–286.

Schwassmann, H.O. (1974). In *Vision in Fishes* (ed. M.A. Ali), pp. 113–135. Plenum Press, New York.

Stell, W.K. and Witkovsky, P. (1973). *J. Comp. Neurol.* **148**, 1–32.

Stell, W.K., Lightfoot, D., Wheeler, T. and Leeper, H. (1975). *Science, NY.* **190**, 989–990.

Sterling, P., Freed, M. and Smith, R.G. (1986). *Trends Neurosci.* **9**, 186–192.

Verrier, M.L. (1928). *Bull. Biol. Fr. Belg., suppl.* **11**, 1–222.

Wagner, H. (1990). In *The Visual System of Fish* (eds R.H. Douglas and M.B.A. Djamgoz), pp. 109–158. Chapman & Hall, New York.

Wagner, H.-J. (1973). *Z. Morph. Tiere.* **72**, 77–130.

Wagner, H.-J. and Ali, M.A. (1978). *Rev. Can. Biol.* **37**, 65–83.

Wagner, H.-J. and Wagner, E. (1988). *Phil. Trans. R. Soc.* **321**, 263–324.

Waterman, T.H. (1954). *Science* **120**, 927–932.

Weiler, R. (1978). *Cell Tissue Res.* **195**, 515–526.

Westheimer, G. (1968). In *Medical Physiology* (ed. V.B. Mountcastle), pp. 1532–1553. Mosby, St Louis, Missouri.

Wunder, W. (1925). *Z. vergl. Physiol.* **3**, 1–63.

Yau, K.-W. and Baylor, D.W. (1989). *Annu. Rev. Neurosci.* **12**, 289–327.

Young, R.W. (1970). *Sci. Am.* **223**, 81–91.

Young, T. (1801). *Phil. Trans. Royal Soc.* **92**, 23–88.

15.2 Mechanosensory Lateral Line: Functional Morphology and Neuroanatomy

Jacqueline F Webb
Department of Biology, Villanova University,
Villanova, Pennsylvania, USA

Introduction

The mechanosensory lateral line system is present in all fishes and in most amphibians, but lost in reptiles and their derivatives (Northcutt, 1989). It consists of a series of sensory organs called neuromasts, which are composed of hair cells and are located in lateral line canals, or on the epithelium of the head and trunk of fishes (Coombs et al., 1988; Webb, 1989b). The structure and function of the mechanosensory lateral line system has been studied by ichthyologists and physiologists for over a century (reviewed by Dijkgraaf, 1962, 1967, 1989; Hensel, 1978; Coombs et al., 1992), and has been the subject of much more intense study in the past 30 years (Cahn, 1967; Atema et al., 1988; Coombs et al., 1989).

In living jawless fishes, the hagfishes and lampreys, the mechanosensory lateral line system is either rudimentary or absent (Northcutt, 1989; Braun, 1996; Braun and Northcutt, 1997). The system is well developed in sharks, skates, and rays (Tester and Kendall, 1967; Northcutt, 1989; Maruska and Tricas, 1998), and in bony fishes (Coombs et al., 1988; Webb, 1989b); it is also present in larval and aquatic adult amphibians (Fritzsch, 1989; Northcutt, 1990; Northcutt et al., 1995). In addition to neuromasts, two other structurally and functionally specialized mechanosensory organs, the vesicles of Savi and the spiracular organs, are only found in some groups of non-teleost fishes (Barry and Bennett, 1989; Northcutt, 1992), but are also considered to be a part of the lateral line system.

Traditionally, the mechanosensory lateral line system was considered to be a component of an 'acousticolateralis system', which included the auditory system and both the mechanosensory and electrosensory lateral line systems (Platt et al., 1989). However, we now know that several fundamental features distinguish these sensory systems from one another. For instance, the mechanosensory lateral line and auditory systems consist of sensory organs composed of hair cells, but develop from different ectodermal **placodes** (several dorsolateral placodes, and a single otic placode, respectively; Northcutt, 1992; see Chapter 27.2) and are innervated by different cranial nerves (a series of lateral line nerves, and the auditory nerve (VIII), respectively, Northcutt, 1989). The electrosensory lateral line system is composed of electroreceptors, which develop from the same placodes as the neuromasts and are innervated by branches of the same lateral line nerves that innervate neuromasts (Northcutt et al., 1995). The auditory system and the mechanosensory and electrosensory lateral line systems have distinct central projection patterns (McCormick, 1989) and respond to different types of stimuli that generally serve different biological roles (Bodznick, 1989; Kalmijn, 1989, 1997; Coombs et al., 1992).

Morphology of the lateral line system on the head

In bony fishes, the lateral line canals are contained in dermal skeletal elements, prominent features of the skull that can be easily observed in dried skeletal material and in cleared and stained specimens (Webb, 1989b; Cubbage and Mabee, 1996; Adriaens et al., 1997; Tarby, 1998; Figure 15.2.1). Typically, these cephalic lateral line canals are present in the neurocranial bones located above the eye (supraorbital canal), and in the circumorbital bones below the eye (infraorbital canal). A canal runs down the length of the preopercular bone (preopercular canal) and into the bones of the lower jaw to the mandibular symphysis (mandibular canal). The head canals generally converge in the otic region of the skull where a canal extends dorsally through the extrascapular bones (= supratemporal commissural canal), and may join the corresponding canal on the opposite side of the head. Finally, a canal generally runs through the post-temporal and supracleithral bones where it joins the trunk canal contained in the lateral line scales. The lateral line canals are lined by a thin epithelium in which neuromast receptor organs are located, and are incorporated into dermal bones, which sit below the basal lamina. In most bony fishes, the lateral line canals are generally well ossified, and are pierced by canal pores, which link the fluid-filled canal lumen with the external environment. The location of the lateral line canals in specific dermal bones of the skull is rather consistent among teleost fishes. Variation in the morphology of the lateral line canals and the bones in which they are contained, and the number and distribution of canal pores on the head, has been used extensively as characters in species descriptions and in phylogenetic reconstructions (Webb, 1989b).

Four general types of cephalic lateral line canal systems have been described among bony fishes: narrow-simple, narrow-branched, reduced, and widened (Figure 15.2.2; reviewed by Webb, 1989b). A narrow-simple canal system is typical of most bony fishes, including experimental model species (Metcalfe,

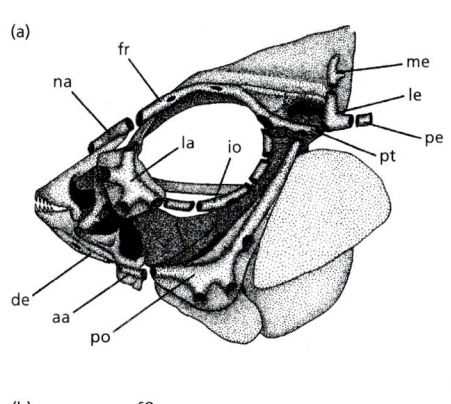

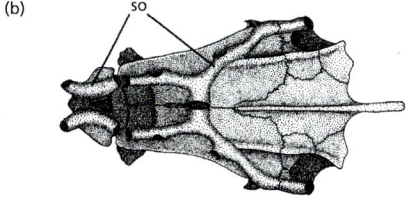

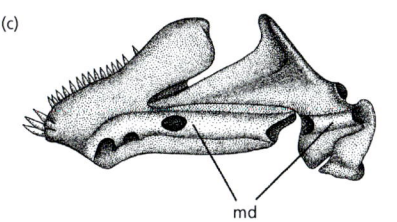

Figure 15.2.1 The skull of the cichlid, *Archocentrus nigrofasciatus* illustrating the location and morphology of the lateral line canals on the head. (a) Lateral view of the skull highlighting the dermal bones containing the lateral line canals: the supraorbital canal in the nasal (na) and frontal (fr) bones, the infraorbital canals in the lacrimal (la) and the infraorbital series (e.g. io), the preopercular canal in the preoperculum (po), the mandibular canal in the dentary (de) and the anguloarticular (aa), the otic canal in the pterotic (pt), the supratemporal commissure in the lateral and medial extrascapulars (le, me), and the post-otic canal in the post-temporal (pe). (b) Dorsal view of neurocranium showing the supraorbital canal (so) in the nasal and frontal bones. (c) Ventrolateral view of mandible showing the mandibular canal (md) in the dentary and anguloarticular bones (from Tarby, 1998).

1989; Puzdrowski, 1989; Cubbage and Mabee, 1996). In these fishes, the canals are well ossified, the canal lumen is generally uniform in diameter, and canal pores appear as holes in the roof of the canal, or as pores at the end of short epithelial tubules that extend from the pore in the canal wall. In a branched canal system (herrings and their relatives, for instance), the bony pores in the canal wall are elongated into bony tubules that are extensively branched and end in numerous terminal pores (Blaxter, 1987). In a reduced canal system, portions of one or more canals are

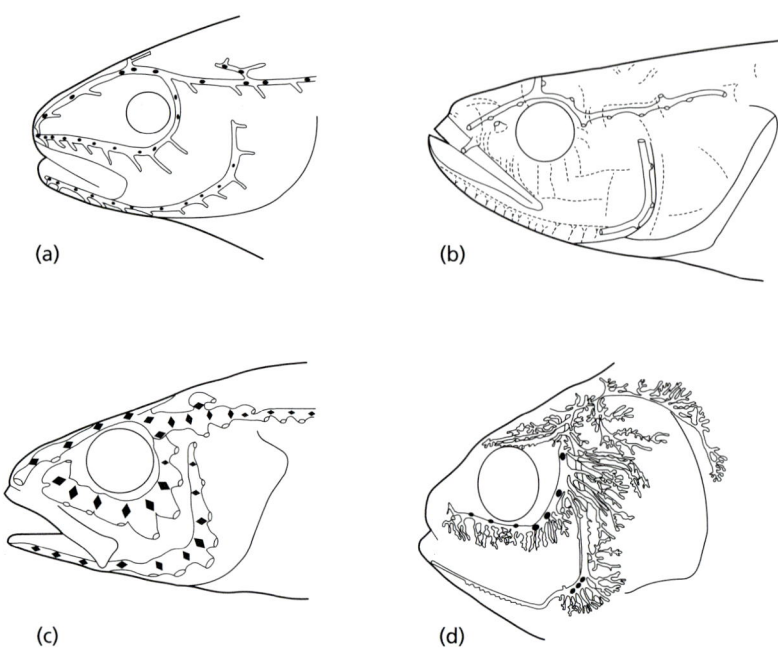

Figure 15.2.2 Four types of head canal systems among teleost fishes: (a) narrow-simple canal system; (b) reduced canal system; (c) widened canal system; (d) branched canal system (from Webb, 1989b, reprinted by permission of Karger).

replaced by superficial neuromasts, which may be greatly proliferated (Tekye, 1990; Wongrat and Miller, 1991). In a widened canal system, canal segments are ballooned out with incompletely ossified walls and a canal roof covered by a tympanum-like epithelium. Small canal pores may be present in the epithelium that covers the canals, but these can be easily overlooked and are generally not noted in the description of these canals (J.F. Webb, personal observation). Widened canals are found among a small number of fish families (Webb, 1989b), including many deep sea fishes (Garman, 1899; Marshall, 1996), some flatfishes (Webb, 1988, 1995) and some freshwater fishes (reviewed by Coombs et al., 1992).

Morphology and distribution of neuromast receptors on the head

Neuromast receptor organs are located in epithelial tissue and consist of a population of directionally polarized hair cells with apical ciliary bundles, surrounded by non sensory supporting cells. The ciliary bundles are embedded in a tall gelatinous **cupula** (Kelly & van Netten, 1991), which is secreted by the nonsensory cells. Displacement of the cupula, and thus the hair cells, by water flow is the basis for neuromast function (see Chapter 27.2). Neuromasts are distributed in predictable locations within the head canals of fishes (Webb and Northcutt, 1997). However, the distribution of canal neuromasts can only be determined if their location is established in larval stages, before they become enclosed by canals (see Chapter 27.2), or if ossified canals of juveniles or adults are sectioned histologically (Tarby, 1998), or dissected for analysis using scanning electron microscopy. In bony fishes with narrow-simple, or narrow-branched canal systems, canal neuromasts are generally oval in shape with their long axis parallel to the long axis of the roughly cylindrical canal (Coombs et al., 1988; Münz, 1989; Tarby, 1998; see Figure 15.2.2). In fishes with a branched canal system, canal neuromasts are found only in the canals, not in the highly branched tubules (Blaxter, 1987) and their morphology and distribution are similar to those in narrow-simple canals. In fishes with widened canals, canal neuromasts are located under bony arches at the constrictions located between ballooned out canal segments. These neuromasts are extremely large (up to 1 mm in diameter, personal observation), are shaped like a diamond or cross, and generally traverse the width of the canal (Garman, 1899; Marshall, 1996; Figure 15.2.2).

The determination of the location of superficial neuromasts in the epithelium of larval or adult fishes requires microscopic analysis using either vital stains (e.g. Blaxter *et al.*, 1983; Tekye, 1990; Mukai and Kobayashi, 1992), histological analysis, or scanning electron microscopy (Tarby, 1998). Alternatively, the location of superficial neuromasts can be determined indirectly by staining the nerves that innervate them (Blaxter *et al.*, 1983; Tekye, 1990). Superficial neuromasts are generally small, and have a round or diamond-shaped outline (Figure 15.2.2). They may occur singly, sitting flush with the epithelium, in small pits, on top of papillae, or on the edge of thin flaps or elongated ridges of skin (Marshall, 1986, 1996; Coombs *et al.*, 1988). Linear series of superficial neuromasts ('pit lines') have been defined on the head of many fishes (Coombs *et al.*, 1988; Northcutt, 1989; Webb and Northcutt, 1997). In some teleosts, the pattern of distribution of superficial neuromasts is extremely complex (Tekye, 1990; Wongrat and Miller, 1991) and their innervation and distribution have been used as the basis for phylogenetic reconstruction (Gill and Bradley, 1992). In most fishes, however, superficial neuromast distributions have not been thoroughly assessed.

Morphology of lateral line canals on the trunk

In bony fishes, the lateral line canal on the trunk ('the lateral line', or more appropriately, the 'trunk canal') is composed of a series of short canal segments, contained in a linear series of overlapping tubed scales, the lateral line scales. Typical lateral line scales appear as a flat plate with a tube superimposed on it (Figure 15.2.3), but lateral line scale morphology varies tremendously among bony fishes (Coombs *et al.*, 1988; Webb, 1990a; Wonsettler and Webb, 1997). Lateral line scales are covered by a thin layer of epithelium and sit in the dermis beneath the basal lamina as do all fish scales. A pore is present at each end of the canal segment in a lateral line scale (suprascalar and infrascalar pores; Coombs *et al.*, 1988; Webb, 1989c). These pores link the canal segments of adjacent overlapping scales which form a continuous epithelium-lined, fluid-filled canal. Additional pores may pierce the wall of the canal segment and provide additional access to the external fluid environment (Webb, 1990a). A neuromast is generally located in the canal segment (or tube) in each lateral line scale in the trunk canals of bony fishes (Suckling, 1967; Webb, 1989b,c; but see Wonsettler and Webb, 1997).

The placement, number, contour and length of the canal on the trunk varies among bony fishes; eight trunk canal patterns have been described among teleost fishes (Figure 15.2.4; Webb, 1989a). **Heterochrony** has been suggested as a developmental/evolutionary mechanism to explain the evolution of these patterns (Webb, 1989a, 1990b), but functional or adaptive explanations for this variation are lacking.

In addition to the neuromast found inside the trunk canal, superficial neuromasts, or 'accessory neuromasts', may be located in the epithelium in close proximity to the trunk canal (Coombs *et al.*, 1988), and may continue onto the caudal fin (Webb, 1989c). In cases where the trunk canal is incomplete and ends short of the caudal peduncle, a series of superficial neuromasts often continues down the trunk in the position where a canal would be present in other species (Webb, 1990b). A limited number of species have a dorsally or ventrally placed canal; other species have a disjunct canal (Webb, 1990b). In some species, the 'main' trunk canal has one or more branches, which may be represented by a linear series of superficial neuromasts (Northcutt, 1989). In other taxa, multiple parallel trunk canals lie along the length of the trunk (Webb, 1989b). In fishes with multiple trunk canals, neuromasts have been found in only one (Wonsettler and Webb, 1997) or two (J.F. Webb, unpublished observation) of the canals. Finally, in some fishes, a trunk canal is absent, but superficial neuromasts may be located in its place (e.g. some cichlids; Webb, 1990b), or are scattered over the trunk (Blaxter *et al.*, 1983; Webb, 1989b).

Neuroanatomy of the mechanosensory lateral line system

The hair cells that compose the canal and superficial neuromasts are innervated by neurons that

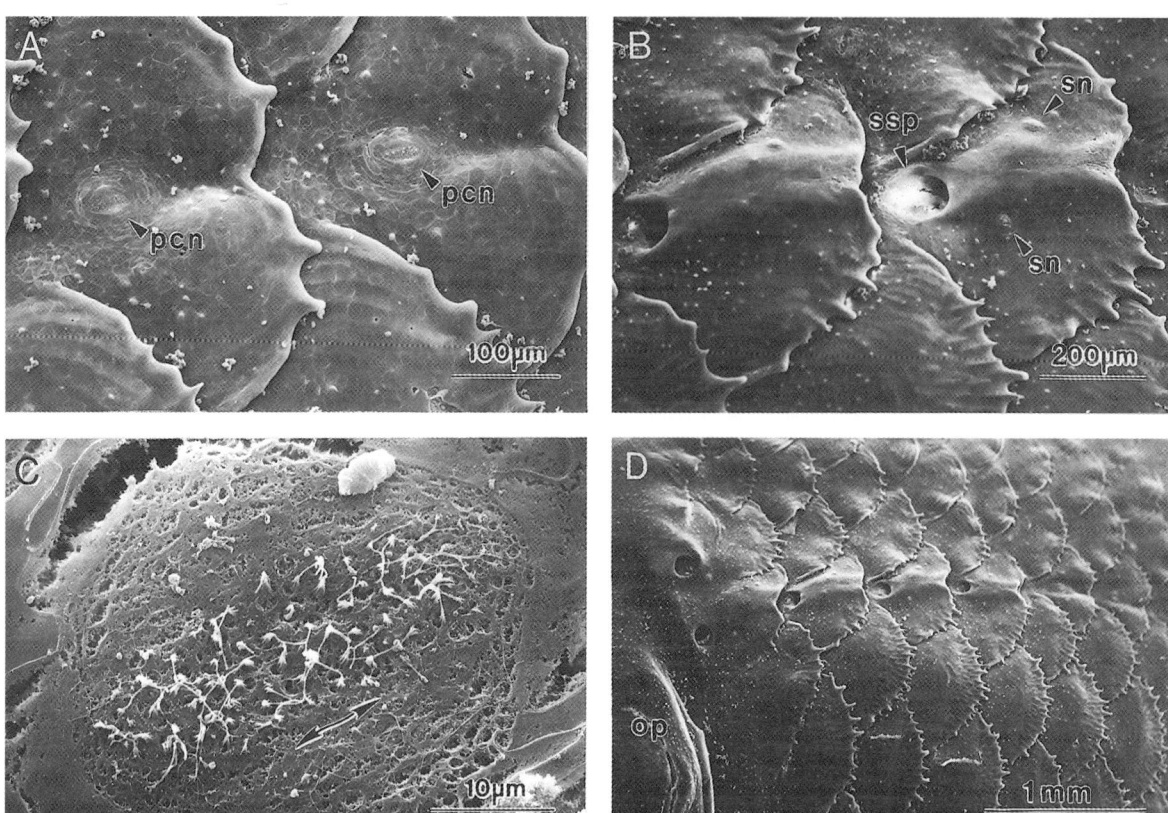

Figure 15.2.3 Scanning electron micrograph of the trunk canal in the cichlid, *Archocentrus nigrofasciatus* (= *Cichlasoma nigrofasciatum*, Webb, 1989a). (A) Two lateral line scales before canal development, showing presumptive canal neuromasts (pcn). (B) Two lateral line scales after canal development showing additional superficial neuromasts (sn) and the suprascalar pore (SSP), which links the canal segment in adjacent scales. (C) Close-up of a presumptive canal neuromast from (A); arrow indicates axis of best physiological sensitivity as determined by hair cell orientation. (D) Anterior end of the trunk canal (just caudal to operculum, op) showing overlapping tubed lateral line scales (from Webb, 1989a, reprinted by permission of Springer-Verlag).

compose a well-defined series of lateral line nerves (Northcutt, 1989; Puzdrowski, 1989; Song and Northcutt, 1991a,b; Figure 15.2.5). These cranial sensory nerves may travel out of the neurocranium with other cranial nerves, but have a series of distinct ganglia, innervate only neuromasts (and electroreceptors in electroreceptive fishes; Northcutt, 1989) and have well-defined central projection sites in the hindbrain. In fishes, several nerves innervate neuromasts on the head. The anterodorsal and anteroventral nerves (considered by some to be branches of the anterior lateral line nerve, ALLN) innervate the neuromast above and below the eye, and the neuromasts on the cheek, opercular series and mandible (including the preopercular and mandibular canals). The middle lateral line nerve (MLLN) innervates neuromasts in the caudal region of the skull. The posterior lateral line nerve (PLLN) innervates another set of neuromasts in the caudal region of the skull and also innervates the neuromasts on the trunk (Northcutt, 1989).

All of the lateral line nerves project to an octavolateralis column in the hindbrain that consists of the medial octavolateralis nucleus (MON), the caudal octavolateralis nucleus, and the magnocellular nucleus, which are distinguished based on cell morphology and the degree of descending input from higher brain centers that is present. Primary projections of the lateral line nerves are also found in the eminentia granularis of the cerebellum. The MON receives the majority of lateral line input to the central nervous system; the magnocellular nucleus also receives input from the inner ear via nerve VIII (McCormick, 1989, 1997; Puzdrowski, 1989; Song and Northcutt, 1991b; Schellart *et al.*, 1992; New and Singh, 1994; New *et al.*, 1996). Central projections from the mechanosensory lateral line system are generally distinct from those of the auditory system (Finger and Tong, 1984; McCormick, 1989) and the electrosensory lateral line system (New and Singh, 1994). Projection sites of the various lateral line nerves

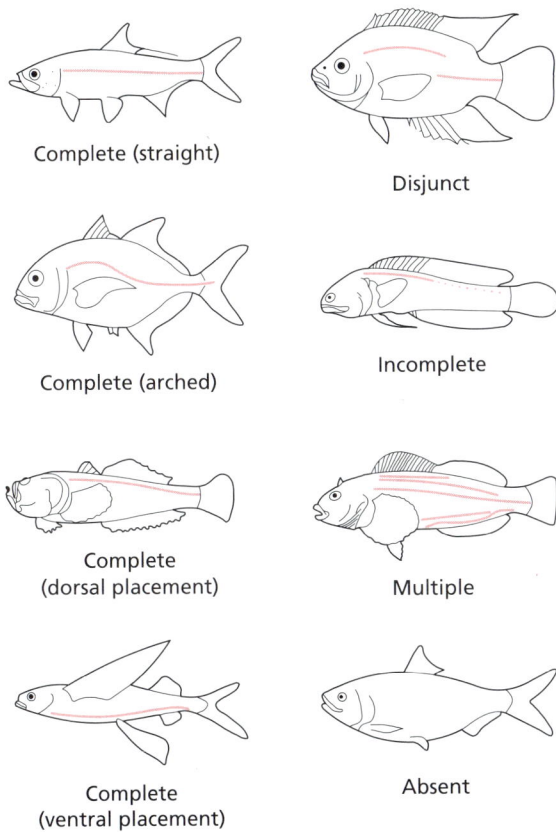

Figure 15.2.4 Eight trunk canal patterns found among teleost fishes (from Webb, 1989b, reprinted by permission of Karger).

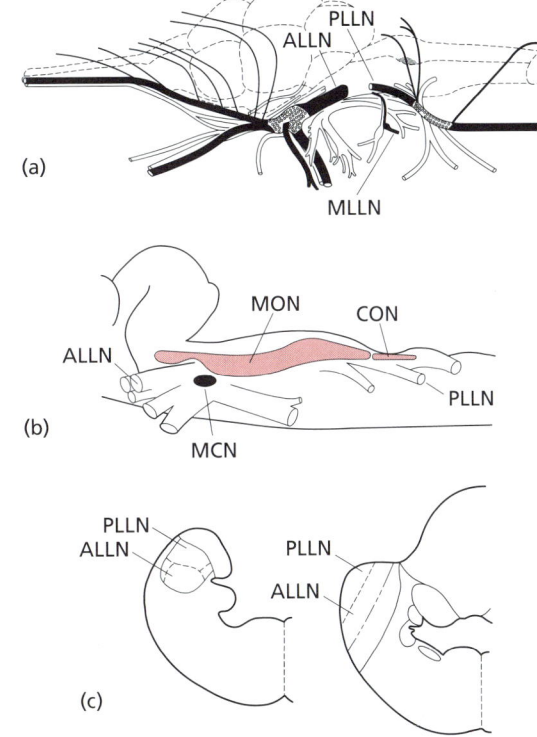

Figure 15.2.5 Lateral line nerves and their central projection in bony fishes. (a) Anterior (ALLN), middle (MLLN) and posterior (PLLN) lateral line nerves (in black) with relationship to the other cranial nerves in the Florida gar, *Lepisosteus osseus* (from Song and Northcutt, 1991a, reprinted by permission of Karger). (b) Lateral view of the central projection areas of the lateral line system in the bowfin, *Amia calva*. The anterior (ALLN) and posterior (PLLN) lateral line nerves project to the medial (MON) and caudal (CON) octavolateralis nuclei (hatched areas), and the magnocellular nucleus (MCN, black) (from McCormick, 1989, reprinted by permission of Springer-Verlag). (c) Cross-sectional diagrams of the primary projections of the anterior (ALLN) and posterior (PLLN) lateral line nerves at caudal (left) and more rostral (right) levels in the bowfin, *Amia calva* (from McCormick, 1989, reprinted by permission of Springer-Verlag).

have been shown to be segregated and demonstrate somatotopy in some species (deRosa and Fine, 1988; Puzdrowski, 1989; Puzdrowski and Leonard, 1993; New and Singh, 1994). The degree to which the projections of different branches of each of these nerves, or inputs from canal versus superficial neuromasts are segregated, is not known. Patterns of central projections vary somewhat among taxa (McCormick, 1982, 1989; Song and Northcutt, 1991b) and there are some differences in projection patterns in fishes that have specialized (McCormick, 1997) and nonspecialized (Schellart et al., 1992) mechanosensory lateral line systems.

Both canal and superficial neuromasts receive efferent input from the rostral and caudal octavolateralis efferent nucleus in the medulla oblongata, as well as from a diencephalic octavolateralis nucleus (Puzdrowski, 1989; Roberts and Meredith, 1989; Song and Northcutt, 1991b; Schellart et al., 1992; New and Singh, 1994; Wagner and Schwartz, 1996). The efferent system develops early and is already apparent in the larvae of zebrafish (Metcalfe et al., 1985).

Functions of the mechanosensory lateral line system

The mechanosensory lateral line system allows fishes to respond to unidirectional (d.c.) or oscillatory water movement (a.c., up to 200 Hz) at relatively short distances (Münz, 1989; Coombs et al., 1998). Effective lateral line stimuli may arise from prey

(Montgomery, 1989; Bleckmann, 1991; Janssen, 1997; Mohr and Bleckmann, 1998; Montgomery and Coombs, 1998), predators (Blaxter and Fuiman, 1990), neighbors in a school (Bleckmann, 1993; Montgomery *et al.*, 1995), or may be generated by water flow over environmental obstacles (Hassan, 1989). Most recently, it has been shown that the lateral line system transduces vibratory reproductive signals in salmon (Satou *et al.*, 1994) and facilitates **rheotaxis** (Montgomery *et al.*, 1997). In addition, efferent input (Tricas and Highstein, 1991) and other inputs (Zottoli and Danielson, 1989; Alborg *et al.*, 1996) contribute to, or modify lateral-line-mediated behavior.

The hydrodynamics of fluid movement in response to a vibrating sphere has been used to study the way in which water in a lateral line canal moves in response to a defined external stimulus (Denton and Gray, 1988, 1989; Bleckmann and Münz, 1990; Coombs, 1994; van Netten and van Maarseveen, 1994; Montgomery and Coombs, 1998). Such studies have shown that the mechanosensory lateral line system responds to noncompressible, local flow in the near field. Canal and superficial neuromasts respond to the acceleration and the velocity components of the stimulus, respectively (Kalmijn, 1989), and differ in their frequency response properties (Münz, 1989).

Several model species have emerged in the study of the functional morphology and physiology of the lateral line system. The morphology, development, and physiology of the lateral line system and lateral-line-mediated behavior have been studied extensively in **clupeoid** fishes (herring and relatives; Blaxter *et al.*, 1983; Denton and Gray, 1983, 1993). The mottled sculpin (*Cottus bairdi*) has been used extensively to study lateral-line-mediated feeding behavior and the neurophysiological responses of canal neuromasts to defined hydrodynamic stimuli (Coombs and Janssen, 1990). The ruffe (*Acerina*, a percid fish), has been used extensively for the experimental analysis of cupular micromechanics and hydrodynamics of fluid movement within its widened head canals (Gray and Best, 1989; van Netten and Khana, 1994). Van Netten and van Maarseveen (1994) have demonstrated that the displacement of the skin overlying the widened canals that occurs in response to a mechanical stimulus accurately predicts the movement of neuromast cupulae and thus predicts the response properties of canal neuromasts. In addition, the functional morphology and neurophysiology of the lateral line have been investigated in fishes with specialized lateral line morphologies including fishes with multiple trunk canals (Bleckmann and Münz, 1990; Wonsettler and Webb, 1997) and surface-feeding fishes (Vogel and Bleckmann, 1997; Mohr and Bleckmann, 1998).

Comparative studies have just begun to provide insights into the adaptive functional evolution of the lateral line system. Recent theoretical and empirical work has shown that there are distinct differences in the functional attributes of narrow and widened head canal systems (Denton and Gray, 1988, 1989). Increased canal width, the presence of a flexible instead of a stiff canal roof, and increased size of neuromasts (typical of widened canals) all contribute to increased neuromast sensitivity, but increased response time of widened canal systems (Coombs *et al.*, 1992). In addition, the response properties of extremely wide canals are similar to that of superficial neuromasts (Coombs *et al.*, 1992). This could account for the evolution of both widened canals and reduced canals (where superficial neuromasts predominate) in deep sea taxa. The ecological significance of functional differences between narrow and widened lateral line systems has been demonstrated recently in two species of percid fishes that use their lateral line systems for prey detection (Janssen, 1997). Interestingly, the nature of structure–function relationships in the lateral line system has been challenged by studies of Antarctic notothenioid fishes in which the frequency response properties of canal neuromasts are not correlated with variation in lateral line canal morphology (Coombs and Montgomery, 1994a,b; Montgomery *et al.*, 1994).

Acknowledgments

I thank Drs Sheryl Coombs and John New for helpful discussions. Melissa Tarby granted permission for the reproduction of several original figures. Supported by NSF grant IBN 9603896.

References

Adriaens, D., Verraes, W. and Taverne, L. (1997). *Eur. J. Morphol.* **35**, 181–208.
Alborg, L., Coombs, S. and New, J.G. (1996). *Soc. Neurosci. Abstr.* **22**, 446.
Atema, J., Fay, R.R., Popper, A.N. and Tavolga, W.N. (eds) (1988). In *Sensory Biology of Aquatic Organisms*. Springer-Verlag, New York.

Barry, M.A. and Bennett, M.V.L. (1989). In *The Mechanosensory Lateral Line – Neurobiology and Evolution* (eds S. Coombs, P. Görner and H. Münz), pp. 591–606. Springer-Verlag, New York.

Blaxter, J.H.S. (1987). *Biol. Rev.* **62**, 471–514.

Blaxter, J.H.S. and Fuiman, L.A. (1990). *J. Mar. Biol. Ass. UK* **70**, 413–427.

Blaxter, J.H.S., Gray, J.A.B. and Best, A.C.G. (1983). *J. Mar. Biol. Ass. UK* **63**, 247–260.

Bleckmann, H. (1991). *Verh. Dtsch. Zool. Ges.* **84**, 105–124.

Bleckmann, H. (1993). In *Behaviour of Teleost Fishes*, 2nd edn (ed. T.J. Pitcher), pp. 201–246. Chapman & Hall, New York.

Bleckmann, H. and Münz, H. (1990). *Brain, Behav. Evol.* **35**, 240–250.

Bodznick, D. (1989). In *The Mechanosensory Lateral Line – Neurobiology and Evolution* (eds S. Coombs, P. Görner and H. Münz), pp. 655–678. Springer-Verlag, New York.

Braun, C.B. (1996). *Brain, Behav. Evol.* **48**, 262–276.

Braun, C.B. and Northcutt, R.G. (1997). *Acta Zool. (Stockholm)* **78**, 247–268.

Cahn, P.H. (ed.) (1967). *Lateral Line Detectors*. Indiana University Press, Bloomington.

Coombs, S. (1994). *J. Exp. Biol.* **190**, 109–129.

Coombs, S. and Janssen, J. (1990). In *Comparative Perception – Volume II: Complex Signals* (eds W.C. Stebbins and M.A. Berkley), pp. 89–123. Wiley, New York.

Coombs, S. and Montgomery, J. (1994a). *Brain, Behav. Evol.* **44**, 287–298.

Coombs, S. and Montgomery, J. (1994b). *Sensory Systems* **8**, 150–156.

Coombs, S., Janssen, J. and Webb, J.F. (1988). In *Sensory Biology of Aquatic Organisms* (eds J. Atema, R.R. Fay, A.N. Popper and W.N. Tavolga), pp. 553–594. Springer-Verlag, New York.

Coombs, S., Görner, P. and Münz, H. (eds) (1989). *The Mechanosensory Lateral Line – Neurobiology and Evolution*. Springer-Verlag, New York.

Coombs, S., Janssen, J. and Montgomery, J. (1992). In *The Evolutionary Biology of Hearing* (eds D.B. Webster, R.R. Fay and A.N. Popper), pp. 267–294. Springer-Verlag, New York.

Coombs, S., Mogdans, J., Halstead, M. and Montgomery, J. (1998). *J. Comp. Physiol. A* **182**, 609–626.

Cubbage, C.C. and Mabee, P.M. (1996). *J. Morphol.* **229**, 121–160.

De Rosa, F. and Fine, M.L. (1988). *Brain, Behav. Evol.* **31**, 312–317.

Denton, E.J. and Gray, J.A.B. (1983). *Proc. Roy. Soc. Lond. B* **218**, 1–26.

Denton, E.J. and Gray, J.A.B. (1988). In *Sensory Biology of Aquatic Organisms* (eds J. Atema, R.R. Fay, A.N. Popper and W.N. Tavolga), pp. 595–618. Springer-Verlag, New York.

Denton, E.J. and Gray, J.A.B. (1989). In *The Mechanosensory Lateral Line – Neurobiology and Evolution* (eds S. Coombs, P. Görner and H. Münz), pp. 229–246. Springer-Verlag, New York.

Denton, E.J. and Gray, J.A.B. (1993). *Phil. Trans. R. Soc. Lond. B* **341**, 113–127.

Dijkgraaf, S. (1962). *Biol. Rev.* **38**, 51–105.

Dijkgraaf, S. (1967). In *Lateral Line Detectors* (ed. P.H. Cahn), pp. 83–95. Indiana University Press, Bloomington.

Dijkgraaf, S. (1989). In *The Mechanosensory Lateral Line – Neurobiology and Evolution* (eds S. Coombs, P. Görner and H. Münz), pp. 7–16. Springer-Verlag, New York.

Finger, T.E. and Tong, S.-L. (1984). *J. Comp. Neurol.* **229**, 129–151.

Fritzsch, B. (1989). In *The Mechanosensory Lateral Line – Neurobiology and Evolution* (eds S. Coombs, P. Görner and H. Münz), pp. 99–114. Springer-Verlag, New York.

Garman, S. (1899). *Mem. Mus. Comp. Zool.* **26**, 1–431.

Gill, H.S. and Bradley, J.S. (1992). *Zool. J. Linn. Soc.* **106**, 97–114.

Gray, J.A.B. and Best, A.C.G. (1989). *J. Mar. Biol. Ass. UK* **69**, 289–306.

Hassan, E.-S. (1989). In *The Mechanosensory Lateral Line – Neurobiology and Evolution* (eds S. Coombs, P. Görner and H. Münz), pp. 217–228. Springer-Verlag, New York.

Hensel, K. (1978). *Acta. Univ. Carolinae – Biologica* 1975–1976, 105–149.

Janssen, J. (1997). *J. Fish Biol.* **51**, 921–930.

Kalmijn, A. (1989). In *The Mechanosensory Lateral Line – Neurobiology and Evolution* (eds S. Coombs, P. Görner and H. Münz), pp. 187–216. Springer-Verlag, New York.

Kalmijn, A. (1997). *Acta. Physiol. Scand.* **161** (Suppl. 638), 25–38.

Kelly, J.P. and van Netten, S.M. (1991). *J. Morphol.* **207**, 23–36.

Marshall, N.J. (1986). *J. Mar. Biol. Ass. UK* **66**, 323–333.

Marshall, N.J. (1996). *J. Fish Biol.* **49** (Suppl. A), 239–258.

Maruska, K.P. and Tricas, T.C. (1998). *J. Morphol.* **238**, 1–22.

McCormick, C.A. (1982). *J. Morphol.* **171**, 159–181.

McCormick, C.A. (1989). In *The Mechanosensory Lateral Line – Neurobiology and Evolution* (eds S. Coombs, P. Görner and H. Münz), pp. 341–364. Springer-Verlag, New York.

McCormick, C.A. (1997). *Hear. Res.* **110**, 39–60.

Metcalfe, W.K. (1989). In *The Mechanosensory Lateral Line – Neurobiology and Evolution* (eds S. Coombs, P. Görner and H. Münz), pp. 147–160. Springer-Verlag, New York.

Metcalfe, W.K., Kimmel, C.B. and Schabtach, E. (1985). *J. Comp. Neurol.* **233**, 377–389.

Mohr, C. and Bleckmann, C. (1998). *Comp. Biochem. Physiol.* **119A**, 807–815.

Montgomery, J.C. (1989). In *The Mechanosensory Lateral Line – Neurobiology and Evolution* (eds S. Coombs, P. Görner and H. Münz), pp. 561–574. Springer-Verlag, New York.

Montgomery, J.C. and Coombs, S. (1998). *J. Exp. Biol.* **201**, 91–102.

Montgomery, J.C., Coombs, S. and Janssen, J. (1994). *Brain, Behav. Evol.* **44**, 299–306.

Montgomery, J.C., Coombs, S. and Halstead, M. (1995). *Rev. Biol. Fish.* **5**, 399–416.

Montgomery, J.C., Baker, C.F. and Carton, A.G. (1997). *Nature* **389**, 960–963.

Mukai, Y. and Kobayashi, H. (1992). *Nippon Suisan Gakkaishi.* **58**, 1849–1853.

Münz, H. (1989). In *The Mechanosensory Lateral Line – Neurobiology and Evolution* (eds S. Coombs, P. Görner and H. Münz), pp. 285–298. Springer-Verlag, New York.

New, J.G. and Singh, S. (1994). *Brain, Behav. Evol.* **43**, 34–50.

New, J.G., Coombs, S., McCormick, C.A. and Oshel, P.E. (1996). *J. Comp. Neurol.* **366**, 534–546.

Northcutt, R.G. (1989). In *The Mechanosensory Lateral Line – Neurobiology and Evolution* (eds S. Coombs, P. Görner and H. Münz), pp. 17–78. Springer-Verlag, New York.

Northcutt, R.G. (1990). *Axolotl Newsletter* **19**, 5–14.

Northcutt, R.G. (1992). In *The Evolutionary Biology of Hearing* (eds D.B. Webster, R.R. Fay and A.N. Popper) pp. 21–48. Springer-Verlag, New York.

Northcutt, R.G., Brandle, K. and Fritzsch, B. (1995). *Dev. Biol.* **168**, 358–373.

Platt, C., Popper, A.N. and Fay, R.R. (1989). In *The Mechanosensory Lateral Line – Neurobiology and Evolution* (eds S. Coombs, P. Görner and H. Münz), pp. 633–654. Springer-Verlag, New York.

Puzdrowski, R.L. (1989). *Brain, Behav. Evol.* **34**, 110–131.

Puzdrowski, R.L. and Leonard, R.B. (1993). *J. Comp. Neurol.* **332**, 21–37.

Roberts, B.L. and Meredith, G.E. (1989). In *The Mechanosensory Lateral Line – Neurobiology and Evolution* (eds S. Coombs, P. Görner and H. Münz), pp. 445–460. Springer-Verlag, New York.

Satou, M., Takeuchi, H.-A., Nishii, J., Tanabe, M., Kitamura, S., Okumoto, N. and Iwata, M. (1994). *J. Comp. Physiol. A* **174**, 539–549.

Schellart, N.A.M., Prins, M. and Kroese, A.B.A. (1992). *Brain, Behav. Evol.* **39**, 371–380.

Song, J. and Northcutt, R.G. (1991a). *Brain, Behav. Evol.* **37**, 10–37.

Song, J. and Northcutt, R.G. (1991b). *Brain, Behav. Evol.* **37**, 38–63.

Suckling, E.E. (1967). In *Lateral Line Detectors* (ed. P.H. Cahn), pp. 45–52. Indiana University Press, Bloomington, Indiana.

Tarby, M.L. (1998). unpublished M.S. thesis, Villanova University, Villanova, PA, USA, 66pp.

Tester, A.L. and Kendall, J.I. (1967) In *Lateral Line Detectors* (ed. P.H. Cahn), pp. 53–69. Indiana University Press, Bloomington, Indiana.

Teyke, T. (1990). *Brain, Behav. Evol.* **35**, 23–30.

Tricas, T.C. and Highstein, S.M. (1991). *J. Comp. Physiol. A* **169**, 25–37.

van Netten, S.M. and Khana, S.M. (1994). *Proc. Natl. Acad. Sci. USA* **91**, 1549–1553.

van Netten, S.M. and van Maarseveen, J.T.P.W. (1994). *Proc. R. Soc. Lond B* **256**, 239–246.

Vogel, D. and Bleckmann, H. (1997). *J. Comp. Physiol. A* **180**, 671–681.

Wagner, T. and Schwartz, E. (1996). *Anat. Embryol.* **194**, 271–278.

Webb, J.F. (1988). *Amer. Zool.* **28**(4), 89A.

Webb, J.F. (1989a). In *The Mechanosensory Lateral Line – Neurobiology and Evolution* (eds S. Coombs, P. Görner and H. Münz), pp. 79–98. Springer-Verlag, New York.

Webb, J.F. (1989b). *Brain, Behav. Evol.* **33**, 34–53.

Webb, J.F. (1989c). *J. Morphol.* **202**, 53–68.

Webb, J.F. (1990a). *Copeia 1990*, 137–146.

Webb, J.F. (1990b). *J. Zool., London* **221**, 405–418.

Webb, J.F. (1995). *Amer. Zool.* **35**(5), 106A.

Webb, J.F. and Northcutt, R.G. (1997). *Brain, Behav. Evol.* **50**, 139–151.

Wongrat, P. and Miller, P.J. (1991). *J. Zool., London* **225**, 27–42.

Wonsettler, A.L. and Webb, J.F. (1997). *J. Morphol.* **233**, 195–214.

Zottoli, S.J. and Danielson, P.D. (1989). In *The Mechanosensory Lateral Line – Neurobiology and Evolution* (eds S. Coombs, P. Görner and H. Münz), pp. 461–480. Springer-Verlag, New York.

15.3 Chemoreception

Toshiaki J Hara
Freshwater Institute, Fisheries and Oceans Canada, Winnipeg, Manitoba, Canada

Peripheral olfactory organ

In vertebrates, the olfactory and gustatory systems comprise the major sensory pathways for detection and identification of chemical stimuli in the environment. Chemical information detected and transmitted directly to the central nervous system (CNS) by neurons of the cranial nerve I is termed olfaction (smell).

The olfactory organ of fish shows considerable diversity, reflecting the degree of development and ecological habitats (e.g. Zeiske *et al*., 1992). In monorhinic cyclostomes (hagfish and lamprey), a single median nostril is located at the rostral part of the dorsal surface of the head (Figure 15.3.1a). During metamorphosis in lampreys, the growth of the enormous upper lip of the mouth gradually displaces the olfactory organ from its original ventral position. The nostril leads to a large olfactory sac (cavity or pit) whose walls are extensively folded (olfactory lamellae). An accessory olfactory organ with tubules leading into the olfactory sac is formed, and a peculiar nasopharyngeal system develops. The nasopharyngeal canal extends from the bottom of the olfactory sac to a pouch. The pouch is a continuation of the nasal tube in a ventral and ventrocaudal direction and lies between the digestive tract and the nasal sac and brain. Its posterior part is enlarged laterally and ends as a blind pouch at the level of the second internal gill opening. Since the pouch is not protected by rigid cartilage, it is subject to the rhythmic contractions and relaxations of the respiratory muscles and body walls, thus sucking in water during the inspiration and expelling it during the expiration phase (Kleerekoper, 1969). The accessory olfactory organ is essentially glandular in character and is neither homologous nor analogous to the **vomeronasal** organ of land vertebrates. Because of its morphological relationships, the nasopharyngeal pouch is considered the forerunner of the nasopharyngeal passageway of higher vertebrates (Leach, 1951).

In elasmobranchs, the paired olfactory pits (nares) are usually situated on the ventral side of the snout (Figure 15.3.1b). Each pit is divided by skin flaps, one medially and one laterally attached into an anterior inlet and a posterior outlet. In some benthic species the posterior opening connects with the mouth cavity, a condition that is lost in teleosts. The flow of water through the olfactory organ is created by the pressure differential between the inlet and outlet nostrils, which is established by forward motion of the animal. In benthic elasmobranchs, the water flow is supported by the respiratory activity.

In teleosts, the paired olfactory organs are usually located on the dorsal side of the head (Figure 15.3.1c and 1d). The eels and morays (Anguilliformes) have large and elongate olfactory pits, extending from the tip of the snout to the orbit of the eye (Figure 15.3.1d), whereas in certain puffers (Tetraodontidae) the nasal sacs are absent. In most teleost fishes, the development of the olfactory organs lies between these two extremes. Many teleosts have an accessory nasal organ that is dorsal, posterior, or ventral to the main nasal cavity. Some species, such as **bathypelagic** *Cyclothone* spp. (Myctophiformes) and **ceratioid** anglerfishes (Lophiiformes), have evolved distinct sexual dimorphism in the olfactory organs (Marshall, 1971). The males have large olfactory organs, whereas they are regressed in the females. Each naris has two openings, anterior inlet and posterior outlet, which are separated by a skin flap. A current of water enters the anterior naris and leaves through the posterior naris either passively by locomotion of fish, or actively

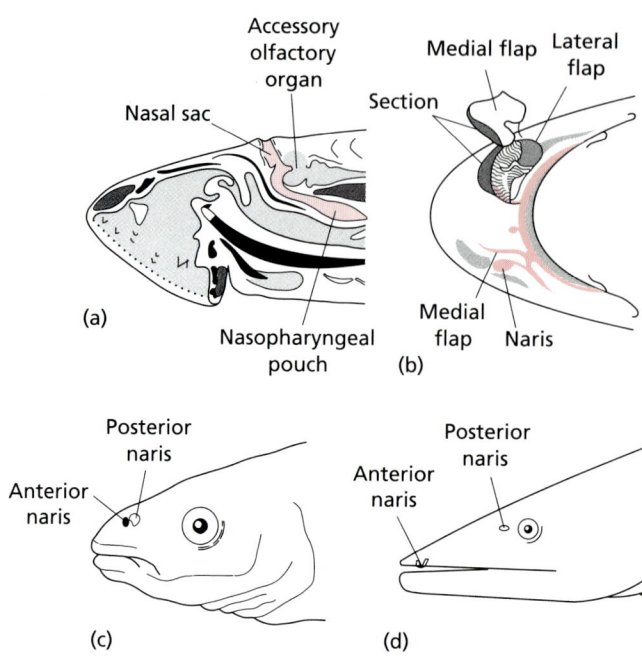

Figure 15.3.1 Diagrams showing positions of the nose in some fishes. (a) Median section of the head showing the olfactory organ and path of water in lamprey. (Modified from Harder, 1975.) (b) Ventral view of the head region of shark. The skin flap on the left side is cut away to show the olfactory organ. (Modified from Kleerekoper, 1969.) (c) and (d) positions of the nares in a salmonid and eel, respectively.

by ciliary action within the cavity or by the pumping action of the cavity generally with the aid of an accessory nasal organ. Unlike some terrestrial vertebrates, there is no contact between the olfactory and respiratory systems in any teleost species. This arrangement came later with the amphibians. Thus, evolutionarily, nasal association with breathing is a secondary function.

The floor of the nasal cavity is lined with the olfactory epithelium (mucosa), which is raised from the floor into a series of folds or lamellae to form a rosette (Figure 15.3.2). The arrangement, shape, and degree of development of the lamellae vary considerably among species. In hagfish the olfactory rosette is composed of about seven free lamellae arranged longitudinally and parallel to one another, whereas in lampreys it consists of a number of longitudinally arranged lamellae radiating from the wall of the olfactory chamber. The olfactory chamber of elesmobranchs is filled with a rosette composed of two rows of olfactory lamellae on each side of a transverse **raphe**. In the majority of the teleosts, the lamellae radiate from a central raphe arising rostrocaudally from the floor. The development of the lamellae begins caudad and continues rostrad, so that the oldest and largest lamella is located most posteriorly. The shape of the rosette is oval (in most fishes), round (e.g. *Esox*) or elongate (e.g. *Anguilla*). In a few species the lamellae are totally absent, however in the majority of species the number of olfactory lamellae varies from a few in sticklebacks to as many as 120 in eels and morays (Yamamoto, 1982; Zeiske *et al.*, 1992). The number of lamellae increases to some extent with growth of an individual, but remains relatively constant after the fish reaches a certain stage in development. The arrangement of the olfactory lamellae is sometimes used as a key to define the genera and subfamilies of fishes (cf. Yamamoto, 1982).

Secondary folding (lamella) of the lamella occurs in some species. Most notable are salmonids; there are 5–10 secondary folds per lamella in adults, but none in the parr and younger stages (Figure 15.3.2). Secondary folding is also found in sharks, Dipnoi, and garfish. The process of secondary folding results in an increase in the surface area of the olfactory epithelium. The total area of the olfactory epithelium has often been correlated with the olfactory sensitivity (e.g. Teichmann, 1954). However, such a simple correlation hardly exists, because, as described below, in no fish species is the sensory epithelium uniformly distributed on the surface of the olfactory lamellae.

In some teleosts (Cyprinidae, Siluridae), the olfactory organ is closely attached to the olfactory bulb (see below); the olfactory nerve is short and a long olfactory tract connects to the telencephalon (pedunculated). In most teleosts, however, the olfactory

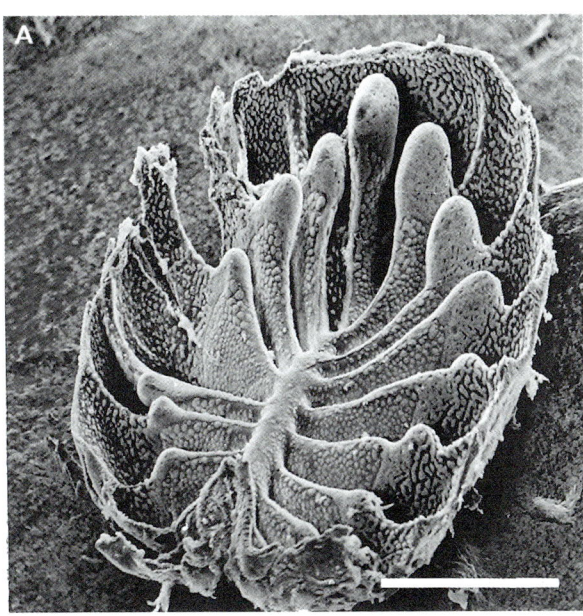 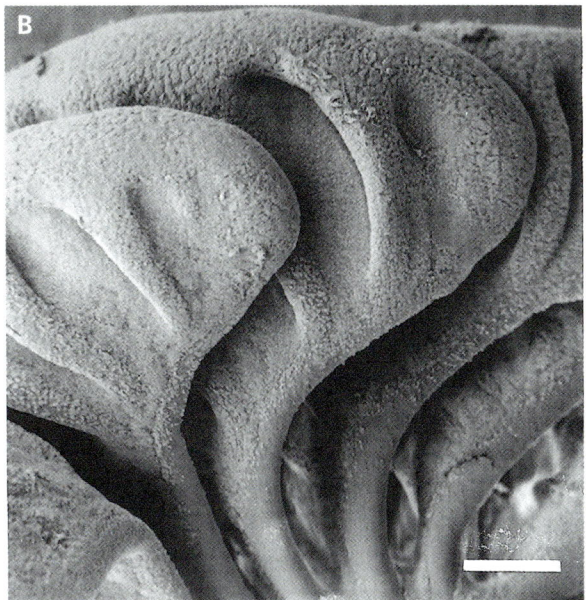

Figure 15.3.2 (A) Scanning electron micrograph of an olfactory rosette of the goldfish, *Carassius auratus*, showing the arrangement of olfactory lamellae and distribution of the sensory epithelium. Scale bar = 1 mm. (B) Scanning electron micrograph of olfactory lamellae of the rainbow trout, *Oncorhynchus mykiss*, showing the development of the secondary folding and distribution of the sensory epithelium. Scale bar = 300 µm.

bulb is close to the telencephalon so that a long olfactory nerve connects with the bulb (sessile).

Gustatory organ

Chemical information detected by specialized epithelial cells (gustatory cells) and transmitted to the CNS by cranial nerve VII (facial), IX (glossopharyngeal), or X (vagal) is termed gustation (taste).

The taste buds constitute the structural basis of the gustatory organ (Figure 15.3.3). In teleosts, taste buds are distributed on the gill rakers and arches, on appendages such as barbels and fins, as well as within the oral cavity. But, unlike humans, they are not usually found on the tongue. In siluroids and cyprinids, taste buds are distributed over the whole body surface. For example, in the yellow bullhead, *Ictalurus natalis*, more than 175 000 taste buds are estimated on the entire body surface alone (Atema, 1971). Some cyprinids have as many as 300 taste buds per square millimeter in the gular regions. The overall taste bud density rank seems to be correlated with the life style from **benthivory** to open water **planktivory** and surface feeding (Gomahr *et al.*, 1992). In salmonids that lack external taste buds, the highest density of 30 per square millimeter is estimated in some areas of the palate, totalling 3000–4000 taste buds on the whole palate (Hara *et al.*, 1993; Figure 15.3.3D). In the amago salmon, *Oncorhynchus rhodurus*, the number of taste buds in the oral cavity increases slowly for 60 days posthatching, then increases markedly over the next 300 days (Komada, 1993). The total number of taste buds in a 360-day-old amago salmon reaches more than 15 000, at which time smolt transformation takes places.

In elasmobranchs, taste buds occur in the mouth and pharynx, and taste-bud-like structures are also present in pits on the body surface. The terminal buds that have extreme similarity with the teleost taste buds are located on the gill arches of larval *Lampetra* (Baatrup, 1983).

Solitary chemosensory cells

In addition to the taste buds, there exists a system of differentiated epithelial sensory cells, which closely resemble gustatory receptor cells but are not organized into discrete end organs (Whitear, 1992). They are termed solitary chemosensory cells (SCCs), and are widespread among the primary aquatic vertebrates, especially fishes. In certain examples, such cells have been shown to be chemosensory, but the

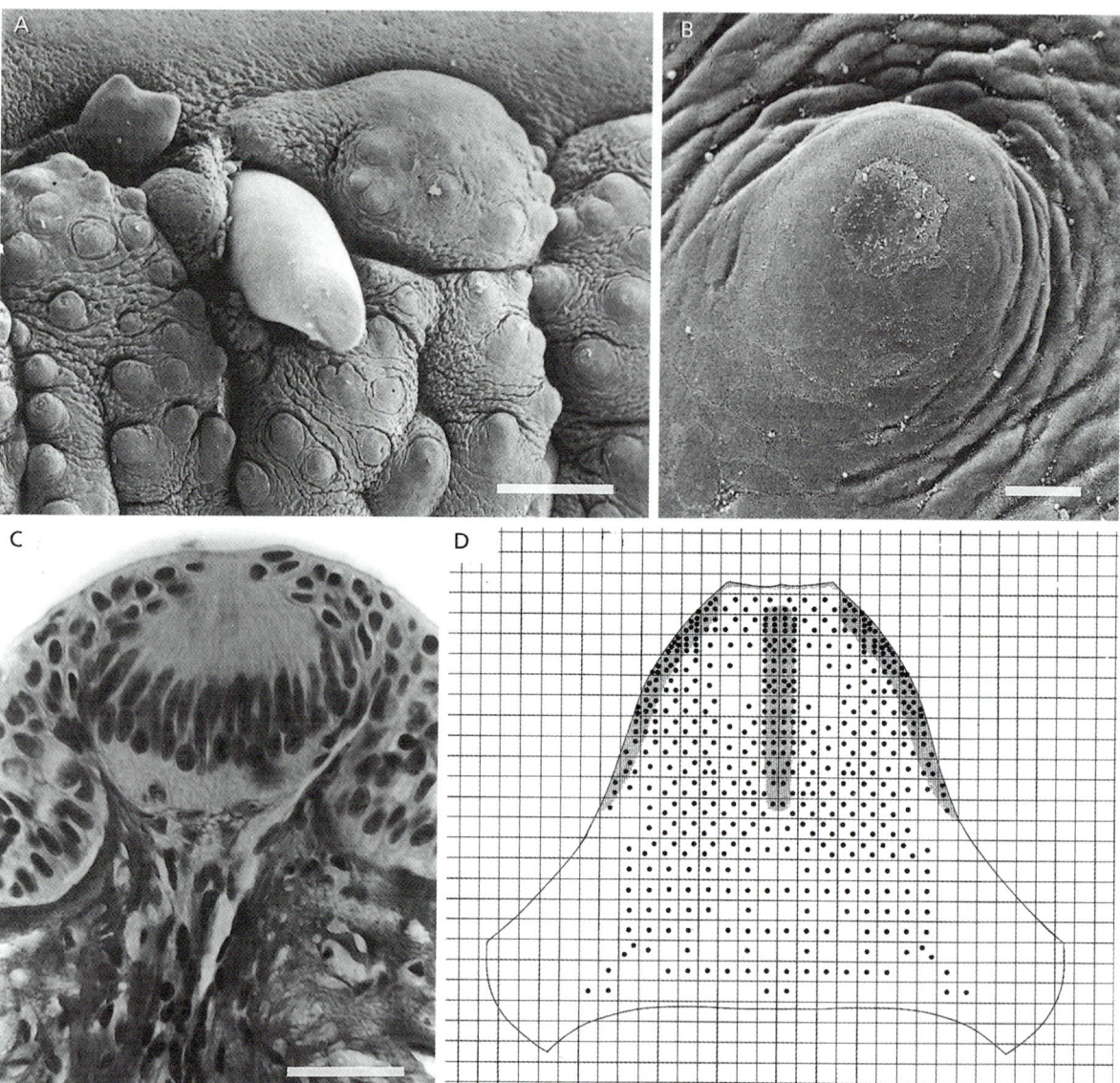

Figure 15.3.3 Anatomy and distribution of taste buds on the palate of the rainbow trout, *Oncorhynchus mykiss*. (A) Scanning electron micrograph of the inside upper lip showing distribution of taste buds. Scale bar = 250 μm. (B) Scanning electron micrograph of a single taste bud located on an elevated papilla. Scale bar = 20 μm. (C) Light micrograph of a longitudinal section through a taste bud on the palate. Scale bar = 30 μm. (D) Schematic illustration of the taste bud distribution on the dorsal oral cavity. Each dot represents five taste buds and each square corresponds to 1 mm². (From Marui *et al.*, 1983.)

physiology, the neural connections, and the distribution of the system are poorly understood. Evidence on central representation of the SCC system is available from only two groups of teleosts, the sea robins (*Prionotus carolinus*) and the rocklings (*Ciliata mustela*). In the former, SCCs in the pectoral fin rays are innervated by spinal nerves. Central pathways devoted to this system are similar to ascending spinal systems in other vertebrates (Finger, 1997). In the rocklings, the anterior dorsal fin vibratile rays contain millions of SCCs (up to 1×10^5 per mm²) but no taste buds; these SCCs are innervated from a component of the recurrent facial nerve (Kotrschal and Finger, 1996).

Development

The olfactory organ in teleosts originates in an **anlage** formed by the ectoderm and the whole organ remains ectodermal throughout its formation. The formation of the nostrils varies in different groups. In rainbow trout *Oncorhynchus mykiss* (Evans *et al.*, 1982), 10–11 days postfertilization, olfactory placodes emerge at a ventrolateral position on the embryo head. They are thickened and begin to form a central depression as the embryo rapidly elongates during the 12–16 days postfertilization. At hatching, approximately 30 days postfertilization, the olfactory groove is enclosed by flaps of skin, forming distinct anterior and posterior nares. The salmonid type of nostril formation is widespread among teleosts, and is considered to be phylogenetically the primary type (Zeiske et al., 1992).

Although an ectodermal origin of the external taste buds is generally accepted, those of the oropharyngeal cavity appear to be of both endodermal and ectodermal origin. Although the first appearance of taste buds varies in different fish species, taste buds generally develop more slowly than the olfactory organ. In rainbow trout, the earliest taste buds can be identified only in larvae 8 days posthatching (Twongo and MacCrimmon, 1977). The olfactory organ, as described above, is already formed 20 days after fertilization and at the day of hatching the olfactory organ is completed by enclosure with skin flaps. The earliest sign of feeding behavior is seen during swim-up (27–29 days posthatching), by which time over 80% of the yolk reserve has been exhausted, and the taste buds have matured (MacCrimmon and Twongo, 1980). In the carp, *Cyprinus carpio*, taste buds mature much earlier, 48 hours posthatching, suggesting earlier initiation of feeding in this species (MacCrimmon and Swee, 1967). In the amago salmon, the first sign of taste buds appears on the third day of hatching, and by 7 days old taste buds are developed on the lips, pharynx, gill arches, and palate (Komada, 1993). Differentiation of taste buds is slower in marine species than in freshwater species and many pelagic larvae do not have taste buds developed at the stage of initial feeding (Iwai, 1980).

References

Atema, J. (1971). *Brain, Behav. Evol.* **4**, 273–294.
Baatrup, E. (1983). *Acta Zool. Stockh.* **64**, 139–147.
Evans, R.E., Zielinski, B. and Hara, T.J. (1982). In *Chemoreception in Fishes* (ed. T.J. Hara), pp. 15–37. Elsevier, Amsterdam.
Finger, T.E. (1997). *Brain, Behav. Evol.* **50**, 234–243.
Gomahr, A., Palzenberger, M. and Kotrschal, K. (1992). *Environ. Biol. Fishes* **33**, 125–134.
Hara, T.J., Sveinsson, T., Evans, R.E. and Klaprat, D.A. (1993). *Can. J. Zool.* **71**, 414–423.
Harder, W. (1975). *Anatomy of Fishes*, Part II. E. Schweizerbart'sche Verlagsbuchhundlung, Stuttgart.
Iwai, T. (1980). In *Fish Behavior and its Use in the Capture and Culture of Fishes* (ed. J.E. Bardach), pp. 124–145. International Center for Living Aquatic Resources Management, Manila, Philippines.
Kleerekoper, H. (1969). *Olfaction in Fishes*. Indiana University Press, Bloomington, Indiana.
Komada, N. (1993). *Jpn. J. Ichthyol.* **40**, 110–116.
Kotrschal, K. and Finger, T.E. (1996). *J. Comp. Neurol.* **370**, 415–426.
Leach, W.J. (1951). *J. Morphol.* **89**, 217–255.
MacCrimmon, H.R. and Swee, U.B. (1967). *J. Fish. Res. Board Can.* **24**, 47–51.
MacCrimmon, H.R. and Twongo, T.K. (1980). *Can. J. Zool.* **58**, 20–26.
Marshall, N.B. (1971). *Exploration in the Life of Fishes*. Harvard University Press, Cambridge, Massachusetts.
Marui, T., Evans, R.E., Zielinski, B. and Hara, T.J. (1983). *J. Comp. Physiol. A* **153**, 423–433.
Teichmann, H. (1954). *Z. Morphol. Oekol. Tiere* **43**, 171–212.
Twongo, T.K. and MacCrimmon, H.R. (1977). *Can. J. Zool.* **55**, 116–128.
Whitear, M. (1992). In *Fish Chemoreception* (ed. T.J. Hara), pp. 103–125. Chapman & Hall, London.
Yamamoto, M. (1982). In *Chemoreception in Fishes* (ed. T.J. Hara), pp. 39–59. Elsevier, Amsterdam.
Zeiske, E., Theisen, B. and Breucker, H. (1992). In *Fish Chemoreception* (ed. T.J. Hara), pp. 13–39. Chapman & Hall, London.

15.4 Hearing

Bernd Fritzsch
Department of Biomedical Sciences, Creighton University, Omaha, Nebraska, USA

Overview

The inner ear of fishes contains three sensory systems. Each of these sensory systems is specialized, based on specific accessory structures, to extract different physical properties of various mechanical stimuli:

1. A system of one (hagfish), two (lampreys) or three (jawed vertebrates) semicircular canals which extract angular acceleration stimuli in the three cardinal planes.
2. A system of one (macula communis in jawless fishes), two (utricle and saccule – some fish) or three **maculae** (utricle, saccule, and lagena – most fish) which extract linear acceleration (gravity) from the environment (Lewis *et al.*, 1985).
3. The auditory system is a highly variable set of sense organs that extracts sound from the environment. Any **otolithic** organ in the aquatic environment can presumably accomplish this by extracting near-field particle motion in a low frequency range (Schellart and Popper, 1992). Alternatively, various sound-pressure transducers are physically coupled to different sensory epithelia of the fish ear to extract sound-pressure energy.
4. In addition, a sensory organ of largely unexplored function, the papilla neglecta (Lewis *et al.*, 1985) exists in many fishes. This sensory epithelium comes as a single patch or two patches. It has been implicated in hearing in elasmobranch fishes (Corwin, 1981) and amphibians (amphibian papilla; Fritzsch and Wake, 1988). However, most recent physiological data in turtles suggest that it may function in most vertebrates as an additional canal crista (Brichta and Goldberg, 1998).

Morphologically, the fish ear can be subdivided into four different types based on a set of uniquely derived features (Figure 15.4.1):

1. The jawless vertebrate ear, which consists of one or two semicircular canals and a more or less well subdivided single otolithic organ, the macula communis.
2. The chondrichthyan ear, which has three semicircular canals, a utricle situated in its own recess, a saccule, a papilla neglecta and, only in elasmobranchs, a lagena. Lungfish, which seem to belong to the sarcopterygian line based on many criteria (Cloutier and Ahlberg, 1996) have an ear very similar to ratfish, a basal chondrichthyan (Figure 15.4.1).
3. The actinopterygian ear. This ear consist of three semicircular canals, a utricle situated in the enlarged crus comunis of the anterior and horizontal semicircular canals, a saccule and, in most derived actinopterygian fish, a lagena. A papilla neglecta is also variably present (Lewis *et al.*, 1985).
4. The sarcopterygian ear. This ear is similar to the actinopterygian ear but has an additional sensory epithelium, the basilar papilla. This sensory epithelium is not present in all members of the sarcopterygian line such as the lungfish and many amphibians. Parsimony argues that the basilar papilla was lost several times in various lines. Alternatively, a basilar papilla evolved several times independently in this lineage alone, always in the same position and always with identical pattern of innervation. Parsimony suggests that this detailed similarity relates to homology of the basilar papilla in sarcopterygians (Fritzsch, 1992).

Overall, the ear reflects in its morphology the consensus of phylogenetic affinities among vertebrates reached based on non-otic morphological criteria and

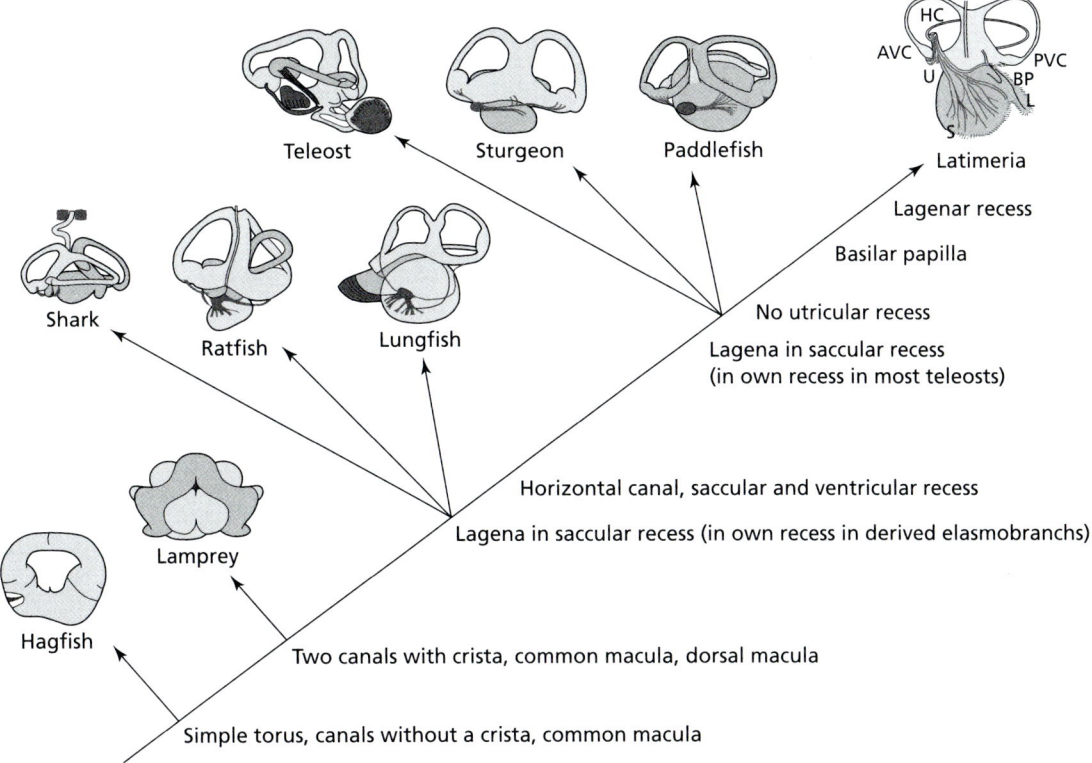

Figure 15.4.1 This image shows the morphological changes in the ear. Note that the cladogram is unrooted and there is no certainty that the hagfish ear displays a primitive feature with the single torus. Cartilaginous fishes and lungfish form a single morphological clade with respect to the organization of the utricle in a separate recess. They share with other jawed vertebrates the horizontal canal. Note that only derived elasmobranchs show a lagena and also have the canals almost completely segregated. Derived actinopterygian fish have a lagena in its own recess that can become fairly large. While superficially formed like a derived actinopterygian fish ear, *Latimeria* displays a separate lagena and a basilar papilla, like tetrapods. AVC, anterior vertical canal; HC, horizontal canal; L, lagena; BP, basilar papilla; S, saccule, U, utricle; PVC, posterior vertical canal. (Modified after Fritzsch, 1992.)

molecular data. Controversial issues, such as the relative position of members of the sarcopterygian line, are aggravated if otic criteria are considered. Besides the importance of otic morphological diversity for evolution of vertebrates in general, the ear presents an excellent model to study structure/function correlation in a tightly constrained system as well as molecular regulation of development (Fritzsch *et al.*, 1998).

The semicircular canals and their function

The semicircular canals consist of two functionally distinct parts, a canal for stimulus acquisition and an ampullary enlargement with the sensory epithelium – the crista – for mechano-electric transduction of the sensory stimulus. Hagfish have a single torus that lies at an angle in the otic capsule (Jørgensen *et al.*, 1998). It is noteworthy that the central lumen of the torus in one family of hagfish is filled with cartilage and in the other with periotic tissue. In lampreys, the single canal has a kink dorsally and curves essentially like an anterior and posterior semicircular canal, respectively, but without a common crus. All jawed vertebrates possess two vertical canals and, in addition, a third, horizontal canal. Fossils indicate that absence of a horizontal canal is primitive, its presence in jawed vertebrates is derived. Recent evidence in mouse mutants suggests that perhaps only one or two genes are relevant for the formation of the horizontal canal. Mutants of the *Otx-1* and *Hmx-3* genes lack a horizontal canal (see Fritzsch and Beisel, 1998, for review). However, it has not been shown that these genes are either absent or not expressed in the ears of lampreys and hagfish, a crucial requirement for the validity of this hypothesis.

Semicircular canals, whether a single torus (hagfish), two canals (lampreys), or three canals (jawed vertebrates) all function using the same principle. A fairly rapid angular acceleration of the head, caused either by self- or by externally generated head movements, lets the endolymph in the canals lag behind the moving canal walls. This lagging endolymph pushes against the moving **cupula** covering the sensory epithelium, the crista. Pressure generated by the lagging endolymph displaces the cupula and causes shearing action on the stereocilia of hair cells. This shearing action ultimately leads to an influx of K^+ through mechanically gated channels at the tips of the stereocilia (see Chapter 27.4, for detail).

The length and width of the semicircular canals vary among fishes. In general, highly mobile fishes seem to have longer and thinner canals. The curvature has a consistent proportion to the body weight of various vertebrates (Lewis *et al.*, 1985). In fish, canals are both longer and thicker than those of other vertebrates of equal body size.

Cartilaginous fish and lungfish have a more pronounced segregation of the semicircular canals from the gravistatic receptors owing to the presence of a separate utricular recess (Figure 15.4.2). In some elasmobranchs there is an almost complete detachment of the canals from the remaining ear, thus likely providing for a more undisturbed detection of specific rotational stimuli (Werner, 1960).

The statolithic organs and their function

The sensory epithelium of the statolithic organ is called the macula. A macula is covered by an otolith/otoconia with a density of about 2.6–3.2 times that of the endolymph (Lewis *et al.*, 1985). Otolith composition varies among bony and cartilaginous fishes (Fermin *et al.*, 1998). All otoconia/otoliths are composed of $CaCO_3$ (Lewis *et al.*, 1985). The high density of the otolith causes it to lag behind in slow, linear acceleration movements or slide with respect to the sensory hair cells when the head is tilted in gravity. Linear acceleration causes a shearing force that either opens or closes ionic channels in the two oppositely polarized hair cell populations of a macula. A fish near a sound source will perceive a comparable stimulus by rocking in the near-field particle motion, essentially working as an accelerometer (Schellart and Popper, 1992; Figure 15.4.3).

The primitive condition could have been a single sensory epithelium, a common macula, covered by a single otolith. This otolithic organ could have rested in the horizontal tube connecting the two semicircular canals (Figure 15.4.1). An ear resembling this hypothetical ear is found in some jawless fishes, such as hagfish. Other jawless fishes such as lampreys show a partial segregation and differential orientation of hair cells in their common macula, interpreted by some to indicate that different sensory epithelia are present in jawed vertebrates. The derived condition found in ratfish, lungfish and paddlefish is the clear segregation of two sensory maculae (the utricle and saccule), each covered by its own otolith. A further derived condition is the formation of a lagena, possibly three times independently among vertebrates (Figure 15.4.1). Whereas the saccule in all jawed vertebrates is in its own recess, the utricle is either in a recess of its own (cartilaginous fish, lungfish) or is in the horizontal common crus connecting the anterior and posterior vertical canal (also called the utricle in actinopterygian and sarcopterygian fishes). Based on comparison with jawless fishes as an outgroup, the position of the utricle in the common crus is possibly the primitive condition whereas the segregation of the utricle into its own recess seems to be a derived character of cartilaginous fishes and lungfishes.

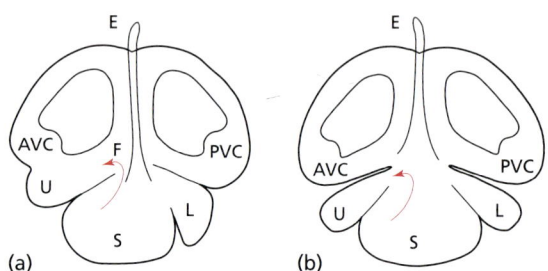

Figure 15.4.2 The distinct differences in the two main types of jawed vertebrate labyrinth is shown (a, b). Horizontal canal is omitted, for clarity. (a) Note that in tetrapods, actinopterygian fish and *Latimeria* there is a single connection through the foramen utriculo-sacculare between the pars inferior (saccule, lagena) and pars superior (canals and utricle) of the ear (arrow). (b) In contrast, in cartilaginous fishes and in lungfish, there is a utricular recess that separates the semicircular canals from the pars inferior (utricle, saccule, lagena). This separation leads in some elasmobranchs to an almost complete isolation of the semicircular canal rings (c). E, endolymphatic duct. For further abbreviations, see Figure 15.4.1. (Modified after Werner, 1960.)

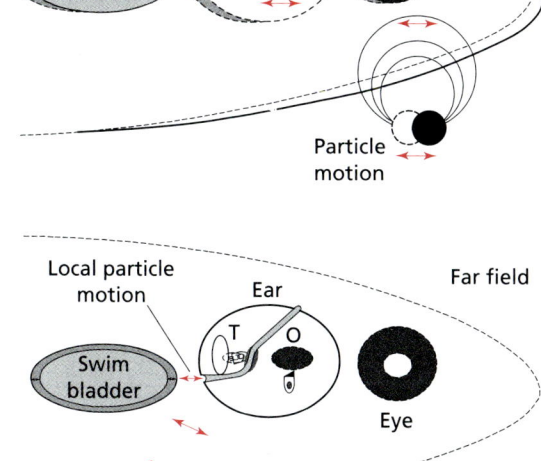

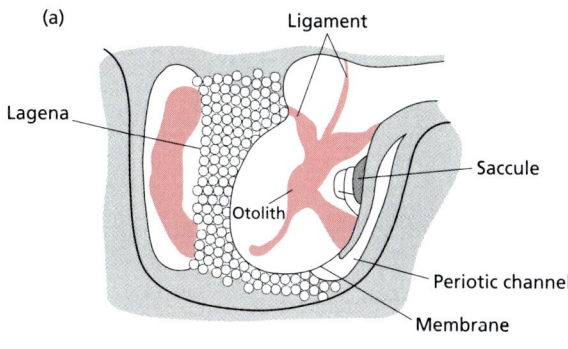

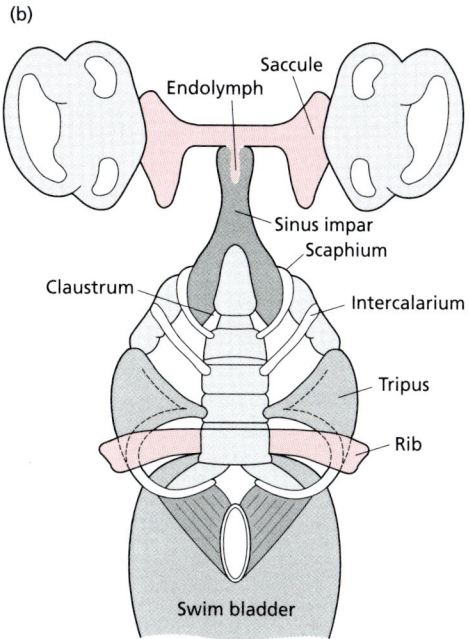

Figure 15.4.3 The distinct effects on otolith movement and sound-pressure-induced particle motion are shown. Note that in the near field, the otolithic organs move more than the otolith (o) due to the higher density of the otolith. There will be local particle motion generated by a gas bubble such as the swim bladder. However, this local particle motion will only exceed the near field particle motion in the far field. (Adapted from Fay and Popper, 1999.)

Figure 15.4.4 The organization of the Weberian ossicles is shown and their connection to the sinus impar which connects to the transverse canal of the endolymphatic duct. Near the saccule the sound-pressure-induced movements of the ossicles will generate fluid movement that will, in turn, generate movements of the saccular macula with respect to the saccular otolith. Notice the winglike protrusion of this otolith compared to the smooth shaped lagenar otolith.
(Modified after Henson, 1974, and Lewis et al., 1985.)

The utricular macula is almost always oriented horizontally (notable exceptions are flatfishes with their secondary change in body orientation, and marine catfish). The utricular macula may be subdivided into three distinct epithelia in herrings, with the middle being related to sound-pressure perception. The sensory epithelium is covered by a calcareous mass, the otolith, which can be smaller or larger than the sensory epithelium (Figure 15.4.4a). This is particularly obvious in the marine catfish *Arias* in which a large round otolith is surrounded by a narrow strip of the utricular macula (Lewis *et al.*, 1985).

The saccular recess is highly diverse in its shape and size. The sensory macula is typically oriented vertically in a more or less sagittal plane. However, it follows the curvature of the saccule and thus is also oriented in a coronal plane at its posterior end. A calcareous otoconial mass or an otolith typically covers the entire sensory epithelium but may occasionally be only half the size (Figure 15.1.4a). The otolith is held in place by specialized attachments to the medial saccular wall (Figure 15.1.4a). In particular in sound-pressure hearing bony fish the saccular otolith can acquire rather complicated shapes. When the saccule and lagena are in a common recess (basic actinopterygian fish, lungfish and some teleosts) or the lagena is only partially segregated (elasmobranchs, some amphibians) there is a common saccular/lagenar otolith/otoconia covering both discontinuous sensory epithelia.

The lagena is a posterior evagination of the saccule. The lagenar recess may be short and its formation is apparently somewhat independent of the presence of a lagenar macula (Fritzsch and Wake, 1988). The lagena is found in many actinopterygian fish and all sarcopterygian fish. It is absent in the ratfish (Fritzsch and Wake, 1988). A lagena with its own recess evolved presumably three times independently – in elasmobranchs, in actinopterygian, and in sarcopterygians (*Latimeria*; Fritzsch, 1992). The lagena was lost secondarily in many deep-sea actinopterygian fish, some amphibians, and all **eutherian** mammals. The macula is mainly oriented vertical but owing to its curvature covers many planes. The lagena is typically smaller than the saccule but can be much larger in some bony fish.

All otolithic organs are sensitive to linear acceleration, generated either by constant tilt in the gravity field or by active or passive movements. The passive movement can be generated in the near field of a sound source and low frequency, direct sound perception is theoretically possible with all otolithic organs but has actually been proven for few (Canfield and Rose, 1996).

The presence of three gravistatic receptors is a somewhat redundant system that permits one of the three receptors to be converted into an organ largely devoted for sound-pressure perception (Fritzsch, 1992; Popper and Fay, 1997). This issue is best exemplified in the tripartite organization of the utricular macula in the herring. It is apparently only the middle part of the utricle that is involved in sound-pressure perception whereas the two lateral parts are dedicated to gravity and linear acceleration perception.

Hearing in water: the role of direct and indirect sound

Any otolithic organ in the aquatic environment can presumably perceive direct or near-field particle motion in a low frequency range (Schellart and Popper, 1992). Alternatively, various sound-pressure transducers are physically coupled to different sensory epithelia of the fish ear to extract sound-pressure energy. Ears are thus converted into sound-pressure receivers for long-range hearing at a higher frequency range (Mann *et al.*, 1998). The sensory epithelia involved in sound-pressure reception are the saccule (many fishes, amphibians) or the utricle (some fishes). In some fish the lagena may respond to low frequency vibrations and at least one sarcopterygian taxon does have a specialized sensory epithelium which could play a role in sound-pressure perception (Fay and Popper, 1999).

A major issue of any auditory receptor is to extract selective frequencies from the stimuli. How this is achieved in fish is unclear. The multiple polarities of hair cell arrangement and their rich variety of various types can help (Popper and Fay, 1997; Fay and Popper, 1999) but details are speculative. Another controversial issue is directional hearing. Recent evidence (Canfield and Rose, 1996) suggests that this is largely accomplished by the direct sound while the indirect sound plays a role in alerting the fish to respond.

Extraction of direct sound in the near field is not necessarily accompanied by any obvious morphological change in the ear and is based only on the relatively higher density of the otoliths. In contrast, extracting sound pressure requires morphological specialization to be able to extract the limited energy available.

Sound-pressure receivers: swim bladders and their connection to the ear

Attenuation of energy transmitted by particle motion (the so-called near field) and sound pressure (the so-called far field) generated from the same source at the same frequency varies over distance (Figure 15.4.3). Thus, energy in the near-field particle motion will dominate closer to the sound source, but further away (in the far field) the sound pressure will dominate (Kalmijn, 1988). Extracting the minute energy present in the sound pressure some distance away requires a specialized mechanical system comparable to the middle ear sound-pressure receiver of terrestrial vertebrates. In essence, such a system will convert sound-pressure energy impacting on a medium of different density into volume changes that generate secondary near-field particle motion of much larger amplitude. This motion, generated by such a sound-pressure converter, will either directly or indirectly drive a nearby part of the ear. In addition, based on associated structures such as specialized perilymphatic conductance pathways, the ear provides little additional resistance to these near-field particle motions. Given that the saccule and lagena are more caudally located in the ear, it is logical that they are more frequently involved in sound-pressure transducers as they are closer to the swim bladder, a sound-pressure converter.

Weberian ossicles

The best known sound-pressure converter among fishes is the system of vertebrae-derived ossicles, commonly known as Weberian ossicles. They are named after the first description by Weber (1820; cited in Henson, 1974). Fishes with this system form a single taxonomically closely associated group called the ostariophysin fishes (Stiassny et al., 1997). Most but not all members of this taxon have this apparatus. It consists of specialized processes of the three first vertebrae, the tripus, intercalarium, scaphium and claustrum (Figure 15.4.4). The latter is in most cases fused to the scaphium. The most specialized ossicle chain exists in certain bottom-dwelling catfish where the ossicles are reduced to one, the tripus. In addition, the swim bladder is encased in bone and probably serves exclusively for sound-pressure reception (Henson, 1974; Bleckmann et al., 1991). Several artesian well-inhabiting catfish have lost their swim bladder. Nevertheless, the tripus is recognizable, albeit fused with the vertebral body. This indicates that proper formation of the ossicles during development requires sound impact or other pressure differences to keep the ossicles mobilized.

The Weberian ossicles are driven by sound pressure or other pressure-induced volume changes. A thin muscle around the swim bladder provides a higher pressure upon contraction, thus ensuring proper apposition of a dorsal slit in the swim bladder envelope with the tip of the tripus at any water pressure. This way the major displacement of the sound-pressure-induced changes is mediated right onto the Weberian ossicles. The claustrum/scaphium impinge on a perilymphatic duct, the sinus impar. The sinus impar ends at the endolymphatic transverse canal that leads to both sacculi. Release windows appear to exist lateral to the saccule (Bleckmann et al., 1991). Thus any movement of the sinus impar is transformed into a fluid movement toward and away from the saccule. In some ostariophysin fishes the otolith of the saccule has winglike projections that lie in the path of the fluid movement (Figure 15.4.4).

The central projections of the saccule have been mapped in a number of ostariophysin fishes and always show segregation from the vestibular projections (Bleckmann et al., 1991; McCormick and Hernandez, 1996). It appears that this specialized area of the hindbrain comes about through segregation from other vestibular nuclei. The higher order projections have been traced to the torus semicircularis of the midbrain (McCormick, 1999).

There is ample evidence that indicates the functional importance of the Weberian ossicles for sound-pressure perception. Tests deflating the swim bladder or disconnecting the ossicles have shown a significant loss in sensitivity and high frequency hearing. Checking the various zebrafish mutants (Whitfield et al., 1996) for absence of swim bladder and, assuming they exist, checking the audiogram in these fish could provide further evidence for the functional role of the Weberian ossicles in sound-pressure perception.

Direct connections: herrings and mormyrids

The swim bladder can also form direct connections with the ear. In the case of the herring this is achieved by two anterior extensions of the swim bladder to the utricle. These bubbles end in an osseous chamber near the utricle (Figure 15.4.5). This specialization may enable some herrings to perceive ultrasound frequencies (Mann et al., 1998). The space between a septum separating the gas bubble from the ear and the utricle is filled with a perilymphatic fluid that appears to be separated from the anterior lateral line canal by a thin membrane (Denton and Gray, 1989). Pressure-induced volume changes will change the curvature of the septum and move perilymphatic fluid toward or away from the lateral line canal, thus predominantly moving the freely suspended middle part of the utricular macula. In fact, it has been shown that liquid flow in this system is directly proportional to pressure changes. Critical tests to prove this hypothesis, like blocking the lateral line exit or changing the viscosity of the perilymph or deflating the gas bubble, need to be done in combination with auditory recordings. The central projections of the different sensory epithelia of the utricle have been mapped and show that the central macula projects distinct from the other maculae (McCormick, 1997).

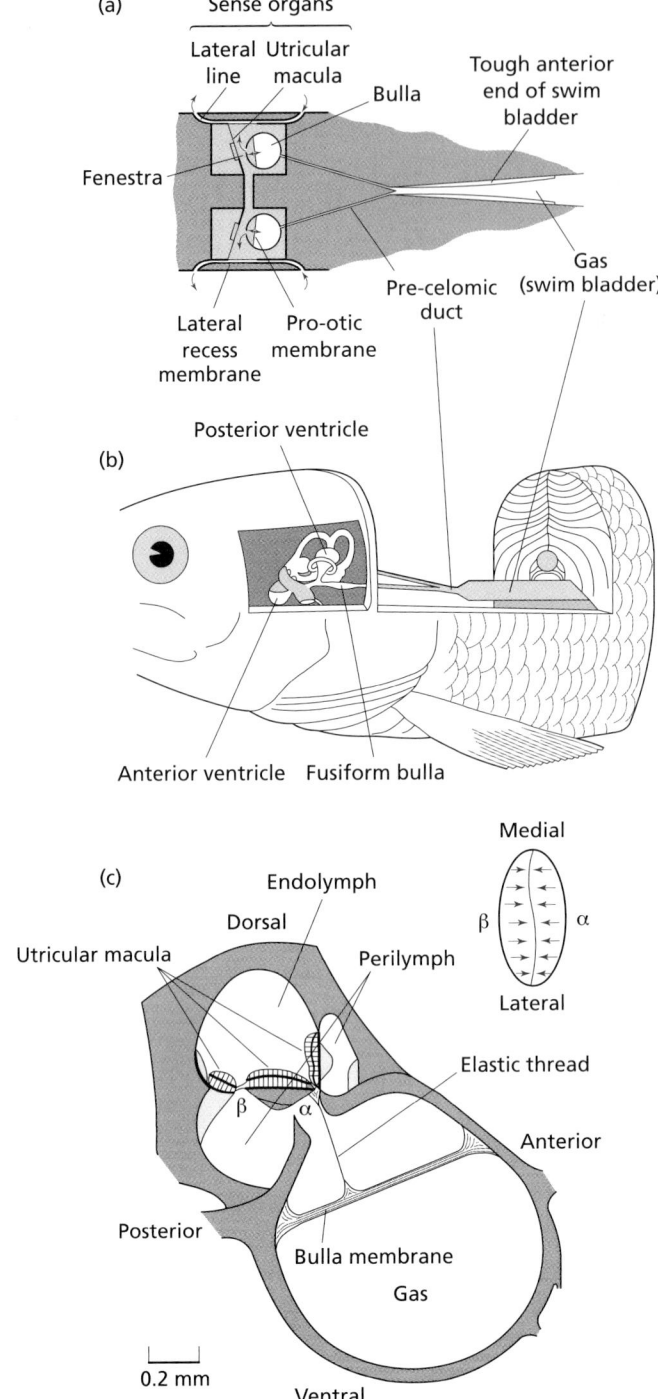

Figure 15.4.5 The morphology (b, c) and the flow of liquid (a) are shown for the herring ear. The swim bladder has two pre-celomic ducts toward the ear (a). One bulla is associated with each utricle. The perilymphatic fluid extends from the bulla to the utricle and the lateral line canal. Pressure changes drive fluid from the bulla across the center part of the tripartite utricle (c) to the lateral line (a).
(Modified after Denton and Gray, 1989; Henson, 1974; and Lewis et al., 1985.)

The ear of mormyrids also has a gas bubble associated with the ear that is ontogenetically derived from the swim bladder, but has lost all connections. This gas bubble is adjusted in its filling by blood capillaries (Henson, 1974). The bubble is next to the saccule and is protected by a loose-fitting bone. Measurements with a laser vibrometer indicate that this bone is free to vibrate when sound pressure impinges on the gas bubble (J.D. Crawford, personal communication). How the sound-induced volume changes drive the saccule and how the pressure changes generated by the volume changes in the gas bubble are released is unclear. Some suggestions for a lateral line connection analogous to herrings or some catfish exists but have not been established rigorously. Irrespective of this unresolved issue, the hearing capacity of mormyrid fishes is beyond question (Crawford *et al.*, 1997). The central projections show that the saccule has distinct territories of segregated projections in the hindbrain, probably devoted to hearing (Bell, 1981).

Anabantids and other solutions for sound-pressure reception

A number of other actinopterygian fish are hearing specialists with various known and unknown adaptations toward sound-pressure reception, such as **anabantid** fishes and **holocentrid** fishes. In the former, a gas bubble from the air breathing chamber is close to the ear and sound-pressure-induced volume changes probably drive the saccule (Henson, 1974, for review). In holocentrids a swim bladder extension reaches the ear (Coombs and Popper, 1979). While it is clear that anabantid fishes qualify as hearing specialists with respect to their frequency range (Ladich and Yan, 1998) the details of sound-pressure perception are not as well worked out as in the cases quoted previously. In addition, the central projection of the presumed 'auditory receptor' has not been mapped in these animals.

Other fishes show a significant capacity for long distance sound perception but do not display any clear sound-pressure reception specialization, for example eels (Schellart and Popper, 1992). How the ears of these animals work is at the moment unclear (Popper and Fay, 1997).

Hearing-related specializations in elasmobranchs

Elasmobranchs show signs of long-range sound perception in behavioral tests but how this is accomplished is unclear. It has been proposed that the papilla neglecta, which can be very large in some elasmobranchs (Corwin, 1981) is the sound receiver. However, there is considerable debate how sound pressure, whose involvement is necessary to explain the long distance sound effects, is actually perceived (Kalmijn, 1988).

Sound production in fish

Among fish, sound production has been found in many taxa and suggests multiple evolutionary origin (Myrberg, 1981; Ladich and Bass, 1998, for review). Two major groups can be identified according to their sound production mechanism: those which use the swim bladder to produce sound and those which use parts of the pectoral girdle to produce sounds. Sonic muscles which insert either directly on the bladder or which vibrate the swim bladder with a tendon or thin bony plates induce the swim bladder sound production. It is possible that fish which also use their swim bladder for sound-pressure reception (catfish) may uncouple the Weberian ossicles through muscle action during sound production (F. Ladich, personal communication). Grinding teeth, snapping tendons or even moving vertebrae can also produce sounds. Sound generated with the pectoral girdle or the swim bladder is typically of low frequency (around 100–500 Hz) whereas other sound can be within the 1–3 kHz range. Even if sound production is performed with a single organ like the swim bladder, there is strong evidence for multiple parallel evolution of this system in unrelated taxa (Ladich and Bass, 1998).

Muscles moving the swim bladder are derived from lateral body wall musculature, are striated and are innervated by sonic motoneurons in the hindbrain (Ladich and Bass, 1998). While the topology of the motoneurons varies somewhat between species, their overall distribution suggests that the neuronal basis for the sonic/vocal circuitry may be conserved among actinopterygian fish and possibly among vertebrates (Ladich and Bass, 1998).

Sounds are produced during various intraspecific interactions such as agonistic behavior, territorial behavior, and mating interactions (Myrberg, 1981; Ladich and Bass, 1998). However, in some species sound production and the appropriate apparatus are better developed in males. Sound can be barely audible or can be so loud that swarms of *Sciana* fish may be mistaken for a submarine or a chorus of *Prochilodus* can be heard several meters outside the water. Sound can be involved in complex social mating interactions which are usually species specific (Crawford et al., 1997). For example, in mormyrid fishes there is interplay between electric organ discharge and sound production. Sounds produced seem to be context-dependent grunts, moans, and growls between 100 and 1000 Hz (Crawford et al., 1997). Thus, at least in mormyrid fish there is a strong correlation with a specialized sound-pressure system of the ear and its use in a species-specific context. However, not all fish that vocalize have specialized sound-pressure receptors (toadfishes) and not all fish with specialized sound-pressure receptors do vocalize (goldfish).

Most intriguing is the development of a large 'rocker' bone that can be moved against the swim bladder in male conger eels (Rose, 1961) and may be the physical basis of sexual dimorphic vocalization (Mann et al., 1997). Further work on such highly dimorphic sound production systems, in particular an analysis of the sensory apparatus, for sexual dimorphism are needed.

Acknowledgments

This work was in part supported by a grant from the NIDCD (2 P01 DC00215-14A1). The author wishes to thank Dr de Caprona for the drawings in Figure 15.4.1 and Drs Bass, Crawford, McCormick, and Ladich for helpful suggestions on the text. Special thanks go to Dr E. Rosa-Molinar for pointing out the literature about the conger eels and their sexual dimorphic swim bladder.

References

Bell, C.C. (1981). *J. Comp. Neurol.* **195**, 391–414.
Bleckmann, H., Niemann, U. and Fritzsch, B. (1991). *J. Comp. Neurol.* **314**, 452–466.
Brichta, A.M. and Goldberg, J.M. (1998). *J. Neurosci.* **18**, 4314–4324.
Canfield, J.G. and Rose, G.J. (1996). *Brain Behav. Evol.* **48**, 137–156.
Cloutier, R. and Ahlberg, P.E. (1996). In *Interrelationship of Fishes* (eds M.L.J. Stiassny, L.R. Parenti and G.D. Johnson), pp. 445–480. Academic Press, San Diego.
Coombs, S. and Popper, A.N. (1979). *J. Comp. Physiol.* **132**, 203–207.
Corwin, J.Y. (1981). In *Hearing and Sound Communication in Fishes* (eds W.N. Tavolga, A.N. Popper and R.R. Fay), pp. 81–105. Springer-Verlag, New York.
Crawford, J.D., Cook, A.P. and Heberlein, A.S. (1997). *J. Acoust. Soc. Am.* **102**, 1200–1212.
Denton, E.J. and Gray, J.A.B. (1989). In *The Mechanosensory Lateral Line Neurobiology and Evolution* (eds S. Coombs, P. Görner and H. Münz), pp. 99–115. Springer-Verlag, New York.
Fay, R.R. and Popper, A.N. (1999). *The Ear of Fishes and Amphibians.* Springer-Verlag, New York.
Fermin, C.D., Lychakov, D., Campos, A., Hara, H., Sondag, E., Jones, T., Jones, S., Taylor, M., Meza-Ruiz, G. and Martin, D.S. (1998). *Histol. Histopathol.* **13**, 1103–1154.
Fritzsch, B. (1992). In *The Evolutionary Biology of Hearing* (eds D.B. Webster, A.N. Popper and R.R. Fay), pp. 351–375. Springer-Verlag, New York.
Fritzsch, B. and Beisel, K. (1998). *Am. J. Hum. Gen.* **63**, 1263–1270.
Fritzsch, B. and Wake, M.H. (1988). *Zoomorphology* **108**, 210–217.
Fritzsch, B., Barald, K. and Lomax, M. (1998). In *Springer Handbook of Auditory Research* (eds E.W. Rubel, A.N. Popper and R.R. Fay), pp. 80–145. Springer, New York.
Henson, O.W. (1974). In *Handbook of Sensory Physiology*, vol. 1. *Auditory System* (eds W.D. Keidel and W.D. Neff), pp. 40–110. Springer, Berlin.
Jørgensen, J.M., Schiri, M. and Geneser, F.A. (1998). *Acta Zool.* **79**, 251–256.
Kalmijn, A.J. (1988). In *Sensory Biology of Aquatic Animals* (eds J. Atema, R.R. Fay, A.N. Popper and W.N. Tavolga), pp. 84–130. Springer-Verlag, New York.
Ladich, F. and Bass, A.H. (1998). *Brain Behav., Evol.* **51**, 315–330.
Ladich, F. and Yan, H.Y. (1998). *J. Comp. Physiol.* **182**, 737–746.
Lewis, E.R., Leverenz, E.L. and Bialek, W.S. (1985). *The Vertebrate Inner Ear.* CRC Press, Boca Raton, Florida.
Mann, D.A., Bowers-Altman, J. and Rountree, R.A. (1997). *Concia* **3**, 610–613.
Mann, D.A., Lu, Z., Hastings, M.C. and Popper, A.N. (1998). *J. Acoust. Soc. Am.* **104**, 562–568.
McCormick, C.A. (1997). *Hear Res.* **110**, 39–60.
McCormick, C.A. (1999). In *The Ear of Fishes and Amphibians* (eds R.R. Fay and A.N. Popper), pp. 155–217. Springer-Verlag, New York.
McCormick, C.A. and Hernandez, D.V. (1996). *Brain Behav. Evol.* **47**, 113–137.
Myberg, A.A. (1981). In *Hearing and Sound Communication in Fishes* (eds W.N. Tavolga, A.N. Popper and R.R. Fay), pp. 395–426. Springer, New York.
Popper, A.N. and Fay, R.R. (1997). *Brain Behav. Evol.* **50**, 213–221.
Rose, J.A. (1961). *Bull. Mar. Sci. Gulf Caribbean.* **11**, 280–308.

Schellart, N.A.M. and Popper, A.N. (1992). In *The Evolutionary Biology of Hearing* (eds D.B. Webster, A.N. Popper and R.R. Fay), pp. 295–321. Springer-Verlag, New York.

Stiassny, M.L.J., Parenti, L.R. and Johnson, G.D. (eds) (1997). *Interrelationship of Fishes*. Academic Press, San Diego.

Werner, G. (1960). *Das Labyrinth der Wirbeltiere*. Fischer Verlag, Jena. 309 pp.

Whitfield, T.T., Granato, M., van Eeden, F.J., Schach, U., Brand, M., Furutani-Seiki, M., Haffter, P., Hammerschmidt, M., Heisenberg, C.P., Jiang, Y.J., Kane, D.A., Kelsh, R.N., Mullins, M.C., Odenthal, J. and Nusslein-Volhard, C. (1996). *Development*, **123**, 241–254.

CHAPTER 16

Reproductive Systems

J M Redding
Department of Biology, Tennessee Technological University, Cookeville, Tennessee, USA

R Patiño
Texas Cooperative Fish and Wildlife Research Unit, Texas Tech University, Lubbock, Texas, USA

Introduction

Fishes are the most numerous and diverse group of vertebrate animals, demonstrating an amazing array of reproductive strategies and processes. Reflecting this functional diversity is a concomitantly diverse set of anatomical specializations that have evolved in the context of the particular genetic heritage and environmental circumstances of a given species. In nearly all fishes, the purpose of these various specializations is to optimize the formation and conjugation of gametes for sexual reproduction; however, a few asexual **gynogenetic** species have been reported in which the genome of the progeny is entirely maternal.

Most species of fish exhibit a **dioecious (gonochoristic) anatomy**, wherein each individual possesses a single sexual phenotype which remains stable after its initial expression. In a male the testes produce sperm, whereas in a female the ovaries produce eggs. However, in some species members of one sex may differ considerably, both anatomically and functionally, in ways that appear to optimize an individual's reproductive success. For example there are two male 'reproductive morphs' of the plainfin midshipman *Porichthys notatus* which differ in body size, relative gonadal weight, color, vocal musculature, neuroanatomy, and reproductive behavior (Bass, 1996).

Some species, representing 14 families of teleostean fishes, are hermaphroditic, a condition in which an individual may express both male and female phenotypes concurrently (synchronous hermaphrodites) or, more commonly, reverse its sex during the course of development (sequential hermaphrodites), often in response to changing environmental circumstances. Examples of sequential hermaphrodism exist among some tropical reef fishes in which a vacancy at the top of a male social dominance hierarchy will induce a subordinate female to reverse sex and assume the top position (Shapiro, 1984).

Production of gametes within the ovary or testes may be synchronous and unitary, as occurs with

Copyright © 2000 Academic Press

one-time spawners like Pacific salmon (genus *Oncorhynchus*), or group synchronous, in which a subset of germ cells produce a cohort of gametes to be released during a limited, seasonal spawning period, typical of most non-tropical species. A third pattern, labeled asynchronous, applies to fish that produce gametes and spawn many times over a protracted period, perhaps during the entire year, as is the case for many tropical species.

The fusion of gametes, i.e. the fertilization of an egg with sperm, occurs externally in the typical oviparous (egg-laying) fish, but in viviparous (live-bearing) species internal fertilization is the rule. After the fusion of gametes, the degree of parental care provided to the offspring varies greatly. Many, perhaps most, species simply abandon their eggs after spawning, whereas others protect them during much of their embryonic and larval development. Indeed, in viviparous species, perhaps 3% of all teleosts, the progeny develop internally, deriving protection and, in some cases, nutrition from the mother (Wourms and Lombardi, 1992).

Description of the remarkable diversity evident in the reproductive biology of fishes would necessarily be encyclopedic, like the *magnum opus* of Breeder and Rosen (1966). Brief review of the scientific literature supports the notion that the vast majority of fishes used for research and aquaculture are classified in the phyletic subdivision Teleostei, a group which includes most of the extant fishes with bony skeletons, such as the Centrachidae, Percidae, Cyprinidae, Salmonidae, and Ictaluridae. Thus, for the purposes of this book we shall confine our discussion to members of that subdivision. Nagahama (1983) provides a good introduction to teleostean reproductive anatomy. More recent and comprehensive reviews of this subject can be found in Redding and Patino (1993) and Patino and Takashima (1995). Those readers who are interested in more evolutionarily primitive fishes may wish to consult Hoar (1969) and Dodd and Dodd (1985), as well as the many good textbooks of comparative vertebrate anatomy.

For practical reasons a few species, such as the rainbow trout (*Oncorhynchus mykiss*), Japanese medaka (*Oryzias latipes*), and the zebrafish (*Danio rerio*) have become popular model organisms. All three of these species are oviparous. The trout, a widely cultured sport and food fish, is easy to propagate and rear artificially, typically in hatchery raceways. Sexual maturation in adults is seasonal and highly synchronous within and between individuals, allowing for efficient gamete collection and spawning procedures. Gametes from individuals can be mixed or used for specific pairwise breeding strategies. Moreover, trout eggs and embryos are relatively large and durable, making them suitable for microinjection and transgenic manipulation. Medaka and zebrafish are small, easily maintained in simple aquaria, and widely used for toxicological and developmental studies. These asynchronous spawners produce small batches of eggs on a schedule that is easily entrained to photoperiod. Mating pairs perform well in groups or in isolation making these two species amenable to pairwise breeding schemes. After adults spawn at the onset of the daily light cycle, medaka embryos remain in a clump attached to the underside of the female from which they can be collected. Zebrafish likewise spawn daily at the onset of the light cycle, but their **demersal eggs** sink to the bottom of the aquarium from whence they are easily collected.

We recognize that virtually every organ system in the body contributes in some way to the processes of reproduction, e.g. vitellogenesis in the liver, olfactory pathways in the nervous system, or even the mouth of a mouth-brooding fish. However, we focus here on the most direct and obvious structural and regulatory elements of reproduction, namely the male and female gonads, associated gonoducts, and hypothalamic and pituitary tissue. Knowledge about the reproductive anatomy of fishes is important both for our basic understanding of their biology and for the practical management of both wild and captive populations.

External anatomy

Some teleostean species exhibit a distinctive sexual dimorphism which allows the observer to separate male and female fish. Such dimorphism can be a permanent feature of the fish's anatomy, such as the **intromittent organ**, a modified anal fin ray, of the male guppy (*Poecillia reticulata*) or the distinctive urogenital papilla of the male channel catfish (*Ictalurus punctatus*). Many female flatfish have an elongated body cavity compared to the male. More often, such dimorphism may become apparent only during sexual maturation or recrudescence. One well-known example of this phenomenon is the hook-like upper jaw and humped-shaped dorsum of male Pacific salmon, both of which are absent in females. Conversely, a distended belly is often a reliable indicator of a mature, egg-bearing female in many species. The urogenital papilla may be enlarged in ovipositors, female fishes specialized for the precise placement of eggs.

Sexually dimorphic coloration, **dichromatism**, is also apparent in some species, the male often assuming brighter, presumably attractive, coloration in advance of courtship. The male three-spined stickleback (*Gasterosteus aculeatus*) in spawning condition has a brilliant red belly unlike the brown or green hue of the female. The male zebrafish has a yellow underside, whereas the female is more silvery. Other sex-specific characteristics may exist, such as the raised and roughened nuptial tubercles and contact organs present in many male fishes (Wiley and Collett, 1970). Body size itself may be correlated to sex, typically the egg-bearing female is larger, as with striped bass (*Morone saxatilis*). However, in some species, such as the channel catfish, the males are larger, typically so if they play a major role in territorial defense and parental care of the developing offspring.

Internal anatomy

In many fishes, particularly those that spawn in masses, the only reliable way to determine the sex of an individual is to observe their internal gonadal structures. If the fish is sexually mature, it is relatively easy to discriminate between the testes and ovary; however, immature specimens may require microscopic examination, and may resist classification altogether if collected before their critical period of sexual differentiation, a developmental event whose timing varies greatly between species. Gonads consist of germ cells, which produce the gametes, and somatic cells, which support, nourish, and regulate the development of germ cells (see Chapter 28). Often the first reliable evidence of sexual dimorphism is the lamellar pattern of the ovarian tissue. Testicular somatic cells organize later into their distinctive lobular patterns. Gonoducts are present in most species to carry the gametes to their appropriate internal or external destinations. The basic organization of the gonads is similar in most vertebrates; however, structural variations among species occur, reflecting phyletic patterns or species-specific adaptations to the environment.

Male structures

The typical teleostean testes are paired structures, bilaterally adjacent to the air bladder and extending nearly the full length of the peritoneal cavity. The paired testes may fuse along their entire length or only along their posterior segments in some species, e.g. poeciliids. The testes are surrounded and suspended by a thin layer of connective tissue, the mesorchium, arising from the dorsal peritoneum and contiguous with that of the air bladder. A testis is usually slightly flattened and has a whitish appearance resulting from the spermatozoa collected inside it. It may be easily confused with the whitish colored fat tissue that is often found around the intestines and cecae. The testis often has an obvious distinct lobular morphology, but in some species it appears smooth and tubular. In some species, the testis may have a non-uniform appearance from anterior to posterior; in the channel catfish, for example, the anterior three-fourths is white, lobate, and actively spermatogenic, whereas the posterior segment is smooth, pink, and lacking spermatozoa (Sneed and Clemens, 1963).

Spermatogenesis within the testes occurs in germinal epithelia organized into lobular divisions composed of blind-ended tubes or sacs (Callard, 1991). In most teleosts sperm are produced throughout the lobule, but in killifish (genus *Fundulus*) and other Atheriniformes sperm production is restricted to the terminal regions of the sacs, farthest from efferent ducts (Grier, 1993). When full with sperm, the mature testes may account for about 10% of the fish's body weight, but in non-spawning condition it regresses to an insignificant and easily overlooked organ in many fishes. The gonado-somatic index (GSI) may be defined as the percentage of total body weight dedicated to the gonad.

$$GSI = 100 \times (gonad\ weight/total\ body\ weight)$$

Some biologists prefer a similar and perhaps more reliable index, the relative gonadal index (RGI) which subtracts gonad weight from the total weight before calculating the percentage.

$$RGI = 100 \times (gonad\ weight/(total\ body\ weight - gonad\ weight))$$

The GSI and RGI are often used to monitor reproductive status of wild and captive fish populations.

The efferent ducts of each spermatogenic lobule are continuous with a coalescing system of ducts which carry mature sperm toward the primary sperm duct associated with each testis. Unlike all other vertebrates, the sperm ducts of teleosts are anatomically independent from the kidney and its ducts. The sperm ducts may be internalized within the testis for much of its length or evident along the dorsal posterior region of each testis. In some species the two sperm ducts may fuse posteriorly. Some fishes exhibit gland-like, lobular outgrowths of the sperm ducts which probably have a secretory function, whereas in other cases, e.g. goldfish (*Carrasius auratus*), modifications of the duct may serve as accessory sperm storage organs, analogous to seminal vesicles in other vertebrates. The ducts terminate at the genital pore which exits independently to the outside between the more anterior anal pore and the more posterior urinary pore (Figure 16.1). Often the genital and urinary pores are associated with an externally prominent urogenital papilla which in some cases is modified for directional release of sperm, like a nozzle on a garden hose, or intromission into the female genital tract. In eels (genus *Anguilla*) and salmonids the sperm ducts are lost secondarily during development so that the sperm is released into the abdominal cavity, from there to be expelled through abdominal pores opening near the anal and urinary pores.

Female structures

The teleostean ovary is generally a bilaterally paired, hollow structure attached to the dorsal peritoneum from which it arose. In some teleosts, e.g. the yellow perch (*Perca flavescens*), the ovary is fused to form a single structure (Figure 16.2), and in others, e.g. the Japanese medaka, only one ovary develops. Oogenesis takes place within the germinal epithelia associated with the lamellar folds of the ovary. With a full complement of mature eggs, the ovaries may account for over 50% of the fish's body weight in some cases; however, in non-spawning condition the ovary regresses to a small fraction of its mature size. As with the male, the GSI and RGI are a practical way to measure reproductive status of female fish, often allowing reasonably accurate predictions of spawning date. However, it should be noted that egg maturity, i.e. readiness for ovulation and fertilization, may not be correlated precisely with gonadal mass. The number of mature eggs contained within the ovary, fecundity, is highly variable between species, ranging from few in the viviparous guppy, to millions in some oviparous pelagic fishes like the Atlantic cod (*Gadus morhua*).

The ovarian tissue is tubular or sac-like and usually clear, pink, yellow or orange in coloration, depending on the pigmentation of the eggs. The mature ovary appears granular; the enclosed eggs may be easily observed with the naked eye or at low power magnification. Typical egg diameter is between 0.4 and 3 mm, but some salmonid species have eggs larger than 5 mm. Marine teleosts tend to have smaller eggs than comparably sized freshwater fishes, resulting in greater fecundity, the millions produced by an Atlantic cod being an extreme example. The external covering of the egg, the chorion, may be smooth or it may be modified for adhesion or attachment to substrate or other eggs as seen in the wiry filaments that protrude from the eggs of Japanese medaka.

Ovulated eggs reside in the ovarian lumen (**ovocoel**) until spawning. In channel catfish as in many other species, ridges of tissue growing from the

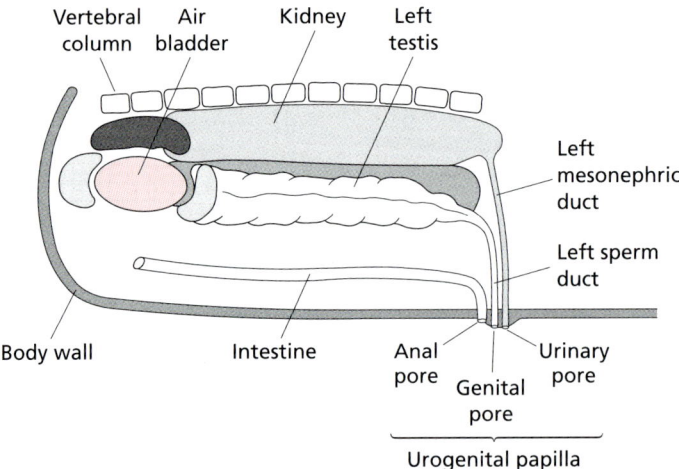

Figure 16.1 Schematic representation of the left testis and sperm duct in a typical teleostean fish in relation to the air bladder to which the testis is connected by a thin layer of connective tissue. The raised urogenital papilla is not present in all species, rather there may be a slit-like invagination in its place.

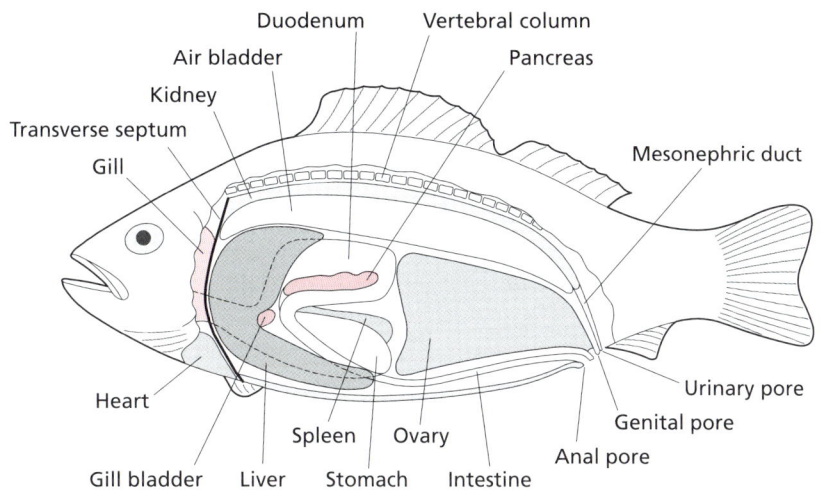

Figure 16.2 Abdominal organs of the female yellow perch, *Perca flavescens*, showing relative size and position of the single fused ovary containing mature eggs. The ovarian duct in this species is evident only as a narrowing at the posterior terminus of the ovary; it can be much longer and more distinct in other fishes. Note the independence of the ovarian duct and genital pore from the mesonephric duct and urinary pore draining the kidney. In fishes with two bilateral ovaries, the ovarian tissue may extend nearly the entire length of the abdominal cavity.

proximal and distal margins of the ovary meet and fuse to form a cavity. In other fishes the lumen forms by the invagination and subsequent fusion of the distal margin of the ovary. In salmonids the lumen appears first as a slit-like fissure in the body of the ovary. The epithelial lining of the lumen may have a secretory function that supplies the nutritional requirements of the stored ovulated eggs. In the case of viviparous fishes, the ovarian lumen may be modified for the storage of sperm and for the physical and nutritional support of developing embryos.

Ovarian ducts are present in most teleosts and are a direct posterior extension of the ovarian wall and lumen rather than a modified Mullarian funnel (oviduct) as seen in other fish taxa and vertebrate groups. Analogous to the sperm ducts, the ovarian ducts penetrate through abdominal pores and terminate independently at the genital pore situated between the anal and urinary pores. In the yellow perch, the single ovarian duct may be apparent only as a narrowing of the posterior ovary which terminates in the abdominal pore (Figure 16.2). In some species, such as the salmonids, ovarian ducts do not develop, so that ovulation involves the deterioration of intraovarian connective tissue, rupture of the ovarian wall, and release of the eggs into the abdominal cavity from which the eggs are expelled via muscular contractions of the body wall through abdominal pores in close proximity to the anal and urinary pores.

Endocrine structures and regulation

Most reproductive functions, from gametogenesis to behavior, are dependent upon the regulatory activity of the brain–pituitary–gonadal axis. Experimental disruption of any part of this axis usually results in profound disturbances of the reproductive process, testifying to its importance. This axis is composed of three primary structural elements: the brain, especially a region called the hypothalamus at the base of the brain near the third ventricle; the pituitary which lies just inferior to the hypothalamus and physically connects to it; and the gonadal tissue, either the testes or the ovary (Figure 16.3).

Linking these structural elements in a regulatory chain are the primary reproductive hormones: gonadotropin-releasing hormone (GnRH) from the hypothalamus, gonadotropins I and II (Gth I, Gth II) from the pituitary, and the sex steroids, androgens or estrogens, from the gonads (see Chapter 28). According to prevailing theory, various internal (e.g. biological clocks) and external (e.g. changing water temperature or photoperiod) stimuli activate neural networks which lead to the release of GnRH which,

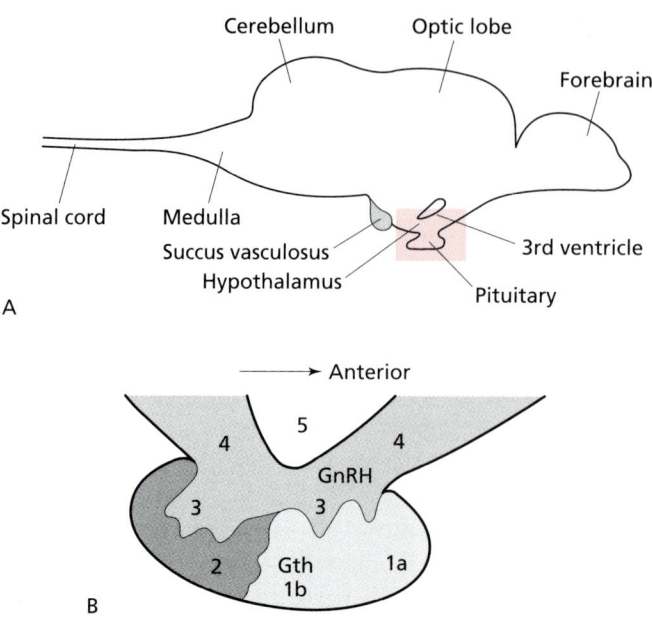

Figure 16.3 (A) Mid-sagittal section of the juvenile trout brain. (B) Enlargement of the boxed area in A, showing the structure of the pituitary gland in relation to the overlying hypothalamus: (1a) rostral pars distalis; (1b) central (proximal) pars distalis, the source of gonadotropic hormones (Gth); (2) pars intermedia; (3) pars nervosa, downward extension of neural tissue into the pituitary; (4) hypothalamic tissue, the source of gonadotropin-releasing hormone (GnRH), and dopamine, an inhibitor of gonadotropin in some species; (5) 3rd ventricle of the brain.

in turn, causes the release of Gth I and Gth II. These pituitary hormones then induce mechanisms in the gonads which promote gamete formation and sex steroid production. Sex steroids coordinate a host of interrelated changes in numerous target tissues, including the gonads, gonoducts, liver and other accessory structures, promoting the maturation and delivery of gametes and accompanying sex behavior.

Conclusion

The fishes, more so than any of the other vertebrate classes, present a fantastically diverse set of reproductive strategies and processes. The anatomical specializations supporting this diversity are likewise highly variable. The generalities presented above apply primarily to the teleostean fishes and should be regarded only as a modest beginning for further study. For those who might become familiar with one or more types of fish in a laboratory setting, beware that the same type of fish collected from wild populations may exhibit both genetic and environmentally induced differences in their anatomy. Recent revelations about the possible effects of endocrine disrupting chemicals in the environment on fish reproductive anatomy and function support this contention and will undoubtedly stimulate much research on this subject for many years into the future.

References

Bass, A.H. (1996). *Am. Scient.* **84**, 352–363.
Breeder, C.M. and Rosen, D.E. (1966). *Modes of Reproduction in Fishes*. Natural History Press, New York.
Callard, G.V. (1991). In *Vertebrate Endocrinology: Fundamentals and Biomedical Implications*, Vol. 4A (eds P.K.T Pang and M.P. Schreibman), pp. 303–341. Academic Press, San Diego.
Dodd, J.M. and Dodd, M.H.I. (1985). In *Evolutionary Biology of Primitive Fishes* (eds R.E. Foreman, A. Gorbman, J.M. Dodd and R. Olsson), pp. 295–319. Plenum Press, New York.
Grier, H.J. (1993). In *The Sertoli Cell* (eds L.D. Russell and M.D. Griswold), pp. 703–739. Cache River Press, Clearwater.
Hoar, W.S. (1969). In *Fish Physiology*, Vol III (eds W.S. Hoar and D.J. Randall), pp. 1–72. Academic Press, New York.
Nagahama, Y. (1983). In *Fish Physiology*, Vol IXa (eds W.S. Hoar, D.J. Randall and E.M. Donaldson), pp. 223–275. Academic Press, New York.

Patino, R. and Takashima, F. (1995). In *An Atlas of Fish Histology: Normal and Pathological Features* (eds F. Takashima and T. Hibiya), pp. 128–153. Kodansha Ltd., Tokyo.

Redding, J.M. and Patino, R. (1993) In *The Physiology of Fishes* (ed. D.H. Evans), pp. 503–534. CRC Press, Boca Raton.

Shapiro, D.Y. (1984). In *Fish Reproduction: Strategies and Tactics* (eds G.W. Potts and R.J. Wootton), pp. 103–118. Academic Press, London.

Sneed, K.E. and Clemens, H.P. (1963). *Copeia* 1963, 606–611.

Wiley, B. and Collett, B. (1970). *Bull. Am. Mus. Nat. Hist.* **143**, 145–216.

Wourms, J.P. and Lombardi, J. (1992). *Am. Zool.* **32**, 276–293.

Plate 1 *Yersinia ruckeri* infected fish exhibit dark coloration, red fins, eyes and mouth.

Plate 2 Skin layers of a jawless fish, a spawning adult Pacific lamprey, *Lampetra tridentata*. The components include a thick epidermis (*ep*) and two layers of dermis, an upper (outer) layer of dense collagenous tissue (*dct*) and a lower hypodermis (*hyp*) consisting largely of adipose tissue. Chromatophores or pigment cells (*ch*) are visible in the upper hypodermis and a small section of muscle (*mu*) is visible beneath the hypodermis. Lampreys are scaleless. Scale bar=100 μm. Hematoxylin and eosin stain.

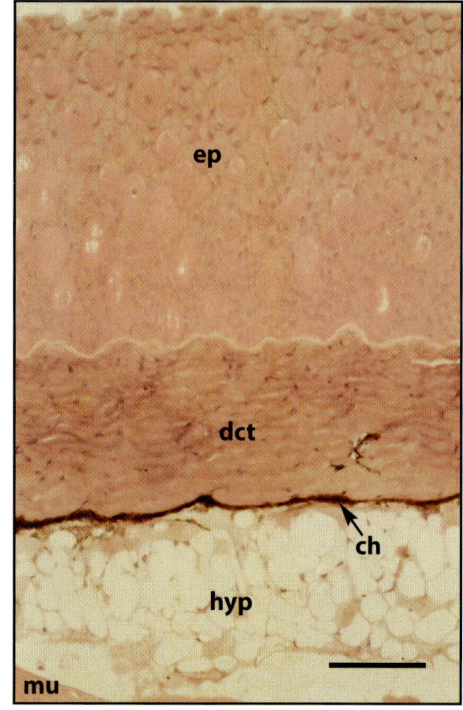

Plate 3 Fin tissue from a chondrichthyan, a smooth dogfish shark, *Mustelus canis*. Tooth-like placoid scales (*sca*) are visible beneath the epidermis (*ep*). The basal plate (*bp*) of each scale is anchored in the dermis, and portions of pulp cavities (*pu*) are visible within the scales. The remnants of epidermis (arrowed), possibly damaged during sampling, suggest that some of the placoid scales of this species may be completely covered by epidermis. (Compare with the partially exposed scales in Figure 5.6b.) The visible dermis of the fin includes an outer layer of loose connective tissue (*lct*) and a layer of dense connective tissue (*dct*). A chromatophore (*ch*) is visible in the dermis. Scale bar=100 μm. Hematoxylin and eosin stain.

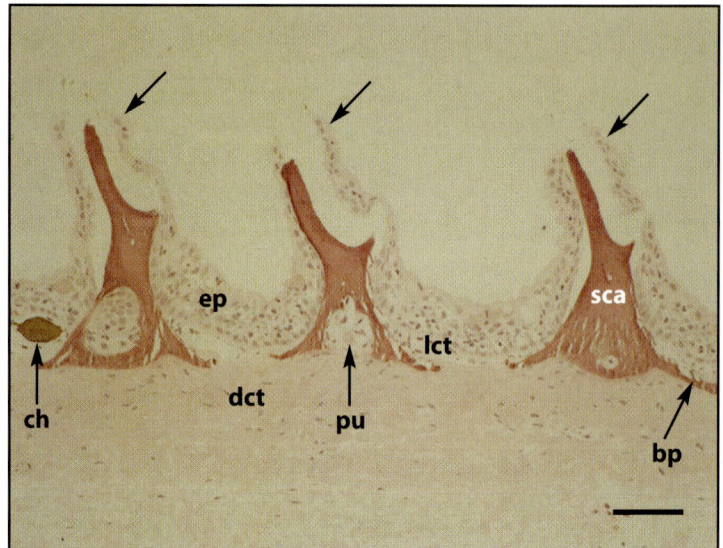

Plate 4 Skin layers of a juvenile steelhead trout, *Oncorhynchus mykiss*. The thin epidermis (*ep*) covers and wraps around the posterior (non-overlapped) portion of each scale (*sca*). The imbricated scales lie in pockets of dermal tissue. The layers of the dermis include an upper layer of loose connective tissue or stratum spongiosum (*lct*), a thick middle layer of dense connective tissue or stratum compactum (*dct*), and a thin lower hypodermis (*hyp*) containing chromatophores (*ch*) and adipose tissue. Subcutaneous muscle (*mu*), including a myoseptum (*my*), is also visible. Scale bar=100 μm. Hematoxylin and eosin stain.

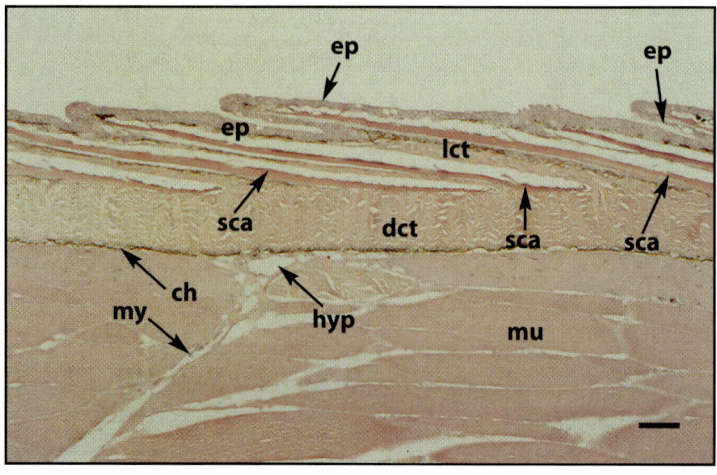

Plate 5 Skin layers of a short-fin eel, *Anguilla australis*. The layers include a thick epidermis (*ep*) and a dermis consisting of an upper layer of loose connective tissue (*lct*), a middle layer of dense connective tissue (*dct*), and a hypodermis (*hyp*) comprised largely of adipose tissue. Chromatophores or pigment cells (*ch*) are visible in the stratum spongiosum (*lct*) and hypodermis. The deeply embedded scales (*sca*) do not overlap. Muscle tissue (*mu*) is visible beneath the skin. Scale bar=100 μm. Hematoxylin and eosin stain. Photomicrograph courtesy of Dr Barbara Nowak.

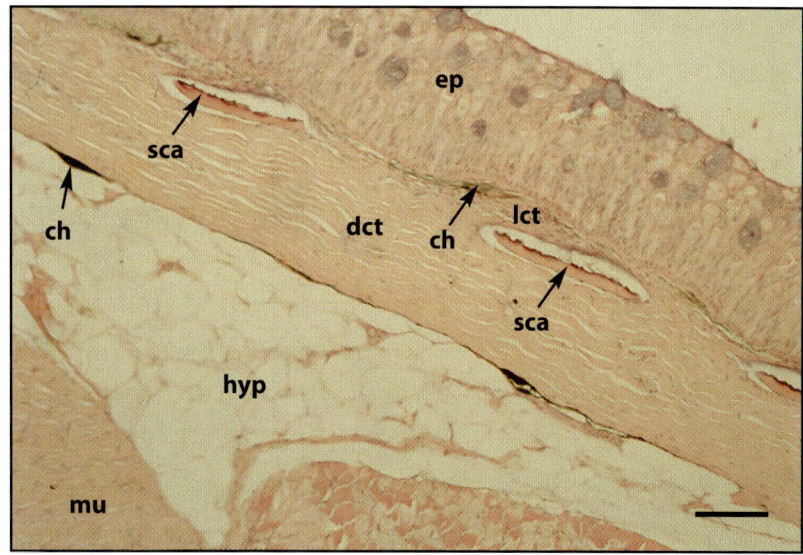

Plate 6 Spinous scale of a toothbrush leatherjacket, *Acanthaluteres vittiger*. This type of scale gives members of the family Monacanthidae their common name of 'filefish.' (a) Spinous scale (*spi*) embedded in the dermis. Adjacent non-spinous scales (*sca*) are also visible. The outer epidermis (*ep*) and two layers of dermis – a layer of loose connective tissue or stratum spongiosum (*lct*) and a layer dense connective tissue or stratum compactum (*dct*) – can be seen. (b) Portion of spinous scale (*spi*) that protrudes from the surface of the skin. Even the protruding portion (spinule) is covered by epidermis (*ep*). Scale bars=100 μm. Hematoxylin and eosin stain. Photomicrographs courtesy of Dr Barbara Nowak.

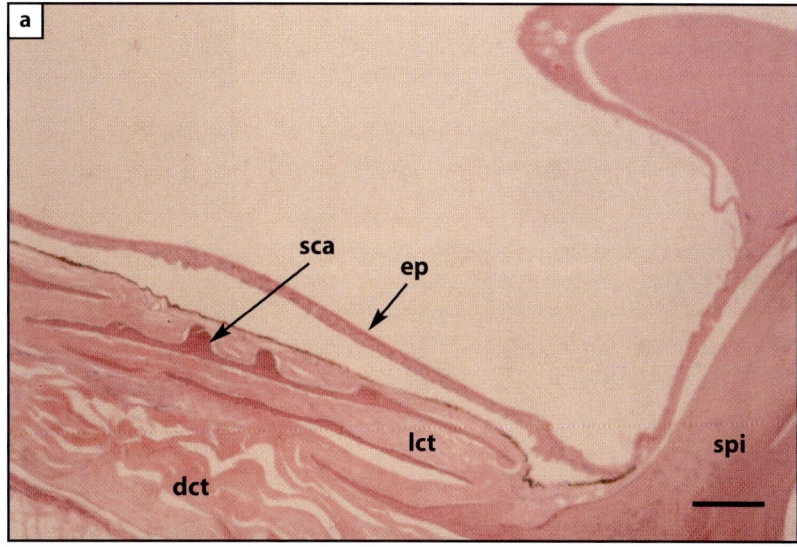

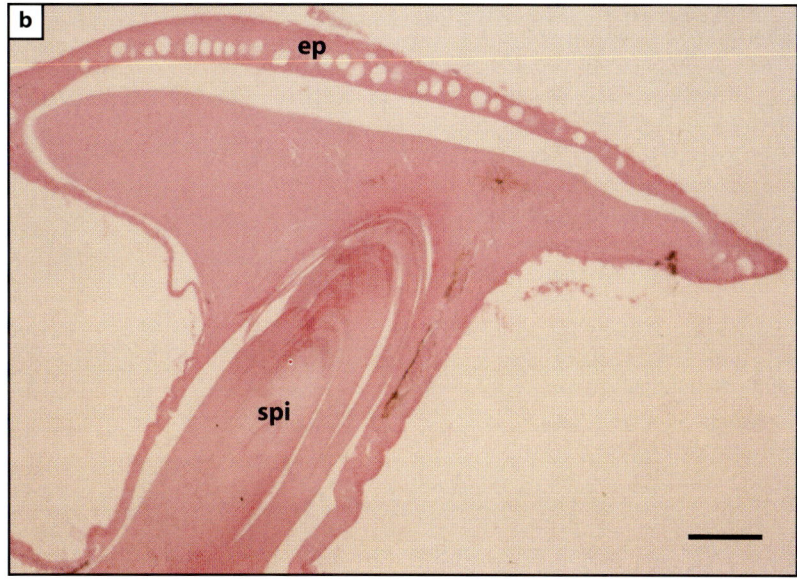

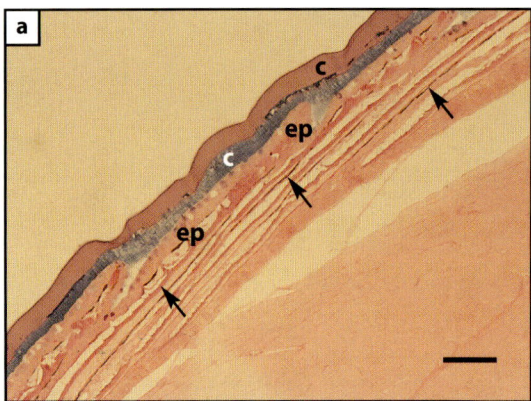

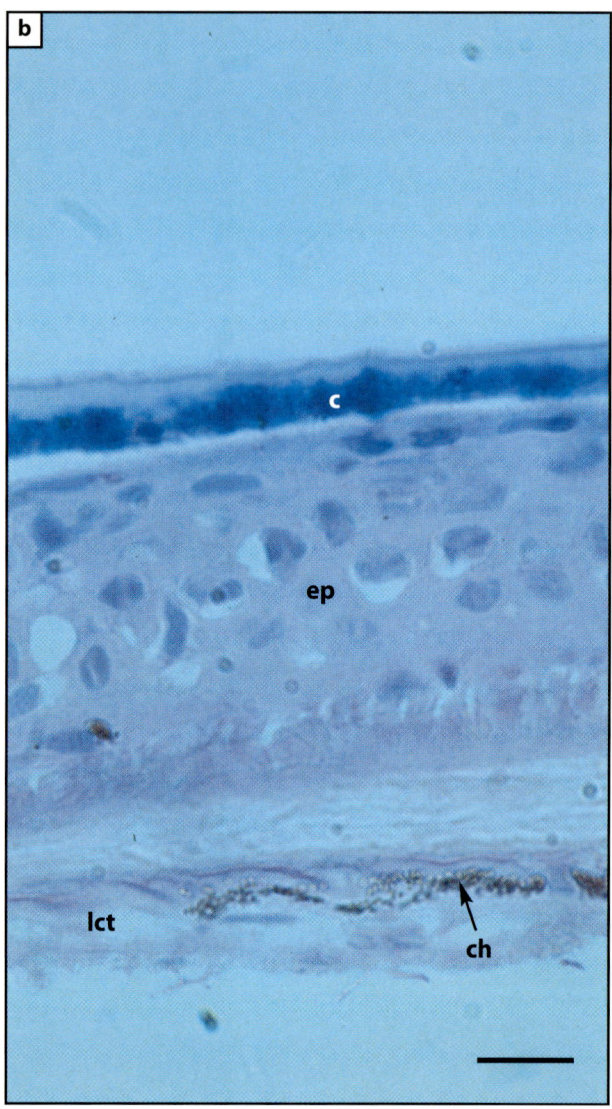

Plate 7 Histological section of skin tissue from a blue gourami, *Trichogaster trichopterus*, fixed to preserve the mucous cuticle and stained with periodic acid-Schiff (PAS) for mucin. This species and other members of the family Anabantidae are adapted for life in freshwater habitats that can be oxygen-poor and dirty. In addition to an accessory respiratory organ that enables this species to obtain atmospheric oxygen, it possesses a thick cuticle that may help to protect the skin from adverse environmental conditions. (a) Thick acellular cuticle (c) overlying the cellular epidermis (ep) and the dermis containing scales (arrows). The cuticle exhibits a bi-layered staining pattern (scale bar=100 μm). (b) Higher-magnification micrograph of an area of skin with a thinner cuticle (c) overlying the epidermis (ep). A chromatophore (ch) is visible in the upper loose connective tissue layer or stratum spongiosum (lct) of the dermis (scale bar= 10 μm). Photomicrographs courtesy of Elena Catap.

Plate 8 Histological section of the epidermis of a jawless fish, a spawning adult Pacific lamprey, *Lampetra tridentata*. The most common epidermal cell types are the mucigenic epithelial cells (ec). Some basal layer epithelial cells (bl) are modified as skein cells (sk) which show characteristic twisted bundles of tonofilaments and are sometimes binucleate (arrow). Granular cells (gr) are present in the middle to upper levels of the epidermis, but the granules are not highly visible in this preparation (see Plate 11). Some of the dense collagenous tissue of the dermis (dct) is visible beneath the epidermis. Scale bar=100 μm. Hematoxylin and eosin stain.

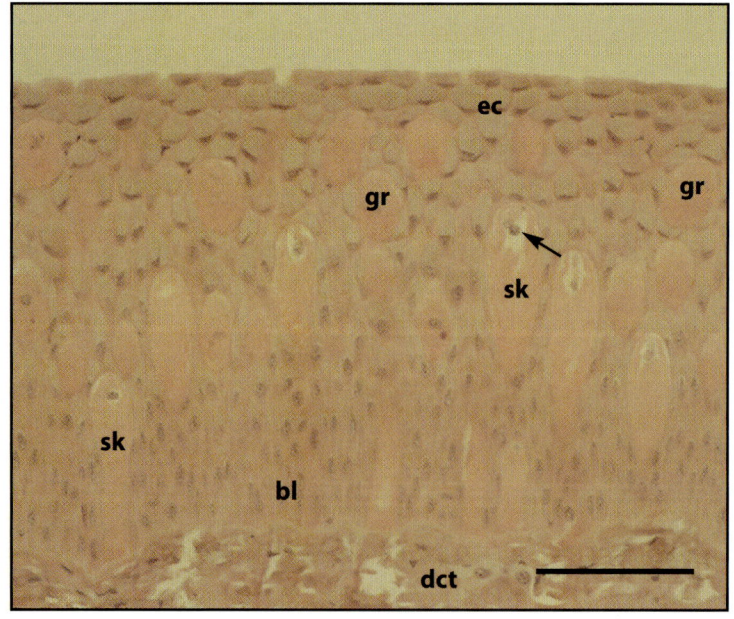

Plate 9 Histological section of the epidermis of a short-fin eel, *Anguilla australis*, showing typical eel-type club cells (cc) with secretory vacuoles (arrows). Other epidermal cell types include epithelial cells (ec) and basal layer epithelial cells *(bl)* as well as mucous goblet cells (gc). Visible dermal elements include loose connective tissue or stratum spongiosum *(lct)*, a scale *(sca)*, and dense connective tissue or stratum compactum *(dct)*. Scale bar=10 µm. Hematoxylin and eosin stain. Photomicrograph courtesy of Dr Barbara Nowak.

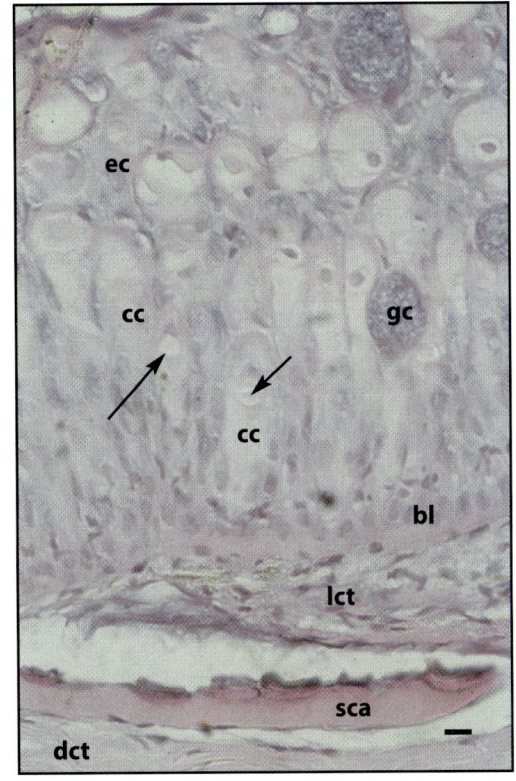

Plate 10 Histological section of the epidermis of a goldfish, *Carassius auratus*, showing typical ostariophysan-type club cells (cc) with a central nucleus (n) but no secretory vacuole. Other epidermal features include flattened surface epithelial cells (sec), mucous goblet cells (gc), and lymphocytes (l). Scale bar=10 µm. Methylene blue-azure II stain.

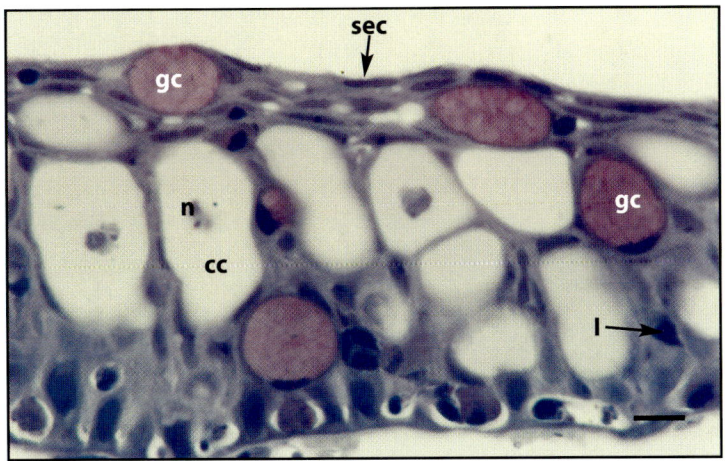

Plate 11 Histological section of the epidermis of a jawless fish, a spawning adult Pacific lamprey, *Lampetra tridentata*, showing characteristic granular cells (gr) with abundant cytoplasmic granules. Nuclei (n) with prominent nucleoli are apparent in some of the granular cells. Skein cells (sk) are also present; two nuclei (arrows) are visible in one skein cell, but the twisted bundles of cytoplasmic tonofilaments are not apparent in this preparation (see Plate 8). Epithelial cells (ec), basal layer epithelial cells (bl) and the basement membrane (bm) are also present. Scale bar=10 µm. Methylene blue-azure II stain.

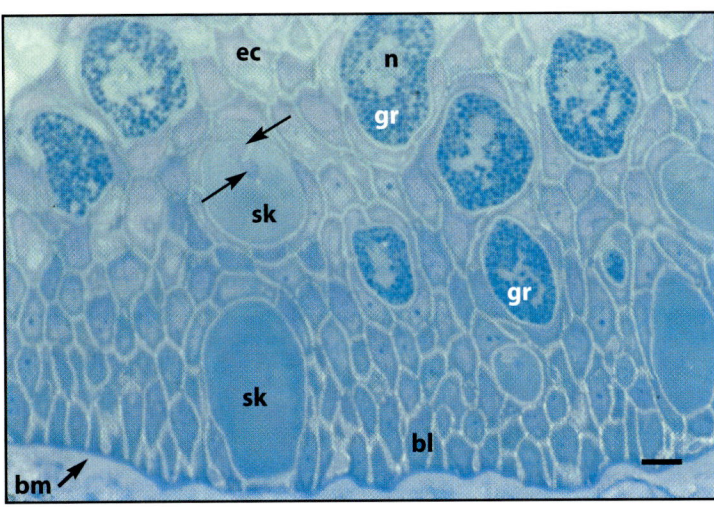

Plate 12 Histological section of fin tissue from a smooth dogfish shark, *Mustelus canis*, showing a melanophore, the most common type of dendritic chromatophore observed in histological sections. A central nucleus (*n*) is visible, as are dendritic processes (*dp*) containing numerous melanosomes, the round to ellipsoid organelles containing black to brownish melanin pigment. A portion of a placoid scale (*sca*) can be seen in the section. Scale bar=10 μm. Hematoxylin and eosin stain.

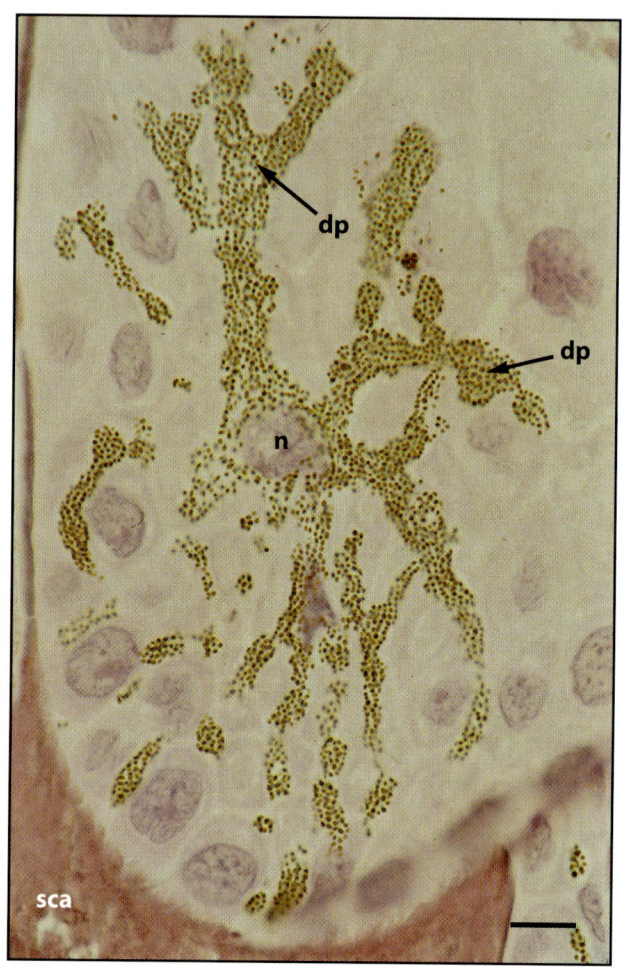

Plate 13 Histological section of the skin of a juvenile chinook salmon, *Oncorhynchus tshawytscha*, showing elements involved in scale growth. The ends of two scales (*sca*) are visible, with associated episquamal scleroblasts (*es*), marginal scleroblasts (*ms*), and hyposquamal scleroblasts (*hs*). The basal plate (*bp*), external layer (*ex*), and circuli (*ci*) of the scales are also apparent. Other elements in the stratum spongiosum or loose connective tissue layer (*lct*) of the dermis include scale pocket lining cells (*spl*), melanophores (*me*), and the basement membrane (*bm*). Cells visible in the epidermis include flattened surface epithelial cells (*sec*), mid-level epithelial cells (*ec*), and cuboidal basal layer epithelial cells (*bl*), as well as a mucous goblet cell (*gc*) and a serous cell (*ser*) that could be a serous goblet cell or sacciform cell. Electron microscopy would be necessary for definitive identification of this cell. Scale bar=10 μm. Hematoxylin and eosin stain.

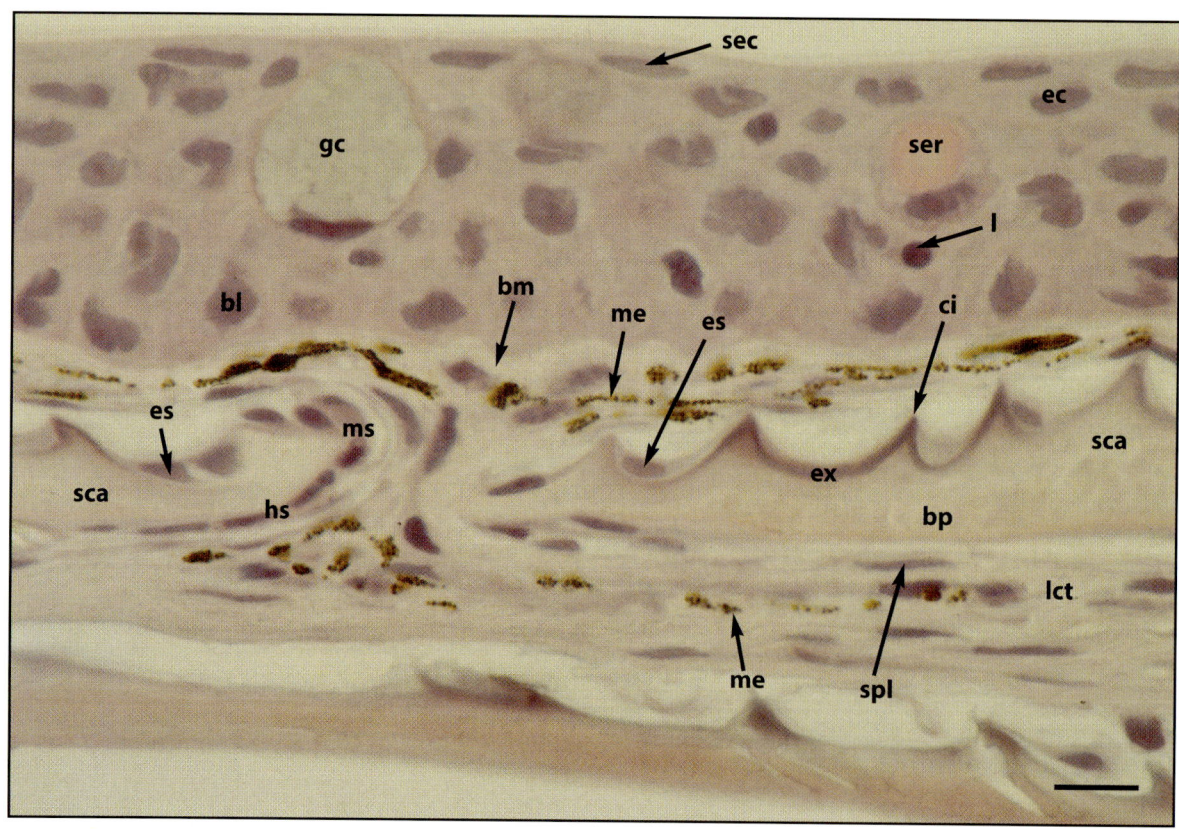

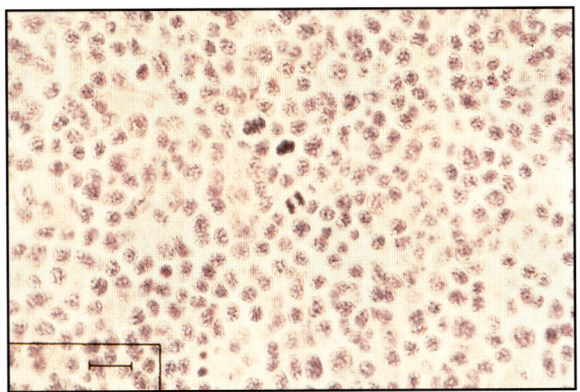

Plate 14 In the thymus of young fish, condensed basophilic nuclear chromatin within mitotic figures of dividing cells can be observed at high magnification. Hematoxylin and eosin. Bar length=10 μm.

Plate 15 The pulp in a spleen is a combination of hematopoietic (red pulp) and lymphopoietic (white pulp) tissue in the rainbow trout. Hematoxylin and eosin. Bar length=100 μm.

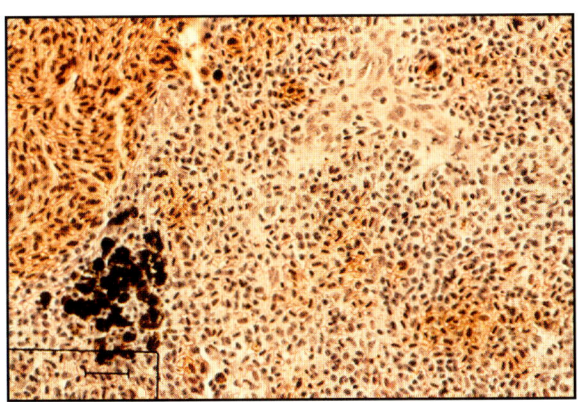

Plate 16 Rainbow trout, splenic tissue loosely organized into areas of pulp, arterioles and macrophages with associated aggregations of dark brown melanin. Hematoxylin and eosin. Bar length=25 μm.

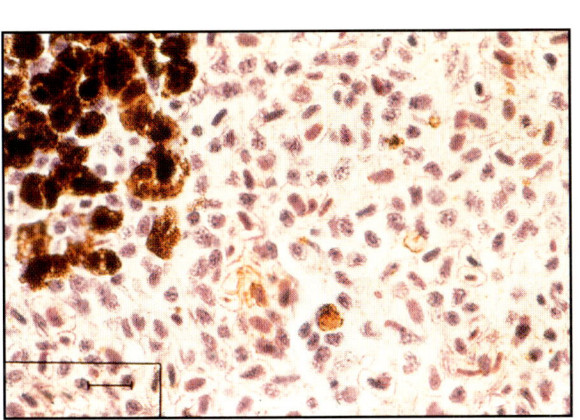

Plate 17 High magnification of melanin granules and phagocytosed material in the cytoplasm of splenic macrophages of rainbow trout. Hematoxylin and eosin. Bar length=10 μm.

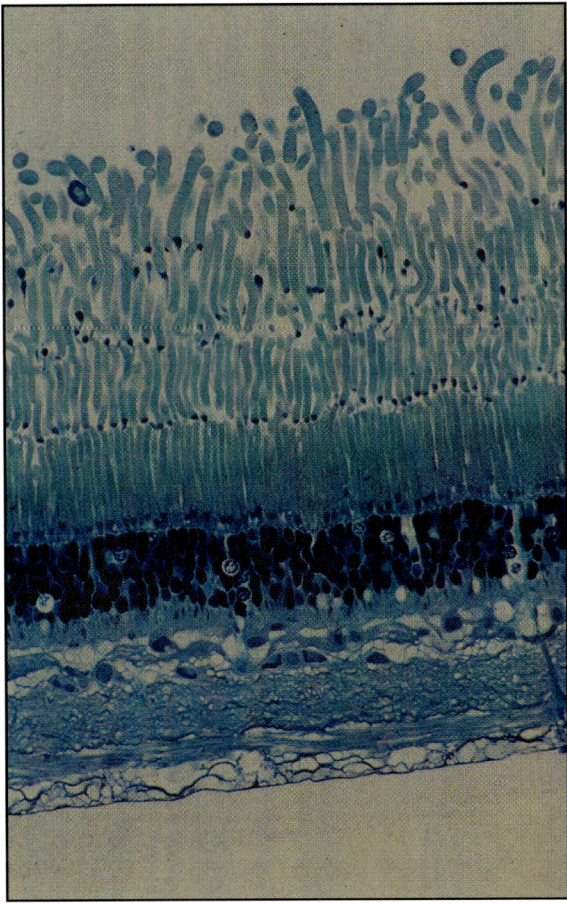

Plate 18 Retinal structures from a deep-sea fish, *Nothacantus chemnitzii*, which shows the characteristic all rod retina with multiple banks of rods with their outer segments stacked. Courtesy of H.J. Wagner.

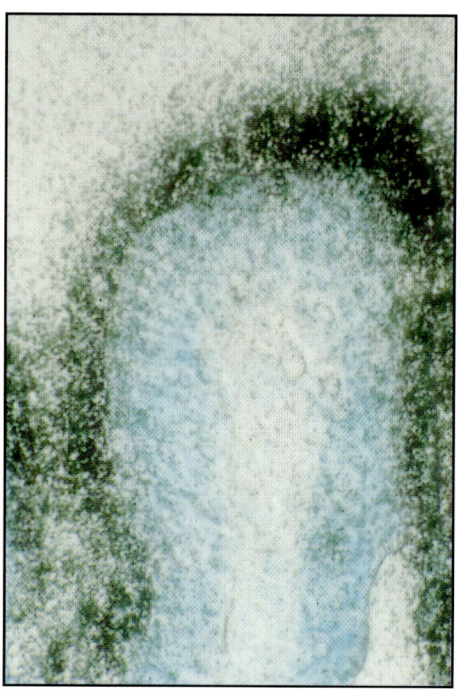

Plate 19 Thaw mounted cryosection of intestinal mucosal fold from catfish exposed to [^{3}H]benzo(a)pyrene. Black signal is silver image resulting from exposure and development of NTB-2 emulsion. Counterstained with methyl green pyronine. (x40).

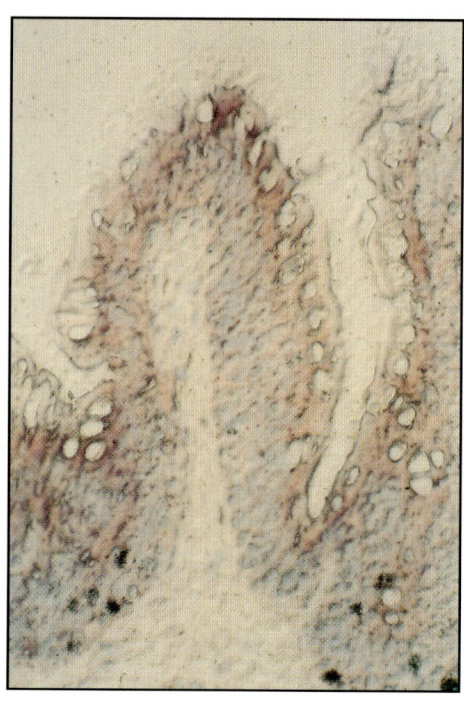

Plate 20 Thaw mounted cryosections of intestinal mucosal fold from catfish exposed to [^{3}H]thymidine in cell turnover experiment. Concentration of black dots in individual cells represent autoradiographic signal. Counterstained with methyl green pyronine. (x40).

Plate 21 Computer enhanced images of intestinal mucosal fold autoradiograph from catfish exposed to [^{3}H]benzo(a)pyrene. Yellow is quantified signal in select region (as outlined in red) as determined by programed size, shape, contrast and threshold parameters. (x40).

PART 4

Microscopic Functional Anatomy

Contents

CHAPTER 17	**Integumentary system**	271
CHAPTER 18	**Skeletal system**	307
CHAPTER 19	**Fish as an experimental model for studying muscle function**	319
CHAPTER 20	**Nervous system**	331
CHAPTER 21	**Respiratory system**	357
CHAPTER 22	**Circulatory system**	369
CHAPTER 23	**Digestive system**	379
CHAPTER 24	**Urinary tract**	385
CHAPTER 25	**Endocrine system**	415
CHAPTER 26	**Immune system**	441
CHAPTER 27	**Sensory systems**	451
CHAPTER 28	**Reproductive system**	489

CHAPTER 17

Integumentary System

Diane G Elliott
Western Fisheries Research Center, Biological Resources Division, US Geological Survey, Seattle, Washington, USA

Introduction

Many of the features of the fish integument can only be observed microscopically. Because there are over 20 000 living fishes, mostly higher bony fishes (teleosts), a great diversity exists in the microscopic anatomy of the integument. A selection of examples from varied taxonomic groups will be used to illustrate the variation in morphological features. The reader is referred to Chapter 5 for a general overview of the structure and functions of the integument.

Epidermis

Fish epidermis is a stratified squamous epithelium (Figure 17.1). The number of cell layers may vary from two in larval fishes to ten or more in adult fishes (Whitear, 1986a). In pelagic fish species, the epidermis is frequently thickest in the dorsal areas of the body, but in **benthic** species, the epidermis covering ventral surfaces is often thicker (Bullock and Roberts, 1974). In some species such as salmonids, the epidermis is frequently thicker in non-scaled areas, such as the top of the head and the fins, than in areas where scales are found (Ferguson, 1989). Epidermal thickness, structure, and the types of cells present also can be influenced by the size, condition, sex, and degree of sexual maturation of a fish (Pickering, 1977; Wilkins and Jancsar, 1979; Pickering and Richards, 1980; Pankhurst, 1982; Nakari et al., 1986; Rydevik, 1988; Burton and Burton, 1989; Ferguson, 1989). Additionally, factors such as changes in environmental conditions or exposure to adverse environmental conditions, handling procedures, dietary deficiencies, and the presence of pathogens or other stressors, can affect the structure and cellular composition of the epidermis (Chapter 4; Pickering and Macey, 1977; Pickering and Richards, 1980; Roberts and Bullock, 1981; Solanki and Benjamin, 1982; Burton et al., 1985; Poston and Wolfe, 1985; Iger et al., 1988, 1994a, 1995; Ferguson, 1989; Taveekijakarn et al., 1994; Berntssen et al., 1997; Blazer et al., 1997; Iger and Abraham, 1997; Paul and Banjeree, 1997; Quiniou et al., 1998). Such changes may be particularly noticeable among fish held in captivity.

Copyright © 2000 Academic Press

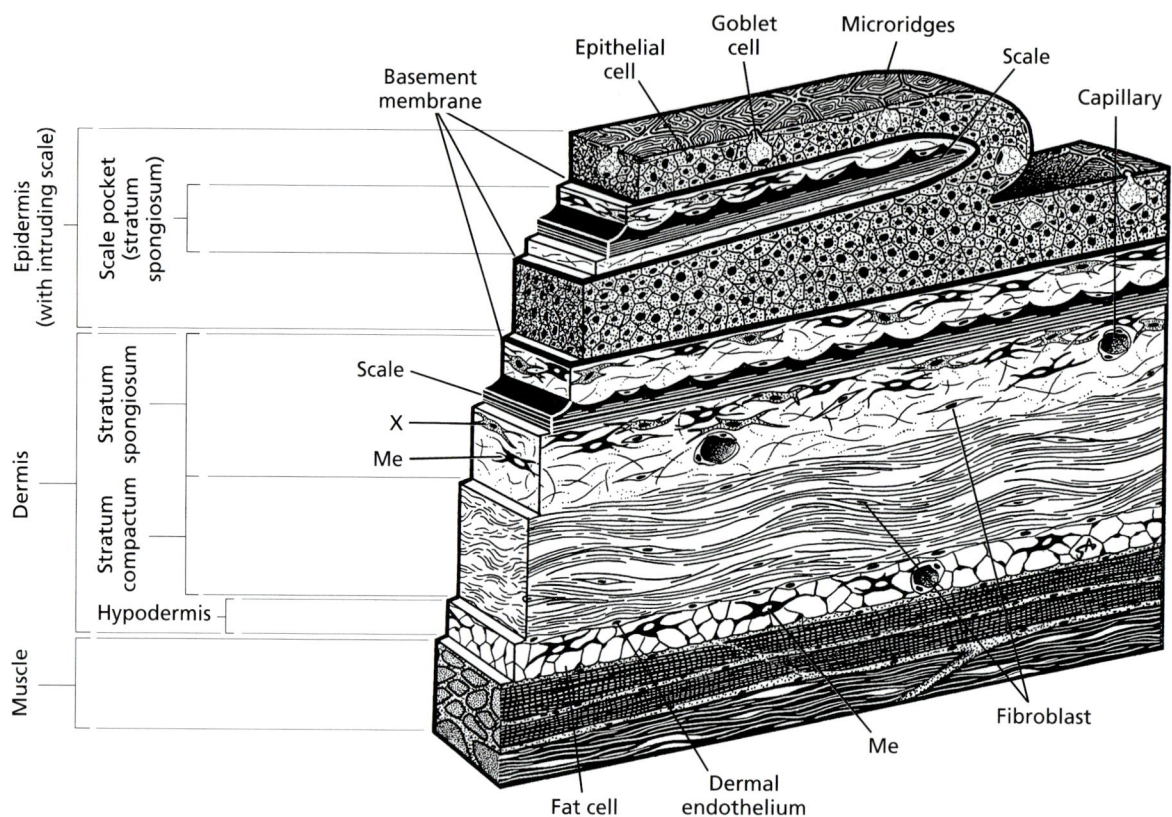

Figure 17.1 Three-dimensional section of the skin of a teleost fish, the coho salmon *Oncorhynchus kisutch*, showing the microscopic structure and some of the major cell types of the epidermis and dermis. Abbreviations: *me* melanophore; *x* xanthophore. (Source: drawing reprinted with permission from Stewart Alcorn; partly modified from Hawkes, 1974b.)

Epidermal cell types and functions

As in all vertebrate epidermis, the fundamental structural unit is the epithelial cell (see below). This is the only constant feature, as a great diversity of cell types exists in the various fish taxa. Some of these include apocrine mucous cells and a variety of other secretory cells, ionocytes, sensory cells, and wandering cells such as leukocytes.

Epithelial cells

The basic cellular element of fish epidermis, the epithelial cell (Figures 17.1–17.11), is known by many names, including: Malpighian cell, epidermal cell, filament-containing cell, filamentous cell, polygonal cell, polyhedral cell, keratocyte, keratinocyte, principal cell, and common cell. In jawless fishes, homologous cells often have been termed 'mucous' cells (see Whitear, 1986a), even though this adjective generally implies a goblet cell (see page 277) in other fishes. Nevertheless, the classical and correct name, according to Whitear (1986a), is epithelial cell.

Unlike the mammalian epidermal epithelial cell, the equivalent cell in teleost fishes is metabolically active and capable of mitotic division throughout all layers of the epidermis, although mitotic activity is most common in the lower layers (Henrickson, 1967; Bullock *et al*., 1978a; Spitzer *et al*., 1982; Tsai, 1996). Among primitive jawless fishes (hagfish and lampreys) epithelial cells are also metabolically active throughout the epidermis, but mitotic activity appears to be confined to cells in the lower layers (Linna *et al*., 1975; Spitzer *et al*., 1979). Dead cells are regularly sloughed from the epidermal surface and replaced by living cells beneath. In a study with the tropical loach *Misgurnus anguillicaudus*, Tsai (1996) found that translocation of epithelial cells from the lower epidermal layers to the surface layer and subsequent sloughing of those cells occurred within a span

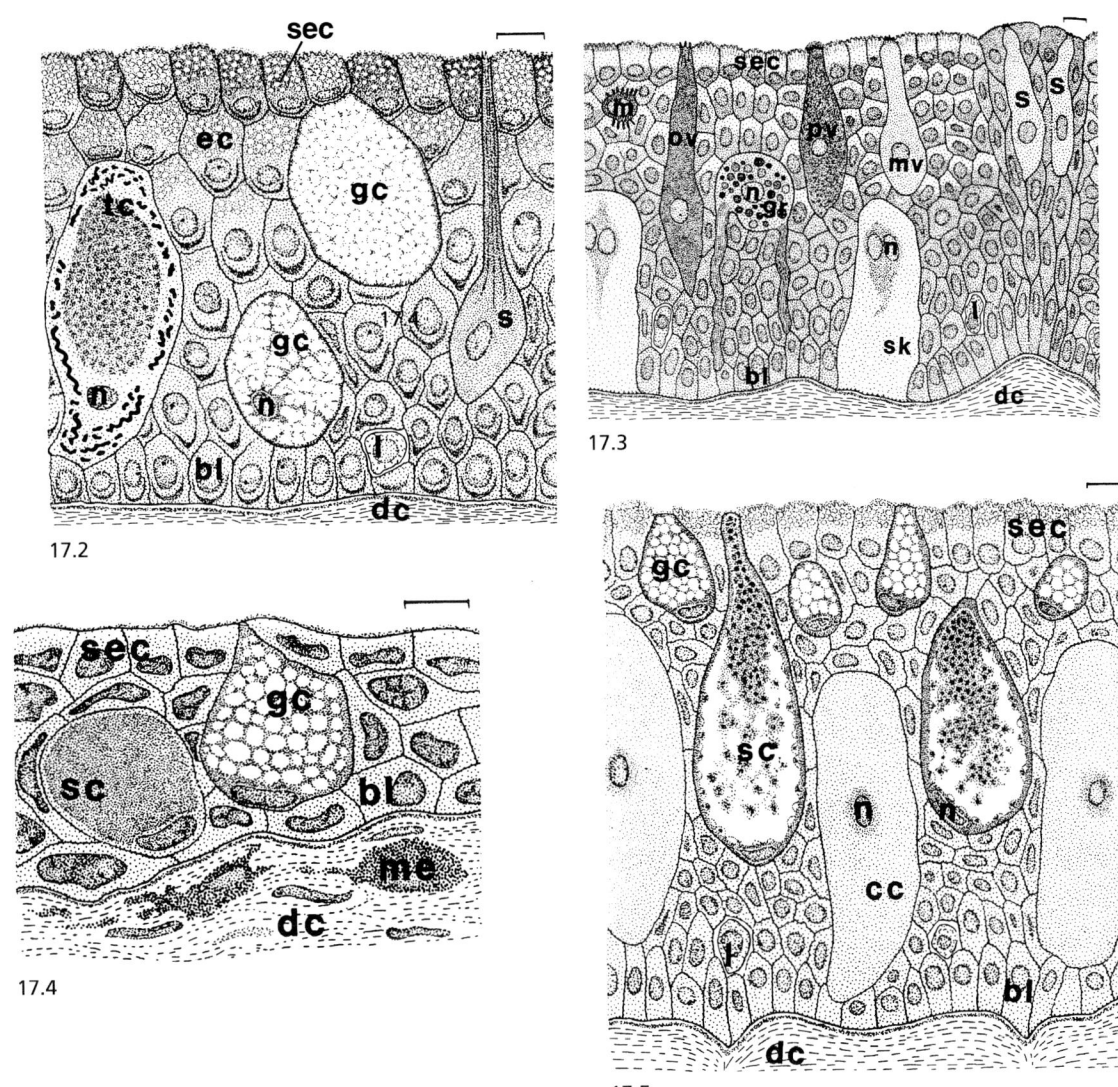

Figures 17.2–17.5 Examples of cell types in the epidermis of non-teleost fishes as seen in sections vertical to the surface. Scale bars = 10 μm. Abbreviations: *bl* basal layer epithelial cell; *cc* club cell; *dc* dermal collagen; *ec* epithelial cell; *gc* mucous goblet cell; *gr* granular cell; *l* lymphocyte; *m* Merkel cell; *me* melanophore; *mv* multivillous cell; *n* nucleus; *ov* oligovillous cell; *pv* polyvillous cell; *s* sensory cell; *sc* sacciform cell; *sec* superficial epithelial cell; *sk* skein cell; *tc* thread cell.

Figure 17.2 The primitive jawless hagfish, *Myxine glutinosa*, showing mucigenic epithelial cells ('small mucous cells' in some literature), a thread cell with a peripheral thread and a central granular mass, goblet cells ('large mucous cells' in some literature), and a sensory cell. (Source: Whitear, 1986a, copyright © 1986 by Springer-Verlag Berlin Heidelberg; sensory cells after Schreiner, 1918.)

Figure 17.3 Composite drawing of a jawless fish (lamprey, *Lampetra* spp.), showing mucigenic epithelial cells, a granular cell, and two basal layer epithelial cells modified as skein cells. The neuroendocrine Merkel cell and the sensory multivillous and oligovillous cells are associated with nerves (not shown), whereas the polyvillous cell (believed to be involved in ion regulation) is not. A sensory end bud (possibly photoreceptive) with two receptor cells and supporting cells is shown at right. (Source: Whitear, 1986a, copyright © 1986 by Springer-Verlag Berlin Heidelberg.)

Figure 17.4 An elasmobranch (ray, *Raja* sp.). Thin epidermis with a goblet cell and sacciform cell. (Source: Rabl, 1931 and Whitear, 1986a.)

Figure 17.5 A polypterid, the reedfish *Erpetoichthys* (*Calamoichthys*) *calabaricus*, showing the four major types of epidermal secretory cells: mucigenic epithelial cells, mucous goblet cells, sacciform cells, and club cells. (Source: Whitear, 1986a, copyright © 1986 by Springer-Verlag Berlin Heidelberg.)

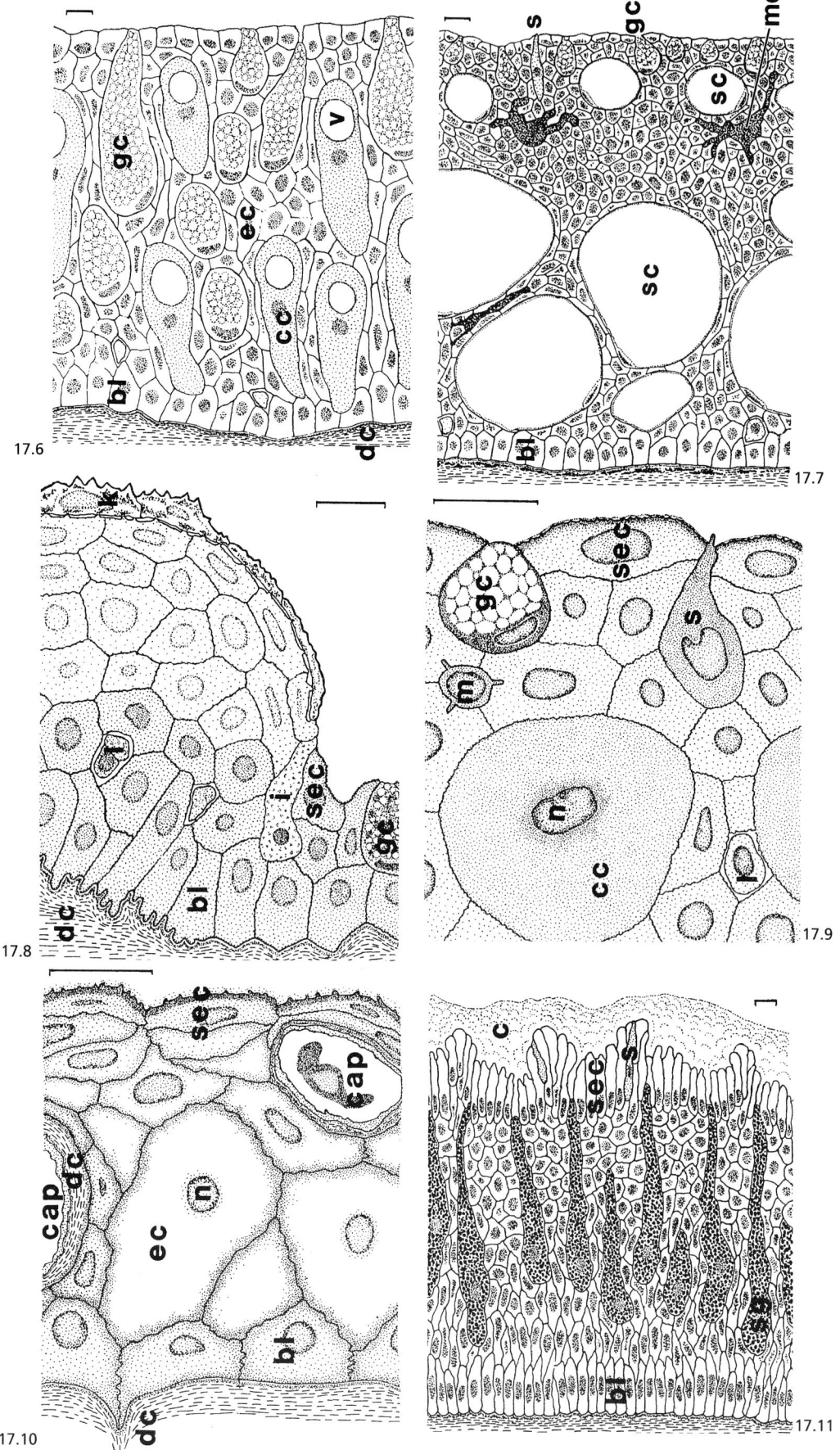

Figures 17.6–17.11 Examples of cell types in the epidermis of teleost fishes as seen in sections vertical to the surface. Scale bars = 100 μm. Abbreviations: *c* cuticle; *cap* capillary blood vessel; *i* ionocyte; *k* keratinized cell; *sg* serous goblet cell; *v* vacuole of eel club cell; see also abbreviations for Figures 17.2–17.5. (Source: all from Whitear, 1986a, copyright © 1986 by Springer-Verlag Berlin Heidelberg.)

Figure 17.6 Eel, *Anguilla anguilla*, with numerous large mucous goblet cells and club cells with large secretory vacuoles.

Figure 17.7 Pollock, *Pollachius virens*, with small epithelial cells and mucous goblet cells but enormous sacciform cells. Melanophores and a solitary sensory cell are also present in the epidermis.

Figure 17.8 Catfish, *Bagarius* sp., showing the side of a keratinized epidermal plaque. Only the superficial layer of epithelial cells is keratinized. Certain other cell types such as goblet cells and ionocytes are confined to non-keratinized areas between the plaques.

Figure 17.9 Outer epidermal layers of a minnow, *Phoxinus phoxinus*, with a mucous goblet cell, solitary sensory cell, lymphocyte, Merkel cell, and a club cell without a secretory vacuole.

Figure 17.10 Thin epidermis of the head of a mudskipper, *Periophthalmus kohlreuteri*, penetrated by blood capillaries accompanied by some dermal collagen. The superficial epithelial cells have a well-developed terminal web (dark shading), whereas the mid-epidermal epithelial cells contain few organelles and are swollen.

Figure 17.11 Distal part of a free pectoral fin ray of a gurnard, *Trigla* sp., with a thick cuticle secreted by columnar superficial epithelial cells. Protruding groups of cells include sensory cells and nerve fibers (latter not shown). The large elongated cells are classified as serous goblet cells.

of about four days. In contrast, a similar process of epithelial cell translocation and shedding in humans requires about 33 to 37 days (Weinstein *et al.*, 1984).

Epithelial cells are typically small relative to many other epidermal cell types (Figures 17.2–17.11), and have basophilic-staining cytoplasm. These cells may vary in shape depending on their position in the epidermis. The basal layer cells adjacent to the acellular basal lamina or basement membrane (page 291) are cuboidal or columnar, whereas the superficial cells are often flattened. Occasional modifications occur in the middle layer of epidermis. For example, among mormyrids and gymnotids, the middle 10 to 30 tiers of epithelial cells are flattened (with a few tiers of rounded cells above and below), forming an insulating layer around the electrosensory organs which traverse the epidermis (Szamier and Wachtel, 1969; Harder, 1971; Srivastava and Szabo, 1974). In contrast, the middle layer epithelial cells are swollen in the horse mackerel *Trachurus* and the mudskipper *Periophthalmus* (Figure 17.10; Whitear, 1986a; Suzuki, 1992).

The shape of the epithelial cell nucleus roughly corresponds to the shape of the cell; the nuclei of basal cells are often elongate and apically located in the cells, whereas the nuclei of more superficial cells are often more round or oval and are centrally (or sometimes basally) located (Figures 17.2–17.11). In histological sections, the nucleus may appear large and pale staining with prominent nucleoli, or smaller and intensely stained; the latter are more common in superficial epithelial cells.

The ultrastructure of fish epithelial cells has been described by several authors (e.g. Henrickson and Matoltsy, 1968a; Roberts *et al.*, 1970; Downing and Novales, 1971a; Bullock and Roberts, 1974; Merrilees, 1974; Harris and Hunt, 1975a; Schwerdtfeger, 1978; Whitear, 1986a; Suzuki, 1992; Suzuki and Hagiwara, 1995). At the ultrastructural level the nucleus is often deeply infolded and closely surrounded by organelles including mitochondria, Golgi systems, ribosomal endoplasmic reticulum, and sometimes lysosomes. In the more superficial cells, accumulations of large secretory vesicles are frequently present in the apical cytoplasm. Secretory vesicles are present in deeper epidermal cells in certain fishes such as dipnoans, eels, some selachians, and especially jawless fishes (Whitear, 1986a). A distinctive feature of the epithelial cell is the presence of **tonofilaments** (intermediate filaments), about 7 to 8 nm in diameter, arranged in bundles (tonofibrils) or randomly distributed. The tonofilaments form an important part of the cytoskeleton of individual cells, and the attachment of tonofibrils to the desmosomal plaques that join adjacent epithelial cells enables the epidermis to respond as a whole to mechanical stress (Matoltsy, 1986). The plasma membranes of neighboring epithelial cells interdigitate extensively, particularly at the lower levels, and desmosomes are present at frequent but often irregular intervals. Hemidesmosomes (half desmosomes) attach the lowermost layer of cells to the basement membrane. Intercellular spaces are also apparent; these may become enlarged in stressed or diseased fish (Whitear, 1986a; Ferguson, 1989).

The superficial epithelial cells, sometimes called pavement cells or squamous cells, always show some modifications compared with the cells below (Schliwa, 1975; Whitear, 1986a). Adjacent surface cells appear to be connected via typical junctional complexes including tight junctions or zonulae occludens (Downing and Novales, 1971a; Roberts et al., 1971). Cytoplasmic organelles such as mitochondria may be less distinct, but tonofilaments are numerous. A network of tonofilament bundles associated with the desmosomes of the junctional complex forms a terminal web at the apex of the cell (Figure 17.10). The outer regions of the web also contain finer filaments identified as actin.

The exterior surface of superficial epithelial cells of teleosts is characterized by microridges or micropapillae, which often form regular fingerprint-like patterns (Figures 17.1 and 17.12a; Hawkes, 1974a; Bereiter-Hahn et al., 1979). The epithelial surface is papillate in dipnoans and pitted like a sponge in primitive actinopterygians and jawless fishes (Whitear, 1986a). The function of the surface microridges is unknown. They may provide some mechanical protection against trauma and aid in holding mucous secretions to the skin surface (Figure 17.12b; Hawkes, 1974a), including the secretions produced by the epithelial cells themselves (see page 288). Microridges also increase the absorptive surface area of epithelial cells (Olson and Fromm, 1973), and may be a factor in enabling the skin to function in gas exchange (Hawkes, 1974a). Bereiter-Hahn (1971) suggested that because microridges can move by contraction of basal actin microfilaments, they might have a role in the initiation of primary wound closure.

In a few fishes, keratinization occurs in the surface epithelial layers (Mittal and Banjeree, 1980). The teeth of lampreys consist of keratinized cones (Mittal and Whitear, 1979), and keratinized oral epithelium has been observed in a number of teleost species (Whitear, 1986a). Breeding males of some species of Salmoniformes, Cypriniformes, and certain other teleost orders display external keratinized nuptial tubercles or contact organs (Chapter 5; Wiley and Collette, 1970). Attachment organs on the lips and fins of fishes in some ostariophysan genera typically have superficial cells bearing horny projections (unculi) of bizarre shapes, which aid in attachment (Roberts, 1982). The skin of the Asiatic catfish *Bagarius* is roughed by the presence of numerous polyhedral keratinized plaques separated by non-keratinized mucogenic epidermis

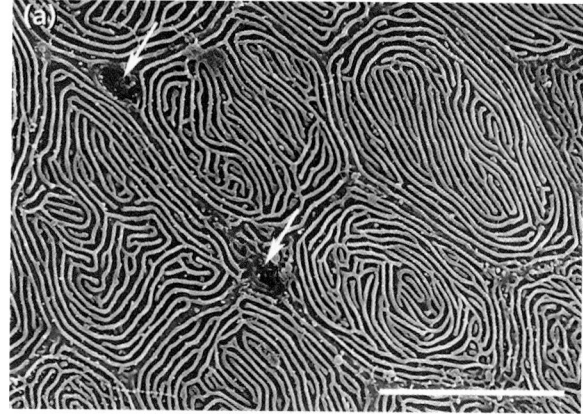

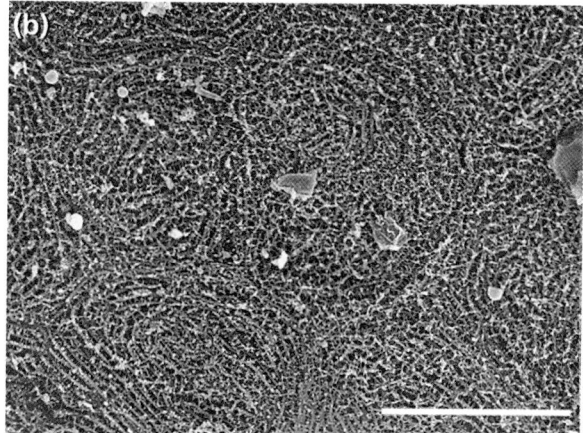

Figure 17.12 Scanning electron micrograph of the epidermal surface of a chinook salmon (*Oncorhynchus tshawytscha*). Scale bars = 10 μm. (a) Individual surface epithelial cells showing microridge patterns; apical ends of superficial goblet cells (arrowed) are visible between epithelial cells. (b) Skin surface fixed to preserve the mucous cuticle external to the superficial epithelial cells. Microridge patterns of superficial epithelial cells are faintly visible beneath the mucus, which has trapped some surface debris.

(Figure 17.8; Mittal and Whitear, 1979; Mittal et al., 1995). As in amphibians, a subcorneal space separates the layer of dead keratinized cells from the replacement cells below. The cells in the outermost replacement layer are joined laterally to one another by tight junctions. The keratinized plaques are believed to function in protecting the skin surface against mechanical damage and potential attack by pathogens (Mittal and Whitear, 1979).

By means of their motility, non-keratinized epithelial cells provide another form of protection for the integument and underlying tissues. The epithelial cell is the key element in the unique wound repair mechanism of fish skin (Phromsuthirak, 1977). Within seconds or minutes after the occurrence of the injury (Bereiter-Hahn, 1986), epithelial cells at the edges of the wound begin migrating toward the wound cavity in compact groups, retaining intercellular contacts as they migrate (Phromsuthirak, 1977; Iger and Abraham, 1990; Sire et al., 1990). These cells rapidly cover the wound and provide a mechanical barrier against infection by opportunistic pathogens from the surrounding water during the early stages of wound repair (Bullock et al., 1978b). Closure of the wound by migrating epithelial cells occurs at a rate as fast as 5–12 µm min^{-1} in teleosts (Bereiter-Hahn, 1986), with complete closure occurring within a few hours to a few days (Mittal and Munshi, 1974; Anderson and Roberts, 1975; Phromsuthirak, 1977; Bullock et al., 1978b; Roubal and Bullock, 1988; Iger and Abraham, 1990; Sire et al., 1990). The actual rate of wound closure depends on such factors as the water temperature, the size and severity of the wound, and perhaps the fish species (Anderson and Roberts, 1975; Phromsuthirak, 1977; Bullock et al., 1978b). Nevertheless, fish epithelial cells are among the fastest migrating cells known from in vitro studies, moving at a speed 10 to 20 times faster than mammalian fibroblasts (Zigmond, 1993). Because of their uniform structure and high motility, fish epithelial cells have been used as an in vitro model for studying the principles of locomotion in simple-shaped cells (Kucik et al., 1990; Heath and Holifield, 1991; Lee et al., 1993).

In addition to wound coverage, migrating epithelial cells may help to protect the skin surface from pathogen invasion via phagocytic activity (Phromsuthirak, 1977; Iger and Abraham, 1990; Peleteiro and Richards, 1990; Åsbakk and Dalmo, 1998). Proteolytic and phagocytic activity to clear debris may facilitate the progression of the migrating epithelial cells toward the center of the wound (Iger and Abraham, 1990). It is also believed that these cells help to eliminate foreign matter from the integument by a process in which cells that become filled with foreign matter subsequently are sloughed from the skin surface (Åsbakk and Dalmo, 1998). In addition, uptake and sequestration of foreign antigens by epithelial cells may play a role in the development of local immunity in the integument (Moore et al., 1998).

Epithelial cell proliferation resulting in epidermal thickening is another protective response frequently observed in fish. This response, epidermal hyperplasia, has been reported in fish infested with skin parasites or exposed to pollutants (Gaines and Rogers, 1975; Iger et al., 1988; Ferguson, 1989). Epithelial cell proliferation and consequent epidermal thickening may also occur at times that fishes are exposed to increased skin abrasion, as during spawning activity (Pickering, 1977; Burton et al., 1985; Nakari et al., 1986).

The mucous secretions produced by superficial epithelial cells also help to protect fish from pathogens. These secretions form a layer called the cuticle on the epidermal surface (Figures 17.11, 17.12b and 17.13; page 290). The continual sloughing and renewal of this layer helps to keep the integument free of bacteria and debris (Roberts and Bullock, 1980; Crouse-Eisnor et al., 1985).

Besides its protective functions, the superficial epithelial cell may be involved in ion exchange between the fish and its environment, at least in freshwater teleosts. Although this role is currently debated, evidence is accruing that the superficial epithelial cell is the site of Na^+ uptake via channels linked electrically to an apical membrane vacuolar H^+-ATPase (proton pump; reviewed by Perry, 1997). The relative roles of superficial epithelial cells and ionocytes (page 285) in Na^+, Cl^-, and Ca^{2+} uptake from the freshwater environment, as well as their roles in acid–base regulation in freshwater fishes, are still uncertain (Perry, 1997).

Goblet cells

The goblet cell is a unicellular exocrine gland common to most animal groups (Whitear, 1986a). This cell, the second category of secretory cell in the fish skin, occurs in the internal epithelia of fish as in other vertebrates, and is almost universally present in the skin, with some exceptions. Lamprey epidermis contains no goblet cells (Figures 17.3 and 17.14); the production of mucus in lampreys is entirely from epithelial cells, which are sometimes called 'mucous cells' (Downing and Novales, 1971a; Whitear, 1986a). In contrast, the morphological features of the 'large mucous cells' of hagfish (Spitzer et al., 1979) qualify them as goblet cells (Figures 17.2 and 17.15; Whitear, 1986a). Goblet cells normally occur in the cartilaginous selachians (Figure 17.4; Whitear, 1986a; Whitear and Moate, 1998). They also occur in relict bony fishes such as the dipnoans and Polypteriformes (Figure 17.5), as well as in most teleosts that have been studied (Figures 17.6–17.9, 17.11; Whitear, 1986a). However, goblet cells are lacking in certain fishes such as the relict bony fish Polydon (Weisel, 1975), and the mudskipper Periophthalmus (Figure 17.10; Suzuki, 1992).

Goblet cells are most frequently recognized in the middle to outer layers of epidermis, although in a

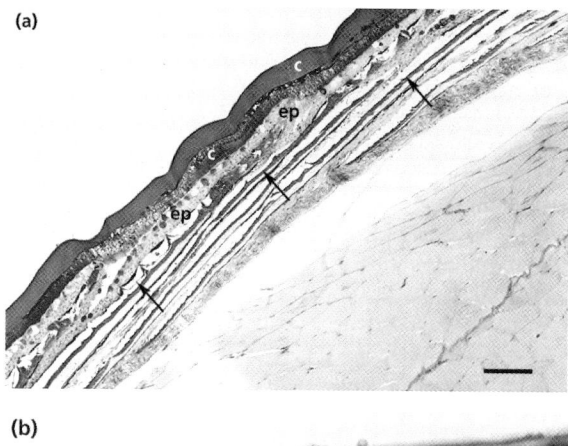

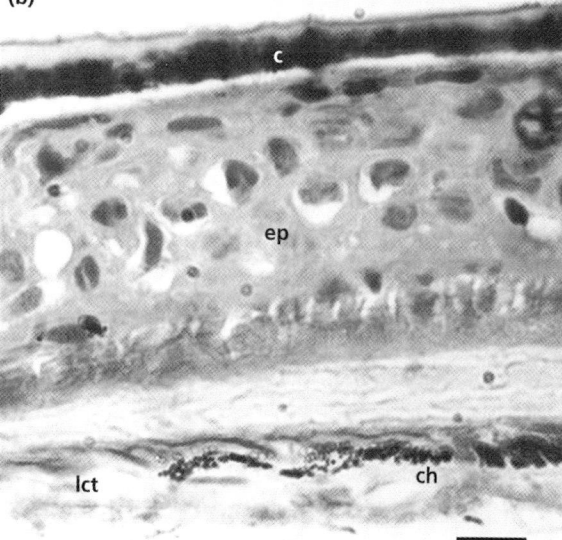

Figure 17.13 Histological section of skin tissue from a blue gourami, *Trichogaster trichopterus*, fixed to preserve the mucous cuticle and stained with periodic acid-Schiff (PAS) for mucin. This species and other members of the family Anabantidae are adapted for life in freshwater habitats that can be oxygen-poor and dirty. In addition to an accessory respiratory organ that enables this species to obtain atmospheric oxygen, it possesses a thick cuticle that may help to protect the skin from adverse environmental conditions. (a) Thick acellular cuticle (c) overlying the cellular epidermis (ep) and the dermis containing scales (arrows). The cuticle exhibits a bi-layered staining pattern (scale bar = 100 μm). (b) Higher-magnification micrograph of an area of skin with a thinner cuticle (c) overlying the epidermis (ep). A chromatophore (ch) is visible in the upper loose connective tissue layer or stratum spongiosum (lct) of the dermis (scale bar = 10 μm). (Source: Photomicrographs courtesy of Elena Catap.) **(See also Colour Plate 7.)**

very thin epidermis, a mature goblet cell may have its base on the basement membrane (Henrickson and Matoltsy, 1968b; Bullock and Roberts, 1974). Evidence suggests that goblet cells differentiate from epithelial cells in the lower layers of the epidermis, and that differentiated goblet cells do not undergo further proliferation (Tsai, 1996).

Immature goblet cells are rounded but become flattened laterally and generally increase in size as they move toward the epidermal surface (e.g. Figure 17.6). As mucous secretions are elaborated, the nucleus and organelles usually become displaced basally. In the hagfish, however, the nucleus of the mature goblet cell is degenerate and centrally located, although it is peripheral in earlier stages (Figures 17.2 and 17.15; Whitear, 1986a). Upon reaching the surface the goblet cell emerges (usually between adjacent epithelial cells), its cell membrane ruptures at the apical point, the cell contents are released, and then the cell dies (Van Oosten, 1957; Harris and Hunt, 1975b). Effete goblet cells that have released their contents are often observed in histological sections.

Ultrastructurally, goblet cell nuclei may not have the extensive infoldings seen in epithelial cell nuclei (Harris and Hunt, 1975b). Although many interdigitations occur between the plasma membranes of goblet cells and neighboring epithelial cells, desmosomes are less common than between adjacent epithelial cells (Bullock and Roberts, 1974; Whitear, 1986a). In general, goblet cells have abundant ribosomal endoplasmic reticulum and prominent Golgi systems (Whitear, 1986a). As the goblet cell progresses toward the epidermal surface, the number of secretory vacuoles increases. When it reaches the surface the cell is packed with secretory vesicles, which differ in shape, electron density and patterning with species (Bullock and Roberts, 1974; Whitear, 1986a). Individual secretory vesicles may be discharged to the surface intact, or vesicular membranes may rupture simultaneously with the cell membrane (Harris and Hunt, 1975b; Whitear, 1986a).

According to Whitear (1986a), goblet cells are of two basic types: mucous goblet cells and serous goblet cells (Figure 17.16). The vesicles of mucous goblet cells contain mucous glycoproteins (page 290) which often show weakly basophilic staining or remain unstained in routine histological sections, whereas the vesicles of serous goblet cells contain basic proteins and thereby exhibit acidophilic staining in routine histological sections (Blackstock and Pickering, 1980). Mucous goblet cells are more common than the serous type, but serous goblet cells have been observed in a variety of fish species, including salmonid, goby, and coelacanth species (see Whitear, 1986a, for a review). Even among fishes in which serous goblet cells occur, however, they may be less numerous than mucous goblet cells in the epidermis (Whitear,

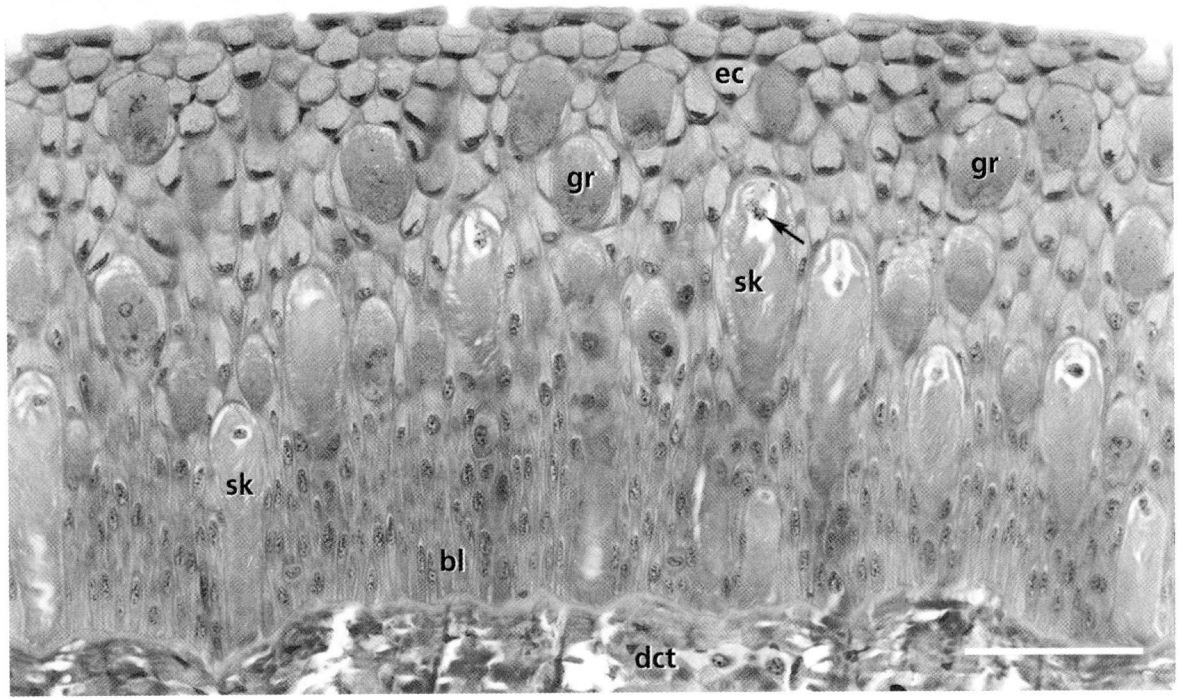

Figure 17.14 Histological section of the epidermis of a jawless fish, a spawning adult Pacific lamprey, *Lampetra tridentata*. The most common epidermal cell types are the mucigenic epithelial cells (*ec*). Some basal layer epithelial cells (*bl*) are modified as skein cells (*sk*) which show characteristic twisted bundles of tonofilaments and are sometimes binucleate (arrow). Granular cells (*gr*) are present in the middle to upper levels of the epidermis, but the granules are not highly visible in this preparation (see Figure 17.19). Some of the dense collagenous tissue of the dermis (*dct*) is visible beneath the epidermis. Scale bar = 100 μm. Hematoxylin and eosin stain. **(See also Colour Plate 8.)**

1986a). The appearance of increased numbers of serous goblet cells has been associated with skin irritation caused by ectoparasite infestation or topical disease treatment with formalin (Blackstock and Pickering, 1980), as well as experimental administration of cortisol, the hormone released during the primary stress response (Chapter 25; Iger *et al*., 1995). The confusing term 'acidophilic granular cell' has been applied to the serous goblet cell in salmonids (Blackstock and Pickering, 1980).

The abundance and size of goblet cells may vary in different regions of the body of a fish (Pickering, 1974; Leonard and Summers, 1976; Burton and Burton, 1989), or even in different areas of the epidermis covering a single scale (Strüssmann *et al*., 1994). A particular distribution of goblet cells may help to ensure an even layer of mucus over the surface of a moving fish (Pickering, 1974), or to reduce friction between overlapping scales during flexion of the body (Strüssmann *et al*., 1994).

Goblet cell numbers may differ between male and female fish of the same species, as in certain salmonid fishes (Pickering, 1977; Nakari *et al*., 1986). The abundance of goblet cells can also change seasonally (Wilkins and Jancsar, 1979; Burton and Fletcher, 1983) and during processes such as larval metamorphosis (Ottesen and Olafsen, 1997), sexual maturation (Pankhurst, 1982; Burton and Burton, 1989) or adaptation to seawater (Solanki and Benjamin, 1982). Some of these changes may be under hormonal control, although many contradictory results have been reported (see Whitear, 1986a, for a review).

Alterations in environmental conditions such as sudden changes in temperature (Quiniou *et al*., 1998), or exposure to ultraviolet radiation (Blazer *et al*., 1997; Kaweewat and Hofer, 1997; Noceda *et al*., 1997), acid water (Zuchelowski *et al*., 1981) or contaminants (Burton *et al*., 1985; Iger *et al*., 1988; Roy, 1988; Berntssen *et al*., 1997) can also affect the number of goblet cells present. For example, acute exposure to an irritant can cause rapid holocrine secretion by goblet cells and a consequent reduction in their numbers (Iger *et al*., 1988; Berntssen *et al*., 1997), whereas chronic exposure to an irritant may result in an increase in goblet cell numbers as new cells differentiate and move to the epidermal surface (Zuchelowski *et al*., 1981; Iger *et al*., 1988). In some instances even a mild stressor, such as a single incidence of handling

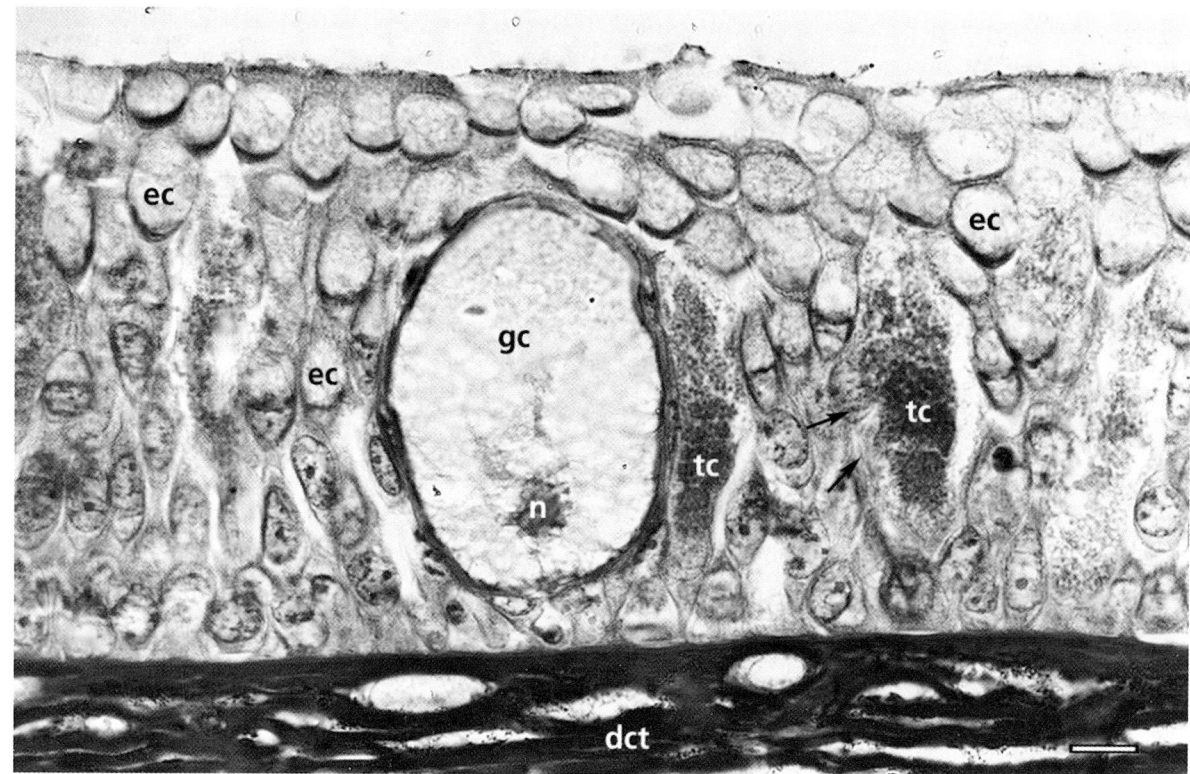

Figure 17.15 Histological section of the epidermis of a jawless fish, the hagfish *Myxine* sp. Epithelial cells (*ec*), also known as 'small mucous cells', are the most numerous cell type. The most prominent cell in the section is a large goblet cell (*gc*) with an eccentric nucleus (*n*); this cell type is sometimes called a 'large mucous cell.' Thread cells (*tc*) are also present. Thread cells contain a central mass of fine granules or globules (visible near *tc* labels) and a dense thread-like component in peripheral areas (arrows). The dense collagenous tissue (*dct*) of the dermis is visible. Scale bar = 10 μm. Hematoxylin and eosin stain.

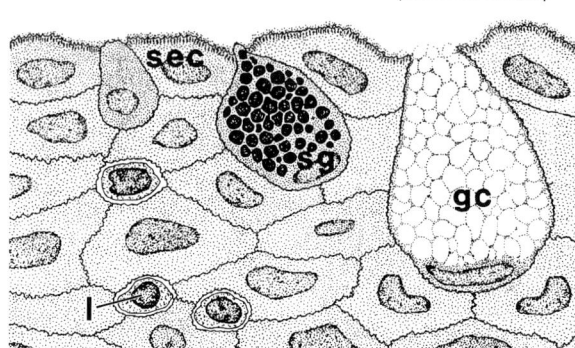

Figure 17.16 Outer layers of the epidermis of a coelacanth, *Latimeria chalumnae*, showing both mucous (*gc*) and serous (*sg*) goblet cells and superficial epithelial cells (*sec*). Several lymphocytes (*l*) are also present in the epidermis. Scale bar = 10 μm. (Source: Whitear, 1986a, copyright © 1986 by Springer-Verlag Berlin Heidelberg.)

of goblet cells (Roy, 1988; Garg and Mittal, 1993; Sanchez *et al.*, 1998), and can result in changes in the chemical composition of the secretion produced (page 290).

In some species such as salmonids, the bulk of the epidermal mucous secretions are from goblet cells (Whitear, 1986a). The chemical composition of goblet cell secretions, which have lubricating and protective functions (Chapter 5), is discussed further on page 290. In addition to secretory functions, goblet cells may have phagocytic abilities and help to remove debris from the integument during wound repair (Iger and Abraham, 1990). Limited evidence suggesting the presence of immunoglobulin in goblet cells led Peleteiro and Richards (1988) to speculate that these cells may have a role in an integumentary-specific immune system involving the production or processing of antibody.

Goblet cells may aggregate in patches on various areas of the body to form multicellular holocrine glands in certain fish species. These and other multicellular glands are reviewed by Whitear (1986a).

(netting), can promote a subsequent increase in epidermal goblet cell numbers (Pickering and Macey, 1977). Exposure to toxicants, disease treatments, or other stressors can also alter the size or morphology

Sacciform cells

Sacciform cells constitute the third category of secretory cells of fish epidermis. These cells have been found in cartilaginous fishes such as rays (Figure 17.4; Whitear, 1986a; Whitear and Moate, 1998) and chimaeras (Whitear, 1986a). They occur in relict bony fishes such as polypterids (Figure 17.5; Whitear, 1981) but have not been reported in dipnoans (Whitear, 1986a). They also appear to be absent from the primitive jawless fishes (Whitear, 1986a). Among the teleosts, sacciform cells are characteristic of the superorders Acanthopterygii, the spiny-rayed fishes, and Paracanthopterygii, a group that includes the Gadiformes or cods (Bullock, 1980; Mittal *et al.*, 1981; Whitear, 1986a). They occur in fishes of other teleost superorders as well, according to reports describing secretory cells in catfish of the genus *Corydoras* in the superorder Ostariophysi (Bhatti, 1938; Henrickson and Matoltsy, 1968c; Mittal *et al.*, 1981), and salmonids in the superorder Protacanthopterygii (Yokoya and Ebina, 1980; Pickering and Fletcher, 1987; López-Dóriga and Martínez, 1993).

By light microscopy, **sacciform cells** have basally located nuclei and acidophilic staining cytoplasm, which may be homogeneous to granular in appearance (Figures 17.4 and 17.5). In many species, sacciform cells cannot be reliably distinguished from serous goblet cells (page 278) in routine histological sections. Indeed, some authors combine these two cell types under the classification 'serous cells' because of the proteinaceous nature of their secretions (page 290; López-Dóriga and Martínez, 1993). However, Whitear (1986a) distinguished sacciform cells from serous goblet cells according to ultrastructural differences, particularly in the packaging of their secretions. The sacciform cell secretion is contained within a single large membrane-bounded vacuole, which is often surrounded by small vesicles ('bubbles') or channels that fuse with the large vacuole and release their contents into its lumen. In contrast, the serous goblet cell secretion is contained within smaller membrane-bounded globules or vesicles that may be dispersed within the cytoplasm and that do not fuse until the secretion is discharged at the epidermal surface. In many species the peripheral vacuoles of sacciform cells appear to arise from endoplasmic reticulum cisternae previously associated with ribosomes, but in others the peripheral cytoplasm is rich in smooth endoplasmic reticulum instead (Whitear, 1986a). Golgi systems are present in sacciform cells but are not as prominent as in goblet cells (Whitear, 1986a).

The contents of the sacciform cell vacuole may be electron-lucent, fibrillar, or have dense (but not membrane-bounded) masses corresponding to the granules seen by light microscopy (Sato, 1978; Whitear, 1986a; Pickering and Fletcher, 1987).

Because of the difficulty in distinguishing sacciform cells by light microscopy and the variation in the ultrastructural appearance of the contents of their secretory vacuoles, these cells have been called by many different names. Some of these include 'granular cells' (Henrickson and Matoltsy, 1968c; Sato, 1978), 'clear mucous cells' (Whitear, 1971b; Phromsuthirak, 1977), 'sacciform granulated cells' (Mittal and Banjeree, 1975), 'granule-sacciform cells' (Bullock, 1980), and 'white cells' (Nishioka *et al.*, 1985).

In the Gadiformes, sacciform cells are enormous and swollen, often appearing cyst-like (Figure 17.7; Bullock, 1980). The distinctive appearance of these cells prompted Bullock *et al.* (1976) to refer to them as 'cystic bullae', but later electron microscopic analyses determined that they were true cells (Mittal *et al.*, 1981). Although the study of Bullock (1980) showed the sacciform cells of many Gadiformes to contain acidophilic granular to colloidal-appearing contents, the cells of some deep-sea species appeared to be empty. He hypothesized that this may have been an artifact resulting from the delay in fixation of the specimens or rapid dissipation of the sacciform cell contents throughout the epidermis.

In some species, sacciform cells appear to differentiate deep in the epidermis, and, like goblet cells, develop their secretion and increase in size as they move toward the epidermal surface (Sato, 1978; Pickering and Fletcher, 1987). Most mature sacciform cells open at the skin surface by an apical pore (Mittal *et al.*, 1981). Among pelagic gadoid fishes, however, the sacciform cells normally appear not to open, whereas among **benthic** gadoids the typical surface openings are present (Bullock, 1980). Some evidence suggests that the usual mode of secretion of these cells is holocrine (Pickering and Fletcher, 1987).

The function(s) of sacciform cells is unknown. The most prevalent hypothesis is that the cells produce 'noxious substances' which may be aimed at repelling or poisoning pathogenic microorganisms as well as potential predators (page 290; Mittal *et al.*, 1981; Whitear, 1986a; Pickering and Fletcher, 1987). A variety of poisonous fishes such as puffers and boxfishes have sacciform cells in the skin (Mittal *et al.*, 1981; Whitear, 1986a). A marked increase in sacciform cell numbers was observed by Pickering and Fletcher (1987) in young brown trout *Salmo trutta* with skin infestations of the protozoan parasite *Ichthyobodo*, prompting the authors to propose that sacciform cell secretions might protect the fish from infestation or damage by skin parasites. Limited evidence suggests that sacciform cells may have an unknown function in the healing of cutaneous wounds, as these cells have been observed in or near areas of regenerating epidermal tissue in brown trout (López-Dóriga and Martínez, 1993) and sticklebacks *Gasterosteus aculeatus* (Phromsuthirak, 1977).

In salmonids, sacciform cell numbers may fluctuate seasonally, and differences in the abundance of these cells have also been observed between male and female fish (Pickering and Fletcher, 1987). However, the sacciform cell is not a consistent feature of the salmonid epidermis (Pickering and Fletcher, 1987; López-Dóriga and Martínez, 1993).

Club cells

The **club cell** is the fourth category of secretory cell in fish epidermis, and is a characteristic feature in the skin of certain taxa of ray-finned (actinopterygian) fishes. Among the fishes with club cells are the primitive Polypteriformes (Figure 17.5) and the eels of the order Anguilliformes (Figure 17.6). The largest group of fishes with club cells is the superorder Ostariophysi, which includes the orders Gonorhynchiformes, Cypriniformes (Figure 17.9), Characiformes, Siluriformes, and Gymnotiformes, and contains over one-quarter of the known fish species and nearly three-fourths of all freshwater species (Whitear, 1986a; Smith, 1992; Moyle and Cech, 1996). The so-called 'club cells' in lampreys (Downing and Novales, 1971b) are actually modified club-shaped binucleate epithelial cells with characteristic twisted bundles of tonofilaments; Lane and Whitear (1980) proposed that these cells be called **'skein cells'** (Figures 17.3 and 17.14) to avoid confusion with actinopterygian club cells.

True club cells are usually large and round or oval to club-shaped, and possess one or two centrally located nuclei with prominent nucleoli (Figures 17.5, 17.6 and 17.9). These cells are generally located in the middle layers of the epidermis, and lack openings to the epidermal surface (Henrickson and Matoltsy, 1968c; Whitear, 1981; Smith, 1982a; Whitear and Mittal, 1983; Whitear and Zaccone, 1984). Club cells interdigitate with adjacent epithelial cells by means of characteristic flattened processes visible by electron microscopy.

Whitear (1986a) classified club cells into two forms based on morphological features. One form, characteristic of eels, has a secretory vacuole (Figures 17.6 and 17.17), whereas the other form, characteristic of ostariophysans, has no vacuole (Figures 17.9 and 17.18). The secretory vacuole of eel club cells resembles that of sacciform cells (page 281) but the secretion is collected in Golgi-derived vesicles. Both forms of club cells have a peculiar cytoplasm that contains discrete coiled filaments associated with the desmosomes in place of the normal bundles of tonofilaments that are seen in epithelial cells (page 275). In eels the helical filaments are regularly arranged and kept in alignment by cross-connectives (Whitear and Zaccone, 1984),

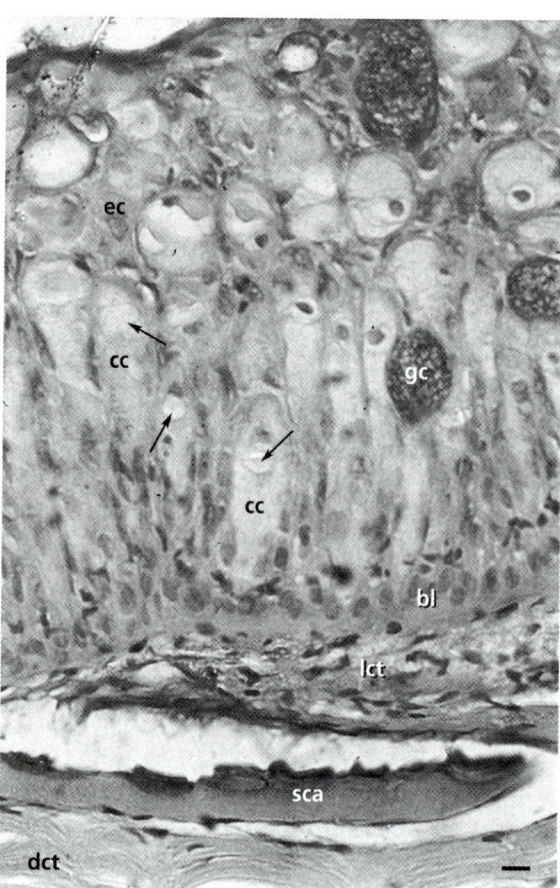

Figure 17.17 Histological section of the epidermis of a short-fin eel, *Anguilla australis*, showing typical eel-type club cells (*cc*) with secretory vacuoles (arrows). Other epidermal cell types include epithelial cells (*ec*) and basal layer epithelial cells (*bl*) as well as mucous goblet cells (*gc*). Visible dermal elements include loose connective tissue or stratum spongiosum (*lct*), a scale (*sca*), and dense connective tissue or stratum compactum (*dct*). Scale bar = 10 µm. Hematoxylin and eosin stain. (Source: photomicrograph courtesy of Dr. Barbara Nowak.) **(See also Colour Plate 9.)**

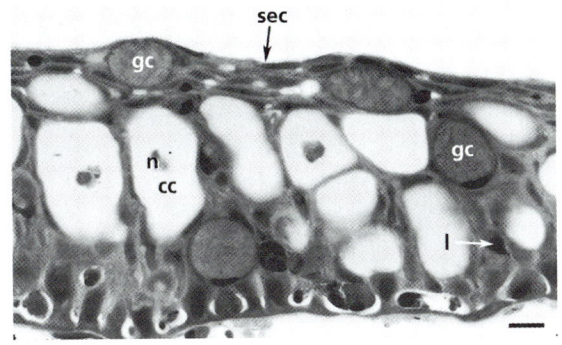

Figure 17.18 Histological section of the epidermis of a goldfish, *Carassius auratus*, showing typical ostariophysan-type club cells (*cc*) with a central nucleus (*n*) but no secretory vacuole. Other epidermal features include flattened surface epithelial cells (*sec*), mucous goblet cells (*gc*), and lymphocytes (*l*). Scale bar = 10 μm. Methylene blue-azure II stain. **(See also Colour Plate 10.)**

but in the ostariophysans the helices are oriented at random (Whitear and Mittal, 1983). Polyribosomes may be directly involved in the formation of the helical filaments, as evidenced by the radiating tracts of ribosomes in the cytoplasm of juvenile club cells (Whitear, 1986a). The morphological features of club cells in polypterids (Figure 17.5) are identical to those in ostariophysans (Whitear, 1981).

In ostariophysans, the club cells function in the alarm substance or **Schreckstoff system**, which has been reviewed by several authors (Pfeiffer, 1977, 1982; Smith, 1977, 1982a, 1986, 1992). The club cells of these fishes contain an alarm pheromone (page 290) that is released into the surrounding water when the epidermis is damaged and the club cells are broken, which frequently happens even with a minor injury to the fragile epidermal cells. Other ostariophysans detect the alarm pheromone by smell, and perform an anti-predator defensive reaction (e.g. gathering into a tight school, diving for cover, or becoming motionless) that is species-specific. Both the cells and the fright reaction develop relatively late in the ontogeny of juvenile fish, and some species or life stages lack either the cells or the reaction. For example, male fathead minnows *Pimephales promelas* lose their club cells (but not the fright reaction) during the spawning season and develop a thickened epidermis with increased numbers of goblet cells (Smith, 1973). In this species and certain other cyprinoids, the seasonal loss of alarm substance cells has been related to abrasive spawning behavior (Smith, 1973, 1976).

Ostariophysan club cells may have functions other than the production of fright pheromones. Damaged club cells of certain ostariophysans release substances that may help to seal skin wounds (page 291; Whitear and Mittal, 1983; Ralphs and Benjamin, 1992).

Among the polypterids and eels, club cells do not appear to be involved in an alarm reaction (Henrickson and Matoltsy, 1968c; Hugie and Smith, 1987). Other functions suggested for the club cells of these fishes include hydrostatic support of the epidermis or secretion of chemicals that serve as predator deterrents or protective agents against parasites, other pathogens, or irritants (Smith, 1977, 1982a, 1986; Shiomi *et al.*, 1988; Mittal and Garg, 1994).

Granular cells

According to Whitear (1986a), the term 'granular cell' should be limited to cells by that name in lamprey epidermis. These cells are located in the middle to upper regions of the epidermis, and are large, with a centrally located nucleus and prominent nucleolus, numerous small granules in the cytoplasm, and appendages penetrating the deeper epidermis (Figure 17.3 and 17.19; Downing and Novales, 1971c; Lethbridge and Potter, 1982). Ultrastructurally, the cytoplasm of immature granular cells is characterized by the presence of many scattered electron-dense granules which are often membrane-bounded, in addition to extensive ribosomal endoplasmic reticulum and moderate numbers of mitochondria (Downing and Novales, 1971c). Numerous 15 nm filaments (thicker than tonofilaments) course through the cytoplasm and form the core of the long, tongue-like appendages. Mature granular cells are largely devoid of cellular organelles, and the cytoplasm is very electron dense. The cytoplasmic granules are also larger than those in immature cells, and some granules may appear paracrystalline or degenerated. Granular cells do not open at the epidermal surface, although the granules are released in damaged skin (Whitear, 1986a).

The numbers of granular cells in the epidermis may vary in different areas of the body, and overall changes in cell abundance may occur during the life cycle of the fish (Lethbridge and Potter, 1982). The function of the cells is not known. Pfeiffer and Pletcher (1964) speculated that granular cells might produce predator-repelling substances, whereas Lethbridge and Potter (1982) hypothesized that the cells might produce antimicrobial substances such as lysozyme, but supporting data for either hypothesis are scarce.

Unfortunately, among teleosts, the term '**granular cell**' has been applied to a variety of distinct cell

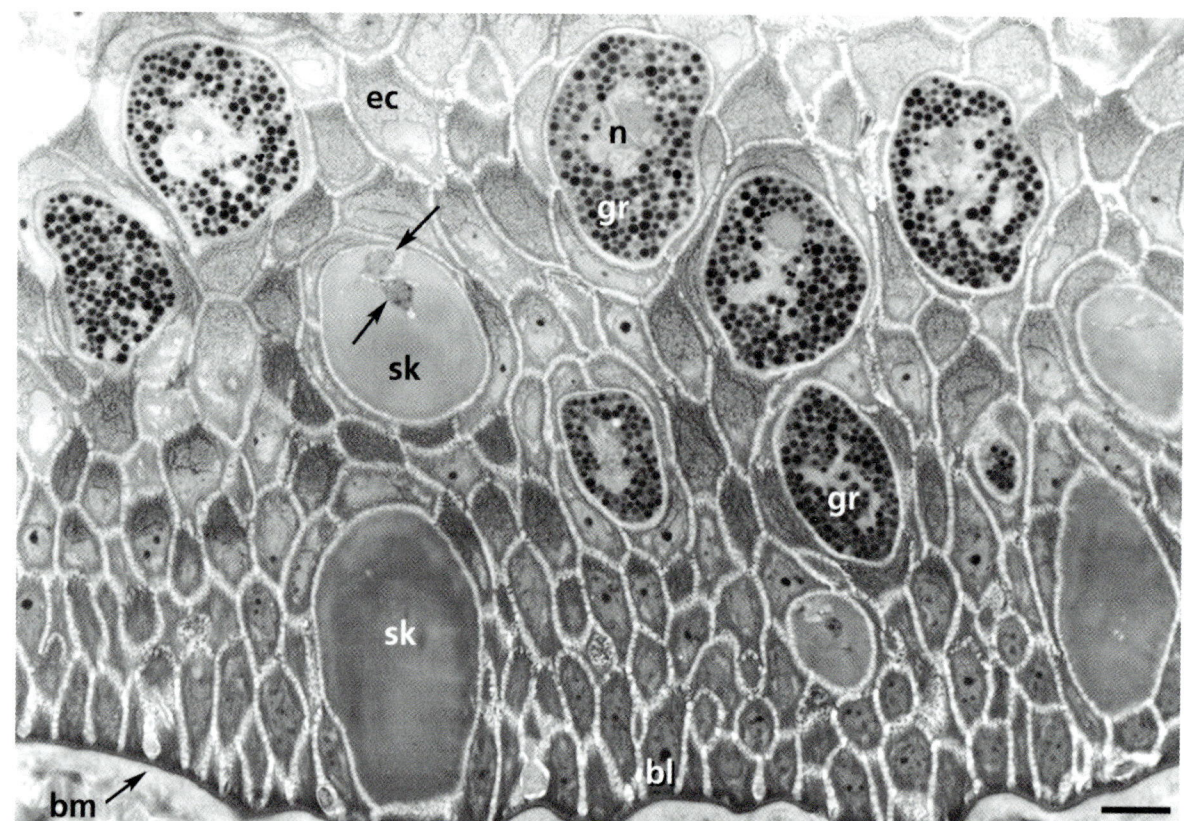

Figure 17.19 Histological section of the epidermis of a jawless fish, a spawning adult Pacific lamprey, *Lampetra tridentata*, showing characteristic granular cells (*gr*) with abundant cytoplasmic granules. Nuclei (*n*) with prominent nucleoli are apparent in some of the granular cells, but the appendages penetrating the deeper epidermis are not visible (see Figure 17.3). Skein cells (*sk*) are also present; two nuclei (arrows) are visible in one skein cell, but the twisted bundles of cytoplasmic tonofilaments are not apparent in this preparation (see Figure 17.14). Epithelial cells (*ec*), basal layer epithelial cells (*bl*) and the basement membrane (*bm*) are also present. Scale bar = 10 μm. Methylene blue-azure II stain. **(See also Colour Plate 11.)**

types, resulting in confusion in the literature (Whitear, 1986a). Teleost cell types sometimes called 'granular cells' include serous goblet cells (page 278; Blackstock and Pickering, 1980), sacciform cells (page 281; Henrickson and Matoltsy, 1968c; Sato, 1978), and cells intrusive in the epidermis (page 288; Barnett *et al.*, 1996). The cytoplasm of all of these cells also exhibits acidophilic staining, adding to the confusion.

Thread cells

The thread cell is an epidermal cell type found only in hagfish (Whitear, 1986a). It is an elongated oval cell with a nucleus near the base of the cell, a dense thread-like component in the peripheral portions of the cell, and a central mass containing fine granules or globules, depending on the species (Figures 17.2 and 17.15). Much of the thick and copious slime discharged by an irritated hagfish (the 'slime eel' of fishing lore) comes from thread cells, which are present along with mucous goblet cells in slime glands along the body (Blackstad, 1963). The 'thread' in each thread cell is an impressive biopolymer (page 291) up to 60 cm long and 3 μm thick, which is coiled in a complex fashion (Koch *et al.*, 1994). These threads are spun from intermediate filaments in association with microtubules and ribosomes (Fernholm, 1981; Downing *et al.*, 1984).

Venom cells

Venomous fishes have multicellular holocrine glands associated with spines to form defensive organs. The structure of venom glands has been described by various authors, including Cameron and Endean (1966, 1970), Russell (1969), Halstead (1988), and Gopalakrishnakone and Gwee (1993). These glands are not truly ducted, but are assemblages of swollen venom cells often interspersed with smaller supporting cells that may help to maintain the cohesiveness of the gland and provide replacements when the venom cells are discharged (Cameron and Endean, 1973; Perriere and Goudey-Perriere, 1989). The cytoplasm of the mature venom cell may be nearly completely filled with secretory granules (Gopalakrishnakone and Gwee, 1993). Although the venom glands may be partially or completely isolated from the epidermis by encapsulation within connective tissue (Perriere and Goudey-Perriere, 1989), the venom cells and supporting cells are believed to be derived from the epidermis (Whitear, 1986a; Perriere and Goudey-Perriere, 1989). Evenomation generally occurs by rupture of the fine epidermal covering of the tip of the spine, and release of the venom into the wound made by the spine.

Ionocytes

Ionocytes and salt glands are auxiliary organs of osmoregulation in vertebrates (Komnick, 1986). The ionocytes in teleosts are commonly called chloride cells or mitochondria-rich cells. They are found in the gill epithelium (Keys and Wilmer, 1932; Komnick, 1986; Perry, 1997) as well as in extrabranchial locations such as the opercular lining skin (Figure 17.20a; Zadunaisky, 1984) and external skin (Figure 17.8; Nonnotte et al., 1979), and are especially prevalent in euryhaline species (Whitear, 1986a). Ionocytes are larger than adjacent epithelial cells, with acidophilic-staining cytoplasm and a nucleus near the base of the cell (Figure 17.20a). The ionocyte often extends through the entire height of the stratified epidermis, with the apex of the cell in direct contact with the external medium and the base resting on the basal lamina (Figure 17.20a; Komnick, 1986). In saltwater-adapted fish, the apical plasma membrane is locally indented to form a small crypt or cavity filled with mucus and frequently subdivided by septa or microplicae (Figure 17.20c); the crypt is often reduced but

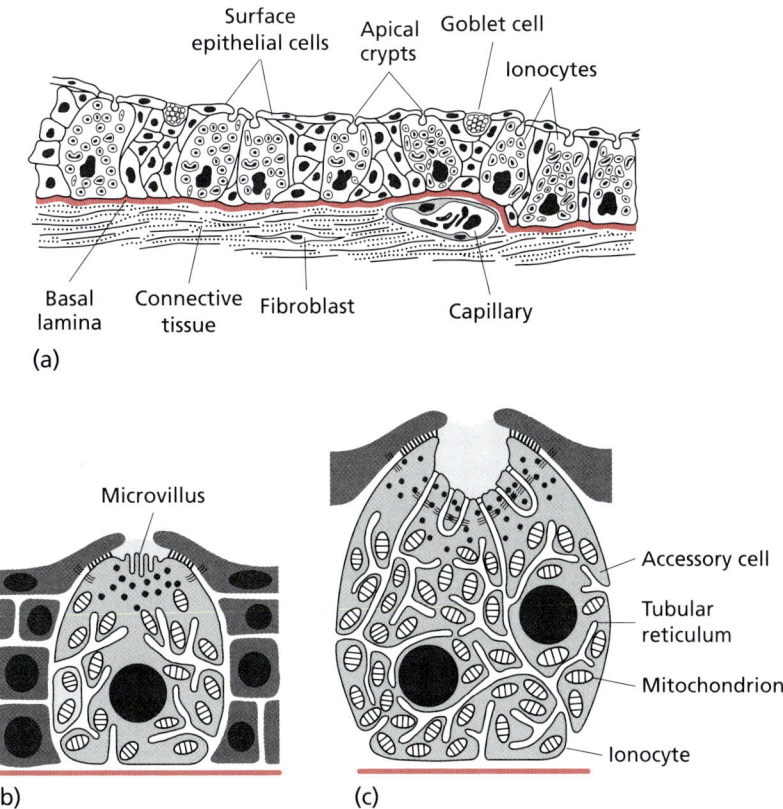

Figure 17.20 Ionocytes, also called chloride cells or mitochondria-rich cells. (a) Abundant ionocytes extending through the entire height of the opercular epithelium of a killifish, *Fundulus heteroclitus*. (b) Diagram of an ionocyte in a freshwater-adapted teleost. (c) Diagram of an ionocyte in a seawater-adapted teleost. In comparison to the ionocyte in the freshwater fish, the ionocyte in the seawater-adapted fish shows an increase in cell size and the number of mitochondria, a more elaborate cytoplasmic tubular reticulum, and a more pronounced apical crypt or cavity (but a reduction in apical microvilli). In seawater, ionocytes and accessory cells (which may represent differentiating ionocytes) form twin cell groups that are interdigitated in the apical region; accessory cells disappear during freshwater adaptation. (Sources: (a) After Degnan et al., 1977. (b) and (c) After Komnick, 1986, copyright ©1986 by Springer-Verlag Berlin Heidelberg.)

the surface microplicae may be more pronounced in freshwater-adapted fish (Figure 17.20b). Ionocytes are joined laterally to adjacent surface epithelial cells via tight junctions. Characteristic ultrastructural features of the cytoplasm include a dense population of mitochondria and an elaborate system of branching and anastomosing tubules, which represent deep infoldings of the basolateral plasma membrane (Komnick, 1986). These tubules reach blind ends in the subapical region of the cell, which contains numerous vesicles.

The functions of ionocytes in teleosts have been reviewed by several authors (Zadunaisky, 1984; Komnick, 1986; Perry, 1997). In general, ionocytes perform an ion-excretory function under the dehydrating condition of seawater, and an ion-absorptive function in fresh water where fish are exposed to hydrating conditions in their habitat. In **anadromous** and **catadromous** species, ionocytes temporarily change between excretory and absorptive functions. Ionocytes have been shown to be responsible for Cl^- secretion in seawater-adapted teleosts (Foskett and Scheffey, 1982), and they are believed to be the principal site of trans-epithelial Ca^{2+} and Cl^- influxes in freshwater teleosts (Perry, 1997). Ionocytes also perform an integral role in acid–base regulation (Heisler, 1993; Perry, 1997). Nevertheless, the relative importance of ionocytes and superficial epithelial cells in certain aspects of ion transfer and acid–base regulation are still unresolved (see page 277).

Ionocytes do not appear to be restricted to teleosts, but are widespread over most taxonomic groups of fishes (Komnick, 1986). Even the primitive hagfish, with a skin highly permeable to water and consequently a blood plasma osmolarity roughly equivalent to that of seawater, possesses mitochondria-rich cells in the gills that are morphologically similar to ionocytes in teleosts (Evans, 1993). Evidence suggests that these cells may be involved in acid–base regulation (Mallat *et al*., 1987; Evans, 1993). Among lampreys, which may migrate between freshwater and marine environments, mitochondria-rich cells in the gills (Peek and Youson, 1979) and the skin (the polyvillous cells of Whitear and Lane, 1983a; Figure 17.3) are believed to function in ionic and acid–base regulation (Evans, 1993). In elasmobranchs, mitochondria-rich cells have been found in the gills (Laurent and Dunel, 1980; Whitear, 1986a) and more recently in the skin (Whitear and Moate, 1998). Evans (1993) speculated that these cells might be involved in ionic or acid–base regulation or both.

Morphological changes occur in ionocytes of euryhaline teleosts during acclimation from fresh water to brackish water, seawater, or concentrated seawater (Figure 17.20b and c; reviewed by Komnick, 1986). Characteristically, these cells increase in number and size during seawater adaptation, whereas adaptation from seawater to fresh water induces degeneration and reduction in cell number. Under certain conditions that challenge ion regulation in fresh water, proliferation of ionocytes may also occur (Perry, 1997).

Epithelial sensory cells and supporting cells

The epidermis of fish is penetrated by nerve fibers (Chapter 20), some of which come to free endings (Whitear, 1971a, 1983), whereas others are associated with differentiated epidermal cells of a sensory nature (Whitear, 1971b, 1977). These sensory cells may be grouped, either as **neuromasts** (mechanoreceptors or electroreceptors), or, in many species of fish, as external taste buds or chemoreceptors (Whitear, 1977; Malinovsky, 1986; Reutter, 1986; Braun, 1998). The sensory cells of these organs are separated from the surrounding non-sensory epithelium by supporting cells, which produce secretions covering the surface of the sensory organs (Whitear, 1986a). The microscopic anatomy of the sensory systems of fishes is discussed further in Chapter 27.

In addition to sensory organs, many fishes from jawless fishes to teleosts have solitary epidermal cells, which appear to be chemosensory in nature, scattered throughout the epidermis (Figures 17.2, 17.3, 17.7, 17.9, and 17.11; Whitear, 1971b, 1986a, 1993; Kotrschal, 1995; Finger, 1997; Kotrschal *et al*., 1998). These cells have been called **oligovillous** cells in lampreys (Figure 17.3) and elasmobranchs (Whitear and Lane, 1983b; Kotrschal *et al*., 1997) and monovillous solitary chemosensory cells in teleosts (Figures 17.7, 17.9, and 17.11; Kotrschal *et al*., 1997); the most common name in current usage is solitary chemosensory cell (Kotrschal, 1995; Finger, 1997; Kotrschal *et al*., 1998). These cells are usually identified as chemoreceptors on the basis of the close resemblance of their fine structure to that of the gustatory cells of the appropriate species (Whitear, 1977; Chapter 27.3). Solitary chemosensory cells generally have one or more sensory villi at the apical end (penetrating to the epidermal surface between epithelial cells) and a nerve fiber at the proximal end. The cells are believed to constitute a water sampling system in the contexts of predator avoidance, habitat recognition, and in some species, finding food (Kotrschal, 1995; Finger, 1997; Kotrschal et al., 1997).

Solitary sensory cells of a non-chemosensory nature may also occur in the epidermis. For example, the multivillous cells of lamprey skin (Figure 17.3; Whitear and Lane, 1983c) may serve as photosensitive receptors in ammocoetes (Steven, 1951).

Merkel cells

Neuroendocrine Merkel cells are found in most vertebrate classes (Hartschuh et al., 1986; Zaccone et al., 1994), and have been reported from lampreys, dipnoans, and teleosts (Figures 17.3 and 17.9; Whitear, 1989; Zaccone et al., 1994). In fishes, Merkel cells are found in the skin and oral epithelium, and occur at various levels in the epidermis (Whitear, 1989; Zaccone et al., 1994). Merkel cells exhibit a range of cytological variation in different vertebrates, but show a number of common structural characteristics indicating that they belong to a homologous series (Whitear, 1989). They can be recognized by the relatively small ratio of cytoplasm to nucleus, the presence of dense-cored cytoplasmic granules, the characteristic finger-like peripheral processes that project between or indent adjacent epithelial cells, the formation of synapses with nerve fibers, and the formation of desmosomal contacts with epithelial cells (Hartschuh et al., 1986; Whitear, 1989; Zaccone et al., 1994). Although Merkel cells were once thought to function primarily in tactile reception (Hartschuh et al., 1986; Whitear, 1989), more recent evidence suggests that they have a multi-functional role as regulatory neuroendocrine cells (Zaccone et al., 1994). Iger et al. (1994b) noted the appearance of Merkel cells throughout the epidermis in carp exposed to cadmium in water, and suggested that this might be a specific response to that toxicant.

Luminescent organs

Luminescent organs or **photophores** (Chapter 5) are present only in marine fishes and are most extensively developed in deep-sea fishes, although they occur in some shallow-water species as well (Lagler et al., 1977). Bioluminescence or 'living light' in fishes is of two types (Harvey, 1957): (i) luminescence which arises from the presence of luminous bacteria living in the luminescent organs in a symbiotic relationship, and (ii) luminescence which arises from self-luminous cells of the fish. Luminous bacteria identified in the light organs of fishes have included *Photobacterium* and *Vibrio* species indistinguishable from free-living species, as well as bacteria which do not belong in any known taxa, but which share phylogenetic origins with *Photobacterium* and *Vibrio* (Haygood and Distel, 1993; Herring, 1993). Self-luminescence can be either light generated by the fish within its own tissues (intracellular luminescence) or by the discharge of a luminous secretion (extracellular luminescence). In some fishes, the bioluminescent substrates (luciferins) are apparently acquired by feeding on luminescent prey (Mallefet and Shimomura, 1995; Mensinger, 1995).

Luminescent organs, whether of the bacterial or self-luminescent type, are epidermal derivatives (Whitear, 1986a), and may occur almost anywhere on the body. Some species have only a few light organs, and others may have hundreds. Bacterial or extrinsic organs are frequently paired and are usually highly vascularized sacs (Bullock and Roberts, 1974). Their light shines continuously, but the light intensity can be controlled by the screening activity of **chromatophores** (page 293) or screen tissues, or by muscular movement. Intrinsic (self-luminescent) light organs may be simple or compound (Bullock and Roberts, 1974). Simple organs consist merely of modified epidermal cells without a lens or reflector, but usually with a pigmentary shield derived from dermal melanophores (page 294). The compound organs are usually embedded within the dermis, and have a structure bearing some similarity to the eye (Chapter 27.1), with an outer melanin layer, a reflecting guanine tapetum, and a central nodule of luminous cells (photocytes) overlaid by a cornea-like epidermis. A detailed description of these organs is beyond the scope of this chapter. However, their structure has been described by Bertin (1958) and Nicol (1969), and their physiology and function have been reviewed by Harvey (1957), Nicol (1969), Herring (1982), and Lombard (1997). Hastings (1996) reviewed the chemistry of bioluminescent systems in fishes and other organisms.

Intrusive cells

Within the epidermis are found several types of cells originating in other tissues. Included among these are chromatophores (page 293), neuromast cells derived from a migratory placode, and neurite processes (Whitear, 1986a). Some of the nerve fibers are connected to the innervated cells discussed on pages 286 and 287, whereas others come to free endings

(Whitear, 1971a, 1983). Nerves passing through the basement membrane into the epidermis lose the Schwann sheath (Chapter 20) and ramify between (or are wrapped in) the epithelial cells.

The epidermis also contains leukocytes (Chapters 22 and 26), which include lymphocytes (Figures 17.2, 17.3, 17.5, 17.8, 17.9, 17.16 and 17.18), macrophages, and various types of granulocytes (Roberts et al., 1971; Bullock and Roberts, 1974; Ferri, 1983; Peleteiro and Richards, 1985, 1990; Ferguson, 1989; Cross and Matthews, 1991; Davidson et al., 1993; Whitear and Moate, 1998). Leukocytes are most abundant between and above the basal layer epidermal cells but can occur at any level of the epidermis and may be exposed at the surface of damaged skin (Whitear, 1986a). Leukocyte numbers in the epidermis may increase greatly during the healing of skin wounds (Phromsithurak, 1977; Iger and Abraham, 1990), certain infectious diseases (Ferguson, 1989) or exposure to pollutants or other stressors (Iger et al., 1988, 1994a,b).

Intrusive cells of foreign origin, such as certain protozoan parasites, may be present in the epidermis (Ferguson, 1989). The origin of one distinctive intrusive cell, the **rodlet cell**, has sparked considerable controversy for more than a century (see e.g. Barber et al., 1979). The rodlet cell has been viewed by some as a sporozoan parasite under the name *Rhabdospora thelohani* (Laguesse, 1895; Anderson et al., 1976; Mayberry et al., 1979a,b), and by others as a normal cellular component of fish (Leino, 1974; Desser and Lester, 1975; Morrison and Odense, 1978; Mattey et al., 1979; Smith et al., 1995; Iger and Abraham, 1997).

Rodlet cells are found in many freshwater and marine teleost species and within many tissues (Leino, 1974; Morrison and Odense, 1978; Mayberry et al., 1979b), but they are usually associated with epithelia (Morrison and Odense, 1978; Barber et al., 1979). These cells have a very characteristic flask-shaped or pear-shaped appearance, with a nucleus at the wider end and acidophilic, often refractile 'rodlets' within membrane-bounded vesicles oriented longitudinally in the cytoplasm (Figure 17.21). A distinctive fibrillar 'capsule' is present in the cytoplasm beneath the plasma membrane. In epithelial tissue, mature rodlet cells are usually oriented parallel to the epithelial cells with the basal end containing the nucleus distal to the lumen or surface, but in connective tissue, no special orientation has been described (Morrison and Odense, 1978).

Proponents of the parasite hypothesis have cited as supporting evidence the rodlet cell's distinctive morphology (Mayberry et al., 1979b), its irregular occurrence within fish species (Barber et al., 1979), the detection of nucleic acid in its rodlets (Barber et al., 1979; Viehberger and Bielek, 1982), its association with certain pathologic conditions (Dawe et al., 1964; Anderson et al., 1976), and the apparent survival of the cell several days past the death of the host (Anderson et al., 1976). Those who believe the rodlet cell to be a normal host cell have noted its wide distribution in fish tissues and in fish species living at remarkably different temperatures and salinities (Leino, 1974; Morrison and Odense, 1978; Ferguson, 1989; Iger and Abraham, 1997). Supporters of the normal cell hypothesis also cite the presence of rodlet cells in larval or very young fish (Leino, 1974; Ferguson, 1989), the lack of a host inflammatory response directed at the cells (Leino, 1974; Desser and Lester, 1975; Mattey et al., 1979; Ferguson, 1989), the presence of desmosomes and junctional complexes between rodlet cells and adjacent epithelial cells (Leino, 1974; Desser and Lester, 1975; Mattey et al., 1979), and a similarity in DNA content between rodlet cell nuclei and teleost host cell nuclei (Barber and Westermann, 1985).

Various functions have been suggested for the rodlet cell, including involvement in ion transport and osmoregulation (Leino, 1974; Morrison and Odense, 1978), lubrication (Leino, 1974; Desser and Lester, 1975), sensory reception (Wilson and Westerman, 1967), and antibiotic effects and pH control (Leino, 1974, 1996). In experiments with trout and carp, Iger and Abraham (1997) found rodlet cells present in the epidermis of fish exposed to a variety of stressors but not in control fish. They concluded that these cells might be involved in the nonspecific defense mechanisms of the skin (and perhaps other epithelia) via the discharge of proteolytic enzymes from the rodlets. Other authors (Leino, 1996; Reite, 1997; Dezfuli et al., 1998) also noted a recruitment of rodlet cells associated with damage to external or internal epithelia by parasites or other insults, and concurred that rodlet cells likely have a defensive function.

Epidermal secretions

The four main sources of secretion in fish skin are epithelial cells (page 272), goblet cells (page 277), sacciform cells (page 281), and club cells (page 282), although other epidermal cells of more limited distribution in fish species (e.g. thread cells of hagfish,

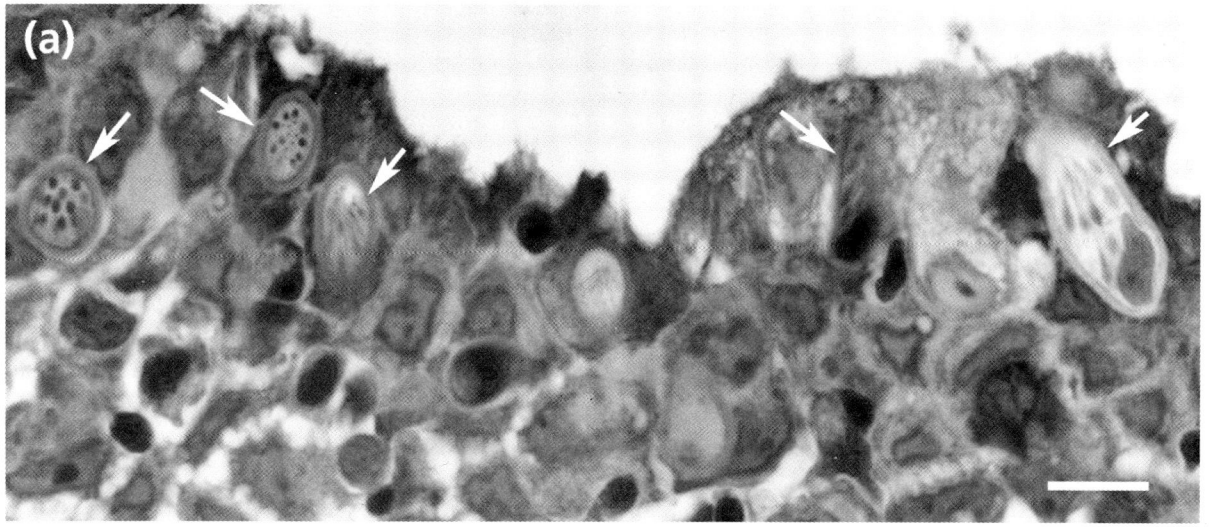

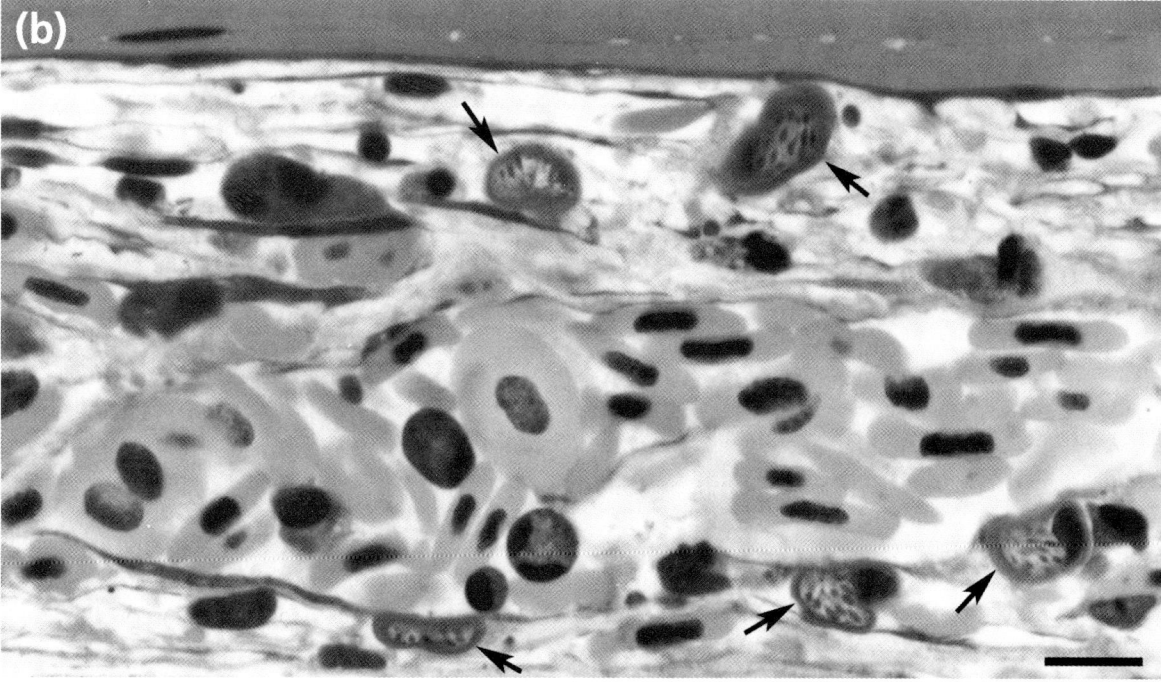

Figure 17.21 Rodlet cells (arrowed) in the skin of a goldfish (*Carassius auratus*) with a skin infection of the bacterium *Aeromonas salmonicida* (bacteria not shown). The fibrillar 'capsule' surrounding the cell beneath the plasma membrane, the longitudinally oriented cytoplasmic vesicles (which contain rodlets visible by electron microscopy), and the nucleus at one end of the cell are characteristics of this cell. Scale bars = 8 μm. Methylene blue-azure II stain. (a) Rodlet cells in the epidermis. In the epidermis, rodlet cells are usually oriented with the end containing the nucleus distal to the surface. (b) Rodlet cells surrounding a blood vessel beneath a scale in the stratum spongiosum of the dermis. The orientation of rodlet cells is variable in connective tissue.

page 284) also contribute. All four principal secretory cell types are represented in the polypterids (Whitear, 1981), whereas the paddlefish *Polydon* possesses only epithelial cells (Weisel, 1975).

The multiple sources of epidermal secretions and the difficulty of collecting them uncontaminated by cellular contents have complicated the interpretation of biochemical analyses. Epidermal secretions are mucous and proteinaceous in varying proportions (Whitear, 1986a). It is convenient to follow the nomenclature of Reid and Clamp (1978), who divided the glycoconjugates (carbohydrate covalently linked to protein) into two broad classes, proteoglycans and glycoproteins, on the basis of their chemical and structural characteristics and their principal locations in the body. The principal component of mucus, and the constituent that gives the secretion its characteristic physicochemical properties, is a glycoprotein of high molecular weight (Reid and Clamp, 1978). Fish mucus glycoproteins may be neutral, but are often made acid by the presence of sialic acid (a carboxylated monosaccharide) or sulfated monosaccharides (Whitear, 1986a; Shephard, 1994). In addition, the surface mucus or 'slime' of fish may contain substances such as crinotoxins (Cameron and Endean, 1973), calmodulin (Flik *et al.*, 1984; Zaccone *et al.*, 1989), pheromones (Hara, 1986), and a variety of antipathogenic substances such as fatty acids, immunoglobulins, complement components, lectins, lysozyme, proteolytic enzymes, and other antimicrobial peptides and proteins (Ingram, 1980; Alexander and Ingram, 1992; Itami, 1993; Shephard, 1994; Lemaitre *et al.*, 1996; Cole *et al.*, 1997; Buchmann and Bresciani, 1998).

Because of the confusion regarding the source(s) of external mucus, histochemical studies of secretory cell contents are generally more rewarding than biochemical studies of surface mucus. According to histochemical analyses, the mucus of goblet cells (page 277) may consist of sialylated, sulfated, or neutral glycoprotein, with taxonomic status or life-style providing few clues regarding the type of mucus produced (Gona, 1979; Whitear, 1986a). In some fishes, goblet cells show a mixed histochemical reaction, suggesting the production of more than one type of glycoprotein (Zaccone, 1983; Whitear and Mittal, 1984). The chemical nature of the mucus produced by goblet cells may shift during larval metamorphosis (Ottesen and Olafsen, 1997), upon exposure of fish to toxicants (Roy, 1988; Garg and Mittal, 1993) or during other environmental changes such as alterations in salinity (Solanki and Benjamin, 1982). Serous goblet cells produce basic protein rather than glycoprotein (Whitear, 1986a).

The external mucous layer known as the cuticle (page 277; Figures 17.11, 17.12b and 17.13; Whitear, 1970) is largely secreted by the surface epithelial cells, but also may be mixed with and modified by secretions from other cells such as goblet cells and sacciform cells (Whitear, 1986a). The cuticle corresponds to the secreted layer found even on internal epithelia that are only exposed indirectly to the environment (Whitear, 1986a). The glycoproteins secreted by surface epithelial cells may be neutral or acid (Carmignani and Zaccone, 1975; Bereiter-Hahn *et al.*, 1980; Mittal *et al.*, 1980; Whitear and Mittal, 1984; Whitear, 1986a; Singh and Mittal, 1990; Ueda *et al.*, 1994). Lipids are also produced by the epithelial cells in some species (Mittal and Banjeree, 1975; Mittal and Whitear, 1979). The cuticle may range from a thin layer barely detectable by electron microscopy to a complex layer 10 μm to 50 μm thick (Whitear, 1986a). In many fishes the cuticle is constantly renewed (Lester, 1972; Whitear, 1986a), but the thick cuticle of the stonefish or gurnard may be retained for many months (Fishelson, 1973). The cuticle may provide protection from microbial colonization (page 277), abrasion (Whitear, 1986a), or, in the case of symbiotic anemone fishes (*Amphiprion*), protection from the stings of the anemones which they inhabit (Lubbock, 1980).

The staining reactions of sacciform cell contents are as varied as their appearance by electron microscopy (page 281), but in general, the secretions are predominantly proteinaceous (Mittal *et al.*, 1981; Whitear, 1986a; Pickering and Fletcher, 1987). The sacciform cells in the epidermis of a few fishes have been reported to show weak reactions for glycoproteins (Mittal and Agarwal, 1977; Bullock, 1980; Mittal *et al.*, 1981, 1994; Whitear, 1986a). Additional histochemical studies have shown varying other constituents such as cholesterol (Mittal and Agarwal, 1977; Zaccone, 1981), nucleic acid (Whitear, 1986a), enzymes (Zaccone, 1982), the bioactive amine serotonin (Zaccone *et al.*, 1986; Fasulo *et al.*, 1993), and bioactive peptides (Zaccone *et al.*, 1987, 1988; Fasulo *et al.*, 1993). Sacciform cells may be analogous to the granular glands of amphibians, which produce crinotoxic or repellent substances (Whitear, 1986a; Fox, 1986), but sacciform cell secretions may have additional protective or regulatory functions as well (page 281; Zaccone, 1981; Fasulo *et al.*, 1993; Mittal *et al.*, 1994).

The club cells of ostariophysans produce an alarm pheromone (page 283) that is a pteridine derivative, probably hypoxanthine-3(N)-oxide (Smith, 1992). Certain non-ostariophysan fishes that lack club cells nevertheless produce an alarm substance that is released from injured skin and recognized by conspecifics and certain other fish species (Smith, 1992; Brown and Smith, 1997). In some fishes lacking club cells, sacciform cells may be the source of the alarm substance (Smith, 1982b).

Club cells produce substances in addition to alarm pheromones, even in ostariophysan species exhibiting an alarm reaction (Ralphs and Benjamin, 1992). The sulfated glycosaminoglycans (chondroitin and keratan sulfate) present in the club cells of some ostariophysans may form a protective 'jelly bandage' to aid in the healing of skin wounds (Ralphs and Benjamin, 1992). Club cells of certain ostariophysans (marine catfishes) produce proteinaceous toxins when the fish are threatened or injured (Al-Hassan *et al.*, 1987; Shiomi *et al.*, 1988); some of these substances may also function to promote wound healing (Al-Hassan *et al.*, 1987). Lectins, which may exhibit antipathogenic activities such as agglutination or precipitation (Alexander and Ingram, 1992), have also been identified from the club cells of some ostariophysans (Danguy *et al.*, 1991). A non-antibody hemagglutinin has been demonstrated in the club cells of a non-ostariophysan, the eel *Anguilla japonica* (Suzuki and Kaneko, 1986), but its function has not been determined. Serotonin has been detected in the club cells of ostariophysan and non-ostariophysan species (Zaccone *et al.*, 1990). The serotonin may have the ability to affect pheromonal or other secretory functions of the cells (Zaccone *et al.*, 1990).

The granular cells of lampreys contain proteinaceous secretions bounded by membranes (Whitear, 1986a). These secretions might have protective functions (page 283).

The two major cell types in hagfish slime glands, thread cells and mucous goblet cells, contribute to the formation of the mass of viscous mucus produced by stressed hagfish (Koch *et al.*, 1991). The cells are released to the epidermal surface by extrusion through a pore in the gland, and rupture on contact with water. Upon release from the cell, the enormous thread cell biopolymer (page 284) uncoils and interacts with goblet cell mucins and seawater, thereby facilitating hydration and modulating the viscoelastic and cohesive properties of the resultant mucus (Koch *et al.*, 1991). The goblet cells contain acid or neutral glycoproteins (Whitear, 1986a), whereas intermediate filament subunits of the thread cell biopolymer appear to be homologues of mammalian epidermal keratin chains (Koch *et al.*, 1995).

Most venoms are complex secretions but usually have a protein or polypeptide component (Russell, 1969; Whitear, 1986a; Halstead, 1988; Gwee *et al.*, 1994; Hopkins and Hodgson, 1998; Lopes-Ferreira *et al.*, 1998). Nevertheless, venoms may be variable in composition even in closely related species (Whitear, 1986a; Halstead, 1988).

Dermis

In the primitive vertebrates, as in worms, the dermis consists essentially of two sets of collagen fibers arranged in opposing geodesic spirals around the body (Whitear, 1986b). This bias sleeve has the ability to bend without wrinkling. Although most fishes require the accommodation of lateral flexions only, the hagfishes are so flexible that they can tie themselves in knots (Strahan, 1963). The overall thickness of the dermis varies with species, position on the body, and life stage (Whitear, 1986b).

The dermis of most fishes is divided into two major layers (Figure 17.1). The upper (outer) layer, the stratum spongiosum or stratum laxum, is a loose network of connective tissue, whereas the lower layer, the stratum compactum, is a dense layer consisting primarily of orthogonal collagen bands. The relative thicknesses of the two layers depends on the presence or absence of scales (Whitear, 1986b); the separation between dermal layers may be indistinct in non-scaled areas (Hawkes, 1974a). The stratum compactum may be reduced in fin tissue (Sharples and Evans, 1996). In lampreys and some primitive actinopterygians, the dermis consists primarily of dense collagenous tissue (Chapter 5, Figure 5.2; Weisel, 1975; Lethbridge and Potter, 1982).

The dermis is separated from the epidermis by an acellular basement membrane (also called the basal lamina or adepidermal membrane). The deep face of the dermal stratum compactum is bounded by a single layer of cells termed the dermal endothelium. Immediately below the dermal endothelium, and separating the dermis from the underlying skeletal musculature, is the hypodermis or subcutis, a layer of well-vascularized loose connective tissue.

As in other vertebrates, collagen is a major component of fish dermal tissue. Although most fish collagen is similar to mammalian type I collagen (Uitto, 1986), some differences in structure and chemical composition have been observed (Whitear, 1986b; Matsui *et al.*, 1990; Nagamalleswari and Joseph, 1991; Kimura, 1992).

Basement membrane

The basement membrane (Figures 17.1 and 17.19) is considered to be a dermal element, although epidermal participation is necessary for its formation (Nadol and Gibbons, 1970; Phromsuthirak, 1977;

Whitear, 1986b). By light microscopy, the presence of the basement membrane is frequently demonstrated with staining by the periodic acid-Schiff (PAS) method (Clark, 1981; Whitear, 1986b), but little structural detail is visible. By electron microscopy, a lucent adepidermal space beneath the basal epidermal cells is crossed by fibrillar material (anchoring filaments) which connect hemidesmosomes to the basement membrane (Whitear, 1986b). The basement membrane is an electron-dense layer, with its thickness depending on the fish species and the position on the body. Additional anchoring filaments or fibers connect the basement membrane to the collagenous tissue beneath. At intervals the basement membrane is penetrated by electron-lucent channels, and it is breached where nerves enter the epidermis. In species with capillaries penetrating the epidermis (e.g. periophthalmids; Figure 17.10), the basement membrane is inflected and accompanies the capillary throughout its course (Whitear, 1986b). As far as is known, the components of the basement membrane complex in fish are the same as those in other vertebrates (Whitear, 1986b; Krieg and Timpl, 1986).

Basement membranes have multiple functions (Krieg and Timpl, 1986). They act as a filtration barrier and control the passage of cells and molecules between different tissues. They are also thought to play a role in the regulation of morphogenesis and wound healing, and they serve as attachment sites for epithelial cells and other cells.

Stratum spongiosum

The supporting tissue of the stratum spongiosum (Figure 17.1) is a loose network of collagen and reticulin fibers. It is contiguous with the basement membrane, and many of its finer fibers insert there. The stratum spongiosum contains the scales (page 296), and a variety of vascular (Chapter 22) and neural (Chapter 20) components. Cellular elements of this layer include fibroblasts, pigment cells (page 293), leukocytes (Chapter 26), and the cells of the scale synthesizing tissues (page 296).

Stratum compactum

The stratum compactum is always present from an early stage in ontogeny (Whitear, 1986b). The major component of this layer (Figure 17.1) is the collagen bundle, which forms a dense matrix above the hypodermis. The collagen fibers are usually highly ordered in a series of layers at right angles to each other. Dermal fibroblasts, which appear as small, darkly staining cells in mature fish, are distributed between the collagen fibers. Few other cells are present in the stratum compactum, although pigment-bearing cells and mast cells may occasionally be seen (Bullock and Roberts, 1974; Hawkes, 1974a). At regular intervals, the stratum compactum is pierced by vertical columns of collagen bearing nerves and blood vessels passing up to the stratum spongiosum.

The stratum compactum is important in locomotor activity (Whitear, 1986b). The 'plywood' structure of this dermal layer provides structural rigidity from stresses impinging on the skin and yet allows flexibility because the sheets of fibers can slip past one another during deformation. As a fish swims, the elasticity in the stretched skin of the convex side of an undulation assists in unbending in the early stages of the next contraction. The stratum compactum of the dermis acts as a tendon in parallel with the muscles.

Dermal endothelium

The single layer of cells lining the deep face of the stratum compactum (Figure 17.1) was termed the dermal endothelium by Whitear *et al.* (1980). It has been found in lampreys and a number of teleost species. This layer consists of modified fibrocytes connected by desmosomes, and may be bounded by a basement membrane on one or both sides. The dermal endothelium forms a complete sheet over most of the head and body except where it is penetrated by nerves or blood vessels, or by the insertions of myosepta (Chapter 5, Figures 5.1 and 5.4; Chapter 19). It apparently does not extend to the fins or opercular valves. The cells have the general character of fibrocytes but usually have less endoplasmic reticulum and possess a large number of caveolae intracellulares. The caveolae sometimes join to form a channel across a cell. Hemidesmosomes may be present on the upper and lower surfaces of the cell. Whitear *et al.* (1980) speculated that the dermal endothelium might function in the regulation of the passage of fluids between the stratum compactum and hypodermis.

Hypodermis

The upper portion of the hypodermis (Figure 17.1; Chapter 5, Figures 5.2, 5.4 and 5.5), immediately below the dermal endothelium, is generally occupied by the deep chromatophore layer (Whitear, 1986b).

Blood vessels, and nerve bundles enclosed in **perineurium**, also lie at this level. Beneath are the swollen cells containing the hypodermal fat, interspersed with loose connective tissue, blood vessels, and thin-walled vessels that might be lymphatic capillaries (Whitear, 1986b). Proximally the loose connective tissue merges into the **myocomma**, or contains dermal bone. The flexibility of the hypodermis allows considerable movement of the integument between the stratum compactum and the superficial myotomes (Bullock and Roberts, 1974).

Specialized dermal elements

Chromatophores

Integumentary colors are primarily dependent on the presence of pigment cells or chromatophores in the skin. In fishes from jawless fishes to teleosts, these cells occur in the dermis (Figure 17.1), where they may be found in the stratum spongiosum, in the hypodermis, or in both (Blackstad, 1963; Bullock and Roberts, 1974; Hawkes, 1974b; Lethbridge and Potter, 1982; Schliwa, 1986; Fujii, 1993a,b). In the stratum spongiosum of non-scaled skin, chromatophores are generally located in the upper portion, separated from the basal lamina by collagen fibrils (Hawkes, 1974b; Fujii, 1993a,b), whereas they are also located beneath the scales in scaled skin (Figure 17.1; Hawkes, 1974b). Epidermal chromatophores are found sporadically (Figure 17.7).

Derived from the neural crest, chromatophores in all vertebrates including fishes have a common embryonic origin. They are now categorized as paraneurons (Fujii, 1993a). They are usually classified according to the color of the pigment they contain. In fishes and other poikilothermic vertebrates, chromatophores are usually categorized into five groups (Fujii, 1993a,b): melanophores (black to brown pigments); erythrophores (red or yellowish pigments); xanthophores (primarily yellow pigments); leucophores (colorless, reflective pigments); and iridophores (colorless, reflective pigments). In addition, a chromatophore with blue pigmentary organelles has been identified in the skin of two callionymid (dragonet) species, *Synchiropus splendidus* and *S. picturatus*; the name 'cyanophore' has been applied to this cell (Goda and Fujii, 1995).

Perhaps because of their shared ontogenetic origin with neurons, most chromatophore types are dendritic cells (Fujii, 1993b), with multiple branched or unbranched processes extending from the cell body (Figure 17.22a). In contrast to the neurons, however, the dendritic processes of chromatophores develop parallel to the plane of the skin, providing effective revelation of the pigmentary colors when the skin is viewed from outside. Dendritic chromatophores include the melanophores, erythrophores, xanthophores, and leucophores. In these cells, pigment-containing organelles (chromatosomes) migrate centripetally (aggregation) or centrifugally (dispersion) in response to various nervous or hormonal signals received by the cells (Figure 17.22b). By contrast, most iridophores lack dendrites and assume a round, oval, or polygonal shape (Figure 17.22a), although some iridophores develop dendritic processes (Fujii, 1993a,b). Chromatophores may vary considerably in diameter, from 10 to 20 μm for the erythrophores in squirrelfish to at least 200 μm for the giant melanophores of the black tetra *Gymnocorymbus ternetzii*, which are easily visible to the naked eye (Schliwa, 1986).

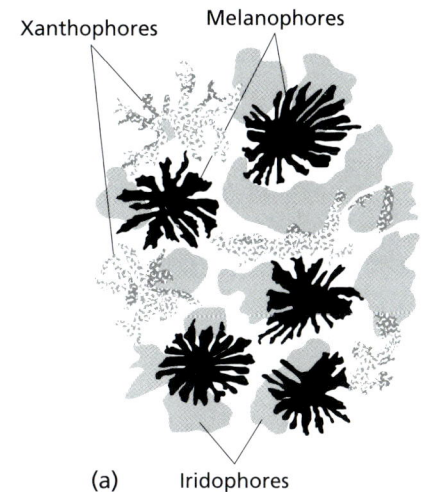

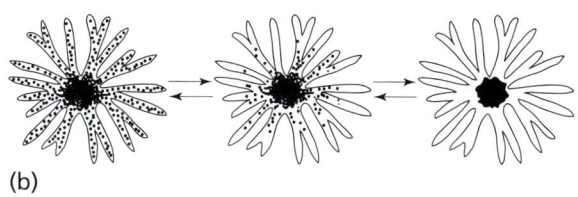

Figure 17.22 (a) Common dendritic and non-dendritic chromatophore types, as viewed from above the plane of the skin of a flounder, *Paralichthys*. (b) Diagram showing the pigment-dispersion response (left) and pigment-aggregation response (right) of chromatosomes in a dendritic chromatophore. (Sources: (a) after Lagler *et al.*, 1977. (b) from Fujii, 1993b.)

Melanophores are the most common dendritic chromatophore in fish. These uninucleate or binucleate cells occur almost everywhere in the skin where any sign of a dark shade is recognizable, and may even be present in the epidermis (Figure 17.7; Imaki and Chavin, 1975; Fujii, 1993a). Melanophores are usually the largest of the dendritic chromatophores. Because of the deposits of melanoprotein inside them, the chromatosomes within melanophores are called melanosomes. Mature melanosomes are round or slightly ellipsoid electron-dense structures with a diameter of about 0.5 µm (Fujii, 1993a) and are bounded by a single membrane; immature melanosomes (premelanosomes) are usually more slender in shape. The dark brown or black coloration makes melanosomes readily visible in routine histological sections (Figure 17.23). In paler mutants or varieties of normally dark species such as the medaka *Oryzias latipes*, the melanosomes are devoid of melanin deposits, although they aggregate and disperse in a manner similar to normal melanosomes (Sugimoto *et al.*, 1985).

Melanophores play a leading role in the rapid changes of color exhibited by many species of fish. Compared with other types of dendritic chromatophores, the melanophores usually display a higher degree of rapid aggregation and dispersion of chromatosomes in response to nervous and hormonal stimuli (Fujii, 1993a).

Xanthophores and erythrophores are the dendritic chromatophores that contribute reddish and yellowish pigments to the coloration of the skin (Figure 17.22a). They are generally smaller than melanophores, but otherwise their morphological features are fundamentally the same (Fujii, 1993a). Usually they are present only in the dermis, although epidermal xanthophores have been observed in an Antarctic blenny, *Trematomus bernacchii* (Obika and Meyer-Rochow, 1990). The chromatosomes, which are limited by a single membrane and contain carotenoids and pteridines, are commonly called **xanthosomes** or **erythrosomes** based on their color, but are sometimes called **carotenoid vesicles** or **pterinosomes** if the chemical identity of their contents is known (Fujii, 1993a). The dimensions of pterinosomes are similar to those of melanosomes, and their contents are visible as a concentric lamellar structure or fine filaments by electron microscopy. Carotenoid vesicles have a barely visible limiting membrane and sometimes look like cytoplasmic oil droplets.

In certain groups of fish within the Osteichthyes such as the Cyprinodontiformes, dendritic chromatophores known as leucophores have been described in the dermis of the skin (Fujii, 1993a). In contrast to the light-absorbing chromatophores (melanophores, xanthophores, and erythrophores), leucophores are light-reflecting chromatophores. Although they are often confused with another light-reflecting chromatophore, the iridophore (see below), they can be distinguished on the basis of the properties of their light-reflecting organelles and their mechanisms of reflecting light (Fujii, 1993b). Leucophores are similar in size to (and sometimes larger than) other dendritic chromatophores, but usually have fewer dendritic processes (Fujii, 1993a). The light-reflecting organelles or leucosomes are globular structures enclosed in a single limiting membrane; by electron microscopy these organelles may appear hollow (Obika, 1988) or contain fibrous or electron-dense material (Menter *et al.*, 1979). Leucosomes scatter light in all directions,

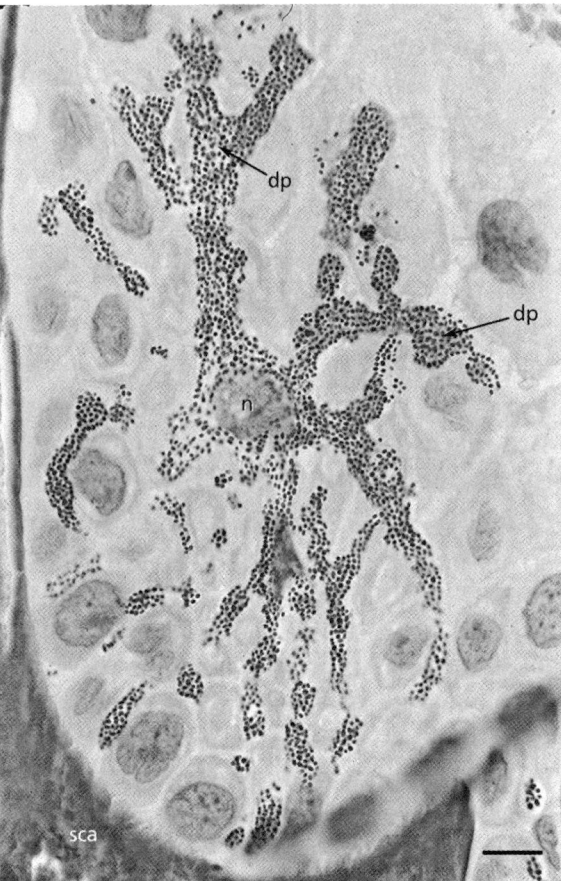

Figure 17.23 Histological section of fin tissue from a smooth dogfish shark, *Mustelus canis*, showing a melanophore, the most common type of dendritic chromatophore observed in histological sections. A central nucleus (*n*) is visible, as are dendritic processes (*dp*) containing numerous melanosomes, the round to ellipsoid organelles containing black to brownish melanin pigment. A portion of a placoid scale (*sca*) can be seen in the section. Scale bar = 10 µm. Hematoxylin and eosin stain. **(See also Colour Plate 12.)**

but the chemical nature of their contents has not been elucidated (Fujii, 1993b). Like the chromatosomes in light-absorbing chromatophores, leucosomes move centripetally or centrifugally within the leucophores. However, during coordinated changes in the overall darkness and lightness of a fish, the response of leucophores to various stimuli understandably may be opposite to that seen in light-absorbing chromatophores (Fujii, 1993a).

Iridophores are non-dendritic or occasionally dendritic chromatophores commonly found in the dermis wherever there are whitish areas of skin (Figure 17.22a); both non-dendritic and dendritic types may be found in the same fish (Hawkes, 1974b). In whitish or silvery skin areas, iridophores may be so densely packed that there is almost no space for other types of chromatophores. Iridophores contain large crystalline platelets that tend to form stacks in the cytoplasm, with uniform spacing between adjacent platelets within a stack (Fujii, 1993a,b). Each platelet is a **monoclinic crystal** composed mainly of guanine, but it may also contain other purines such as hypoxanthine and uric acid (Fujii, 1993b). Although the platelets are transparent, the presence of these highly refractive structures (with refractive indices ≥ 1.83) in the watery cytoplasm renders the iridophores highly reflective (Fujii, 1993b). Despite the fact that iridophore platelets of many fishes do not exhibit motile responses of any kind to stimulation, iridophores play an integral part in the high reflectivity which is apparent in the whitish or metallic appearance of the sides or belly skin in many species (Denton, 1970; Land, 1972).

Some of the non-dendritic iridophores in certain teleostean species have been found to be motile, with the essence of the motility being a relatively rapid change in the distance between adjacent platelets in a pile, or a change in the angle of inclination of the platelets (Oshima and Nagaishi, 1992; Fujii, 1993a,b). Some dendritic iridophores are also motile; in these cells the platelets aggregate and disperse in much the same manner as the chromatosomes in other dendritic chromatophores, but the movement of the platelets is slow (Iga et al., 1987; Fujii et al., 1991).

The great variety of skin hues observed macroscopically in fishes is due in large part to the arrangement of chromatophores (Fujii, 1993a). When chromatophores of different types overlap one another, a subtractive mixture of component pigments is created, akin to the mixing of paints. When differently colored chromatophores are placed side by side, colors are produced by an additive mixture of pigments, as in offset printing. The body colors of many species result from the integration of these effects. A specialized interrelationship between chromatophores has been identified in coho salmon *Oncorhynchus kisutch*, in which iridophores are encompassed by the dendritic processes of subjacent melanophores (Hawkes, 1974b). By movement of the melanosomes in and out of the melanophore processes, the melanophore can modify the bright colors of the iridophore above it by obscuring its light-reflecting capacity. This is similar to the dermal chromatophore unit involved in rapid color changes in amphibians (Bagnara et al., 1968).

Rapid changes in hue and pattern of coloration due to motile activities of chromatophores, called physiological color changes, are primarily controlled by the nervous (Chapter 20) and endocrine (Chapter 25) systems (Fujii, 1993a,b). In at least some dendritic chromatophores, the mechanism of chromatophore motility involves the microtubules that radiate from the cell center toward the periphery of the dendritic processes. Although the microtubules apparently function as cytoskeletal elements that maintain the stellate shape of the cells (Fujii, 1993a,b), they also provide a framework for the movement of the pigment granules (Chen and Wang, 1993; Fujii, 1993a,b). In fish melanophores, a **kinesin**-related end motor drives the granules along a radial array of uniformly polarized microtubules away from the cell center during dispersion (Rodionov et al., 1991), and a **dynein**-related motor drives the granules in the opposite direction during aggregation (Nilsson and Wallin, 1997). A microtubule-independent motility system involving actin filaments has also been identified in fish melanophores; this system is required for achieving uniform distribution of dispersed pigment granules and is functionally coordinated with the microtubule system (Rodionov et al., 1998). Cytoskeletal systems other than microtubules and actin may also be involved in the movement of pigment granules in certain dendritic chromatophores (Wang et al., 1997).

Morphological color changes, accompanied by changes in the net amount of pigmentary materials in the skin, generally proceed more slowly than physiological color changes (Fujii, 1993a). Examples of morphological changes include the remarkable alterations in color and pattern that occur during maturation of some marine angelfishes (Pomacanthidae), and the more subtle changes that occur in the medaka *Oryzias latipes* during prolonged adaptation to a dark or light background. These phenomena may result from changes in the number and size of pigment cells, and changes in the amount of

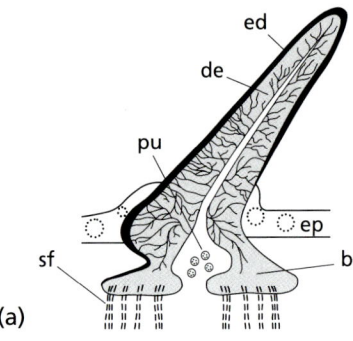
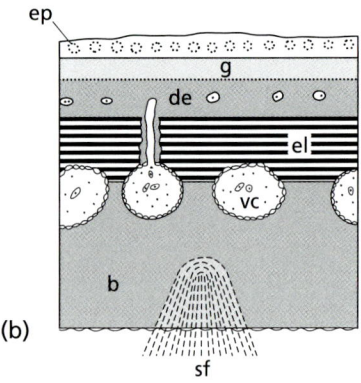
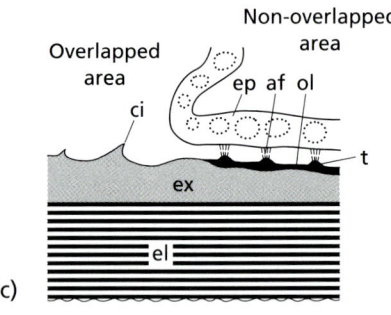
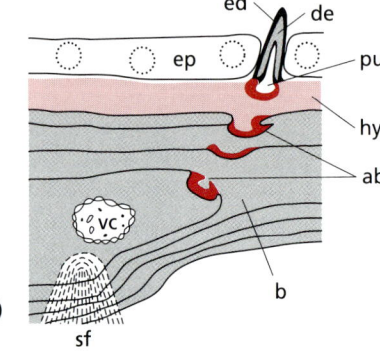

Figure 17.24 Schematic drawings comparing major structural layers and components of fish scales and scutes shown in cross section (not drawn to scale). (a) Placoid scale (odontode) of a chondrichthyan. (b) Portion of a ganoid scale of a polypterid. (c) Portion of an elasmoid scale of a teleost (cichlid). (d) Portion of a scute of a teleost (the armored catfish *Corydoras*) with an odontode projecting from the surface. Abbreviations: *ab* attachment bone; *af* anchoring collagen fibers; *b* bone; *ci* circulus; *de*

pigment in the cells (Fujii, 1993a; Sugimoto, 1993; Sugimoto and Oshima, 1995). When present, epidermal melanophores can play a large role in physiological color changes through the transfer of melanin to other cells in the epidermis (Fujii, 1993a). Pigmentation due to the deposition of carotenoids in erythrophores or xanthophores can be influenced by dietary conditions, because fish cannot synthesize carotenoids (Fujii, 1993a).

Scales

The literature on the morphology, ontogeny, and regeneration of fish scales is extensive, and has been reviewed by Van Oosten (1957), Whitear (1986b), Sire (1990), Raschi and Tabit (1992), Bereiter-Hahn and Zylberberg (1993), and Huysseune and Sire (1998), among others. The general structure and orientation of the various fish scale types are described in Chapter 5.

The tooth-like placoid scale (dermal denticle or odontode) of chondrichthyans (Figure 17.24a; Chapter 5, Figures 5.3 and 5.6; see also Chapter 18) is fixed directly in the dermis by anchoring collagen fibers (called Sharpey's fibers) attached to the acellular bony basal plate (Reif, 1980). The spine, which erupts through the epidermis in first-generation scales shortly before hatching (Reif, 1980), consists of dentine covered by enameloid (Van Oosten, 1957), with perhaps a thin exterior layer of true enamel (Huysseune and Sire, 1998). The pulp cavity enclosed by the spine may be subdivided into branching canals from which radiate numerous branching canaliculi called dentinal tubules (Van Oosten, 1957). The pulp cavity maintains contact with the superficial layer of the dermis via nerves, blood vessels, and lymph channels, which enter through openings in the basal plate (Van Oosten, 1957).

The development of first-generation scales in the shark *Scyliorhinus canicula* was described by Reif (1980). The basal layer of epidermal epithelial cells ('dental epithelial cells') becomes cuboidal, and mesenchymal cells accumulate immediately below the epidermis at particular sites. Each accumulation of mesenchymal cells organizes into a dermal papilla, which then invaginates slightly into the basal layer of

dentine; *ed* enameloid; *el* elasmodine; *ep* epidermis; *ex* external layer; *g* ganoine; *hy* hyaloine; *ol* outer limiting layer; *pu* pulp cavity; *sf* Sharpey's fibers; *t* tubercle; *vc* vascular canal. (Sources: (a) Modified from Bertin, 1958, (b) After Sire, 1990. (c) Modified from Sire (1988, 1990). (d) After Sire and Huysseune, 1996; reproduced with permission from Blackwell Science Ltd.)

the epidermis, forming a dental papilla surrounded by a dental organ comprised of basal epidermal cells. The dental papilla and dental (enamel) organ sink deeper into the dermis, and enameloid and dentine are deposited and mineralize. The basal plate of the scale containing the anchoring fibers is formed by osteoblasts, which are derived from the fibrous layer of the dermis. Subsequently, the scale erupts from the epidermis. During the life of the shark, placoid scales are continually shed (by resorption of the anchoring collagen bundles in the dermis) and replaced. Scales that regenerate after skin injury may differ from the original scales in size, shape, and orientation (Reif, 1978).

Ganoid scales are thick juxtaposed osseous plates anchored into the dermis by numerous large collagenous Sharpey's fibers (Figure 17.24b; Chapter 5, Figure 5.7; Chapter 18). The classical ganoid scale of polypterids is composed of two fused parts (Sire, 1990). The superficial part is a vascular layer of dentine (Chapter 18) covered by a hypermineralized layer of ganoine (Chapter 18), and is considered to be derived from dental units (odontodes) that have fused together. The lower part is a thick vascular osseous basal plate beneath a stack of thin collagenous layers; this plate constitutes a pseudo-lamellar bone and represents the dermal bone component of the scale. In lepisosteids, the dentine layer is absent, and the ganoine is in direct contact with the osseous basal plate (Kerr, 1952).

The development of ganoid scales in polypterids has been well studied (see e.g. Sire et al., 1987; Sire, 1989, 1990, 1995; Huysseune and Sire, 1998). The first elements of the ganoid scale matrix are deposited in dermal papillae close to the epidermis. The structure of this matrix, which consists of small patches of woven-fibered material, resembles that of the external layer in teleost elasmoid scales (see below). Below this tissue anlage, successive layers of collagen fibers are deposited in a plywood-like arrangement to form **elasmodine** (Figure 17.24b) similar to that in teleost scales (see below); the elasmodine later mineralizes incompletely from the surface downward. Subsequent to elasmodine formation, the tissues characteristic of the ganoid scale are deposited. The dentine layer develops at the upper surface of the elasmodine layer. This layer comprises a vascular network surrounded by woven-fibered osseous material. When the dentine layer reaches the epidermis, ganoine is deposited on its surface by epidermal cells. At about the same time that ganoine deposition begins, the thick bony basal plate starts to form around vascular cavities on the lower surface of the elasmodine layer. The basal plate rapidly thickens by the deposition of pseudo-lamellar fibered bone. In regenerating ganoid scales, no dentine and elasmodine layers are formed, and ganoine is deposited directly on the surface of the regenerated bony plate (Huysseune and Sire, 1998).

Despite a great diversity in size and shape, the elasmoid scales (Chapter 5, Figures 5.8 and 5.9) of teleosts and certain relict bony fishes are relatively homogeneous in orientation and microscopic structure (Sire, 1990). Elasmoid scales are inserted into oblique pockets (Figure 17.25a) in the stratum spongiosum of the dermis and usually overlap each other (exceptions noted in Chapter 5), with the non-overlapped posterior region covered by a fold of epidermis. The structure of the teleost elasmoid scale is characterized by two distinct parts: a thin superficial layer, termed the external layer or osseous layer, and a thick layer of elasmodine usually called the basal plate or fibrillary plate (Figures 17.24c, 17.25b and 17.26; Fouda, 1979a; Schönbörner et al., 1979; Yamada and Watabe, 1979; Sire, 1990).

The external layer averages 3 μm in thickness (Sire, 1990) and has regular elevations (Figures 17.25b and 17.26) that constitute the circuli (Chapter 5, Figure 5.8) at the scale surface. The well-mineralized matrix of the external layer resembles woven-fibered bone (Sire, 1990). Ontogenic studies have indicated that the external layer is laid down at one time during scale development and does not thicken thereafter (Sire and Géraudie, 1983). In primitive teleost fishes (osteoglossids, elopids, clupeids, mormyrids, and gymnarchids), cells are trapped in the matrix of the external layer, whereas in more advanced acellular-boned teleosts, no cells are trapped in this layer (Meunier, 1987).

The basal plate is deposited below the external layer and thickens during the entire life of the fish, reaching a thickness of more than 150 μm in adult specimens (Sire, 1990). The basal plate is partially mineralized and its organic matrix consists of a stack of thick sheets or layers of collagen fibrils that are regularly disposed (Figure 17.25d). In each layer, the fibrils have the same orientation, but the orientation changes from one layer to another, resulting in plywood-like elasmodine (formerly isopedine; Chapter 18). Cells have been observed within these collagenous layers in osteoglossids and some characoids, but the elasmodine is acellular in other teleosts examined (Meunier, 1987). A few cyprinids such as the goldfish *Carassius auratus* have sheets of vertically oriented collagen in

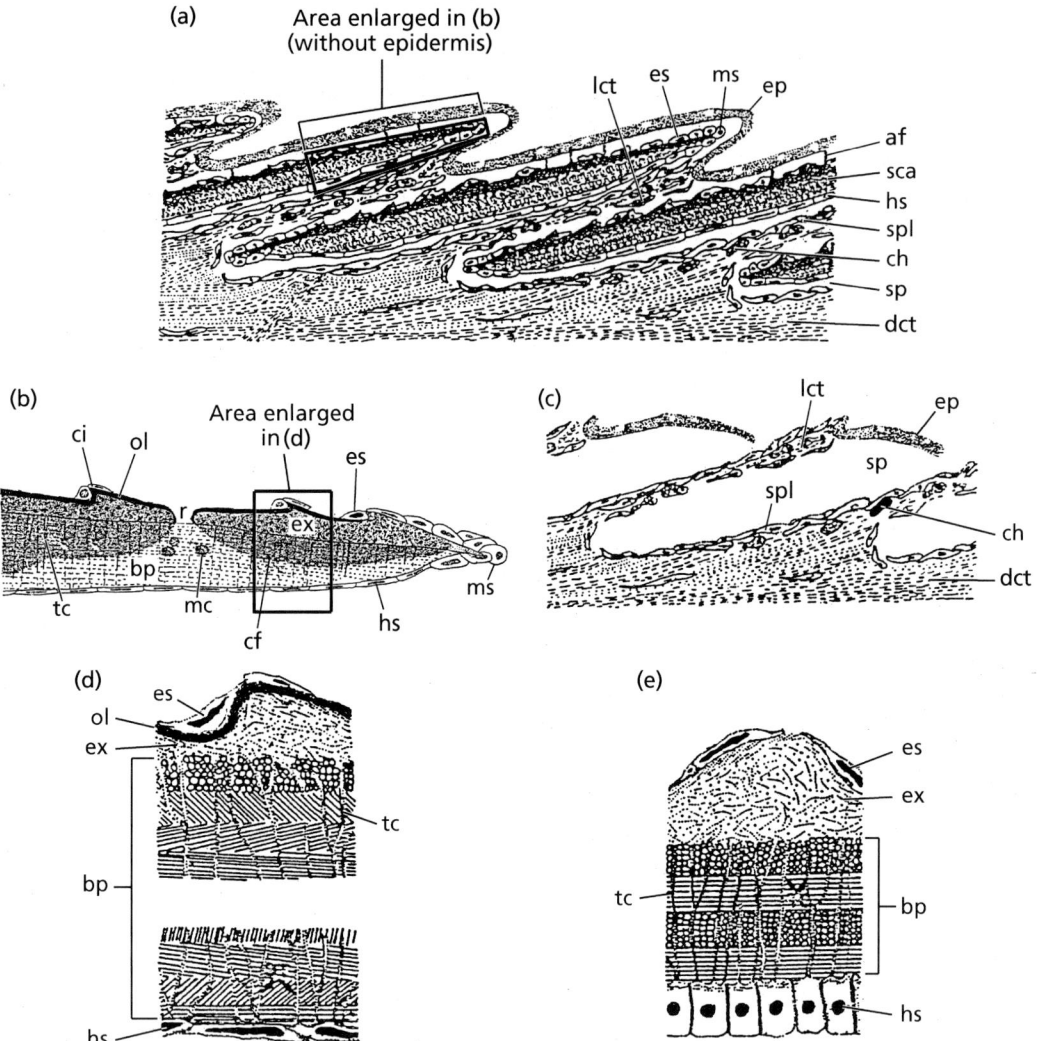

Figure 17.25 Orientation and microscopic structure of ontogenic and regenerating scales of a teleost, the goldfish *Carassius auratus*. (a) Ontogenic scales in pockets, showing anchoring bundles of collagen fibers, scale forming cells (scleroblasts), and scale pocket lining cells. (b) Enlargement of the posterior (non-overlapped) region of an ontogenic scale, showing elements involved in growth and mineralization of the scale. (c) The skin just after scale removal, with empty scale pockets shown. The scleroblasts (along with epidermal and dermal tissue from the posterior area of the pocket) have been removed with the scale, but scale pocket lining cells remain. (d) Enlargement of an ontogenic scale showing the twisted plywood-like structure of the basal plate elasmodine, crossed by sheets of vertical thin collagen fibers, that is characteristic of this species. (e) Enlargement of a regenerating scale showing the orthogonal plywood-like structure of the basal plate elasmodine that is typical of regenerating scales in teleosts, and characteristic of ontogenic scales in some species. The hyposquamal scleroblasts which produce the collagen fibrils forming the basal plate are cuboidal in regenerating scales in contrast to the flattened cells in ontogenic scales. The outer limiting layer is not present in newly regenerating scales. Abbreviations: *af* anchoring bundles of collagen fibers; *bp* basal plate; *cf* calcification front; *ch* chromatophore; *ci* circulus; *dct* dense connective tissue (stratum compactum); *ep* epidermis; *es* episquamal scleroblast; *ex* external layer; *hs* hyposquamal scleroblast; *lct* loose connective tissue (stratum spongiosum); *mc* Mandl's corpuscle; *ms* marginal scleroblast; *ol* outer limiting layer; *r* radius; *sca* scale; *sp* scale pocket; *spl* scale pocket lining cells; *tc* vertical thin collagen fibers. (Sources: (a) and (c) After Bereiter-Hahn and Zylberberg, copyright ©1993, page 626, with permission from Elsevier Science. (b) After Schönbörner et al., 1979, Zylberberg and Nicholas, 1982 and Whitear, 1986b. (d) and (e) After Bereiter-Hahn and Zylberberg, copyright ©1993, page 632, with permission from Elsevier Science.)

the basal plate (Figure 17.25b and d; Zylberberg and Nicholas, 1982; Whitear, 1986b).

During the growth of most elasmoid scales, a third well-mineralized layer, termed the outer limiting layer (Figure 17.25b; Schönbörner et al., 1979), is deposited at the surface of the external layer. However, deposition of the outer limiting layer only occurs in the posterior non-overlapped regions of the

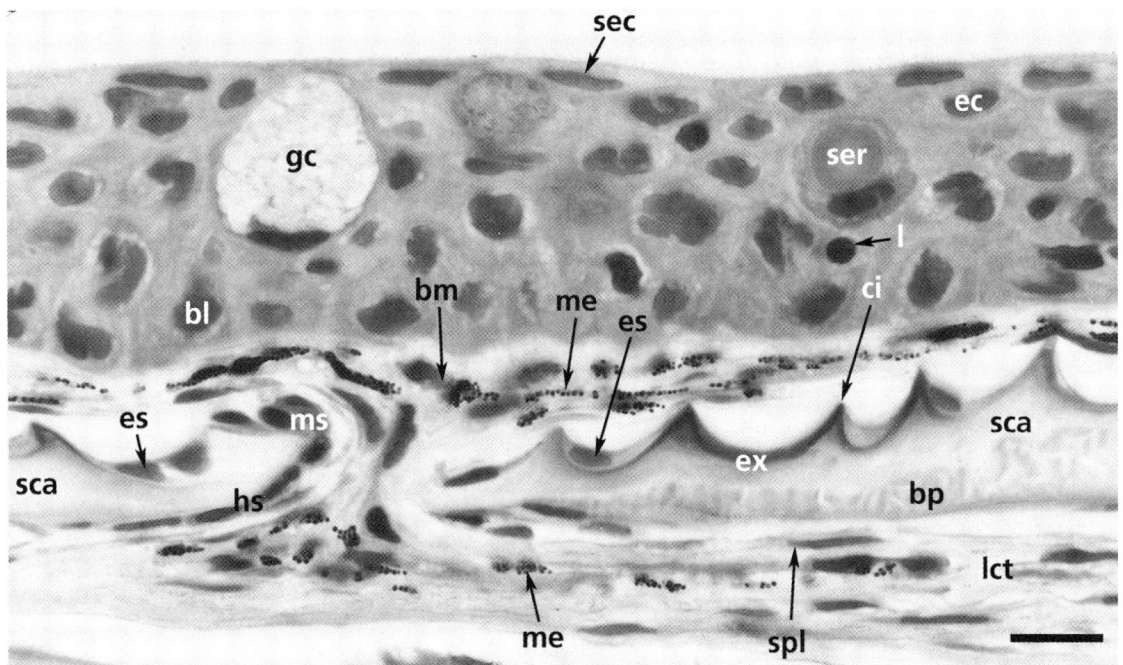

Figure 17.26 Histological section of the skin of a juvenile chinook salmon, Oncorhynchus tshawytscha, showing elements involved in scale growth. The ends of two scales (sca) are visible, with associated episquamal scleroblasts (es), marginal scleroblasts (ms), and hyposquamal scleroblasts (hs). The basal plate (bp), external layer (ex), and circuli (ci) of the scales are also apparent. Other elements in the stratum spongiosum or loose connective tissue layer (lct) of the dermis include scale pocket lining cells (spl), melanophores (me), and the basement membrane (bm) separating the dermis from the epidermis. Cells visible in the epidermis include flattened surface epithelial cells (sec), mid-level epithelial cells (ec), and cuboidal basal layer epithelial cells (bl), as well as a mucous goblet cell (gc) and a serous cell (ser) that could be a serous goblet cell or sacciform cell. Electron microscopy would be necessary for definitive identification of this cell. Scale bar = 10 μm. Hematoxylin and eosin stain. **(See also Colour Plate 13.)**

scale close to (or in direct contact with) the basement membrane of the dermis (Figure 17.24c; Sire, 1985, 1990; Sire et al., 1997). The matrix of the outer limiting layer contains few or no collagen fibrils (Sire, 1985). This layer thickens throughout the life of the fish, reaching a thickness of more than 50 μm in old adult cichlids (Sire, 1985), and it may provide additional protection for the non-overlapped portion of the scale (Sire et al., 1997).

The development of elasmoid scales is similar to the early stages of development of ganoid scales in polypterids (Huysseune and Sire, 1998). Mesenchymal cells initially accumulate under the epidermis at precise loci, with initiation presumably controlled by signals from epithelial cells. These cells proliferate to form dermal papillae that invaginate slightly into the epidermis. The mesenchymal cells of each dermal papilla differentiate first to produce the woven-fibered initial scale matrix (the external layer). These differentiated cells persist at the scale margins, allowing the scale to increase in size laterally, without thickening of the external layer. The mesenchymal cells at the upper surface of the developing scale (closest to the epidermis) then stop or slow their activity, whereas those located below (facing the dermis) differentiate into elasmoblasts and synthesize the elasmodine. The outer limiting layer is later deposited on the surface of the external layer in the posterior areas of the scale. Evidence suggests that the basal epithelial cells in the epidermis participate in the deposition of the outer limiting layer (Sire et al., 1997).

The external layer of the elasmoid scale mineralizes as soon as it is deposited, with the mineral crystals oriented along the collagen fibrils of the woven-fibered matrix (Huysseune and Sire, 1998). Mineralization of the elasmodine matrix is considerably delayed and limited to the superficial collagen layers (Brown and Wellings, 1969; Lanzing and Wright, 1976; Yamada and Watanabe, 1979). The mineralization front advances downward from the external layer, preceded by small corpuscle-like patches of mineral, Mandl's corpuscles (Figure 17.25b; Schönbörner et al., 1981; Meunier, 1984). Mineralization of the outer limiting layer occurs as soon as it is formed, probably under the influence of the basal epidermal cells (Sire, 1988; Sire et al., 1997).

Mature elasmoid scales are anchored in the scale pockets by sparse bundles of collagen fibers arising from the posterior superficial surface of the scale (Figure 17.25a; Junquiera et al., 1970; Mittal and Banjeree, 1974; Zylberberg and Meunier, 1981; Sire, 1986). Each scale is surrounded by a layer of osteoblast-like cells, the **scleroblasts** or scale-forming cells, constituting the scale-sac or scale-bag (Figure 17.25a; Sire et al., 1990). The deep surface of each pocket is lined by a single-layered or bi-layered cellular sheet, the scale pocket lining, which separates the pocket from the subjacent stratum compactum (Figure 17.25a; Whitear et al., 1980). The scale pocket lining is composed of fibrocytes joined by numerous desmosomes, and a basement membrane is present on both the upper and lower faces of the lining. When a scale is removed from its pocket, the soft tissues (epidermis and dermis) located around the posterior non-overlapped portion of the scale are removed with it (Figure 17.25c; Sire and Géraudie, 1984). The scale-forming cells of the scale-bag also remain attached to the scale and are removed (Sire and Géraudie, 1984; Sire, 1989). Following closure of the epidermal wound (page 276), the upper layer of cells of the scale pocket lining in the cichlid *Hemichromis bimaculatus* has been shown to differentiate and form a new scale papilla, which then functions in the formation of the new scale (Sire, 1989).

The restoration of the ontogenic scale pattern in regenerated scales depends on the presence of the scale pocket lining cells; if they are damaged, abnormal scalation patterns and abnormal scales may be produced (Neave, 1940). Although the chronology of events in scale regeneration is almost identical to that occurring during ontogeny (Sire and Géraudie, 1984), the early events may occur more rapidly during scale regeneration (Bereiter-Hahn and Zylberberg, 1993). Differences in structure and superficial ornamentation between ontogenic and regenerated scales are apparent at both light microscopic and ultrastructural levels (Figure 17.25d and e; Neave, 1940; Blair, 1942; Fouda, 1979b; Sire and Géraudie, 1984). These structural variations may be related to the differences in the rates of the initial stages of scale formation (Bereiter-Hahn and Zylberberg, 1993).

In contrast to elasmoid scales, the osseous dermal plates or **scutes** which occur in certain teleosts (Figure 17.24d; Chapter 5, Figure 5.11) are not localized in scale pockets, are firmly anchored in the dermis by large Sharpey's fibers, and are juxtaposed rather than overlapped (Sire, 1990). In the armored catfish *Corydoras*, the scutes consist of a thick basal plate covered by **hyaloine**, a non-osseous, non-collagenous, hypermineralized tissue (Chapter 18; Sire, 1993; Huysseune and Sire, 1998). In more advanced teleosts such as sticklebacks, thunnids, and tetraodontids, scutes show a wide diversity of shapes but are only composed of bone (Huysseune and Sire, 1998). The dermal plates of certain teleosts are ornamented with spines (Chapter 5; Whitear, 1986b). The scutes of *Corydoras* are noteworthy because they have attached odontodes (dermal denticles) that are similar in structure to placoid scales in sharks, with a dentine cap enclosing a simple pulp cavity and covered with a thin layer of enameloid (Figure 17.24d; Sire and Huysseune, 1996).

Scute development has been most completely studied in *Corydoras* (Bhatti, 1938; Sire, 1993). Initially, condensations of mesenchymal cells in the dermis form loosely organized 'dermal papillae'. The scute anlage is formed by incorporation of existing dermal collagen bundles and deposition of new bony material, resulting in a woven-fibered region. The scute then enlarges by apposition of parallel-fibered bone. Sharpey's fibers are incorporated into this bone in the basal region of the scute. After the scute is well developed, a layer of hyaloine is deposited on its surface.

Fin rays

The fins of fishes are supported by dermal rays or spines. In lampreys and hagfishes, the fin supports are unsegmented cartilaginous rods that provide only weak support (Lagler et al., 1977). The dermal fin rays of elasmobranchs and bony fishes are known collectively as dermatotrichia (Lagler et al., 1977). These include several kinds of collagenous structures with varying degrees and types of mineralization; dermatotrichia may be segmented or unsegmented, branched or unbranched, biserial (two lateral components paired on the midline) or uniserial (see e.g. Lagler et al., 1977; Whitear, 1986b). The microscopic structural and phylogenetic relationships among fin rays, spines, and other dermal elements such as scales are discussed in Chapter 18.

Acknowledgments

I thank Carla Aiwohi for extensive technical assistance and library research for this chapter. I am indebted to Dr Barbara Nowak and Elena Catap of the School of Aquaculture, University of Tasmania, Launceston, Tasmania, Australia, for some of the photomicrographs used in this chapter. I am grateful to Dr Susan Gutenberger, Lower Columbia River Fish Health Center, US Fish and Wildlife Service, Underwood, Washington, USA for providing the lamprey skin tissues, and to Dr George Sanders, Department of Comparative Medicine, University of Washington, Seattle, Washington, USA for providing the shark fin tissues. I thank Stewart Alcorn, School of Fisheries, University of Washington, for the drawing for Figure 17.1.

References

Alexander, J.B. and Ingram, G.A. (1992). *Annu. Rev. Fish Dis.* **2**, 249–279.

Al-Hassan, J.M., Thomson, M., Summers, B. and Criddle, R.S. (1987). *Comp. Biochem. Physiol. B Comp. Biochem.* **88**, 813–822.

Anderson, C.D. and Roberts, R.J. (1975). *J. Fish Biol.* **7**, 173–182.

Anderson, C.D., Roberts, R.J., MacKenzie, K. and McVicar, A.H. (1976). *J. Fish Biol.* **8**, 331–341.

Åsbakk, K. and Dalmo, R.A. (1998). *J. Mar. Biotechnol.* **6**, 30–34.

Bagnara, J.T., Taylor, J.D. and Hadley, M.E. (1968). *J. Cell Biol.* **38**, 67–79.

Barber, D.L. and Westermann, J.E.M. (1985). *J. Fish Biol.* **27**, 817–826.

Barber, D.L., Westermann, J.E.M. and Jensen, D.N. (1979). *J. Fish Biol.* **14**, 277–284.

Barnett, R.R., Akindele, T., Orte, C. and Shephard, K.L. (1996). *J. Fish Biol.* **49**, 148–156.

Bereiter-Hahn, J. (1971). *Cytobiologie* **4**, 73–102.

Bereiter-Hahn, J. (1986). In *Biology of the Integument – 2. Vertebrates* (eds J. Bereiter-Hahn, A.G. Matoltsy and K.S. Richards), pp. 443–471. Springer-Verlag, Berlin.

Bereiter-Hahn, J. and Zylberberg, L. (1993). *Comp. Biochem. Physiol. A Comp. Physiol.* **105**, 625–641.

Bereiter-Hahn, J., Osborn, M., Weber, K. and Vöth, M. (1979). *J. Ultrastruct. Res.* **69**, 316–330.

Bereiter-Hahn, J., Richards, K.S., Elsner, L. and Vöth, M. (1980). *Proc. R. Soc. Edinb. Sect. B (Biol.)* **79**, 105–111.

Berntssen, M.H.G., Kroglund, F., Rosseland, B.O. and Wendalaar Bonga, S.E. (1997). *Can. J. Fish. Aquat. Sci.* **54**, 1039–1045.

Bertin, L. (1958). In *Traité de Zoologie*, Vol. XIII (ed. P.-P. Grassé), pp. 433–550. Masson, Paris.

Bhatti, H.K. (1938). *Trans. Zool. Soc. Lond.* **24**, 1–102.

Blackstad, T.W. (1963). In *The Biology of* Myxine (eds A. Brodal and R. Fänge), pp. 195–230. University Press, Oslo.

Blackstock, N. and Pickering, A.D. (1980). *Cell Tiss. Res.* **210**, 359–369.

Blair, A.A. (1942). *J. Fish. Res.* **5**, 440–447.

Blazer, V.S., Fabacher, D.L., Little, E.E., Ewing, M.S. and Kocan, K.M. (1997). *J. Aquat. Anim. Health* **9**, 132–143.

Braun, C.B. (1998). *J. Comp. Neurol.* **392**, 135–163.

Brown, G.A. and Wellings, S.R. (1969). *Z. Zellforsch. Mikrosk. Anat.* **93**, 571–582.

Brown, G.E. and Smith, R.J.F. (1997). *Can. J. Zool.* **75**, 1916–1922.

Buchmann, K. and Bresciani, J. (1998). *Parasitol. Res.* **84**, 17–24.

Bullock, A.M. (1980). *Ocean. Mar. Biol. Annu. Rev.* **18**, 251–315.

Bullock, A.M. and Roberts, R.J. (1974). *Oceanogr. Mar. Biol. Ann. Rev.* **13**, 383–411.

Bullock, A.M., Roberts, R.J. and Gordon, J.D.M. (1976). *J. Mar. Biol. Assoc. UK* **56**, 213–226.

Bullock, A.M., Marks, R. and Roberts, R.J. (1978a). *J. Zool. (Lond.)* **184**, 423–428.

Bullock, A.M., Marks, R. and Roberts, R.J. (1978b). *J. Zool. (Lond.)* **185**, 197–204.

Burton, D. and Burton, M.P.M. (1989). *J. Fish Biol.* **35**, 845–853.

Burton, D. and Fletcher, G.L. (1983). *J. Mar. Biol. Assoc. UK* **63**, 273–287.

Burton, D., Burton, M.P., Truscott, B. and Idler, D.R. (1985). *Proc. R. Soc. Lond. Ser. B Biol. Sci.* **225**, 121–128.

Cameron, A.M. and Endean, R. (1966). *Toxicon* **4**, 111–121.

Cameron, A.M. and Endean, R. (1970). *Toxicon* **8**, 171–178.

Cameron, A.M. and Endean, R. (1973). *Toxicon* **11**, 401–410.

Carmignani, M.P.A. and Zaccone, G. (1975). *Acta Histochem.* **52**, 100–110.

Chen, J.S. and Wang, S.M. (1993). *Arch. Histol. Cytol.* **56**, 451–458.

Clark, G. (1981). In *Staining Procedures*, 4th edn (ed. G. Clark), pp. 171–215. Williams and Wilkins, Baltimore.

Cole, A.M., Weis, P. and Diamond, G. (1997). *J. Biol. Chem.* **1997**, 12008–12013.

Cross, M.L. and Matthews, R.A. (1991). *J. Fish Biol.* **39**, 279–283.

Crouse-Eisnor, R.A., Cone, D.K. and Odense, P.H. (1985). *J. Fish Biol.* **27**, 395–402.

Danguy, A., Genten, F. and Gabius, H.J. (1991). *Eur. J. Basic Appl. Histochem.* **35**, 341–357.

Davidson, G.A., Ellis, A.E. and Secombes, C.J. (1993). *J. Fish Biol.* **42**, 301–306.

Dawe, C.J., Stanton, M.F. and Schwartz, F.J. (1964). *Cancer Res.* **24**, 1194–1201.

Degnan, K.J., Karnaky, K.J., Jr. and Zadunaisky, J.A. (1977). *J. Physiol. (Lond.)* **271**, 155–191.

Denton, E.J. (1970). *Phil. Trans. R. Soc. Lond. B Biol. Sci.* **258**, 285–313.

Desser, S.S. and Lester, R. (1975). *Can. J. Zool.* **53**, 1483–1494.

Dezfuli, B.S., Capuano, S. and Manera, M. (1998). *J. Fish Biol.* **53**, 1084–1095.

Downing, S.W. and Novales, R.R. (1971a). *J. Ultrastruct. Res.* **35**, 282–294.

Downing, S.W. and Novales, R.R. (1971b). *J. Ultrastruct. Res.* **35**, 295–303.

Downing, S.W. and Novales, R.R. (1971c). *J. Ultrastruct. Res.* **35**, 304–313.

Downing, S.W., Spitzer, R.H., Koch, E.A. and Salo, W.L. (1984). *J. Cell Biol.* **98**, 653–669.

Evans, D.H. (1993). In *The Physiology of Fishes* (ed. D.H. Evans), pp. 315–341. CRC Press, Boca Raton, Florida.

Fasulo, S., Tagliafierro, G., Contini, A., Ainis, L., Ricca, M.B., Yanaihara, N. and Zaccone, G. (1993). *Arch. Histol. Cytol.* **56**, 117–125.

Ferguson, H.W. (1989). *Systemic Pathology of Fish*. Iowa State University Press, Ames.

Fernholm, B. (1981). *Acta Zool. (Stockh.)* **62**, 137–145.

Ferri, S. (1983). *Anat. Anz.* **154**, 161–168.

Finger, T.E. (1997). *Brain Behav. Evol.* **50**, 234–243.

Fishelson, L. (1973). *Z. Zellforsch. Mikrosk. Anat.* **140**, 497–508.

Flik, G., van Rijs, J.H. and Wendelaar Bonga, S.E. (1984). *Eur. J. Biochem.* **138**, 651–654.

Foskett, J.K. and Scheffey, C. (1982). *Science* **215**, 164–166.

Fouda, M.M. (1979a). *J. Fish Biol.* **15**, 173–183.

Fouda, M.M. (1979b). *J. Zool (Lond.)* **189**, 503–509.

Fox, H. (1986). In *Biology of the Integument – 2. Vertebrates* (eds J. Bereiter-Hahn, A.G. Matoltsy and K.S. Richards), pp. 116–135. Springer-Verlag, Berlin.

Fujii, R. (1993a). In *The Physiology of Fishes* (ed. D.H. Evans), pp. 535–562. CRC Press, Boca Raton, Florida.

Fujii, R. (1993b). *Int. Rev. Cytol.* **143**, 191–255.

Fujii, R., Hayashi, H., Toyohara, J. and Nishi, H. (1991). *Zool. Sci.* **8**, 461–470.

Gaines, J.L., Jr and Rogers, W.A. (1975). In *The Pathology of Fishes* (eds W.E. Ribelin and G. Migaki), pp. 429–441. University of Wisconsin Press, Madison.

Garg, T.K. and Mittal, A.K. (1993). *Biomed. Environ. Sci.* **6**, 119–133.

Goda, M. and Fujii, R. (1995). *Zool. Sci.* **12**, 811–813.

Gona, O. (1979). *Histochem. J.* **11**, 709–718.

Gopalakrishnakone, P. and Gwee, M.C.E. (1993). *Toxicon* **31**, 979–988.

Gwee, M.C.E., Gopalakrishnakone, P., Yuen, R., Khoo, H.E. and Low, K.S.Y. (1994). *Pharmacol. Ther.* **64**, 509–528.

Halstead, B.W. (1988). *Poisonous and Venomous Marine Animals of the World*. Darwin Press, Princeton, New Jersey.

Hara, T.J. (1986). In *The Behavior of Teleost Fishes* (ed. T.J. Pitcher), pp. 152–176. Croom Helm, London.

Harder, W. (1971). *Z. Zellforsch. Mikrosk. Anat.* **114**, 262–270.

Harris, J.E. and Hunt, S. (1975a). *Cell Tiss. Res.* **163**, 535–543.

Harris, J.E. and Hunt, S. (1975b). *Cell Tiss. Res.* **157**, 553–565.

Hartschuh, W., Weihe, E. and Reinecke, M. (1986). In *Biology of the Integument – 2. Vertebrates* (eds J. Bereiter-Hahn, A.G. Matoltsy and K.S. Richards), pp. 605–620. Springer-Verlag, Berlin.

Harvey, E.N. (1957). In *The Physiology of Fishes*, Vol. 2 (ed. M.E. Brown), pp. 345–366. Academic Press, New York.

Hastings, J.W. (1996). *Gene* 173, 5–11.

Hawkes, J.W. (1974a). *Cell Tiss. Res.* **149**, 147–158.

Hawkes, J.W. (1974b). *Cell Tiss. Res.* **149**, 159–172.

Haygood, M.G. and Distel, D.L. (1993). *Nature* **363**, 154–156.

Heath, J. and Holifield, B. (1991). *Nature* **352**, 107–108.

Heisler, N. (1993). In *The Physiology of Fishes* (ed. D.H. Evans), pp. 343–378. CRC Press, Boca Raton, Florida.

Henrickson, R.C. (1967). *Experientia* **23**, 357–358.

Henrickson, R.C. and Matoltsy, A.G. (1968a). *J. Ultrastruct. Res.* **21**, 194–212.

Henrickson, R.C. and Matoltsy, A.G. (1968b). *J. Ultrastruct. Res.* **21**, 213–221.

Henrickson, R.C. and Matoltsy, A.G. (1968c). *J. Ultrastruct. Res.* **21**, 222–232.

Herring, P.J. (1982). *Oceanogr. Mar. Biol. Ann. Rev.* **20**, 415–470.

Herring, P.J. (1993). *Nature* **363**, 110–111.

Hopkins, B.J. and Hodgson, W.C. (1998). *Toxicon* **36**, 791–793.

Hugie, D.M. and Smith, R.J.F. (1987). *Can. J. Zool.* **65**, 2057–2061.

Huysseune, A. and Sire, J.-Y. (1998). *Eur. J. Oral Sci.* **106** (suppl. 1), 437–481.

Iga, T., Takabatake, I. and Watanabe, S. (1987). *Comp. Biochem. Physiol. C Comp. Pharmacol.* **8**, 319–324.

Iger, Y. and Abraham, M. (1990). *J. Fish Biol.* **36**, 421–437.

Iger, Y. and Abraham, M. (1997). *Tiss. Cell* **29**, 431–438.

Iger, Y., Abraham, M., Dotan, A., Fattal, B. and Rahamim, E. (1988). *J. Fish Biol.* **33**, 711–720.

Iger, Y., Jenner, H.A. and Wendelaar Bonga, S.E. (1994a). *J. Fish Biol.* **44**, 921–935.

Iger, Y., Lock, R.A.C., Van der Meij, J.C.A. and Wendelaar Bonga, S.E. (1994b). *Arch. Environ. Contam. Toxicol.* **26**, 342–350.

Iger, Y., Balm, P.H.M., Jenner, H.A. and Wendelaar Bonga, S.E. (1995). *Gen. Comp. Endocrinol.* **97**, 188–198.

Imaki, H. and Chavin, W. (1975). *Cell Tiss. Res.* **158**, 363–373.

Ingram, G.A. (1980). *J. Fish Biol.* **16**, 23–60.

Itami, T. (1993). *J. Shimonoseki Univ. Fish.* **42**, 1–71.

Junquiera, L.C.U., Toledo, A.M.S. and Porter, K.R. (1970). *Arch. Histol. Jpn.* **32**, 1–15.
Kaweewat, K. and Hofer, R. (1997). *J. Photochem. Photobiol. B Biol.* **41**, 222–226.
Kerr, T. (1952). *Proc. Zool. Soc. Lond.* **122**, 55–78.
Keys, A.B. and Wilmer, E.N. (1932). *J. Physiol.* **76**, 368–378.
Kimura, S. (1992). *Comp. Biochem. Physiol. B Comp. Biochem.* **102**, 225–260.
Koch, E.A., Spitzer, R.H., Pithawalla, R.B. and Downing, S.W. (1991). *Cell Tiss. Res.* **264**, 79–86.
Koch, E.A., Spitzer, R.H., Pithawalla, R.B. and Parry, D.A. (1994). *J. Cell Sci.* **107**, 3133–3144.
Koch, E.A., Spitzer, R.H., Pithawalla, R.B., Castillos, F.A. III and Parry, D.A. (1995). *Int. J. Biol. Macromol.* **17**, 283–292.
Komnick, H. (1986). In *Biology of the Integument – 2. Vertebrates* (eds J. Bereiter-Hahn, A.G. Matoltsy and K.S. Richards), pp. 499–516. Springer-Verlag, Berlin.
Kotrschal, K. (1995). *Environ. Biol. Fishes* **44**, 143–155.
Kotrschal, K., Krautgartner, W.-D. and Hansen, A. (1997). *Chem. Senses* **22**, 111–118.
Kotrschal, K., Royer, S. and Kinnamon, J.C. (1998). *J. Struct. Biol.* **124**, 59–69.
Krieg, T. and Timpl, R. (1986). In *Biology of the Integument – 2. Vertebrates* (eds J. Bereiter-Hahn, A.G. Matoltsy and K.S. Richards), pp. 788–799. Springer-Verlag, Berlin.
Kucik, D.F., Elson, E.L. and Sheetz, M.P. (1990). *J. Cell Biol.* **111**, 1617–1622.
Lagler, K.F., Bardich, J.E., Miller, R.R. and Passino, D.M. (1977). *Ichthyology*, 2nd edn. John Wiley, New York.
Laguesse, E. (1895). *Rev. Biol. Nord France* **7**, 343–363.
Land, M.F. (1972). *Prog. Biophys. Mol. Biol.* **24**, 75–106.
Lane, E.B. and Whitear, M. (1980). *Can. J. Zool.* **58**, 450–455.
Lanzing, W.J.R. and Wright, R.G. (1976). *Cell Tiss. Res.* **167**, 37–47.
Laurent, P. and Dunel, S. (1980). *Am. J. Physiol.* **238**, 147–159.
Lee, J., Ishihara, A., Theriot, J.A. and Jacobson, K. (1993). *Nature* **362**, 167–171.
Leino, R.L. (1974). *Cell Tiss. Res.* **155**, 367–381.
Leino, R.L. (1996). *Can. J. Zool.* **74**, 217–225.
Lemaitre, C., Orange, N., Saglio, P., Saint, N., Gagnon, J. and Molle, G. (1996). *Eur. J. Biochem.* **15**, 143–149.
Leonard, J.B. and Summers, R.G. (1976). *Cell Tiss. Res.* **171**, 1–30.
Lester, R.J.G. (1972). *J. Parasitol.* **53**, 717–722.
Lethbridge, R.C. and Potter, I.C. (1982). In *The Biology of Lampreys*, Vol. 3 (eds M.W. Hardisty and I.C. Potter), pp. 377–448. Academic Press, London.
Linna, T.J., Finstad, J. and Good, R.A. (1975). *Am. Zool.* **15**, 29–38.
Lombard, N. (1997). *Bioluminescence in Insects and Fishes: Mechanisms, Functions and Applications*. École Nationale Vétérinaire, Lyon, France.
Lopes-Ferreira, M., Barbaro, K.C., Cardoso, D.F., Moura-Da-Silva, A.M. and Mota, I. (1998). *Toxicon* **36**, 405–410.
López-Dóriga, M.V. and Martínez, J.L. (1993). *J. Zool. (Lond.)* **230**, 425–432.
Lubbock, R. (1980). *Proc. R. Soc. Lond. Ser. B Biol. Sci.* **207**, 35–61.
Malinovsky, L. (1986). In *Biology of the Integument – 2. Vertebrates* (eds J. Bereiter-Hahn, A.G. Matoltsy and K.S. Richards), pp. 535–560. Springer-Verlag, Berlin.
Mallatt, J., Conley, D.M. and Ridgeway, R.L. (1987). *Can. J. Zool.* **65**, 1956–1965.
Mallefet, J. and Shimomura, O. (1995). *Mar. Biol.* **124**, 381–385.
Matoltsy, A.G. (1986). In *Biology of the Integument – 2. Vertebrates* (eds J. Bereiter-Hahn, A.G. Matoltsy and K.S. Richards), pp. 255–277. Springer-Verlag, Berlin.
Matsui, R., Ishida, M. and Kimura, S. (1990). *Comp. Biochem. Physiol. B Comp. Biochem.* **95**, 669–675.
Mattey, D.L., Morgan, M. and Wright, D.E. (1979). *J. Fish Biol.* **15**, 363–370.
Mayberry, L.F., Bristol, J.R., Sulimanovic, D., Fijan, N. and Petrenic, Z. (1979a). *Fish Pathol.* **21**, 145–150.
Mayberry, L.F., Marchiondo, A.A., Ubelaker, J.E. and Kazic, D. (1979b). *J. Protozool.* **26**, 168–178.
Mensinger, A.F. (1995). *Environ. Biol. Fish.* **44**, 133–142.
Menter, D.G., Obika, M., Tchen, T.T. and Taylor, J.D. (1979). *J. Morphol.* **160**, 103–120.
Merrilees, M.J. (1974). *J. Ultrastruct. Res.* **47**, 272–283.
Meunier, F.J. (1984). *Am. Zool.* **24**, 953–964.
Meunier, F.J. (1987). *Ann. Biol.* **26**, 201–233.
Mittal, A.K. and Agarwal, S.K. (1977). *J. Zool. (Lond.)* **182**, 429–439.
Mittal, A.K. and Banjeree, T.K. (1974). *J. Zool. (Lond.)* **174**, 341–355.
Mittal, A.K. and Banjeree, T.K. (1975). *Can. J. Zool.* **53**, 833–843.
Mittal, A.K. and Banjeree, T.K. (1980). In *The Skin of Vertebrates* (eds R.I.C. Spearman and P.A. Riley), pp. 1–12. Linn. Soc. Symp. No. 9. Academic Press, London.
Mittal, A.K. and Garg, T.K. (1994). *J. Fish Biol.* **44**, 857–875.
Mittal, A.K. and Munshi, J.S.D. (1974). *Acta Anat.* **88**, 424–442.
Mittal, A.K. and Whitear, M. (1979). *Cell Tiss. Res.* **202**, 213–230.
Mittal, A.K., Whitear, M. and Agarwal, S.K. (1980). *J. Zool. (Lond.)* **191**, 107–125.
Mittal, A.K., Whitear, M. and Bullock, A.M. (1981). *Z. Mikrosk. Anat. Forsch. (Leipz.)* **95**, 559–585.
Mittal, A.K., Ueda, T., Fujimori, O. and Yamada, K. (1994). *Histochem. J.* **26**, 666–677.
Mittal, A.K., Garg, T.K. and Verma, M. (1995). *Jpn. J. Ichthyol.* **42**, 187–191.
Moore, J.D., Ototake, M. and Nakanishi, T. (1998). *Fish Shellfish Immunol.* **8**, 393–407.

Morrison, C.M. and Odense, P.H. (1978). *J. Fish. Res. Board Can.* **35**, 101–116.

Moyle, P.B. and Cech, J.J., Jr (1996). *Fishes: An Introduction to Ichthyology*, 3rd edn. Prentice-Hall, Upper Saddle River, New Jersey.

Nadol, J.B. and Gibbons, J.R. (1970). *Z. Zellforsch. Mikrosk. Anat.* **106**, 398–413.

Nagamalleswari, D. and Joseph, K.T. (1991). *Biochem. Int.* **24**, 1063–1073.

Nakari, T., Soivio, A. and Pesonen, S. (1986). *J. Fish Biol.* **29**, 451–457.

Neave, F. (1940). *Q. J. Microsc. Sci.* **81**, 541–568.

Nicol, J.A.C. (1969). In *Fish Physiology*, Vol. III (eds W.S. Hoar and D.J. Randall), pp. 355–400. Academic Press, New York.

Nilsson, H. and Wallin, M. (1997). *Cell Motil. Cytoskeleton* **38**, 397–409.

Nishioka, R.S., Lai, K.V. and Bern, H.A. (1985). *Aquaculture* **45**, 393–394.

Noceda, C., Sierra, S.G. and Martínez, J.L. (1997). *Dis. Aquat. Org.* **31**, 103–108.

Nonnotte, G., Nonnotte, L. and Kirsch, R. (1979). *Cell Tiss. Res.* **17**, 387–396.

Obika, M. (1988). *Zool. Sci.* **5**, 311–321.

Obika, M. and Meyer-Rochow, V.B. (1990). *Pigment Cell Res.* **3**, 33–37.

Olson, K.R. and Fromm, P.O. (1973). *Z. Zellforsch.* **143**, 439–449.

Oshima, N. and Nagaishi, H. (1992). *Comp. Biochem. Physiol. A Comp. Physiol.* **102**, 273–278.

Ottesen, O.H. and Olafsen, J.A. (1997). *J. Fish Biol.* **50**, 620–633.

Pankhurst, N.W. (1982). *J. Fish Biol.* **21**, 549–561.

Paul, V.I. and Banjeree, T.K. (1997). *Dis. Aquat. Org.* **28**, 151–161.

Peek, W.D. and Youson, J.H. (1979). *J. Morphol.* **160**, 143–164.

Peleteiro, M.C. and Richards, R.H. (1985). *J. Fish Dis.* **8**, 161–172.

Peleteiro, M.C. and Richards, R.H. (1988). *J. Fish Biol.* **32**, 845–858.

Peleteiro, M.C. and Richards, R.H. (1990). *J. Fish Dis.* **13**, 225–232.

Perriere, C. and Goudey-Perriere, F. (1989). *Toxicon* **27**, 287–295.

Perry, S.F. (1997). *Annu. Rev. Physiol.* **59**, 325–347.

Pfeiffer, W. (1977). *Copeia* **1977**, 653–655.

Pfeiffer, W. (1982). In *Chemoreception in Fishes* (ed. T.J. Hara), pp. 307–325. Elsevier, Amsterdam.

Pfeiffer, W. and Pletcher, T.F. (1964). *J. Fish. Res. Board Can.* **21**, 1083–1088.

Phromsuthirak, P. (1977). *J. Fish Biol.* **11**, 193–206.

Pickering, A.D. (1974). *J. Fish Biol.* **6**, 111–118.

Pickering, A.D. (1977). *J. Fish Biol.* **10**, 561–566.

Pickering, A.D. and Fletcher, J.M. (1987). *Cell Tiss. Res.* **247**, 259–265.

Pickering, A.D. and Macey, D.J. (1977). *J. Fish Biol.* **10**, 505–512.

Pickering, A.D. and Richards, R.H. (1980). *Proc. R. Soc. Edinb. Sect. B Biol. Sci.* **79**, 93–104.

Poston, H.A. and Wolfe, M.J. (1985). *J. Fish Dis.* **8**, 451–460.

Quiniou, S.M.-A., Bigler, S., Clem, L.W. and Bly, J.E. (1998). *Fish Shellfish Immunol.* **8**, 1–11.

Rabl, H. (1931). In *Anatomie der Wirbeltiere*, Bd. I. (eds L. Bolk, E. Göppert, E. Kallius and W. Lubosch), pp. 280–306. Urban and Schwarzenberg, Berlin.

Ralphs, J.R. and Benjamin, M. (1992). *J. Fish Biol.* **40**, 473–475.

Raschi, W. and Tabit, C. (1992). *Aust. J. Mar. Freshwat. Res.* **43**, 123–147.

Reid, L. and Clamp, J.R. (1978). *Br. Med. J.* **34**, 5–8.

Reif, W.-E. (1978). *Zoomorphologie* **90**, 101–111.

Reif, W.-E. (1980). *J. Morphol.* **166**, 275–288.

Reite, O.B. (1997). *Fish Shellfish Immunol.* **7**, 567–584.

Reutter, K. (1986). In *Biology of the Integument – 2 Vertebrates* (eds J. Bereiter-Hahn, A.G. Matoltsy and K.S. Richards), pp. 586–604. Springer-Verlag, Berlin.

Roberts, R.J. and Bullock, A.M. (1980). *Proc. R. Soc. Edinb. Sect. B.* **79**, 87–91.

Roberts, R.J. and Bullock, A.M. (1981). *Fish Pathol.* **15**, 237–239.

Roberts, R.J., Shearer, W.M., Elson, K.G.R. and Munro, A.L.S. (1970). *J. Fish Biol.* **2**, 223–229.

Roberts, R.J., Young, H. and Milne, J.A. (1971). *J. Fish Biol.* **4**, 87–98.

Roberts, T.R. (1982). *Zool. Scr.* **11**, 55–76.

Rodionov, V.I., Gyoeva, F.K. and Gelfand, V.I. (1991). *Proc. Natl. Acad. Sci. USA* **88**, 4956–4960.

Rodionov, V.I., Hope, A.J., Svitkina, T.M. and Borisy, G.G. (1998). *Curr. Biol.* **8**, 165–168.

Roubal, F.R. and Bullock, A.M. (1988). *J. Fish Biol.* **32**, 545–555.

Roy, D. (1988). *Ecotoxicol. Environ. Saf.* **15**, 260–271.

Russell, F.E. (1969). In *Fish Physiology*, Vol. III (eds W.S. Hoar and D.J. Randall), pp. 401–449. Academic Press, New York.

Rydevik, M. (1988). *J. Fish Biol.* **33**, 941–944.

Sanchez, J.G., Speare, D.J., Sims, D.E. and Johnson, G.J. (1998). *J. Comp. Pathol.* **118**, 81–87.

Sato, M. (1978). *Jpn. J. Ichthyol.* **24**, 231–238.

Schliwa, M. (1975). *J. Ultrastruct. Res.* **52**, 377–386.

Schliwa, M. (1986). In *Biology of the Integument – 2 Vertebrates* (eds J. Bereiter-Hahn, A.G. Matoltsy and K.S. Richards), pp. 65–77. Springer-Verlag, Berlin.

Schönbörner, A.A., Biovin, G. and Baud, C.A. (1979). *Cell Tiss. Res.* **202**, 203–212.

Schönbörner, A.A., Meunier, F.J. and Castanet, J. (1981). *Tiss. Cell* **13**, 589–597.

Schreiner, K.E. (1918). *Arch. Mikrosk. Anat.* **92**, 1–63.

Schwerdfeger, W.K. (1978). *Z. Mikrosk. Anat. Forsch.* **92**, 193–205.

Sharples, A.D. and Evans, C.W. (1996). *Dis. Aquat. Org.* **24**, 81–91.

Shephard, K.L. (1994). *Rev. Fish Biol. Fish.* **4**, 401–409.

Shiomi, K., Takamiya, M., Yamanaka, H., Kikuchi, T. and Suzuki, Y. (1988). *Toxicon* **26**, 353–361.

Singh, S.K. and Mittal, A.K. (1990). *J. Fish Biol.* **36**, 9–19.

Sire, J.-Y. (1985). *Ann. Sci. Nat. (Zool.) Paris* **13**, 163–180.

Sire, J.-Y. (1986). *J. Fish Biol.* **7**, 713–724.

Sire, J.-Y. (1988). *Cell Tiss. Res.* **253**, 165–172.

Sire, J.-Y. (1989). *Am. J. Anat.* **186**, 315–323.

Sire, J.-Y. (1990). *Neth. J. Zool.* **40**, 75–92.

Sire, J.-Y. (1993). *J. Morphol.* **215**, 225–244.

Sire, J.-Y. (1995). *Connect. Tiss. Res.* **33**, 213–222.

Sire, J.-Y. and Géraudie, J. (1983). *Acta Zool. (Stockh.)*, **64**, 1–8.

Sire, J.-Y. and Géraudie, J. (1984). *Cell Tiss. Res.* **237**, 537–547.

Sire, J.-Y. and Huysseune, A. (1996). *Acta Zool. (Stockh.)*, **77**, 51–72.

Sire, J.-Y., Géraudie, J., Meunier, F.J. and Zylberberg, L. (1987). *Am. J. Anat.* **180**, 391–402.

Sire, J.-Y., Boulekbache, H. and Joly, C. (1990). *Biol. Cell.* **68**, 147–158.

Sire, J.-Y., Quilhac, A., Bourguignon, J. and Allizard, F. (1997). *J. Morphol.* **231**, 161–174.

Smith, R.J.F. (1973). *Can. J. Zool.* **51**, 875–876.

Smith, R.J.F. (1976). *Can. J. Zool.* **54**, 1172–1182.

Smith, R.J.F. (1977). In *Chemical Signals in Vertebrates* (eds D. Müller-Schwarz and M.M. Mozell), pp. 303–320. Plenum, New York.

Smith, R.J.F. (1982a). In *Chemoreception in Fishes* (eds T.J. Hara), pp. 327–342. Elsevier, Amsterdam.

Smith, R.J.F. (1982b). *Can. J. Zool.* **60**, 1067–1072.

Smith, R.J.F. (1986). In *Chemical Signals in Vertebrates 4* (eds D. Duvall, D. Müller Schwarz and R.M. Silverstein), pp. 99–115. Plenum, New York.

Smith, R.J.F. (1992). *Rev. Fish Biol. Fish.* **2**, 33–63.

Smith, S.A., Caceci, H., Marei, E.-S. and El-Habback, H.A. (1995). *J. Fish Biol.* **46**, 241–254.

Solanki, T.G. and Benjamin, M. (1982). *J. Fish Biol.* **21**, 563–575.

Spitzer, R.H., Downing, S.W. and Koch, E.A. (1979). *Cell Tiss. Res.* **197**, 235–255.

Spitzer, R.H., Koch, E.A., Reid, R.B. and Downing, S.W. (1982). *Cell Tiss. Res.* **222**, 339–357.

Srivastava, C.B.L. and Szabo, T. (1974). *J. Ultrastruct. Res.* **48**, 69–91.

Steven, D.M. (1951). *J. Microsc. Sci.* **92**, 233–247.

Strahan, R. (1963). In *The Biology of Myxine* (eds A. Brodal and R. Fänge), pp. 22–32. University Press, Oslo.

Strüssmann, C.A., Nin, F. and Takashima, F. (1994). *Copeia* **1994**, 956–961.

Sugimoto, M. (1993). *Comp. Biochem. Physiol. A Comp. Physiol.* **104**, 513–518.

Sugimoto, M. and Oshima, N. (1995). *Pigment Cell Res.* **8**, 37–45.

Sugimoto, M., Oshima, N. and Fujii, R. (1985). *Zool. Sci.* **2**, 317–322.

Suzuki, N. (1992). *Jpn. J. Ichthyol.* **38**, 379–396.

Suzuki, N. and Hagiwara, K. (1995). *Bull. Nansei Natl. Fish. Res. Inst.* **28**, 33–41.

Suzuki, Y. and Kaneko, N. (1986). *Dev. Comp. Immunol.* **10**, 509–518.

Szamier, R.B. and Wachtel, A.W. (1969). *J. Morphol.* **128**, 261–289.

Taveekijakarn, P., Miyazaki, T., Matsumoto, M. and Arai, S. (1994). *J. Aquat. Anim. Health* **6**, 251–259.

Tsai, J-C. (1996). *J. Zool.* **239**, 591–599.

Ueda, T., Mittal, A.K., Fujimori, O. and Yamada, K. (1994). *Acta Histochem. Cytochem.* **27**, 613–617.

Uitto, J. (1986). In *Biology of the Integument – 2. Vertebrates* (eds J. Bereiter-Hahn, A.G. Matoltsy and K.S. Richards), pp. 800–809. Springer-Verlag, Berlin.

Van Oosten, J. (1957). In *The Physiology of Fishes*, Vol. 1 (ed. M.E. Brown), pp. 207–244. Academic Press, New York.

Viehberger, G. and Bielek, E. (1982). *Experientia* **38**, 1216–1218.

Wang, S.M., Chen, J.S., Fong, T.H., Hsu, S.Y. and Lim, S.S. (1997). *Cell Motil. Cytoskeleton* **36**, 216–227.

Weinstein, G.D., McCollough, J.L. and Ross, P. (1984). *J. Invest. Dermatol.* **82**, 623–628.

Weisel, G.F. (1975). *J. Morphol.* **145**, 143–150.

Whitear, M. (1970). *J. Zool. (Lond.)* **160**, 437–454.

Whitear, M. (1971a). *J. Zool. (Lond.)* **163**, 231–236.

Whitear, M. (1971b). *J. Zool. (Lond.)* **163**, 237–264.

Whitear, M. (1977). *Symp. Zool. Soc. Lond.* **39**, 291–313.

Whitear, M. (1981). *Z. Mikrosk. Anat. Forsch. (Leipz.)* **95**, 531–543.

Whitear, M. (1983). *Acta Biol. Hung.* **34**, 303–319.

Whitear, M. (1986a). In *Biology of the Integument – 2. Vertebrates* (eds J. Bereiter-Hahn, A.G. Matoltsy and K.S. Richards), pp. 8–38. Springer-Verlag, Berlin.

Whitear, M. (1986b). In *Biology of the Integument – 2. Vertebrates* (eds J. Bereiter-Hahn, A.G. Matoltsy and K.S. Richards), pp. 39–64. Springer-Verlag, Berlin.

Whitear, M. (1989). *Arch. Histol. Cytol.* **52**(suppl.), 415–422.

Whitear, M. (1993). In *Advances in Fish Research*, Vol. 1 (ed. B.R. Singh), pp. 169–184. Narendra Publishing House, Delhi.

Whitear, M. and Lane, E.B. (1983a). *J. Zool.* **199**, 345–358.

Whitear, M. and Lane, E.B. (1983b). *J. Zool.* **199**, 359–384.

Whitear, M. and Lane, E.B. (1983c). *J. Zool.* **201**, 259–272.

Whitear, M. and Mittal, A.K. (1983). *Z. Mikrosk. Anat. Forsch. (Leipz.)* **97**, 141–157.

Whitear, M. and Mittal, A.K. (1984). *J. Fish Biol.* **25**, 317–331.

Whitear, M. and Moate, R. (1998). *J. Zool.* **246**, 275–285.

Whitear, M. and Zaccone, G. (1984). *Z. Mikrosk. Anat. Forsch (Leipz.)* **98**, 481–501.

Whitear, M., Mittal, A.K. and Lane, E.B. (1980). *J. Fish Biol.* **17**, 43–65.

Wiley, M.L. and Collette, B.B. (1970). *Bull. Am. Mus. Nat. Hist.* **143**, 143–216.

Wilkins, N.P. and Jancsar, S. (1979). *J. Fish Biol.* **15**, 299–307.

Wilson, J.A.F. and Westerman, R.A. (1967). *Z. Zellforsch. Mikrosk. Anat.* **83**, 196–206.

Yamada, J. and Watabe, N. (1979). *Cell Tiss. Res.* **201**, 409–422.

Yokoya, S. and Ebina, Y. (1980). *Zool. Mag. (Tokyo)* **89**, 58–60.

Zaccone, G. (1981). *Z. Mikrosk. Anat. Forsch. (Leipz.)* **95**, 809–826.

Zaccone, G. (1982). *Basic Appl. Histochem.* **26**, 199–207.

Zaccone, G. (1983). *Histochemistry* **78**, 163–175.

Zaccone, G., Fasulo, S., Lo Cascio, P. and Licata, A. (1986). *Cell Tiss. Res.* **246**, 679–682.

Zaccone, G., Fasulo, S., Lo Cascio, P., Licata, A. and Ainis, L. (1987). *Basic Appl. Histochem.* **31**, 455–464.

Zaccone, G., Fasulo, S., Licata, A., Ainis, L. and Lo Cascio, P. (1988). *Ann. NY Acad. Sci.* **547**, 464–465.

Zaccone, G., Fasulo, S., Ainis, L. and Contini, A. (1989). *Histochemistry* **91**, 13–16.

Zaccone, G., Tagliafierro, G., Fasulo, S., Contini, A., Ainis, L. and Ricca, M.B. (1990). *Histochemistry* **93**, 355–357.

Zaccone, G., Fasulo, S. and Ainis, L. (1994). *Histochem. J.* **26**, 609–629.

Zadunaisky, J.A. (1984). In *Fish Physiology*, Vol. XB (eds W.S. Hoar and D.J. Randall), pp. 129–176. Academic Press, New York.

Zigmond, S.H. (1993). *Cell Motil. Cytoskeleton* **25**, 309–316.

Zuchelowski, E.M., Lantz, R.C. and Hinton, D.E. (1981). *Anat. Rec.* **200**, 33–39.

Zylberberg, L. and Meunier, F.J. (1981). *J. Zool (Lond.)* **195**, 459–471.

Zylberberg, L. and Nicholas, G. (1982). *Cell Tissue Res.* **223**, 349–367.

CHAPTER 18

Skeletal System

A Huysseune
Biology Department, Ghent University, Ghent, Belgium

Introduction

The skeleton serves functions for support, protection, locomotion, buoyancy, feeding and respiration. Some of these functions are apparent at the gross anatomical level, while others are tuned on a scale that requires study at the microscopic anatomical level. In addition, the skeleton fulfills several physiological functions, including roles in growth and mineral homeostasis. Growth, pathologic conditions, or seasonal activity can be registered, temporarily or permanently, in the skeleton and provide the investigator with useful tools to reconstruct the animal's life.

From an evolutionary perspective, a distinction has to be made between endoskeleton and dermal skeleton.[1] The endoskeleton includes cartilages and their ossifications that usually develop in deeper parts of the body. The dermal skeleton includes elements that are originally formed in the skin dermis without a cartilaginous template. They comprise dermal and membrane bones,[2] teeth, odontodes, ganoid and elasmoid scales, scutes, bony fin spines (spiny rays) and the different types of dermotrichia (fin rays).[3]

Although the skeleton is intuitively associated with bone, many more tissues make up a functional skeleton. Chondrichthyans and osteichthyans present an astonishing diversity of skeletal tissues unequalled in any other vertebrate class and which is even more pronounced in fossil than in extant species (for general accounts see Stéphan, 1900; Schaffer, 1930; Ørvig, 1951, 1967; Blanc, 1953; Schaeffer, 1977; Hall, 1978; Meunier, 1983; Francillon-Vieillot *et al.*,

---

[1] We will follow the argumentation of Francillon-Vieillot *et al.* (1990) and use the term dermal skeleton rather than exoskeleton, its obvious antonym.

[2] The terms membrane bone and dermal bone are usually considered synonyms in terms of their histogenesis (direct ossification in the mesenchyme). According to Patterson (1977) they should be distinguished because of different evolutionary origin and membrane bones should be used exclusively for endoskeletal bones which are no longer preformed in cartilage.

[3] The spiny rays, and the so-called lepidotrichia found in all osteichthyans except for the dipnoans, are of bony nature. Dipnoan camptotrichia consist of a subepidermal layer of mineralized bone, and of a deeper layer of non-mineralized matrix considered to be of pre-osseous nature; chondrichthyan ceratotrichia and osteichthyan actinotrichia are unmineralized fin rays, built from elastoidin, and assumed to be homologous (see Francillon-Vieillot *et al.*, 1990, for further reading and references).

1990; Ricqlès et al., 1991; Beresford, 1993). Rather than viewing them as discrete categories, these skeletal tissues should be considered as a continuous spectrum produced by cytodifferentiating modulation of common stem cell scleroblasts (Moss and Moss-Salentijn, 1983). It is nevertheless important to attempt to classify the different skeletal tissues as a workable tool for studying form and function in the fish skeleton.

Skeletal tissue diversity is most pronounced in the craniofacial skeleton. This is not surprising given its highly kinetic nature, the numerous elements from which it is built up, hence the number of joints and other types of connections between skeletal elements, and their degrees of freedom (see Chapter 6).

Cartilage

Structural diversity

In general terms cartilages are resilient stiff tissues which resist compression (Beresford, 1993). According to classical histology textbooks (usually based on mammalian, if not human histology), there are three major types of cartilage, each of which fulfils specific roles in the body. Even a brief glimpse at some sections of a few fish species reveals that such a classification is untenable for fish and that teleosts, the majority of the living fishes, present an astounding diversity of cartilages, both in young stages and adults. Given that there are no clear-cut categories and that cartilage types can grade into one another, both in time and in space, this points to the difficulty of analyzing their function. In a series of papers, Benjamin (1989a,b, 1990) and Benjamin and Ralphs (1991) made a valuable attempt to characterize many of the teleostean cartilages; these papers provide the most recent important source on their distribution, structure and function. Below, only some general trends will be presented.

Many of the biomechanical properties of cartilage depend upon the amount and composition of the extracellular matrix (ECM) (Beresford, 1993). The ECM can be anything between a thin seam surrounding abundant cells, or abundant with few cells. Cell-rich cartilages often support lips and labial folds, or gill filaments (Benjamin, 1990). A common and well-studied type is a tissue called hyaline-cell cartilage (HCC). Hyaline-cell cartilage has little matrix, a mild affinity for alcian blue, and houses closely-packed, hyaline cells (characterized in routine histological sections by abundant, chromophobic cytoplasm) (Figure 18.1). Acting as a damper against mechanical stresses, and providing the mouth region of bottom-dwelling species with a flexible support, it is a particularly important tissue in cyprinids (Benjamin, 1989a). In certain instances, as in the black molly, *Poecilia sphenops*, it develops as a secondary cartilage from the periosteum of perichondral or dermal bone, probably in response to mechanical stimulation, and becomes a permanent feature of the adult skeleton (Benjamin, 1989b). Its frequent merging with other cartilage types suggests that HCC is the most labile of the cartilaginous tissues and occupies a key position in teleosts (Benjamin, 1990). A widespread example of matrix-rich cartilage is the hyaline cartilage found in the neurocranium of many species. Matrix-rich hyaline cartilage (the 'true' hyaline cartilage of birds and mammals) probably serves much the same function as mammalian hyaline cartilage in protection, support and growth.

The histological diversity of at least the cranial cartilages is accompanied by variation in their matrix macromolecules. Collagen type II (cartilage-type collagen) is often absent in cell-rich cartilages; chondroitin sulfate is ubiquitous, but keratan sulfate is sometimes absent (Benjamin and Ralphs, 1991). Elastic fibers are often found in the matrix of certain cell-rich cartilages, as in barbels, or, rather surprisingly, at the joints between adjacent branchial elements (Benjamin, 1990; personal observations).

The shape, size and cytoplasmic characteristics of the cells contained within the cartilage matrix, the chondrocytes, can be extremely variable. There is little doubt that this reflects their metabolic activities and determines the type of cartilage in which they eventually come to lie. Cells are responsible for growth of the cartilage; in appositional growth, chondroblasts are recruited from the perichondrium at the periphery; in interstitial growth chondrocytes divide, or increase in size, or secrete matrix, or a combination of these. Regional variations in the extent to which these mechanisms contribute to the overall size increase of a cartilage, determine its shape. True **hyaline** cartilage often shows distinct maturation gradients, either unidirectional, at the extremity of a cartilage, thus producing an epiphysis-like area (cf. Haines, 1934) (Figure 18.2), or, more often, bidirectional, as commonly seen in the cartilaginous braincase. An epiphysis has a zone of undifferentiated cells, a proliferative zone of flattened cells, and a zone of hypertrophied cells, derived from the flattened cells, yet not arranged in columns (Figure 18.2).

Cartilage calcification is common in chondrichthyans, where the mineralized cartilage functionally replaces a bony skeleton (bone is present at the base of the dermal denticles, or placoid scales (Moss, 1977), and has also been demonstrated in the endoskeleton, notably in shark vertebrae (Bordat, 1987)). The endoskeletal, peripheral cartilaginous mineralizations, or tesserae (Kemp and Westrin, 1979; Clement, 1992), remain separate, enabling the cartilage to grow. In teleosts, cartilage calcification is probably limited to true hyaline cartilages and not widespread. Cartilage calcifies through spheritic mineralization (Figure 18.3); mineral crystals are deposited concentrically around a mineralization centre, independent of the spatial organization of the collagen. This coalescence of globular clusters of mineral crystals creates a characteristic pattern of so-called Liesegang waves (Meunier, 1983). Unlike endothermic vertebrates, normal calcification is not always dependent upon preliminary chondrocyte hypertrophy. In sharks, e.g., calcification of cartilage is associated with vital non-hypertrophic chondrocytes (Moss and Moss-Salentijn, 1983).

Many of the cartilages in fish skeletons, especially the cell-rich cartilages, persist in adults. Matrix-rich hyaline cartilage is partially or entirely resorbed in many taxa. Prior to this process, the cartilage becomes surrounded by a perichondral bone rim (see below). This appears to be necessary for the cartilage to be resorbed. Resorption may or may not be associated with cartilage calcification, but apparently is always associated with perichondral bone deposition (Figures 18.2 and 18.3). The mechanism of cartilage resorption is not well understood. Clastic cells involved in resorption of the perichondral bone rim have been observed to affect the cartilage as well (Sire et al., 1990). In view of their small size and their frequent mononuclear condition, it may well be that such clasts have been frequently overlooked and that they are much more common than hitherto assumed in the process of cartilage destruction.

Vascular elements seem to penetrate the medullary cavity only after the cartilage has been resorbed by other cells. The significance of such blood vessels is not well understood. Nutrition of the cartilage is probably by diffusion, as in mammalian cartilage. Cartilage canals, which may occur in large cartilaginous elements, probably serve to supply tissues other than the cartilage through which they run.

Distribution and function of cartilaginous tissues

In general terms, cartilage provides support and protection; it builds different types of joints and menisci where it meets specific biomechanical demands, or it acts as a template for subsequent bone formation and thus plays a pivotal role in the animal's growth. Beyond this generalization, the specific functions of the many different cartilage types remain largely unexplored, with some exceptions. Anker (1987), e.g., used an inductive approach to derive the function from the shape and structure of the different cartilages and other connective tissues in the jaw region of a teleost fish. Only such detailed studies will allow us to understand the functional significance of the different types of cartilage so abundantly present in certain taxa.

Many types of cartilage are permanent components of the adult skeleton. As Beresford (1993) has pointed out, this wide range of cartilages does not reflect unstable, primitive or poorly differentiated variants or forerunners of more 'typical' cartilage, but tissues adapted to specific, and fine-tuned mechanical and physiological functions, the nature of which remains largely unanswered.

Bone

Structure and function at the tissue level

Classically, bone is composed of three constituents: cells (osteocytes); an organic matrix; and a mineral phase. Bone, as a tissue, is the result of an ossification process, i.e. the secretion of an unmineralized precursor of the bone matrix (osteoid), in most cases followed by its mineralization. Bone matrix is deposited either directly in connective tissue (resulting in dermal and membrane bones), or on the surface of or within a cartilage model (producing perichondral and endochondral bones, respectively). In some cases, parts of tendons may mineralize to form so-called sesamoid bones. The teleostean urohyal is a well-known example (but see Patterson, 1977).

Dermal or membrane bones having no topographic relationship with any cartilage are said to be **achondral** (Figure 18.4). In many cases however the

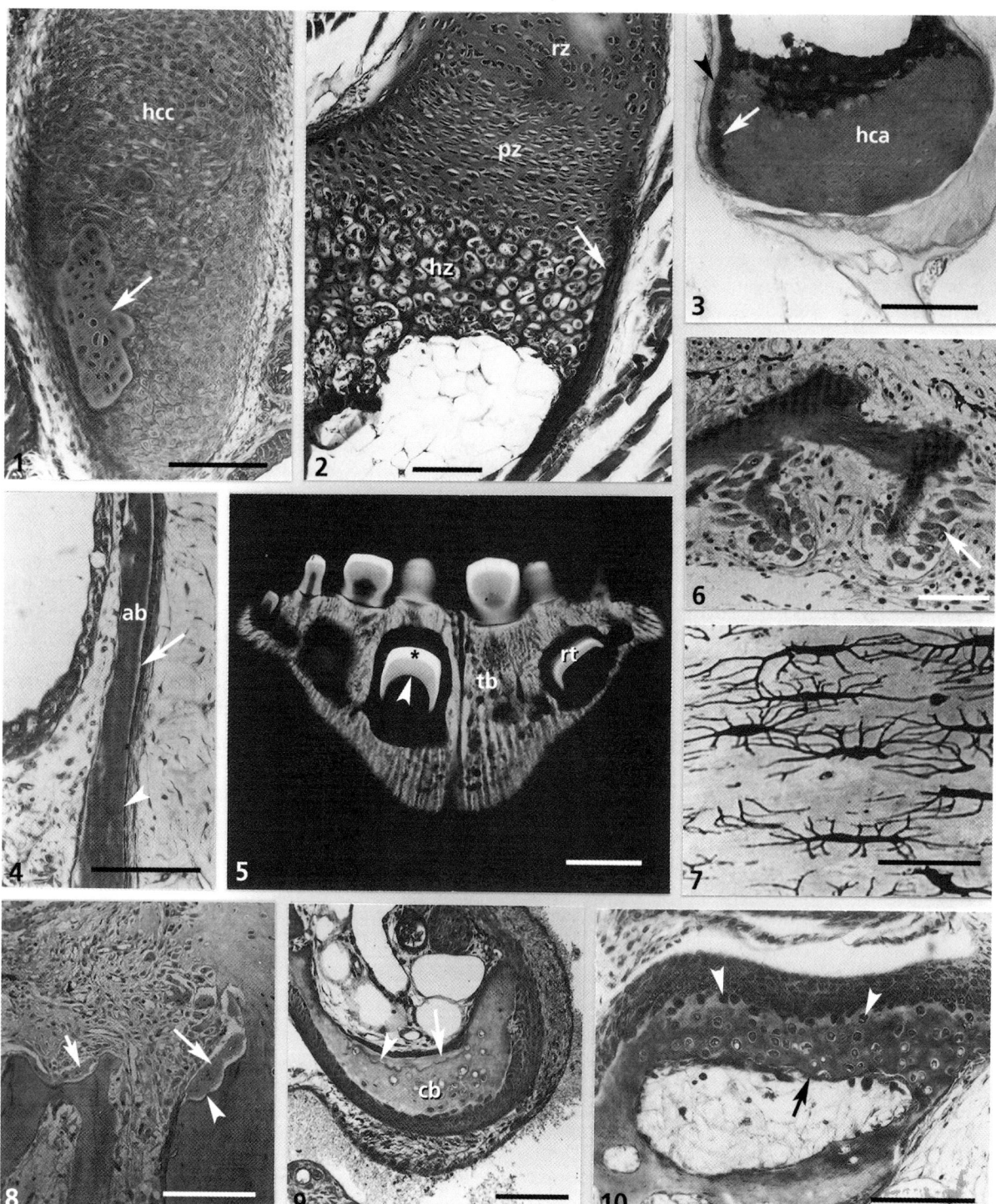

Figure 18.1 Lip cartilage of a nearly adult specimen of the cichlid *Astatotilapia elegans* (55 mm standard length, SL), composed of hyaline cell cartilage (hcc), locally differentiating into hyaline cartilage (arrow). Bouin-fixed, decalcified, 4 μm plastic section, stained with toluidine blue. Scale bar = 100 μm.

Figure 18.2 Branchial cartilage in a juvenile specimen of *Oncorhynchus mykiss* (approx. 150 mm SL). Cell-rich hyaline cartilage undergoing hypertrophy and resorption. Note the resting zone (rz), a zone of proliferation (pz) and a zone of (non-columnar) hypertrophic chondrocytes (hz). Cellular perichondral bone with sparse osteocytes (arrow) covers the hypertrophic cartilage. Bouin-fixed, decalcified, 10 μm paraffin section, stained with PAS-hemalum-picro-indigo-carmine. Scale bar = 100 μm.

Figure 18.3 Hyosymplectic cartilage in a juvenile specimen of the cichlid *Astatotilapia elegans* (25 mm SL) at its joint with the opercular bone. Hyaline cartilage undergoing spheritic calcification. Note the layer of (heavily stained) calcified cartilage (arrow) adjacent to a layer of (unstained) acellular perichondral bone (arrowhead). Formalin-fixed, undecalcified, 10 μm frozen section, stained with toluidine blue. Scale bar = 100 μm.

Figure 18.4 Infraorbital bone in a nearly adult specimen of the cichlid *Astatotilapia elegans* (55 mm SL). Compact acellular bone (ab) covered with a thin periosteum. Resting cells (non-active osteoblasts) (arrowhead) grade into cuboidal, more active osteoblasts arranged in a pseudoepithelial lining (arrow). Bouin-fixed, decalcified, 4 μm plastic section, stained with toluidine blue. Scale bar = 100 μm.

Figure 18.5 Pharyngeal jaw of the molluscivorous cichlid *Astatoreochromis alluaudi* (70 mm SL) composed of trabecular bone (tb). Trabeculae are oriented to withstand the pressure exerted on the jaw during snail crushing. Replacement teeth (rt) develop in the medullary cavity. Note distinct difference in mineralization between enameloid cap (asterisk) and underlying dentine (arrowhead). Microradiograph of transverse ground section, approximately 100 μm thick. Scale bar = 1 mm.

Figure 18.6 Osteoblasts (arrow) around the dentary bone of a 52 mm SL specimen of the cichlid *Eretmodus* cf. *cyanostictus*. Alcohol-fixed, decalcified, 2 μm plastic section stained with toluidine blue. Scale bar = 50 μm.

Figure 18.7 Osteocytes from a spiny ray in the dorsal fin of a carp *(Cyprinus carpio)*. Note extensive processes housed in a canalicular system. Silver impregnation according to Von Ebner. Courtesy of F.J. Meunier. Scale bar = 50 μm.

Figure 18.8 Osteoclasts (arrows) around the dentary bone of a 52 mm specimen of the cichlid *Eretmodus* cf. *cyanostictus*. One osteoclast lies within a distinct Howship lacuna (arrowhead). Same preparation as in Figure 18.6. Scale bar = 100 μm.

Figure 18.9 Cement line (arrowhead) in an articular area of the premaxillary bone of a nearly adult specimen of the cichlid *Astatotilapia elegans* (55 mm SL). The scalloped surface indicates that it is a reversal line; new, acellular, bone (arrow) has been deposited against the eroded surface of sparsely cellular chondroid bone (cb). Same preparation as in Figure 18.1. Scale bar = 100 μm.

Figure 18.10 Same preparation as in Figure 18.1. Chondroid bone in an articular area of the infraorbital bone. Note ongoing inclusion of osteoblast-like cells on the articular side (arrowheads), and breakdown at the opposite side (arrow). Scale bar = 100 μm.

shape of these bones mimics the outline of a cartilaginous element, though remaining separated from it by a few cell layers (often limited to a thin perichondrium and a thin periosteum). There is a clear functional advantage related to this so-called **parachondral** position of bones. Not only does it add strength to the cartilaginous skeleton but this is done in a way that precisely matches the mechanical requirements first imposed upon the cartilage. Parachondral bones are frequently found along the skull floor and in the skull roof.

Perichondral bone is very common in fishes. Nearly all the matrix-rich hyaline cartilages are covered at some stage during ontogeny by a layer of perichondral bone, except at their articular surfaces. The bone matrix is laid down at the immediate contact of the outer layers of the cartilage, and an admixture of both matrices can occur (Figure 18.3). Perichondral bone often persists as a thin layer without further (periosteal) ossification. The periosteum is usually thin, and not distinctly subdivided into an outer fibrous and an inner osteogenic layer.

In general, fish possess rather few 'long bones' with growth plate-like arrangements exhibiting typical endochondral (replacement) bone formation as seen in higher vertebrates. Resorption of cartilage is usually along a straight front, and no interterritorial bars of (calcified) cartilage persist as surfaces for endochondral bone formation (Figures 18.2 and 18.3). In some instances, e.g. in rod-like branchial elements, a thin seam of bone (closing plate) is deposited periodically along the straight resorption front, delimiting a new compartment of the medullary cavity, thus creating a ladder-like architecture of the bone (cf. Haines, 1934; personal observations). The rarity of endochondral (endosteal) trabeculae is suggested by Moss (1963) to be related to the non-weight-bearing role of the bones.

A vast array of cell types and tissue constituents can colonize the medullary cavity, which arises after resorption of the cartilage. These include blood vessels, nerve tracts or thin nerve fibers, undifferentiated mesenchyme and different progenitor cells. Unlike in tetrapods, the medullary cavity of fish bones is never hematopoietic. The endosteum is usually hardly more than a single layer of either resting cells, osteoblasts, or osteoclasts, or combinations of these. In many cases, the medullary cavity is completely filled with fat. These fat-filled cavities appear to be far more common in some taxa than in others, and their distribution may be related to an assumed role in buoyancy.

The architecture of bone is all-important from a functional viewpoint. The architecture deals with the fine structural organization of bones as organs, in particular with features such as porosity, vascularization and fine spatial arrangement of the major components. Compact bone has little or no cavities and few vascular elements. Such a condition is commonly seen in flat dermal bones of the cranial roof, or of the opercular series (Figure 18.4). Cancellous bone has a loose structure being composed of trabeculae separated by numerous extensions of the medullary cavity with blood vessels and loose connective tissue. **Trabecular** bone is a special category of cancellous bone in which the bone trabeculae take a preferential orientation in response to functional demands. A specific loading regime can entail the deposition of trabecular bone in the jaw medullary cavity in a cichlid fish (Huysseune et al., 1994) (Figure 18.5). Again, as with cartilage, the listed categories are stereotypes and several intermediate types can be distinguished.

When a vascular canal running through the bone of a vertebrate is surrounded by centripetally deposited layers of bone, a primary **osteon** is formed. New blood vessels and accompanying cells can penetrate such bone to form secondary osteons, thereby destroying the primary osteons. The tissue built up in this way is called Haversian bone. Scattered secondary osteons are well known from the skeleton of many fish (Ricqlès et al., 1991), but secondary bone reconstruction to produce dense Haversian bone has been reported only in large fish with a high metabolism such as tunafish (Amprino and Godina, 1956).

The bony skeleton experiences a continuous process of localized resorption and rebuilding, called remodeling, which is related to three factors (Ricqlès et al., 1991): (i) the need during ontogeny for bone to reshape to the appropriate adult morphology (size increase of individual skeletal elements is rarely isometric); (ii) the adaptation to special needs imposed by the environment, in particular certain physical conditions (physiological adaptation); and (iii) the response to physiological demands related to calcium metabolism. The challenge is to know how changes in shape and size of the bone are realized while the bone continuously fulfills all its mechanical and physiological functions. Remodeling is considerable in teleosts with a high metabolic activity, such as Thunnidae. Numerous species with less elevated metabolism exhibit much less remodeling (Meunier and François, 1992). Remodeling is often pronounced in dentigerous bones which need to accommodate successive generations of teeth.

Structure and function at the cytological and molecular level

The major organic constituent of the matrix is its fibrillar component, collagen. Minor constituents of vertebrate bone are glycoproteins, proteoglycans and lipids (Francillon-Vieillot et al., 1990) and there is no reason to believe that osteichthyan bone would be different in this aspect. The smallest unit visible at the transmission electron-microscopical (TEM) level is the collagen fibril, with a diameter of between 20 and 50 nm, and exhibiting cross-striation (Meunier, 1983). The normal collagen of bone is type I, but in some species bone has been shown to be positive for type II collagen (Benjamin and Ralphs, 1991). Collagen is a highly tension-resistant material and the predominant orientation of the fibrils, along with the way in which they are organized into fibers and fiber bundles, is informative of the nature of the tensile stresses to which the bone is submitted. In addition, there is a clear relationship between tissue typology and rate of bone deposition (Francillon-Vieillot et al., 1990; Ricqlès et al., 1991). Woven-fibered bone shows no preferential fibril orientation, is usually the first matrix laid down in a bone, and is adapted to quick embryonic growth (Ricqlès et al., 1991). In ontogeny, it is usually complemented by parallel-fibered bone, where overall fibril orientation is similar, and fibrils often lie parallel to the long axis of the bone. The majority of teleost bones are of this type. This type of bone is deposited more slowly than woven-fibered bone (Ricqlès et al., 1991). In lamellar bone matrix, subsequent lamellae, each with a distinct fibril orientation, are laid down on the inside of previously deposited lamellae around a vascular canal, thereby producing primary and secondary osteons. Subsequent lamellae need not take a concentric arrangement; they can become piled up in a stack. Such plywood-like matrices occur in the elasmodine (formerly called isopedine) of the basal plate of elasmoid scales, the scale type that covers the majority of the living bony fishes (Meunier, 1983; Sire, 1990; Zylberberg et al., 1992). Elasmodine is unusual in other aspects and it will be dealt with separately. Sharpey's fibers, fiber bundles penetrating into the bone, and which serve an anchoring function (e.g. for ligaments or tendons), are commonly found in fish primary bone (Meunier, 1983).

The process of matrix deposition is not continuous and the matrix usually carries obvious signs of periods of growth arrest or of resorption. These landmarks, or cement lines, persist in the tissue as long as the matrix carrying them is not itself resorbed. Cement lines are characterized by their strong tinctorial affinities and have been shown to be slightly more mineralized and contain more ground substance than the matrix between the lines (Castanet et al., 1993). When there is a mere arrest of growth, the cement line (rest line) is usually smooth, but when it represents the outline of a surface previously submitted to resorption, the cement line (reversal line) often has a scalloped appearance. Cement lines are persistent structures telling past histories; they offer precious information to those studying the dynamics of growth and remodeling of the skeleton at an individual scale. In some skull bones and in fin spines, cement lines may reflect cyclical growth and provide valuable information on the life history of the animal or, at a larger scale, on the age structure of the population (Castanet et al., 1993). One should nevertheless be aware of possible pitfalls since non-cyclical events can affect the growth of an individual and generate additional growth marks (Castanet et al., 1993).

The cells secreting the unmineralized precursor of bone (osteoid) are **osteoblasts** (Figure 18.6). They derive from osteoprogenitor cells, the origin of which has not been identified in fish. Depending on their secretory activity, but also their location on a bone, osteoblasts can take various shapes: pear-shaped around the free edge of a growing bone; plump and spindle-shaped along a bone surface; or cuboidal or even columnar in a pseudo-epithelial lining (compare Figures 18.4 and 18.6). Apart from their shape and arrangement, secretory osteoblasts can be distinguished in ordinary preparations by their intense basophilia; their distribution provides an easy way to evaluate the dynamic state of the bone (whether or not it is at rest, or in a phase of construction or resorption). Resting osteoblasts constitute the so-called lining cells (Meunier, 1983).

When osteoblasts become enclosed in osteoid, they become **osteocytes**. Teleosts present a considerable variation in both osteocyte concentration and morphology, according to the species considered and the location within the bone (Stéphan, 1900; Blanc, 1953; Moss, 1963; Meunier, 1983) (Figure 18.7). The number and length of the cytoplasmic processes housed in canaliculi can vary to a great extent (Stéphan, 1900; Meunier, 1983). There is little doubt that, as in endothermic animals, these cellular processes allow communication between osteocytes and exchange of metabolites.

The presence or absence of encapsulated osteoblasts distinguishes the two major categories of bone encountered in fishes: cellular (osteocytic) bone, with osteocytes enclosed in lacunar spaces in the bone matrix; and acellular (anosteocytic) bone, totally devoid of osteocytes. During acellular bone formation, the cells retract from the surface in much the same way as odontoblasts (the cells producing dentine) retract from the dentine surface, yet without leaving behind cellular processes as odontoblasts normally do (exceptions are the cellular prolongations in holostean bone, e.g. in the scales of *Lepisosteus* (Sire and Meunier, 1994) or the osteoblastic prolongations sometimes found in acellular bone of higher teleosts (Meunier, 1983). Young cellular bone in fishes is often acellular; only once the matrix has gained a certain thickness do cells become embedded. Cellular bone may be so sparsely populated by osteocytes as to be seemingly acellular in some areas; also, low-diameter vascular and non-vascular canals can easily be mistaken for osteocytic lacunae when cut in cross-section (Moss, 1965). Despite the absence of osteocytes, acellular bone is as capable of remodeling as is cellular bone (Witten and Villwock, 1997). Given the fact that acellular bone occurs in the so-called higher orders of teleosts which are the most numerous, evolved, diversified and widespread, the adaptive significance of acellular over cellular bone has long been questioned (cf. Parenti, 1986). Several authors have proposed that cells may have been disposed of during evolution because fish may not normally need bone as a mineral reservoir or osteocytes to mobilize this calcium (Moss, 1963). Families tend to have either cellular or acellular bone (that is, if one disregards the nature of the scales), and in only one case do both tissue types seem to be present in a single species (*Albula vulpes*, Moss, 1961b). There is no correlation between bone histology (cellular or acellular bone) and aquatic environment (marine or freshwater) (Moss, 1965). Clearly, comparative studies on teleostean cellular and acellular bone offer an exquisite opportunity to elucidate the role of osteocytes in the physiology of bone, yet few bone physiologists seem to have taken advantage of this possibility.

Resorption of both cellular and acellular bone is largely effected by resorbing cells which are often mononuclear and therefore easily overlooked in histological sections. Despite claims to the contrary, such cells present a characteristic ruffled border and clear zones, and deserve the term of osteoclasts (Sire et al., 1990). In other instances, they resemble the giant, multinuclear cells known from mammals (Figure 18.8). Lopez (1973) suggested two more mechanisms for bone resorption: osteocytic osteolysis, and **halastasis**, the latter involving removal of the unstable phase of calcium phosphate without lysis of the matrix.

Bone tissues eventually mineralize. As in other vertebrates, the inorganic phase is calcium phosphate, largely in the form of hydroxyapatite (Meunier, 1983). The degree of mineralization of both acellular and cellular bone in osteichthyans is similar to that of mammalian bone, yet the range of values is slightly higher (Meunier and Huysseune, 1992). Acellular bone appears to be slightly more mineralized than cellular bone, and bone in marine teleosts slightly more than in freshwater species (Meunier and François, 1992). Mineralization of osteoid is achieved by inotropic calcification, characterized by crystal deposition along the collagen fibrils (Meunier, 1983). Matrix vesicles, considered to play a significant role in the nucleation of hydroxyapatite crystals in endothermic vertebrates, have also been reported in fish bone (Lopez et al., 1978; Weiss and Watabe, 1979).

Teleosts are generally able to maintain constant serum calcium levels, of the same order of magnitude as those of mammals (Moss, 1965). The source of mineral ions for fish is environmental water, rather than food (Meunier, 1983; Takagi and Yamada, 1992). The maintenance of homeostatically regulated values of calcium and phosphate appears to be independent of bone type and marine or freshwater environment (Moss, 1965). The role of bone, in particular acellular bone, in calcium homeostasis has been disputed. Despite claims to the contrary, acellular bone is metabolically active and can mobilize calcium, especially in stress situations (Weiss and Watabe, 1978; Takagi and Yamada, 1992), just like cellular bone (Kacem et al., 1998).

The hormones involved in calcium metabolism in fish differ from those in terrestrial vertebrates. Calcium regulation in fish is dominated by the hypercalcemic action of prolactin, a pituitary hormone (fish lack a parathyroid gland) and the hypocalcemic action of hypocalcin, a hormone produced by the corpuscles of Stannius (these are absent, however, in chondrichthyans; Clement, 1992). The ultimobranchial body secretes a hormone that is close to mammalian calcitonin but that is probably not significantly involved in controlling the exchange between bone and body fluids (Wendelaar Bonga and Lammers, 1982; Clement, 1992).

Chondroid bone

Chondroid bone stands for tissues which combine a mineralized, bone-like matrix, with cells which are reminiscent of cartilage cells by their size, shape, cytological characteristics, and lack of canalicular system (Huysseune and Sire, 1990) (Figures 18.9 and 18.10). Chondroid bone can be anything between almost acellular and highly cellular, multilayered to nearly single-layered (Meunier and Huysseune, 1992). It is unlikely, though, that chondroid bone can turn into either acellular or cellular bone, in the way suggested by Moss (1961a, 1963).

Chondroid bone is overwhelmingly present in teleosts. It is found in joints and other sites subjected to mechanical stress where it seems to meet two demands: the need for an accelerated local growth rate and the need for a shear-resistant support (Huysseune and Sire, 1990; Meunier and Huysseune, 1992).

Teeth and dental tissues

Teeth, composed of dentine overlain by a cap of hypermineralized tissue, are present in chondrichthyans and osteichthyans (actinopterygians as well as sarcopterygians).[4] Again, the spectrum of tissues building up these structures is broad, especially when fossils are considered. **Odontodes** (placoid scales in chondrichthyans, and dermal denticles on the body surface of some osteichthyans), although physically related to the integument, are considered homologous to teeth (cf. Huysseune and Sire, 1998; see Smith and Coates, 1998 for a challenging view on the evolutionary origin of teeth). It is assumed that odontodes have been modified during evolution to give rise to odontocomplexes and that some dermal skeletal elements contain odontode-derived tissues (e.g. the tissues covering the cranial bones, ganoid scales and fin rays in

---

[4] Agnathans (hagfish and lampreys) possess horny teeth.

polypterids and in lepisosteids) or tissues suspected to be odontode-derived (e.g. in the elasmoid scales, and in the surface of **scutes**) (see Huysseune and Sire, 1998, for a circumstantial discussion on these issues).

The hypermineralized cap of teeth consists of enamel or enameloid (Figure 18.5). Once completed, both resemble each other, but the developmental pathways producing them are quite different: enameloid is of mixed epithelial–mesenchymal origin, enamel is purely epithelial (Smith, 1995). Enamel is assumed to occur exclusively in *Latimeria*, dipnoans and basal actinopterygians; chondrichthyans and advanced actinopterygians would possess enameloid only (but see the discussion by Smith, 1995). The crystallites in enameloid are organized into parallel groups or crystal bundles, the pattern of which appears to be linked to the cutting or crushing function of the tooth (Smith, 1995). Crystallite arrangement in enamel, though more ordered than in enameloid, does not reach the highly ordered patterns observed in mammals. A hypermineralized collar surrounds the shaft in many actinopterygians and is considered by Smith (1995) to be enamel.

Dentine is a hard bone-like tissue composed of largely the same major constituents as bone. Usually dentine contains tubules housing cytoplasmic processes persisting after retraction of the odontoblasts (but see Ørvig, 1951, for a comprehensive discussion on the nature of dentine in fish). When the dentine is atubular however, as in many first-generation teeth, it is hardly distinguishable from acellular bone or even cellular bone with low levels of osteocytic incorporation. Some proteins appear to be dentine-specific, at least in mammals, notably dentine phosphoprotein (DPP) and dentine sialoprotein (DSP) (Butler, 1998); if such would be the case in teleosts, their detection would greatly help to analyze the developmental and evolutionary relationships of bone and dentine in fish.

Teeth in chondrichthyans are attached to the cartilage by means of fibrous tissues. In osteichthyans a wide range of structures serve to attach the tooth to its support, some of which appear to be dentinous in nature, others bony. The variable nature of the attachment structure has led to much confusion in its terminology (see Huysseune and Sire, 1998). The attachment structure firmly **ankyloses** the tooth if there is continuity in mineralization between tooth and dentigerous bone, or allows mobility in areas where unmineralized fibers connect tooth and bone (see Fink, 1981, for a phylogenetic interpretation).

Non-osseous tissues of the dermal skeleton

Other elements of the dermal skeleton contain tissues which cannot be easily classified as either bone or dental tissue. This is the case for the tissue covering the cranial bones, ganoid scales and fin rays in polypterids and lepisosteids (ganoine), for tissues building up the elasmoid scale (notably the elasmodine of the basal plate), or for the superficial tissue on the scutes of armored catfish (hyaloine). It has been argued that these tissues are probably derived from tissues building ancestral odontocomplexes (but see the review by Huysseune and Sire, 1998).

Ganoine is a hypermineralized, non-collagenous, stratified tissue of epidermal origin and homologous to true enamel (Ørvig, 1951; Sire *et al.*, 1987; Huysseune and Sire, 1998). Hyaloine is another highly mineralized, non-collagenous tissue sharing structural characteristics with other superficial tissues of the dermal skeleton (Sire, 1993). Elasmodine has a plywood-like structure; its strata contain collagen fibrils that range in diameter from 60 to 100 nm. Elasmodine remains largely unmineralized and mineralized strata have a lower degree of mineralization compared to the bone tissue of the same species (Meunier, 1983). In addition to an incomplete mineralization, cells are usually absent from elasmodine (Meunier and Huysseune, 1992). Evolutionary loss of mineralization not only affects the basal plate of elasmoid scales; it is also observed in the fin rays of dipnoans and involves a local decrease in the amount of mineralized tissue rather than an overall decrease in the level of mineralization (Meunier and François, 1992).

Conclusion

The aquatic environment imposes different physical and physiological conditions upon the organism as compared to terrestrial life. Within this environment, fish have been able to experiment widely with the building materials of the skeleton, resulting in an astonishing range of combinations of the three basic components of skeletal material (cells, matrix and inorganic constituents), organized into an equally astonishing range of skeletal tissues. Each of these tissues represents the outcome of specific developmental pathways and functional adaptations, the nature of which we still largely ignore. The invaluable advantage of such a tissue richness resides in the potential explanatory power that lies hidden in comparative studies of the structure and function of the fish skeleton.

Acknowledgments

Professor B.K. Hall and Dr T. Miyake (Dalhousie University, Halifax, Canada), Prof. F.J. Meunier (MNHN, Paris, France) and Dr J.-Y. Sire (CNRS, Université Paris 7, France) are gratefully acknowledged for their helpful comments on the manuscript.

References

Amprino, R. and Godina, G. (1956). *Publ. Staz. Zool. Napoli* **28**, 62–71.
Anker, G.C. (1987). *Neth. J. Zool.* **37**, 394–427.
Benjamin, M. (1989a). *Anat. Embryol.* **179**, 285–303.
Benjamin, M. (1989b). *J. Anat.* **164**, 145–154.
Benjamin, M. (1990). *J. Anat.* **169**, 153–172.
Benjamin, M. and Ralphs, J.R. (1991). *J. Anat.* **179**, 137–148.
Beresford, W.A. (1993). In *The Skull. Vol. 2. Patterns of Structural and Systematic Diversity* (eds J. Hanken and B.K. Hall), pp. 69–130. The University of Chicago Press, Chicago.
Blanc, M. (1953). *Mém. Mus. natn. Hist. nat., Série A* **7**, 1–146.
Bordat, C. (1987). *Can. J. Zool.* **65**, 1435–1444.
Butler, W.T. (1998). *Eur. J. Oral Sci.* **106** (Suppl. 1), 204–210.
Castanet, J., Francillon-Vieillot, H., Meunier, F.J. and de Ricqlès, A. (1993). In *Bone, Vol. 7B* (ed. B.K. Hall), pp. 245–283. CRC Press, Boca Raton.
Clement, J.G. (1992). *Aust. J. Mar. Freshwater Res.* **43**, 157–181.
Fink, W.L. (1981). *J. Morphol.* **167**, 167–184.
Francillon-Vieillot, H., de Buffrénil, V., Castanet, J., Géraudie, J., Meunier, F.J., Sire, J.Y., Zylberberg, L. and de Ricqlès, A. (1990). In *Skeletal Biomineralization: Patterns, Processes and Evolutionary Trends* (ed. J.G. Carter), pp. 471–530. Van Nostrand Reinhold, New York.
Haines, R.W. (1934). *Quart. J. Microsc. Sci.* **77**, 77–97.
Hall, B.K. (1978). *Developmental and Cellular Skeletal Biology*. Academic Press, New York. 304pp.
Huysseune, A. and Sire, J.-Y. (1990). *Tissue Cell* **22**, 371–383.
Huysseune, A. and Sire, J.-Y. (1998). *Eur. J. Oral Sci.* **106**, 437–481.
Huysseune, A., Sire, J.-Y. and Meunier, F.J. (1994). *J. Morphol.* **221**, 25–43.
Kacem, A., Meunier, F.J. and Baglinière, J.L. (1998). *J. Fish Biol.* **53**, 1096–1109.
Kemp, N.E. and Westrin, S.K. (1979). *J. Morphol.* **160**, 75–102.
Lopez, E. (1973). *Mém. Mus. Nat. Hist. nat.* **80**, 1–90.
Lopez, E., Baud, C.A., Boivin, G. and Lallier, F. (1978). *Ann. Biol. anim. Bioch. Biophys.* **18**, 105–117.
Meunier, F.J. (1983). Les tissus osseux des Ostéichthyens. Structure, genèse, croissance et évolution. Archives et Documents, Micro-Edition, Inst. Ethnol., S.N. 82-600-**328**, 1–200.
Meunier, F.J. and François, Y. (1992). *Ann. Biol.* **31**, 169–184.
Meunier, F.J. and Huysseune, A. (1992). *Neth. J. Zool.* **42**, 445–458.
Moss, M.L. (1961a). *Am. J. Anat.* **108**, 99–109.
Moss, M.L. (1961b). *Acta Anat.* **46**, 343–362.
Moss, M.L. (1963). *Ann. NY Acad. Sci.* **109**, 337–350.
Moss, M.L. (1965). *Acta Anat.* **60**, 262–276.
Moss, M.L. (1977). *Amer. Zool.* **17**, 335–342.
Moss, M.L. and Moss-Salentijn, L. (1983). In *Cartilage. Vol. I. Structure, Function and Biochemistry* (ed. B.K. Hall), pp. 1–30. Academic Press, New York, London.
Ørvig, T. (1951). *Ark. Zool.* **2**, 321–454.
Ørvig, T. (1967). In *Structural and Chemical Organization of Teeth, vol. 1.* (ed. A.E.W. Miles), pp. 45–110. Academic Press, London.
Parenti, L.R. (1986). *Zool. J. Linn. Soc.* **87**, 37–51.
Patterson, C. (1977). In *Problems in Vertebrate Evolution* (eds S. Andrews, R. Miles and A. Walker), pp. 77–121. Academic Press, London.
Ricqlès, A. de, Meunier, F.J., Castanet, J. and Francillon-Vieillot, H. (1991). In *Bone Vol. 3* (ed. B.K. Hall), pp. 1–78. CRC Press, Boca Raton.
Schaeffer, B. (1977). In *Problems in Vertebrate Evolution.* Linnean Society Symposium Series, vol. 4 (eds S.M. Andrews, R.S. Miles and A.D. Walker), pp. 25–52.
Schaffer, J. (1930). In *Handbuch der mikroskopischen Anatomie des Menschen. Bd. 2, Vol. 2* (ed. W. von Möllendorf), pp. 1–390. Springer Verlag, Berlin.
Sire, J.-Y. (1990). *Neth. J. Zool.* **40**, 75–92.
Sire, J.-Y. (1993). *J. Morphol.* **215**, 225–244.
Sire, J.-Y. and Meunier, F.J. (1994). *Acta Zool., Stockh.* **75**, 235–247.
Sire, J.-Y., Géraudie, J., Meunier, F.J. and Zylberberg, L. (1987). *Am. J. Anat.* **180**, 391–402.
Sire, J.-Y., Huysseune, A. and Meunier, F.J. (1990). *Cell Tissue Res.* **260**, 85–94.

Smith, M.M. (1995). In *Evolutionary Change and Heterochrony* (ed. K.J. McNamara), pp. 125–150. John Wiley and Sons, New York.

Smith, M.M. and Coates, M. (1998). *Eur. J. Oral. Sci.* **106** (Suppl. 1), 482–500.

Stéphan, P. (1900). *Bull. Sci. Fr. Belg.* **33**, 281–429.

Takagi, Y. and Yamada, J. (1992). *Comp. Biochem. Physiol.* **102A**, 481–485.

Weiss, R.E. and Watabe, N. (1978). *Comp. Biochem. Physiol.* **60A**, 207–211.

Weiss, R.E. and Watabe, N. (1979). *Calcif. Tissue Int.* **28**, 43–56.

Wendelaar Bonga, S.E. and Lammers, P.I. (1982). *Gen. Comp. Endocrinol.* **48**, 60–70.

Witten, P.E. and Villwock, W. (1997). *J. Appl. Ichthyol.* **13**, 149–158.

Zylberberg, L., Géraudie, J., Meunier, F.J. and Sire, J.-Y. (1992). In *Bone Vol. 4* (ed. B.K. Hall), pp. 171–224. CRC Press, Boca Raton.

CHAPTER 19

Fish as an Experimental Model for Studying Muscle Function

Lawrence C Rome
Department of Biology, University of Pennsylvania,
Philadelphia, Pennsylvania, USA

Introduction

Animals perform a tremendous variety of motor tasks, ranging from making explosive movements taking less than ~50 ms, to swimming hundreds of miles in the ocean, to producing sound at several hundred hertz for communication. These activities require very different outputs from the muscles (Rome, 1997; Rome and Lindstedt, 1997). Accordingly, research over the past several decades has revealed that muscle tissue has tremendous diversity, perhaps more than any other tissue.

Recently, with the advent of new integrative muscle physiology techniques, research has focused on comparing these diverse muscle properties to the function muscles must actually perform *in vivo* (Rome *et al.*, 1993; Lutz and Rome, 1994). This has led to two important conclusions: first, there is an excellent matching between the properties of the muscles and the mechanical performance that is required of the muscles *in vivo*; second, no one muscle can perform all required motor activities effectively – rather, different types of muscles (along with different muscular anatomies) are required to perform the full repertoire of motor activities in which animals engage (Rome

Copyright © 2000 Academic Press

and Lindstedt, 1998). These facts no doubt underlie the large diversity of muscle properties observed.

Why fish provide a superior experimental model for exploring muscle function

These breakthroughs in the understanding of muscle function have been fueled in large part by the use of fish as an experimental model. Although some investigators have had a primary interest in fish, many have used fish to exploit their experimental advantages (Curtin and Woledge, 1988; Rome *et al.*, 1988; Altringham and Johnston, 1990). Thus, in this chapter, we will not only discuss the muscular system of fish, but we will emphasize the uniqueness of the system, and the special role it has played (and continues to play) in the field of muscle design.

In all vertebrates there is great diversity of properties of muscle fibers within a given animal to perform different functions, and these can be usefully categorized as different fiber types. However, characterizing the properties of these different fiber types and how they are used to perform different functions was problematic prior to the use of fish. Mammalian muscle is typically heterogeneous, that is slow-twitch muscle fibers are interspersed with fast-twitch muscle fibers (Figure 19.1). Experimentally, this makes it very difficult to identify which fibers are active during particular motor activities (i.e. electromyography electrodes implanted in the muscles will pick up signals from fibers of all types, making it impossible to discriminate which type(s) are powering a particular movement). In addition, any given bundle of fibers dissected from a muscle will necessarily contain more than one fiber type, making it difficult to differentiate the properties of one fiber type from another (Rome, 1998). Speculation about the properties of different fiber types, and how they are used during activities such as running and weight lifting is rampant in the sports industry. However, in the past these conclusions have been based on only indirect evidence.

In fish, by contrast, the different muscle fiber types are organized into large, homogeneous, and ana-

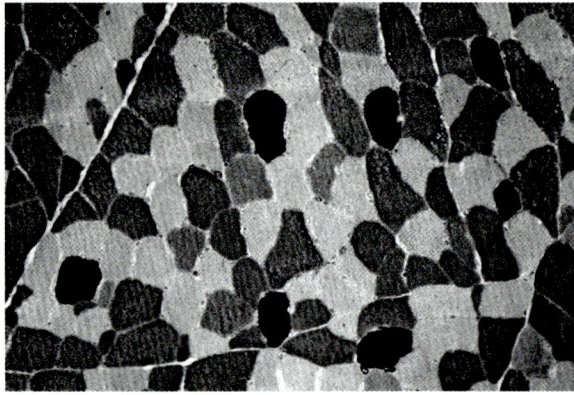

Figure 19.1 Cross-section of typical mammalian muscle histochemically stained to show different fiber types. Darkly stained fibers are type I, unstained fibers are type IIa, and intermediate stained fibers are type IIb. Note the fiber type can vary from one cell to the next.

tomically separated regions which are obvious to the naked eye (Figure 19.2c). This enables researchers to overcome both obstacles mentioned above (Rome *et al.*, 1988). First, electromyography (EMG) electrodes can be implanted in these anatomically separate regions to monitor which swimming activities are powered by each fiber type (see Figure 19.2c) (Rome *et al.*, 1984). Second, one can dissect bundles of fibers, which are all of the same type, and hence one can, in a relatively straightforward manner, determine the mechanical, biochemical and ultrastructural properties of each muscle fiber type (Figure 19.2a,b). Because of these advantages, the fish muscular system has become widely used by those trying to understand comparative muscle design (Rome, 1994; Rome and Lindstedt, 1997). It should be noted that in some species (e.g. salmonids), this separation may not be complete (i.e. red muscle fibers are interspersed with white muscle fibers). Consequently, prior to using a particular species as an experimental model, its fiber type distribution has to be determined.

Muscle fiber types

In this section, four different fish muscle fiber types are briefly defined and how they are used *in vivo* is discussed. In the rest of the chapter, the differences in properties and usage are explored in greater detail. We focus here on three trunk muscle fiber types used for swimming and one muscle fiber type associated with the swimbladder used for sound production. It should be noted that additional fiber types are found

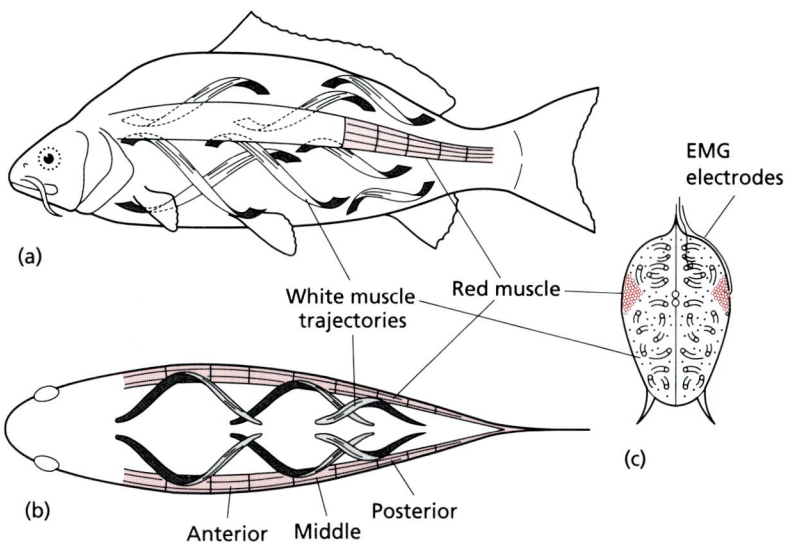

Figure 19.2 Three views of carp. The red muscle forms a thin sheet just under the skin (its thickness is exaggerated for illustrative purposes in part (c). The red fibers run parallel to the body axis (a,b). Note, the pink muscle (not shown) also forms a thin sheet of parallel fibers that lies just medial to the red. The white fibers run in a helical fashion with respect to the backbone (c). Consequently, white fibers need shorten by only about one-quarter as much as the red ones to produce a given curvature change of the body. By placing electromyography (EMG) electrodes as shown, one can monitor the activity of different fiber types during swimming. EMGs reveal that at low swimming speeds only the red fibers are used, whereas at high swimming speeds and the escape response, the white muscle fibers are used as well. (Adapted from Rome et al., 1988.)

in the jaws, attached to the eyes, and attached to the fins. In addition, some muscles have lost their ability to generate force altogether, and have evolved into heater organs (Block, 1991). Discussion of these additional fiber types goes beyond the coverage here.

Mechanical properties

Muscle fiber types are defined by both their mechanical properties and their metabolic properties. The mechanical properties can be expressed by measurements of the maximum velocity of shortening (V_{max}), the rate of activation and force generation, and the rate of muscle relaxation. The rate at which muscle uses energy is also generally proportional to the overall speed of the muscle (Woledge et al., 1985). Muscles used for slow motor activities (e.g. maintaining posture or performing slow movements), typically activate and relax slowly, have a low V_{max} and use energy slowly. These are called slow-twitch muscle fibers. Muscles used for fast and powerful movements, activate and relax quickly, have a high V_{max} and use energy rapidly. These are called fast-twitch muscles. Muscles used for rapid eye movements or to generate sound for communication activate and relax extremely rapidly, have a high V_{max} and use energy extremely rapidly. These are called 'superfast-twitch muscles'.

Metabolic properties

Slow-twitch fibers are thought to be recruited even during low intensity activity, and hence are used a large percentage of the time. They must therefore be highly aerobic to prevent build-up of anaerobic by-products (e.g. lactic acid) which cause fatigue. In mammals, slow-twitch fibers with high aerobic capacities are called 'slow-twitch oxidative fibers' (SO, type I). In fish, they are called 'slow-twitch red fibers' (the red color in fish muscle is due to the presence of high densities of myoglobin, mitochondria and capillaries necessary to support a high oxygen consumption rate). At the other end of the spectrum, the 'fast-twitch glycolytic' muscles (FG or type IIb in mammals), are the last fiber types to be recruited and thus are used only occasionally (for rapid acceleration or escape). They use energy very rapidly – at a rate which cannot be maintained aerobically. Thus they use glycolytic metabolism almost exclusively which results in rapid fatigue. In most fish, this muscle is white (due to very low densities of myoglobin, mitochondria and capillaries), and hence is referred to as 'fast-twitch white muscle'. There is also an intermediate fiber type, that is both fast and used aerobically (e.g. for long distance fish swimming and bird flying). In mammals, these are called 'fast-twitch oxidative glycolytic' (FOG) or type IIa fibers. In fish, they are called 'pink muscle' (due to intermediate densities of myoglobin, mitochondria and blood supply). In fish, the pink muscle forms a thin sheet which sits between the red and white muscle. Finally, superfast-twitch muscles can be either highly aerobic or anaerobic depending on their function. Two aerobic muscles are the shaker muscle of rattlesnakes, which shakes its tail for many hours, and the swimbladder muscle of the midshipman (*Porichthys notatus*), which produces a long and continuous mating call. The swimbladder muscle of the oyster toadfish (*Opsanus tau*), in contrast, has a low aerobic capacity, reflecting that its mating call is intermittent.

Recruitment of different fiber types

In the preceding paragraph we discussed the general usage of different fiber types. Prior to work on fish, the recruitment pattern of different fiber types was rarely measured during locomotion. However, taking advantage of the anatomical separation of the muscle fiber types in fish has made it relatively straightforward to measure recruitment. In experiments on swimming carp (*Cyprinus carpio*), it was found that at relatively slow steady swimming speeds only slow-twitch red fibers were recruited, and at high speeds the fast-twitch white muscles were additionally recruited (Rome *et al.*, 1984). Interestingly, when fish recruited the white muscle they shifted from 'steady' swimming to 'burst and coast' swimming (Rome *et al.*, 1990). Subsequent work on the marine scup (*Stenotomus chrysops*) showed that the intermediate pink muscle was additionally recruited at intermediate steady swimming speeds (Coughlin and Rome, 1999). Thus as the fish swims faster, faster muscle fiber types are additionally recruited, but the slower types are not turned off. The swimming speed at which faster muscle types are recruited seems to be well explained by the biomechanics of the muscle and swimming (see below).

Muscle structure

Muscle structure can be viewed at three levels: the gross, the cellular, and the molecular level. There is a great deal of variation in muscles at these levels, some of which is associated with different fiber types.

Gross level

The gross structure sets the gearing of muscles, that is, how much body movement a given shortening of muscle can produce. This gearing is set by both the orientation of the fibers within the muscle and the placement of the origin and insertion of the muscle. One fiber arrangement is 'parallel', in which the fibers run parallel to the long axis of the body – this is evident in the red and pink muscle of fish. Another orientation is a complex helical arrangement found in the white muscle of fish (Figure 19.2) (Rome *et al.*, 1988; Rome and Sosnicki, 1991). Different fiber arrangements can have a marked effect on the function of muscle. In fish, for instance, parallel orientation to the long axis endows the red and pink muscle with a relatively low gear ratio, whereas the helical structure of the white muscle endows it with a high gear ratio. The different gear ratios are equivalent to low and high gear on a 10-speed bicycle. This difference, together with differences in V_{max}, has a dramatic effect on muscle function in fish (see below).

Cellular level

At the cellular level, the space within a muscle fiber is largely occupied by three components in differing proportions (force-generating myofibrils, the Ca^{2+}-pumping sarcoplasmic reticulum (SR), and oxygen-consuming mitochondria). As these different structures are necessary for different functions, increase in the space occupied by one structure (and its function) must be at the expense of another. This competition for space puts some limitations on the properties of different fiber types (Rome and Lindstedt, 1998). A fourth component, stored metabolic substrates such as glycogen and fat, is generally small and is ignored here.

Aerobic muscle fibers, red and pink, must have a high density of mitochondria. Indeed, the red muscle of fish generally contains 25–30% mitochondria (Johnston, 1983). The volume of mitochondria, in turn, reduces the space available for the force-generating myofibrils and, hence, red fibers generate less force than white fibers. Superfast fibers need to have high volumes of SR to cycle Ca^{2+} very rapidly, and thus myofibrillar volume is accordingly reduced which contributes to a reduction in force (see below).

Ultrastructural/molecular level

The force-generating myofilaments consist of myosin thick filaments and actin thin filaments. The thick filaments are 1.6 μm long in vertebrates (Page and Huxley, 1963). Because the distribution of myosin heads along thick filaments is similar, all vertebrates have the same number of myosin heads. However, there are different types (isoforms) of myosin, and these different types (slow, fast and superfast) confer many of the observed differences in mechanical properties (e.g. different V_{max}s). The thin filaments are made primarily of actin monomers which are identical in different fiber types, but the number of monomers in a filament, and hence its length can vary. Whereas thin filaments are generally 1.0 μm long in fish and most other vertebrates (Sosnicki et al., 1991), mammalian thin filaments can be longer (1.2–1.3 μm long) which leads to a change in the sarcomere length–tension relationship.

The thin filament also contains the regulatory proteins which control the activation and relaxation of muscle. These consist of the troponin complex (troponin C, troponin T and troponin I) as well as the tropomyosin. Again, different muscle fiber types have different isoforms of these proteins, which alter their performance (see muscle activation, force generation and muscle relaxation, below).

Design of the fish muscular system

Having described some of the basic elements of the fish muscular system, we will now examine more functional aspects and point out along the way the insights that the fish muscular system has provided to the field of muscle design in general. Fish have been a very helpful model in elucidating how muscular systems are designed. The reason is threefold. First, fish undergo a wide range of motor movements from swimming (see Figure 19.3) to producing sound at 200 Hz, and these can be easily elicited and quantitatively analyzed. Second, these motor movements are powered by different muscle fiber types permitting exploration of the function of different fiber types. And most importantly, one can experimentally determine: (i) which fiber types are powering a particular

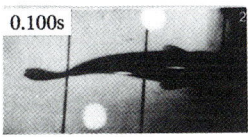
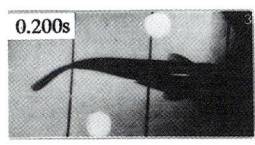
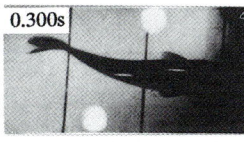
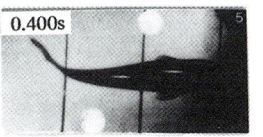

Figure 19.3 High speed motion picture frames during steady swimming with red muscle (left, five film frames separated by 0.1 s) and escape response with white muscle (right, six frames at 5 ms intervals). When fish swim, the muscle on the concave side *actively* shortens and the muscle on the convex side is lengthened (while relaxed) by active shortening of the contralateral muscle. The sarcomere length change is proportional to the curvature of the backbone. Note the extreme curvature of the backbone during the escape response compared to very little curvature during steady swimming. Note also the relative time scales: one full tail beat of steady swimming takes 0.4 s whereas it takes only 0.025 s for the fish to complete bending during the escape response.

activity; (ii) what length change and stimulation pattern they undergo; and (iii) the mechanical properties of the different fiber types (Rome, 1998).

Fish produce a wide range of movements (Rome et al., 1988). The sarcomere length (SL) change of the muscle depends largely on the curvature of the spine. High speed motion pictures of carp swimming at 20 cm s^{-1} reveal that there is very little curvature in the backbone (Figure 19.3). In contrast, during the escape response, there is a large amount of curvature in the backbone (Figure 19.3). Note also the difference in time scales. During steady swimming, one tailbeat takes 400 ms (i.e. 2.5 Hz), whereas in the escape response, the fish goes from straight to highly curved in only 25 ms. *How has the muscular system evolved so that it can produce both these types of movements effectively?*

Myofilament overlap and V/V_{max} – two design 'goals' of the muscular system

From cell physiology, we may anticipate that there are some rules (Rome, 1994; Rome and Lindstedt, 1997) that are followed when an animal muscular system evolves. During full activation, the force generated depends on the amount of overlap between the myosin thick filament and actin thin filament or, more precisely, the number of myosin crossbridges which can interact with actin sites (Gordon et al., 1966). Force is maximal when there is optimal overlap between the filaments so that the maximum number of myosin heads can attach to actin. Force falls as the sarcomere length increases, and the overlap and number of myosin heads that can interact declines. Force also declines when the sarcomeres become too short, reflecting possible interference in force generation by myosin heads (Figure 19.4).

It would seem sensible that during evolution the gear ratio of the animal's muscle fibers and their myofilament lengths are varied, so that no matter what movements the animal makes, the muscle would operate at optimal myofilament overlap (i.e. where the muscle generates near maximal force; see Figure 19.4). As such, gear ratio and myofilament lengths can be viewed as the *design parameters* (those components that can be varied during evolution). Myofilament overlap can be viewed as a *design constraint* or *design goal* (i.e. the rule by which the variation in parameters is adjusted). As both design parameters are anatomical

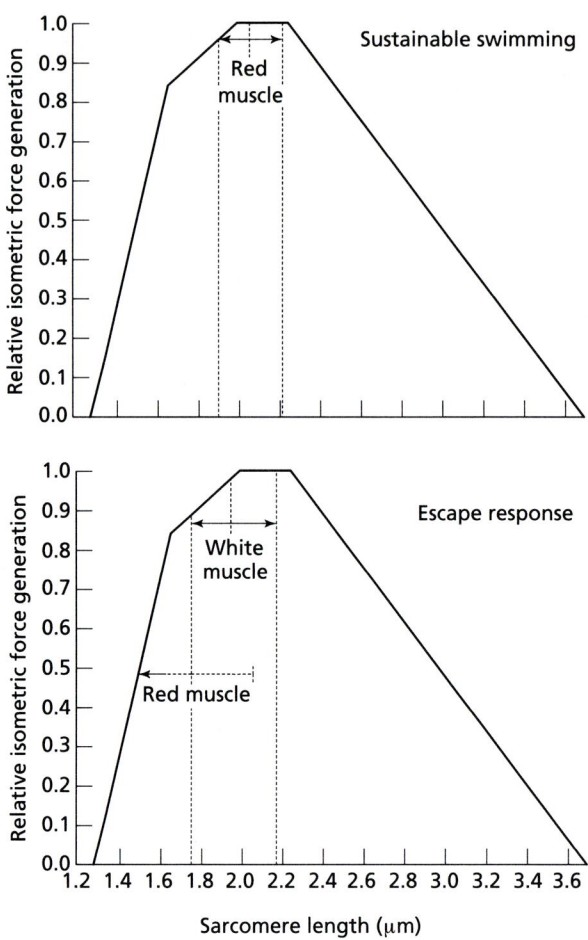

Figure 19.4 Myofilament overlap in swimming carp. During sustained swimming (top) only the red muscle is active. The dotted lines and arrows show the SL excursion (1.91–2.23 μm). Note, that the force does not drop below 96%. If the red muscle had to power the more extreme escape response (bottom), it would have to shorten to 1.4 μm where it generates little tension. Instead, the white muscle, which has a fourfold greater gear ratio, is used. In the posterior region of the fish, the white muscle shortens to only 1.75 μm (arrows and dotted line), where at least 85% maximal tension is generated. In the rest of the fish the excursion is smaller and the force higher. Note because carp myofilament lengths are the same as frog, the frog SL–tension curve is used to describe the SL–tension curves for carp red and white muscles.

features of the muscle, one being at the organ level and the other at the molecular level, this can be viewed as a *structural* design consideration.

There is also a dynamic design consideration which takes into account the fact that muscle shortens during locomotion. When a muscle is isometric, it generates its maximal force. However, to move limbs or bend the body, muscles must shorten. Muscle shortening is also necessary to perform mechanical work – a requirement of increasing the mechanical energy of a body (e.g. during jumping) and for overcoming drag (e.g. during flying or swimming).

The relationship between the force and velocity is described by the force–velocity curve (Figure 19.5). The force muscle generates is a function of the velocity when it shortens (V), or more precisely a function of V/V_{max} (Hill, 1950). The faster the muscle shortens, the less force it generates, until it generates zero force when shortening at its maximum velocity of shortening, V_{max}. In addition and more importantly, the mechanical power that a muscle generates and the efficiency (mechanical power output/rate of metabolic energy usage) with which it generates mechanical power are functions of V/V_{max} as well (Figure 19.5). As mentioned previously, animals have different fiber types (i.e. different V_{max}s within the same animal). Hence we might anticipate that the muscular system of animals who need to generate power (swimming fish, jumping frogs, flying birds) would be designed so that no matter what movement the animal makes, the muscle fibers operate over a range of V/V_{max} values (0.15–0.40) where the fibers generate maximal power with near maximal efficiency. Thus, the design parameters V_{max} and fiber gear ratio are varied in such a way that they operate under the design constraint of V/V_{max}.

It is important to emphasize that myofilament overlap and V/V_{max} are *potential* constraints, which are derived exclusively from experiments on *isolated* muscle. It was necessary to determine whether animals *actually use their muscles* over this narrow range of values during their full range of locomotion, which had never been previously determined.

Myofilament overlap

Experiments on fish have provided considerable insight into myofilament overlap in vertebrates. During caudal fin propulsion, most fish bend their backbone. By a combination of high speed motion pictures, and anatomical and mathematical approaches which relate SL to backbone curvature, it was found that at low swimming speeds in carp (Figure 19.3), the red muscle, which powers this movement, undergoes cyclical SL excursions between 1.91 and 2.23 μm centered around a sarcomere length of 2.07 μm (Figure 19.4A) (Rome *et al.*, 1988; 1990; Rome and Sosnicki, 1991). Further, using quantitative electron microscopy, it was found that the thick and thin filament lengths of the red (1.52 μm and 0.96 μm) and white (1.56 μm and 0.99 μm) muscle in carp are similar to that in frog (Sosnicki *et al.*, 1991). Using the frog SL–tension relationship to approximate that of the red and white muscle, shows that the red muscle operates over a range of SLs where no less than 96% maximal tension is generated (Figure 19.4A) (Rome *et al.*, 1988; 1990; Rome and Sosnicki, 1991).

The most extreme movement that carp make, the escape response (pictured in Figure 19.3B), involves a far greater curvature of the backbone than steady swimming. *If* the red muscle were powering this movement, it would have to shorten to a SL of 1.4 μm where low forces and even irreversible damage can occur (Figure 19.4B) (Rome *et al.*, 1988; Rome and Sosnicki, 1991). Rather, it is the white muscle which performs the movement because the white muscle has a different fiber orientation than the red. The red muscle fibers run parallel to the long axis of the fish (Figure 19.2) just beneath the skin. The white muscle fibers, by contrast, run in a helical orientation with respect to the long axis of the fish.

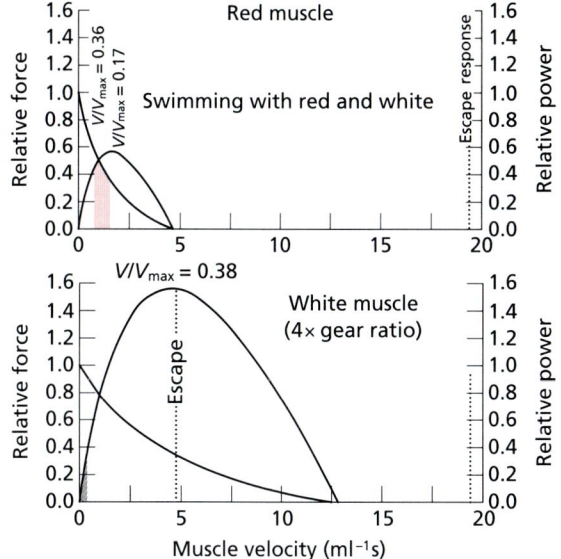

Figure 19.5 V/V_{max} in swimming carp. Force–velocity (dashed) and power–velocity (solid) curves of the red muscle (upper) and white muscle (lower). During steady swimming (upper) the red muscle shortens at a V of 0.7–1.5 muscle lengths (ML)/s (shaded region) corresponding to a V/V_{max} of 0.17–0.38 where maximum power and efficiency are generated. The red fibers cannot power the escape response because they would have to shorten at 20 ML/s (upper) or four times their V_{max}. The escape response is powered by the white muscle which, because of its fourfold higher gear ratio, needs shorten at only 5 ML/s, which corresponds to a $V/V_{max} = 0.38$ where maximum power is generated (lower). The white muscle would not be well suited to power steady swimming movements (shaded, lower panel), as it would have to shorten at a V/V_{max} of 0.01–0.03, where power and efficiency are low.

As predicted by Alexander (1969) and experimentally verified by Rome and Sosnicki (1991), the helical orientation endows the white fibers with a fourfold higher gear ratio than the red fibers (i.e. the white muscle can produce a given backbone curvature while undergoing only one-quarter of the SL excursion). Thus to power this most extreme movement of fish, on average the white muscle must shorten to a SL of 1.82 µm, and at this SL the muscle generates about 94% maximal force (Figure 19.4B).

As shown above, the myofilament overlap is never far from its optimal level even in the most extreme movements. It appears, therefore, that animals are designed in such a way that *no matter what the movement*, the muscles used generate nearly optimal forces. As such, myofilament overlap can be considered a design constraint (i.e. a part of the system that is kept constant). Given the movements that fish need to make, two design parameters (fiber gear ratio and myofilament lengths) are adjusted during evolution such that the muscle fibers being used always operate at near maximal myofilament overlap and force generation (Rome and Sosnicki, 1991; Rome and Lindstedt, 1997).

V/V_{max}

The two main fiber types in fish, red muscle and white muscle, have different V_{max} values in addition to the aforementioned different gear ratios. (NB Pink muscle is abundant in some species, has intermediate properties and appears to follow the same set of design rules (Coughlin *et al.*, 1996).) The first question which was asked is why do animals have different fiber types and are the faster fibers used to power faster movement (higher Vs) while operating at the same V/V_{max}?

As illustrated in Figure 19.5, Rome and colleagues found that the V_{max} of carp red muscle was 4.65 muscle lengths per second (ML/s) and the V_{max} of carp white muscle was 2.5 times higher, 12.8 ML/s (Rome *et al.*, 1988). During steady swimming the red muscle is used over ranges of velocities of about 0.7 to 1.5 ML/s (Figure 19.5A, shaded part of the curve (Rome *et al.*, 1988; 1990; Rome and Sosnicki, 1990)). This corresponds to a V/V_{max} of 0.17–0.36, which is where maximum power is generated. At higher swimming speeds (higher Vs) the fish recruited their white muscle because the mechanical power output of the red muscle actually declines.

It is clear from Figure 19.5A, that the red muscle cannot possibly power the escape response. To power the escape response, the red muscle would have to shorten at 20 ML/s, which it clearly cannot do, as this is four times its V_{max}. Even if white muscle were placed in the same orientation occupied by the red (i.e. same gear ratio), it could not power the escape response either, because its V_{max} is only about 13 ML/s. However, because of its fourfold higher gear ratio, the white muscle need shorten at only 5 ML/s to power the escape response (Figure 19.5B). This corresponds to a V/V_{max} of about 0.38, which is where white muscle generates maximum power (Rome *et al.*, 1988; Rome and Sosnicki, 1991).

If the white muscle does so well at producing fast movements, then why don't fish have only one fiber type and let the white muscle power the slow swimming movements as well? The white muscle could certainly power slow swimming, but it is not used because its high V_{max} and fourfold higher gear ratio would make its V/V_{max} at slow swimming speeds so low (i.e. 0.01–0.03; shaded portion of Figure 19.5B) that the muscle's efficiency would be nearly 0.

Thus the red and white muscle forms a two gear system. To achieve a wide repertoire of movements fish must use different fiber types with different V_{max}s and different gear ratios to power different movements. The red muscle powers slow movements, while the white muscle powers very fast movements, both while working at the appropriate V/V_{max} (0.17–0.38). The effectiveness of white muscle to power fast movements depends on the product of its gear ratio and V_{max}. In terms of backbone curvature, the white muscle can produce 10-fold faster movements (2.5-fold higher V_{max} × fourfold higher gear ratio) than the red muscle. If the red muscle had the same gear ratio as the white, it could not produce the escape response, nor could the white muscle if it had the gear ratio of the red. What is needed is both the correct V_{max} and the correct gear ratio to produce the full repertoire of movements (Rome *et al.*, 1988; Rome and Sosnicki, 1991).

Additional studies have revealed that fast and slow swimming species of fish (Rome *et al.*, 1992a,b), fish swimming at low and high temperatures (Rome and Sosnicki, 1990; Rome *et al.*, 1990; 1992a,b), and even jumping frogs (Lutz and Rome, 1994) also use their muscle at optimal V/V_{max} (0.17–0.36). This provides additional evidence that V/V_{max} is an important design constraint.

It appears from these examples that many animals use their muscles *over a narrow range of myofilament overlap and over a narrow range of V/V_{max}* where muscle generates maximum force and maximum power with optimal efficiency. Therefore, during evolution, three design parameters (gear ratio, V_{max} and myofilament lengths), appear to have been adjusted so as to obey these design constraints. Hence, these design constraints appear to constitute two of the rules by which muscular systems have evolved (Rome, 1997; Rome and Lindstedt, 1997).

Muscle activation, force generation and relaxation

Different muscle fiber types are designed to activate, generate force and relax over a wide range of frequencies. Here we examine the modifications of muscles to operate at different frequencies. Muscle is turned on by the release of calcium (Ca^{2+}) from the sarcoplasmic reticulum (SR, which stores Ca^{2+}) into the myoplasm. Ca^{2+} in turn binds to troponin removing the inhibition from the thin filament, and thereby allowing the myosin crossbridges to attach and generate force (Figure 19.6).

For the muscle to relax, the process must be reversed; Ca^{2+} must unbind from troponin, so that inhibition can be returned to the thin filament. Thus the myoplasmic calcium must be lowered and this is done by Ca^{2+} being pumped into the SR. Ca^{2+} must then unbind from troponin, and finally the crossbridges must detach.

To reveal how these processes are adjusted to allow muscle to operate over a wide range of frequencies, fish again have been advantageous because in a single species the properties of superfast muscles used to produce sound could be compared to slow and fast-twitch locomotory muscles used for swimming (Rome et al., 1996). The male toadfish (*Opsanus tau*) produces a 'boatwhistle' mating call 10–12 times every minute for many hours to attract females to its nest. A tone is generated by oscillatory contractions of the muscles encircling the fish's gas-filled swimbladder at 200 times per second (Fine, 1978; Bass and Baker, 1991). Such high frequency stimulation of typical locomotory muscles (which relax relatively slowly) would produce a completely fused (i.e., constant force) tetanus because they would be unable to relax between stimuli. Because maintained tension would simply compress the bladder and prevent it from vibrating, sonic muscles must be specifically modified to turn on and off rapidly.

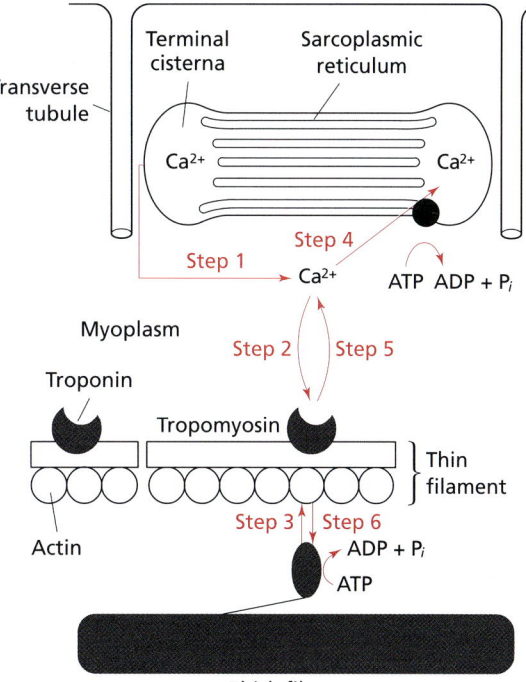

Figure 19.6 Major kinetic steps in muscle activation and relaxation. Activation: step 1, Ca^{2+} is released from the SR into the myoplasm; step 2, Ca^{2+} binds to troponin, releasing inhibition of the thin filament; step 3, crossbridges then attach. Relaxation: step 4, Ca^{2+} is resequestered from the myoplasm by the Ca^{2+} pumps; step 5, Ca^{2+} comes off troponin, thereby preventing further crossbridge attachment; step 6, crossbridges then detach. For a muscle to relax rapidly, steps 4–6 must all be very fast.

The muscle activation–relaxation rates vary by ~50-fold between locomotory and sonic fiber types in toadfish (Rome et al., 1996). Toadfish red muscle (used for slow steady swimming at ~2 Hz) has a twitch half-width (the time duration at the 50% force level) of about 500 ms, compared to approximately 200 ms for white muscle (used for burst swimming at ~5 Hz), and about 10 ms for the swimbladder muscle (upper panel of Figure 19.7).

For any muscle to activate and relax rapidly, two conditions must be met. First, calcium, the trigger for muscle contraction, must enter the myoplasm rapidly and be removed rapidly (Figure 19.6, steps 1, 4). Second, myosin crossbridges must attach to actin and generate force soon after the Ca^{2+} level rises and then detach and stop generating force soon after the Ca^{2+} level falls (Figure 19.6, steps 2, 3, 5, 6).

The time course of Ca^{2+} release and re-uptake during contractions can be determined by injecting muscle cells with a fluorescent Ca^{2+}-sensitive dye, and tracking fluorescence with time (Figure 19.7, bottom).

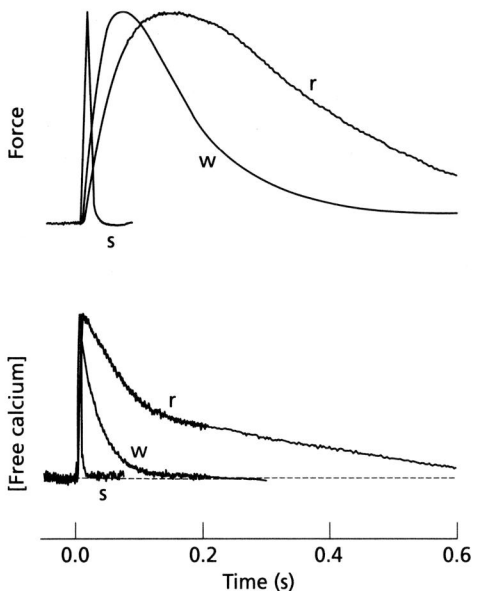

Figure 19.7 Twitch tension (upper) and calcium transients (lower) of three fiber types from toadfish at 15°C. In each case, the force and the calcium records have been normalized to their maximum value. The twitch and calcium transient become briefer going from the slow-twitch red fiber (r), to the fast-twitch white fiber (w) to the superfast-twitch swimbladder fiber (s). (Adapted from Rome et al., 1996.)

The Ca^{2+} transient in the sonic muscles is the fastest ever measured for any fiber type (a half-width of ~3.4 ms at 16°C and 1.5 ms at 25°C) (Rome et al., 1996). The importance of the Ca^{2+} transient duration in setting the twitch duration can be seen in Figure 19.7 which shows that between the slow-twitch red fibers and the superfast-twitch swimbladder fibers, the half-widths of the Ca^{2+} transient and the twitch speeded up in parallel (by ~50-fold).

The significance of a fast Ca^{2+} transient is most apparent during repetitive stimulation. During stimulation of slow red muscle at a modest 3.5 Hz (Figure 19.8), the time course of Ca^{2+} uptake is so slow that $[Ca^{2+}]$ does not have time to return to baseline between stimuli. Even the lowest myoplasmic $[Ca^{2+}]$ between stimuli was above the threshold required for force generation in this fiber type, thus resulting in a partially fused tetanus. By contrast, the swimbladder's Ca^{2+} transient is so rapid that even with 67 Hz stimulation, the $[Ca^{2+}]$ return to baseline between stimuli is complete (Figure 19.8B; note the 50× faster time base). In addition, $[Ca^{2+}]$ is below the threshold for force generation for more than half of the time in all but the first stimulus. Hence, the Ca^{2+} transient is sufficiently rapid to permit the oscillation in force required for sound production.

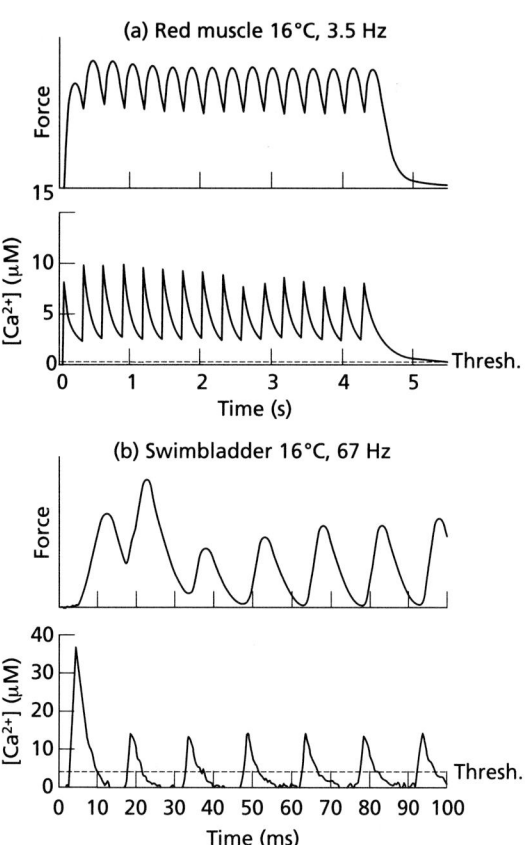

Figure 19.8 Force production and calcium transients during repetitive stimulation. (A) Slow twitch red fiber stimulated at 3.5 Hz. The threshold $[Ca^{2+}]$ for force generation was derived by force-pCa experiment on skinned fibers and is shown with a dotted line. (B) Swimbladder fiber stimulated at 67 Hz. Note that the $[Ca^{2+}]$ threshold for force production is much higher for the swimbladder than for the red fiber and the different scale for the ordinate. (Adapted from Rome et al., 1996.)

Even though $[Ca^{2+}]$ returns rapidly to baseline, the swimbladder fiber could not relax quickly unless its troponin rapidly released bound Ca^{2+} (Figure 19.6, step 5). Indeed, kinetic modeling indicates that if the swimbladder troponin had the same off-rate for Ca^{2+} (k_{off}) as fast-twitch fibers, then occupancy of its troponin sites with Ca^{2+} would not decline sufficiently rapidly to permit the observed rapid fall in force. Experiments on the Ca^{2+} sensitivity of troponin suggest that swimbladder muscle has a threefold lower affinity troponin than fast-twitch muscle, and hence a threefold faster k_{off}. With this higher k_{off}, the modeled rate of troponin deactivation no longer appears limiting. Thus the less sensitive troponin with its more rapid k_{off} may be another modification for faster relaxation (Rome et al., 1996).

The final requirement for force to drop quickly following the dissociation of Ca^{2+} from troponin is a fast crossbridge detachment rate (Figure 19.6, step 6). Indeed, the maximum velocity of shortening (V_{max}; which is thought to be affected by crossbridge detachment rate) of swimbladder muscle (~ 12 ML/s) is exceptionally fast, fivefold and 2.5-fold faster than toadfish red and white muscle respectively. Further, direct measurement of the crossbridge detachment rate shows it to be exceptionally fast (~ 110 s^{-1} or about 50 times faster than the toadfish red fibers and the well studied rabbit fast fibers). This fast detachment rate would permit the rapid relaxation rate observed (Rome et al., 1999).

The toadfish experiments have thus identified three kinetic variables that change progressively as twitch speed increases from the slow-twitch of red fibers to the superfast-twitch of swimbladder fibers: (i) the duration of the Ca^{2+} transient must become shorter, which in turn requires more rapid calcium release and re-uptake which is achieved principally by an increased density of SR Ca^{2+} pumps; (ii) troponin needs a faster off-rate for Ca^{2+}, which requires molecular modification of troponin to a lower affinity type; (iii) crossbridges must detach more rapidly, which involves molecular modification of myosin.

Changes in all of these parameters in concert enable swimbladder fibers to perform mechanical work at high operating frequencies (100–200 Hz) (Rome et al., 1996; Rome and Lindstedt, 1998). In contrast, vertebrate locomotory muscles lack these properties, and thus the red and white swimming muscles of toadfish cannot perform work beyond 15 Hz, and even the fastest vertebrate locomotory muscle measured (i.e. lizard hindlimb muscle) cannot perform work beyond about 30–40 Hz.

Muscle energetics

As we have seen, muscle can be built to be very fast, however, the faster the muscle, the more energy is used. Here muscle energy usage is briefly examined. The energetic cost of contractions is set by two components – the ATP used by Ca^{2+} pumps, which accounts for about 30% of the total, and the ATP used by the crossbridges, which accounts for the remainder (Woledge et al., 1985) (Figure 19.6). Fibers used for high frequencies have extraordinarily high ATP utilization rates when active because the rate of ATP utilization by both of these components is extremely rapid. For instance, swimbladder fibers appear to pump Ca^{2+} back into the SR at about 50 times the rate as red muscle fibers. Because the Ca^{2+} must be pumped up a concentration gradient into the SR, this requires ATP. Thus the swimbladder fibers' Ca^{2+} pumps use ATP 50-fold faster than those of the red muscle. It has been further found that the crossbridges of swimbladder fibers use ATP about six times faster than red muscle fibers. Thus during sound production (where Ca^{2+} is being continuously pumped and crossbridges are continuously generating force and splitting ATP), the swimbladder will use ATP at a far greater rate than the red muscle.

Interestingly, because of the swimbladder's high SR volume and high crossbridge detachment rate constant, it generates only a small fraction ($\sim$ one-tenth) of the force of the red muscle. The fact it also uses energy at a much faster speed, results in the cost of force generation being about 20 times higher in swimbladder than that in the red muscle. This makes it far too costly for this superfast muscle to be used to power relatively slow movements such as swimming. By contrast, the locomotory red and white muscles are far too slow to power high frequency sound production. Hence, these different muscles represent mutually exclusive designs to power different activities (Rome and Lindstedt, 1998; Rome et al., 1999).

Summary

Animals perform a very wide range of motor activities. This requires a great diversity of different muscle properties. Comparison of the muscle properties and their usage shows that each component of the muscular system appears to be well adjusted to its *in vivo* function. Because increased muscle speed results in increased energetic cost, a basic tenet in the design of muscular systems is that muscle speed is set to be sufficiently fast to perform a given activity, but not faster than necessary, as this would entail a large waste of energy.

Importantly, fish, because of the anatomical separation of their muscles, and the extraordinary range of movements their muscle power, have provided the experimental model by which much of what we know about muscle design has been determined. It is likely that their preeminence in elucidating principles of muscle design will continue in the future.

Acknowledgments

This work was supported by grants from the NIH (AR38404 & AR46125) and NSF (IBN-9514383).

References

Alexander, R.M. (1969). *J. Mar. Biol. Assoc.* **49**, 263–290.

Altringham, J.D. and Johnston, I.A. (1990). *J. Exp. Biol.* **148**, 395–402.

Bass, A.B. and Baker, R. (1991). *Brain, Behav. Evol.* **37**, 1–15.

Block, B. (1991). *Amer. Zool.* **31**, 726–742.

Coughlin, D.J. and Rome, L.C. (1999). *Biol. Bull.* **196**, 145–152.

Coughlin, D.J., Zhang, G. and Rome, L.C. (1996). *J. Exp. Biol.* **199**, 2703–2712.

Curtin, N.A. and Woledge, R.C. (1988). *J. Exp. Biol.* **140**, 187–197.

Fine, M.L. (1978). *Oecologia* **36**, 45–57.

Gordon, A.M., Huxley, A.F. and Julian, F.J. (1966). *J. Physiol.* (*London*) **184**, 170–192.

Hill, A.V. (1950). *Sci. Prog.* **38**, 209–229.

Johnston, I.A. (1983). In *Fish Biomechanics* (eds P.W. Webb and D. Weihs), pp. 36–67. Praeger, New York.

Lutz, G. and Rome, L.C. (1994). *Science* **263**, 370–372.

Page, S.G. and Huxley, H.E. (1963). *J. Cell Biol.* **19**, 369–390.

Rome, L.C. (1994). In *Comparative Vertebrate Exercise Physiology* (ed. J.H. Jones), pp. 125–179, Academic Press, Orlando.

Rome, L.C. (1997). *Am. Sci.* **85**, 356–363.

Rome, L.C. (1998). *Comp. Biochem. Physiol. B* **120**, 51–72.

Rome, L.C. and Lindstedt, S.L. (1997). In *Handbook of Physiology. Section 13, Comparative Physiology* (ed. W. Dantzler), pp. 1587–1651. Oxford University Press, New York.

Rome, L.C. and Lindstedt, S.L. (1998). *News Physiol. Sci.* **13**, 261–268.

Rome, L.C. and Sosnicki, A.A. (1990). *J. Physiol.* (*London*) **427**, 151–169.

Rome, L.C. and Sosnicki, A.A. (1991). *Am. J. Physiol.* (*Cell Physiol.*) **260**, C289–C296.

Rome, L.C., Loughna, P.T. and Goldspink, G. (1984). *Am. J. Physiol.* **247**, r272–r279.

Rome, L.C., Funke, R.P., Alexander, R.M., Lutz, G., Aldridge, H.D.J.N., Scott, F. and Freadman, M. (1988). *Nature* **355**, 824–827.

Rome, L.C., Funke, R.P. and Alexander, R.M. (1990). *J. Exp. Biol.* **154**, 163–178.

Rome, L.C., Choi, I., Lutz, G. and Sosnicki, A.A. (1992a). *J. Exp. Biol.* **163**, 259–279.

Rome, L.C., Sosnicki, A.A. and Choi, I. (1992b). *J. Exp. Biol.* **163**, 281–295.

Rome, L.C., Swank, D. and Corda, D. (1993). *Science* **261**, 340–343.

Rome, L.C., Syme, D.A., Hollingworth, S., Lindstedt, S.L. and Baylor, S.M. (1996). *Proc. Natl. Acad. Sci.* **93**, 8095–8100.

Rome, L.C., Cook, C., Syme, D., Connaughton, M., Ashley-Ross, M., Klimov, A., Tikunov, B. and Goldman, Y.E. (1999). *PNAS* **96**, 5826–5831.

Sosnicki, A.A., Loesser, K. and Rome, L.C. (1991). *Am. J. Physiol.* (*Cell Physiol.*) **260**, C283–C288.

Woledge, R.C., Curtin, N.A. and Homsher, E. (1985). *Energetic Aspects of Muscle Contraction*. Academic Press, New York.

CHAPTER 20

Nervous System

Ann B Butler
Krasnow Institute for Advanced Study and
Department of Psychology,
George Mason University, Fairfax, Virginia, USA

Abbreviations

III	oculomotor nucleus
A	nucleus anterior
AO	accessory optic nucleus
C	central zone of optic tectum
CE	nucleus centralis of the inferior lobe of the hypothalamus
CN	central nucleus
CP	central posterior nucleus
D	dorsal pallium
DAO	dorsal accessory optic nucleus
Dc	central zone of area dorsalis telencephali
Dc-1,2,3,4	parts of central zone of area dorsalis telencephali
Dd	dorsal zone of area dorsalis telencephali
DF	nucleus diffusus of the inferior lobe of the hypothalamus
Dl-d	dorsal part of lateral zone of area dorsalis telencephali
Dl-p	posterior part of lateral zone of area dorsalis telencephali
Dl-v	ventral part of lateral zone of area dorsalis telencephali
Dm-1,2,3,4	parts of medial zone of area dorsalis telencephali
DM	dorsomedial neuropil
DOT	dorsal optic tract
Dp	posterior zone of area dorsalis telencephali
DP	dorsal posterior nucleus
E	entopeduncular nucleus
ECL	external cellular layer of olfactory bulb
ET	eminentia thalami
FR	fasciculus retroflexus
G	periventricular gray zone of optic tectum
GL	glomerular layer of olfactory bulb
H	habenula
Hy	hypothalamus
I	nucleus intermedius
ICL	internal cellular layer of olfactory bulb
Ld	pars dorsalis of the lateral pallium
Lv	pars ventralis of the lateral pallium
M	medial pallium
MaOT	marginal optic tract

Copyright © 2000 Academic Press

MTP	nucleus medianus of the posterior tuberculum
NC	nucleus corticalis
nMLF	nucleus of the medial longitudinal fasciculus
nPVO	nucleus of the paraventricular organ
NT	nucleus taenia
OL	olfactory tract
ON	olfactory nerve
P1-v	ventral part of first pallial zone
P1-d	dorsal part of first pallial zone
P2	second pallial zone
P2mc	magocellular part of pallial area 2 of hagfish
P3 (Figure 20.2)	pallial area 3 of hagfish
P3 (Figure 20.3)	third pallial zone
P4l	lateral part of pallial area 4 of hagfish
P4m	medial part of pallial area 4 of hagfish
P5	pallial area 5 of hagfish
PC	central pretectal nucleus
PCo	posterior commissure
PG	preglomerular nuclear complex
PO	posterior pretectal nucleus
PPd	nucleus pretectalis periventricularis pars dorsalis
PPm	magnocellular part of periventricular preoptic nucleus
PPp	parvicellular part of periventricular preoptic nucleus
PPv	nucleus pretectalis periventricularis pars ventralis
PSm	nucleus pretectalis superficialis pars magnocellularis
PSp	nucleus pretectalis superficialis pars parvocellularis
PT	nucleus posterior thalami
PVO	paraventricular organ
S	superficial zone of optic tectum
SAC	stratum album centrale
SC	suprachiasmatic nucleus
SCO	subcommissural organ
SFGS	stratum fibrosum et griseum superficiale
SGC	stratum griseum centrale
SM	stratum marginale
SO	stratum opticum
SOF	secondary olfactory fiber layer of olfactory bulb
SPV	stratum periventriculare
tc	tela choroidea (ependymal lining of overlying lateral ventricle)
TCo	tectal commissure
Tel	telencephalon
TL	torus longitudinalis
TLa	torus lateralis
TN	terminal nerve
TP	nucleus tuberis posterior
TPp	periventricular nucleus of posterior tuberculum
trolp	deep part of lateral olfactory tract
trols	superficial part of lateral olfactory tract
trov	ventral olfactory tract
TS	torus semicircularis
VAO	ventral accessory optic nucleus
Vc	commissural nucleus of area ventralis telencephali
Vd	dorsal nucleus of area ventralis telencephali
Vi	intermediate nucleus of area ventralis telencephali
Vl	lateral nucleus of area ventralis telencephali
VL	nucleus ventrolateralis
VMc	caudal part of nucleus ventromedialis
Vn	'nother nucleus of area ventralis telencephali
VOT	ventral optic tract
Vp	postcommissural nucleus of area ventralis telencephali
Vs	supracommissural nucleus of area ventralis telencephali
Vv	ventral nucleus of area ventralis telencephali
ZL	zona limitans diencephali

Introduction

The nervous system in all vertebrates is composed of neurons, glial cells, and ependymal cells. The latter line the ventricular surfaces, while the glia are interspersed with neurons. Neurons are the active, conductive elements of the nervous system (Figure 20.1). This introduction contains basic material on neurons and how they are organized and interconnected

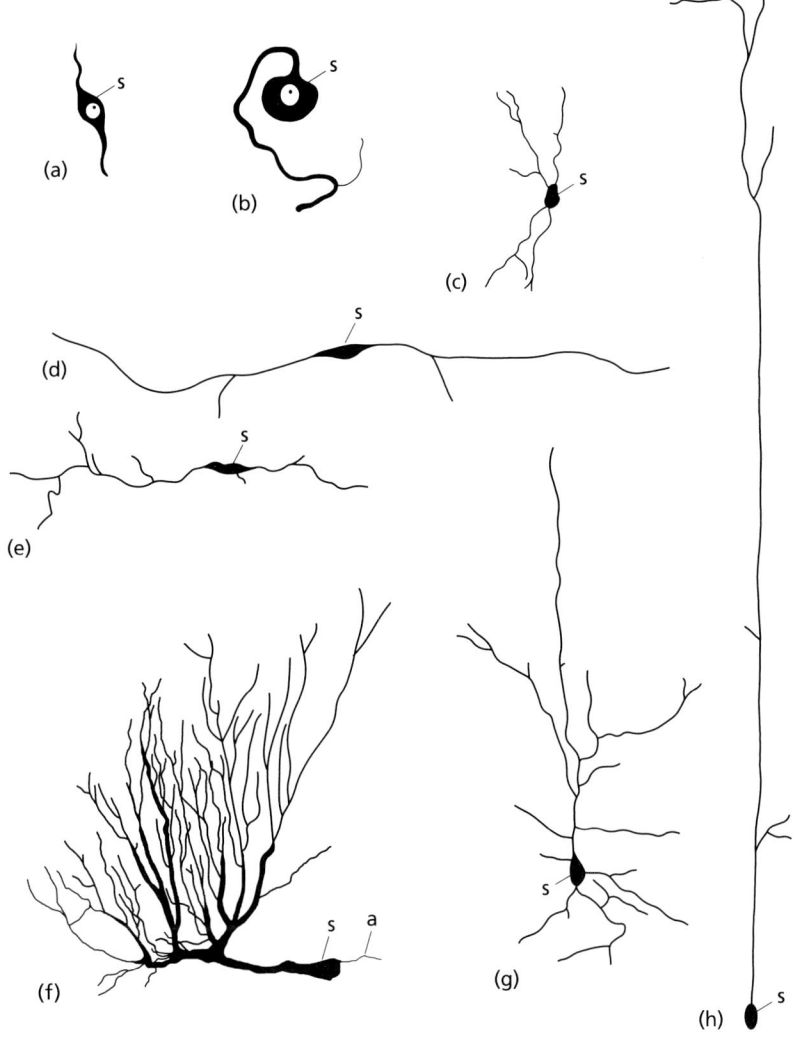

Figure 20.1 Drawings of neurons: (a) bipolar and (b) unipolar neurons with processes that extend peripherally and centrally, after Parent (1996); (c and g) multipolar cells and (d and e) horizontal cells from the optic tectum of the dogfish shark, *Scyliorhinus canicula*, after Manso and Anadón (1991); (f) Purkinje cell from the cerebellum of the trout *Salmo gairdneri*, after Pouwels (1978); (h) piriform cell from the optic tectum of the perciform fish *Eugerres plumieri*, after Vanegas et al. (1974). Abbreviations: a, axon; s, soma.

within the nervous system. Basic terminology for discussing groups of neurons and neuronal pathways is introduced. Those already familiar with this material may wish to bypass it in favor of the remainder of the chapter.

Most neurons are defined by the presence of long, thin processes that arise from the cell body (soma), extend for some distance from the cell body, and are able to conduct impulses, i.e. waves of transient changes in the membrane potential, along their lengths either toward or away from the cell body. The processes that generally conduct impulses away, or efferent, from the cell body are called axons; they usually give rise to a number of branches in the region where they terminate and may also give rise to collateral branches along their course. The terminal branches of axons form specialized endings called synapses that abut the membranes of the target cell processes. Synapses are the sites of communication between neurons, which occurs most often by release of chemical neurotransmitters. The processes that generally receive the incoming, or afferent, input from axons are called dendrites, which means 'branches.' Dendrites generally conduct changes in membrane potential towards the cell body. The dendrites frequently have a number of branches, and the resultant formation is referred to as the dendritic arborization or the dentritic tree. The great majority of synapses occur between axons and dentrites (axodendritic), but rare exceptions occur and include synapses between axons and axons (axoaxonic), axons and cell bodies (axosomatic), and dendrites and dendrites (dendrodendritic).

Neurons are frequently classified based on the shape and/or size of the cell bodies or the shape and extent of their dendritic processes (Figure 20.1). Neurons that have small cell bodies are sometimes described as parvicellular, while those with large cell bodies may be termed magnocellular, for example. The terminology may alternatively describe cell body shape, e.g. **piriform** or **fusiform**; the number of sites, or poles, where dendritic processes arise from the cell body, e.g. unipolar, bipolar, multipolar; the scientist(s) who first identified a particular cell type, e.g. Purkinje, Cajal-Retzius; various features of the dentritic spread, e.g. horizontal, reticular; or particular unique features, e.g. chandelier, **amacrine**.

Sets of neuron cell bodies that have the same afferent and efferent connections and that perform the same function usually occur in anatomically discrete groups. Such a neuronal group may form a localized aggregation that in a particular plane of section appears spherical or ovoidal in shape. An aggregation, or cell group, of this sort is most often called a nucleus, which means 'nut,' but may be called a variety of other terms, such as ganglion (often but not only for neuronal aggregations in the peripheral nervous system), locus (e.g. locus coeruleus), area (e.g. area dorsalis), gray (e.g. periaqueductal gray), tuberculum (e.g. posterior tuberculum), or lobe (e.g. inferior hypothalamic lobe). Neuronal aggregations that are elongated, which usually occurs along the rostrocaudal axis, may be called a column. The defining feature of nuclei and these other similar aggregations is that the dentritic arborization does not extend beyond the border of the aggregation. Some cell groups are subdivided into laminae (layers), and in such cases, the dendrites of the neurons in one lamina do not extend beyond it into the region of any other lamina. Due to this architectural feature, afferent axons to the cell group (or to one laminar subdivision of it) terminate over most or all of the dendritic tree rather than on only limited parts of the dendrites.

The local dendritic arborizations that characterize nuclei are in contrast to the dentritic arborizations of cortices. Cortex consists of a minimum of two to three layers formed by cell bodies and/or their processes with the defining characteristic that the dendrites of the layered cell bodies extend beyond the cell body layer of origin into at least one other layer. The elongation of the dendrites of the individual neurons in the same direction allows various afferent inputs to be layered in their pattern of termination and thus uniformly segregated to a particular portion of the dendrites at a uniform distance from all of the cell bodies within a particular lamina. Such an arrangement gives precedence to the more proximally ending fibers that have the greater influence on the cell bodies. Cortices generally also contain more diverse cell types than nuclei, which results in an increased capacity for more complex local processing of inputs. Further, interactions occur vertically across the neuronal cell layers in cortex, and the term column is also used to apply to vertical (i.e. deep to superficial across all layers), functionally coherent units within cortex.

The axons of groups of neurons, whether located within nuclei or cortices, may have local connections but most frequently form bundles that pass between various other cell groups to reach their target areas for termination. Axons are frequently referred to as fibers, and the bundles that they form are generally referred to as tracts or pathways and named in a variety of ways, including bundles (e.g. medial forebrain bundle), **fascicles** (e.g. fasciculus gracilis), **funiculi** (e.g. dorsal funiculus), or capsules (e.g. internal capsule). Some axonal pathways terminate in sites that lie on the same side of the brain as their site of origin and are referred to as ipsilateral, while those that cross the midline and terminate in sites that lie on the opposite side are referred to as contralateral, and those that terminate partly on the ipsilateral side and partly on the contralateral side are referred to as bilateral. For pathways that cross the midline, those that terminate in a site that is in a mirror-image position to the site of origin are called **commissures**, whereas those that terminate in a site that is at different location or level of the neuraxis are called **decussations**.

The terms afferent and efferent, as mentioned above, refer to the direction of the normal axonal conduction with respect to the nuclei of origin and termination: a single axon that originates from a neuron in nucleus 'A' and terminates on the dendrites of a neuron in nucleus 'B' is efferent from nucleus 'A' and afferent to nucleus 'B'. The term projection refers to any set of axons that arises from one site and terminates in another, e.g. nucleus 'A' projects to nucleus 'B', and the pathway itself can be referred to as a projection. An alternate use of the terms afferent and efferent concerns the sensory and motor neurons of the peripheral nervous system, respectively, in the sense that peripheral sensory fibers are afferent to the central nervous system, and the axons of motor neurons that innervate muscles or glands are efferent from it. In that sense, the terms afferent and sensory are sometimes interchangable as are the terms efferent and motor.

The ganglia, cranial nerves, and peripheral nerves of the peripheral nervous system were discussed in Chapter 8, in addition to the gross functional anatomy of the central nervous system. The central nervous system, which is the focus of this chapter, comprises the brain and spinal cord, the axial length of which can be referred to as the neuraxis. As discussed in Chapter 8, the brain can be divided into three major regions – forebrain, midbrain, and hindbrain – while the spinal cord is more uniform in structure throughout its extent. The developing neural tube has a ventral part, the basal plate, and a dorsal part, the alar plate. As development proceeds, the tube gains flexures and various evaginations (out-pocketings) of its walls. Between and within the major groups of fishes, extensive variation occurs in the degree to which various parts of the neuraxis, particularly those of the alar plate, are enlarged and elaborated.

Principles of sensory and motor system organization

Basic patterns can be recognized in the sets of connections for the relay of sensory inputs from all the different sensory receptor systems up to the forebrain and for the relay of motor outputs from various sites in the brain down to motor neuronal cell groups in the brainstem and spinal cord. These basic patterns can serve as templates for the details of the particular relay cell groups specific to each system. The same basic patterns characterize cell groups of fishes and other vertebrates as well.

Among sensory systems, a remarkable degree of uniformity exists in the organization of central pathways across systems despite extensive differences in the type of physical stimuli and in their transduction mechanisms (Hodos and Butler, 1997, 2000). For all spinal nerves and most cranial nerves, the sensory, bipolar neuron cell bodies lie within peripherally located ganglia and project to multipolar neurons within the central nervous system (as discussed in Chapter 8). The multipolar neurons that receive input from the sensory bipolar neurons are the first in a chain of several multipolar neurons that relay the sensory information rostrally along the neuraxis and can thus be referred to as first-order multipolar neurons. Bipolar and first-order multipolar neurons can be identified in comparable positions in the sensory chain for all sensory systems. In most cases, the first-order multipolar neurons project rostrally to one of two sites: either to a nucleus within the diencephalon or to part of the midbrain roof. In the latter case, the midbrain roof then projects rostrally to a nucleus within the diencephalon. Diencephalic cell groups subsequently project to the telencephalon. A number of telencephalic regions receive these various sensory inputs and are extensively interconneted with each other.

The basic templates for motor pathways follow a similar multi-synaptic strategy. Various sites within the telencephalon project either to the midbrain roof directly or to diencephalic nuclei, which in turn project to more caudal levels, including the midbrain roof. The latter structure and numerous other sites relay descending motor information to the reticular formation in the brainstem, which in anamniotes is the major source of descending projections to the motor nuclei of cranial nerves and to motor neurons within the spinal cord.

Regional anatomy

Numerous cell groups are present in each major division of the brain and in the spinal cord, and their interconnections are often extensive and complex. Many of the cell groups are part of the relay systems for sensory and motor pathways, while others are concerned with analysis and integration of multiple inputs or with maintenance of brain and/or body functions. A selection of some of the nuclei, areas, and cortical structures in each region are surveyed here as a basis for appreciating the specific pathways for particular sensory and motor systems discussed below. While the connections of cell groups vary considerably as to whether they are only ipsilateral, only contralateral, or bilateral to various degrees, comments on the laterality of projection systems for the most part are omitted here, since they would add an additional layer of complexity to the information presented. Most commissural and reciprocal connections are likewise not included. For appreciating the neuroanatomy of a teleost brain, an excellent atlas of the brain of the zebrafish *Danio rerio* has recently been published by Wullimann *et al.* (1996), and for cartilaginous fishes, the reader is referred to Smeets *et al.* (1983). A textbook on comparative vertebrate neuroanatomy by Butler and Hodos (1996) provides a more detailed account of most of the material covered here. A recently published extensive and comprehensive work on the vertebrate central nervous system by Nieuwenhuys *et al.* (1998) provides a much greater depth and breadth of coverage and extensive references.

Forebrain: telencephalon

The telencephalon has a dorsal, pallial part and a ventral, subpallial part. The latter includes a laterally located striatum related to motor functions and a medial septal area related to the medial pallium, hypothalamus, and related visceral functions. In jawless fishes (agnathans) and cartilaginous fishes (chondrichthyans), which have evaginated telencephalons, topologic relationships of various regions of the pallium are generally more straightforward to interpret than is the case in the everted telencephalons (Nieuwenhuys, 1962, 1963; and see Chapter 8) of ray-finned fishes (actinopterygians). In hagfishes, however, the enlargement of telencephalic pallial regions occurs to the degree that the ventricular space is secondarily obliterated, obscuring topologic relationships (Wicht and Northcutt, 1992), and in some sharks, skates, and rays (Group II cartilaginous fishes: see Chapter 8), similarly exuberant telencephalic development also obscures the basic topography of the pallium.

Identification of particular pallial regions by their afferent and efferent connections and by histochemical characteristics has helped to clarify the identity of a number of areas (see Northcutt, 1995; Braford, 1995). Use of as many criteria as possible rather than only one or a few criteria is of course an essential strategy for making neural comparisons across species, given the diversity of morphology and the degree of phylogenetic divergence within fishes as well as across vertebrates. Localization of the main olfactory pallial area(s) by tracing the central projections of the olfactory bulb only has been of arguable help; such olfactory pallial area(s) could serve as a topologic landmark for the developmentally original lateral part of the pallial mantle, but the independent evolution of the pallium in various groups of fishes complicates the comparison, particularly in fishes with everted brains (see Braford, 1995). In hagfishes olfactory bulb projections (Figure 20.2b) are massive and distribute to most of the pallium (Wicht and Northcutt, 1993), where the pallium is not only enlarged but also attains a dramatically laminated structure. In lampreys, which have a comparatively small telencephalon, olfactory projections (Figure 20.2a) are also distributed to most of the pallium (Northcutt and Puzdrowski, 1988). The telencephalons of jawed fishes have large pallial regions that are free of olfactory input, however. Olfactory projections are predominantly restricted to a small medial area and a more extensive lateral area of the pallium (Figure 20.2c) in cartilaginous fishes (Ebbesson and Heimer, 1970) and to the topologic lateral pallial region in ray-finned fishes. The latter region lies in a medial position (Figure 20.3a) due to eversion (see Chapter 8) in Group I fishes such as the cladistian *Polypterus* (Braford and Northcutt, 1983) and in a secondarily-achieved ventrolateral position (Figure 20.3b) in Group II ray-finned fishes. In jawless and

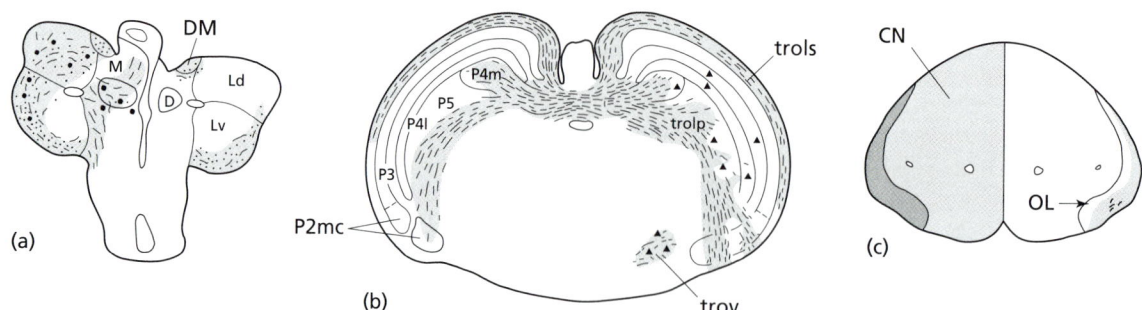

Figure 20.2 Drawings of transverse sections through the telencephalons of (a) the silver lamprey *Ichthyomyzon unicuspis*, after Northcutt and Puzdrowski (1988); (b) the hagfish *Eptatretus stouti*, after Wicht and Northcutt (1993); and (c) the nurse shark *Ginglymostoma cirratum*, after Ebbesson and Heimer (1970). The olfactory bulb projections to the telencephalon are indicated by dashed lines and the shaded regions on both sides in (a) and (b) and on the right side in (c). (The larger dots and triangles represent neuron cell bodies that are retrogradely filled by the olfactory bulb injection and hence project to the olfactory bulb.)

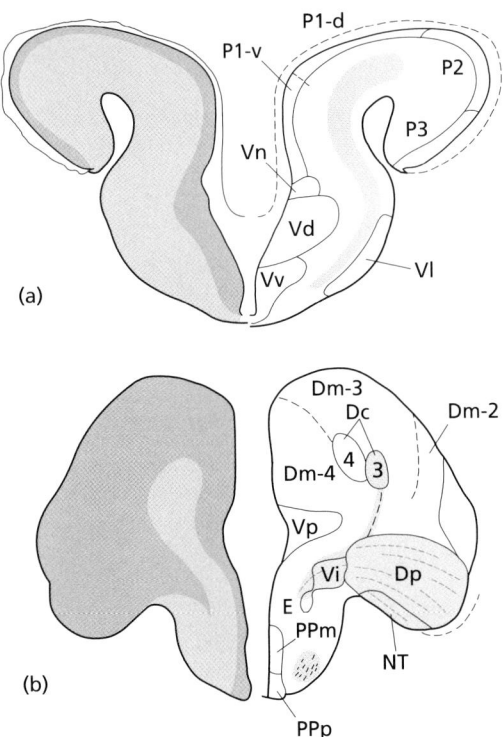

Figure 20.3 Transverse sections through the telencephalon of (a) the cladistian fish *Polypterus palmas* and (b) the green sunfish *Lepomis cyanellus*, after Northcutt and Davis (1983), with olfactory projections indicated by the shaded regions on the right side.

cartilaginous fishes, nomenclature for telencephalic cell groups varies but is straightforward for both pallial and subpallial regions (Nieuwenhuys, 1977; Northcutt, 1978; Wicht and Northcutt, 1992), as shown in Figure 20.2. In *Polypterus*, a simple nomenclature (Braford and Northcutt, 1983) is also established (Figure 20.3a). In teleosts (Figure 20.4), nomenclature is based on neutral topographic terms (Northcutt and Davis, 1983). Identification of regions that correspond to specific regions of the telencephalon in vertebrates other than ray-finned fishes remains tentative at best (Braford, 1995), even in differentiating between pallium and subpallium (Northcutt and Braford, 1980; Nieuwenhuys et al., 1998). An area dorsalis (D) and an area ventralis (V) have some but perhaps not all parts that are pallial and subpallial, respectively.

Area dorsalis can be divided into medial (Dm), dorsal (Dd), lateral (Dl), and central (Dc) parts (Figure 20.4). Dc is markedly heterogeneous (Braford, 1995; Lopes Corrêa et al., 1998), with some of its parts connectionally related to each of the other dorsal parts – Dm, Dd, and Dl. A caudal (posterior) part of Dl, referred to as Dp, and its adjoining part of Dc, is the major olfactory bulb target within the dorsal area. In general, Dm, Dd, Dl, and Dc all receive major sensory inputs relayed from various sites in the diencephalon. Dc is a particularly complex and pivotal part

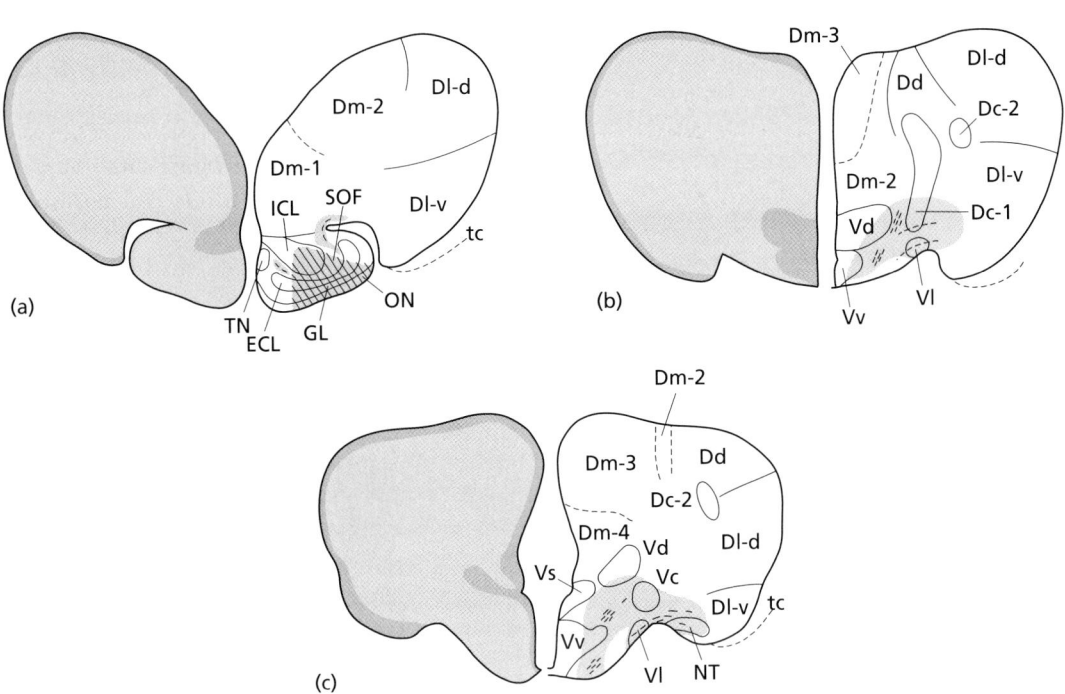

Figure 20.4 Transverse sections through the telencephalon of the green sunfish *Lepomis cyanellus*, after Northcutt and Davis (1983). Projections of the olfactory bulb throughout the telencephalon are shown in the line drawings, represented by dashes and the shaded regions on the right side. The section shown in (f) is the same as that in Figure 20.3b.

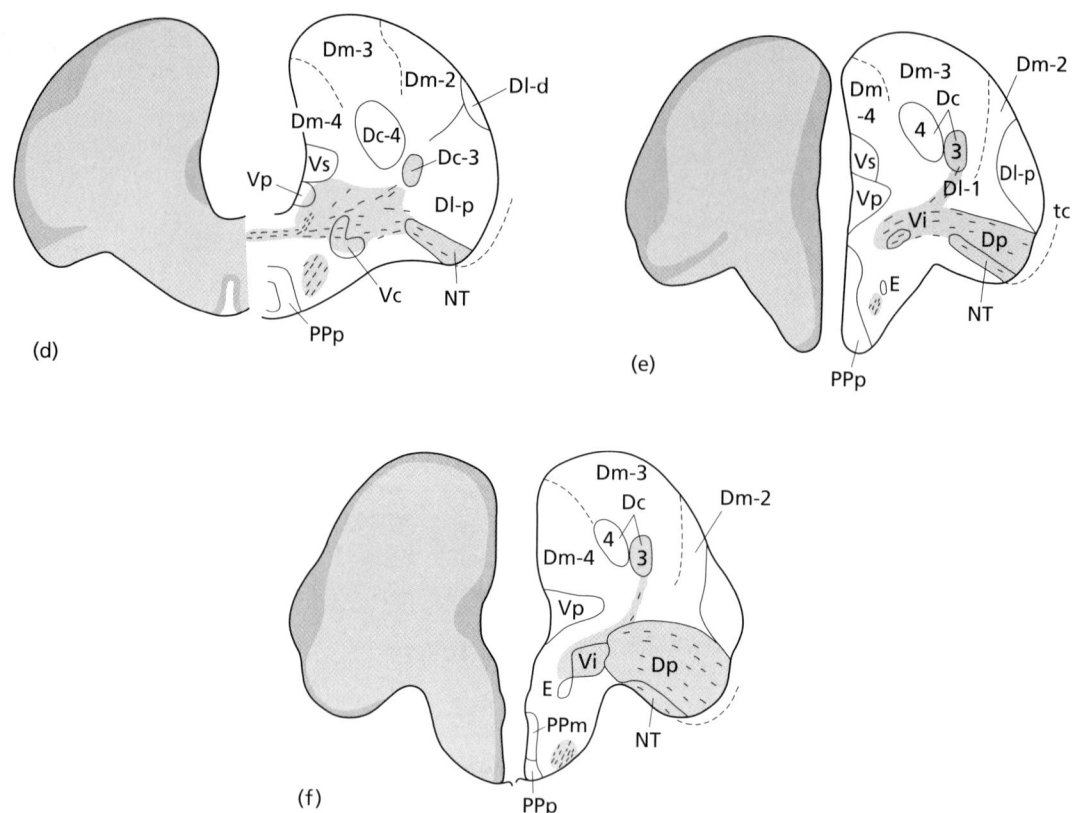

Figure 20.4 (continued)

of telencephalic circuitry. It has major reciprocal connections with other parts of the dorsal area (Murakami et al., 1983) and also receives substantial inputs from parts of the ventral area of the telencephalon (Lopes Corrêa et al., 1998). It is a major source of extratelencephalic projections to the more caudal parts of the brain (Lopes Corrêa et al., 1998), and is thus a key component of descending motor pathways.

Unlike the dorsal area, the ventral area of the telencephalon does not undergo eversion during embryological development (Nieuwenhuys, 1962, 1963). It can be subdivided into a number of parts, again with topographic nomenclature (Figure 20.4). The various parts of the ventral area receive olfactory input from the olfactory bulb (e.g. Finger, 1975; Bass, 1981a; Levine and Dethier, 1985). Efferent projections from various ventral parts include reciprocal ones to the olfactory bulb (e.g. Oka, 1980; von Bartheld et al., 1984; Murakami et al., 1983) and those to Dc (Lopes Corrêa et al., 1998) as well as to more caudal parts of the brain such as the habenula, hypothalamus, and midbrain tegmentum (Shiga et al., 1985b).

Based mainly on immunohistochemical criteria, a dorsal striatum (homologous to the caudate-putamen of mammals and often referred to in fishes simply as the striatum) has been at least tentatively identified in jawless, cartilaginous, and ray-finned fishes (see Northcutt, 1995). The dorsal part of the ventral area, Vd, of ray-finned fishes has many substance P (SP)-containing and enkephalin (ENK)-containing neurons and fibers and is richly innervated by dopamine (DA)-containing fibers, and areas with a similar histochemical profile occur in other fishes in the ventrolateral part of the telencephalon (see Reiner et al., 1998). In contrast, a good candidate for a homologue of the amniote dorsal pallidum (the globus pallidus) – as characterized by rich innervation by SP+ and ENK+ woolly fibers[1] with a lack of DA+ fiber innervation that borders the striatal field – has been identified in jawless and cartilaginous fishes but not in ray-finned fishes (Reiner et al., 1998). While a population of pallidal neurons may be present in ray-finned fishes, they may not be organized in a discrete field or in the same manner as in other vertebrates. As with the various parts of the area dorsalis, correspondences to particular parts of the

---

[1] The term woolly fiber was used by Haber and Nauta (1983) to characterize the appearance of ENK+ and SP+ terminal endings on the dendrites of pallidal neurons.

area ventralis of the brains of amniotes remain largely enigmatic.

Forebrain: diencephalon

The diencephalon comprises six regions in jawed fishes, each of which contain multiple nuclei. Rostrally, four dorsal-to-ventral parts are present – the epithalamus, dorsal thalamus, ventral thalamus, and the hypothalamus and preoptic area. More caudally, the pretectal area and the posterior tuberculum are present. Additionally, the retinae of the paired lateral eyes arise from the diencephalon during embryological development. In mammals, the nuclei of the dorsal thalamus are the only diencephalic nuclei that project to the telencephalon, and many relay ascending sensory information to various areas of the neocortex (Jones, 1985). In jawed fishes at least some nuclei in four of the six diencephalic regions are involved in the relay of sensory information to the telencephalon: the dorsal thalamus, the ventral thalamus, the hypothalamus, and the posterior tuberculum (or tubercle). In jawless fishes sensory-relay components of the posterior tuberculum have not been identified, and in hagfishes the anatomical relationships of various diencephalic nuclei and regions are somewhat obscured by evagination and exhuberent enlargement during development (Wicht and Northcutt, 1992; and see Nieuwenhuys *et al.*, 1998).

While some differences occur in particular diencephalic components, the basic organization of the diencephalon is relatively constant across teleosts and can be summarized here. The epithalamus includes the pineal gland, which projects to multiple sites, most of which are in the diencephalon (Ekström and van Veen, 1983; Ekström, 1984) and include the other part of the epithalamus, the habenular nuclei (Figure 20.5). The latter generally relay telencephalic inputs (Shiga *et al.*, 1985a) to a ventromedial midbrain site, the interpeduncular nucleus, but whether the latter projects to the topologic medial pallium as it does in mammals is unknown.

The dorsal thalamus consists of three nuclei (Figure 20.5). The most rostral, nucleus anterior, receives a substantial input from the retina (e.g. Braford and Northcutt, 1983; Butler and Northcutt, 1992). The two more caudal nuclei, the dorsal posterior and central posterior nuclei, receive various sensory inputs predominantly from the midbrain roof, or tectum. This basic organization is in accord with dorsal thalamic organization across all vertebrates: a rostral, or **lemnothalamic** (Butler, 1994) part, that predominantly receives retinal and other **lemniscal** (i.e. direct) input not relayed through the midbrain roof and a caudal part, or **collothalamic** (Butler, 1994) part, that receives inputs relayed through the midbrain roof.

The ventral thalamus in teleosts consists of several nuclei, including nucleus intermedius and nuclei ventrolateralis and ventromedialis (Figure 20.5). The latter nucleus has a substantial set of connections and is involved in the relay of visual, somatosensory, auditory, lateral line, and cerebellar inputs to several parts of the dorsal telencephalon, as well as to sites in the ventral telencephalon, dorsal thalamus, posterior tuberculum, midbrain, hindbrain, and spinal cord (Ito *et al.*, 1986; Striedter, 1991; and see Nieuwenhuys *et al.*, 1998). Nuclei intermedius and ventrolateralis have less global connections; they predominantly relay retinal and other inputs to the midbrain tectum (Striedter, 1990b). These ventral thalamic connections differ markedly from the connections of motor system-related nuclei assigned to the ventral thalamus in mammals and other amniotes, and the respective sets of nuclei may thus be independently derived and unrelated phylogenetically.

The hypothalamus has a number of different nuclei and/or regions (Braford and Northcutt, 1983). In teleosts and most other fishes it is characterized by large, evaginated inferior lobes (Figure 20.5). The paraventricular part of the hypothalamus contains numerous neurons, collectively called the paraventricular organ, that make contact with the cerebrospinal fluid (CSF) of the third ventricle (Fryer *et al.*, 1985) and project to the pituitary gland, or hypophysis, where they influence the release of hormones (Peter and Fryer, 1983). The more laterally lying tuberal region includes nucleus lateralis tuberis, which is functionally related to the pituitary gland, as is the case for similar nuclei in other vertebrates, but via direct projections rather than via a hypophyseal portal system (Peter and Fryer, 1983). In contrast, the more rostrally located nucleus anterior tuberis serves as a sensory relay for lateral line system input via the torus semicircularis (Finger, 1980) to Dm in the telencephalon (Striedter, 1990b, 1991) and other sites. The large inferior lobe (one on each side) includes several nuclei. Stimulation of the inferior lobe causes feeding and aggressive biting (Demski and Knigge, 1971), and its connections are consistent with a pivotal role in feeding behavior. The inferior lobe receives visual input from pretectal nuclei (discussed below) as well as gustatory inputs (e.g. Kanwal *et al.*, 1988); it has reciprocal

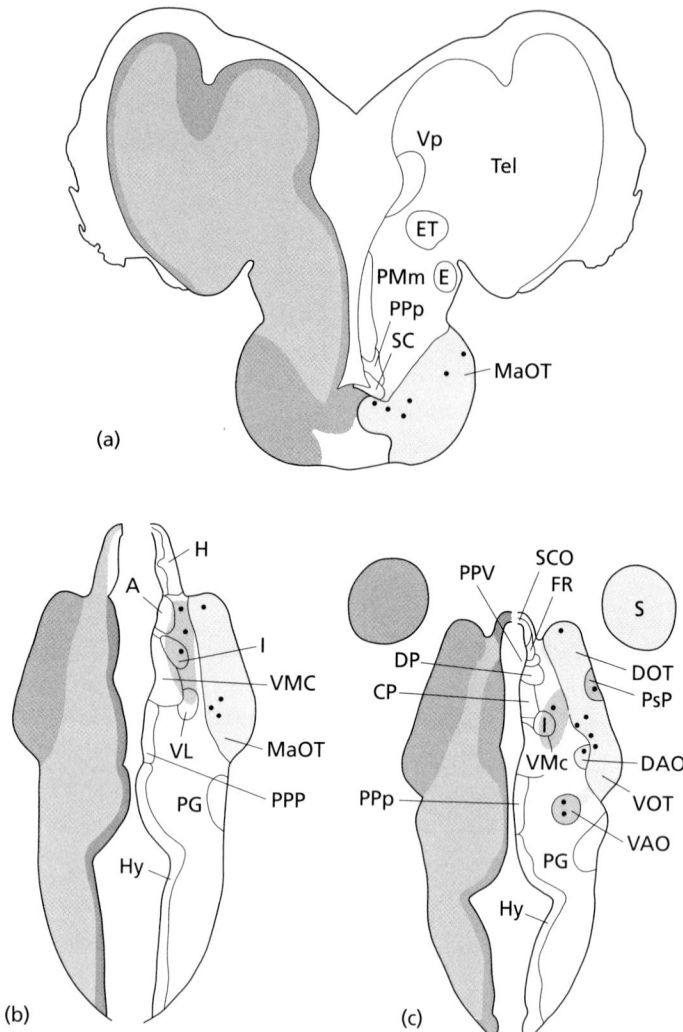

Figure 20.5 Transverse sections through the diencephalon and midbrain of the bowfin *Amia calva*, after Butler and Northcutt (1992). Projections of the retinal ganglion cells throughout these regions are shown in the line drawings, represented by shaded areas on the right side. Large dots represent retrogradely filled neuron cell bodies.

telencephalic connections (Murakami *et al.*, 1983; Wullimann *et al.*, 1991) and projects to trigeminal and facial motor nuclei (Luiten and van der Pers, 1977). In addition to these various nuclei and the inferior lobe, the hypothalamus has a rostral-most part called the preoptic area, in reference to its proximity to the optic chiasm. The preoptic area includes several nuclei and is functionally related to the periventricular and tuberal parts of the hypothalamus (see Nieuwenhuys *et al.*, 1998).

The posterior tuberculum lies caudal to the tubercular hypothalamus but generally is considered to be a separate component of the diencephalon (Braford and Northcutt, 1983; Northcutt, 1995). In most fishes, it comprises periventricular and migrated portions; the latter includes a number of sensory relay nuclei that mimic a number of the dorsal thalamic nuclei of mammals and other amniotes in their connections. The migrated portion includes a group of nuclei called the preglomerular nuclear complex (Figure 20.5). Some of these nuclei relay ascending visual, auditory, lateral line, and gustatory inputs to parts of the telencephalon (Murakami *et al.*, 1986a; Kanwal *et al.*, 1988; Wullimann, 1988). The periventricular part of the posterior tuberculum includes CSF-contacting neurons that are part of the paraventricular organ (as discussed above in the hypothalamus) and a population of DA-containing neurons that project to the telencephalon – particularly to the striatum – and widely to other regions of the brain as well (Meek, 1994; Smeets and Reiner, 1994). In ray-finned fishes, these DA-containing neurons are restricted to the periventricular zone, but in cartilaginous fishes, they are migrated into two cell groups – a medially lying

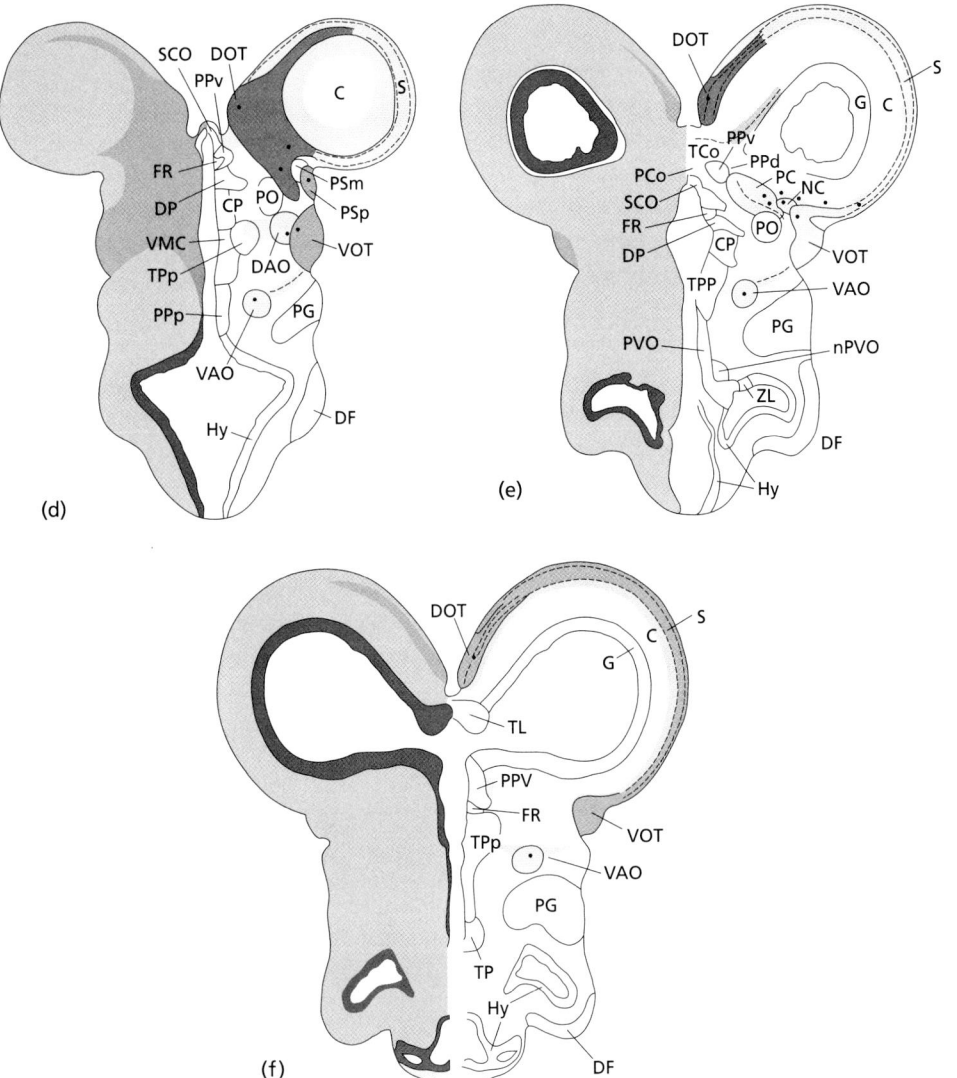

Figure 20.5 (continued)

ventral tegmental area and a more laterally positioned substantia nigra (Meredith and Smeets, 1987) – which likely correspond to the same-named nuclei in amniotes. These DA-containing cell groups occur in amniotes in the midbrain rather than at the slightly more rostral level of the caudal hypothalamus-midbrain transition area, which would correspond to the position of the posterior tuberculum in fishes (see Smeets and Reiner, 1994; Butler and Hodos, 1996; Reiner et al., 1998).

The pretectal region rostrally precedes the midbrain tectum and, like the tectum, has superficial, central, and periventricular zones (Braford and Northcutt, 1983). Within the diencephalon, the pretectum shares with the epithalamus the distinction of not projecting to the telencephalon (see Butler and Hodos, 1996; Nieuwenhuys et al., 1998). The major sensory input to the pretectum is visual (e.g. Braford and Northcutt, 1983; Northcutt and Butler, 1991, 1993). The periventricular pretectum receives visual inputs and also cerebellar input and projects to the optic tectum (Meek et al., 1986b; Striedter, 1990b). The superficial and central pretectal nuclei (Butler et al., 1991) (Figure 20.5) are involved in visually-guided feeding behavior. They relay visual inputs through one nucleus, the posterior pretectal nucleus (PO, as shown in Figure 20.5) in some teleosts or via two nuclei – pretectalis superficialis pars intermedius (PSi) and glomerulosus (Sakamoto and Ito, 1982) – in acanthopterygians and paracanthopterygians to the inferior lobe of the hypothalamus (Wullimann and Meyer, 1990) as part of the circuitry involved in feeding that was discussed above. The variation in the nuclei involved in this circuit across fishes is one of many examples of the plasticity of the brain over evolution, as are features of the pretectum in catfishes and gymnotids. In these fishes, the central part of the pretectum contains an additional nucleus, nucleus electrosensorius, which relays electrosensory input from the midbrain to part of the preglomerular nuclear complex in catfish (Striedter, 1991) or to a nucleus involved with the jamming avoidance response in gymnotids called the prepacemaker nucleus (Bastian and Yuthas, 1984 and see Chapters 15.1 and 27.1). The prepacemaker nucleus is a specialized, lateroventral subdivision of the central posterior nucleus (of the dorsal thalamus) and is discussed further below in the context of motor system regulation of electromotor behavior. A further unusual variation in catfishes is that they entirely lack a superficial part of the pretectum (Striedter, 1990a).

Midbrain (mesencephalon)

The midbrain roof contains structures involved in sensory relay and integrative functions, and the midbrain tegmentum contains a variety of cranial nerve and other nuclei. The cranial nerve nuclei (Figure 20.6) of the tegmentum lie in a medial position and are those of the oculomotor (III) and trochlear (IV) nerves (see Chapter 8). The former includes preganglionic parasympathetic neurons of the Edinger-Westphal nucleus (Wathey, 1988), as well as the groups of somatic motor neurons for the extraocular muscles, including the specialized group of somatic motor neurons in *Astroscopus* that innervate its electric organ (Leonard and Willis, 1979). The tegmentum also includes the midbrain reticular formation, which is a diffusely arranged set of cell columns with extensive connections and multiple integrative functions, and various nuclei that project to the spinal cord (Miller and Kriebel, 1986), cerebellum, tectum, and other midbrain regions (Meek *et al.*, 1986a, 1986b; Wullimann and Northcutt, 1990; Wullimann and Roth, 1994), and to other brainstem sites. A small red nucleus, which is partially defined by its receipt of cerebellar projections, is present in ray-finned fishes (Wullimann and Northcutt, 1988), while cartilaginous fishes have a larger red nucleus in the tegmentum (Smeets and Nieuwenhuys, 1976; Smeets *et al.*, 1983) that occurs at the same level as their DA-containing cell groups, the ventral tegmental area and substantia nigra (Meredith and Smeets, 1987). A tegmental structure that is unique to ray-finned fishes is the torus lateralis (Figure 20.6), which lies along the lateral aspect of the tegmentum; it is known to project to the telencephalon in the cladistian fish *Polypterus* (Northcutt, 1981b). The roof structures of the midbrain are the optic tectum, which predominantly receives visual input and in ray-finned fishes has an associated, medially lying structure, the torus longitudinalis, and the torus semicircularis, which predominantly receives auditory and lateral line inputs. The hallmark of the midbrain roof structures is their preservation of the topologic maps of sensory space from the various sensory afferent projection systems. In midbrain roof structures that are laminated, the sensory maps of the various systems are vertically in register with each other for each location in external physical space. For example, fibers that carry auditory information from a particular point in space that is relayed to the optic tectum from the torus semicircularis synapse on neurons that lie directly deep to neurons that receive visual inputs from the same point in space. Further, motor-related outputs of the optic tectum likewise spatially correspond with the sensory inputs for the same point in space.

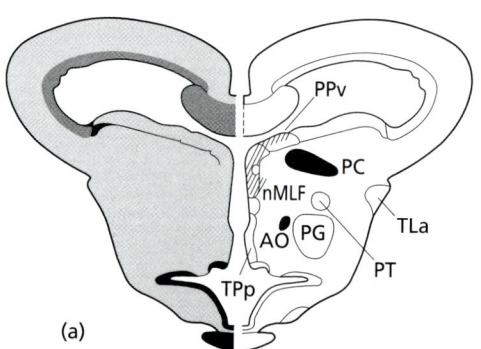 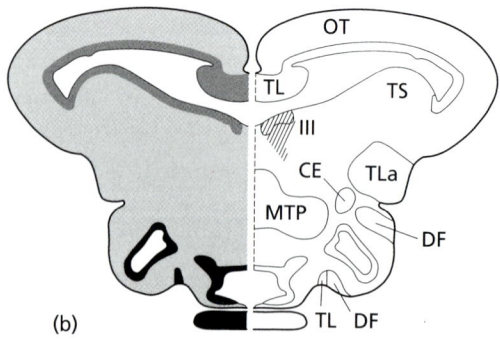

Figure 20.6 Transverse sections through the midbrain of the longnose gar *Lepisosteus osseus*, after Braford and Northcutt (1983). On the right side, retinal input is indicated by solid black and cerebellar input by diagonal lines. Retinal input also extends throughout the optic tectum.

The torus semicircularis is the homologue of the inferior colliculus of mammals, which in mammals (and other amniotes) is a purely auditory structure in the absence of the lateral line system. A torus semicircularis has not been identified in hagfishes, but in lampreys and jawed fishes, it forms a ridge along the ventral part of the tectal ventricle (Figure 20.6). In cartilaginous fishes, the torus semicircularis is called the lateral mesencephalic nuclear complex, or nucleus; it has auditory, mechanosensory, and electrosensory divisions, with the latter positioned in its dorsolateral part (Boord and Northcutt, 1988). In non-electroreceptive teleosts, the torus semicircularis is of moderate size. It contains a medial (central) nucleus that relays auditory inputs to multiple sites, including the anterior tuberal nucleus of the hypothalamus and the central posterior nucleus of the dorsal thalamus (Echteler, 1984).

The lateral line parts of the torus semicircularis are much more dramatically developed in electrosensory teleosts. In passively electrosensory teleosts – ictalurids (catfishes) and the notopterid xenomystines (featherbacks, or knifefishes) – the torus semicircularis is large and has a medial acoustic relay zone, a lateral electrosensory relay zone, and a mechanosensory relay zone in an intermediate position (Knudsen, 1977, 1978; Finger, 1986; Braford, 1986). In actively electrosensory teleosts, the torus semicircularis is even more enlarged and differs somewhat between the gymnotids (electric eels and knifefishes) and mormyrids, reflecting its independently achieved elaboration in these two groups. In the latter, at least six distinct nuclei are present in the torus semicircularis (Bell et al., 1981; Bell and Szabo, 1986). In gymnotids, the torus semicircularis reaches its apogee; it is massively enlarged and fused across the midline, and its extensive dorsal part, which receives the electrosensory lateral line input, is laminated (Carr et al., 1981). At least twelve layers and sublayers can be distinguished in *Eigenmannia*, and the various electrosensory and other inputs (including cerebellar, somatosensory, and tectal) are precisely restricted among them, as are the origins of the various efferent components (Carr et al., 1981; Carr and Maler, 1985, 1986). This elaborate laminar structure allows for integration of inputs that is used to produce the jamming avoidance response (Heiligenberg, 1988) and is one of numerous examples of specialized alar plate elaboration and lamination for exploitation of a particular sensory system. The optic tectum is likewise an example of alar plate lamination. It is laminated in all vertebrates with multiple laminae in each of the superficial, central, and deep (or periventricular) zones, and its degree of lamination varies with the development of this part of the visual system (see Butler and Hodos, 1996). The optic tectum of fishes and most other vertebrate groups is the homologue of the superior colliculus in mammals. In lampreys (Kennedy and Rubinson, 1984) and hagfishes the optic tectum is not dramatically enlarged, but it does achieve significant expansion in many cartilatigous fishes (Northcutt, 1977, 1978; Ebbesson, 1984). In ray-finned fishes the optic tectum has multiple layers within each zone and a considerable variety of cell types (Figures 20.5d–f and 20.7) (Vanegas et al., 1974, 1984). Its most massive input is from retinal ganglion cells via the optic nerve (see Chapter 8), and the precise **retinotopic map** of visual space is preserved across the tectal landscape (e.g. Rusoff and Easter, 1980; Fraley and Sharma, 1986; Presson and Fernald, 1986; Springer and Mednick, 1986). While the optic tectum is predominantly a visual system structure, it also receives auditory and somatosensory inputs and carries out multimodal integration functions. The

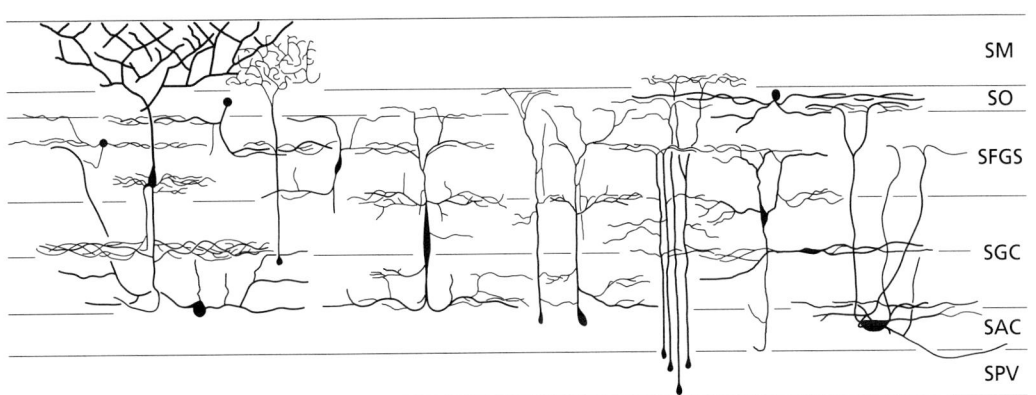

Figure 20.7 Drawings of Golgi-stained neurons from the optic tectum of a perciform teleost, *Eugerres plumieri*, after Vanegas et al. (1974).

auditory and somatosensory inputs are relayed into the optic tectum in precise register with the visual map of space. Further, the retina and optic tectum continue to grow throughout the life of teleost fishes, requiring continuous revision of the retinotopic map and its registration with the other sensory maps, which the maintained plasticity of this system permits (Cook *et al.*, 1983; Easter and Stuermer, 1984).

In ray-finned fishes a unique associated structure, the torus longitudinalis, lies at the medial end of the tectal hemisphere. This structure is densely packed with small neurons that resemble the small granule cells of the cerebellum. It receives an input from the valvula cerebelli (Ito and Kishida, 1978) and telencephalic input via tegmental and preglomerular nuclei (Wullimann and Roth, 1994) and projects to the optic tectum (Ito and Kishida, 1978; Northmore *et al.*, 1983). The telencephalic area Dc also sends direct projections to the optic tectum (Lopes Corrêa *et al.*, 1998). Efferent tectal projections are not only to numerous more rostral sites, including the dorsal posterior nucleus of the dorsal thalamus and part of the preglomerular nuclear complex (e.g. Northcutt and Butler, 1980; Wullimann and Northcutt, 1990), but also to more caudal brainstem targets, including the reticular formation and the neighborhood of the oculomotor nuclei (Sharma *et al.*, 1985).

Hindbrain (rhombencephalon)

The hindbrain contains some elaborated alar plate structures, cranial nerve nuclei (as introduced in Chapter 8), and the hindbrain reticular formation (see Butler and Hodos, 1996; Nieuwenhuys *et al.*, 1998). The latter is the caudal continuation of the midbrain reticular formation and extends throughout the central part of most of the hindbrain; it is a set of diffusely arranged neurons with extensive dendritic trees that receive and integrate numerous inputs. The reticular formation serves to coordinate the activities of many different cell groups throughout most of the brain and spinal cord. It includes a set of norepinephrine-containing neurons, the locus coeruleus, and various groups of serotonin-containing neurons, the nuclei of the raphe, all of which have extensive projections that innervate numerous sites. It is also a major source of descending projections to the spinal cord, and the set of neurons that give rise to these projections include some specialized giant neurons – Müller and Mauthner cells – that mediate rapid evasive and escape responses (Fetcho and Faber, 1988; and see Butler and Hodos, 1996).

In most fishes, the hindbrain roof comprises several major laminated structures, including the cerebellum, electrosensory lateral line lobes (ELLL) where present, and the facial and vagal lobes. Lampreys have only a rudimentary cerebellar region that lies at the rostral end of the fourth ventricle, and hagfishes apparently lack any cerebellar tissue (Nieuwenhuys *et al.*, 1998). A well developed cerebellar region is present in all jawed fishes and has several parts – a corpus – which has many cytoarchitectonic similarities to the cerebellum in amniotes and will be considered first, and other parts more related to the lateral line system.

Both ray-finned and cartilaginous fishes have a corpus cerebelli that consists of an outer molecular layer, granule cells, Purkinje cells, and other related cell types (Nieuwenhuys, 1967; Smeets *et al.*, 1983). In most cartilaginous fishes – as in amniotes – the Purkinje cells occur in a regular layer and are the efferent cells of the cerebellar cortex, projecting to a deep cerebellar nucleus, which in turn gives rise to the major efferent cerebellar tracts. In ray-finned fishes, in contrast, Purkinje cells project to other cells, called **eurydendroid cells**, that give rise to cerebellar efferents (Nieuwenhuys *et al.*, 1974; Finger, 1983a). In teleosts the cerebellar corpus receives input from the dorsal preglomerular nucleus, which relays electrosensory and telencephalic inputs (Wullimann and Northcutt, 1990), and from octaval and mechanosensory nuclei as well as numerous other brainstem structures (Meek *et al.*, 1986a,b; Wullimann and Northcutt, 1988). The cerebellum also receives visual input (Wullimann and Northcutt, 1988) that derives from various nuclei, including the accessory optic nuclei and, via relays, nucleus pretectalis superficialis pars magnocellularis (PSm), which receives tectal input (Northcutt, 1983b). The cerebellar efferent projections from the eurydendroid cells are to a wide variety of targets, including the reticular formation and the red nucleus, which lies in the midbrain, and tectal, pretectal, thalamic, and hypothalamic sites, as well as the valvula of the cerebellum (Wullimann and Northcutt, 1988).

In cartilaginous fishes lateral line-related cerebellar regions have not received much study. In ray-finned fishes the cerebellar valvula, which is a structure that is unique to ray-finned fishes, and the granular eminences of the caudal cerebellar lobe are extensively involved with the lateral line system. These structures are both involved in feedback loops from the lateral line part of the torus semicircularis via the pre-eminential nucleus (Carr *et al.*, 1981;

Maler *et al.*, 1982; Bastian and Bratton, 1990), which lies at the cranial end of the midbrain and also receives inputs from the electrosensory lateral line lobe (Maler *et al.*, 1982). The valvula and the caudal granular eminences project back to the ELLL (Carr and Maler, 1986). The valvula also has extensive other inputs and outputs (see Nieuwenhuys *et al.*, 1998); it reaches an extremely large size in electrosensory fishes, particularly in the mormyrid *Gnathonemus* (Nieuwenhuys and Nicholson, 1967; Meek, 1992).

The electrosensory lateral line lobe is likewise a large, laminated structure in electrosensory teleosts, in contrast to the mechanosensory region, which has a more modest size and simpler structure in all ray-finned fishes (McCormick, 1982, 1989). These electrosensory and mechanosensory regions are the site of termination of the respective components of the lateral line nerves (see Chapter 8). Their major rostral projections are to parts of the torus semicircularis in the midbrain (see Finger, 1986; McCormick, 1989).

The hindbrain roof additionally has two or three elaborated lobes related to the gustatory system – the facial and vagal lobes and in some fishes a glossopharyngeal lobe. These lobes receive gustatory input (see Finger, 1983b) from the ventrolateral components of the respective cranial nerves (VII, X, and IX: see Chapter 8). Like the cerebellar corpus, valvula, and ELLL, these lobes (Figure 20.8) are enlarged and laminated (Morita *et al.*, 1983). These lobes project rostrally to multiple sites that form several ascending gustatory pathways (Finger, 1983b).

The cranial nerve nuclei of the hindbrain include these lateral line and gustatory lobes. They also include the various sensory and motor nuclei associated with the trigeminal (V) nerve, the motor nucleus of the abducens (VI) nerve, the motor nuclei of the facial and glossopharyngeal nerves, the octaval (VIII) nuclear region, and the motor and non-gustatory sensory nuclei of the vagus nerve (see Chapter 8, Table 1). The motor components of these nuclei were discussed in Chapter 8, and the major projections of the sensory nuclei will be considered below in the context of their respective ascending systems.

Spinal cord

The tendency of the alar plate to elaborate into lobes extends to the spinal cord in a least two cases. The sea robin (*Prionotus carolinus*) uses specialized pectoral fin rays to search for food, and their sensory fibers

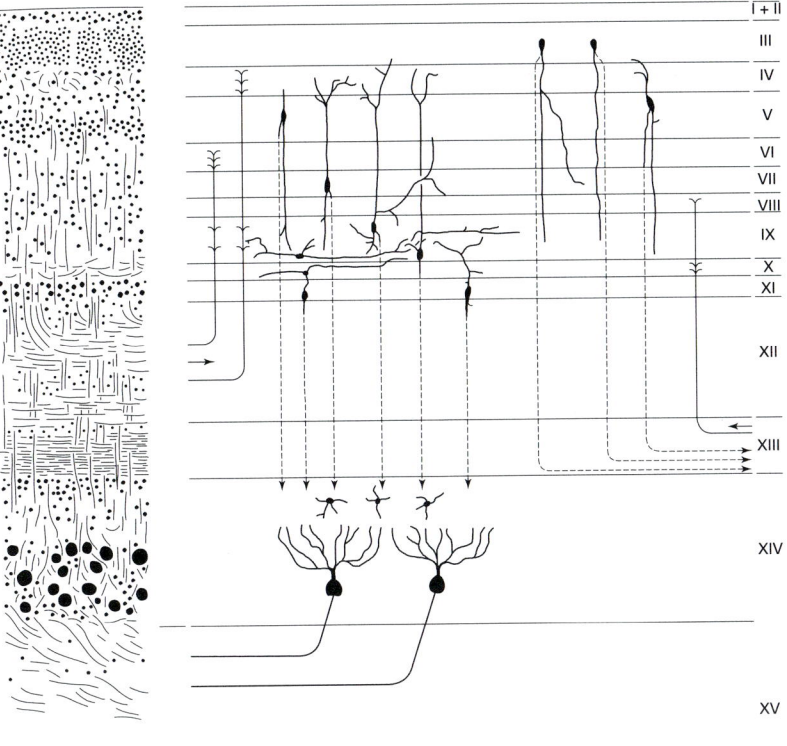

Figure 20.8 Laminar organization of the cyprinid vagal lobe, as reconstructed from Nissl-stained and Golgi material after Morita *et al.* (1983), and after Nieuwenhuys *et al.* (1998). Roman numerals indicate layers.

project into four paired sets of accessory lobes in the rostral, dorsal spinal cord that in turn project to the motor neurons that innervate these rays (Finger and Kalil, 1985). A similar specialization with accessory spinal lobes also occurs in gurnards (Nieuwenhuys and Meek, 1985).

The overall organization of the spinal cord in fishes is similar to that in other vertebrates. Neuron cell bodies lie in a central position and form an alar plate-derived dorsal horn and a basal plate-derived ventral horn. These horns contain the sensory-recipient and motor neurons, respectively. Fiber systems are located in the peripheral region. Ascending spinal cord projections include those to brainstem areas such as the reticular formation, various cranial nerve motor nuclei, the valvula and granular eminences of the cerebellum, and the torus semicircularis (Oka et al., 1986b). Descending pathways to the spinal cord originate in large part from the reticular formation (Oka et al., 1986a) and are discussed below in the context of motor systems.

Sensory and motor systems

Most sensory and motor pathways are described here only for ray-finned fishes, since considerably less information is available for other fishes. The sensory systems are generally organized in the basic patterns discussed above, with relay to the diencephalon either directly or via the midbrain roof and relay from the diencephalon to the telencephalon. Only the major systems are surveyed here. For the interest of those with some familiarity with mammalian neuroanatomy, brief comparisons with sensory and motor system pathways in mammals (see Butler and Hodos, 1996; Parent, 1996; Nieuwenhuys et al., 1998), as generally representative of amniotes, are noted but with the caveat that due to the independent evolution of some of the components of these pathways, such comparisons are at least in part similes rather than intimations of homology and must be treated with considerable caution, particularly regarding the various areas of the telencephalon.

Olfactory system

Olfactory bipolar receptor cells lie in the olfactory epithelium (see Chapter 27.3) and project to first-order multipolar cells, the mitral cells, in the olfactory bulb. Bipolar neurons terminate on the distal ends of mitral cell dendrites in complex formations called glomeruli. The mitral cells project centrally to multiple sites in the telencephalon (Figure 20.4). In jawless fishes, these projections are widespread (Northcutt and Puzdrowski, 1988; Wicht and Northcutt, 1993), whereas in jawed fishes they are restricted to specific areas; in the pallium, afferent olfactory fibers innervate only its topologically lateral part. In cartilaginous fishes, such as squalomorph sharks, olfactory projections are to a medial as well as a lateral pallial area, which are interpreted as subdivisions of the topologic lateral pallium (Ebbesson and Heimer, 1970; Northcutt, 1978). In ray-finned fishes with simply organized pallia, olfactory projections are restricted to the medial, i.e. the topologic lateral part (Braford and Northcutt, 1983). Pallial projections in teleosts termiante in Dp, which is presumably secondarily migrated to a lateral position, and they are also widespread to all of the ventral areas of the telencephalon (von Bartheld et al., 1984; Levine and Dethier, 1985). The olfactory pathway to the topologic lateral pallium is similar to that of mammals, where the olfactory bulb projects to lateral (piriform) cortex. Likewise, at least some components of olfactory projections to ventral telencephalic areas in fish may correspond to (or mimic) olfactory projections to parts of the amygdala in amniotes.

Visual system

In the retina receptor cells project to bipolar cells, which in turn project to ganglion cells (see Chapter 27.1). The retinal ganglion cells are the first-order multipolar neurons of the visual system and project centrally via the optic nerve to the optic chiasm, where most or all decussate to the contralateral side and continue as the optic tract. Retinal axons terminate (Figure 20.5) in nuclei in the preoptic area, dorsal thalamus, ventral thalamus, and pretectum and in the optic tectum (e.g. Braford and Northcutt, 1983; Striedter, 1990b; Northcutt and Butler, 1991; Butler and Northcutt, 1992).

In the dorsal thalamus a dense terminal field lies on the dendrites of the cells that form nucleus anterior, and in the ventral thalamus projections are distributed to nucleus intermedius, ventromedialis, and ventrolateralis. The retina projects to a number of pretectal nuclei (many of which have names descriptive of their location and/or cell type), including nucleus pretectalis superficialis pars parvocellularis (PSp), corticalis (NC), pretectalis centralis (PC), and the dorsal (PPd) and ventral (PPv) parts of nucleus pretectalis periventricularis. PSp and NC are involved in the circuitry for feeding behavior mediated through the inferior lobe (IL) of the hypothalamus (Wullimann and Meyer, 1990); they project to the IL via relay through the posterior pretectal nucleus or through nucleus pretectalis superficialis pars intermedius (PSi) and nucleus glomerulosus, as discussed above. Two accessory optic nuclei that lie ventral to the main pretectal area also receive retinal input. None of the retinorecipient pretectal nuclei or the accessory optic nuclei project to the telencephalon.

Retinal projections to the optic tectum preserve the retinotopic map of visual space as they distribute to its upper layers (see Vanegas et al., 1984). In the freshwater butterfly fish *Pantodon buchholzi*, for example, the ventral part of the optic tract arises from the dorsal hemiretina and distributes to the ventrolateral half of the optic tectum, while the dorsal optic tract arises from the ventral hemiretina and distributes to the dorsomedial half of the optic tectum (Saidel and Butler, 1991). Tectal efferent projections likewise arise in a spatially ordered manner. An ascending tectothalamic projection terminates in the dorsal posterior nucleus of the dorsal thalamus (e.g. Northcutt and Butler, 1980; Braford and Northcutt, 1983; Striedter, 1990b), and the dorsal posterior nucleus in turn projects to Dm in the telencephalon (Echteler and Saidel, 1981).

Direct thalamotelencephalic visual projections from nucleus anterior occur in at least some fishes. In the cladistian fish *Polypterus*, nucleus anterior projects to the telencephalon (Northcutt, 1981b); in goldfishes it is known to project specifically to Dl (Echteler and Saidel, 1981). However, in catfishes, nucleus anterior has no telencephalic projections (Striedter, 1990b).

In mammals two major visual pathways to the telencephalon are present – one with direct retinal input to the dorsal lateral geniculate nucleus of the dorsal thalamus, and from there to striate cortex, and the other with relay through the optic tectum (superior colliculus) and then via the lateral posterior/pulvinar complex of the dorsal thalamus to extrastriate cortices. The nucleus anterior of fishes appears to constitute a **lemnothalamic** structure (Butler, 1994) that relays sensory input directly to the telencephalic pallium, in a manner similar to that of the dorsal lateral geniculate nucleus (and other nonvisual lemnothalamic nuclei) of mammals. The tecto-dorsal posterior-telencephalic pathway of fishes likewise resembles the corresponding collothalamic pathway (Butler, 1994) of mammals.

Gustatory system

The facial and vagal lobes (and the glossopharyngeal lobe where present) receive gustatory input via cranial nerves VII, IX, and X and contain the first-order multipolar neurons of this system. Output from these lobes is relayed rostrally to a secondary gustatory nucleus in the pons (Finger, 1983b). This nucleus gives rise to an ascending system that projects to one or more nuclei that lie within or in the vicinity of the preglomerular nuclear complex,[2] as well as to the inferior lobe of the hypothalamus (Morita et al., 1980; Finger, 1983b; Kanwall et al., 1988; Wullimann, 1988). These diencephalic sites project rostrally to areas within the telencephalon including Dm and its related part of Dc (Kanwall et al., 1988). This ascending gustatory system differs from its mammalian counterpart, which derives from the same three cranial nerve components but involves a modestly-sized brainstem gustatory nucleus, which projects to a dorsal thalamic nucleus (part of the ventral posteromedial nucleus) for subsequent relay to gustatory neocortical regions.

Acousticolateral system

The acousticolateral system comprises vestibulocochlear, mechanosensory lateral line, and electrosensory lateral line components. Since the microscopic anatomy of electrosensory receptors is not considered separately in this volume, a brief overview is given here (Zakon, 1986, 1988). The two major types of electroreceptive organs are ampullary organs and tuberous organs (see Chapter 8). The former contain receptor cells that have a kinocilium, microvilli, or both on their apical surfaces and lie within an epithelium of support cells, to which they are coupled with tight junctions. The ampullary receptor cells are sensitive to low-frequency electric fields. Tuberous organs occur only in two groups of actively electroreceptive teleosts – mormyrids and gymnotids. They contain receptor cells that are typically covered with

---

[2] These nuclei are termed lateralis thalami and lobobulbaris in catfishes (Kanwal et al., 1988) and nucleus glomerulosus in cyprinids (Morita et al., 1980). The latter nucleus is not the same as the nucleus glomerulosus (NG) discussed above in the pretectum that is involved in a relay from PSp to PSi to NG to the inferior lobe of the hypothalamus.

microvilli and are attached to supporting epithelial cells by tight junctions. Tuberous receptor cells are tuned to frequencies in the range of the electric organ discharges (EODs) of these fishes and are sensitive to various aspects of these discharges.

Due to the common embryological derivation of bipolar sensory neurons from the dorsolateral series of placodes (see Chapter 8) and the common midbrain relay site of the torus semicircularis, the several components of the acousticolateral system are considered here together. The octavolateral area of the brainstem contains the first-order multipolar neurons for these systems and includes the various auditory and mechanosensory lateral line nuclei in all fishes, as present in the bowfin *Amia* (Figure 20.9a), for example (McCormick, 1989). The lateralis column of nuclei lies dorsal to the octavus (auditory and vestibular) column. In electrosensory fishes, the electrosensory lateral line column of nuclei lies lateral to the mechanosensory column (Figure 20.9b and c). The mechanosensory column is thus consistently referred to as the medial lateral line column (or nucleus) whether or not a more lateral, electrosensory column is present (see Nieuwenhuys *et al.*, 1998).

The torus semicircularis comprises separate auditory, mechanosensory, and – where present – electrosensory nuclei, with the auditory region generally the most medioventral (although it is termed the central nucleus) and the electrosensory most dorsolateral (Figure 20.9d–f). In cyprinids the central nucleus receives octaval input and other inputs, including those from the anterior tuberal nucleus of the hypothalamus, the central posterior nucleus of the dorsal thalamus, and Dc of the telencephalon (Echteler, 1984). Some variation in afferent input occurs across species, but the major acoustic input from octaval nuclei is consistent (Striedter, 1991). The auditory central nucleus and related nuclei of the torus semicircularis project rostrally to targets that include the central posterior nucleus, the anterior tuberal nucleus, and the lateral preglomerular nucleus (Echteler, 1984; Striedter, 1991) for subsequent relay to multiple telencephalic pallial areas. The mechanosensory lateral line part of the torus semicircularis

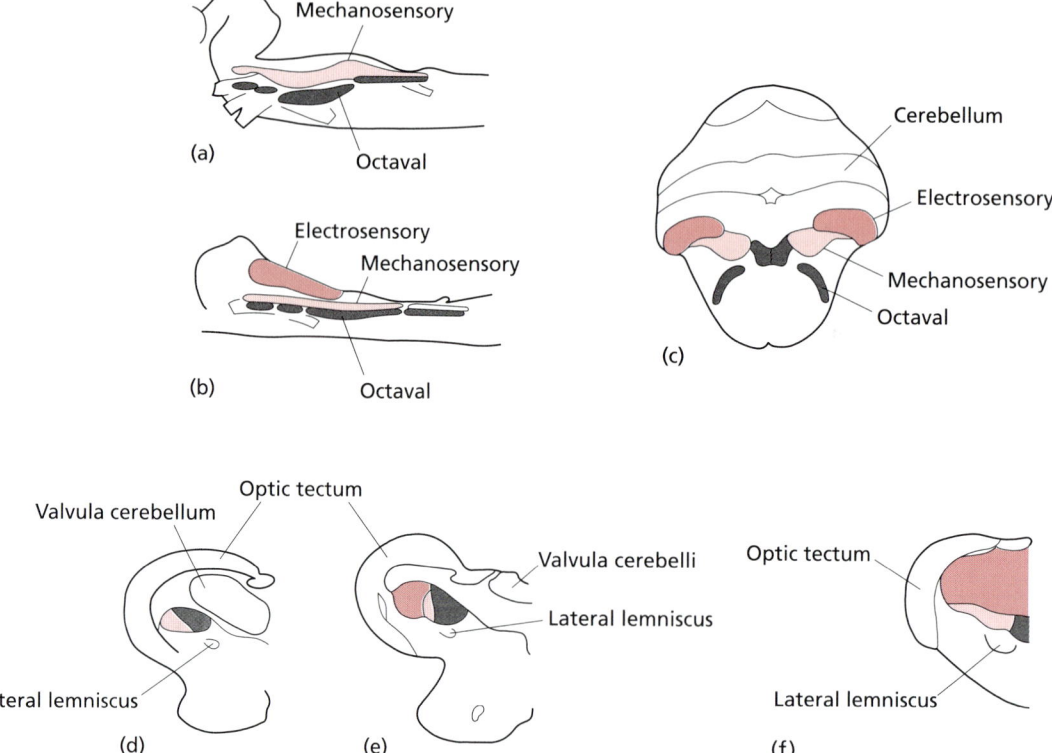

Figure 20.9 Octaval (grey), mechanosensory (light pink), and electrosensory (darker pink) areas in the brainstem of fishes: (a) projection of octaval and mechanosensory nuclei in the bowfin *Amia* onto a lateral view of the brainstem with rostral towards the left; (b) a similar projection in the sturgeon *Acipenser* including the electrosensory region; (c) a transverse section through the octavolateral area in the osteoglossomorph fish *Xenomystus*; (d–f) transverse hemisections through the torus semicircularis in three ostariophysan fishes – *Cyprinus*, *Ictalurus*, and *Eigenmannia*, respectively. All parts of this figure were adapted from McCormick (1989) and are after Butler and Hodos (1996).

receives input from the respective brainstem nuclei (Finger and Tong, 1984) and projects rostrally predominantly to the anterior tuberal and lateral preglomerular nuclei (Murakami *et al.*, 1986a, 1986b; Striedter, 1991) for relay to the telencephalon. The electrosensory region of the torus is highly elaborated in the two groups of actively electroreceptive teleosts – mormyrids and gymnotids. In the latter (Carr *et al.*, 1981; Carr and Maler, 1986; Heiligenberg, 1990) the dorsal electrosensory part of the torus has twelve layers (Figure 20.10), is fused across the midline, and receives its predominant input from the electrosensory lateral line lobe, with additional inputs from structures including the optic tectum, cerebellum, and the descending trigeminal nucleus (Carr *et al.*, 1981). Efferent projections from the electrosensory dorsal torus are to nucleus electrosensorius in the pretectum, the optic tectum, and the pre-eminential nucleus (Carr *et al.*, 1981). The latter nucleus is involved with feedback connections to the electrosensory lateral line lobe and with the valvula and caudal cerebellar granular eminences as discussed above. The ascending pathways to the telencephalon are relayed from nucleus electrosensorius via the lateral preglomerular nucleus.

Ascending auditory pathways in mammals are similar to the pathway via the central posterior nucleus in ray-finned fishes. Cochlear nuclei in the brainstem receive eighth nerve afferents that innervate hair cells in the cochlea. These nuclei project rostrally to the auditory part of the midbrain roof (inferior colliculus), which in turn project to the several nuclei of the medial geniculate body in the dorsal thalamus. Medial geniculate projections are to auditory cortical areas. No mammals (or other amniotes) have a lateral line system, since it was lost with

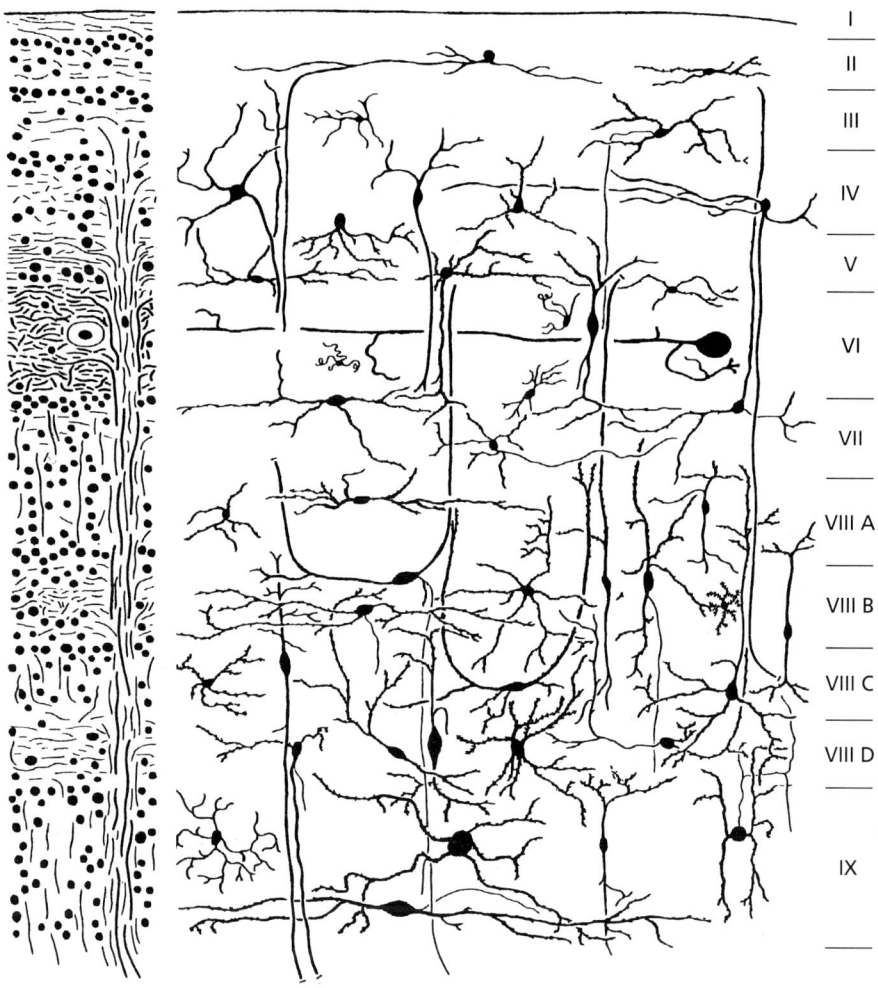

Figure 20.10 Laminar organization of the torus semicircularis of the actively electrosensory fish *Eigenmannia*, as reconstructed from Nissl-stained and Golgi material after Carr *et al.* (1981) and Carr and Maler (1985), and after Nieuwenhuys *et al.* (1998). Roman numerals indicate layers.

the transition to the terrestrial environment. Interestingly, however, platypuses have independently re-evolved mechanoreception and electroreception via receptors on the bill (Gregory et al., 1988) for use in their aquatic habitat. These receptors are not innervated by lateral line nerves but rather by the trigeminal nerve, and the inputs are relayed by the trigeminal system to the somatosensory part of the neocortex (Scheich et al., 1986; Manger et al., 1996). In fishes, the basic sensory system template of ascending relay of inputs through brainstem nuclei (or laminated lobes) to the midbrain roof and thence to diencephalic nuclei for subsequent projection to the telencephalon is common to these various acousticolateral systems. The plastic potential for elaboration of the neuraxis is strikingly apparent, however, in the various non-dorsal thalamic nuclei that are involved in these systems. Likewise, at the telencephalic level, the projections are to multiple pallial regions rather than to just one region for a particular modality. The results of independent evolution in alar plate structures given basic sensory inputs demonstrate both the constraints and the degrees of freedom in vertebrate sensory system design.

Other sensory systems

The ventromedial nucleus of the ventral thalamus is an important relay nucleus for somatosensory and motor-related feedback pathways to the telencephalon. Its connections are reminiscent of some of the components of the ventral tier nuclei (of the dorsal thalamus) of mammals, which relay ascending somatosensory inputs and inputs from the cerebellum, telencephalic basal ganglia, and substantia nigra to somatomotor cortices. In the teleost *Sebastiscus* nucleus ventromedialis receives visual and other inputs from the optic tectum, auditory and lateral line inputs from the torus semicircularis, motor-related inputs from the cerebellum and reticular formation, and direct somatosensory inputs from the primary sensory trigeminal nucleus and the spinal cord; it projects to sites in the telencephalon that include Dm, Dd, Dc, and part of area ventralis (Ito et al., 1986).

Motor systems

Major descending projections to the spinal cord in fishes originate in the reticular formation. The reticular formation interconnects cranial nerve motor nuclei so that motor functions of head musculature are coordinated, and it gives rise to spinal cord projections (Roberts, 1992; Timerick et al., 1992; Lee et al., 1993; Prasada Rao et al., 1993) that likewise direct body motor behaviors. The reticular formation includes the giant Müller and Mauthner cells for very rapid avoidance and escape behaviors as discussed above. In some ray-finned fishes a small red nucleus (nucleus ruber) that receives afferent projections from the corpus cerebelli and projects to the spinal cord has been identified in the midbrain tegmentum (Northcutt, 1983a; Wullimann and Northcutt, 1988), but the rubrospinal system is much better developed in at least some cartilaginous fishes (Ebbesson and Campbell, 1973; Smeets and Timerick, 1981).

One of the major influences on the reticular formation is the optic tectum via its descending tectobulbar projections. The optic tectum is influenced by more rostral levels of the neuraxis via descending projections from the telencephalon, some of which are direct and others of which are relayed via diencephalic nuclei. Major descending systems from the telencephalon originate in Dc, which projects to multiple sites including the periventricular part of the pretectum, the torus semicircularis, and the optic tectum, through which Dc could thus affect the reticular formation via the descending tectal pathways (Bass, 1981b; Lopes Corrêa et al., 1998). Similarly, efferents to numerous brainstem sites arise from other parts of area dorsalis, including Dm and Dl (Murakami et al., 1983 and see Nieuwenhuys and Meek, 1990). The projections from dorsal telencephalic regions to the pretectum are reminiscent of the major descending system in birds and reptiles from the dorsal pallidum to the pretectum and from thence to the optic tectum, a pathway which is also present in mammals but minor in comparison to their pallidothalamocortical system (Reiner et al., 1984; Reiner et al., 1998).

The motor-related basal plate exhibits some capacity for specialization, such as the electromotor nucleus (within the oculomotor nuclear complex) in *Astroscopus* mentioned above, if not as frequently as the alar plate. In the actively electrosensory gymnotid and mormyrid teleosts, a specialized part of the reticular formation serves an electromotor function, affecting the output of signals by the system. In gymnotids, such as *Eigenmannia* and *Apteronotus*, the prepacemaker nucleus of the diencephalon (as discussed above) projects to a pacemaker nucleus located in the medullary reticular formation (Heiligenberg et al., 1981). The latter nucleus has small neurons that generate a constant firing pattern and are thus referred to as pacemaker cells; they project to larger neurons within the same nucleus, which in turn project to electromotor neurons in the spinal cord (see Dye and Meyer, 1986). The small neurons regulate the jamming avoidance response via the larger neurons, while direct prepacemaker projections to the larger neurons affect the production of the 'chirps' and 'yodels' that are used for conspecific communication (Dye and Heiligenberg, 1987). In the mormyrid *Gnathonemus* a similar but independently evolved electromotor nucleus is present in the medullary reticular formation (Elekes *et al.*, 1985), and its output is regulated by a small group of electromotor command neurons that lie ventral to it (Bell *et al.*, 1983). These command neurons receive descending input from a precommand nucleus located in the midbrain (Bell *et al.*, 1983).

A different sort of motor specialization that has also arisen multiple times independently is that for generating sounds via muscles associated with the swim bladder or the pectoral girdle/fin. Fishes that have evolved such sound-producing systems include siluriforms (catfish), scorpaeniforms (rockfish), and batrachoidiforms (midshipman and toadfishes) (Bass, 1985; Ladich and Bass, 1998). The muscles that generate sounds by various mechanisms are innervated by a group of specialized neurons in the sonic motor nucleus, which is located at variable positions in the caudal medullary brainstem and extending into the upper spinal cord depending upon the species (Ladich and Bass, 1998). Bass and Baker (1997) have proposed that homologous regions of the caudal part of the developing hindbrain give rise to these various sets of sonic motor neurons in fishes as well as to the neural circuitry used for sound-producing behaviors in tetrapods.

Comparative perspective

Living jawless fishes have adopted specialized parasitic/scavenger strategies, and, as in all species, exhibit a mix of specialized and primitive characters. The widespread olfactory projections to the telencephalon, the absence of a significant cerebellum, and the lack of most alar plate elaborations in these fishes may have also been features of the earliest vertebrates, but given the lack of additional living vertebrate outgroups or fossil record, models of ancestral vertebrate brains must remain tentative. The gain of migratory neural crest and placodal tissues (Northcutt and Gans, 1983; Gans and Northcutt, 1983; Noden, 1991; Northcutt, 1996 and see Chapter 8) was a seminal event in the earliest vertebrates as was the gain of a central nervous system with definitive rostrocaudal parts of the forebrain, midbrain, hindbrain, and spinal cord and with definitive alar and basal plates.

Many variations on the theme of alar plate elaboration at various levels of the neuraxis – from the telencephalon to the spinal cord – have occurred. Every vertebrate from mammals to hagfishes has at least one alar plate elaboration in the midbrain – the optic tectum – and eyes (even if secondarily reduced in some species), which can be considered an alar plate elaboration of the diencephalon. Lamination is an often achieved cytoarchitecture with presumed advantages for the processing and integration of inputs, but elaboration of nuclear regions also occurs and is particularly dramatic in the telencephalons of many cartilaginous and ray-finned fishes (Northcutt, 1981a) as well as in reptiles and birds (see Karten and Shimizu, 1989; Karten, 1991). The sites of alar plate elaboration along the neuraxis are stunningly variable among fishes – from the accessory spinal lobes in sea robins and gurnards to the massive telencephalons in hagfishes and in many jawed fishes to the huge electrosensory lateral line lobes and valvulae in actively electrosensory teleosts to the impressive facial and vagal lobes in gustatory specialists. In amniotes alar plate elaboration likewise occurs at most levels of the neuraxis, as with the neocerebellar hemispheres of mammals, the exceptionally enlarged and complex optic tecta of birds, and the telencephalic pallial regions in mammals, some lizards, crocodiles, and birds.

In comparing the organization of the brain across jawed vertebrates, one can note that amniotes share many more fundamental similarities with cartilaginous fishes than they do with ray-finned fishes. These similarities occur across a variety of systems and locations and contrast with a number of more specialized, derived features that characterize the brains of ray-finned fishes (Northcutt, 1995). In the telencephalon, the developmental eversion of the pallium in ray-finned fishes (and partially in the coelacanth fish *Latimeria*) is in contrast to the evagination of the pallium in all other vertebrate groups. In the subpallium, cartilaginous fishes and amniotes both have an identifiable dorsal pallidum, identified by the presence of SP+ fibers and ENK+ woolly fibers in a DA fiber-poor field that neighbors a DA fiber-rich striatal field, while such a field has not yet been identified in ray-finned fishes. In the midbrain, cartilaginous fishes and amniotes share similar migrated DA-containing cell groups, the substantia nigra and ventral tegmental area, that project to the striatum, in contrast to the non-migrated posterior tubercular population of DA-containing cells in ray-finned fishes. Cartilaginous fishes and amniotes likewise share a red nucleus of significant size, versus the small red nucleus identified in ray-finned fishes. In the cerebellum features shared by cartilaginous fishes and amniotes include the Purkinje cells being the efferent cells of the cortex and the presence of a deep nucleus that is the gateway to extracerebellar targets, versus the intracortical projections of Purkinje cells and the eurydendroid efferent cells of ray-finned fishes. The common features of the brains in cartilaginous fishes and amniotes would appear to give us the best glimpse of what the brains of ancestral jawed vertebrates were like.

Not all features are common among cartilaginous fishes and amniotes of course; abundant differences exist, such as projections to the telencephalon from probable posterior tubercular homologues in cartilaginous fishes. Nonetheless, the brains of members of the actinopterygian radiation are as unique as they are diverse. Ray-finned fishes are a treasure-trove of brain diversity that can be mined extensively for insights into the full range of capacities and constraints for nervous system evolution.

Acknowledgments

In writing this chapter the author depended heavily upon the recently published, comprehensive, and definitive statement of our time on the organization of the central nervous system in fishes and other vertebrates, *The Central Nervous System of Vertebrates*, by R. Nieuwenhuys, H.J. ten Donkelaar, and C. Nicholson, 1998. The reader is referred to this work of astonishing breadth and depth for further information on all of the areas covered here. This work was partially supported by NSF grant IBN 9728155.

References

Bass, A.H. (1981a). *J. Morphol.*, **169**, 91–111.
Bass, A.H. (1981b). *Brain, Behav. Evol.*, **19**, 1–16.
Bass, A.H. (1985). *Brain, Behav. Evol.*, **27**, 115–131.
Bass, A.H. and Baker, R. (1997). *Brain, Behav. Evol.*, **50**, 3–16.
Bastian, J. and Bratton, B. (1990). *J. Neurosci.*, **10**, 1226–1240.
Bastian, J. and Yuthas, J. (1984). *J. Comp. Physiol.*, **154**, 895–908.
Bell, C.C. and Szabo, T. (1986). In: *Electroreception* (eds T.H. Bullock and W. Heiligenberg), John Wiley & Sons, New York, pp. 375–421.
Bell, C.C., Finger, T.E. and Russell, C.J. (1981). *Exp. Brain Res.*, **42**, 9–22.
Bell, C.C., Libouban, S. and Szabo, T. (1983). *J. Comp. Neurol.*, **216**, 327–338.
Boord, R.L. and Northcutt, R.G. (1988). *Brain, Behav. Evol.*, **32**, 76–88.
Braford, M.R., Jr. (1995). *Brain, Behav. Evol.*, **46**, 259–274.
Braford, M.R., Jr. (1986). In: *Electroreception* (eds T.H. Bullock and W. Heiligenberg), pp. 453–464. John Wiley & Sons, New York.
Braford, M.R., Jr. and Northcutt, R.G. (1974). *J. Comp. Neurol.*, **156**, 165–178.
Braford, M.R., Jr. and Northcutt, R.G. (1983). In: *Fish Neurobiology, Vol. 2, Higher Brain Areas and Functions* (eds R.E. Davis and R.G. Northcutt), pp. 117–163. University of Michigan Press, Ann Arbor, MI.
Butler, A.B. (1994). *Brain Res. Rev.*, **19**, 29–65.
Butler, A.B. and Hodos, W. (1996). *Comparative Vertebrate Neuroanatomy: Evolution and Adaptation*. Wiley-Liss, New York, 514 pp.
Butler, A.B. and Northcutt, R.G. (1992). *Brain, Behav. Evol.*, **39**, 169–194.
Butler, A.B., Wullimann, M.F. and Northcutt, R.G. (1991). *Brain, Behav. Evol.*, **38**, 92–114.
Carr, C.E., Maler, L., Heiligenberg, W. and Sas, E. (1981). *J. Comp. Neurol.*, **203**, 649–670.
Carr, C.E. and Maler, L. (1985). *J. Comp. Neurol.*, **235**, 207–240.
Carr, C.E. and Maler, L. (1986). In: *Electroreception* (eds T.H. Bullock and W. Heiligenberg), pp. 319–374. John Wiley & Sons, New York.
Cook, J.E., Rankin, E.C.C. and Stevens, H.P. (1983). *Exp. Brain Res.*, **52**, 147–151.
Demski, L.S. and Knigge, K.M. (1971). *J. Comp. Neurol.*, **143**, 1–16.
Dye, J. and Heiligenberg, W. (1987). *J. Comp. Physiol.*, **161**, 187–200.
Dye, J.C. and Meyer, J.H. (1986). In: *Electroreception* (eds T.H. Bullock and W. Heiligenberg), pp. 71–102. John Wiley & Sons, New York.
Easter, S.S. and Stuermer, C.A.O. (1984). *J. Neurosci.*, **4**, 1052–1063.
Ebbesson, S.O.E. (1984). In: *Comparative Neurology of the Optic Tectum* (ed. H. Vanegas), pp. 33–46. Plenum Press, New York.
Ebbesson, S.O.E. and Campbell, C.B.G. (1973). *J. Comp. Neurol.*, **152**, 233–254.
Ebbesson, S.O.E. and Heimer, L. (1970). *Brain Res.*, **17**, 47–55.

Echteler, S.M. (1984). *J. Comp. Neurol.*, **230**, 536–551.

Echteler, S.M. and Saidel, W.M. (1981). *Science*, **212**, 683–685.

Ekström, P. (1984). *J. Comp. Neurol.*, **226**, 321–335.

Ekström, P. and van Veen, T. (1983). *Cell Tiss. Res.*, **232**, 141–155.

Elekes, K., Ravaille, M., Bell, C.C., Libouban, S. and Szabo, T. (1985). *Neurosci.*, **15**, 417–429.

Fetcho, J.R. and Faber, D.S. (1988). *J. Neurosci.*, **8**, 4192–4213.

Finger, T.E. (1975). *J. Comp. Neurol.*, **161**, 125–141.

Finger, T.E. (1980). *Science*, **210**, 671–673.

Finger, T.E. (1983a). In: *Fish Neurobiology, Vol. 1, Brain Stem and Sense Organs* (eds R.G. Northcutt and R.E. Davis), pp. 261–284. University of Michigan Press, Ann Arbor, MI.

Finger, T.E. (1983b). In: *Fish Neurobiology, Vol. 1: Brain Stem and Sense Organs* (eds R.G. Northcutt and R.E. Davis), pp. 285–309. University of Michigan Press, Ann Arbor, MI.

Finger, T.E. (1986). In: *Electroreception* (eds T.H. Bullock and W. Heiligenberg), John Wiley & Sons, New York, pp. 287–317.

Finger, T.E. and Kalil, K. (1985). *J. Comp. Neurol.*, **239**, 384–390.

Finger, T.E. and Tong, S.-L. (1984). *J. Comp. Neurol.*, **229**, 129–151.

Fraley, S.M. and Sharma, S.C. (1986). *Brain Res.*, **385**, 179–184.

Fryer, J.N., Boudreault-Chateauvert, C. and Kirby, R.P. (1985). *J. Comp. Neurol.*, **242**, 475–484.

Gans, C. and Northcutt, R.G. (1983). *Science*, **220**, 268–274.

Gregory, J.E., Iggo, A., McIntyre, A.K. and Proske, U. (1988). *J. Physiol.*, **400**, 349–366.

Haber, S.N. and Nauta, W.J.H. (1983). *Neurosci.*, **9**, 245–260.

Heiligenberg, W. (1988). *Brain, Behav. Evol.*, **31**, 6–16.

Heiligenberg, W. (1990). *Synapse*, **6**, 196–206.

Heiligenberg, W. Finger, T., Matsubara, J., Carr, R. (1981). *Brain Res.*, **211**, 418–423.

Hodos, W. and Butler, A.B. (1997). *Brain, Behav. Evol.*, **50**, 189–197.

Hodos, W. and Butler, A.B. (2000). In: *Brain Evolution and Cognition* (eds G. Roth and M.F. Wulliamann). Spektrum Akademischer Verlag, Heidelberg, in press.

Ito, H. and Kishida, R. (1978). *J. Comp. Neurol.*, **181**, 465–476.

Ito, H., Murakami, T., Fukuota, T., Kishida, R. (1986). *J. Comp. Neurol.*, **250**, 215–227.

Jones, E.G. (1985). *The Thalamus*. Plenum Press, New York, 935 pp.

Kanwal, J.S., Finger, T.E. and Caprio, J. (1988). *J. Comp. Neurol.*, **278**, 353–376.

Karten, H.J. (1991). *Brain, Behav. Evol.*, **38**, 264–272.

Karten, H.J. and Shimizu, T. (1989). *J. Cognit. Neurosci.*, **1**, 291–301.

Kennedy, M.C. and Rubinson, K. (1984). In: *Comparative Neurology of the Optic Tectum* (ed. H. Vanegas), pp. 1–13. Plenum Press, New York.

Knudsen, E.I. (1977). *J. Comp. Neurol.*, **173**, 417–432.

Knudsen, E.I. (1978). *J. Neurophysiol.*, **41**, 350–364.

Ladich, F. and Bass, A.H. (1998). *Brain, Behav. Evol.*, **51**, 315–330.

Lee, R.K.K., Eaton, R.C. and Zottoli, S.J. (1993). *J. Comp. Neurol.*, **329**, 539–556.

Leonard, R.B. and Willis, W.D. (1979). *J. Comp. Neurol.*, **183**, 397–414.

Levine, R.I. and Dethier, S. (1985). *J. Comp. Neurol.*, **237**, 427–444.

Lopes Corrêa, S.A., Grant, K. and Hoffmann, A. (1998). *Brain, Behav. Evol.*, **52**, 81–98.

Luiten, P.G.M. and van der Pers, J.N.C. (1977). *J. Comp. Neurol.*, **174**, 575–590.

Maler, L., Sas, E., Carr, C.E. and Matsubara, J. (1982). *J. Comp. Neurol.*, **211**, 154–164.

Manger, P.R., Calford, M.B. and Pettigrew, J.D. (1996). *Proc. Roy. Soc. Lond. B.*, **263**, 611–617.

Manso, M.J. and Anadón, R. (1991). *J. Comp. Neurol.*, **307**, 335–349.

McCormick, C.A. (1982). *J. Morphol.*, **171**, 159–181.

McCormick, C.A. (1989). In: *The Mechanosensory Lateral Line: Neurobiology and Evolution* (eds S. Coombs, P. Görnera and H. Munz), pp. 341–364. Springer-Verlag, Berlin.

Meek, J. (1992). *Europ. J. Morphol.*, **30**, 37–51.

Meek, J. (1994). In: *Phylogeny and Development of the Catecholamine Systems in the CNS of Vertebrates* (eds W.J.A.J. Smeets and A. Reiner), pp. 49–76. Cambridge University Press, Great Britain.

Meek, J., Nieuwenhuys, R. and Elsevier, D. (1986a). *J. Comp. Neurol.*, **245**, 319–341.

Meek, J., Nieuwenhuys, R. and Elsevier, D. (1986b). *J. Comp. Neurol.*, **245**, 342–358.

Meredith, G.E. and Smeets, W.J.A.J. (1987). *J. Comp. Neurol.*, **265**, 530–548.

Miller, K.E. and Kriebel, R.M. (1986). *Brain Res. Bull.*, **16**, 183–188.

Morita, Y., Ito, H. and Masai, H. (1980). *J. Comp. Neurol.*, **191**, 119–132.

Morita, Y., Murakami, T. and Ito, H. (1983). *J. Comp. Neurol.*, **218**, 378–394.

Murakami, T., Morita, Y. and Ito, H. (1983). *J. Comp. Neurol.*, **216**, 115–131.

Murakami, T., Fukuoka, T. and Ito, H. (1986a). *J. Comp. Neurol.*, **247**, 383–397.

Murakami, T., Ito, H. and Morita, Y. (1986b). *Brain Res.*, **382**, 97–103.

Nieuwenhuys, R. (1967). *Progr. Brain Res.*, **25**, 1–93.

Nieuwenhuys, R. (1962). *J. Morphol.*, **111**, 69–88.

Nieuwenhuys, R. (1963). *J. Hirnforsch.*, **6**, 171–200.

Nieuwenhuys, R. (1977). *Ann. N.Y. Acad. Sci.*, **299**, 97–145.

Nieuwenhuys, R. and Meek, J. (1985). In: *Functional Morphology of Vertebrates* (eds H.-R. Duncker and G. Fleischer), pp. 515–528. Gustav Fischer Verlag, Stuttgart.

Nieuwenhuys, R. and Meek, J. (1990). In: *Cerebral Cortex, Vol. 8A: Comparative Structure and Evolution of Cerebral Cortex, Part I* (eds E.G. Jones and A. Peters), pp. 31–73. Plenum Press, New York.

Nieuwenhuys, R. and Nicholson, C. (1967). *Nature*, **215**, 764–765.

Nieuwenhuys, R., Pouwels, E. and Smulders-Kersten, E. (1974). *Z. Anat. Entwickl. Gesch.*, **144**, 315–336.

Nieuwenhuys, R., ten Donkelaar, H.J. and Nicholson, C. (1998). *The Central Nervous System of Vertebrates.* Springer-Verlag, Berlin, 2219 pp.

Noden, D.M. (1991). *Brain, Behav. Evol.*, **38**, 190–225.

Northcutt, R.G. (1977). *Am. Zool.*, **17**, 411–429.

Northcutt, R.G. (1978). In: *Sensory Biology of Sharks, Skates, and Rays* (eds E.S. Hodgson and R.F. Mathewson), pp 117–193. Department of the Navy, Arlington, VA.

Northcutt, R.G. (1981a). *Ann. Rev. Neurosci.*, **4**, 301–350.

Northcutt, R.G. (1981b). *Neurosci. Lett.*, **22**, 219–222.

Northcutt, R.G. (1983a). *Anat. Rec.*, **205**, 144A.

Northcutt, R.G. (1983b). In: *Fish Neurobiology, Vol. 2: Higher Brain Areas and Functions* (eds R.E. Davis and R.G. Northcutt), pp. 1–42. University of Michigan Press, Ann Arbor, MI.

Northcutt, R.G. (1995). *Brain, Behav. Evol.*, **46**, 275–318.

Northcutt, R.G. (1996). *Israel J. Zool.*, **42**, S273–S313.

Northcutt, R.G. and Braford, M.R., Jr. (1980). In: *Comparative Neurology of the Telencephalon* (ed. S.O.E. Ebbesson), pp. 41–98. Plenum Press, New York.

Northcutt, R.G. and Butler, A.B. (1980). *Brain Res.*, **190**, 333–346.

Northcutt, R.G. and Butler, A.B. (1991). *Brain, Behav. Evol.*, **37**, 333–354.

Northcutt, R.G. and Butler, A.B. (1993). *J. Comp. Neurol.*, **328**, 547–561.

Northcutt, R.G. and Davis, R.E. (1983). In: *Fish Neurobiology, Vol. 2: Higher Brain Areas and Functions* (eds R.E. Davis and R.G. Northcutt), pp. 203–236. University of Michigan Press, Ann Arbor, MI.

Northcutt, R.G. and Gans, C. (1983). *Quart. Rev. Biol.*, **58**, 1–28.

Northcutt, R.G. and Puzdrowski, R.L. (1988). *Brain, Behav. Evol.*, **32**, 96–107.

Northmore, D.P.M., Williams, B. and Vanegas, H. (1983). *J. Comp. Physiol.*, **150**, 39–50.

Oka, Y., (1980). *Brain Res.*, **185**, 215–225.

Oka, Y., Satou, M. and Ueda, K. (1986a). *J. Comp. Neurol.*, **254**, 91–103.

Oka, Y., Satou, M. and Ueda, K. (1986b). *J. Comp. Neurol.*, **254**, 104–112.

Parent, A. (1996). *Carpenter's Human Neuroanatomy*, 9th Edition. Williams & Wilkins, Baltimore, 1011 pp.

Peter, R.E. and Fryer, J.N. (1983). In: *Fish Neurobiology, Vol. 2: Higher Brain Areas and Functions* (eds R.E. Davis and R.G. Northcutt), pp. 165–201. University of Michigan Press, Ann Arbor, MI.

Pouwels, E. (1978). *Anat. Embryol.*, **153**, 37–54.

Prasada Rao, P.D., Jadhao, A.G. and Sharma, S.C. (1993). *Brain Res.*, **620**, 211–220.

Presson, J.C. and Fernald, R.D. (1986). *Devel. Brain Res.*, **26**, 179–186.

Reiner, A., Brauth, S.E. and Karten, H.J. (1984). *Trends Neurosci.*, **7**, 320–325.

Reiner, A., Medina, L. and Veenman, C.L. (1998). *Brain Res. Rev.*, **28**, 235–285.

Roberts, B.L. (1992). *Rev. Fish Biol.* **2**, 243–266.

Rusoff, A.C. and Easter, S.S. (1980). *Science*, **208**, 311–312.

Saidel, W.M. and Butler, A.B. (1991). *Brain, Behav. Evol.*, **38**, 154–168.

Sakamoto, N. and Ito, H. (1982). *J. Comp. Neurol.*, **205**, 291–298.

Scheich, H., Langner, G., Tidemann, C., Coles, R.B. and Guppy, A. (1986). *Nature*, **319**, 401–402.

Sharma, S.C., Dunn-Meynell, A.A. and Kobylack, M.A. (1985). *Neurosci. Lett.*, **59**, 265–270.

Shiga, T., Oka, Y., Satou, M., Okumoto, N. and Ueda, K. (1985a). *Brain Res. Bull.*, **14**, 55–61.

Shiga, T., Oka, Y., Satou, M., Okumoto, N. and Ueda, K. (1985b). *Brain Res.*, **364**, 162–177.

Smeets, W.J.A.J. and Nieuwenhuys, R. (1976). *J. Comp. Neurol.*, **165**, 333–368.

Smeets, W.J.A.J. and Reiner, A. (eds.) (1994). *Phylogeny and Development of Catecholamine Systems in the CNS of Vertebrates.* Cambridge University Press, Great Britain, 488 pp.

Smeets, W.J.A.J. and Timerick, S.J.B. (1981). *J. Comp. Neurol.*, **202**, 473–491.

Smeets, W.J.A.J., Nieuwenhuys, R. and Roberts, B.L. (1983). *The Central Nervous System of Cartilaginous Fishes. Structure and Functional Correlations.* Springer-Verlag, Berlin, 266 pp.

Springer, A.D. and Mednick, A.S. (1986). *J. Comp. Neurol.*, **247**, 221–232.

Striedter, G.F. (1990a). *Brain, Behav. Evol.*, **36**, 329–354.

Striedter, G.F. (1990b). *Brain, Behav. Evol.*, **36**, 355–377.

Striedter, G.F. (1991). *J. Comp. Neurol.*, **312**, 311–331.

Timerick, S.J.B., Roberts, B.L. and Paul, D.H. (1992). *Brain, Behav. Evol.*, **39**, 93–100.

Vanegas, H., Laufer, M. and Amat, J. (1974). *J. Comp. Neurol.*, **154**, 43–60.

Vanegas, H., Ebbesson, S.O.E. and Laufer, M. (1984). In: *Comparative Neurology of the Optic Tectum* (ed. H. Vanegas), pp. 93–120. Plenum Press, New York.

von Bartheld, C.S., Meyer, D.L., Fiebig, E. and Ebbesson, S.O.E. (1984). *Cell Tissue Res.*, **238**, 475–487.

Wathey, J.C. (1988). *J. Comp. Physiol.*, **162**, 511–524.

Wicht, H. and Northcutt, R.G. (1990). *Brain, Behav. Evol.*, **36**, 315–328.

Wicht, H. and Northcutt, R.G. (1992). *Brain, Behav. Evol.*, **40**, 25–64.

Wicht, H. and Northcutt, R.G. (1993). *J. Comp. Neurol.*, **337**, 529–542.

Wullimann, M.F. (1988). *Neurosci. Lett.*, **86**, 6–10.

Wullimann, M.F. and Meyer, D.L. (1990). *Brain, Behav. Evol.*, **36**, 14–29.

Wullimann, M.F. and Northcutt, R.G. (1988). *Brain, Behav. Evol.*, **32**, 293–316.

Wullimann, M.F. and Northcutt, R.G. (1990). *J. Comp. Neurol.*, **297**, 537–552.

Wullimann, M.F. and Roth, G. (1994). *Brain, Behav. Evol.*, **44**, 338–352.

Wullimann, M.F., Meyer, D.L. and Northcutt, R.G. (1991). *J. Comp. Neurol.*, **312**, 415–435.

Wullimann, M.F., Rupp, B. and Reichert, H. (1996). *Neuroanatomy of the Zebrafish Brain: A Topological Atlas*. Birkhaüser Verlag, Basel, 144 pp.

Zakon, H.H. (1986). In: *Electroreception* (eds T.H. Bullock and W. Heiligenberg), pp. 103–156. John Wiley & Sons, New York.

Zakon, H.H. (1988). In: *Sensory Biology of Aquatic Animals* (eds J. Atema, R.R. Fay, A.N. Popper and W.N. Tavolga), pp. 813–850. Springer-Verlag, New York.

CHAPTER 21

Respiratory System

Kenneth R Olson
Indiana University School of Medicine,
South Bend Center for Medical Education,
University of Notre Dame,
Notre Dame, Indiana, USA

Abbreviations

AC	accessory cells
AF	afferent filamental artery
ALA	afferent lamellar arteriole
BC	basal cell
BL	basal lamina
C	collagenous columns
CC	chloride cells
CR	cartilaginous support rod
DC	dark cell
DG	dense-cored granules
EC	endothelial cells
EF	efferent filamental artery
ELA	efferent lamellar arteriole
FB	fibrous border
IFE	interlamellar filamental epithelium
IL	interlamellar vasculature
LC	light cell
M	mitochondria
MC	mucous cell
MF	myosin filaments
N	nucleus
NEC	neuroepithelial cell
NF	nerve fibers
NP	nerve plexus
NV	nutrient vessel
OMC	outer marginal channel
PC	pillar cells
PE	pavement epithelial cell
PF	pillar cell flange
PM	plasma membrane
R	red blood cell
RL	respiratory lamellae
RS	rodlet sacs
TS	tubular system
UC	undifferentiated cell
VS	vascular space
VT	vesicotubular system
W	environmental water
WF	water flow

Copyright © 2000 Academic Press

Introduction

The general anatomy of the gill arch and the relationship between the arch support tissues and the gill filaments was described in Chapter 9. From a functional perspective, the gill arch proper probably does little more than provide structural support for the filaments. The anatomy of the tissues that comprise the arch, e.g. cartilage, bone, skeletal and smooth muscle, and other connective tissues, is generally similar to that found in all vertebrates and will not be considered here. The emphasis of this chapter is on the gill filament whose unique structure has become highly specialized in order to perform the variety of diverse activities requisite for an aquatic existence.

The gill filament is the functional unit of the gill and this is where essentially all of the homeostatic activities of the gill arch are carried out. Oxygen is taken up by simple diffusion across the epithelium of the respiratory lamellae. Carbon dioxide is excreted either as a dissolved gas across the lamellae, or in the form of bicarbonate by specialized ion transporting cells found on the filament body and occasionally on the lamellae. These and other ion transporting cells also transport sodium, calcium, hydrogen and chloride ions and thereby are vital in acid–base balance, calcium metabolism and osmoregulation. Other epithelial cells secrete mucus or provide a physiological barrier from the environment. Gill vessels not only serve as conduits for blood, they also metabolically activate or inactivate numerous hormones and other molecules in the plasma and in response to neural and endocrine signals they provide a means for selective control of tissue perfusion. Additional cells provide structural, metabolic or other forms of support which, although vital, is of lesser concern and in the interest of space will be largely omitted.

Gill filament

The gross anatomy of the gill filament has been the subject of a number of reviews (Hughes, 1984; Laurent; 1984; Olson, 1991, 1996) and the orientation planes and general anatomy can be found in Chapter 9.

Figure 21.1 shows the functional elements of the gill filament. The ion transporting cells (chloride cells or **ionocytes**) are found on the body, or alamellar, portion of the filament. These cells are especially prevalent along the afferent margin of the filament and in the interlamellar filamental epithelium, i.e. the filament epithelium between adjacent lamellae (Figure 21.2). If chloride cells are present on the lamellar epithelium they are most commonly located near the body of the filament. Chloride cells are rarely observed on the more lateral areas of the lamella, although they may appear here as well under certain conditions, such as excessively low ambient sodium and chloride (Laurent and Hebibi, 1989; Perry and Laurent, 1989; Laurent and Perry, 1990). Chloride cells are rarely found along the efferent margin of the filament. Perhaps this is because the inspired water initially comes into contact with the gill along this surface and the shear forces here are probably the most severe.

Mucous cells cover the filament and extend out on to the lamellae. They are most prevalent on the efferent edge of the filament where their protective functions are probably most required. **Neuroepithelial cells** are also thinly scattered along the efferent margin beneath the epithelium. The majority of the lamellar and filamental surface is covered by simple squamous 'pavement' epithelial cells.

Interstitial and undifferentiated cells are found throughout the body of the filament. Cells in the interlamellar area are precursor cells for new chloride cells. Cells embedded in the area of the inner margin of the lamella differentiate into lamellar pillar cells and the lamellae essentially grow out from their base as the fish grows (Hughes, 1984).

The filament body also contains the blood vessels supplying the respiratory lamella (afferent filamental artery, AF, and afferent lamellar arteriole, ALA), vessels draining the lamella (efferent filamental artery, EF, and efferent lamellar arteriole, ELA) as well as the extensive and complex filamental circulation. Support for these tissues is provided by the cartilaginous support rod (CR) and a variety of other supportive and interstitial cells.

Epithelium

The anatomy of the gill epithelium has been summarized in several reviews (Laurent and Dunel, 1980; Laurent, 1984; Olson, 1996). Laurent and Dunel (1980) and Laurent (1984) suggested that there are at least two general types of epithelium, one covering the filament (and arch support tissue) and the other on the lamellae. Both areas have the three major cell

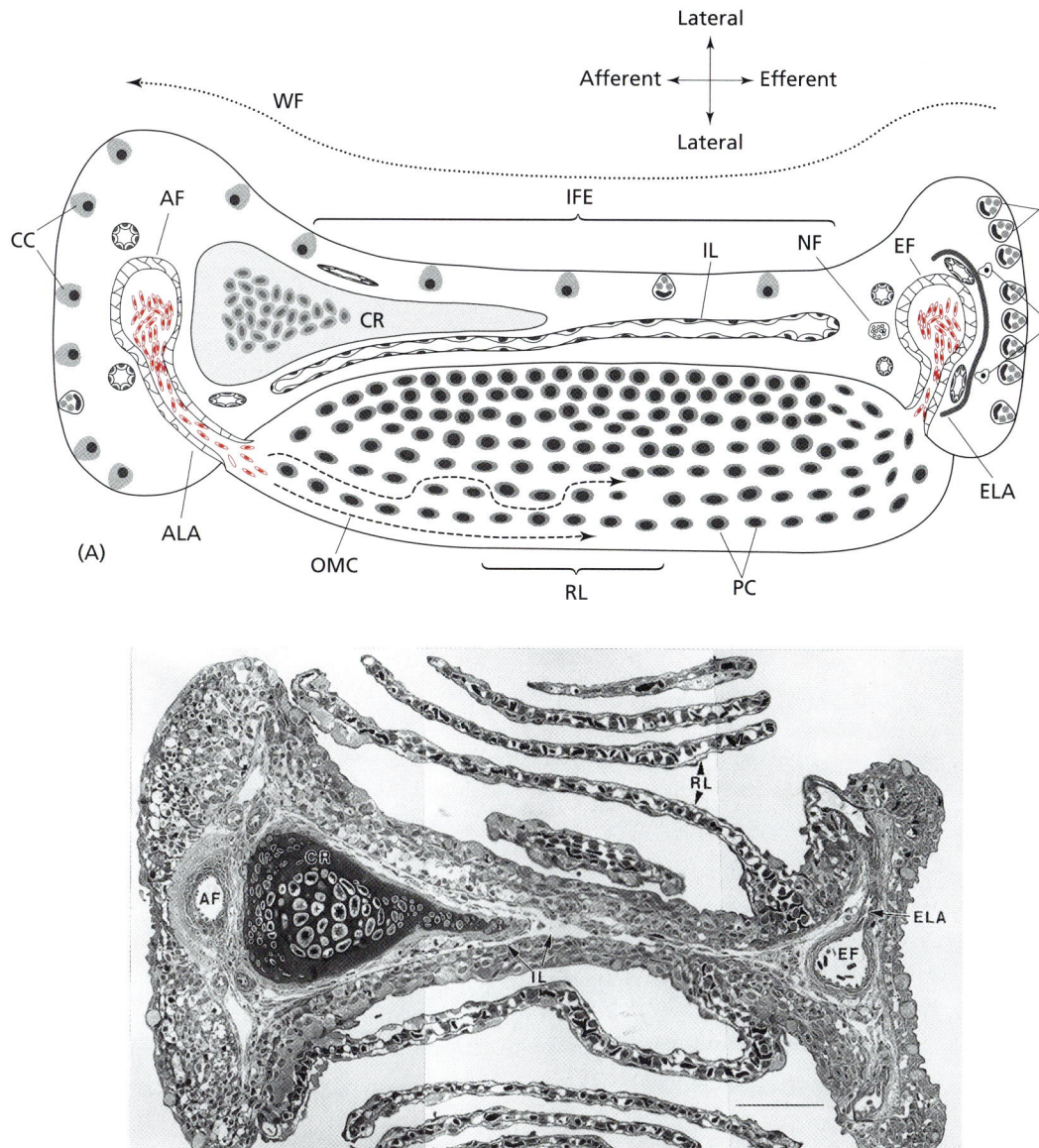

Figure 21.1 (A) Drawing of an idealized gill filament in a typical bony fish. In this section the barbell-shaped body of the filament is cut in cross-section and a planar section is cut through one respiratory lamella (RL; bottom). Chloride cells (CC) are most prevalent in the epithelium along the afferent edge of the filament and in the interlamellar filamental epithelium (IFE), i.e. the filamental epithelium between lamellae. Mucous cells (MC) and an occasional sensory neuroepithelial cell (NEC) are more common along the efferent edge of the filament, although the MC can be found throughout the filament. A cartilaginous rod (CR) runs the length of the filament for structural support. The afferent filamental artery (AF) travels the length of the filament and gives rise to the afferent lamellar arterioles (ALA) that supply the RL. In some species (as shown) the AF is closely apposed to the CR and here the vessel wall is considerably reduced. Blood percolates around the pillar cells (PC) or along the outer marginal channel (OMC) on its way across the lamella and exits via the efferent lamellar arteriole (ELA) and efferent filamental artery (EF). Water flow (WF; dashed arrow) is countercurrent to lamellar blood flow. Only a portion of the extensive interlamellar vasculature (IL) is shown. One or several nerve fibers (NF) are often found near the EF. Figure not drawn to scale. (B) Light micrograph of a section through a trout gill filament. (Courtesy of Dr S. Dunel-Erb, unpublished.) Abbreviations as in Figure 21.1A. Bar = 100 μm.

types: pavement, chloride and mucous. Other cells, such as neuroepithelial cells, granular cells, taste buds and **rodlet cells**, are less common and their distribution is often quite limited and highly variable.

Bony fish maintain osmolarity of their body fluids relatively constant (300–350 mosm L^{-1}) whether they live in fresh water, where osmolarity is often below 1 mosm L^{-1}, or salt water (sea water) with an

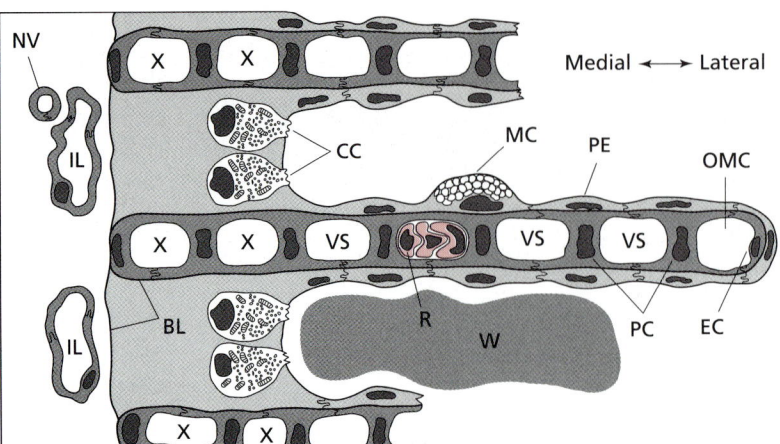

Figure 21.2 Drawing of a section perpendicular to three respiratory lamellae and parallel to the axis of the filament. The lamella is comprised of two layers of pavement cell epithelium (PE) that cover a planar sheet of pillar cells (PC). Cytoplasmic extensions between adjacent PC form the vascular space (VS) which is occupied by three contorted red blood cells (R) in the center of the figure. The thick nuclear region of the PE is aligned with the pillar cell body, thereby minimizing the diffusion distance between the environmental water (W) and VS. True endothelial cells (EC) are only found around the peripheral margin of the lamella, i.e. the outer marginal channel (OMC). Chloride cells (CC) are common in the interlamellar filamental epithelium but may also be found on the lamella (not shown) along with an occasional mucous cell (MC). A sheet of basal lamina (BL) lines the filament beneath the inner margin of the lamella and extends out between the PE and PC to envelop the lamellae. Filamental nutrient (NV) and interlamellar (IL) vessels lie beneath the BL. In most fish, from 10 to 20% of the medial VS may be embedded in the filament and unavailable for gas exchange (X). Figure not drawn to scale. (Modified from Olson, 1991.)

osmolarity of around 1000 mosm L^{-1}. Freshwater fish counter the osmotic gain of water and loss of salts by excreting a dilute urine and actively absorbing salts at the gill. Saltwater fish have just the opposite problem and they counter dehydration by drinking sea water and excreting the excess salt load across the gill.

The identities of specific gill cells involved in ion transport have not been completely resolved although recent studies have been summarized to form a working hypothesis (Filk *et al.*, 1995; Goss *et al.*, 1995; Lin and Randall, 1995; McCormick, 1995). Pavement cells, especially those on the lamella (and perhaps chloride cells), contain a vacuolar-type H^+-ATPase in the apical (water side) membrane that transports H^+ into the environment and this establishes an electrochemical gradient that drives Na^+ absorption from the water into the cell. A Na^+–K^+-ATPase in the basolateral (interstitial side) membrane delivers Na^+ to the plasma. Less is known of the site of chloride absorption, although it is thought to occur primarily in chloride cells. Calcium may also be actively absorbed from the freshwater environment by chloride cells. Chloride cells of saltwater fish actively secrete chloride and sodium follows passively down an electrochemical gradient, probably via a paracellular pathway between the epithelial cells and through the tight junctions.

Pavement cells

Pavement cells are by far the most prevalent and may account for over 95% of all gill epithelial cells (Goss *et al.*, 1995). In cross-section, these simple squamous cells are fairly nondescript and contain the usual compliment of intracellular organelles. Their most notable feature is the external membrane, whose topography varies with location and species. In cross-section the outer membrane appears to form microvilli (Figure 21.3), however, as viewed from the surface this may vary from microvillous to microridged (Figure 21.4). Microridges are commonly arranged in large fingerprint whorls on pavement cells from the arch support tissue and some areas on the filament (Figure 21.4A). They become smaller and more elaborate on the interlamellar filamental epithelium and can be quite variable, from extensive to non-existent, on the lamellae (Figure 21.4B–D). One nearly universal characteristic of pavement cells is the presence of a continuous **microridge** along the cell junction, even if the cell does not have any other microridges or microvilli (Figure 21.4D).

The function of microridges is not known. They may trap, help hold, or distribute mucus, provide an unstirred boundary layer, add mechanical strength, permit rapid changes in cell volume, or even enhance

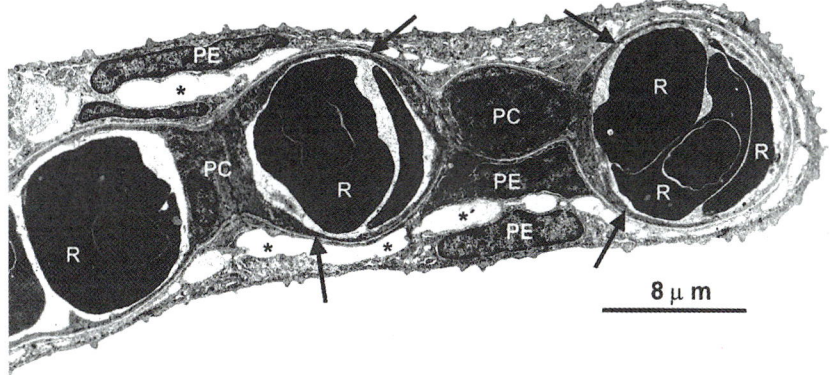

Figure 21.3 Transmission electron micrograph of a trout respiratory lamella. Cytoplasmic flanges (arrows) extend from pillar cells (PC) and delimit the vascular space. The basal lamina appears as a thin white band above and beneath the PC. The outer marginal channel (far right) is filled with three red blood cells (R). The surface of pavement epithelial cells (PE) contains microridges and appears irregular in cross-section. A sinus space (∗) is common beneath the epithelium. Two epithelial cell nuclei are present above the PC on the left and beneath the PC on the right.

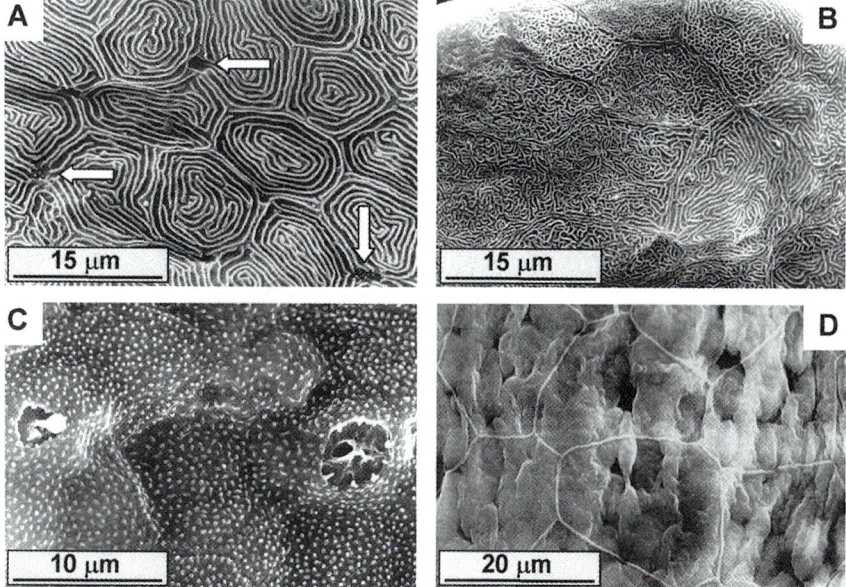

Figure 21.4 Scanning electron micrograph of gill epithelial cells. A. Whorled pavement cells common on arch support tissue and filament body with recessed mucous cells (arrows); from walking catfish, *Clarias batrachus*. B. Complex microridges on lamellar surface of rainbow trout, *Oncorhynchus mykiss*. C. Microvillous lamellar surface of walking catfish with two surface mucous cells. A mucus droplet is being secreted from the cell on left. D. Lamellar pavement cells from the bluefish, *Pomatomus saltatrix*, are devoid of microridges or microvilli. (A and C modified from Olson *et al.*, 1995; B and D modified from Olson, 1996.)

flexibility of the gill (reviewed in Olson, 1996). They may also increase the membrane surface to accommodate more ion pumps. Nevertheless, the high degree of variability suggests that form is somehow coupled to function.

Chloride cells

Chloride cells, also called ionocytes or mitochondria-rich cells, have long been implicated in ion transport. As shown in Figures 21.5 and 21.6, chloride cells can be broadly classified into two categories: (i) surface cells with an apical (exposed to the environment) membrane that is flush with the pavement cells; or (ii) recessed cells with the apical membrane opening into a pit that is partially covered by overlying pavement cells. Surface cells are more common in freshwater fish, whereas saltwater fish have recessed chloride cells (there are numerous exceptions in both instances). A number of subtypes of chloride cells

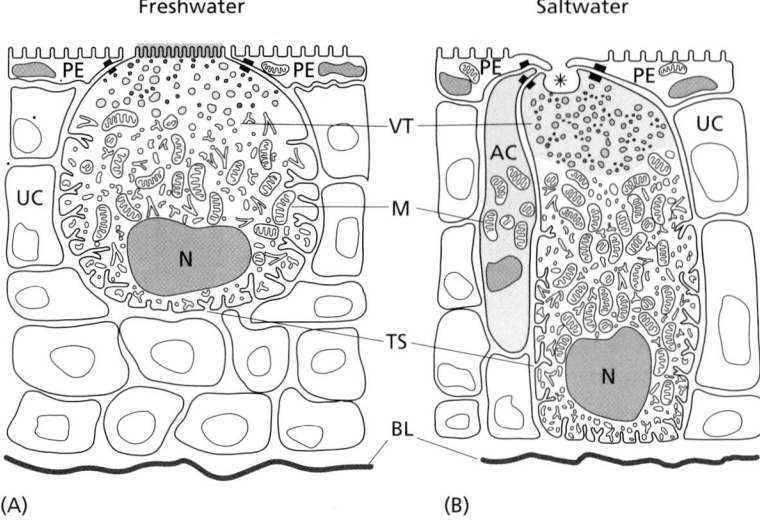

Figure 21.5 Drawing of chloride cells from freshwater and saltwater fish. Chloride cells are round to oval in freshwater fish and the microvillous apical membrane is even with the surface membranes of pavement epithelial cells (PE). The apical membrane in chloride cells from saltwater fish forms an apical pit (*) and is nearly covered by PE. The saltwater chloride cell is elongated and usually extends to the basal lamina (BL). It also has a greater density of mitochondria (M) and the tubular system (TS) and vesicotubular system (VT) are more elaborate. Accessory cells (AC) are associated with saltwater, but not freshwater chloride cells, undifferentiated cells (UC) are common in both environments. Tight junctions between the chloride and pavement (and accessory) cells are found in the apical region (black bars). Figure not drawn to scale.

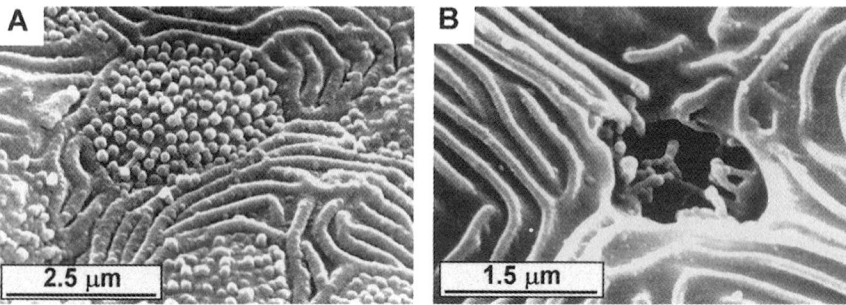

Figure 21.6 Scanning electron micrographs of (A) microvilli on the apical surface of a chloride cell from a freshwater climbing perch, *Anabas testudineus* and (B) the recessed apical pit of a saltwater mullet, *Mugil cephalus*. (A from Olson, 1996, with permission; B from Hossler et al., 1979, with permission.)

have been described, e.g. light cells and dark cells, microvillous and smooth surface cells, A cells and B cells, **α and β cells**, large and small cells, accessory cells, and so on (Pisam et al., 1987, 1988). It is likely that these cells perform distinct functions, however, the specifics have not been established and the present discussion will be limited to a general description of the surface and recessed cells.

All chloride cells have two striking features: abundant mitochondria and a **basolateral membrane** that is so extensively invaginated that all but the apical region of the cell appears to be filled with a tubular reticulum. This elaborate reticulum is the site of Na^+–K^+-ATPase and both the surface area of the reticulum and enzyme activity increase dramatically when fish are adapted to salt water (Karnaky et al., 1976; Pisam, 1981). The high density of mitochondria is consistent with the elevated energy requirements of highly metabolically active cells. Vesicles and short tubules, the vesiculotubular system (Pisam, 1981), replace the mitochondria and tubular reticulum in the apical region of the cell. This system is derived from the endoplasmic reticulum and it is not associated with the basolateral membrane (Pisam et al., 1983). These vesicles appear to fuse with the apical membrane and perhaps they are involved in delivery of glycoproteins to the cell surface (Pisam et al., 1983) or in rapid insertion/removal of ion channels or transporters

(Lin and Randall, 1995). The vesiculotubular system also becomes more developed during saltwater adaptation (Pisam, 1981). Nuclei are nearly always basally positioned.

Chloride cells whose apical surfaces are flush with the pavement epithelium (freshwater type) are typically round to oval and they are separated from the basal lamina of the filament by one or more undifferentiated cells (Figure 21.5A). The apical surface is generally studded with microvilli (Figure 21.6A) and the junction between chloride and pavement cells only has a single ridge contributed by the latter (Olson, 1996). Adaptation to very low external sodium chloride concentrations increases the surface area of apical membrane that is exposed to the environment (Perry and Laurent, 1989). Tight junctions with multiple sealing strands are found between these chloride cells and pavement cells (Sardet et al., 1979).

Recessed chloride cells, where the apical membrane is reduced in area, depressed into a shallow pit, and nearly covered by adjacent pavement cells, are common in saltwater species (Figures 21.5B and 21.6B). These cells are elongated and generally extend all the way across the epithelium and contact the basal lamina. Accessory cells are closely associated with recessed chloride cells. Although accessory cells are structurally similar to chloride cells, they do not extend to the basal lamina. Several accessory cells and one or more chloride cells usually share a common apical pit (Sardet et al., 1979). The apical membranes of accessory and chloride cells are highly interdigitated and the surface contact between the cells is elaborated (Sardet et al., 1979; Dunel-Erb and Laurent, 1980). Unlike pavement–pavement or pavement–chloride cell junctions, the tight junctions between chloride cells and between chloride and accessory cells have a single sealing strand and appear to be quite leaky (Sardet et al., 1979; Dunel-Erb and Laurent, 1980). This appears to be the paracellular pathway for passive sodium excretion that accompanies active transcellular chloride secretion by the chloride cell. Not surprisingly, accessory cells are found in saltwater fish, but are absent from freshwater fish.

It should be noted that comparisons of the effects of salinity adaptation on chloride cell numbers and anatomy have been done on relatively few species of fish and the above descriptions have been highly generalized. Exceptions to these have been observed (see Olson, 1996) and caution should be exercised when initial observations are made on novel species.

Mucous cells and granular cells

Mucus is primarily secreted by mucous cells (also called goblet cells), although other cells, such as granular cells may contribute to the total secretory product (Hidalgo et al., 1987). Mucous cells are found scattered over the arch, filament, and basal regions of the lamella (Figures 21.2, 21.4A and C). These cells have a basal nucleus and are filled with large mucus droplets. Mucous cells may contain acidic glycoproteins, neutral glycoproteins, or a combination of the two (Fletcher et al., 1976), suggesting that there are different populations of mucous cells on the gill. The number and/or size of mucous cells may be increased by adaptation to high salinity (sea water) or to ion-poor water (Olson, 1996). Mechanical abrasion and a variety of water-borne contaminants are potent stimuli for mucus secretion.

Less is known of granular cell secretions and their function or control. These cells are embedded in the epithelium and may only secrete mucus intermittently (Hidalgo et al., 1987). Their secretory product, glycoproteins with abundant sialic acid residues, appears similar to that of mucous cells.

Neuroepithelial cells

Neuroepithelial cells belong to the group of amine precursor uptake and decarboxylation (APUD) endocrine cells and are believed to serve as oxygen sensors in the gill (Dunel-Erb et al., 1982, 1994; Bailly et al., 1992). They are generally restricted to the efferent border of the filament and they are most prevalent near the tip of the filament. Because they rest on the basal lamina of the filament and are thus 50–100 μm beneath the epithelium (Figure 21.1A), they are actually closer to the blood vessels communicating with the nutrient and interlamellar system, than with either the efferent lamellar arteriole, efferent filamental artery, or epithelial surface. Dense-cored vesicles (80–100 nm) containing serotonin are especially abundant in the basal region of the neuroepithelial cell (Figure 21.7) and the cells become degranulated when the fish is exposed to hypoxia. Several different types of nerve fibers penetrate through the basal lamina and they are in close contact with the basal area of the neuroepithelial cells where the dense-cored granules are concentrated.

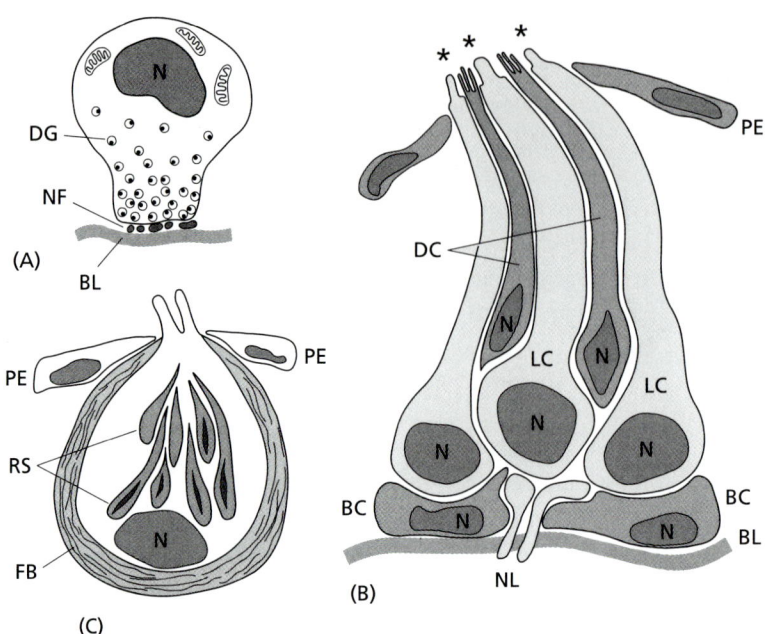

Figure 21.7 Epithelial cell types. (A) Neuroepithelial cells rest on the basal lamina (BL). Dense-cored granules (DG) containing serotonin are concentrated in this region and are closely associated with nerve fibers (NF) that penetrate the BL. (B) Taste buds contain light (LC) and dark (DC) sensory cells and bulge out through the pavement epithelial cells (PE). LC have a single thick apical microvillous (*), DC have up to 20 small thin microvilli. These cells rest on flattened basal cells (BC) and a dense nerve plexus (NP) passes through the basal lamina (BL) and around the BC to the light and dark cells. (C) Rodlet cells have a thick fibrous border (FB) and numerous electron-dense spindle- or club-shaped rodlet sacs (RS) with a dense core. Several microvilli protrude above the pavement epithelium (PE). Figures not drawn to scale.

Other cells – taste bud and rodlet

Taste buds (Figure 21.7B) are generally restricted to the arch support tissue and gill rakers; they are rare to non-existent on filaments and are absent from lamellae. Taste buds have two types of sensory cells, light and dark (Walker et al., 1981; Hossler and Merchant, 1983), and a single taste bud may have 10–25 of each. Both are long spindle-shaped cells with basal nuclei and either a single thick microvillus (light cell) or up to 20 thin microvilli (dark cell) protrude above the epithelium. A dense nerve plexus passes through the basal lamina and around the somewhat flattened basal cells to innervate the light and dark cells.

Rodlet cells (Figure 21.7C) are thinly dispersed on the gill arch, body of the filament, interlamellar filamental epithelium and basal areas of the lamellae (Mattey et al., 1979; Leino, 1982; Karlsson, 1983). They have an unusually thick fibrous border immediately beneath the cell membrane that may have contractile properties. Much of the remaining cytoplasm is occupied by numerous electron-dense spindle- or club-shaped rodlet sacs and most sacs contain an additional dense core. Several thin elongated microvilli project above the pavement epithelium. Rodlet cells appear to be secretory, however, their specific function is unknown.

Blood vessels

The three main vascular networks in the gill filament – respiratory, nutrient and interlamellar – have been described in Chapter 9. In most instances these vessels are indistinguishable from vessels found in other vertebrates and their fine structure is not particularly noteworthy. There are, however, three exceptions – the pillar cells of the lamellae, the interlamellar vessels in the filament, and the small vessels in the filament and arch that give rise to the secondary circulation. The following sections describe the anatomy of pillar cells and interlamellar vessels. Because secondary vessels are found in both the gill and systemic circulations, their anatomy is described in Chapter 22.

Pillar cells

Pillar cells are spool-shaped cells with a cylindrical body and extensive cytoplasmic flanges, the latter extending between adjacent pillar cells and thereby delimiting the lamellar vascular compartment (Figures 21.2, 21.3 and 21.8; reviewed in Olson, 1991). Pillar cells are often arranged in rows in the outer margin of the lamella which provides an uninterrupted path for red blood cells traversing the lamella (Figures 21.1 and 21.9A). Pillar cells are more densely packed in the inner margin of the lamella and red blood cells are often excluded from this region. The inner margin is embedded in the body of the filament (Figure 21.2) and is probably of little value in gas exchange. However, this area may be important in metabolism of circulating biomolecules, e.g. hormones, because it provides the greatest pillar cell to plasma surface area. The metabolic capabilities of pillar cells have been well established (Olson, 1998).

The lamellar basement membrane is comprised of two parallel sheets of basal lamina separated by pillar cells. Blood pressure in the gill lamellae is higher than in any other capillary bed in the body and in order to keep the lamellae from ballooning out under this pressure the two sheets of basal lamina are connected to each other by numerous collagenous columns (Figure 21.8). Because collagen is thrombogenic when exposed to plasma, the pillar cell plasma membrane envelops the column (Figure 21.8), much like a tree growing around a wire fence. Thus even though the column remains outside the cell, it is effectively separated from the plasma. Typically there are 4–10 columns per pillar cell. Myosin filaments run along the pillar cell axis and may provide additional structural support or even exhibit contractile activity.

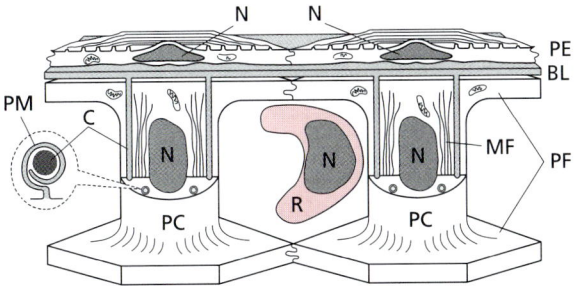

Figure 21.8 Relationship between pillar cells (PC), basal lamina (BL) and pavement epithelium (PE) in lamella. The BL is a thin sheet above and below (not shown) the PC. Thin columns of BL material (C) connect the two parallel sheets and the PC plasma membrane (PM) envelops the columns to prevent thrombogenesis. Strands of myosin filaments (MF) traverse the PC longitudinal axis. PF, pillar cell flange; R, red blood cell. Figure not drawn to scale.

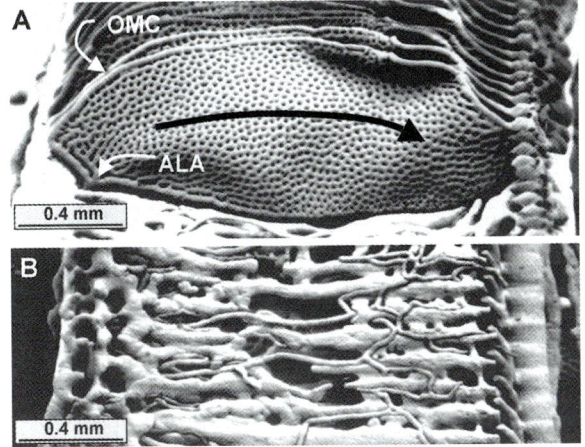

Figure 21.9 Scanning electron micrograph of vascular corrosion replicas of a respiratory lamella (A) and the interlamellar filamental vasculature (B). With this method the blood vessels are filled with a liquid resin that then polymerizes forming a rigid replica of the vascular space. The surrounding tissue is then digested away and the replica can be examined. (A) Blood enters the lamella from the lower left via the afferent lamellar arteriole (ALA) and exits on the right. The outer marginal channels (OMC) of several lamellae are visible at the top. (B) The lamella have been removed to expose the interlamellar filamental vasculature (large diameter) and nutrient vessels (narrow diameter) below.

Interlamellar and nutrient vessels

The body of the filament is perfused by two circulations, an extensive interlamellar system and nutrient vessels (reviewed in Olson, 1991, 1996). Both systems lie medial to the basal lamina that envelops the filament (Figure 21.2).

Nutrient vessels arise from the base of the efferent filamental artery and are a part of the secondary circulation. Nutrient capillaries (Figure 21.10) have a relatively thick endothelium and may be associated with pericytes. The plasma appears quite electron opaque when viewed with the electron microscope suggesting an abundance of protein. Nutrient vessels may be drained by a separate venous system or they may empty into the interlamellar vessels.

Interlamellar vessels and their longitudinal connecting vessels give the appearance of a ladder running the length of the filament with the interlamellar vessels as the rungs (Figure 21.9B). Interlamellar vessels are thin-walled large-bore vessels (Figure 21.10) reminiscent of mammalian lymphatics. In fact, adjacent endothelial cells appear to overlap each other forming

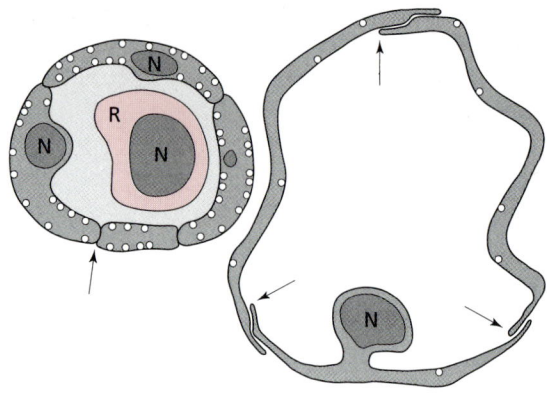

Figure 21.10 Two types of capillaries are present in the body of the filament. Nutrient capillaries (left) have relatively thick-walled endothelial cells and cell junctions (arrow) rarely overlap. Pericytes (not shown) are relatively common as are red blood cells (R). The plasma is electron opaque, indicative of protein. Interlamellar capillaries are thin-walled and cell junctions usually overlap and appear valve-like. Interlamellar vessels are highly distensible and partially collapse under low pressure, pushing the nucleus (N) into the lumen. Red cells are uncommon and plasma appears to have less protein. Figure not drawn to scale.

valve-like junctions similar to those typical of lymphatic capillaries. Interlamellar vessels also appear quite distensible and under low pressure the vessel partially collapses, forcing the endothelial cell nucleus to bulge out into the lumen of the vessel. This vasculature appears to be perfused mainly from very narrow-bore vessels arising from the medial wall of the efferent filamental artery, which may account for the relative paucity of red blood cells in this system. Because the interlamellar vessels course beneath the area of the filament epithelium rich in chloride cells, it has been suggested that this vascular network is important in gill ion transport. However, as shown in Figure 21.2, the chloride cells are actually closer to the inner margin of the lamella, leaving the function of the interlamellar system still speculative.

Acknowledgments

The author is grateful to the National Science Foundation for past and current support through NSF Grant Nos PCM 79-23703, PCM 84-048897, DCB 86-16028, INT 83-00721, INT 86-02965, INT 86-18881, IBN 91-05247 and IBN 97-23306.

References

Bailly, Y., Dunel-Erb, S. and Laurent, P. (1992). *Anat. Rec.* **233**, 143–161.

Dunel-Erb, S. and Laurent, P. (1980). *J. Morphol.* **165**, 175–186.

Dunel-Erb, S., Bailly, Y. and Laurent, P. (1982). *J. Appl. Physiol.* **53**, 1342–1353.

Dunel-Erb, S., Chevalier, C. and Laurent, P. (1994). *Can. J. Zool.* **72**, 1794–1799.

Filk, G., Verbost, P.M. and Wendelaar Bonga, S.E. (1995). In *Cellular and Molecular Approaches to Fish Ionic Regulation* (eds C.M. Wood and T.J. Shuttleworth), pp. 317–342. Academic Press, San Diego.

Fletcher, T.C., Jones, R. and Reid, L. (1976). *Histochem. J.* **8**, 597–608.

Goss, G., Perry, S. and Laurent, P. (1995). In *Cellular and Molecular Approaches to Fish Ionic Regulation* (eds C.M. Wood and T.J. Shuttleworth), pp. 257–284. Academic Press, San Diego.

Hidalgo, J., Velasco, A., Sanchez Aguayo, I. and Amores, P. (1987). *Histochemistry* **88**, 65–73.

Hossler, F.E. and Merchant, L.H. (1983). *Am. J. Anat.* **166**, 299–312.

Hossler, F.E., Ruby, J.R. and McIlwain, T.D. (1979). *J. Exp. Zool.* **208**, 399–406.

Hughes, G.M. (1984). In *Fish Physiology Vol. X Gills, Part A, Anatomy, Gas Transfer, and Acid-Base Regulation* (eds W.S. Hoar and D.J. Randall), pp. 1–72, Academic Press, New York.

Karlsson, L. (1983). *J. Fish. Biol.* **23**, 511–524.

Karnaky, K.J., Jr., Kinter, L.B., Kinter, W.B. and Stirling, C.E. (1976). *J. Cell. Biol.* **70**, 157–177.

Leino, R. (1982). *Can. J. Zool.* **60**, 2768–2782.

Laurent, P. (1984). In *Fish Physiology, Vol. X Part A, Anatomy, Gas Transfer, and Acid-Base Regulation* (eds W.S. Hoar and D.J. Randall), pp. 73–183. Academic Press, New York.

Laurent, P. and Dunel, S. (1980). *Am. J. Physiol.* **238**, R147–R159.

Laurent, P. and Hebibi, N. (1989). *Can. J. Zool.* **67**, 3055–3063.

Laurent, P. and Perry, S.F. (1990). *Cell Tissue Res.* **259**, 429–442.

Lin, H. and Randall, D. (1995). In *Cellular and Molecular Approaches to Fish Ionic Regulation* (eds C.M. Wood and T.J. Shuttleworth), pp. 229–255. Academic Press, San Diego.

Mattey, D.L., Morgan, M. and Wright, D.E. (1979). *J. Fish Biol.* **15**, 363–370.

McCormick, S.D. (1995). In *Cellular and Molecular Approaches to Fish Ionic Regulation* (eds C.M. Wood and T.J. Shuttleworth), pp. 285–315. Academic Press, San Diego.

Olson, K.R. (1991). *J. Electron Microsc. Technique* **19**, 389–405.

Olson, K.R. (1996). In *Fish Morphology: Horizons of New Research* (eds H.M. Dutta and J.S.D. Munshi), pp. 31–45. Science Publishers, Lebanon, NH.

Olson, K.R. (1998). *Comp. Biochem. Physiol.* **119A**, 55–65.

Olson, K.R., Ghosh, T.K., Roy, P.K. and Munshi, J.S.D. (1995). *Anat. Rec.* **242**, 383–399.

Perry, S.F. and Laurent, P. (1989). *J. Exp. Biol.* **147**, 147–168.

Pisam, M. (1981). *Anat. Rec.* **200**, 401–414.

Pisam, M., Chretien, M., Rambourg, A. and Clermont, Y. (1983). *Anat. Rec.* **207**, 385–397.

Pisam, M., Caroff, A. and Rambourg, A. (1987). *Am. J. Anat.* **179**, 40–50.

Pisam, M., Prunet, P., Boeuf, G. and Rambourg, A. (1988). *Am. J. Anat.* **183**, 235–244.

Sardet, C., Pisam, M. and Maetz, J. (1979). *J. Cell Biol.* **80**, 96–117.

Walker, E.R., Fidler, S.F. and Hinton, D.E. (1981). *Anat. Rec.* **200**, 67–81.

CHAPTER 22

Circulatory System

Kenneth R Olson
Indiana University School of Medicine,
South Bend Center for Medical Education,
University of Notre Dame,
Notre Dame, Indiana, USA

Abbreviations

A	artery
AE	arterial endothelial cell
APC	anterior wall of pericardial cavity
AV	atrioventricular valve/orifice
BA	bulbus arteriosus
BAA	bulbus arteriosus adventita
BL	basal lamina
BV	bulboventricular ring
CL	compact layer with radial fibers
CT	circular tubules
CV	cover cell
DES	desmosome
E	endothelial cell
EBA	efferent branchial artery
EF	efferent filamental artery
EL	elongated filament
EO	osmophilic endothelial organelles
FT	filamentous tube-like structures
G	glycogen granules
ICP	interstitial cover cell process
IEP	interstitial endothelial cell process
IL	interlamellar vascular system
LE	longitudinal element
LT	longitudinal tubule
MF	myofibril
NF	nerve fibers
P	pericyte
R	red blood cell
RE	radial element
RSR	reticular sarcoplasmic reticulum
S	secondary vessel
SC	Schwann cell
SM	smooth muscle
SSC	subsarcolemmal cisterna
V	vein
VB	ventriculo-bulbar valve
VW	ventricular wall
VE	villuous endothelial cell
WG	whorled pattern with central glycogen
WP	Weibel Palade bodies

Copyright © 2000 Academic Press

Overview

The fish cardiovascular system is in many respects an evolutionary prototype and its design employs numerous features common to all vertebrates. Cardiac muscle provides the energy for blood flow, vascular smooth muscle is used to regulate blood distribution and return and capillary endothelial cells are the perm-selective barriers across which nutrients are exchanged. However, with 400 million years of evolution it is not surprising that there are a few obvious, and perhaps fundamental, differences between the cells and tissues that make up the fish and mammalian systems.

The emphasis of this chapter will be on the more unique features of the fish cardiovascular system. The fish heart has a single atrium and ventricle and the latter is seldom required to develop pressures in excess of 40 mmHg (one-third that of mammals), moreover, it is rarely perfused with oxygenated blood and the myocardium is accordingly different. Fish blood vessels can be functionally (and structurally) divided into compliance, conductance, resistance, exchange and capacitance vessels. These descriptors are similar to those used in mammals, although anatomical solutions to a common problem are not always identical and the differences will be stressed. Fish apparently lack a lymphatic system, yet they possess another, perhaps related, vascular network, the secondary circulation. This circulation is both an anatomical curiosity and a physiological enigma. Finally, elaborate countercurrent arteriole/capillary systems, the retia mirabilia, are uniquely designed to concentrate gases in the swimbladder, oxygen in the eye, and in tuna the retia help retain heat in select tissues.

The heart

The overall shape and arrangement of the ventricle and ventricular myocardial fibers in teleost fish hearts has been correlated with their life style (Santer and Greer Walker, 1980; Santer et al., 1983; Santer, 1985; Farrell and Jones, 1992; Agnisola and Tota, 1994; see also Chapter 10). Most fish are relatively sedentary and they have either saccular or tubular hearts with only a spongy (trabecular) myocardium. Active fish have pyramidal-shaped ventricles and both a spongy myocardium and an outer layer of compact myocardium. In teleosts with both spongy and compact myocardium the percent of ventricular mass occupied by the compact layer varies between 16% (conger eel, *Conger conger*) and 74% (bigeye tuna, *Thunnus obesus*); 20–45% being the most common (Farrell and Jones, 1992; Sánchez-Quintana et al., 1996). The spongy and compact layers in teleosts are separated by a layer of connective tissue in all but the region immediately surrounding the atrioventricular and bulboventricular valves, where there is a direct connection between **fascicles** in the two layers (Sánchez-Quintana et al., 1996). Hearts from elasmobranchs (sharks, skates and rays) always have both spongy and compact layers and there is direct continuity between the layers.

There is less variation in the anatomy of the teleost atrium and a definitive compact layer is lacking. Trabeculae are common and important in atrial function (see Chapter 10).

Myocardial fibers

Myocardial fibers in the spongy layer are commonly organized into fascicles in the shape of bundles or thin sheets that branch frequently and are interconnected with each other (Figure 22.1). Fascicles may extend several hundred micrometers but they are generally less than 25–35 μm in diameter. This increased surface area is essential because the fibers obtain their oxygen directly from venous blood in the ventricular lumen. There is no coronary circulation in the spongy myocardium of bony fish with the exception of a very

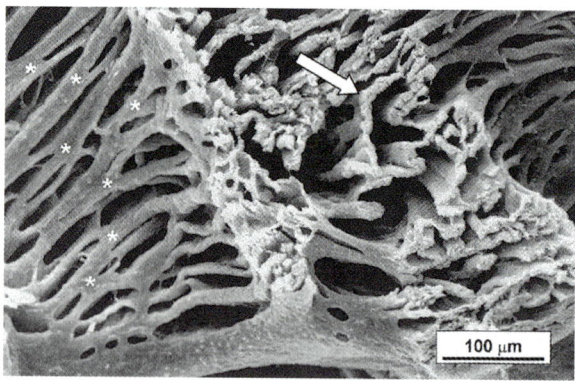

Figure 22.1 Scanning electron micrograph of the trabeculated ventricular myocardium in a sedentary fish, *Heteropneustes fossilis*. Fibers in center of micrograph have been cut nearly perpendicular to long axis. Cardiac myocytes are arranged in branched interconnected bundles (∗) or sheets (arrow) thereby exposing large areas of the plasma membrane to blood in the ventricular lumen. (Micrograph from Olson and Datta Munshi, unpublished observation.)

few of the most active species. The fascicles are collectively arranged to form branching **lacunae** in the ventricular wall. The lacunae enhance the mechanical efficiency of the fascicles (see Chapter 10).

Myocardial fibers in the compact myocardium are arranged into tightly packed fascicles. This, plus the connective tissue layer separating the spongy and compact layers, excludes blood from the ventricular lumen and these cells must be nourished by a coronary circulation. Myocardial fascicles in the compact myocardium of pyramidal hearts are arranged in two layers, or less commonly one or three layers, the number of which depends on the extent of the compact layer (Farrell and Jones, 1992; Sánchez-Quintana et al., 1996). In the outer layer, the fascicles usually run longitudinally and in some species they insert into the bulboventricular fibrous ring (Figure 22.2a). Contraction of the outer layer during ventricular systole produces an axial shortening of the ventricle. Fascicles in the inner layer encircle the vertices of the pyramidal ventricle (Figure 22.2b). This effectively subdivides the lumen into smaller chambers and presumably provides greater mechanical advantage. The fascicles usually encircle the atrioventricular and bulboventricular junctions and they may insert directly into the fibrous bulboventricular valvular ring. Contraction of the inner layer decreases the radial dimensions of the vertices of the pyramidal ventricle.

Cardiocytes

The structure of fish cardiocytes has been summarized by Satchell (1991), Farrell and Jones (1992) and Tibbits et al. (1992) and is shown in Figure 22.3. Fish cardiocytes are smaller in diameter than their mammalian counterparts (1–12.5 µm versus 10–25 µm). Myofibrils may occupy half of the volume of the cell and in many fish they are restricted to the periphery and lie just beneath the sarcolemma. A, I and Z bands are present and six actin-containing thin filaments surround each myosin-containing thick filament. Spherical or oval mitochondria and a single nucleus (not shown in Figure 22.3) are often centrally located. Cardiocytes lack typical T tubules, although **caveolae**, 0.15 µm diameter flask-shaped infoldings of the sarcolemma, may serve this function. The sarcoplasmic reticulum (SR) is reduced in size and four types of SR have been described (Figure 22.3b). Subsarcolemmal **cisternae** lie beneath the cell membrane and communicate with circular tubules that encircle the myofibrils. A few longitudinal tubules connect adjacent circular tubules and both communicate with an often sparsely distributed reticular lattice.

The contractile proteins of mammalian cardiocytes are activated almost exclusively by release of calcium from intracellular stores, whereas activation of the contractile apparatus of fish cardiocytes is dependent primarily on external calcium. Thus in fish cardiocytes the greater surface to volume ratio and peripheral orientation of myofibrils enhances delivery of extracellular calcium to the myofibrils and an extensive system of transverse tubules and sarcoplasmic reticulum is probably not necessary.

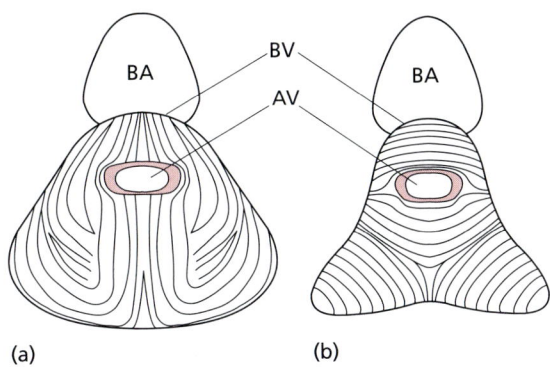

Figure 22.2 Dorsal view of the compact myocardium and bulbus arteriosus (BA) of a pyramidal ventricle illustrating the orientation of myocardial fascicles (solid lines). a. Fascicles in the outer layer often run longitudinally and may insert into the bulboventricular fibrous ring (BV). Their contraction decreases the longitudinal axis of the ventricle. b. Fascicles in the inner layer encircle the vertices of the pyramid and their contraction decreases the radial dimension. Increased mechanical efficiency is obtained by subdividing the ventricular lumen into several smaller-diameter chambers. AV, orifice of atrioventricular valve. (Figures were drawn from micrographs and description by Sánchez-Quintana et al., 1996.)

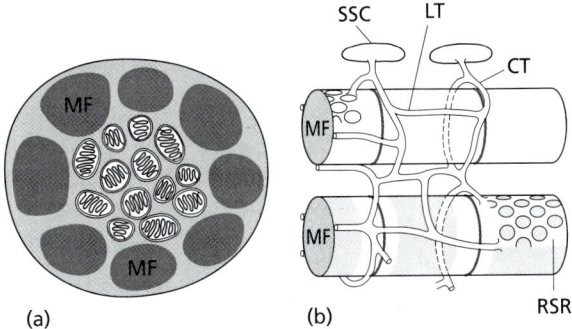

Figure 22.3 Structure of cardiocytes. a. Cross-section through a cardiocyte showing the peripheral arrangement of the myofibrils (MF) and centrally distributed mitochondria. b. Relationship between two myofibrils and subsarcolemmal cisterna (SSC), circular tubules (CT), longitudinal tubules (LT) and reticular sarcoplasmic reticulum (RSR). (b adapted from Farrell and Jones, 1992, with permission.)

Connective tissue

Connective tissue in the fish heart, as in mammals, consists of a hierarchical arrangement of epimysium, perimysium and endomysium (Sánchez-Quintana *et al.*, 1996). The **epimysium** forms a thick epicardial layer and a thinner endocardial layer in all hearts. Thick coiled **perimysial** sheaths form an extensive irregular framework that penetrates the compact myocardium and supports the fascicles. These are especially pronounced in very active fish and they may store energy near the end of systole. Perimysial fibers also radiate from the coronary vessels, perhaps helping maintain patency of the vessel during systole. The perimysium in trabeculated myocardium provides anchoring points for the fascicles. Individual and bundled collagen fibrils in the **endomysium** ensheath the myocytes. In trabeculated hearts these fibers are either oblique or transverse to the long axis of the cell.

Endothelium

At least two types of endothelial cells line the atrium and cover the trabeculae in the ventricle (Sætersdal *et al.*, 1974). They range from flat to cuboidal. One type of endothelial cell has catecholamine-containing granules. The other endothelial cells are highly phagocytic (atrial even more so than ventricular) and they actively remove considerable amounts of foreign matter such as ferritin particles (Leknes, 1987) and colloidal carbon (Ellis *et al.*, 1976; although these may be macrophages). These cells have numerous bristle-coated vesicles (probably clathrin coated), moderately dense-bodied inclusions and lysosomes are abundant (Leknes, 1982). The phagocytic-type cells appear unique to the heart and they are not present in the adjacent bulbus arteriosus.

Innervation

Distribution of autonomic nerves to the teleost heart has been reviewed by Morris and Nilsson (1994). Inhibitory parasympathetic (cholinergic) innervation is supplied by paired vagus (X cranial) nerves that travel along the ductus Cuvier. The **parasympathetic** ganglion lies near the sinoatrial border and the post-ganglionic fibers travel to (in order of decreasing nerve density) sinoatrial nerve plexus (pacemaker), atrioventricular border, atrium and ventricle. Excitatory **sympathetic** (adrenergic) nerves may travel one of three routes, either with the vagus nerves (forming the **vagosympathetic trunk**), along the anterior pair of spinal nerves, or along the coronary arteries. There are a few exceptions, such as pleuronectids, which lack adrenergic innervation.

Peripheral circulation

Blood vessels are anatomically divided into arteries, capillaries and veins, whereas they are often functionally characterized as compliance, conductance, resistance, exchange and capacitance vessels (reviewed in Olson, 1997). Compliance vessels dampen the ventricular pulse. This reduces the ventricular load by decreasing systolic pressure, protects the delicate capillaries from excessive pressure, and maintains blood flow even when the ventricle is at rest. Compliance vessels are arteries and by far the most important and unique compliance vessel in fish is the bulbus arteriosus. The ventral aorta, dorsal aorta and larger systemic arteries also contribute to compliance, however they mainly serve as low resistance conductance pathways to distribute the blood. Most of the vascular resistance is in small arteries and arterioles ($< 500\,\mu m$ diameter) and here is where nerves, hormones and local factors regulate blood flow distribution between and within organs. Capillaries, and probably small venules, are sites for exchange between blood and tissues. The capacitance vessels, venules and veins, store blood and through them the return of blood to the heart is regulated.

Arteries

Compliance vessels

Bulbus arteriosus

The bulbus arteriosus is a tear-drop shaped vessel that serves as major **depulsator** (aka windkessel vessel; reviewed in Satchell, 1991; Bushnell *et al.*, 1992). The vessel wall is rich in elastic tissue and vascular smooth muscle and is innervated by autonomic nerves. Elastin in fish has fewer cross links than mammalian elastin and appears more elastic. Thus the bulbus is not only highly compliant but it is possible

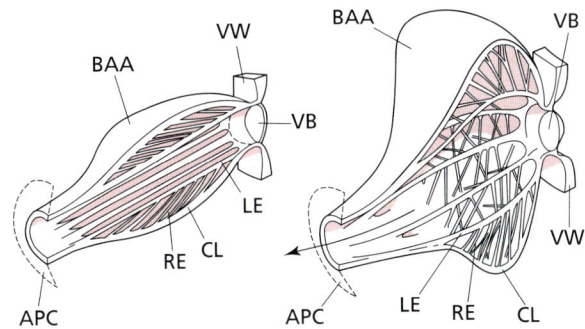

Figure 22.4 The bulbus arteriosus of a number of fish contains both longitudinal (LE) and radial (RE) elements. During ventricular ejection (systole), blood flows into the bulbus, and energy is stored during distension of the bulbar wall and RE. The LE prevent elongation of the bulbus and serve as the attachment site for the RE. Elastic recoil of the bulbus helps maintain arterial pressure during diastole. Other abbreviations on page 369 (From Priede, 1976, with permission.)

that compliance is actively regulated. The bulbus of many fish is also structurally reinforced with trabecular longitudinal (axial) and radial elements (Figure 22.4; Priede, 1976) to withstand large volume/pressure oscillations. During ventricular ejection, potential energy is stored in the elastic walls and perhaps radial trabeculae. This energy is then dissipated during diastole to maintain arterial pressure.

Conductance and resistance vessels

The general anatomical features of fish blood vessels have been described by Satchell (1991) and Bushnell *et al.* (1992). With the exception of a few minor variations, the anatomy of fish and mammalian arteries (and veins) is similar and the mammalian nomenclature is used for fish as well.

The blood vessel wall has three layers (from inside outward): the tunica intima, tunica media and tunica adventitia. The relative thickness of each layer varies with the vessel's function.

The tunica intima is bounded on the blood (luminal) side by a single layer of endothelial cells and on the exterior (abluminal) side by a thin, fenestrated elastic sheet, the internal elastic lamina. In vessels where flow rate and shear forces are high, such as conducting arteries, endothelial cells are shaped like elongated diamonds with their long axis parallel to the direction of blood flow; whereas venous endothelial cells are more irregularly shaped and show no flow-directed orientation (Olson and Villa, 1991). The endothelial surface also appears to be vessel specific, i.e. endothelial cells lining the ventral aorta are smooth, scattered endothelial projections, similar to those found in mammals (Smith *et al.*, 1971) protrude from the celiacomesenteric artery and endothelial projections are especially abundant on the surface of the anterior cardinal vein (Olson and Villa, 1991). Prostaglandin-like endothelium-derived relaxing factors (EDRF) are released by both arterial and venous endothelia in fish (Olson and Villa, 1991).

The tunica media consists mainly of smooth muscle and in the more muscular vessels the cells may be oriented circumferentially and longitudinally (Bushnell *et al.*, 1992) or helically (Packer and Olson, unpublished observation). Elastic fibers may be randomly interspersed between the smooth muscle cells, or in vessels with higher blood pressure, e.g. the ventral aortas of trout, carp, eel and tuna, the tunica media may be organized into distinct laminae. In some vessels dense elastic fibers are condensed into a sheet forming an outer elastic lamina.

The tunica adventitia is commonly a thin layer of collagen with some elastin. In some vessels, such as the dorsal aorta (see below) and afferent filamental artery of the gill, the adventitia is affixed to surrounding skeletal or cartilaginous tissue.

Ventral and dorsal aortas

The ventral aorta is a thick elastic artery and in the carp as many as 20 superimposed elastic laminae have been identified. Smooth muscle is also abundant in the ventral aorta and this vessel has been the subject of numerous *in vitro* **myographic** studies of fish vascular physiology and pharmacology.

The dorsal aorta is the longest artery and in most fish its structure is unlike that of other arteries. It is a relatively thin-walled vessel with a limited amount of smooth muscle and elastin and abundant collagen. The dorsal wall of the aorta is especially thin and here it is attached to the ventral surface of the spinal column by transverse bands of collagen fibers. The ventral ligament of the spinal cord may project into the lumen of the dorsal aorta and serve as a mechanical booster pump (see Chapter 8 and Priede, 1975).

Other conducting arteries contain various mixes of elastin, smooth muscle and connective tissue. Wall thickness decreases with vessel diameter and there is a progressive reduction in the amount of adventitia and smooth muscle. Many of these vessels are innervated by sympathetic adrenergic nerves and to a lesser extent by nerves containing neuropeptides such as vasoactive intestinal peptide (Morris and Nilsson, 1994). A few endothelial cells in arteries and larger arterioles have electron-dense granules that appear to contain catecholamines (Davison, 1987). These may also play a role in regulating blood flow.

Arterioles

Arterioles are short ($\leq$500 μm), narrow ($\leq$300 μm) and their muscular medial layer is important for regulating vascular resistance in response to **neurocrine**, **endocrine** and **paracrine** stimuli (Satchell, 1991; Olson, 1997). The smallest arterioles have a single layer of smooth muscle (Figure 22.5). Arteriolar endothelial cells may be squamous but in many fish they are cuboidal (Davison, 1987; Satchell, 1991). The vessel lumen may be the size of a single red blood cell. In a detailed study of arterioles in swimming muscles of the leatherjacket, *Parika scaber*, Davison (1987) observed longitudinally oriented contractile-like fibrils in the endothelial cells and circumferentially oriented fibrils in the smooth muscles. He proposed that contraction of the former would shorten the vessel and contraction of the latter would decrease its radius. Davison (1987) also observed that fish arteriolar smooth muscle has considerably less fibrillar material than similar size mammalian arterioles, which may reflect the lower blood pressures commonly found in fish.

Capillaries

Much of the interest in fish capillaries has been directed toward understanding the factors that affect capillary density (cf. Egginton, 1992), and there has not been a systemic examination of capillary ultrastructure. However, from the available information it appears that fish capillaries are anatomically similar to those found in mammals (Satchell, 1991) and consist of a single layer of squamous endothelial cells surrounded by a basement membrane. Typically they are 4–10 μm in diameter and 500–1000 μm long. Most capillaries are continuous, especially in skeletal muscle, although some fenestrated and discontinuous capillaries are present in other tissues. Pinocytotic vesicles are abundant on both luminal and abluminal surfaces and to a lesser extent in the cytoplasm. Contractile-like filaments are also present.

Periendothelial cells (pericytes) have also been found attached to the abluminal surface of arterioles, capillaries and venules (Figure 22.6; Couch, 1990). Fish pericytes are similar to those in mammals and have common identifying characteristics: (i) they are encompassed within a basal lamina that is continuous with the basal lamina of the adjacent vessel; (ii) they contain many plasmalemmal vesicles; and (iii) they possess an apparent 'sole' region, an area where they are in closest contact with the endothelium. The sole region also contains contractile actin- and myosin-like filaments but plasmalemmal vesicles are absent.

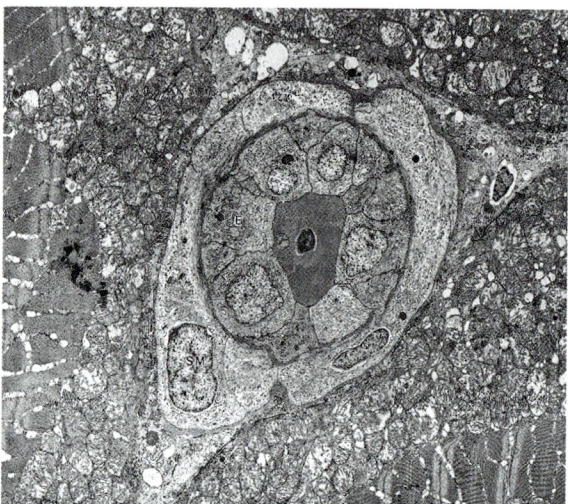

Figure 22.5 Transmission electron micrograph of a terminal arteriole in the swimming muscle of the leatherjacket, *Parika scaber*. This 14 μm-diameter vessel has a single layer of smooth muscle (SM) and cuboidal endothelial cells (E). A single red blood cell fills the lumen. (From Davison, 1987, with permission.)

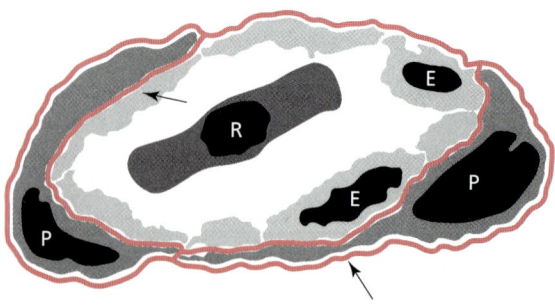

Figure 22.6 Diagram of capillary and two pericytes (P) in the gut of the sheepshead minnow. Pericyte cell bodies are surrounded by a basal lamina (arrows), overlie endothelial cell (E) junctions and contain contractile filaments. (After Couch, 1990.)

Veins

Fish veins and venules are structurally similar to those in mammals, although they have proportionally thinner walls and less smooth muscle.

Secondary circulation and lymphatics

The early literature describing an extensive lymphatic system in teleosts (summarized in Kampmeier, 1969) has been seriously challenged by modern anatomical studies and it is now thought that fish not only lack true lymphatics, but instead they possess a unique vasculature, the secondary circulation. The secondary circulation derives its name from the fact that it is a vascular network that originates from the main circulation (primary circulation) and forms a second artery-capillary-vein circuit in-parallel with the primary system (reviewed in: Vogel, 1985; Olson, 1992, 1996; Steffensen and Lomholt, 1992; see also Chapter 10). The anatomy and physiology of this circulation is unquestionably the most perplexing question in the fish cardiovascular system.

The secondary system has several distinguishing characteristics:

1. It only originates from postlamellar gill arteries (efferent filamental and efferent branchial), the dorsal aorta, or systemic segmental arteries.
2. It appears to be primarily associated with epithelial tissues and is found in the gill, skin, fins, oral mucosa and peritoneal lining (although it has also been observed in the tuna heat-exchanger). It does not appear to be present in brain, skeletal muscle, coronary or splanchnic circulations.
3. Parent vessels that form the secondary system have an unusually narrow diameter (2–20 μm) and follow a relatively long (200–300 μm) and very tortuous path before they anastomose with each other to form progressively larger arteries.
4. Secondary capillaries have very thin walls, appear to lack a basement membrane, and are generally similar to mammalian lymphatic capillaries. Secondary veins are also thin-walled, but they are not otherwise notable. They drain into primary veins of the systemic circulation.

Little is known of the physiological attributes of the secondary system. The narrow lumen of the parent arterioles restricts access of blood cells and hematocrit may be as low as 1 or 2%. Initial indirect estimates of the volume of the secondary system suggested it was an astounding 1.5 times that of the primary circulation (Steffensen and Lomholt, 1992), although more recent studies indicate that it may be only 10–20% the volume of the primary system (Bushnell *et al.*, 1998). Either way, it is a significant system.

Gill secondary system

The body of the gill filament contains two circulatory systems, the nutrient and interlamellar (Chapters 9 and 21) that have many, but not always all, of the characteristics of secondary vessels.

Nutrient circulation

Narrow-bore, tortuous arterioles arise from the gill efferent branchial and the efferent filamental arteries in the basal region of the filament near the anastomosis of the efferent filamental and branchial arteries (Figure 22.7). Curiously, numerous long microvilli stud the endothelial cell surface in both the lumen of

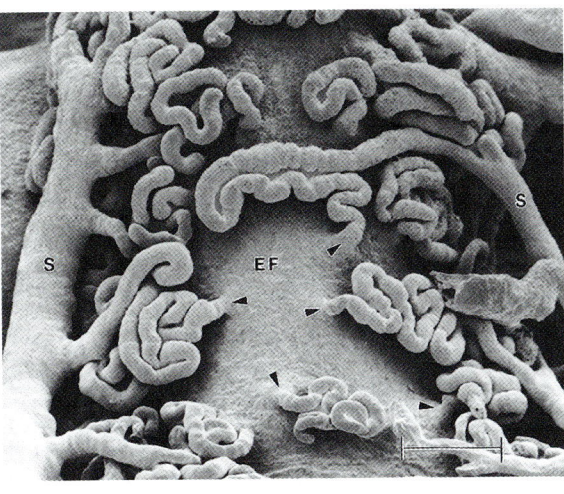

Figure 22.7 Scanning electron micrograph of a vascular corrosion replica of secondary vessels in the gill of the climbing perch (*Anabas testudineus*). Numerous narrow-bore, tortuous arterioles (arrowheads) arise from the gill efferent filamental artery (EF) and subsequently anastomose to form secondary arteries (S). Bar = 40 μm. (From Olson *et al.*, 1986, with permission.)

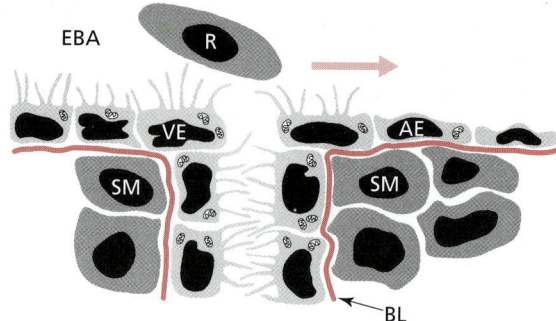

Figure 22.8 Diagram of a longitudinal section through the lumen of the efferent branchial artery (EBA) illustrating the origin of the secondary vessels shown in Figure 22.7. Villous endothelial cells (VE) guard the lumen of the secondary vessel and are also preferentially located on the upstream lumen of the EBA. AE, typical arterial endothelial cell; BL, basal lamina; SM, smooth muscle; R, red blood cell; arrow shows direction of blood flow. (Drawn from descriptions by Vogel, 1978.)

the parent artery (efferent branchial or filamental) near the origin of the secondary arteriole and in the secondary arteriole itself (Figure 22.8). The combined effect of a narrow lumen and the bristly microvilli appears to prevent red blood cells from entering this circulation, although Steffensen and Lomholt (1992) proposed that the microvilli may actually enhance red cell entry.

The tortuous secondary arterioles anastomose with each other to form progressively larger arteries that supply what resembles a nutrient system for the arch support tissue and the body of the filament. In some fish, such as the catfish, this system drains from the filament through its own venous network. In many other species the nutrient vessels drain directly into (or supply?) the interlamellar system of the filament.

Interlamellar circulation

The interlamellar system appears to have its origin via short arterioles that originate from the medial wall of the efferent filamental artery in the distal 90% of the filament (Figure 22.9). The lumen of these vessels is also nearly occluded and the endothelium frequently has microvillar projections and whorled cytoplasmic structures (Vogel *et al.*, 1974). The interlamellar system has a thin-walled endothelium with overlapping cell junctions and it is readily distended by even a slight increase in blood pressure.

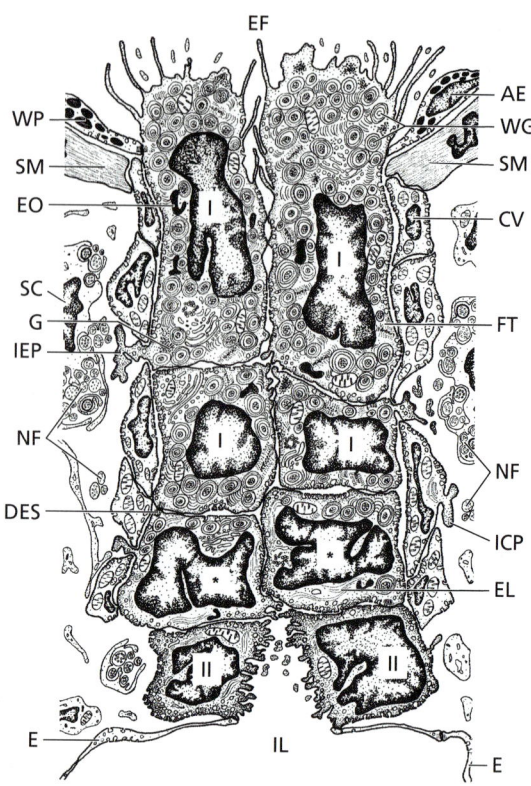

Figure 22.9 Diagram showing the narrow arteriovenous anastomoses from the efferent filamental artery (EF) to the interlamellar system (IL) in the gill of *Tilapia mossambica*. Type I endothelial cells contain numerous filaments in concentric whorl-like patterns with centrally located glycogen granules (WG). Type II (II) and intermediate endothelial cells (∗) lack the whorled patterns. Other abbreviations: AE, arterial endothelial cell; DES, desmosome; E, endothelial cell; EL, elongated filament; EO, osmophilic endothelial organelles; FT, filamentous tube-like structures; G, glycogen granules; ICP, interstitial cover cell process; IEP, interstitial endothelial cell process; NF, nerve fibers; SC, Schwann cell; SM, smooth muscle; WG, whorled pattern with central glycogen; WP, Weibel Palade bodies. (From Vogel *et al.*, 1974, with permission.)

Systemic secondary circulation

The systemic secondary system is similar to that in the gill, although it has received considerably less attention. Steffensen *et al.* (1986) described long secondary vessels in the fins of the glass catfish that drain into a large diameter longitudinal lateral vein and Vogel (1985) has shown that secondary vessels form a planar meshwork covering the surface of each scale. However, Lahnsteiner *et al.* (1990) reported that secondary vessels from dorsal segmental arteries anastomose directly with a dorsal cutaneous vein and blood returns directly to the heart without an intervening capillary bed. Clearly the systemic secondary system needs additional study.

Retia mirabilia

The close physical association between arteries and veins of similar size is a common feature in vertebrate tissues. Usually this is either the most anatomically or hemodynamically efficient route and countercurrent exchange between the arteries and veins is usually neither intended nor desired. In fish, however, there are at least three instances where small blood vessels are specifically arranged to make use of this countercurrent flow. These countercurrent vascular networks, called **retia mirabilia** (wonderful web), are found in most fish in the swimbladder and in the choroid gland (behind the eye) where they are important in concentrating gases, especially oxygen. They are also found in muscle and near the brain, eye, and gut in tuna and a few sharks, where they are important in regulating temperature of these organs. Fish retia are summarized in Satchell (1991) and Bushnell et al. (1992).

All retia in fish share common anatomical features that optimize exchange between the inflow and outflow vessels (Figure 22.10):

1. Retial vessels are arteriole to capillary size in diameter and the supply and return vessels form a dense network of parallel tubes separated by only a sparse interstitial matrix.
2. In most instances retial vessels are considerably longer than typical capillaries, in a 1.9 kg tuna vessels of the heat exchanging rete are 1 cm long (Stevens et al., 1974).
3. An additional capillary bed is located between the inlet (afferent) and outlet (efferent) retial vessels. These capillaries supply the secretory epithelium of the swimbladder, the choriocapillarias behind the eye, and muscle, brain, eye, and gut tissue in tunas.

Retia may be involved in either countercurrent exchange or countercurrent multiplication. Countercurrent exchange, such as occurs in the tuna heat exchanger, is a passive process. Heat generated through tissue metabolism warms the blood and as this warm blood passes into the efferent retial vessels the heat is transferred to blood in the afferent vessels thereby returning heat to the tissue. This enables

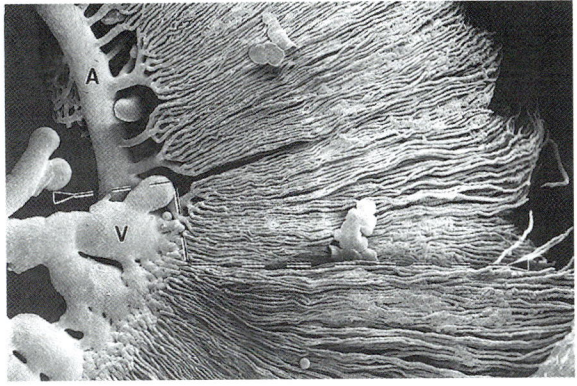

Figure 22.10 Vascular corrosion replica of the vessels comprising the choroid rete mirable of *Channa gachua* (viewed from behind the eye). A horseshoe-shaped artery (A) supplies smaller arterioles that radiate out to form long parallel capillaries that in turn supply the choriocapillarias (not shown). Closely apposed venous capillaries (below dashed line and arrowheads) complete the countercurrent loop and return blood to a central vein (V). Capillaries from this 75 g fish are approximately 1 mm long. (Adapted from Munshi et al., 1994, with permission.)

tissues associated with the rete to operate at temperatures 10–20°C warmer than the rest of the body.

Countercurrent multipliers in the swimbladder and eye require external energy and they can produce a standing gradient that is greater than that which can be produced by simple diffusion. The net effect of this complex process is to force the unloading of gases from blood entering the organ and to prevent their loss in the effluent blood. For example, the choroid gland can increase the partial pressure of oxygen (pO_2) to over 500 mmHg in the fish eye. This is five times the pO_2 of arterial blood and it is necessary to supply oxygen to the highly active retina, which, unlike mammals, is avascular. Somewhat similar processes are employed in the swimbladder for rapid secretion of oxygen, nitrogen and carbon dioxide. This rapidly changes the density of the fish and enables the now neutrally buoyant fish to move up or down the water column.

Acknowledgment

The author is grateful to the National Science Foundation for past and current support through NSF Grant Nos PCM 79-23703, PCM 84-048897, DCB 86-16028, INT 83-00721, INT 86-02965, INT 86-18881, IBN 91-05247 and IBN 97-23306.

References

Agnisola, C. and Tota, B. (1994). *Cardioscience* **5**, 145–153.

Bushnell, P.G., Jones, D.R. and Farrell, A.P. (1992). In *Fish Physiology Vol. XII, Part A. The Cardiovascular System* (eds W.S. Hoar, D.J. Randall and A.P. Farrell), pp. 89–120, Academic Press, San Diego.

Bushnell, P.G., Conklin, D.J., Duff, D.W. and Olson, K.R. (1998). *J. Exp. Biol.* **201**, 1381–1391.

Couch, J.A. (1990). *Anat. Rec.* **228**, 7–14.

Davison, W. (1987). *Cell Tissue Res.* **248**, 703–708.

Egginton, S. (1992). *J. Zool.* **226**, 691–698.

Ellis, A.E. and Munroe, A.L.S. (1976). *J. Fish Biol.* **8**, 67–78.

Farrell, A.P., Jones, D.R. and Roberts, R.J. (1992). In *Fish Physiology Vol. XII, Part A. The Cardiovascular System* (eds W.S. Hoar, D.J. Randall and A.P. Farrell), pp. 1–73. Academic Press, San Diego.

Kampmeier, O.F. (1969). *Evolution and Comparative Morphology of the Lymphatic System.* Charles C. Thomas, Springfield, IL.

Lahnsteiner, A., Lametschwandtner, A. and Patzner, R.A. (1990). *Scan. Microsc.* **4**, 111–124.

Leknes, I.L. (1982). *Anat. Anaz., Jena* **152**, 125–128.

Leknes, I.L. (1987). *Acta Histochem. (Jena)* **81**, 175–182.

Morris, J.L. and Nilsson, S. (1994). In *Comparative Physiology and Evolution of the Autonomic Nervous System* (eds S. Nilsson and S. Holmgren), pp. 193–246. Harwood Academic Publishers, Chur.

Munshi, J.S.D., Roy, P.K., Ghosh, T.K. and Olson, K.R. (1994). *Anat. Rec.* **238**, 77–91.

Olson, K.R. (1992). In *Fish Physiology Vol. XII, Part B. The Cardiovascular System* (eds W.S. Hoar, D.J. Randall and A.P. Farrell), pp. 136–232. Academic Press, San Diego.

Olson, K.R. (1996). *J. Exp. Zool.* **275**, 172–185.

Olson, K.R. (1997). In *The Physiology of Fishes* (ed. D.H. Evans), pp. 129–154. CRC Press, Boca Raton.

Olson, K.R. and Villa, J. (1991). *Am. J. Physiol. Regul. Integr. Comp. Physiol.* **260**, R925–R933.

Olson, K.R., Munshi, J.S.D., Ghosh, T.K. and Ojha, J. (1986). *Am. J. Anat.* **176**, 305–320.

Priede, I.G. (1975). *J. Zool., Lond.* **175**, 39–52.

Priede, I.G. (1976). *J. Fish Biol.* **9**, 209–216.

Sætersdal, T.S., Justesen, N.-P. and Krohnstad, A.W. (1974). *J. Mol. Cell. Cardiol.* **6**, 415–437.

Sánchez-Quintana, D., García-Martínez, V., Climent, V. and Hurlé, J.M. (1996). *J. Exp. Biol.* **275**, 112–124.

Santer, R.M. (1985). *Adv. Anat., Embryol. Cell Biol.* **89**, 1–102.

Santer, R.M. and Greer Walker, M. (1980). *J. Zool., Lond.* **190**, 259–272.

Santer, R.M., Greer Walker, M.G., Emerson, L. and Witthames, P.R. (1983). *Comp. Biochem. Physiol.* **76A**, 453–457.

Satchell, G.H. (1991). *Physiology and Form of Fish Circulation.* Cambridge University Press, New York.

Smith, U., Ryan, J.W., Michie, D.D. and Smith, D.S. (1971). *Science* **173**, 925–927.

Steffensen, J.F. and Lomholt, J.P. (1992). In *Fish Physiology Vol. XII, Part A. The Cardiovascular System* (eds W.S. Hoar, D.J. Randall and A.P. Farrell), pp. 185–213. Academic Press, San Diego.

Steffensen, J.F., Lomholt, J.P. and Vogel, W.O.P. (1986). *Acta. Zool. (Stockh.)* **67**, 193–200.

Stevens, E.D., Lam, H.M. and Kendall, J. (1974). *J. Exp. Biol.* **61**, 145–153.

Tibbits, G.F., Moyes, C.D. and Hove-Madsen, L. (1992). In *Fish Physiology Volume XII, Part A. The Cardiovascular System* (eds W.S. Hoar, D.J. Randall and A.P. Farrell), pp. 267–297. Academic Press, San Diego.

Vogel, W.O.P. (1978). *Z. Mikrosk. Anat. Forsch.* **92**, 565–570.

Vogel, W.O.P. (1985). In *Cardiovascular Shunts. Phylogenetic, Ontogenetic and Clinical Aspects* (eds K. Johansen and W. Burggren), pp. 143–159, Munksgaard, Copenhagen.

Vogel, W., Vogel, V. and Schlote, W. (1974). *Cell Tissue Res.* **155**, 491–512.

Chapter 23

Digestive System

Randal K Buddington
Department of Biological Sciences, Mississippi State University, Mississippi, USA

Victoria Kuz'mina
Institute for the Biology of Inland Waters, Borok, Yaroslavl, Russia

Introduction

Even though the gross anatomy of the digestive tract, particularly the alimentary canal, is highly variable among fish, the types of cells and tissues (i.e. the 'building blocks') are conserved. An appropriate analogy is how houses built for different environments may vary in structure, but the same basic materials are used for their construction.

Similar to all other vertebrates the different regions of the alimentary canal consist of the same four basic tissue layers (Figure 23.1). The internal mucosal layer consists of two components. The first is the epithelium which serves as a selectively permeable membrane that separates the body from the outside (lumen). The epithelium lining the two ends of the digestive system develops from two invaginations: the **stomodeum** forms the mouth and the **proctodeum** forms the anus. The epithelium lining the remainder of the alimentary canal is derived from endoderm, with evaginations of this tissue forming the accessory organs, including the liver and gall bladder, pancreas, and air bladder. Directly under the epithelium is the lamina propria, which is largely connective tissue. Capillaries are present in the lamina propria and they allow for the exchange of materials between the host and the outside environment, using the epithelium as the site of exchange. Specifically, nutrients absorbed from food are transferred to the blood in the capillaries, and conversely, the blood can deliver other materials or by-products of metabolism

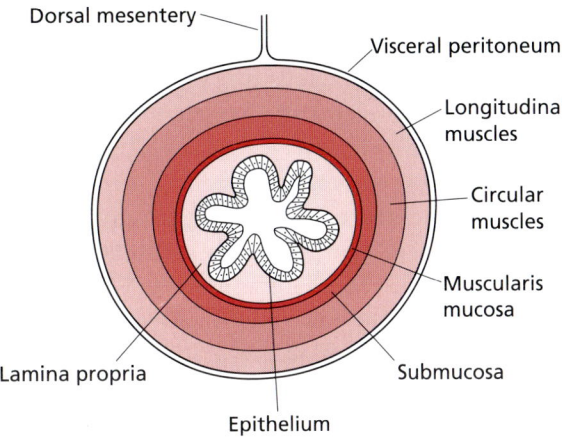

Figure 23.1 The organization of the tissue layers in the fish alimentary canal.

Copyright © 2000 Academic Press

for secretion into the intestine. In the intestine lymph vessels can be seen and these are involved in the absorption of fat, other nutrients and fluids. A thin layer of smooth muscle known as the muscularis mucosa can sometimes be seen between the lamina propria and the deeper tissue layers, but it is less developed in fish compared to that of mammals. Extensions of the muscularis mucosa, when present, penetrate up into the villi and provide the villi with the ability to move, which effectively 'stirs' the layer of fluid overlying the epithelium. This reduces the thickness of the boundary layer that covers the epithelium and limits diffusion of nutrients down to the absorptive cells, and by so doing enhances the absorption of the luminal contents.

The second tissue layer is called the submucosa. This layer supports the mucosa and consists mainly of connective tissue. The submucosa is penetrated by nerves and blood vessels that extend up to the mucosa.

The third tissue layer is known as the muscularis. This layer is actually composed of two coats of smooth muscle. The inner circular layer is arranged such that the fibers loop around the lumen. Fibers of the second, or longitudinal, layer are oriented in a proximal to distal direction. Contractions of the muscle layers are important for the segmental contractions that mix food items and the peristaltic contractions that propel the **chyme** in a distal direction.

The fourth, and outermost layer, is known as the **serosa**. In reality, the serosa is the visceral peritoneum that covers the alimentary canal and other viscera lying within the peritoneal cavity.

Mouth and Pharynx

The microscopic anatomy of the fish mouth and pharynx is basically the same as that seen in other vertebrates. Only the macroscopic structure varies widely, corresponding with the wide diversity of food habits. Mucous cells are interspersed among the epithelial cells lining the oral cavity and pharynx. Salivary gland cells are present, but unlike other vertebrates they exist as solitary cells, not multicellular glands. The secretions from the mucous and salivary gland cells serve to lubricate the food, and in some filter feeders help to 'trap' small particles.

Taste buds, which consist of elongate sensory cells with accessory cells, can be found on the lips, tongue and throughout the oral cavity, but can also be located outside of the mouth, such as in the sawfish. They can also be found on the body surface and fins of some species, with the catfish as an excellent example. The external taste buds allow for recognition of food items without ingestion and are particularly well developed in species that live in low light, poor visibility environments.

Esophagus

The epithelium of the esophagus is stratified and consists mainly of cuboidal cells with numerous mucus-secreting goblet cells and occasional lymphocytes. The underlying submucosa is thick because of abundant connective tissue. Striated muscle fibers can be seen in the two muscle layers. The only other region of the alimentary canal where striated muscle is seen is the anal sphincter. In some species the striated muscle is gradually replaced by smooth muscle as the esophagus extends caudally. However, in other species striated muscle can be seen throughout the esophagus, and can be even more developed in the caudal end of the esophagus. Exemplary of this is the expanded terminal region of the esophagus in carp.

Stomach

The stomach is commonly separated into three regions (cardiac, fundic and pyloric) based on location. However, the three regions can also be distinguished by histological features. The cardiac region, which receives and stores food entering from the esophagus, is usually nonsecretory, or has only a few gastric glands. This region is lined by a stratified epithelium. **Rugae**, which are ridges with extensive and well developed muscle fibers in the mucosa, allow the cardiac region to distend and accommodate large meals.

The fundic and pyloric regions have secretory functions in fish with true stomachs. Both are lined by a simple epithelium with a mixture of columnar cells that lack microvilli and numerous goblet cells that secrete a mucus with a basic pH which protects the gastric epithelium from autodigestion. In fish with true stomachs, simple and branched tubular gastric glands penetrate deep into the lamina propria of the mucosa, with the abundance and depth of the glands greater in the fundic region. Goblet cells are present along the neck of the gastric glands and in the pyloric region are even more numerous in the necks of the glands. At the base of the glands is a population of oxyntopeptic cells that secrete both acid (HCl) and pepsin, which initiate chemical digestion of dietary inputs. The ultrastructural features of the oxyntopeptic cells is consistent with the dual functions of acid and enzyme secretion. The apical domain of the secretory cells has an extensive network of tubules that is involved with the secretion of acid. Zymogen granules that contain the digestive enzymes are located more at the base of the **oxyntopeptic** cells where they are associated with a well developed rough endoplasmic reticulum that is needed to support an intense level of protein synthesis. The granules are transported to the apical membrane and released by the process of exocytosis.

The submucosa under the gastric epithelium consists of connective tissue with interspersed smooth muscle fibers. The next layer, the muscularis, consists of the inner circular and outer longitudinal smooth muscle layers. In some species striated muscle can be seen in the cardiac region of the stomach, and this is thought to represent an extension of the striated muscle present in the esophagus. In some species the two muscle layers of the pyloric region are highly developed forming a '**gizzard**', which is used for grinding food items (e.g. the sturgeons and gizzard shad). A third muscle layer can be seen in the gizzard of some **species**.

At the terminus of the stomach the smooth muscle layers are both thicker and constricted, forming a structure known as the pyloric sphincter. This structure regulates the passage of food items from the stomach into the intestine. The sphincter is either absent or poorly developed in stomachless fishes.

Intestine

The partly digested food entering the intestine from the stomach is referred to as **chyme**. Enzymes, electrolytes, water and other solutes (e.g. bile) secreted by the intestine itself and associated accessory organs (pancreas and gall bladder) are added to the chyme and continue the chemical digestion started in the stomach. To a large extent, the capacity of the intestine to hydrolyze and absorb food items is directly related to the amount of surface area. Similar to other vertebrates, the mucosa of the intestine has a complex architecture that greatly increases the amount of surface area compared to that of a smooth bore cylinder. In most species the mucosa is arranged into complex folds called villi that can give the intestinal lining the appearance of velvet. The finger-like villi seen in mammals effectively increase the surface area by about 10-fold over that of a smooth-walled tube. Although the specific size and shape of the villi vary among the different species of fish, they effectively increase digestive surface area by several-fold. The depth and complexity of the mucosa, hence villus architecture, also varies among regions of the intestine, being greatest in the proximal intestine and decreasing distally.

The different strategies used to increase the gross digestive surface area (longer intestines, more complex mucosal architecture, presence of pyloric ceca, or development of a **spiral valve**) are simply modifications of the existing intestine and do not involve the development of new tissues. For example, the ceca are simply evaginations of the adjacent intestine and they only differ from the proximal intestine in being thinner. The spiral valve present in the distal intestine of some species is formed from an extension of the mucosa and submucosa.

The intestine is perhaps the organ best characterized by histology. A simple epithelium is dominated by columnar absorptive cells (**enterocytes**). The enterocytes are usually elongated, but can be cuboidal. In some primitive species a portion of the enterocytes are ciliated. Enterocytes are considered to be 'polarized' cells. The outer, or apical, membrane domain that is exposed to the luminal contents of the intestine is invested with enzymes that complete the final stages of hydrolyzing water-soluble nutrients and another set of proteins known as **transporters** that are responsible for absorbing the constituent building blocks. The other domain, known as the basolateral membrane, has another set of functional proteins, including a different group of carriers that transport absorbed nutrients into the blood. Nutrients that are water soluble are dependent on the transporters in both sets of membranes to reach the blood. Lipid-soluble nutrients are able to traverse the enterocytes passively by diffusing through both sets of membranes down concentration gradients and specialized transporters are not needed.

The individual enterocytes are linked together by tight junctions forming an epithelium that effectively forms a barrier that prevents large molecules and organisms from gaining entrance to the body. However, the tight junctions are not so well developed that they prevent the movement of water and monovalent ions. This has led to the intestinal epithelium being considered as 'leaky'.

The enterocytes that constitute the intestinal epithelium are constantly being replaced. This is accomplished in an orderly fashion. New enterocytes are produced at the base of the villi by proliferating stem cells. As the new enterocytes migrate up the villi they differentiate and acquire functional capabilities. The enterocytes eventually migrate to the tips of the villi where they are shed. Villi grow in length when rates of enterocyte proliferation, hence replacement, exceed the rate of shedding at the villus tip, and will shorten when the rate of shedding is greater than that for replacement. In mammals the lifespan of an enterocyte is about 2–3 days, coinciding with complete replacement of the villus epithelium. The replacement process in fish is slower compared to mammals. The lifespan is even longer at colder temperatures, and in some species complete replacement of the enterocyte population can take several weeks.

Interspersed among the enterocytes are goblet cells which are filled with secretory granules. The mucus secreted by the goblet cells and other digestive secretions form a boundary layer that covers the epithelium and is referred to as the **glycocalyx**.

Other cell types that can be seen in the epithelium include immune associated cells (e.g. macrophages) and cells that appear to have endocrine secretory functions based on the contents of cytoplasmic granules. Much less is known about these other cell types despite the obvious importance they have in protecting against invasion and regulating digestive and metabolic processes.

The underlying submucosa is thinner than that seen in the stomach and contains connective tissue, blood and lymph vessels, smooth muscle fibers and nerves. The muscularis has both the circular and longitudinal layers seen in all regions of the alimentary canal. The serosa can be highly pigmented in some species.

Functional anatomy of the intestine

The concept of membrane digestion has been used to describe the functional activities of the enterocytes that dominate the epithelium in conjunction with enzymes and features of the overlying mucous layer. Therefore, the greater the surface area, the greater the digestive capacities. In addition to the gross anatomical strategies that increase the amount of surface area available for digestion, the apical membrane of each enterocyte is lined by microvilli. These are extensions of the cell membrane that surround an inner framework composed of actin filaments that are organized by 'linker' proteins. The microvilli are estimated to increase the surface area available for hydrolysis and absorption by about 200-fold more. The length and number of microvilli vary among intestinal regions, species of fish, feeding state and environmental conditions. In contrast the diameter of the microvilli is much more consistent and ranges from about 10 to 14 μm. Although other ultrastructure characteristics are shared by enterocytes, there can be differences, such as variation of the number, degree of development, and distribution of enterocyte organelles.

The enzymes associated with the apical membrane of enterocytes include disaccharidases (e.g. maltase and trehalase), and a diversity of peptidases that release smaller peptides and free amino acids. There are several classes of transporters for absorbing the different types of solutes (sugars, amino acids, vitamins, bile acids, etc.). The specific proportions of the different functional proteins vary among regions of the intestine as well as among species.

Absorption of proteins and carbohydrates is largely dependent on the luminal and brush border enzymes that release the constituent amino acids, peptides and sugars, which are subsequently absorbed. The process by which fats are absorbed and transported to the blood of fish differs from that for the other nutrients, but appears to be similar to that known for mammals and other vertebrates. Briefly, fatty acids and monoglycerides released from dietary fats by the action of luminal lipases are absorbed, re-esterified in the enterocyte, and then transferred to the blood as a component of particles synthesized by the endoplasmic reticulum.

The enterocytes of many, if not most, fish are also capable of macromolecular absorption. This ability has been demonstrated numerous times using intraluminal injections of horseradish peroxidase or other macromolecules and then being able to detect them inside the enterocytes using cytochemical methods. The capacity to take in macromolecules is particularly well developed in larval fish that do not yet have a 'mature' digestive system and in adults of other species that lack true secretory stomachs.

The proportions of the different enzymes and transporters present in the brush border membrane are matched to the composition of the natural diet. For example, the activities of carbohydrases and rates of glucose transport are higher in omnivorous and herbivorous species compared to strict carnivores.

Despite the diversity of feeding habits, fish must eat sufficient quantities of food to meet requirements for protein and essential amino acids. Because the requirements do not differ widely among fish, the activities of the peptidases and amino acid transporters are not as variable among fish as those seen for the carbohydrases and glucose transport.

The epithelium lining the intestine can be damaged by some components that are used to formulate some production diets. For example, feeding salmonids diets that contain soybean meal causes an inflammation (enteritis) in the distal intestine that is associated with a reduction in the amount of mucosa and a decrease in the activities of certain digestive functions. Although the causes for the enteritis are not yet known, they are suggestive of an allergic reaction to a component of soybeans.

Although the majority of research has been directed at understanding the responses of the brush border functions to the composition of the diet, it is now recognized that the densities, types, and functional characteristics of the different enzymes and transporters present in the membranes of enterocytes are also responsive to changes in environmental conditions such as temperature and salinity.

Multicellular glands are not present in the mucosa, but throughout the stomach and intestine there are large individual cells that are granulated and have been shown to have secretory functions. A number of different types of endocrine cells have been identified in the intestine with most characterized as being or similar to **enterochromaffin cells**. Although many of the secretory cells are located in the mucosa, endocrine and other secretory cell types can be detected throughout other tissues of the digestive system. Most of the secretory cells synthesize polypeptides that can act locally (**paracrine**) or systemically (**endocrine**). Exemplary are the cells in the proximal intestine that synthesize cholecystokinin and secretin to regulate exocrine pancreatic functions in response to the chemical composition of the chyme in the intestine. The best known examples of digestive system endocrine cells are those of the pancreas. The endocrine pancreas produces several hormones, and the four best known are insulin, glucagon, somatostatin and pancreatic polypeptide. Each hormone is synthesized by a specific type of cell. When the numerous different secretory cell types are considered collectively, the digestive system represents the largest endocrine organ in the body and the associated secretions are critical for regulating physiological and metabolic processes throughout the body.

Although endocrine cells of the intestine are known to be important for regulating digestion and metabolism in mammals, the specific roles they play in fish have not been adequately studied and defined. In addition to the influence of endocrine secretions, intestinal functions are also subject to regulation by the multitude of neurons that comprise the 'enteric nervous system'.

The intestinal epithelium is also an important tissue for osmoregulation. This is particularly true for marine species that must drink sea water to compensate for the osmotic loss of body fluids to the environment. This is accomplished by intestinal absorption of monovalent ions present in sea water (e.g. Na and Cl), which drives the movement of water out of the gut into the blood. The excess ions are then disposed of by the gills. Corresponding with the changes in functional demands placed on the intestine during adaptation to different salinities, there are changes in hydrolytic and transport functions.

Another key function of the intestine is the defense against invasion. This is accomplished by both non-specific and specific defense mechanisms. The non-specific mechanisms include the tight junctions that link the enterocytes together and effectively form a barrier. The mucus secreted by the goblet cells also provides a barrier that reduces the ability of pathogens to invade. Additionally, phagocytic cells in the epithelium and underlying tissue layers are able to respond to organisms and particles that manage to pass the epithelial barrier. The lymphocytes present in the epithelium are known as intraepithelial lymphocytes (IEL), with additional lymphocytes found in the underlying tissue layers. The lymphocytes are able to recognize and react to specific antigens present in the intestines of fish. The collection of lymphocytes and associated tissues present in the tissues of the alimentary canal is often referred to as the mucosal associated lymphoid tissue (MALT).

Accessory organs

The pancreas develops as an extension of the alimentary canal epithelium just distal to the future stomach and the ducts that link the pancreas to the intestine (or **ceca**) are the remnants of the embryological origin. The pancreas has both endocrine and exocrine functions. In many, if not most species of fish, the pancreas is a diffuse organ, but is a compact organ in some fish as it is in other vertebrates. In most species the pancreatic tissue is hard to detect macroscopically. Patches of exocrine and endocrine pancreatic tissue can be seen in several sites, including the intestine, the visceral mesenteries, and alongside blood vessels near the intestine. Pancreatic tissue has also been reported in the liver and spleen.

The exocrine pancreatic cells are more abundant than the endocrine cells and are detected by staining for the presence of zymogen granules, which represent the digestive enzymes, or by using antibodies that are selective for specific enzymes. The digestive secretions enter the proximal intestine through a single duct or multiple ducts. In addition to secreting a multitude of digestive enzymes (e.g. several proteases, amylase, lipase, deoxy and ribonucleases), the acinar cells of the pancreas produce bicarbonate to neutralize gastric acid and other electrolytes.

The liver also develops as an outgrowth of the epithelium of the developing intestine and at about the same level as the presumptive pancreas. The bile duct, much like the pancreatic duct, is a remnant of the embryologic origin of the liver and it provides a connection with the intestine.

Most hepatic cells, like the enterocytes, are 'polarized' and have two distinct membrane domains. One domain is exposed to the blood whereas the other membrane domain faces the caniculi that collect secretions that eventually constitute the bile and are secreted into the intestine.

References

Bæverfjord, G. and Krogdahl, A. (1996). Development and regression of soybean meal induced enteritis in Atlantic salmon, *Salmo salar* L., distal intestine: a comparison with the intestines of fasted fish. *J. Fish Dis.* **19**, 375–387.

Buddington, R.K., Krogdahl, A. and Bakke-McKellep, A.M. (1997). The intestines of carnivorous fish: structure and functions and the relations with diet. *Acta Physiol. Scand.* **161 (Suppl. 638)**, 67–80.

Fänge, R. and Grove, D. (1979). Digestion. In *Fish Physiology*, vol. 8 (eds W.S. Hoar, D.J. Randall and J.R. Brett), pp. 161–260. Academic Press, New York.

Kuperman, B.I. and Kuz'mina, V.V. (1994). The ultrastructure of the intestinal epithelium in fishes with different types of feeding. *J. Fish Biol.* **44**, 181–193.

Kuz'mina, V.V. and Gelman, A.G. (1997). Membrane-linked digestion in fish. *Rev. Fish. Sci.* **5**, 99–129.

Iwama, G. and Nakamishi, T. (eds) (1996). *The Fish Immune System: Organism, Pathogen, and Environment*. Academic Press, New York.

Reinecke, M., Müller, C. and Segner, H. (1997). An immunohistochemical analysis of the ontogeny, distribution, and coexistence of 12 regulatory peptides and serotonin in endocrine cells and nerve fibers of the digestive tract of the turbot, *Scophthalmus maximus* (Teleostie). *Anat. Embryol.* **195**, 87–102.

Romer, A.S. (1970). *The Vertebrate Body*. W.B. Saunders, Philadelphia.

Suyehiro, Y. (1941). A study on the digestive system and feeding habits of fish. *Jap. J. Zool.* **10**, 1–303.

Yasutake, W.T. and Wales, F.H. (1983). *Microscopic Anatomy of Salmonids: An Atlas*. United States Department of the Interior, Resource Publication 150, Washington, DC.

Zapata, A.G. and Cooper, E.L. (eds) (1990). *The Immune System: Comparative Histopathology*. John Wiley, New York.

CHAPTER 24

Urinary Tract

Marlies Elger
Institut fur Anatomie und Zellbiologie I, Universität Heidelberg, Heidelberg, Germany

Hartmut Hentschel
Max-Planck-Institut fur molekulare Physiologie, Dortmund, Germany

Margaret Dawson and J Larry Renfro
Department of Physiology and Neurobiology, University of Connecticut, Connecticut, USA

Introduction

Renal corpuscles similar to those found in the glomerular kidneys of higher vertebrates are present in the majority of fish kidneys. However, a general tendency toward reduction in the number of nephron segments is observed in more recently evolved teleosts. In lower teleosts, such as eels, salmonids and ostariophysids (e.g. carp and catfish), multisegmental nephrons prevail. In advanced marine teleosts from different orders, such as percids, killifish, flounder and sticklebacks, the distal segment is generally absent. The greatest parsimony in the development of renal structures is seen in the aglomerular kidneys of several different orders that include toadfish, anglerfish, sea horses and needlefish. In these, mostly marine fishes, the glomeruli and the proximal tubule segment I never develop from the nephron anlage, and the distal segment is absent.

Renal vascular system

Architecture of the vessel wall

Intrarenal arteries and arterioles

Renal vessels are similar in structure to those of similar size found elsewhere in the body. The lumen is lined with squamous endothelium. The endothelial cells, whose nuclei bulge into the lumen, are arranged

Copyright © 2000 Academic Press

longitudinally. The endothelial cells lie on a thin basement membrane, which is generally continuous; small gaps are present where endothelium contacts underlying smooth muscle cells. The next layer is a tunica media consisting of several layers of smooth muscle cells arranged circularly in the intrarenal arteries and one muscular layer in arterioles. A continuous basement membrane surrounds each muscle cell. Numerous caveolae are formed on the cell membranes of both endothelial and smooth muscle cells. The cytoplasm of the endothelial cells often contains specific granules, rough endoplasmic reticulum and small mitochondria. The cytoplasm of the smooth muscle cells is densely packed with myofilaments, a few strands of rough endoplasmic reticulum and mitochondria with cristae in a moderately dense matrix. Smooth muscle cells are connected by gap junctions (nexus). A thin tunica elastica interna with a few elastic fibrils is present. The adventitia beneath the muscular layers consists of fibrocytes and bundles of collagen fibrils. The efferent arterioles differ from the afferent arterioles in having a smaller diameter and in the lack of a continuous muscular layer. Granulated cells are present in the terminal portions of the walls of intrarenal arteries of teleosts (Elger *et al.*, 1984b). Their morphology and staining properties suggest that the granulated cells correspond to the renin-producing cells of the juxtaglomerular apparatus of higher vertebrates (Nishimura and Bailey, 1982). These cells are found in various teleosts from fresh water and salt water including aglomerular fish.

Preglomerular sphincters

Preglomerular sphincters, accompanied by **varicosities** of unmyelinated adrenergic nerves, have been described near the origin of the afferent arterioles in the trout (Elger *et al.*, 1984b). Similar sphincters were observed in the shark, *Scyliorhinus caniculus*. These sphincters were not observed in **stenohaline** freshwater fish such as the Prussian carp, *Carassius auratus gibeli*.

Intrarenal veins

The walls of the intrarenal veins and peritubular venous sinusoids have a very thin endothelium of squamous cells. Pores bridged by a membrane are frequently observed. A thin discontinuous basement membrane separates the endothelium from bundles of collagen fibrils and fibrocytes.

Nephron and collecting duct system

Definition and evolutionary aspects

Morphological investigations have revealed renal corpuscles (glomeruli) and renal tubules (uriniferous tubules) in most fishes. The nephron, i.e. the epithelia of the renal corpuscles and the major portions of the renal tubules, can be divided into two regions – the proximal and the distal nephron. By definition, the proximal nephron derives from that end of the S-shaped anlage, which grows in the direction of the arterial blood vessels. Further differentiation of the proximal end of the nephron leads to the formation of the glomerulus (glomerular nephron). Besides the glomerular epithelia, the following portions belong to the proximal nephron: the neck segment, the proximal tubule and the intermediate segment (Figure 24.1). The distal nephron originates from the other end of the anlage. It consists of the distal tubule with early and late segments. The collecting tubule is the end portion of the unbranched renal tubule, however it does not belong to the nephron proper. It connects the nephron with the arborescent system of collecting ducts that in turn coalesce to form the archinephric duct. This glomerular, multisegmental nephron may be considered a general or basic nephron type, from which other nephron types can be derived. This generalized nephron is found in archaic bony fish, such as Polypteriformes and Dipnoi, whereas other types of nephron are present in the kidneys of the Agnatha, the Elasmobranchiomorpha and the majority of the Actinopterygii including teleosts.

In all fishes, the borderline between the late distal tubule and the collecting tubule-collecting duct system is indistinct. In teleosts, major parts of the collecting tubule-collecting duct system develop from distal portions of the nephron. In several teleost groups, the proximal end of the tubule fails to contact the arterial system to form glomeruli. These aglomerular (as well

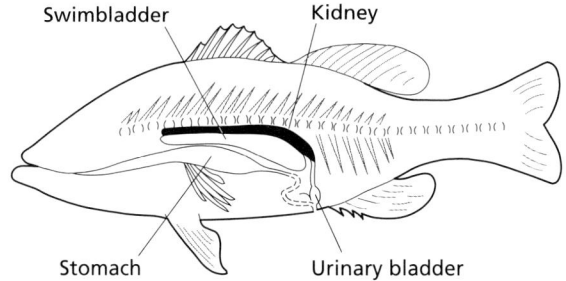

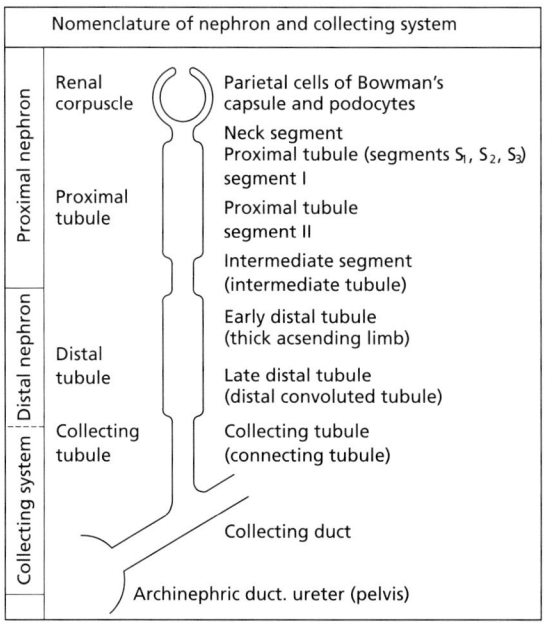
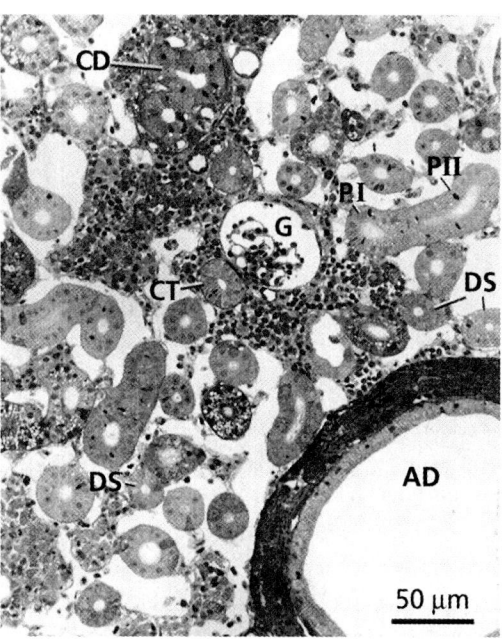

Figure 24.1 Nomenclature of nephron and collecting system. The general retroperitoneal location of the kidneys in a teleost is shown relative to the internal and external anatomy in the top panel. The terminology of the corresponding nephron segments of the mammalian kidney is indicated in parentheses. (From Hentschel and Elger, 1987, with permission.) A light micrograph in the right panel shows the various nephron segments of a teleost (rainbow trout, *Oncorhyndrus mykiss*) in cross-section. Note that the renal tissue of teleosts has no apparent zonation. The renal tubules are partly embedded in interstitial tissue and large sinus capillaries of the renal portal systems. AD, archinephric duct; CD, collecting duct; CT, collecting tubule; DS, distal segment; G, glomerulus; PI, proximal segment PI; PII, proximal segment PII. Bar: 50 μm.

as pauciglomerular) kidneys are found predominantly in marine species of advanced teleost orders.

In the Myxinoidea of the class of Agnatha, the nephron is rudimentary or 'atubular'. This nephron type contrasts sharply with the multisegmental nephron of the other agnathan subgroup, the Petromyzontia, which comprises recent lampreys. The renal tubules of the latter group superficially resemble those of higher vertebrates, because they display a countercurrent arrangement of distal portions with a nephron loop in a distinct zone. Lampreys and the hypothetical ancient freshwater fish presumably evolved from a common ancestral vertebrate. It is a matter of debate whether this extinct, unknown vertebrate had a nephron like that of recent myxinids, or whether the latter represents regressive evolution from a more elaborate type of kidney.

The recent cartilaginous and bony fishes, as well as the tetrapods most probably had as a common ancestor an ancient freshwater fish (Figure 24.2) (Hentschel and Elger, 1987) based on the following observations. First, the exceptionally long nephron of elasmobranchs can theoretically be derived from a multisegmental nephron that is similar to that of the freshwater-dwelling Polypteridae. Unlike the segmentation of the polypterid nephron, the proximal nephron of elasmobranchs has an additional second major portion (referred to as proximal tubule segment

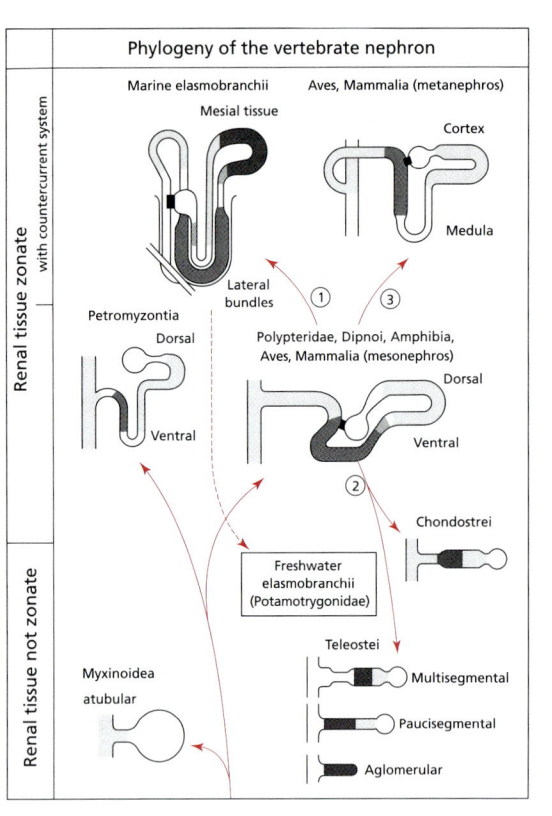
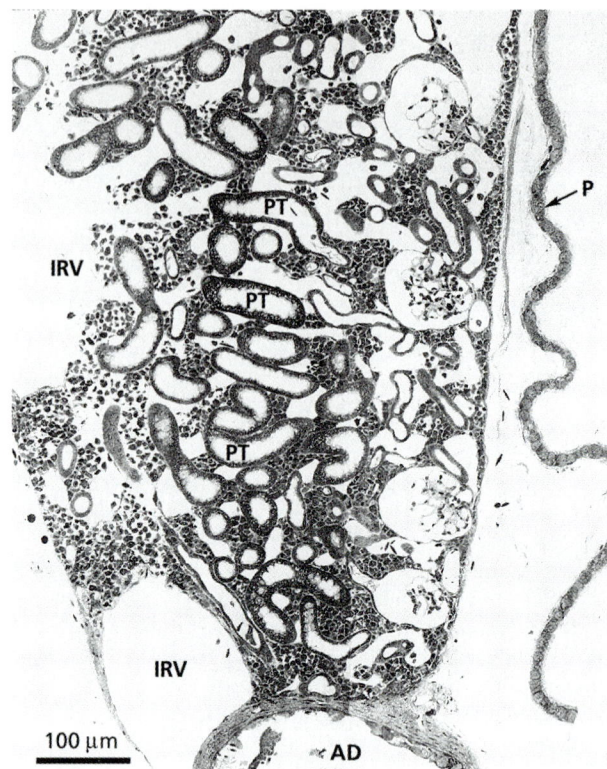

Figure 24.2 Organization of the renal tissue and phylogeny of the vertebrate nephron. Right panel: Three levels of renal organization have evolved in the vertebrates. Kidneys can be (a) without zonation, (b) zonate without countercurrent system, and (c) zonate with countercurrent system. The kidneys of reptiles are similar to those of polypterids and amphibians, with regard to the absence of a functional countercurrent system. Deviating from this scheme, in several species of Petromyzontia the course of the nephron may be reversed, with the loop running dorsally. Multiciliary cells are not present in the avian and mammalian mesonephros; in the mammalian-type nephrons of domestic fowl, the entire loop of Henle consists of early distal tubule cells. Note that the aglomerular teleost nephron, although placed at the bottom of this scheme because of its simple architecture is a highly advanced form with extensive reduction of nephron structures. (From Hentschel and Elger, 1989, with permission.) The light micrograph in the right panel shows a cross section of the zonate kidney of *Polyterus*. Glomeruli and distal segments are located ventrally (on the right side), and proximal tubules (PT) and collecting tubules (small profiles) are located dorsally. On the dorsal side run large intrarenal veins (IRV) which draw blood from the sinus capillaries of the renal portal system. AD, archinephric duct; P, ciliated peritoneum. Bar: 100 μm.

II, PII). The nephron runs through two kidney zones, a zone of mesial tissue and a zone of lateral countercurrent bundles (Figures 24.4 and 24.5). The mesial tissue is similar to the dorsal zone of polypterids, lungfish and amphibians, and it superficially resembles the renal cortex of birds and mammals in that the convoluted parts of the majority of proximal tubules are located in this zone. The lateral bundle zone contains a highly organized countercurrent system. Because of the presence of a separate arterial blood supply to the lateral bundle zone, and the intricate arrangement of single nephron segments with two

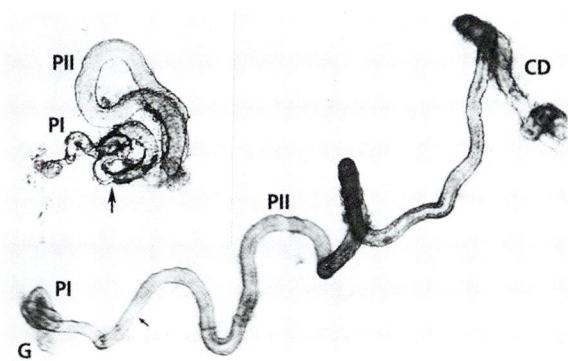

Figure 24.3 Microdissected renal tubules of the winter flounder, *Pleuronectes americanus*. Long nephron and a short coiled nephron (top, left) are shown. The paucisegmental nephron in this advanced teleost has a small glomerulus (G) which leads via a narrow neck segment to the very short first brush border segment, the proximal segment PI, which has two to four bends in the vicinity of the glomerulus. The elaborate second proximal segment PII is distinctly thicker than PI and joins directly the collecting tubule/collecting duct system (CD). A distal segment is not developed in this species. Arrows mark the transition between PI and PII. Light micrograph (LM), ×100. (From Elger *et al.*, 1998, with permission.)

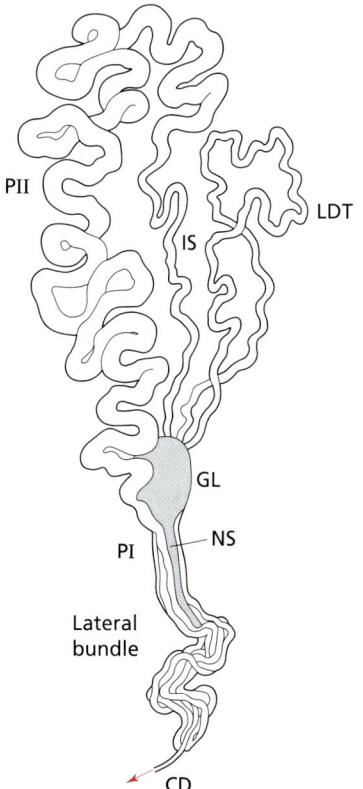

Figure 24.4 Young nephron of dogfish, *Scyliorhinus*, according to microdissections of E. Borghese. The glomerular nephron runs alternating between mesial tissue zone and lateral bundle zone. GL, glomerulus; CD, collecting duct; IS, intermediate segment; LDT, late distal tubule; NS, neck segment; PI, proximal tubule I; PII, proximal tubule II. (After Borghese, 1966, modified. From Hentschel and Elger, 1987, with permission.)

hairpin loops and a peculiar central vessel, these lateral bundles appear even more complex than the countercurrent systems in the avian and mammalian renal medulla. Second, the Chondrostei, Holostei and lower Teleostei exhibit a multisegmental nephron with a PII. The nephron of advanced teleosts is rather short and consists of a few segments only (Figure 24.3). The phylogenetic trend towards a reduction in number of nephron segments culminates in the aglomerular kidneys that lack the renal corpuscle portions of the proximal tubule and the distal nephron segments (Figure 24.2).

Glomerulus

Gross anatomy

The generalized glomerulus type

The renal corpuscle consists of a globular cluster of specialized branching and anastomosing capillaries (the glomerulus) extending into a pouch-like expansion of the proximal nephron, the visceral and parietal layers of Bowman's capsule. The glomerular capillaries form several lobules emanating from the afferent arteriole. Capillaries of this glomerular tuft coalesce to form the efferent arteriole that exits the glomerulus near the entry point of the afferent arteriole (vascular pole). The mesangium lies between the two arterioles and follows the capillaries into the central regions of the glomerulus providing an attachment of all glomerular capillaries to the mesangial cell layer throughout their entire length, but not always by their entire circumference (see below). The **parietal** layer of Bowman's capsule has an epithelium of squamous cells that lie on a thick basement membrane. At one end of the glomerulus (urinary pole), the parietal epithelium and the basement membrane are continuous with the neck region of the proximal tubule. At the vascular pole, the parietal layer approaches the arterioles and folds backwards, becoming the visceral layer which, together with the glomerular basement membrane, closely follows the external contours of the capillary tuft and the more centrally located mesangium. The visceral layer consists of an epithelium of highly specialized cells, the podocytes, which face the urinary space. An important site of filtration is the peripheral part of the glomerular capillaries where the podocytes and the glomerular basement membrane overlay the fenestrated capillary endothelium. In contrast, the central region of the lobulus, which contains most of the mesangium, is not involved directly in ultrafiltration.

The earliest vertebrate group with the aforementioned generalized glomerulus type is in the recent Polypteridae. This glomerular design, which we have labelled the polypterus-type, is found in various classes of fish from archaic Actinopterygii (such as ganoid fish, including sturgeons, polypterids, bowfin) to advanced Actinopterygii including stenohaline freshwater Teleostei and Dipnoi, and this type of glomerulus closely resembles the glomeruli of other, higher, vertebrate classes (see Figure 24.11b). Other types of glomeruli with a substantially different structural organization from this polypterus-type glomerulus have evolved in the Agnatha (hagfish and lampreys), Elasmobranchiomorphii (sharks and skates), and in most marine and euryhaline Teleostei.

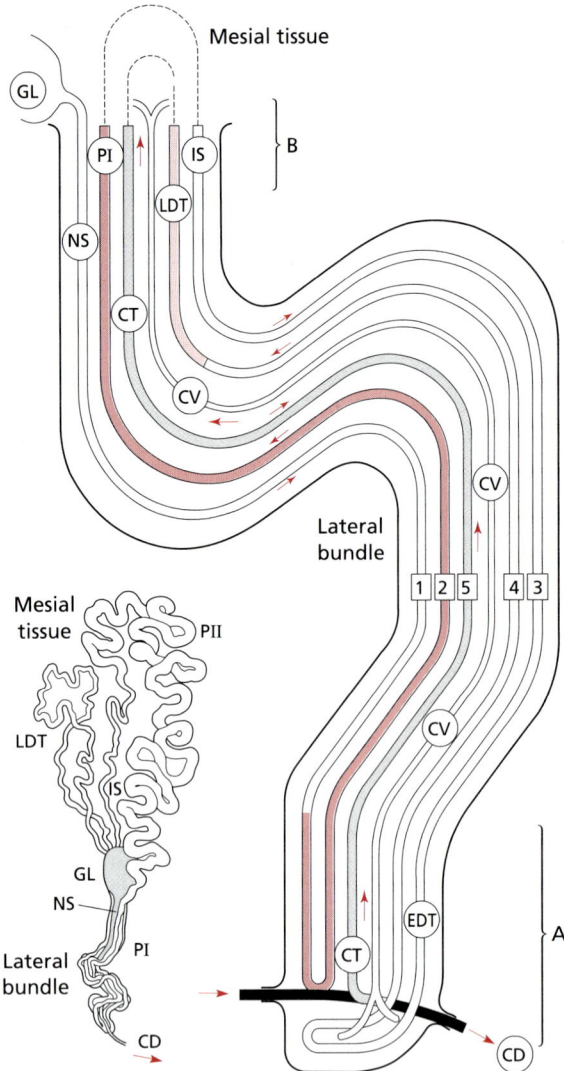

Figure 24.5 Schematic drawing of the arrangement of nephron segments and portions of the collecting tubule-collecting duct system inside the countercurrent bundle of marine elasmobranchs. Five tubular profiles and a central vessel are enclosed in the bundle sheath (thick line). The bundle sheath is discontinuous at the transition of the lateral bundle zone to the mesial tissue zone. The course of the nephron in the mesial zone is indicated by broken lines (compare, however, inset for presentation of nephron segments in the mesial tissue zone). Thick arrows show the direction of supposed flow in the central vessel (CV), thin arrows indicate the direction of urine flow inside the renal tubule segments. For simplicity, the actual thickness of the different tubule segments is not drawn to scale. The color code of the segments: red = proximal tubule segment PIa; light green = early distal tubule EDT; dark green = late distal tubule LDT; light blue = collecting tubule CT; dark blue = collecting duct CD. The portion within the brace A refers to the distal end of the bundle, GL, glomerulus; IS, intermediate segment; NS, neck segment. (From Hentschel et al., 1998b).

Myxine type

The renal corpuscles of the hagfish, such as *Myxine glutinosa*, *Paramyxine atami* and *Eptatretus stouti*, are exceptionally large (500 to 1500 μm in diameter). The parietal layer of Bowman's capsule is continuous at the urinary pole with a rather short neck region that in turn is confluent with the archinephric duct. One, rarely two, afferent arterioles supply a large glomerular tuft with several lobules. Multiple efferent arterioles leave the glomerular tuft from various locations. Therefore, a single vascular pole is not present. The mesangium extends from the central region of the lobules to the peripheries of the capillary loops. Numerous cell processes of mesangial cells are interposed between the glomerular basement membrane and the endothelium. The mesangial matrix blends with the subendothelial space that contains collagen fibrils and microfibrils (see Figure 24.11a).

Lamprey type

Several species of lamprey (e.g. *Lampetra fluviatilis*, *Entosphenus japonica* and *Petromyzon marinus*) have

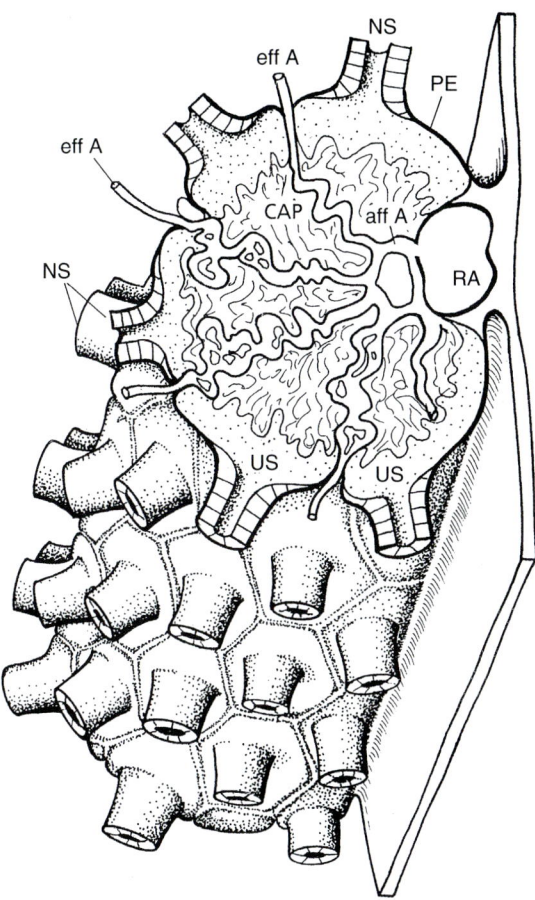

Figure 24.6 Glomus of a lamprey. Three-dimensional schematic drawing of a part of the glomus of one kidney. The glomus is composed of multiple renal corpuscles. The renal artery (RA) which runs in parallel to the aorta, gives rise to very short afferent arterioles (aff A). Branches of the aff A split into glomerular capillaries (CAP), which run in septae and give rise to efferent arterioles (eff A). The septae are bordered by podocytes. The urinary spaces (US) of adjacent nephrons are completely separated from each other by the septae, however, the capillaries of one septum filter into several nephrons. NS, neck segment; US, urinary spaces of two different nephrons. (After Youson and Ooi, 1979. In Hentschel and Elger, 1989, modified, with permission.)

and closely follow the capillary loops with a layer of podocytes. The parietal layer of Bowman's capsule consists of squamous cells and is restricted to the small polygonal sectors that are contributed by each single nephron to the external wall of the glomus. The gross architecture of the glomus with its close arrangement of capillary beds has several functional consequences which distinguish the glomus from the glomeruli in other vertebrate classes. As in single glomeruli in other vertebrates, the urinary spaces of all nephrons are separate from each other, as demonstrated by casts of single nephrons with Microfil (Logan et al., 1980a). Thus the podocytes with their foot processes and filtration slits form a structurally continuous layer that encloses the urinary space of one nephron. However, each capillary septum is bordered by the podocytes of more than two adjacent nephrons; hence the filtrate generated from the blood deriving from one afferent arteriole appears in the lumina of several renal tubules. Consequently, differences in function of single nephrons, which are related to the hemodynamics of the arterial blood supply (intermittency, recruitment), and which are generally found in other fish, have not been observed in lampreys.

Marine elasmobranch type

The **opisthonephric** kidneys of elasmobranchs grow by the continuous formation of new nephrons (Hentschel, 1991). Accordingly, even in adult fish, the renal corpuscles display a great variability in size and differentiation depending on their respective age. Mature glomeruli are rather large ovoid to ellipsoid structures (approximately $350 \times 150\,\mu m$) (Figure 24.7). One afferent arteriole, rarely two arterioles, supplies the large glomerular tuft that is drained by multiple (up to six and occasionally even more) efferent arterioles (Figure 24.8). The efferent arterioles leave Bowman's capsule at various locations along the ellipsoid. Thus, no typical vascular pole is present, but the region may be called a 'vascular field'. The parietal layer of Bowman's capsule consists of squamous cells that gradually increase in height at the urinary pole. Here the cells are equipped with numerous cilia and resemble the epithelial cells of the neck segment. The visceral layer of Bowman's capsule consists of podocytes. The space between the glomerular basement membrane and the endothelium is filled by processes of mesangial cells and fibular material (collagen fibers type I, microfibrils) (Figure 24.11c).

no individual glomeruli. One large glomus with multiple capillary convolutions is present in each of the paired elongated kidneys (Figure 24.6) (Youson, 1981). Renal arteries branching from the aorta supply the elongate glomus with blood via very short, orderly afferent arterioles. Efferent arterioles exit the glomus toward the tubular mass of the kidney. The capillary convolutions connect the afferent and efferent arterioles forming voluminous septum-like structures. The most proximal portions of the nephrons, the visceral epithelia, are inserted between the septa

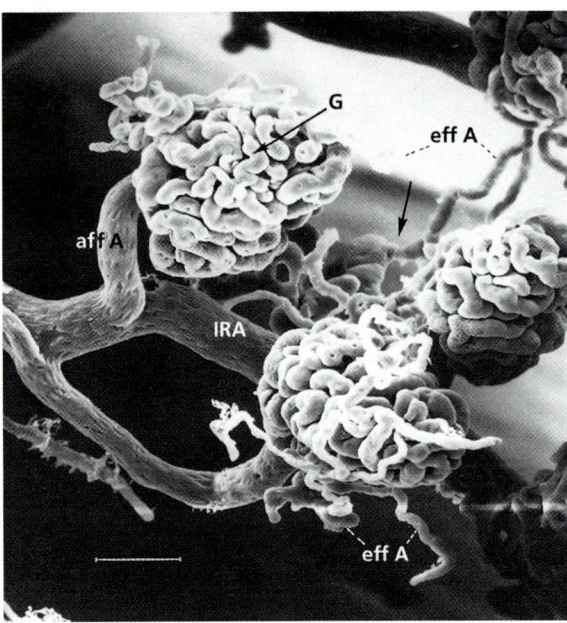

Figure 24.7 Glomeruli of the skate, *Raja erinacea*. Scanning electron micrograph of a corrosion cast with Mercox showing the microvasculature of three glomeruli. One afferent arteriole (aff A) arising from an intrarenal arteriole (IRA) enters the capillary tuft of a glomerulus (G). Four to six small efferent arterioles (eff A) originate from each glomerulus. The transition of an efferent arteriole to a large venous sinus of the fenestrated peritubular circulation is shown (arrow). The surface of the casts of the arterioles and the glomerular capillaries is indented by bulging endothelial cells. Bar = 100 μm. (After Hentschel, 1988. From Hentschel and Elger, 1989, with permission.)

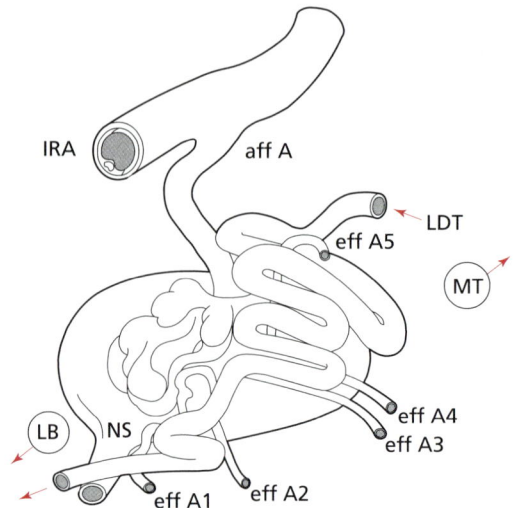

Figure 24.8 Renal corpuscle of the dogfish, *Scyliorhinus caniculus*. Reconstruction from a series of 10-μm sections. The arterial blood, coming from the intrarenal artery (IRA), via preglomerular sphincter and afferent arteriole (aff A), perfuses the glomerulus and leaves by several small efferent arterioles (eff A1–eff A5). The efferent arterioles originate along a furrow-like vascular field of the renal corpuscle. A convolute of the renal tubule meanders between the efferent arterioles, and has a close contact to the vascular field. The transition from the late distal tubule (LDT), arriving from the mesial tissue (MT), to the collecting tubule running into the lateral bundle (LB) occurs in this region. NS = Neck segment. (From Hentschel and Elger, 1989, with permission.)

Marine teleost type

The glomeruli of teleosts exhibit wide interspecific variation in size and structural organization of the filtration barrier. In contrast to the glomeruli of freshwater species of all classes of fish, the glomeruli of marine teleosts are generally very small (sometimes less than 30 μm in diameter), although there may be exceptions to this rule. The glomeruli of marine fish (and euryhaline fish in sea water) differ from the generalized vertebrate model by the presence of an extended mesangium that surrounds the entire periphery of the capillaries (Figure 24.11d). It has been suggested that the reduction in size occurring in glomeruli of **euryhaline** teleosts during adaptation to sea water is dependent on the action of vasoactive hormones. A similar diminution of glomerular size occurs in sticklebacks in fresh water under the influence of sexual hormones. The glomerular structure may be greatly altered in pauciglomerular marine fish, corresponding to degenerated glomeruli in **stenohaline** freshwater teleosts during adaptation to brackish water.

Zonation of the renal tissue

In most classes of lower vertebrates the arrangement of nephrons and blood vessels leads to a higher renal organization which is apparent in a clear zonation of the renal tissue. Thereby the location of the glomeruli and of several nephron segments is confined to special zones. Despite the high interspecies diversity in the number of nephron segments, the earliest portion of the distal nephron (that is the distal segment in fish such as Petromyzontia, Polypteridae and Dipnoi, which is equivalent to the early distal tubule in Amphibia and Reptilia, or the thick ascending limb of Henle in Aves and Mammalia) is generally located in a zone apart (mostly ventrally) from the glomerular population. The majority of the proximal segments together with late portions of the distal nephron are located in a zone dorsal to their parent glomeruli

(Figure 24.2). An area of close contact between the distal nephron and the vascular pole of its parent glomerulus is formed in zonate kidneys, with the exception of the **cyclostomes**.

In the highly organized zonate kidneys of marine elasmobranchs, a countercurrent arrangement of selected tubular segments (Figure 24.5) allows for additional important renal functions, such as renal retention of urea, but unlike the loop of Henle in Aves and Mammalia, it does not allow formation of urine more osmotically concentrated than plasma.

In Chondrostei and Teleostei a clear zonation is not apparent. However, renal structures apparently are not distributed at random. Frequently, the glomeruli are located in the vicinity of the collecting duct system even though a close contact point between the distal segment and the vascular pole is lacking in the differentiated nephron. Because the teleost kidney is presumably phylogenetically younger than the zonate kidney of the ancient Polypteridae, the lack of zonation and the lack of contact points, as well as the disappearance of nephron segments in the advanced fishes, clearly gives evidence of regressive processes in the evolutionary line leading to modern teleosts.

Cytological organization

In contrast to the diversity of glomerular types among the classes of vertebrates, the fine structural organization of cells in the filtration barrier is quite similar. Four cell types are regularly present in the renal corpuscles of all vertebrate classes studied so far: (i) endothelial cells; (ii) mesangial cells; (iii) podocytes; and (iv) parietal cells.

Endothelial cells

Capillaries consist of large, flat cells with attenuated cytosolic regions. The perinuclear region contains the cell organelles and bulges into the capillary lumen. The attenuated cell portions are fenestrated by regularly spaced, large pores (40–70 nm in diameter) in freshwater bony fish (Figure 24.10) and elasmobranchs. Fenestrations are scarce in *Myxine* and sticklebacks. The pores are open in flounder, trout and carp, but a thin diaphragm spanning several or all pores has been reported from *Raja*, *Gasterosteus* and *Sparus*. Micropinocytotic vesicles are numerous in **agnathan** kidneys and in several marine and euryhaline teleosts. A thin glycocalyx has been demonstrated by staining with Alcian blue in carp. In *Gasterosteus aculeatus*, tight junctions and desmosomes connect endothelial cells.

Mesangial cells

These cells are of irregular shape, frequently stellate with cytoplasmic processes. These cells subserve structural support, contraction and phagocytosis, properties that have been described in lampreys. Mesangial cell processes extend peripherally, occasionally completely around the capillary loops, and insinuate themselves between the glomerular basement membrane and the underlying layer of endothelium in Agnatha, Elasmobranchii, several marine Teleostei such as flounder, and euryhaline Teleostei in sea water (Figure 24.11d). The mesangial cells have peculiar tubular invaginations in Myxinoidea. Collagen fibrils are regularly present in the mesangial matrix that fills the subendothelial space in Agnatha and Elasmobranchii. Microfibrils (tubular structures, 12 to 15 nm thick) have been found in the matrix of agnathans, skates and flounder.

Podocytes

The visceral layer of the Bowman's capsule consists of a highly specialized cell type, the podocytes, and a well-defined basement membrane, the glomerular basement membrane (GBM). The most important specialization of the epithelium is the filtration slit, which represents the extracellular route for the filtration process (Figures 24.9 and 24.10). During differentiation of the epithelium, the simple cuboidal cells undergo several major architectural changes transforming them into the podocyte phenotype. These changes include a profound alteration of the nature of cell–cell adhesion and elaborate cell shape changes involving the three major cytoskeletal components: microfilaments, microtubules and intermediate filaments. The cell–cell junctions (zonula occludens, zonula adherens) are translocated toward the base of the epithelium during differentiation. The typical structure of the junctional complex is lost; in its place a thin extracellular membrane, the slit diaphragm, is formed, which interconnects the cells close to the GBM. The gap between the cells (filtration slit) at the level of the slit diaphragm has a constant width of 30 to 45 nm. In fish, occasionally two or more slit membranes span the filtration slit. Focal tight junctions, adherens junctions and desmosomes have been

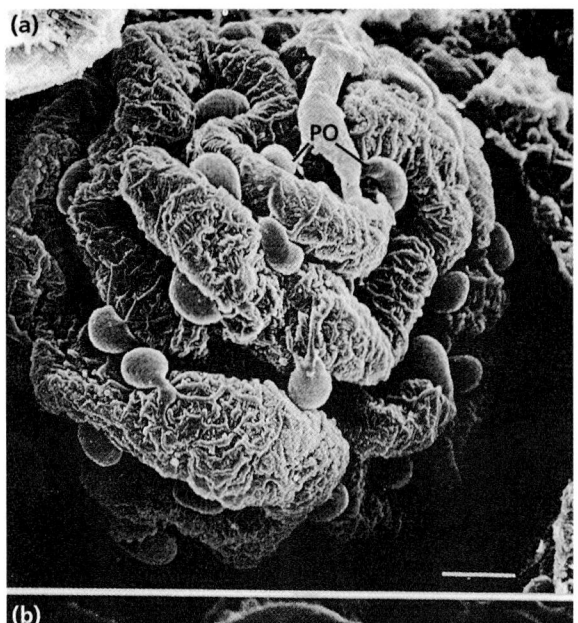

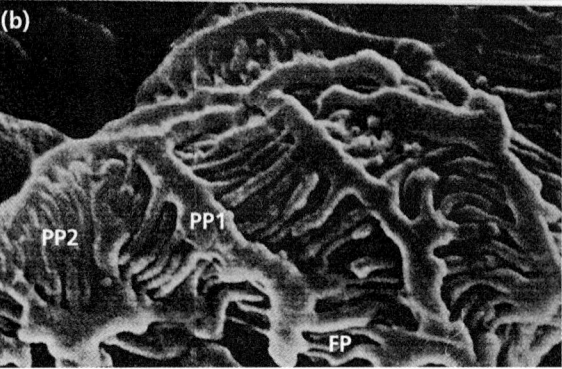

Figure 24.9 Glomerulus of the Prussian carp, *Carassius auratus gibelio*. Scanning electron micrograph of critical-point dried tissue demonstrating the surface architecture of the epithelial layer on the capillary loops. (a) Overview. PO, podocyte cell bodies. Bar = 10 μm. (b) Arrangement of cell processes. The foot processes (FP) of adjacent cells interdigitate between major or primary processes (PP1 and PP2) of the two cells. Note the high degree of interdigitation in this freshwater fish with a high glomerular filtration rate. ×13 400. (From Elger and Hentschel, 1981; in Hentschel and Elger, 1989.)

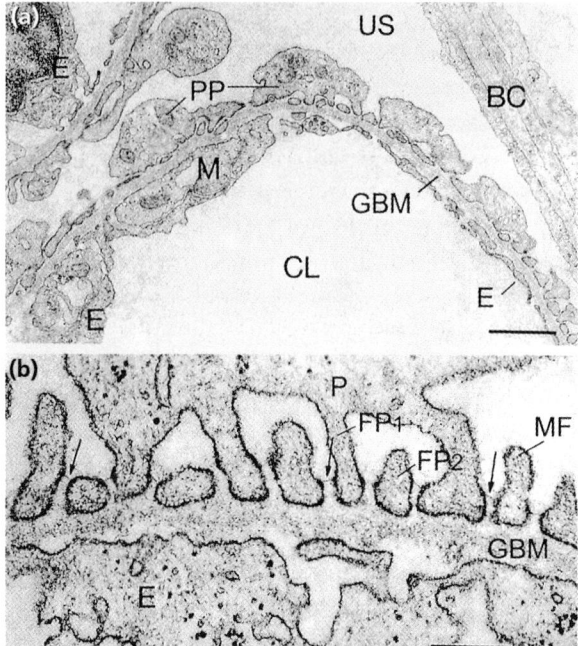

Figure 24.10 Cross-section of peripheral wall of a glomerular capillary of *Carassius auratus gibelio*. (a) The filtration barrier consists of fenestrated endothelial cells (E), glomerular basement membrane (GBM) and epithelial cells of the visceral layer (podocytes). Several primary processes (PP) of podocytes are seen. Cytoplasmic processes of mesangial cells (M) are infrequent in the subendothelial region. BC, parietal epithelium of Bowman's capsule; US, urinary space; CL, capillary lumen. Bar = 1 μm. (b) Filtration barrier at a higher magnification. Foot processes (FP1) originating from an elevated podocyte cell body (P) alternate with foot processes (FP2) deriving from a neighbor cell. The filtration slit between the foot processes of these two cells are cut several times and they are spanned by a slit diaphragm (arrows). The foot processes contain microfilament bundles. Bar = 0.5 μm. (From Elger and Hentschel, 1981; in Hentschel and Elger, 1989).

observed in well-developed glomeruli of hagfish, lampreys, skates, dogfish, *Polypterus*, carp and stickleback. The elongation of the filtration slit to an appropriate length parallel to the GBM occurs by extensive interdigitation of adjacent cells. Two major types of cell processes are formed during interdigitation of the cells: (i) foot processes and (ii) primary processes (Figure 24.9). The foot processes are comprised of relatively short finger-like extensions (up to 1 μm), that are attached to the GBM along their entire length. Their shape is defined by bundles of microfilaments, which run along the cell membrane parallel to the longitudinal axis of the processes, form a bend and continue into the next foot process of the same cell. Foot processes of adjoining cells alternate in a regular manner, lining the meandering filtration slit. In cross-sections through the interdigitating system of foot processes, a regular alternating pattern of the foot processes of the two neighboring cells can be observed (Figure 24.10). The system of interdigitating foot processes can fold, thereby providing further space for an elongation of the filtration slits. Thus, large cell processes, the so-called primary processes, develop, which are contiguous with the GBM and which are lined by foot processes. These cell processes (primary processes) may extend deeply into their neighbor cell. Frequently the cell body of the podocyte is elevated from the GBM, because the interdigitating primary processes with their interdigitating foot processes reach beneath the nuclear region. This extreme amplification of the cell membrane allows almost the entire surface of the capillary tuft (and the mesangium) to be covered by a continuous system of interdigitating foot processes interconnected by an extremely long filtration slit membrane.

Because the slit membrane is considered to be the borderline between the apical and the basal cell membrane, all cell membrane proteins lining the cell processes and the elevated portions of the perikaryon above the level of the slit membrane belong to the luminal domain facing the urinary space. The luminal membrane as well as the slit membrane are covered by a thick alcianophilic glycocalyx in carp. Coated pits abound along the basal cell membrane. The perikaryon contains a conspicuously well-developed Golgi apparatus, free ribosome, profiles of granular and smooth endoplasmic reticulum lysosomes and microfilaments. Numerous microtubules and intermediate filaments extend from the perikaryon into the primary processes.

Glomerular basement membrane (GBM)

The GBM is mainly produced by the podocyte layer and covers the entire capillary tuft. In the more central regions of the glomerular lobules the GBM faces the mesangium. In its pericapillary course the GBM may closely approach the fenestrated endothelium. However, in many species the mesangium (matrix and often also cell processes) invades the dilated subendothelial space (Figure 24.10). Therefore, the GBM and the endothelial basement membrane never fuse, even in mature glomeruli. In these glomeruli a discontinuous endothelial basement membrane may be seen. Major constituents of the GBM are collagen type IV, the glycoproteins laminin and nidogen/entactin, and heparin sulfate proteoglycans. Not all of these proteins, which distinguish the GBM from other basement membranes, have been identified in fish; the expression of distinct changes may reveal evolutionary aspects of GBM development.

Parietal cells

The squamous cells of the parietal epithelium of Bowman's capsule have a simple shape without interdigitation and contain few cell organelles. A single apical cilium may be present. Intracellular bundles of microfilaments run in virtually all directions around the glomerulus.

Aspects of structural–functional correlation of the glomerulus

The glomerulus is the location of primary urine formation by ultrafiltration of the plasma (Yokota *et al.*, 1985). The glomerular capillary wall restricts the passage of plasma proteins and allows rapid filtration of water and small solutes. The amount of filtered fluid depends largely on the available filtration surface determined by the number of glomeruli, size and differentiation of the glomerular tufts, and the total filtration slit area. Other factors determining the glomerular filtration are the colloid osmotic pressure of the blood and the hydraulic permeability of the filtration barrier. The hydrostatic pressure difference between the glomerular capillaries and the urinary space is lower in **ectothermic** species than in mammals. The filtration of macromolecules depends on their size, shape and electrical charge. The fixed negative charges, which are abundant in the glycocalyx of the epithelial cells and endothelial cells (e.g. podocalyxin in the rat) and in the GBM (proteoglycan side chains), account for the charge selectivity of the glomerulus. Anionic sites in the glomerular filtration barrier of fish have been localized by cationic probes in cyclostomes and goldfish (Reale *et al.*, 1981; Elger *et al.*, 1984a).

Quantitative morphological data taking into account intraspecific heterogeneity and interspecific differences are not available in the lower vertebrates. A given fish kidney contains a wide size range of glomeruli. This is the result of ongoing nephrogenesis in the adult kidney. In teleost kidneys, the newest and smallest glomeruli are scattered throughout the kidneys. In several other fishes, including *Polypterus*, marine elasmobranchs and lungfish, new glomeruli are added in a lateral position, whereas the oldest, largest glomeruli are located near the midline of the body, resulting in a zonate appearance for the kidney. Variations in size that are independent of glomerular age occur during adaptation to water of different salinities. A change in the available filtration surface

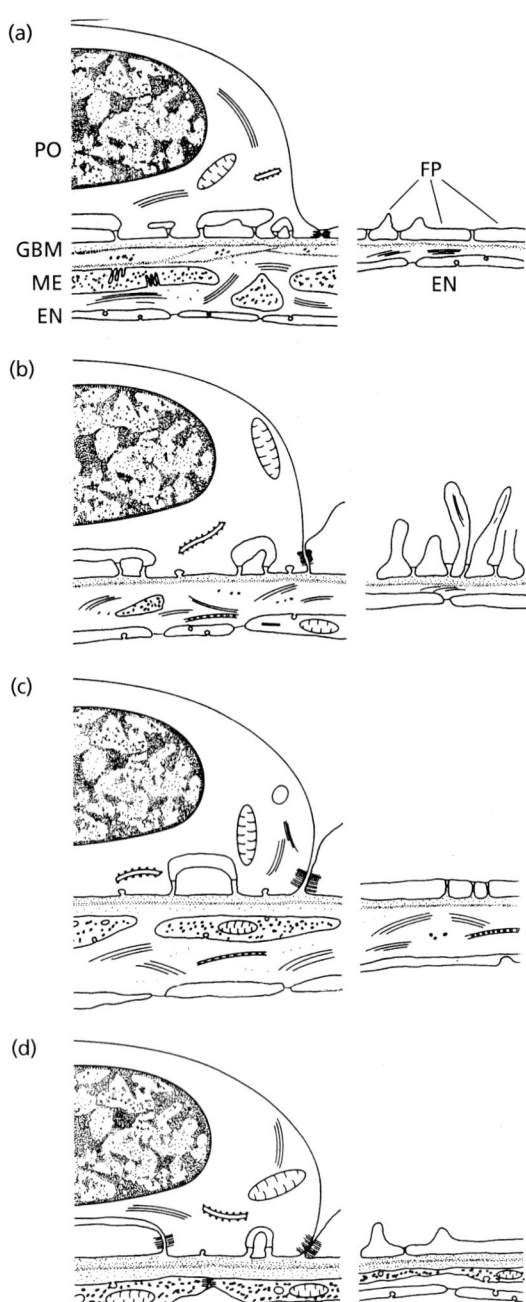

Figure 24.11 Architecture of the filtration barrier. Four types, which were described in lower vertebrates, are schematically drawn. Two different modifications of each type are shown, which are usually present in one single glomerulus. (a) *Myxine* type. The cell body of the podocyte (PO) rests with broad foot processes on the glomerular basement membrane (GBM). The lamina densa frequently splits up into two to three anastomosing layers. Large processes of mesangial cells and mesangial matrix as well as collagen fibers extend into the subendothelial space. The endothelium (EN) is irregularly fenestrated. FP, foot processes; ME, mesangium. (b) *Polypterus* type. Large areas of the barrier show regularly alternating foot processes. The lamina densa is comparatively thin. The filtration barrier of other freshwater fish (lungfish, freshwater teleosts) is very similar. The high degree of interdigitation of the podocytes in this type resembles that of birds and mammals. (c) Elasmobranch type. This type is characterized by a very irregular formation of foot processes and filtration slits. Beneath the lamina densa lies an extended subendothelial space, which contains microfibrils, collagen fibrils (type I) and processes of mesangial cells. (d) Marine teleost type. The basement membrane is rather thick and dense. Processes of mesangial cells may be present as a continuous layer, frequently encompassing the periphery of the capillaries. Foot processes and endothelial fenestrations are poorly developed. (From Hentschel and Elger, 1989, with permission.)

and/or alteration of the properties of the filtration barrier would be of advantage in euryhaline freshwater fish during adaptation to saline water. The hypothesis has been advanced that mesangial cell processes penetrate into the pericapillary region, thereby diminishing the available filtration surface and the hydraulic conductivity. By these modifications in the glomerular capillary wall, the glomerular filtration rate could be lowered (Hickman and Trump, 1969). From the comparative morphology of the filtration barrier it becomes evident that those animals which possess a rather thin basement membrane, extensive interdigitation with a high number of foot processes and filtration slit profiles, rare peripheral mesangium, and a high number of endothelial pores have rather high glomerular filtration rates. This type of glomerular filtration barrier is typically found in freshwater fish (e.g. *Polypterus*, sturgeons, stenohaline freshwater elasmobranchs and teleosts) (Figure 24.11b). Freshwater fish rely on a copious urine flow to eliminate water, resulting in a considerably reduced ultrafiltration coefficient, K_f. The filtration barrier of several marine fish (Figure 24.11a,c) is considerably thickened by an extended mesangium, which is also located in the peripheral capillary wall, and the filtration area is diminished by the smallness of the capillary tuft (marine teleosts such as flounder with less than 50 μm diameter) and the paucity of the filtration slit profiles. Accordingly, physiological studies generally find very low glomerular filtration rates in marine teleosts compared to freshwater teleosts.

Long-term adaptation of freshwater carp to salt water resulted in a 90% reduction in the number of glomeruli (Elger and Hentshel, 1981). In the remaining glomeruli, the ultrafiltration coefficient seems to be lowered by extensive folding of the glomerular basement membrane. This may indicate that glomeruli of freshwater fish degenerate in sea water. For short-term adaptation of the glomerular filtration rate, intermittent function of glomeruli by the action of the arterial preglomerular vasculature appears to be an important mechanism in fish (Elger et al., 1984a). Preglomerular sphincters, in addition to a dense innervation of the arterial smooth muscle layer, have been found in several fish (Elger et al., 1984b).

In summary, the glomeruli are at the center of an extremely flexible ultrafiltration system. Fish are exposed to a host of environmental changes that may disrupt fluid and electrolyte balance. Neural and humoral regulation mediate both long- and short-term adaptations of individual animals. Appropriate changes in glomerular filtration can be affected in the short term by such physiological processes as hemodynamic alterations at the preglomerular vasculature, changing glomerular intermittency, and by changes in filtration slit characteristics. Longer term changes may be accomplished by nephrogenesis and glomerular recruitment.

Renal tubule

Neck segment

The neck segment is a characteristic feature of the renal tubule in lampreys, elasmobranchs, polypterids, holosteans, several teleosts and lungfish (Hentschel and Elger, 1989). The epithelium of the neck segment consists of columnar to cuboidal cells with multiple cilia extending into the tubular lumen (about 30 per cell in *Polypterus*) (Figure 24.12). These cilia have the commonly found arrangement of 9 + 2 microtubular structures. The cytoplasm of the multiciliary cells contains the striated rootlets of the **kinocilia**, many mitochondria, a well-developed Golgi complex, lipid droplets, lysosomes and glycogen granules. The apical cell membrane may have a few short microvilli. The basal and lateral cell membranes usually have no infoldings or interdigitations. Several species have no distinct neck segment; only a few cells that differ from those of the adjoining segments are found in the corresponding area of the tubules of the sterlet and of several teleosts including goldfish, rainbow trout, sticklebacks and the cotttids. These cells may contain apical mucous vacuoles or may appear largely undifferentiated.

The multiciliary cell, although typical of the neck segment, is not restricted to this segment. Single multiciliary cells are present in the brush-border segments of the proximal tubule in bony fish (carp, zebrafish, trout, flounder, killifish and lungfish); clusters of these cells are found in the proximal tubule of elasmobranchs, and in the intermediate segment of elasmobranchs, polypterids and lungfish (see below). The so-called neck segment of *Myxine* consists of proximal segment I cells, and, hence, it is not comparable to the neck segments of other species.

Functional considerations

Cells similar to the multiciliary cells of the neck segment are found in the renal organs of invertebrates and function in the axial propulsion of fluid in the tubule; this suggests a similar function for these cells in the neck segment of fishes. In addition, transepithelial transport of solutes and fluid may occur in the exceptionally long neck segment of marine elasmobranchs. Net reabsorption of urea has been observed in isolated perfused ciliated segments of the lateral bundle, presumably the neck segment. This mechanism *in vivo* could shift urea from the neck segment to the adjacent PIa. Countercurrent multiplication could create a urea-free zone at the tip of the hairpin loop. Urea analyses in various zones of the kidney of the skate *Raja erinacea* are consistent with this mechanism.

Proximal tubule

General remarks

A luminal brush border is universally present in the proximal tubule. The only other renal structures with similar dense and regular brush borders are the archinephric duct of Myxinidae and the most cranial portion of the archinephric duct in teleosts. A convoluted portion (pars convoluta) and a straight portion (pars recta) of the proximal tubule in lampreys and lungfish have been described. It appears that in most Osteichthyes the entire proximal tubule is convoluted. In marine elasmobranchs, the initial portion of the proximal tubule runs countercurrent to the neck segment (first hairpin loop) in the lateral bundles. The following portions are convoluted in the region of the glomeruli and thereafter in the mesial zone. Two major portions of the proximal tubule can be discerned in elasmobranchs (*Scyliorhinus caniculus*, *Raja erinacea*) (Figure 24.5), Chondrostei (e.g.

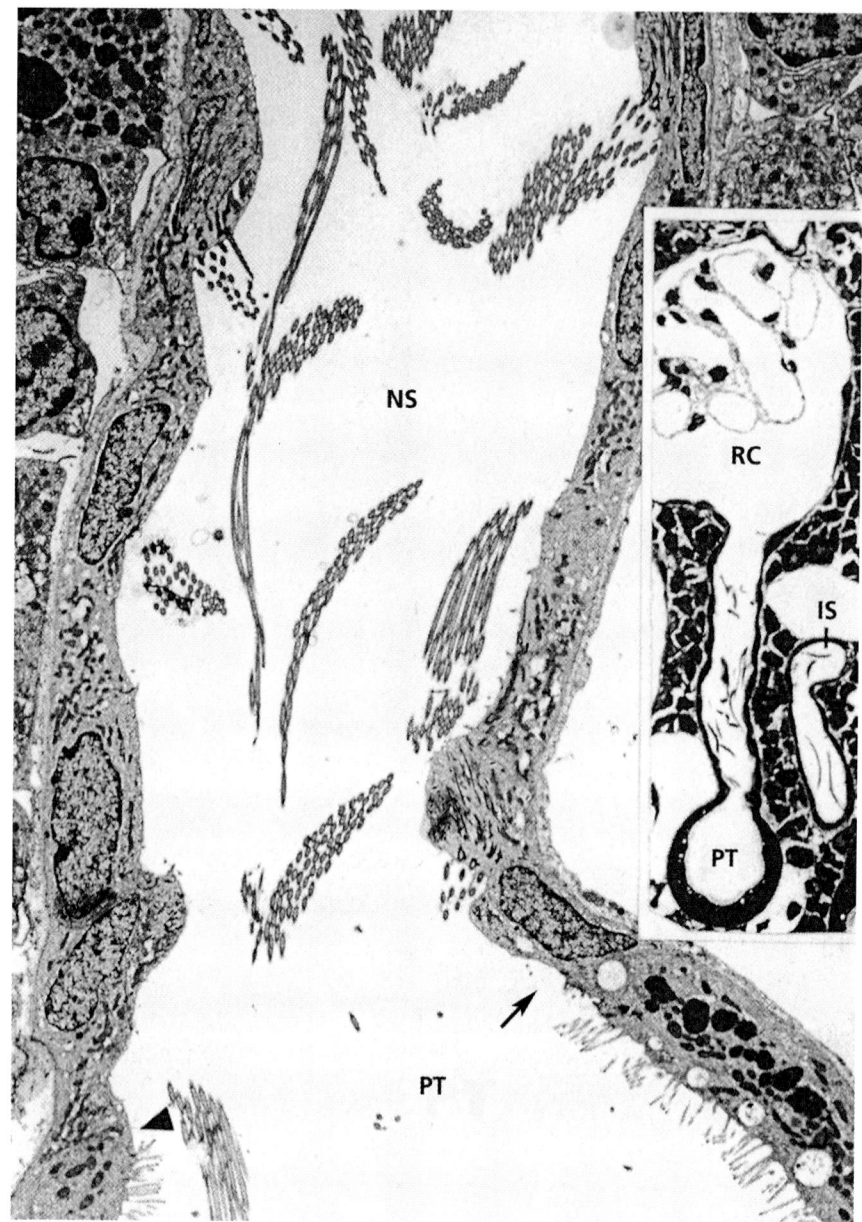

Figure 24.12 Neck segment of the archaic bony fish, *Polypterus senegalus*. The neck segment (NS) consists of multiciliary cells, from which flames of cilia extend into the lumen. The abrupt transition to the proximal tubule (PT) is shown (arrowhead). Inset: Light micrograph (LM) showing an overview of the renal corpuscle (RC), the neck segment and the proximal tubule (PT). The ciliated intermediate segment (IS) runs in a direction opposite to the NS. TEM, ×4200; inset: Semithin section; LM, ×350. (From Hentschel and Elger, 1989, with permission.)

Acipenser ruthenus), Holostei (*Amia calva*) and glomerular teleosts (Figure 24.3). These are termed proximal tubule segment I (PI) and proximal tubule segment II (PII); the distinction is based on morphological differences and tracer studies with peroxidase (Hentschel and Elger, 1989). PI epithelium has an apical endocytic apparatus connected to the elaborate lysosomal system. This morphological indicator of protein reabsorption is lacking in PII. With reference to this definition, the proximal tubule of lampreys and polypterids can be regarded as belonging to PI in its entirety. The proximal tubule (principal segment) of aglomerular teleosts belongs to PII.

The end of the proximal tubule is marked by the abrupt disappearance of the brush border, irrespective of whether the subsequent segment is an intermediate segment (in polypterids and elasmobranchs), a distal tubule (most freshwater teleosts), or a collecting tubule (most marine teleosts). A peculiar renal organization is present in the hagfish, the most primitive of recent vertebrates, where the epithelium of the archinephric duct has morphological and functional

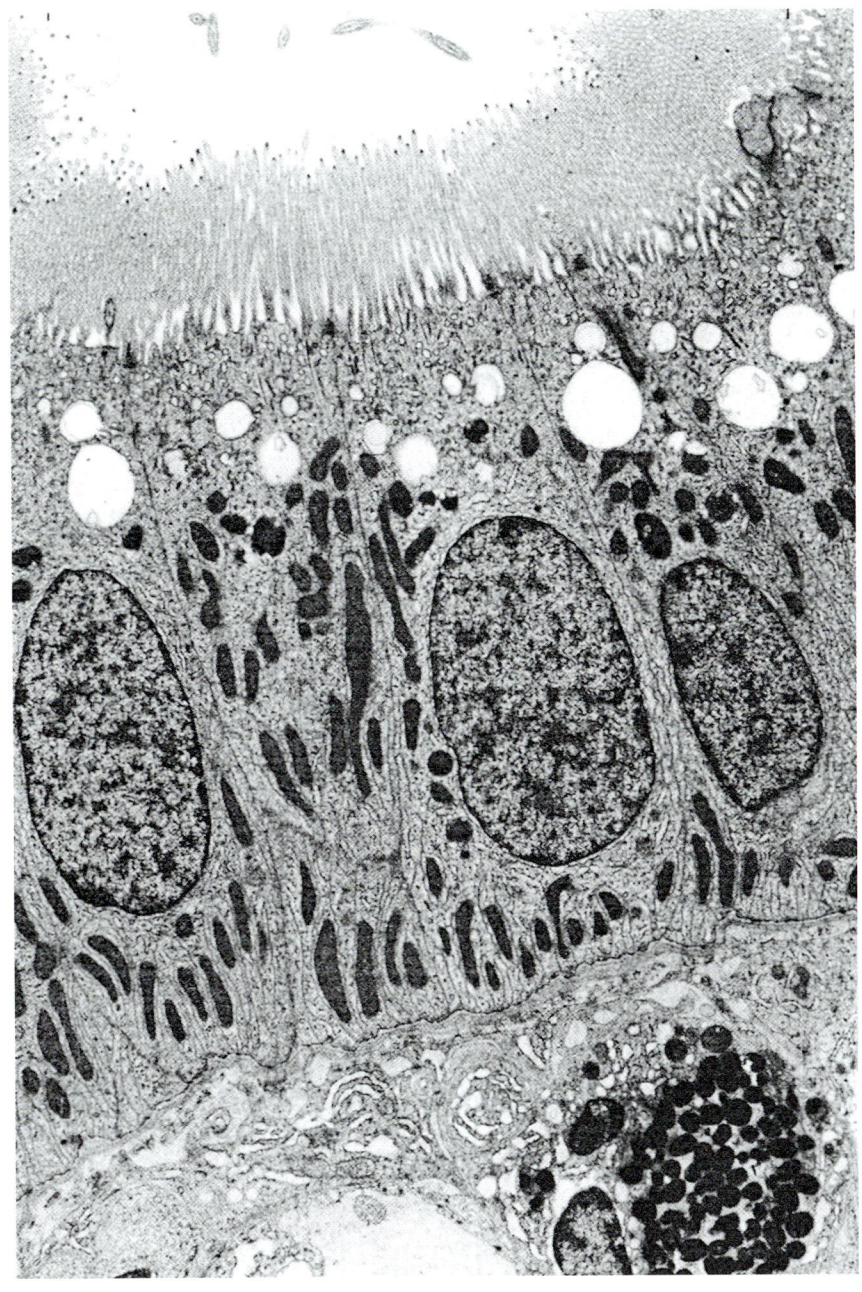

Figure 24.13 Proximal tubule segment PI of the rainbow trout, *Salmo gairdneri*. The apex displays a dense brush border. The lateral cell membranes are straight, which is characteristic for teleost renal tubule cells. Amplifications of the basal and lateral cell membrane are provided by invaginations, which are associated with numerous mitochondria. The apical endocytic apparatus consists of many small vesicles. Several large vacuoles representing early endosomes are present. TEM, ×8000. (From Hentschel and Elger, 1989, with permission.)

analogy with PI. An extended apical endocytic apparatus has also been demonstrated.

Proximal tubule segment I

This portion of the proximal nephron is present in all vertebrates from the Agnatha to the Mammalia. It is only lacking in the kidneys of aglomerular teleosts. In the PI of all species, a brush border of microvilli is typically present on the apical surface of the epithelial cells (PI cells) (Figure 24.13). Very small spindle-shaped cells without a brush border are intercalated between the PI cells in many teleosts. They may be amoeboid. In addition, single multiciliary cells are interspersed with PI cells. Thus, up to three cell types may be present in this segment and, similarly, in PII.

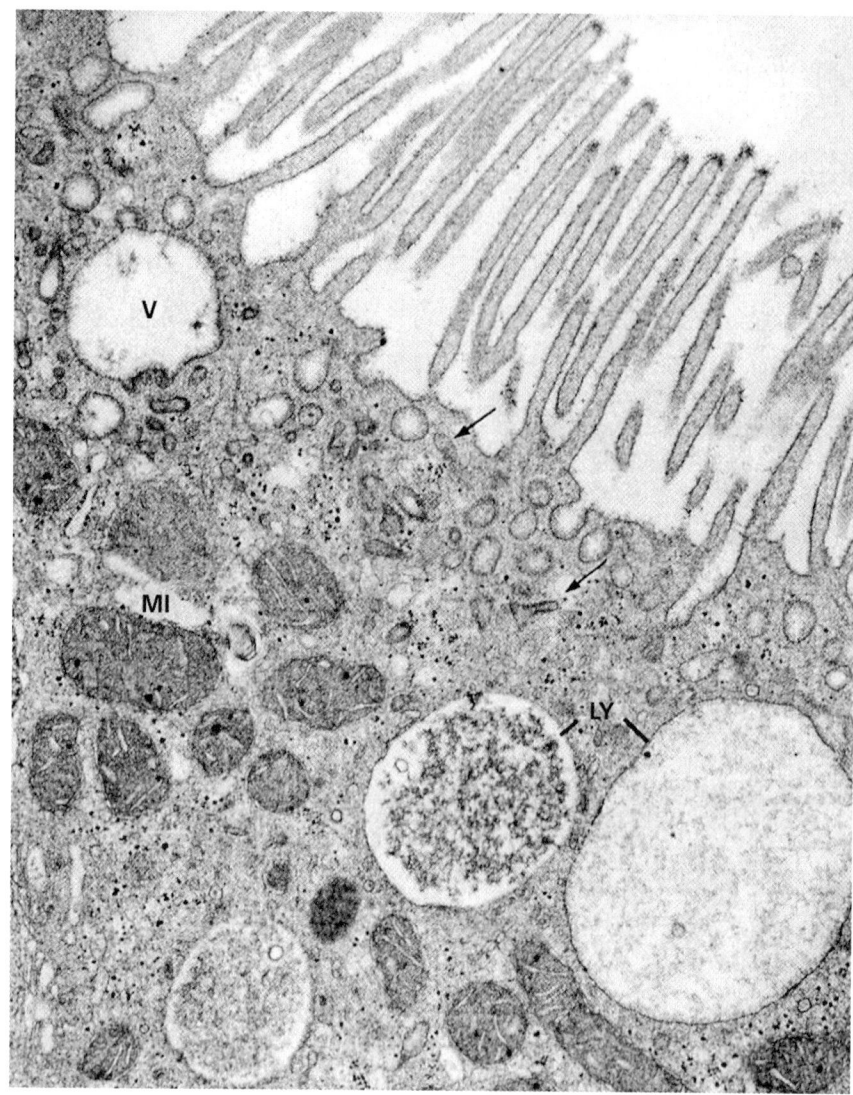

Figure 24.14 Apical region of proximal segment PI cell of *Polypterus senegalus*. The apical cytoplasm contains endocytic vesicles, dense tubule (arrows) and early endosomes (V). LY, lysosomes; MI, mitochondria. TEM, ×24 000. (From Hentschel and Elger, 1989, with permission.)

The epithelium of PI is comparatively uniform in teleosts such as trout, carp, sticklebacks and flounder. In contrast, in lampreys, elasmobranchs and polypterids a gradual transition between the cells of early and late portions of PI is observed. The length and density of the microvilli exhibit considerable differences in the species with a long PI. Generally, the early portion has longer and more regular microvilli than the late portion. In marine elasmobranchs, the microvilli are rather small and irregular in the first portion of the PI (PIa) which is located in the lateral countercurrent bundles. A brush border of tall microvilli is present on the epithelial cells of PIb outside the bundles. The microvilli are covered with a glycocalyx. An axial bundle of actin-like filaments has been observed in actinopterygid fish such as *Amia calva*, carp and flounder. Microtubules have been described in the microvilli of sticklebacks. The apical cell membranes of the PI cells are characterized by numerous clathrin-coated pits between the microvilli (Figure 24.14). The apical cytoplasm contains small round vesicles, dense tubules and vacuoles. These organelles, corresponding to endosomes, and the coated pits are involved in endocytosis and membrane recycling. The morphological equivalent of the apical endocytic system is present in all glomerular kidneys described.

A great variety of membrane-bound inclusions have been described which are similar to peroxisomes. Most of the large inclusions appear to be secondary lysosomes. Besides the large basal nucleus, these columnar cells have a well-developed Golgi complex, large mitochondria, cisternae of the rough and smooth endoplasmic reticulum, and free ribosomes. Large mitochondria frequently lie in the vicinity of the basolateral amplifications (interdigitations, invaginations). In the apical zone, the cells are connected by junctional complexes with a zonula occludens (tight junctions) and a zonula adherens. Desmosomes are frequently present. Unlike those found in the mammalian proximal tubule, the tight junctions of the proximal tubule are rather deep in all species of lower vertebrates including fish. In trout, for instance, they consist of three to seven anastomosing plus two to six parallel strands. Gap junctions indicate electrical coupling of PI cells comparable to mammalian proximal tubules (Elger and Hentschel, 1986). In elasmobranchs, the lateral cell membranes of adjacent cells enclose a narrow intercellular space of nearly constant width. Extensive folding of these parallel lateral cell membranes results in lateral interdigitation. In teleosts, the polygonal columnar cells exhibit a rather straight lateral cell membrane. The intercellular space is narrow. Instead of a basal labyrinth of cellular interdigitations, an intracellular system of branched membranes lining flat saccular spaces is formed. These spaces are connected with the exterior by rows of small oval pores located in the basal and lateral parts of the cell membranes. Thus, the membranous intracellular system is an invaginated branched system of the basolateral cell membrane (Figure 24.13) (see also the distal tubule below).

In the lungfish, the lateral intercellular spaces are organized differently. They are narrow beneath the junctional complex, becoming progressively wider in the direction of the cell base. Partial interdigitation of adjacent cells is observed by the presence of small cytoplasmic folds projecting into the enlarged intercellular space. The lateral intercellular space is continuous with a basal extracellular labyrinth.

Proximal tubule segment II

A second major part of the proximal tubule can be discerned in the Elasmobranchiomorphii and the Actinopterygii. Marked differences in the fine structure of the epithelial cells of this segment (PII cells) compared to the PI cells and the proximal tubule cells of other vertebrates suggest that it may have no homologous counterpart in other vertebrate classes (Hentschel and Elger, 1989).

Elasmobranchiomorphii

The PII follows its highly convoluted course exclusively in the mesial zone of dogfish, such as *Scyliorhinus caniculus*, and marine skates, such as *Raja erinacea* (Figure 24.5). This kidney zone is irrigated mainly with venous blood of the portal system that runs in the extended sinus capillaries between the tubules. The epithelium consists of very large, high columnar cells that become progressively smaller in the direction of the intermediate segment. Multiciliary cells are intermingled singly, or in small groups, especially at the end of the segment. The apical cell membrane of PII cells is greatly amplified by a dense brush border of tall and slender microvilli (Figure 24.17). Beneath the brush border lies a cytoplasmic zone with numerous small clear vesicles. Mitochondria are lacking in this apical zone. The nucleus is located centrally. Large mitochondria are packed densely in the perinuclear region. The lateral cell membranes are rather straight in *Scyliorhinus* with occasional folds that provide some interdigitation of adjacent cells. In the skate the intercellular

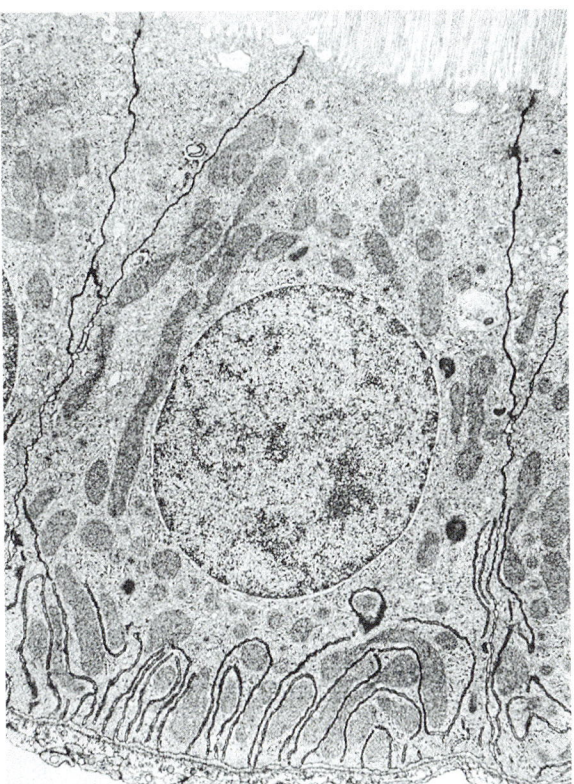

Figure 24.15 Proximal segment PII of *Salmo gairdneri*. The basal cell membrane is amplified by numerous laminate invaginations, which are filled by material that stains densely with tannic acid. TEM, ×4000. (From Hentschel and Elger, 1989, with permission.)

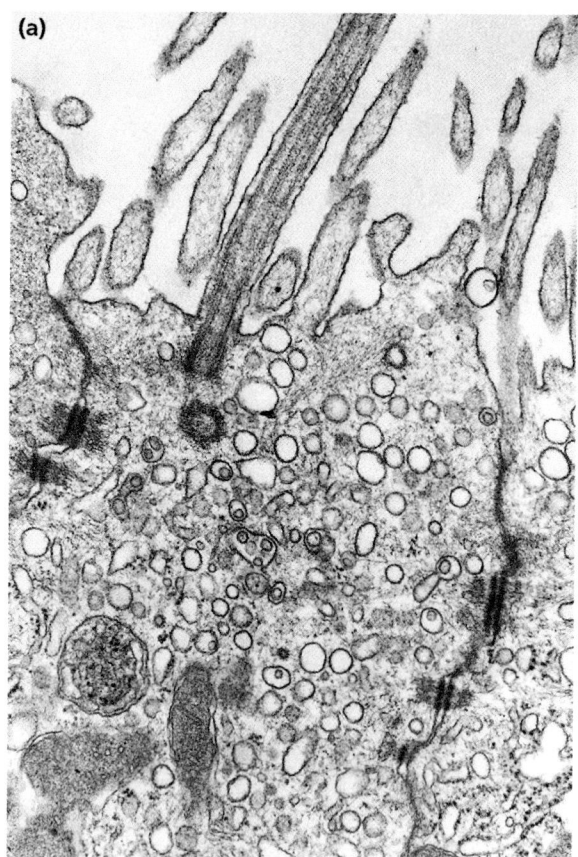

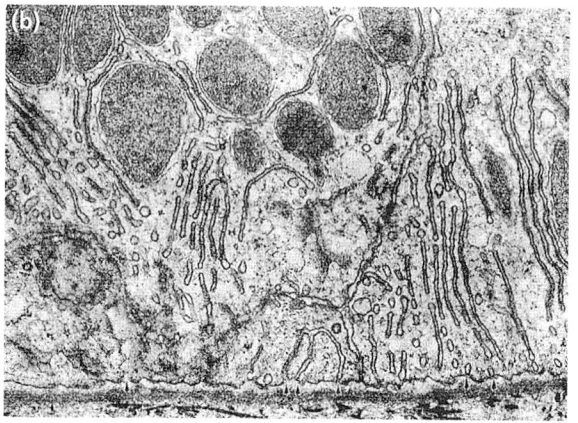

Figure 24.16 Proximal segment PII cell of the winter flounder, *Pleuronectes americanus*. (a) Apical region of PII cell. Smooth clear vesicles abound in the zone beneath the brush border, which consists of long slender and loosely arranged microvilli. A single cilium projects from the brush border cell. Mitochondria display only few cristae in a dense matrix, thereby differing largely from mitochondria in flounder PI cells. Endocytic organelles, such as dense tubules and large early endosomes, which are characteristically found in PI cells, are lacking in PII. The tight junction (zonula occludens) displays a large apicobasal depth. TEM, ×32 000 (from Elger *et al.* 1998, with permission). (b) Basal region of PII cell. Numerous infoldings of the basolateral cell membrane associated with mitochondria are present. These folds have, at their origin (arrowheads) from the peripheral straight basolateral cell membrane, a tubular appearance. TEM, ×29 000 (from Elger *et al.*, 1998, with permission).

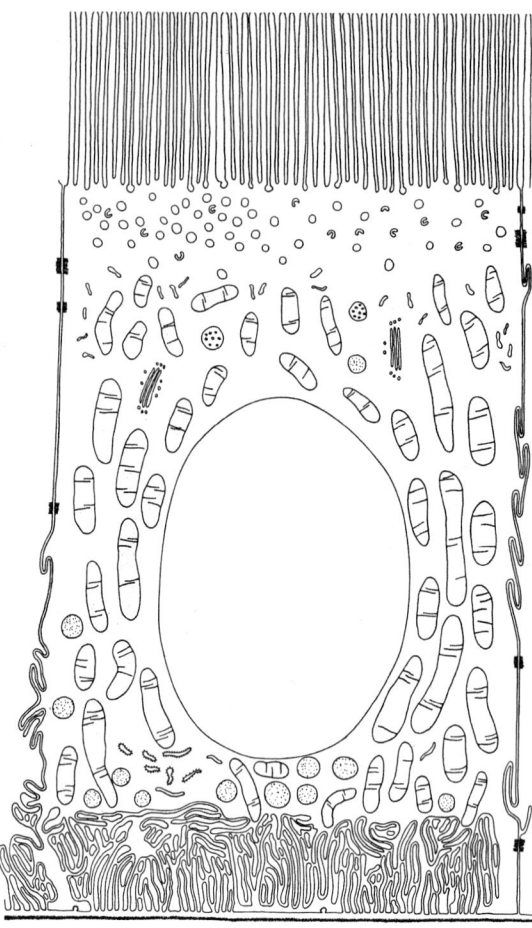

Figure 24.17 Schematic drawing of proximal segment PII cell of a marine elasmobranch, the dogfish *Scyliorhinus caniculus*. The high brush border consists of long regular microvilli. The basal cell membrane forms an extended basal labyrinth of densely packed meandering folds without mitochondria in between. At variance to the PI the apical cytoplasmic zone in PII contains numerous small vesicles with a smooth surface. (From Hentschel and Elger, 1989, with permission.)

space is narrow in the region beneath the apical junctional complex and widens in the direction of the cell base. The lateral cell membranes are greatly enlarged by meandering folds that protrude into the intercellular space. The basal region of the epithelium contains an elaborate basal labyrinth. The densely packed folds of the labyrinth and the lateral amplifications do not contain mitochondria. The cytoplasm of the cell body is complemented by a Golgi system, small cisternae of the smooth and rough endoplasmic reticulum and lysosome-like bodies.

Actinopterygii

The transition from PI to PII occurs abruptly in several teleosts, such as trout, sticklebacks and killifish, *Fundulus heteroclitus*. In other species, such as plaice, *Pleuronectes platessa*, the transition is gradual. It appears that the PII is the longest segment within the teleost nephron. The epithelial cells are generally tall columnar with a centrally located nucleus (Figure 24.15). The microvilli of the brush border are frequently shorter and fewer at the end of the PII. In the winter flounder PII cells often bear only a few long and slender microvilli, which may predominate near the lateral cell margin. The apical cytoplasm of PII cells is generally filled by small vesicles, which may be round (flounder) (Figure 24.16a) or saccular (trout). The zonula occludens is made up of a broad band of anastomosing strands as seen in the trout. These tight junctions are impermeable to lanthanum as observed in the plaice (Ottosen 1978). A few desmosomes can be found. Numerous gap junctional plaques provide electrical coupling of the PII cells. Lateral and basal cell membranes are amplified by a system of intracellular saccular invaginations as in PI cells (Figure 24.16b). Large mitochondria are found adjacent to the basolateral invaginations as well as throughout the cell, except for the apical region. Similar to PI, PII also contains interspersed multiciliary cells and migrating cells.

Sexual dimorphism

Profound differences in the PII in relation to the sexual cycle of males have been well-documented in several teleost species. The structure of the epithelium of PII in sticklebacks, for example, is influenced by testosterone and probably by other male sexual hormones. During the reproductive cycle of *Spinachia spinachia* and other sticklebacks, the cells grow conspicuously and the cytoplasm is filled with secretory granules that contain a neutral glycoprotein. The Golgi complex is very elaborate at this stage and the cisternae of the rough endoplasmic reticulum are greatly enlarged and filled with moderately dense material. Mitochondria are scarce and basolateral invaginations are lacking. The nucleus is large and may contain several densely staining nucleoli. After the breeding season these nephrons degenerate and are partly replaced by newly differentiating tubules. Sexual dimorphism in the end portion of the proximal tubule has also been reported in males of a lungfish species with nest-building behavior.

Aglomerular teleosts

Electron-microscopic histology of kidneys from several aglomerular teleost species indicates that the principal portion of their renal tubule is homologous to the PII of other teleosts. An apical endocytic apparatus, similar to that in the proximal tubule of other vertebrates, is lacking in aglomerular species; this suggests that the PI vesicles function in the reabsorption of filtered proteins. The apical cytoplasmic zone contains an abundance of smooth-surfaced vesicular or saccular elements. This zone resembles the apical cytoplasmic zone of the PII cells in glomerular teleosts. Structural differences exist among the species with respect to the density of basolateral amplifications and associated mitochondria, the height and number of microvilli in the brush border, and the number of lysosomes and multivesicular bodies.

Functional considerations of the proximal tubule

The proximal tubule is the principal site of reabsorption of filtrate. Reabsorption of low-molecular-weight proteins (less than 40 kD) takes place by endocytosis into epithelial cells in glomerular kidneys (Youson, 1981; Ottosen and Maunsbach, 1973). Several sugars and most amino acids are reabsorbed via sodium-dependent transport mechanisms in the luminal membrane. For example, studies with brush border membrane vesicles revealed a sodium-dependent transport system for D-glucose in the archinephric duct of the hagfish and the proximal tubule of flounder and trout (Freire *et al.*, 1995; Kanli and Terreros, 1997). Brush border membrane vesicles of flounder and of the aglomerular toadfish *Opsanus tau* (PII) translocate L-alanine and L-glutamic acid in a sodium-dependent manner (Eveloff *et al.*, 1980; Wolff *et al.*, 1997). A characteristic feature of the brush border is the high activity of alkaline phosphatase, which was demonstrated in the archinephric duct of the Atlantic hagfish and in the proximal tubule of fish, such as dogfish and carp (for literature see Hentschel and Elger, 1989).

Secretion of ions and organic substances also takes place mainly in the proximal tubule (Beyenbach and Baustian, 1989; Beyenbach and Liu, 1996; Miller et al., 1997; Renfro, 1997). Secretion is the only mechanism available for the initial formation of urine in aglomerular fishes. In the aglomerular fish, sinusoid capillaries of the renal portal system, the only peritubular circulation, surround the early portions of the renal tubule (PII). Hence urine production clearly depends on transepithelial transport of substances and fluid from the venous portal system to the tubular lumen of the PII. Secretion of divalent ions in the PII has been shown in marine teleosts and elasmobranchs by micropuncture and in vitro (Renfro, 1980; Stolte et al., 1977; Beyenbach and Frömter, 1985). These data, and the absence of an endocytic apparatus in the PII, indicate that the apical vesicular apparatus of PII cells in glomerular and aglomerular teleosts and elasmobranchs is functionally different from the vesicular compartment in the PI and that the PII vesicles are involved in secretory processes.

The length of the PI and PII in glomerular teleosts varies considerably between the species. In stenohaline freshwater teleosts such as goldfish and zebrafish, sections of PI segments are seen with high frequency in histological samples, whereas in the nearly stenohaline marine winter flounder more than 90% of the proximal tubule consists of the secretory PII (Elger et al., 1998) (Figure 24.3). A rudimentary PI segment is often associated with very few and very small glomeruli suggesting that the length of the PI is related to the function of the glomeruli. Killifish tubules and flounder primary cultures are extensively used to study important transport processes such as organic anion secretion (Miller and Pritchard, 1994) and P-glycoprotein mediated secretion of substances, e.g. xenobiotics (Miller, Sussman, Renfro, 1998, Schramm et al., 1995). **P-glycoprotein**, which was localized to the brush border of flounder proximal tubule cells (Sussman-Turner and Renfro, 1995), is a highly conserved integral membrane protein functioning as an energy-dependent efflux pump for certain lipophilic aromatic compounds entering the cell by diffusion.

Recently, a sodium-phosphate (P_i) cotransport system was cloned from winter flounder and localized mainly to the basolateral region of PII cells (Werner et al., 1994; Elger et al., 1998). Unlike the mammalian proximal tubule, teleost kidneys may either reabsorb filtered P_i or secrete excess P_i into the urine (Gupta and Renfro, 1991; Lu et al., 1994, Renfro 1997). This localization suggests that PII is the major site for secretion of phosphate.

Intermediate segment

Multiciliary cells are typically present in the well-defined intermediate segment of archaic freshwater fish, such as *Polypterus senegalus* and *Erpetoichthys calabaricus*, and in several lungfish species (Figure 24.12). In these species, the intermediate segment is clearly different from the end of the proximal tubule and the beginning of the distal tubule. Furthermore, the intermediate segment greatly resembles the neck segment. Both segments can be observed in the same region, between the dorsal and ventral zone of the zonate kidneys. However, they run in opposite directions.

The intermediate segment of lampreys contains no multiciliary cells, although such cells are found in the neck segment. Between the columnar cells of the proximal tubule and the cuboidal to low columnar cells of the distal tubule is a section of tubule composed of flattened cells.

Elasmobranchs have an intermediate segment with multiciliary cells comparable to that of archaic bony fishes (Figure 24.5). The neck segment and intermediate segment in the complex kidneys of skates and dogfish resemble one another histologically. In the past this has frequently been a source of confusion. The intermediate segment begins in the mesial tissue at the end of PII and runs into the lateral bundles. Here it is connected rather abruptly with the first portion of the distal tubule.

In teleosts, the intermediate segment is generally limited to a short section of tubule with a mixture of cell types, rather than a distinct segment. The species studied to date with electron microscopy have generally had a mosaic of brush border cells and multiciliary cells at the terminal end of PII. Antarctic aglomerular teleosts of the family Nototheniidae have a short transitional zone between PII and the collecting tubule-collecting duct system containing multiciliary cells intermingled with mucous cells. A distinct intermediate segment is also lacking in the sturgeon and the bowfin. As a general rule, multiciliary cells have not been found in the epithelium of tubular portions distal to the end of the intermediate segment of the proximal tubule.

Functional considerations

Presumably, the multiciliary cells of the intermediate segment assist in the propulsion of fluid through the renal tubule, as has been suggested in the neck segment and brush border. In addition, the movement of the cilia may keep the tubular fluid well mixed and, hence, prevent the development of local areas of high salt concentration. Particularly in marine teleosts, which combine a very low urine flow with the secretion of divalent cations in PII, such mixing could be an important mechanism in the prevention of stone formation.

Distal tubule

The distal tubule consists of at least two portions that are referred to as early distal tubule which is homologous to the early distal tubule (EDT) of Amphibia and the mammalian thick ascending limb of Henle, and the late distal tubule. These portions are characteristically present in the generalized vertebrate nephron (a multisegmental nephron) in archaic bony fish (*Polypterus senegalus, Protopterus dolloi, Amia calva*). In higher teleosts the distal nephron and collecting duct system consist of a reduced number of segments and cell types (see below). Based on the study of marine teleosts as compared with freshwater teleosts the distal tubule was thought to be absent in the former. In fact, this segment is present in many marine species including marine lampreys, elasmobranchs, sturgeons and several marine teleosts, e.g. the catfish *Plotosus lineatus*. Conversely, distal segments are lacking in several freshwater teleosts, such as the stickleback and the bullhead, *Cottus gobio*. Aglomerular teleost species also lack a distal segment (Figure 24.2).

Early distal tubule

The beginning of the EDT is marked by an abrupt change of the epithelium from the preceding segment (intermediate tubule or PII). The cells of the EDT in fishes are structurally similar to those of the EDT of amphibians, reptiles, birds and mammals (Hentschel and Elger, 1987). Electron microscopy has revealed that the EDT consists of one epithelial cell type with the following characteristics (Figure 24.18):

1. The basolateral cell membrane displays elaborate amplification, which frequently extends almost to the apical cell membrane. The mode of membrane amplification is characteristic of the respective class of vertebrates (see below).
2. Numerous large mitochondria are arranged between or in the vicinity of the basolateral amplifications.
3. The apical cell membrane in most cases has only a few short and stubby microvilli.
4. Round or elongated vesicles are found in a narrow apical cytoplasmic zone.
5. The zonula occludens of the apical junctional complex is narrow and consists of tightly packed parallel strands, unlike those of the subsequent segments of the distal nephron and the CT/CD system (Elger and Hentschel, 1986; Hentschel et al., 1993).
6. Apical interdigitations and, therefore, elongations of the zonula occludens are usually seen in species which display basolateral interdigitations (Figure 24.19).

Two major types of amplification of basolateral cell membranes can be discerned in the EDT of different vertebrate classes, namely (i) basolateral invaginations and (ii) basolateral or lateral interdigitation.

Basolateral invaginations

In lampreys, sturgeons and teleosts, the EDT cells, like the proximal tubule cells, are polygonal, and the intercellular spaces are narrow and of equal width. In the lampreys, an intracellular dense, highly branched tubular system is in continuity with the basolateral cell membrane. The profiles of the tubular system have a constant diameter. In sturgeons and teleosts, tubular and laminate invaginations extend into the cell body and are connected with the straight outer basolateral cell membrane by short tubular portions. The extracellular space of the invaginations is narrow. A structural specialization of the basolateral membrane portions was found in teleosts and lampreys in thin sections and freeze-fracture replicas: a thick extracellular coat, and a special arrangement and high number of intramembrane particles were found in the invaginated membrane portions, compared to the straight membranes that face the intercellular space (Wendelaar Bonga and Veenhuis, 1974).

Basolateral interdigitation

In the EDT of elasmobranchs and archaic bony fishes, there are basolateral or lateral interdigitations between adjacent cells. The structure and location of these interdigitations varies from species to species. In the elasmobranch EDT, the basal cell membrane is

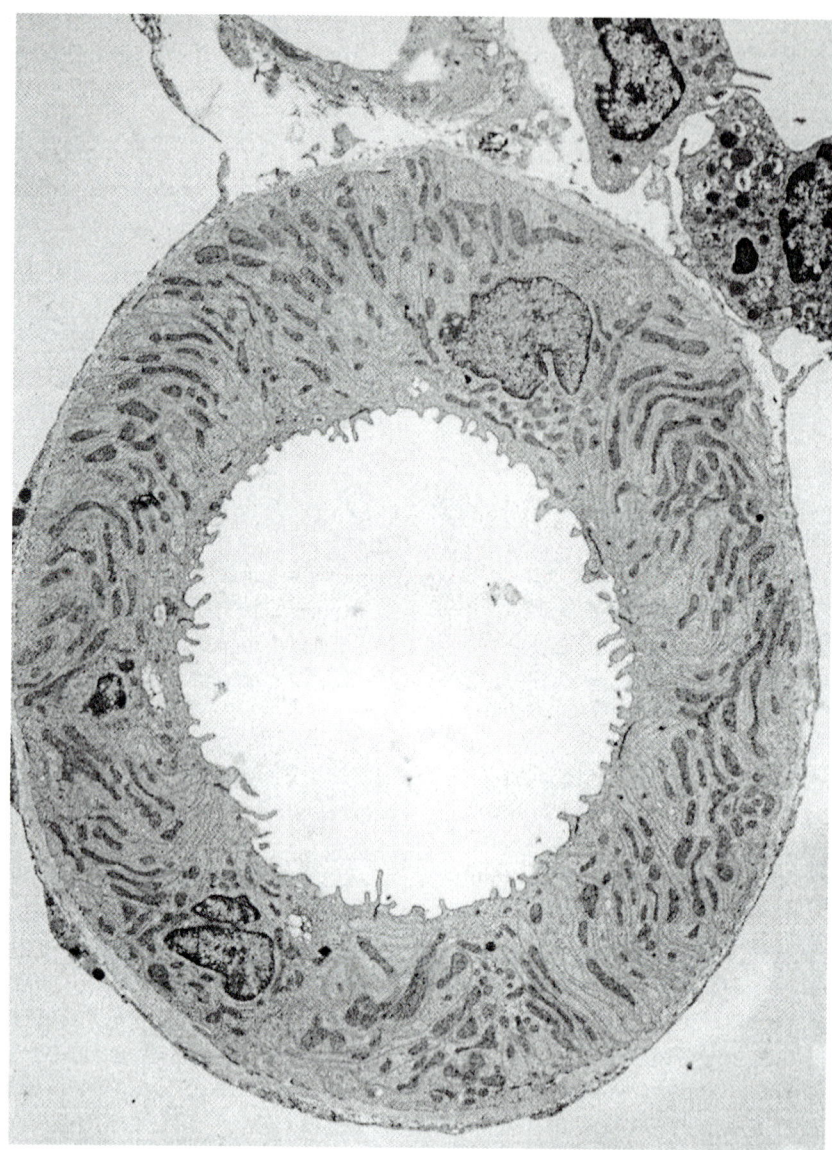

Figure 24.18 Early distal tubule (distal segment) of *Polypterus senegalus*. The epithelium consists of one cell type. The apical cell membrane shows a few short microvilli. Elaborate interdigitations of the basolateral cell membranes are closely associated with mitochondria. TEM, ×4200. (From Hentschel and Elger, 1989, with permission.)

usually smooth. Shallow folds cover the lateral cell membrane and interdigitate with those of neighboring cells. The intercellular space is narrow and uniform in width. A cytoplasmic zone adjacent to but not within these folds contains the mitochondria. The basal cell membrane is usually smooth. In *Polypterus*, cell processes which are oriented perpendicularly to the cell base and extend from the luminal surface to the base of the epithelium interdigitate with those of neighboring cells. (Figure 24.18). The intercellular space is narrow. In addition, laminate invaginations of the basal plasma membrane may extend to below the apical cell surface. The elongated mitochondria are located within the interlocking cell processes. In *Protopterus*, the overall course of the basolateral cell membrane is similar to that of *Polypterus*. However, the intercellular spaces are more or less distended and the lateral cell membrane is further enlarged by microplicae and microvilli devoid of mitochondria.

Functional considerations

The EDT (distal segment) is a diluting segment, in which sodium chloride is reabsorbed from the tubular lumen (driven by the basolateral Na,K,ATPase) and

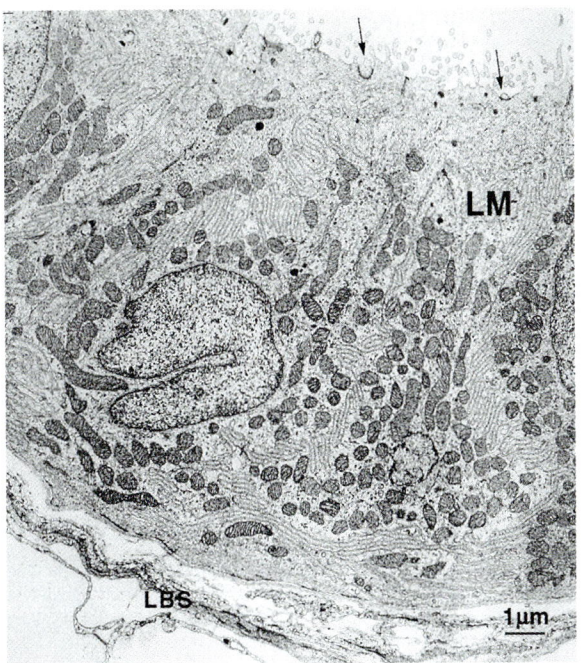

Figure 24.19 Early distal tubule cell of *Scyliorhinus caniculus*, located in the countercurrent system of the lateral bundle. Lateral (LM) and basal membranes are extensively folded. Several bent profiles of the tight junctional belt are seen (arrows), which, due to apical interdigitation of the lateral cell borders, is linearly elongated around the cell. LBS, lateral bundle sheath. TEM, Bar = 1 µm. (From Hentschel and Elger, 1989, with permission.)

water is left behind, as was shown in lampreys (Logan et al., 1980b), marine elasmobranchs (Friedman and Hebert, 1990; Hebert and Friedman, 1990) and teleosts (trout) (Nishimura et al., 1983). The epithelium displays a lumen-positive potential difference. The Na–K–Cl cotransport protein of the shark rectal gland (sNKCCl) mediating cellular uptake of NaCl is localized to the apical plasma membrane of the EDT in the countercurrent bundle of this marine elasmobranch (Biemesderfer et al., 1996). As there is no need to prevent any excretory loss of NaCl in their hyperosmolar environment, the EDT is lacking in the axial nephron of most but not all marine teleosts (see above). In the nephrons with countercurrent arrangement of distal nephron segments and parts of the CT/CD system, i.e. in marine elasmobranchs, the diluting function appears to be involved in urea retention. Reabsorbed NaCl accumulates in the interstitium of the EDT lateral bundle, probably drawing water from the neck segment and proximal tubule. During passage of the fluid along the bundle via the central vessel directed to the circulatory system of the mesial tissue, urea may be taken up from the collecting tubule by countercurrent exchange.

Late distal tubule

In archaic bony fish (Polypteridae, *Protopterus*) a late distal tubule segment (LDT) is well developed arising gradually from the preceding segment. The cytological organization of the LDT cells superficially resembles that of the cells in the EDT. However, the cells are higher than in the preceding segment, and the elaborate basolateral amplifications reach below the apical cell membrane. The apical zonula occludens is deeper and consists of a higher number of anastomosing strands. In the end portion of the LDT intercalated cells are interspersed between the late distal tubule cells (see below).

In marine elasmobranchs, a peculiar structural organization of the LDT was observed. As the early distal tubule returns from the tip of the lateral bundle to the glomerulus, the epithelial cells gradually decrease in height, becoming LDT cells. In the beginning these cells have an apical mucous coat. The LDT runs into the mesial tissue, where it coils extensively (thin tubules) between the coilings of the proximal segment PII (Figure 24.5). Here, the epithelial cells lack an apical mucous coat. Clusters of cells were observed that have the important characteristics of intercalated cells (studs on the cytoplasmic side of apical vesicles and the apical cell membrane, intramembranous rod-shaped particles) (Hentschel et al., 1993). The presence of H^+–K^+-ATPase was demonstrated in this segment by immunocytochemistry (Hentschel et al., 1993; Swenson et al., 1994). An important function of the LDT in marine elasmobranchs appears to be the recognition of calcium and magnesium in the tubular fluid. The divalent cation-sensing receptor cloned from dogfish kidney, SCaR, has been localized to the apex of this epithelium (Hentschel et al., 1998a). In addition to its presence in the EDT, the apical Na^+–K^+–Cl^--cotransporter is also found along the LDT (Biemesderfer et al., 1996). The LDT returns to the glomerulus and, after an elaborate contact with the vascular field of the glomerulus, becomes the collecting tubule that enters the bundle.

In teleosts, a distinct LDT segment is regularly not present. It appears that many transport functions of the LDT are performed by the CT/CD system, the archinephric duct and the urinary bladder (see below).

Collecting tubule–collecting duct (CT/CD) system

The CT/CD system begins as an unbranched CT. The system connects each nephron to the archinephric ducts or the secondary ureters in the mature kidney of fishes. The transition in epithelium type from the end of the distal tubule is gradual.

Ontogenetic considerations

The development of the CT, and supposedly its functions, varies greatly among vertebrates. The diversity may be partly a result of the different ontogenetic development of the kidneys. In the development of the definitive kidney of lampreys (Petromyzontia, agnathan fish), a tree-like system of CTs differentiates from the nephrogenic material, however, separately from the S-shaped anlagen. Five to seven secondary and tertiary branches (the initial CTs) are present with a common trunk (the large CT or CD) that joins the archinephric duct (Youson, 1981).

In gnathostomous fish, prolonged growth of the kidney, presumably lifelong, parallels the growth of the body. In addition to well-differentiated nephrons that lie in the renal tissue, developing nephrons are present in sexually mature elasmobranchs, such as dogfish and skates (Hentschel, 1991), and in actinopterygians, such as sterlets, trout, carp and sticklebacks (Hentschel and Elger, 1987). In elasmobranchs, the collecting duct tree seems to grow by the differentiation of new branches (CTs) that are connected at their blind ends with S-shaped bodies. During the differentiation of the S-shaped body into the nephron loops, the lumen of the latter becomes continuous with the CT/CD system. In many teleosts, the remarkable postnatal growth of the kidney is accomplished by development of new S-shaped bodies in the vicinity of the CT/CD system. The developing tubules insert into the distal nephron portions (CTs) of existing nephrons.

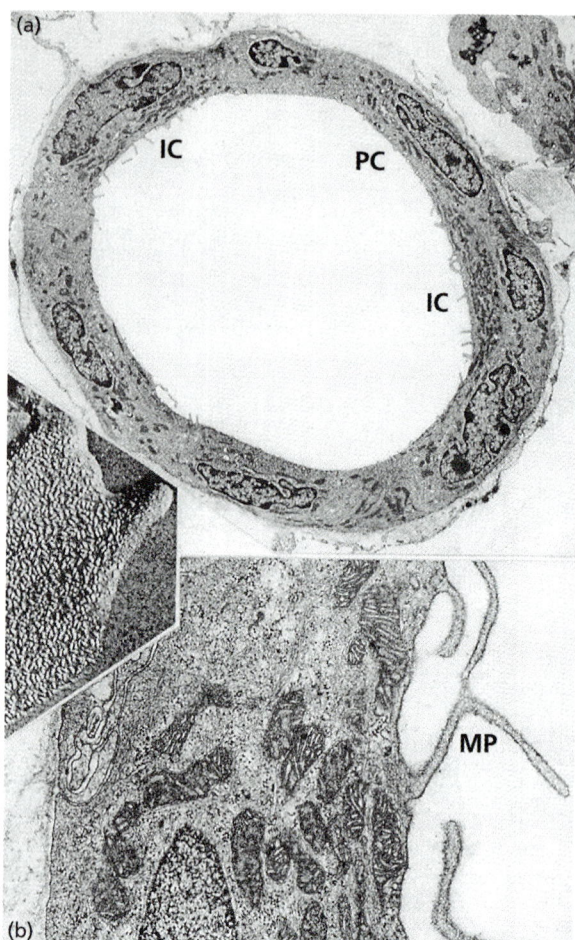

Figure 24.20 Collecting tubule of *Polypterus senegalus*. (a) Cross-section demonstrating the flat epithelium consisting of two epithelial cell types, namely collecting tubule cells (principal cells, PC) displaying a smooth cell surface, and intercalated cells (IC) with microplicae. (b) Part of intercalated cell. Numerous mitochondria accumulate in the upper part of the cell. MP, microplicae. Inset: Freeze-fracture replica demonstrating densely aggregated rod-shaped particles on the protoplasmic face of the apical plasma membrane of an intercalated cell. TEM, (a) ×3500; (b) ×26 000; (inset) ×65 000. (From Hentschel and Elger, 1989, with permission.)

Cytological organization

In most groups of fish (Petromyzontia, archaic bony fish such as the Polypteridae, *Protopterus*), the epithelium of the CT and (or) the CD consists of segment-specific principal cells (CT cells, CD cells) and a variable number of intercalated cells (Figure 24.20). In marine elasmobranchs, intercalated cells are located in the thin coiling of the mesial tissue zone, named late distal tubule. Intercalated cells have not been detected in teleost kidney.

Petromyzontia

In lampreys, the CT and the small CDs consist of one cell type with elaborate basolateral infoldings and numerous mitochondria. Generally, CD cells are taller than those of the CT. Intercalated cells with prominent apical microplicae associated with intracellular stud-like structures are found in the CD.

Teleostei

In most teleosts the CT and small CDs, like those of lampreys, are composed of one cell type displaying the features of a high transport activity. In several species (e.g. flounder, trout), the epithelium overlays a continuously increasing layer of smooth muscle cells. The secretion of mucus by the CD cells is frequently observed in teleosts, such as flounder (Figure 24.21), trout and cyprinids (goldfish, carp). Larger CDs may display additional cell types, i.e. **ionocytes** (trout) or goblet cells (cyprinids). In the sterlet and the marine catfish, *Plotosus*, more than one cell type is present in the CT. Many stenohaline marine glomerular teleosts (e.g. winter flounder) and aglomerular teleosts lack the early distal tubule, and the CT/CD system is directly connected to the end of the proximal segment PII (Figure 24.3).

Elasmobranchii

In *Scyliorhinus caniculus*, the CT (portion in the bundle) (Figure 24.5) contains one epithelial cell type. The apical cytoplasmic zone of these cells contains granules which are often in contact with the plasma membrane. These granules probably contain mucus and are responsible for the positive reaction with Alcian blue. The cell apex has short microvilli. An active Golgi apparatus, distended cisternae of the rough endoplasmic reticulum, multivesicular bodies and lysosomes are present, and, in the supranuclear zone thick bundles of microfilaments are attached to the terminal web. The zonula occludens exhibits a large apicobasal depth. The lateral cell membrane is amplified in a manner similar to the distal tubule.

Polypteridae and Dipnoi

The CT and CD epithelia are lower than those of the late distal tubule, and contain intercalated cells. The zonula occludens are deep with multiple strands. Alcianophilic mucous granules accumulate in the apical cytoplasmic zone.

In *Polypterus senegalus* (Polypteridae), the elaborate basolateral interdigitations in the CT cells are

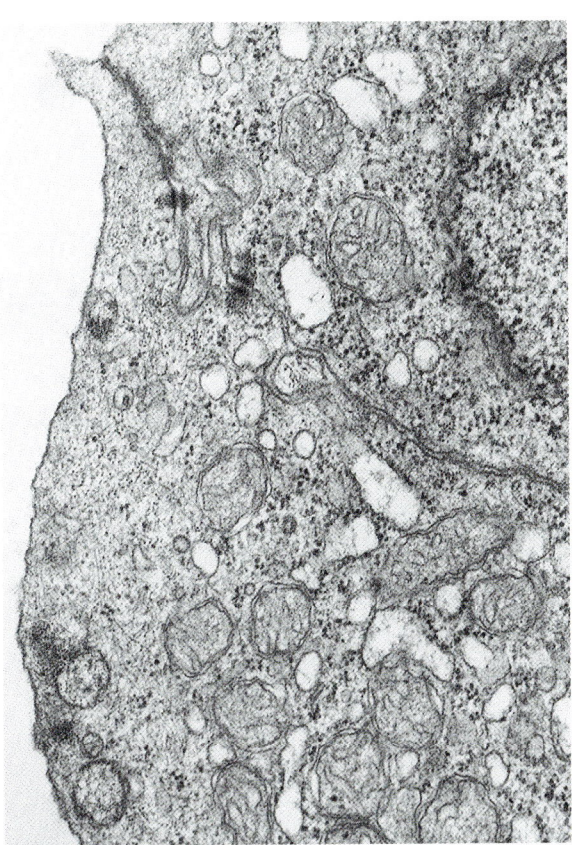

Figure 24.21 Apical region of collecting tubule cell of *Pleuronectes americanus*. The collecting tubule of teleosts consists of one epithelial cell type only. Short stubby microvilli are occasionally present at the apical cell pole. Dense mucous vesicles (arrows) containing fibrous material are located at the apical cell membrane. Mitochondria are numerous and contain many cristae in a light matrix. TEM, ×40 000. (From Elger *et al.*, 1998, with permission.)

arranged in randomly oriented stacks without mitochondria between the folds (Figure 24.20). The numerous short mitochondria lie in the perinuclear cytoplasm together with bundles of microfilaments, Golgi fields, short cisternae of the rough endoplasmic reticulum and free ribosomes (polysomes). The apical cell surface is devoid of microvilli. In *Protopterus dolloi* (Dipnoi), the lateral plasma membrane displays microplicae and microvilli and is associated with mitochondria, similar to the distal tubule. However, only a few short basal invaginations are present. Prominent Golgi fields and glycogen deposits are numerous, as are tubular profiles of mostly smooth endoplasmic reticulum. Microfilaments abound between the granules beneath the apical cell membrane which displays a few stubby microvilli.

Intercalated cells

Intercalated cells may be interspersed between LDT cells in the end portion of the late distal tubule and between CT/CD cells. They are found in the kidneys of representative species of all classes or subgroups of fish except teleosts (Hentschel and Elger, 1987). The cytoplasm of these cells is frequently more dense than that of the surrounding segment-specific cells. The cells have abundant carbonic anhydrase and mitochondria. The latter are usually not associated with basolateral amplifications. The apical cell membrane may be greatly amplified by microplicae or microvilli.

In the urinary bladder of toads and turtles and in the collecting duct system of mammals (Kriz and Kaissling, 1999), at least two types of intercalated cells are found, named alpha and beta cells or type A and B cells, along with many transitional types. The main differences between type A and B cells are in the structural and functional polarity of the cells. In type A cells, the apical cell membrane and associated vesicular and tubular structures in the cytoplasm are equipped with intramembranous rod-shaped particles as revealed on freeze-fracture replicas, and, on their cytoplasmic side, with stud-like structures. In type B cells these structures are confined to the basolateral cell membrane and basal vesiculotubular profiles. All other cytological features vary greatly among the classes and among the intercalated cells of a given species. In type A cells, the apical studs were shown to be associated with an electrogenic proton ATPase of the type involved in proton secretion in the mammalian kidney. The basolateral membrane contains an erythrocyte anion exchanger ('band 3' protein). In contrast, type B cells are thought to secrete protons and reabsorb bicarbonate. Type A intercalated cells were observed in lampreys and polypterids (Hentschel and Elger, 1987). Clusters of cells in the late distal tubule of marine elasmobranchs are also similar to the A type (see above in paragraph LDT). Cells which may represent a type B cell were found in the sterlet, *Acipenser ruthenus*. A morphologically distinct cell type, called **flask cell**, is present in lungfish.

Polypteridae and Dipnoi

Two to three type A intercalated cells are usually seen in cross-sections through the CT and CD in *Polypterus* (Figure 24.20). They are characterized by conspicuous microplicae of the apical cell membrane.

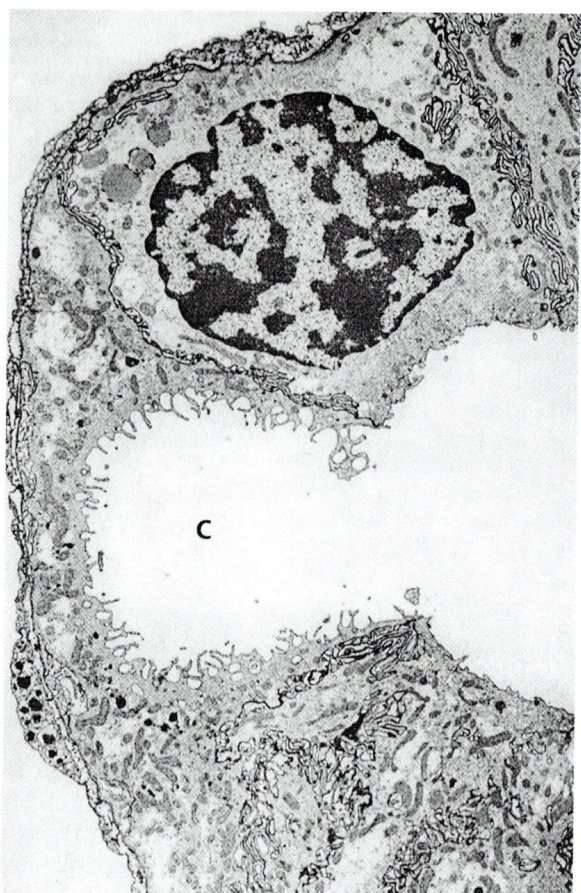

Figure 24.22 Intercalated (flask) cell of collecting tubule of the lungfish, *Protopterus dolloi*. The large cavity (C) of the flask cell is delineated by the invaginated apical cell membrane displaying branching microplicae. Numerous mitochondria are distributed throughout the cytoplasm. Tannic acid staining; TEM, ×4550. (From Hentschel and Elger, 1989, with permission.)

The perinuclear zone is enriched by numerous small elongated mitochondria. The cytoplasm contains many polysomes, bundles of microfilaments and Golgi fields. Amplifications of the basolateral cell membrane are rare. Freeze-fracture replicas reveal a high number of rod-shaped particles in the luminal membrane. The most prominent feature of the intercalated cell of the lungfish (*Protopterus*) is the large extracellular cavity delimited by the apical plasma membrane that invaginates deeply to the basal quarter of the cell (flask cell) (Figure 24.22). The basal cell border is straight. The cytoplasm is characterized by numerous elongated mitochondria and ribosomes. Small vesicles abound beneath the apical cell membrane and are frequently seen in contact with the latter. The invaginated apical cell membrane is further enlarged by ramifying folds. At the base of these folds, regularly spaced studs are revealed on the cytoplasmic side of the membrane. Similar studs can be seen on elongated flat membranous profiles in the cytoplasm. Rod-shaped particles were revealed in the membranes by freeze-fracture studies.

Functional considerations of CT/CD system

The functions of the segments of the CT/CD system are not as well studied as those of the EDT in non-mammalian species. Sodium may be reabsorbed and net secretion of potassium might occur in archaic fish, such as *Polypterus*, sturgeons and lungfish as it does in mammals. Exposure to potassium results in equivalent structural responses (enlargement of basolateral cell membrane area) in amphibians and mammals. In animals that possess intercalated cells (all vertebrate groups, with the exception of teleosts), a contribution of these cells to excretion of protons (type A cells) or of bicarbonate (type B cells) can be assumed.

In addition to the peculiarities in the ontogenetic development and the absence of morphological structures indicative of intercalated cells, it appears that several functions of the CT/CD system, the archinephric duct and urinary bladder of teleosts, are more reminiscent of the distal tubule than of the collecting duct system of mammals. NaCl can be reabsorbed along the entire CT/CD system in flounder. The electroneutral thiazide-sensitive Na/Cl transport system which, in the rat, is restricted to the distal convoluted tubule, was first described and cloned from the flounder urinary bladder (Gamba *et al.*, 1993) (Figure 24.23). Moreover, the divalent ion sensing receptor, CaR, is detectable in considerable amounts in the urinary bladder and in a subpopulation of cells in the CT/CD system in flounder and killifish (Baum *et al.*, 1996). Recent studies using ion microscopy or immunocytochemistry suggest functions of the CT/CD system in reabsorption of magnesium (killifish) (Chandra *et al.*, 1997) and phosphate (flounder) (Elger *et al.*, 1998).

Glomerulo-tubular contact

A contact between the distal nephron and its glomerular arterioles is commonly found in the vertebrates and apparently has its roots in ontogenetic development. In the zonate kidneys of Polypteridae and the

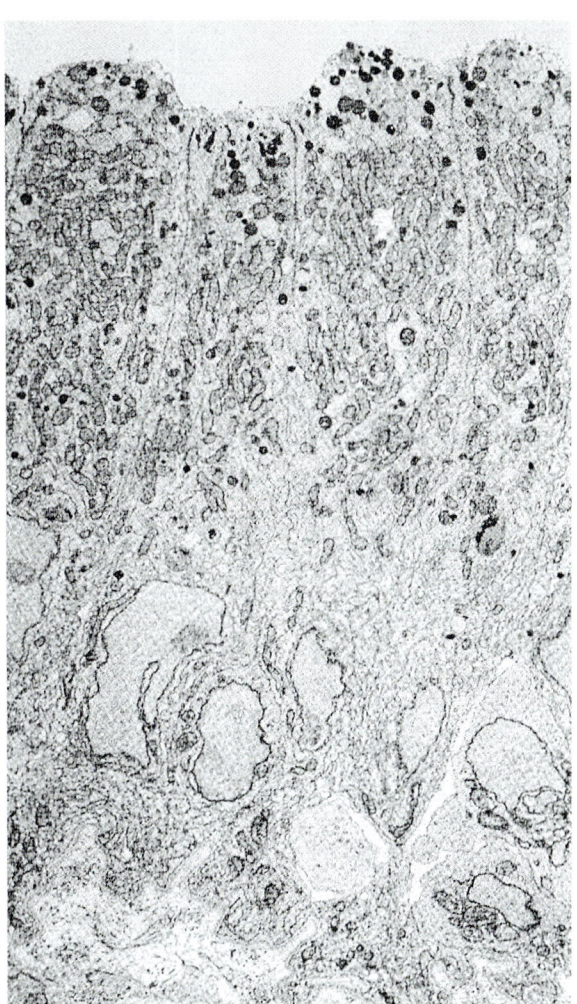

Figure 24.23 Urinary bladder epithelium of *Pleuronectes americanus*. (From Renfro *et al.*, 1976, with permission.)

lungfish, the early distal tubule forms an association with the glomerular vascular pole similar to that seen in the amphibian and mammalian kidney. In the zonate kidney with the peculiar countercurrent system in marine elasmobranchs, a close contact is formed between the glomerular vascular field and the transition between the late distal tubule and collecting tubule (Figures 24.5 and 24.8). Cyclostomes and teleosts do not form a close contact between the vascular pole and the EDT. In teleosts, with their non-zonated kidneys, the collecting tubule generally passes near the glomerulus. A role for this contact in the regulation of the glomerular filtration rate in relation to the NaCl concentration in the tubular fluid at this site has not yet been established in lower vertebrates. In the mammalian kidney the macula densa salt concentration has two effects: (i) the angiotensin-II-dependent regulation of glomerular arteriolar resistance through tubuloglomerular feedback (TGF); and

(ii) control of renin secretion (Schnermann *et al.*, 1998). Granulated cells have been described in the renal vasculature in elasmobranchs and teleosts (both glomerular and aglomerular) regardless of whether such a close contact exists (Nishimura and Bailey, 1982; Elger *et al.*, 1984b; Brown *et al.*, 1990; Lacy and Reale, 1995). The occurrence of angiotensin and angiotensin receptors has been shown in several fish species (Uva *et al.*, 1992; Takei *et al.*, 1998).

References

Baum, M., Hentschel, H., Elger, M., Brown, E.M., Hebert, S.C. and Harris, H.W. (1996). *Bull. Mount Desert Island Biol. Lab.* **35**, 31–32.

Beyenbach, K.W. and Baustian, M.D. (1989). In *Structure and Function of the Kidney. Comparative Physiology*, vol. 1 (ed. R.K.H. Kinne), p. 103. Karger, Basel.

Beyenbach, K.W. and Frömter, E. (1985). *Am. J. Physiol.* **234**, F282–F295.

Beyenbach, K.W. and Liu, P.L. (1996). *Kidney Int.* **49**, 1543–1548.

Bielek, E. (1981). *Cell Tissue Res.* **220**, 163–180.

Biemesderfer, D., Payne, J.A., Lytle, C.Y. and Forbush, B. (1996). *Am. J. Physiol.* **270**, F927–F936.

Brown, J.A., Gray, C.J. and Taylor, S.M. (1990). *Prog. Clin. Biol. Res.* **342**, 528–533.

Chandra, S., Morrison, G.H. and Beyenbach, K.W. (1997). *Am. J. Physiol.* **273**, F939–F948.

Elger, M. and Hentschel, H. (1981). *Cell Tissue Res.* **220**, 73–85.

Elger, M. and Hentschel, H. (1986). *Cell Tissue Res.* **244**, 395–401.

Elger, M., Kaune, R. and Hentschel, H. (1984a). *J. Comp. Physiol. B* **154**, 225–231.

Elger, M., Wahlqvist, I. and Hentschel, H. (1984b). *Cell Tissue Res.* **237**, 451–458.

Elger, M., Sakai, T., Winkler, D. and Kriz, W. (1991). *Contrib. Nephrol.* **95**, 22–33.

Elger, M., Werner, A., Herter, P., Kohl, B., Kinne, R.K. and Henstchel, H. (1998). *Am. J. Physiol.* **274**, F374–F383.

Eveloff, J., Field, M., Kinne, R. and Murer, H. (1980). *J. Comp. Physiol.* **135**, 175–182.

Freire, C.A., Kinne-Saffran, E., Beyenbach, K.W. and Kinne, R.K. (1995). *Am. J. Physiol.* **269**, R592–R602.

Friedman, P.A. and Hebert, S.C. (1990). *Am. J. Physiol.* **258**, R398–R408.

Gamba, G., Saltzberg, S.N., Lombardi, M., Miyanoshita, A., Lytton, J., Hediger, M.A., Brenner, B.M. and Hebert, S.C. (1993). *Proc. Natl. Acad. Sci. USA* **90**, 2749–2753.

Gupta, A. and Renfro, J.L. (1991). *Am. J. Physiol.* **260**, R704–R711.

Hebert, S.C. and Friedman, P.A. (1990). *Am. J. Physiol.* **258**, R409–R417.

Henderson, I.W., O'Toole, L.B. and Hazon, N. (1988). In *Physiology of Elasmobranch Fishes* (ed. T.J. Shuttleworth), p. 201. Springer, New York.

Hentschel, H. (1988). *Am. J. Anat.* **183**, 130–147.

Hentschel, H. (1991). *Am. J. Anat.* **190**, 309–333.

Hentschel, H. and Elger, M. (1987). *Adv. Anat. Embryol. Cell Biol.* **108**, 1–151.

Hentschel, H. and Elger, M. (1989). In *Comparative Physiology. Structure and Function of the Kidney* (ed. R.K.H. Kinne), p. 1. Karger, Basel.

Hentschel, H., Mahler, S., Herter, P. and Elger, M. (1993). *Anat. Rec.* **235**, 511–532.

Hentschel, H., Nearing, J., Harris, H.W. and Elger, M. (1998a). *J. Am. Soc. Nephrol.* **9**, 554A–555A.

Hentschel, H., Storb, U., Teckhaus, L. and Elger, M. (1998b). *Anat. Embryol. (Berl.)* **198**, 73–89.

Hickman, C.P. and Trump, B.F. (1969). *Fish Physiology* (eds W.S. Hoar and D.J. Randall), p. 91. Academic Press, New York.

Kanli, H. and Terreros, D.A. (1997). *Acta Physiol. Scand.* **160**, 267–276.

Kriz, W. and Kaissling, B. (1999). In *The Kidney: Physiology and Pathophysiology*, 3rd edn. (eds D.W. Seldin and G. Giebisch). Raven Press, New York.

Lacy, E.R. and Reale, E. (1995). In *Fish Physiology*, vol. 14, *Cellular and Molecular Approaches to Fish Ionic Regulation* (eds W.S. Hoar, D.J. Randall and A.P. Farrell), p. 107. Academic Press, New York.

Lacy, E.R., Luciano, L. and Reale, E. (1991). *Anat. Embryol. (Berl.)* **183**, 475–481.

Logan, A.G., Moriarty, R.J., Morris, R. and Rankin, J.C. (1980a). *Anat. Embryol. (Berl.)* **158**, 245–252.

Logan, A.G., Morris, R. and Rankin, J.C. (1980b). *J. Exp. Biol.* **88**, 239–247.

Lu, M., Barber, L.E. and Renfro, J.L. (1994). *Am. J. Physiol.* **267**, F624–F631.

Miller, D.S. and Pritchard, J.B. (1994). *Am. J. Physiol.* **267**, R695–R704.

Miller, D.S., Fricker, G. and Drewe, J. (1997). *J. Pharmacol. Exp. Ther.* **282**, 440–444.

Miller, D.S., Sussman, C.R. and Renfro, J.L. (1998). *Am. J. Physiol.* **275**, F785–F795.

Nishimura, H., Bailey, J.R. (1982). *Kidney Int. Suppl.* **12**, S185–S192.

Nishimura, H., Imai, M. and Ogawa, M. (1983). *Am. J. Physiol.* **244**, F247–F254.

Ottosen, P.D. (1987). *Cell Tissue Res.* **194**, 207–218.

Ottosen, P.D. and Maunsbach, A. (1973). *Kidney Intern'l* **3**, 315–326.

Reale, E., Luciano, L., Kuhn, K., Stolte, H. and Brod, J. (1981). *Ren. Physiol. Basel* **4**, 85–89.

Renfro, J.L. (1980). Relationship between renal fluid and Mg secretion in a glomerular marine teleost. *Am. J. Physiol.* **238**, F92–F98.

Renfro, J.L. (1997). *Fish Physiol. Biochem.* **17**, 377–383.

Renfro, J.L., Miller, D.S., Karnaky, K.J. Jr. and Kinter, W.B. (1976). *Am. J. Physiol.* **231**, 1735–1743.

Schnermann, J., Traynor, T., Yang, T., Arend, L., Huang, Y.G., Smart, A. and Briggs, J.P. (1998). *Kidney Int. Suppl.* **67**, S40–S45.

Schramm, U., Fricker, G., Wenger, R. and Miller, D.S. (1995). *Am. J. Physiol* **268**, F46–F52.

Stolte, H., Galaske, R.G., Eisenbach, G.M., Lechene, C., Schmidt-Nielsen, B. and Boylan, J.W. (1977). *J. Exp. Zool.* **199**, 403–410.

Sussman-Turner, C. and Renfro, J.L. (1995). *Am. J. Physiol.* **268**, F135–F144.

Swenson, E.R., Fine, A.D., Maren, T.H., Reale, E., Lacy, E.R. and Smolka, A.J. (1994). *Am. J. Physiol.* **267**, F639–F645.

Takei, Y., Itahara, Y., Butler, D.G., Watanabe, T.X. and Oudit, G.Y. (1998). *Gen. Comp. Endocrinol.* **110**, 140–146.

Uva, B., Masini, M.A., Hazon, N., O'Toole, L.B., Henderson, I.W. and Ghiani, P. (1992). *Gen. Comp. Endocrinol.* **86**, 407–412.

Wendelaar Bonga, S.E. and Veenhuis, M. (1974). *J. Cell Sci.* **14**, 587–609.

Werner, A., Murer, H. and Kinne, R.K. (1994). *Am. J. Physiol.* **267**, F311–F317.

Wolff, N.A., Werner, A., Burkhardt, S. and Burckhardt, G. (1997). *FEBS Lett.* **417**, 287–291.

Yokota, S.D., Benyajati, S. and Dantzler, W.H. (1985). *Ren. Physiol. Basel* **8**, 193–221.

Youson, J.H. (1981). In *The Biology of Lampreys* (eds M.W. Hardisty and I.C. Potter), p. 191. Academic Press, New York.

CHAPTER 25

Endocrine System

David M Janz and Lynn P Weber
Department of Zoology, Oklahoma State University, Stillwater, Oklahoma, USA

Introduction

The recent advances in molecular biology have allowed an explosion of new information in endocrinology and opened the door to understanding evolution on a much different level than before. Although traditional techniques such as removal/replacement are still very important as corroborative evidence and to put newly discovered gene products into a physiological context, the pace of discovery in fields like endocrinology would be slower without these technological advances. The next few decades will be exciting in endocrinology.

The basis of endocrine and neural function is communication between cells via a chemical messenger. At one time these two systems were considered distinct, but we now know that drawing a definite line between the two can be very difficult. The traditional definitions of a hormone versus a neurotransmitter are too rigid to fit what we know about endocrinology today, particularly in teleosts since they lack a well-developed hypophysial portal system and rely on neurohemal secretion, both in the brain and in the urophysis, much more than higher vertebrates.

Molecular structure

There are four major categories of messenger molecules: amines, steroids, peptides and lipids. The scope of this chapter does not allow discussion of the first and last categories. The big explosion in knowledge has been in the peptide hormone category, thanks to molecular biology and improved protein chemistry techniques. These hormones will be discussed, with reference to papers containing sequence data as much as possible throughout this chapter. This is intended to aid the reader in finding information necessary for using molecular tools in endocrinology research, not to bog the reader down in molecular detail. Because peptide hormones diverge in structure to a great extent compared with the other three classes of hormones, researchers have realized that knowledge of specific amino acid or genetic sequences in every species is necessary to draw appropriate conclusions about localization and function. Thus, in order to conduct good comparative research in this day and age, a researcher must find sources of homologous antibodies or generate his or her own, as well as design appropriate **mRNA** or DNA probes, all of which is based on molecular knowledge. Although by no means

Copyright © 2000 Academic Press

comprehensive, hopefully this chapter will give the reader an idea of where to start looking in the literature for existing knowledge in molecular endocrinology of fish. For more detailed comparative molecular endocrinology references, the reader is referred to Bentley (1998) or Sherwood and Hew (1994). Whenever possible, review articles or current literature will be cited, both of which contain landmark references.

Gene expression and regulation

It is important to study the tissues where hormones are expressed and released from in order to understand how a hormone is regulated and integrated with physiological processes to maintain homeostasis. However, an emerging area of research in endocrinology is the study of specific enzymes that produce or process hormones, their regulation and their location. Steroidogenic enzymes and enzymes that produce amine or lipid hormones are fairly well characterized. On the other hand, this is not true for peptide hormones. Peptide hormones are often synthesized as prohormones or even preprohormones that must undergo a series of specific proteolytic steps or other post-translational modification such as glycosylation or amidation before they are biologically active. Thus, the highly specific and diverse proteolytic enzymes responsible for producing active peptide hormones, the '**prohormone convertases**' represent a rapidly expanding area of research.

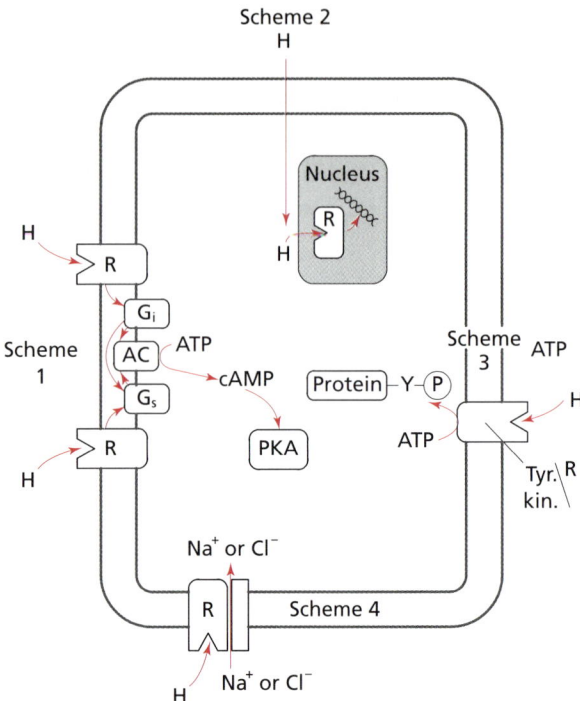

Figure 25.1 Schematic of the four general schemes of endocrine receptor signal transduction in cells. Scheme 1 shows an example of a receptor system that couples to G proteins and effectors. In this figure, two different receptors are shown to regulate activity of the effector, adenylyl cyclase, which controls production of the second messenger, cAMP. Scheme 2 shows the basic signaling of steroid hormones. Scheme 3 shows an example of a receptor that possesses intrinsic enzymatic (tyrosine kinase) activity. Scheme 4 shows a receptor that is also an ion channel with ion permeability controlled by hormone binding. H, hormone; R, receptor; G_i, G protein that inhibits adenylyl cyclase; G_s, G protein that stimulates adenylyl cyclase; AC, adenylyl cyclase; PKA, cAMP-dependent protein kinase. Protein Y-P is a protein phosphorylated on a tyrosine residue and Tyr. Kin. is tyrosine kinase.

Hormone effectors in target tissues

Hormones work by activating specific receptors on their target tissues. Knowledge of the specific receptor type involved in a given hormone response, its location and its regulation are necessary pieces of information in understanding endocrine processes. Once a hormone binds to its receptor, the message is 'transduced' to the inside of the cell via the signaling system to which the receptor couples. Although there are more receptor types than there are hormones, representing a huge diversity of receptors, there are surprisingly few signal transduction systems.

In general, there are four major types of signaling schemes utilized by receptors (see Figure 25.1 for schematic):

1. The receptor is inserted in the plasma membrane and couples to 'effector' enzymes via heterotrimeric G proteins that catalyze production of intracellular second messengers.
2. The receptor binds directly to target DNA in the nucleus of cells and acts as a **transcription factor** once the hormone is bound.
3. The receptor itself possesses enzymatic activity that is activated upon hormone binding.
4. The receptor itself forms an ion channel that is opened or closed upon hormone binding.

Many growth hormones or hormones with mitogenic activity use the third scheme since the receptors possess intrinsic tyrosine kinase activity, while the fourth scheme is limited to a few neurotransmitters, mostly in the brain. The second scheme which is referred to as 'nuclear receptors' is utilized by most steroid hormones. The first scheme, **G protein**/effector enzyme coupling, represents the majority of receptors and is complicated. The best characterized of these signaling cascades is **adenylyl cyclase** (formerly called adenylate cyclase) which produces the second messenger, cyclic AMP (**cAMP**), from **ATP**. The cAMP in turn activates another enzyme, cAMP-dependent protein kinase (**PKA**), which then phosphorylates specific target proteins on serine or threonine residues to produce the response associated with the hormone. The activity of adenylyl cyclase can either be suppressed or enhanced by a given receptor, depending on which G protein couples with it. Receptors that stimulate adenylyl cyclase activity couple to G_s, while receptors that inhibit its activity couple to G_i. Another commonly used signaling effector is phospholipase C (**PLC**), often coupled to receptors via G_q. PLC catalyzes the hydrolysis of the membrane phospholipid, phosphatidylinositol-1,4,5-bisphosphate (**PIP$_2$**) into two second messenger products: inositol-1,4,5-trisphosphate (**IP$_3$**) and diacylglycerol (**DAG**). IP$_3$ causes release of calcium from intracellular stores in the endoplasmic reticulum while DAG stimulates Ca^{2+}/lipid-dependent protein kinase (**PKC**), both of which activate a series of other cascades to mediate the responses associated with the hormone.

Hypothalamic and pituitary hormones

Growth hormone-releasing hormone and PACAP

Molecular structure

Growth hormone-releasing hormone (**GHRH**; also referred to as growth hormone-releasing factor (GRF) or somatocrinin) is a peptide hormone of 45 amino acid residues in teleost and chondrostean species that belongs to the secretin-glucagon family of hormones. The C-terminal of the GHRH peptide is post-translationally modified, namely by amidation of the C-terminus. Other processing, such as cleavage to form smaller GHRH-like products of 10, 19 and 28–29 amino acid residues, has also been shown to occur in salmon (*Oncorhynchus* spp.) and carp (*Cyprinus carpio*) (Sherwood *et al.*, 1994). The salmon and catfish (*Clarias macrocephalus*) GHRH gene codes not only for GHRH, but also for another 38 amino acid peptide hormone called PACAP (*p*ituitary *a*denylate *c*yclase-*a*ctivating *p*olypeptide). In fish, GHRH and PACAP are located on the same gene, while in mammals these two proteins are products of two separate genes. Piscine GHRH peptides are highly related to each other, however when compared to human, only 41–62% of the amino acid sequence is conserved (Sherwood *et al.*, 1994). On the other hand, the PACAP sequence is much more highly conserved than the GHRH sequence with 89–92% homology in deduced sequence between human and teleost PACAP.

Gene expression and regulation

GHRH-like and PACAP gene expression have been found in many catfish tissues, including brain, stomach, testes and ovary, but not in the pancreas, pituitary, muscle and liver (Sherwood *et al.*, 1994). Although GHRH and PACAP are located on the same gene, the expression of these two hormones is differentially regulated via exon skipping in salmon (Parker *et al.*, 1997). PACAP mRNA was found to be expressed in brain, heart and pituitary of goldfish (*Carassius auratus*) (Wong *et al.*, 1998a).

Hormone effectors in target tissues

The target of GHRH is anterior pituitary cells and it has been shown to cause **GH** release from cultured rainbow trout (*Oncorhynchus mykiss*) pituitary cells (Luo *et al.*, 1990) and from goldfish pituitaries *in vitro* or *in vivo* (Vaughan *et al.*, 1992). However, there appears to be species-specificity in the ability of GHRH to stimulate release of GH. For example, human GHRH cannot stimulate GH release in various other species, including many fish (Sherwood *et al.*, 1994). Therefore studies using GHRH should use the hormone from the species being studied or a very closely related species. GHRH receptors are coupled to adenylyl cyclase and increase somatotrope cAMP levels which in turn leads to GH secretion. In fact, any agents which lead to increases in intracellular cAMP in somatotropes are potent stimulators of GH secretion (reviewed in Melamed *et al.*, 1998). Long-term effects of GHRH on somatotropes include GH gene transcription and translation via cAMP binding to cAMP responsive elements upstream of the GH gene and stabilization of GH mRNA.

PACAP, as its name implies, stimulates adenylyl cyclase in pituitary cells to produce cAMP and cause GH release (Wong *et al.*, 1998a; Miyata *et al.*, 1989; Chartrel *et al.*, 1991). PACAP may play a role not only in GH release, but also in **LH**, **FSH** and **ACTH** release. In fact, PACAP may be a more potent stimulator of GH secretion than GHRH in salmonids (Parker *et al.*, 1997). PACAP elicits its effects via two different receptor subtypes, type I and type II PACAP receptors which are both G protein-coupled receptors, also known as **PVR1** and **VIP** receptors, respectively. The recently cloned goldfish type I PACAP receptor was shown to stimulate increases in cAMP production when expressed in **COS** cells (Wong *et al.*, 1998a).

Growth hormone

Molecular structure

Growth hormone (GH) is a single 21 kDa polypeptide hormone containing two disulfide bonds and belongs to a hormone superfamily that includes prolactin and somatolactin. The complete amino acid sequence has been determined from many teleost fish, as well as GH cloning (Chen *et al.*, 1994). There are four highly conserved α-helical domains in all vertebrate GHs that form bundles, A_{GH}, B_{GH}, C_{GH} and D_{GH}, with highly variable regions in between each conserved region (Chen *et al.*, 1994). Overall, the predicted structure of teleost GHs is likely to be similar to the porcine GH structure deduced by X-ray crystallography (Abdel-Meguid *et al.*, 1987).

Gene expression and regulation

Release of GH is regulated by many factors and varies seasonally with somatic growth (reviewed in Peter and Marchant, 1995). GH release is stimulated by GHRH, dopamine, thyrotropin-releasing hormone, neuropeptide Y (**NPY**), gonadotropin-releasing hormone, and cholecystokinin (**CCK**), but is inhibited by somatostatin (also referred to as somatotropin release-inhibiting factor (**SRIF**)), insulin-like growth factor, GH, glutamate, norepinephrine, and serotonin (Peter and Marchant, 1995; Trudeau *et al.*, 1996).

Hormone effectors in target tissues

A primary function of GH is to promote somatic growth in fish, however this function is accomplished indirectly. When GH is released, it increases transcription and release of insulin-like growth factors (IGF-I or IGF-II). The mitogenic effects of GH are thus mediated by **IGFs** in fish. Radioreceptor assays have been developed for GH in coho salmon (*Oncorhynchus kisutch*) and rainbow trout using radioactively labeled GH. A single population of GH receptors were found, with the majority of receptors found in liver (Gray *et al.*, 1990; Sakamoto and Hirano, 1991; Yao *et al.*, 1991), but receptors were also found in gill, intestine and posterior kidney (Sakamoto and Hirano, 1991). Also, the number of GH receptors increase in the liver of trout after transfer to salt water from fresh water (Sakamoto and Hirano, 1991) and are thought to mediate necessary metabolic changes required for the saltwater transition.

In mammals, after GH binds to GH receptors in target tissue the GH receptor/ligand complex dimerizes (Calduch-Giner *et al.*, 1997). Once the GH receptors are activated, a tyrosine kinase called **JAK** (*ja*nus *k*inase) is activated and transduces the GH signal in target tissues. Activated JAK phosphorylates specific proteins that translocate to the nucleus and transcriptionally activate GH target genes (Carter-Su and Smit, 1998).

Gonadotropin-releasing hormone

Molecular structure

Gonadotropin-releasing hormone (**GnRH**) is found in multiple forms in fishes. GnRH is a decapeptide hormone that is phosphorylated at the *N*-terminus and amidated at the *C*-terminus (see Table 25.1). The amino acid sequence is highly conserved among teleost fish sharing 70–90% identical amino acids (Sherwood *et al.*, 1994). Generally, two or more GnRH forms can be detected in each fish species. For example, both salmon-type (**sGnRH**) and chicken-type GnRH-II (**cGnRH-II**) exist in the brain of goldfish and salmonids (Yu *et al.*, 1988; King and Millar, 1992). The gene for GnRH encodes not only for GnRH, but also for a GnRH-associated peptide (**GAP**) and a signal peptide for processing of the pre-prohormone peptide. Unlike the GnRH hormone moiety that is highly conserved between fish and

TABLE 25.1: Amino acid sequence of GnRH hormones from several fish species and a mammal for comparison. Bold letters indicate amino acid residue differences compared to the salmon sequence

Species	1 2 3 4 5 6 7 8 9 10
Salmon	pGlu-His-Trp-Ser-Tyr-Gly-Trp-Leu-Pro-Gly-NH$_2$
Dogfish	pGlu-His-Trp-Ser-**His**-Gly-Trp-Leu-Pro-Gly-NH$_2$
Catfish	pGlu-His-Trp-Ser-**His**-Gly-**Leu-Asn**-Pro-Gly-NH$_2$
Lamprey III	pGlu-His-Trp-Ser-**His-Asp**-Trp-**Lys**-Pro-Gly-NH$_2$
Lamprey I	pGlu-His-**Tyr**-Ser-**Leu-Glu**-Trp-**Lys**-Pro-Gly-NH$_2$
Chicken II (same as form in ratfish, dogfish and catfish)	pGlu-His-Trp-Ser-**His**-Gly-Trp-**Tyr**-Pro-Gly-NH$_2$
Mammal (same as sturgeon form)	pGlu-His-Trp-Ser-Tyr-Gly-**Leu-Arg**-Pro-Gly-NH$_2$

Source: Adapted from Sherwood *et al.*, 1994.

higher vertebrates, GAP is not well conserved (Sherwood *et al.*, 1994). The function of the GAP peptide is not known.

Gene expression and regulation

GnRH is found primarily in the hypothalamus and released into the anterior pituitary, although sGnRH and a novel GnRH-like protein were found recently in the ovary of goldfish (Lin and Peter, 1996; Pati and Habibi, 1998). In hermaphroditic dusky anemonefish (*Amphiprion melanopus*) and bluehead wrasse (*Thalassoma bifasciatum*), the location of GnRH was found to vary according to sex and the stage of sexual development. Higher levels of GnRH-immunoreactive nerve fibers were observed in the preoptic area of males compared to females or juveniles (Grober and Bass, 1991; Elofsson *et al.*, 1997). This observation may indicate that high levels of preoptic area GnRH are required for masculinization of hermaphroditic fish.

Hormone effectors in target tissues

The primary role of GnRH is to cause release of gonadotropins from the pituitary, but only sGnRH, not **cGnRH**, is thought to regulate this release in salmonids (Demski, 1987; Amano *et al.*, 1998). GnRH receptors are found on gonadotropes of all fish, as well as somatotropes of goldfish, but not catfish (Cook *et al.*, 1991; Bosma *et al.*, 1997). Prolonged treatment with a salmon GnRH analog leads to increased body growth and GH release in goldfish, carp and tilapia (*Oreochromis aureus*) (reviewed in Melamed *et al.*, 1998). However, this is not consistently observed and GnRH may not stimulate GH release in all fish species. Also, cGnRH-II-immunoreactive fibers have been found to terminate near somatolactin cells in sockeye salmon (*Oncorhynchus nerka*) (Parhar *et al.*, 1996), suggesting a possible role of GnRH in prolactin release.

Receptor binding assays for GnRH have been developed in African catfish (*Clarias gariepinus*) and goldfish, revealing the presence of a single and multiple GnRH receptors, respectively (Habibi *et al.*, 1987; De Leeuw *et al.*, 1988). These G-protein-coupled receptors are expressed in the pituitary, other parts of the brain and in the gonads. Stimulation of GnRH receptors on target cells elicits increases in calcium influx via voltage-sensitive calcium channels, sodium influx, activation of cAMP-dependent protein kinase (**PKA**) and activation of protein kinase C (PKC) which ultimately leads to increased gonadotropin or GH expression and secretion (Bosma *et al.*, 1997; Chang *et al.*, 1997; Melamed *et al.*, 1998).

Gonadotropins

Molecular structure

Gonadotropin (**GTH**) is found in two major forms in most teleosts (follicle-stimulating hormone (FSH)-like and luteinizing hormone (LH)-like), called GTH-I and GTH-II, respectively (Li and Ford, 1998; Melamed *et al.*, 1998). However, some fish species such as European eel (*Anguilla anguilla*) and catfish appear to possess only GTH-II since many attempts to purify GTH-I from these species have not been successful. The GTH hormones belong to a heterodimeric glycoprotein superfamily that includes thyroid-stimulating hormone (**TSH**), with each hormone comprised of a common α-subunit and differing β-subunits (Gharib *et al.*, 1990). GTH-I and GTH-II peptides have been purified and characterized from several fish species, including chum salmon (*Oncorhynchus keta*), coho salmon, and carp (Itoh *et al.*, 1988; Suzuki *et al.*, 1988a,b; Swanson *et al.*, 1991; Van Der Kraak *et al.*, 1992). Also, cDNAs have been isolated from several fish species for both GTH hormones (Trinh *et al.*, 1986; Sekine *et al.*, 1989; Lin *et al.*, 1992). The GTH hormone α- and β-subunits are non-covalently associated, 113 and 119 amino acid residues long, respectively, and encoded by separate genes. Because GTH-I and GTH-II differ only in their β-subunits, β-subunits determine biochemical and physiological specificity. Although the α-subunit is the same for both GTH hormones, an additional α-subunit variant (α_1) that is minor in abundance and that only associates with GTH-Iβ has been detected in salmon and carp (Suzuki *et al.*, 1988b; Lo *et al.*, 1991).

Gene expression and regulation

GTH-α is expressed in both gonadotropes and thyrotropes because this subunit is common to both TSH and the two GTHs. However, based on studies in salmonids and bluefin tuna (*Thunnus thynnus*) pituitaries, GTH-Iβ- and GTH-IIβ-subunits are likely expressed only in gonadotropes and not in other pituitary cell types (Naito *et al.*, 1993; Kagawa *et al.*, 1998). The two different GTH-β-subunits are not co-localized in the same gonadotrope cells, except perhaps in platyfish, but rather in two distinct populations of pituitary gonadotropes of most fish (Melamed *et al.*, 1998).

Hormone effectors in target tissues

The main targets of GTH are ovary and testis. Specifically, in fish ovaries, the targets are granulosa and thecal cells that surround the oocyte, not the oocyte itself (Van Der Kraak *et al.*, 1998). In fish testes, the targets of GTH are Sertoli and Leydig cells (Swanson *et al.*, 1997). Receptor binding studies in the amago salmon (*Oncorhynchus rhodurus*) suggest that only one type of GTH receptor is present on both thecal and granulosa cells (Kanamori and Nagahama, 1988), while two different GTH receptors are present in coho salmon ovaries (Yan *et al.*, 1992). The first GTH receptor subtype, GTH-RI, binds both **GTH-I** and **GTH-II** with a slight preference for GTH-I, while the second GTH receptor subtype, **GTH-RII**, specifically binds GTH-II (see Figure 25.2). **GTH-RI** is found on both thecal and granulosa cells while GTH-RII is found on only granulosa cells (Nagahama *et al.*, 1994). Only GTH-RI are expressed in coho salmon ovarian follicles during early stages of vitellogenesis, while during final oocyte maturation, GTH-RII receptors appear in granulosa cells along with GTH-RI in both ovarian cell types (Swanson *et al.*, 1997). Similarly, in fish testes GTH-RI is found during early spermatogenesis on Sertoli cells while during final spermiation, GTH-RII receptors appear on Leydig cells (Swanson *et al.*, 1997).

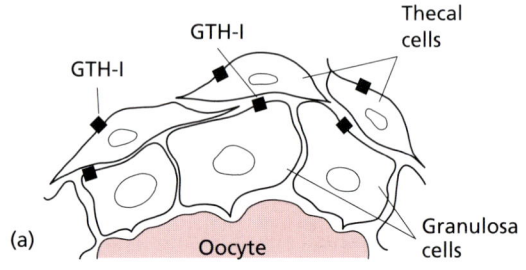

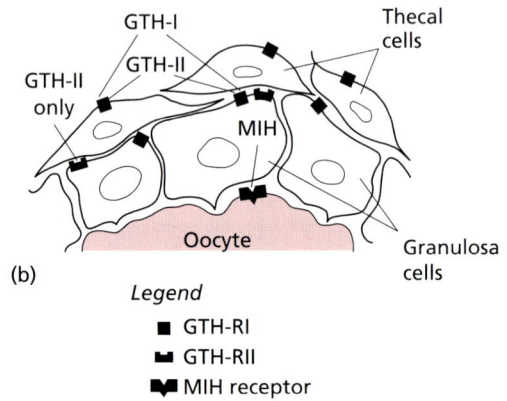

Figure 25.2 Schematic of coho salmon ovarian cells with the cell-type-specific receptors at two stages of development: (a) vitellogenesis and (b) oocyte maturation. During vitellogenesis, GTH-I is the major endocrine input to the ovary and GTH-RI receptors occur on both somatic cells, granulosa and thecal cells. During oocyte maturation, both GTH-I and GTH-II are important endocrine signals to the ovarian somatic cells. The shift to maturation occurs via GTH-RI on both granulosa and thecal cells, but also via the appearance of GTH-RII on the granulosa cells. Finally, the oocyte begins to respond to maturation signals via MIH receptors on its plasma membrane.

Both GTH receptor types are G-protein-coupled receptors that stimulate adenylyl cyclase and cAMP production in testes and ovary (Wade and Van Der Kraak, 1991; Nagahama et al., 1994). In addition to the primary role played by cAMP, however, IP$_3$ production, increases in intracellular calcium, protein kinase C activity and arachidonic acid metabolites may all play roles in signaling of GTH receptors (Van Der Kraak and Wade, 1994). Thecal cells at all stages of development produce and release testosterone in response to GTH. However, thecal cells in preovulatory follicles mainly produce 17α-hydroxyprogesterone (17α-P) in response to GTH. In granulosa cells from vitellogenic follicles, GTH stimulates production of 17β-estradiol. In contrast, granulosa cells from preovulatory follicles produce mainly maturation-inducing hormone (MIH or 17α,20β-dihydroxy-4-pregnen-3-one or 17α,20β-DP) in response to GTH. Increasing intracellular calcium using the calcium ionophore A23187 or stimulating PKC with phorbol esters prevents production of 17α,20β-DP regardless of the presence of GTH or cAMP (Van Der Kraak, 1991; Nagahama et al., 1994). Thus, calcium and PKC activity appear to oppose the actions of GTH or cAMP in fish granulosa cells. In fish testes, production of cAMP due to GTH stimulation leads to production of both 11-ketotestosterone and testosterone (Wade and Van Der Kraak, 1991).

The steroidogenic enzymes in the gonadal cells are expressed differentially, depending on the stage of gonadal maturity. Regulation of synthetic enzyme levels contributes to the switchover from estradiol to 17α,20β-DP production by the piscine granulosa cells (see Chapter 13, Figures 13.2 and 13.3 for oocyte physiology schematic). This steroidogenic switchover is accomplished by an increase in granulosa cell expression of 20β-hydroxysteroid dehydrogenase (20β-HSD) which converts 17α-P into 17α,20β-DP. Production of 20β-HSD may be a direct result of GTH-II production and the presence of GTH-RII on granulosa cells late in oocyte development. Also, activity and mRNA expression of aromatase in granulosa cells falls just prior to oocyte maturation, ensuring a large precursor availability for granulosa cell production of 17α,20β-DP (Kanamori et al., 1988; Tanaka et al., 1992). The precise role of GTH-RI versus GTH-RII in this switchover, as well as the change from GTHI to GTHII release from the pituitary, in mediating this steroidogenic switchover is not clear.

Thyrotropin-releasing hormone and thyrotropin

Molecular structure

Thyrotropin-releasing hormone (**TRH** or **TRF**) has not been characterized from fish, but is likely to be close or identical in structure to the simple mammalian TRH peptide, pGlu-His-Pro-NH$_2$ (Morley, 1981). Thyrotropin (or thyroid-stimulating hormone or TSH) protein has been isolated from carp, European eel, tilapia and coho salmon (see Moriyama et al., 1997). TSH, being a member of the glycoprotein hormone family, is composed of two peptides, α- and β-subunits, with the β-subunits conferring TSH properties. Also, **cDNA** has been obtained for the TSH β-subunit from rainbow trout and European eel. A large degree of species specificity of the TSH hormone is consistent with the low degree of homology of the TSH β-subunits between species and has presented difficulties in directly measuring TSH levels in fish until recently (Moriyama et al., 1997).

Gene expression and regulation

TRH is expressed in the hypothalamus (Jackson and Reichlin, 1974), while TSH is expressed in thyrotrope pituitary cells in fish. The regulation of TRH expression in fish is not known. On the other hand, steady-state levels of TSH mRNA and secretion have been shown to be negatively regulated by thyroid hormones in coho salmon (Larsen et al., 1997; Moriyama et al., 1997).

Hormone effectors in target tissues

The presumed target of TRH is pituitary thyrotropes where it is known to stimulate TSH secretion in mammals. However, TRH could not stimulate TSH secretion in coho salmon pituitary cells, while urotensin I and other members of the corticotropin-releasing factor (**CRF**) family were potent TSH-releasers (Larsen et al., 1998). Although the physiological role of TRH (if any) in the release of TSH in fish is unclear, binding of TRH to its receptors in dogfish (*Scyliorhinus canicula*) pituitary has been shown to increase adenylyl cyclase activity and increase cAMP levels (Deering and Jones, 1975). Knowledge of TSH action in fish is not much clearer than TRH, although presumably the TSH receptor on thyroid follicular cells is similar to mammalian receptors which stimulate increased adenylyl cyclase and PLC activity to cause thyroid hormone synthesis and release (Roger et al., 1997).

Pro-opiomelanocortin, endorphin, α-melanocyte-stimulating hormone and adrenocorticotrophic hormone

Molecular structure

Pro-opiomelanocortin (**POMC**) is a peptide prohormone which, when processed, produces several different hormones with divergent physiological roles. The hormone products, which differ depending on post-translational processing, are adrenocorticotrophic hormone (ACTH or corticotropin), β-endorphin, β-lipotropin (**β-LPH**) and melanotropin (melanocyte-stimulating hormone or MSH). In fish, cDNA for two isoforms of POMC, POMC-A and POMC-B, have been obtained from trout and sockeye salmon, both of which have highly conserved sequences compared with higher vertebrates (Salbert et al., 1992; Okuta et al., 1996). In mammals and elasmobranchs, there are α, β and γ forms of **MSH** produced, but teleost fish appear to lack the γ-MSH sequence. Also, piscine POMC-A is longer than that of mammals by 25 amino acid residues. The POMC-A extension appears to have several protease processing sites, suggesting that fish express two novel POMC peptide derivatives. One of these derivatives may be amidated on the C-terminal (Tollemer et al., 1999). Cyclostomes such as sea lamprey (*Petromyzon marinus*) also possess two POMC genes, but they differ markedly from other vertebrates with one isoform lacking MSH and the other isoform containing two novel MSH peptides (Heinig et al., 1995; Takahashi et al., 1995).

Initial processing of POMC begins in all POMC-expressing cells with proteolytic cleavage to form **β-LPH** and ACTH (see Figure 25.3 for processing schematic; mammalian processing reviewed in Rouille et al., 1995). ACTH is a mature hormone produced from this cleavage, while β-LPH is considered to be only a prohormone in fish. However, ACTH can also be cleaved further to produce α-MSH and corticotropin-like intermediate lobe protein (**CLIP**). The overlapping biological activity of ACTH and α-MSH arises from the fact that α-MSH is a fragment of ACTH. The prohormone β-LPH can be cleaved to produce γ-LPH and β-endorphin, both of which can be further processed to β-MSH and metenkephalin, respectively. In corticotrope cells, the principal product secreted is ACTH, while melanotropes secrete α-MSH and CLIP (derived from ACTH), as well as β-MSH (derived from β-LPH) (Vallarino et al., 1989).

ACTH is a highly conserved 39 amino acid peptide, particularly in the amino terminal 24 residues where it is usually acetylated. α-MSH is a 13 amino acid peptide, while β-MSH has 16–22 residues depending on the species. Diacetylated α-MSH is the most common and most potent form in teleosts, but un- and monoacetylated α-MSH are also detectable (Stouthart et al., 1998). The amino acid sequence of α-MSH is relatively invariant among fish, while that of β-MSH is not as well conserved. Endorphins will not be discussed further in this section.

Gene expression and regulation

The pars intermedia of the pituitary contains high levels of POMC-expressing neurons in corticotrope and melanotrope cells (Naito et al., 1984). POMC mRNA was found to decrease in trout exposed to stress or increase in trout reared in dark-colored tanks (Suzuki et al., 1997). However, regulation of the prohormone convertases may be more critical than regulation of POMC expression. If the type of proteolytic enzyme isoform or expression level changed,

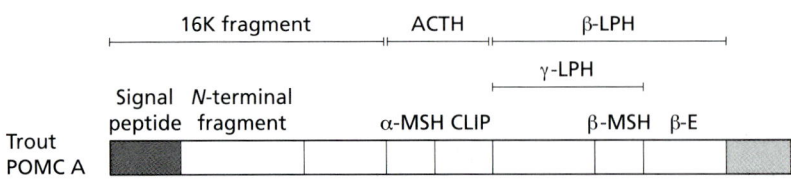

Figure 25.3 Proteolytic processing scheme of rainbow trout pro-opiomelanocortin (POMC-A) to produce the wide variety of hormone products. Hatched bars at the C-terminus indicate the unique fish extension to POMC-A. β-E is β-endorphin. (Modified from Tollemer et al., 1999.)

POMC processing and the amount/types of POMC hormones produced would also change. However, very little is known regarding the regulation of prohormone convertases in fish and represents an area in need of investigation (Rouille et al., 1995).

Hormone effectors in target tissues

The primary role of ACTH is to cause release of cortisol from inter-renal tissues in fish, although α-MSH also mediates cortisol release (Wendelaar Bonga, 1997). Both α-MSH and β-MSH mediate dispersion of the pigment, melanin, in fish scales which darkens the skin of some fish in response to dark backgrounds (Baker et al., 1984).

There is little information on ACTH receptors in fish, but the ability of ACTH to stimulate cortisol release suggests that they must be located on inter-renal cells. ACTH receptors are likely to be coupled to adenylyl cyclase via G_s which stimulates increases in cAMP and cortisol release, similar to that in mammals. Information on the mechanism of action of MSH is a little more abundant. The MSH receptors themselves are not well characterized, but are known to stimulate increases in cAMP production via stimulation of adenylyl cyclase activity in melanocytes to mediate the dispersion of melanin (Sammak et al., 1992), although inositol-1,4,5-trisphosphate (IP_3)-mediated increases in intracellular calcium may also play a role (Fuji et al., 1991).

Melanin-concentrating hormone

Molecular structure

Melanin-concentrating hormone (**MCH**) is a cyclic heptadecapeptide hormone that has been purified or cloned from chum salmon, tilapia and rainbow trout (see Balm and Gröneveld, 1998). The mature MCH sequence is highly conserved or identical among fish such as tilapia, salmon, eel and bonito (*Katsuwonus pelamis*), but diverges from mammalian MCH. The MCH prohormone contains not only the MCH moiety, but two other peptides in tandem. In trout, one of these other peptides, **NEV**, is 13 amino acids long and may be related to **NEI** (neuropeptide-E-I) which is co-released with MCH in mammals. However, in tilapia the equivalent peptide, called **Mgrp**, may be 22 amino acids (Balm and Gröneveld, 1998).

Gene expression and regulation

MCH is located in the hypothalamus of vertebrates and only in fish is it released into the general circulation via the neurohypophysis. In mammals it appears to be restricted to a neural role. The hypothalamic MCH is located in two distinct regions, the **NRL** and **NLT** neuronal nuclei in fish, with the former serving mainly neurotransmitter functions and the latter involved in endocrine functions (Gröneveld et al., 1995). In support of the regional differences in MCH roles, hypothalamic MCH mRNA expression in the NLT, but not the NRL region was found to increase in response to stress in tilapia (Gröneveld et al., 1995). MCH immunoreactivity has also been found in fish spleen, intestine and testis (Kawauchi, 1989).

Hormone effectors in target tissues

MCH was first noted in fish because it causes pallor, accomplished by stimulating aggregation of fish scale melanin. This occurs in opposition to the effects of α-MSH both acutely in response to stress and chronically as an adaptation to a light-colored background (Baker, 1991). MCH inhibits α-MSH and CRF secretion, leading to suppression of the stress response in fish, but not in mammals. MCH may also be involved in osmoregulation and somatolactin release, although this observation is inconsistent (Francis et al., 1997). Little is known about MCH receptors and their signaling.

Vasotocin and isotocin

Molecular structure

Vasotocin and isotocin are neural nonapeptides found in fish. They are related to the mammalian vasopressin and oxytocin, respectively (see Table 25.2). A variety of other very similar neural peptides have been isolated from elasmobranchs. The members of this highly conserved neural peptide family are amidated on their C-terminal and tend to differ from each other by only one or two amino acid residues. The hormone peptide forms a cyclic structure via a disulfide bond between two conserved cysteine residues (Urano et al., 1994).

Vasotocin and isotocin cDNA have been obtained from many fish species, including white sucker (*Catostomus commersoni*), chum salmon, masu

TABLE 25.2: Amino acid sequence of vasotocin and isotocin peptides in fish compared to mammalian homologues. Bold letters indicate amino acid residue differences compared to the mammalian sequence

Hormone	Amino acid sequence	Species
Vasopressin family		
	1 2 3 4 5 6 7 8 9	
Arg-vasopressin	Cys-Tyr-Phe-Gln-Asn-Cys-Pro-Arg-Gly-NH$_2$	Mammals
Vasotocin	Cys-Tyr-**Ile**-Gln-Asn-Cys-Pro-Arg-Gly-NH$_2$	Teleosts, dogfish and lungfish
Oxytocin familiy		
	1 2 3 4 5 6 7 8 9	
Oxytocin	Cys-Tyr-Ile-Gln-Asn-Cys-Pro-Leu-Gly-NH$_2$	Mammals
Mesotocin	Cys-Tyr-Ile-Gln-Asn-Cys-Pro-**Ile**-Gly-NH$_2$	Lungfish, non-mammals
Isotocin	Cys-Tyr-Ile-**Ser**-Asn-Cys-Pro-**Ile**-Gly-NH$_2$	Teleosts
Glumitocin	Cys-Tyr-Ile-**Ser**-Asn-Cys-Pro-**Glu**-Gly-NH$_2$	Ray
Aspargtocin	Cys-Tyr-Ile-**Asn**-Asn-Cys-Pro-Leu-Gly-NH$_2$	*Squalus acanthias*
Valitocin	Cys-Tyr-Ile-Gln-Asn-Cys-Pro-**Val**-Gly-NH$_2$	*Squalus acanthias*
Asvatocin	Cys-Tyr-Ile-**Asn**-Asn-Cys-Pro-**Val**-Gly-NH$_2$	*Scyliorhinus caniculus*
Phasvatocin	Cys-Tyr-**Phe**-**Asn**-Asn-Cys-Pro-**Val**-Gly-NH$_2$	*Scyliorhinus caniculus*

Source: Adapted from Urano *et al.*, 1994.

salmon (*Oncorhynchus masu*), sockeye salmon, hagfish and lamprey (see Urano *et al.*, 1994; Kuno *et al.*, 1996). Probably due to tetraploidy, salmonids and castomids possess two gene copies each of vasotocin and isotocin which differ slightly in nucleotide sequence. The vasotocin and isotocin precursor proteins contain three regions: a signal peptide, a hormone moiety and a cysteine-rich peptide referred to as neurophysin. Another peptide, **copeptin**, can be produced by processing of the neuropeptide hormones in salmonids. **Neurophysin** may act as a chaperone during processing of vasotocin or isotocin, while the function of copeptin is not clear.

Gene expression and regulation

Generally, each fish species has one form of vasotocin and one form of oxytocin-like hormones, except for the cyclostomes, such as lamprey, which appear to express only a vasotocin peptide (Lane *et al.*, 1988). Vasotocin and isotocin-like peptides are produced in the hypothalamus of fish and secreted from separate neuronal populations in the neurohypophysis (Hyodo and Urano, 1991). Hypothalamic neurosecretory cells from eel are stimulated by glutamic acid, acetylcholine (via muscarinic receptors), dopamine, noradrenaline, vasotocin, isotocin, and angiotensin II *in vitro* to release these hormones, as well as increase their expression (Urano *et al.*, 1994). Expression and release of vasotocin and isotocin were also found to increase at the onset of spawning in pink salmon (*Oncorhynchus gorbuscha*) (Garlov and Khutinaev, 1991). Furthermore, transition from fresh water to salt water decreased mRNA/protein expression and release of both vasotocin and isotocin in diverse fish species (Urano *et al.*, 1994).

Hormone effectors in target tissues

Vasotocin and isotocin stimulate the release of ACTH from corticotropes, although not with the same efficacy as CRF or urotensin (Fryer *et al.*, 1985). This effect, in conjunction with direct effects on gill epithelium, confirms that the primary role of vasotocin and isotocin is osmoregulation in fish. These neuropeptides have also been shown to modify blood pressure, with an initial drop followed by an increase in rainbow trout blood pressure, although isotocin had less of a vasoregulatory effect than vasotocin (Warne and Balment, 1998).

Radiolabeled mammalian arginine–vasopressin revealed binding sites in the preoptic area and pituitary of sea bass, likely representing vasotocin receptors (Moons *et al.*, 1989). Additional vasotocin binding sites have been found in gills, from both freshwater- and seawater-adapted eels (Guibbolini et al., 1988). A vasotocin receptor has been cloned from several teleost species and most closely resembles the mammalian V_{1a} receptor (Mahlmann et al., 1994). The vasotocin receptor interacts with vasotocin, but not isotocin, suggesting that another receptor has yet to be characterized in fish (Guibbolini and Lahlou, 1990). Although isotocin receptors have not been characterized in fish, both vasotocin and isotocin inhibit adenylyl cyclase activity in rainbow trout gill epithelium (Guibbolini and Lahlou, 1992).

Gonadal hormones

Sex steroid hormones

Molecular structure

The sex steroid hormones are organic fused ring structures that are derived from cholesterol. The main sex steroids in fish are 17β-estradiol (E_2) and testosterone (T), but both 17α,20β-dihydroxy-4-pregnen-3-one (17α,20β-DP) and 11-ketotestosterone may play specific yet key roles in fish. The progesterone family of hormones, including 17α,20β-DP have 21 carbons and are referred to as C_{21} steroids, while androgens are C_{19} and estrogens are C_{18} steroid hormones.

Expression and regulation

Control of sex steroid hormone production occurs mainly by regulation of key enzymes in the biosynthetic pathway. In fish, the key biosynthetic enzymes are members of the cytochrome P450 (CYP450) enzyme family. $P450_{scc}$, which mediates side chain cleavage of cholesterol, is the rate-limiting step in all steroid hormone biosynthesis (see Figure 25.4 for synthesis schematic). Other biosynthetic P450 enzymes important for sex steroids are 3β-hydroxysteroid dehydrogenase (3β-HSD), 17β-hydroxysteroid dehydrogenase (17β-HSD), 17α-hydroxylase and aromatase. In fish ovaries, another important enzyme is 20β-hydroxysteroid dehydrogenase (20β-HSD) which is responsible for producing the maturation-inducing hormone (MIH), 17α,20β-DP. Sex steroid biosynthesis occurs mostly in the gonads, but extragonadal steroidogenesis is found in the brain, pituitary and spinal cord of fish (Pasmanik and Callard, 1985; Andersson et al., 1988). However, aromatase has recently been detected not only in kidneys, but also in muscle, stomach and intestine of rainbow trout (Belvedere et al., 1998). Also, 20β-HSD activity was detected in trout gill and the equivalent cyprinid enzyme, 20α-HSD, was found in goldfish eye, gills, heart and blood (Ebrahimi and Kime, 1998).

Levels of these steroidogenic enzymes may be regulated in a coordinate manner or individual enzymes may be induced at particular times in a specific tissue (Parker and Schimmer, 1993; see above section on GTH). Brain aromatase is high in goldfish compared with most other vertebrates and expression of its mRNA was found to fluctuate seasonally, with forebrain levels peaking in the spring in both sexes (Gelinas et al., 1998).

Hormone effectors in target tissues

The target of sex steroid hormones is largely gonadal, although the liver and brain are important targets as well. Estrogen receptors, which are members of a large nuclear receptor superfamily that act as transcription factors, have been separated into two different subtypes (ERα and ERβ) in mammals. Estrogen receptor cDNA, similar to ERα, has been cloned from rainbow trout and tilapia (Pakdel et al., 1994; Tan et al., 1996). ERα does not appear to be involved in somatic growth and development, but is responsible for mediating gonadal development and maturation in fish. All nuclear receptor family members, including ER, have six conserved domains (A to F domains), the most highly conserved of which is the C domain (LeDréan et al., 1994). The C domain contains two copies of a conserved DNA binding motif called zinc fingers. The largest domain is the E domain which is well conserved among steroid hormone receptors and is responsible for both hormone and heat shock protein (HSP) binding, as well as hormone-dependent transactivation of specific gene transcription (LeDréan et al., 1994). Androgen receptors are also part of the nuclear receptor superfamily, although relatively little is known about piscine androgen receptors. Recently, binding studies have shown that rainbow trout or goldfish brain, testis and ovary androgen receptor do not bind androgen ligands in the same way as mammalian androgen receptors, suggesting that there are large species differences in these receptors (Wells and Van Der Kraak, 1998).

When inactive, the estrogen and androgen receptors are located in the nucleus in fish and higher

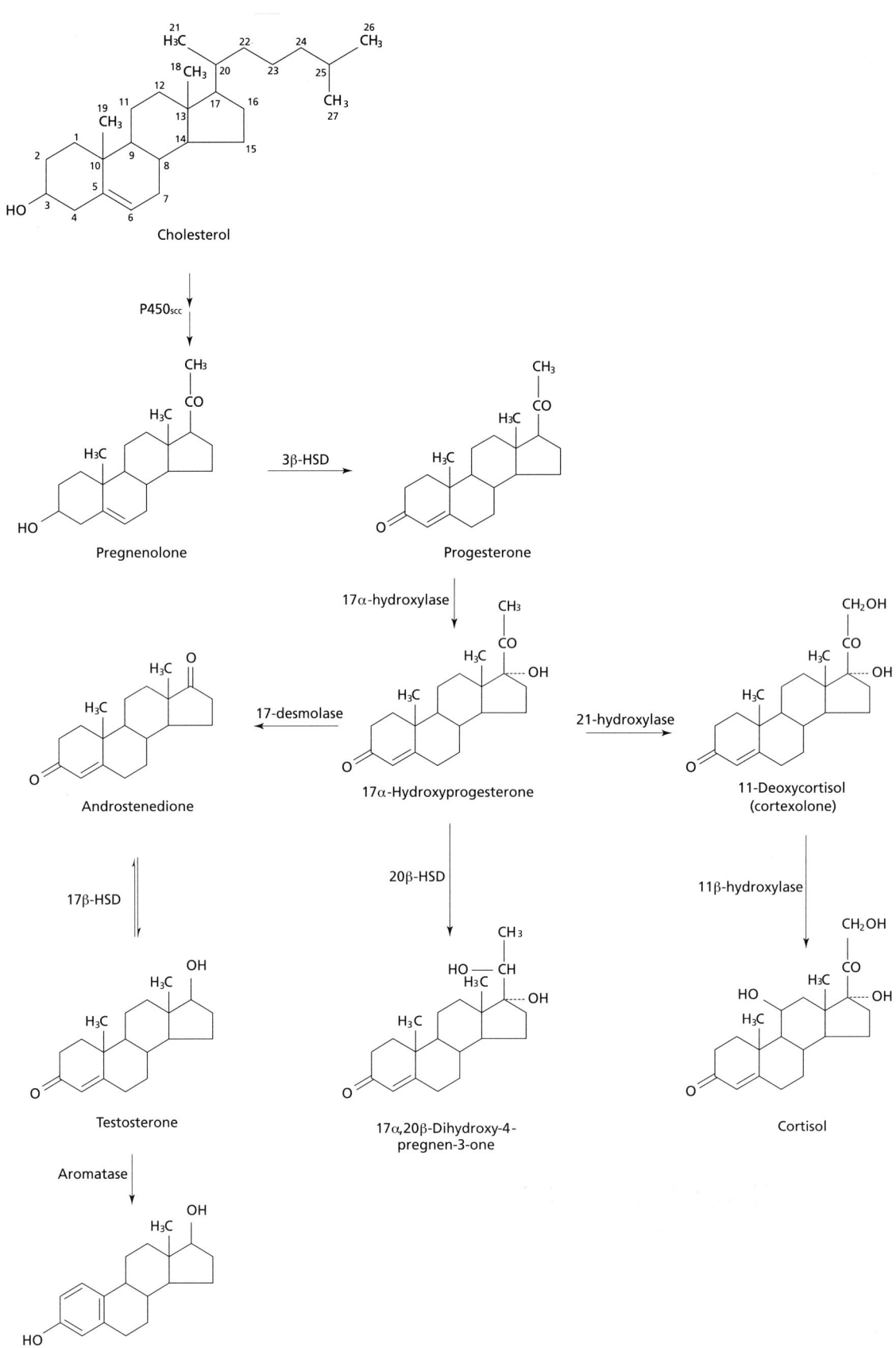

Figure 25.4 Biosynthesis of important piscine steroid hormones. Synthetic enzymes include P450$_{scc}$ (P450 side chain cleavage), 3β-HSD (3β-hydroxysteroid dehydrogenase) and 20β-HSD (20β-hydroxysteroid dehydrogenase).

vertebrates (Welshons et al., 1984; Gelinas and Callard, 1997). In fish, both receptor types are probably complexed with several pairs of **HSPs** which stabilize the receptor in its inactive form and prevent its association with DNA. For ER in mammals, these are hsp56, hsp70 and hsp90. For the mammalian androgen receptor, only hsp90 is thought to be associated. The sex steroid then diffuses across the plasma membrane of target cells, binds to ER which promotes dimerization and phosphorylation of receptor/hormone complexes, as well as loss of the HSPs. The steroid receptor is now active and binds to hormone-responsive elements (**HREs**) of the target cell DNA that are specific for each steroid hormone receptor, resulting in transcriptional activation of the gene immediately downstream of the HRE (Beato and Sanchez-Pacheco, 1996). A well-known target of ER is the vitellogenin gene in oviparous vertebrate liver, thus estrogen stimulation leads to increased expression of the vitellogenin protein (Pelissero et al., 1993). Another ER target is the ER gene itself in liver, thus stimulation of ER leads to increased expression of ER protein in fish liver (LeDréan et al., 1994).

The $17\alpha,20\beta$-DP receptor is located on the plasma membrane of fish oocytes, making this receptor very different from other steroid hormone receptors (Nagahama et al., 1994). The $17\alpha,20\beta$-DP receptors stimulate production of one subunit of a cytosolic second messenger complex called maturation-promoting factor or **MPF**. MPF has been purified from carp and is a complex of the cell cycle proteins, cdc2-kinase and cyclin B (Yamashita et al., 1992). The identity of the two **MPF** subunits has been confirmed with cDNA from goldfish (Nagahama et al., 1994). $17\alpha,20\beta$-DP probably elicits its effect by increasing levels of cyclin B which complexes with cdc2-kinase and promotes autophosphorylation of cdc2-kinase. This active complex then mediates germinal vesicle breakdown and oocyte maturation.

Thyroid hormones

T_4 and T_3

Molecular structure

Thyroid hormones are iodinated derivatives of the amino acid, tyrosine. In fish, the first type is thyroxine or $3,5,3',5'$-tetraiodothyronine (T_4), whereas the second type is its metabolite, $3,5,3'$-triiodothyronine (T_3). Active T_3 should not be confused with an inactive metabolite of T_4 that is also present in vertebrates, $3,3',5'$-triiodothyronine (reverse T_3 or rT_3). Thyroid hormones form spontaneously from tyrosine in cells where iodine is concentrated, such as in thyroid follicles. Thyroglobulins are large iodine-binding proteins found throughout vertebrates, including fish (Suzuki et al., 1975) and act as intermediates during iodination of tyrosine.

Expression and regulation

TSH stimulates T_4 release from thyroid follicular cells in all vertebrates. The majority of T_3 is produced peripherally after conversion of circulating T_4 by activating deiodinases, namely type I and II deiodinases (**D1** and **D2**; outer ring deiodinases). D2 deiodinases are highly expressed in liver and kidney of tilapia, African catfish, rainbow trout and turbot (*Scophthalmus maximus*) (Darras et al., 1998). However, brain contains high levels of **D3** deiodinase (an inner ring deiodinase) which converts T_4 to the inactive rT_3. Type I and type III deiodinases have been cloned from tilapia (Van Der Geyten et al., 1998), while type II has been cloned from killifish (*Heroclitus fundulus*) (Valverde-R et al., 1997). Food deprivation has been shown to decrease the levels of D2 deiodinase in tilapia, particularly in the liver, which correlated with decreased T_3 levels (Darras et al., 1998).

Hormone effectors in target tissues

In higher vertebrates T_4 is somewhat active and T_3 is the most active thyroid hormone, while T_3 may be the only active thyroid hormone in fish. The target of T_3 is mainly peripheral tissue where it stimulates increases in metabolism, mobilizes lipids and increases protein synthesis. T_3 also stimulates increases in GH mRNA expression and GH release from the pituitary (Moav and McKeown, 1992). Thyroid receptors belong to the superfamily of nuclear receptors that includes the steroid hormone receptors discussed above. In mammals, inactive thyroid receptors are located within the nucleus of target cells already bound to specific DNA sequences (hormone-responsive elements or HREs) where they act as gene repressors, along with a co-repressor protein called **SMRT** (silencing mediator of retinoic acid and thyroid hormone receptor) (Chen and Evans, 1995). T₃ diffuses across the plasma membrane, binds to the thyroid receptor and causes release of SMRT. This alleviates the repression of transcription of the downstream gene, allowing production of its mRNA and subsequent translation into protein. No heat shock proteins (HSPs) are known to be associated with the thyroid receptor.

Gastroenteropancreatic hormones

Insulin and insulin-like growth factors

Molecular structure

Insulin is a two-chain peptide hormone produced from one linear preprohormone, linked by two disulfide bonds. The B chain of insulin is about 30 and the A chain 21 amino acids long (reviewed in Mommsen and Plisetskaya, 1991). Mature insulin is produced first by cleavage of the signal peptide from the preprohormone, to produce proinsulin, followed by cleavage of the C-peptide from between the A and B chains to form mature insulin (Rouille et al., 1995). The insulin gene has been cloned from several species of fish, including anglerfish (*Lophius americanus*), hagfish and salmon. In some teleosts such as chum salmon, there are two different genes that produce different insulin isoforms (Mommsen and Plisetskaya, 1991). All members of the insulin family exhibit high degrees of homology between fish and mammalian forms (about 60–90%).

Insulin-like growth factors (IGF-I and IGF-II) are related both structurally and functionally to insulin. IGF-I and -II are two-chain peptide hormones that originate from preprohormones processed similarly to insulin. IGF-I genes have been cloned from many salmonid species, including Atlantic salmon (*Salmo salar*), chinook salmon, chum salmon and rainbow trout (see Chan and Steiner, 1994). The IGF-II gene has been cloned from rainbow trout (Shamblott and Chen, 1992).

Gene expression and regulation

Insulin is expressed and released from pancreatic tissue in fish. Amino acids and to some extent glucose are potent secretagogues in fish which differs from mammals where glucose is the most potent (Mommsen and Plisetskaya, 1991).

In coho salmon, IGF-I mRNA expression was highest in the liver, but was also found in adipose tissue, brain, heart, kidney, liver, muscle, ovaries, testes and spleen (Duguay et al., 1992). On the other hand, IGF-II appears to be expressed only during embryogenesis and juvenile development, except perhaps in females during vitellogenesis (Le Bail et al., 1998).

Hormone effectors in target tissues

The targets of insulin include most tissues, particularly skeletal muscle and lipocytes, where it stimulates nutrient uptake, increases in protein synthesis and deposition of fatty acids into lipids (Mommsen and Plisetskaya, 1991). The main role of IGF-I may be to mediate effects of growth hormone, namely somatic growth and/or osmoregulatory adaptation to fresh water, but evidence for such roles is inconsistent (Chan and Steiner, 1994). Also, gonadal IGFs stimulate steroidogenesis and may mediate the steroidogenic effect of growth hormone (Le Bail et al., 1998).

Insulin and IGF receptors are very similar and are considered to belong to the same receptor family. The effects of IGF-I and IGF-II are largely mediated by the type I IGF receptor which is similar to the insulin receptor, both being tetrameric plasma membrane receptors composed of two α- and two β-subunits linked by disulfide bonds. A unique feature of this superfamily of receptors is that both insulin and IGFs can bind to their own and each other's receptors, albeit with slightly different potencies. The α-subunits are extracellular and bind IGF, while the β-subunits cross the membrane and possess tyrosine kinase activity that is activited upon ligand binding. A type I IGF receptor has been cloned recently from turbot and is expressed in extremely high levels in fish skeletal muscle (Elies et al., 1998). This illustrates an important difference between mammals and fish because mammals have more insulin receptors than IGF receptors throughout the body, while the reverse is true in fish.

Glucagon

Molecular structure

Glucagon is a 29 amino acid peptide belonging to a superfamily that also includes secretin, vasoactive intestinal peptide (VIP), gastric inhibitory peptide (GIP), growth hormone-releasing hormone, and PACAP. Secretin, VIP and GIP will not be discussed in this chapter due to limited information and space. Glucagon is a preprohormone that also encodes several other hormones, including glucagon-like peptide (GLP), depending on post-translational processing (Rouille et al., 1995). Glucagon has been purified and sequenced from many fish species, including dogfish, anglerfish, channel catfish (*Ictalurus punctatus*), coho salmon, European eel and flounder (*Platichthys flesus*), with several species expressing two different isoforms (see Duguay and Mommsen, 1994). Very little is known about piscine glucagon, except that peptide sequences of the hormone moiety, but not intervening sequences of the prohormone, are reasonably well conserved with higher vertebrate glucagons.

Gene expression and regulation

Glucagon expression has been found in the intestine and Brockmann bodies of many teleosts, the latter being the major source. The brain may also express glucagon, but the role of this peptide there is not clear. Secretion of glucagon is inhibited by glucose while amino acids may stimulate secretion (Duguay and Mommsen, 1994).

Hormone effectors in target tissues

The main target of glucagon is the liver where it acts as an anti-insulin factor by stimulating gluconeogenesis and glycogenolysis. The glucagon receptors in American eel (*Anguilla rostrata*), rainbow trout and brown bullhead (*Ictalurus nebulosus*) may not couple to adenylyl cyclase like mammalian receptors since cAMP levels do not change after stimulation with physiological levels of glucagon. The fish hepatic glucagon receptor may couple to phospholipase C and increases in intracellular calcium instead (Duguay and Mommsen, 1994).

Somatostatin

Molecular structure

There are at least two somatostatin (SST) or SRIF (somatotropin release-inhibitory factor) precursors, preprosomatostatin I (PPSS I) and II (PPSS II) which are encoded by two separate genes in teleosts, similar to mammals (reviewed in Conlon et al., 1997). After post-translational processing of the PPSS I and II forms, two different products are formed, SST-14 and SST-22, respectively. The core of all somatostatins is SST-14, a highly conserved cyclic peptide linked by a disulfide bridge between cysteine^4 and cysteine^{14}. A larger variant, SST-28, may have a specific role in endocrine pancreatic cells or may serve only as a circulating precursor for active SST-14. Also, SST-25 has been found in eels and salmonids while SST-34 and -37 have been found in cyclostomes (Conlon et al., 1997). The different forms arise from either of the two precursor hormones and represent differences in the proteolytic processing of somatostatin.

Gene expression and regulation

Although both PPSS-I and PPSS-II are expressed in the Brockmann bodies of fish, they do not appear to be co-localized, but rather are found in distinct cell types (Sevarino et al., 1989). SST-14 is released from rainbow trout Brockmann body D-cells primarily in response to glucose, although cholecystokinin and neuropeptide Y also stimulate its release to some extent (Eilertson et al., 1996). SST (or SRIF) is also found in the hypothalamus of fish (Power et al., 1996).

Hormone effectors in target tissues

The main targets of somatostatin released from D-cells are adjacent cells (i.e. paracrine action) where it inhibits release of insulin and glucagon. Another target of SST-14 is the pituitary where it inhibits GH release in many teleosts, including grass carp (*Cyprinus idellus*) (Wong et al., 1998b). However, the longer somatostatin forms, SST-22 of catfish and SST-25 of salmonids, do not appear to be involved in inhibition of GH release (Peter and Marchant, 1995). SRIF receptor stimulation by somatostatin in tilapia or rainbow trout pituitaries reduces GH expression and release by decreasing cAMP levels (Melamed et al., 1998).

Osmoregulatory hormones

Cortisol

Molecular structure and regulation

Cortisol is a C_{21} steroid hormone that is an organic fused ring structure derived from cholesterol, similar to the sex steroid hormones. Cortisol is synthesized via 21-hydroxylase and 11β-hydroxylase (see Figure 25.4), then released from the inter-renal tissue of fish head kidney in response to a variety of stimuli, but predominantly after ACTH stimulation (Wendelaar Bonga, 1997).

Hormone effectors in target tissues

In fish, both mineralocorticoid (i.e. osmoregulatory) and glucocorticoid (i.e. metabolic) effects are mediated by cortisol, unlike mammals where separate steroid hormones (aldosterone and corticosterone, respectively) mediate each effect (Hanke and Kloas, 1995). The primary cortisol target for osmoregulation is the gills, while for metabolic effects it is the liver. Other physiological processes such as reproduction also affect cortisol release. For instance, cortisol release is increased and decreased by 17α,20β-DP (MIH) and 11-ketotestosterone, respectively, in rainbow trout (Young et al., 1996; Barry et al., 1997).

Consistent with the fact that fish use only cortisol to regulate dual functions, only glucocorticoid (**GR**) receptors and not mineralocorticoid (**MR**) receptors have been found in fish. GR cloned from rainbow trout is similar to mammalian GR, particularly in the DNA binding C domain, but with an extra insert between the two zinc fingers (Ducouret et al., 1995). GR binds cortisol with high affinity, but lacks any affinity for binding aldosterone (the mineralocorticoid of higher vertebrates). GR is expressed in intestine, chloride cells of the gills, liver, leukocytes and hypothalamus (Knoebl et al., 1996; Teitsma et al., 1997; Uchida et al., 1998).

GR is a member of the nuclear receptor superfamily (see sex steroid hormone section of this chapter). However, unlike other steroid hormone receptors that are strictly nuclear, cortisol receptors are cytoplasmic when inactive and translocate to the nucleus upon hormone binding (Allison and Omeljaniuk, 1998). Piscine GR is likely to be associated with hsp90, hsp70, and hsp56 similar to mammalian GR. Analogous to ER and the estrogen gene, a target of the cortisol/GR transcriptional promoter complex is the GR gene itself (Maule and Schreck, 1991). Another GR target may be the sodium pump or Na^+/K^+-ATPase of gill chloride cells (Madsen, 1990).

Urotensins and corticotropin-releasing factors

Molecular structure

Several proteins that stimulate release of ACTH (andrenocorticotrophic hormone), similar to mammalian corticotropin-releasing factor (CRF or **CRH**), have been characterized in fish (reviewed in Lederis et al., 1994). Both piscine isoforms, CRF1 and CRF2, are close in structure to the 41 amino acid mammalian CRF. Another uniquely piscine CRF homologue was found in the urophysis and has been called urotensin I. Thus, the current piscine CRF family is composed of CRF1, CRF2 and urotensin I.

Carp pro-urotensin I has been cloned and is a 145 amino acid peptide (Ishida et al., 1986). CRF precursor proteins characterized from the white sucker have 46–54% conservation in amino acid sequence compared to mammalian CRF. The similarity in mature hormone structure between CRF and urotensin I allows antibodies raised to ovine CRF to cross-react with urotensin I (Cioni et al., 1998). Both CRF and urotensin I precursor proteins consist of three domains: a signal peptide, a highly conserved peptide of unknown function and the hormone moiety (Ishida et al., 1986; Morley et al., 1991). Post-translational processing of CRF and urotensin I involves cleavage of the signal peptide from the precursor, then amidation of the C-terminal of the hormone peptide to produce a mature α-helical CRF or urotensin I hormone.

Another fish urophysial protein is urotensin II (see Table 25.3), but it is structurally distinct from urotensin I or CRF since it is a dodecapeptide with a ring formed by a disulfide bond between two cysteines (Conlon et al., 1996). In fact, urotensin II structurally resembles somatostatin more than it resembles CRF. Multiple urotensin II isoforms arising from multiple genes have been found in white sucker (two) and carp (four) (Ohsako et al., 1986).

TABLE 25.3: Amino acid sequences of urotensin II from different species of fish. Bold letters indicate amino acid residue differences compared to the lamprey and elasmobranch sequence

Species	1 2 3 4 5 6 7 8 9 10 11 12
Lamprey, dogfish and rays	Asn-Asn-Phe-Ser-Asp-Cys-Phe-Trp-Lys-Tyr-Cys-Val
Sturgeon, paddlefish	**Gly-Ser**-Thr-Ser-**Glu**-Cys-Phe-Trp-Lys-Tyr-Cys-Val
Sucker (IIB)	**Gly-Ser-Asn-Thr-Glu**-Cys-Phe-Trp-Lys-Tyr-Cys-Val
Trout	**Gly-Gly-Asn**-Ser-**Glu**-Cys-Phe-Trp-Lys-Tyr-Cys-Val
Carp (IIβ)	**Gly-Gly-Asn-Thr-Glu**-Cys-Phe-Trp-Lys-Tyr-Cys-Val
Flounder	**Ala-Gly-Thr**-Thr-**Glu**-Cys-Phe-Trp-Lys-Tyr-Cys-Val
Goby	**Ala-Gly-Thr-Ala**-Asp-Cys-Phe-Trp-Lys-Tyr-Cys-Val

Source: Adapted from Conlon *et al.*, 1996.

Gene expression and regulation

CRFs are present in the hypothalamus of fish and their expression is inhibited by corticosteroids, representing a negative feedback loop (Morley *et al.*, 1991). Both urotensin I and urotensin II are found in fish hypothalamus, but they are also co-localized in Dahlgren cells that project from the spinal cord to the neurohemal (teleosts) or neurosecretory (elasmobranchs) urophysis (Cioni *et al.*, 1998). Urophysectomy leads to increases in urotensin I expression in one of two areas of the brain of white sucker where urotensin I is found, suggesting that there is tissue-specific regulation of urotensin I levels (Morley *et al.*, 1991).

Hormone effectors in target tissues

CRF and urotensins stimulate release of ACTH from corticotropes in the pituitary and inter-renal cells, although they can also stimulate release of ACTH and α-MSH from melanotropes in goldfish (Lederis *et al.*, 1994). In addition to a role in stress/osmoregulation, the urotensins, but not CRF, are noted for their vasoactive actions with urotensin I mediating vasorelaxation and urotensin II mediating vasoconstriction of a variety of vascular tissues (Conlon *et al.*, 1996). Finally, CRF and urotensin I may play a major role in release of thyrotropin (TSH) from pituitary thyrotropes (Larsen *et al.*, 1998).

CRF receptors have been tentatively separated into two different subtypes based on functional differences. Type I CRF receptors exhibit a high potency for urotensin I, and are found in goldfish pituitary corticotropes and in mammalian arterial tissue. Type II receptors have a similar potency for CRF and urotensin I, and are found in mammalian pituitary corticotropes. The CRF receptor signaling mechanism is not known in fish, but may be similar to mammalian receptors that involve cAMP, calcium and protein kinase C (Lederis *et al.*, 1994).

Urotensin II receptors stimulate adenylyl cyclase activity and increases in target cell cAMP (Perrott *et al.*, 1993). Responsiveness of the kidney to urotensin II stimulation has been observed in seawater-, but not freshwater-adapted flounder, suggesting that salinity may regulate levels or responsiveness of the urotensin II receptors (Stenhouse and Balment, 1998).

Angiotensins

Molecular structure

Angiotensins are highly conserved peptides present throughout all vertebrates. The preprohormone of angiotensins is angiotensinogen that is produced in the liver and present in the general blood circulation. Angiotensinogen is cleaved by the kidney enzyme 'renin' to form the decapeptide angiotensin I (Ang I; see Table 25.4). Ang I is then converted by angiotensin-converting enzyme (**ACE**) to the octapeptide angiotensin II (**Ang II**) which is considered to be physiologically the most important angiotensin (reviewed in Wilson, 1984). In fish, the proteolytic products of Ang II do not appear to have any physiological roles, unlike in mammals.

Gene expression and regulation

Angiotensins are expressed in blood, brain and kidney in teleosts, as well as the rectal gland of elasmobranchs

TABLE 25.4: Amino acid sequence of angiotensin I from several fish species and from human for comparison. Bold letters indicate amino acid residue differences compared to the human sequence. Arrow indicates the point of cleavage by angiotensin-converting enzyme (ACE) that produces angiotensin II

Species	1 2 3 4 5 6 7 8 ↓ 9 10
Human	Asp-Arg-Val-Tyr-Ile-His - Pro -Phe-His-Leu
Salmon	**Asn**-Arg-Val-Tyr-**Val**-His-Pro-Phe-**Asn**-Leu
Eel	**Asn**-Arg-Val-Tyr-**Val**-His-Pro-Phe-**Gly**-Leu
Goosefish	**Asn**-Arg-Val-Tyr-**Val**-His-Pro-Phe-**His**-Leu
Dogfish	**Asn**-Arg-**Pro**-Tyr-Ile-His-Pro-Phe-**Gln**-Leu

Source: Adapted from Takei *et al.*, 1993.

(Galli and Phillips, 1996). More important than tissue expression of angiotensins is the location of renin and ACE since angiotensinogen is present in the general circulation and therefore available to most tissues. Renin is stored in the kidney with prorenin and both are released into the plasma. Isoforms of this enzyme have been identified in non-renal tissues, including brain, uterus and corpuscles of Stannius. In dogfish, ACE was found in gills, spleen, kidney and brain (Uva *et al.*, 1992).

Hormone effectors in target tissues

The main targets of angiotensins are vascular and renal smooth muscle cells. Angiotensin receptor stimulation leads to vasoconstriction and increased blood pressure, as well as anti-natriuresis, decreased prolactin secretion and increased drinking in seawater fish. Both **Ang I** and II are also known to stimulate ACTH release from pituitary corticotropes in fish, unlike mammals where only Ang II is capable of ACTH release (Lederis *et al.*, 1994).

Angiotensin receptors have been separated into several major subtypes (AT_1–AT_4) in vertebrates, but only AT_1 and AT_2 have functions clearly attributed to them. In Japanese dogfish (*Triakis scyllia*), **AT** receptors were found in highest levels in inter-renal tissue and gills, followed by rectal gland, intestine, heart, liver and kidney (Tierney *et al.*, 1997). However, the dogfish AT receptors, as well as receptors that cause ACTH release in other fish, do not resemble mammalian AT_1 or AT_2 receptors pharmacologically (Lederis *et al.*, 1994; Tierney *et al.*, 1997). In rainbow trout and European eel, AT receptors were found in brain, smooth muscle, kidney, gill, skin, intestine and heart, but again were unlike either mammalian AT receptor (Cobb and Brown, 1992; Marsigliante *et al.*, 1996).

AT receptors couple to G proteins and phospholipase C (PLC) in mammalian systems, although the transducers are unknown in fish. Stimulation of gill AT receptors stimulates increases in Na^+/K^+-ATPase activity in both freshwater- and saltwater-adapted eel (Marsigliante *et al.*, 1997).

Natriuretic peptides

Molecular structure

Atrial natriuretic peptide (**ANP** or atrial natriuretic factor or **ANF** or **atriopeptin**) hormones identified in eel and rainbow trout atria are 27 and 29 amino acids long, respectively, while mammalian ANP is 28 residues (Takei *et al.*, 1997). ANP belongs to the type A family of natriuretic peptides. Type C natriuretic peptides are present mostly in fish brain, but are also found in hearts of some fish species. This 22 amino acid fish brain natriuretic peptide is identical to the mammalian type C natriuretic peptide (**CNP**) (Galli and Phillips, 1996). A third type of natriuretic peptide found in rainbow trout and eel ventricles, **VNP**, is 36 or 37 amino acids long, respectively and is not sufficiently similar to belong to any mammalian natriuretic peptide group (Takei *et al.*, 1997). All natriuretic peptides have a disulfide bond between their two cysteine residues, forming a peptide loop that is necessary for their biological activity.

In mammals, ANP prohormone contains the ANP moiety at the C-terminal (proANP residues 99–126), as well as two other peptides that exhibit similar biological activity to ANP, but with longer durations of action (Vesely et al., 1994). The rainbow trout ANP prohormone is a 126 amino acid peptide and is similar in structure to the mammalian ANP (Cousins et al., 1997). The two other ANF peptides have been named based on their most potent effects: long-acting natriuretic peptide (proANP residues 1–30) and vessel dilator (proANP residues 31–67).

Gene expression and regulation

ANP and VNP are expressed in the atria and ventricles, respectively, of fish while CNP is expressed in the brain of fish. The release of ANP is higher in seawater than in freshwater fish, reflecting a major role in osmoregulation (Westenfelder et al., 1988), but it is not known whether atrial/ventricular expression increases in response to a saline environment. Recently, proANP 1–30 and proANP 31–67 were found circulating at three times higher concentrations than ANP itself in the plasma of both seawater- and freshwater-adapted rainbow trout (Cousins et al., 1997). The release of the proANP hormones increased to a greater extent in seawater-adapted trout in response to increased cardiac filling pressure, suggesting that these other proANP hormones may also be important in osmoregulation.

Hormone effectors in target tissues

In general, natriuretic peptides antagonize all physiological actions of angiotensins. Specifically, natriuretic peptides cause vasorelaxation, natriuresis, increase prolactin secretion, and inhibit ACTH secretion. Natriuretic peptide receptors are present in the gills, vascular smooth muscle and renal tubular cells in fish (Evans et al., 1989). Mammalian natriuretic peptide receptors have been divided into two subtypes: type A which prefers to bind ANP and type B which prefers CNP. Both type A and type B ANP receptors have been cloned from eel (see Takei et al., 1997), but other natriuretic peptide receptor(s) are likely to be present since fish VNP is fairly different from mammalian natriuretic peptides. The known ANP receptors possess an intrinsic **guanylyl cyclase** activity. If guanylyl cyclase activity is stimulated by hormone binding, **cGMP** production increases which in turn stimulates cGMP-dependent protein kinase (**PKG**) activity. Potent stimulation of guanylyl cyclase activity by CNP has been found in spiny dogfish (*Squalus acanthias*) rectal gland, likely mediating CNP-stimulated rectal gland salt secretion (Gunning et al., 1997).

Somatolactin

Molecular structure

Somatolactin is a unique glycosylated peptide hormone from fish pituitary that is structurally related to growth hormone and prolactin. Somatolactin was purified and sequenced originally from Atlantic cod (*Gadus morhua*) (Rand-Weaver et al., 1991), but cDNA has since been obtained from several other teleosts (see Kaneko, 1996). The somatolactin gene codes for a prohormone composed of a 24–26 amino acid signal peptide followed by a 207 (flounder) or 209 (salmon/cod) amino acid hormone moiety. The amino acid sequences of the three piscine somatolactin isoforms are highly conserved with each other and are approximately 24–26 kDa in weight. In fact, the amino acid sequence conservation among the piscine somatolactin isoforms is greater than that among growth hormone. There are three characteristic disulfide bonds formed by a series of cysteine residues that are unique to somatolactins.

Gene expression and regulation

Somatolactin mRNA is expressed exclusively in the pituitary, particularly the pars intermedia, and not in other tissues such as stomach, liver, heart, intestine, spleen or brain (Johnson et al., 1997). The pars intermedia is an area that is also associated with high levels of POMC-related hormones (Naito et al., 1984). Both plasma and pituitary levels of somatolactin increased as Atlantic salmon matured sexually and decreased as a result of castration, suggesting that somatolactin is regulated by gonadal hormones (Mayer et al., 1998).

Hormone effectors in target tissues

The target of somatolactin has, until recently, remained relatively unknown (Kaneko, 1996). A recent study in flounder suggests that the kidney may be a major target (Lu et al., 1995). Somatolactin was found to stimulate inorganic phosphate resorption in kidney proximal tubule epithelial cells via stimulation of cAMP production, suggesting that the somatolactin receptor couples positively to **adenylyl cyclase**. However, further research in this area and into identifying somatolactin receptors themselves is needed.

Prolactin

Molecular structure

Prolactin, being a member of the growth hormone and somatolactin family, is an approximately 200 amino acid glycosylated hormone that likely forms the characteristic bundle of α helices of this family. Teleost prolactins are single chain polypeptides with two cross-linking disulfide bridges. Two forms of prolactin have been isolated from several fish species, including salmon, carp, eel and tilapia, which are likely to be products of separate genes (see Suzuki et al., 1991). These prolactin isoforms show strong homology to each other (60–80%), but low homology with mammalian prolactins (20–30%). cDNA clones have also been obtained for the two prolactin isoforms in tilapia, named **tiPRL$_1$** and tiPRL$_2$, which are 188 and 177 amino acids long, respectively (Rentier-Delrue et al., 1989).

Gene expression and regulation

Prolactin is expressed and released from lactotrope cells of the teleost pituitary, with both isoforms being co-expressed in the same cells (Specker et al., 1993).

Hormone effectors in target tissues

The target of prolactin is mainly the kidney, but also the gills and intestine, consistent with a major role in osmoregulation and freshwater adaptation (Prunet et al., 1990). However, the mechanism by which osmoregulation is accomplished, either through Na^+ retention or Cl^- reabsorption or both, appears to vary greatly among different fish species. Early receptor binding studies may have led to erroneous conclusions that multiple receptors are present in liver, gills and kidney. The early studies used ovine prolactin and this was later determined to bind non-discriminately to both growth hormone and prolactin receptors in fish species such as tilapia (reviewed in Prunet and Auperin, 1994). Thus, caution must be exercised in interpreting results from studies using heterologous hormones. More recently, using homologous mRNA probes, single transcripts of prolactin receptor mRNA were found in rainbow trout and tilipia kidney, gills, gut, liver, skin, head, kidney, spleen and gonads, but not liver (Prunet et al., 1997). This suggests that a single prolactin receptor may be present in each of these piscine tissues, contradicting earlier studies which suggested the presence of multiple prolactin receptors. However, the number and affinity of gill prolactin receptors for both isoforms of prolactin are regulated by the osmotic environment, with both parameters increasing in tilapia transferred from fresh to brackish water. The mechanism(s) by which this occurs is not known (Prunet and Auperin, 1994).

Similar to the growth hormone receptor, prolactin receptors may dimerize upon prolactin binding (Sandra and Prunet, 1998). Once the prolactin receptors are activated, a tyrosine kinase called JAK (*ja*nus *k*inase) is activated and transduces the prolactin signal in target tissues, but this needs to be confirmed in fish tissues (Carter-Su and Smit, 1998).

Calcitonin

Molecular structure

Calcitonin is a 32 amino acid long peptide hormone and has been purified from salmonids. Three different isoforms of calcitonin have been identified from salmon, with all species expressing calcitonin I, but variable expression of calcitonin II or calcitonin III (Potts et al., 1972). Calcitonin has also been isolated from a cartilaginous fish, the stingray (*Dasyatis akatei*) (Takei et al., 1991). Through alternative splicing of the calcitonin mRNA, another peptide product of 37 amino acids called calcitonin gene-related peptide (**CGRP**) can be produced and has been detected in salmon (Jansz and Zanberg, 1992).

Gene expression and regulation

In salmonids, calcitonin gene expression has been found mainly in the gills (Martial et al., 1994). However, calcitonin immunoreactivity has been found in nervous tissue and in the pituitary (Deftos et al., 1980; Girgis et al., 1980). CGRP is more of a neuropeptide than calcitonin, but may also be released into the general circulation to perform endocrine functions (Lamharzi and Fouchereau-Peron, 1996).

Hormone effectors in target tissues

The role of calcitonin in fish is contentious, but has been shown to inhibit calcium transport in isolated, perfused gills (reviewed in Wendelaar Bonga and Pang, 1991). Recently, this was corroborated *in vivo* when calcitonin was shown to inhibit whole body uptake of ^{45}Ca in fingerling rainbow trout (Wagner et al., 1997). In support of a direct action of calcitonin on fish gills is the observation of specific calcitonin receptors in trout gill cells (Fouchereau-Peron et al., 1981). Also, calcitonin increased the expression of stanniocalcin mRNA in fingerling rainbow trout, suggesting a concerted action between these two hypocalcemic hormones in fish. Release of calcitonin from the gills produces relatively low concentrations in the general circulation, but because calcitonin is released and acts locally in the gills, concentrations may be sufficiently high there to have an effect on calcium transport (Wagner et al., 1997). However, work is needed to distinguish the role, if any, calcitonin has in osmoregulation in fish, as well as characterization of its receptor and signaling mechanism.

The role of CGRP is sketchy in fish, but it was shown to be released at higher concentrations and bind specifically to its receptors in the gills when rainbow trout were transferred from fresh to sea water (Lamharzi and Fouchereau-Peron, 1996).

Stanniocalcin

Molecular structure

Stanniocalcin (**hypocalcin** or formerly **teleocalcin**) is a glycoprotein that appears to be unrelated to any other vertebrate hormone. Amino acid sequences and/or cDNA sequences for stanniocalcin have been obtained from Australian eel (*Anguilla australis*), coho salmon and chum salmon, ranging from 231 to 233 amino acid residues in size (Butkus et al., 1987; Wagner et al., 1992; Yamashita et al., 1995). Wide variations in the apparent size of stanniocalcin, 32–60 kDa, may be a result of disulfide-linked dimerization of the hormone or variable carbohydrate content.

Gene expression and regulation

Stanniocalcin protein is expressed in the organ that it was named for, the corpuscles of Stannius, located near the kidney of fish. Stanniocalcin mRNA was thought to be exclusively expressed in the corpuscles of Stannius, but recent work suggests it may also be located in the renal tubules of certain teleosts (Wagner, 1994; Marra et al., 1997). Release of stanniocalcin is stimulated by hypercalcemia and likely arises by some unknown calcium sensor in the stanniocalcin cells themselves (Wagner, 1994). Also, increasing length of exposure or increasing hypercalcemia both stimulate increases in stanniocalcin mRNA expression *in vitro* (Wagner and Jaworski, 1994).

Hormone effectors in target tissues

The targets of stanniocalcin are the gills, intestine and kidney where it has anti-hypercalcemic action by affecting Ca^{2+} and phosphate transport (Wagner, 1994). Virtually nothing about stanniocalcin receptors or their signaling mechanisms is known.

Conclusion

Knowledge of molecular endocrinology in fish is rapidly expanding and will continue to do so in the future. This information is invaluable, particularly to comparative endocrinologists and molecular evolutionary biologists. However, as more genes are cloned for peptide hormones or hormone processing enzymes, endocrinologists must take the responsibility of putting them into a physiological context. This represents the real challenge of the future.

References

Abdel-Meguid, S.S., Shieh, H.S., Smith, W.W., Dayringer, H.E., Violand, B.N. and Bentle, L.A. (1987). *Proc. Natl. Acad. Sci. USA* **84**, 6434–6437.

Allison, C.M. and Omeljaniuk, R.J. (1998). *Gen. Comp. Endocrinol.* **110**, 2–10.

Amano, M., Ashihara, M.O., Yoshiura, Y., Kitamura, S., Ikuta, K. and Aida, K. (1998). *Cell Tissue Res.* **292**, 267–273.

Andersson, E., Borg, B. and Lambert, J.G.D. (1988). *Gen. Comp. Endocrinol.* **72**, 394–401.

Baker, B.I. (1991). *Int. Rev. Cytol.* **126**, 1–47.

Baker, B.I., Wilson, J.F. and Bowley, T.J. (1984). *Gen. Comp. Endocrinol.* **55**, 142–149.

Balm, P.H.M. and Gröneveld, D. (1998). *Ann. NY Acad. Sci.* **839**, 205–209.

Barry, T.P., Riebe, J.D., Parrish, J.J. and Malison, J.A. (1997). *Gen. Comp. Endocrinol.* **107**, 172–181.

Beato, M. and Sánchez-Pacheco, A. (1996). *Endocrin. Rev.* **17**, 587–609.

Belvedere, P., Dalla Valle, L., Lucchetti, A., Ramina, A., Vianello, S. and Colombo, L. (1998). *Ann. NY Acad. Sci.* **839**, 589–591.

Bentley, P.J. (1998). *Comparative Vertebrate Endocrinology*, 3rd edn. Cambridge University Press, Cambridge, UK.

Bosma, P.T., Kolk, S.M., Rebers, F.E.M., Lescroart, O., Roelants, I., Willems, P.H.G.M. and Schulz, R.W. (1997). *J. Endocrinol.* **152**, 437–446.

Butkus, A., Roche, P.J., Fernley, R.T., Haralambidis, J., Penschow, J.D., Ryan, G.B., Trahair, J.F., Tredear, G.W. and Coghlan, J.P. (1987). *Mol. Cell Endocrinol.* **54**, 123–133.

Calduch-Giner, J.A., Sitjà-Bobadilla, A., Alvarez-Pellitero, P. and Pérez-Sánchez, J. (1997). *Cell Tissue Res.* **287**, 535–540.

Carter-Su, C. and Smit, L.S. (1998). *Recent Prog. Horm. Res.* **53**, 61–82.

Chan, S.J. and Steiner, D.F. (1994). In *Fish Physiology*, Vol. 13 (eds N.M. Sherwood and C.L. Hew), pp. 213–224. Academic Press, San Diego, CA.

Chang, J.P., Wong, A.O.L., Van Goor, F., Goldberg, J.I., Jobin, R.M. and Johnson, J.D. (1997). *Advances in the Molecular Endocrinology of Fish*. 2nd IUBS Toronto Symposium, S7.

Chartrel, N., Tonon, M.C., Vaudry, H. and Conlon, J.M. (1991). *Endocrinology* **129**, 3367–3371.

Chen, J.D. and Evans, R.M. (1995). *Nature* **377**, 454–457.

Chen, T.T., Marsh, A., Shamblott, M., Chan, K.M., Tang, Y.L., Cheng, C.M. and Yang, B.Y. (1994). In *Fish Physiology*, Vol. 13 (eds N.M. Sherwood and C.L. Hew), pp. 3–66. Academic Press, San Diego, CA.

Cioni, C., De Vito, L., Greco, A. and Pepe, A. (1998). *J. Morphol.* **235**, 59–76.

Cobb, C.S. and Brown, J.A. (1992). *J. Comp. Physiol.* **162**, 197–202.

Conlon, J.M., Yano, Y., Waugh, D. and Hazon, N. (1996). *J. Exp. Zool.* **275**, 226–238.

Conlon, J.M., Tostivint, H. and Vaudry, H. (1997). *Regul. Pept.* **69**, 95–106.

Cook, H., Berkenbosch, J.W., Fernhout, M.J., Yu, K.L., Peter, R.E., Chang, J.P. and Rivier, J.E. (1991). *Regul. Pept.* **36**, 369–378.

Cousins, K.L., Farrell, A.P., Sweeting, R.M., Vesely, D.L. and Keen, J.E. (1997). *J. Exp. Biol.* **200**, 1351–1362.

Darras, V.M., Mol, K.A., Van Der Geyten, S. and Kühn, E.R. (1998). *Ann. NY Acad. Sci.* **839**, 80–86.

Deering, D.J. and Jones, A.C. (1975). *J. Endocrinol.* **64**, 49–57.

Deftos, L.J., Burton, D.W., Watkins, W.B. and Catherwood, B.C. (1980). *Gen. Comp. Endocrinol.* **42**, 9–18.

De Leeuw, R., Conn, M.P., Van T Veer, C., Goos, H.J. and Van Oordt, P.G.W.J. (1988). *Fish Physiol. Biochem.* **5**, 99–107.

Demski, L.S. (1987). In *Psychobiology of Reproductive Behavior: An Evolutionary Perspective* (ed. D. Crews), pp. 1–27. Prince-Hall, Englewood Cliffs, NJ.

Ducouret, B., Tujague, M., Ashra, J., Mouchel, N., Servel, N., Valotaire, Y. and Thomson, E.B. (1995). **136**, 3774–3783.

Duguay, S.J. and Mommsen, T.P. (1994). In *Fish Physiology*, Vol. 13 (eds N.M. Sherwood and C.L. Hew), pp. 225–271. Academic Press, San Diego, CA.

Duguay, S.J., Park, L.K., Samadpour, M. and Dickhoff, W.W. (1992). *Mol. Endocrinol.* **12**, 25–37.

Ebrahimi, M. and Kime, D.E. (1998). *Ann NY Acad. Sci.* **839**, 581–583.

Eilertson, C.D., Carneiro, N.M., Kittilson, J.D., Comley, C. and Sheridan, M.A. (1996). *Regul. Pept.* **63**, 105–112.

Elies, G., Groigno, L., Wolff, J., Boeuf, G. and Boujard, D. (1998). *Ann. NY Acad. Sci.* **839**, 513–514.

Elofsson, U., Winberg, S. and Francis, R.C. (1997). *J. Comp. Physiol. A* **181**, 484–492.

Evans, D.H., Chipouras, E. and Payne, J.A. (1989). *Am. J. Physiol.* **257**, R939–R945.

Fouchereau-Peron, M., Moukhtar, M.S., Le Gal, Y. and Milhaud, G. (1981). *Comp. Biochem. Physiol.* **68A**, 417–421.

Francis, K., Suzuki, M. and Baker, B.I. (1997). *Neuroendocrinology* **66**, 195–202.

Fryer, J.N., Lederis, K. and Rivier, J. (1985). *Regul. Pept.* **11**, 11–15.

Fuji, R., Wakatabi, H. and Oshima, N. (1991). *J. Exp. Zool.* **259**, 9–17.

Galli, S.M. and Phillips, M.I. (1996). *Proc. Soc. Exp. Biol. Med.* **213**, 128–137.

Garlov, P.E. and Khutinaev, A.S. (1991). *Zh. Evol. Biokhim. Fiziol.* **27**, 41–48.

Gelinas, D. and Callard, G.V. (1997). *Gen. Comp. Endocrinol.* **106**, 155–168.

Gelinas, D., Pitoc, G.A. and Callard, G.V. (1998). *Mol. Cell. Endocrinol.* **138**, 81–93.

Gharib, D.D., Wierman, M.E., Shupnik, M.A. and Chin, W.W. (1990). *Endocrin. Rev.* **11**, 177–199.

Girgis, S.I., Galan, F.G., Arnett, T.R., Rogers, R.M., Bone, Q., Ravazzola, M. and MacIntyre, I. (1980). *J. Endocrinol.* **87**, 375–382.

Gray, E.S., Young, G. and Bern, H.A. (1990). *J. Exp. Zool.* **256**, 290–296.

Grober, M.S. and Bass, A.H. (1991). *Brain Behav. Evol.* **38**, 302–312.

Gröneveld, D., Balm, P., Martens, G. and Wendelaar Bonga, S. (1995). *J. Neuroendocrinol.* **7**, 527–533.

Guibbolini, M.E. and Lahlou, B. (1990). *Am. J. Physiol.* **258**, R3–R9.

Guibbolini, M.E. and Lahlou, B. (1992). *Peptides* **13**, 865–871.

Guibbolini, M.E., Henderson, I.W., Mosley, W. and Lahlou, B. (1988). *J. Mol. Endocrinol.* **1**, 125–130.

Gunning, M., Solomon, R.J., Epstein, F.H. and Silva, P. (1997). *Am. J. Physiol.* **273**, R1400–R1406.

Habibi, H.R., Peter, R.E., Sokolowska, M., Rivier, J.E. and Vale, W.W. (1987). *Biol. Reprod.* **36**, 844–853.

Hanke, W. and Kloas, W. (1995). *Horm. Metab. Res.* **27**, 389–397.

Heinig, J.A., Keeley, F.W., Robson, P., Sower, S.A. and Youson, J.H. (1995). *Gen. Comp. Endocrinol.* **99**, 137–144.

Hyodo, S. and Urano, A. (1991). *Zool. Sci.* **8**, 1005–1022.

Ishida, I., Ichikawa, T. and Deguchi, T. (1986). *Proc. Natl. Acad. Sci. USA* **83**, 308–312.

Itoh, H., Suzuki, K. and Kawauchi, H. (1988). *Gen. Comp. Endocrinol.* **71**, 438–451.

Jackson, I.M.D. and Reichlin, S. (1974). *Endocrinology* **95**, 854–862.

Jansz, H.S. and Zandberg, J. (1992). *Ann. NY Acad. Sci.* **657**, 63–69.

Johnson, L.L., Norberg, B., Willis, M.L., Zebroski, H. and Swanson, P. (1997). *Gen. Comp. Endocrinol.* **105**, 194–209.

Kagawa, H., Kawazoe, I., Tanaka, H. and Okuzawa, K. (1998). *Gen. Comp. Endocrinol.* **110**, 11–18.

Kanamori, A., Adachi, S. and Nagahama, Y. (1988). *Gen. Comp. Endocrinol.* **72**, 13–24.

Kaneko, T. (1996). *Int. Rev. Cytol.* **169**, 1–24.

Kawauchi, H. (1989). *Life Sci.* **45**, 1132–1140.

King, J.A. and Millar, R.P. (1992). *Trends Endocrinol. Metab.* **3**, 339–346.

Knoebl, I., Fitzpatrick, M.S. and Schreck, C.B. (1996). *Gen. Comp. Endocrinol.* **101**, 195–204.

Kuno, S., Kubokawa, K., Nagasawa, H., Urano, A. and Ishii, S. (1996). In *Environmental & Conservation Endocrinology* (ed. J. Joss), pp. 211–212. Macquarie University, Sydney.

Lamharzi, N. and Fouchereau-Peron, M. (1996). *Gen. Comp. Endocrinol.* **102**, 274–280.

Lane, T.F., Sower, S.A. and Kawauchi, H. (1988). *Gen. Comp. Endocrinol.* **70**, 152–157.

Larsen, D.A., Dickey, J.T. and Dickhoff, W.W. (1997). *Gen. Comp. Endocrinol.* **107**, 98–108.

Larsen, D.A., Swanson, P., Dickey, J.T., Rivier, J. and Dickhoff, W.W. (1998). *Gen Comp. Endocrinol.* **109**, 276–285.

LeBail, P.Y., Gentil, V., Noel, O., Gomez, J.M., Carre, F., Le Goff, P. and Weil, C. (1998). *Ann. NY Acad. Sci.* **839**, 157–161.

Lederis, K., Fryer, J.N., Okawara, Y., Schönrock, C. and Richter, D. (1994). In *Fish Physiology*, vol. 13 (eds N.M. Sherwood and C.L. Hew), pp. 67–100, Academic Press, San Diego, CA.

LeDréan, Y., Pakdel, F. and Valotaire, Y. (1994). In *Fish Physiology*, vol. 13 (eds N.M. Sherwood and C.L. Hew), pp. 331–366. Academic Press, San Diego, CA.

Li, M.D. and Ford, J.J. (1998). *J. Endocrinol.* **156**, 529–542.

Lin, X.W. and Peter, R.E. (1996). *Gen. Comp. Endocrinol.* **101**, 282–296.

Lin, X.W., Lin, H.R. and Peter, R.E. (1993). *Gen. Comp. Endocrinol.* **89**, 62–71.

Lin, Y.W.P., Rupnow, B.A., Price, D.A., Greenberg, R.M. and Wallace, R.A. (1992). *Mol. Cell. Endocrinol.* **85**, 127–139.

Lo, T.B., Huang, F.L., Liu, C.S., Chang, Y.S., Chang, G.D. and Huang, C.J. (1991). *Bull. Inst. Zool. Acad. Sin. Mongr.* **16**, 37–59.

Lu, M., Swanson, P. and Renfro, J.L. (1995). *Am. J. Physiol.* R577–R582.

Luo, D., McKeown, B.A., Rivier, J. and Vale, W. (1990). *Gen. Comp. Endocrinol.* **80**, 288–298.

Madsen, S.S. (1990). *Comp. Biochem. Physiol.* **95**, 171–175.

Mahlmann, S., Meyerhof, W., Hausmann, H., Heierhorst, J., Schönrock, C., Zwiers, H., Lederis, K. and Richter, D. (1994). *Proc. Natl. Acad. Sci. USA* **91**, 1342–1345.

Marra, L.E., Paramsothy, K., Kawauchi, H. and Youson, J.H. (1997). In *Advances in the Molecular Endocrinology of Fish*. 2nd IUBS Toronto Symposium, p.13.

Marsigliante, S., Muscella, A., Vilella, S., Nicolardi, G., Ingrosso, L., Ciardo, V., Zonno, V., Vinson, G.P., Ho, M.M. and Storelli, C. (1996). *J. Mol. Endocrinol.* **16**, 45–56.

Marsigliante, S., Muscella, A., Vinson, G.P. and Storelli, C. (1997). *J. Mol. Endocrinol.* **18**, 67–76.

Martial, K., Maubras, L., Taboulet, J., Jullienne, A., Berry, M., Milhaud, G., Benson, A.A., Moukhtar, M.S. and Cressant, M. (1994). *Proc. Natl. Acad. Sci. USA* **91**, 4912–4914.

Maule, A.G. and Schreck, C.B. (1991). *Gen. Comp. Endocrinol.* **84**, 83–93.

Mayer, I., Rand-Weaver, M. and Borg, B. (1998). *Gen. Comp. Endocrinol.* **109**, 223–231.

Melamed, P., Rosenfeld, H., Elizur, A. and Yaron, Z. (1998). *Comp. Biochem. Physiol.* **119C**, 325–338.

Miyata, A., Arimura, A., Dahl, D.H., Minamino, N., Uehara, A., Jiang, L., Culler, M.D. and Coy, D.H. (1989). *Biochem. Biophys. Res. Comm.* **164**, 567–574.

Moav, B. and McKeown, B.A. (1992). *Horm. Metab. Res.* **24**, 10–14.

Mommsen, T.P. and Plisetskaya, E.M. (1991). *Rev. Aquat. Sci.* **4**, 225–259.

Moons, L., Cambre, M., Batten, T.F.C. and Vandesande, F. (1989). *Neurosci. Lett.* **100**, 11–16.

Moriyama, S., Swanson, P., Larsen, D.A., Miwa, S., Kawauchi, H. and Dickoff, W.W. (1997). *Gen. Comp. Endocrinol.* **108**, 457–471.

Morley, J.E. (1981). *Endocrin Rev.* **2**, 396–436.

Morley, S.D., Schönrock, C., Richter, D., Okawara, Y. and Lederis, K. (1991). *Mol. Mar. Biol. Biotechnol.* **1**, 48–57.

Nagahama, Y., Yoshikuni, M., Yamashita, M. and Tanaka, M. (1994). In *Fish Physiology*, vol. 13 (eds

Naito, N., Takahashi, A., Nakai, Y. and Kawauchi, H. (1984). *Gen. Comp. Endocrinol.* **56**, 185–192.

Naito, N., Suzuki, K., Nozaki, M., Swanson, P., Kawauchi, H. and Nakai, Y. (1993). *Fish Physiol. Biochem.* **11**, 241–246.

Ohsako, S., Ishida, I., Ichikawa, T. and Deguchi, T. (1986). *J. Neurosci.* **6**, 2730–2735.

Okuta, A., Ando, H., Ueda, H. and Urano, A. (1996). *Zool. Sci.* **13**, 421–427.

Pakdel, F., Petit, F., Anglade, I., Kah, O., Delaunay, F., Bailhache, T. and Valotaire, Y. (1994). *Mol. Cell Endocrinol.* **104**, 81–93.

Parhar, I.S., Pfaff, D.W. and Schwanzel-Fukuda, M.S. (1996). *Mol. Brain Res.* **41**, 216–227.

Parker, D.B., Power, M.E., Swanson, P., Rivier, J. and Sherwood, N.M. (1997). *Endocrinology* **138**, 414–423.

Parker, K.L. and Schimmer, B.P. (1993). *Trend Endocrinol. Metab.* **4**, 46–50.

Pasmanik, M. and Callard, G.V. (1985). *Gen. Comp. Endocrinol.* **60**, 244–251.

Pati, D. and Habibi, H.R. (1998). *Endocrinology* **139**, 2015–2024.

Pelissero, C., Flouriot, G., Foucher, J.L., Bennetau, B., Dunogues, J. and Sumpter, J.P. (1993). *J. Steroid Biochem. Mol. Biol.* **44**, 263–272.

Perrott, M.N., Sainsbury, R.J. and Balment, R.J. (1993). *Gen. Comp. Endocrinol.* **89**, 387–395.

Peter, R.E. and Marchant, T.A. (1995). *Aquaculture* **129**, 299–321.

Potts, J.T., Keutman, H.T., Niall, H.D., Habener, J.F. and Tregear, G.W. (1972). *Gen. Comp. Biochem. Suppl.* **3**, 405–410.

Power, D.M., Canario, A.V.M. and Ingleton, P.M. (1996). *Gen. Comp. Endocrinol.* **101**, 264–274.

Prunet, P. and Auperin, B. (1994). In *Fish Physiology*, vol. 13 (eds N.M. Sherwood and C.L. Hew), pp. 367–391. Academic Press, San Diego, CA.

Prunet, P., Avella, M., Fostier, A., Björnsson, B., Boeuf, G. and Haux, C. (1990). In *Progress in Comparative Endocrinology* (eds A. Epple, C.G. Scanes and M.T. Stetson), pp. 547–552. Wiley-Liss, New York.

Prunet, P., Sandra, O. and Le Rouzic, P. (1997). In *Advances in the Molecular Endocrinology of Fish*. 2nd IUBS Toronto Symposium, S9.

Rand-Weaver, M., Noso, T., Muramoto, K. and Kawauchi, H. (1991). *Biochemistry* **30**, 1509–1515.

Rentier-Delrue, F., Swennen, D., Prunet, P., Lion, M. and Martial, J.A. (1989). *DNA* **8**, 261–270.

Roger, P.P., Christophe, D., Dumont, J.E. and Pirson, I. (1997). *Eur. J. Endocrinol.* **137**, 579–598.

Rouille, Y., Duguay, S.J., Lund, K., Furuta, M., Gong, Q., Lipkind, G., Oliva, A.A., Chan, S.J. and Steiner, D.F. (1995). *Front. Neuroendocrinol.* **16**, 322–361.

Sakamoto, T. and Hirano, T. (1991). *J. Endocrinol.* **130**, 425–433.

Salbert, G., Chauveau, I., Bonnec, G., Valotaire, Y. and Jego, P. (1992). *Mol. Endocrinol.* **6**, 1605–1613.

Sammak, P.J., Adams, S.R., Harootunian, A.T., Shliwa, M. and Tsien, R.Y. (1992). *J. Cell. Biol.* **117**, 57–72.

Sandra, O. and Prunet, P. (1998). *Ann. NY Acad. Sci.* **839**, 172–176.

Sekine, S., Saito, A., Itoh, H., Kawauchi, H. and Itoh, S. (1989). *Proc. Natl. Acad. Sci. USA* **86**, 8645–8649.

Sevarino, K.A., Stork, P., Ventimiglia, R., Mandel, G. and Goodman, R.H. (1989). *Cell* **57**, 11–19.

Shamblott, M.J. and Chen, T.T. (1992). *Proc. Natl. Acad. Sci. USA* **89**, 8913–8917.

Sherwood, N.M. and Hew, C.L. (eds) (1994). *Fish Physiology*, vol. 13. Academic Press, San Diego, CA. 513pp.

Sherwood, N.M., Parker, D.B., McRory, J.E. and Lescheid, D.W. (1994). In *Fish Physiology*, vol. 13 (eds N.M. Sherwood and C.L. Hew), pp. 3–66. Academic Press, San Diego, CA.

Specker, J.L., Kishida, M., Huang, L., King, D.S., Nagahama, Y., Ueda, H. and Anderson, T.R. (1993). *Gen. Comp. Endocrinol.* **89**, 29–38.

Stenhouse, A.A. and Balment, R.J. (1998). *Ann. NY Acad. Sci.* **839**, 389–391.

Stouthart, A.J., Lucassen, E.C., van Strien, F.J., Balm, P.H., Lock, R.A. and Wendelaar Bonga, S.E. (1998). *J. Endocrinol.* **157**, 127–137.

Suzuki, K., Kawauchi, H. and Nagahama, Y. (1998a). *Gen. Comp. Endocrinol.* **71**, 292–301.

Suzuki, K., Kawauchi, H. and Nagahama, Y. (1988b). *Gen. Comp. Endocrinol.* **71**, 302–306.

Suzuki, M., Bennett, P., Levy, A. and Baker, B.I. (1997). *Gen. Comp. Endocrinol.* **107**, 341–350.

Suzuki, R., Yasuda, A., Kondo, J., Kawauchi, H. and Hirano, T. (1991). *Gen. Comp. Endocrinol.* **81**, 391–402.

Suzuki, S., Gorbman, A., Rolland, M., Montfort, M.F. and Lissitzky, S. (1975). *Gen. Comp. Endocrinol.* **26**, 59–69.

Swanson, P., Suzuki, K., Kawauchi, H. and Dickhoff, W.W. (1991). *Biol. Reprod.* **44**, 29–38.

Swanson, P., Planas, J. and Dittman, A. (1997). In *Advances in the Molecular Endocrinology of Fish*. 2nd IUBS Toronto Symposium, S8.

Takahashi, A., Amemiya, Y., Sarashi, M., Sower, S.A. and Kawauchi, H. (1995). *Biochem. Biophys. Res. Comm.* **213**, 490–498.

Takei, Y., Takahashi, A., Watanabe, T.X., Nakajima, K., Sakakibara, S., Sasayama, Y., Suzuki, M. and Oguru, C. (1991). *Biol. Bull.* **180**, 485–488.

Takei, Y., Hasegawa, Y., Watanabe, T.X., Nakajima, K. and Hazon, N. (1993). *J. Endocrinol.* **139**, 281–285.

Takei, Y., Fukuzawa, A., Itahara, Y., Watanabe, T.X., Yoshizawa-Kumagaye, Y., Nakajima, K., Yasuda, A.,

Smith, M.P., Duff, D.W. and Olson, K.R. (1997). *FEBS Lett.* **414**, 377–380.

Tan, N.S., Lam, T.J. and Ding, J.L. (1996). *Mol. Cell Endocrinol.* **120**, 177–192.

Tanaka, M., Telecky, T.M., Fukada, S., Adachi, S., Chen, S. and Nagahama, Y. (1992). *J. Mol. Endocrinol.* **8**, 53–61.

Teitsma, C.A., Bailhache, T., Tujague, M., Balment, R.J., Ducouret, B. and Kah, O. (1997). *Neuroendocrinology* **66**, 294–304.

Tierney, M., Takei, Y. and Hazon, N. (1997). *Gen. Comp. Endocrinol.* **105**, 9–17.

Tollemer, H., Teitsma, C., Leprince, J., Bailhache, T., Vandesande, F., Kah, O., Tonon, M. and Vaudry, H. (1999). *Cell Tissue Res* **295**, 409–417.

Trinh, K.Y., Wang, N.C., Hew, C.L. and Crim, L.W. (1986). *Eur. J. Biochem.* **159**, 619–624.

Trudeau, V.L., Sloley, B.D., Kah, O., Mons, N., Dulka, J.G. and Peter, R.E. (1996). *Gen. Comp. Endocrinol.* **103**, 129–137.

Uchida, K., Kaneko, T., Tagawa, M. and Hirano, T. (1998). *Gen. Comp. Endocrinol.* **109**, 175–185.

Urano, A., Kubokawa, K. and Hiraoka, S. (1994). In *Fish Physiology*, vol. 13 (eds N.M. Sherwood and C.L. Hew), pp. 101–132. Academic Press, San Diego, CA.

Uva, B., Masini, M.A., Hazon, N., O'Toole, L.B., Henderson, I.W. and Ghiani, P. (1992). *Gen. Comp. Endocrinol.* **86**, 407–412.

Vallarino, M., Delbende, C., Bunel, D.T., Ottonello, I. and Vaudry, H. (1989). *Peptides* **10**, 1223–1230.

Valverde-R, C., Croteau, W., Lafleur, G.J., Orozco, A. and St Germain, D.L. (1997). *Endocrinology* **138**, 642–648.

Van Der Geyten, S., Sanders, J.P., Darras, V.M., Kühn, E.R., Leonard, J.L. and Visser, T.J. (1998). *Ann. NY Acad. Sci.* **839**, 498–499.

Van Der Kraak, G. (1991). *Gen. Comp. Endocrinol.* **81**, 268–275.

Van Der Kraak, G. and Wade, M.G. (1994). In *Perspectives in Comparative Endocrinology* (eds K.G. Davey, R.E. Peter and S.S. Tobe), pp. 59–63. National Research Council of Canada, Ottawa, ON.

Van Der Kraak, G., Suzuki, K., Peter, R.E., Itoh, H. and Kawauchi, H. (1992). *Gen. Comp. Endocrinol.* **85**, 217–229.

Van Der Kraak, G., Chang, J.P. and Janz, D.M. (1998). In *The Physiology of Fishes*, 2nd edn (ed. D.H. Evans), pp. 465–488. CRC Press, Boca Raton.

Vaughan, J.M., Rivier, J.E., Spiess, J., Peng, C., Chang, J.P., Peter, R.E. and Vale, W.W. (1992). *Neuroendocrinology* **56**, 539–549.

Vesely, D.L., Douglass, M.A., Dietz, J.R., Gower, W.R., McCormick, M.T., Rodriguez-Paz, G. and Schocken, D.D. (1994). *Circulation* **90**, 1129–1140.

Wade, M.G. and Van Der Kraak, G. (1991). *Gen. Comp. Endocrinol.* **83**, 337–344.

Wagner, G.F. (1994). In *Fish Physiology*, vol. 13 (eds N.M. Sherwood and C.L. Hew), pp. 273–306. Academic Press, San Diego, CA.

Wagner, G.F. and Jaworski, E.M. (1994). *Mol. Cell. Endocrinol.* **99**, 315–322.

Wagner, G.F., Dimattia, G.E., Davie, J.R., Copp, D.H. and Friezen, H.G. (1992). *Mol. Cell. Endocrinol.* **90**, 7–15.

Wagner, G.F., Jaworski, E.M. and Radman, D.P. (1997). *J. Endocrinol.* **155**, 459–465.

Warne, J.M. and Balment, R.J. (1998). *Ann. NY Acad. Sci.* **839**, 412–413.

Wells, K. and Van Der Kraak, G. (1998). Soc. Environ. Toxicol. Chem. 19th Annual Meeting, p. 59. SETAC Press, Pensacola, FL.

Welshons, W.V., Lieberman, M.E. and Gorski, J. (1984). *Nature* **307**, 747–749.

Wendelaar, Bonga, S.E. (1997). *Physiol. Rev.* **77**, 591–625.

Wendelaar Bonga, S.E. and Pang, P.K.T. (1991). *Int. Rev. Cytol.* **128**, 139–213.

Westenfelder, C.F., Birch, M., Baranowski, R.L., Rosenfeld, M.J., Shiozawa, D.K. and Kablitz, C. (1988). *Am. J. Physiol.* **255**, F1281–F1286.

Wilson, J.X. (1984). *Endocrin. Rev.* **5**, 45–61.

Wong, A.O.L., Leung, M.Y., Shea, W.L.C., Tse, L.Y., Chang, J.P. and Chow, B.K.C. (1998a). *Endocrinology* **139**, 3465–3479.

Wong, A.O.L., Ng, S., Lee, E.K.Y., Leung, R.C.Y. and Ho, W.K.K. (1998b). *Gen. Comp. Endocrinol.* **110**, 29–45.

Yamashita, M., Fukada, S., Yoshikuni, M., Bulet, P., Hirai, T., Yamaguchi, A., Lou, Y.H., Zhao, A. and Nagahama, Y. (1992). *Dev. Biol.* **149**, 8–15.

Yamashita, K., Koide, Y., Itoh, H., Kawada, N. and Kawauchi, H. (1995). *Mol. Cell. Endocrinol.* **112**, 159–167.

Yan, L., Swanson, P. and Dickoff, W.W. (1992). *Biol. Reprod.* **47**, 144–153.

Yao, K., Niu, P.D., Le Gac, F. and Le Bail, P.Y. (1991). *Gen. Comp. Endocrinol.* **81**, 72–82.

Young, G., Thorarensen, H. and Davie, P.S. (1996). *Gen. Comp. Endocrinol.* **103**, 301–307.

Yu, K.L., Sherwood, N.M. and Peter, R.E. (1988). *Peptides* **9**, 625–630.

CHAPTER 26

Immune System

David B Powell
ProFISHent, Inc., Redmond, Washington State, USA

Blood and lymphatic vessels

The distribution network of the lymphatic and blood vessels provides a rapid, targeted response to fight the invasion of foreign organisms with a combination of both humoral and cellular defense systems. Successful destruction and elimination of minor infections results in the production of serum antibodies and populations of activated cells which can more quickly and effectively combat recurrent invasions. The duration of immunity will depend on many factors including the age of the fish, type of pathogen and environmental conditions.

Piscine oxygen-transporting erythrocytes contain basophilic nuclei resulting in a more darkly staining blood smear (Figure 26.1). These cells are also larger in size and more oblong in shape (13–16 μm long) than their mammalian counterparts. The total concentration of leukocytes is less than that of mammals, ranging between 8000 and 21 000 cells per cubic millimetre (Yasutake and Wales, 1983). Lymphocytes make up the majority of the white blood cell population (greater than 90%) followed by lower concentrations of neutrophilic granulocytes, thrombocytes and monocytes. Our limited knowledge of the functions of these cells makes it difficult to correlate these cell types with those described in the mammalian scientific literature. The fact that different fish species have different subpopulations of granulocytes complicates generalizations about cell function and morphology. As new studies are done, evidence is accumulating that most fish (more advanced that agnathans) possess many of the same basic immunological mechanisms as higher vertebrates including: functional interactions among T lymphocytes, B lymphocytes, a major histocompatibility complex (**MHC**), immunoglobins and accessory molecules (cytokines, etc.) (Warr, 1997).

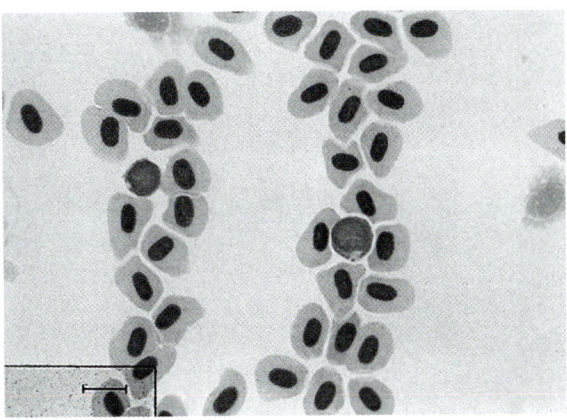

Figure 26.1 Nucleated erythrocytes and two lymphocytes in the blood of rainbow trout. Leishman and Giemsa stain. Bar length = 10 μm.

Copyright © 2000 Academic Press

A typical circulating neutrophilic granulocyte can be distinguished by its multilobed nucleus in rainbow trout (Figure 26.2). The corresponding cell type in catfish is quite different in appearance with a rounder eccentric oval nucleus and **azurophilic** cytoplasmic granules (Ainsworth, 1992). Granulocytes can also be distinguished from other cells by their characteristic positive staining with myeloperoxidase (Yasutake and Wales, 1983). The small pseudopodia of circulating leukocytes are also commonly observed in stained smears and *in vitro* primary cell cultures of rainbow trout blood (Figure 26.2).

Many pathogens are able to infect the blood vessels and hematopoietic tissue of fish. Monocyte-like cells and neutrophilic granulocytes can be activated to move into infected tissue and engulf and destroy bacterial invaders. Neutrophilic granulocytes from rainbow trout, rockfish, catfish, carp and goldfish have all been shown to actively phagocytose bacteria by light microscopy (Weinreb, 1963; Suzuki, 1984; Siwicki *et al.*, 1985; Finco-Kent and Thune, 1987). Wounds exposing collagen trigger thrombocytes to release clotting factors to precipitate fibrin polymers that help wall off damaged areas adjacent to blood vessels and capillaries. B lymphocytes (plasma cells) respond to foreign invaders by dividing and producing IgM-like antibody (tetrameric immunoglobin in teleosts) that specifically binds the surface of pathogens resulting in their agglutination and **opsonization**. Bacterial aggregates formed this way are more efficiently engulfed and destroyed by the host cellular defenses. Experimental evidence for this process in rainbow trout and coho salmon was obtained by demonstrating significantly increased phagocytosis of *Aeromonas salmonicida* following treatment of these bacteria with either antiserum or complement (Sakai, 1984).

Complement in fish serum has been shown to enhance the immune response and inactivation of a variety of pathogenic bacteria, parasites and viruses. For example, complement neutralizes *A. salmonicida* extracellular proteases, *Vibrio* bacterial species, hemoflagellates, IPN, IHN and VHS viruses (Dorson *et al.*, 1979; Dorson, 1981; Sakai, 1984, Jones and Woo, 1987). In addition to this direct action, complement influences cellular responses by stimulating leukocyte migration (Griffin, 1984).

Teleost fish possess a complement (C) system similar to that of mammals with a functional antibody-mediated pathway (using C1, C2 and C4) and an alternate (properdin) pathway for complement activation. This system is heat labile and Ca^{2+}/Mg^{2+} dependent but active at lower temperature ranges (down to 4°C) in fish compared to mammals (Roitt *et al.*, 1993). A complement fixation test using sheep red blood cells and hyperimmune serum from fish (exposed to specific pathogens or antigens) provides the means to evaluate titers of lytic antibodies or complement activity (Ingram, 1990).

Eicosanoids are another important group of signaling molecules produced by blood leukocytes (and other tissues) to stimulate non-specific immune responses. In particular, prostaglandin E_2, leukotriene B_4 and lipoxins A_4 and B_4 are eicosanoids that may play significant roles in regulating fish inflammatory and defense systems (Rowely, 1992). Prostaglandin and leukotriene molecules have been isolated from white blood cells of rainbow trout and marine fish (Matsumoto *et al.*, 1989; Pettitt *et al.*, 1989). Prostaglandin E2 is known to promote vasodilation allowing the migration of monocytes in circulation to penetrate damaged or infected tissues (Roitt *et al.*, 1993). Lipoxins and leukotriene B4 induce directed locomotion or migration of fish leukocytes (Hunt and Rowely, 1986; Rowely, 1992).

Thymus

The thymus supplies lymphocytes to various organs during early development to assist in immunological responses to disease organisms (Romano *et al.*, 1997). However, compared to mammalian systems, our understanding of the precise functions of these cells

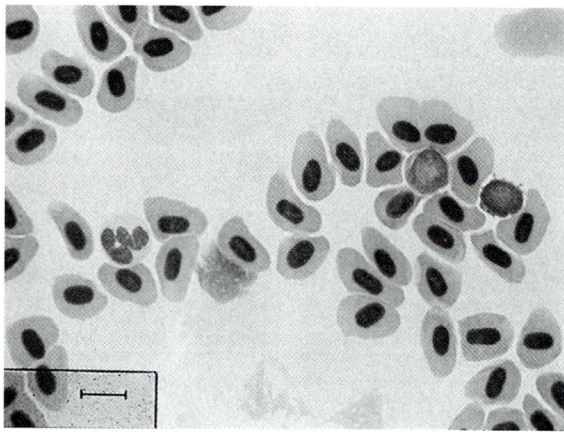

Figure 26.2 Neutrophilic granulocyte characterized by its multilobed basophilic nucleus (left center of photo). The small pseudopodium of a circulating leukocyte is also visible on the right side of this microphotograph. Leishman and Giemsa stain. Bar length = 10 μm.

is very limited in fish. Surgical removal of the thymus from rainbow trout during early development results in a depletion of lymphocytes in the spleen (Grace and Manning, 1980). A second effect of this procedure is to delay the rejection of **allographs** (Manning *et al.*, 1982). However, if thymic ablation is done on mature fish, there is no delay in tissue rejection (Sailendri and Muthukkaruppan, 1975). Such experiments support the concept that the rejection of allographs is transferred to other tissues via the migration of 'T lymphocytes' out of the thymus prior to maturation (Manning *et al.*, 1982).

The thymus is the first organ to become infiltrated with lymphocytes (thymocytes) in early development. In cross-section, epithelial tissue projections that extend in from the outer capsule have sometimes been described as being similar to Hassall's corpuscles found in mammals (Belova, 1976). However, this comparison is still very controversial, and to date, no distinct function has been demonstrated in fish. It is known that a framework of reticuloendothelial cells loosely supports a high concentration of lymphocytes. A mature thymus typically exhibits an indistinct boundary between more basophilic outer cortical tissue and inner medullary tissue. In rainbow trout, the basophilic regions are not restricted to the cortical tissue and are generally visible at relatively low magnification (Figure 26.3). Histologically, the tissue appears homogeneous due to the predominance of thymocytes which have very little cytoplasm surrounding the prominent nuclei of these cells. The inner surface is characterized by invaginations of connective tissue to support the stroma of this organ (Figure 26.4).

In the thymus of young fish, mitotic figures of this rapidly dividing tissue can be observed under high magnification (Figure 26.5). A monoclonal antibody (WCL9) was developed against immature thymocytes in carp (*Cyprinus carpio*) that does not bind with lymphocytes in the spleen or kidney (Rombout *et al.*, 1997). This finding is an indication that significant changes in the expression of surface antigens occur following the exit of these cells from the thymus.

The movement of thymocytes has been traced in rainbow trout using radioactive isotopes. When injected into the thymus, thymocytes were found to migrate in greatest numbers to the spleen (up to 45%) and, in lesser numbers (up to 17%), to the kidney (Tatner, 1985). When blood lymphocytes are labeled and injected, it was found that surface immunoglobulin-negative lymphocytes tended to localize in the thymus (Tatner and Findley, 1991). It has also been

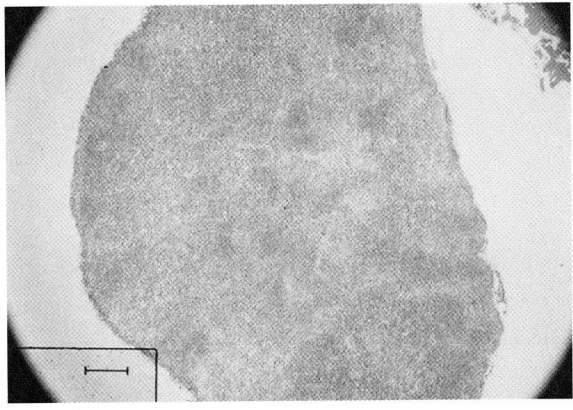

Figure 26.3 Low magnification of a cross-section of a rainbow trout thymus. Note the basophilic regions are not restricted to the cortical tissue. Hematoxylin and eosin. Bar length = 200 μm.

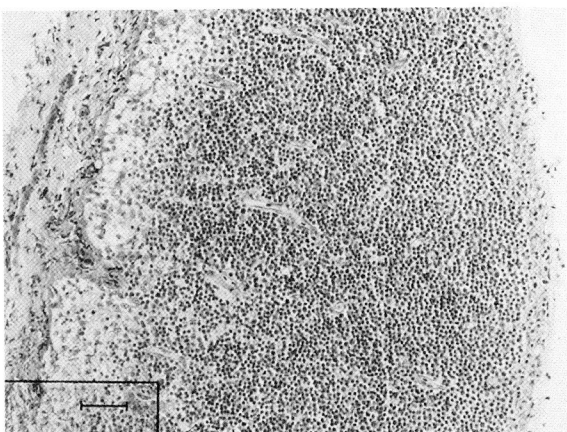

Figure 26.4 Enlarged cross-section of a rainbow trout thymus. Note the connective tissue supporting the stroma of the thymus. Hematoxylin and eosin. Bar length = 50 μm.

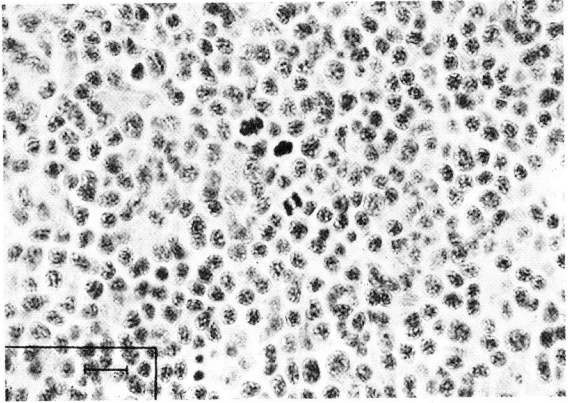

Figure 26.5 In the thymus of young fish, condensed basophilic nuclear chromatin within mitotic figures of dividing cells can be observed at high magnification. Hematoxylin and eosin. Bar length = 10 μm. **(See also Colour Plate 14.)**

suggested that the capillaries in the thymus of teleost fish are selectively permeable (Ellis *et al.*, 1976). This would differ from mammalian systems in which the thymus contains a capillary barrier that isolates the developing thymocytes from foreign antigens and prevents the entry of particulates. The consequences of such differences are unknown at this time.

Kidney

In teleost fish, the kidney is a multifunctional organ. The kidney not only fulfills excretory and osmoregulatory functions, it contains high numbers of hematopoietic and phagocytic cells (Figure 26.6). The anterior kidney is primarily composed of hematopoietic tissue and the posterior portion contains the tubules and glomeruli of the excretory system. Evidence of lymphopoiesis in the anterior section comes from observations of multiple mitotic figures and larger immature lymphocytes in the anterior kidney. The removal of aging blood cells and particulate matter is carried out by the reticular endothelial system of circulating and tissue-associated macrophages that are concentrated in the spleen and kidney (Figure 26.7). These functions appear to be analogous to those of the mammalian bone marrow, kidney and lymph node tissues.

Phagocytosis in the kidney is an essential element in the functioning of the immune system of lower vertebrates. Toxicants, wounds, or infective agents can elicit a strong response of phagocytic cells to wall off damaged tissue and isolate the site to start the healing process. Large particles or chronic inflammation stimulate the formation of giant syncytial cells from the fusion of macrophage cell aggregates. The kidney also contains antibody-producing B cells for the development of humoral immunity against foreign materials and infective agents. Particles in intraperitoneally injected vaccines containing bacteria are commonly phagocytosed and transported to the kidney and spleen for further processing. Like the spleen, enlargement or gross swelling of the kidney indicates a possible infection, toxicity or osmoregulatory problem.

Phagocytes from the anterior kidney of rainbow trout have been broadly classified based on characteristics of morphology (extent of granulation, melanin, etc.), phagocytic capacity and adherence to glass or charged materials (Braun-Nesje, 1982). Macrophages and sinusoidal endothelial cells in rainbow trout appear to express 'scavenger receptors' because they

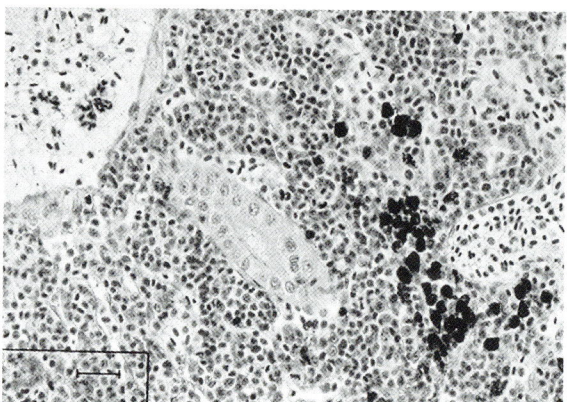

Figure 26.6 Rainbow trout kidney with osmoregulatory tubuoles, hematopoietic tissue and phagocytic cells. Hematoxylin and eosin. Bar length = 25 μm.

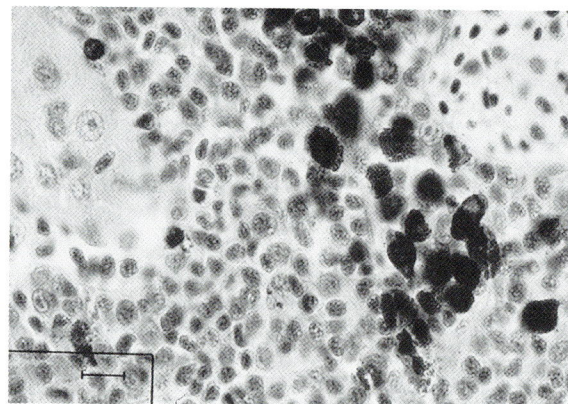

Figure 26.7 Reticular endothelial system of rainbow trout kidney tissue as demonstrated by aggregates of macrophages. Hematoxylin and eosin. Bar length = 10 μm.

engulf and concentrate LdL (low density lipoproteins) similar to mammalian macrophages (Goldstein *et al.*, 1979; Dannevig *et al.*, 1994; Ganassin and Bols, 1999).

Phagocytes are also able to generate a 'respiratory burst' of highly reactive oxygen intermediates (H_2O_2, etc.) to inactivate virally infected cells or engulfed bacteria. Bayne and Levy (1991) demonstrated that adrenocorticotropin could increase the respiratory burst activity. He determined this relationship using *in vitro* measurements of luminol chemiluminescence and associated cytochrome C reductions in the presence of catecholamine agonists.

Non-specific immune reactions can be affected by stressful conditions. For example, extensive handling (30 min) or confinement stressors have been associated with increases in serum lysozyme levels in rainbow trout (Mock and Peters, 1990; Fevolden and Roed, 1993). There are multiple conflicting reports describing the effect of stress on the rate of phagocytosis. Many experiments indicate that macrophages in stressed rainbow trout can exhibit marked hypertrophy, increased pseudopodia and higher rates of phagocytosis (Nuessgen et al., 1991). Other researchers observed the opposite effect, notably reduced phagocytosis, in samples taken from corticosteroid-injected catfish or salmon (Ainsworth et al., 1991; Narnaware et al., 1994). Narnaware et al. (1997) observed a decline in the rate of phagocytosis, as measured by yeast cell engulfment in the kidney and spleen, temporarily following periods of stress. The reasons for such contradictory observations probably include differences in duration and severity of stress and different assay systems and sampling effects caused by leukocyte redistribution within the various organs (Moynihan et al., 1994; Weyts et al., 1999).

The anterior kidney of fish may modulate the interactions of the immune system with the hypothalamo-pituitary-interrenal (HPI) axis since both functions are carried out by this organ. This modulation could occur by the exchange of immune and endocrine signal molecules to receptors of both systems. Corticosteroid receptors have been identified in carp and salmon leukocytes (Maule and Schreck, 1990; Weyts et al., 1998). Carp lymphocytes and neutrophils demonstrated a high affinity for cortisol with apparently the same receptor molecule (Weyts et al., 1998). Similarly, coho cells bound triamcinolone acetonide, a corticosteroid analog, suggesting a possible mechanism for linkage of the immune response to endocrine homeostatic systems (Maule and Schreck, 1991). Weyts et al. (1999) suggested that cortisol may act on the kidney to regulate immunity during periods of stress by suppressing specific immune responses and enhancing short-term non-specific responses.

Interleukin-like leukocyte signaling molecules, known to promote specific immunological responses in mammalian systems, have been identified in a variety of fish species. Macrophage supernatants from carp express functional molecules similar to mammalian macrophage activating factor and interleukin IL-1 (Verburg-van Kemenade et al., 1995, 1996). Interleukin-2 activity was detected in carp lymphoblasts in response to mitogen stimulation (Caspi and Avtalion, 1984; Grondel and Harmsen, 1984). Interestingly, with the exception of tumor necrosis factor (TNFα), tests with recombinant mammalian cytokines did not produce effects in either carp or catfish cells (Ellsaesser and Clem, 1994; Jang et al., 1995; Verburg-van Kemenade et al., 1995). This host specificity may be partially explained by DNA sequence analyses of the IL-1B gene. Comparisons of the rainbow trout genome revealed a less than a 50% homology with human and mouse IL-1B genes and undergoes incomplete splicing (Zou et al., 1999).

Although not directly antiviral, interferons are cytokines released from virally infected cells (type I) or properly triggered T lymphocytes (type II) to reduce viral infection. In fish, interferon activity can be isolated from kidney, blood, spleen, heart and liver (Secombes, 1994). Injection of interferon produced against one virus can passively protect fish against experimental challenge with other pathogenic viruses. In rainbow trout fry, the timing of the highest concentrations of interferon in the blood occurs about two weeks following immersion challenge with either IPNV or IHNV (Renault et al., 1991; Dorson et al., 1992).

Intestine

The lower (descending) intestine of fish does not contain the large organized concentrations of lymphocytes, granulocytes and macrophages arranged like the Peyer's patches of mammals. However, these cell types are loosely arranged at lower concentrations in the intestinal mucosa and lamina propria beneath the lamina epithelialis (Figure 26.8). This gut-associated lymphoid tissue (GALT) includes immunoglobulin-positive macrophages and B lymphocytes, as well as putative immunoglobulin-negative T cells and natural killer cells (Rombout et al., 1993). The development of a monoclonal antibody using intestinal lymphocyte

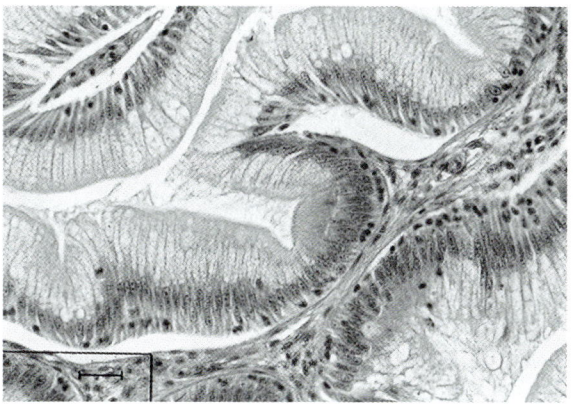

Figure 26.8 Descending intestine of rainbow trout with parallel absorbent cells arranged to form the lamina epithelialis and underlying laminar propria containing a mixture of connective tissue and phagocytic cells. Hematoxylin and eosin. Bar length = 25 μm.

membranes revealed cross-reactivity with skin and gill lymphocytes but not to immunoglobin B cells, macrophages or non-specific cytotoxic lymphocytes (Rombout *et al.*, 1998). This work provides further evidence that T-cell populations exist in the lamina propria of the intestine and may assist with mucosal immunity.

Intestinal macrophages are thought to have important antigen-presenting functions for the development of specific immunity against pathogens in the digestive tract (Rombout *et al.*, 1985, 1989). Intestinal antigen uptake and oral vaccination efficacy appear to be maximized when antigens are protected from gastric secretions either by encapsulation or anal intubation (Joosten *et al.*, 1977; Quentel and Vigneulle, 1997). Although still rudimentary, our understanding of the immunological mechanisms of piscine intestinal tissue has increased with the advent of immunohistochemistry. Using this technique, the transport of soluble antigens across the intestinal epithelium appears to occur to a much greater extent than particulate antigens.

Liver

The liver in fish is normally a uniform reddish tan in color and smooth in texture. Microscopically, the parenchyma can be seen as a three-dimensional network of sinusoids composed of cuboidal hepatocytes with narrow spaces lined with endothelial cells (Figure 26.9). In some fish species, basophilic exocrine pancreatic tissue can be observed surrounding the portal veins (Kendall and Hawkins, 1975). Phagocytic mononuclear cells residing in the liver sinusoids are known as Kupffer cells. These cells phagocytose particulate antigens such as bacteria and viruses in the blood. If the function of Kupffer cells in fish is similar to the liver phagocytes in mammals, then these cells are capable of presenting antigens to T lymphocytes (Roitt *et al.*, 1993). Chronically infected fish often show multiple aggregations of lymphocytes adjacent to major blood vessels in the liver.

The liver may be responsible for the production of acute phase serum proteins during an infection. In particular, **C-reactive protein** levels (named for its binding to C-protein of pneumococci) have been shown to increase two to threefold in the serum and liver of rainbow trout 48 hours after exposure to endotoxin from *Salmonella typhimurium* or other stressors (Samad, 1990). In mammals, this acute phase protein has been

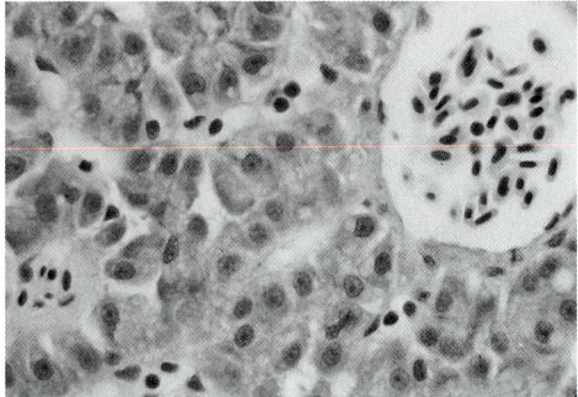

Figure 26.9 Liver parenchyma can be seen as a cross-section through a three-dimensional network of sinusoids composed of cuboidal hepatocytes with narrow spaces lined with endothelial cells. Hematoxylin and eosin stains, high magnification.

shown to attach in a calcium-dependent reaction to the surface of a wide variety of fungi and bacteria (Roitt *et al.*, 1993). Once bound, C-reactive protein assists the binding of complement to opsonize the foreign material for enhanced phagocytosis. An analysis of various rainbow trout tissues revealed C-reactive protein associated with the membranes of peripheral blood lymphocytes, polymorphonuclear cells and the lymphoid cells of the thymus, spleen and anterior kidney (Samad, 1990).

Spleen

The spleen may be a primary site for the identification and removal of cells weakened by age or infection. Such cells are targets for phagocytosis and deposition in the spleen. Enlargement of the spleen (splenomeglia) is an indicator of possible systemic bacterial or viral infection. The spleen has multiple functions in teleost fish. In addition to hematopoiesis, the spleen may have immunological functions similar to lymph nodes in mammals. The outer capsule of the spleen is composed of connective tissue forming a thin outer wall and scattered invaginating connective tissue projections. The pulp matrix is held together by a diffuse network of argyrophilic fibers (Roberts, 1978). The spleen also retains a substantial number of mature red blood cells that can be released into circulation during periods of stress or extended physical activity. Thrombocytes, responsible for regulating blood clotting (presumably by secreting thromboplastin-like substances), are predominantly produced in the spleen (Romano et al., 1997).

Splenic tissue is typically composed of a mixture of three main tissue types: pulp, ellipsoids and melanomacrophage centers. The pulp is a combination of hematopoietic red pulp and lymphopoietic white pulp tissue (Figure 26.10). The separation of the two tissue groups is much less distinct in fish (Osteichthyes) than in mammals. In rainbow trout, the splenic tissue appears to remain mixed until the age of 6 months to a year when focal areas of lymphocytic aggregations appear (Castillo, 1991). These aggregates have been hypothesized to function as sites for antigen trapping and processing as part of an immune response to pathogens or vaccines (Zapata et al., 1997).

Splenic ellipsoids are divided arterioles forming dense-walled capillaries that are capable of collecting enormous quantities of small particulate antigens. Roberts (1978) demonstrated this filtering capacity by intravenously injecting Indian ink into plaice fish. The dark ink clearly lined the ellipsoid basement membranes in histological sections of the spleen. Trapped material is then engulfed by macrophage cells and transported to melanomacrophage centers for processing. These centers are typically adjacent to minor blood vessels in teleost species. In rainbow trout, splenic tissue appears loosely organized into areas of pulp, arterioles and macrophages with associated aggregations of dark brown melanin granules (Figure 26.11). The number and concentration of these aggregates increase with age. At high magnification, melanin granules and phagocytosed material can be seen in the cytoplasm of splenic macrophages (Figure 26.12).

Traditionally immunological functions of specific fish tissues or cell types have been studied *in vitro* using short-lived primary cultures from individual fish. Recently, new fish cell lines with discrete immunological characteristics have become available for experiments that can be replicated in multiple laboratories. For example, a monocyte/macrophage-like cell line (RTS11) has been cultivated from rainbow trout spleen tissue that can be used to evaluate cellular differentiation and mechanisms of phagocytosis (Ganassin and Bols 1996, 1999). This cell line is characterized by three basic morphologic types: round smooth cells with a kidney shaped nucleus (the predominant form); transitional suspended granular cells; and adherent macrophage-like cells.

The round monocyte-like cells can be stimulated by LPS (lipopolysacharide) or PMA (phorbol ester)

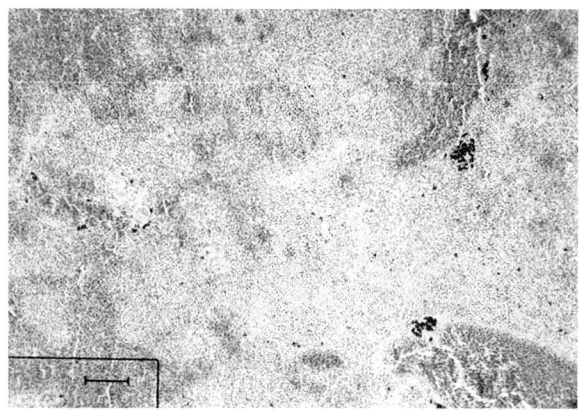

Figure 26.10 The pulp in a spleen is a combination of hematopoietic (red pulp) and lymphopoietic (white pulp) tissue in the rainbow trout. Hematoxylin and eosin. Bar length = 100 μm. **(See also Colour Plate 15.)**

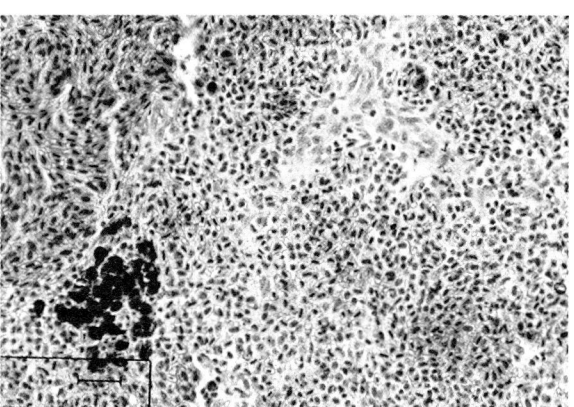

Figure 26.11 Rainbow trout, splenic tissue loosely organized into areas of pulp, arterioles and macrophages with associated aggregations of dark brown melanin. Hematoxylin and eosin. Bar length = 25 μm. **(See also Colour Plate 16.)**

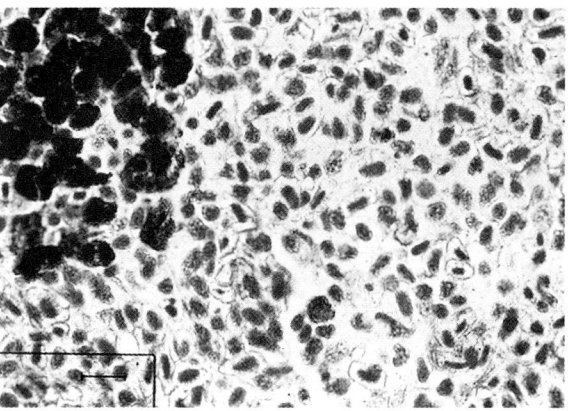

Figure 26.12 High magnification of melanin granules and phagocytosed material in the cytoplasm of splenic macrophages of rainbow trout. Hematoxylin and eosin. Bar length = 10 μm. **(See also Colour Plate 17.)**

to differentiate, become granular, adherent and actively phagocytic (Ganassin and Bols, 1999). The production of lysozyme-like enzymes into the culture medium by the trout RTS11 cells is another characteristic shared by mammalian macrophages and established macrophage cell lines (Polansky et al., 1985). As new cell lines and monoclonal surface marker libraries are developed for the study of the spleen's function in fish, we will gain a better understanding of the cellular and molecular basis of these responses.

References

Ainsworth, A.J. (1992). *Ann. Rev. Fish Dis.* 123–148.
Ainsworth, A.J., Dexiang, C. and Waterstrat, P.R. (1991). *J. Aquat. Anim. Health* 3, 41–47.
Bayne, C.J. and Levy, S. (1991). *J. Fish. Biol.* 38(4), 609–619.
Belova, A.V. (1976). *J. Ichthy.* 16, 112–119.
Braun-Nesje, R. (1982). *Dev. Comp. Immunol.* 6(2), 281–291.
Caspi, R. and Avtalion, R. (1984). *Dev. Comp. Immunol.* 8(1), 51–60.
Castillo, A. (1991). Doctoral Thesis, University of Leon, Spain.
Dannevig, B.H., Lauve, A., Press, C.M. and Landsverk, T. (1994). *Fish Shell. Immunol.* 4, 3–18.
Dorson, M. (1981). *Dev. Biol. Stand.* 49, 307–319.
Dorson, M., Torchy, C. and Michel, C. (1979). *Ann Recher. Vet.* 10, 529–534.
Dorson, M., DeKinklelin, P. and Torchy, C. (1992). *Fish Shell. Immunol.* 2, 311–313.
Ellis, A.E., Munro, A.L.S. and Roberts, R.J. (1976). *J. Fish Biol.* 8, 67–78.
Fevolden, S.E. and Roed, K.H. (1993). *J. Fish Biol.* 43, 919–930.
Finco-Kent, D. and Thune, R.L. (1987). *J. Fish Biol.* 31 (Suppl. A), 41–49.
Ganassin, R.C. and Bols, N.C. (1996). *Fish Shell. Immunol.* 6, 17–34.
Ganassin, R.C. and Bols, N.C. (1999). *Fish Shell. Immunol.* 8, 457–476.
Goldstein, J.L., Ho, Y.K., Basu, S.K. and Brown, M.S. (1979). *Proc. Natl. Acad. Sci.* 76, 335–337.
Grace, M.F. and Manning, M.J. (1980). *Dev. Comp. Immunol.* 4, 255–264.
Griffin, B.R. (1984). *Dev. Comp. Immunol.* 8, 589–597.
Grondel, J. and Harmson, E. (1985). In *Fish Immunology* (Manning, M. and Tatner, M., eds) 261–262.
Hunt, T.C. and Rowely, A.F. (1986). *Immunology* 59, 563–568.
Ingram, G.A. (1990). In *Techniques in Fish Immunology* (eds J.S. Stolen, T.C. Fletcher, D.P. Anderson, B.S. Roberson and W.B. van Muiswinkel), pp. 25–44. Fair Haven, New Jersey.
Jang, I., Hardie, J. and Secombes, C. (1995). *J. Leukocyte Biol.* 57(6), 943–947.
Jones, S.R. and Woo, P.T. (1987). *J. Fish Dis.* 10, 395–402.
Joosten, E., Caumartin-Dhieux, C., van Muiswinkel, W. and Rombout, J. (1997). In *Fish Vaccinology. Developments in Biological Standards*, vol. 90 (eds R. Gudding, A. Lillehaug, P.J. Midtlyng, F. Brown), p. 439. Karger, Basel.
Kendall, M.W. and Hawkins, W.E. (1975). *J. Fish. Res. Board Can.* 32, 1459–1464.
Manning, M.J., Grace, M.F. and Secombes, C.J. (1982). *Dev. Comp. Immunol.* 2, 75–82.
Matsumoto, H. Iijima, N. and Kayama, M. (1989). *Comp. Biochem. Physiol.* 93b, 397–402.
Maule, A.G. and Schreck, C.B. (1990). *Gen. Comp. Endocrinol.* 77, 448–455.
Maule, A.G. and Schreck, C.B. (1991). *Gen. Comp. Endocrinol.* 84, 83–93.
Mock, A. and Peters, G. (1990). *J. Fish Biol.* 37, 873–885.
Moynihan, J.A., Ader, R. and Cohen, N. (1994). In *Neuropeptides and Immunoregulation* (eds B. Scharrer, E.M. Smith, G.B. Stephano), pp. 120–138. Springer Verlag, Berlin.
Narnaware, Y.K., Baker, B.I. and Tomlinson, M.G. (1994). *Fish Physiol Biochem.* 13(1), 31–40.
Narnaware, K., Kelly, S. and Woo, N. (1997). *Fish Shellfish Immunol.* 7(7), 515–517.
Nuessgen, A., Raabe, A. and Moeck, A. (1991). *Fish Shell. Immunol.* 1(1), 17–31.
Pettitt, T.R., Barrow, S.E. and Rowely, A.F. (1989). *Biochem. Biophys. Acta* 1003, 1–8.
Polansky, D.A., Yang, A., Brader, K. and Miller, D.M. (1985). *Br. J. Hematol.* 60, 7–17.
Quentel, C. and Vigneulle, M. (1997). In *Fish Vaccinology. Developments in Biological Standards*, vol. 90 (eds R. Gudding, A. Lillehaug, P.J. Midtlyng, F. Brown), pp. 69–78. Karger, Basel.
Renault, T., Torchy, C. and DeKinkelin, P. (1991). *Dis. Aquat. Org.* 10, 23–29.
Roberts, R.J. (1978). In *Fish Pathology*. Baillière Tindall, Cassell Ltd & Macmillan Publishing Co. Inc., pp. 13–54. London, UK.
Roitt, I.M., Brostoff, J. and Male, D.K. (1993). In *Immunology*, 3rd edn. Mosby-Year Book Europe Ltd., pp. 1–12.
Romano, N., Taverne-Thiele, J.J., Van-Maanen, J.C. and Rombout, J.H.M.W. (1997). *Fish Shell. Immunol.* 7(7), 439–453.
Rombout, J.H.W.M., Joosten, P.H.M., Engelsma, M.Y., Vos, A.P., Taverne, N. and Taverne-Thiele, J.J. (1998). *Dev. Comp. Immunol.* 22(1), 63–77.
Rombout, J.M., Lamers, C.J., Helfrich, M.H., Decker, H. and Taverne-Thiele, J.J. (1985). *Cell Tissue Res.* 239, 519–530.

Rombout, J.M., Taverne-Thiele, A.J. and Villena, M.I. (1993). *Dev. Comp. Immunol.* **17**, 55–66.

Rombout, J.M., Van Den Berg, A.A., Van Den Berg, C.A., Witte, P. and Egberts, E. (1989). *J. Fish Biol.* **35**, 179–186.

Rombout, J.H., Van-De-Waal, J.W., Companjen, A., Taverne, N. and Taverne-Thiele, J-J. (1997). *Dev. Comp. Immunol.* **21**(1), 35–46.

Rowely, A.F. (1992). In *Techniques in Immunology – 2* (eds J.S. Stolen, T.C. Fletcher, D.P. Anderson, S.L. Kaatari and A.F. Rowley), pp. 143–156. Fair Haven, New Jersey.

Sailendri, K. and Muthukkaruppan, V.R. (1975). *J. Exp. Zool.* **191**, 371–383.

Sakai, D.K. (1984). *J. Fish Dis* **7**, 329–338.

Samad, F. (1990). Diss. Abst. Int. Pt. B Sci. and Eng., Diss. Ph.D No. DA9011681 **50**(12), 186.

Secombes, C.J. (1994). In *Techniques in Fish Immunology – 3* (eds J.S. Stolen, T.C. Fletcher, A.F. Rowley, J.T. Zelikoff, S.L. Kaatari and S.A. Smith). Fair Haven, New Jersey.

Siwicki, A., Studnicka, M. and Ryka, B. (1985). *L. Bamidgeh.* **37**, 123–128.

Suzuki, K. (1984). *Bull. Japan. Soc. Sci. Fish.* **50**, 1305–1315.

Tatner, M.F. (1985). *Dev. Comp. Immunol.* **9**, 85–91.

Tatner, M.F. and Findley, C. (1991). *Fish Shell. Immunol.* **1**, 107–117.

Verburg-van Kemenade, B.M., Weyts, F.A., Debets, R. and Flik, G. (1995). *Dev. Comp. Immunol.* Jan–Feb, **19**(1), 59–70.

Verburg-van Kemenade, B., Daly, J., Groeneveld, A. and Wiegertjes, G. (1996). *Vet. Immunol. Immunopathol.* **51**(1–2), 189–200.

Warr, G.W. (1997). In *Fish Vaccinology. Developments in Biological Standards*, vol. 90 (eds R. Gudding, A. Lillehaug, P.J. Midtlyng, F. Brown), pp. 15–21. Karger, Basel.

Weinreb, E.L. (1963). *Anat. Rec* **147**, 219–238.

Weyts, F.A.A., Verburg-van Kemenade, B.M.L. and Flik, G. (1998). *Gen. Comp. Endocrinol.* **111**, 1–8.

Weyts, F.A.A., Cohen, N., Flik, G. and Verburg-van Kemenade, B.M.L. (1999). *Fish Shell. Immunol.* **9**, 1–20.

Yasutake, W.T. and Wales, J.H. (1983). In *Microscopic Anatomy of Salmonids: An Atlas*, US Dept. Interior, Fish and Wildlife Service, Resource Publication 150, pp. 82–83. Washington DC.

Zapata, A.G., Torroba, M., Varas, A. and Jimenez, E. (1997). In *Fish Vaccinology. Developments in Biological Standards*, vol. 90 (eds R. Gudding, A. Lillehaug, P.J. Midtlyng and F. Brown), pp. 23–32. Karger, Basel.

Zou, J., Cunningham, C. and Secombes, C. (1999). *European J. Biochem*, **259**(3), 901–908.

CHAPTER 27

Sensory Systems

27.1 Vision

Russell D Fernald
Department of Psychology, Stanford University,
Pal Alto, California, USA

Introduction

Of all the structures in the eye, the retina is certainly the most important. Images in the world are focused by the cornea and lens on to the retina at the very back of the eye. This thin, transparent bit of nervous tissue serves the critically important function of transforming physical energy in the form of photons, into neural energy in the form of ionic currents. The details of this exquisite transformation are now known in remarkable detail. How photon energy results in seeing, however, is not well understood, though a great deal is known about the structure, connections and function of the retina. The events that lead from photon absorption to processing of visual information in the retina are the topics of this chapter.

The retinal structure of teleost fish does not differ fundamentally from that found in other vertebrates as expected given the conservative nature of evolution. Nonetheless, it has proven more difficult in teleosts than in other vertebrate classes to identify a general model. This is because teleosts comprise the largest number of vertebrate species (> 25 000), and their enormous range and diversity of habitats have resulted in a wide variety of retinal types. Teleosts range from fish that live in shallow water and feed on the surface to bioluminescent deep-sea fishes and include spectacular examples such as the flashlight fish (*Photoblepheron* sp.) that provides its own illumination for some activities.

In this chapter the following features of teleost retinas will be presented at the microscopic level including: (i) the process of phototransduction; (ii) the flow of information through the neural retina; (iii) information about the five retinal cell types; (iv) adaptations to different visual environments including the consequences of continuous retinal growth in teleosts.

Phototransduction

All known vertebrates with eyes transform collected light into neural signals using the same fundamental

Copyright © 2000 Academic Press

mechanism: two molecules which, in combination, absorb photons and produce an ionic current proportional to the number of photons absorbed. The combination of molecules, together called the 'visual pigment', consists of **opsin**, a glycoprotein that enfolds a light-sensitive chromophore, called retinal. The visual pigment is physically located in specialized photoreceptor cells called rods and cones. All vertebrates have rods, cones or both rods and cones arrayed across the retinal surface. The photoreceptive opsin/retinal combination is located at the very back of the eye, near the sclera, in a specialized portion of the photoreceptor cells, called the outer segment (Figure 27.1.1 and Table 27.1.1).

Every photoreceptor type can be characterized by its sensitivity to the two defining parameters of light: wavelength and intensity. Ultimately, transduction of these two variables together with their spatial location identifies events in the visual world. The output of an individual photoreceptor is proportional to the number of photons that have activated it. Thus, information about wavelength depends on knowing the characteristics of individual photoreceptor wavelength sensitivity. Since wavelength response is critically important, visual pigments in all kinds of photoreceptors are characterized by their spectral sensitivity which is measured by the wavelength at which there is maximum absorption (λ_{max}). Peak wavelength absorption of rod photoreceptors in most vertebrates is typically centered at or near 500 nm and most vertebrates, including fish, have only a single kind of rod photoreceptor. In contrast, many vertebrates have two or more kinds of cone photoreceptors. Cone photoreceptors can differ morphologically as well as in their peak wavelength response (λ_{max}). Since individual photoreceptors can only report on the number of photons received, perception of colors requires comparison of information from cone photoreceptors that have different wavelength specificities. At the phenomenological level, color perception arises from the comparison of photon information coming from cone photoreceptors which have different characteristic λ_{max} values. Which mechanisms produce the differences in peak wavelength absorbance that distinguish different photoreceptor types?

Opsins are seven transmembrane domain proteins comprised of *ca.* 348 amino acids of approximately 40 kD molecular mass (Figure 27.1.1B). The transmembrane domains are alpha helices that comprise about 50% of the molecular mass. Enfolded in opsin within the span of the membrane, is retinal, a light-absorbing chromophore based on vitamin A.

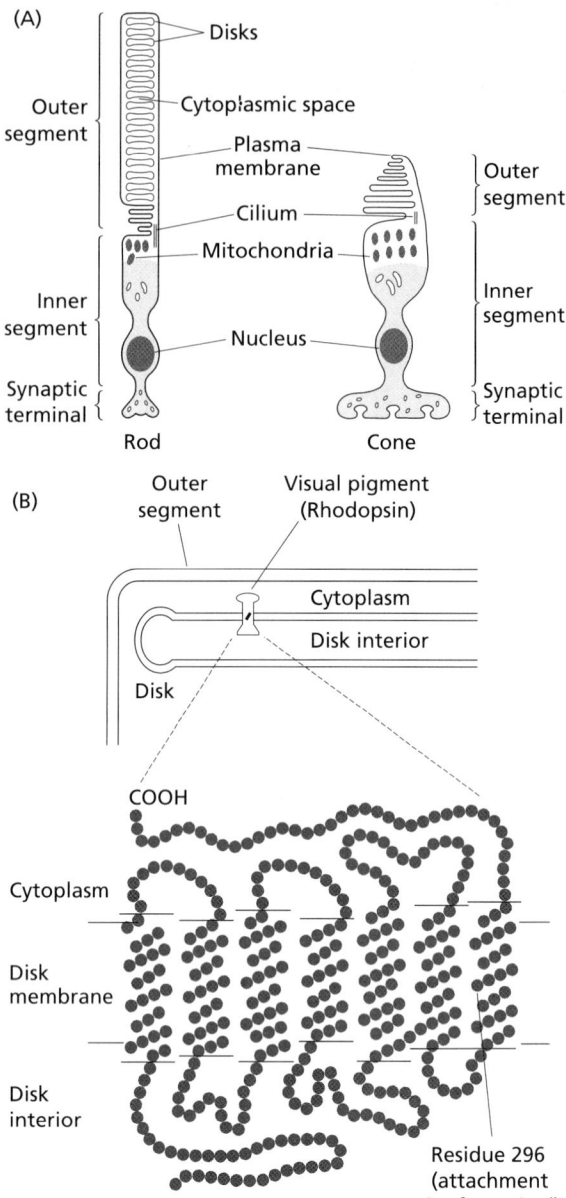

Figure 27.1.1 (A) Schematic illustration of typical rod and cone photoreceptors. Light enters photoreceptors from below and passes through the whole cell before reaching the light-sensitive portion called the outer segment. Photoreceptors are unusual in that their synaptic connections require the postsynaptic input lie within the synaptic terminal of the photoreceptor. (B) Within the outer segments of rods and cones, the light-sensitive photopigment opsin is located in a membrane. Opsin enfolds a light-absorbing chromophore, retinal, that is linked to opsin at residue 296 via a Schiff base. (Adapted from Kandel et al., 1991.)

Fish have two kinds of retinal, one based on vitamin A_1 (11-*cis*-retinal) and the other based on vitamin A_2 (11-*cis*-3-dehydroretinal). Most other vertebrates have only 11-*cis*-retinal. By convention, when opsin

TABLE 27.1.1: Rod and cone photoreceptor comparison		
Characteristics	**Rod photoreceptors**	**Cone photoreceptors**
Role	Nocturnal, scotopic vision	Diurnal, phototopic vision
Sensitivity	High (more photopigment)	Low (less photopigment)
	Single photon detection	Much higher threshold
Acuity	Low – many rods converge	High – less convergence
Color	No – single photoreceptor type	Yes – multiple cone types with different wavelength sensitivity

combines with vitamin A_1 based retinal, the combination is called rhodopsin. This form is found in all vertebrates, including some teleosts. When opsin combines with vitamin A_2 based retinal, the combination is called **porphyropsin**, and this form is found in teleosts, amphibians and aquatic reptiles.

Retinal is surrounded by the seven transmembrane domains of opsin, and held to opsin by a Schiff base linkage to a lysine at position 296 in the polypeptide chain, lying midway in the seventh helix. The wavelength specificity of a particular opsin is set by the interaction between the opsin molecule and retinal. Since retinal alone absorbs light optimally at *ca.* 240 nm, the wavelength specificity of particular opsins (both rhodopsins and porphyropsins) is determined by the nature of particular amino acids in the vicinity of the retinal (e.g. Chang *et al.*, 1995). It is thought that key differences in the properties of amino acids near retinal in opsin molecules (e.g. charge, polarity, etc.) cause different species of opsins to absorb light optimally at different wavelengths, ranging from about 410 nm to 590 nm. Indeed, five key amino acid locations appear to be of primary importance (Neitz *et al.*, 1991; Asenjo *et al.*, 1994) but these alone cannot explain the complete range of λ_{max} values measured in a range of vertebrates suggesting that additional molecular mechanisms are yet to be discovered (Yokoyama and Radlwimmer, 1998).

The DNA sequence of opsin is known for a large number of species (e.g. Applebury, 1994) and comparison of these has shown that all opsins are derived from a common ancestor. Rhodopsins show considerably more homology than do the cone opsins. It seems probable that an ancestral gene duplicated and then was subjected to further selection leading to the current diversity of opsin types (Nathans *et al.*, 1986).

Absorption of light by retinal causes it to twist around a double bond and lengthen by 5 Å which, in turn, alters the structure of the opsin molecule enfolding it. This dimensional change activates a biochemical cascade that amplifies the effect and ultimately changes the flow of current through the outer segment membrane. This fundamental transduction event causes activated photopigment molecules to become inactive, requiring regeneration through several intermediates.

Through this transduction event in the outer segment of photoreceptors, the energy in absorbed photons becomes known to the nervous system as a change in inward sodium current across the photoreceptor outer segment. This results in a change in the flow of neurotransmitters from the photoreceptors, proportional to the intensity of light. The photoreceptors communicate with the bipolar cells of the retina that connect to ganglion cells for transmission of the information to the brain.

Information flow in the retina

The retina is a neuroepithelial structure segregated into layers, each of which contains neurons responsible for collecting, transforming and finally transmitting information to the brain. There are five classes of neurons and two basic stages of synaptic connection. It is useful to think of these arrays of cells in their distinct layers as processing information in two orthogonal directions (Figure 27.1.2). In the 'through' or vertical direction, excitation of a photoreceptor by light initiates a signal that passes from the photoreceptors to bipolar cells and then to ganglion cells. In the ganglion cells, visual information is encoded in action potentials that travel to the brain via the optic nerve. The idea that there is a direct pathway from photon

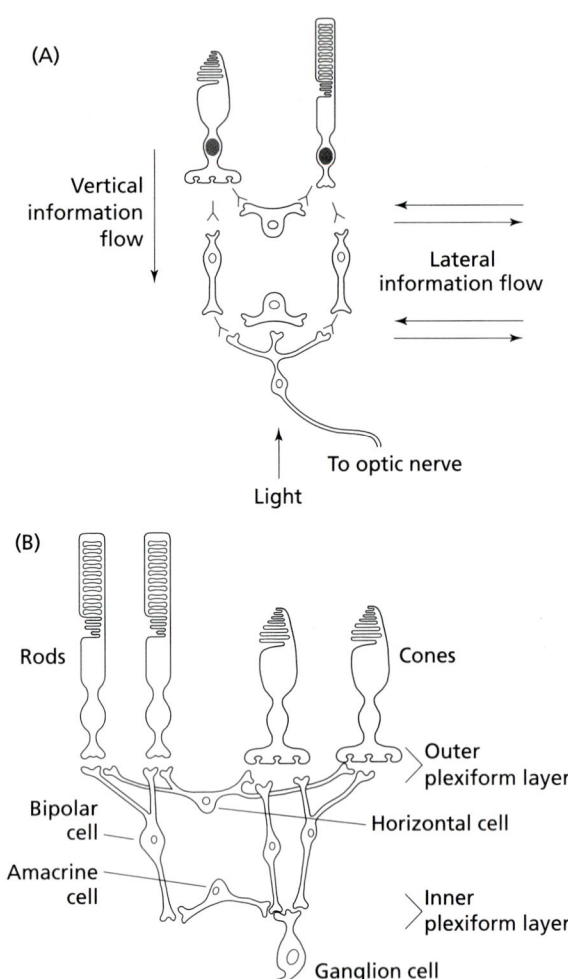

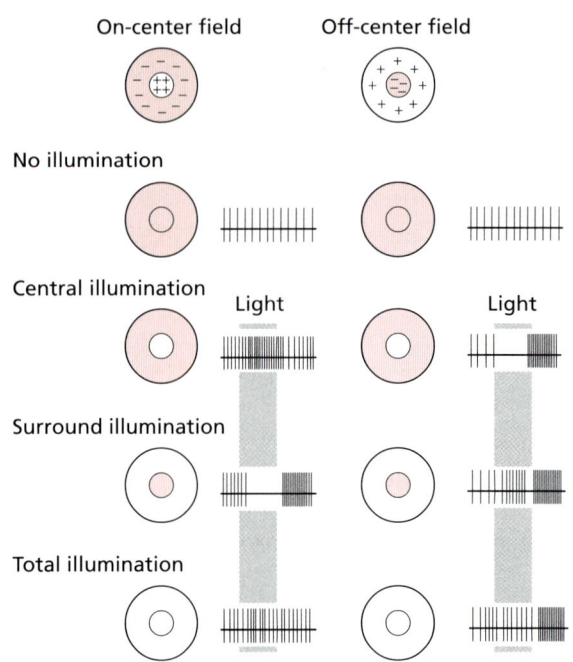

Figure 27.1.2 (A) Schematic illustration comparing the direct or 'vertical' flow of information from photoreceptors to bipolar cells to ganglion cells and the interactions or 'lateral' flow of information mediated by horizontal cells and amacrine cells. Photic stimulation initiates activity in both of these pathways so that the information delivered as a distribution of light in space from the visual world is transformed to activation of receptive fields within ganglion cells. This fundamental processing step is common to all vertebrates. (B) Retinal cell types arranged in the laminar organization characteristic of all vertebrate retinas. There are three layers containing cell nuclei and two layers (outer and inner plexiform layers) where the synaptic contacts are located. Identification of the cell types responsible for both the vertical and horizontal information flow in the retina are depicted in Figure 27.1.2A.

Figure 27.1.3 Receptive field types as represented in the firing patterns of ganglion cell axons. Ganglion cells respond best when there is a high contrast in their receptive fields. Ganglion cell receptive fields are circular and there are two kinds of receptive fields: on-center cells and off-center cells. As shown in the figure, on-center cells increase their activation when stimulated in the central circular area and are inhibited when the concentric surround field is stimulated. Off-center cells have the opposite polarity of responses. In the figure, the pattern of action potentials fired by the ganglion cell being studied is shown by vertical bar. (Adapted from Kandel *et al.*, 1991.)

detection to the brain is appealing in some respects. However, information arriving at the photoreceptors is modified substantially by interactions that occur orthogonal to this direct flow of information. Lateral interactions modulate the direct retinal pathway at two levels. First, in the outer plexiform layer, interactions occur via horizontal cells which interconnect photoreceptors. Second, in the inner plexiform layer, interactions occur via amacrine cells which connect with bipolar and ganglion cells. Through these lateral interactions the original photoreceptor signals are modified substantially, both in time and space before arriving at the ganglion cell for transmission to the brain.

The connections resulting from these lateral interactions produce the first level of analysis or image extraction of a visual scene. Imagine a scene falling on the retina. At the level of the photoreceptor, each point in space is represented by a value equivalent to the wavelength, as indicated by which photoreceptor is active, and intensity, as indicated by the activity of that photoreceptor. How is this map of photon capture transformed into visual information? The first step is the conversion of **photic information** into receptive fields (Figure 27.1.3). This is the best studied and perhaps most important transformation in the processing of visual information. In fact, the information contained in the firing patterns of retinal ganglion cells must provide a good representation of the visual scene because it is all the brain receives.

Individual ganglion cells respond only when light falls on a restricted region of the retina, known as the receptive field of that ganglion cell. The receptive fields of ganglion cells are circular and consist of two parts: a center and a surround in which light has opposite effects. These different responses define the cell type. 'On-center' ganglion cells, as the name implies, respond with increased rate of action potentials when the center circle of the receptive field is illuminated. In these cells, light in the surrounding receptive field inhibits the rate of action potentials. 'Off-center' ganglion cells have receptive fields organized in the opposite fashion so that light in the center of the receptive field is inhibitory and light in the surround is excitatory. Based on these receptive fields, ganglion cells respond best to changes in the visual stimulus, in time or in space or in spectral composition. Ganglion cells representing visual information from the entire field of view send signals to the brain in parallel, meaning that different attributes of the visual world can be characterized simultaneously as a function of their location in space.

Retinal structure

Information processing within the retina produces the receptive field organization described above. This functional organization depends on particular characteristics of retinal neurons and their interconnections. This section examines the phenotypes of individual cells and how they are wired together to produce the observed receptive fields.

Studying the anatomical substrate of fish retina has a legendary history beginning with the work of Cajal (1892) and continued by Wunder (1925) who compared retinal structures of 24 fish freshwater species using histological methods. Subsequently, Verrier (1928), Dathe (1969), Bathelt (1970) and Wagner (1972) all added descriptive analyses of fish retinas. This early work could not provide definitive answers to the key question of how the retina is wired but did provide baseline information about retinal structure and cell types. The unusual specializations of deep-sea fishes have also been described because of their novel adaptations to life in low photon levels (Brauer, 1908; Munk 1966; Locket 1971a,b and see Figure 27.1.4).

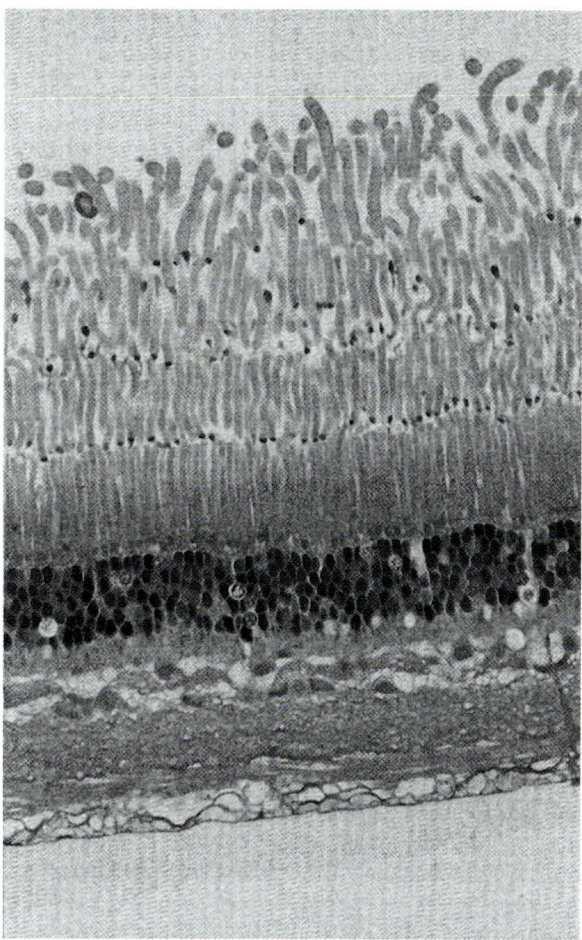

Figure 27.1.4 Retinal structures from a deep-sea fish, *Nothacantus chemnitzii*, which shows the characteristic all rod retina with multiple banks of rods with their outer segments stacked. (Courtesy of H.J. Wagner.) **(See also Colour Plate 18.)**

Ahlbert (1969) first recognized that there was an important relationship between habitat and retinal structure by comparing retinal structure in teleosts with distinctly different habitats. In 1976 Ali and Anctil produced a descriptive catalog of the retinas of fishes extending and elaborating that relationship. As exemplified therein, there is a vast array of fish retinal types. In addition to these original survey works, there has been a substantial investigation of individual species with the intention of discovering the direct causal relationship between structure and function (Fernald, 1990b).

One general principle to emerge from analysis of teleost fish retinas is that relative retinal thickness is directly related to the functionality of the retina (Wagner, 1990). For example, diurnal species with the need for good vision generally have a 1:1 ratio of the photoreceptor layer thickness (including the pigmented epithelium) relative to the neural retina. In contrast, fishes from a low light level habitat have an inner retina that is only *ca*. 15–25% of the photoreceptor layer. In deep-sea fishes typically the lowest values are found (Munk, 1964, 1966). Retinas differ primarily because there are far fewer cells used or perhaps needed for the lateral processing discussed above.

The particular retinal morphology of a fish species is not strictly tied to its particular taxonomic position but rather depends significantly on its ecological and **ethological** context. Figure 27.1.4 shows a striking example of this principle. The primary constraints on the evolution of particular retinal structures is that the eyes must collect and focus sufficient light to provide the requisite spatial resolution for behavior. Ultimately, however, each species uses the same retinal cells to meet their particular needs. Understanding the different cell types can help illuminate just how distinct cell types subserve particular perceptual needs.

Retinal cell types and connections

Within the retina proper, there are five principal neuronal cell types. In addition, there are two important additional cell types: glial cells and pigmented epithelial cells. The predominant glial cell in the retina is the Müller cell, which spans the entire neural retina. The spreading endfeet of these cells form the internal limiting membrane at the **vitread** retinal surface and the external limiting membrane at the **sclerad** end of the retina. Müller cell bodies lie in the inner nuclear layer and lamellar processes ensheath cell bodies of retinal neurons. These cells serve to maintain ionic balance within the retina (Newman and Zahs, 1997) and are also structurally important. Whether glial cells also play some role in visual processing remains to be seen.

The other main non-neuronal cell type is the pigmented epithelial cell which constitutes the outermost portion of the retina. These simple cuboidal cells comprise the basal lamina of the retina. Where the cells meet retinal tissue, there are extended processes enveloping the inner and outer segments of the photoreceptor cells. PE cells contain spherical or rod-shaped pigment granules, i.e. melanosomes that shift in location depending on the state of light adaptation. In the light, they are apical and effectively isolate cone outer segments from each other while shielding rod outer segments from the intense daylight (Douglas, 1982a). In the dark, the melanosomes move to the cell bodies. Interestingly, for species that occupy habitats with low light levels, the pigment granule density is reduced and reflecting substances replace them which increase the chances that an outer segment will catch photons.

Pigmented epithelial cells are perhaps best known because they phagocytose the tips of rods and cones. Since rods and cones constantly renew their outer segments, the regulation of production and phagocytosis must be exquisitely balanced. In addition to these supporting players, the main cells of the retina are described in detail below.

Photoreceptors

Photoreceptors are the primary input to the retina. Since these cells collect and transduce light, their structure and arrangement reflect adaptations to maximize photon capture in the rods and acuity in the cones. The most effective method for increasing sensitivity in rods is to increase the photon collection surface area. This translates into maximizing the number of the opsin-containing membranes. Indeed, photoreceptors with increased outer segment area achieved via large or numerous outer segments in rod photoreceptors are found in fish living in low light levels. In some fish living in murky water, the rods are grouped in bundles and many deep-sea species have their rods arranged in multiple banks (Munk, 1966). Both of these solutions allow increased sensitivity.

The adaptations producing more acute vision require organizing cone photoreceptors with respect to each other. In teleost fish that rely on vision for behavioral interactions, the cone photoreceptors are organized in remarkably precise arrays as first noted by Ryder (1895). These mosaics appear organized either in alternating rows of single and double cones ('row pattern') or in squares with four cone pairs set around a single, central cone ('square pattern') (e.g. Eigenmann and Shafer, 1900; Wagner, 1990). Fish with the potential of color vision have cones of different wavelength specificity and the chromatic organization of their cone mosaic optimizes chromatic detection (Fernald, 1981). Transitions from one mosaic type to another associated with a change in life habits have been reported in salmon and trout (Lyall, 1957) and in perch (Ahlbert, 1969). In elasmobranchs, cone photoreceptors are rare and no such mosaics exist. Indeed, there is evidence that there is only one cone type in elasmobranchs (Cohen, 1990).

Photoreceptors and the pigmented epithelium exhibit some remarkable transformations in synchrony with the light/dark cycle. Phagocytosis and renewal of the photoreceptor outer segments occurs in a regular rhythm (O'Day and Young, 1978). The disks containing opsin are produced every day, with a strong diurnal rhythm as reflected in opsin abundance and in opsin mRNA production (Korenbrot and Fernald, 1989). In contrast, phagocytosis of a single photoreceptor occurs only once every few days, but is linked to the light/dark cycle as well.

Although photoreceptor outer segment renewal is found in all vertebrates, additional retinal transformations related to the light/dark cycle enhance retinal function in teleosts. The most dramatic of these are the retinomotor movements which are the wholesale daily rearrangement of photoreceptor positions. In the light-adapted retina, cone photoreceptors move vitread so they are adjacent to the outer limiting membrane, pigment granules migrate vitread into the processes around the cone outer segments forming a dense absorbing layer sclerad to the cone outer segment boundary and the rod photoreceptors extend sclerad through elongation of the myoids beyond the pigment layer so they no longer receive photic input. Retinomotor movements occur in some other cold-blooded vertebrates as well as in some bird species (Walls, 1942). In teleosts, these movements can be as much as 40–80 µm and the pigment may migrate *ca.* 100 µm. The consequence of this positioning is that only the cones receive light input, and the cone outer segments are optically isolated from each other. In darkness, these processes reverse so that the rod myoids contract bringing the rod photoreceptors adjacent to the outer limiting membrane, the pigment granules migrate to the **perikarya** of the pigmented epithelial cells and the cone myoids elongate, placing the cones sclerad to the outer limiting membrane but vitread to the pigmented epithelial band. Consequently, the cone photoreceptors can receive light and may participate in vision at low light levels. The regulation of retinomotor responses and cellular events which underlie them have been reviewed (Burnside and Nagle, 1983) as has the distribution among teleosts (Ali and Wagner, 1975). It has been reported that the cone mosaic changes form during retinomotor movements (Kunz, 1980), although this does not appear to be the case (Fernald, 1982).

There are three generally accepted notions about the selective advantages of such movement. First, retinomotor movements result in essentially two separate functional retinas: cones for photopic vision and rods for scotopic vision (Herzog, 1905). Second, the dispersed pigment granules contribute to photopic acuity by absorbing scattered rays refracted out of the cone outer segments (Garten, 1907). Third, the pigmented epithelium shields rod outer segments from bleaching during bright light (Garten, 1907). A fourth advantage is to increase retinal acuity, especially in animals with small eyes (Fernald, 1988). In small fish, the retinal acuity of an all cone retina resulting from retinomotor movements is twice what it would be if rod photoreceptors were also in place (Fernald, 1988). Conversely, the uniform field of rod photoreceptors during periods of low light intensity would increase threshold detection since no photons fall on low sensitivity cones. Thus retinomotor movements offer a neat solution to maximally packing cone photoreceptors.

Horizontal cells

Horizontal cells comprise a large, distinct cell class located just vitread of the external limited membrane. These cells are also organized in a mosaic pattern, particularly visible in young animals (Hagedorn and Fernald, 1992). There are three horizontal cells with long axons (*ca.* 500 µm) connected to cones (H1, H2, H3) and one connected to rods (H4) which has no axon (Stell and Lightfoot, 1975; Weiler, 1978; Downing, 1983). Following the initial proposals of Stell *et al.* (1975), four interrelated roles for horizontal cells in transforming photic information can be stated (Wagner, 1990): first they mediate chromatic interactions between different spectral cone types; second, they generate the antagonistic surround of bipolar cell receptive fields; third, they modulate spatially summed signals in the outer plexiform layer via gap junctions; and fourth, they constitute an additional pathway between the outer plexiform and inner nuclear layers. The H1 cells use GABA as a neurotransmitter (Lam and Steinman, 1971) but the neurotransmitters in the remaining horizontal cell types are unknown (e.g. Lasater, 1990).

The schematic connections of horizontal cells is shown in Figure 27.1.6 and the role of horizontal cells in defining the receptive field is shown in Figure 27.1.5.

Bipolar cells

Information about photon capture in photoreceptors is transmitted to the brain via bipolar and then ganglion cells in the retina. Bipolar cells in the inner nuclear

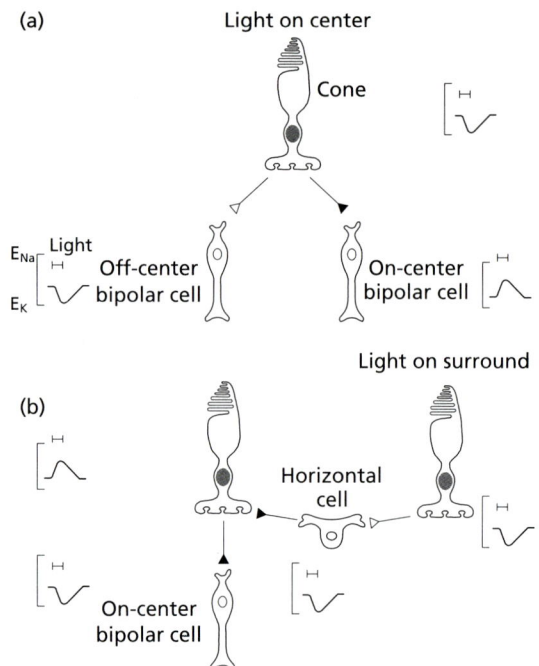

Figure 27.1.5 (A) Generation of receptive field types by wiring of photoreceptors. The responses of bipolar cells depend directly on their input from photoreceptors and horizontal cells. Here, a cone photoreceptor is shown schematically as connected to two bipolar cells: one with off-center response properties and one with on-center response properties. When light is absorbed by the photoreceptor, the on-center bipolar cell is excited and the off-center bipolar cell is inhibited. The different response results from different neurotransmitters.
(B) Retinal circuitry leading to the center-surround organization of the receptive fields. An on-center bipolar cell is shown receiving input from a photoreceptor in the surround of its visual field. These lateral interactions between neighbor bipolar cells is mediated by horizontal cells. (Adapted from Kandel et al., 1991.)

layer, which serve as the main conduit from photoreceptors to ganglion cells, also have synapses from horizontal cells which regulate the lateral interactions across the retina and from dopaminergic interplexiform cells. In a pioneering work, Scholes (1975) examined teleost bipolar cell connections identifying at least 10 types. He found: (i) that selective bipolar cells connect only to various spectral cone types (cone bipolar cells); (ii) that some bipolar cells connect predominantly to cones but also receive some rod input (mixed bipolar cells); and (iii) that still other bipolar cells have massive input, primarily from rods with a few red-sensitive cones included. This linking of the scotopic/rod pathway to the red cone bipolars is different from mammalian retinas which have an exclusive cone (scotopic) pathway (e.g. Sterling et al., 1986). Although serotonin has been indicated as a neurotransmitter in bipolar cells of elasmobranchs (Brunn et al., 1984), in teleosts it is unknown what bipolar cells use to communicate with their partners.

The bipolar cells are organized with antagonistic receptive fields with the 'OFF' center cells located distal to the 'ON' center cells in the inner plexiform layer. The details of how the bipolar cell types are connected to produce the receptive field types observed in physiological recordings are shown in Figures 27.1.5 and 27.1.6.

Interplexiform cells

Interplexiform cells were first described by Cajal (*Box salpa*, 1892) but only quite recently have their

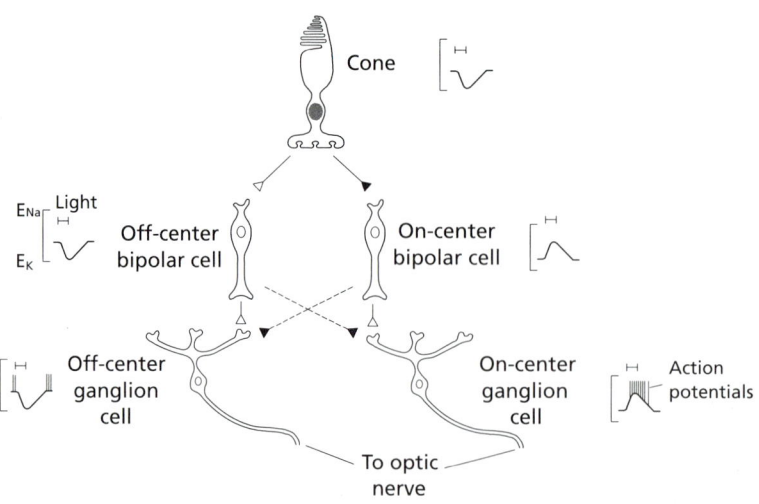

Figure 27.1.6 Schematic illustration showing that cones contribute to the receptive fields of many ganglion cells. Their connections determine how this input is distributed. (Adapted from Kandel et al., 1991.)

characteristics been examined in attempts to understand their function. These cells contain either dopamine (Dowling and Ehinger, 1975, 1978) or accumulate glycine (Marc *et al.*, 1979). Those dopaminergic cells receive input from horizontal cell axons (Marshak and Dowling, 1987), from amacrine dendrites (Dowling and Ehinger, 1978) and from the nervus terminalis via peptidergic transmitters (Zucker and Dowling, 1987). These terminate primarily on horizontal cells but also on some bipolar cells (Wagner and Wulle, 1992). The glycine accumulating cells receive input from H1 horizontal cells (Marc and Liu, 1984) and from horizontal cell axons (Marshak and Dowling, 1987). The latter cells show light evoked slow responses similar to those of amacrine and bipolar cells (Hashimoto *et al.*, 1980) suggesting that they might comprise a separate functional cell class. Taken together, the interplexiform cells appear to be important in the efferent control of retinal response, acting by the release of dopamine into the outer plexiform layer.

Amacrine cells

Amacrine cells are local circuit neurons which make contact with every type of retinal cell except photoreceptors (Dowling, 1979). These cells may be the most diverse cell class in the CNS with a record number of 48 distinct subtypes identified based on structure in one teleost (roach) (Wagner and Wagner, 1988). Matching this diversity is that found in amacrine neurotransmitter types (Lasater, 1990). GABA, glycine and acetylcholine have been found in amacrine cells as have numerous neuropeptides. Given the enormous diversity, amacrine cell function is difficult to characterize. Wagner (1990) suggests three possible principles of operation: first, populations of amacrine cells may act as multicellular aggregates via gap junction coupling; second, individual amacrine cells might function alone under certain conditions (Miller, 1979); third, parts of the dendritic field of amacrine cells may be functional microcircuits under certain conditions (Bloomfield and Miller, 1982).

Ganglion cells

Ganglion cells collect information about the visual scene and relay it to the brain via action potentials. The structural and functional analysis of ganglion cells have revealed both clear identifying features and remarkable complexity. As in other sensory systems, stimulus information is represented in ganglion cells as receptive fields. For all ganglion cells, these receptive fields are concentric rings containing two distinctive regions: a center and a surround (see Figure 27.1.3). The receptive fields can be classified in two ways: on center/off surround and off center/on surround. On-center cells respond with increased firing of action potentials when stimulated in the center of their field and inhibited when stimulated in the surround of their receptive field. Off-center cells have the opposite response type. Taken together, these two cell types are thought to have evolved to make ganglion cell responses highly sensitive to edges but relatively insensitive to changes in ambient light levels. The potential importance of different kinds of receptive field type is suggested in species where these are differentially distributed across the retina. Fish that use silhouetting for prey capture have more off-center receptive fields in the ventral retina and more on-center receptive fields in the dorsal region of the retina (Lythgoe, 1979).

Clearly fish live in and evaluate a complex visual environment. It is relatively easy to demonstrate that receptive field structures accentuate the contrast of natural edges. It is this property that is thought to have led to their evolution.

Retinal growth in teleosts

Unlike warm-blooded vertebrates, the central nervous system of fish grows by adding neurons throughout life. This is not a small effect, since in rapidly growing animals, as much as 90% of the retinal cells present at one year will have been added after hatching (Fernald, 1984). Growth continues while the animal behaves normally, feeding and breeding without interruption. Because of this capability, fish have become extremely useful model systems for the analysis of growth and development spawning numerous recent reviews (Fernald, 1989, 1990b; Powers and Raymond, 1990).

As the eye grows, one immediate question is whether the field of view changes. It does not. Several studies have shown that the visual angle subtended by the optics is independent of eye size (Easter *et al.*, 1977; Fernald, 1984). In addition, neither the lens quality (Fernald and Wright, 1983, 1985) nor the accommodative mechanism (Fernald and Wright, 1985) is compromised during growth. Thus as the animal grows, the image delivered to the retina is constant in quality. Regulation of the growth of fish is complex, but in some species this is socially regulated (Fraley and Fernald, 1982). How these social signals result in metabolic changes is still unknown, but social effects on some physiological systems are beginning to be understood (Davis and Fernald, 1990; Francis *et al.*, 1993; Hofmann *et al.*, 1999).

Retinal development has been studied in several teleost species (Müller, 1952; Hollyfield, 1972; Grün, 1975; Sandy and Blaxter, 1980; Sharma and Ungar, 1980; Kunz, 1983; Hagedorn and Fernald, 1992) and all of these studies confirm two general principles. First, there is a sequential production of retinal cell phenotypes which is essentially the same in all vertebrates, beginning with ganglion cells and ending with rod photoreceptors; second, there are two distinct developmental gradients: one which extends vitread to sclerad and the other which extends from the center to the periphery of the developing eye. Fish which metamorphose exhibit an extreme form of the first developmental principle, adding rod photoreceptors at metamorphosis which can be as much as 4 months after the animal is living independently (review in Evans and Fernald, 1990).

Although the phenomenology of retinal development is known, the controlling factors are not. Recent work in other vertebrate species has shown that the final cell phenotype is not related to cell lineage (Turner and Cepko, 1987; Holt *et al.*, 1988) but rather that unspecified local factors govern cell fate. An interesting difference between fish retinal development and that of mammals is that although cell death is an important factor in shaping mammalian retinas it is not so in teleosts (Hoke and Fernald, 1998). Cell specification in teleosts appears to be accurate enough during development not to require subsequent adjustment through pruning.

Perhaps the most unusual feature of teleost retinal development is that it does not stop at hatching. Instead, there is continued growth of the retina from a germinal zone located at the margin of the eye. In addition to the cell addition, retinal stretching also occurs (e.g. Fernald, 1984). First described by Müller (1952), this germinal zone is a compressed version of the central to peripheral gradient present in embryogenesis. At the outermost edge, a single layer of pseudo-stratified epithelial cells represent the beginning of cell addition and about 150μm centralwards fully differentiated, laminated retinal structure exists. Consistent with the late addition of rod photoreceptors in embryogenesis (e.g. Raymond, 1985; Hagedorn and Fernald, 1992), there is a secondary zone of rod neurogenesis, central to the marginal germinal zone (Fernald, 1989). Even more remarkable, the teleost retina adds rod photoreceptors throughout the central retina (Fernald and Johns, 1980; Johns and Fernald, 1981). These cells arise from a set of precursor cells of unknown origin which divide in the outer nuclear layer amidst the differentiated layer of rod nuclei. Discovery and details of this phenomenon have recently been reviewed (Fernald, 1989). The role of these new rods appears to be to preserve the retinal sensitivity in the face of massive stretching of the extant retinal tissue (Fernald, 1985; Powers and Raymond, 1990).

In fish requiring excellent vision for social interactions, the eye growth must be achieved while maintaining specific areas of specialization. This means that the higher concentration of cone photoreceptors in the temporal pole must be maintained if the animal is to preserve high resolution vision. Recently, it has been shown that this is achieved by differential retinal stretching (Zygar *et al.*, 1999) rather than differential cell addition. How this is achieved is unknown, however.

As the retina grows, many more ganglion cells are added which must be integrated into the primary projection to the **optic tectum**. The cells are added to the optic nerve in an unusual fashion first recognized and described by Scholes (1979). The problem is that the teleost optic nerve in visually oriented animals has a retinotopic order at all ages, despite the fact that it contains an increasing number of optic fibers. Scholes (1979) found that in the ribbon shaped optic nerve, the new fibers are all added at one edge, preserving **retinotopy** in the nerve but allowing for the dramatic increase in cell fiber number.

New axons reaching the optic tectum pose another topological problem during adult growth because the tectum adds cells in a pattern different from that of the retina. Whereas the retinal growth is symmetrical around the margin of the eye (Müller, 1952), the tectal growth pattern is asymmetric, appearing roughly in a horseshoe shape with the major growth zone at the rear (caudal) margin (Meyer, 1978). Despite the non-congruence of retinal and tectal growth patterns, the visual field is represented on the surface of the tectum in a consistent fashion in all sizes of animals. This requires the continuous remodeling of the terminal arbors of optic fibers during growth.

The dynamic nature of visual system growth in adulthood is all the more remarkable because animals remain socially interactive without any signs that the nervous system is being enlarged and rearranged. This suggests that neural remodeling could be a very rapid process, possibly in phase with the light/dark cycle similar to that seen in dynamic changes of retinal cell connectivity (see above). Because of this continued visual system modification, teleosts have become favored models for studies of nervous system development.

Summary

The connections among cell types in the retina provide a remarkable preliminary analysis of the visual scene before any information gets to the brain. Because fish species are so numerous, much of what we know is derived from careful studies of fish visual systems. In particular, the idea that there is a flow of information through the retina and also across the retina at two levels has provided significant insight into how all vertebrae retinas analyze information. This chapter covers information from its arrival as photons to its coding into ganglion cell firing patterns. Neither how these firing patterns represent the information nor how this leads to action is known.

Acknowledgments

Work supported by grant NIH NEI 05051 to RDF.

References

Ahlbert, I. (1969). *Arkiv. für Zoologie* **22**, 445–480.
Ali, M.A. and Anctil, M. (1976). *Retinas in Fish: An Atlas*. Springer-Verlag, Berlin, Heidelberg, New York.
Ali, M.A. and Wagner, H.-J. (1975). In *Vision in Fishes* (ed. M.A. Ali), pp. 369–396. Plenum Press, New York.
Applebury, M.L. (1994). In *Molecular Evolution of Physiological Processes* (ed. D.M. Fambrough), pp. 235–248. Rockefeller University Press, New York.
Asenjo, A.B., Rim, J. and Oprian, D.D. (1994). *Neuron* **12**, 1131–1138.
Bathelt, D. (1970). *Zool. Jb. Abt. Anat.* **87**, 402–470.
Bloomfield, S.A. and Miller, R.F. (1982). **208**, 288–303.
Brauer, A. (1908). *Die Tiefseefische. II Anatomische Teil. B. Augen*. G. Fischer Verlag, Jena.
Brunn, A., Ehinger, B. and Sytsma, V.M. (1984). *Brain Res.* **295**, 233.
Burnside, B. and Nagle, W. (1983). In *Progress in Retinal Research*, vol. 2 (eds N. Osborne and G. Chader), pp. 67–109. Pergamon Press, Oxford.
Cajal, S.R. (1892). *Cellule* **9**, 121–225.
Chang, B.S., Crandall, K.A., Carulli, J.P. and Hartl, D.L. (1995). *Mol. Phylogenet. Evol.* **4**, 31–43.
Cohen, J.L. (1990). In *The Visual System of Fish* (eds R.H. Douglas and M.B.A. Djamgoz), 465pp. Springer-Verlag, New York.
Dathe, H.H. (1969). *Z. mikrosk.-anat. Forsh.* **80**, 269–319.
Davis, M.R. and Fernald, R.D. (1990). *J. Neurobiol.* **21**, 1180–1188.
Douglas, R.H. and Wagner, H.J. (1982). Annual Spring Meeting of the Association for Research in Vision and Ophthalmology Incorporated, Sarasota, FLA., USA, 2–7 May 1982. *Invest. Ophthal. Vis. Sci.* **22**, 140.
Dowling, J.E. (1979). In *The Neurosciences. Fourth Study Program* (eds F.O. Schmitt, and F.G. Warden), pp. 163–181. MIT Press, Cambridge, Mass.
Dowling, J.E. and Ehinger, B. (1975). *Science (Washington DC)* **188**, 270–273.
Dowling, J.E. and Ehinger, B. (1978). *Proc. Roy. Soci. Lond. – B. Biol. Sci.* **201**, 7–26.
Downing, J.E.G. (1983). University of London.
Easter, S.S. Jr., Johns, P.A. and Baumann, L.R. (1977). *Vis. Res.* **17**, 469–476.
Eigenmann, C. and Shafer, G. (1900). *Amer. Natur.* **XXXIV**, 109–118.
Evans, B. and Fernald, R. (1990). *Neurobiology* **21**, 1037–1052.
Fernald, R. (1982). *Comp. Neurol.* **206**, 379–389.
Fernald, R.D. (1983). *Nature* **301**, 618–620.
Fernald, R. (1984). *Amer. Sci.* **72**, 58–65.
Fernald, R. (1985). *Environ. Biol. Fish.* **13**, 113–123.
Fernald, R.D. (1988). In *Sensory Biology of Aquatic Animals* (eds J. Atema, R.R. Fay, A.N. Popper and W.N. Tavolga), pp. 435–466. Springer-Verlag, Berlin.
Fernald, R. (1989). In *Development of the Vertebrate Retina* (eds B.L. Finlay and D.R. Sengelaub), pp. 247–265. Plenum Press, New York and London.
Fernald, R.D. (1990a). In *The Visual System of Fish* (eds R.H. Douglas and M.B.A. Djamgoz), pp. 45–62. Chapman & Hall, New York.
Fernald, R.D. (1990b). In *The Visual System of Fish* (eds R.H. Douglas and M.B.A. Djamgoz), pp. 443–464. Chapman & Hall, New York.
Fernald, R.D. and Johns, P. (1980). *Invest. Ophthal Vis. Sci.* **19**, 69.
Fernald, R.D. and Johns, P.R. (1981). *Invest. Ophthal. Vis. Sci.* **20**, 77.
Fernald, R.D. and Wright, S. (1983). *Nature* **301**, 618–620.
Fernald, R.D. and Wright, S. (1985). *Vis. Res.* **25**, 163–170.

Fraley, N.B. and Fernald, R.D. (1982). *Z. Tierpsychol.* **60**, 66–82.

Francis, R.C., Soma, K. and Fernald, R.D. (1993). *Proc. Natl. Acad. Sci. USA* **90**, 7794–7798.

Garten, S. (1907). In *Graefe-Saemisch Handbuch der Gesamten Augenheilkunde*. Leipzig.

Grün, G. (1975). *J. Embryol. Exp. Morphol.* **33**, 243–257.

Hagedorn, M. and Fernald, R.D. (1992). *J. Comp. Neurol.* **321**, 193–208.

Hashimoto, Y., Abe, M. and Inokuchi, M. (1980). *Brain Res.* **197**, 331–340.

Herzog, H. (1905). *Arch. Anat. Physiol. (Physiol. Abstr.)* **516**, 413.

Hofmann, H.A., Benson, M.E. and Fernald, R.D. (1999). *Proc. Natl. Acad. Sci.* **96**, 14171–14176.

Hoke, K. and Fernald, R.D. (1998). *Dev. Brain Res.* **111**, 143–146.

Hollyfield, J.G. (1972). *J. Comp. Neurol.* **144**, 373–380.

Holt, C.E., Bertsch, T.W., Ellis, H.M. and Harris, W.A. (1988). *Neuron* **1**, 15–26.

Johns, P.R. and Fernald, R.D. (1981). *Nature* **293**, 141.

Kandel, E.R., Schwartz, J.H. and Jessel, T.M. (1991). *Principles of Neural Science*. Elsevier, New York.

Korenbrot, J.I. and Fernald, R.D. (1989). *Nature* **337**, 454–457.

Kunz, Y.W. (1980). *Experientia* **36**, 1371–1374.

Kunz, Y.W. (1983). *Experientia* **39**, 1049–1050.

Lam, D.M.K. and Steinman, L. (1971). *Proc. Natl. Acad. Sci. USA* **66**, 2777.

Lasater, E.M. (1990). In *The Visual System of Fish* (eds R.H. Douglas and M.B.A. Djamgoz), 211 pp. Chapman & Hall, New York.

Locket, N.A. (1971a). *J. Mar. Biol. Ass. UK* **51**, 79–91.

Locket, N.A. (1971b). *Proc. Roy. Soc., B* **178**, 161–184.

Lyall, A. (1957). *Quart. J. Microsc. Sci.* **98**, 101–110.

Lythgoe, J.N. (1979). *The Ecology of Vision*. Clarendon Press, Oxford.

Marc, R.E. and Liu, W.-L.S. (1984). *Nature, Lond.* **312**, 266–269.

Marc, R.E., Lam, D.M.K. and Stell, W.K. (1979). *Invest. Ophthalmol. Vis. Sci.* **18**, 34.

Marschak, D.W. and Dowling, J.E. (1987). *J. Comp. Neurol.* **256**, 430–443.

Meyer, R. (1978). *Exp. Neurol.* **59**, 99–111.

Miller, R.F. (1979). In *The Neurosciences: Fourth Study Program* (eds F.O. Schmitt and F.G. Warden), pp. 227–245. MIT Press, Cambridge, Mass.

Müller, H. (1952). *Zool. Jb. Abt. allg. Zool. Physiol.* **63**, 275–324.

Munk, O. (1964). *Galathea Rep.* **7**, 137–149.

Munk, O. (1966). *Dana Rep.* **70**, 1–62.

Nathans, J., Piantanida, T., Eddy, R., Shows, T. and Hogness, D. (1986). *Science* **232**, 203–252.

Neitz, M., Neitz, J. and Jacobs, G. (1991). *Science* **252**, 971–974.

Newman, E.A. and Zahs, K.R. (1997). *Science* **275**, 844–847.

O'Day, W.T. and Young, R.W. (1978). *J. Cell. Biol.* **76**, 593–604.

Powers, M.K. and Raymond, P.A. (1990). In *The Visual System of Fish* (eds R.H. Douglas and M.B.A. Djamgoz), pp. 419–442. Chapman & Hall, New York.

Raymond, P. (1985). *J. Comp. Neurol.* **236**, 90–105.

Ryder, J. (1895). *Proc. Acad. Nat. Sci.*, 161–167.

Sandy, J. and Blaxter, J. (1980). *J. Mar. Biol. Ass. UK* **60**, 59–71.

Scholes, J. (1975). *Phil. Trans. Roy. Soc. Lond., B* **270**, 61–118.

Scholes, J. (1979). *Nature* **278**, 620–624.

Sharma, S.C. and Ungar, F. (1980). *J. Comp. Neurol.* **191**, 373–382.

Stell, W.K. and Lightfoot, D.O. (1975). *J. Comp. Neurol.* **159**, 473–502.

Stell, W.K., Lightfoot, D., Wheeler, T. and Leeper, H. (1975). *Science* **190**, 989–990.

Sterling, P., Freed, M. and Smith, R.G. (1986). *Trends Neurosci.* **9**, 186–192.

Turner, D.L. and Cepko, C.L. (1987). *Nature* **328**, 131–136.

Verrier, M.L. (1928). *Bull. Biol. Fr. Belg.*, Supp. **11**, 1–222.

Wagner, H.-J. (1972). *Z. Morph. Tiere*, **72**, 77–130.

Wagner, H.-J. (1990). In *The Visual System of Fish*, vol. 345 (eds R.H. Douglas and M.B.A. Djamgoz), pp. 109–157. Chapman & Hall, New York.

Wagner, H.-J. and Wagner, E. (1988). *Phil. Trans. Roy. Soc.* **321**, 263–324.

Wagner, H.J. and Wulle, I. (1992). *Vis. Neurosci.* **9**, 325–333.

Walls, G.L. (1942). *The Vertebrate Eye and its Adaptive Radiation*. Hafner, New York.

Weiler, R. (1978). *Cell Tissue Res.* **195**, 515–526.

Wulle, I. and Wagner, H.-J. (1990). *Verh. Anat. Ges., Jena* **261**, 359–366.

Wunder, W. (1925). *Z. vergl. Physiol.* **4**.

Yokoyama, S. and Radlwimmer, F.B. (1998). *Mol. Biol. Evol.* **15**, 560–567.

Zucker, C.L. and Dowling, J.E. (1987). *Nature, Lond.* **330**, 166–168.

Zygar, C.A., Lee, M.L. and Fernald, R.D. (1999). *J. Neurobiol.* **41**, 435–442.

27.2 Mechanosensory Lateral Line: Microscopic Anatomy and Development

Jacqueline F Webb
Department of Biology, Villanova University, Villanova, Pennsylvania, USA

The mechanosensory lateral line system of bony fishes is composed of a series of receptor organs called neuromasts, which are located on the epithelium or in lateral line canals on the head and trunk (Figure 27.2.1), and are innervated by several lateral line nerves, which project to the hindbrain (see Chapter 15.2).

Structure and function of neuromast receptor organs

Neuromast receptor organs are epithelial structures composed of a population of sensory hair cells (which are sensitive to the displacement of their apical ciliary bundles), and nonsensory supporting cells and mantle cells (e.g. Hama and Yamada, 1977; Rouse and Pickles, 1991a,b).

The apical ciliary bundle of each sensory hair cell is composed of one long kinocilium (with a 9 + 2 microtubule configuration) and a cluster of shorter stereocilia (composed of actin), which are graded in length and located to one side of the kinocilium (Flock and Wersall, 1962; Flock and Duvall, 1965; Rouse and Pickles, 1991b; Cernuda-Cernuda and Garcia-Fernandez, 1992). Tip links between rows of stereocilia, which have been described in the inner ear of vertebrates, are present in the neuromasts of two species of fishes (Rouse and Pickles, 1991b). The morphology of the kinocilium and stereocilia and

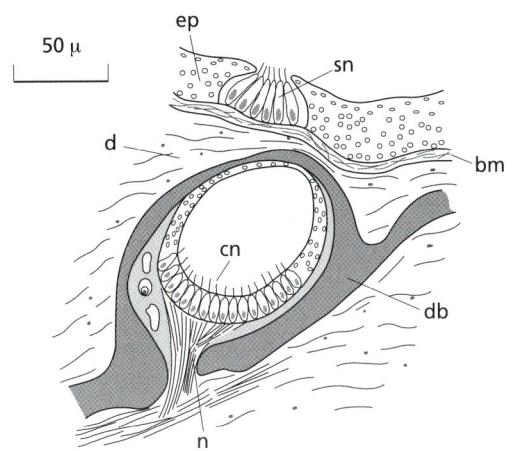

Figure 27.2.1 Diagrammatic representation of a superficial neuromast (sn) in the epidermis (ep) and a canal neuromast (cn) inside a bony canal (db) in the cyprinid, *Hybopsis aestivalis*: bm, basement membrane; d, dermis; n, nerve. (From Reno, 1969, reprinted with permission by The American Society of Ichthyologists and Herpetologists.)

the number of stereocilia in a hair cell vary within and among neuromasts within an individual (e.g. Yamada, 1973; Cernuda-Cernuda and Garcia-Fernandez, 1992), and among species (Popper and Platt, 1993). Hair cell heterogeneity, similar to that found in the auditory system of mammals (e.g. Type I vs. Type II hair cells), has been demonstrated in fishes, based on the differential susceptibility of hair cells in canal and superficial neuromasts to destruction by gentamicin sulfate (Song et al., 1995).

Each hair cell is morphologically polarized as a result of the relative positions of the single kinocilium and the cluster of stereocilia on its apical surface. The morphological polarization of a hair cell defines its axis of best physiological sensitivity to water movement (see Chapter 15.2). The physiological properties of mechanosensory hair cells in the lateral line system have only been studied in a few fish species (e.g. Flock and Wersall, 1962; Kroese and van Netten, 1989; van Netten, 1997; Wiersinga-Post and van Netten, 1998). Our knowledge of hair cell physiology is based on extensive work in the auditory system of vertebrates (e.g. Hudspeth, 1983; Corwin and Warchol, 1991; Guth et al., 1998; see Chapter 27.4).

Within a neuromast, hair cells are oriented 180° to one another and occur in an approximately 50:50 ratio (Hama, 1972; Rouse and Pickles, 1991a; Figure 27.2.2), so that each neuromast has a single axis of best physiological sensitivity (but see Shardo, 1996). This is in contrast to the maculae of the otolithic organs of the ear of fishes, which consist of patches of hair cells with either one or two different axes of best physiological sensitivity (Popper and Platt, 1993; see Chapter 15.4). The hair cells of canal neuromasts are oriented parallel to the long axis of the canal so that movement of fluid along the length of the canal can provide an effective mechanical stimulus for the canal neuromast. The axis of best physiological sensitivity of superficial neuromasts that occur in linear series, including those that are homologues of canal neuromasts (e.g. pedomorphic canal neuromasts, Webb, 1989b; Northcutt, 1992, 1997; Webb and Northcutt, 1997), tends to be parallel to the line of neuromasts. In contrast, the orientation of superficial neuromasts that are accessory to a canal tends to be perpendicular to the axis of the canal (e.g. Marshall, 1986; Webb, 1989c), thus adding another axis of best physiological sensitivity to the population of neuromasts. As a result, the lateral line system can respond to water movements arising from several directions. Superficial neuromasts that do not appear to be accessory to a canal occur on both the head and trunk of

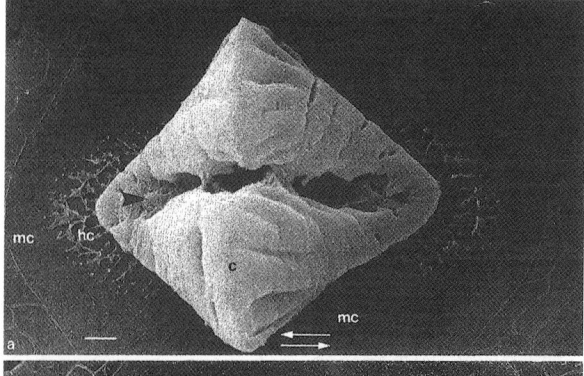
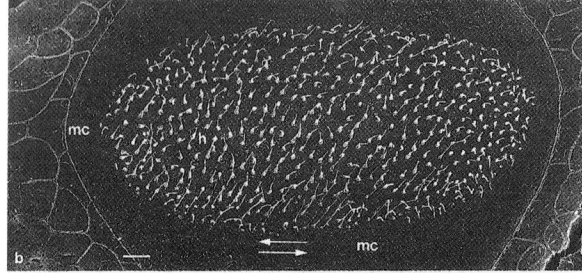
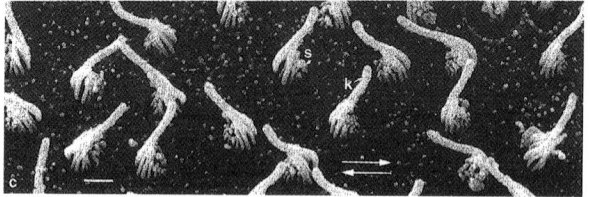

Figure 27.2.2 Hair cell distribution and morphology within a neuromast of the coral reef fish, *Apogon cyanosoma*. (a) Cupula (c) partially covering the neuromast; mantle cells (mc) and hair cells (hc) are evident (scale bar = 10 μm). (b) Close-up of the surface of the neuromast showing the oval sensory strip with its population of hair cells, and the mantle cells (mc) that surround the sensory strip and define the outer perimeter of the neuromast (scale bar = 10 μm). (c) Close-up of the hair cells of the sensory strip showing the kinocilium (k) and stereocilia (s) of each hair cell. The overall axis of polarization of the hair cell population is represented by arrows (scale bar = 1 μm). (From Rouse and Pickles, 1991b, reprinted by permission of Wiley-Liss, Inc.)

fishes (e.g. Northcutt, 1989; Teyke, 1990; Webb and Northcutt, 1997).

Hair cells are located in a portion of the neuromast called the sensory strip, which is typically round, oval or elongated (Figures 27.2.2 and 27.2.3). The sensory strip is surrounded by mantle cells that secrete the cupula (Kelly and van Netten, 1991; Mukai et al., 1991; Rouse and Pickles, 1991b; Mukai and Kobayashi, 1992) in which the ciliary bundles of all of the hair cells are embedded (Figure 27.2.2). The cupula provides a mechanical linkage between the hair cells and the external hydrodynamic environment. The functional significance of heterogeneity

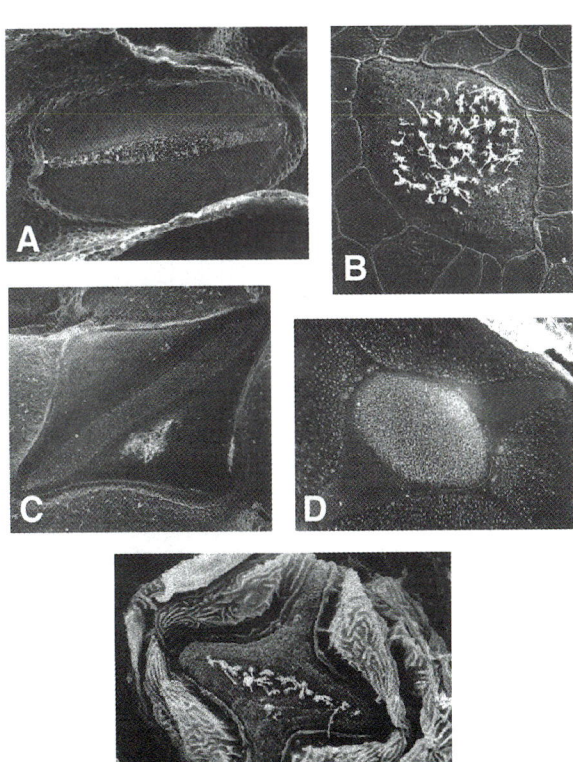

Figure 27.2.3 Scanning electron micrographs demonstrating morphological variation in neuromasts among teleost fishes. (A) Canal neuromast from the narrow head canal of the sculpin, *Cottus bairdi* (courtesy of S. Coombs). (B) A superficial neuromast from the trunk of the flatfish, *Scophthalmus aquosus* (Newman, Chambers and Webb, unpublished data). (C) Canal neuromast from a widened head canal of the blind side of the head in the rex sole, *Glyptocephalus* sp. (D) Canal neuromast from a widened head canal of the clown knifefish, *Notopterus chitala* (courtesy of S. Coombs). (E) Superficial neuromast from the blind side of the head of the California tongue sole, *Symphurus atricauda*.

in cupular structure is still unclear (Kelly and van Netten, 1991; Rouse and Pickles, 1991b). The cupula of superficial neuromasts grows continuously (Mukai and Kobayashi, 1992); the height of the cupula of canal neuromasts is limited by canal diameter. The shape of the neuromast generally defines the contour of the perimeter of the cupular base (but preparation artifact is common and cupular morphology is generally hard to discern.

The morphology of canal neuromasts is correlated with canal morphology. Neuromasts in narrow canals tend to be oval, with the long axis parallel to the axis of the canal (Figure 27.2.3A). Neuromasts in widened canals are much larger (in some cases, an order of magnitude larger); their hair cells are located in a round or oval sensory strip, but the mantle cells that surround the sensory strip and define the outer perimeter of the neuromast, span the width of the canal (Coombs *et al.*, 1988; Webb, 1989a; Figure 27.2.3C and D). Superficial neuromasts (= pit organs, Northcutt and Bleckmann, 1993; Webb and Northcutt, 1997) tend to be smaller than canal neuromasts within a given individual (Münz, 1989; Wonsettler and Webb, 1997). They are generally round or oval, and have a round or narrow and elongated sensory strip (Figure 27.2.3B and E). Recently, Marshall (1996) described two new types of superficial neuromasts in two species of deep-sea fishes, which appear to consist only of a very long, thin sensory strip that is located either flat on the skin surface, or on the edge of a flap of skin.

The length of kinocilia and stereocilia, number and density of hair cells, the shape and stiffness of the cupula, and the size and shape of a neuromast all have functional consequences (Denton and Gray, 1989; van Netten and Kroese, 1989; Kelly and van Netten, 1991; van Netten and Khanna, 1994). Interspecific and ontogenetic variation in these parameters need to be investigated in order to understand their potential adaptive and evolutionary implications.

Development of the mechanosensory lateral line system

The development of the mechanosensory lateral line system occurs in three stages: (i) embryonic patterning of neuromast receptors and sensory neuron outgrowth; (ii) differentiation and growth of neuromasts; and (iii) morphogenesis of lateral line canals on the head and trunk (reviewed by Blaxter, 1987; Webb, 1989b, 1999; Northcutt, 1992).

Like those of the sensory maculae of the inner ear, the hair cells in neuromasts of the lateral line system in both fishes and amphibians develop from cranial ectodermal **placodes** (reviewed by Northcutt, 1992; Webb and Noden, 1993; Northcutt *et al.*, 1994; Figure 27.2.4). Several older studies have provided detailed descriptions of the differentiation of neuromasts and sensory neurons from cranial ectodermal placodes (reviewed by Northcutt, 1992, 1997). More recently, it has been clearly demonstrated in both fishes and especially in amphibians (Fritzsch and Neary, 1998), that several lateral line placodes arise in cranial

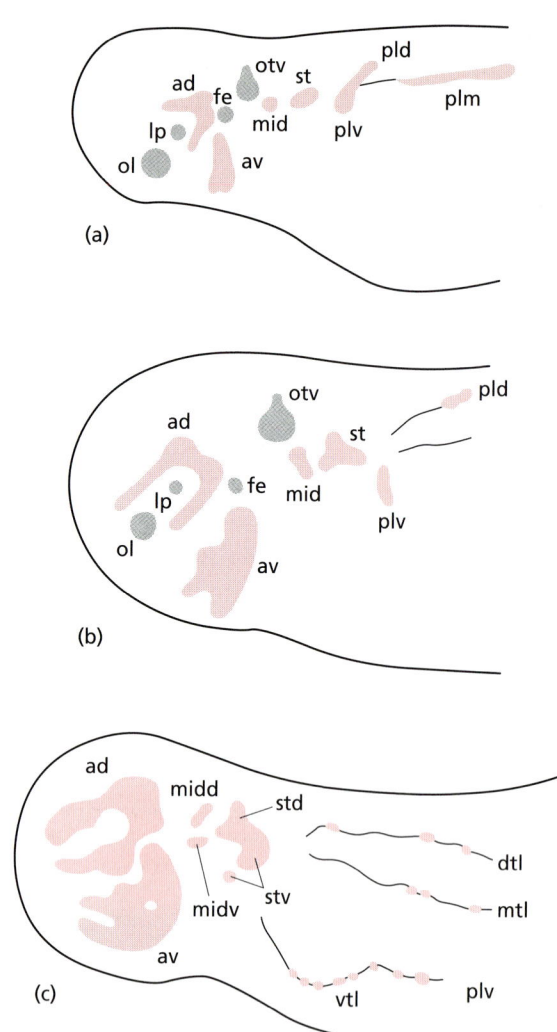

Figure 27.2.4 Ectodermal placodes on the head of the axolotl, an amphibian model system in which lateral line placodes are well known. (a) and (b) The lateral line placodes (black), and other sensory and neurogenic placodes (gray) in stage 35 and stage 37 axolotl embryos. (c) The lateral line placodes in a stage 38 larval axolotl, showing the elongation of placodes on the head and trunk that delineate lines of neuromasts that will differentiate from them: ad, anterodorsal lateral line placode; dtl, dorsal trunk lateral line; fe, facial epibranchial placode; lp, lens placode; mid, middle lateral line placode; midd, dorsal subdivision of middle lateral line placode, midv, ventral subdivision of middle lateral line placode; mtl, main trunk lateral line; ol, olfactory placode; otv, otic (octaval) vesicle; pld, dorsal subdivision of posterior lateral line placode; plm, main subdivision of posterior lateral line placode; plv, ventral subdivision of posterior lateral line placode; std, dorsal subdivision of supratemporal lateral line placode; stv, ventral subdivision of supratemporal lateral line placode; st, supratemporal lateral line placode. (From Northcutt et al., 1994, reprinted by permission of Wiley-Liss, Inc.)

ectoderm and elongate over the head (Northcutt et al., 1994, 1995; Northcutt, 1997) and down the trunk (Metcalfe et al., 1985; Vischer, 1989; reviewed by Northcutt, 1992, 1997), delineating the lines of neuromasts that subsequently differentiate in situ (Otsuka and Nagai, 1997; Figure 27.2.4). One recent paper has suggested that, in addition to placodes, neuromasts are partially derived from neural crest cells (Collazo et al., 1994). The sensory neurons that compose the lateral line nerves and innervate the neuromasts (Northcutt et al., 1994; Northcutt and Brandle, 1995; see Chapter 15.2), and the electroreceptor organs (in those species that have them), also differentiate from lateral line placodes (Northcutt et al., 1995; Fritzsch and Neary, 1998).

At hatching, neuromasts generally are present in the epithelium on both the head and trunk (e.g. Blaxter et al., 1983; Blaxter, 1987; Metcalfe, 1989; Blaxter and Fuiman, 1990; Otsuka and Nagai, 1997; Figure 27.2.5). Hair cells of differentiated neuromasts that are present in embryos and larvae are innervated, and are considered functional (Otsuka and Nagai, 1997). Neuromast number (e.g. Metcalfe, 1989; Vischer, 1989; Harvey et al., 1992) and neuromast size (Münz, 1989; Wonsettler and Webb, 1997; Tarby, 1998) increase, and neuromast and cupular shape change, ontogenetically (e.g. Blaxter et al., 1983; Münz, 1986, 1989; Webb, 1989b; Harvey et al., 1992; Wonsettler and Webb, 1997). Evidence from amphibians demonstrates that neuromast size increases with the addition of hair cells that differentiate from support cells, and that hair cells are capable of regeneration (Balak et al., 1990; Corwin and Warchol, 1991; Jones and Corwin, 1996). Hair cells in various stages of development are present in neuromasts, and new hair cells are produced in pairs, with opposite polarities, which may be the result of a single mitotic event (Rouse and Pickles, 1991a). Hair cells have been shown to exhibit turnover within a neuromast (Rouse and Pickles, 1991a), but with one known exception (Shardo, 1996), the axis of orientation of the hair cells in a neuromast, and therefore the neuromast's axis of best physiological sensitivity, appears not to change ontogenetically (Münz, 1989; Vischer, 1989; Webb, 1989b).

Late during the larval period, or at transformation to the juvenile stage, the lateral line canals form as epithelial ridges begin to rise on either side of a subset of individual neuromasts on the head (the presumptive canal neuromasts; Münz, 1986), making it appear that these neuromasts are sinking between these ridges into epithelial grooves. Intramembranous

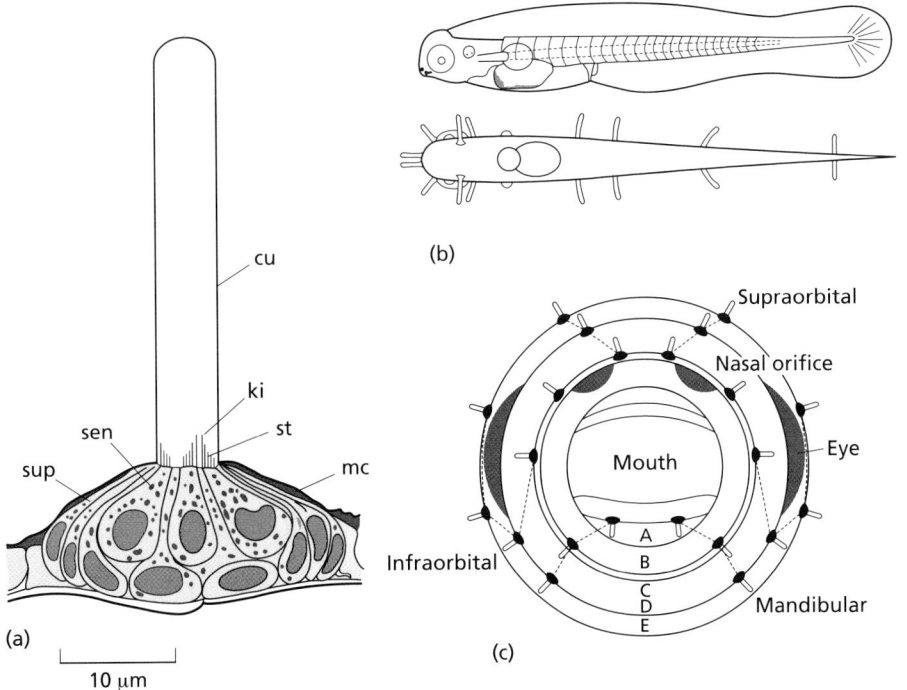

Figure 27.2.5 Superficial neuromasts in larval fishes. (a) Schematic of a section of a superficial neuromast: cu, cupula; ki, kinocilium; mc, mantle cell; sen, sensory cell; st, stereocilium; sup, supporting cell. (b) Dorsal view of a larva of a rabbitfish, *Siganus guttatus* (2.8 mm TL) indicating the location of the cupulae of superficial neuromasts. (c) Anterior view of the arrangement of superficial neuromasts on the head of a goldfish larva at different levels: A, rostral to olfactory sac; B, at level of olfactory sac; C, rostral to eye; D, caudal to eye; E, halfway between eye and auditory vesicle. (From Webb, 1999.)

ossification within these ridges forms the canal walls. The ridges fuse over the neuromast and the canal walls extend medially and fuse to form the canal roof (e.g. Webb, 1989c; Tarby, 1998; Figure 27.2.6). The epithelial tissue surrounding and lining adjacent canal segments fuse leaving a common epithelial pore between adjacent canal segments. Intramembranous ossification of canal segments that compose a lateral line canal (e.g. the mandibular canal, see Chapter 15.2) is non-synchronous and non-sequential (Kapoor, 1961; Webb, 1989b,c; Tarby, 1998). After canal segment ossification, the bony walls of adjacent segments fuse and the lateral line canals become integrated into the cranial dermal bones forming the 'lateral line bones', which are prominent components of the adult fish skull (e.g. Lekander, 1949; Disler, 1960; Branson and Moore, 1962; Cubbage and Mabee, 1996; Adriaens *et al.*, 1997; Tarby, 1998; see Chapter 15.2). After canals are formed and integrated into dermal bones, neuromast size and canal diameter continue to increase with fish size (Tarby, 1998). Neuromasts that remain superficial stay relatively small throughout life (Münz, 1989) and additional superficial neuromasts may continue to differentiate and proliferate throughout the larval and juvenile periods.

On the trunk, superficial neuromasts occur in one or more linear series in the epithelium overlying the myomeres of the trunk musculature (Blaxter *et al.*, 1983; Metcalfe, 1989), and increase in number before a final number of neuromasts is established. The lateral line scales develop in the dermis beneath each of the trunk neuromasts, late during the larval stage. Canal morphogenesis is similar to that on the head. Each neuromast becomes enclosed in a canal segment formed by the fusion of two epithelial ridges that rise on either side of the neuromast. A canal segment ossifies intramembranously in association with the flat scale beneath it, forming a tubed lateral line scale (Webb, 1989c; Wonsettler and Webb, 1997). Adjacent canal segments are associated with individual scales; the canal lumen is linked by a common epithelial pore, but the ossified walls and roof of adjacent canal segments do not fuse. Canal morphogenesis generally proceeds in a rostral to caudal direction (Vischer, 1989; Webb, 1989a, 1999; Wonsettler and Webb, 1997), resulting in the development of a pored trunk canal that consists of a series of overlapping, tubed

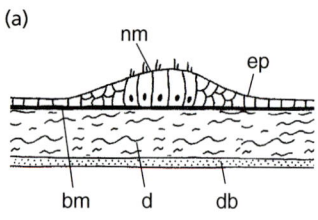
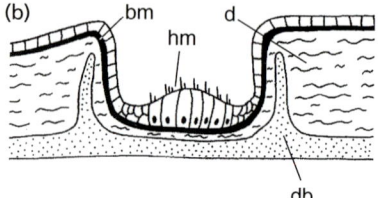
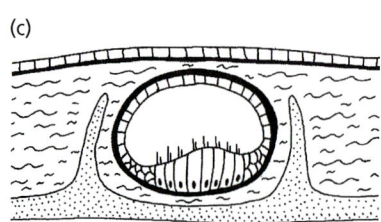
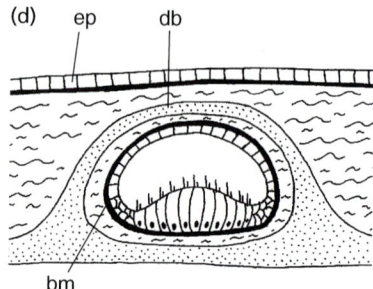

Figure 27.2.6 Diagrammatic representation of the morphogenesis of the lateral line canals of the head in the cichlid, *Archocentrus nigrofasciatus*. (a) Stage I – A superficial neuromast (nm) is located in the epidermis (ep) that sits above the basement membrane (bm); underlying dermal bone is present in the dermis. (b) Bony canal walls ossify intramembranously and extend upward from the underlying dermal bone (db) on either side of a neuromast, which appears to sink into an epithelial groove. (c) The epithelium fuses over the neuromast (nm) forming an epithelial canal. (d) The dermal canal bone continues to ossify intramembranously and the canal walls fuse medially to form the canal roof: nm, neuromast; bm, basement membrane; d, dermis; dm, dermal bone; ep, epithelium. (From Tarby, 1998.)

lateral line scales (see Chapter 15.2). Superficial neuromasts, which are generally smaller than canal neuromasts may remain in the epithelium in association with the canal (Webb, 1989b,c; Wonsettler and Webb, 1997).

Mechanisms responsible for patterns of development in the lateral line system of both fishes and amphibians are now being explored using modern approaches in a limited number of model species (see Fritzsch and Neary, 1998). Perturbations of the lateral line system have been observed in zebrafish mutants (e.g. Whitfield *et al.*, 1996; Nicholson *et al.*, 1998). In addition, the distribution pattern of neuromast-specific antigens (Kornblum *et al.*, 1990), expression patterns of **HOX genes** (Ekker *et al.*, 1997; Metscher *et al.*, 1997), and the role of neural crest in patterning and morphogenesis of the lateral line system (Smith *et al.*, 1988, 1990; Parichy, 1996a,b) have been examined in zebrafish and amphibian model systems. The examination of patterns of gene expression, as well as the nature of tissue interactions during embryogenesis will shed light on the mechanisms underlying the patterning of neuromast receptors on both the head and trunk of fishes. The nature of the interaction between neuromasts and dermal bone that can account for the pattern of distribution of the lateral line canals in dermal skeletal elements (see Devillers, 1947; DeBeer, 1985) and the mechanisms underlying the morphogenesis of lateral line canals (see Wonsettler and Webb, 1997, for discussion) are still not fully understood.

Acknowledgments

Dr Sheryl Coombs kindly provided two original photos. Ms Melissa Tarby granted permission for the reproduction of an original illustration from her unpublished MS thesis, and Leo Smith assisted in the preparation of figures. Supported by NSF grant IBN 9603896.

References

Adriaens, D., Verraes, W. and Taverne, L. (1997). *Eur. J. Morphol.* **35**, 181–208.
Balak, K.J., Corwin, J.T. and Jones, J.E. (1990). *J. Neurosci.* **10**, 2502–2512.
Blaxter, J.H.S. (1987). *Biol. Rev.* **62**, 471–514.
Blaxter, J.H.S. and Fuiman, L.A. (1990). *J. Mar. Biol. Ass. UK* **70**, 413–427.
Blaxter, J.H.S., Gray, J.A.B. and Best, A.C.G. (1983). *J. Mar. Biol. Ass. UK* **63**, 247–260.

Branson, B.A. and Moore, G.A. (1962). *Copeia* **1962**, 1–108.
Cernuda-Cernuda, R. and Garcia-Fernandez, J.M. (1992). *J. Hirnforsch.* **33**, 107–113.
Collazo, A., Fraser, S.E. and Mabee, P.M. (1994). *Science* **264**, 426–430.
Coombs, S., Janssen, J. and Webb, J.F. (1988). In *Sensory Biology of Aquatic Organisms* (eds J. Atema, R.R. Fay, A.N. Popper and W.N. Tavolga), pp. 553–594. Springer-Verlag, New York.
Corwin, J.T. and Warchol, M.E. (1991). *Ann. Rev. Neurosci.* **14**, 301–303.
Cubbage, C.C. and Mabee, P.M. (1996). *J. Morphol.* **229**, 121–160.
de Beer, G.R. (1985). *The Development of the Vertebrate Skull*. University of Chicago Press, Chicago.
Denton, E.J. and Gray, J.A.B. (1989). In *The Mechanosensory Lateral Line: Neurobiology and Evolution* (eds S. Coombs, P. Görner and H. Münz), pp. 229–246. Springer-Verlag, New York.
Devillers, C. (1947). *Ann. Palaeontol. (Paris)* **33**, 1–94.
Disler, N.N. (1960). *Lateral Line Sense Organs and their Importance in Fish Behavior*. Israel Program for Scientific Translation, Translated from Russian, Jerusalem, 1971, 328pp.
Ekker, M., Akimenko, M.-A., Allende, M.L., Smith, R., Drouin, G., Langille, R.M., Weinberg, E.S. and Westerfield, M. (1997). *Mol. Biol. Evol.* **14**, 1008–1022.
Flock, A. and Duvall, A.J. (1965). *J. Cell Biol.* **25**, 1–8.
Flock, A. and Wersall, J. (1962). *J. Cell Biol.* **15**, 19–27.
Fritzsch, B. and Neary, T.J. (1998). In *Amphibian Biology*, Vol. 3 (eds H. Heatwole and E.M. Dawley) pp. 878–922. Surrey Beatty and Sons, Australia.
Guth, P.S., Perin, P., Norris, C.H. and Valli, P. (1998). *Prog. Neurobiol.* **54**, 193–247.
Hama, K. (1972). *Acta Histochem. Cytochem.* **5**, 258–260.
Hama, K. and Yamada, Y. (1977). *Cell Tissue Res.* **176**, 23–36.
Harvey, R., Blaxter, J.H.S. and Hoyt, R.D. (1992). *J. Mar. Biol. Assoc. UK* **72**, 651–668.
Hudspeth, A.J. (1983). *Ann. Rev. Neurosci.* **6**, 187–215.
Jones, J.E. and Corwin, J.T. (1996). *J. Neurosci.* **16**, 649–662.
Kapoor, A.S. (1961). *Copeia* **1961**, 176–181.
Kelly, J.P. and van Netten, S.M. (1991). *J. Morphol.* **207**, 23–36.
Kornblum, H., Corwin, J.T. and Trevarrow, B. (1990). *J. Comp. Neurol.* **301**, 162–170.
Kroese, A.B.A. and van Netten, S.M. (1989). In *The Mechanosensory Lateral Line: Neurobiology and Evolution* (eds S. Coombs, P. Görner, and H. Münz) pp. 265–284, Springer-Verlag, New York.
Lekander, B. (1949). *Acta Zool. (Stockholm)*. **30**, 1–131.
Marshall, N.J. (1986). *J. Mar. Biol. Assoc. UK* **66**, 323–333.
Marshall, N.J. (1996). *J. Fish Biol.* **49** (Suppl. A), 239–258.
Metcalfe, W.K. (1985). *J. Comp. Neurol.* **238**, 218–224.
Metcalfe, W.K. (1989). In *The Mechanosensory Lateral Line: Neurobiology and Evolution* (eds S. Coombs, P. Görner and H. Münz) pp. 147–160, Springer-Verlag, New York.
Metcalfe, W.K., Kimmel, C.B. and Schabtach, E. (1985). *J. Comp. Neurol.* **233**, 377–389.
Metscher, B.D., Northcutt, R.G., Gardiner, D.M. and Bryant, S.V. (1997). *Dev. Genes Evol.* **207**, 287–295.
Mukai, Y. and Kobayashi, H. (1992). *Nippon Suisan Gakkaishi.* **58**, 1849–1853.
Mukai, Y., Kobayashi, H. and Yoshikawa, H. (1992). *Jap. J. Ichthyol.* **38**, 411–417.
Münz, H. (1986). *Ann. Mus. Roy. Afr. Centr. Sc. Zool.* **251**, 85–89.
Münz, H. (1989). In *The Mechanosensory Lateral Line: Neurobiology and Evolution* (eds S. Coombs, P. Görner and H. Münz), pp. 285–298. Springer-Verlag, New York.
Nicholson, T., Rusch, A., Friedrich, R.W., Granato, M., Ruppersberg, J.P. and Nusslein-Volhard, C. (1998). *Neuron* **20**, 271–283.
Northcutt, R.G. (1989). In *The Mechanosensory Lateral Line: Neurobiology and Evolution* (eds S. Coombs, P. Görner and H. Münz), pp. 17–78. Springer-Verlag, New York.
Northcutt, R.G. (1992). In *The Evolutionary Biology of Hearing* (eds D.B. Webster, R.R. Fay, A.N. Popper), pp. 21–48. Springer-Verlag, New York.
Northcutt, R.G. (1997). *Brain, Behav., Evol.* **50**, 25–37.
Northcutt, R.G. and Bleckmann, H. (1993). *J. Comp. Physiol. A.* **172**, 439–446.
Northcutt, R.G. and Brandle, K. (1995). *J. Comp. Neurol.* **355**, 427–454.
Northcutt, R.G., Catania, K.C. and Criley, B.B. (1994). *J. Comp. Neurol.* **340**, 480–514.
Northcutt, R.G., Brandle, K. and Fritzsch, B. (1995). *Dev. Biol.* **168**, 358–373.
Otsuka, M. and Nagai, S. (1997). *Zool. Sci.* **14**, 475–481.
Parichy, D.M. (1996a). *Devel. Biol.* **175**, 265–282.
Parichy, D.M. (1996b). *Devel. Biol.* **175**, 283–300.
Popper, A.N. and Platt, C. (1993). In *Physiology of Fishes* (ed. D.H. Evans), pp. 99–136. CRC Press, Boca Raton.
Reno, H.W. (1969). *Copeia* **1969**, 736–773.
Rouse, G.W. and Pickles, J.O. (1991a). *Proc. Roy. Soc. Lond. B.* **246**, 123–128.
Rouse, G.W. and Pickles, J.O. (1991b). *J. Morphol.* **209**, 111–120.
Shardo, J.D. (1996). *Copeia* **1996**, 226–228.
Smith, S.C., Lannoo, M.J. and Armstrong, J.B. (1988). *J. Morphol.* **198**, 367–379.
Smith, S.C., Lannoo, M.J. and Armstrong, J.B. (1990). *Anat. Embryol.* **182**, 171–180.
Song, J., Yan, H.Y. and Popper, A.N. (1995). *Hear. Res.* **91**, 63–71.
Tarby, M.L. (1998). Unpub. MS. Thesis, Villanova University, Villanova, PA, USA, 64pp.

Teyke, T. (1990). *Brain, Behav., Evol.* **35**, 23–30.

Van Netten, S.M. (1997). *Biophys. Chem.* **68**, 43–52.

Van Netten, S.M. and Khanna, S.M. (1994). *Proc. Natl. Acad. Sci.* **91**, 1549–1553.

Van Netten, S.M. and Kroese, A.B.A. (1989). In *The Mechanosensory Lateral Line: Neurobiology and Evolution* (eds S. Coombs, P. Görner and H. Münz), pp. 247–264. Springer-Verlag, New York.

Van Netten, S.M. and van Maarseveen, J.T.P.W. (1994). *Proc. R. Soc. Lond. B* **256**, 239–246.

Vischer, H. (1989). *Brain, Behav., Evol.* **33**, 205–222.

Webb, J.F. (1989a). *Brain, Behav., Evol.* **33**, 34–53.

Webb, J.F. (1989b). In *The Mechanosensory Lateral Line: Neurobiology and Evolution* (eds S. Coombs, P. Görner and H. Münz), pp. 79–98. Springer-Verlag, New York.

Webb, J.F. (1989c). *J. Morphol.* **202**, 53–68.

Webb, J.F. (1999). In *The Origin and Evolution of Larval Forms* (eds B.K. Hall and M.H. Wake), pp. 109–158. Academic Press, San Diego.

Webb, J.F. and Noden, D.M. (1993). *Amer. Zool.* **33**, 434–447.

Webb, J.F. and Northcutt, R.G. (1997). *Brain, Behav., Evol.* **50**, 139–151.

Whitfield, T.T., Granato, M., Van Eeden, F.J.M., Schack, U., Brand, M., Furutani-Seidi, M., Haffter, P., Hammerschmidt, M., Heisenbert, C.-P., Jiang, Y.-J., Kane, D.A., Kelsh, R.N., Mullins, M.C., Odenthal, J. and Nusslein-Volhard, C. (1996). *Development* **123**, 241–254.

Wiersinga-Post, J.E.C. and van Netten, S.M. (1998). *Proc. Roy Soc. Lond. B.* **265**, 615–623.

Wonsettler, A.L. and Webb, J.F. (1997). *J. Morphol.* **233**, 195–214.

Yamada, Y. (1973). *J. Ultrastruc. Res.* **43**, 1–17.

27.3 Chemoreception

Toshiaki J Hara
Freshwater Institute, Fisheries and Oceans Canada, Winnipeg, Manitoba, Canada

Olfactory epithelium

The olfactory lamella is composed of two layers of epithelium enclosing a thin stromal sheet. The stromal sheet that is separated from the epithelium by a basal lamina is filled with loose fibers and connective tissues. It also contains blood vessels and nerve axon bundles arising from the receptor cells (see below). Occasionally free nerve endings, presumably of the terminal nerve and trigeminal nerve, are also present in the olfactory epithelium (Figure 27.3.1D).

The epithelium is separated into two regions, sensory and non-sensory (Figure 27.3.1; cf. Figure 15.3.2). The sensory epithelium, a pseudostratified columnar epithelium, consists of receptor cells, supporting cells, ciliated non-sensory cells and basal cells (Figure 27.3.1B). Its thickness ranges from 35 to 75 µm. The non-sensory epithelium is of a stratified squamous type and usually thinner than the sensory epithelium. It consists of ciliated and non-ciliated surfaces (Figure 27.3.1A). The non-ciliated area has a general structure identical to that of the external epidermis from which it is derived. The surface has microridges arranged in fingerprint-like patterns.

The sensory epithelium exhibits various distribution patterns within the lamella. The following four types can be identified (Yamamoto, 1982): (i) continuous except for the lamella margin (*Auguilla*, *Ictalurus*); (ii) separated regularly by the non-sensory epithelium (Salmoniformes; cf. Figure 15.3.2B); (iii) interspersed irregularly with the non-sensory epithelium (*Gasterosteus*, *Thunnus*); and (iv) scattered in islets (*Phoxinus*, *Cyprinus*, *Carassius*; cf. Figure 15.3.2A).

Receptor cells (olfactory neurons)

The olfactory receptor cell is a bipolar primary neuron with a cylindrical dendrite that terminates at the free surface of the epithelium, i.e. directly exposed to the external environment (Figure 27.3.1B). This anatomical position contrasts strikingly with visual and auditory receptor cells, which are protected by

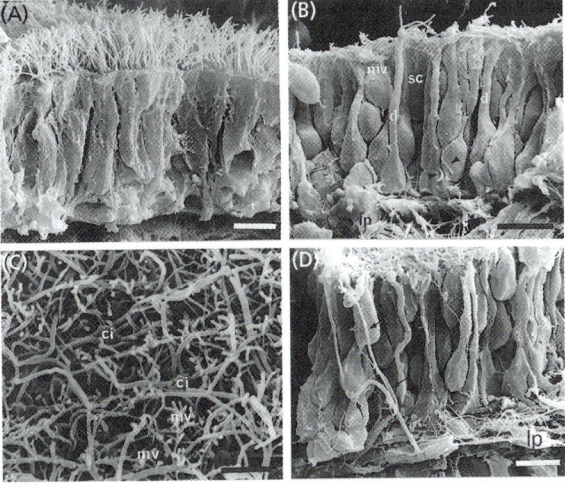

Figure 27.3.1 Scanning electron micrographs of sensory and non-sensory olfactory epithelium of the zebrafish, *Danio rerio*. Scale bars = 5 µm. (A) The non-sensory epithelium showing ciliated non-sensory cells with numerous kinocilia. (B) The sensory epithelium showing ciliated and microvillar (mv) receptor cell with long dendrites (d) separated by the supporting cells (sc). Arrowheads, olfactory axons; lp, lamina propria. (C) The surface of the sensory epithelium showing olfactory cilia (ci) and microvilli (mv) protruding from the olfactory knobs. (D) The sensory epithelium showing free nerve endings (arrows) leaving the lamina propria and reaching the apical portion of the epithelium. (From Hansen and Zeiske, 1998.)

membranes, fluid baths, bones and other structures serving to transduce and process the signals before they are perceived. The distal end of the dendrite forms a swelling (olfactory knob) protruding slightly above the epithelial surface. The proximal part of the perikaryon tapers to form an axon. The axons pass through the basement membrane, become grouped in the submucosa, and form the olfactory nerve fascicles, which run posteriorly to end in the olfactory bulb (see below).

In teleosts, at least two morphologically and ontogenetically distinct receptor cell types generally exist: ciliated and microvillar (Figure 27.3.2). The basic cytological features of these cell types are fundamentally the same (Zielinski and Hara, 1988). Both receptor cell types are present in hagfish, but only ciliated receptor cells are known in lampreys. Elasmobranchs possess only microvillar receptor cells. The ciliated receptor cell has four to eight cilia radiating from an olfactory knob. Each cilium measures 2–7 µm in length and 0.2–0.3 µm in diameter. The cilia show the 9 + 2 arrangement of microtubules, which is identical with that of common kinocilia. The microvillar receptor cell bears from 30 to 80 microvilli (2–5 µm long and about 0.1 µm wide) depending upon species. Approximately 5–10 million olfactory neurons comprise the sensory epithelium on each side of the nose in an average teleost. No clear functional differentiation appears to exist between these two receptor cell types. In the Arctic char (*Salvelinus alpinus*) and brook char (*Salvelinus fontinalis*), ciliated receptor cells are specific for bile-acid-like substances whereas the microvillar are specific for amino acids (Thommesen, 1983). Ciliated receptor cells respond to amino acids in the rainbow trout embryos at their early developmental stages (Zielinski and Hara, 1988). Furthermore, both sharks having only microvillar receptor cells and sea lamprey having only ciliated receptor cells respond to amino acids (Zeiske et al., 1986; Li et al., 1995).

A unique feature of the olfactory system is the regenerative capacity of the receptor neurons. Unlike other neurons of the vertebrate CNS, the receptor cells are continuously renewed in adults. Experimental severance of the olfactory nerve or application of aquatic toxicants causes degeneration of the sensory neurons followed by reconstitution of a new population of functional neurons (Evans et al., 1982; Evans and Hara, 1985; Zielinski and Hara, 1992). In rainbow trout, olfactory neurons gradually degenerate within 2 weeks of axotomy and virtually disappear in 2–4 weeks. Regeneration of the receptor neurons starts 7–8 weeks postaxotomy, reaching approximately normal structural configuration at about 12 weeks. Restoration of olfactory activity coincides with morphological repopulation of receptor neurons after 12 weeks. The regenerative process in rainbow trout is slower than that in other teleost fishes (e.g. Ichikawa and Ueda, 1977) and mammals (e.g. Graziadei and Monti-Graziadei, 1979). Most interesting is the fact that ciliated receptor cells appear in the olfactory epithelium before microvillar receptor cells as in embryonic rainbow trout (Zielinski and Hara, 1992). Similar differential morphological and functional development of ciliated and microvillar receptor cells is seen in the olfactory epithelium of goldfish (Zippel et al., 1997). These authors suggest that microvillar receptor cells mediate responses to pheromones, whereas ciliated cells mediate responses to amino acids. In the mouse, the vomeronasal organ and its microvillar receptor cells form after the olfactory organ (Cuschieri and Bannister, 1975). Thus, the consistent delayed formation of the trout microvillar receptor cells, combined with their functional differentiation, supports the hypothesis that the vomeronasal organ may be present in some groups of fishes in a form that has not been recognized (Eisthen, 1992).

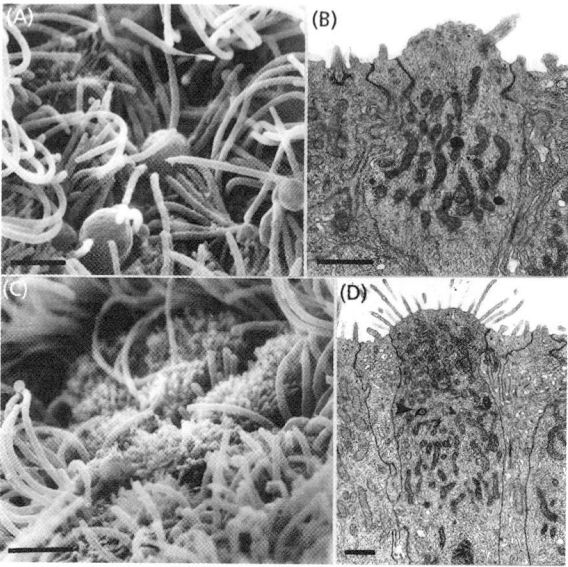

Figure 27.3.2 (A) Scanning and (B) transmission electron micrographs of ciliated olfactory receptor cells of the rainbow trout, *Oncorhynchus mykiss*. Scale bars = 3 and 0.3 µm, respectively. (C) Scanning and (D) transmission electron micrographs of microvillar olfactory receptor cells of rainbow trout. Scale bars = 4 and 1 µm, respectively. (From Zielinski and Hara, 1988).

Non-receptor cells

The supporting cells are columnar epithelial cells extending vertically from the epithelial surface to the basal lamina, forming a mosaic interspersed with receptor cells. The free surface is flat, with a few microvilli (cf. Figure 27.3.1). Supporting cells adjoin receptor cells, ciliated non-sensory cells, and other supporting cells at the free surface by an apical tight junction. The function of the supporting cells in fish olfaction is not defined, but in higher vertebrates they may have a role in the binding of odour molecules as they dissociate from olfactory receptor sites (Zielinski *et al.*, 1988, and see below).

Ciliated non-sensory cells are typical columnar epithelial cells with a wide flat surface, from which long (20–30 μm) kinocilia extend (cf. Figure 27.3.1). These cilia create weak currents over the olfactory lamellae, presumably assisting in water renewal and the transport of stimulant molecules in the olfactory organ. The basal cells are small and undifferentiated cells lying adjacent to the basal lamina and having no cytoplasmic processes reaching the free surface.

The basal cells in the sensory epithelium are assumed to be the progenitors of the receptor and supporting cells. Increased mitotic figures are seen in the basal region in a reconstituting epithelium after degeneration (Evans *et al.*, 1982). In addition, mucous cells (goblet cells) are scattered in the non-sensory epithelium. The mucous droplets are released through the stomatum of the cell surface of the epithelium. Secretions from the supporting cells also contribute to the mucous layer in fish, which, unlike other vertebrates, lack the mucous glands in the lamina propria of the olfactory epithelium.

In addition, gonadotropin-releasing hormone (GnRH) neurons originate in the medial olfactory placode during embryonic development and migrate into the basal forebrain in salmonids (Parhar *et al.*, 1995). The intracerebral expression of GnRH is not detected until the peak of **smoltification**, which coincides with a peak in thyroid hormones, and with downstream migratory behaviour (Parhar and Iwata, 1996). The terminal nerve (cranial nerve 0; TN) is another neuronal system associated with the olfactory system through their development from the olfactory placode (Sheldon, 1912; Demski, 1993). The TN is also GnRH-immunoreactive, and sends its axons to the olfactory epithelium. Its involvement in chemosensory function related to reproduction has been suggested (Demski and Northcutt, 1983), however recent electrophysiological studies do not support this hypothesis and its function still remains unclear (Fujita *et al.*, 1991). Recently, in rainbow trout, magnetoreceptor cells have been identified in the lamina propria of the olfactory epithelium, which is linked to the brain via the superficial ophthalmic ramus of the trigeminal nerve (Walker *et al.*, 1997). The cells contain magnetite crystals and respond to changes in the intensity of an imposed magnetic field, which is implicated in the magnetic orientation in salmonid navigation.

Taste buds

The taste bud is bulbiform in shape and varies in size (45–75 μm in height and 30–50 μm in width; Figure 27.3.3). Gustatory (receptor), supporting, and basal cells constitute the taste bud. The gustatory cells have a single or occasionally two apical processes (1.5–3 μm in length and 0.5 μm in diameter). Just below the processes, many electron-dense vesicles (50–70 nm) are seen in the peripheral cytoplasm. Synapses between gustatory cells and nerves are small and located by increased density of the membranes and postsynaptic dense layer under the nerve membrane. The gustatory synapse is not always associated with vesicles. The number of gustatory cells in a taste bud varies considerably, e.g. from five in *Pomatoschistus* (Gobiidae) to 67 in *Corydoras* (Callichthyidae) (Jakubowski and Whitear, 1990). Gustatory cells within the taste bud are also continually replaced. The life span of an individual gustatory cell in catfish is only 12–42 days (Raderman-Little, 1979). As a consequence, gustatory cells at different stages of development are present within a single taste bud, which may be represented by the often-referred to dark, light and intermediate gustatory cells.

The supporting cells have a number of small (100–200 nm in diameter and 0.5–1.0 μm in length) microvilli at their apical surfaces. The distal cytoplasm contains secretory vesicles that open at the surface to secrete mucus. The basal cells, typically one to five per taste bud, are situated at the bottom of the taste bud. They are attached to the gustatory and supporting cells by desmosomes. Unmyelinated nerve fibers packed closely together are intermingled with the processes of gustatory and basal cells, forming the nerve plexus (cf. Figure 15.3.3C). This is the terminal structure of the gustatory nerve.

Solitary chemosensory cells

The SCCs are bipolar, with an apical process, or processes, at the surface of the epidermis, and the proximal region associated with a neurite profile (Whitear, 1992). They may be found in the external skin and oropharyngeal epithelium, including the gills. The apical process is usually a single stout projection (1–2 μm in length), but occasionally a cluster of microvilli set on the apex of the cells. The cytoplasm contains numerous mitochondoria, Golgi systems giving rise to characteristic vesicles, endoplasmic reticulum, coated vesicles, and a few multivesicular bodies and lysosomes. There are also well-developed, vertically set microtubules. Vesicles are present within the apical projection and may open to the surface. These vesicles (50–70 nm in diameter) have electron-dense contents, which correspond to the distal vesicles of gustatory cells. Slender nerve fibers make passing contact with the SCCs, usually near the base and sometimes indenting the membrane. In the rocklings, the fine facial nerve fibers supplying the vibratile fin originate from a small population of **geniculate** ganglion sensory neurons. These fibers terminate within a dorsal part of the medullary facial lobe (Kotrschal and Whitear, 1988). Synaptic specializations are rarely seen, except in favorable situations such as the rockling vibratile fin. Like those on the gustatory cells, synapses consist of membrane densities, with a postsynaptic density under the neurite membrane; occasionally the base of the cell is full of vesicles.

Molecular basis of signal transduction

Olfaction

The initial event of olfaction is binding of an odorant molecule with a specific membrane receptor located in the cilia or microvilli of the olfactory neuron. Direct evidence for the existence of such receptors was provided by studies that showed the binding of radiolabeled amino acids to olfactory cilia preparations in rainbow trout (e.g. Rhein and Cagan, 1983). Recently, a large multigene family has been identified in a variety of vertebrates including catfish and zebrafish that probably encode for hundreds of different odorant receptors expressed by olfactory neurons (Buck and Axel, 1991; Ngai et al., 1993a; Barth et al., 1996). The size of the receptor repertoire of catfish and zebrafish is far smaller than that of human and rodent (100 vs. 500–1000 genes). These genes are expressed only in the olfactory sensory neurons. *In situ* hybridization studies also show that individual olfactory neurons express different complements of odorant receptors and are therefore functionally distinct (Ngai et al., 1993b). The odorant receptors are a member of a large superfamily of G-protein-coupled receptors, exhibiting seven hydrophobic domains (Buck and Axel, 1991).

The binding of odor molecules to odorant receptors induces a transduction cascade that ultimately leads to depolarization of the neuron and action potential generation in the olfactory axon. A variety of receptor-mediated transduction events are known (Figure 27.3.3). Chemoreception in fishes uses at least two processes: one that alters the concentration of second messengers, mediated most probably by intervening GTP (guanosine 5′-triphosphate)-binding regulatory proteins (G-proteins), and another that affects intracellular ionic activity via stimulus-gated ion channels (Brand and Bruch, 1992). Stimulus (amino acids) binding to receptor (R) in the ciliary membrane activates G-proteins that stimulate the activity of adenylyl cyclase (AC). AC generates cyclic AMP (cAMP) from ATP, which directly gates an ion channel in the ciliary plasma membrane, allowing influx of sodium and calcium ions. This leads to change in membrane potential that results in action potential generation in the axons. This transduction scheme is widely distributed among vertebrates. In catfish, two other second messengers, inositol triphosphate (IP_3) and diacylglycerol (DAG), are also involved in the transduction. In teleosts, the third transduction scheme, i.e. stimulus binding to receptor (R), directly opens an ion channel, which allows influx of sodium and calcium ions leading to a generation of receptor potential. Although multiple transduction mechanisms are identified in teleosts, details on which particular transduction pathway is involved in which stimulus group remain largely unknown. The transduction pathways described above are based exclusively on data obtained for odorant amino acids in channel catfish, *Ictalurus punctatus*. Different transduction mechanisms may function in other fish species, and for other odorants including reproductive pheromones, gonadal steroids and prostaglandins.

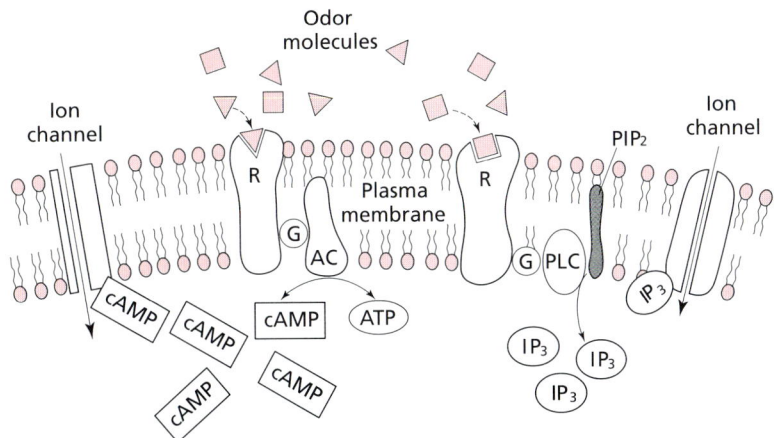

Figure 27.3.3 Schematic diagram to summarize mechanisms of olfactory and gustatory transduction. Stimulus binding to receptor (R) in the ciliary membrane activates GTP-binding regulatory protein (G), which, in turn, stimulates the activity of a second messenger producing enzyme adenylyl cyclase (AC). AC generates cyclic AMP (cAMP) from ATP. cAMP directly gates an ion channel in the ciliary membrane, allowing influx of Na and Ca ions to create a generator potential. In gustatory transduction, G protein activates a second messenger producing enzyme phospholipase C (PLC). PLC metabolizes phosphatidylinositol bisphosphate (PIP_2) into two second messengers, inositol triphosphate (IP_3) and diacylglycerol (DAG). IP_3 releases Ca from internal stores, which triggers neurotransmitter release.

Gustation

Two possible transduction sequences are suggested for gustation of amino acids in channel catfish (Brand and Bruch, 1992; Caprio *et al.*, 1993). Binding of L-alanine to receptor (R) triggers a G-protein-dependent increase in the production of the second messengers IP_3 and cAMP, while binding of L-arginine and L-proline to their respective receptors is directly coupled to the activation of non-selective cation channels. Again, it should be pointed out that the models of gustatory signal transduction sequences depicted above are derived from the gustatory system of channel catfish that has unique physiological response characteristics for amino acids.

Neural projections and central olfactory pathways

Olfactory nerve

The olfactory nerve, unmyelinated axons of the receptor neurons, course to the first relay station, the olfactory bulb. A topographical relationship can be found between the olfactory epithelium, nerve and bulb. According to Sheldon's (1912) classical work with carp, the olfactory nerve consists of medial and lateral bundles. The former, originating mainly from the more rostral olfactory lamellae, projects to the medial part of the olfactory bulb, while the latter originating mainly from the more caudal projects to the lateral part. As mentioned above, molecular biological studies with catfish indicate that individual receptors or receptor subfamilies are found in cells distributed throughout the olfactory epithelium, with no segregation along anterior–posterior, dorsal–ventral, or medial–lateral axes (Ngai *et al.*, 1993a). Also, in zebrafish neurons projecting into a single identified glomerulus are randomly dispersed, but regularly spaced over the olfactory epithelium (Baier *et al.*, 1994). Neurons expressing specific receptors therefore appear to be randomly distributed within the olfactory epithelium in teleosts. If spatial segregation is employed to decode olfactory information, then neurons expressing a given receptor must project to discrete loci within the olfactory bulb.

Functional significance of this scheme is supported by electrophysiological experiments (Thommesen, 1978; Hara and Zhang, 1996). In salmonid fishes, olfactory neurons responsive to two distinct odorant groups, amino acids and the bile acids, are distributed randomly throughout the olfactory rosette and yet project to spatially segregated regions of the bulb. Figure 27.3.4 schematically illustrates this

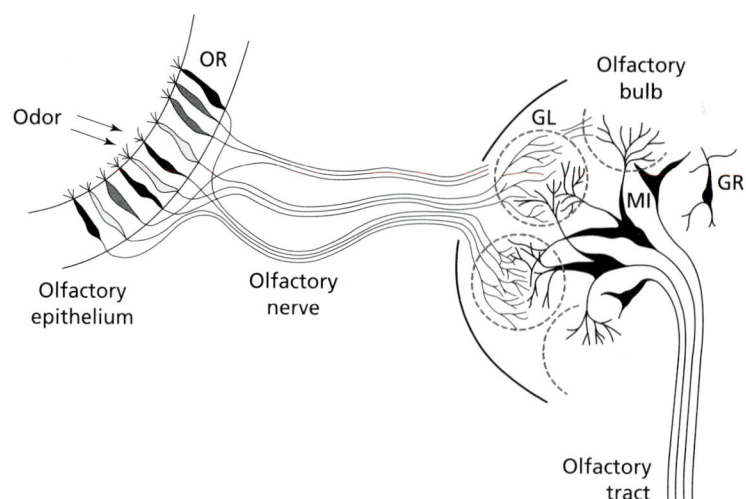

Figure 27.3.4 Schematic representation of the cellular anatomy of the peripheral olfactory system and the neural organization in the olfactory bulb of a teleost. Olfactory receptor cells (OR) are randomly distributed in the olfactory epithelium. Dispersed olfactory receptor cells with common receptors converge on to common olfactory glomeruli (GL) located in segregated regions of the olfactory bulb and transmit information through common mitral cells (MI). GR, granule cells. (Modified from Hara, 1993.)

neural organization. The axons of the olfactory neurons come together to form the olfactory nerve **fasciculi** which, after penetrating the bone underlying the epithelium, run caudally to terminate in the superficial layers of the bulb. Whether the olfactory nerve fibers are organized in some fashion as in the carp or are intermingled, is not known. After entering the bulb, the primary axons are organized in a very consistent fashion and terminate at a distinctly positioned glomerular layer of the bulb in the rainbow trout (Riddle and Oakley, 1992). Because the individual nerve fibers do not branch before entering the glomeruli (Allison, 1953), the signal transduced by a neuron is targeted exclusively to a single specific glomerulus. In mammals, functionally distinct classes of neurons are segregated in several discrete zones of the olfactory epithelium. Neurons in different spatial zones, or domains, of the epithelium project axons to different regions of the bulb. Thus, it appears that the entire fish olfactory epithelium may approximate a single compartment of the mammalian olfactory epithelium (Ngai et al., 1993c).

Olfactory bulb

The microscopic structure of the olfactory bulb is generally similar in all vertebrates. In fishes the olfactory bulb is poorly differentiated and lamination is not as distinct as in higher vertebrates. Generally, four layers are distinguishable, from the outside toward the inside: (i) olfactory nerve layer; (ii) the glomerular layer, (iii) the mitral cell layer, and (iv) the internal cell layer. The dominant feature of the bulb is the synaptic contact between the primary olfactory nerve fibers and dendrites of secondary neurons, mitral and tufted cells. Because dispersed olfactory neurons with common odorant receptors converge on common glomeruli located in segregated regions of the olfactory bulb and transmit information through common mitral cells, glomeruli appear to be functional units that respond differentially to different odorants.

In the glomeruli the highly important synaptic contact between the primary and the secondary olfactory neurons takes place, and this alone must indicate that they have considerable physiological significance. The incoming olfactory nerve fibers do not divide before entering the glomeruli, although they branch freely within them (Allison, 1953). Hence, it is significant that the axon of a receptor neuron does not terminate in more than one glomerulus, and each glomerulus receives impulses from a segregated and independent collection of olfactory neurons (Figure 27.3.4). Synapses between the primary olfactory nerve terminals and the mitral cell dendrites are excitatory and asymmetrical, characterized by a dense thickening of the postsynaptic membranes (Oka et al., 1982; Satou, 1992). The mitral cell has a relatively large cell body and several main dendrites reaching out to branch in two or more glomeruli, which receive numerous synaptic inputs from the olfactory nerve on their spiny appendages. In addition, their somata and dendritic shafts make numerous dendro-dendritic, reciprocal synapses with the peripheral dendrites of granule cells, thus receiving excitatory and inhibitory influences. Centrifugal fibers from the telencephalic hemisphere terminate in the internal cell layer, making synapses with dendrites and **somata** of granule cells, through which the mitral cell activity is modulated by influence from more central parts of the brain (Hara and Gorbman, 1967).

In rainbow trout, most primary nerve axons terminate in nine discrete terminal fields in the glomerular layer (Riddle and Oakley, 1992). The size and positioning of the nine fields are constant across individuals. Axonal fascicles end cohesively in typical spheroidal glomeruli in most fields, but in two lateroposterior fields axons end in a diffuse brush pattern. In zebrafish, approximately 80 glomeruli are present in a typical olfactory bulb (Baier and Korsching, 1994). They are arranged in a stereotyped pattern and bilaterally symmetrical. Twenty-two glomeruli can be identified in different individuals by their characteristic positional and morphological features. The remaining glomeruli are embedded in the glomerular plexus or densely clustered in the dorsal olfactory bulb. The latter might correspond with the diffuse brush pattern seen in the rainbow trout. In mammals, the total number of glomeruli is estimated to be approximately 2000 (Allison, 1953). If each neuron expresses only one receptor type and each type of neuron connects to a characteristic glomerulus, the number of glomeruli should be roughly the same as the number of receptor types.

Evidence for a functional organization in the bulb comes from electrophysiological studies of olfactory rosette and bulb in salmonids (Thommesen, 1978; Hara and Zhang, 1996, 1998). Amino acids and bile acids induce electrical responses in two segregated regions – the lateroposterior and mid-olfactory bulb in salmonids. When the olfactory rosette is subjected to a series of partial lamellectomy (removal of the lamellae), bulbar responses to these odorants are reduced uniformly, independent of the rosette region removed. These data indicate that in salmonids olfactory neurons responsive to amino acids and bile acids are randomly distributed throughout the olfactory rosette, and yet project to spatially segregated regions. Furthermore, when two odorants are applied in mixture, generated signals are encoded simultaneously and independently.

Olfactory tract and its central projections

The axons of the mitral cells form the majority of the afferent fibers of the olfactory tract through which information from the olfactory bulb is conveyed to the telencephalic hemispheres. The olfactory tract consists of two main bundles, lateral (LOT) and medial (MOT). In the carp, the axons from the mitral cells in the lateral part of the olfactory bulb run mainly through the LOT, whereas those in the medial part of the bulb run mainly through the MOT (Sheldon, 1912). The LOT and MOT fibers terminate bilaterally in several regions in the telencephalon and diencephalon. Generally, in cyprinids three terminal fields in the telencephalon and one in the diencephalon are identified (Oka *et al*., 1982). The projection pattern is essentially the same in rainbow trout (Northcutt and Davis, 1983). Some MOT fibers continue into the preoptic region, where they may synapse on dendrites of the preoptic neurons. Electrical and chemical stimulation of the olfactory tract of goldfish induce depletion of stainable neurosecretory granules and trigger action potentials in the preoptic neurons (Kandel, 1964; Jasinski *et al*., 1967). Electrical stimulation and lesion experiments also show that telencephalic and preoptic areas contribute to aspects of reproductive activities of several teleost fishes (Demski and Hornby, 1982; Kyle *et al*., 1982; Satou *et al*., 1982; Koyama *et al*., 1984). Most significant is the fact that both LOT and MOT fibers reach the contralateral hemispheres via the anterior and **habenular** commissures. Centrifugal fibers originating in both ipsilateral and contralateral telencephalic hemispheres run through the MOT. It thus appears that extensive interaction takes place here at the telencephalic level between the lateral and medial olfactory subsystems (Satou, 1992).

Functional separation of the olfactory system is demonstrated in several species. Electrical stimulation of each olfactory tract bundle elicits distinct behavioral patterns in free-swimming cod (*Gadus morhua*) (Döving and Selset, 1980). In goldfish, section of the MOT reduces courtship behavior, while feeding response is affected more by section of the LOT than of the MOT (Stacey and Kyle, 1983). In support of this, responses to identified pheromones (gonadal steroids and prostaglandins) are mainly mediated by the MOT, but responses to amino acids and crude food odors are mediated by both the LOT and MOT (Sorensen *et al*., 1991).

Extrabulbar projection of primary olfactory nerve

Not all primary olfactory nerves make contact with the secondary olfactory neurons in the olfactory bulb. Recent immunohistochemical studies in teleosts indicate that some primary olfactory fibers reach extrabulbar targets directly (Bazar et al., 1987; Honkanen and Ekström, 1990; Szabo et al., 1991; Riddle and Oakley, 1992; Becerra et al., 1994). These fibers project through the MOT and reach the ventral telencephalic and preoptic regions believed to be involved in sexual behavior (Becerra et al., 1994; Anadón et al., 1995).

Interestingly, in salmonids, olfactory neurons responsive to amino acids and bile acids project to spatially segregated regions of the olfactory bulb, and thereby generated signals are encoded spatially and temporally. However, stimulation by presumed hormonal pheromones, gonadal steroids and prostaglandins did not induce any response in the olfactory bulb. These data suggest that olfactory signals due to pheromones are processed in a manner distinct from those for amino acids and bile acids, and may be mediated by extrabulbar primary olfactory fibers bypassing the bulb (Hara and Zhang, 1998).

Gustatory nerves and their central projections

Each taste bud, depending upon its location, is innervated by one of three gustatory nerves by which peripheral gustatory inputs are transmitted to the CNS: (i) facial (cranial nerve VII); (ii) glossopharyngeal (cranial nerve IX); and (iii) vagal (cranial nerve X). Normally, the facial nerve innervates taste buds on the external body, lips, and rostral palate, which are important in appetitive (food search) feeding activity in some species (e.g. catfish). The glossopharyngeal nerve innervates taste buds along the floor of the oral cavity and gill rakers, while the vagal nerve innervates taste buds in the pharynx. These latter two nerves are believed to be important in ingestive and swallowing behavior. The facial and vagal nerves respectively terminate in the facial and vagal lobes of the medulla. The glossopharyngeal nerve terminates in a dorsal medullary region between the facial and vagal lobes (Kanwal and Finger, 1992). The relative complexity of the gustatory nerve termination in the medulla depends upon the development of appetitive and ingestive behavior in a particular fish species. In species such as catfish, the facial lobe is highly developed, whereas in species such as goldfish and carp the vagal lobe is highly developed.

Little is known about how the nerve endings synapse with the gustatory cell in fishes. Thus our knowledge of how individual fibers relay gustatory information to the CNS in teleosts is extremely limited. In higher vertebrates, a single gustatory nerve fiber innervates several taste buds and each gustatory cell forms more than one synapse, suggesting a high degree of convergence of input on to gustatory nerve fibers (Kinnamon, 1987). A limited body of data suggests that different fiber types exist in the same fish species that are widely different in their sensitivities to members of the same class of chemicals. At the same time, the chemical specificity of different fiber types across species can differ radically (Caprio et al., 1993).

References

Allison, A.C. (1953). *Biol. Rev. Cambridge Phil. Soc.* **28**, 195–244.

Anadón, R., Manso, M.J., Rodríguez-Moldes, I. and Becerra, M. (1995). *Neurosci. Lett.* **191**, 157–160.

Baier, H. and Korsching, S. (1994). *J. Neurosci.* **14**, 219–230.

Baier, H., Rotter, S. and Korsching, S. (1994). *Proc. Natl. Acad. Sci. USA* **91**, 11646–11650.

Barth, A.L., Justice, N.J. and Ngai, J. (1996). *Neuron* **16**, 23–34.

Bazer, G.T., Ebbesson, S.O.E., Reynolds, J.B. and Bailey, R.P. (1987). *Cell Tissue Res.* **248**, 499–503.

Becerra, M., Manso, M.J., Rodríguez-Moldes, I. and Anadón, R. (1994). *J. Comp. Neurol.* **342**, 131–143.

Brand, J.G. and Bruch, R.C. (1992). In *Fish Chemoreception* (ed. T.J. Hara), pp. 126–149. Chapman & Hall, London.

Buck, L. and Axel, R. (1991). *Cell* **65**, 175–187.

Caprio, J., Brand, J.G., Teeter, J.H., Valentinčič, T., Kalinoski, D.L., Kohbara, J., Kumazawa, T. and Wegert, S. (1993). *Trends Neurosci.* **16**, 192–197.

Cuschieri, A. and Bannister, L.H. (1975). *J. Anat.* **119**, 471–498.

Demski, L.S. (1993). *Acta Anat.* **148**, 81–95.

Demski, L.S. and Hornby, P.J. (1982). *Can. J. Fish. Aquat. Sci.* **39**, 36–47.

Demski, L.S. and Northcutt, R.G. (1983). *Science* **220**, 435–437.

Döving, K.B. and Selset, R. (1980). *Science* **207**, 559–560.

Eisthen, H.L. (1992). *Microsc. Res. Tech.* **23**, 1–21.

Evans, R.E. and Hara, T.J. (1985). *Brain Res.* **330**, 65–75.

Evans, R.E., Zielinski, B. and Hara, T.J. (1982). In *Chemoreception in Fishes* (ed. T.J. Hara), pp. 15–37. Elsevier, Amsterdam.

Fujita, I., Sorensen, P.W., Stacey, N.E. and Hara, T.J. (1991). *Brain Behav. Evol.* **38**, 313–321.

Graziadei, P.P.C. and Monti Graziadei, G.A. (1979). *J. Neurocytol.* **8**, 1–17.

Hansen, A. and Zeiske, E. (1998). *Chem. Senses* **23**, 39–48.

Hara, T.J. (1993). In *Behaviour of Teleost Fishes* (ed. T.J. Pitcher), pp. 171–199. Chapman & Hall, London.

Hara, T.J. and Gorbman, A. (1967). *Comp. Biochem. Physiol.* **21**, 185–200

Hara, T.J. and Zhang, C. (1996). *Neurosci. Res.* **26**, 65–74.

Hara, T.J. and Zhang, C. (1998). *Neuroscience* **82**, 301–313.

Honkanen, T. and Ekström, P. (1990). *J. Comp. Neurol.* **292**, 65–72.

Ichikawa, M. and Ueda, K. (1977). *Cell Tissue Res.* **183**, 445–455.

Jakubowski, M. and Whitear, M. (1990). *Z. Mikrosk.-Anat. Forsch.* **104**, 529–560.

Jasinski, A., Gorbman, A. and Hara, T. (1967). In *Neurosecretion* (ed. F. Stutinsky), pp. 106–123. Springer-Verlag, Berlin.

Kandel, E.R. (1964). *J. Gen. Physiol.* **47**, 691–717.

Kanwal, J.S. and Finger, T.E. (1992). In *Fish Chemoreception* (ed. T.J. Hara), pp. 79–102. Chapman & Hall, London.

Kinnamon, J.C. (1987). In *Neurobiology of Taste and Smell* (eds T.E. Finger and W.L. Silver), pp. 277–297. Wiley-Interscience, New York.

Kotrschal, K. and Whitear, M. (1988). *J. Comp. Neurol.* **268**, 109–120.

Koyama, Y., Satou, M., Oka, Y. and Ueda, K. (1984). *Behav. Neural Biol.* **36**, 229–241

Kyle, A.L., Stacey, N.E. and Peter, R.E. (1982). *Behav. Neural Biol.* **36**, 229–241.

Li, W., Sorensen, P.W. and Gallaher, D. (1995). *J. Gen. Physiol.* **105**, 569–589.

Ngai, J., Dowling, M.M., Buck, L., Axel, R. and Cgess, A. (1993a). *Cell* **72**, 657–666.

Ngai, J., Chess, A., Dowling, M.M., Necles, N., Macagno, E.R. and Axel, R. (1993b). *Cell* **72**, 667–680.

Ngai, J., Chess, A., Dowling, M.M. and Axel, R. (1993c). *Chem. Senses* **18**, 209–216.

Northcutt, R.G. and Davis, R.E. (1983). In *Fish Neurobiology*, vol. 2 (eds R.G. Northcutt and R.E. Davis), pp. 203–236. University of Michigan Press, Ann Arbor, MI.

Oka, Y., Ichikawa, M. and Ueda, K. (1982). In *Chemoreception in Fishes* (ed. T.J. Hara), pp. 61–75. Elsevier, Amsterdam.

Parhar, I.S. and Iwata, M. (1996). *Neurosci. Res.* **26**, 299–308.

Parhar, I.S., Iwata, M., Pfaff, D.W. and Schwanzel-Fukuda, M. (1995). *J. Comp. Neurol.* **362**, 256–270.

Raderman-Little, R. (1979). *Cell Tissue Kinet.* **12**, 269–280.

Rhein, L.D. and Cagan, R.H. (1983). *J. Neurochem.* **41**, 569–577.

Riddle, D.R. and Oakley, B. (1992). *J. Comp. Neurol.* **324**, 575–589.

Satou, M. (1992). In *Fish Chemoreception* (ed. T.J. Hara), pp. 40–59. Chapman & Hall, London.

Satou, M., Oka, Y., Fujita, I., Koyama, Y., Shiga, T., Kusunoki, M., Matsushima, T. and Ueda, K. (1982). *Zool. Mag.* **91**, 459.

Sheldon, R.E. (1912). *J. Comp. Neurol.* **22**, 177–339.

Sorensen, P.W., Hara, T.J. and Stacey, N.E. (1991). *Brain Res.* **558**, 343–347.

Stacey, N.E. and Kyle, A.L. (1983). *Physiol. Behav.* **30**, 621–628.

Szabo, T., Blähser, S., Denizot, J.-P. and Ravaille-Véron, M. (1991). *C.R. Hebd. Sésnc. Acad. Sci., Paris, Ser. III* **312**, 555–560.

Thommesen, G. (1978). *Acta Physiol. Scand.* **102**, 205–217.

Thommesen, G. (1983). *Acta Physiol. Scand.* **117**, 241–249.

Walker, M.M., Diebel, C.E., Haugh, C.V., Pankhurst, P.M., Montgomery, J.C. and Green, C.R. (1997). *Nature* **390**, 371–376.

Whitear, M. (1992). In: *Fish Chemoreception* (ed. T.J. Hara), pp. 103–125. Chapman & Hall, London.

Yamamoto, M. (1982). In *Chemoreception in Fishes* (ed. T.J. Hara), pp. 39–59. Elsevier, Amsterdam.

Zeiske, E., Caprio, J. and Gruber, S.H. (1986). In *Indo-Pacific Fish Biology: Proceedings of the 2nd International Conference on Indo-Pacific Fishes*, pp. 381–391. Ichthyological Society of Japan, Tokyo.

Zielinski, B. and Hara, T.J. (1988). *J. Comp. Neurol.* **271**, 300–311.

Zielinski, B.S. and Hara, T.J. (1992). *Microsc. Res. Tech.* **23**, 22–27.

Zielinski, B.S., Getchell, M.L. and Getchell, T.V. (1988). *Anat. Rec.* **221**, 769–779.

Zippel, H.P., Sorensen, P.W. and Hansen, A. (1997). *J. Comp. Physiol. A* **180**, 39–52.

27.4 Hearing

Bernd Fritzsch
Department of Biomedical Sciences, Creighton University, Omaha, Nebraska, USA

Introduction

For functional analysis the fish ear can be viewed as two distinct compartments, each dedicated to a specific set of functions. At a macroscopic level (presented in Chapter 15.4) the morphology of the ear dictates which stimulus will be preferentially perceived by a given sensory organ. For example, semicircular canals acquire, based on their morphology, angular acceleration in various directions and those stimuli will be perceived by the sensory epithelia associated with them (Figure 27.4.1). In contrast, the existence of an **otolith** shifts the preferred stimulus to linear acceleration in the otolithic organs (Figure 27.4.1). This physical association with the otolith allows hair cells to perceive relative movements, such as gravity, between the otolith and the fish. Other structures associated with the ear, such as the swimbladder, shift the preferred stimulus to sound pressure, conducted by specialized, perilymphatic inner ear pathways to the appropriate sensory epithelium for perception.

In contrast to the multitude of stimuli that can be perceived by the ear, is the largely similar mechano-electric transduction found in all hair cells. Nevertheless, specific stimuli can be perceived based on the macroscopic structures of the ear which are designed to extract specific physical properties of the various stimuli and conduct them to areas of the brainstem specialized for the appropriate information processing. Thus, each of the macroscopically distinct parts of the ear (semicircular canals, otolithic organs) is associated with sensory receptor epithelia dedicated to perceive predominantly the stimuli conducted to them by the associated structures. The perception of a stimulus ultimately requires a mechano-electric transduction of the stimulus in specialized sensory cells, called hair cells, and its conductance to the brain by afferents for further information processing.

This chapter deals with the mechanosensory transducers of the various sensory organs, the hair cells, as well as the afferent and efferent connections of these sensory epithelia that provide the basis for acquisition of distinct stimuli by the brain.

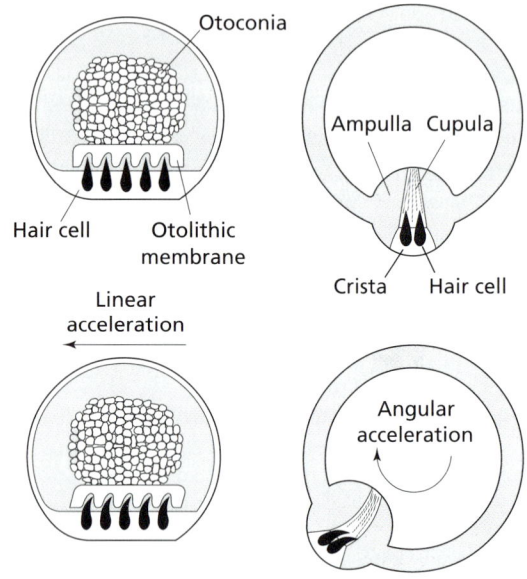

Figure 27.4.1 The different effects of angular and linear acceleration are shown. Angular acceleration uses the lagging of the fluid behind the rotating canal walls to generate a push against the cupula covering the sensory epithelium, the crista. In contrast, linear acceleration uses the inertia of the otolith to generate a shearing force at the apex of the hair cells. This linear acceleration can be generated by movement or by gravity when the fish tilts in space.

The hair cell as a mechanosensory transducer: correlating structure and function

Hair cells are the universal mechano-electric transducers of the inner ear and the lateral line of vertebrates. All hair cells share an apical specialization consisting of numerous protrusions of variable length, hence the name hair cells (Figure 27.4.2). During development, all mechanosensory hair cells form from specialized embryonic tissue, the placodes (Fritzsch et al., 1999, for review). Hair cells have no axon of their own to obtain contact with the brainstem. Instead, ganglion cells, derived from the same placodes, connect them with the brainstem. Formation of hair cells and ganglion cells depends on pro-neuronal genes of the basic helix–loop–helix type and both cell types may be clonally related (Fritzsch et al., 1999). Hair cells also provide neurotrophic support for the developing sensory neurons and the sensory neurons degenerate in the absence of this support in mutant mice (Fritzsch et al., 1999, for review.)

The apical specializations of hair cells consist of a single, asymmetrically arranged kinocilium and a variable number of stereocilia. The stereocilia of hair cells are modified microvilli, the kinocilium is a modified flagellum present in epidermal cells. Microvilli are filled with actin and form rather stiff rods, kinocilia contain the typical $9+2$ arrangement of microtubules. Microvilli are constricted near their insertion into the hair cell (Figure 27.4.2). The central filaments extend into a dense intracellular network at the apex of the hair cell, the cuticular plate. Stereocilia can pivot around their constriction point and will do so if a mechanical stimulus deflects them. Stereocilia are organized like organ pipes with the tallest near the kinocilium and the shortest the farthest away. The hair cell is thus polarized in the direction of increasing and decreasing length of stereocilia (Figure 27.4.2).

Stereocilia are linked at their tips by threads of extracellular material. These threads connect the tip of the shorter stereocilia with the side of the taller (Gillespie and Corey, 1997). The tip links are probably inserted on both sides into cation channels. These as yet uncharacterized channels will open when the bundle is deflected in the direction of the taller stereocilia (more tension on the tip links) and they will close if the movement is in the direction of the shorter stereocilia (Figure 27.4.2). Myosin-mediated movements can adjust the tension of the tip links to guarantee optimal stimulus response under a wide variety of deflections of the hair cell apex (Steyger et al., 1998).

Supporting cells surround hair cells. The tight junctions between them serve to completely separate the extracellular environment, surrounding the hair cell apex, from the extracellular environment along the side and the base of each hair cell. The stereocilia extend into the inner compartment of the ear which is filled with a fluid rich in K^+ and called the endolymph. In contrast, a different extracellular fluid, rich in Na^+ and called the perilymph, surrounds the basal and lateral areas of a hair cell. Consequently, an electrochemical potential exists between the endolymph and the perilymph (Lewis et al., 1985). Thus, when a mechanical stimulus opens the channels at the tip

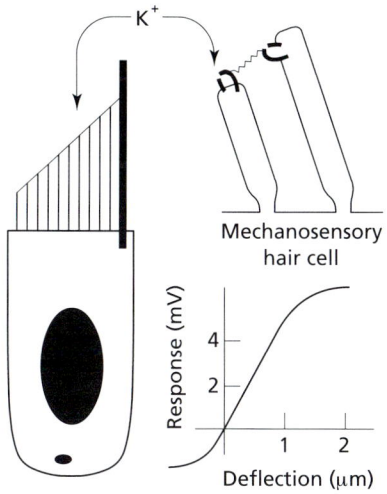

Figure 27.4.2 The basic organization of a mechanosensory hair cell is shown. Note that the kinocilium (black bar) is connected to the ever shorter hair cells by apical links. These links (shown on the top right) connect two mechanosensitive channels with each other. When movement is pushing the bundle to the right, the tip lengths tighten and open the channels. Thus deflection to the right over a fraction of a micrometer will cause changes in the electric potential of a few millivolts. In contrast, deflection to the left will cause hyperpolarization of the resting potential of the hair cell. These changes will cause release of synaptic transmitter at the afferent synapse. (Modified after Lewis et al., 1985.)

links, K^+ will enter into the hair cell and cause depolarization of the resting potential. This will lead to influx of Ca^{2+} through voltage gated Ca^{2+} channels which, in turn, will result in transmitter release proportional to the change in resting potential. The transmitter, probably glutamate, will diffuse across the synaptic cleft to the afferent nerve ending and elicit here a change in resting potential by opening glutamatergic channels, such as NMDA (Pickels, 1996).

This way, a mechanical stimulus causes a graded change in the resting potential of hair cells and, through the synaptic transmission, an excitation of afferents. Afferents will conduct this excitation as action potential frequency to the brain for further interpretation. As outlined above, the relevant stimulus for a given hair cell will be determined by the associated structures (i.e. whether a hair cell is in a semicircular canal or has an otolith covering it) rather than by the hair cell itself. However, some evidence exists for frequency selectivity of hair cells and hair cells embedded in the same endorgan do not respond equally to the same stimulus. Thus while the overall stimuli such as linear or angular acceleration are determined by associated structures, hair cells can respond differently to the intensity of these stimuli, largely owing to variation in the ventrolateral channels. Many of these channel variations have been analyzed with patch clamps in terrestrial vertebrates but less is known for fishes.

The semicircular canals

The sensory epithelia of the semicircular canals, called crista ampullaris, are an aggregation of hair cells situated in an enlargement of each canal, the ampulla. A crista is typically covered by a gelatinous **cupula** and is composed of various forms of hair cells that are all polarized in one direction only. Variation of the crista hair cells concerns the length and size of the apical stereocilia which provide the physical basis for hair cells of variable sensitivity (Lewis et al., 1985).

In hagfish, the cristae form a ring or annulus of hair cells with very long cilia extending into a barely noticeable ampullary enlargement of the single torus (Figure 27.4.3). Hagfish appear to lack a cupula entirely. All hair cells are polarized away from the common **macula**. Some data suggest that several hair cells have a central kinocilium surrounded by microvilli. However, since the turnover rate of the hair cells in the hagfish is unknown it cannot be excluded that these cells are in fact developing hair cells. The anterior crista is innervated by the anterior vestibular nerve and the posterior crista is innervated by a separate posterior nerve (Amemiya et al., 1985).

Lampreys have unique sensory epithelia for their two canals. Each 'crista' is composed of two (Hagelin, 1974) or three arms (Lewis et al., 1985) extending along the ampullary walls in roughly orthogonal planes. Hagelin (1974) claims that only two arms exist and these are possible homologues of the two hemicristae found in tetrapods. A single cupula covers all sensory epithelia. Polarity of all hair cells is toward the canals. A third sensory epithelium, sometimes referred to as the papilla neglecta (Lewis et al., 1985), is present in the dorsal confluence of the two semicircular canals. Based on its development and pattern of innervation it has been suggested that it is the sensory epithelium for the absent horizontal canal (Hagelin, 1974; Fritzsch, 1998).

All jawed fishes have three semicircular canals with an essentially identical organization of the sensory epithelia, which rides on an elevated ridge of connective tissue, hence called crista. Only some tetrapods have a horizontal canal crista that appears to be a hemicrista compared with those present in the vertical canals. The more centrally located hair cells with longer, thicker stereocilia respond quickly to rotation whereas the more peripheral cells with shorter and thinner stereocilia respond more slowly. Nerve fibers seem to innervate either central or peripheral hair cells and show a similar characteristic response to stimuli. The anterior branch of the vestibular nerve innervates the anterior and horizontal canal cristae whereas the posterior branch of the vestibular nerve innervates the posterior crista. The central projection of these fibers is such that the crista organs terminate adjacent to each other and somewhat distinct from the gravistatic sensory organs (Mensinger et al., 1997).

It is noteworthy that the organization of the sensory epithelia of jawless vertebrates varies substantially whereas the sensory epithelia of the semicircular canal system are extremely constant in jawed vertebrates. This suggests that the organization of the semicircular canal with elevated cristae has not been changed. In contrast, the organization of the semicircular sensory epithelia of hagfish and lampreys

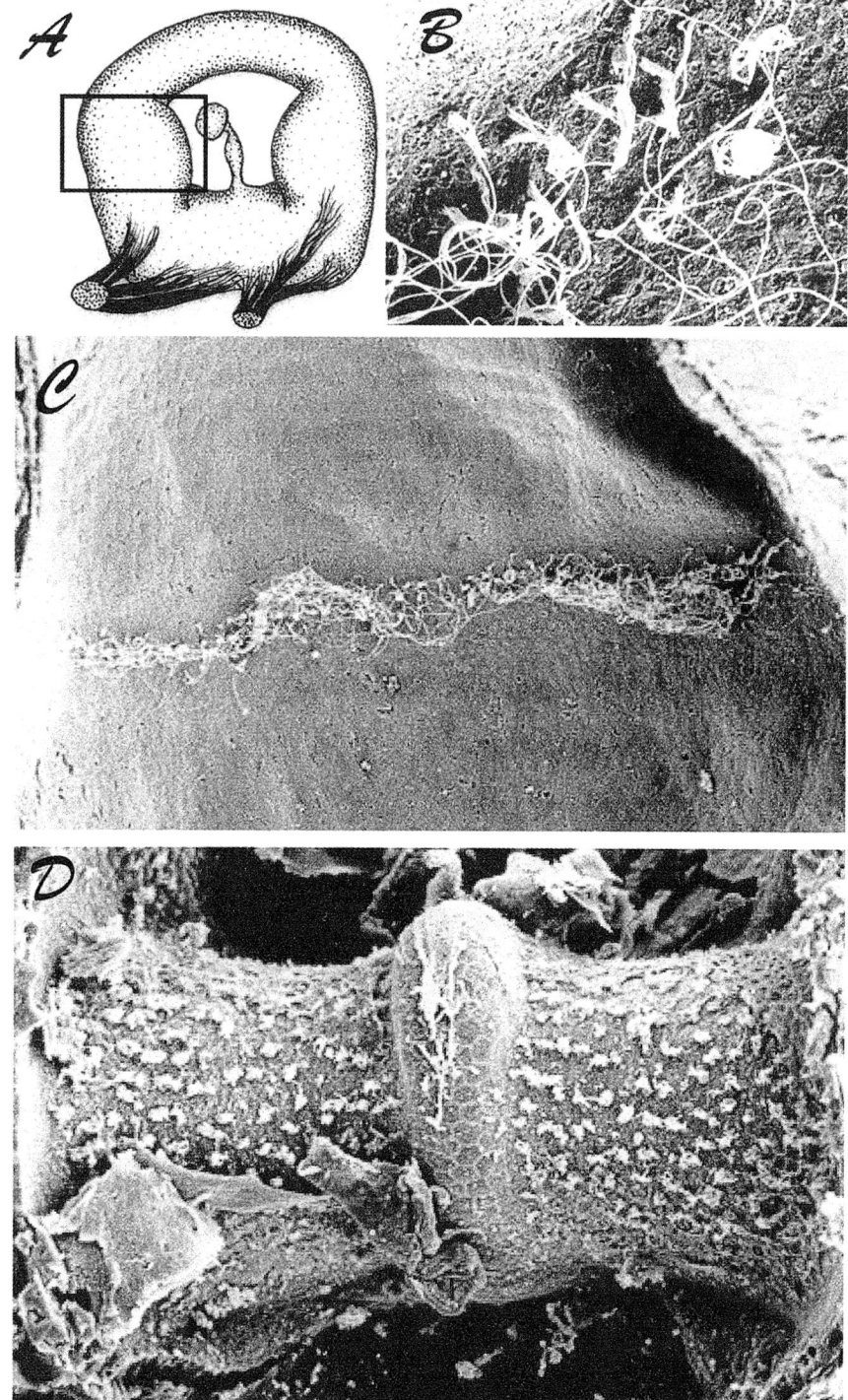

Figure 27.4.3 The semicircular canal organization in hagfish (A–C) and a tetrapod (D) is shown. All jawed vertebrates have a crista studded with hair cells in the ampullary enlargements (D). In life, all hair cells of jawed vertebrate cristae are covered by a gelatinous cupula (removed in D). In contrast, hagfish have hardly any ampullary enlargement and the hair cells form a ring around each canal. Hair cells are all polarized toward the canal and have very long kinocilia that extend directly into the lumen (A–C). No cupula is formed in hagfish.

deviates significantly from each other and from jawed vertebrates. Together with the presence of only two canals in fossil jawless vertebrates this suggests that hagfish and lampreys present a progressive change in the semicircular sensory epithelia. More detailed physiology is needed to be able to specify the potential driving force for this reorganization of the sensory epithelia.

The statolithic organs: opposing polarity in different ways

Each macula of the statolithic organs (saccule, utricle and lagena) is composed of two or more variations of hair cells that are organized in two or more opposing polarities (Lewis et al., 1985). All hair cells in fish maculae belong to a single type without a complete calyx. However, they may have a partial calyx, at least in some bony fish (Popper and Fay, 1997).

Hair cells can be classified based on their apical specialization. For example hair cells have a variable number of stereocilia (between 10 and 250) and the relative length of stereocilia with respect to kinocilia can vary. This variable length has been suggested as playing a role in frequency detection (micromechanical frequency tuning), based on an analogy of variation of apical specialization along the mammalian cochlea. In fact, there is evidence suggesting that electrical tuning of hair cells exists in some fish and some crude frequency mapping has been established in the goldfish saccule (Popper and Fay, 1997; Fay and Popper, 1999). In contrast, a macromechanical tuning with frequency dispersal along an organ comparable to the mammalian cochlea has not been demonstrated in fish.

As outlined above, hair cells are polarized. This polarization varies systematically among the different sensory epithelia of the ear. It can be extremely simple as in the semicircular canal cristae where all hair cells show a single polarity, away from the gravistatic sensors. Similarly, the mammalian cochlea has also only one axis of polarity for all hair cells (Lewis et al., 1985). In contrast, the otolithic sensors have at least two or more axes of polarity for the hair cells. With respect to variation of this general theme between animals, there is again a level of complexity among the otolithic sensory epithelia.

The utricle shows a simple opposing polarity of two populations of hair cells. Opposing polarity means that the kinocilia are facing each other where the polarity reverses in the sensory epithelium. This line of polarity reversal is within a curved area of the epithelium characterized by large hair cells with larger and taller stereociliary bundles, referred to as striola (Figure 27.4.4). This general pattern of two opposing polarities of hair cells is found in the utricle of all vertebrates, irrespective to its orientation (horizontally or vertically) and whether or not it is subdivided into several patches of sensory epithelia (Lewis et al., 1985).

The saccule typically has two patches of hair cells of opposing polarity (Figure 27.4.4). In contrast to the utricle, the hair cells are oriented with the kinocilia facing away from each other at the dividing line. In actinopterygian fishes, the saccular macula shows four or more distinct groups of uniform hair cell polarity forming patches of opposing polarity with each other. Typically, the more posterior and central parts of the macula are dorsoventrally polarized whereas the anterior component is anterioposteriorly polarized. A similar change in orientation has also been found in elasmobranchs (Lewis et al., 1985). In fact, in many actinopterygian fishes the orientation of hair cells in the saccule can be rather complex (Lu et al., 1998). This complexity may be related to extraction of various vectorial components of sound stimulation for directional hearing (Lu et al., 1998). Recent electrophysiological data suggest that various hair cells with different orientation respond preferentially to stimuli in specific directions. The fish ear could thus represent a directional pressure-gradient receiver, like some ears in anurans and birds (Popper and Fay, 1997; Fay and Popper, 1999). In this context it is noteworthy that many sound pressure receiving fish, such as ostariophysin fish and mormyrids, have a very simple sensory epithelium characterized by two opposing polarities of hair cells oriented dorsoventrally (Figure 27.4.4). It would be important to show that the polarity of the hair cells is in line with near field particle motion induced by the sound pressure receiver coupled to this sensory epithelium. Some catfish have duplications of the anterior macula thus showing a change in polarity twice across the macula (Fay and Popper, 1999). The significance of these duplications is largely unknown but could relate to an increased resolution of sound pressure reception.

The most variable of polarity patterns among vertebrates is shown by the lagena (Figure 27.4.4). Three major patterns can be characterized: (i) a patchy alteration of polarity without any distinct dividing lines (most elasmobranchs); (ii) two oppositely polarized hair cell patches facing away from each other at a dividing line (land vertebrates); (iii) two or more sensory patches facing toward each other at a dividing line or displaying gradual changes in orientation along a dividing line (actinopterygian fishes; Figure 27.4.4). Unfortunately, no data other then abstracts

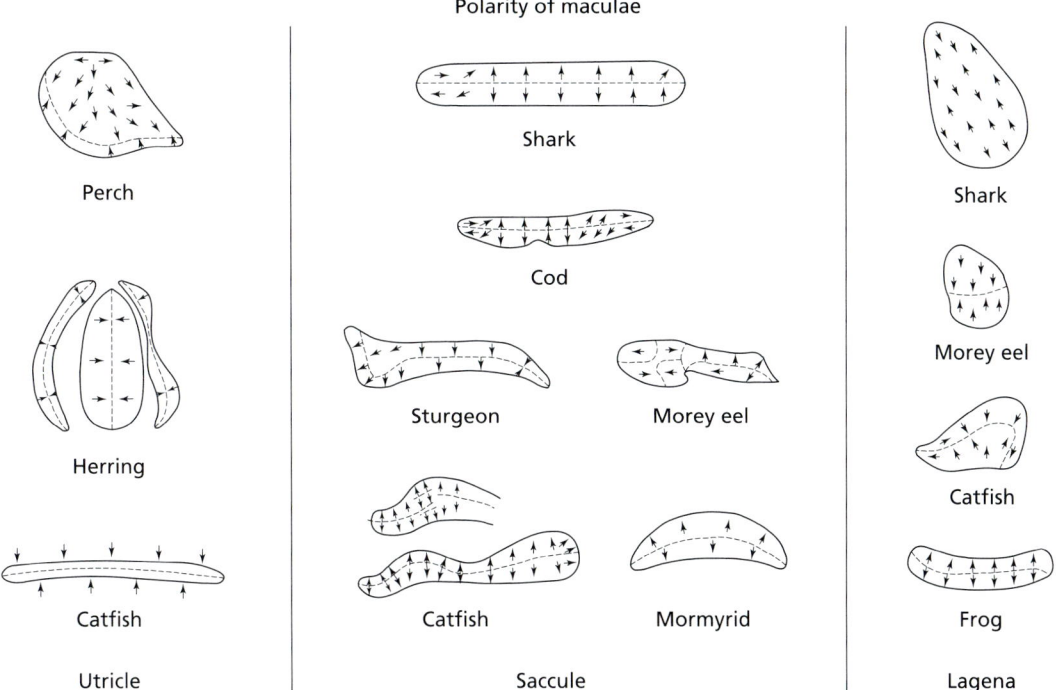

Figure 27.4.4 The polarity of hair cells within various sensory epithelia is shown by arrows. Note that the utricle in all vertebrates is always organized into two patches of opposite polarity facing each other, irrespective of its shape or multiplication. Many saccular hair cell polarity patterns are simply bidirectional, facing away from each other. However, among actinopterygian fishes there are increasing levels of complexity with some species showing multiple patches of hair cell polarity (morey eel). Some sound pressure hearing fish have a very simple organization of two populations of hair cells with opposing polarity, such as mormyrid fish. Others, like some catfish, show a more complicated pattern with multiple hair cell patches of different polarity. The lagena shows three basic patterns of hair cell polarity. A random opposing polarity is found in sharks, hair cell polarity facing each other (as in the utricle) is found in actinopterygian fish and polarity facing away from each other (as in the saccule) is found in tetrapods. These data suggest that the lagena evolved three times, each with a different polarity pattern. (Modified after Lewis et al., 1985.)

are published as yet on the organization in lungfish and *Latimeria*, crucial groups for the evolution of land vertebrates. These data reinforce the view presented in Chapter 15.4 that the lagena evolved at least three times independently among vertebrates – once in elasmobranchs, once in actinopterygian fishes and once in land vertebrates.

Afferent fiber connections of the fish ear

Afferent fibers from the ear enter the brainstem via the VIIIth cranial nerve (the stato-acoustic nerve of terrestrial vertebrates). Afferents from the anterior vertical canal, horizontal canal (or dorsal macula in lampreys; Fritzsch, 1998) and utricle are within the anterior branch of the VIIIth nerve. Fibers coming from the posterior vertical canal, lagena and papilla neglecta (where present) are within the posterior branch. The saccule receives fibers from a variable mix of both the anterior or posterior branch and, in some species even exclusively from either the anterior or the posterior branch. Within the sensory epithelia, fibers tend to innervate hair cells of a single polarity. Unusually for vertebrates, myelinated fibers enter into the sensory epithelia in actinopterygian fishes. Specialized areas of the sensory maculae, e.g. the striola, may be innervated by fibers dedicated exclusively to these areas. Moreover, it appears that cells responding uniformly to a given stimulus may be connected by a single fiber (Popper and Fay, 1997). Among vertebrates, the ear of the hagfish is unique in that it is supplied by two distinct anterior and posterior nerves, each with its own ganglion (Amemiya *et al.*, 1985).

Upon entering the brainstem, fibers bifurcate to form ascending and descending **fascicles**. These fascicles run through an assembly of neurons in the brainstem called the vestibular nuclei. In most fish five discrete cellular aggregations have been identified among these cells (McCormick, 1997, 1999). Within these nuclei, fibers run in discrete, sensory-organ-specific fascicles (Mensinger et al., 1997). Bulk labeling (Bleckmann et al., 1991) and detailed fiber labeling (Mensinger et al., 1997) show that each endorgan has both distinct and overlapping projections with other organs to the vestibular nuclei (McCormick, 1999). Interestingly, there appears to be a correlation between anatomy and the response dynamics of a given fiber (Mensinger et al., 1997).

As in other vertebrates, the semicircular canals project more medially and the gravistatic sensory system projects more laterally in a given vestibular nucleus (Dickman and Fang, 1996; McCormick, 1997). However, at the level of individual fibers this segregation is less clear (Mensinger et al., 1997).

The vestibular nuclei connect at least to three targets within the brain: the ocular motor nuclei, the motor neurons for the muscles of the neck, and the motor neurons for the muscles of the fins. The sound pressure receiving auditory sensors typically project into a segregated area that may be called 'auditory nuclei'. These 'auditory nuclei' are either medially (McCormick, 1997), dorsolaterally (Will and Fritzsch, 1988) or dorsally (amniotes; Carr, 1992) of the vestibular projection. How this segregation of auditory and vestibular fibers develops is unknown (Fritzsch, 1997; McCormick, 1999).

Higher order projections of auditory parts have been analyzed electrophysiologically in the torus semicircularis and show a temporal coding of acoustic stimuli comparable to other vertebrates (Bodnar and Bass, 1999; Crawford and Huang, 1999). Thus, despite a very different auditory periphery, the central properties of signal analysis appear to be rather similar between auditory systems and even between auditory and electric signals (Bodnar and Bass, 1999; Fay and Popper, 1999).

The efferent system of the ear and the lateral line

An additional system common to all mechanosensory hair cells are neurons in the brainstem which terminate with their axons either on the hair cells or on afferent fibers which innervate hair cells. Since these

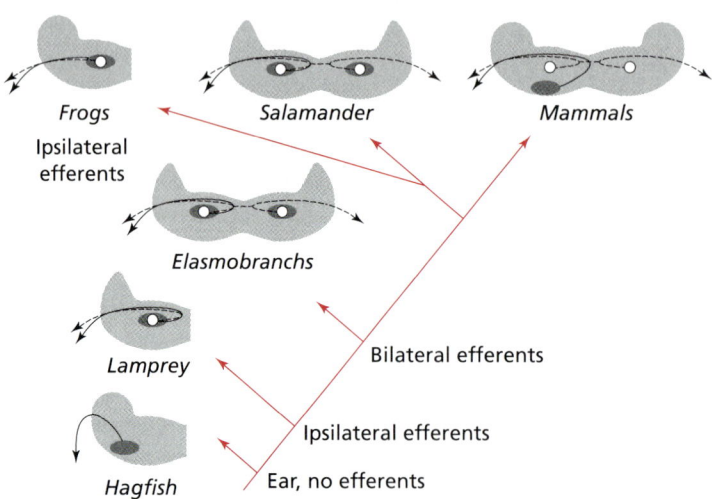

Figure 27.4.5 This scheme shows the evolution of efferents in craniate vertebrates. Hagfish have a facial motor nucleus (black) but are the only craniates without efferents to the ear. Lampreys have efferents to the ear (gray), the cell bodies of which are coextensive with facial motoneurons (black). In elasmobranchs there is some segregation and a bilateral distribution and projection of efferents. This bilateral distribution of efferents and their fibers is likely primitive for jawed vertebrates. Frogs and actinopterygian fish (not shown) have ipsilateral efferents coextensive with facial motoneurons whereas salamanders show bilateral distribution of efferents. In mammals, vestibular and cochlear efferents are segregated from each other and from the facial motor nucleus. It appears that absence of efferents is primitive for craniates, unilateral efferents to the ear alone is primitive for lampreys and bilateral efferents which are also common to the inner ear and the lateral line are primitive for jawed vertebrates. Phylogenetic and ontogenetic evidence supports the view that efferents to the ear are redirected facial motoneurons. (After Fay and Popper 1999.)

fibers are projecting from the brainstem to the ear they have been named the efferent fibers to the inner ear and lateral line. The cells that give rise to these fibers come in a puzzling array of different positions inside the hindbrain and form bilateral populations in many vertebrates. Even if the efferent cells are unilateral, individual cells may have bilateral distribution of their dendrites (Highstein, 1992). Both comparative (Roberts and Meredith, 1992) and developmental (Bruce et al., 1997) data strongly suggest that efferent neurons are derived from facial branchial motoneurons. Efferent neurons segregate from facial branchial motoneurons through differential migration, typically after their axon has reached through unknown navigational processes a different target, the inner ear or the lateral line (Fritzsch, 1998). Either axonal processes, or migrating efferent neurons, or their dendrites can cross the floor plate of the developing brain stem to reach the contralateral side (Figure 27.4.5). Individual efferent cells may even branch in aquatic vertebrates to supply both lateral line and inner ear hair cells (Highstein, 1992; Hellmann and Fritzsch, 1996). Efferent cells can also have axonal branches to reach both inner ears (Fritzsch and Wahnschaffe, 1987; Warr, 1992) and they may branch profusely to supply a variable set of endorgans in the ear.

Consistent with the motor neuron origin of efferent neurons is that efferent neurons are predominantly cholinergic (Eybalin, 1993). The cholinergic receptor on hair cells has recently been identified (Elgoyhen et al., 1994). Mutants for this gene are currently analyzed and show lack of function attributed to efferent fiber action (Liberman et al., 1999). Other transmitters exist also in the efferent system, in particular to the inner hair cells of the cochlea (Eybalin, 1993). Electrophysiological evidence suggests that efferents can influence the activity profile of afferents differentially, with the most pronounced effects achieved on the more sensitive afferents (Highstein, 1992). These effects may even play a role in arousal toward specific stimuli (Highstein, 1992).

References

Amemiya, F., Kishida, R., Goris, R.C., Onishi, H. and Kusunoki, T. (1985). *Brain Res.* **337**, 73–79.
Bleckmann, H., Niemann, U. and Fritzsch, B. (1991). *J. Comp. Neurol.* **314**, 452–466.
Bodnor, D.A. and Bass, A.H. (1999). *J. Neurophysiol.* **81**, 552–563.
Bruce, L.L., Kingsley, J., Nichols, D.H. and Fritzsch, B. (1997). *Int. J. Dev. Neurosci.* **15**, 671–692.
Carr, C.E. (1992). In *The Evolutionary Biology of Hearing* (eds D.B. Webster, A.N. Popper and R.R. Fay), pp. 351–375. Springer-Verlag, New York.
Crawford, J.D. and Huang, X. (1999). *J. Exp. Biol.* **202**, 1417–1426.
Dickman, J.D. and Fang, Q. (1996). *J. Comp. Neurol.* **367**, 110–131.
Elgoyhen, A.B., Johnson, D.S., Boulter, J., Vetter, D.E. and Heinemann, S. (1994). *Cell* **79**, 705–715.
Eybalin, M. (1993). *Physiol. Rev.* **73**, 309–373.
Fay, R.R. and Popper, A.N. (1999). *The Ear of Fishes and Amphibians*. Springer-Verlag, New York.
Fritzsch, B. (1997). *Brain, Behav. Evol.* **50**, 38–49.
Fritzsch, B. (1998). *Brain, Behav. Evol.* **52**, 207–217.
Fritzsch, B. and Wahnschaffe, U. (1987). *Neurosci. Lett.* **81**, 48–52.
Fritzsch, B., Pirvola, U. and Ylikoski, J. (1999). *Cell Tissue Res.* **295**, 369–382
Gillespie, P.G. and Corey, D.P. (1997). *Neuron* **19**, 955–958.
Hagelin, L.-O. (1974). *Acta Zool. Suppl.* 1–218.
Hellmann, B. and Fritzsch, B. (1996). *Brain, Behav. Evol.* **47**, 185–194.
Highstein, S.M. (1992). *Ann. NY Acad. Sci.* **656**, 108–123.
Lewis, E.R., Leverenz, E.L. and Bialek, W.S. (1985). *The Vertebrate Inner Ear*. CRC Press, Boca Raton.
Liberman, C., Vetter, D.E. and Heinemann, S.F. (1999). *ARO Abstract* **21**, 137.
Lu, Z., Song, J. and Popper, A.N. (1998). *J. Comp. Physiol.* **182**, 805–815.
McCormick, C.A. (1997). *Hear. Res.* **110**, 39–60.
McCormick, C.A. (1999). In *The Ear of Fishes and Amphibians* (eds R.R. Fay and A.N. Popper). Springer-Verlag, New York.
Mensinger, A.F., Carey, J.P., Boyle, R. and Highstein, S.M. (1997). *J. Comp. Neurol.* **384**, 71–85.
Pickles (1996). *An Introduction to the Physiology of Hearing*. Academic Press, London.
Popper, A.N. and Fay, R.R. (1997). *Brain, Behav. Evol.* **50**, 213–221.
Roberts, B.L. and Meredith, G.E. (1992). In *The Evolutionary Biology of Hearing* (eds D.B. Webster, A.N. Popper and R.R. Fay), pp. 182–210. Springer-Verlag, New York.
Steyger, P.S., Gillespie, P.G. and Baird, R.A. (1998). *J. Neurosci.* **18**, 4603–4615.
Warr, W.B. (1992). In *The Anatomy of the Mammalian Auditory Pathways* (eds R.R. Fay, A.N. Popper and D.B. Webster), pp. 410–448. Springer, New York.
Will, U. and Fritzsch, B. (1998). In *The Evolution of the Amphibian Auditory System* (eds B. Fritzsch, M. Ryan, W. Wilczynski, T. Hetherington and W. Walkowiak) pp. 159–184. Wiley, Chichester.

CHAPTER 28

Reproductive Systems

R Patiño
Texas Cooperative Fish & Wildlife Research Unit, Texas Tech University, Lubbock, Texas, USA

J M Redding
Department of Biology, Tennessee Technological University, Cookeville, Tennessee, USA

Introduction

The focus of this chapter is the brain–pituitary–gonadal axis. The emphasis is on teleosts, since most species used in basic and applied research are from this phyletic group. Because of space limitations, our presentation relies more on recent literature reviews than on the primary literature. An attempt has been made to minimize overlap with material covered in Part 3 of this volume, specifically in Chapters 13 and 16.

The functional structure of the teleost gonad and its control by the brain–pituitary axis have been intensively studied. These studies have been undertaken partly because of basic scientific curiosity and partly because of applied interests. For example, many of the commercially important teleosts are difficult to spawn in captivity or do not spawn at all. Thus, much of the applied research on fish reproductive biology aims to develop improved techniques of artificial spawning. The ability to manipulate the sexual phenotype of fishes also can be of benefit to aquacultural operations since it allows the selection of desirable sex-linked traits such as growth rate or egg production. Consequently, a considerable amount of research has been dedicated to defining the developmental organization and sexual differentiation of the teleost gonad. It is noteworthy that most of the research on teleost reproductive physiology has been directed toward females. One reason for this female-biased attention is that, typically, it is more difficult to maintain captive females in reproductive condition than it is to maintain males. Increasing attention has been placed recently on issues related to reproductive toxicology and environmental endocrine disruption. These various applied interests have helped establish an overall knowledge of teleost reproductive physiology that is, as a group, more advanced than in other fish taxa and perhaps other poikilothermic vertebrates.

Copyright © 2000 Academic Press

Brain–Pituitary

The functional structure of the brain–pituitary axis in relation to reproduction has been recently reviewed for fishes (Aida et al., 1995; Schreibman and Magliulo-Cepriano, 1999). The pituitary gland of teleosts is composed of two parts, the adenohypophysis and the neurohypophysis (see Chapter 16, Figure 16.3B). The embryonic origin of the adenohypophysis is the buccal epithelium and that of the neurohypophysis is the diencephalon (brain). Unlike most other vertebrates, vascular communication between the neurohypophysis and the adenohypophysis in teleosts is minimal or non-existent. Instead, the cells of the adenohypophysis are directly innervated by or in close association with nerve fibers from the neurohypophysis. These nerve fibers originate in the hypothalamus and other parts of the brain, and their secretory products control hormone production and release from the adenohypophyseal cells.

The hormone-producing cells of the teleost adenohypophysis are typically segregated according to type. This segregation helps define three distinct adenohypophyseal regions: rostral pars distalis, central (proximal) pars distalis, and pars intermedia. The two teleost gonadotropins, GtH-I and GtH-II are produced by separate cells (gonadotrops) within the central pars distalis (Figure 28.1). Gonadotrop activity is regulated by stimulatory and inhibitory factors (Khan and Thomas, 1999). The primary stimulatory system consists of gonadotropin-releasing hormone (GnRH) neurons. Several types and patterns of distribution of GnRH have been reported in the teleost brain, but it is generally accepted that GnRH neurons from the preoptic and anterior hypothalamic brain areas are important for the regulation of reproductive activity. In some species such as carps and catfishes, gonadotrop activity is negatively controlled by dopamine. However, negative dopaminergic control is minimal or even absent in other species such as salmonids and Atlantic croaker.

Gonads

This section discusses structural aspects of gonadal sex differentiation, the adult ovary and the adult testis. The focus is on **gonochoristic** species, where the two sexes are found in separate individuals. Hermaphroditism is mentioned only within a

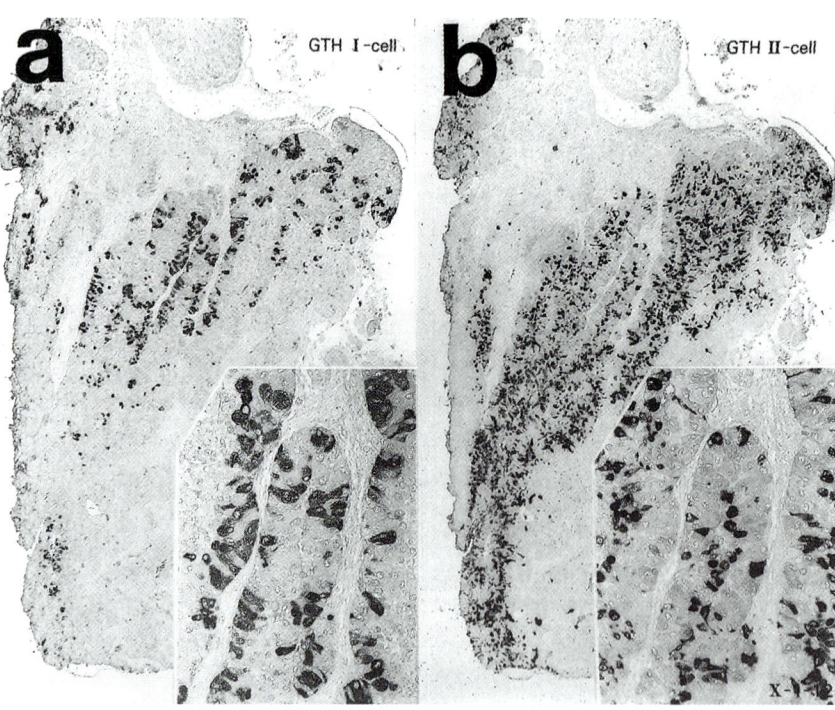

Figure 28.1 Immunocytochemical localization of gonadotrops producing GtH-I (a) and GtH-II (b) in the central pars distalis of the amago salmon (*Oncorhynchus rhodurus*) pituitary. Darkly stained cells indicate a positive reaction to the respective antisera. Adjacent sections were used for this analysis. Note that the GtH-I and GtH-II antisera did not stain the same cells. Magnification, ×47 or ×146 (inset). (From Aida et al., 1995. Reprinted with permission.)

histopathological context. A recent description of the differentiation and structure of the teleost gonad was provided by Patiño and Takashima (1995). Other recent reviews focusing on the structure of adult gonads include those by Callard (1991) and Grier (1993) for the testis and Wallace and Selman (1990) for the ovary.

Gonadal sex differentiation

Sex determination refers to the genetic or environmental prescription of sex in gonochoristic individuals. Genotypic sex determination occurs at fertilization whereas environmental sex determination occurs during embryonic or larval development (see also later discussion). Sex differentiation is the developmental expression of phenotypic sex. Primary sex differentiation pertains to the gonads (gonadal sex differentiation) and secondary sex differentiation refers to all other tissues (Strüssmann and Patiño, 1999). The discussion that follows is limited to gonadal sex differentiation.

Germ plasm elements (e.g. **vasa protein**) have been recognized in the teleost embryo as early as the two-cell stage of development (Yoon et al., 1997). Following their segregation from somatic cell lines, primordial germ cells (PGCs) migrate to the site of the germinal ridges (primordial gonads) where they populate and contribute to the establishment of sexually indifferent gonads (Figure 28.2a). The primordial gonads consist of two parallel cords situated along the upper peritoneal wall on either side of the dorsal mesentery. At this stage of development, the cords have a simple architecture consisting of PGCs and stroma.

Primordial gonads remain sexually indifferent from days (e.g. medaka, Satoh, 1974), to weeks (e.g. fathead minnow, Figure 28.2; tilapias, Nakamura and Takahashi, 1973; channel catfish, Patiño et al., 1996), to months (e.g. common carp, Davies and Takashima, 1980; rainbow trout, Takashima et al., 1980) or even years (e.g. black carp, Shelton and Rothbard, 1995) after fertilization. Ovarian differentiation in teleosts generally precedes testicular differentiation and can be recognized by an enhanced mitotic activity of the germ cells (now called oogonia) and the enlargement of putative ovaries (compare Figures 28.2b and 28.3a). Some of the oogonia quickly enter the meiotic process and become oocytes (Figure 28.2b). Because it is easily recognized, the onset of oocyte meiosis is a commonly used histological index of ovarian differentiation.

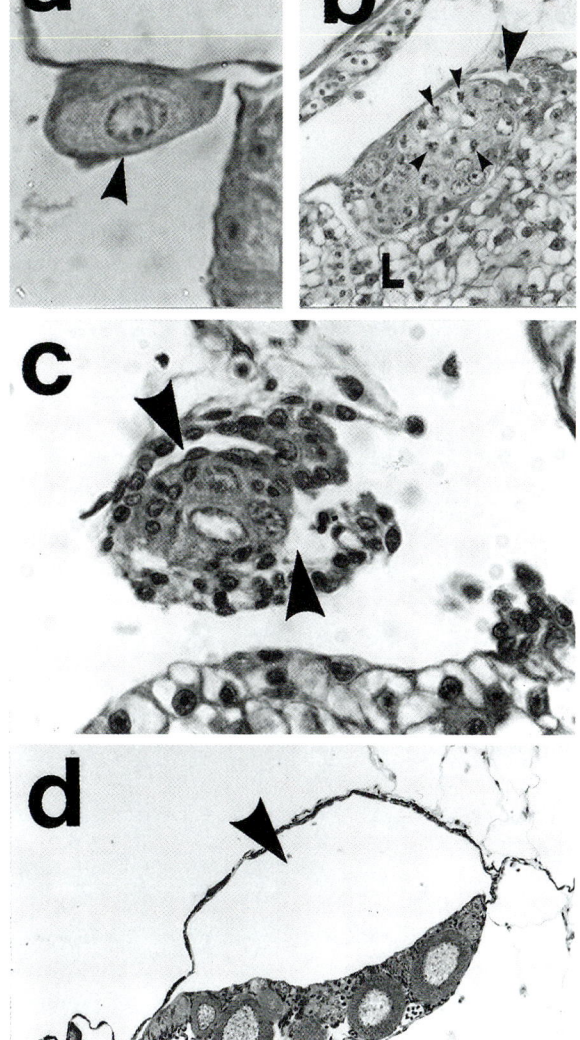

Figure 28.2 Ovarian differentiation in fathead minnow (*Pimephales promelas*) reared at 24°C. (a) Cross-section of indifferent gonad at 7 days post-fertilization (dpf). Note PGC (arrowhead) and thin layer of somatic cells at the periphery. (b) Cross-section of ovary at 16 dpf (sandwiched between liver (L) and air bladder). Note nest of meiotic germ cells at bouquet stage of development (small arrowheads), and a dorsal cavity (large arrowhead). (c) Cross-section of anterior region of ovary at 22 dpf. Note dorsal cavity (upper arrowhead) and formation of ventral cavity (lower arrowhead). This particular section of the ovary (in the anterior region of the gonad) contained few germ cells thus facilitating the definition of somatic tissue patterns. (d) Cross-section of ovary at 36 dpf. Note the relatively large perinucleolar (pre-vitellogenic) oocytes with lightly stained germinal vesicles, the dorsal cavity (upper arrowhead), and the ventral cavity (lower arrowhead). Magnification, ×165 (d), ×330 (b), or ×825 (a, c). Hematoxylin–eosin stain.

However, in some species such as certain tilapias (Nakamura and Takahashi, 1973; Yoshikawa and Oguri, 1978) and channel catfish (Patiño et al., 1996), ovarian stroma differentiation may precede germ cell differentiation. Following the onset of meiosis, oocyte growth continues at a relatively fast pace, but the meiotic process is arrested at prophase of the first division (diplotene stage) through most of oogenesis until the time of oocyte maturation and ovulation (see Ovary).

Teleost ovaries contain an ovarian lumen (**ovocoel**). The ovocoel typically forms either by the simple fusion of gonadal tissue outgrowths (e.g. channel catfish; Patiño et al., 1996) or by internal fission (e.g. rainbow trout; Takashima et al., 1980). A curious pattern of ovocoel development occurs in fathead minnow. In this species, the distal end of the developing ovary fuses laterally with the abdominal wall to form a dorsal cavity (Figure 28.2b–d). This pattern of cavity formation is similar to the case of the guppy, where the single ovary forms an ovocoel by fusion of tissue outgrowths with the peritoneal wall (Takahashi, 1974). However, in fathead minnow a secondary cavity also forms by the ventral fusion of a separate pair of tissue outgrowths (Figure 28.2c and d). The developing ovarian lamellae appear to extend into the ventral cavity (Figure 28.2d), suggesting that the ovocoel of adult minnow ovaries originates from this cavity. On gross inspection, the oviducts of female fry are essentially germ-cell-free, posterior extensions of the ovarian stroma. Typically, the two oviducts fuse into a single duct before reaching the genital pore. With further development, the lumen of each oviduct forms a continuous cavity with its respective ovocoel.

An early sign of testicular differentiation is the organization of germ cells into nests resembling testicular lobules (Figure 28.3b and c; see also Testis). However, an unequivocal sign of testicular differentiation in teleosts is the appearance of efferent sperm ducts (Figure 28.3d). The PGCs of the differentiating testis are now termed spermatogonia. Unlike meiosis in oocytes, meiosis in spermatocytes is a continuous, uninterrupted process. Despite the observation that ovaries normally differentiate earlier, it is not uncommon among teleosts for males to reach sexual maturity before females. On gross inspection, the main sperm ducts of male fry appear as germ-cell-free, posterior extensions of the testicular stroma. Like the case of oviducts, the two sperm ducts typically fuse into a single duct before reaching the genital pore. The lumina of the efferent duct system of the developing and adult testis (see Testis) are continuous with the lumina of the sperm ducts. In some species, such as catfishes (Grizzle and Rogers, 1976; van den Hurk et al., 1987), the posterior region of the testis does not develop a germinal epithelium. This region is known as the seminal vesicle, and its function in the adult fish may be related to pheromone production (Schoonen and Lambert, 1987) and control of sperm motility (van den Hurk et al., 1987).

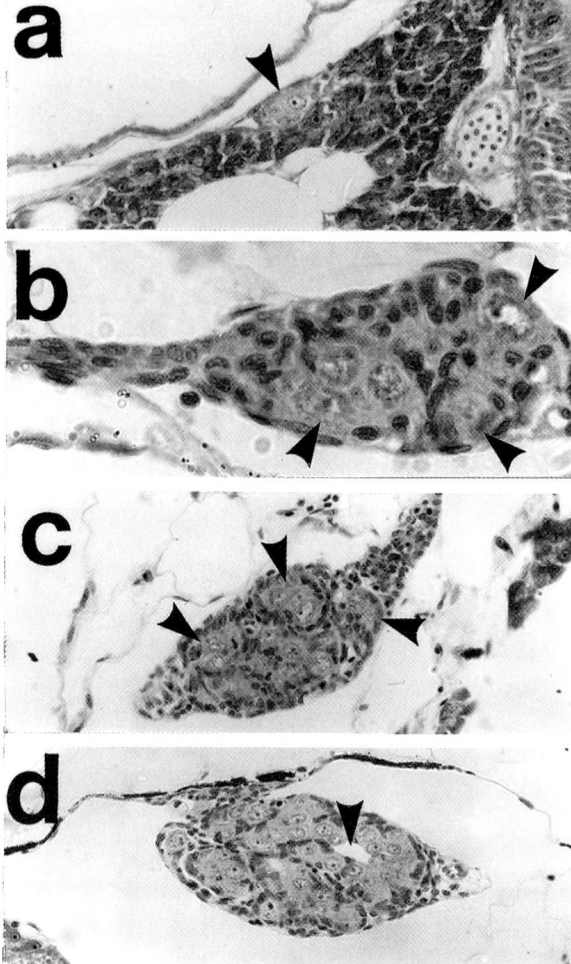

Figure 28.3 Testicular differentiation in fathead minnow (*Pimephales promelas*) reared at 24°C. (a) Cross-section of indifferent, putative testis (arrowhead) at 16 days post-fertilization (dpf). The testis is sandwiched between the air bladder above and basophilic (possibly pancreatic) tissue below. The nucleus of one PGC is clearly visible. Note the small size of the putative testis compared to the ovaries of fish of the same age (see Figure 28.2b). Sex-linked differences in gonad size were first apparent as early as 10–13 dpf (not shown). (b) Cross-section of putative testis at 28 dpf. Note the organization of primitive lobules (arrowheads) containing small nests of germ cells. (c) Cross-section of testis at 36 dpf. Testicular lobules are more easily recognized at this stage (arrowheads). (d) Cross-section of testis at 44 dpf. Note the presence of the efferent duct (arrowhead). Magnification, ×330 (a, c, d) or ×825 (b). Hematoxylin–eosin stain.

Sex steroids play an important role in the process of gonadal sex differentiation in fishes (Patiño, 1997; Nakamura et al., 1998). There is evidence indicating that the site of sex steroid production is the morphologically indifferent gonad, but the genetic mechanisms that control steroid hormone production as well as the specific cellular and physiological responses of the indifferent gonad to steroids are uncertain.

If applied at the appropriate dose and time (usually around the time of gonadal sex differentiation), exogenous estrogens and androgens can override the genotypic sex and yield female-skewed or male-skewed populations, respectively. The sex-reversed individuals are functionally indistinguishable from normal individuals. However, exposure to exogenous sex steroids or sex-steroid-like compounds (e.g. xenoestrogens) can also have abnormal effects on sex determination, such as the induction of hermaphroditic gonads (Figure 28.4). This phenomenon can occur when the dose of steroids used for sex reversal is too high or too low, or when the period of exposure is inadequate. Paradoxical effects of steroids on gonadal sex determination also have been reported. In channel catfish, for example, treatment with several androgens causes functional gonadal feminization even when using androgens that cannot be converted endogenously to estrogens (Davis et al., 1990).

Environmental factors such as temperature can affect not only the rate of gonadal sex differentiation but also the process of sex determination (Conover and Kynard, 1981; Strüssmann and Patiño, 1999). Manipulations of water temperature at the appropriate time in development (typically also around the period of morphological sex differentiation) have been shown to override the genetically prescribed sex in species such as channel catfish (Patiño et al., 1996) with well-established chromosomal mechanisms of sex determination (Davis et al., 1990). Moreover, in some species such as the South American atherinid, *Odonthestes bonariensis*, water temperature may be the main and perhaps only determinant of sex (Strüssmann and Patiño, 1995, 1999). Temperature can also affect gonadal processes other than sex determination and differentiation. For example, high sublethal temperatures have been shown to induce germ cell degeneration in the gonads of male and female teleosts in a manner similar to that reported during spontaneous or experimentally-induced cryptorchidism in certain male mammals (Strüssmann and Patiño, 1999).

Ovary

The teleost ovary is of the cystovarian type where ovulation occurs inside the ovary into the ovocoel (Hoar, 1969). In some teleosts such as salmonids, the oviduct may be secondarily lost during development and the ovulated eggs are released into the abdominal cavity.

The basic unit of the ovary is the ovarian follicle, which contains the oocyte and the follicle cells.

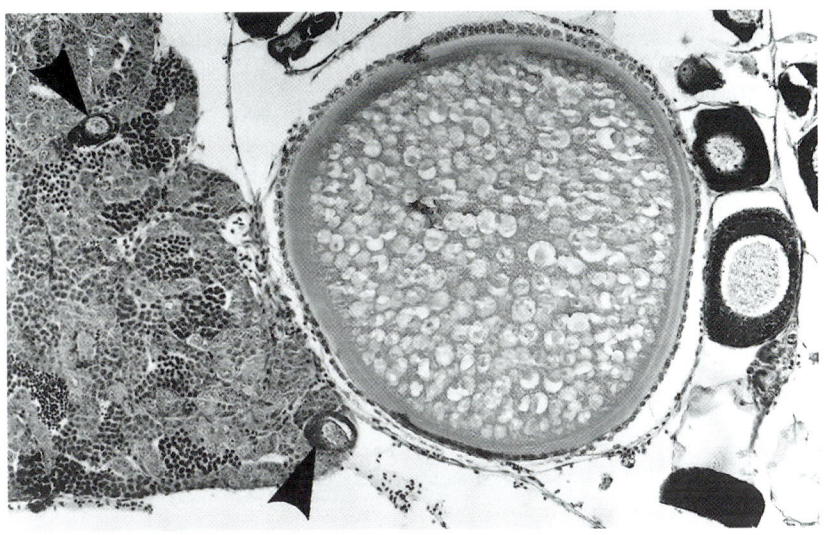

Figure 28.4 Sagittal section of ovotestis in 5-month-old fathead minnow (*Pimephales promelas*) treated with 1 μM estradiol-17β (in water) from 5 to 25 days post–fertilization at 24°C. Note the clear separation of seemingly normal ovarian tissue (right; showing an early vitellogenic oocyte and several perinucleolar oocytes marked by their basophilic cytoplasm) and testicular tissue (left; showing collection of spermatocysts at various stages of development). Small perinucleolar oocytes (arrowheads) were also observed embedded within the testicular tissue. Magnification, ×165. Hematoxylin–eosin stain.

Oocyte growth is typically marked by a pre-**vitellogenic** phase followed by a vitellogenic phase. Prominent cytological stages of pre-vitellogenic oocytes include bouquet (e.g. Figure 28.2b), chromatin-nucleolus, perinucleolar (e.g. Figures 28.2d, 28.4 and 28.5c), and cortical alveoli Figure 28.5a (see also Patiño and Takashima, 1995). The common use of the term 'yolk vesicle' to describe the cortical alveoli seems inappropriate since the contents of these structures do not contribute to embryonic development (Wallace and Selman, 1990). The hormonal control of pre-vitellogenic growth is poorly understood. However, the vitellogenic phase is to a large extent under the control of estrogens, which regulate the production of hepatic vitellogenin (yolk precursor) and other factors necessary for this phase of follicular growth (see also Chapter 13).

In cross-section, vitellogenic and full-grown follicles are composed of the centrally located oocyte, an acellular vitelline envelope (zona radiata), a monolayer of granulosa cells, an acellular basement membrane, and a heterogeneous thecal cell layer containing steroidogenic cells, blood vessels and cells, and other cell types (Figure 28.6). Thecal and granulosa cells are major sites of steroid hormone production in the ovary (Nagahama, 1997; see also Chapter 13 in this volume). The vitelline envelope is laid down between the oocyte and the granulosa cell layer during late pre-vitellogenesis (Patiño and Takashima, 1995). Recent studies have reported that

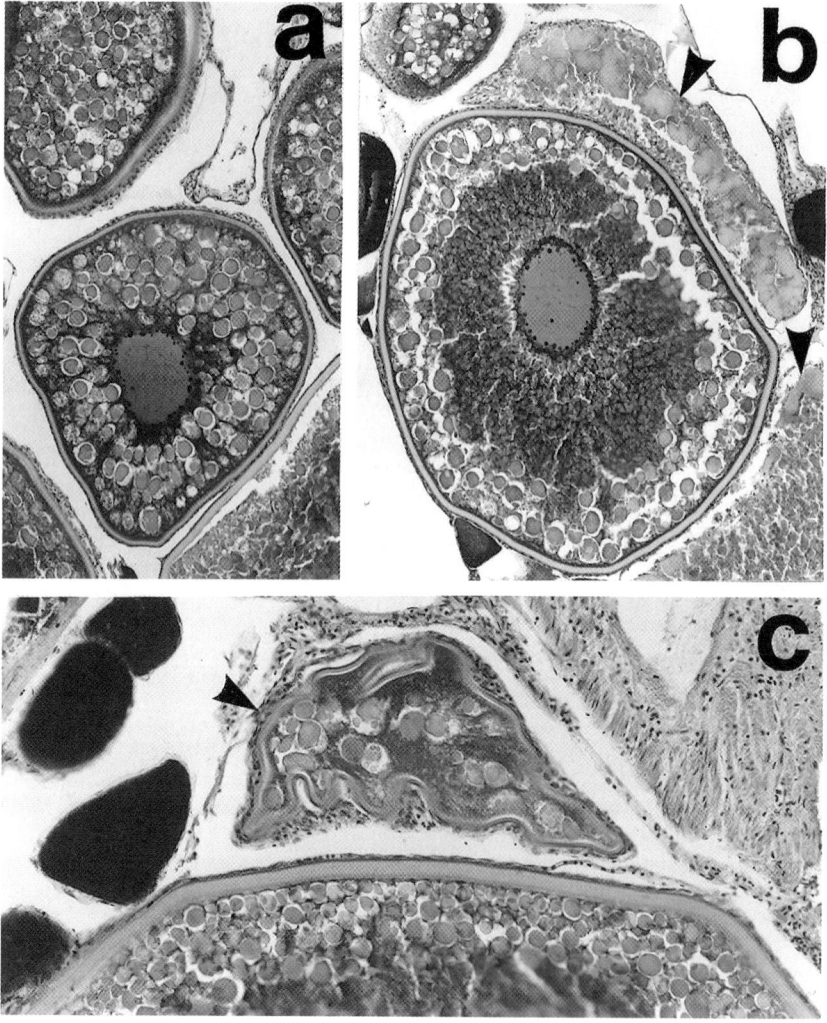

Figure 28.5 Growth and atresia of ovarian follicles in carp (*Cyprinus carpio*). (a) Cross-section of follicle at the interphase between the end of cortical alveoli stage and the beginning of vitellogenic growth. Note the presence of cortical alveoli occupying most of the oocyte's volume around the germinal vesicle. (b) Cross-section of follicle at mid-vitellogenic growth. Note the accumulation of yolk around the germinal vesicle while the cortical alveoli are displaced to the periphery of the cytoplasm. Atretic follicles (arrowheads) are also seen flanking the vitellogenic follicle. (c) Atretic follicle (arrowhead). Note the late-vitellogenic follicle at the bottom of the panel and the strongly basophilic cytoplasm of several small perinucleolar oocytes on the left. Magnification, ×82.5 (a, b) or ×165 (c). Hematoxylin–eosin stain.

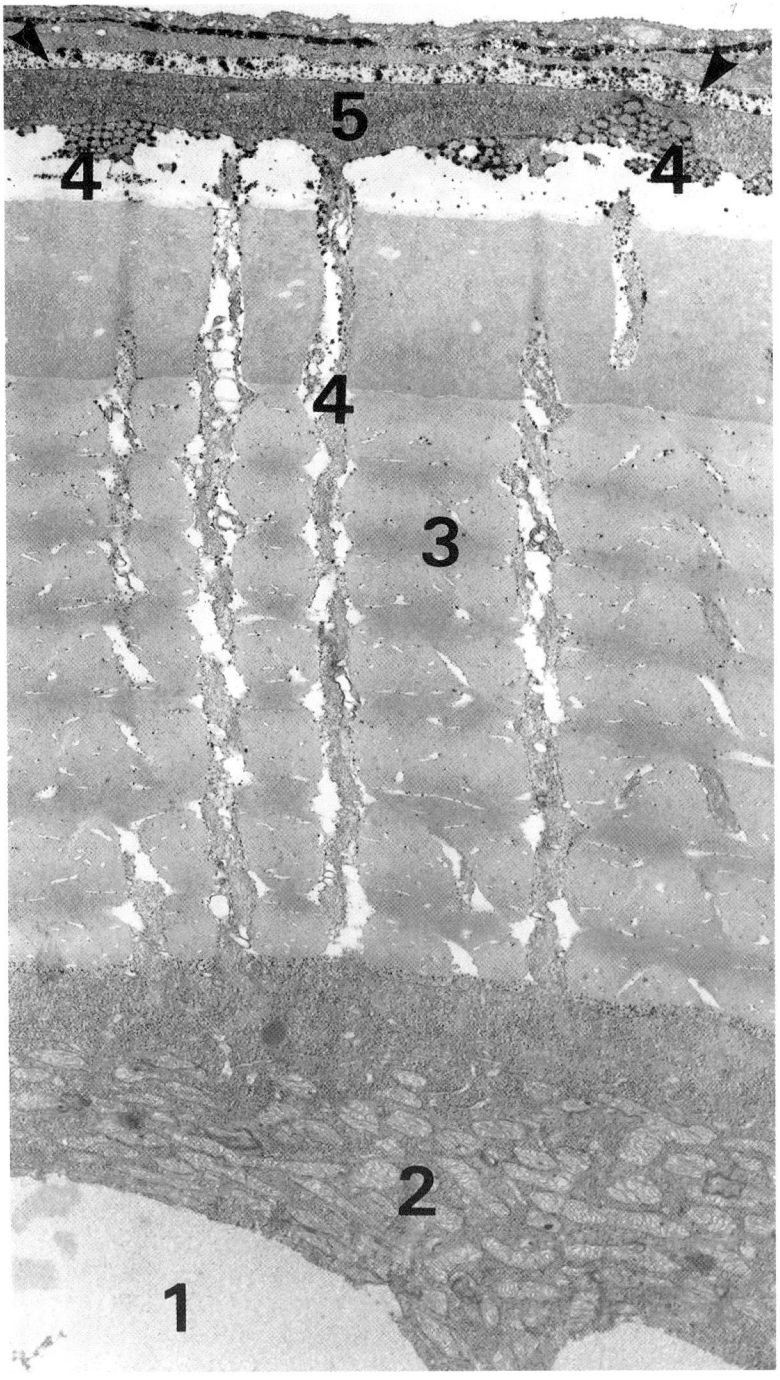

Figure 28.6 Electron photomicrograph of a cross-section of an ovarian follicle of Atlantic croaker (*Micropogonias undulatus*). The specimen was a full-grown follicle sampled after the acquisition of oocyte maturational competence. Note the presence of cortical alveoli (1) and mitochondrial masses (2) in the oocyte's peripheral cytoplasm. The vitelline envelope (3) is perforated by pore canals, which allow oocyte microvilli (4) access to the granulosa cell layer (5). The oocyte microvilli are seen forming 'bundles' (4) that are in close contact with the granulosa cell layer. A basement membrane (arrowheads) separates the granulosa cell layer from the outer, heterogeneous thecal cell layer. Magnification, ×13 750.

vitelline envelope proteins are detected in blood before the onset of vitellogenin production, confirming that the start of vitelline envelope formation precedes the vitellogenic growth phase of the follicle (e.g. Hyllner *et al.*, 1994; Celius and Walther, 1998). Vitelline envelope proteins are produced in the liver and are also under the control of estradiol-17β.

The oocyte remains in close contact with the granulosa cells throughout vitellogenic growth via numerous microvilli that extend from the oocyte surface and cross the vitelline envelope through pore canals. The oocyte microvilli make contact with granulosa cells and form gap junctions (GJ) at these points of contact. Gap junctions are aggregates of intercellular channels often found between neighboring cells. They allow the direct cytoplasmic exchange of small molecules up to 1000 Da in molecular mass (Beyer, 1993). Homologous GJ occur between granulosa cells of vitellogenic follicles (Figure 28.7a), although the stage of development at which they first appear has not been clearly established. Heterologous GJ between oocyte microvilli and granulosa cells (Figure 28.7b) are first recognized in cortical alveoli follicles (Iwamatsu *et al.*, 1988). In general, GJ of the vertebrate ovarian follicle are considered to be important for the regulation of follicular growth. In teleost fishes, heterologous and homologous GJ of full-grown follicles are markedly stimulated by gonadotropin (York *et al.*, 1993; Patiño and Kagawa, 1999) and insulin-like growth factor I (Patiño and Kagawa, 1999) immediately preceding the resumption of the first meiotic division. One proposed role for the enhanced levels of heterologous GJ is that they are an essential component of the signaling mechanism for maturational hormones (Patiño, 1997; see also below). Homologous GJ also exist among thecal cells, but their role in follicular physiology and their hormonal regulation have not been fully explored.

The hormone-dependent enhancement of GJ in full-grown follicles coincides with the acquisition of oocyte maturational competence (OMC), an event defined as the ability of the intrafollicular oocyte to respond to steroidal maturation-inducing hormone (MIH) (Patiño, 1997). The induction of OMC seems to be under the specific control of GtH-II since GtH-I is without effect (Kagawa *et al.*, 1998). However, OMC development is independent of steroid production (Patiño and Thomas, 1990a). Following the acquisition of OMC, GtH-II stimulates the production of MIH by the ovarian follicle. The MIH binds to high affinity receptors on the surface of the oocyte (e.g. Patiño and Thomas, 1990b) and triggers the resumption and completion of the first meiotic division.

The germinal vesicle (nucleus) is located at the center of the oocyte throughout growth of the ovarian follicle. Typically, the germinal vesicle starts to migrate toward the animal pole during late vitellogenesis or during the acquisition of OMC. Exposure to MIH causes the dissolution of the germinal vesicle as part of the process of meiotic cell division. The disappearance of the germinal vesicle is commonly referred to as germinal vesicle breakdown, and this event has become a standard index to monitor the resumption of the first meiotic division. Expulsion of the oocyte from the follicle (ovulation) normally follows maturation (Goetz *et al.*, 1991). Heterologous GJ are lost just before the onset of ovulation (York *et al.*, 1993; Patiño and Kagawa, 1999). Unlike the eggs of mammals, the ovulated eggs of teleost fishes are devoid of surrounding granulosa cells. Meiosis in the mature, ovulated oocyte is arrested once again, this time at metaphase of the second division. The trigger for the resumption of the second meiotic division is the fertilization of the egg by sperm. The post-ovulatory (oocyte-free) follicle may continue to produce steroids (Nagahama *et al.*, 1982).

Follicular **atresia** is a common occurrence in the teleost ovary (Figure 28.5b and c). Stress and other environmental insults can increase the incidence of atresia, but this phenomenon is also observed in otherwise healthy individuals. Its precise causes are unclear although a mechanism involving apoptotic cell death has been proposed (see Chapter 13).

Unlike adult ovaries of mammals which contain only oocytes and no oogonia, adult ovaries of teleosts contain a permanent pool of oogonia which can continue to proliferate and replenish the oocyte population. At least some of the oogonia seem to retain the sexual bipotentiality typical of PGCs, since treatment with androgenic steroids can cause **ovotestis** formation in adult ovaries of some species (Nakamura *et al.*, 1998).

Testis

Fish testes have been classified into three types, lobular, anastomosing tubular and tubular, of which only the first two are found in teleosts (Grier, 1993). The lobular type essentially consists of a collection of blind-ended sacs (lobules) which drain into the efferent (sperm) duct system. The lobules can be hollow (most species) or solid (e.g. Atheriniformes). Hollow lobules have a fairly homogeneous distribution of germ cells at various stages of development along their walls, and the mature sperm is released into the lumen (which is connected to the efferent duct system). Solid lobules have the germ cell units (spermatocysts; see below) arranged in increasing developmental stage from the blind end of the lobule to the efferent duct end, and spermatogonia are found only at the blind end. Anastomosing tubular testes have a branching network of tubules and are found in species such as gar, catfish (*Ictalurus* sp.),

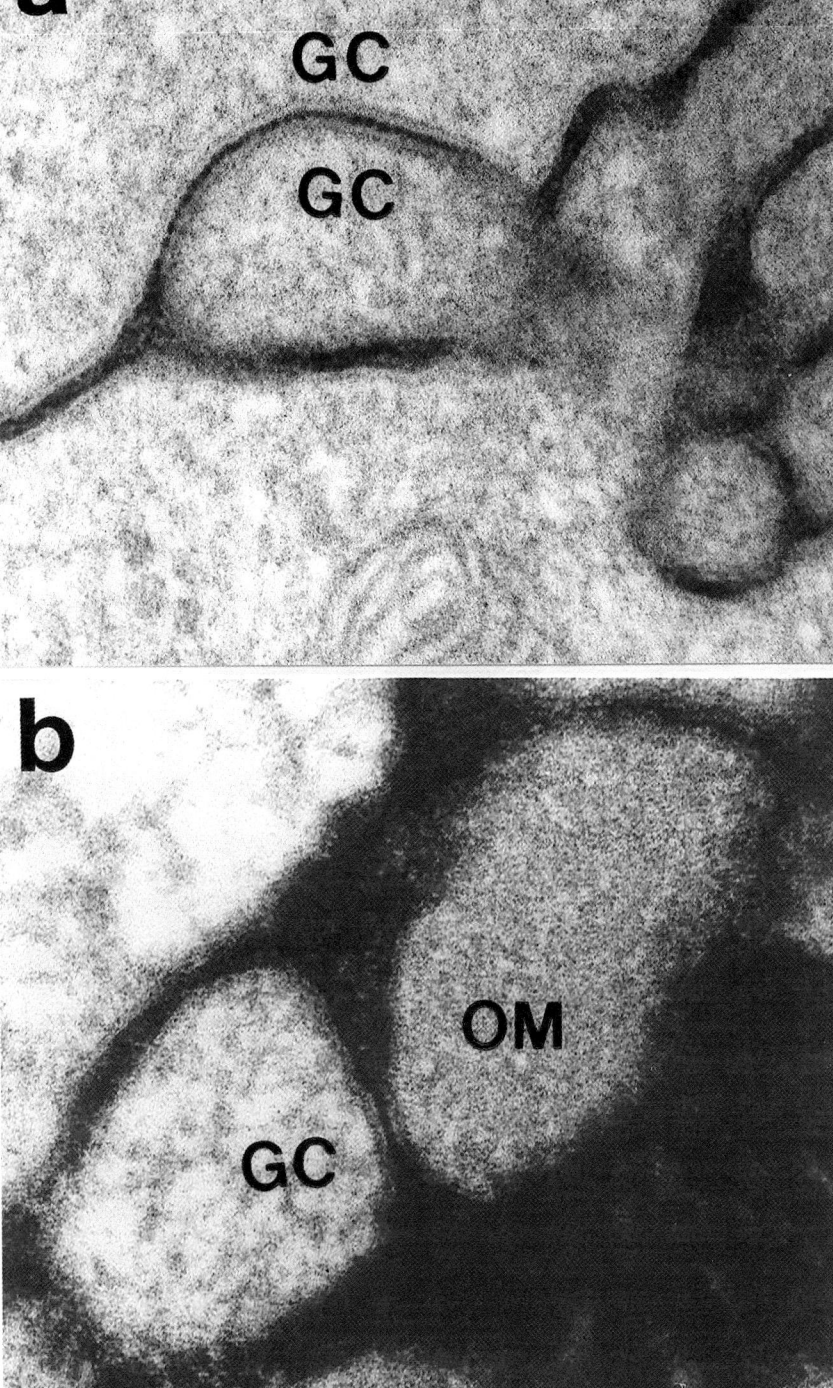

Figure 28.7 Electron photomicrograph of gap junctions (GJ) in full-grown ovarian follicles of Atlantic croaker (*Micropogonias undulatus*). Gap junctions between cells are recognized by their seven-layered organization. The ovarian follicles were stimulated *in vitro* with human chorionic gonadotropin to induce oocyte maturational competence, and subsequently exposed to maturation-inducing hormone to induce the resumption of meiosis. The follicles shown here had not yet undergone germinal vesicle breakdown. (a) Homologous GJ between two granulosa cells (GC); note that the cell on the bottom of the picture had extended a microvillus or finger-like projection to make GJ contact with the neighboring cell. (b) Heterologous GJ between granulosa cell (GC) microvillus and oocyte microvillus (OM). Magnification, ×220 000. Gap junctions were highlighted with lanthanum. (Reprinted with modification from York *et al.*, 1993.)

herring, shad, and others (Grier, 1993). The general organization of their germinal tissue is similar to that of the hollow lobular testes. Among fishes, a tubular testis type has been described only for the lungfish (Grier, 1993).

The basic germinal unit of lobular and anastomosing tubular testes is the spermatocyst, which is a spherical structure composed of germ cells and Sertoli cells (Figure 28.8). Sertoli cells support and regulate the development of spermatocytes. As in most other vertebrates, the Sertoli cells in teleost fishes form a cell barrier that is developmentally regulated (Patiño and Takashima, 1995). The purpose of the Sertoli cell barrier is to regulate precisely the microenvironment necessary for proper germ cell differentiation. One prominent feature of the barrier is the presence of tight junctions between Sertoli cells. However, the junctional complex of the Sertoli cell barrier in the mammalian testis includes not only tight junctions but also GJ (e.g. Pelletier and Byers, 1992). The existence of GJ between Sertoli cells in the teleost testis has not been demonstrated but the presence of testicular connexin mRNA, which encodes the proteins that form GJ channels, was recently documented for Atlantic croaker (X. Chang, R. Patiño, P. Thomas and V. Lee, unpublished data). Clusters of interstitial (Leydig) cells are typically found in the interstices between the testicular lobules. These cells are steroidogenic and produce androgens. Homologous GJ have been demonstrated between Leydig cells in the mammalian testis (e.g. Varanda and de Carvalho, 1994) but have not yet been described in the teleost testis.

Two types of spermatogonia are typically recognized, primary and secondary (Figure 28.9). Large germ cells (stem cells) also can be seen occasionally (Figures 28.8 and 28.9). These cells may be the source of primary spermatogonia and may retain sexual bipotentiality, since the adult teleost testis can form an ovotestis after exogenous estrogen treatment (Nakamura *et al.*, 1998). Pre-vitellogenic oocytes have been observed, although rarely, embedded in the testes of carp collected from feral populations (Goodbred *et al.*, 1997). Stem cells and spermatogonia proliferate mitotically, but secondary spermatogonia become irreversibly committed to enter meiosis (Callard, 1991). Spermatogenesis occurs in three phases: meiosis, spermiogenesis and **spermiation**. During meiosis, primary spermatocytes divide to form secondary spermatocytes, and the latter divide again to form haploid spermatids (Figure 28.9). Within a spermatocyst, the mitotic divisions of secondary spermatogonia and the meiotic divisions of spermatocytes are incomplete and form a syncytium of germ cells joined by intercellular cytoplasmic bridges. All cells of the syncytium develop in a synchronous manner (Figure 28.9). Spermiogenesis refers to the transformation of spermatids into spermatozoa (sperm). Prominent morphological changes during spermiogenesis include the transformation of the

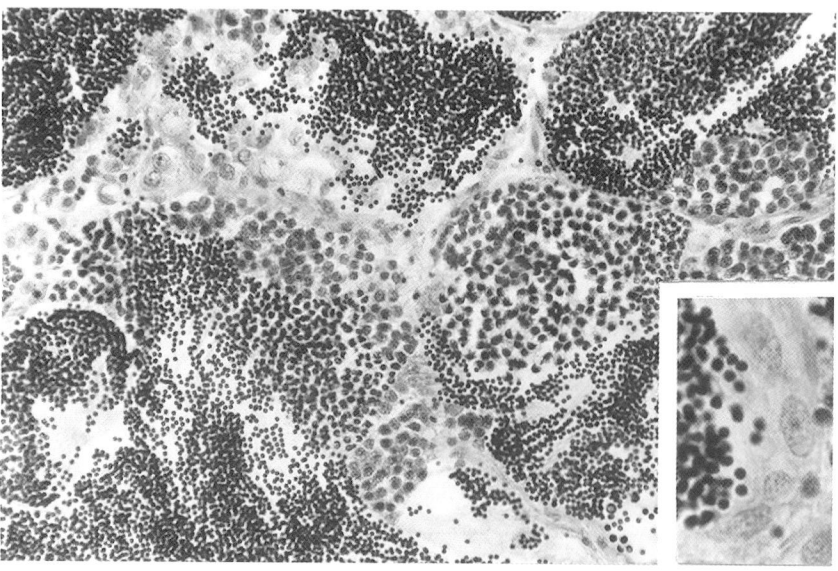

Figure 28.8 Cross-section of carp (*Cyprinus carpio*) testis. Groups of synchronously developing germ cells correspond to distinct spermatocysts. Although somewhat difficult to distinguish on the picture, lobules can be defined by their light-staining borders consisting of fibroblasts, spermatogonia and Sertoli cells. Inset: Sertoli cells along the wall of a lobule after spermiation (the Sertoli cells contain a nucleus with single nucleolus); free sperm is observed on the left side of the inset. Magnification, ×330 or ×825 (inset). Hematoxylin–eosin stain.

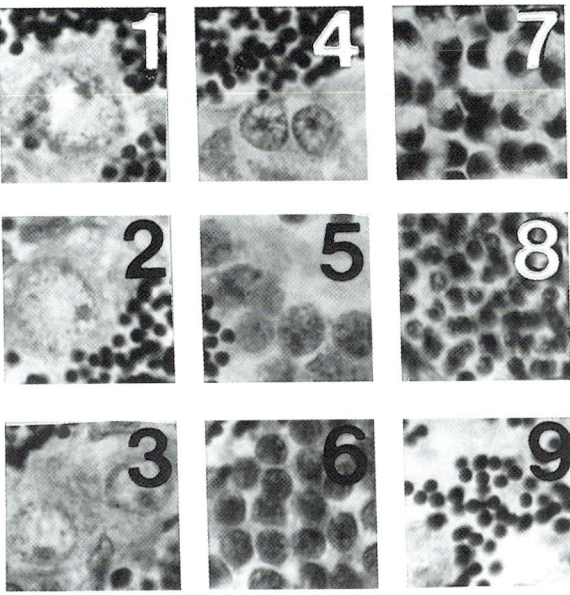

Figure 28.9 Male germ cell development in carp (*Cyprinus carpio*). (1) and (2), single stem cells; (3), pair of primary spermatogonia; (4), pair of secondary spermatogonia; (5), nest of primary spermatocytes; (6), nest of secondary spermatocytes; (7), nest of secondary spermatocytes in meiotic division; (8), nest of spermatids; and (9), free sperm. Magnification, ×825. Hematoxylin–eosin stain.

nucleus from a round to an elongated form (with some exemptions such as carp, in which it remains round; Figures 28.8 and 28.9), the development of a flagellum, and the elimination of most of the spermatid's cytoplasm, which is discarded as residual body and phagocytosed by Sertoli cells (Patiño and Takashima, 1995). Teleost sperm do not have a true acrosome. Spermiation refers to the release of sperm from the spermatocyst into the efferent duct system. While in the efferent duct system the sperm are not motile. In African catfish, one function of the fluid produced by the seminal vesicle may be to maintain sperm immotility (van den Hurk et al., 1987). Sperm motility is usually triggered by changes in the surrounding medium which accompany spawning.

Conclusion

This brief presentation cannot do justice to the wealth of knowledge available concerning the functional structure of the teleost reproductive system. However, we have attempted to highlight common themes of current interest that seem evident among many teleosts. For more complete descriptions, the reader is referred to the list of references provided and to the numerous original research papers available in the literature.

Acknowledgments

Mr Scott York assisted with the electron micrography of Atlantic croaker specimens; Mr Juan Zalapa and Mr Cevdet Uguz prepared the histological specimens for the fathead minnow; and Mr Zalapa assisted with the carp histological preparations. This work was performed in part under the auspices of the Texas Higher Education Coordinating Board Advanced Technology Program Grant No. 0205-44-2612, and Enhancing Animal Reproductive Efficiency Program, US Department of Agriculture grant numbers 96-35203-3465 and 97-35203-4805. The Texas Cooperative Fish & Wildlife Research Unit is jointly sponsored by the United States Geological Survey's Division of Biological Resources, Texas Tech University, Texas Parks & Wildlife Department, and The Wildlife Management Institute.

References

Aida, K., Kaneko, T., Oguri, M. and Sato, H. (1995). In *An Atlas of Fish Histology*, 2nd edn (eds F. Takashima and T. Hibiya), pp. 154–185. Kodansha-Gustav Fischer Verlag, Tokyo–Stuttgart New York.

Beyer, E.C. (1993). *Intl. Rev. Cytol.* **137**, 1–37.

Callard, G.V. (1991). In *Vertebrate Endocrinology: Fundamentals and Biomedical Implications*, vol. 4A (eds P.K.T. Pang and M.P. Schreibman), pp. 303–341. Academic Press, New York.

Celius, T. and Walther, B.T. (1998). *J. Endocrinol.* **158**, 259–266.

Conover, D.O. and Kynard, B.E. (1981). *Science* **213**, 577–579.

Davies, P.R. and Takashima, F. (1980). *J. Tokyo Univ. Fish.* **66**, 191–199.

Davis, K.B., Simco, B.A., Goudie, C.A., Parker, N.C., Cauldwell, W. and Snellgrove, R. (1990). *Gen. Comp. Endocrinol.* **78**, 218–223.

Goetz, F.W., Berndtson, A.K. and Ranjan, M. (1991). In *Vertebrate Endocrinology: Fundamentals and Biomedical Implications*, vol. 4A (eds P.K.T. Pang and M.P. Schreibman), pp. 127–203. Academic Press, New York.

Goodbred, S.L., Gilliom, R.J., Gross, T.S., Denslow, N.P., Bryant, W.L. and Schoeb, T.R. (1997). United States Geological Survey, Open-File Report 96–627, 48pp.

Grier, H.J. (1993). In *The Sertoli Cell* (eds L.D. Russell and M.D. Griswold), pp. 703–739. Cache River Press, Clearwater.

Grizzle, J.M. and Rogers, W.A. (1976). *Anatomy and Histology of the Channel Catfish*, 94 pp. Auburn University Agricultural Experiment Station, Auburn Alabama.

Hoar, W.S. (1969). In *Fish Physiology*, vol. 3 (eds W.S. Hoar and D.J. Randall), pp. 1–72. Academic Press, New York.

Hyllner, S.J., Silversand, C. and Haux, C. (1994). *Mol. Reprod. Dev.* **39**, 166–175.

Iwamatsu, T., Ohta, T., Oshima, E. and Sakai, N. (1988). *Zool. Sci.* **5**, 353–573.

Kagawa, H., Tanaka, H., Okuzawa, K. and Kobayashi, M. (1998). *Gen. Comp. Endocrinol.* **112**, 80–88.

Khan, I.A. and Thomas, P. (1999). In *Encyclopedia of Reproduction*, vol. 3 (eds E. Knobil and J.D. Neill), pp. 552–564. Academic Press, San Diego.

Nagahama, Y. (1997). *Steroids* **62**, 190–196.

Nagahama, Y., Kagawa, H. and Young, G. (1982). *Can. J. Fish. Aquat. Sci.* **39**, 56–64.

Nakamura, M. and Takahashi, H. (1973). *Bull. Fac. Fish. Hokkaido Univ.* **24**, 1–13.

Nakamura, M., Kobayashi, T., Chang, X. and Nagahama, X. (1998). *J. Exp. Zool.* **281**, 362–372.

Patiño, R. (1997). *Progr. Fish Cult.* **59**, 118–128.

Patiño, R. and Kagawa, H. (1999). *Gen. Comp. Endocrinol.* **115**, 454–462.

Patiño, R. and Takashima F. (1995). In *An Atlas of Fish Histology*, 2nd edn (eds F. Takashima and T. Hibiya), pp. 128–153. Kodansha-Gustav Fischer Verlag, Tokyo, Stuttgart, New York.

Patiño, R. and Thomas, P. (1990a). *Biol. Reprod.* **43**, 818–827.

Patiño, R. and Thomas, P. (1990b). *Gen. Comp. Endocrinol.* **78**, 204–217.

Patiño, R., Davis, K.B., Schoore, J.E., Uguz, C., Strüssmann, C.A., Parker, N.C., Simco, B.A. and Goudie, C.A. (1996). *J. Exp. Zool.* **276**, 209–218.

Pelletier, R.M. and Byers, S.W. (1992). *Microsc. Res. Tech.* **20**, 3–33.

Satoh, N. (1974). *J. Embryol. Exp. Morphol.* **32**, 195–215.

Schoonen, W.G. and Lambert, J.G. (1987). *Gen. Comp. Endocrinol.* **68**, 375–386.

Schreibman, M.P. and Magliulo-Cepriano, L. (1999). In *Encyclopedia of Reproduction*, vol. 3 (eds E. Knobil and J.D. Neill), pp. 812–822. Academic Press, San Diego.

Shelton, W.L. and Rothbard, S. (1995). In *Proceedings of the Fifth International Symposium on the Reproductive Physiology of Fish* (eds F.W. Goetz and P. Thomas), p. 143. FishSymp95, Austin, Texas.

Strüssmann, C.A. and Patiño, R. (1995). In *Proceedings of the Fifth International Symposium on the Reproductive Physiology of Fish* (eds F.W. Goetz and P. Thomas), pp. 153–157. FishSymp95, Austin, Texas.

Strüssmann, C.A. and Patiño, R. (1999). In *Encyclopedia of Reproduction*, vol. 4 (eds E. Knobil and J.D. Neill), pp. 402–409. Academic Press, San Diego.

Takahashi, H. (1974). *Bull. Fac. Fish. Hokkaido Univ.* **25**, 174–199.

Takashima, F., Patiño, R. and Nomura, M. (1980). *Bull. Japan. Soc. Sci. Fish.* **46**, 1317–1322.

van den Hurk, R., Resink, J.W. and Peute, J. (1987). *Cell Tissue Res.* **247**, 573–582.

Varanda, W.A. and de Carvalho, A.C. (1994). **267**, C563–C569.

Wallace, R.A. and Selman, K. (1990). *J. Electron Microsc. Tech.* **16**, 175–201.

Yoon, C., Kawakami, K. and Hopkins, N. (1997). *Development* **124**, 3157–3165.

York, W.S., Patiño, R. and Thomas, P. (1993). *Gen. Comp. Endocrinol.* **92**, 418–420.

Yoshikawa, H. and Oguri, M. (1978). *Bull. Japan. Soc. Sci. Fish.* **44**, 313–318.

PART 5

Procedures

Contents

CHAPTER 29	**Stress and anesthesia**	**503**
CHAPTER 30	**Collection of body fluids**	**513**
CHAPTER 31	**Routes of administration for chemical agents**	**529**
CHAPTER 32	**Necropsy**	**543**
CHAPTER 33	**Surgical techniques**	**557**
CHAPTER 34	**Fixation of fish tissues**	**569**
CHAPTER 35	**Autoradiography of fishes**	**579**

CHAPTER 29

Stress and Anesthesia

Henrik Kreiberg
Canada Department of Fisheries & Oceans, Pacific Biological Station, Nanaimo, British Columbia, Canada

Introduction

Many researchers, technicians and students have some basic familiarity with live fish for use in experimentation. The goldfish (*Carassius auratus*) and the killifish (*Fundulus heteroclitus*) for example have been laboratory standards for many decades, in large part due to their ease of maintenance and tolerance for disruptive circumstances. This chapter is not about how to work with goldfish (although use of the suggestions offered will improve such work).

Rather, this chapter recognizes that there are many other species of freshwater and saltwater fish whose predatory habit or ecological specialization makes them well suited to research protocols where a more sensitive organism provides a clearer result. These species include the salmonids, percids, centrarchids and the great majority of marine fish species dwelling in temperate waters. Keeping and using such species in peak condition presents greater challenges, and requires deeper insight into their functioning and responses as animals in order to serve scientific as well as humanitarian principles.

Nearly all fish laboratory protocols share some requirement for physically handling the animals. It is during such events that the majority of problems occur in the keeping of fish. It can be said that a handling event is a compression of the entire gamut of threats and difficulties a fish may face during captivity. My purpose here is to review some of the main handling considerations in research procedures using fish less tolerant of abrupt change, and to provide some lead-in references for further exploration. The concepts presented are eminently applicable to everyday concerns in the design and management of fish-holding facilities, inasmuch as reducing stress to a minimum at all times will always be beneficial to fish. If the following comments serve also to heighten the reader's awareness of fish as sentient beings who share considerable physiology and biochemistry with humans, despite lacking soft fur and appealing eyes, a further progression will have occurred.

Copyright © 2000 Academic Press

The stress response

Fish perceive handling as a general and potentially dangerous stress. Their stress response is practically speaking like our own. Their nervous system detects the threat, and almost instantly, adrenaline is released into the bloodstream. This hormone is followed closely by others such as cortisol, all of which go to work preparing the fish for its reaction, e.g. escape (Mazeaud and Mazeaud, 1981). The result is that blood glucose, red blood cell counts, heartbeat and ventilation rate all increase, and digestive processes may cease temporarily. Once triggered, even by a very transient stress such as being dipnetted in air from one tank to another, the sequence plays through, although in rough proportion to the severity and duration of the stressor.

While a fish may not suffer disadvantage from briefly increased blood glucose or red blood cell counts (both of which serve primarily to enhance the fish's ability to escape from danger), other stress response components have significant drawbacks. Loss of appetite can persist following a handling stress (Wedemeyer, 1976), and normal reproductive functions may be suppressed for a considerable time (Carragher and Sumpter, 1990). Adrenaline disturbs ion transport at the gill membrane, where only a scant few cell layers separate the world within from the surrounding environment. Fish maintain their blood chemistry markedly different from the water which surrounds them, both in fresh water and in sea water. Adrenaline and cortisol both cause temporary changes in gill permeability which in fresh water allows dilution of the blood by excessive entry of water, and vice versa in normal sea water (Mazeaud et al., 1977; Folmar and Dickhoff, 1980). Consequently, blood levels of calcium, magnesium, sodium and other vital electrolytes are pushed out of normal operating ranges for as much as 24 hours after a brief stress such as dipnetting (Wedemeyer, 1972). The burden of trying to restore physiological and metabolic order consumes valuable stored energy, which may leave the fish less capable of fighting pathogens or adapting to new temperatures or reduced oxygen availability. Cortisol elevation itself suppresses immune system function (Barton et al., 1987; Maule et al., 1987) and a frequent sign of poor handling practices in fish is a subsequent infection.

Mitigating the stress response

From the foregoing, it would appear that if the handler can keep the fish from perceiving a disturbance, many problems can be averted. This may not be as difficult as it sounds. In some instances, conditioning to handling has been accomplished by associating presentation of a dipnet with daily feeding. Actual netting out, when it occurs, is thus only a minor departure from routine. A variation of this is to use a flashing light to signal feeding time; operation of the light preceding any other essential procedures then takes advantage of the conditioning, and is not limited to a particular action. Trapping in smaller swim-in enclosures, with bait as necessary, also provides an option for low-stress gathering of fish.

Fish are generally speaking more comfortable under lower light and exposure conditions. Therefore, if they must be held interim in a smaller container, simply covering it to exclude light can be helpful (Hubbs et al., 1988). In chinook salmon undergoing simulated transport in small containers, exclusion of light reduced the hormonal response to stress by about 25% over unprotected fish (Wedemeyer, 1985). Many who work with fish find domestic plasticware such as pails and storage containers very useful; simply putting the lid on when fish are inside will make an appreciable difference to their condition. Choosing a darker color for these accessories, many of which are available in dark browns, greens and blues, will also be valuable. Where recycled metal containers are used, painting the interior a dark color is a good option. This philosophy is also applicable to the fishes' home tanks, where a dark color at least on the floor will improve acceptability. Paints used should be inert or non-leaching when cured.

Physical factors other than light and shelter can also be useful in reducing stress. Fish take their body temperature from the water they live in, and studies have shown that coolwater fish species perceive rapid temperature changes as a stress, responding with elevated cortisol levels and ensuing impacts (Wedemeyer, 1969; Strange et al., 1977; Barton and Peter, 1982). Synchronizing the air and water temperatures of each location a fish will occupy during a short protocol will do much to avoid stress from this source, even if it may inconvenience the researcher. However attractive it may appear to 'slow' a fish's metabolism using short-term temperature manipulation, and thereby reduce stress, it should be apparent from the foregoing that the opposite can occur. In general, many of the other stress mitigation methods described herein offer equivalent or better ways to minimize fish discomfort than does temperature manipulation.

Other key factors of habitat quality, particularly oxygen content, should also be kept appropriate to the particular fish species' needs in all waters to which they will be exposed (reviewed by Davis, 1975). In fresh water, pH and mineral content can vary widely with the water source, and care should be taken both to know the natural ranges for these variables and to ensure that they are not exceeded when making temporary arrangements. Full strength sea water in contrast is normally consistent in its mineral content, pH and buffering capacity. In either type of water, however, use of 'conditioned' or 'seasoned' water (in which the fish have previously spent some time) should be regarded as an option to reduce stress, provided the water is well aerated and free of waste products. If change is unavoidable, then reducing the rate of change will in nearly all cases be a better choice, giving the animals the most time in which to adapt, and minimizing the chance that change is perceived as a threat. For water temperature, a broad rule of thumb in fish is to restrict temperature changes to one or two degrees centigrade per day, which allows physiological compensation to take place.

Further relief from handling stress can be achieved by keeping fish in their home element as much as possible. It may seem intuitive to use a dipnet to catch and then move the fish in air when transferring fish between containers. With some fish species, however, there can be a heavy price for this convenience. Use of in-water transfers reduced 96-h mortality from 82% (for 'dry' transfers) to 15% in American shad (Murai et al., 1979) and reduced long-term mortality from 52% to 15% in Atlantic salmon (Flagg and Harrell, 1990). Principal causes of mortality, which may ultimately be attributed to a pathogen, are the loss of the protective layers of mucus and scales which serve as a fish's primary defense against pathogens and osmoregulatory difficulty.

Survival after dipnetting in air is a complex matter depending on factors which will vary from one instance to the next. Nevertheless, there is clear advantage to using in-water transfers even if the benefit is less dramatic than reduced mortality. Nets may still be used to crowd and guide the fish into a cloth-lined scoop or other device allowing them to be lifted and moved without exposure to air, as the damage clearly occurs when the fish are no longer supported by their home element.

Fishes' home element can also be improved upon for purposes of handling. Like us, fish maintain their blood and tissues at a salt composition intermediate between fresh water and normal sea water. As they pass large volumes of water over their gills, they must constantly expend some energy to maintain the imbalance, except when in brackish water of osmotic pressure similar to their blood (i.e. 12 to 15 parts per thousand salinity). The provision of brackish water surroundings when fish are likely to be stressed has definite advantages, both in conserving energy for physiological emergency purposes and in countering the electrolyte disturbances associated with stress-elevated adrenaline and cortisol levels described above. The presence of salts softens the side effects of the hormones released during the initial stress response, and in the case of chloride and other monovalent anions, can also directly suppress cortisol secretion (Tomasso et al., 1980).

The utility of providing isosmotic surroundings has been demonstrated in various fish species dwelling in fresh or salt water, as well as those which can inhabit both, including threadfin and American shad (Collins and Hulsey, 1963; Murai et al., 1979), blueback herring (Guest and Prentice, 1982), yellowfish and bass (Hattingh et al., 1975), muskellunge (Miles et al., 1974), rainbow and brown trout (Haswell and Thorpe, 1982; Nikinmaa et al., 1983), chinook salmon (Long et al., 1977), and steelhead trout and coho salmon (Wedemeyer, 1972). In some cases, the immediate improvements were striking, e.g. greatly improved survival following a transport procedure, while in others the benefits were most apparent in reduced disturbance of electrolyte balance. In all cases, however, fish undergoing a stressful procedure with a coincident mild salt exposure came through the ordeal in superior physiological condition for coping with subsequent challenges. Undoubtedly, the experience itself was also less traumatic.

Use of moderate salinity water to mitigate stressful procedures with fish is effective regardless of whether the fish are coming from salt or fresh water, and likewise independent of the destination's salinity. In cases where fish will be returned to fresh water, it will be beneficial if they can be given a recovery period of 1–2 days in brackish water. This serves to reduce osmotic stress and provides a reservoir for essential ions which the fish may need to replace in its blood and tissues. One should be careful not to take too simplistic an approach and attempt to create salinity with a single salt, e.g. pure sodium chloride. Natural sea water relies on several ions for its characteristic salinity, and advantage is gained by providing as full a spectrum as possible. In fact, higher salinity created by only one ion may be harmful. Natural or artificial sea water provides the best mixture (a general recipe may be found in Spotte, 1970). Alternatively, concentrations of single ions or salts reflecting their abundance in natural sea water may serve (for sodium chloride, this will be of the order 0.5 to 1% in aqueous solution, to achieve a final salinity of about half that of normal sea water). The use of brackish water in the manner described above may have one further advantage, that of killing or restraining some pathogenic organisms which may otherwise exploit a fish's temporary distress to invade and establish an infection.

Anesthesia

A considerable number of chemical anesthetics, as well as a few alternative approaches, have been used in fish culture and research to induce various planes of anesthesia (reviewed in Bell, 1967; McFarland and Klontz, 1969; and Summerfelt and Smith, 1990). In the practical sense, most anesthetic applications are intended to maximize survival while a fish is removed from water and/or being transported, or to prevent pain and immobilize fish during a procedure such as measuring, blood sampling or surgery. While numerous investigators have undertaken to describe the progressive stages of anesthesia in their own terms and scales, there is a broad congruence among them: fish either retain or lose their balance, and if balance is lost, they may retain some ability to move or they may be so heavily affected as to be inert. Beyond this stage is death (brain failure), but full and rapid recoveries are possible from any preceding stage. It may be convenient to think of these incremental stages as light sedation (fish remain upright), heavy sedation (fish lose their balance) and complete anesthesia (fish become inert). The following discussion addresses mainly the use of chemical anesthetics.

Practical aspects of anesthetizing fish

Anesthetics are selected using a variety of criteria, including efficacy, availability, regulatory and cost considerations, and unfortunately, sometimes merely from habit. Reviews of fish anesthetics such as McFarland (1960), Bell (1967), McFarland and Klontz (1969) and Ross and Ross (1984) provide a wide variety of options, often with species toxicity and safety-margin information which will help in preventing needless stress and mortality to fish. Careful consultation should be part of the choice of an anesthetic, as few drugs are identical in their powers. Some outstanding differences which must be considered include induction times, recovery times, safety margins (i.e. the ratio between an effective anesthetic dose and a lethal dose), effect on reflex or spontaneous motor activity, and the health risk to both fish and researcher (several anesthetics have in recent years been identified as possible carcinogens).

The great majority of procedures with fish where anesthetics are indicated are relatively brief, lasting only minutes. The field of extended surgical interventions, requiring supplemental gill irrigation and multistage anesthetic applications, is beyond the scope of this chapter. In most circumstances, the drug is applied to the fish via uptake at the gill filaments from the surrounding water (sometimes termed 'inhalation anesthesia'). Fish are usually transferred as quickly and unobtrusively as possible from their home tank to a smaller water container or bath containing the drug; it is rare that chemical anesthetics are delivered by injection. Reasons for this include the ease with which dosing can be controlled via adjustment of diluting volume in the bath or the time exposed to the drug. In comparison, it is often difficult to achieve these objectives via injection without subjecting the fish to further handling to obtain accurate weight information or to access suitable injection sites, and this method is often confined to special cases where lengthy periods of anesthesia are desired.

Most fish will survive anesthesia better on an emptied stomach, which may require withholding food for 48 hours or longer. Freshly prepared anesthetic solutions are the basis of accuracy and reliability. Working solutions need to be clearly labeled with content and strength, and original dosing calculations should be verified at least once before commencing a protocol. If considerable numbers of fish are expected to transit the anesthetic bath, provision of supplemental aeration is advisable. Furthermore, a small number of fish should be anesthetized and observed to make a full recovery to alertness again before the anesthetic bath is deemed ready for use. It may be worth while, particularly in handling sensitive species, to consider addition of a mucus-replacing polymer such as polyvinyl pyrrolidone (Wedemeyer, 1992). Such compounds, used in anesthetic baths at rates of 50 to 200 parts per million, may help to avoid osmotic distress and secondary infections occurring after handling.

Induction of anesthesia proceeds in a largely similar manner for most non-sedentary fish. Upon introduction to the drug, most fish settle rapidly, perhaps displaying several 'coughs' or flaring of their opercula. Excessive activity of this nature is a sign that the drug is noxious, which is not advantageous in the long run. Sometimes preceded by several minutes of moderate to relaxed activity, fish will descend to the floor of the bath and begin to lose balance. After some efforts to regain their upright position, they will lie quietly on their sides and continue to ventilate, usually with a somewhat shallower and more rapid rhythm than during the upright phase. At this point, they may be handled for less exacting procedures such as routine weighing. Deeper anesthesia is appropriate for processes where some pain may be expected, e.g. blood sampling or minor incisions, in which case cessation of opercular movement should be noted before handling. In bottom-dwelling and flatfish species, the principal guide to onset of anesthesia is slowing of the opercular rhythm. It is vital to retain a clear awareness of each fish's drug exposure time, particularly when working with groups of fish, in order to avoid overexposures.

Recovery from anesthesia is often as simple as returning fish to drug-free water. If care is taken to avoid pile-ups of inert fish in the recovery area, which can interfere with recovery, most fish will regain their balance and normal responses within no more than twice the time needed for induction. In overdose situations, it may be necessary to stimulate the fish by causing waterflow over the gills, or by gentle hand massage of the heart area while the fish remains in water. Frequently repeated anesthesia with no apparent consequence has been reported for some species and drug combinations (Schoettger and Julin, 1967; McFarland and Klontz, 1969), but in general, fish should be given a period of several days between exposures to anesthetic.

Pre-handling applications

Sedation is a useful option for reducing sensory awareness, and thus for mitigating short-term stress. Light anesthesia, which allows fish to maintain equilibrium, swimming and breathing, has been used in some cases to decrease stress in fish transports (reviewed in Wedemeyer, 1996), and has been suggested as a means to reduce metabolic rates and conserve oxygen over unsedated fish (Baudin, 1932; Piper *et al.*, 1982). It can also be extremely valuable when there is a requirement for frequent handling of fish, or for species such as mackerel and tuna which are difficult to keep under laboratory conditions.

Deriving the most value from sedation however is a matter of timing and chemistry. Studies have shown that in a procedure involving crowding, capture, dipnetting and deep anesthesia, it is the initial crowding stage during which most of the fish's stress response is invoked (Strange and Schreck, 1978; Barton *et al.*, 1980; Matthews *et al.*, 1986). Simply using a sedative in the water in a transport container is therefore something of a catch-up effort, which may be completely ineffective in preventing the physiological stress reaction (Kreiberg, 1992). As well, other research has shown that while many anesthetics are effective for rapid induction of deep anesthesia, some of those most widely used, e.g. **tricaine methane sulfonate** (**TMS** or **MS-222**) and 2-phenoxy ethanol, have an excitatory effect during initial absorption (Barton and Peter, 1982; Davis *et al.*, 1982), which largely defeats the purpose of calming fish which are to remain conscious under weaker doses of a drug.

A few fish anesthetics however are classed as hypnotics, and do not have an excitatory impact. Chiefly these are etomidate and its analogue metomidate. Initial studies with these compounds indicated their suitability for use in fish in terms of wide safety margins, lack of electrolyte disturbance and minimal elevation of cortisol in striped bass and channel catfish (Davis *et al.*, 1982; Limsuwan *et al.*, 1983). Kreiberg and Powell (1991) subsequently showed that if a lightly-sedating dose of metomidate was added unobtrusively to the home tank of some chinook salmon, the plasma cortisol response to stress of mild crowding and dipnetting in air was negligible relative to unsedated control fish, whose cortisols became strongly elevated with the same handling. They also noted that complete anesthesia with MS-222 following the initial sedation with metomidate was efficient. This two-step handling practice has been adopted by other researchers striving to minimize the stress impacts of research sampling.

MS-222

This drug, whose suitability for cold-blooded animals was originally noted in 1920, has subsequently found wide application in fish hatcheries and countless research applications throughout the world. It was known originally as tricaine methane sulfonate, but the terms TMS, tricaine and MS-222 have been used interchangeably in more recent times. There is an extensive literature devoted to its efficacy, probably the greatest for any fish anesthetic. It is similar in many respects to cocaine, novocaine and benzocaine, all of which are familiar from human uses (and which have also been used as fish anesthetics, especially the latter).

MS-222 is directly water soluble (unlike benzocaine), has a good shelf-life stored in the powder form, clears rapidly from the tissues of recovered fish, and has been found to function predictably in a great variety of fish species. While it shows somewhat reduced efficacy with higher temperatures and has numerous documented physiological consequences in treated fish (Ross and Ross, 1984), these factors do not detract seriously from its general suitability for rendering complete anesthesia, and to a lesser degree, for sedation applications. Although it is comparatively expensive to purchase, it remains the drug of choice for many applications. Table 29.1 provides a laboratory user's overview of some better-known fish anesthetics.

MS-222 has received regulatory clearance in several countries for use in fish destined for human consumption, usually with pre-harvest withdrawal times of between 5 and 21 days, and in this respect the drug may continue to find relatively wide use in fish. Most reports suggest a concentration of 40 to 100 parts per million as an effective range for complete anesthesia and recovery within 2 to 4 minutes. It is worth noting that stock solutions of MS-222 are strongly acidic in soft fresh water, with the potential for gill irritation to fish. In the absence of natural buffering, such as provided by normal sea water, it is valuable and considerate to buffer solutions, e.g. with sodium bicarbonate (Allen and Harman, 1970) or sodium hydroxide (Wedemeyer, 1970). Other toxicity problems can arise with MS-222 if the stock solution is exposed to sunlight, hence this should be avoided.

Clove oil compounds

The aromatic oils from the clove plant have shown good prospect as fish anesthetics (Endo *et al.*, 1972; Hikasa *et al.*, 1986). Clove oil and its main effective ingredients eugenol and isoeugenol are widely used as food and fragrance additives. Due to this familiarity factor, the existence of a considerable body of human testing for health consequences, and the generally low cost of usage, these compounds may find wider application in research with fish in the near future. It is important however for the researcher to be aware that the physiological effects of clove oil compounds on fish are undocumented at this time, and that they may therefore have unspecified impacts on biological variables being measured.

Carbon dioxide

One other anesthetic currently in vogue due to its environmental acceptability is carbon dioxide. It may

TABLE 29.1: Overview of some chemical fish anesthetics. 'T' indicates sedative dosage for transport, 'A' gives dosage for complete anesthesia

Compound	Dosage	Notes
Benzocaine	10–30 ppm (T), 40–100 ppm (A)	Widely used and researched
Carbon dioxide	200–800 ppm (A)	Widely used (smaller fish); scant research; induction stressful
Clove oil (eugenol)	10–30 ppm (T), 50–100 ppm (A)	Effective, inexpensive; not widely studied
MS-222 (TMS)	5–30 ppm (T), 40–100 ppm (A),	Widely used and researched
Methyl pentynol	1–2 mL L^{-1} (T), 3 mL L^{-1} (A)	'methyl parafynol'; not widely used; slow recovery
Metomidate	0.1–0.5 ppm (T), 3–5 ppm (A)	Minimal stress response; better as sedative
2-phenoxy ethanol	100–300 ppm (A)	Widely used and researched; health safety concerns
Propanidid	0.5–3 mL L^{-1} (A)	Not widely used; expensive
Quinaldine	2–5 ppm (T), 5–25 ppm (A)	Considerable usage; inexpensive
Tertiary amyl alcohol	0.2–0.3 mL L^{-1} (T), 2–5 mL L^{-1} (A)	Not widely used; slow induction, recovery

be created cheaply and readily using vinegar and baking soda, sublimated dry ice, or obtained pure from commercial gas suppliers. It is probably one of the crudest and least elegant methods of immobilizing fish. At high concentrations the gas is toxic and completely inhibits oxygen consumption (Shelton, 1970). Its anesthetic effect at lower concentrations arises from its power to reduce the animal's capacity to extract oxygen from the environment (Black *et al.*, 1954), and is not dissimilar to applying a choke-hold and causing a black-out. There is no documented reduction of nervous sensation, hence the induction phase is in most fish species characterized by strong physical avoidance reactions. Coupled with inherent difficulties in maintaining control of dosing due to carbon dioxide's volatility, it may serve humanitarian and more mundane purposes equally to explore other options, even though they may require greater effort initially.

Electronarcosis

An electrical current is widely used to stun and capture wild fish, particularly so in fresh water. It has been applied successfully for temporary immobilization of captive fish from 15 cm to over 100 cm in length (Hartley, 1967; Orsi and Short, 1987). Its physiological stress impacts have not been well investigated, but it is generally considered that only in a few discrete forms has it any analgesic value. Effective electronarcosis is sensitive to numerous electrical and biological variables, and is perhaps more demanding of the operator than are most chemical anesthesia methods. It can be the optimum methodology for certain very brief and repetitive procedures such as high-seas tagging programs, but it has not found wide application in laboratory work with fish.

Regulatory

Prospective users of fish culture chemicals and drugs should always familiarize themselves with local regulatory conditions and occupational health information before using any compound. The mere fact that a drug can be purchased cannot be taken to mean that it is approved for use in fish, and serious legal consequences may result from improper usage. Choice of anesthetic is often determined by a researcher based on the nature of the procedure for which a fish is destined. More recently however, regulatory acceptance of a drug may be the pivotal criterion for its selection, a fact which may compromise a laboratory procedure in terms of fish stress. Under such circumstances, it will be especially valuable to recognize and apply additional methods of stress mitigation.

Sentience, analgesia and euthanasia

Fish possess several of the characteristics associated with the capacity of animals to feel pain, including the necessary brain structure, opioid compound receptors and the capacity to associate neutral events with noxious stimuli (Bateson, 1992). Lacking the ability to vocalize, they express discomfort by listlessness, loss of appetite and condition, abnormal behavior such as withdrawal from the main group of their companions in a tank, and abnormal coloration, frequently darker.

Newborn humans show elevated blood cortisol, glucose and lactate in response to pain from surgery, all of which are reduced under anesthesia (Anand *et al.*, 1985). Fish, as we have seen, also possess these responses. The laboratory user of fish thus should have no doubt as to the sensibility and sentience of fish, and if a procedure is deemed at all likely to cause pain, it is humane and responsible to take the most appropriate measures to minimize pain and distress.

Euthanasia, which in the strict sense connotes a peaceful painless death, is too often employed in a euphemistic sense to cover any instance where an animal is put to death. The humane killing of fish relies on one or more of the following: oxygen starvation, irreversible depression of vital nervous system parts (e.g. by anesthetics), or physical destruction of the brain. The onus remains on the animal handler to be as well informed as possible about the pharmacology and physiological impacts of a method of 'euthanasia', which may mean going beyond blind adoption of a previously reported method. Use of carbon dioxide, for example, would not, in the light of foregoing comment, rate as a good method of euthanasia due to its violent induction and mode of action.

By contrast, use of established lethal levels of central nervous system depressants such as MS-222 and benzocaine would be preferable. These compounds are recommended for fish in guides issued by the Canadian Council on Animal Care (Olfert *et al.*,

1993) and the American Veterinary Medical Association (Andrews *et al.*, 1993). Research with MS-222 has shown that stress, as indicated by elevation of plasma cortisol prior to death, did not reach levels warranting humanitarian concern if an established lethal dose was used (Strange and Schreck, 1978). As with reversible induction, use of appropriate buffering described above would avoid pH-related trauma.

A rapid stunning blow to the head is also considered to be a humane method of killing fish, although it is most suitable for larger fish, and is usually preceded by anesthesia to quieten the fish. If the physical stunning method is used, common sense would suggest that the task be placed in the most experienced hands available. Suffocation, either by draining a tank or removing fish to a dry tub or pail, is not a humane alternative. In general, if fish are small, a lethal dose of appropriate anesthetic is best. Euthanasia of fish should be treated with the same level of preparation applied to any other critical phase of a study or project, both for the sake of the animals and to minimize distress to staff who are taking part.

In summary, fish represent such a wealth of evolutionary diversity and specialization that it is difficult to go beyond generalities in addressing the subject of stress while they are in captivity. Many fish species do however perceive and respond to disruptive influences in a similar manner physiologically, which opens the door to making such experiences less stressful for them. None of the processes and mitigative options discussed in this chapter are especially new, and most of them are within the reach of practically anyone who keeps and looks after fish. Unfortunately, such information is not always found under one heading. The present survey is necessarily brief, but it may prove helpful in increasing awareness of stress-related considerations in research and other uses of fish, and thereby improve the lot both of the fish and of the researchers who use and depend upon them.

References

Allen, J.L. and Harman, P.D. (1970). *Progr. Fish Cult.* **32**(2), 100.
Anand, K., Brown, M.J., Causon, R.C., Christofides, N.D., Bloom, S.R. and Aynsley-Green, A. (1985). *J. Pediatr. Surg.* **20**, 41–48.
Andrews, E.J., Bennett, B.T., Clark, J.D., Houpt, K.A., Pascoe, P.J., Robinson, G.W. and Boyce, J.R. (1993). *J. Am. Vet. Med. Assoc.* **202**(2), 231–249.
Barton, B.A. and Peter, R.E. (1982). *J. Fish Biol.* **20**, 39–51.
Barton, B.A., Peter, R.E. and Paulencu, C.R. (1980). *Can. J. Fish. Aquat. Sci.* **37**, 805–811.
Barton, B.A., Schreck, C.B. and Barton, L.D. (1987). *Dis. Aquat. Organisms* **2**, 173–185.
Bateson, P. (1992). *New Scientist* **134**(1818), 30–33.
Baudin, L. (1932). *Compt. Rend. Soc. Biol.* **109**, 731.
Bell, G.R. (1967). *Fish. Res. Bd. Can. Bull.* **148**. 11pp.
Black, E.C., Fry, F.E.J. and Black, V.S. (1954). *Can. J. Zool.* **32**, 408–420.
Carragher, J.F. and Sumpter, J.P. (1990). *Gen. Comp. Endocrinol.* **77**, 403–407.
Collins, J.L. and Hulsey, A.H. (1963). *Progr. Fish Cult.* **25**, 105–106.
Davis, J.C. (1975). *J. Fish. Res. Bd. Can.* **32**, 2295–2332.
Davis, K.B., Parker, N.C. and Suttle, M.A. (1982). *Progr. Fish Cult.* **44**(4), 205–207.
Endo, T., Ogishima, K., Tanaka, H. and Ohshima, S. (1972). *Bull. Jap. Soc. Sci. Fish.* **38**(7), 761–767.
Flagg, T.A. and Harrell, L.W. (1990). *Progr. Fish Cult.* **52**, 127–129.
Folmar, L.C. and Dickhoff, W.W. (1980). *Aquaculture* **21**, 1–37.
Guest, W.C. and Prentice, J.A. (1982). *Progr. Fish Cult.* **44**(4), 183–185.
Hartley, W.G. (1967). In *Fishing with Electricity* (ed. R. Vibert). EIFAC.
Haswell, M.S. and Thorpe, G.J. (1982). *Progr. Fish Cult.* **44**(4), 179–183.
Hattingh, J., Fourie, F.L. and Van Vuren, J.H. (1975). *J. Fish Biol.* **7**, 447–449.
Hikasa, Y., Takase, K., Ogasawara, T. and Ogasawara, S. (1986). *Jap. J. Vet. Sci.* **48**(2), 341–351.
Hubbs, C., Nickum, J.G. and Hunter, J.R. (1988). *Fisheries* **13**(2), 16–22.
Kreiberg, H. (1992). *Bull. Aquacult. Assoc. Can.* **92**(3), 52–54.
Kreiberg, H. and Powell, J. (1991). *World Aquacult.* **22**(4), 58–59.
Limsuwan, C., Limsuwan, T., Grizzle, J.M. and Plumb, J.A. (1983). *Can. J. Fish. Aquat. Sci.* **40**, 2105–2112.
Long, C.W., McComas, J.R. and Monk, B.H. (1977). *Mar. Fish Rev.* **39**(7), 6–9.
McFarland, W.N. (1960). *Calif. Fish Game* **46**(4), 407–431.
McFarland, W.N. and Klontz, G.W. (1969). *Fed. Proc.* **28**(4), 1535–1540.
Matthews, G.M., Park, D.L., Achord, S. and Kaattari, T.E. (1986). *Trans. Am. Fish. Soc.* **115**, 236–244.
Maule, A.G., Schreck, C.B. and Kaattari, S.L. (1987). *Can. J. Fish. Aquat. Sci.* **44**, 161–165.
Mazeaud, M.M. and Mazeaud, F. (1981). In *Stress in Fish* (ed. A.D. Pickering), pp. 47–68. Academic Press, London.

Mazeaud, M.M., Mazeaud, F. and Donaldson, E.M. (1977). *Trans. Am. Fish. Soc.* **106**(3), 201–212.

Miles, H.M., Loehner, S.M., Michaud, D.T. and Salivar, S.L. (1974). *Trans. Am. Fish. Soc.* **103**(2), 336–342.

Murai, T., Andrews, J.W. and Muller, J.W. (1979). *Progr. Fish Cult.* **41**(1), 27–29.

Nikinmaa, M., Soivio, A., Nakari, T. and Lindgren, S. (1983). *Aquaculture* **34**, 93–99.

Olfert, E.D., Cross, B.M. and McWilliam, A.A. (1993). *Guide to the Care and Use of Experimental Animals*, vol. 1. Canadian Council on Animal Care, Ottawa. 211pp.

Orsi, J.A. and Short, J.W. (1987). *Progr. Fish Cult.* **49**, 144–146.

Piper, R.G., McElwain, I.B., Orme, L.E., McCraren, J.P., Fowler, L.G. and Leonard, J.R. (1982). *Fish Hatchery Management*. US Fish Wildlife Service, Washington, DC. 517pp.

Ross, L.G. and Ross, B. (1984). *Anesthetic and Sedative Techniques for Fish*. Institute of Aquaculture, University of Stirling, Stirling. 35pp.

Schoettger, R.A. and Julin, A.M. (1967). US Fish & Wildlife Service Investigation in Fish Control 13. 15pp.

Shelton, G. (1970). In *Fish Physiology*, vol. 4 (eds W.S. Hoar and D.J. Randall), pp. 293–352. Academic Press, New York.

Spotte, S. (1970). *Fish and Invertebrate Culture*. Wiley Interscience, New York. 145pp.

Strange, R.J. and Schreck, C.B. (1978). *J. Fish. Res. Bd. Can.* **35**, 345–349.

Strange, R.J., Schreck, C.B. and Golden, J.T. (1977). *Trans. Am. Fish. Soc.* **106**(3), 213–218.

Summerfelt, R.C. and Smith, L. (1990). In *Methods in Fish Biology* (eds C.B. Schreck and P.B. Moyle), pp. 213–272. Am. Fish. Soc., Bethesda.

Tomasso, J.R., Wright, M.I., Simco, B.A. and Davis, K.B. (1980). *Progr. Fish Cult.* **42**, 144–146.

Wedemeyer, G. (1969). *Comp. Biochem. Physiol.* **29**, 1247–1251.

Wedemeyer, G. (1970). *J. Fish. Res. Bd. Can.* **27**(5), 909–914.

Wedemeyer, G. (1972). *J. Fish. Res. Bd. Can.* **29**(12), 1780–1783.

Wedemeyer, G. (1976). *J. Fish. Res. Bd. Can.* **33**(12), 2699–2702.

Wedemeyer, G. (1985). Contract report to Bonneville Power Administration, #82-19. Portland. 70pp.

Wedemeyer, G. (1992). *World Aquacult.* **23**(4), 47–50.

Wedemeyer, G. (1996). In *Principles of Salmonid Culture* (eds W. Pennell and B.A. Barton), pp. 727–758. Elsevier Science, Amsterdam.

Chapter 30

Collection of Body Fluids

Marsha C Black
College of Agricultural and Environmental Sciences,
The University of Georgia, Athens, Georgia, USA

Introduction

As fish are employed in increasing numbers in physiological, pharmacological and toxicological research there is a critical need for detailed procedures for sampling their excreta. This chapter provides details of multiple approaches for collecting blood, urine, feces and sperm from fish, including construction of cannulas and catheters. Special considerations for sample treatment and preservation are also presented within each section.

Blood collection

Anticoagulants

Anticoagulants are employed in blood collection to prevent blood clotting during and after collection. If a blood vessel or the heart is cannulated for repeated blood sampling or physiological measurements (e.g. blood pressure, cardiac output, blood volume), the cannula is flushed with an anticoagulant solution following sample collections and small amounts of anticoagulant are often injected into the cannulated vessel or heart to further retard clotting (e.g. Conte et al., 1964; Smith and Bell, 1964; Soivio et al., 1975; Barron et al., 1987). Moreover, many biochemical analyses are performed on anticoagulant-treated whole blood or plasma. Alternatively, if blood coagulation times are to be measured or if serum is desired for a biochemical analysis, whole blood may be collected into a tube containing no anticoagulant. Following complete coagulation serum is further separated from clotted cells by centrifugation.

Heparin dissolved in saline (usually 0.9 or 1 mg L^{-1} NaCl) is the most commonly used anticoagulant in fish studies (Axelsson and Fritsche, 1994). Heparin is available in several salt forms, including sodium, calcium, ammonium or lithium heparin. The concentration of heparin in heparinized saline varies mostly depending on the fish species, weight or application, although blood pressure, heart rate, temperature and cannula material are also cited as important factors (Axelsson and Fritsche, 1994). If calcium ions are to be measured in heparinized plasma, heparin concentrations should not exceed 100 U mL^{-1} (Sachs et al., 1989, in Axelsson and Fritsche, 1994). Specific examples of heparin use in cannulation experiments are given in Table 30.1.

Copyright © 2000 Academic Press

TABLE 30.1: Examples of anticoagulants used in vessel cannulation experiments

Species	Age or weight	Anticoagulant	Concentration	Vessel cannulated	Reference
Oncorhynchus mykiss	553–1346 g	heparin	540 U mL^{-1} in saline	dorsal aorta, celiac artery	Gilmour et al., 1994
O. mykiss	4 yr	sodium heparinate	1 mg L^{-1} in 1% saline	dorsal aorta	Soivio et al., 1975
O. mykiss	150–450 g	heparin	10 U mL^{-1} in 0.9% saline	dorsal aorta	Barron et al., 1987
Salmo gairdneri (Steelhead trout)	135–800 g	heparin	140 000 U mL^{-1}	dorsal aorta	Conte et al., 1964
Cyprinus carpio	—	heparin	10 U mL^{-1} in saline	dorsal aorta	Garey, 1969
Salmo salar	580–1160 g	heparin	1000 U Hepalin® solution	dorsal aorta	Kiessling et al., 1995

Oxalate or citrate salts are also used as anticoagulants in fish experiments (Smith, 1929; Axelsson and Fritsche, 1994). The anticoagulant activity of these salts results from reductions in blood Ca^{2+}, causing inhibition of thrombin formation. Blood is usually collected into a tube containing added oxalate or citrate salts (Axelsson and Fritsche, 1994) or commercially available tubes containing premeasured amounts of oxalate or citrate salts (e.g. Vacutainer® tubes, Becton Dickinson). No studies were found that used these salts as an anticoagulant in cannulation procedures, probably due to their inhibition of Ca^{2+}, which may cause unwanted vascular or systemic effects.

Ethylenediamine tetra acetic acid (EDTA) also provides anticoagulant activity through reducing free Ca^{2+} concentrations. It has been identified as the best anticoagulant to use when measuring hematological parameters in fish (Hesser, 1960).

Single sample collection

Caudal puncture

If a single sample of blood is desired, caudal puncture is the most common method used by investigators. Caudal puncture will yield whole blood volumes of 0.2 mL to greater than 10 mL from most fish weighing more than 25 g. Prior to drawing blood, fish should be anesthetized in a dilute solution of MS222 (3-aminobenzoic acid ethyl ester methane sulfonate) or a similar anesthetic agent until the fish loses equilibrium. (See the review by Iwama and Ackerman (1994) for a description of fish anesthetics and dosages.) The anesthetized fish is placed ventral side up on a stable surface. (Note: A piece of styrofoam with a V-shaped groove cut into it will stabilize the fish nicely.) Blood is drawn into a 1 or 5 mL glass or plastic syringe (depending on the fish size and/or desired quantity of blood), using a 1 to 1.5 inch (2.5 to 3.75 cm) 22 gauge needle. A Vacutainer® syringe may be used for drawing blood from large fish; however, because the amount of suction cannot be controlled, this method may cause collapsed vessels in some fish. The syringe or Vacutainer® tube may contain an anticoagulant, if desired. The needle is inserted approximately 5 mm posterior to the anal fin, along the midline of the fish's body (see Figure 30.1, caudal puncture). Gently insert the needle (with the bevel facing the fish's posterior end) straight down into the caudal region of the fish. Slowly guide the needle downward until the needle just touches the spine, then back off the spine approximately 1 mm to position the needle in the caudal vein. Blood will pool into the needle hub when the needle is in the vein. Holding the syringe in place by the needle, gently apply light suction with the syringe until the desired amount of blood has been withdrawn. Take care not to apply too much suction, as the vessel may collapse. After the blood is drawn, remove the needle and gently invert the syringe several times to mix the blood (and anticoagulant, if added). The blood may now be discharged from the syringe into another container. (Note: Never discharge blood through the needle, as this may cause hemolysis of the blood sample.) Return the fish to the water, irrigating its gills until **opercular** movement is resumed.

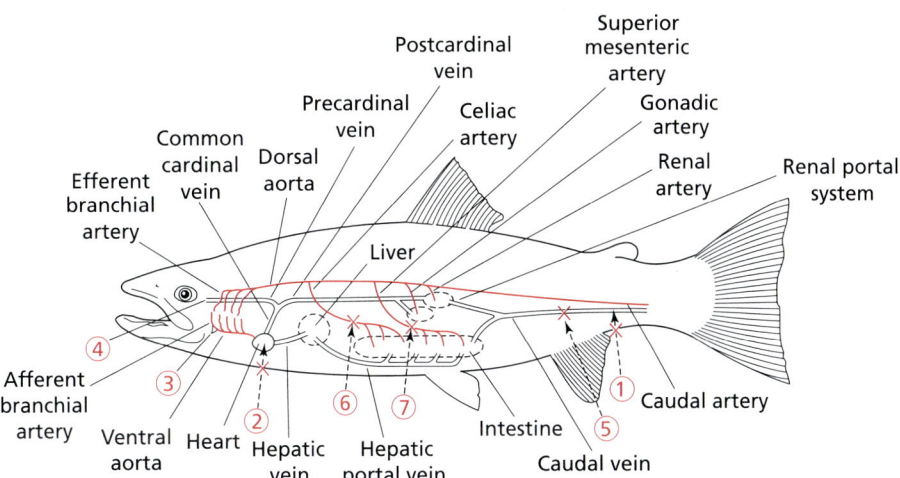

Figure 30.1 Diagram of a fish showing the major blood vessels and locations for drawing blood. 1, Caudal puncture; 2, cardiac puncture, cannulation; 3, ventral aorta puncture, cannulation (see Figure 30.2 for detail); 4, dorsal aorta cannulation (see Figure 30.2); 5, caudal vein cannulation; 6, celiac artery cannulation; 7, mesenteric artery cannulation.

Severance of the caudal peduncle

Another method of quickly collecting blood from anesthetized fish is by severing the caudal peduncle, posterior to the anal fin with a sharp knife or scalpel (Smith, 1929; Cockell et al., 1992). Small quantities of blood can be collected from the severed body into microcapillary tubes and larger amounts can be drained into a test tube, with or without an anticoagulant added to the receiving vessel. The major disadvantage of this method is that the fish is sacrificed by the collection method.

For very small fish (e.g. mosquitofish, *Gambusia* sp. or medaka, *Oryzias latipes*) with minute blood volumes, red blood cells (RBCs) can be collected using a modification of this method (personal communication, C. Theodorakis, TIEHH, Texas Tech University, Lubbock, TX). Fish are anesthetized to loss of equilibrium with MS222 (Iwama and Ackerman, 1994) and doused with 70% ethanol. The caudal peduncle is severed with sterile scissors and the body of the fish is immediately placed head-up into a 10 × 75 mm PE tube containing 1 mL of a sterile fish physiological saline solution (Table 30.2). As the tube is gently swirled, a thin stream of blood will slowly flow out of the severed body of the fish and the red blood cells (RBCs) will settle to the bottom of the tube. When the blood flow ceases, remove the fish with sterile forceps, resuspend the RBCs in the physiological saline, transfer the saline to a conical microcentrifuge tube and centrifuge at 16 000 rpm or less for 1 min. After centrifugation, the physiological saline is decanted and RBCs are resuspended in an additional milliliter of physiological saline and centrifuged as before for 1 min. Once centrifuged and decanted, the washed RBCs can be used for biochemical analyses, such as DNA strand length determinations.

TABLE 30.2: Directions for preparing fish physiological saline solution with EDTA added as an anticoagulant

6.44 g NaCl
0.011 g KCl
0.022 g $CaCl_2$
0.012 g $MgSO_4$
0.007 g KH_2PO_4
0.010 g $NaHCO_3$
0.0186 g EDTA
Weigh the salts and place in a 1 L volumetric flask. Dilute to 1.0 L with deionized water. Autoclave the solution before use.

Cardiac puncture

Ventricular cardiac puncture can be used for a single blood collection of large volumes. Fish are anesthetized and placed ventral side up in a stable position, as previously described. The point of insertion is located on the linea alba, an indentation along the ventral midline of the fish as it passes between the pectoral fins (see Figure 30.1, cardiac puncture). A 23 gauge needle attached to a syringe is inserted slowly, moving straight down at this location until blood appears in the needle hub, indicating that the ventricle has been pierced. Sampling catfish and other fishes with a bony pectoral girdle is best accomplished by puncturing the midline below the pectoral fins, and then angling the needle slightly toward the anterior end of the fish, rather than straight down (Klonz and Smith, 1968).

Blood may also be collected from the ventral aorta, although this procedure is only used on large fish (> 1.5 kg) and is more difficult than a ventricular puncture (Klonz and Smith, 1968). In an anesthetized fish, the ventral aorta is sampled with a 22 gauge, 1.5 inch needle. The fish is positioned dorsal side up, with the mouth held open. The needle is guided through the mouth to the base of the lingual frenum (see Figure 30.1, ventral aorta puncture). The tissue under the tongue is punctured in a downward and slightly posterior direction until the cartilage is just pierced. If a slight negative pressure is applied to the syringe, blood will pool in the needle hub, indicating that the aorta has been pierced (Klonz and Smith, 1968).

Serial sampling techniques

Cannula designs

Most cannulas used to collect blood from fishes are constructed of medical-grade polyethylene (PE) tubing, a flexible, yet highly manipulative material that is non-toxic and non-reactive with fish tissues (Axelsson and Fritsche, 1994). One notable exception to the use of PE tubing is the use of polyvinylchloride (PVC) tubing for blood collected for blood-gas measurements (Garey, 1969). The permeability of PVC tubing to gases, such as O_2 and CO_2, is much lower than PE tubing (Lebovits, 1966). Regardless of the tubing material the inside diameter of the tubing at its minimum diameter should be as large as possible, approaching the same internal diameter as the vessel to be cannulated. This will reduce the suction needed

TABLE 30.3: Size specifications for medical-grade PE tubing (ClayAdams Intramedic, Becton Dickinson, Parsippany, NJ) used for cannula and catheter fabrication

Tubing	ID (mm)	OD (mm)	Needle gauge (to fit into tubing)
PE 10	0.28	0.60	30
PE 20	0.38	1.09	27, 26
PE 50	0.58	0.96	23
PE 90	0.86	1.27	20
PE 160	1.14	1.57	18
PE 190	1.19	1.70	18
PE 200	1.40	1.90	17
PE 204	1.57	2.09	16
PE 240	1.67	2.42	15

for blood collection into the syringe and minimize adverse effects, such as blood cell damage and hemolysis (Axelsson and Fritsche, 1994). Table 30.3 presents the size specifications of commercially available PE tubing (Clay Adams, Bectin Dickinson Co., Parsippany, NJ).

Early versions of cannulas were constructed of PE tubing attached to needle tips (Conte *et al.*, 1964; Smith and Bell, 1964). More recently, blood cannulas have been constructed entirely from synthetic tubing, such as PE, PVC, silicon elastomer or polyurethane (e.g. Soivio *et al.*, 1975; Thomas and Le Ruz, 1982; Barron *et al.*, 1987; Axelsson and Fritsche, 1994). Insertion of PE tubing into the vessel will minimize the potential vascular stress associated with an indwelling needle. Cannulas may be fabricated from two different sizes or types of tubing (e.g. PE joined with PVC, silicon elastomer, or polyurethane tubing) bonded with cyanoacrylate glue or fused after softening with heat or a solvent such as benzene (Axelsson and Fritsche, 1994). This allows the cannula to be composed of a short length of smaller diameter tubing (e.g. PE 10 or PE 50) for easy vessel insertion, coupled with a longer length of tubing having a larger diameter or other properties. As an alternative, PE tubing with a wide diameter may be tapered by gently heating and drawing out the tubing, creating an area with a smaller diameter that can be cut with a scalpel and used as the insertion end of the cannula (Axelsson and Fritsche, 1994). One or more heat-formed bubbles may be added to a PE cannula for stabilizing it within the buccal cavity.

For the majority of vessel cannulations in fish (e.g. dorsal and ventral aorta, caudal, and branchial cannulations), the vessel cannot be isolated and the surrounding tissue must be pierced *en route* to the vessel. Several methods exist allowing easy insertion of most cannulas through tissue into a vessel. For cannulas constructed of tubing attached to a needle, the needle end is inserted through the tissue and directly into the vessel, where it remains for the duration of the experiment (Conte *et al.*, 1964; Smith and Bell, 1964).

Another method employs the use of a **trocar** or larger bore needle, which is initially inserted into the vessel, but withdrawn after a PE catheter of smaller diameter has been threaded through it into the vessel (Garey, 1969; Axelsson and Fritsch, 1994). A variation of the threaded needle method uses a stiff, sharpened wire (e.g. steel piano wire) threaded inside the cannula, which creates a smaller insertion hole in the vessel (Soivio *et al.*, 1975; Axelsson and Fritsch, 1994). Additionally, the snug-fitting wire will function as a low pressure syringe plunger. The wired cannula is inserted into the vessel and the wire withdrawn slowly (up to 10 mm past the insertion point) until blood is observed in the cannula. If no blood follows the wire withdrawal, it may be repositioned and withdrawn again; however, care must be exercised as excess wire manipulation may cause vessel damage or constriction. Once blood enters the cannula, the wire is completely removed and the cannula is immediately filled with heparinized saline and plugged (Soivio *et al.*, 1975).

Blood collection from restrained and free-swimming fish

Dorsal aorta cannulation

The dorsal aorta is the most frequently cannulated vessel in studies requiring serial blood collections from restrained and free-swimming fish. The cannulation procedure is easily performed on most large fish (> 200 g), including many salmonids (rainbow trout, pink salmon, sockeye salmon and Atlantic salmon), channel catfish, common carp, and dogfish shark (e.g. Conte et al., 1964; Smith and Bell, 1964; Garey, 1969; Soivio et al., 1975; Opdyke et al., 1982; Axelsson and Fritsche, 1994; Gilmour et al., 1994; Black et al., 1995; Kiessling et al., 1995; McKim et al., 1996). The method has been used for making sequential measurements of hematological and blood chemistry parameters (Soivio et al., 1975), blood acid–base status (Currie and Tufts, 1993; Gilmour et al., 1994), blood gas concentrations (Garey 1969; Barron et al., 1987; Currie and Tufts, 1993); blood volume determinations (Conte et al., 1964; Smith and Bell, 1964), systemic blood pressure measurements (Barron et al., 1987; Axelsson and Fritsche, 1994) and for monitoring concentrations of chemicals in blood (Black et al., 1995; Kiessling et al., 1995).

Cannulation of the dorsal aorta can be conducted quickly (usually in less than 5 min) in the absence of gill irrigation after deeply anesthetizing the fish with a high concentration of MS222. For example, exposure to 100 mg L^{-1} of MS222 for approximately 5 min was sufficient to anesthetize a 500 g adult rainbow trout for cannulation (Black et al., 1995). Once the fish is fully anesthetized, a 16 gauge needle is inserted through the fleshy part of the snout into the fish's buccal cavity. A 20 mm piece of heat-flared PE 160 tubing is inserted through the needle. The needle is retracted, allowing the flared end of the tubing to rest against the buccal lining, with the tubing protruding from the surface of the fish's snout. This tube provides an unrestrained passage through which the cannula will exit the buccal cavity (Figure 30.2).

The fish is then wedged on its back into a V-shaped holder and the lower jaw is retracted with a suspended hook or loop of wire, so that the mouth is held open. The cannula used for adult rainbow trout is made from tapered PE 50 tubing with a bubble located approximately 30–50 mm from the insertion tip (Black et al., 1995, modified from Soivio et al., 1975). The cannula is filled with heparinized saline (Table 30.1), fitted with a sharpened steel wire (previously described) and inserted at a shallow angle into the midline of the roof of the buccal cavity at the second pharyngeal arch (see Figure 30.1, dorsal aorta cannulation). A slight resistance will be felt as the tip of the wire pierces the vessel and blood should be observed filling the cannula. The wire is withdrawn and the cannula is pinched to halt the flow of blood, while the cannula's free end is threaded through the PE 160 exit tube to the fish's exterior. The bubble of the cannula should rest against the flared end of the tube to stabilize the cannula and prevent it from being pulled out of the vessel. Then the cannula is quickly flushed with heparinized saline (being careful not to inject more than 0.3 mL into the vessel) and plugged with a hedgehog quill or straight pin or heat sealed (Soivio et al., 1975). The cannula should be anchored at the first pharyngeal gill arch with a single 4-0 silk suture, which is wrapped twice around the cannula before it is knotted. The fish is then resuscitated by carefully forcing a stream of oxygenated water over its gills for several minutes. This basic technique may be employed to allow serial blood collections from transected (Black et al., 1995), restrained (Soivio et al., 1975) or free-swimming fish (Soivio et al., 1975; Kiessling et al., 1995) and has allowed repeated blood collections for several weeks.

Ventral aorta cannulation

Cannulation of the ventral aorta of adult rainbow trout is also performed on deeply anesthetized fish, without gill irrigation. The cannula is made from PE 50 tubing, bubbled at 2.5 and 5 cm from the cannula tip (Gamperl an Boutilier, unpublished – described by Currie and Tufts, 1993; Axelsson and Fritsche, 1994). Prior to cannulation a 16 gauge needle is used to form an exit hole in the trout's mouth near the juncture of the upper and lower jaw. A flared piece of PE 160 tubing is threaded through the needle, as previously described. The fish is positioned dorsal side up on a V-shaped wedge and the upper jaw is held open. As the ventral aorta lies just below cartilage located along the ventral midline of the floor of the buccal cavity, an insertion hole into the cartilage must first be made with a 20 gauge needle. The needle is positioned at the second gill arch and inserted at a

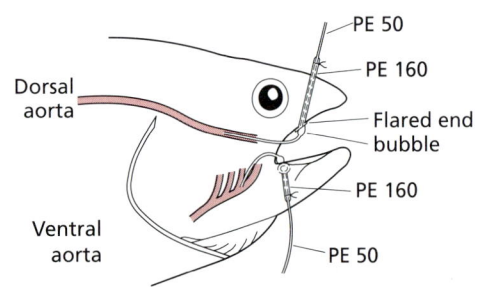

Figure 30.2 Diagram showing the position of cannulas inserted into the dorsal aorta and ventral aorta. (Modified from Axelsson and Fritsche, 1994.)

slightly downward angle until the cartilage is barely pierced (see Figure 30.1, ventral aorta cannulation). Once the cartilage is pierced, the needle is replaced with a wired cannula (with an unsharpened wire) and the wire is withdrawn, causing blood to flow into the cannula (Figure 30.2). The cannula is positioned for optimal blood flow, filled with heparinized saline, plugged and secured with a suture at the first bubble. The plugged end of the cannula is then passed to the outside of the buccal cavity through the PE 160 tube, so that the second bubble rests against the flared end of the exit tube (Gamperl an Boutilier, unpublished – described by Currie and Tufts, 1993; Axelsson and Fritsche, 1994). Clotting, cited as a potential problem with this technique (Axelsson and Fritsche, 1994), is remedied with frequent flushing of the cannula with heparinized saline. This cannula may be used with restrained and free-swimming fish (Currie and Tufts, 1993).

Caudal vein cannulation

Serial blood samples can also be collected from the caudal vein of large fish. The fish is anesthesized and placed on its side on a flat surface. The caudal vein is approached from the side of the fish just ventral to the lateral line at the level of the anal fin (see Figure 30.1, caudal vein cannulation). A thin-walled 18 gauge needle is inserted at this point, pointing slightly toward the head, until it penetrates the vein. The cannula, constructed of unaltered PE 50 tubing filled with heparinized saline, is inserted through the needle and into the vessel (Hickman, unpublished – described in Klonz and Smith, 1968). The cannula is secured to the fish's body with a suture (4-0 silk), which is wrapped twice around the tubing before knotting.

Cannulation of isolated vessels

If the vessel to be cannulated can be dissected and isolated from surrounding tissue (e.g. celiac or mesenteric artery), it can be cannulated directly through a hole cut in the vessel (see Figure 30.1, celiac artery; mesenteric artery). Upon isolation the vessel is clamped with a small bulldog clamp or jeweler's forceps above the cannula insertion point and occluded by a suture (4-0 silk) knotted below the insertion point. A small transverse cut is made in the vessel, through which a heparin-filled cannula is inserted and immediately tied in place with another loop of suture. The clamp is then removed, restoring blood flow to the cannulated vessel (Black, unpublished; Axelsson and Fritsche, 1994).

Urine collection

Studies of fish renal function and urinary excretion of salts, hormones, drugs, xenobiotics or their metabolites by fish require the collection of urine from restrained or free-swimming fish. Fish urine can be collected from living fish using urethral or urinary bladder catheters (Wood and Randall, 1973; Kakuta et al., 1986; Wood and Patrick, 1994; Black et al., 1995) or by collecting naturally-vented urine into an isolated chamber. Others describe a simple procedure for collecting urine from anesthetized or freshly killed fish without the use of a cannula, by palpating the fish's abdominal region (Rall and Burger, 1967; Rocha and Reis-Henriques, 1996).

Urinary bladder catheters

Urinary bladder catheters (sometimes referred to as cannulas) are inserted through the urinary papilla and into the urinary bladder. Because of the immense variability in the urogenital anatomy of different fish species, catheters must be custom-fitted. Before a catheter is designed, it is advised that the investigator dissect the urogenital region of a member of the species to be used. Information collected from the dissection will enable the investigator to custom-fit the catheter and avoid catheter insertion errors, which can result in the collection of undesirable substances, such as milt (seminal fluid and sperm), feces or other body fluids (Wood and Patrick, 1994).

In an early experiment in which the common carp (*Ciprinus carpio*) and goldfish (*Carassius auratus*) were catheterized, Smith (1929) used an L-shaped glass catheter sewn to the urinary papilla and attached to a rubber bag to collect urine from the urinary bladder. More recently most catheters are constructed from PE tubing, although some employ other materials attached to the tubing, such as the perforated end of an infant feeding tube. Other catheters are fabricated from several sizes of PE tubing bonded together with glue. Several designs of catheters custom fitted for adult rainbow trout and common carp are discussed in detail.

Most catheters are described for use with adult rainbow trout and are based on a similar design. In general, the insertion tip that will reside in the bladder must be smooth and the tubing sides may be perforated with several small holes, to avoid total occlusion by mucous plugs. The distal end (approximately 200 mm) cut from an infant feeding tube (e.g. $3\frac{1}{2} \times 12$ inch Argyle® Feeding Tube, Sherwood Medical, St. Louis, MO) is perforated and works well as the insertion tip. It can be easily attached to additional PE tubing to extend the catheter to the desired length (A. Oikari, personal communication; Black et al., 1995). If PE tubing is used to form the insertion tip, the end must be heat-smoothed and perforations can be made with a heated needle. Tubing sizes may vary from PE 60 to PE 190, depending on the fish size and the desired tightness of the catheter in the urinary papilla. The insertion tip is then attached to additional lengths of PE tubing, as desired.

Wood and Randall (1973) describe a catheter designed for use in free-swimming trout. It is constructed from 45 cm of PE 60 tubing threaded through a 10 to 15 mm PE 190 cuff (with slightly flared ends), which is secured with glue at approximately 10–15 mm from the smoothed and perforated tip of the PE 60 tubing. The PE 190 cuff blocks the flow of urine around the catheter and helps to anchor the catheter within the urinary papilla. Curtis and Wood (1991) flared the catheter tip to 1.5 × the original diameter instead of using the cuff. Another modification of this design was used to catheterize adult common carp (Kakuta et al., 1986). This catheter was constructed of PE tubing (0.3 to 1 mm ID). Heat-formed bubbles were placed 3 and 20 mm from the perforated tip and were used as internal and external anchors for the catheter.

Urinary bladder catheterization of restrained and free-swimming fish

Fish are lightly anesthetized with MS222 or another agent (Iwama and Ackerman, 1994) and placed ventral side up on a V-shaped block. As urinary bladder catheterization is often coupled with other surgical procedures, the gills are continuously bathed with flowing, aerated anesthetic (e.g. MS222) during this procedure. Using a syringe, the urinary catheter is completely filled with either physiological saline or deionized water. Wood and Patrick (1994) recommend using saline for marine fish and water for freshwater fish, although most others report using saline for freshwater fish (Kakuta et al., 1986; Scott and Liley, 1994; Black et al., 1995). Care must be taken not to leave bubbles in the catheter, as they will impede urine flow and are difficult to remove once the catheter is in place. The fluid-filled catheter is inserted into the urinary papilla and gently directed into the urinary bladder (see discussion of the importance of prior dissection of this area) (Figure 30.3A). If the catheter has added cuffs or bubbles, make sure that these are in place within the papilla. The catheter should then be tested to ensure urine flow by removing the syringe and observing the flow of urine from the catheter end, making sure no air bubbles are detected at the insertion end or along the length of the catheter. If urine flow and no air bubbles are detected the catheter should be injected with 0.5–1 mL of saline or water (Curtis and Wood, 1991; Wood and Patrick, 1994) and plugged until all surgical procedures are completed. The catheter is then secured to the urinary papilla or the anal fin with a 4.0 silk suture, wrapped twice around the catheter before knotting (Wood and

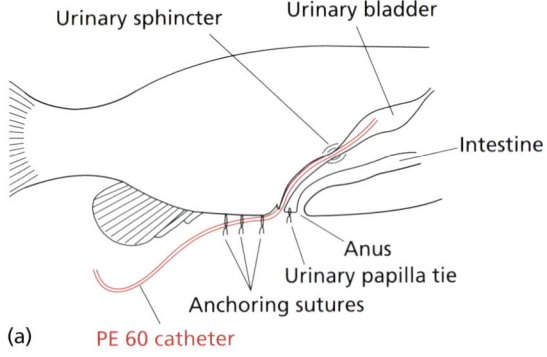

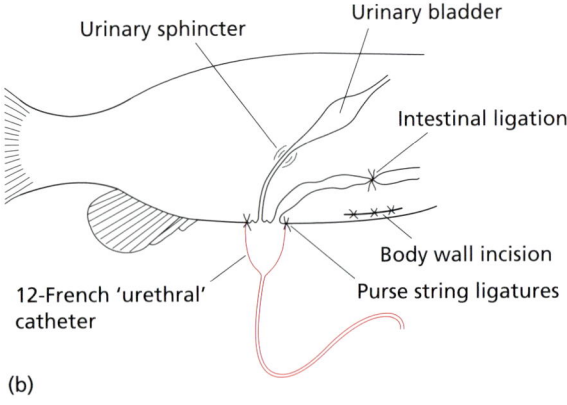

Figure 30.3 Catheters for collecting urine from fish. The internal urinary catheter (A) drains the urinary bladder and the external catheter (B) collects naturally-vented urine. (Modified from Wood and Patrick, 1994.)

Randall, 1973; Scott and Liley, 1994; Black *et al.*, 1995). If desired, other sutures may also be used to further anchor the catheter to the fish, but are necessary only if the fish will be free-swimming after catheterization.

A flexible collecting bag (e.g. balloon or condom) can be attached to the end of a well anchored short catheter to collect urine from free-swimming fish (Smith, 1929; Goldstein *et al.*, 1964; Wood and Patrick, 1994). Alternatively, the catheter of free-swimming fish may be plugged and the cannula drained periodically, to empty the bladder. Both methods have their drawbacks. Periodically draining the bag or plugged catheter requires extra handling and stress to the fish. Morover, the restricted urine flow of both methods may alter renal function (Wood and Patrick, 1994).

If fish are placed into a physiological chamber (Smith, 1929; McKim and Goeden, 1982; Black *et al.*, 1995) or restraining system (Soivio *et al.*, 1975) after catheterization, urine may be continuously collected into a tared container (Duff *et al.*, 1997) or collected at timed intervals using a fraction collector (Black *et al.*, 1995). For these applications, the open end of the catheter should be positioned in the collecting vessel or fraction collector at approximately 2–8 cm below the level of the water in the chamber. This level will ensure an adequate hydraulic head (1–2 mmHg) for draining the bladder (Duff *et al.*, 1997) without introducing air or creating excess suction, which may cause the bladder to collapse against the catheter (Wood and Patrick, 1994).

Laboratory diuresis, a temporary increase in urine production, may result from stress associated with handling, anesthesia or invasive surgical procedures. This effect may last for several hours or even days (Hunn and Willford, 1970). As a result of laboratory diuresis, ion or chemical concentrations measured in urine shortly after catheterizing may be diluted. Thus, it is recommended that urine collection for experimental measurements be delayed for 24 to 72 h after catheterization (Wood and Patrick, 1994).

Collecting normally vented urine

External catheter

In most fish urinary output and composition are known to be altered by an in-dwelling bladder catheter, primarily through elimination of bladder storage and modification of the urine (Wood and Patrick, 1994). To avoid this potential interference, yet maintain the ability to collect undiluted urine, Curtis and Wood (1991, 1992) designed an external catheter to sample naturally vented urine from rainbow trout. As the external catheter is fitted over the entire urogenital opening, fish must be starved for one week prior to the experiment and the posterior intestine must be sutured shut to prevent the passage of intestinal fluids or fecal casts into the catheter. Additionally, fish should not be catheterized during their breeding season to prevent contamination of urine by milt or other reproductive fluids. The external catheter is fabricated by cutting off the flared end of a 40 cm Bard all-purpose urethral catheter (size 12 French, elastic rubber; Davol Inc., Cranston, RI), forming a wide base that is sutured (2-0 silk) in a purse-string to the flesh around the urogenital opening (Figure 30.3B). The catheter is further adhered to the fish with cyanoacrylate cement and tested for leaks by filling with water. The fish is then resuscitated and placed in a holding chamber. The catheter is allowed to drain into a container placed ≤3 cm below the surface of the water (Wood and Patrick, 1994).

Post-bladder catheter

A post-bladder catheter was also designed by Wood *et al.* (1995) to collect naturally vented urine from toadfish (*Opsanus beta*). This catheter is fabricated from a short length (10 mm) of heat-smoothed PE10 attached to a longer length (40 cm) of PE50 tubing, attached to a syringe. The catheter is filled with saline and the PE10 end is inserted less than halfway into the urinary papilla. The catheter is secured within the papilla with a loop of 4-0 silk tied around it and the papilla. The free end of the catheter is further secured to the fish's body by threading through attachments made from latex strips glued to the fish with cyanoacrylate cement. Upon resuscitating the fish, naturally-vented urine is collected through the catheter into a vessel placed not more than 3 cm below the water level of the holding chamber (Wood *et al.*, 1995).

Divided physiological chamber

Fish physiological chambers are used to isolate multiple excretion routes in fish for physiological and toxicokinetic studies (e.g. McKim and Goeden, 1982; Black *et al.*, 1995; McKim *et al.*, 1996). Chamber designs range from simple two-sectioned boxes with inlet and outlet ports for separating gill and urinary elimination (e.g. Smith, 1929) to more complex chambers designed for use with cannulated and catheterized fish to separate multiple elimination routes (e.g. McKim and Goeden, 1982; Black *et al.*, 1995; McKim *et al.*, 1996).

The simplest physiological chamber is a box or cylinder, with a flexible membrane that divides it into two chambers (e.g. Smith, 1929). An oval, fish-sized hole is cut into the membrane and a fish is placed into the chamber so that the membrane fits snugly around its body, just posterior to the gills. The membrane provides separation between substances excreted by the gills and those excreted posterior to the barrier by the urinary bladder and/or intestine.

More recent adaptations of the original wooden chamber (Smith, 1929) are fabricated from clear or darkened plexiglass and the dividing membrane can be secured to the fish with sutures or cyanoacrylate adhesive (Figure 30.4) (Post *et al.*, 1965; Wood *et al.*, 1995). Water levels in each chamber are balanced with standpipes and the water is aerated and, in some cases, recirculated (Post *et al.*, 1965; Black *et al.* 1995). Prior to experimentation the barrier must be tested with a dye to ensure the absence of leaks. Barrier integrity can be further aided by rendering the fish inactive through **spinal transection** (McKim and Goeden, 1982; Black *et al.*, 1995), continuous treatment with anesthesia (Smith, 1929) or by using restraining devices to hold the fish in place (Wood *et al.*, 1995). To prevent fecal contamination of urinary output, the posterior intestine can be **ligated** in starved fish (Wood and Patrick, 1994).

Physiological chambers are advantageous for collecting urine from fish that are too small to catheterize and are also used as a relatively simple method to screen renal excretion. Disadvantages of using a physiological chamber (versus catheters) for collecting urine include: dilution of urine by the collection chamber water; potential contamination of urinary output with fecal material, dermal products or bacterial colonies; leakage of the barrier; stress induced by chamber confinement or transection; and if fish are anesthetized while in the chamber, altered urine flow and composition resulting from prolonged anesthesia (Smith, 1929; Wood and Patrick, 1994; Black, unpublished).

Collection of fecal materials

Many chemicals with high molecular masses (>300 Da) are mostly excreted by the gastrointestinal route and may be measured in feces following elimination (Parkinson, 1996). Fish feces have been monitored to measure the availability and digestibility of various components (vitamins, minerals, amino acids, etc.) of fish diets (Satoh *et al.*, 1992; Hajen *et al.*, 1993; Anderson *et al.*, 1995; Sugiura *et al.*, 1998). Inert markers not absorbed by the digestive tract, such as yttrium or chromium oxide (Y_2O_3, Cr_2O_3), may be

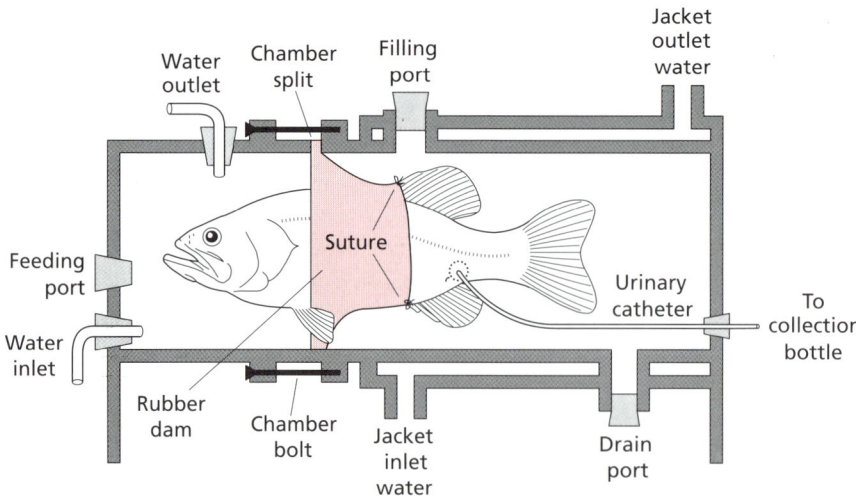

Figure 30.4 Diagram of a physiological chamber. The posterior section is fitted with a drain for collecting fecal materials from fish fitted with a urinary catheter. However, naturally-vented urine could also be collected if the fish has been starved and its intestine ligated. (Modified from Post *et al.*, 1965.)

co-administered for calculation of absorption efficiencies of diet components. Feces may also be collected to study the elimination of endogenous and xenobiotic substances from fish (Black et al., 1995; Wood et al., 1995). Feces may be collected from individuals or pooled from multiple fish, depending on the experimental goal. The most commonly used methods will be discussed in detail.

Collection of gravity-settled feces

Manual collection

The easiest method of feces collection is by manually siphoning, pipetting or netting the particulate fecal material from the bottom of a tank. These manual collection methods are time-consuming and labor-intensive and must be done on a daily basis, to reduce leaching of materials from feces and potential bacterial contamination and degradation of fecal components (Anderson et al., 1995).

Guelph system

The Guelph system combines a stationary drainpipe with a gravity settling column to enable periodic collection of feces with no fish disturbance. Originally developed at the University of Guelph (Guelph, Ontario, Canada), it is most often used in studies quantifying diet assimilation efficiencies in fish. The system is composed of a fish tank with a slanted or conical bottom that drains into a horizontal collection tube connected to a vertical settling column (Cho et al., 1982; Satoh et al., 1992; Hajen et al., 1993). Two versions of the original Guelph system are presented in Figure 30.5 (Cho et al., 1982; Hajen et al., 1993). As water and fecal material from the tank pass from the collection tube into the settling column, fecal material settles to the conical bottom and can be removed via a collection port at the bottom of the column. Water flows out of the outlet at the top of the column. Flow rates through the system (usually 2 to 8 L min^{-1}) are adjusted to maximize settling conditions in the column, but not in the collecting tube (Satoh et al., 1992; Hajen et al., 1993). The major advantage of the Guelph system is that it is a quick and simple method for collecting feces from many fish at once. The primary disadvantage is that sampling is still manually controlled, although it should be feasible to automate

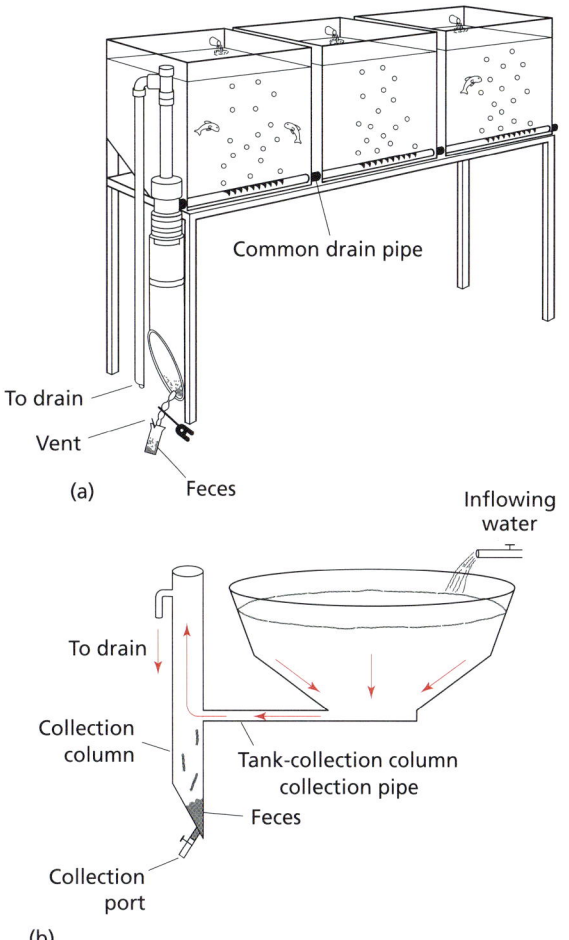

Figure 30.5 Two modifications of the Guelph system for collecting feces from multiple fish. (Modified from Cho et al., 1982; Hajen et al., 1993.)

the system with computer-controlled solenoid valves on inlet and outlet pipes.

TUF column system

Similar to the Guelph system, the TUF column system also employs a settling column (Satoh et al., 1992). It is composed of a 5 × 30 cm stoppered tube with a 2.5 cm inlet siphon tube that extends from near the bottom of the tank to the column, where it enters through the stopper (Figure 30.6). A smaller exit tube drains water from the top of the settling column and when positioned to siphon continuously, the rate of water movement through the column is determined by the incoming flow rate to the fish tank. Satoh et al. (1992) used an incoming flow rate of 0.6 L min^{-1}. Experiments comparing protein and lipid digestibilities of rainbow trout found the TUF and Guelph systems to be reliable and comparable fecal collection systems (Satoh et al., 1992).

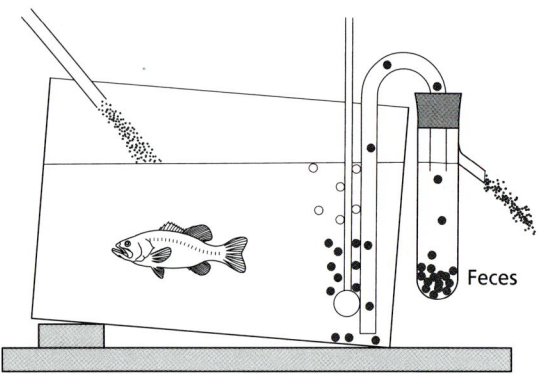

Figure 30.6 The TUF column for collecting feces from multiple fish. (Modified from Satoh *et al.*, 1992.)

Divided physiological chambers

Feces may also be collected into the posterior chamber of a divided physiological chamber if the urinary bladder has been catheterized (Post *et al.*, 1965). Water containing fecal material is easily sampled from the fecal collection chamber through a drain port, without interrupting the experiment or disturbing the fish (Figure 30.4). The major disadvantages of this method are similar to those of manual collection of feces from a tank: the possibility of bacterial contamination and degradation coupled with the difficulties of sampling the particulate feces from chamber C. The major advantage is that the feces of a single fish is collected in a relatively small amount of water (approximately 1 L), allowing easy quantitation of fecal excretion products, especially if the material to be assayed is radiolabeled.

Manual stripping

Anderson and colleagues (1995) describe a manual stripping method to collect feces from Atlantic salmon (*Salmo salar*). Fish are lightly anesthetized with 2-phenoxyethanol. Feces are stripped by applying gentle pressure to the ventral abdominal area, beginning just posterior to the pelvic fins and moving posteriorly to the anal opening. Feces are collected directly from the anus into an open vessel. This method is especially useful when collections must be made from large numbers of fish, and fecal dilution in tank water is not desired.

Fecal catheters

Two fecal catheter designs were described in the literature. Black and colleagues (1995) constructed a fecal catheter for 200–500 g rainbow trout from flexible latex tubing of the following dimensions: 5 mm ID, 6 mm OD by 100 cm in length. The cannula is marked at 5 cm intervals with a black permanent marker. (Smaller intervals may also be marked, if desired.) Marks should extend completely around the catheter. The tip of the catheter is cut at an angle and the resulting point rounded to prevent tearing of the intestinal lining as it is inserted. An anchoring collar, made of a 5 mm piece of 5.5–6 mm ID tubing, is slipped over the catheter and positioned 5 cm from the insertion end.

The fish is anesthetized with MS222 and placed ventral side up on a V-shaped support. The fish gills should be continuously bathed in a dilute solution of MS222 (20–30 mg L^{-1}) while catheters are installed. If both urinary and fecal catheters are to be installed, the urinary catheter should be installed first, then followed by the fecal catheter. In preparation for the fecal catheter, a purse-string suture (4-0 silk) is stitched around the anus and left unknotted. The fecal catheter is inserted so that the anchoring collar is positioned just outside the anus. (Note that the fecal catheter is not filled with any liquid.) The purse-string suture is tightened around the catheter, knotted and then secured to the anchoring collar with a stitch, taking care to pierce only the collar. The fish is then placed in a physiological chamber, with the fecal catheter exiting the chamber and emptying into a beaker or flask. Fish are allowed to recover overnight from surgical manipulations.

Fecal samples may be collected over the entire exposure duration or collections may occur over timed intervals. Samples are 'timed' by noting the length of passage of the feces in the catheter at specified times. These measurements are made using the 5 cm increment lines. Once the exposure is concluded, the catheter is removed and both ends are clamped with hemostats. The net distance travelled by the feces in the catheter during each timed sampling interval is calculated. Sampling begins at the external end of the catheter. The first sampling distance (time 0 to the first sample time) is clamped off with hemostats and the catheter is cut. Feces are expelled from the cut section into a sampling container. Sampling continues until all intervals have been cut from the catheter (Black *et al.*, 1995). Sampled feces may then be digested or extracted for further analysis.

This method is especially useful for collecting timed samples for calculation of elimination rates for excreted substances. Using this method Black and colleagues (1995) calculated fecal excretion rates for pentachlorophenol (PCP) over 24 and 36 h collection periods. Disadvantages of this method include the manual observations needed for timed collections and sample 'bleeding' if feces are liquefied. However, if the fecal material consists of solid material, little or no sample 'bleeding' is expected between adjacent collecting locations. This was demonstrated in excretion experiments using i.p. injected radiolabeled PCP (Black, unpublished). Fecal samples measured during the overnight acclimation period prior to PCP injection remained uncontaminated with the radiolabel, even when adjacent samples contained high levels of radioactivity.

Wood and colleagues (1995) constructed a catheter for collecting 'rectal fluid' from adult toadfish from a 40 cm Bard urethral catheter. Prior to insertion the catheter was fitted with a flared rigid plastic insertion piece (2.5 mm ID × 5 mm OD (flare) × 10 mm long) and filled with a 0.9% saline solution. The rigid insertion piece was gently placed into the anus approximately 5 mm and secured with a purse-string suture (as described above). The catheter was anchored with latex strips sutured to the fish body. Because the catheter was filled with fluid (saline) the collecting vessel must be placed approximately 3 cm below the water surface to facilitate collection. Rectal fluid collected with this catheter was successfully analyzed for nitrogen excretion products (urea and ammonia). Combined urinary and rectal fluid N-excretion values were similar to concurrent measurements made from free-swimming fish. The primary advantage of catheterization was the ability to separate urinary and fecal contributions to N-excretion. Furthermore, the intermittent nature of urinary and fecal elimination could only be observed upon catheterization (Wood et al., 1995).

Collection of gametes

Fish gametes (eggs and sperm) are collected routinely by hatcheries for external fertilization techniques. Gametes are also collected for aquaculture and basic research applications, ranging from optimization of gamete storage conditions (e.g. Bart et al., 1995; Rana, 1995a,b; Weismann et al., 1995) to measuring biochemical parameters of seminal plasma and fluid (Viljoen and Van Vuren, 1992; Rana, 1995a), hormone manipulation experiments (Breton et al., 1995; Goren et al., 1995); and studies of reproductive behavior (Olsén and Liley, 1993; Scott et al., 1994).

Egg stripping

The easiest method for collecting unfertilized eggs from ripe female fish is by manual stripping. A slit is cut in the female's abdomen with a scalpel, the abdomen is gently pressed, and the eggs are expressed from the abdominal slit into a suitable container. If fertilized eggs are desired, milt may be hand stripped from a spermiating male (see section below) directly on to the eggs. Milt and eggs are gently mixed with a small quantity of spring water, rinsed with flowing water and placed in a suitable container for hardening (Pillay, 1993).

Fertilized egg collection

Fertilized eggs can be collected from pond cultured fish (e.g. channel catfish, carp) in spawning containers, cylindrical breeding refugia placed on the bottom of the broodstock pond or enclosure during the spawning season. Old milk cans or any cylindrical container with a small opening (20 to 30 cm ID) are suitable for use as spawning containers and are placed in the pond with the open end oriented toward the pond's center (Stickney, 1994). Female fish deposit gelatinous egg masses into these containers and the eggs are subsequently fertilized, guarded and oxygenated by a single male fish. Containers are checked daily and removed when eggs are detected. After collection, fertilized eggs are usually disinfected in a 10% **betadine** solution, rinsed, and hatched in a specialized trough in well-oxygenated water at 25–28°C (Johnson, 1993).

Fish can also be induced to spawn in lab aquaria, allowing the process to be more easily controlled and monitored. During the spawning season, mature female fish are injected with a hormone (e.g. fish pituitary extract, partially purified fish gonadotropin, human chorionic gonadotropin or luteinizing hormone) to induce ovulation (Pillay, 1993; Stickney, 1994). Hormone-injected females are paired with a single male fish and placed in an aquarium under flow-through conditions. Spawning should occur within 24 h of injection. If spawning does not occur, female fish may be reinjected with hormone. Brood fish should be removed immediately following egg fertilization, allowing eggs to be removed (Stickney, 1994).

TABLE 30.4: Selected milt cryopreservation methods

Fish species	Diluent	Cryoprotectant	Reference
Rainbow trout	0.6 M sucrose	Me$_2$SO	Rana, 1995b
Rainbow trout	0.6 M sucrose	20% DMSO	Steinberg et al., 1995
Blue catfish	Hanks soln.	DMSO	Bart et al., 1995
Ictaluridae sp.	Hanks soln.	5% MeOH	Tiersch, 1995
Cyprinid sp.	Ca^{2+}-free Hanks	10% DMSO	Tiersch, 1995
Sparidae sp.	Hanks soln.	10% DMSO	Tiersch, 1995

Milt collection by stripping

Milt (seminal fluid and sperm) can be collected from male fish by stripping, achieved by applying gentle pressure on the fish's abdomen until milt is expressed. Stripping is an easy technique that is achieved without anesthesia or specialized collection materials. However, urine contamination of milt is nearly unavoidable with this method and can comprise up to 80% of the volume collected (Rana, 1995a).

Milt collection by syringe and catheter

Alternatively, milt may be collected from ripe male fish using a 2.5 ml syringe (with the needle removed) inserted into the genital opening (Viljoen and Van Vuren, 1992). Gentle pressure is applied to the abdomen while collecting the milt sample and care should be taken to prevent contact of the syringe tip with urine, feces or mucus. Using this method small amounts of uncontaminated milt (0.2 mL) may be collected without anesthesia (Viljoen and Van Vuren, 1992). Contamination can also be avoided by collecting milt with a catheter temporarily inserted into the genital opening, while applying gentle abdominal pressure (Rana, 1995a).

Sample preservation

Frequently milt is collected for fertilization experiments, and must be preserved for future use. **Cryopreservation** appears to be the most used method, although the chemical additives used in preserving sperm samples are still under investigation. Table 30.4 provides a summary of several cryopreservation methods for sperm. No long-term storage methods are currently available for eggs or embryos (Bromage, 1995).

References

Anderson, J.S., Lall, S.P., Anderson, D.M. and McNiven, M.A. (1995). *Aquaculture* **138**, 291–301.

Axelsson, M. and Fritsche, R. (1994). In *Biochemistry and Molecular Biology of Fishes*, vol. 3 (eds P.W. Hochachka and T.P. Mommsen), pp. 17–36. Elsevier, Amsterdam.

Barron, M.G., Tarr, B.D. and Hayton, W.L. (1987). *J. Fish Biol.* **31**, 735–744.

Bart, A.N., Wolfe, D.F. and Dunham, R.A. (1995). In *Proceedings of the Fifth International Symposium of the Reproductive Physiology of Fish* (eds F.W. Goetz and P. Thomas), p. 111. Fish Symposium 95, Austin, Texas.

Black, M.C., Björkroth, K. and Oikari, A. (1995). In *Environmental Toxicology and Risk Assessment – Third Volume, ASTM STP 1218* (eds J.S. Hughes, G.R. Biddinger and E. Mones), pp. 351–364. American Society for Testing and Materials, Philadelphia.

Breton, B., Roelants, I., Mikolajczyk, T., Epler, P. and Ollevier, F. (1995). In *Proceedings of the Fifth International Symposium of the Reproductive Physiology of Fish* (eds F.W. Goetz and P. Thomas), pp. 102–104. Fish Symposium 95, Austin, Texas.

Bromage, N. (1995). In *Broodstock Management and Egg and Larval Quality* (eds N.R. Bromage and R.J. Roberts), pp. 1–24. Blackwell, Oxford.

Cho, C.Y., Slinger, S.J. and Bayley, H.S. (1982). *Comp. Biochem. Physiol.* **73B**, 25–41.

Cockell, K.A., Hilton, J.W. and Bettger, W.J. (1992). *Comp. Biochem. Physiol.* **103C**, 453–458.

Conte, F.P., Wagner, H.H. and Harris, T.O. (1964). *Am. J. Physiol.* **205**, 533–540.

Currie, S. and Tufts, B.L. (1993). *Fish Physiol. Biochem.* **12**, 183–192.

Curtis, B.J. and Wood, C.M. (1991). *J. Exp. Biol.* **155**, 567–583.

Curtis, B.J. and Wood, C.M. (1992). *J. Exp. Biol.* **173**, 181–203.

Duff, D.W., Conklin, D.J. and Olson, K.R. (1997). *J. Exp. Zool.* **278**, 215–220.

Garey, W.F. (1969). *J. Appl. Physiol.* **27**, 756–757.

Gilmour, K.M., Randall, D.J. and Perry, S.F. (1994). *Respir. Physiol.* **96**, 259–272.

Goldstein, L., Forster, R.P. and Fanelli, G.M., Jr. (1964). *Comp. Biochem. Physiol.* **12**, 489–499.

Goren, A., Gustafson, H. and Doering, D. (1995). In *Proceedings of the Fifth International Symposium of the Reproductive Physiology of Fish* (eds F.W. Goetz and P. Thomas), pp. 99–101. Fish Symposium 95, Austin, Texas.

Hajen, W.E., Beames, R.M., Higgs, D.A. and Dosanjh, B.S. (1993). *Aquaculture* **112**, 321–332.

Hesser, E.F. (1960). *Prog. Fish Cult.* **22**, 164–171.

Hunn, J.B. and Willford, W.A. (1970). *Comp. Biochem. Physiol.* **33**, 805–812.

Iwama, G.K. and Ackerman, P.A. (1994). In *Biochemistry and Molecular Biology of Fishes*, vol. 3 (eds P.W. Hochachka and T.P. Mommsen), pp. 1–15. Elsevier, Amsterdam.

Johnson, M. (1993). In *Aquaculture for Veterinarians: Fish Husbandry and Culture* (ed. L. Brown), pp. 249–270. Pergamon Press, Oxford.

Kakuta, I., Nanba, K., Uematsu, K. and Murachi, K. (1986). *Bull. Jap. Soc. Sci. Fish.* **52**, 2079–2089.

Kiessling, A., Dosanjh, B., Higgs, B., Deacon, G. and Rowshandeli, N. (1995). *Aquacult. Nutr.* **1**, 43–50.

Klonz, G.W. and Smith, L.S. (1968). In *Methods of Animal Experimentation* (ed. W.I. Gay), pp. 323–385. Academic Press, New York.

Lebovits, A. (1966). *Mod. Plastics* **43**, 139–146.

McKim, J.M. and Goeden, H.M. (1982). *Comp. Biochem. Physiol.* **72C**, 65–74.

McKim, J.M., Nichols, J.W., Lien, G.J., Hoffman, A.D., Gallinat, C.A. and Stokes, G.N. (1996). *Fund. Appl. Toxicol.* **31**, 218–228.

Olsén, K.H. and Liley, N.R. (1993). *Gen. Comp. Endocrinol.* **89**, 107–118.

Opdyke, D.F., Carroll, R.G. and Keller, N.E. (1982). *Am. J. Physiol.* **242**, R306–R310.

Parkinson, A. (1996). In *Casarett and Doulls's Toxicology: The Basic Science of Poisons* (eds C.D. Klaassen, M.O. Amdur, J. Doull), pp. 113–186. McGraw-Hill, New York.

Pillay, T.V.R. (1993). In *Aquaculture Principles and Practices*, pp. 156–173. Fishing News Books, Blackwell, Oxford.

Post, G., Shanks, W.E. and Smith, R.R. (1965). *Prog. Fish Cult.* **27**, 108–111.

Rall, D.P. and Burger, J.W. (1967). *Am. J. Physiol.* **212**, 354–356.

Rana, K. (1995a). In *Broodstock Management and Egg and Larval Quality* (eds N.R. Bromage and R.J.R. Roberts), pp. 53–75. Oxford, Boston.

Rana, K. (1995b). In *Proceedings of the Fifth International Symposium of the Reproductive Physiology of Fish* (eds F.W. Goetz and P. Thomas), pp. 85–89. Fish Symposium 95, Austin, Texas.

Rocha, M.J. and Reis-Henriques, M.A. (1996). *Comp. Biochem. Physiol.* **115C**, 257–264.

Sachs, C., Rabouine, P., Chaneac, M., Kindermans, C. and Deechaux, M. (1989). *Scand. J. Clin. Invest.* **49**, 647–651.

Satoh, S., Cho, C.Y. and Watanabe, T. (1992). *Nippon Suisan Gakkaishi* **58**, 1123–1127.

Scott, A.P. and Liley, N.R. (1994). *J. Fish Biol.* **44**, 117–129.

Scott, A.P., Liley, N.R. and Vermeirssen, E.L.M. (1994). *J. Fish Biol.* **44**, 131–147.

Smith, H.W. (1929). *J. Biol. Chem.* **81**, 727–742.

Smith, L.S. and Bell, G.R. (1964). *J. Fish. Res. Bd. Canada* **21**, 711–717.

Soivio, A., Nynolm, K. and Westman, K. (1975). *J. Exp. Biol.* **62**, 207–217.

Steinberg, H., Hedder, H., Baulain, R. and Holtz, W. (1995). In *Proceedings of the Fifth International Symposium of the Reproductive Physiology of Fish* (eds F.W. Goetz and P. Thomas), p. 146. Fish Symposium 95, Austin, Texas.

Stickney, R.R. (1994). In *Principles of Aquaculture*, pp. 352–414. John Wiley, New York.

Sugiura, S.H., Dong, F.M. and Hardy, R.W. (1998). *Aquaculture* **160**, 283–303.

Thomas, S. and Le Ruz, H. (1982). *J. Comp. Physiol.* **48**, 123–130.

Tiersch, T. (1995). In *Proceedings of the Fifth International Symposium of the Reproductive Physiology of Fish* (eds F.W. Goetz and P. Thomas), p. 147. Fish Symposium 95, Austin, Texas.

Viljoen, B.C.S. and Van Vuren, J.H.J. (1992). *Comp. Biochem. Physiol.* **103A**, 253–257.

Weismann, T., Lahnsteiner, F. and Patzner, R.A. (1995). In *Proceedings of the Fifth International Symposium of the Reproductive Physiology of Fish* (eds F.W. Goetz and P. Thomas), pp. 90–92. Fish Symposium 95, Austin, Texas.

Wood, C.M. and Patrick, M.L. (1994). In *Biochemistry and Molecular Biology of Fishes*, vol. 3 (eds P.W. Hochachka and T.P. Mommsen), pp. 127–143. Elsevier, Amsterdam.

Wood, C.M. and Randall, D.J. (1973). *J. Comp. Physiol.* **82**, 257–276.

Wood, C.M., Hopkins, T.E., Hogstrand, C. and Walsh, P.J. (1995). *J. Exp. Biol.* **198**, 1729–1741.

CHAPTER 31

Routes of Administration for Chemical Agents

Marsha C Black
College of Agricultural and Environmental Sciences,
The University of Georgia, Athens, Georgia, USA

Introduction

Choice of an appropriate route of administration for chemical agents is a crucially important aspect of the experimental design of fish exposures. As with mammals, use of different administration routes in exposures with fish may yield different peak concentrations, bioavailabilities, biotransformation pathways and half-lives of the administered chemical. Some exposure routes are more similar to natural, environmental exposures than others and are better choices for use in ecological research. Alternatively, other routes of administration may be more suited for pharmacokinetic or physiological studies. This chapter details many approaches for administering chemical agents to fish and describes their potential research applications.

Water-borne exposures

Uptake of chemicals by fish gills

The most prevalent route of exposure of fish for chemical agents is via the gill. Fish gills have an enormous surface area, approximately 50% of the entire surface area of the fish (Laurén, 1991). Gill secondary lamellae, flattened ridges protruding perpendicularly from the primary lamellae, provide an effective and extensive surface for gas exchange. The epithelium of the secondary lamellae is typically two cells thick, occasionally studded with chloride and mucous cells. It encloses a thin basement membrane covering the pillar cells, which support the inner vascular spaces of the secondary lamellae (Heath, 1995; Gilmour, 1998). Thus the diffusion distance for respiratory gases (O_2, CO_2) and other substances entering or exiting the gill is quite small – ranging from 1 to 20 µm, depending on the fish species (Hughes, 1979; Gilmour, 1998). Moreover, countercurrent exchange provided by opposing flows of blood within the gill and water pumped over lamellar surfaces, helps maintain a favorable concentration gradient for gas diffusion (Piiper and Scheid, 1984).

Although designed to facilitate diffusion of respiratory gases, fish gills also provide routes for other molecules to be accumulated by fish. Small hydrophilic molecules (e.g. NH_3, CO_2 and urea) can pass through small aqueous pores or gaps between cells in the gill epithelium (reviewed in Walsh, 1998). Larger neutral hydrophobic molecules, including many drugs and toxic organic chemicals, readily diffuse across the gill epithelium into the vascular space. Diffusion or uptake efficiency of these chemicals by the gills depends primarily on their hydrophobicity and molecular size. Chemicals having a moderately high hydrophobicity (log K_{OW} ranging from 3 to 6) (McKim et al., 1985) and a molecular cross-section of less than 9.5 Å (Opperhuizen et al., 1985) are efficiently accumulated by diffusion across gill epithelia. In addition, free metal ions can bind to negatively charged sites on fish gills. Once bound to the gill epithelium many metals use existing ion transfer mechanisms, such as calcium channels or protein-mediated endocytosis for entry into the gill (reviewed in Simkiss, 1996).

Laboratory exposures

Numerous methods exist for exposing fish to water-borne chemicals. Exposure chambers can be simple containers such as beakers or fish tanks, using solvent-rinsed glass containers for organic chemical exposures and acid-washed (2% nitric acid) plastic containers for metal exposures. Fish are usually exposed to water-borne chemicals under one of the following possible scenarios: (i) static (no exposure water change); (ii) static/renewal (periodic change of exposure water); and (iii) flow-through (continuous flow of exposure water through the tank) (Rand and Petrocelli, 1985). Static systems are appropriate only for short exposures (less than 48 h), while static/renewal systems can be used for juvenile fish exposure durations of up to 8 days (USEPA, 1993; ASTM

1998). Flow-through systems are the best system to use to eliminate possible effects due to metabolic waste accumulation and to ensure a constant exposure concentration, especially in exposures lasting longer than 96 h. However, flow-through systems require larger volumes of exposure water, additional equipment (e.g. pumps and/or proportional diluters) and generate more waste for disposal than other methods (Rand and Petrocelli, 1985).

Chemicals used in laboratory exposures can be dissolved directly in water or may be solubilized with the aid of a co-solvent or vehicle, such as triethylene glycol, ethanol, methanol or acetone. The use of surfactants (e.g. TWEEN 80) or dimethyl sulfoxide (DMSO) is not recommended as these chemicals may alter fish gill membrane properties. If a vehicle is used the same amount should be added to all test vessels and the amount added should not exceed 0.05% (0.5 mL L^{-1}) of the final test volume. At this concentration the vehicle should not alter the toxicity or chemical form of the agent that is solubilized or produce any toxicity itself. To verify this, a vehicle control should be included in the experimental design for comparison with the negative control water (ASTM, 1998).

Specialized exposure chambers that provide constant, gentle agitation and aeration are required for exposing fish eggs to water-borne chemicals. Diamond and colleagues (1995) describe a chamber for simultaneously hatching and exposing fathead minnow eggs to water-borne toxicants (Figure 31.1). Chambers are constructed from a 1 L beaker fitted with an aerated egg incubation chamber, which will incubate up to 30 eggs. Chambers can be suspended in

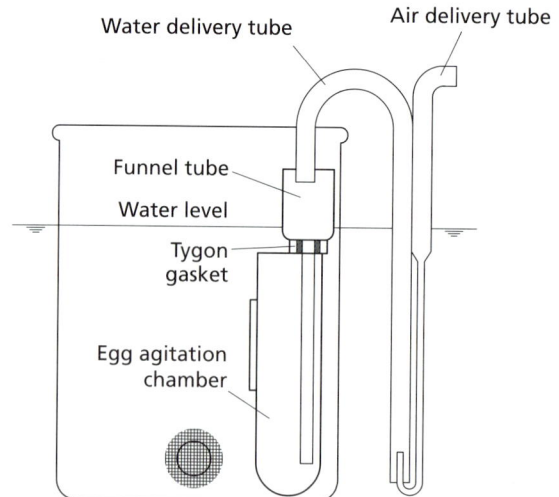

Figure 31.1 Exposure/hatching chamber for fish eggs. (Modified from Diamond et al., 1995.)

aquaria or modified for flow-through exposures without submersion. With suitable supports and battery powered pumps these chambers may also be adapted for *in-situ* field exposures.

If smaller hatch/exposure chambers are desired, Helmstetter *et al.* (1996) describe a chamber fabricated from glass scintillation vials. Each vial is fitted with a glass Pasteur pipette through which purified compressed air is bubbled to maintain >4 mg L^{-1} dissolved oxygen concentrations in 20 mL of solution. The solution requires no renewal over the 21 day hatching period, as hatch time was found to increase with daily water changes. Although the scintillation vial chambers were designed to incubate topically-exposed medaka eggs, this chamber could also be used for exposure to water-borne exposures, possibly with the addition of a limited number of solution changes.

Physiological chambers (Figure 31.2) can be used to separate routes of uptake (e.g. gill vs. skin) or when it is desirable to measure compound uptake in the absence of elimination from renal or fecal routes (McKim and Goeden, 1982; Black and McCarthy, 1988, 1990). Fish placed in these chambers may be transected, anesthetized or restrained to restrict their movements while in the chamber. A latex membrane sewn to the fish's mouth and attached to the sides of the chamber effectively separates incoming water (chamber A) from water that has passed over the fish gills (chamber B). By measuring the chemical concentrations in water collected from chambers A and B, the gill extraction efficiency for the chemical can be calculated. Compound dose (mg kg^{-1} day^{-1}) can be easily calculated from measurements of exposure concentration and extraction efficiency coupled with physiological measurements of ventilation volumes and rates by the fish, simultaneously measured with compound exposure in the physiological chamber.

These extraction efficiency and dose measurements have been most useful in mechanistic studies, including experiments detecting the influence of hypoxia, sorption, gill tissue damage and temperature change on contaminant uptake by fish gills (McKim and Goeden, 1982; Black and McCarthy, 1988, 1990; Black *et al.*, 1991). Gill extraction efficiencies coupled with physiological measurements have also been used to detect the mechanism of action for toxicants (McKim *et al.*, 1987a,b) and to parameterize physiologically-based pharmacokinetic models for toxicants (McKim and Nichols, 1992).

Field exposures

Many types of *in-situ* cages are available for fish exposure to contaminants in the natural environment. For adult or juvenile fish, cages are usually constructed of aquaculture mesh stretched over a cylindrical or rectangular frame fabricated from metal or PVC pipe (McKee *et al.*, 1989; Bugley and Shepard, 1991; Budhabhatti and Maughan, 1994). Cages may be attached to flotation devices, allowing only contact with water-borne contaminants (e.g. Webster *et al.*, 1992, 1993). Martin and Black (1995) describe a cylindrical cage design that can be easily modified for water-only or sediment exposures by altering the position of the cage floor (Figure 31.3). The cage floor can be

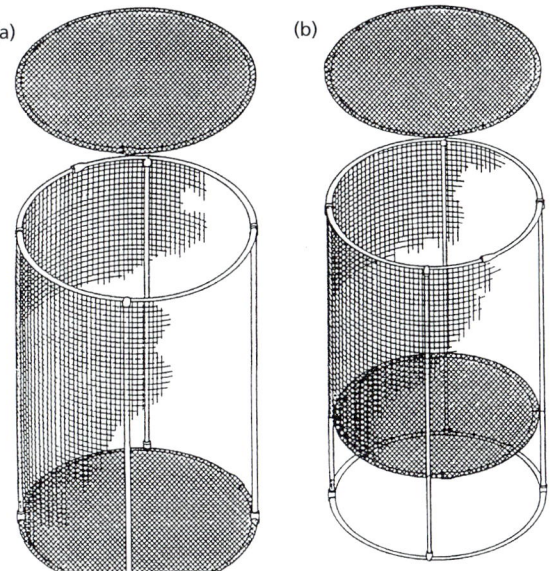

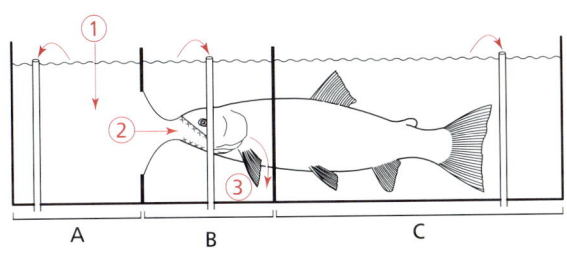

Figure 31.2 Diagram of a fish physiological chamber. (1) Exposure water enters chamber A; (2) water is pumped over the gills; and (3) water expelled into chamber B. (Modified from Black and McCarthy, 1988.)

Figure 31.3 *In-situ* exposure cages for adult fish constructed from PVC pipe and plastic mesh. Cage A provides exposure to water and sediment, while cage B provides exposure only to water. (Modified from Martin and Black, 1995.)

directly embedded in the sediment or suspended 18 inches (45 cm) above a support frame embedded in the sediments.

Sasson-Brickson and Burton (1991) designed an exposure chamber for *in-situ* exposure of *Ceriodaphnia dubia* neonates, which could be easily adapted for use with larval fish (Figure 31.4). The cylindrical chambers are constructed of clear acrylic pipe with a removable lid with mesh-covered inlet and outlet ports and a removable mesh base. The chamber is designed to be placed in a flowing water environment, as site water circulates through the chamber by entering and exiting the lid ports. The fine mesh (149 μm) used for *C. dubia* experiments could be replaced with a more coarse mesh for larval fish exposures to increase water circulation and contact with sediment particles.

Oral administration

Routine feeding with impregnated food

Although used less frequently in research applications with fish, dietary administration is often the method chosen for administration of therapeutic agents in aquaculture. There are several benefits to administering drugs or chemicals with food. The ease of applying the drug to the food and the ability to treat many fish at one time make this an easy method to use, if exact doses are not needed. Chemical agents can be mixed with a solvent (e.g. ethanol) and sprayed on pelleted fish chow (e.g. Fagerland and McBride, 1978). The chemical may also be applied to the chow in a solution of

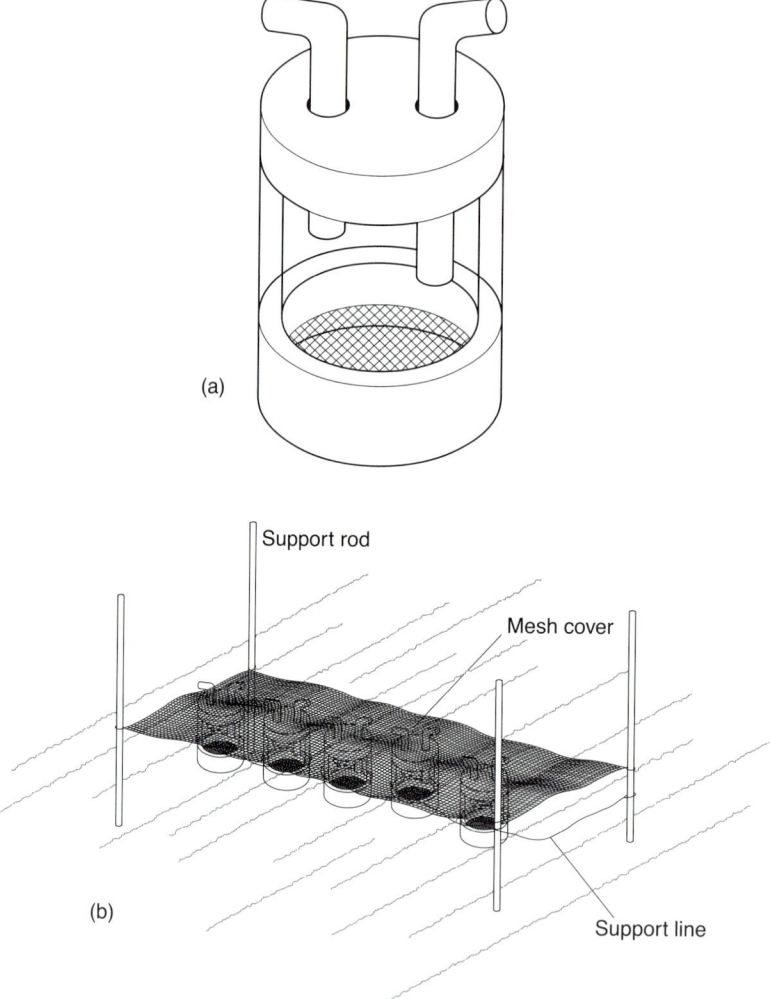

Figure 31.4 (a) *In-situ* exposure cage for larval fish constructed from acrylic pipe and fine mesh (149 μm). (b) Diagram showing *in-situ* placement of exposure cages in a stream. (Modified from Sasson-Brickson and Burton, 1991.)

corn oil (up to 200 mL/5 kg pelleted chow), 5% agar or 5% gelatin to aid in adherence to the chow (Scott, 1993). Small amounts of radiolabeled chemical can be mixed with the nonlabeled chemical to enable tracking the chemical and its metabolites in blood or other tissues without complicated extraction and purification steps. Treated pellets can be administered to the water by hand or by an automatic feeder in large aquaculture applications. Although some of the applied chemical may be lost by dissolution in water, especially for water-soluble chemicals, most is consumed along with the food.

Systemic absorption of the administered chemical is relatively rapid. Significant accumulation of radiolabeled methyltestosterone was detected in fish tissues 2 h after feeding sprayed chow (Fagerland and Dye, 1979). Tissue levels of testosterone appeared to reach equilibrium concentrations 24 h after feeding testosterone sprayed on food to coho salmon (*Oncorhynchus kisutch*) (Fagerland and McBride, 1978). In a similar experiment with carp (*Cyprinus carpio*), 4 days' feeding was required for testosterone to reach equilibrium concentrations in fish tissues (Lone and Matty, 1981). Chemical absorption from impregnated food depends on the rate of chemical dissolution from the food, its absorption efficiency in the stomach and intestine and influences of other factors, such as chemical and microbial degradation in the gut and binding to tissues (Levine, 1990). Although in both experiments the chow was sprayed with nearly identical concentrations of testosterone, chemical absorption may have been altered by the type of chow used in each experiment. Fagerland and McBride (1978) specified the use of a moist chow, whereas Lone and Matty (1981) did not identify whether the chow was moist or dry. Also, physiological or metabolic differences between the two fish species may have caused differences in absorption, distribution and metabolism of the testosterone. Thus absorption efficiencies and time needed to attain equilibrium concentrations in fish will vary, depending on chemical properties, chow properties and the fish species.

Routine feeding with contaminated prey

Fish can also be exposed to many types of chemical agents through consumption of contaminated prey. This method works best with contaminants that are easily **bioaccumulated** by the prey species, including many hydrophobic organic compounds and heavy metals. Organic contaminants may be radiolabeled to facilitate measuring accumulated contaminants in prey and the fish that consume them. Prey organisms are first exposed to sublethal concentrations of the contaminant (or multiple contaminants) for a sufficient duration for bioaccumulation to occur. Once the desired amount of chemical has been accumulated by prey organisms, they are presented as food items to the fish.

The major benefit of this method of administering chemicals is its realism. This method mimics actual food chain transfer in the field, and is often the preferred method of contaminant exposure for studies of chemical assimilation efficiencies and bioaccumulation potential for compounds and mixtures. A limitation of this method involves the additional toxicant exposure needed to produce contaminated prey. More chemical is needed to ensure sufficient accumulation in live prey through water-borne exposures than would be used in an oral bolus dose (see subsequent sections on oral bolus doses). As **trophic** transfer of chemicals is often less than 10%, lengthy exposures to contaminated prey may be necessary to attain the desired chemical concentration or effect in consumer fish. In addition, this method offers little control over the chemical dose or form, especially if the chemical is **biotransformed** by the prey organism. Inconsistencies in dose may also result from variable chemical accumulation by prey organisms and prey selection by the consumer fish. To identify and quantify all forms of the chemical, including all biotransformation products, chemical extraction and analysis of a subset of prey organisms can be performed, but may be time-consuming and costly. The need for this extra chemical analysis of prey species may be avoided by selecting test chemicals that either are not biotransformed by that species or biotransform very slowly relative to the duration of the experiment.

A variation of this technique is the use of food made from contaminated fish, rather than using live contaminated prey. Cleland and colleagues (1988) fed rainbow trout (*Oncorhynchus mykiss*) a diet of halogenated hydrocarbon contaminated coho salmon harvested from Lake Michigan and Lake Ontario. The contaminated diet was prepared from salmon flesh that was eviscerated, ground in a meat grinder, and supplemented with vitamins. Sufficient quantities of the contaminated salmon diet for the exposure were prepared and frozen at $-20°C$. Trout fed the contaminated salmon diet for 20 weeks accumulated similar quantities of halogenated hydrocarbons as contained in the salmon flesh. This method is subject to similar limitations as the use of live prey.

Gavage with impregnated food or gelatin capsules

If more exact dosages for individual fish are required, the chemical agent may be dissolved in a vehicle (e.g. water, ethanol, glycerol, fish oil, corn oil) and either mixed with a slurry of crushed and/or dampened chow or added to a gelatin capsule. This enables a precisely measured dose to be administered along with the capsule or food. Fish should be lightly anesthetized with MS222 or a similar agent (Iwama and Ackerman, 1994). The food or capsule is inserted directly into the stomach of the anesthetized fish with a plunger or modified syringe. Following resuscitation the fish should be carefully observed to be certain that it retains the capsule or all of the administered food.

Collier and colleagues (1978) found that distribution patterns and metabolism products of naphthalene administered to coho salmon in a gelatin capsule were similar to those measured following water exposures and IP injections. Regardless of the administration route used, hydrocarbon accumulation varied widely among individuals. However, capsule administration yielded individual variability that was comparable with other methods of administration (Collier et al., 1978).

Injection techniques

The most common method of administering a bolus dose to fish is by injection. Ideally the chemical agent is dissolved directly in physiological saline (refer to Chapter 30, Table 30.2). However, hydrophobic chemicals may be dissolved in very small quantities of a co-solvent (e.g. ethanol, methanol, or DMSO) prior to dilution with saline. For chemicals not soluble or stable at a neutral pH, the pH of the injection solution may be adjusted with acid or base (Perry and Reid, 1994). The final injection volumes should be as small as possible to minimize physiological alterations to the fish. Additionally, vehicle and/or sham-injected fish should always be used in parallel with chemically injected fish, to control for any effects of injection or the vehicle.

Chemical agents may be injected directly into veins or arteries (intravascular), into the peritoneal cavity (intraperitoneal) or into the muscle (intramuscular) of adult fish. Fish eggs or embryos may be injected into the perivitelline space or yolk sac with a micro syringe (Black et al., 1985). In general, injection techniques provide a high internal dose with rapid distribution to the tissues. Injection methods are especially useful for acute studies where high chemical concentrations are desired.

Intravascular injection

Chemical agents dissolved in an appropriate vehicle can be injected directly into a vessel (blind puncture) or into an in-dwelling vascular cannula. Injections directly into veins or arteries, intravascular (IV) injections, provide the most rapid and accurate administration of a chemical dose. The chemical is placed directly into the systemic circulation and because there are no physical barriers to cross, systemic distribution of the entire dose can take place immediately upon injection (Levine, 1990).

Blind puncture IV injections are usually administered into the caudal vein or artery (Perry and Reid, 1994). The procedure is nearly identical to the method of obtaining a blood sample by caudal puncture, described in Chapter 30. The fish is anesthetized with MS222 or a similar agent (Iwama and Ackerman, 1994) and placed ventral side up on a stable surface. A 1 mL syringe with a 1 to 1.5 inch (2.5–3.75 cm) 22 or 23 gauge needle (depending on the size of the fish) is loaded with the dissolved chemical. Air is evacuated from the syringe and needle. The needle is slowly inserted along the ventral midline of the fish approximately 5 mm posterior to the anal fin until the tip of the needle touches the spine. The needle is retracted slightly (≈ 1 mm) until blood pools into the needle hub, indicating that the needle is positioned inside the vessel (Figure 31.5). At this time the agent is injected slowly into the vessel. The needle is withdrawn and the fish is resuscitated in flowing water. This method may be preferable for single injections of tracer dyes or chemicals when no further blood sampling will take place. Otherwise, the use of an in-dwelling cannula for injections and taking subsequent multiple blood samples may be more convenient.

Arterial or venous cannulas provide perfect locations for injecting a chemical agent. (See Chapter 30 for descriptions of cannulas and instructions for their installation.) Prior to injection a solution of the

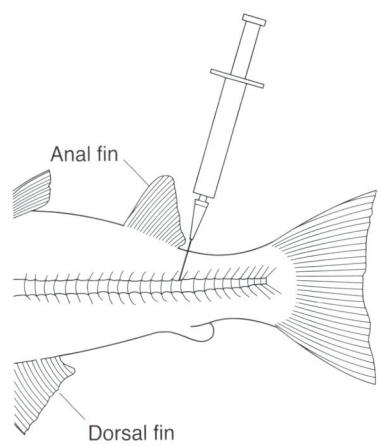

Figure 31.5 Illustration of the position for the syringe and needle during caudal puncture for injecting chemical agents. (Modified from the cover illustration of Brown, 1993.)

chemical agent is loaded into a syringe. The needle used should be of an appropriate size to fit inside the cannula (see Chapter 30, Table 30.3). Assuming the cannula is already in place, fish restrained in physiological chambers need not be anesthetized during the injection process. Free-swimming fish fitted with cannulas may be removed from the water, anesthetized (Iwama and Ackerman, 1994) and placed in a stable V-shaped holder for cannula injection. Alternatively, the fish may be lightly anesthetized and placed into a submerged holding cell for injection. A holding cell can be constructed of a piece of PVC pipe with a diameter slightly wider than the fish, cut to the length of the fish and sealed on each end with a wide mesh (one end must be removable). A groove is cut into the top of the cell for access to the cannula.

Prior to attaching the syringe needle to the cannula, the cannula plug should be removed and the heparin drained out. When blood appears at the end of the cannula, the needle should be attached and the chemical agent slowly injected manually or using a syringe pump. Once completely injected the cannula can be pinched to stop the outward flow of blood, the injection syringe is replaced with one filled with heparin (see Chapter 30, Table 30.1, for a range of possible concentrations) and the cannula is refilled with heparin and plugged. Free-swimming fish can then be resuscitated and/or released.

Intraperitoneal injection

Intraperitoneal (IP) injections employ injection of a chemical agent into the intraperitoneal cavity of the fish. Although care must be taken to avoid puncturing internal organs, this method usually requires less skill than most IV methods. Single IP injections can be made directly into the fish or through an IP cannula; multiple injections of chemicals dissolved in non-viscous vehicles are usually made via an IP cannula (Black et al., 1995). High blood concentrations of IP-injected chemicals are usually attained quite rapidly following injection, with maximum blood levels appearing within 1 to 2 h (e.g. Pankhurst et al., 1986; Black et al., 1995). The percentage of the IP dose absorbed to the systemic circulation will likely be less than the dose absorbed from an IV injection, but should be equal to or greater than absorption from intramuscular (IM) or oral doses.

Prior to injection the chemical agent is dissolved in saline or an appropriate vehicle. Hydrophobic chemicals are often dissolved in oil (fish, corn, peanut or mineral oil) (Varanasi et al., 1979; Pankhurst et al., 1986; Vijayan et al., 1997). If a slower release time is desired the chemical can be dissolved and injected in warmed cocoa butter, which solidifies post-injection inside fish held at or below 20°C (Pankhurst et al., 1986). Chemicals may also be solubilized in ethanol or another co-solvent prior to diluting in saline or oil (Black et al., 1995). The choice of vehicle may alter the rate and percentage of chemical absorbed by the systemic circulation. For example, in goldfish (*Carassius auratus*) that were IP-injected with 17β-estradiol (E_2) dissolved in saline, peanut oil and cocoa butter, the highest plasma E_2 concentrations were attained with the saline or oil vehicle. However, a more sustained blood level was observed in fish injected with E_2 dissolved in warmed cocoa butter (Pankhurst et al., 1986).

Direct IP injections are made with a syringe fitted with an appropriately sized needle (1 to 1.5 inches (2.5–3.75 cm) long; 18 to 23 gauge needle, depending on the viscosity of the vehicle). If the fish can be restrained and stabilized during injection, no anesthesia is necessary for making IP injections, although for large fish light anesthesia may be necessary. Fish should be laid on its side on a stable surface. IP injections are made into the peritoneal cavity, with the needle entering the area just anterior to the pelvic fins and along the midline of the fish (Figure 31.6) (Wall, 1993). A slight resistance may be felt as the peritoneum is penetrated by the needle. Once the needle is inside the peritoneum, the chemical can be slowly injected. Then the syringe needle is withdrawn and the fish is resuscitated in flowing water.

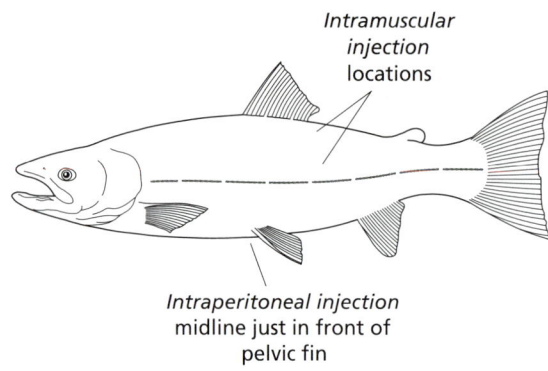

Figure 31.6 IP and IM injection locations on adult fish. (Modified from Wall, 1993.)

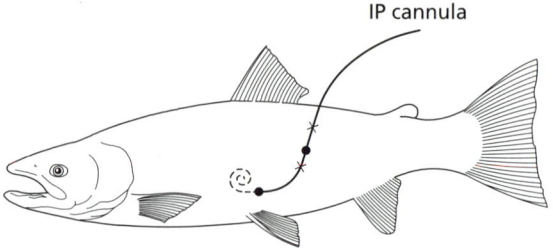

Figure 31.7 Diagram illustrating the position of an IP cannula installed in an adult fish. The dashed line indicates the position of the cannula inside the peritoneal cavity of the fish. Suture locations are indicated with an 'X'.

Fish can also be injected via an IP cannula. With a cannula in place it is easy to perform multiple injections over time in free-swimming (temporarily restrained or anesthetized during administration) or restrained fish. A prolonged dose can also be administered via a cannula to restrained fish using a syringe pump.

Black and co-workers (1995) used an IP cannula constructed from 60 cm of PE50 tubing for injecting ^{14}C-labeled pentachlorophenol into 400–500 g rainbow trout. The end of the tubing is first drawn over a flame and cut with a scalpel to form a tapered end. Then the first 10 cm of the tubing is heat-shaped in hot water to form a loose spiral. This spiral is followed by 5 cm of straight tubing, followed by a heat-shaped elbow bend. The rest of the tubing is left straight. Two anchor bubbles are formed within 2 cm of the elbow bend on both sides to help prevent the cannula from slipping through the anchoring sutures (Figure 31.7). Prior to insertion, the cannula is filled with physiological saline. A 70 cm piece of steel piano wire (0.5 mm ID) is sharpened at one end and carefully threaded through the cannula, beginning at the straight end and progressing toward the spiral end. To avoid piercing the cannula wall, the spiral end must be uncoiled as the wire is slowly pushed through it. The cannula is ready to be inserted when the sharpened tip of the steel wire has been completely threaded through the cannula so that it protrudes slightly from the tapered end of the elongated spiral.

The fish should be anesthetized and laid on its side as described for direct IP injections. To aid in the cannula insertion, the skin of the fish can be pierced with a 20 gauge needle at the IP injection location previously described. (On a 400–500 g rainbow trout this location is on the side of the fish, approximately 2 cm above the pelvic fins.) The wired cannula is inserted into the pierced hole at a shallow angle headed toward the anterior end of the fish. Then, the wired cannula should be gently pushed down through the peritoneum and just into the peritoneal cavity. Once the cannula is inside the cavity, the wire is withdrawn, while the spiral end of the cannula is threaded completely into the cavity. Once inside the cavity the spiral will help anchor the cannula in the fish. Then the cannula is affixed to the side of the fish with two or three sutures, sewn on both sides of the anchor bubbles (Figure 31.7). Once the cannula is in place and secured, it can be used immediately for injecting the chemical agent or be filled with saline and plugged. The cannula should be filled with physiological saline and plugged following all chemical injections (Black *et al.*, 1995).

Intramuscular injection

Perhaps the least used among injection methods for pharmacokinetic studies, intramuscular (IM) injection has wide use for parenteral administration of therapeutic agents. Intramuscular and IP injections are especially useful for administering chemicals that are unpalatable, thus precluding administration in food. Intramuscular injection is an easy injection method to use, and in most cases, can be used without anesthetizing the fish. Distribution rates for a chemical agent administered by IM injection are slower than with IV injection and hinge on the rate of systemic absorption. Systemic absorption rates of IM-injected chemicals are largely dependent on the degree of vascularization of the muscle injected, blood flow rates (usually proportional to the activity of the organism) and the rate of chemical diffusion across the capillary endothelium (Levine, 1990).

Chemicals are usually solubilized in an appropriate vehicle (e.g. physiological saline, ethanol, or oil) prior to IM injection. However, depot injections of the chemical in a semisolid matrix (e.g. cocoa butter) are possible if a more prolonged chemical administration is desired (Levine, 1990). Injections of the chemical are made in the muscle either at the base of the dorsal fin (Scott, 1993) or midway between the dorsal fin and lateral line (Wall, 1993) (Figure 31.6), using a 20–23 gauge needle. A slightly larger needle (e.g. 16 or 18 gauge) may be used if viscous or semisolid vehicles are used. An appropriate number of sham-injected fish should always be included in the experimental design to control for possible effects due to injection and the vehicle used.

Microinjection of fish eggs

As the use of fish in **carcinogenesis** assays increased, methods were developed to administer high doses of carcinogens to eggs and embryonic fish. Because many of these chemicals were only sparingly soluble in water, a microinjection technique was developed to safely inject small quantities of chemicals directly into fish eggs (Maccubbin and Black, 1986). This technique has been used successfully for injecting chemicals into large fish eggs (e.g. rainbow trout) and has increased the application of this species in carcinogenesis testing.

A microinjection apparatus developed by Black and colleagues (1985), shown in Figure 31.8a, is fabricated from a syringe microburet (e.g. Micromatic Instruments, Model SB2) connected with PE tubing to the non-injection end of a 31 gauge stainless steel hypodermic needle. The hypodermic needle is attached to a micromanipulator (Brinkmann Instruments) to facilitate uniform, precise injections into the perivitelline space or yolk sac of fish eggs.

Prior to injection the chemical agent is solubilized in a vehicle (e.g. glass-distilled DMSO or physiological saline). The hypodermic needle and microburet are loaded with the chemical and air is expelled. Injections are usually made in eggs at the eyed stage and 0.1 to 1 μl are typically injected. Just before injection 10 to 15 eggs are removed from the culture solution and placed on a grooved plexiglass block. Using

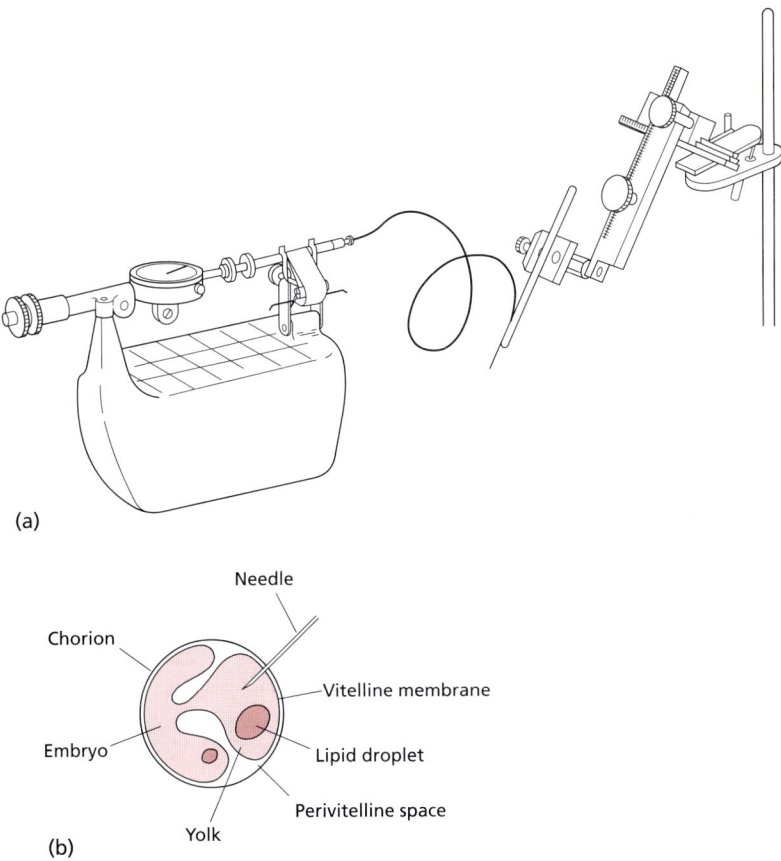

Figure 31.8 (a) Apparatus designed for making microinjections into fish eggs. The apparatus may also be used for topical applications of chemicals to eggs. (b) Locations for microinjection into fish eggs. (Modified from Black *et al.*, 1985.)

the micromanipulator and observing an egg through a stereo microscope, the injection needle is lowered into the egg until it pierces the chorion and enters the perivitelline space (Fig. 31.8b). The chemical can be injected into the perivitelline space or the needle can be further directed into the yolk sac for injection (Metcalfe and Sonstegard, 1984; Black et al., 1985). Once all eggs have been injected in this manner they are returned to containers with aerated culture water. Injection of 10–15 eggs usually takes less than 1 min. Black and colleagues (1985) found survival of injected embryos to be age dependent, with increased survival (up to 80–90% of noninjected control survival) when injections were made in older eggs that were within 2–3 days of hatching.

Implants

Pellet implants

Chemical agents may be encapsulated in a solid matrix, which is pelletized and surgically implanted within the body of a fish. Implants provide the best method for attaining a sustained release of a chemical agent to a fish without multiple applications. They have been used most frequently to administer analogues of gonadotropin-releasing hormones (GnRH) and reproductive hormones to fish (e.g. Trudeau et al., 1991; Hodson and Sullivan, 1993; Woods and Sullivan, 1993), and have potential applications in aquaculture for inducing sexual maturation and/or spawning in broodstock fish. Implants are also under investigation as a method to administer toxic substances to fish. Recently a synthetic, chemically-impregnated pellet was tested as a method to deliver a sustained, sublethal methyl mercury dose to adult Nile tilapia (*Oreochromis niloticus*) (Arnold et al., 1999).

The rate of chemical release from implants depends mostly on the chemical and physical properties of the matrix, although properties of the chemical agent may also be involved. The most frequently used matrix ingredients are various combinations of cellulose and cholesterol (e.g. Sherwood et al., 1988; Hodson and Sullivan, 1993; Woods and Sullivan, 1993) and **silastic elastomer** (e.g. Pankhurst et al., 1986; Trudeau et al., 1991; Zou et al., 1997). Both matrices have delivered sustained doses of impregnated steroid hormones in research trials, although no studies were found that directly compared the release properties of these two matrices.

Research trials using *in-vitro* perfusion chambers (mechanical simulations of fish implantations) found that release rates for GnRH were altered by regulating the relative percentages of cholesterol and cellulose in the pellet (Sherwood et al., 1988). Pellets containing 95–100% cholesterol (0–5% cellulose) yielded a slow, sustained release of hormone with 40% of the dose released gradually over 3 weeks. By increasing the cellulose percentage to 25–100%, a more rapid hormone release was observed. Woods and Sullivan (1993) found differences in spawning success in female striped bass (*Morone saxatilis*) implanted with a GnRH analogue in pellets containing 95% cholesterol versus pellets with 80% cholesterol. Differences in spawning success were partially attributed to differences in timing and duration of GnRH analogue release.

Silastic, a synthetic elastomer material, is also used in formulation of solid pellet implants and capsules for implants. This material is molded into the desired shape, rather than pressed and does not require a pellet-making apparatus. It is less brittle than cellulose/cholesterol pellets and can be cut into a desired length to customize the dose for each fish (Pankhurst et al., 1986). The only major disadvantage of silastic pellets is that they may release less chemical over time than cellulose/cholesterol pellets (Sherwood et al., 1988).

Cellulose/cholesterol pellet implants are formulated by first mixing the matrix materials (cellulose and cholesterol) in the desired percentages. The powdered chemical agent is then added to the matrix materials and mixed well. If the matrix contains 95–100% cholesterol the matrix–chemical mixture must be warmed (30°C oven, 1 h) prior to making pellets to reduce the brittleness of the pressed pellet. Cellulose/cholesterol implants are press-formed from the powdered material in a pellet maker (e.g. Parr Instrument Co., Moline, IL) (Sherwood et al., 1988). The finished pellet size depends on the size of the press mold, with a cylinder of 3 mm diameter and 3 mm height being a representative size (Sherwood et al., 1988; Hodson and Sullivan, 1993; Woods and Sullivan, 1993).

Silastic pellets are prepared from unpolymerized elastomer (Silastic 382 medical grade elastomer, Dow Corning Corporation). The dry, powdered chemical agent is first combined with the dry elastomer material. A liquid accelerator is added to the elastomer mixture at a ratio of 1 g elastomer mixture to 5 μL accelerator. This mixture is molded into pellets (e.g. $2 \times 2 \times 30$ mm) and allowed to solidify. Chemical dosages for individual fish may be cut from these pellets with a scalpel.

Prior to implant insertion, fish are lightly anesthetized with MS222 or a similar agent. Pellets can be surgically implanted into the peritoneal cavity of fish through small incisions (2–3 mm) made in the body wall (Trudeau et al., 1991; Zou et al., 1997). Another method of implantation is to load the pellet into a 10 or 12 g hypodermic needle or **trocar** and 'inject' it into the fish. A common injection location for pellets is the peritoneal cavity (e.g. Pankhurst et al., 1986) (see previous descriptions of IP injections). Others have injected silastic pellets into the dorsal lymphatic sinus of fish (Hodson and Sullivan, 1993; Woods and Sullivan, 1993). In adult striped bass (weights ranging from 3.1 to 8.1 kg), this sinus is accessed by inserting the needle or trocar into the region just posterior to the second dorsal fin (Figure 31.9). The authors suggest injecting the pellets approximately 2 mm into the sinus. If a trocar is used instead of a hypodermic needle, a small incision must be made through the fish scales prior to injection (Hodson and Sullivan, 1993; Woods and Sullivan, 1993).

Although pellet implantation provides an excellent means for administering sustained doses of hormonal agents, the method is still under development for other applications. Insufficient data presently exist for the use of this method for chemicals other than reproductive hormones, without conducting range-finding studies. Formulation of the appropriate dose of chemicals can also take several experimental trials to determine the bioavailability of the chemical in each pellet matrix (Arnold et al., 1999). Finally, like all injection methods this administration route sacrifices realism for convenience; thus it is more appropriate for use in pharmacokinetic or physiological research, rather than in ecotoxicology studies.

Osmotic pumps

Although designed to deliver sustained doses of hormones, proteins and drugs to mammals (e.g. Dalia et al., 1998; Kraeling et al., 1998; Wang and Du, 1998), osmotic pumps may provide another option for chemical administration to fish. Two general types of osmotic pumps are currently in use: an implantable, refillable pump and an oral tablet form. Of these two forms implantable osmotic pumps have the most potential for applications in fish research, although no studies were found to support this use.

Osmotic pumps use osmotic pressure to deliver drugs and other chemical agents over periods of time ranging from 24 h to 4 weeks. They are designed to deliver these substances at a constant rate (zero order kinetics), independent of the chemical's molecular mass or solution viscosity. The basic design for implantable osmotic pumps consists of an osmotic pressure compartment surrounding a refillable reservoir compartment, with a flexible interface separating the two compartments (Figure 31.10). The exterior surface of the osmotic pressure compartment is

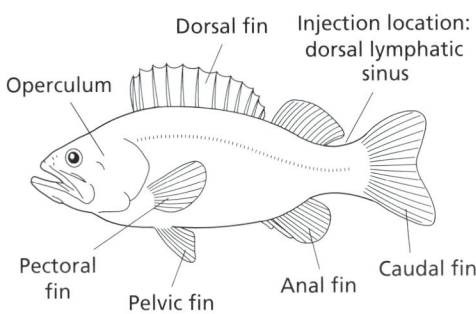

Figure 31.9 Diagram of a fish showing the injection site for implanting pellets into the dorsal lymphatic sinus. (Modified from Branson, 1993.)

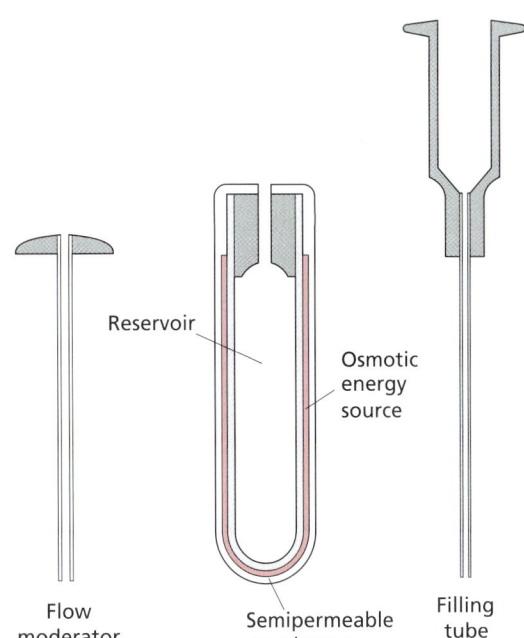

Figure 31.10 Diagram of an implantable mini-osmotic pump. (Modified from Theeuwes, 1980.)

covered with a biocompatible, semipermeable material that is permeable only to water. Its interior section is filled with a salt (osmotic energy source) that will dissolve in the inflowing water, creating a hypertonic solution. The expanding salt solution exerts hydrostatic pressure on the flexible membrane, which forces the chemical out of the reservoir through an orifice or flow modulator.

In rodents and small mammals the osmotic pump is usually implanted subcutaneously or intraperitoneally. Implanted pumps may also be connected to a cannula for direct delivery of the chemical to the vascular system (Jerzewski and Chien, 1992). Mini-osmotic pumps used in animal research have a typical reservoir volume of 0.2 mL with a release rate of 5 to 10 $\mu L\ h^{-1}$. Typical dimensions for these pumps are 2.5 cm length × 0.7 cm width (Eckenhoff and Yum, 1981, cited in Jerzewski and Chien, 1992). Implants of these dimensions should be sufficiently small for intraperitoneal implantation in larger fish (e.g. adult salmonids) via a shallow incision in the peritoneum of the fish. Refer to the surgical procedure for IP insertion of pellets (described previously in this chapter) for guidance on pump insertion. After surgically implanting the osmotic pump an antibiotic may be applied at the incision site to prevent infection prior to closing the incision with sutures or surgical glue.

Topical exposure

Although topical administration is not used as a route of administration for juvenile and adult fish, it can be a good method for exposing fish eggs to chemical agents. Topical application of chemicals is most useful for exposing small eggs (e.g. Japanese medaka, *Oryzias latipes*) to high concentrations of chemicals, as these eggs are not large enough for injection (Helmstetter and Alden, 1995a). However, this method has also been used successfully to deliver toxicants to larger eggs such as salmonid eggs (Maccubbin and Black, 1986).

Prior to application, the chemical agent is dissolved in a vehicle. Glass-distilled DMSO is usually used because of its membrane-penetrating qualities and low toxicity to fish embryos (e.g. Metcalfe and Sonstegard, 1984; Black *et al*., 1985; Maccubbin and Black, 1986; Helmstetter and Alden, 1995a,b). The apparatus described for microinjecting eggs can be used to deliver an accurate topical dose to fish eggs (Maccubbin and Black, 1986) (see Figure 31.8 and apparatus description in section describing microinjection of fish eggs). Alternatively, a push-button repeating dispenser (Hamilton Model PB-600-1 with 5 μL syringe), Chaney repeating dispenser, or digital syringe (e.g. Hamilton, Model 7000.5) fitted with a 10 cm × 0.17 mm OD fused silica capillary on-column needle (Supelco, Belefonte, MA) can also be used (Helmstetter and Alden, 1995b; Helmstetter *et al*., 1996). The chemical solution is drawn into the dispensing apparatus and air is flushed from the system. Eyed eggs (rainbow trout) or 48-h-old medaka eggs are removed from the culture solution, placed on a glass surface, and blotted dry with a glass fiber filter paper. The application needle is lowered to just above each egg surface and the desired amount (0.1 to 1 μL) slowly added to the surface of each egg (Maccubbin and Black, 1986). The topically applied chemical is allowed to be absorbed for 1 min before the egg is gently rinsed and returned to the culture media (Helmstetter and Alden, 1995b).

Egg viability using this method is excellent and generally exceeds 90% for DMSO controls. This method appears to be very useful for safe delivery of measured chemical doses to fish eggs. However, some chemicals are absorbed poorly after topical application. Helmstetter and Alden (1995a) tested the efficacy of topical application of seven chemicals with various chemical properties. The efficiency of chemical absorption was determined to be directly related to the chemical's log K_{OW}, with hydrophobic chemicals being absorbed to a greater extent than hydrophilic chemicals. Based on their findings, Helmstetter and Alden (1995a) present an equation that can be used to calculate the topical dose needed to simulate a desired external aqueous exposure concentration.

References

Arnold, B.S., Jagoe, C.H. and Gross, T.S. (1999). In *Environmental Toxicology and Risk Assessment – Eighth Volume. ASTM STP 1364* (eds D.S. Henshel, M.C. Black, and M.C. Harrass), pp. 413–422. American Society for Testing and Materials, West Conshohocken, PA.

ASTM (1998). In *Annual Book of ASTM Standards*, vol. 11.05. *Biological Effects and Environmental Fate; Biotechnology; Pesticides*. E729–96, pp. 218–238. American Society for Testing and Materials, West Conshohocken, PA.

Black, J.J., Maccubbin, A.E. and Schiffert, M. (1985). *J. Natl. Canc. Inst.* **75**, 1123–1128.

Black, M.C. and McCarthy, J.F. (1988). *Environ. Toxicol. Chem.* **7**, 593–600.

Black, M.C. and McCarthy, J.F. (1990). *Aquat. Toxicol.* **17**, 275–290.

Black, M.C., Millsap, D.S. and McCarthy, J.F. (1991). *Physiol. Zool.* **64**, 145–168.

Black, M.C., Björkroth, K. and Oikari, A. (1995). In *Environmental Toxicology and Risk Assessment – Third Volume, ASTM STP 1218* (eds J.S. Hughes, G.R. Biddinger and E. Mones), pp. 351–364. American Society for Testing and Materials, Philadelphia.

Branson, E. (1993). In *Aquaculture for Veterinarians: Fish Husbandry and Medicine* (ed. L. Brown), pp. 1–30. Pergamon Press, Oxford.

Brown, L. (1993). *Aquaculture for Veterinarians: Fish Husbandry and Medicine*. Pergamon Press, Oxford.

Budhabhatti, J. and Maughan, O.E. (1994). *Prog. Fish Cult.* **56**, 147–148.

Bugley, K. and Shepard, G. (1991). *N. Am. J. Fish. Manage.* **11**, 468–471

Cleland, G.B., Oliver, B.G. and Sonstegard, R.A. (1988). *Aquat. Toxicol.* **13**, 281–290.

Collier, T.K., Thomas, L.C. and Malins, D.C. (1978). *Comp. Biochem. Physiol.* **61C**, 23–28.

Dalia, A.D., Norman, M.K., Tabet, M.R., Schleuter, K.T., Tsibulsky, V.L. and Norman, A.B. (1998). *Brain Res.* **797**, 29–34.

Diamond, S.A., Oris, J.T. and Guttman, S.I. (1995). *Environ. Toxicol. Chem.* **14**, 1387–1388.

Fagerland, U.H.M. and Dye, H.M. (1979). *Aquaculture* **18**, 303–315.

Fagerland, U.H.M. and McBride, J.R. (1978). *J. Fish. Res. Bd. Can.* **35**, 893–900.

Gilmour, K.M. (1998). In *The Physiology of Fishes*, 2nd edn (ed. D.H. Evans), pp. 101–127. CRC Press, Boca Raton, Florida.

Heath, A.G. (1995). *Water Pollution and Fish Physiology*, 2nd edn. Lewis Publishers, Boca Raton, Florida.

Helmstetter, M.F. and Alden III, R.W. (1995a). *Aquat. Toxicol.* **32**, 1–13.

Helmstetter, M.F. and Alden III, R.W. (1995b). *Aquat. Toxicol.* **32**, 15–29.

Helmstetter, M.F., Maccubbin, A.E. and Alden III, R.F. (1996). In *Techniques in Aquatic Toxicology* (ed. G.K. Ostrander), pp. 93–124. Lewis Publishers, Boca Raton, Florida.

Hodson, R.G. and Sullivan, C.V. (1993). *Aquacult. Fish. Manage.* **24**, 389–398.

Hughes, G.M. (1979). In *Environmental Physiology of Fishes* (ed. M.A. Ali), pp. 33–56. Plenum Press, New York.

Iwama, G.K. and Ackerman, P.A. (1994). In *Biochemistry and Molecular Biology of Fishes*, vol. 3 (eds P.W. Hochachka and T.P. Mommsen), pp. 1–15. Elsevier, Amsterdam.

Jerzewski, R.L. and Chien, Y.W. (1992). In *Treatise on Controlled Drug Delivery: Fundamentals, Optimization, Applications* (ed. A. Kydonieus), pp. 225–253. Marcel Dekker, New York.

Kraeling, R.R., Johnson, B., Barg, C.R. and Rampacek, G.B. (1998). *Biol. Repro.* **58**, 1199–1205.

Laurén, D.J. (1991). In *Aquatic Toxicology and Risk Assessment, 14th Volume, ASTM STP 1124* (eds M.A. Mayes and M.G. Barron), pp. 223–244. American Society of Testing and Materials, Philadelphia.

Levine, R.R. (1990). *Pharmacology: Drug Actions and Reactions*, 4th edn. Little, Brown and Company, Boston.

Lone, K.P. and Matty, A.J. (1981). *Aquaculture* **24**, 315–326.

Maccubbin, A.E. and Black, J.J. (1986). In *Aquatic Toxicology and Environmental Fate, 9th Volume STP 921* (eds T.M. Poston and R. Purdy), pp. 277–286. American Society for Testing and Materials, Philadelphia.

McKee, D.A., Ordner, M.T., Houston, D. and Lawrence, A. (1989). *Prog. Fish Cult.* **51**, 166–167.

McKim, J.M. and Goeden, H.M. (1982). *Comp. Biochem. and Physiol.* **72C**, 65–74.

McKim, J.M. and Nichols, J.W. (1992). In *Aquatic Toxicology: Molecular, Biochemical and Cellular Perspectives* (eds D.C. Malins and G.K. Ostrander), pp. 469–519. Lewis Publishers, Boca Raton, Florida.

McKim, J., Schmieder, P. and Veith, G. (1985). *Toxicol. Appl. Pharmacol.* **77**, 1–10.

McKim, J.M., Schmieder, P.K., Carlson, R.W., Hunt, E.P. and Niemi, G.J. (1987a). *Environ. Toxicol. Chem.* **6**, 295–312.

McKim, J.M., Schmieder, P.K., Niemi, G.J., Carlson, R.W. and Henry, T.R. (1987b). *Environ. Toxicol. Chem.* **6**, 313–328.

Martin, L.K. and Black, M.C. (1995). *Prog. Fish Cult.* **57**, 323–325.

Metcalfe, C.D. and Sonstegard, R.A. (1984). *J. Natl. Canc. Inst.* **73**, 1125–1132.

Opperhuizen, A., van der Veld, E., Gobas, F.A.P.C., Llem, D.A.K. and van der Steen, J.M.D. (1985). *Chemosphere* **14**, 1871–1896.

Pankhurst, N.W., Stacey, N.E. and Peter, R.E. (1986). *Aquaculture* **52**, 145–155.

Perry, S.F. and Reid, S.G. (1994). In *Biochemistry and Molecular Biology of Fishes*, vol. 3 (eds P.W. Hochachka and T.P. Mommsen), pp. 85–92. Elsevier, Amsterdam.

Piiper, J. and Scheid, P. (1984). In *Fish Physiology*, vol. XA (eds W.S. Hoar and D.J. Randall), pp. 229–262. Academic Press, London.

Rand, G.M. and Petrocelli, S.R. (1985). In *Fundamentals of Aquatic Toxicology* (eds G.M. Rand and S.R.

Petrocelli), pp. 1–28. Hemisphere Publishing, Washington.

Sasson-Brickson, G. and Burton, G.A., Jr. (1991). *Environ. Toxicol. Chem.* **10**, 201–207.

Scott, P. (1993). In *Aquaculture for Veterinarians: Fish Husbandry and Medicine* (ed. L. Brown), pp. 131–152. Pergamon Press, Oxford.

Sherwood, N.M., Crim, L.W., Carolsfield, J. and Walters, S.M. (1988). *Aquaculture* **74**, 75–86.

Simkiss, K. (1996). In *Ecotoxicology: A Hierarchical Treatment* (eds M.C. Newman and C.H. Jagoe), pp. 59–83. Lewis Publishers, Boca Raton, Florida.

Theeuwes, F. (1980). In *Controlled Release Technologies* (ed. A.F. Kydonieus), Ch. 10. CRC Press, Boca Raton, Florida.

Trudeau, V.L., Peter, R.E. and Sloley, B.D. (1991). *Biol. Repro.* **44**, 951–960.

USEPA (1993). Methods for measuring the acute toxicity of effluents and receiving waters to freshwater and marine organisms (ed. C.I. Weber), pp. 45–56. US Environmental Protection Agency EPA/600/4-90/027F.

Varanasi, U., Gmur, D.J. and Tresler, P.A. (1979). *Arch. Environ. Contam. Toxicol.* **8**, 673–692.

Vijayan, M.M., Feist, G., Otto, D.M.E., Schreck, C.B. and Moon, T.W. (1997). *Aquat. Toxicol.* **37**, 87–98.

Wall, T. (1993). In *Aquaculture for Veterinarians: Fish Husbandry and Medicine* (ed. L. Brown), pp. 193–221. Pergamon Press, Oxford.

Walsh, P.J. (1998). In *The Physiology of Fishes*, 2nd edn (ed. D.H. Evans), pp. 199–214. CRC Press, Boca Raton.

Wang, D.H. and Du, Y. (1998). *J. Hypertens.* **16**, 467–475.

Webster, C.D., Tidwell, J.H. and Yancey, D.H. (1992). *Prog. Fish Cult.* **54**, 92–96.

Webster, C.D., Tidwell, J.H., Goodgame, L.S. and Johnsen, P.B. (1993). *Prog. Fish Cult.* **55**, 95–100.

Woods, L.C. III and Sullivan, C.V. (1993). *Aquacult. Fish. Manage.* **24**, 211–222.

Zou, J.J., Trudeau, V.L., Cui, Z., Brechin, J., Mackenzie, K., Zhu, Z., Houlihan, D.F. and Peter, R.E. (1997). *Gen. Comp. Endocrinol.* **106**, 102–112.

CHAPTER 32

Fish Necropsy

Jeffrey P Fisher
Pentec Environmental, Inc., Edmonds, Washington, USA

Mark S Myers
National Marine Fisheries Service, Northwest Fisheries Science Center, Seattle, Washington, USA

Introduction

Careful necropsy technique is critical to establish reproducibility and credibility in fish anatomy (Harder, 1975; Rich and Smith, 1978; Yasutake and Wales, 1983; Roberts and Sheperd, 1986; Chiasson and Radke, 1996), pathology, toxicology, immunology, neurobiology and taxonomy. In pathology, fish necropsy is employed for the purposes of diagnosing **enzootic** and **epizootic** infection, for characterizing **idiopathic** or **toxicopathic** disease, and for satisfying requirements of routine fish health inspections required for the marketing of food fish cultured throughout many parts of the world (Strange, 1983; Phillips, 1988; Hunn and Schnick, 1990; Reimschuessel, 1993; AFS, 1994). Aquaculture research has additional needs where necropsy is required, such as studies to evaluate the effects of dietary constituents on body composition and food conversion efficiency as reflected by somatic indices such as liver-to-body-weight ratios (Brown and Gratzek, 1980; Anderson and Gutreuter, 1983).

Although this text is devoted to the use of the fish as a laboratory animal, fish are often used as sentinels of chemical contaminant exposure in the environment. As such, field necropsy may be required to investigate fish kills or to establish ecological risks from sublethal levels of toxicant exposure by assessing aspects of their internal condition. Such studies could involve tissue resection for various chemical contaminant analyses (Stehr et al., 1993; Hunn and Schnick, 1990), biopsy for histopathology (Ostrander et al., 1993; Wooster et al., 1993), or a broad assessment of somatic indices that collectively may indicate population stress when compared against unexposed populations (Adams et al., 1993).

Somatic and stress indices obtainable through field necropsy are also possible with focused laboratory-based toxicological research, where, for example, necropsy may be required to identify systemic lesions resultant from acute or chronic toxicant exposure. The advantage provided by the laboratory setting is that more sophisticated techniques can be applied and more control of the conditions under which the necropsy occurs can be maintained. Neurochemistry

Copyright © 2000 Academic Press

and behavioral toxicology of fishes may require the dissection of the fish brain, a procedure more amenable to laboratory facilities where appropriate control over necropsy conditions can be established.

Chapter 32 illustrates techniques that may be used to perform fish necropsies in fish of the salmoniform and pleuronectiform body types. The techniques can be applied to fishes of different morphology, with only minor variations. The techniques illustrated here do not assume the necessity of a sterile field, although such conditions are often required for fish disease diagnosis or surgery. (Specific procedures for those protocols are detailed elsewhere in the handbook).

Salmoniform fish type

Characterize the epidermal integrity and condition of all fins. Record the location, size, and severity of ulcerations, abrasion, erosion, hemorrhage, other gross lesions, and parasitism. Extend all fins for proper inspection (Figure 32.1). Inspect the dorsal and ventral surface of the specimen. Distend the pelvic, pectoral and anal fins; record fin lengths and condition, and excise with scissors or scalpel if needed for study (e.g. histopathology). Examine vent (anus) for signs of hemorrhage, discharge, and/or parasitism (Figure 32.2).

Identify any ocular and dermal abnormalities and anomalies. For example, the specimen pictured exhibits periorbital hemorrhage (Figure 32.3). In a healthy fish, the lens and overlying cornea should be clear with no abrasions or discoloration.

Note obstructions or abnormalities within the buccal cavity (Figure 32.4). Gill rakers should be clearly separated and fully developed.

Deflect the operculum (gill cover) toward the head and make a cut with scissors along the preopercular margin, beginning at the ventral margin of the preoperculum near the base of the gill arches (Figure 32.5).

Examine the pseudobranch on the ventral surface of the preoperculum (not pictured). Examine the gill filaments, gill arches and gill rakers. Record external gill architecture, noting color, erosion, necrosis, hemorrhage, exudate and typical signs of parasitism (e.g. cysts) (Figure 32.6).

Using scissors, clip the outer gill arch with attached gill filaments at the ventral and dorsal connections. Underlying gill arches can be removed sequentially. The gills pictured exhibit mild erosion along the outer margin of filaments.

Make an incision immediately rostral to vent (Figure 32.7). Avoid perforation of intestine or rectum. If sterility is required (e.g. for bacteriological or virological samples), swab external surface with disinfectant such as an iodophore or alcohol prior to making incision. Use blunt-nosed scissors and cut in an anterodorsal direction through the **hypaxial** musculature, arcing slightly above and around the internal organs within the peritoneal cavity. Make a second anterior (rostral) cut along the ventral midline. The anterodorsal cut should eventually be below the lateral line approximately 4–6 scale widths. Avoid contact with internal organs. Reflect the hypaxial musculature overlying the body cavity, and at the dorsal opercular margin make a delicate anteroventral cut around the gills and pericardium, taking care to avoid puncturing the heart (Figure 32.8). Examine the condition of the visceral milieu (Figure 32.9). For example, the stomach should be elastic, the spleen should be dark red and firm, the liver should be deep pink to red and soft, but not friable. The **pyloric ceca** should be filamentous and fully extended with intact mesentery between individual ceca. Record signs of abnormalities, such as the following: size, shape and numeric anomalies; density, consistency and textural alterations; color anomalies; fluid and hemodynamic anomalies (e.g. ascites, edema, congestion, hemorrhage, hematoma); cysts and nodules; and more general anomalies such as exudate, erosion, distention, rupture, ulcerations, and grossly visible parasites.

To remove the visceral mass, begin by grasping the anterior portion of the esophagus with forceps, and while gently pulling the esophagus caudally, transect the esophagus with scissors at the esophageal sphincter (Figure 32.10). Be careful to avoid perforation of the gall bladder in this step. Displace the alimentary tract and other associated viscera from the trunk cavity from the rostral position and, with scissors, transect the rectum immediately rostral to the anus (Figure 32.11). Cut and tease away mesenteric attachments (connective tissue) between the alimentary tract and liver (Figure 32.12). Extend the components of the alimentary tract for determination of architectural integrity (Figure 32.13). To harvest the spleen, clasp it by its attached mesentery and cut it away from remaining attached viscera.

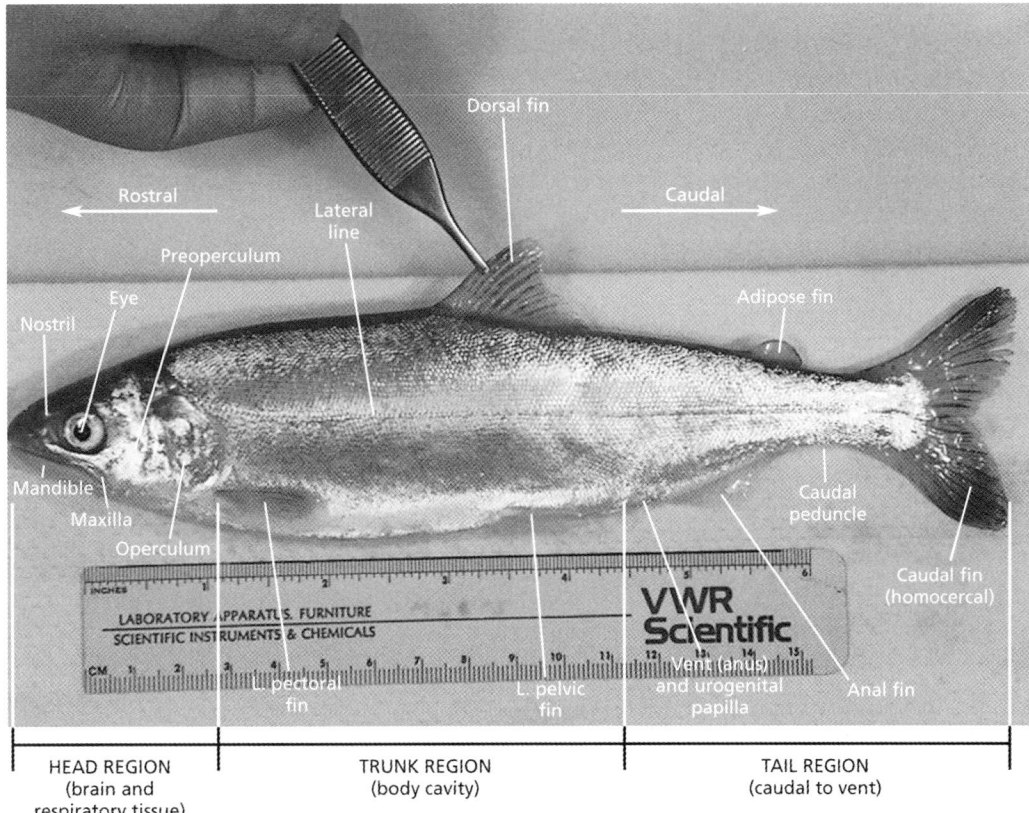

Figure 32.1 External examination: lateral view.

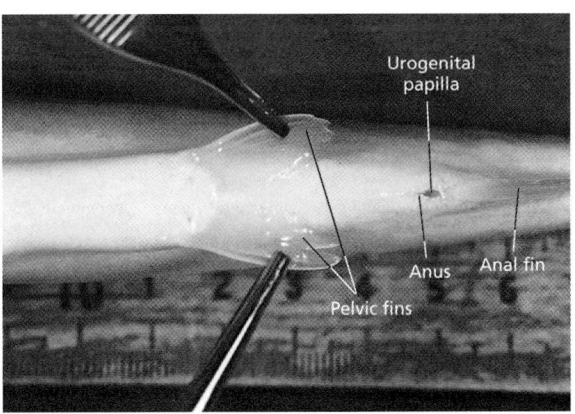

Figure 32.2 External examination: ventral view.

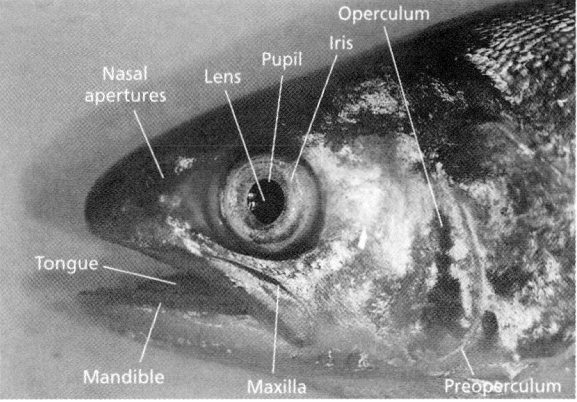

Figure 32.3 External examination: head region.

Separate mesenteric attachments connecting the gall bladder to liver and upper intestines. Clasp bile duct with forceps and cut bile duct with scissors, taking care to avoid spillage of bile from bladder on to the liver (Figure 32.14).

The bile can be subsequently harvested by perforating the bladder with a No. 11 scalpel blade while suspended over the mouth of an amber glass vial (Figure 32.15). Such a procedure is effective for analyzing a variety of fluorescent aromatic compounds, primarily representing metabolites of polycyclic aromatic hydrocarbons (Krahn et al., 1986).

Cut caudally through the esophageal and stomach walls to expose the esophageal and gastric mucosa (Figure 32.16). Record the presence or absence of food, the quantity and quality of that observed, and any irregularities in the mucosal architecture (parasitism, hemorrhage, ulceration, etc.). Proceed caudally to reveal the pyloric sphincter (separating the stomach from the upper intestine), and the mucosa of the upper and lower intestine (Figure 32.16).

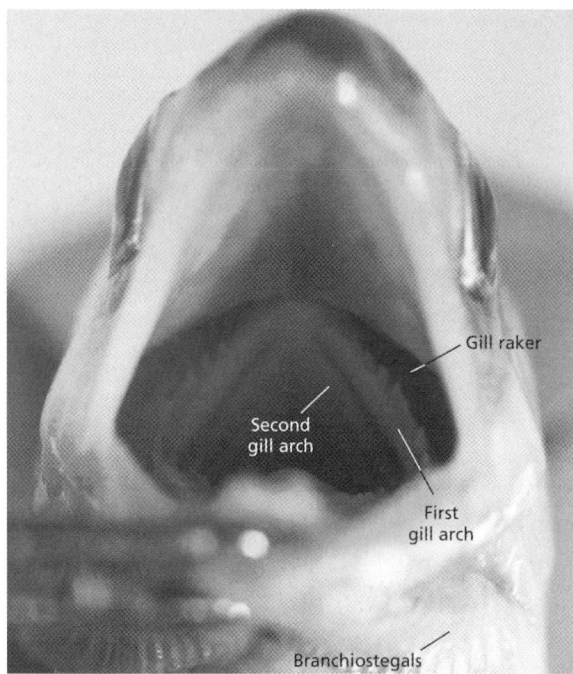

Figure 32.4 Buccal cavity examination.

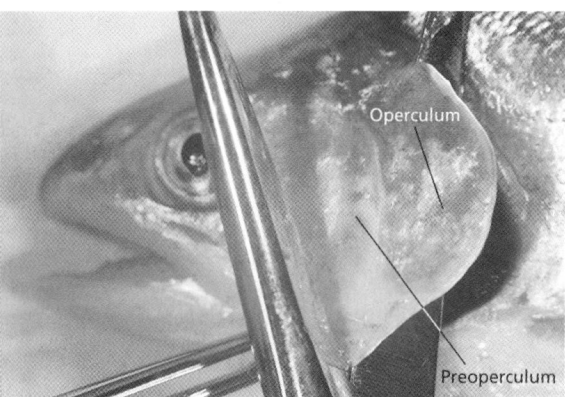

Figure 32.5 Gill examination: operculum removal.

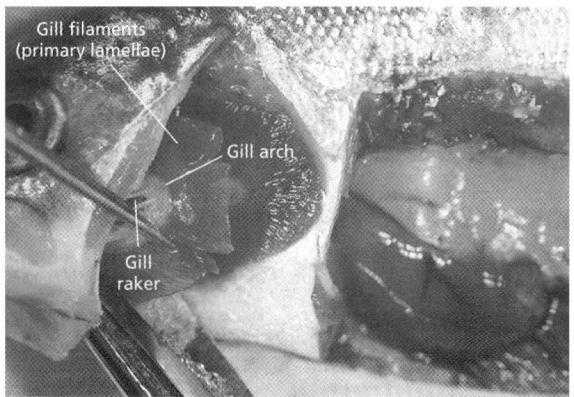

Figure 32.6 Gill examination: tissue excision.

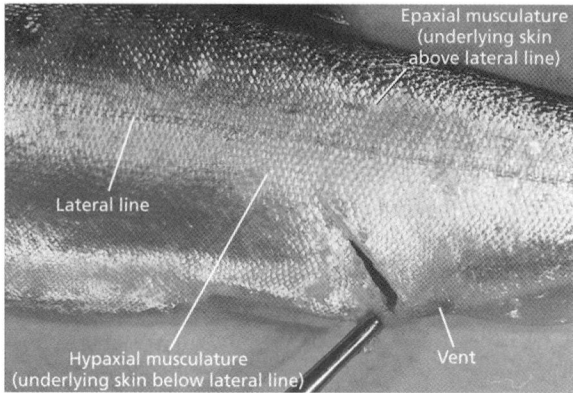

Figure 32.7 Internal examination: entry incision.

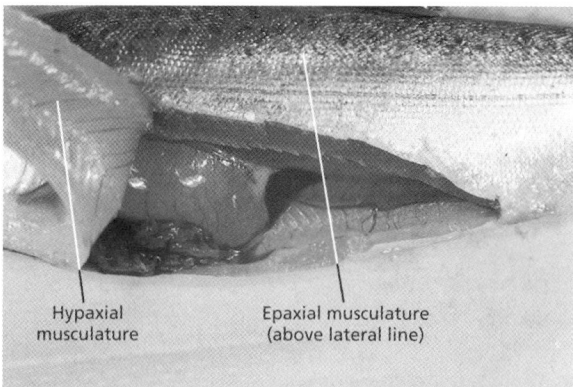

Figure 32.8 Internal examination: entry.

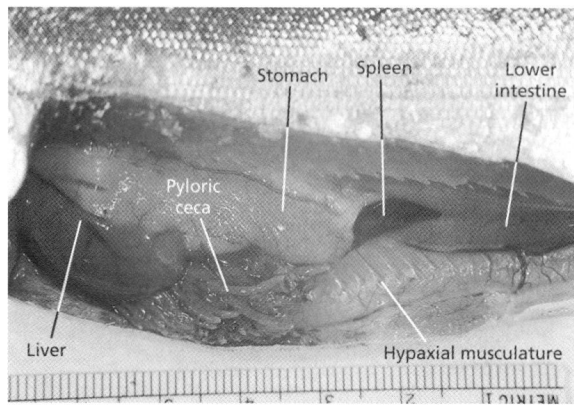

Figure 32.9 Internal examination: characterizing organ condition.

Expose the pericardial cavity by carefully cutting through the pericardium (Figure 32.17). This procedure can be conducted either before or after the removal of the viscera. Clasp bulbus arteriosus with forceps and transect at the rostral end to remove the heart from the pericardial cavity. Examine the structural integrity of the heart chambers. The bulbus should be elastic, non-friable and cream colored. The conical-shaped ventricle should be firm and light red;

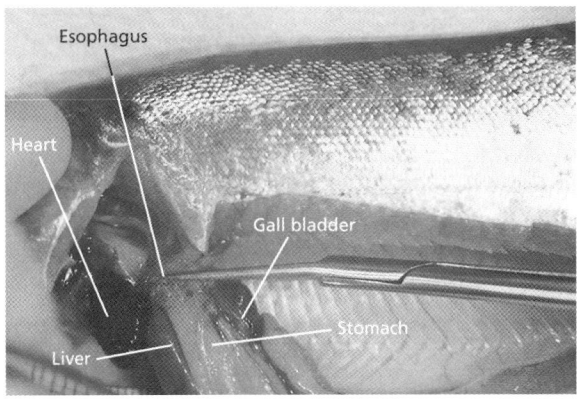

Figure 32.10 Internal examination: resection of alimentary tract (rostral).

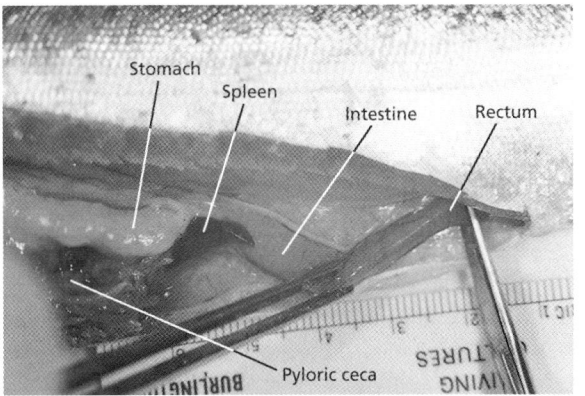

Figure 32.11 Internal examination: resection of alimentary tract (caudal).

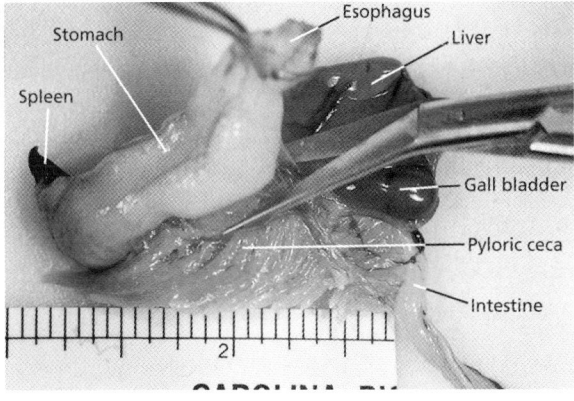

Figure 32.12 Alimentary tract: separation of internal organs.

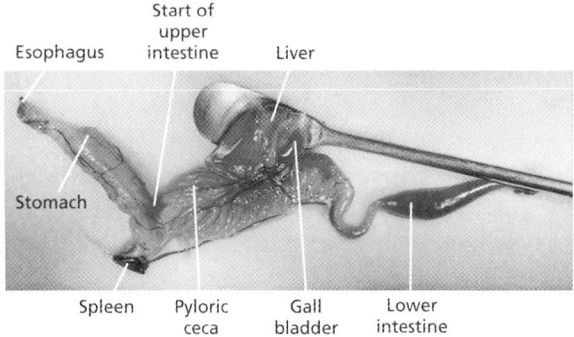

Figure 32.13 Alimentary tract: gross inspection.

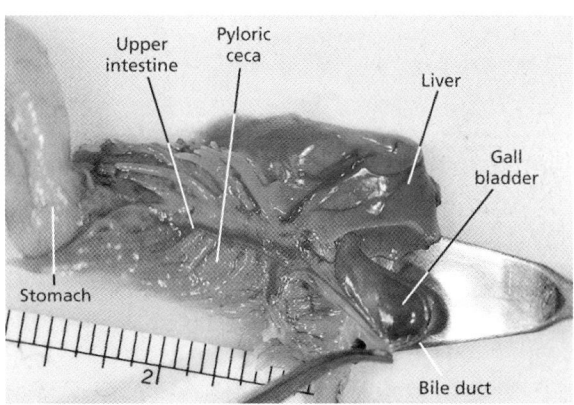

Figure 32.14 Alimentary tract: resection of gall bladder.

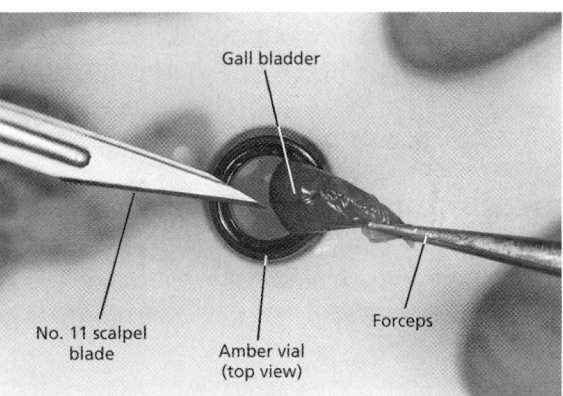

Figure 32.15 Alimentary tract: bile harvest.

the atrium and sinus venosus should be loose and sac-like, with a deep, ruby red color (Figure 32.18). In smaller fish, as pictured, the sinus venosus will be indiscernible without microscopy.

Note the condition of the dorsally positioned gonads (Figure 32.19). In immature fish or fish captured outside of their reproductive cycle, they may

Figure 32.16 Alimentary tract: stomach and intestines inspection.

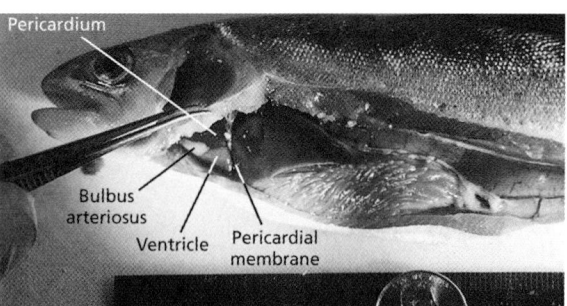

Figure 32.17 Heart: exposure of the pericardial cavity.

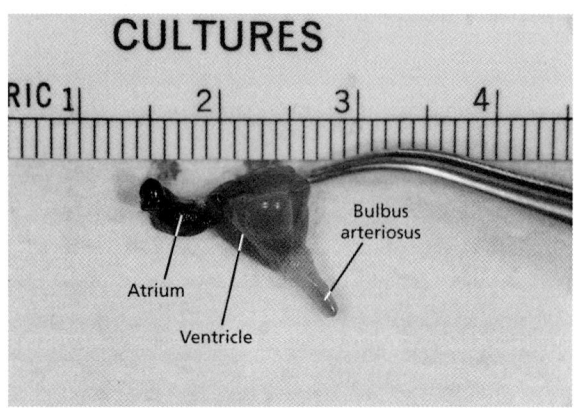

Figure 32.18 Heart: inspection and resection.

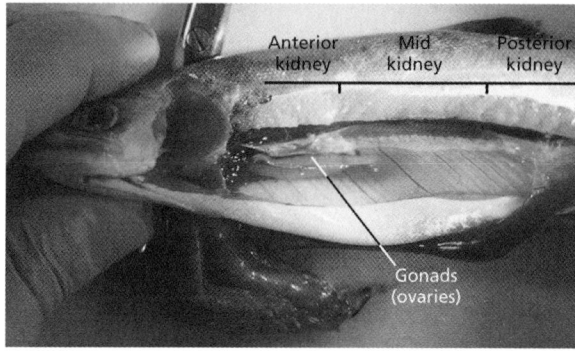

Figure 32.19 Gonads (testes and ovaries): inspection and resection.

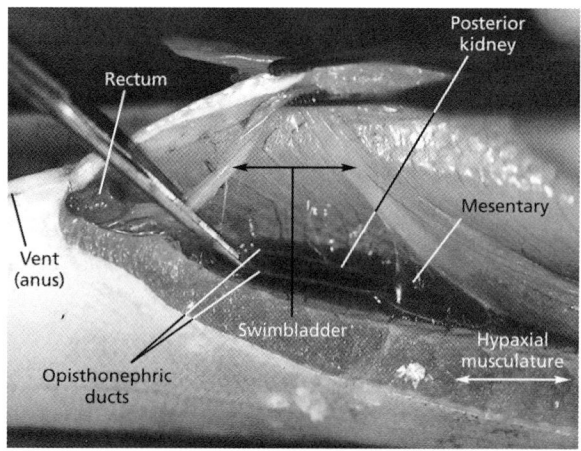

Figure 32.20 Swimbladder: inspection and resection.

not be grossly visible. Color is highly variable based on maturity and species. Those pictured are immature ovaries. Gonads in mature fish may occupy a more ventral position and represent the largest component of the visceral mass.

The swimbladder is dorsally situated above the visceral mass overlaying the kidney. In the Salmonidae it connects rostrally to the esophagus through a pneumatic duct (not pictured). Transect the swimbladder immediately rostral to its caudal connection (Figure 32.20). Excise the swimbladder by reflecting the membranous sac rostrally and transecting at the pneumatic duct.

Upon removal of the swimbladder, the anterior, middle and posterior kidney are clearly observed in their retroperitoneal position, immediately ventral to the spinal column. The paired **opisthonephric** ducts and urinary bladder can be seen overlying the posterior kidney (Figure 32.21). To resect a kidney section (e.g. for histopathology), make a longitudinal incision through the overlying peritoneal membrane in the caudal direction. Transect the kidney tissue at the rostral end of a desired tissue block and cut dorsally down to the vertebral column. Proceed with two oblique and caudal cuts along both sides of the kidney in similar fashion and remove the underlying tissue block (Figure 32.22). Gently lift the kidney section away from the spinal column and transect it at the caudal end (Figure 32.23).

To expose the brain, use a sharp scalpel to shave the overlying skin, cartilage, and bone overlying the brain beginning just caudal to the nostrils (Figure 32.24). The cerebellum (metencephalon) and optic

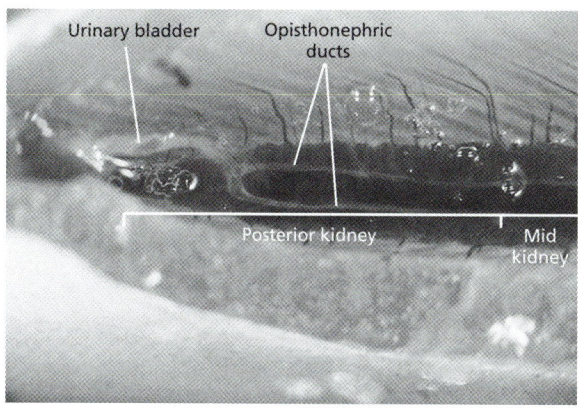

Figure 32.21 Kidney: examination.

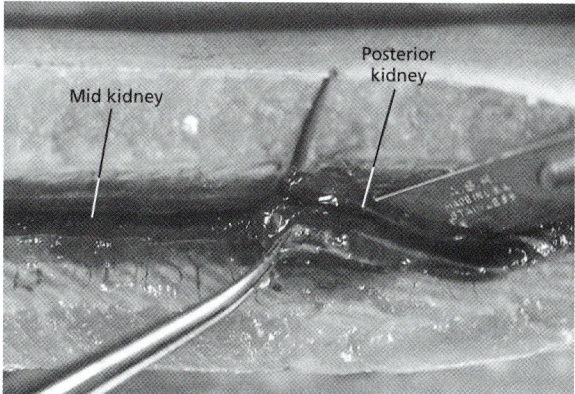

Figure 32.22 Kidney resection: initiation.

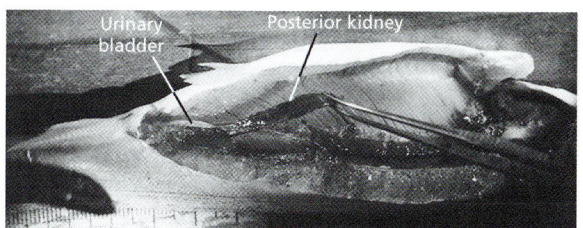

Figure 32.23 Kidney resection: completion.

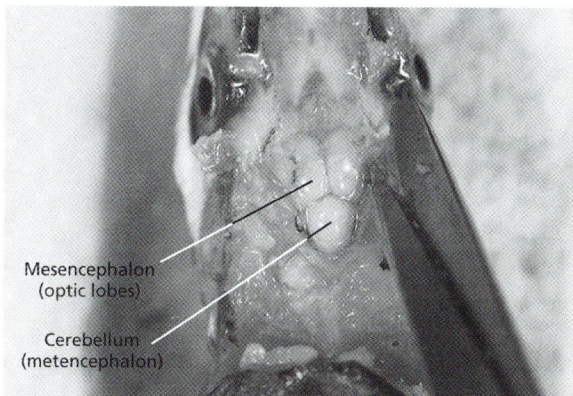

Figure 32.24 Brain dissection: initiation.

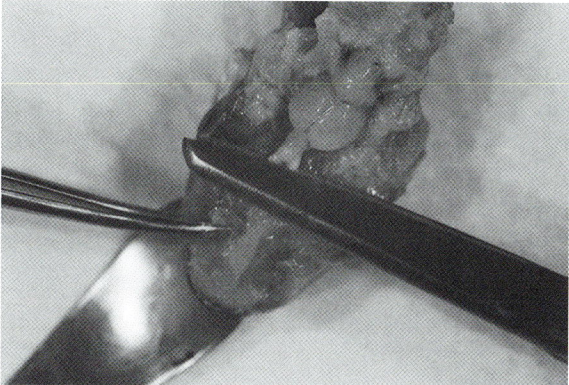

Figure 32.25 Brain resection: transection of optic nerves.

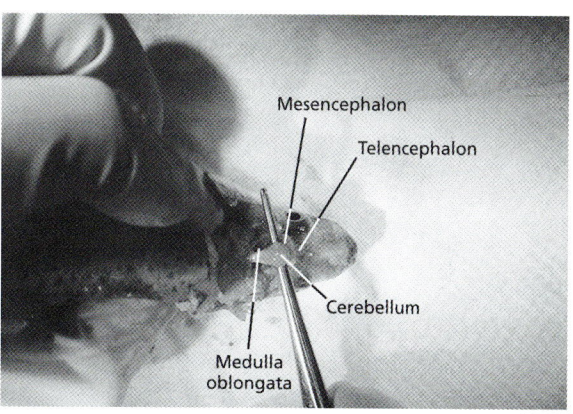

Figure 32.26 Brain resection: transection of medulla.

lobes (mesencephalon) are generally prominent in predatory fish like salmonids compared to the forebrain (telencephalon). Cut laterally through the brain case on either side of the cerebellum and optic lobes. Transect the optic and olfactory nerves (Figure 32.25) and medulla oblongata (myelencephalon) (Figure 32.26). The brain may then be lifted out of the skull. Procedures for the removal and examination of the eye are similar to that described for pleuronectiform fish (see pages 554–556).

Pleuronectiform fish type

The necropsy procedure for a pleuronectiform fish (i.e. flatfish) is essentially the same as that described above for salmonid fishes, with some important exceptions based on the lack of bilateral symmetry in flatfishes. The procedures described below for English sole (*Pleuronectes vetulus*), a member of the family Pleuronectidae (right-eyed flounders), are applicable

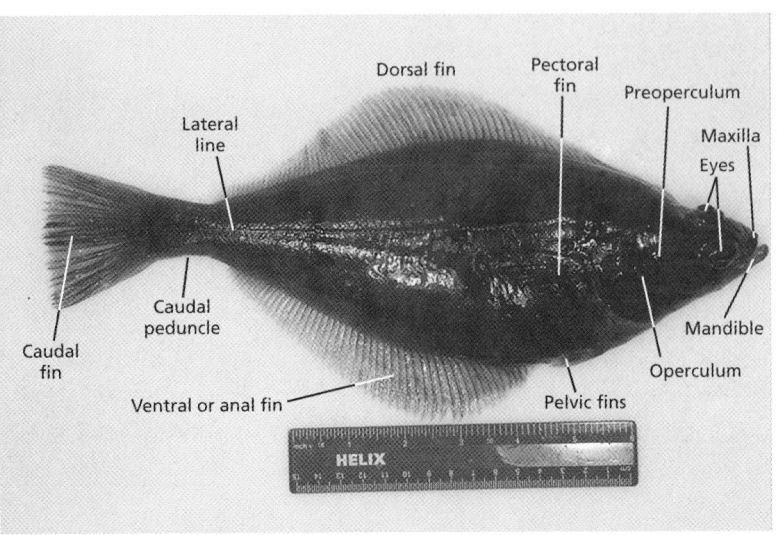

Figure 32.27 External examination: eyed-side view of English sole.

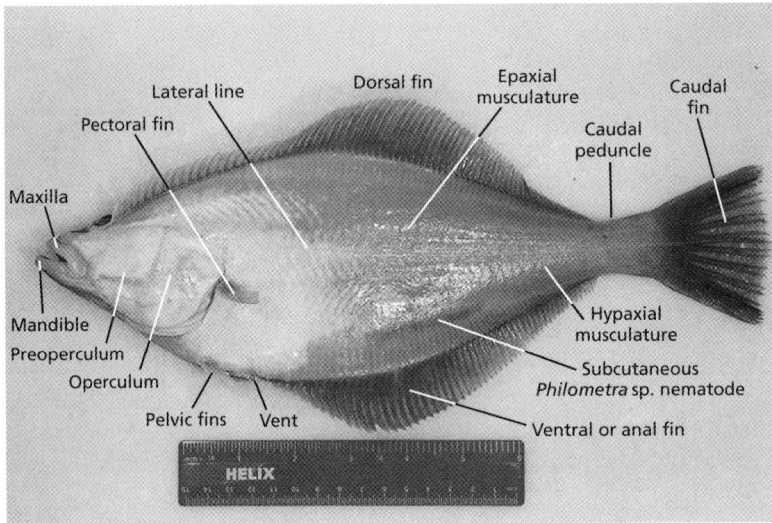

Figure 32.28 External examination: blind-side view.

to members of this family as well as those of the Bothidae (left-eyed flounders).

Begin by inspecting the condition and color of epidermis, eyes, dermis and all extended fins, and vent, viewing the specimen from both the eyed side (Figure 32.27) and blind side (Figure 32.28), recording all pertinent information, including parasitism, as described above for the salmonid example.

Note that flatfish possess paired pectoral and pelvic fins, and single elongate dorsal and ventral (anal) fins, and a typical caudal fin. The specimen shown in Figure 32.28 displays a subcutaneous *Philometra* sp. nematode worm in the hypaxial musculature overlying the ovarian cavity.

To examine the gills, place the flatfish with its blind side facing up, deflect the operculum toward the head, and make a cut with scissors along the preopercular margin, beginning at the ventral margin of the preoperculum near the base of the gill arches (Figure 32.29); proceed dorsally near the preopercular–opercular junction (Figure 32.30). Examine the prominent pseudobranch on the ventral surface of the preoperculum (not pictured). Using the same method described for the salmonid, examine the gill filaments, gill arches and gill rakers; record anomalies and excise the outer gill arch (Figure 32.31). In most cases, the entire gill arch can be preserved for histology by placement into a tissue cassette and then into an appropriate fixative.

To expose the organs in the peritoneal cavity, with the blind side of the fish still facing upward, use scissors to make an incision immediately caudal to the

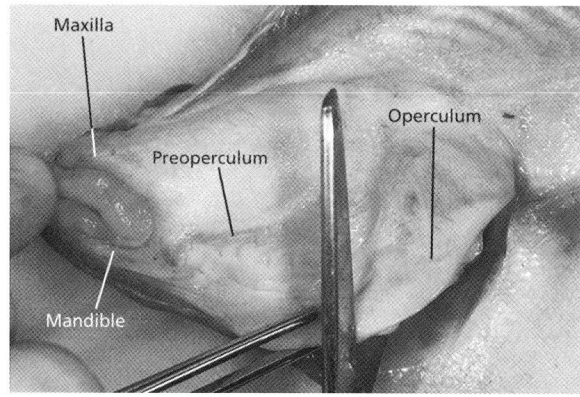

Figure 32.29 External examination, head and opercular regions: preparation for gill tissue excision.

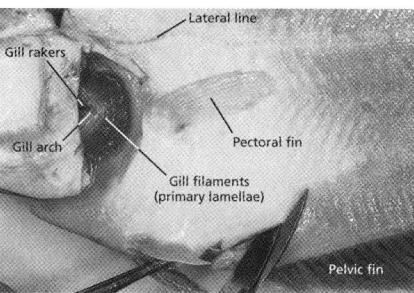

Figure 32.30 Gill examination, gross, and gross internal examination: entry incision into peritoneal cavity.

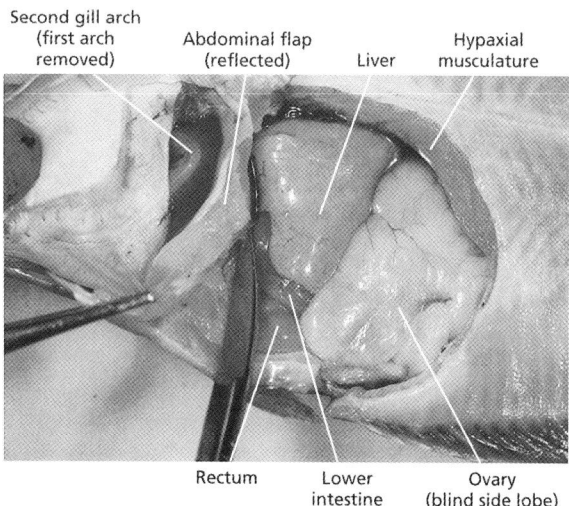

Figure 32.31 Internal examination: entry into and exposure of peritoneal cavity.

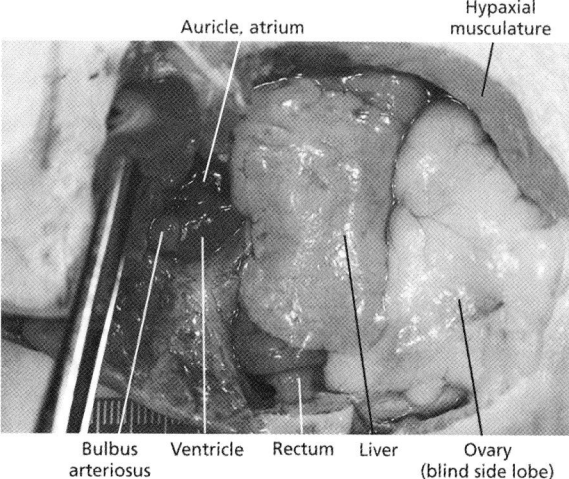

Figure 32.32 Internal examination: exposure of heart, examination of internal organs.

vent (Figure 32.30); proceed posteriorly and dorsally along the caudal margin of the peritoneal cavity, being careful to avoid perforation of lower intestine or rectum, or damaging the underlying ovary or testis. If necessary, use sterile technique as described for the salmonid necropsy. Continue this cut through the **hypaxial** musculature overlying the abdomen, following a roughly semicircular pattern (Figure 32.31), sequentially following the caudal and dorsal margins of the peritoneal cavity. Make a second anterior (rostral) cut along the ventral midline between and past the bases of the pelvic fins, and finish the exposure of the visceral organs by making a cut dorsally along the anterior margin of the peritoneal cavity; reflect the main abdominal flap dorsally and rostrally (Figure 32.31). Complete this process by carefully trimming any remaining hypaxial muscle near the anterior margin of the peritoneal cavity, posterior to the gills and pericardium, and anterior to the base of the blind side pectoral fin. Avoid contact with internal organs when making these cuts, especially the final one where the pericardium can easily be punctured. Examine the condition of the visceral organs and record any anomalies (Figures 32.31 and 32.32), as in the method described for salmon.

To expose the pericardial cavity, carefully make a scissors cut through the pectoral girdle and musculature associated with the cleithrum and coracoid bones, and through the pericardium (Figure 32.32); discard the blind-side half of the pectoral girdle. This procedure can be conducted either before or after the removal of the viscera. In this chapter, this procedure was done for the purposes of accurately displaying the main components of the heart *in situ* in a flatfish species (Figures 32.32 and 32.33). In practice in the field, the heart is typically collected by first cutting the exposed pericardium (just rostral to the anterior margin of the peritoneal cavity) with a scalpel blade. Next, reach anteriorly into the pericardial cavity with forceps, and clasp the anterior portion of the bulbus arteriosus with forceps and transect it at the

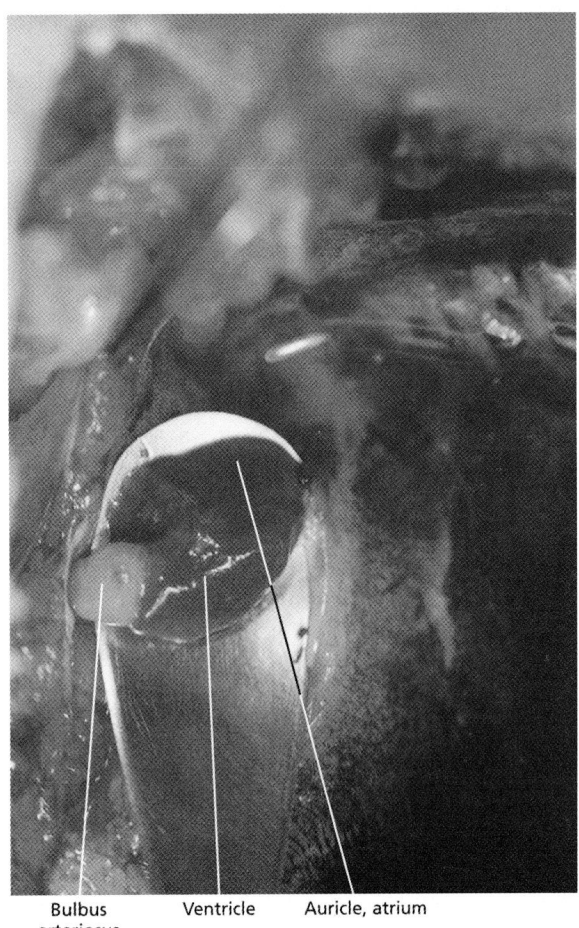

Figure 32.33 Internal examination: heart, examination and tissue collection.

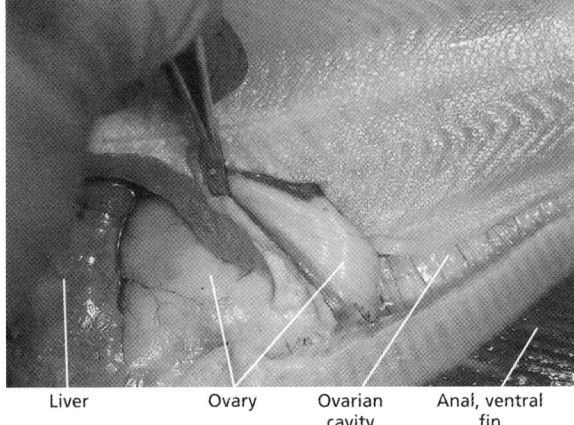

Figure 32.34 Internal examination: gonads (ovary), examination and tissue collection.

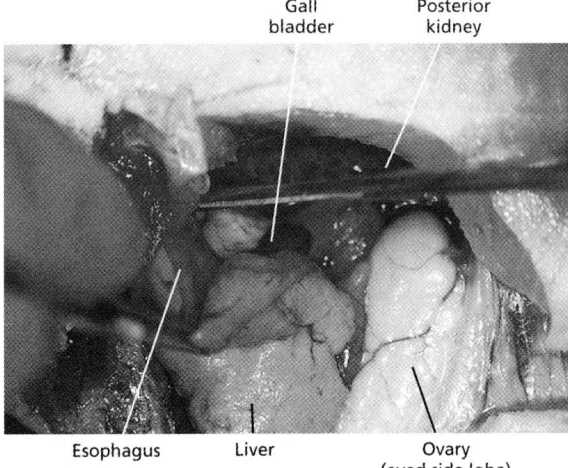

Figure 32.35 Internal examination: resection of visceral mass, transection of esophagus.

rostral end to remove the heart from the pericardial cavity. Examine the condition of the heart chambers (Figure 32.33), as is done for the salmon. In most cases, the heart can be preserved whole in a tissue cassette and placed in fixative; for heart tissue samples > 3 mm in thickness, portions of the heart may have to be bisected with a scalpel blade.

Gonadal tissues in adult flatfish are present bilaterally as two lobes at the posterior margin of the peritoneal cavity, and in the case of the ovary, extend posteriorly and bilaterally into two cone-shaped cavities that are extensions of the peritoneal cavity (Figure 32.34). Ovarian tissue is most easily collected by first making a cut with scissors posteriorly along the ventral margin of the entire length of the ovarian cavity on the blind side, followed by grasping the exposed posterior end of the ovary and gently pulling the ovary anteriorly. During this process, trim any fascia or connective tissue attachments between ovary and peritoneum with fine scissors, proceeding from the posterior to anterior end of the ovarian cavity (Figure 32.34). The entire lobe of the ovary can then be removed, placed flat on a necropsy board and examined routinely. The same process can also be conducted for the eyed-side lobe of the ovary by turning the fish eyed side up and conducting the same procedure. Sections for histology can be excised as a 3 mm transverse section cut from the middle region of the conically shaped ovarian lobe. Testes (not shown) are also situated in the posterior margin of the peritoneal cavity, but do not extend into the posterior musculature as the ovaries do, and are easily excised with scissors.

The visceral mass of flatfish is removed using identical methods to those for salmonid and other bilaterally symmetrical fishes, involving initial transection of the anterior esophagus (Figure 32.35) followed by transection of the rectum (Figure 32.36). Take special care to avoid perforation of the gall bladder in this

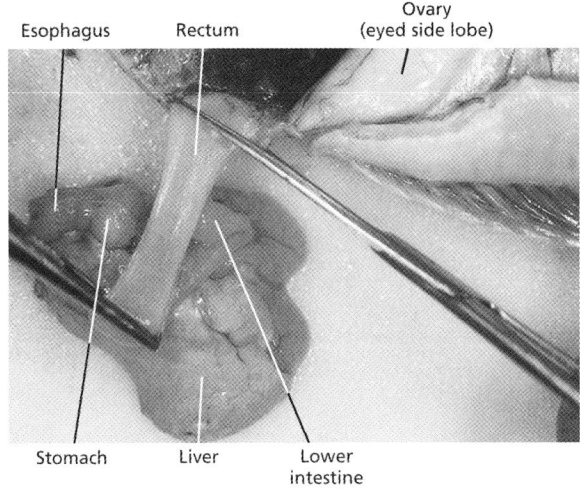

Figure 32.36 Internal examination: resection of visceral mass, transection of rectum.

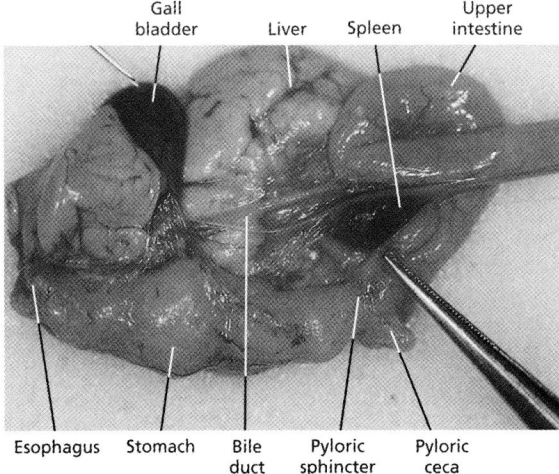

Figure 32.37 Internal examination: examination of organs of visceral mass after resection; eyed-side view.

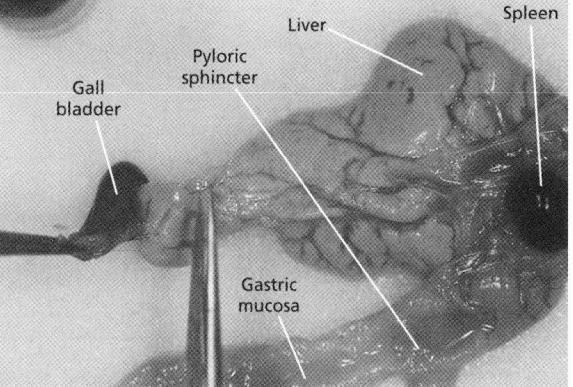

Figure 32.38 Internal examination: resection of gall bladder for bile collection.

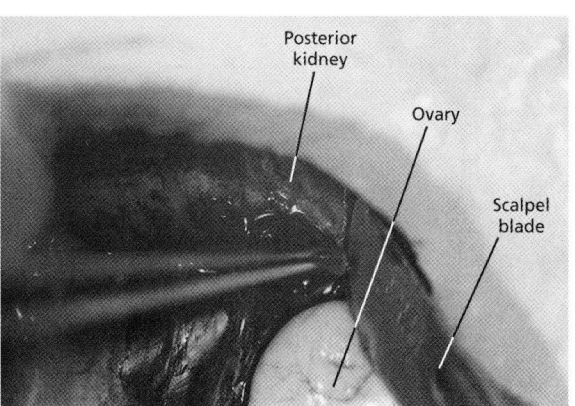

Figure 32.39 Internal examination: resection of posterior kidney.

step. After removal from the abdominal cavity, the entire visceral mass is then rotated 180° along its long axis and placed on a clean, flat surface to reveal all organs and the mesenteric attachments among them (Figure 32.37). Cut and tease away mesenteric attachments between the alimentary tract and liver (Figure 32.38). Bile may be then collected from the gall bladder using the same method as described for salmonids, beginning first by separating mesenteric attachments connecting the gall bladder to liver and upper intestine (Figure 32.38). The spleen (Figure 32.38) is collected by clasping it by its attached mesentery, cutting it away from remaining attached viscera, and then bisecting it with a scalpel to view the cut surface (not shown). If the thickness of the bisected spleen is > 3 mm, it is then placed into a tissue cassette and pre-

served in fixative – otherwise it must be trimmed further. The liver is now separated from the alimentary tract and examined for any anomalies. Sections of liver for histology are cut through its entire depth from either the longitudinal or vertical axis at < 3 mm using very sharp scissors; the sections are placed in a tissue cassette and fixed routinely. Examination of the alimentary tract (e.g. stomach, Figure 32.38) is conducted exactly as described previously for salmon.

The kidney in flatfish is located retroperitoneally at the dorsal margin of the peritoneal cavity (Figure 32.39) and ventral to the vertebral column; it is not separable grossly into anterior and posterior kidney. Resection of kidney tissue in flatfish (Figures 32.39 and 32.40) follows the same procedure as that for salmonids, although tissue collection is somewhat more problematic because of the difficult angle from which one must approach the tissue collection. Sections of kidney for histopathological examination should not exceed 3 mm in thickness, for optimal fixation.

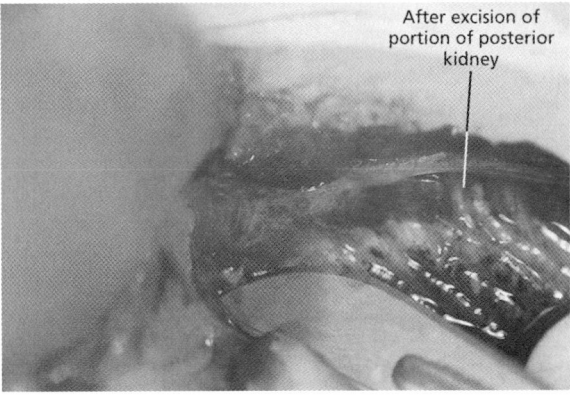

Figure 32.40 Internal examination: posterior kidney resected.

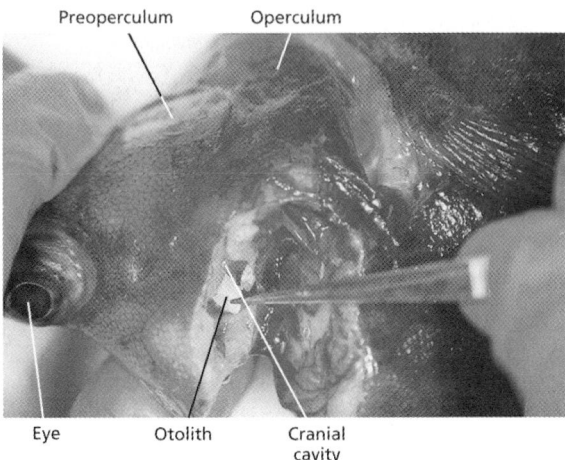

Figure 32.42 Collection of otoliths for fish age determination: extracting an otolith.

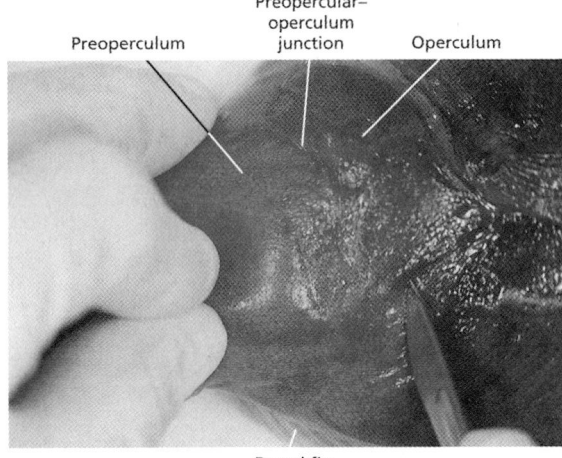

Figure 32.41 Collection of otoliths for fish age determination: lining up the incision.

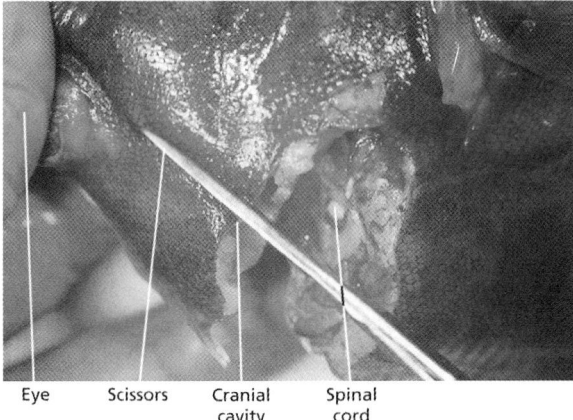

Figure 32.43 Brain dissection: initial cut along dorsal margin of cranial cavity.

In the necropsy of flatfish for research in any discipline (e.g. histopathology, toxicology, immunology, epizootiology, stock assessment), it is highly advisable to collect **otoliths** ('earbones') from the labyrinths or cavities lateral to the medulla oblongata, for determination of fish age. There are three otoliths, each within a sac of the labyrinth organ, but in practice it is rarely possible to collect more than two. In English sole and similar flatfish species, otoliths are routinely collected by placing the fish eyed side up, and making a vertical incision (with a sharp knife or scalpel blade) at a slightly anterior angle through the depth of the cranium at a level approximately midway between the posterior edge of the preoperculum and the posterior edge of the operculum (Figure 32.41), on the dorsal side of the head. Note position of scalpel blade in Figure 32.41, which represents the position at which this cut is made. After opening the incision further by forcing down on either side of the cut, forceps are then used to extract one or two otoliths from the otolith organs lateral to the medulla (Figure 32.42). In practice in the field, otoliths are typically collected just after collection of routine biological information (length, weight, sex) and collection of blood for serological, immunological, or endocrinological studies. This cut also serves to sever the spinal column and effectively sacrifice the fish so that necropsy techniques can be carried out more easily and humanely.

To collect the brain and associated organs (e.g. pituitary gland) from a flatfish, we have been successful in using the following method. After making the cut to collect otoliths, make another cut with blunt scissors along the dorsal margin of the brain cavity to near the base of the eye, being very careful to avoid contact with the fragile brain tissue (Figure 32.43). Make a similar and parallel cut along the ventral margin of

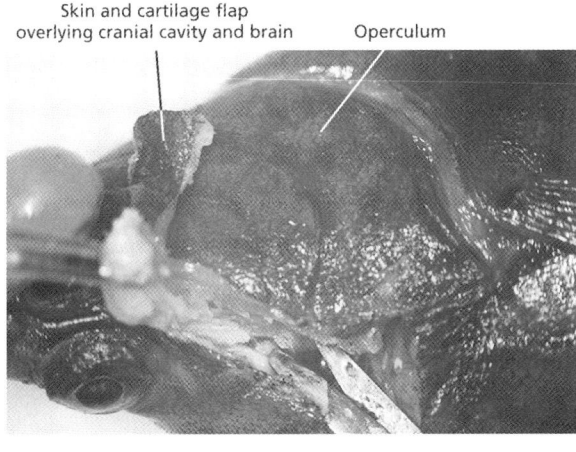

Figure 32.44 Brain dissection: second cut and removal of cranium.

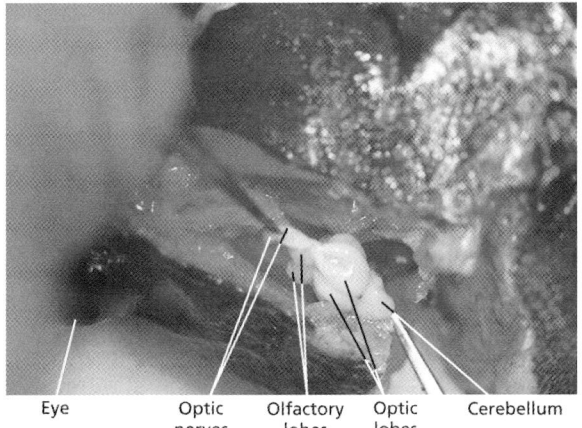

Figure 32.45 Brain dissection: cutting of optic nerves and removal of brain.

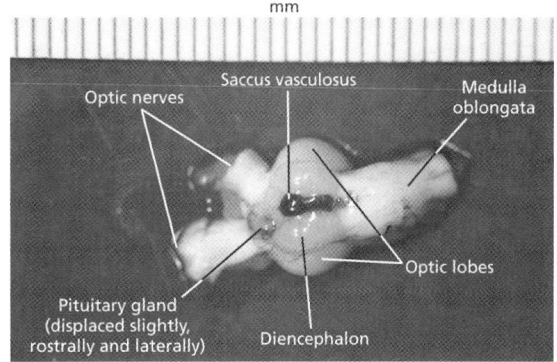

Figure 32.46 Brain examination: blind-side view.

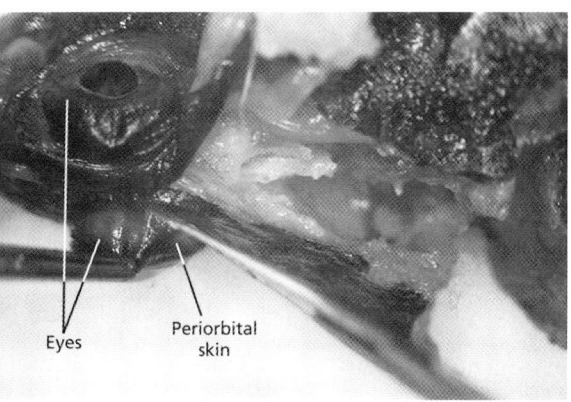

Figure 32.47 Eye resection: cutting of periorbital skin.

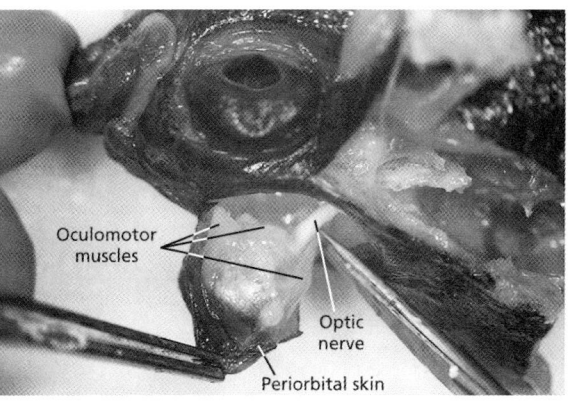

Figure 32.48 Eye resection: cutting of oculomotor muscles and optic nerve.

the cranial cavity, avoiding contact with the ventral surfaces of the brain (Figure 32.44). Grasp the posterior end of this cranial flap with a stout set of forceps and pull the flap upward and rostrally, while exerting downward pressure on either side of the cranium. This cut exposes the brain, and usually permits removal of the intact brain. However, additional scissor cuts along either side of the brain may be necessary to expose it sufficiently to permit intact removal. After the brain is fully exposed from a top view, gently grasp the medulla and pull the brain posteriorly to expose the tough, stringy optic nerves, which then may be cut with a sharp pair of scissors (Figure 32.45). The brain can now be removed from the cranial cavity for examination, including examination from the ventral view (Figure 32.46) to expose the pituitary.

To remove the eye, cut through the periobital skin (Figure 32.47) and oculomotor muscles (Figure 32.48) around the orbit of the eye. Transect the optic nerve, if not already completed from brain dissection (Figure 32.48). Invert the eye with optic nerve facing upward, support the eye laterally on both sides with the inside edges of forceps, and bisect the eye with a sharp scalpel (Figure 32.49). At a position near the optic nerve, transect eye through sclera, retina, vitreous body, and down through cornea to resect lens (Figure 32.50).

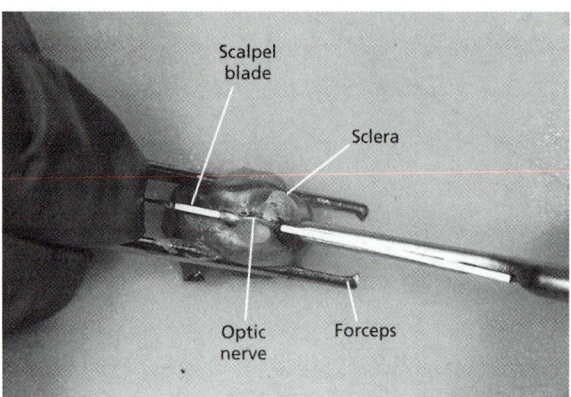

Figure 32.49 Bisection of eye: preparation for cut, optic nerve facing upward.

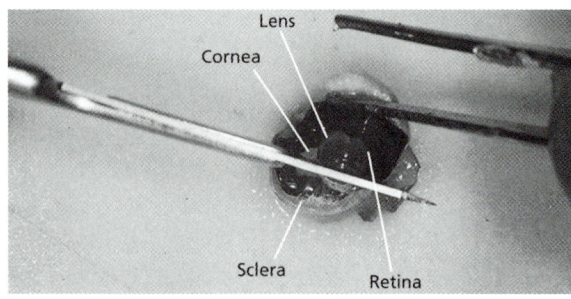

Figure 32.50 Bisection of eye: cut through optic nerve, through sclera and down through cornea.

For histological preparations, the bisected halves of the eye should be placed in the tissue cassette, and eventually embedded, with the cut side down so that all tissue layers of the eye can be examined in one section.

References

Adams, S.M., Brown, A.M. and Goede, R.W. (1993). *Trans. Am. Fish. Soc.* **122**, 63–73.

AFS (American Fisheries Society) (1994). In *Blue Book. Suggested Procedures for Identification of Certain Finfish and Shellfish Pathogens*, 4th edn (ed. J.C. Thoesen), pp. 1–14. AFS Fish Health Section, Bethedsda, Maryland.

Anderson, R.O. and S.J. Gutreuter (1983). In *Fisheries Techniques* (eds L.A. Nielsen and D.L. Johnson), pp. 283–300. American Fisheries Society, Bethesda, Maryland.

Brown, E.E. and Gratzek, J.B. (1980). In *Fish Farming Handbook*, pp. 237–337. AVI Publishing Company, Westport, Connecticut.

Chiasson, R.B. and Radke, W.J. (1966). *Laboratory Anatomy of the Perch*, 4th edn. Wm. C. Brown, Dubuque, Iowa.

Harder, W. (1975). *Anatomy of Fishes. Part II: Figures and Plates*. E. Schweizerbart/sche Verlagsbuchhandlung (Nägele u. Obermiller), Germany.

Hunn, J.B. and Schnick, R.A. (1990). In *Field Manual for the Investigation of Fish Kills* (eds F.P. Meyer and L.A. Barclay), pp. 30–40. US Fish and Wildlife Service, Resource Publication 177, Washington, DC.

Krahn, M.M., Rhodes, L.D., Myers, M.S., MacLeod, and Malins, D.C. (1986). *Arch. Environ. Contam. Toxicol.* **15**, 61–67.

Ostrander, G.K., Blair, J.B., Hurst, J. and Stark, B. (1993). *Aquat. Toxicol.* **25**, 31.

Phillips, P.H. (1988). In *Fish Diseases*, pp. 95–99. Post Graduate Committee in Veterinary Science, University of Sydney, Australia.

Reimschuessel, R. (1993). In *Fish Medicine* (ed. M.K. Stoskopf), pp. 160–165. W.B. Saunders Co., Harcourt, Brace, Jovanovich, Philadelphia.

Rich, A.A. and Smith, L.S. (1978). *Smith's Introductory Anatomy and Biology of Selected Fish and Shellfish*. College of Fisheries, University of Washington, Seattle.

Roberts, R.J. and Shepherd, C.J. (1986). *Handbook of Trout and Salmon Diseases*, 2nd edn. Fishing News Books, Farnham, Surrey, UK.

Stehr, C.M., Myers, M.S. and Willis, M.L. (1993). In *Sampling and Analytical Methods of the National Status and Trends Program National Benthic Surveillance and Mussel Watch Projects 1984–1992. Volume II: Comprehensive Descriptions of Complementary Measurements*, pp. 63–69. NOAA Technical Memorandum NOS ORCA 71, Silver Spring, Maryland.

Strange, R.J. (1983). In *Fisheries Techniques* (eds L.A. Nielsen and D.L. Johnson), pp. 337–346. American Fisheries Society, Bethesda, Maryland.

Wooster, G.A., Hsu, H.M. and Bowser, P.R. (1993). *J. Aquat. An. Health.* **5**, 157–164.

Yasutake, W.T. and Wales, J.H. (1983). *Microscopic Anatomy of Salmonids: An Atlas*. US Department of the Interior, Fish and Wildlife Service, Resource Publication 150, Washington, DC.

CHAPTER 33

Surgical Techniques

Gerald R Johnson
Department of Veterinary Pathology and
Microbiology, Atlantic Veterinary College,
University of Prince Edward Island, Charlottetown,
Prince Edward Island, Canada

Introduction

The principles of surgery, when adapted to fish, are similar to those of terrestrial species. Aquatic **ectothermic** animals require changes in operating room set-up, surgical techniques and selection of materials, which may not be readily apparent to those approaching fish surgery for the first time. Very specific housing and maintenance requirements before, during and after surgery are equal to surgical expertise to ensure the survival and the return to normal physiologic function of the fish patient. For fish, water provides a total life-support system supplying oxygen, controlling body temperature, contributing to buoyancy and shelter; and removing all metabolic wastes (Casebolt et al., 1998). Provision of a proper aquatic environment during all stages of any surgical procedure while still allowing clean and practical conditions for the surgeon requires considerable planning. A variety of surgical techniques for fish have been developed: (i) to facilitate research on body functions of fish (Klontz and Smith, 1968); (ii) to provide non-lethal biopsy procedures (Wooster et al., 1993b); (iii) for removal of tumors (Harms et al., 1995); and (iv) to facilitate implantation of transmitters and other location devices for remote study of fish behavior (Bidgood, 1980). Increasing numbers of fish are held in private or public aquaria which allow them to live longer and, more importantly, remain visible and valuable. More than ever, geriatric diseases in fish require surgical attention including removal of tumors (Reid and Backman, 1986), unsightly injuries and correction of failed vision (Harms et al., 1995; Nadelstein et al., 1997). Valuable fish, such as prize koi carp, form strong human–animal bonds and some of their diseases, once considered fatal, are being addressed by surgical intervention (Lewbart et al., 1995, 1998). In commercial aquaculture, new species of fish are constantly investigated for their suitability for intensive production. Brood stock may be captured from the wild by hook and line causing cutaneous lesions and fractures that require surgical repair to allow survival in captivity. Research experience in applied and basic science requiring accurate data from catheterized fish held in free-swimming states can be obtained by surgical intervention (Klontz and Smith, 1968; Sohlberg et al., 1997). There are numerous indications for fish surgery in veterinary practice, in the laboratory and in the field. These procedures can be carried out

Copyright © 2000 Academic Press

humanely and effectively with little or no mortality when basic principles and precautions are followed.

Books have been written on the details of veterinary surgical technique. Almost every common mammalian species has a surgery textbook describing the anatomic and physiologic peculiarities of that species as well as aberrant responses to various anesthetics, and surgical or implanted materials. Texts describing surgery in fish have been around for some time (Klontz and Smith, 1968), and descriptions of techniques and suitable materials continue to develop (Reinecker and Ruddell, 1974; Tytler and Hawkins, 1981; Ostrander et al., 1993; Stoskopf, 1993b; Wooster et al., 1993a). Surgery in fish is complex and intricate. Anyone attempting any invasive surgery for research should be properly trained in surgical aseptic technique or obtain the services of a veterinary surgeon.

Animal care

Worldwide, the use of animals for experimentation is open to increasing scrutiny and accountability regarding experimental procedures and research design. A global trend to reduce animal use in testing and to change testing from mammals to species lower on the evolutionary scale has increased interest in fish as research animals. The numbers of fish used experimentally has risen dramatically in the last decade (Canadian Council of Animal Care, 1996). In many countries, fish are classified with other vertebrates and fall under the same regulations as mammals while in others they remain unclassified. Concern for fish welfare is increasing (Verheijen and Flight, 1992). Surgical intervention for any experimentation requires careful consideration of the fish patient's well-being and must concur with animal care and use guidelines as they apply to surgical level of invasiveness (Olfert et al., 1993). All protocols for fish surgery should be reviewed by independent animal care committees or their equivalent. Evaluation should include adequate housing, appropriate water conditions, approved statistical analysis to ensure minimal use of animals, assessment of surgical expertise, operating system evaluation, and post-operative care including clear guidelines as to experimental end points. A continual and thorough assessment of surgical outcomes should be routine to ensure that modifications in surgical techniques will not compromise the fish's well-being. Appropriate attention to pre-surgical planning, personnel training, aseptic surgical techniques and animal well-being will provide the best possible outcome of surgical intervention and the most reliable research data. Surgery on pet fish should be carried out by a licensed veterinarian.

Principles of surgery

Asepsis, the prevention of contact with microorganisms, is essential for any procedure which penetrates the skin, enters a body cavity or exposes internal tissues. Integral asepsis involving sterile materials, drapes, gowns, gloves, as well as a bacteria-free operating room and air supply, are required for mammalian surgery but are almost impossible for fish surgery. However, every attempt should be made to keep surgery conditions as close to mammalian standards as possible. Knecht et al. (1987) provide a review of basic techniques, instruments, and sterile procedures. The concept of delivering anesthetic in free flowing water over the gills to a fish on the surgery table, often within a centimeter of the incision, may be a difficult adjustment for a surgeon and represents one of the challenges to aseptic technique. To maintain the integrity of external mucus and skin, the fish must be kept moist during the entire surgery and at the same temperature as the water from which the fish was removed (see section below on operating equipment). Only the immediate surgical site is maintained dry. The isolation and preparation of an aseptic incision site, though essential, differs from the mammalian norm for a variety of reasons.

With all of the requirements for moisture and water flow during surgery, one might expect an increase in contaminant infections. However, most fish have a lower fluctuating body temperature than do mammals. The bacterial flora to which the fish are sensitive differs from the commonly encountered contaminants from air, fomites, and humans so microbial agents are less likely to survive at the lower temperature or in the foreign environment provided by the fish. Ectothermia provides some advantage in avoiding wound contamination in piscine species with lower natural temperatures (Wagner et al., 1999), while those fish that live at temperatures closer to the mammal, can expect to have a greater risk of wound contamination from common environmental contaminants. Regardless of the relative risk, vigilance should be maintained and surgery under aseptic conditions will provide better success and more humane conditions for the fish.

Anesthesia for surgery

The use of anesthetics in fish has been discussed in Chapter 29 and should be reviewed carefully to select an effective approved compound for the particular surgical procedure being contemplated. The reactions to anesthesia are different with various taxonomic groups. Any fish being considered for surgery should be well understood and trials anesthetizing healthy fish are recommended to ensure proper dosage and accurate calculation of time to reach the necessary plane of anesthesia (Schramm and Black, 1984). Many canulation and transmitter insertions take less than 4 min so fish can be induced to the proper plane of anesthesia, operated on and returned to the recovery tank without additional sedation. Surgical interventions longer than a few minutes run the risk of inducing severe hypoxia (Iwama et al., 1987) and therefore require constant irrigation of the gills with oxygen-saturated water or well oxygenated water containing a low level of anesthetic to prolong the time of anesthesia. If water is recirculated, care must be taken to ensure oxygen levels are not depleted and pH remains neutral, as irrigation of gills will consume oxygen from the water and raise the carbon dioxide level by normal respiratory exchange (Iwama et al., 1987). Whenever possible water used for anesthetic and recovery should be the same as the water from which the fish was selected and to which it will return. Gregarious species of fish, accustomed to close social interaction, should be returned to the same tank to which they were acclimated before surgery and with the same cohort population. Fulfilling the behavioral needs of fish may determine success of post-surgical recovery. The social hierarchy in salmon for instance is important to recover normal swimming activity and return to normal feeding. Fishes that live a more solitary existence such as pike, need to have sufficient places to hide during the recovery stage. Close observation may be necessary to avoid attack of a patient by other fish that perceive anesthetic recovery as aggressive or abnormal behavior. Fish involved in field surgery and returned to their natural field environment directly afterward, have a less severe problem with social behavior than those in tanks as the environment provides familiarity, space and sufficient hiding during recovery (Bidgood, 1980).

Anesthesia in fish is normally delivered in water. The most common is tricaine methane sulfonate (trade names TMS, Syndel International or MS222, Argent Laboratories). Although other compounds are available, few have received formal regulatory approval. For surgery, only approved products should be used. When anesthesia is long or the level of anesthesia adjusted by irrigating the gills, special care must be taken to avoid any contamination of the surgical site.

Healing of surgical incisions

The components of healing in fish are similar to mammals though there are some crucial differences necessary to adapt to the aquatic environment and an ectothermic nature (Ferguson, 1989). The epithelium and associated cuticle and mucous layers provide physical protection, as well as defense against the osmoregulatory imbalance which would ensue if water was allowed to gain entry into the body in the freshwater environment or to leave the body in the saltwater context (Ferguson, 1989). Fish skin is non-keratinized squamous epithelium able to proliferate from any layer of the epidermis (pleuripotential). In addition, several authors have demonstrated the rapid migration of epithelial cells to cover an incision or other injury in the absence of mitotic activity. This unique epithelial migration protects the fish from the effects of the aquatic environment (Ozaki et al., 1973; Bullock et al., 1978; Ferguson, 1989) (Santos Ruiz et al., 1996; Quilhac and Sire, 1999).

Role of ambient temperature

All aspects of healing are influenced by temperature to some degree but epithelial migration seems least affected. Consistently, at a variety of temperatures, in different species of fish, epithelial covering of injured sites occurs within hours (Bullock et al., 1978; Santos Ruiz et al., 1996; Quilhac and Sire, 1999). The epidermal mitotic activity and underlying fibrosis begin after about 3 days but can be delayed and have reduced epithelial thickness when ambient temperatures are lower (Finn and Nielsen, 1971a; Roberts et al., 1973; Bullock et al., 1978). Quilhac and Sire determined that wound healing in a cichlid proceeded as fast as $500\,\mu m\,h^{-1}$ for re-epithelialization, while Santos Ruiz et al. (1996) documented wound healing in the amputated fin of goldfish within 12 h for epithelium and cellular proliferation by 60 h (Santos Ruiz et al., 1996; Quilhac and Sire, 1999).

Inflammatory cells which are essential to wound healing, are more affected by temperature. In rainbow trout and Atlantic salmon, inflammatory cells begin to move into the area to clean up foreign debris and inhibit infection in about 12 h at 15°C. However, salmonids may have little or no inflammatory cell response as late as 4 days if temperature is maintained at 5°C (Finn and Nielsen, 1971a). Fibrous invasion of the wound begins about 3 days post-operatively when temperatures are optimal. Fibroplasia in trout appears advanced at 8 days at 15°C but is just beginning at 16 days at very low temperatures (Finn and Nielsen, 1971a,b). The speed at which the fibrous proliferation closes the internal wound and provides strength sufficient to take over from the sutures varies with the temperature and becomes a factor in the selection of suture material.

Scales, if removed to facilitate surgery, leave deep lesions extending through the dermis. The descaling lesions are usually covered quickly by mobile cutaneous epithelium, providing they were removed cleanly and carefully. Replacement of scales takes considerable time (Sire *et al.*, 1997) but does not affect integrity of the incision.

The absolute temperature at which fish are held is not the only factor that will affect healing. Each species of fish has an optimal temperature range and will function best within that range. Held at room temperature, salmonids are above their optimal temperature range while tilapia are below theirs. Any fish should be acclimated within their optimal temperature range for 14 days before elective or experimental surgery and returned to that system (preferably to the original tank for social and familiarity reasons) after surgery to ensure optimal healing rates. As a general rule, healing mechanisms remain consistent when judged by optimal temperatures and vary little between species within that temperature range.

Complications

In the event of bacterial contamination, or physical disruption, the skin will continue to migrate but will float off the incision site and appear as white, cotton-like projections, sometimes confused with fungal contamination. Examination of the strands by direct smear using a light microscope often reveals only epithelial cells as opposed to fungal contamination. Control of any microbial contamination is essential to timely healing of incisions. The best control is strict aseptic surgical technique. Antibiotics can be used locally and systemically without interfering with healing but they are not usually required if operative conditions are aseptic.

Operating equipment

Surgical suite safety

Detailed planning is essential whether researchers are developing a specific operating room within an experimental unit or modifying mammalian surgery suites to suit an ectothermic aquatic species. Only general differences are dealt with here.

Water will be a part of every consideration. Holding facilities for the fish may be in the same room as the patient. Water will be flowing for respiration if anesthesia is prolonged by intention or because of complications. Water that ends up on the floor can be very slippery causing some occupational and health safety concerns. The smooth easily sterilized floors of a normal surgery can become a liability when fish surgery is being done. Floors may need to be rough enough for traction or temporarily covered to create a non-slip surface. Well-sloped drains should be available if large fish are being considered as patients.

Electricity for special lighting, for water pumps to deliver anesthetic, and for monitoring equipment is likely to be present. To ensure human safety all electrical sources, either on the main line or at the receptacle, should contain **ground fault interrupters (GFI)**. These electrical devices detect any rapid draw of electricity and shut down instantly to optimize human safety. Another alternative is to reduce all electrical sources to 12 volt systems. Both add some expense to the process but no surgery in wet conditions should be conducted without considering the implications of electrical accidents. GFI are required near sinks or other water sources in most electrical codes.

Surgical tables

Every operating table has to meet some basic criteria. The aquatic patient must be able to maintain all essential body functions including adequate respiration, cardiac sufficiency, normal physiologic body temperatures and external body moisture. For the surgeon, the table must provide a stable operating area and suitable ergonomic comfort in terms of height, slope and lighting. Despite differences in conditions and complications of water flowing nearby, the incision area must remain aseptic, free of contaminating water and accessible to the surgeon.

Stability of the patient is paramount to good surgical intervention but the range of species characteristics in fish requires some innovation to meet the challenges of each species. Fish that are elliptical or round in body type are best accommodated by a 'V'-shaped trough. These have been constructed out of many materials including Styrofoam, plexiglass, wood, etc. (Klontz and Smith, 1968; Reinecker and Ruddell, 1974; Goetz, et al., 1977; Stoskopf, 1993b; Tytler and Hawkins, 1981; Lewbart et al., 1995). One of the more robust and adjustable designs is a 4 inch (10 cm) plastic PVC pipe cut in half longitudinally. Length can vary depending on the size range of fish anticipated. A central bolt permanently attached through the pipe allows it to be attached to a tripod (camera, video, or heavier survey type tripods) for easy adjustment of slope for drainage, height for the surgeon, or so it can be leant to either side for unusual lighting complications. The pipe table will accommodate a variety of sizes of fish, by adjusting the thickness with foam sponge inserted as both a bed and a positioning support. The foam offers cushioning but can also help maintain temperature with the addition of appropriate water, either by soaking before surgery or by adding water to the foam during the entire procedure. Foam sponge comes in several types and one must check to ensure that open cell foam is used if water absorption and retention is desired. Straps are not usually required but are easily attached when needed. Care must be taken to ensure that any compression of the fish is absolutely minimal as they tend not to have the heavy skeletal support of mammals.

Obviously fish species present more shapes than will conform to any one table. Flatfish require a flat soft surface of appropriate moisture and temperature levels (Reinecker and Ruddell, 1974). Broad, soft, moist straps may be required for stability and restraint. Laterally flattened fish, especially those that are small, may require tables developed for that particular surgery. Fortunately soft absorbent materials are common. Whenever foam sponge or other latex or Styrofoam materials are improvised in fish surgery, they should be well rinsed to avoid the possibility of any chemical contaminants.

The design of the temperature control systems will determine how the table is sloped. For longer procedures, a flow-through system allowing water to run slowly through the foam bed keeps optimal moisture and temperature. A separate drainage system must be installed for the anesthesia that is flowing over the gills to avoid cross-contamination of the two water sources.

Equipment

Sterile surgical packs

All surgical equipment should be properly sterilized and packed. There are several methods to accomplish this but autoclaving is preferred with the proper addition of scalpel blades, sutures and needles after the pack is opened. If wet sterilization has been used, disinfectant residues may complicate and slow surgical procedures. Equipment lists should be well thought out before beginning. Several generic equipment lists are available (Wooster et al., 1993a).

Drapes

Draping isolates the incision site eliminating external contamination of the wound. The moist body of the fish complicates traditional drape material which tends to be absorbent. The material used to pack surgical instruments (Sterilization Wrap, Source Medical) provides an alternative that is readily available, autoclaves well and is relatively waterproof while still remaining dry on the upper side. It is strong, cheap and easily cut to suit the shape of the incision site. A variety of other materials with a plastic backing stick (Kaydry Table Soaker, VWR Scientific Products) well to the fish mucus and remain steady eliminating the need to attach the material with gels or clamps. If necessary, absorbent materials can be placed under theses drapes to maintain skin moisture. Except in very large fish, clamps are generally avoided due to interference of the scales, trauma to the friable skin and disruption of myofibers which are easily traumatized.

Sutures and needles

The choice of sutures and needles must be defined by the procedure being considered. A **laparotomy** requires a much different choice of instruments than repair of a corneal ulcer. As noted earlier ectothermic animals vary in the rate at which they heal. Absorbable suture materials will dissolve in fish (Stoskopf, 1993b; Gilliland, 1994), however, the time taken may vary considerably due to temperature-induced changes in inflammatory cell and macrophage infiltration. In many experimental systems, permanent stitches are used if fish can be easily captured and stitches removed with light anesthetization. In general, fish with lower optimal temperatures and those at suboptimal temperatures require suture material that will last longer, up to 3 to 4 weeks. Since no internal suturing of each layer is usually done due to the tendency of the peritoneum and other fascia to tear, sutures must hold until fibrous invasion is complete (Bart and Dunham, 1990; Gilliland, 1994). Although permanent sutures provide the least risk to premature release of the incision, they require additional capture and anesthetic for removal.

Steel staples have been successfully used and provide fast closure of incisions suitable for some procedures (Mulford, 1984). Surgical glues work well for small incisions and avoid the need to remove stitches (Olson and Bruce, 1987; Nemetz and Macmillan, 1988). Experimental protocols may be required to determine if a specific type of suture material suits the type of surgery being requested.

Fish surgical procedures

General concepts

Fish surgery is indicated in a wide number of circumstances. Pet fish requiring emergency repair or valuable display fish needing cosmetic surgery are becoming more common within veterinary practice. In scientific experimentation, miniaturization of sensors allows accurate real-time information in cohabited fish rather than studies in close confinement. In the field, transmitters and **transponders** can be inserted under fish skin or into the abdominal cavity to provide months of tracking information to study migration, behavior, or link to reproductive records from previous years. In all cases, fish require humane handling, expedient intervention and rapid recovery to a normal physiologic state. The extent to which those objectives are met, determines the accuracy and reliability of the research results. Many procedures have been described and investigated but only those dealing with normal recovery are dealt with here. Procedures developed for terminal experiments are not considered.

Surgical techniques

Before surgery, food should be withheld from fish patients for at least 24 h to reduce the possibility of regurgitation and subsequent gill damage. Reduction in gut content makes handling of the viscera easier and reduces the risk of contamination (Klontz and Smith, 1968).

Surgical preparation begins when the patient is stable, well anesthetized and respiring in a normal manner. Surgical site disinfection should be done quickly and gently to avoid irritating the skin in a manner that may delay healing. The incision site is dried and disinfected with a suitable compound. Alcohol-based surgical scrubs are not recommended (Stoskopf, 1993b). Povidone iodine-based disinfectants, especially those which do not include acids and detergent compounds (i.e. Ovidine, Syndel Laboratories International) can be gently applied to the operative area. No scrubbing *per se* is required as disinfectants will flow into the remaining mucus and cuticle. Simple drying using sterile gauze or towels followed by an application of the disinfectant will usually suffice. Opinions vary on the necessity of disinfection and its effect on healing (Wagner *et al.*, 1999) but preparing an aseptic surface for incision is a safeguard to introducing bacteria or other biologic contaminants systemically. On some fish, scales may have to be removed after initial preparation. Scale removal leaves a wound through the epidermis and dermis. Scales should be removed by grasping each individual scale and pulling posteriorly to remove it cleanly from the scale pocket. Appropriately sized hemostats provide positive clamping and minimize the trauma of removal. When there are too many small scales to contemplate individual removal, a scalpel can be used to scrape the scales off toward the posterior aspect of the fish. Descaling in a reverse direction as practiced in food preparation leaves a large rough poor healing lesion. Only remove the scales necessary to create the incision as the remainder provide protection and stability to the site. A second drying and disinfection of the surgical site is required after scale removal to ensure that no microorganisms contaminate the scale pockets or enter the body by this route.

Suitable drapes are applied so that only the incision site is exposed. All surgical equipment must be sterile and properly packed. Instruments are generally laid out on a mobile waist high table to be easily accessible to the surgeon. Separate instruments are required for the external and internal incisions. Surgeons should be properly masked and gloved before proceeding to open the skin with one set of instruments, switching to the second set of instruments for internal use, and carrying out the required surgical procedures (Knecht *et al.*, 1987).

Though the epidermis and peritoneum tend to be friable and tear easily, the overall integument is tough and resilient due to a compact dermal layer of dense collagen in most species. Surgeons require experience to obtain the optimal scalpel pressure to maintain control when the blade breaks through this stratum compactum. Scalpel blade size and shape should be chosen to allow sufficient pressure to be exerted while maintaining ample safety to avoid damaging underlying organs. One clean incision creates less trauma, promoting rapid first intention healing than do multiple attempts to incise the skin. Experienced surgeons make a considerable difference in the ability of the fish to heal rapidly without complications.

The integument and muscle tissue of fish are less vascularized and the blood pressure is significantly lower than mammalian counterparts. Incisions can be made in the mid-ventral or paramedial abdomen, for instance, with little or no effort directed toward tying off bleeding vessels. Normal fish blood clots extremely rapidly. As a result, many procedures can be completed in minutes and often do not require continuous anesthesia to provide sufficient time to complete the surgery (see Chapter 30). Closing of individual layers is recommended only for very large fish. In most circumstances, single interrupted sutures to appose all layers at once are sufficient due to epidermal cell mobility discussed above in the section on healing of surgical incisions.

Laparotomy

The most common surgical interventions in fish are cannulations previously described in Chapter 30. Laparotomy is most commonly used for internal surgery. Both the ventral midline laparotomy and paramedian abdominal laparotomy have been successful (Klontz and Smith, 1968; Wagner et al., 1999). Depending on species, the former is preferred (Stoskopf, 1993b), especially in round fish like salmonids with lighter scales, a strong 'linea alba' and a relatively small vascular component along the ventral midline. The incision can be created quickly and easily opened with forceps or retractors for adequate exposure to the abdominal cavity. A laparotomy is a common procedure for biopsy (Wooster et al., 1993a), tumor excision (Reid and Backman, 1986; Harms et al., 1995), exploratory surgery for diagnosis (Lewbart et al., 1998), gonadectomy (Brown and Richards, 1979; Underwood et al., 1986) and for implantations of various devices (Hart and Summerfelt, 1975; Bidgood, 1980). Incisions sites are species dependent and normally chosen by the area of the abdomen to which access is required. Protocols for each species should be reviewed or developed before beginning a major project. The pelvic girdle migrated cranially as fish evolved so position varies between species and may represent an inconvenient obstacle. Incisions cranial or caudal to the pelvis are preferred though the pelvic bones can be split and rejoined if necessary. A detailed post-mortem examination of the species involved may be required to determine the optimal incision sites. Carnivorous fish have simple alimentary systems and there is usually sufficient space available to work quickly and safely. Herbivorous fish have much more intestinal mass which is tightly bound in mesentery and may complicate laparotomies.

After the skin is excised and the peritoneum exposed and punctured, the incision can be lengthened using a scalpel or scissors using the blunt portion toward the interior while lifting the abdominal wall to avoid any internal organs. Any internal bleeding on the viscera must be attended to immediately before proceeding. Fish blood clots rapidly and small bleeders are of no consequence but internal bleeding, even with the low blood pressure found in fish, represents an important lesion.

As described earlier, closure on most fish consists of simple interrupted stitches through the entire body wall. Close apposition without excessive tightening is essential to provide a framework for epithelial migration and subsequent fibrous proliferation and to prevent water leakage. Both the skin and peritoneum will cover the incision rapidly if nothing obstructs their migration so every attempt should be made to ensure proper alignment of apposing tissue layers. Surgeons must be careful not to inadvertently include mesenteric fat or other internal tissues into the incision during closure. Such disruption will prevent healing. Infections must also be avoided and some authors recommend antibiotic powder sprinkled into the incision site. There is no evidence that this provides any advantage.

Elective and emergency surgeries

Debilitating wounds can occur as a result of capture, routine handling, accidents within display tanks, or adverse social interaction in community tanks.

Hierarchical battles during breeding seasons may leave a valuable fish with life threatening injuries. As fish age in captivity, diseases, rarely seen in wild populations, have appeared, including ocular problems (Nadelstein et al., 1997), swimbladder dysfunction (Lewbart et al., 1995), neoplasia (Stoskopf, 1993b; Harms et al., 1995) and fractures (Bakal et al., 1997).

In aquaculture, Arctic char have a well-described syndrome of swimbladder hyperinflation but the problem occurs in many species periodically. Swimbladders in fish become distended and cause fish to float at the surface during rest. Eventually the sustained swimming against this increased buoyancy causes fatigue and death. Many types of repair have been attempted. A hypodermic needle inserted laterally through a prepared site on the abdominal wall or dorsocranially through the cloaca work most effectively for short- to medium-term relief. Evaluation of the reproductive state of valuable brood fish can be enhanced by simple surgical intervention with videolaparoscopy via the genital pore (Ortenburger et al., 1996). Examinations can ensure the sex of the fish, the reproductive state and estimate the time to spawning.

Field surgery and tagging

Fisheries biologists use a variety of internal and external tags from fin clips to subcutaneous and abdominal transponders to identify a segment of a large subset of fish to study population dynamics. Some tagging systems cause virtually no adverse reactions such as visual implant tags while others cause considerable cutaneous damage. Roberts et al. (1973) studied the effects of tagging procedures and described both short term and long-term pathologic reactions.

Surgical implantation of radio transmitters evolved with the miniaturization of telemetry devices in the early 1970s (Bidgood, 1980). Hydrosonic or acoustic transmitters and their power supply became small enough to fit in the abdomen of sport fish in order to assess migration routes, study social behavior or localize reproductive sites within a water body (Johnsen and Hasler, 1977; Prince and Maughan, 1978).

Fish captured from the wild may be subject to field-based surgery. All of the principles of surgery discussed previously for laboratory-based surgery apply when attempting a laparotomy for permanent transmitter insertion into the abdomen in a field situation. Procedures need to ensure adequate anesthesia, temperature control, expedite well-planned procedures, asepsis and evaluation of after-effects of the surgery. Experimental fish may have to be held until recovery is complete before return to their natural environment. However, many species find confinement the most stressful and are best released. Good research data depend on good recovery from surgery so transmitters should be retrieved immediately to assess surgery failures if fish do not continue to move.

Surgical complications

The most common cause of surgical failure is due to inappropriate tension on the sutures. Sutures that are placed too tightly will tend to tear through the tissue while sutures that are loose allow contamination with water and water-borne microorganisms, causing dehiscence of the incision or inhibiting epithelial migration. Surprisingly, breakdown of the incision site may not be immediately fatal. In rainbow trout, the omental adipose tissue can adhere or block the incision site sufficiently to protect against leakage of water. Osmoregulatory capability of the fish may remain intact until the fish can be captured, anesthetized and the incision debrided and closed a second time.

Experiments evaluating surgical implantation of transmitters have noted some interesting complications. Within hours of the surgery, rather large transmitters have been evacuated from gonadal pores assumed to be too small to physically accommodate the transmitter or from erosion through the lower intestine (Schramm and Black, 1984; Chisholm and Hubert, 1985). Some implants have rested against the incision and with the constant force of gravity caused dehiscence of the incision. Inflammation due to foreign body response to the transmitter covering normally encases the transmitter holding it firmly in place. Once walled off within the abdomen, a neutral density transmitter has virtually no adverse effect on most fish. However, some authors have noted transmitter evacuation several weeks after implantation due to 'intestinalization' of transmitters (Summerfelt and Mosier, 1984). Encapsulation of the transmitter adjacent to the intestinal wall encompasses the transmitter and eventually becomes integrated into the intestine which envelops the transmitter. Defecation of the transmitter occurs several weeks after the surgery despite complete uncomplicated healing of the incision (Chisholm and Hubert, 1985). With the use of more inert materials, the fibrous encapsulation of implants has been reduced.

Managing convalescents

Holding facilities

Post-operatively fish have to be held in the best facilities possible. Each species has its own requirements as reviewed in Chapters 1 and 2 of this text. In addition to optimal general husbandry, extra precautions in the first 3 days of recovery include an ability to observe patients closely, maintenance of a low microbiologic challenge, absence of cutaneous parasites, and sufficient protection to overcome territorial aggression yet allow appropriate social interaction with cohorts to help restore normal exercise and feeding habits. Fish tanks collect suspended solids originating from excess feed, fecal material, or abundant sloughed mucus, all of which can act as a nidus for bacterial proliferation which can contaminate the incision, consume oxygen from hospital tank water, and add to ammonia production (Casebolt et al., 1998). Re-establishing normal feeding habits helps to ensure sufficient energy stores to optimize healing. Post-surgical appetite varies with species. Any uneaten food, even live diets such as *Artemia*, contaminate recovery tanks. Many diets contain a significant amount of oil which is difficult to remove from the water. Dry diets should be inspected to ensure that no dusty or crumbled feed is included in the pellets fed. The effects of such precautions are not readily apparent but will affect water quality precipitously when they are allowed to build up. Post-surgical containment should have extra filtering capability to remove any waterborne complications including uneaten food, oil lost from the food or mortalities and other suspended solids. Ultraviolet sterilization of any recirculating water as it re-enters the recovery tank helps to control build up of microbials during post-operative care.

Holding facilities should be monitored more often for adequate oxygen supply and appropriate water chemistry, especially watching for elevated levels of ammonia, nitrite, nitrate and carbon dioxide. Currents should be adjusted in recovery tanks to allow for mild mandatory exercise, being careful not to cause any fatigue due to excessive exercise. Care must be taken to ensure that post-operative fish are not exposed to exogenous toxins such as chlorine from domestic water supplies, cleaning agents, pesticides like fly sprays, etc., or herbicides, especially those used to control mildew. Any silicone used to repair tanks, fix hoses, or prevent leaks must not contain any mildewicides as these are very toxic to fish.

Importance of stress reduction

Returning fish to the same cohort group in the same tank with optimal environmental conditions helps to reduce stress in most species. In all cases, adequate safe shelter should be available in case adverse social reactions are directed toward fish that may swim abnormally several hours after recovery from anesthesia (Stoskopf, 1993a). Shelter may be as simple as lengths of plastic pipe of appropriate size for a single fish to swim into, or more complex arrangements such as stiff plastic netting that allows water and visual contact without the possibility of physical interaction.

The importance of this extra effort shows in the consistency (or lack thereof if ignored) of the results of the research that follows. Everything from behavior to drug metabolism is affected by chronic stresses on the fish (Iwama et al., 1987). Diseases commonly follow stressful situations and fish that return to health earlier will be better able to avoid diseases and resume healthy normal metabolism. For instance, many parasites that live on fish in experimental systems without consequence and often unnoticed, will proliferate rapidly in apparently 'clear water' following surgical intervention. Fish demonstrating any irritation following surgery should be carefully examined for superficial skin or gill protozoa and monogenians.

Lighting

Often overlooked, lighting inconsistencies are very stressful for some species of fish. Post-operatively, the human tendency to check fish often using bright room lighting systems has to be avoided. Some fish, like freshwater tropicals, tend not to be concerned about bright lights or the change from light to dark. Salmonids, however, express measurable stress to rapid change from light to dark and vice versa (Rostant, 1996). Gradual light increases imitating dawn or dusk are preferred for diurnal light controls and maintenance of appropriate low light conditions during the day minimizes the stress of post-operative living. Examinations with a pencil beam flashlight allow adequate scrutiny of incisions without disturbing the fish. Fish are sensitive to photoperiod and should remain on the correct photoperiod control throughout the surgery and afterward so no hormonal complications to healing occur.

Summary

Good surgery is a team effort. All persons involved in an experimental protocol requiring surgery on fish should be well trained in aseptic technique and principles of surgery. Additional challenges involve maintenance of a clean and practical interface between the aquatic environment and the site of surgery. Because surgical procedures add a significant stress to the fish, the risk of disease rises requiring the utmost sanitation during procedures and the best of conditions for convalescent care. Understanding of the specific needs of the species of fish being used, including pre- and post-operative housing and care, is essential to humane treatment and a high level of survivability.

References

Bakal, R.S., Love, N.E., Lewbart, G.A. and Berry, C.R. (1997). *Vet. Radiol. Ultrasound* **39**(4), 318–321.

Bart, A.N. and Dunham, R.A. (1990). *Prog. Fish Cult.* **52**, 241–246.

Bidgood, B.F. (1980). *Fisheries Report 20. Field Surgical Procedure for Implantation of Radio Tags in Fish*. Alberta Fish and Wildlife Division, Edmonton, Canada.

Brown, L.A. and Richards, R.H. (1979). *Vet. Rec.* **104**, 215.

Bullock, A.M., Marks, R. and Roberts, R.J. (1978). *J. Zool., Lond.* **185**, 197–204.

Canadian, Council of Animal Care (1996). *Animal Utilization Survey*, pp. 1–4. Canadian Council on Animal Care, Ottawa, Canada.

Casebolt, D.B., Speare, D.J. and Horney, B.S. (1998). *Lab. Anim. Sci.* **48**(2), 124–136.

Chisholm, I.M. and Hubert, W.A. (1985). *Trans. Am. Fish. Soc.* **114**, 766–767.

Ferguson, H.W. (1989). In *Skin, in Systemic Pathology of Fish* (ed. H.W. Ferguson), pp. 41–42. Iowa State University Press, Ames, Iowa.

Finn, J.P. and Nielsen, N.O. (1971a). *J. Path.* **105**, 257–268.

Finn, J.P. and Nielsen, N.O. (1971b). *J. Fish Biol.* **3**, 463–478.

Gilliland, E.R. (1994). *Prog. Fish Cult.* **56**, 60–61.

Goetz, F.W., Jr., Hoffman, R.A. and Pancoe, W.L. (1977). *J. Fish Biol.* **10**, 287–290.

Harms, C.A., Bakal, R.S., Khoo, L.H., Spaulding, K.A. and Lewbart, G.A. (1995). *J. Am. Vet. Med. Assoc.* **207**, 1215–1217.

Hart, L.G. and Summerfelt, R.C. (1975). *Trans. Am. Fish. Soc.* **1**, 56–59.

Iwama, G.K., Boutilier, R.G., Heming, T.A., Randall, D.J. and Mazeaud, M. (1987). *Can. J. Zool.* **65**, 2466–2470.

Johnsen, P.B. and Hasler, A.D. (1977). *Trans. Am. Fish. Soc.* **6**, 556–559.

Klontz, G.W. and Smith, L.S. (1968). In *Methods in Animal Experimentation*, vol. 3, 3rd edn (ed. W. Gay), pp. 323–385. Academic Press, London.

Knecht, C.D., Algernon, R.A., Williams, D.J. and Johnson, J.H. (1987). *Fundamental Techniques in Veterinary Surgery*, 3rd edn. W.B. Saunders, Philadelphia.

Lewbart, G.A., Stone, E.A. and Love, N.E. (1995). *J. Am. Vet. Med. Assoc.* **207**, 319–321.

Lewbart, G.A., Spodnick, G., Barlow, N., Love, N.E., Geoly, F. and Bakal, R.S. (1998). *Vet. Rec.* **143**, 556–558.

Mulford, C.J. (1984). *N. Am. J. Fish. Manage.* **4**, 571–573.

Nadelstein, B., Bakal, R. and Lewbart, G.A. (1997). *J. Am. Vet. Med. Assoc.* **211**(5), 603–606.

Nemetz, T.G. and Macmillan, J.R. (1988). *Trans. Am. Fish. Soc.* **117**, 190–195.

Olfert, E.D., Cross, B.M. and McWilliam, A.A. (1993). *Guide to the Care and Use of Experimental Animals*, vol. 1. pp. 1–11. Canadian Council on Animal Care, Ottawa, Canada.

Olson, M.E. and Bruce, J.B. (1987). *Lab Anim.* **16**(2), 27–30.

Ortenburger, A.I., Jansen, M.E. and Whyte, S.K. (1996). *Can. Vet. J.* **37**, 96–100.

Ostrander, G.K., Blair, J.B., Stark, B.A. and Hurst, J.G. (1993). *Aqua. Toxicol.* **25**, 31–41.

Ozaki, H., Akima, R. and Harada, M. (1973). *J. Tokyo Univ. Fish.* **59**, 69–78.

Prince, E.D. and Maughan, O.E. (1978). *Prog. Fish Cult.* **40**(3), 90–93.

Quilhac, A. and Sire, J.Y. (1999). *Anat. Rec.* **254**, 435–451.

Reid, S. and Backman, S. (1986). *Can. Vet. J.* **29**, 172.

Reinecker, R.H. and Ruddell, M.O. (1974). *Prog. Fish Cult.* **36**(2), 111–112.

Roberts, R.J., MacQueen, A., Shearer, W.M. and Young, H. (1973). *J. Fish Biol.* **5**, 497–503.

Rostant, E.S. (1996). Biochemical and Behavioral Indicators of Sensory Perception in Atlantic Salmon (SALMO SALAR). MSc Thesis, University of Prince Edward Island, Canada.

Santos Ruiz, L., Santa Maria, J.A. and Becerra, J. (1996). *Int. J. Dev. Biol.* Suppl 1, 183S–184S.

Schramm, H.L. Jr. and Black, D.J. (1984). *Prog. Fish Cult.* **46**, 185–190.

Sire, J.Y., Allizard, F., Babiar, O., Bourguignon, J. and Quilhac, A. (1997). *J. Anat.* **190**(Pt 4), 545–561.

Sohlberg, S., Martinsen, B., Horsberg, T.E. and Soli, N.E. (1997). *J. Vet. Pharmacol. Therap.* **20**, 493–495.

Stoskopf, M.K. (1993a). In *Fish Medicine* (ed. M.K. Stoskopf), pp. 98–112. W. B. Saunders, Philadelphia.

Stoskopf, M.K. (1993b). In *Fish Medicine* (ed. M.K. Stoskopf), pp. 91–97. W. B. Saunders, Philadelphia.

Summerfelt, R.C. and Mosier, D. (1984). *Trans. Am. Fish. Soc.* **113**, 760–766.

Tytler, P. and Hawkins, A.D. (1981). *Vivisection, Anaesthetics and Minor Surgery*, pp. 247–278. Academic Press, London.

Underwood, J.L., Hestand, R.S. and Thompson, B.Z. (1986). *Prog. Fish Cult.* **48**, 54–56.

Verheijen, F.J. and Flight, W.G.F. (1992). *What We May and May Not Do to Fish*, pp. 1–6. UBI, Neuroethology Group, Comparative Physiology, Utrecht.

Wagner, G.N., Stevens, E.D. and Harvey-Clark, C. (1999). *J. Aqua. Anim. Hlth.* **11**, 373–382.

Wooster, G.A., Hsu, H.M. and Bowser, P.R. (1993a). *A Manual for Nonlethal Surgical Procedures to Obtain Tissue Samples for Use in Fish Health Inspections.* NRAC Publication No. 112-1993, pp. 1–27. Northeastern Regional Aquaculture Center, Massachusetts.

Wooster, G.A., Hsu, H.M. and Bowser, P.R. (1993b). *J. Aqua. Anim. Hlth.* **5**, 157–164.

CHAPTER 34

Fixation of Fish Tissues

John W Fournie
US Environmental Protection Agency, National Health and Environmental Effects Research Laboratory, Gulf Ecology Division, Gulf Breeze, Florida, USA

Rena M Krol and William E Hawkins
Institute of Marine Sciences, The University of Southern Mississippi, Ocean Springs, Mississippi, USA

Introduction

This chapter deals with the fixation of fish tissues and whole fish. Traditionally, fixation has been applied to animal tissues mainly for histological or pathological studies. Development of new molecular and immunologic tools now allows tissue and cellular localization of nucleotide sequences and antigens, even from archived specimens. With fishes, the need for conducting anatomical pathology analyses has increased in recent years due to the incorporation of fish health condition as a component in state and federal monitoring programs, the need to monitor the health of various types of finfish in aquaculture, and the increased use of fish models in aquatic toxicology, carcinogenicity testing, and biomedical research. Molecular and immunologic applications involving fish have lagged behind those using mammals but, for the most part, reagents and procedures used for mammals are directly applicable to fishes. Regardless of application, specimens must be properly processed so that structure is maintained and reliable histological evaluations can be performed. The process begins with fixation, the most important step in producing good histologic specimens. Realizing that no one fixative is ideal for all situations, careful consideration should be given to study objectives and potential future uses of the specimens.

In this chapter, we provide information dealing with the principal facets of fixing fish tissues, beginning with a brief discussion of the purpose of fixation and the basic elements of the chemistry of fixation. We describe the various types of fixatives available, provide recipes for the most commonly used fixative

Copyright © 2000 Academic Press

mixtures, and briefly discuss the advantages and disadvantages of the fixatives. The choice of fixatives for specific procedures such as preparation of liver tissue for light microscopy, the preparation of whole small fish for histological evaluation, and the use of molecular and immunohistochemical techniques are discussed. As modern communication technologies increase the speed of exchange and availability of information, the Internet becomes more important. Here, we list and briefly describe some currently accessible websites and servers that might be useful when considering studies that deal with fixation of fish tissues.

Fixation protocols for fish in general are found in Hinton (1990). Protocols for several individual species include the following: channel catfish, *Ictalurus punctatus* (Grizzle and Rogers, 1976); striped bass, *Morone saxatilis* (Groman, 1982); several salmonids (Yasutake and Wales, 1983); guppy, *Poecilia reticulata*, and Japanese medaka, *Oryzias latipes* (Fournie et al., 1996).

Purpose and principles of fixation

The objective of fixation is to preserve cell constituents and the morphology of tissues in a condition as near as possible to that existing during life. This is accomplished by the prevention or inhibition of autolysis and putrefaction. Autolysis occurs when proteolytic enzymes split proteins into amino acids making them incapable of coagulation, the principal fixation mechanism of some fixatives. Putrefaction is postmortem decomposition caused by non-pathogenic bacteria. Fixation also serves to harden naturally soft tissue which allows for easy manipulation during subsequent processing and sectioning and aids in visualizing structures when chemicals and biological dyes are applied (Sheehan and Hrapchak, 1980).

To achieve these objectives, some principles must be adhered to regardless of study objective or choice of fixative. Detailed instructions on fixation procedures are found in several classical references on histotechnique including the following: Gray, 1952; Baker, 1958; Davenport, 1960; Lillie, 1965; Luna, 1968; Pearse, 1968; Humason, 1972; Sheehan and Hrapchak, 1980; Presnell and Schreibman, 1997. Tissue samples or whole animals *must* be placed into an appropriate volume of fixative immediately upon removal or as soon after death as possible. The optimal size for tissues to be fixed for light microscopy is about 2 cm^2 in greatest dimension and about 3–4 mm thick but these values vary with the density of the tissue. For example, compared with a compact tissue such as liver, much larger pieces of spongy tissues such as fish heart can be adequately fixed. Blood should always be washed away as it retards penetration of the fixative. Heat and desiccation are the only physical processes involved in fixation. Heat causes rapid coagulation and both heat and desiccation cause extreme distortion of tissues. Generally, fixation for 6–48 h at room temperature is adequate. Good fixation brings out the refractive index of tissues (approximately 1.54) which is then modified by staining (Sheehan and Hrapchak, 1980). Sometimes it is necessary to store tissues for prolonged periods before they are processed. Regardless of the fixative used, most references recommend that tissues be stored either in 70% ethanol or in 10% neutral buffered formalin. Storage in ethanol restores some of the color lost due to formalin fixation and does not affect staining quality. Storage in buffered formalin results in some gradual loss of basophilic staining properties and there is concern that the formation of polymers in formalin mask some antigenic properties. Improper storage can result in severe damage due to aldehydes (when stored in formalin) which inhibit enzyme activity and also to oxidation which can affect staining. Although ethanol storage is best, most laboratories use buffered formalin because of cost (Bochenski, 1984). Thompson (1966), however, recommends that tissues fixed in buffered formalin should be stored in buffered formalin.

Chemistry of fixation

Fixatives are chemical substances (i.e. liquids, solids, or combinations of them) that act as coagulants or non-coagulants of proteins in cells and tissues (Putt, 1972). At their isoelectric pH, proteins are neutral. At that point, they show minimal solubility and viscosity, the lowest osmotic pressure and the least swelling. Differences in isoelectric pH caused by fixing solutions affect how well tissues stain. In living cells, natural proteins occur as colloidal solutions or as gels. Fixation renders most of the protein of tissues insoluble and converts the living, natural jelly-like condition into a fine-ground spongy mass. Coagulating fixatives such as picric acid, mercuric chloride and ethanol precipitate protein to form a clot of finely shredded strands which stabilize the proteins, permitting rapid penetration of dehydrating, clearing and embedding agents. As they penetrate tissues, coagulating fixatives produce complete fixation by strengthening protein linkages against disruption by subsequent processing. However, coagulating fixatives can introduce artifacts into tissues such as alteration or disruption of nuclear and cytoplasmic membranes, displaced mitochondrial cristae, and disrupted, unevenly distributed or damaged cellular ground substances. Non-coagulating fixatives such as formaldehyde and osmium tetroxide do not transform the protoplasm in that way, resulting in minimal distortion of cells and tissues and reduction in artifacts. Most commonly used fixatives are composed of three or four chemical substances formulated to balance the disadvantages of one component with the advantages of another (Putt, 1972). Therefore, many fixatives contain both coagulants and non-coagulants. Coagulants facilitate embedding processes whereas non-coagulants control both the texture and the staining reactions of the tissue.

Because formaldehyde is the most commonly used simple fixative and is a component of the majority of fixative mixtures used in fish pathology, we provide some specific information on the chemistry of formalin fixation. Formalin is a good polymerizing fixative because of its ability to form links between adjacent protein chains. The most frequently encountered reaction of formaldehyde is its addition to a compound containing a reactive hydrogen atom with the resultant formation of a hydroxymethyl compound (Pearse, 1968). This compound is usually reactive and it may condense with an additional hydrogen atom to form a methylene bridge. The conversion of free amino groups to methylene derivatives is the most marked change in the process (Davenport, 1960).

Types of fixatives

Several criteria should be considered in choosing a fixative including: (i) the kind of tissue; (ii) the purpose of the study; (iii) the type of lesion; (iv) whether fish tissues or whole fish are being prepared; and (v) the staining or special techniques to be employed. These and other considerations will play a role in determining whether a simple fixative, a coagulant or non-coagulant fixative, or a fixative mixture is used.

Most studies dealing with the suitability of various fixatives have been conducted on mammalian tissues. Several studies have examined various fixatives for fish tissues. Some of those are Hinton *et al.* (1984), who examined fixatives for use with specific tissues and techniques on several fishes; Speilberg *et al.* (1993), who evaluated different fixatives for light microscopic studies of salmonid liver tissues; and Fournie *et al.* (1996), who reported techniques for the preparation of whole small fish for histological evaluation. Ortego *et al.* (1994) and Hemmer *et al.* (1998) investigated fixation procedures related to the immunoreactivity of mammalian antibodies used in immunohistochemical studies with fishes. These and some other references on fixatives are summarized later.

Simple coagulant fixatives

This group of fixatives includes ethanol, picric acid, mercuric chloride and chromium trioxide. Below, we provide detailed information on two of these that are used alone but can also be combined with other fixatives. Even though these chemicals are rarely used individually as fixatives, it is important to realize that several chemical substances can serve as fixatives and to be aware of their fixative properties.

Ethanol

Ethanol is a non-additive, coagulant fixative that is occasionally used for fixing enzymes. It is a strong cytoplasmic coagulant that penetrates rapidly, does not fix chromatin well and is seldom used alone. Absolute ethanol is the standard concentration for fixation. It shifts the isoelectric points of proteins less than other fixatives (Baker, 1958). Ethanol shrinks and hardens tissues, lyses red blood cells and leukocytes, and renders glycogen insoluble. Ethanol is the recommended fixative for glycogen. It is compatible with picric acid, mercuric chloride, formaldehyde and acetic acid.

Picric acid (trinitrophenol)

Picric acid penetrates slowly and fixes rapidly. A saturated aqueous solution (0.9–1.2% aqueous) is the standard concentration for fixation (Presnell and Schreibman, 1997). It is the only compound that is used both as a fixative and as a dye. Picric acid is an excellent protein coagulant and forms protein picrates with a strong affinity for acid dyes (Humason, 1972). Although it causes up to 50% shrinkage, it leaves the tissues soft. It does not dissolve lipids or fix carbohydrates well but, nevertheless, is recommended as a fixative for glycogen. Picric acid is also very tolerant of mixture with other fixatives (see the section on Bouin's fixative below).

Simple non-coagulant fixatives

This group includes formaldehyde, osmium tetroxide, potassium dichromate and acetic acid. These fixatives act by breaking down cross-linkages between protein molecules and by releasing radicals that associate with water molecules. Formaldehyde is the most frequently utilized fixing agent in histological laboratories.

Formaldehyde

Formaldehyde is a gas which is soluble to approximately 37–40% by weight in water and is commonly known as 40% formalin (Putt, 1972). Formalin preserves the structure of the living cell better than any other primary fixative except osmium tetroxide (Baker, 1958). It has a relatively rapid rate of penetration due to its low molecular weight. Formalin is a good preservative of lipids, hardens tissues well but does not fix soluble carbohydrates. Formalin fixation is compatible with a wide variety of staining procedures and specimens can be stored in it for many years without excessive deterioration. Formalin is inexpensive, easy to prepare and maintains cell morphology, antigenicity and genetic material well. It is a component of most of the commonly used fixative mixtures.

NEUTRAL BUFFERED FORMALIN (NBF)
Formaldehyde (37–40%) 100.0 ml
Distilled water 900.0 ml
Sodium phosphate monobasic 4.0 g
Sodium phosphate dibasic
 (anhydrous) 6.5 g

Tissue may remain in fixative indefinitely; action is progressive. NBF is often referred to as the 'universal' fixative.

Acetic acid

Acetic acid is one of the oldest known fixatives (Humason, 1972). It causes cell constituents, particularly protein gels, to swell far more than any fixative (Baker, 1958). Therefore, it is seldom used alone but combined with a reagent that has shrinking characteristics. It penetrates rapidly and gives lifelike preservation to nuclei. However, it precipitates nucleoproteins and destroys mitochondria, Golgi bodies and calcium deposits (Humason, 1972). It does not fix lipids or carbohydrates. Tissues should be transferred to alcohol rather than water to minimize swelling. Acetic acid leaves tissues much softer than any other fixative.

Fixative mixtures

Fixative mixtures consist of two or more substances, each of which may act as a fixative when used alone. So far as possible, the components should complement one another and compensate for their respective deficiencies. Therefore, a trio of ingredients is frequently used that consists of acetic acid, a coagulant fixative and a non-coagulant fixative. In mixtures, acetic acid antagonizes the shrinking effect of other fixatives while non-coagulant fixatives stabilize proteins. Tissues are not readily infiltrated with paraffin unless they have been made spongy to some extent by the inclusion of a coagulant fixative. If a coagulant is used alone, the cytoplasm and nucleoplasm tend to be transformed into a coarse meshwork. However, if formaldehyde or osmium tetroxide is included, a compromise is reached. The protoplasm is more smoothly fixed but the paraffin can still penetrate. Below, we describe the most commonly used fixative mixtures, the recipes for preparing these mixtures and briefly discuss the advantages and disadvantages of the mixtures.

Bouin's and Lillie's fixatives

Bouin's fixative consists of formalin, acetic acid and picric acid. In Lillie's fixative, which is closely related to Bouin's, the acetic acid is replaced by formic acid. Both Bouin's and Lillie's penetrate rapidly and evenly without shrinkage, are excellent for muscle and connective tissue, and are recommended for soft and friable tissues and for the demonstration of glycogen. Bouin's and Lillie's should not be used for tissues to be stained by the Feulgen reaction due to the potential for over-hydrolysis of nucleic acid, i.e. the coagulation of nucleoprotein caused by the picric acid. Because Bouin's and Lillie's contain picric acid, they form protein picrates which are water-soluble. Therefore tissues should be transferred directly to 50–70% alcohol to remove excess picric acid. Fixation for 24 h to several weeks is recommended. Fixation for long periods, several months or more, leads to poor nuclear staining. The advantage of Lillie's over Bouin's as a fixative for whole small fish is that the stronger formic acid of Lillie's decalcifies better than the acetic acid of Bouin's fixative.

As a fixative mixture, the components of Bouin's complement one another. The formalin fixes cytoplasm and nucleoplasm and the picric acid coagulates cytoplasm to readily admit paraffin. Acetic acid causes swelling of tissues and breaks salt linkages between protein chains. It does not harden tissue and actually prevents some hardening caused by alcoholic treatment. It dissolves out certain cell inclusions such as mitochondria, Golgi bodies and some minerals such as calcium (Presnell and Schreibman, 1997). Acetic acid counteracts shrinkage caused by picric acid. It compensates for the deficiencies of both formaldehyde and picric acid. Picric acid offsets disadvantages of formalin such as the excessive hardening of tissue and delayed paraffin infiltration and renders cytoplasm basophilic.

BOUIN'S FIXATIVE
Picric acid (saturated aqueous)	750 ml
Formaldehyde (37–40%)	250 ml
Glacial acetic acid	50 ml

Because of its acetic acid content, do not use this fixative for cytoplasmic inclusions. Large vacuoles often form in tissues. Wash in 70% alcohol plus several drops of ammonium hydroxide for several days until no more yellow comes out. Most of the yellow color should be removed before sections are stained. Otherwise, treat slides in 70% alcohol plus a few drops of saturated lithium carbonate to further remove the yellow stain. Some of the yellow color will be lost in the series of alcohols used in processing.

LILLIE'S FIXATIVE
Picric acid, 1–2% aqueous (1–2 g/100 ml water)	850 ml
Formaldehyde (37–40%)	100 ml
Formic acid, 90–95% aqueous	50 ml

This fixative is used for 1 to 2 days. To extract some of the yellow stain from the picric acid, place in 70–80% alcohol for 2–3 days.

Dietrich's and Davidson's fixatives

Dietrich's fixative consists of formaldehyde, ethanol, glacial acetic acid and distilled water (Gray, 1954; Yevich and Yevich, 1994). It provides good cellular detail with little shrinkage or distortion. Specimens can be stored in Dietrich's for two to three months if necessary without excessively hardening the tissues. The use of Dietrich's also shortens the time necessary in decalcifying fluid because the glacial acetic acid acts as a decalcifying agent. Fish larvae and very small fish that have been in Dietrich's for 3 to 4 days need little if any decalcification (Yevich and Yevich, 1994). Tissues that have been in Dietrich's fixative for an extended period must be washed in running water for 24 h in order to prevent the formation of a black formalin precipitate on the tissue sections. Overall, this is an excellent fixative mixture for whole fish and fish tissues and is used by the US Environmental Protection Agency and the National Oceanic and Atmospheric Administration for their extensive field monitoring and sampling programs (Macauley and Summers, 1991).

DIETRICH'S FIXATIVE
Distilled water	750 ml
Ethanol (95%)	375 ml
Formaldehyde (37–40%)	125 ml
Glacial acetic acid	25 ml

Davidson's fixative is similar to Dietrich's and consists of formaldehyde, ethanol, glycerin, glacial acetic acid and distilled water. Davidson's fixative was introduced by W.H. Hartmann and named for W.M. Davidson, a British hematologist (Richmond, 1997). Use of this fixative was publicized by Moore et al. (1953) in their studies on sex chromatin. In that paper, they indicated that 10 parts glycerin was omitted from the original Davidson's recipe. Davidson's fixative with sea water replacing distilled water was chosen by Shaw and Battle (1957) as the best fixative for marine bivalve molluscs. Davidson's is frequently used for marine fishes and was recently evaluated on Atlantic salmon liver by Speilberg *et al.* (1993). They found it to be one of two fixatives that provided the best maintenance of liver morphology at the light microscopic level. Davidson's is also an excellent fixative for fish gonads.

DAVIDSON'S FIXATIVE

Formaldehyde (37–40%)	200 ml
Glycerol	100 ml
Glacial acetic acid	100 ml
Absolute alcohol	300 ml
Distilled water	300 ml

Fixation for electron microscopy

Fixation methods for both transmission electron microscopy (TEM) and scanning electron microscopy (SEM) are similar. However, fixation is far more critical for electron microscopy than for light microscopy. As with all fixatives, the goal is to preserve the tissue quickly with minimum swelling or shrinkage. Electron microscopy presents other physical factors such as exposure of the tissue to a vacuum and to an electron beam. The speed of preservation depends on the rate of penetration and fixative reactivity which, in turn, are dependent on pH, osmolarity, temperature, duration of fixation, fixative concentration and specimen size. Treatments of these aspects of chemical fixation are provided by Dawes (1988) and Hayat (1989).

Coagulative fixatives are inappropriate for electron microscopy because they distort cellular detail too severely. Non-coagulative additive fixatives commonly used for electron microscopy are glutaraldehyde, osmium tetroxide and formaldehyde. These fixatives can be additive to cell constituents other than proteins. For example, because osmium tetroxide is a heavy metal, it can stabilize some lipids or stain tissue because it increases the tissue density. Generally, formaldehyde penetrates rapidly but fixes slowly, glutaraldehyde penetrates slowly but fixes rapidly, and osmium tetroxide penetrates and fixes slowly. A double fixation method with initial fixation with glutaraldehyde or Karnovsky's fixative, a mixture of glutaraldehyde and formaldehyde (Karnovsky, 1965), followed by post-fixation with osmium tetroxide is most effective for preservation (Hayat, 1989). Glutaraldehyde causes more nuclear shrinkage but less cytoplasmic shrinkage than osmium tetroxide or formaldehyde. Formaldehyde is advantageous because it cross-links both nucleoprotein and DNA, while glutaraldehyde cross-links only nucleoprotein.

Although fixatives penetrate more rapidly at room temperature, tissues are fixed for electron microscopy at around 4°C because there is less extraction of cellular components and autolysis is slowed. Tissues must be obtained within minutes of death to preserve cellular constituents. This seems to be especially true for fish, particularly those taken from warm waters, perhaps because fish are **poikilothermic**. The optimal size of tissues is 0.5–1.0 mm^3. If tissues cannot be cubed because the area of interest would be lost, pieces with at least one dimension of 1 mm thickness or less may be used.

The pH of the vehicle buffer for solutions of glutaraldehyde and osmium tetroxide is a critical factor in good fixation. Buffers influence the appearance and stainability of organelles. The molecular mass of proteins is related to pH. Sodium cacodylate, which is widely used, is effective between pH 6.4 and 7.4 (Dawes, 1988). The pH of most animal tissues is between 6.0 and 8.0 (Hayat, 1989). Sodium cacodylate does not affect enzymatic localization and electrolytes such as calcium can be added as stabilizers without causing precipitation as can occur with phosphate buffers. The same buffer should be used for both initial and post-fixation solutions.

The total osmolarity of a fixative solution affects cell size and shape. Each cell type has an optimal osmolarity, and since tissues are composed of multiple cell types, it is important to find an average osmolarity to produce the most satisfactory preservation. Osmolarity has minimal effect on dense tissues but a stronger effect on less compact, more hydrated tissues. The osmolality range of fixative and vehicle for mammalian tissues is 500–700 mosm. For fish, glutaraldehyde fixatives in cacodylate buffer which fall within the osmolality range for mammalian tissues are routinely used. Karnovsky's fixative (about 2000 mosm) also is often used for marine specimens. The addition of non-electrolytes (sucrose, glucose, dextran) and electrolytes ($CaCl_2$ and $NaCl$) stabilize osmolarity and decrease cell swelling and extraction of cellular components. They help maintain cell shape and stabilize membrane and cytoskeletal elements (Hayat, 1989).

An optimal concentration is important in fixation. If the concentration is too low, the time for fixation must be lengthened, causing extraction of cellular components. If concentrations are too high, fine cellular structure may be damaged. Osmium is an effective cross-linking agent at the appropriate concentration (about 1%), but can cleave protein molecules at high concentrations.

Because a wide variety of factors must be considered when choosing a fixative for electron microscopy, processed specimens should be evaluated for evidence of satisfactory preservation. Membranes should be continuous, without distortion or breakage. The spaces between membranes should be granular and there should be no enlarged intercellular spaces in healthy cells. Hayat (1989) provides a detailed summary of cellular components and their satisfactory appearance.

2.5% GLUTARALDEHYDE IN 0.1 M SODIUM CACODYLATE BUFFER WITH 0.6 M SUCROSE

0.2 M Sodium cacodylate buffer, pH 7.4	250.0 ml
Glutaraldehyde (25%)	50.0 ml
Sucrose	100.0 g
Distilled water	500.0 g

0.2 M SODIUM CACODYLATE BUFFER, pH 7.4

Dissolve 42.8 g of cacodylic acid or 31.99 g of cacodylic acid (anhydrous) in 990 ml of distilled water. Adjust pH to 7.4 with 1 mol l^{-1} hydrochloric acid.

KARNOVSKY'S SOLUTION (Karnovsky, 1965)

Dissolve 2 g paraformaldehyde in 25 ml water by heating to 60–70°C. Add 1–3 drops 1 mol l^{-1} NaOH with stirring until the solution clears. The solution is cooled, 5 ml of 50% glutaraldehyde is added, and the volume is made to 50 ml with 0.2 M cacodylate buffer, pH 7.4–7.6. The final pH is 7.2. If cacodylate is used, 25 mg $CaCl_2$ anhydrous is added.

Fixation and special procedures

Liver tissue

Speilberg et al. (1993) evaluated five different aldehyde-containing fixatives for light microscopic studies of liver tissue from Atlantic salmon (*Salmo salar*). Fixatives evaluated included Bouin's, Davidson's, 10% formalin with modified Millonig's phosphate buffer, 10% formalin + 1% glutaraldehyde in 0.1 M phosphate buffer, and modified half strength Karnovsky's. The study indicated that phosphate-buffered formalin and Davidson's provided the best maintenance of liver morphology. The principal criteria they evaluated were the morphological appearance of hepatocyte nuclei and cytoplasm, liver vasculature, bile ducts, connective tissue and artifacts in hematoxylin and eosin-stained sections. Hepatocyte cytoplasm and nuclei were homogeneous, well outlined, and exhibited only moderate shrinkage. This study at least partially confirmed previous reports that phosphate buffered formalin and Bouin's were the most widely used and efficacious fixatives in fish pathology (Hinton et al., 1984; Roberts, 1989). Most authors consider phosphate buffered formalin the fixative of choice for general use (Bullock, 1989; Ferguson, 1989).

Whole small fish

Fournie et al. (1996) evaluated several histology protocols for whole small fish specimens. High quality sections can be difficult to produce because small fish are composed of a variety of tissues that range in consistency from soft visceral organs to hard bones and scales. In a test using 840 one-year-old medaka to compare various fixation and decalcification parameters, one group of medaka was scaled, a second group was not scaled, and a third group was not scaled but later decalcified for 4, 8, or 12 h in TBD-1, a 14% hydrochloric acid solution. Specimens from each of these five groups were placed into three different fixatives (Lillie's, Bouin's, or 10% neutral buffered formalin) for varying time periods. After fixation, specimens were placed in 10% neutral buffered formalin (NBF) for storage. In the hematoxylin- and eosin-stained sections of the whole fish, the liver was considered to be the most important indicator of fixation quality. The best overall protocols were fixation in Bouin's or Lillie's for 2 to 4 days, followed by storage in 10% NBF, then an 8- or 12-h decalcification step in TBD-1. Since there was no definitive combination that produced clear results above all others, a fixation time of 3 days in Bouin's with an 8-h decalcification in TBD-1 was recommended.

Immunohistochemical techniques

Generally, it appears that protocols developed for immunohistochemical studies using mammalian models are suitable for fish. Formalin fixation seems to work well, and with rodent tissues, it has been shown that specimens fixed and stored in formalin for over 24 months can yield acceptable results with staining for proliferating cell nuclear antigen (PCNA) (Greenwell et al., 1991). Because formalin cross-links proteins, epitopes of some antigens become concealed and require enzymatic digestion or antigen retrieval techniques to unmask them. Studies on PCNA detection in the medaka (*Oryzias latipes*) fixed in Lillie's showed that fixative was less desirable for PCNA than was formalin (Ortego et al., 1994). However, application of antigen retrieval methodologies helped overcome the fixation problems and allowed PCNA detection in medaka fixed in other fixatives such as Davidson's, but not in medaka fixed in acetone (Ortego, unpublished data). Hemmer et al. (1998) described the effects of Bouin's, Dietrich's, and Lillie's on the immunostaining patterns, intensity, and tissue specificity of mammalian P-glycoprotein antibody in sheepshead minnow (*Cyprinodon variegatus*). Immunoreactivity in liver, intestine, exocrine pancreas and kidney was found to be fixative-dependent. With Bouin's fixative lower overall immunostaining intensities and quenching were observed in some tissue/antibody combinations. Good immunostaining was observed in tissues fixed in Dietrich's and Lillie's fixatives. Overall, their data indicated that tissue fixation has a significant impact on P-glycoprotein antibody immunoreactivity in fish tissues and must be considered in the comparison and interpretation of results.

Molecular techniques

The application of the polymerase chain reaction (PCR) to paraffin-embedded histologic specimens allows the tissue and cellular localization of oligonucleotide sequences. For *in situ* PCR, fixation should yield optimal cellular morphology and preserve nucleic acid integrity (Muro-Cacho, 1997). Whereas with immunohistochemistry, methodologies have been adapted to formalin-fixed tissues, aldehyde-based fixatives are actually preferred for preservation of nucleic acids. Nevertheless, the cross-linking properties of aldehyde fixatives sometimes require special proteolytic digestion conditions to be applied. Some fixatives such as those containing picric acid (Bouin's and Lillie's) or that contain mercury (Zenker's solution) degrade nucleic acids and tissues fixed in those solutions generally are not useful for molecular studies (Ben-Ezara et al., 1991; Greer et al., 1991).

Safety

The process of fixation obviously involves cell death and therefore substances used in the process must be handled carefully. Manufacturers' Material Safety Data Sheets (MSDS) contain detailed information on individual chemicals, compounds and mixtures. The MSDS sheets that accompany each chemical product when it is purchased or that are available from the supplier should be the definitive resource for safety and handling data. Below, safety and handling information is given on some of the more commonly used substances in histological fixation.

Aldehydes (glutaraldehyde, formaldehyde)

These can cause dermatitis if permitted to wet the skin and fumes can be noxious. Therefore, they should be handled in a fume hood while wearing gloves. Breathing the vapors is dangerous and formaldehyde is considered to be a carcinogen. Because of the hazards associated with the use and disposal of formaldehyde, non-formaldehyde fixative solutions are under development but at the present time are not frequently employed.

Sodium cacodylate

Sodium cacodylate contains arsenic. If inhaled or absorbed through the skin, it can cause dermatitis and liver and kidney inflammation. Gloves should be worn and the chemical should be handled under a fume hood. Arsenic gas can be produced if sodium cacodylate comes in contact with acids or drain cleaners in the plumbing.

Osmium tetroxide

Osmium tetroxide is volatile at room temperature and the fumes are caustic to the eyes, nose and throat. Exposure must be limited and the chemical should be handled in a fume hood while wearing gas-tight goggles.

Ethanol

Ethanol is highly flammable and volatile and can produce toxic fumes.

World Wide Web access to histological information

Vast amounts of information are accessible on the Internet. Listed below are several websites and servers where information on chemical fixation and other information related to histotechnique can be found.

1. 'Histology Resources' contains a set of histotechnology tutorials:
   http://www.pharm.arizona.edu/centers/tox_center/swehsc/exp_path/w3-histo.html
2. The Histotech's Homepage contains links to many histology sites:
   http://www.histology.to
3. The Immunohistochemistry Home Page includes a variety of information about various immunohistochemistry topics:
   http://immuno.hypermart.net
4. The Histonet is a list server which provides a forum for questions and answers related to histology:
   Histonet@Pathology.swmed.edu
5. A site with information on types of fixatives, factors that influence fixation, and choice of fixatives is:
   http://www-medlib.med.utah.edu/WebPath/HISTHTML/HISTOTCH/HISTOTCH.HTML

Summary and conclusions

Fixation of tissues is one of the oldest procedures in biomedicine. Fixation is a simple process to execute if a few principles are given attention such as the rapid acquisition of the tissue to avoid autolysis and making certain the size of the tissue and the volume of fixative are appropriate for good histologic specimens. Development of molecular and immunohistochemical procedures that can be carried out on histologic slides is bringing about a resurgence of interest in basic histology and facilitates investigating structure and function simultaneously. New special and general purpose fixation fluids are becoming available but many of the fluids developed earlier in the twentieth century are still appropriate. Because of the empirical nature of studies on comparative effects of fixatives and the need to communicate technological advances rapidly, the Internet will become a more important source for these types of data.

Acknowledgments

We thank Lee Courtney for his critical review and Val Coseo for formatting this manuscript. This paper is number 1083 of the US Environmental Protection Agency, National Health and Environmental Effects Research Laboratory, Gulf Ecology Division, Gulf Breeze, Florida 32561.

References

Baker, J.R. (1958). *Principles of Biological Microtechnique*. 3rd edn, John Wiley, New York.

Ben-Ezara, J., Johnson, D.A., Rossi, J., Cook, N. and Wu, A. (1991). *J. Histochem. Cytochem.* **39**, 351–354.

Bochenski, G.M. (1984). *National Society for Histotechnology Self-Assessment Examinations: Fixation*. National Society for Histotechnology, Lanham, Maryland. 45pp.

Bullock, A.M. (1989). In *Fish Pathology*, 2nd edn (ed. R.J. Roberts), pp. 374–405, Baillière, London.

Davenport, H.A. (1960). *Histological and Histochemical Technics*. W.B. Saunders, Philadelphia.

Dawes, C.J. (1988). *Introduction to Biological Electron Microscopy*, pp. 73–115. Ladd Research Industries, Burlington, Vermont.

Ferguson, H.W. (1989). *Systemic Pathology of Fish*. Iowa State University Press, Ames, Iowa.

Fournie, J.W., Hawkins, W.E., Krol, R.M. and Wolfe, M.J. (1996). *Techniques in Aquatic Toxicology* (ed. G.K. Ostrander). pp. 577–587. Lewis Publishers, Boca Raton, Florida.

Gray, P. (1952). *Handbook of Basic Microtechnique*. McGraw-Hill, New York.

Gray, P. (1954). *The Microtomist's Formulary and Guide*. Blakiston, New York. 794pp.

Greenwell, A., Foley, J.F. and Maronpot, R.R. (1991). *Canc. Lett.* **59**, 251–256.

Greer, C.E., Petterson, S.L., Kiviat, N.B. and Manos, M.M. (1991). *Am. J. Clin. Pathol.* **32**, 85–112.

Grizzle, J.M. and Rogers, W.A. (1976). *Anatomy and Histology of the Channel Catfish*. Auburn University Agricultural Experiment Station, Auburn, Alabama, 94pp.

Groman, D.B. (1982). *Histology of the Striped Bass*. Bethesda, Maryland. *Am. Fish. Soc. Monogr.* **3**. 116pp.

Hayat, M.A. (1989). *Principles and Techniques of Electron Microscopy*, 3rd edn, 78pp. CRC Press, Boca Raton, Florida.

Hemmer, M.J., Courtney, L.A. and Benson, W.H. (1998). *J. Exp. Zool.* **281**, 251–259.

Hinton, D.E. (1990). In *Methods for Fish Biology* (eds C.B. Schreck and P.B. Moyle), pp. 191–209. American Fisheries Society, Bethesda, Maryland.

Hinton, D.E., Walker, E.R., Pinkstaff, C.A. and Zuchelkowski, E.M. (1984). *Natl. Canc. Inst. Monogr.* **65**, 291–320.

Humason, G.L. (1972). *Animal Tissue Techniques*, W.H. Freeman, San Francisco. 641pp.

Karnovsky, M.J. (1965). *J. Cell Biol.* **27**, 137A.

Lillie, R.D. (1965). *Histopathologic Technic and Practical Histochemistry*, 3rd edn. McGraw-Hill, New York.

Luna, L.G. (ed.) (1968). *Manual of Histologic Staining Methods*. McGraw-Hill, New York. 258pp.

Macauley, J.M. and Summers, J.K. (1991). *Near Coastal Louisianian 1991 Province Demonstration Project – Field Operations Manual*. Gulf Breeze (Florida): US Environmental Protection Agency, Office of Research and Development, Environmental Research Laboratory. Report ERL-GB-NOSR-119.

Moore, K.L., Graham, M.A. and Barr, M.L. (1953). *Surg. Gynecol. Obstet.* **96**, 641–648.

Muro-Cacho, C.A. (1997). *Front. Biosci.* **2**, c15–29.

Ortego, L.S., Hawkins, W.E., Walker, W.W., Krol, R.M. and Benson, W.H. (1994). *Biotech. Histochem.* **69**, 317–323.

Pearse, A.G.E. (1968). *Histochemistry*, 3rd edn, vol. 1. Williams & Wilkins, Baltimore, Maryland.

Presnell, J.K. and Schreibman, M.P. (1997). *Humason's Animal Tissue Techniques*, 5th edn, pp. 17–44. Johns Hopkins University Press, Baltimore, Maryland.

Putt, F.A. (1972). *Manual of Histopathological Staining Methods*, John Wiley, New York. 335pp.

Richmond, R.S. (1997). http://members.aol.com/rsrichmond/histology.html

Roberts, R.J. (ed.) (1989). *Fish Pathology*, 2nd edn. Baillière, London.

Shaw, B.L. and Battle, H.I. (1957). *Can. J. Zool.* **35**, 325–347.

Sheehan, D.C. and Hrapchak, B.B. (1980). *Theory and Practice of Histotechnology*, 2nd edn, pp. 40–58. Reissued in 1987 by Battelle Press, Columbus, Ohio.

Speilberg, L., Evensen, Ø., Bratberg, B. and Skjerve, E. (1993). *Dis. Aquat. Org.* **17**, 47–55.

Thompson, S.W. (1966). *Selected Histochemical and Histological Methods*. Charles C. Thomas, Springfield, Illinois.

Yasutake, W.T. and Wales, J.H. (1983). *Microscopic Anatomy of Salmonids: An Atlas*. United States Department of the Interior. Resource Publication 150, Washington, DC. 189pp.

Yevich, P.P. and Yevich, C.A. (1994). In *Biomonitoring of Coastal Waters and Estuaries* (eds I. Kees and M. Kramer), pp. 179–204. Lewis Publishers, Boca Raton, Florida.

CHAPTER 35

Autoradiography of Fishes

Kevin M Kleinow
School of Veterinary Medicine, Louisiana State University, Baton Rouge, Louisiana, USA

Introduction

Autoradiography as a methodology utilizes radiation originating from radioactive material contained within a specimen, to expose a nuclear or photographic emulsion. In so doing, autoradiography produces an image reflective of the distribution of the source of radioactivity within the sample. The essence of autoradiography is that it permits the distribution of radioactivity to be related to the detailed structure of the specimen without disturbing local architecture. Since each crystal of the emulsion is an independent detector, the definition of the record resulting from this technique may range from the organ to the cellular or organelle level depending on the techniques employed. Its greatest utility is for specimens which exhibit heterogeneity of signal and source distribution on a small scale. In this realm it outperforms other techniques including conventional radioisotopic techniques and chemical analysis where procurement of very small samples, detection sensitivity and background considerations may be problematic.

Autoradiography is a technique applied to many facets of the experimental biology of fishes. Radioisotopes of biologically important compounds can be administered to fish such that chemical location and persistence can be observed and in some cases quantified. Many compounds of interest to biologists have elemental isotopes (^{3}H, ^{14}C, ^{35}S, ^{125}I) which are suitable for autoradiography. In addition, other radioisotopes of lesser scientific utility and diminished technical suitability have found usage with specific autoradiographic approaches in studies with fishes.

Basic considerations

A number of books and reviews discuss the application of autoradiography (Caro *et al.*, 1962; Miller *et al.*, 1964; Stumpf and Sar, 1975; Stumpf, 1976; Rogers, 1979). These sources provide background

Copyright © 2000 Academic Press

regarding the basic principles and artistry of autoradiography. Whatever the approach used, autoradiography requires three basic components: the specimen, radioisotope in the specimen and a recording modality such as a photographic or nuclear emulsion. A variety of formats, as discussed in this chapter, exist to provide a successful interaction between these components (Figure 35.1). Integral to autoradiography as a technique are a series of sequential processing steps each of which can be modified for a particular application (Table 35.1). Within the context of this interaction and these processing steps are the phenomena of resolution, efficiency, and background which ultimately influences the outcome and interpretability of the autoradiographs.

Radioisotopic considerations

Ionizing radiation (alpha-α, beta-β or gamma-γ) imparts energy to silver halogen crystals in nuclear or photographic emulsions. As this occurs an electron is trapped at a sensitivity speck converting silver ion to elemental silver thus producing a latent image. Each of the radiation types interacts differently with the emulsion. Alpha particles with a large mass, a high initial energy characteristic of the isotope (all are > 4 MeV) and a 2^+ charge which attracts orbital electrons dissipate energy to the emulsion very rapidly resulting in a dense latent image. Beta particles with a single negative charge and a characteristic isotope spectrum of often lower energies lose energy through the repulsion of emulsion electrons. Because of their size and energy β-particles lose energy at a slower rate than α-particles thus resulting in less dense latent images. Gamma rays of electromagnetic radiation must collide directly with electrons or nuclei in order to lose energy, resulting in widely spaced interactions with the emulsion. Of the ionizing radiation which may interact with an emulsion α- and β-particles are of the greatest utility for autoradiography.

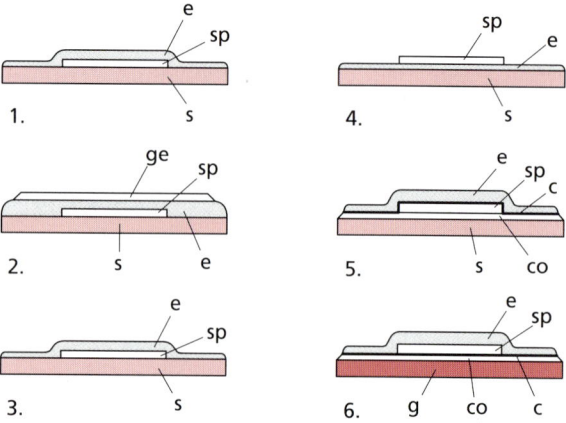

Figure 35.1 Common autoradiographic exposure arrangements between supporting structure, specimen and photographic or nuclear emulsion: (1) Conventional approach; (2) stripping film technique; (3) cryosection thaw mounted on slide; (4) cryosection thaw mounted on emulsion; (5) flat substrate approach for electron microscope; (6) loop method for electron microscope. Key: e, emulsion or film; sp, specimen; s, slide; g, grid; ge, gelatin; c, carbon; co, collodion.

Resolution

Alpha and beta particles may be emitted in any direction from the isotope. For α- particles the effect is discrete as all particles are of the same energy. Conversely, the greater the maximum energy of the β-particle spectrum the greater the interim distance from the source in which silver grains may be formed in the emulsion. Grain density in the emulsion is greatest directly over the source while dropping off to background levels as the distance from the source increases. Beta-emitting isotopes with lower energies will exhibit greater resolution as the distance traveled from the source by the β- particle will be less. Beta particles from tritium, for example, have a maximal energy of 18.5 keV and rarely travel more than 1 μm in nuclear emulsion while ^{14}C with a maximal energy of 0.156 MeV may travel a fourfold greater distance. Higher energy β-emitting isotopes such as ^{32}P (1.709 MeV), ^{35}S (0.167 MeV) and ^{131}I (0.806 MeV)

TABLE 35.1: Experimental steps common to autoradiographic procedures
Radioisotope administration[a]
Sample collection
Sample preparation
Sectioning
Pre-exposure staining[b]
Film or emulsion application
Exposure
Emulsion development
Post-exposure staining[b]
Assessment
[a]Radioisotopes may be administered *in vivo* or after sectioning *in vitro*; [b]Either pre- or post-exposure staining.

have poorer resolution values. Resolution is also influenced by exposure time, distance between emulsion and source, emulsion thickness, emulsion crystal size and source thickness. Longer exposure times generally result in lower resolution as an emulsion crystal is a developed silver grain only once. The greater the time the greater the chance that a crystal directly adjacent to the source will be hit without further development while distal hits are recorded. As the distance between the source and the emulsion is decreased the better the resolution. Emulsion grains over the source have a greater probability of being exposed the closer the distance. Thicker emulsions will generally lower resolution while smaller emulsion crystals will increase resolution. In the former case developed grain densities will superimpose with thicker emulsions and for the latter the developed grain will correspond more closely to the particle trajectory when smaller crystal emulsions are used. If the source is thicker the resolution may also be poorer as a series of source points are available with each overlaying the effect of the other in the emulsion.

Efficiency

The lower the initial β-particle energy the greater the energy loss to the emulsion and thus the greater the silver deposits of the latent image. Conversely, greater penetration through the specimen and emulsion occurs with greater energy. For many isotopes the radiation in the specimen contributing to the autoradiograph is limited by the travel distance of the particle, which in turn is related to the particle's energy. Some particles which originate in the specimen do not have sufficient energy to pass out of the specimen and are said to undergo self-absorption. Thus autoradiographs of low energy isotopes can only be related to the region of the specimen closest to the emulsion. The particles contributing to the autoradiograph are greater for higher energy β-emitters, but the energy loss and latent image formed per particle is less. Other factors which influence efficiencies include thickness of the emulsion, sensitivity of the emulsion, temperature and duration of exposure, developer selection and conditions of development. Although any of these features may be problematic the most significant factors for low energy isotopes are self-absorption and source–emulsion distance. For higher energy particles emulsion thickness and the sensitivity of the emulsion take on greater importance.

Background

Factors other than radiation from the experimental source may result in silver grains in the developed emulsion. Background may result with exposure of the emulsion to light, contamination of emulsion with metallic ions, exposure of emulsion to extraneous sources of radiation, pressure-induced artifacts in drying or handling the emulsion, interaction of emulsion with reactive groups in the biological specimen (positive and negative chemography), and overdevelopment of the nuclear emulsion. Comparison of grain density in autoradiographs of unlabeled samples with those containing an experimental source must be carried out in order to recognize true signal. Background may vary substantially from areas over or under the specimen to distant areas as well as between regions of a specimen. Chemography is the most insidious form of background as chemical groups may be tissue- or region-specific giving rise to what is an apparently realistic autoradiograph. Controls for positive (increased signal) and negative (fading of signal) chemography are an absolute necessity to validate any autoradiograph (especially cryosections) due to the plausibility of the non-signal.

Methodologies

An important consideration regarding the autoradiographic methodology to be used is the level of organization to which the investigation is directed. Different techniques are utilized for macroscopic specimens as compared to those samples intended for either light microscopy (LM) or autoradiography with the electron microscope (EM). On the one hand the presence of radioactivity is recognized as a gross visual picture by the generalized blackening of a film while for LM and EM techniques the signal is recognized as a higher density of individual silver grains which may be observed on a cellular or ultrastructural scale. Due to differences in resolution the ease of application varies dramatically from the relatively straightforward techniques for macroscopic applications to more demanding LM and EM procedures.

Macroscopic autoradiography

Macroscopic or whole body autoradiography provides a record of the distribution of radioactivity for viewing by the unaided eye. As a technique it surveys all tissues, organs or sites impartially without sampling bias providing a total picture of the compound's distribution. This procedure has been performed with a variety of fish species including northern pike (Balk et al., 1984), cod (Ingebrigtsen and Solbakken, 1985; Ingebrigtsen et al., 1988, 1990, 1992), flounder (Ingebrigtsen and Solbakken, 1985), bluegill (Lucas et al., 1979), carp (Till, 1978), fathead minnow (Till, 1978), and rainbow trout (Bergsjo et al., 1979; Ingebrigtsen et al., 1988; Czerwinska et al., 1990; Ingebrigtsen et al., 1990; Waddell et al., 1990; Ngethe et al., 1992). The method has been used for examining the distribution of environmental contaminants (Balk et al., 1984; Ingebrigtsen and Solbakken, 1985; Ingebrigtsen et al., 1990, 1992; Waddell et al., 1990), naturally occurring isotopes (Lucas et al., 1979), aquaculture drugs (Bergsjo et al., 1979; Ingebrigtsen et al., 1988; Czerwinska et al., 1990), mycotoxins in food products (Ngethe et al., 1992) and nutrients (Skeen and Northmore, 1984) in fishes.

Although there are variants of the procedure most experiments described with fish as the species of interest use a close adaptation of the original method first described by Sven Ullberg (1954). At the appropriate interval following dosing, fish are anesthetized with either an overdose of MS-222 or benzocaine. These preparations generally use small sized fish with most ranging from 7 to 140 g although some as large as 235 g have been used. Smaller fish are easier to work with and handle as the resulting microtome sections are quite fragile. Once the fish is euthanized the fish are frozen. The most common approach to freezing is to embed the fish in carboxymethyl cellulose (1%) and rapidly immerse the embedded fish in hexane cooled to $-80°C$ by use of solid CO_2. Another variant of this theme is to freeze the fish separately, then later embed and freeze in the carboxymethyl cellulose gel. Freezing may also be accomplished with isopentane cooled with liquid nitrogen (liquid nitrogen alone causes cracking). After the fish is embedded and frozen it is attached to a microtome stage. Carboxymethyl cellulose gel is utilized as the adhesive by placing a layer between the stage and the fish and freezing. At this point it is important to maintain temperature at or about $-20°C$ for the remainder of the experiment. The exact temperature may be varied in a narrow range to facilitate quality sectioning. Once attached to the stage the block may be sectioned on a variety of cryomicrotomes or a heavy microtome maintained in a cold room at $-20°C$. The most popular cryomicrotome specific for this purpose is the PMV cryomicrotome (PMV 450 MP, Palmstierna Mekaniska Verkstad, Stockholm, Sweden). Once attached to the microtome the embedded fish block is sectioned at a slow steady cutting speed, removing material until the level of interest is reached within the sample. In order to obtain complete sections of a comparatively large specimen a sheet of specialized cold flexible tape (No. 821, 3M Co., St Paul, Mn, USA) is applied to the 'soon to be cut surface' of the frozen squared block. As the block is sectioned (optimum thickness 20–40 μm) the tape is pressed to the block ahead of the knife while the tape and attached section is gently lifted clear of the knife and block distal to the blade. Before exposing the film to the specimen the sections are freeze dried on the tape to prevent translocation of the label and to minimize chemographics between the specimen and the film. Freeze drying may be carried out in a cold room ($-20°C$) or by other means such as with the cryostat. Times for freeze drying vary between 24 and 168 h depending on the thickness of the sections and local conditions.

A variety of X-ray films have been utilized for macroscopic autoradiography. The major determinant in the film selected is the isotope utilized in the study. Tritium has an extremely low efficiency relative to conventional X-ray film. This is due to the fact that tritium is particularly sensitive to self-absorption in the specimen and conventional film often has an anti-abrasion coat which blocks tritium. Tritium-sensitive films without the anti-abrasion coat and composed of large emulsion crystals are available and have been used with fish studies (Balk et al., 1984; Czerwinska et al., 1990; Ngethe et al., 1992). Two such commercially available films are Ultrafilm 3-H (LKB, Bromma Sweden) and Hyperfilm 3-H (Amersham, England). Other X-ray films commonly used with ^{14}C or ^{35}S in fish studies are Kodirex (Kodak) (Bergsjo et al., 1979), Kodak type AA X-ray film (Waddell et al., 1990) and Structurix D7R (Agfa-Gevaert, Antwerp, Belgium) (Ingebrigtsen and Solbakken, 1985; Ingebrigtsen et al., 1988, 1992). The exposure of the film to the specimen occurs in direct opposition either in a sandwich format with blotting paper or in an X-ray cassette. In either case exposure is accomplished in a light-tight encasement generally at $-20°C$. According to the literature the period of exposure varies with the amount of radiolabel present ranging from 4 weeks to 1 year. Time selection is based on adequate time for signal development without over-exposure. Often several exposure times are developed to provide the proper results in a timely fashion. After exposure films are developed conventionally usually with Kodak D-19 developer according to manufacturer's instructions.

Autoradiographic approaches for the light microscope

Numerous investigational studies have utilized nuclear and photographic emulsion-based light microscope autoradiographic techniques with fish. Among other areas and investigators the greatest applications of these techniques have occurred in the study of the fish visual system (Reperant et al., 1981, 1982; Cohen, 1985; Hals et al., 1986; O'Connor et al., 1987; Saidel and Butler, 1991; Medina et al., 1993), neural system (Kim et al., 1979a,b; Galeo et al., 1987; Fine et al., 1990, 1996; Dietl and Palacios, 1991; Tong et al., 1992; Smith et al., 1996), endocrinological system (Holtzman and Schreibman, 1975; Davis et al., 1977; Kim et al., 1979a, b; Woodhead et al., 1979, Van Eys et al., 1983; Wabuke-Bunoti and Firling, 1983; Bosma et al., 1997) and for applications such as cell turnover studies (Hyodo-Taguchi, 1970; Garcia and Johnson, 1972; Stroband and Debets, 1978; Raderman-Little, 1979; Kiss et al., 1988), studies of meiosis and spermiogenesis (Sinha et al., 1979, 1983) and toxicant localization on a cellular scale (Kleinow et al., 1996). Since these investigations make use of a variety of radiolabeled molecules, tissue types and emulsions/films with differing characteristics, many variants of the technique are presented in the literature. Each is a customization of the technique for the specific application. The techniques used in the processing of the tissues for sectioning are based in part upon the diffusibility of the radiolabeled molecule of interest. The following are the basic autoradiographic approaches which have found use with fish.

Conventional processing

Radiolabeled precursors such as amino acids and nucleosides may be used to demonstrate the synthesis, transport tract and location of biological macromolecules: proteins and nucleic acids. Many of the macromolecules synthesized from radiolabeled precursors are retained by fixation procedures of conventional histological processing while the unincorporated precursor is often lost. Thus radiolabeled proteins and nucleic acids are available to the autoradiographic process. Among the fixation techniques available, the use of Bouin's fixation (Stroband and Debets, 1978; Corwin, 1981; Rao and Sharma, 1982; Gona, 1984; Galeo et al., 1987; Kiss et al., 1988; Saidel and Butler, 1991; Becerra et al., 1996), formaldehyde-based fixation (Hyodo-Taguchi, 1970; Finger, 1978a, b; Meyer and Ebbesson, 1981; Reperant et al., 1981, 1982; Medina et al., 1993) or a modification of glutaraldehyde fixation (Raderman-Little, 1979; Hals et al., 1986; Szamier and Ripps, 1983) are the most widely used for this application. Zenkers and Carnoys fixatives have also been used successfully (Hyodo-Taguchi, 1970; Van Eys et al., 1983). Tissues, once fixed, are typically dehydrated in a graded alcohol series and embedded in conventional paraffin. Use of other embedding materials such as Epon 812 (Szamier and Ripps, 1983; Hals et al., 1986) and Araldite 506 (Raderman-Little, 1979) and JB-4 (Van Eys et al., 1983) have been associated with either glutaraldehyde or Carnoy's based fixation and higher resolution-based techniques. Once embedded, tissues are sectioned at thicknesses ranging from 1 to 15 μm with most studies cutting sections from 4 to 12 μm. Sections attached to slides and in the case of paraffin, dewaxed, may be subjected to counter staining before or after the application of a nuclear or photographic emulsion. Typical counter stains which are used to visualize tissues in context with the emulsion are cresyl violet (Northcutt and Wathey, 1980; Meyer and Ebbesson, 1981; Reperant et al., 1981; Saidel and Butler, 1991; Medina et al., 1993), toluidine blue (Rao and Sharma, 1982), 1% azure II methylene blue (Szamier and Ripps, 1983), hematoxylin and eosin (Gona, 1984; Becerra et al., 1996), giemsa (Garcia and Johnson, 1972) and hemalum eosin-based stains (Hyodo, 1965; Hyodo-Taguchi, 1970).

Slides are dipped into the nuclear emulsion covering the specimen under safe light conditions appropriate for the specific emulsion. Emulsions commonly used include Kodak NTB-2 or NTB-3 and Ilford K-5. A variety of stripping films have also been used for these applications. These products are composed of emulsion attached to a gelatin matrix. The stripping film, upside down and hydrated, is floated on to a slide with the attached specimen. The emulsion is then dried prior to exposure. Exposure for all techniques is timed specifically for the application (typically 14 to 35 days) in light-tight boxes with desiccant at 4–5°C. Following exposure slides are commonly developed with Kodak D-19 or Dektol developer.

Another circumstance for conventional processing is where the radiolabeled compound is not a precursor, but is bound. Various toxicants, drugs, hormones and neurotransmitters may act in this manner. While covalent binding may survive conventional processing non-covalent binding often does not. Label may be removed by fixation-induced stereochemical changes in the binding site, large volumes of aqueous solutions may remove compound in early stages of fixation and the binding site may not be precipitated by fixation. All such applications require test runs to examine the retention characteristics of the system being considered. If conventional processing results in compound loss cryosection-based techniques may provide an alternative strategy.

Cryosection-based techniques

Conventional histological preparations using liquid fixatives, dehydration fluids, embedding media and other fluids may be a source of translocation and leaching of diffusible radiolabeled substances. Because of these concerns alternative procedures such as cryosectioning and freeze drying (Stumpf and Roth, 1966) are used to maintain the compound in place. The selection of a technique may be based on the tissues of interest. Tissues with different density components and with high water contents may disintegrate upon freeze drying and may benefit from a cryosectioning and thaw mount approach. Specimens which have significant chemographic effects may benefit by freeze drying before application of an emulsion. A variety of applications with diffusible hormones (Davis *et al.*, 1977; Kim *et al.*, 1979a,b; Fine *et al.*, 1990, 1996) and **xenobiotics** (Kleinow *et al.*, 1996) have utilized cryosectioned fish tissues. Most of these techniques are an adaptation of original methods presented by Miller *et al.* (1964), Stumpf (1976) and Stumpf and Sar (1981).

Fish exposed to the radiolabeled ligand *in vivo* are euthanized and tissues of interest are immediately subsectioned into small blocks on an ice cold tray. Small tissue size ensures almost instantaneous freezing, reducing the potential for tissue artifacts and compound movement. Tissues are frozen by immersion in a variety of media supercooled with liquid nitrogen including liquid freon (Fine *et al.*, 1990; Kleinow *et al.*, 1996), isopentane (Fine *et al.*, 1996), and liquid propane (Kim *et al.*, 1979a,b). Once frozen, tissues are kept frozen until and through sectioning. Tissues are attached to a stage of a cryostat using media such as OTC (Miles, Inc., Elkhart, IN) and cut into sections usually 4 μm in thickness. Frozen sections are either thaw mounted on emulsion precoated (Kodak NTB-2 or NTB-3, New Haven, CT) slides under safelight conditions or mounted on slides then covered with emulsion. Thin sections have minimal amounts of diffusible fluid which is rapidly lost during drying and fixation of the section to the slide. Because potential interaction of the tissues with the emulsion is greater for cryosections than with other approaches, control slides for positive and negative chemography as well as background are a necessary requirement. Slides should be stored for the appropriate exposure interval at 4 to $-15°C$ in light-proof boxes containing a desiccant. Exposure intervals in published studies range from 14 to 600 days. Following exposure, slides are warmed to room temperature and developed using conventional techniques with products such as Kodak D-19 developer and Kodak fixer. Since the sections are thaw mounted, care must be exercised to prevent loss of the section from the slide during the development process. Following development, slides are gently washed in a water rinse and air dried. Tissues may be stained with methyl green pyronin (Stumpf and Sar, 1981), cresyl violet (Davis *et al.*, 1977), or methylene blue–basic fuchsin (Kim *et al.*, 1979b) to provide visualization of cellular and subcellular detail without obscuring the autoradiographic signal (Figure 35.2).

Other applications of autoradiography with cryosectioned tissues have utilized an *in vitro* approach to the use of radiolabeled compounds. The use of this

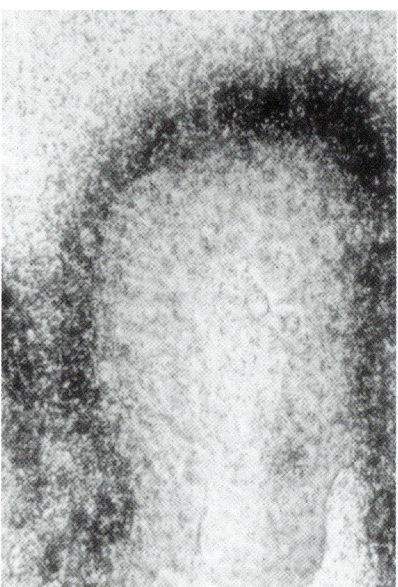

Figure 35.2 Thaw mounted cryosection of intestinal mucosal fold from catfish exposed to [^{3}H]benzo(a)pyrene. Black signal is silver image resulting from exposure and development of NTB-2 emulsion. Counterstained with methyl green pyronine. (×40). **(See also Colour Plate 19.)**

technique in fish has been confined largely to the examination of receptor or neurotransmitter location in the brain (Phan and Maler, 1983; Pack et al., 1989; Dietl and Palacios, 1991; Tong et al., 1992; Smith et al., 1996). In these procedures the brains are removed and instantly frozen. Once frozen, tissues were cryosectioned producing sections ranging from 10 to 32 μm. At this step the sections are either thaw mounted on gelatin-coated slides and stored at −20°C until future use, or air dried. For the frozen sections, following thawing, tissues are generally preincubated with a Tris-based buffer and air dried or the sections are dried first then preincubated. Following preincubation the sections are incubated with the radioligand of interest usually in a Tris or phosphate buffer. After incubation with the radioligand for an application-specific interval the slides are copiously washed in ice cold buffer, rinsed in distilled water and air dried. Several approaches to the autoradiographic procedure may be utilized at this point. Sections may be opposed to Ultrofilm (LKB Inc., Gaithersburg, MD) or its equivalent (Pack et al., 1989; Dietl and Palacios, 1991; Tong et al., 1992), Kodak X-Omat AR film (Smith et al., 1996) or dipped in liquid emulsion (Kodak NTB-2) (Phan and Maler, 1983). After development, sections may be counterstained with cresyl violet, luxol blue, toluidine blue or other similar stains.

Autoradiographic approaches for the electron microscope

A variety of high resolution autoradiographic approaches have been utilized in the experimental biology of fishes (Bunt and Klock, 1980; Sealock and Kavookjian, 1980; Szamier and Ripps, 1983; Peyrichoux et al., 1986; Noe et al., 1988). Autoradiographic techniques geared towards visualization by the electron microscope require specialized methods regarding the handling of tissues, emulsions and staining. Most of the recent studies utilizing the EM are modifications or adaptations of older techniques (Caro et al., 1962; Salpeter and Bachmann, 1964; Larra and Droz, 1970; Bok and Young, 1972). It is fair to say that the full potential of EM resolution capabilities is not realized in use with autoradiography, as limitations are evident with the emulsion sensitivity and silver grain size.

Tissue preparation is based on electron microscopy techniques. One fixation method as used for fish is described by Sealock and Kavookjian (1980). In this method teleost Ringer's supplemented with 2% glutaraldehyde and 1% paraformaldehyde is used for an initial 5 min perfusion. Tissues are transferred to an ice cold tannic acid fixative (275 mM NaCl, 7.6 mM KCl, 415 mM urea, 2 mM sodium phosphate, 2% glutaraldehyde, 0.3% tannic acid pH 7.1) for 4–6 h. Following an overnight wash in Ringer's the tissues were post-fixed for 6–9 h with 0.2% osmium tetroxide. Tissues were subsequently washed in 100 mM sodium cacodylate pH 7.0, then treated with 1% tannic acid in cacodylate buffer, washed twice with cacodylate buffer and rinsed with water. Samples are dehydrated in a graded alcohol series and embedded in Epon 812 media.

Autoradiographic preparations for the electron microscope require special considerations in emulsion selection and use. Emulsions with the greatest sensitivity and resolution must be selected to complement most suitably the use of an electron microscope. The most popular emulsion for use with fish is Ilford L-4 (Sealock and Kavookjian, 1980; Szamier and Ripps, 1983; Peyrichoux et al., 1986; Noe et al., 1988). Samples must be mounted on a grid for viewing. Two basic approaches have been developed to accommodate the use of a grid in autoradiography. Either the sample is coated with emulsion elsewhere and moved to the grid or the grid is coated with emulsion in some fashion. Both methods have advantages and disadvantages. While the flat substrate method which coats a sample on a microscope slide via a dipping technique allows for uniform thickness of emulsion and better autoradiographic technique, removing a completed autoradiograph to an EM grid for viewing can be problematic (Saltpeter and Bachmann, 1972). The loop method, first proposed by Caro and Tubergen (1962), on the other hand places emulsion on the sample mounted on the grid via a loop. No transfer is necessary for viewing with this method, but layer thickness can vary. Both techniques have been reported for use with fish (Sealock and Kavookjian, 1980; Noe et al., 1988).

For the loop method, ultrathin sections (< 1 μm) cut on a diamond or glass knife are mounted on a grid coated with a collodion film overlaid by an evaporated layer of carbon. After drying the grid is attached to a microscope slide. A platinum wire is inserted in emulsion of proper consistency to fill the loop. On touching the grids the emulsion falls off the loop and covers the grid. Samples covered with emulsion are desiccated overnight. Slides are placed in light-proof boxes with desiccant and are purged with nitrogen, argon or

CO_2 before sealing. The boxes are placed in a refrigerator at 4°C for the exposure duration. Development of emulsion for all applications is often accomplished using phenidon (Szamier and Ripps, 1983; Peyrichoux et al., 1986) or Microdol X (Noe et al., 1988). After development autoradiographs are post-stained with either 2% uranyl acetate or lead citrate (Sealock and Kavookjian, 1980; Szamier and Ripps, 1983; Falcon and Collin, 1985).

Quantitation

Quantitative approaches to autoradiography range from relative measurements to absolute measurements. Stringency in each step of the autoradiographic technique is necessary to produce meaningful results. This is especially true in cross comparisons of data collected or processed at different times or under different conditions. Factors such as section thickness and consistency, emulsion thickness, emulsion dryness, chemography, exposure times and developer use can contribute to variability. Because of the nature of the recording media the relationship between radiation dose and silver grains developed is not linear throughout an open-ended exposure regime. When 10% or less of the emulsion is developed the relationship follows a linear progression. As more crystals are developed further multiple hits are not recorded in a linear fashion. In a similar way, as recording proceeds adjacent grains overlap and even fuse making it difficult to assess and quantitate individual events. Within a sample of a particular exposure duration certain areas may be amenable to quantitative analysis whereas other areas may yield less than optimal quantitative conditions because of over-development. Controls for background, negative and positive chemography, latent fading and internal reference standards are integral for stringent quantitative assessment.

However, quantitative analysis may be as simple as the presence or absence of radioactive signal or gauged relative to a quantitative ladder of radioactive sources. Studies examining cell turnover kinetics with ^{3}H[thymidine] make use of the presence or absence of developed silver grains within cells to quantitate cells which are proliferating (Figure 35.3). Within this context quantitation is predicated by controlling and factoring out background which may be misconstrued as an abundance of cell turnover. Whole body autoradiography of fishes on the other

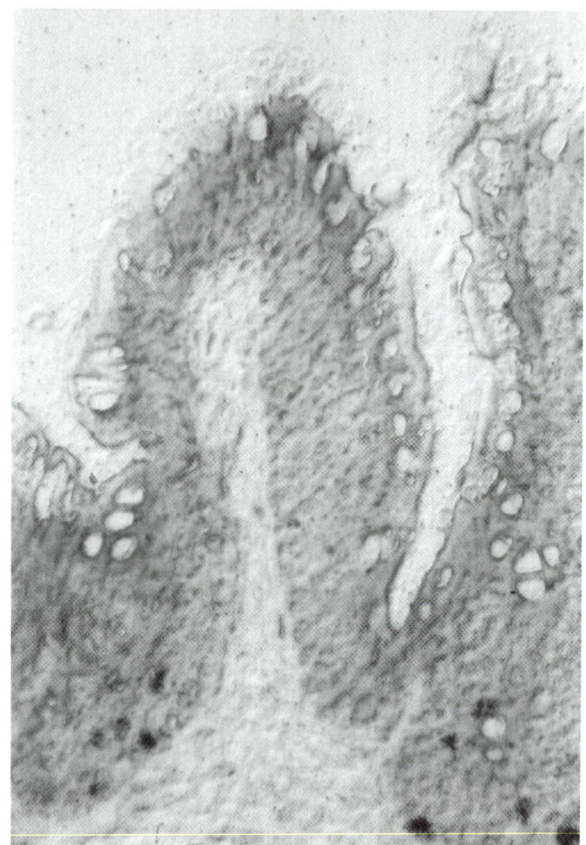

Figure 35.3 Thaw mounted cryosections of intestinal mucosal fold from catfish exposed to [^{3}H]thymidine in cell turnover experiment. Concentration of black dots in individual cells represent autoradiographic signal. Counterstained with methyl green pyronine. (×40). **(See also Colour Plate 20.)**

hand has made use of radiation standard grey scales (Ingebrigtsen et al., 1988; Czerwinska et al., 1990). These scales provide a relative pictorial index of radioactivity in individual organs. Other relative comparisons with fish whole body autoradiographic films have used densitometric evaluations to compare differences in organ signal using the blood as the reference tissue (Czerwinska et al., 1990).

In addition to these methods, quantitation has also been accomplished by counting either individual silver grains or the area in which the silver grains occupy. Counting may be accomplished manually using grids and visual discrimination (Reperant et al., 1981) or by image analysis systems (Pack et al., 1989; Dietl and Palacios, 1991; Kleinow et al., 1996). One approach for computer-assisted image analysis and relative quantitation was accomplished using Image-1/AT, Noran version, 4.0 3A software (Kleinow et al., 1996). The signal of interest was highlighted using threshold range and contrast features within the analysis program. Silver grain particles were further discriminated from tissue structure and confluent signal by use of signal size and shape discriminators (Figure 35.4). Internal area calibrations were installed in order to quantitate silver signal per unit area. As enhanced microscopic images were viewed on a monitor, regions for quantitation were outlined electronically. The total area (μm^2) occupied by silver grains in each respective tissue region was calculated by using the background corrected counts of silver grain particles contained inside the delinated area. Signal density was obtained by dividing the signal object area value by the regional area.

Quantitative assessment throughout a time course study presents a number of challenges not the least of which is the question of measurable signal development. Measurable and comparable signal development is required in all samples throughout the time course. Early sampling times with higher radiation levels may exhibit over-development of signal while later times with lower radiation levels may demonstrate under-exposure. One experimental approach is the selection of a single exposure time which provides the linear grain development for the area of interest throughout the time course. Through the use of multiple exposure times for all samples one is able to select microscopically a single optimum autoradiographic exposure time which allows comparable quantitation for all samples (Kleinow et al., 1996). Strict controls and standards are an integral component to these types of analysis.

References

Balk, L., Meijer, J., DePierre, J.W. and Appelgren, L. (1984). *Toxicol. Appl. Pharmacol.* **74**, 430–449.

Becerra, J., Junqueira, L.C.U., Bechara, I.J. and Montes, G.S. (1996). *Arch. Histol. Cytol.* **59**, 15–35.

Bergsjo, T., Nafstad, I. and Ingebrigtsen, K. (1979). *Acta Vet. Scand.* **20**, 25–37.

Bok, D. and Young, R.W. (1972). *Vision Res.* **12**, 161–168.

Bosma, P.T., Kolk, S.M., Rebers, F.E.M., Lescroart, O., Roelants, I., Willems, P.H.G.M. and Schulz, R.W. (1997). *J. Endocrinol.* **152**, 437–446.

Bunt, A.H. and Klock, I.B. (1980). *Invest. Ophthalmol. Vis. Sci.* **19**, 707–719.

Caro, L.G., Van Tubergen, R.P. (1962). *J. Cell Biol.* **15**, 173–188.

Cohen, J.L. (1985). *Brain Res.* **332**, 169–173.

Corwin, J.T. (1981). *J. Comp. Neurol.* **201**, 541–553.

Czerwinska, K., Sund, R.B. and Sognen, E. (1990). *Acta Pharm. Nord.* **2**, 377–386.

Davis, R.E., Morrell, J.I. and Pfaff, D.W. (1977). *Gen. Comp. Endocrinol.* **33**, 496–505.

Dietl, M.M. and Palacios, J.M. (1991). *Brain Res.* **539**, 211–222.

Falcon, J. and Collin, J.P. (1985). *J. Pineal Res.* **2**, 357–373.

Fine, M.L., Keefer, D.A. and Russel-Mergenthal, H. (1990). *Brain Res.* **536**, 207–219.

Fine, M.L., Chen, F.A. and Keefer, D.A. (1996). *Brain Res.* **709**, 65–80.

Finger, T.E. (1978a). *Brain Res.* **153**, 608–614.

Finger, T.E. (1978b). *J. Comp. Neurol.* **180**, 691–706.

Galeo, A.J., Fine, M.L. and Stevenson, J.A. (1987). *J. Neurobiol.* **18**, 359–373.

Garcia, M.N. and Johnson, H.A. (1972). *Cell Tissue Kinetics* **5**, 331–339.

Gona, O. (1984). *Gen. Comp. Endocrinol.* **55**, 289–294.

Hals, G., Christensen, B.N., O'Dell, T., Christensen, M. and Shingai, R. (1986). *J. Neurophysiol.*, **56**, 19–31.

Holtzman, S. and Schreibman, M.P. (1975). *Gen. Comp. Endocrinol.* **25**, 447–455.

Hyodo, Y. (1965). *Radiation Res.* **26**, 383–394.

Hyodo-Taguchi, Y. (1970). *Radiation Res.* **41**, 568–578.

Ingebrigtsen, K. and Solbakken, J.E. (1985). *J. Toxicol. Environ. Hlth.* **16**, 197–205.

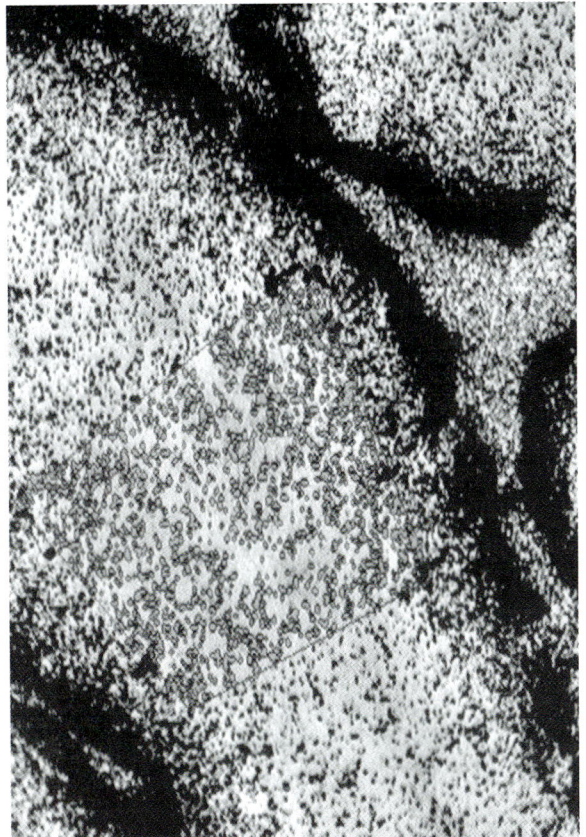

Figure 35.4 Computer enhanced images of intestinal mucosal fold autoradiograph from catfish exposed to [^{3}H]benzo(a)pyrene. Yellow is quantified signal in select region (as outlined in red) as determined by programed size, shape, contrast and threshold parameters. (×40). **(See also Colour Plate 21.)**

Ingebrigtsen, K., Solbakken, J.E., Norheim, G. and Nafstad (1988). *J. Toxicol. Environ. Hlth.* **25**, 361–372.

Ingebrigtsen, K., Hektoen, H., Andersson, T., Bergman, A. and Brandt, I. (1990). *Pharmacol. Toxicol.* **67**, 344–345.

Ingebrigtsen, K., Hektoen, H., Andersson, T., Wehler, E.K., Bergman, A. and Brandt, I. (1992). *Pharmacol. Toxicol.* **71**, 420–425.

Kim, Y.S., Sar, M. and Stumpf, W.E. (1979a). *Cell Tissue Res.* **198**, 435–440.

Kim, Y.S., Stumpf, W.E. and Sar, M. (1979b). *Brain Res.* **170**, 43–59.

Kiss, R., deLaunoit, Y., Lenglet, G. and Danguy, A. (1988). *Acta Zool.* **69**, 225–230.

Kleinow, K.M., Smith, A.A., McElroy, A.E. and Wiles, J.E. (1996). *Marine Environ. Res.* **42**, 65–73.

Larra, F. and Droz, B. (1970). *J. Microsc.* **9**, 845–880.

Lucas, H.F., Jr, Simmons, D., Markun, F., Farnham, J. and Keenan, M. (1979). *Hlth Physics* **36**, 147–152.

Medina, M., Reperant, J., Ward, R., Rio, J.P. and Lemire, M. (1993). *Anat. Embryol.* **187**, 167–191.

Meyer, D.L. and Ebbesson, S.O.E. (1981). *Cell Tissue Res.* **218**, 389–401.

Miller, O.L., Stone, G.E. and Prescott, D.M. (1964). *J. Cell Biol.* **23**, 654–658.

Ngethe, S., Horsberg, T.E. and Ingebrigtsen, K. (1992). *Aquaculture* **108**, 323–332.

Noe, B.D., Amherdt, M., Perrelet, A. and Orci, L. (1988). *Pancreas* **3**, 700–713.

Northcutt, R.G. and Wathey, J.C. (1980). *Neurosci. Lett.* **20**, 237–242.

O'Conner, P.M., Zucker, C.L. and Dowling, J.E. (1987). *J. Neurochem.* **49**, 916–920.

Pack, A.M., Caine, S.B., Winokur, A., Manaker, S. and Fishman, A.P. (1989). *J. Comp. Neurol.* **287**, 19–27.

Peyrichoux, J., Pierre, J., Reperant, J. and Rio, J.P. (1986). *J. Comp. Neurol.*, **246**, 364–381.

Phan, M. and Maler, L. (1983). *Neurosci. Lett.* **42**, 137–143.

Raderman-Little, R. (1979). *Cell Tissue Kinet.* **12**, 269–280.

Rao, P.D.P. and Sharma, S.C. (1982). *J. Comp. Neurol.* **210**, 37–48.

Reperant, J., Vesselkin, N.P., Ermakova, T.V., Rustamov, E.K., Rio, J.P., Palatnikov, G.K., Peyrichoux, J. and Kasimov, R.V. (1982). *Brain Res.* **251**, 1–23.

Reperant, J., Rio, J.P., Miceli, D., Amouzou, M. and Peyrichoux, J. (1981). *Brain Res.* **217**, 225–243.

Rogers, A.W. (1979). *Techniques of Autoradiography*, pp. 1–429, Elsevier/North Holland Biomedical Press, Amsterdam.

Saidel, W.M. and Butler, A.B. (1991). *Brain Behav. Evol.* **38**, 154–168.

Salpeter, M.M. and Bachmann, L. (1964). *J. Cell Biol.* **22**, 469.

Salpeter, M.M. and Bachmann, L. (1972). In *Principles and Techniques of Electron Microscopy*, vol. II (ed. M.A. Hayat). Van Nostrand Reinhold, New York.

Schmatolla, E. and Erdmann, G. (1973). *J. Embryol. Exp. Morph.* **29**, 697–712.

Sealock, R. and Kavookjian, A. (1980). *Brain Res.* **190**, 81–93.

Sinha, G.M., Mondal, S.K., Midya, T. and Ghosal, S.K. (1979). *Leipzig* **93**, 442–448.

Sinha, G.M., Mondal, S.K. and Ghosal, S.K. (1983). *Cytologia* **48**, 87–93.

Skeen, L.C. and Northmore, P.M. (1984). *Neurosci. Lett.* **52**, 191–197.

Smith, A., Trudeau, V.L., Williams, L.M., Martinoli, M.G. and Priede, I.G. (1996). *J. Neuroendocrinol.* **8**, 655–658.

Stroband, H.W.J. and Debets, F.M.H. (1978). *Cell Tissue Res.* **187**, 181–200.

Stumpf, W. (1976). In *Methods in Cell Biology*, vol. XII (ed. D.M. Prescott) pp. 171–193, Academic Press, New York.

Stumpf, W.E. and Roth, L.J. (1966). *J. Histochem. Cytochem.* **14**, 274–287.

Stumpf, W.E. and Sar, M. (1975). In *Methods in Enzymology*, vol. XXXVI. Hormone Action, Part A, Steroid Hormones (eds B.W. O'Malley and J.G. Hardman), pp. 135–156. Academic Press, New York.

Stumpf, W.E. and Sar, M. (1981). In *Techniques in Neuroanatomical Research* (eds C.H. Heym and Forssmann, W.G.), Ch. 13. pp. 247–254. Springer, New York.

Szamier, R.B. and Ripps, H. (1983). *J. Comp. Neurol.* **215**, 51–62.

Till, J.E. (1978). *Hlth. Phys.* **34**, 333–343.

Tong, C.K., Pan, M.P. and Chang, Y.C. (1992). *Neuroscience* **49**, 237–246.

Ullberg, S. (1954). *Acta Radiol. Suppl.* **118**, 1–110.

Van Eys, G.J.J.M., Lowik, C.W.G.M. and Wendelaar Bonga, S.E. (1983). *Gen. Comp. Endocrinol.* **49**, 277–285.

Wabuke-Bunoti, M.A.N. and Firling, C.E. (1983). *Gen. Comp. Endocrinol.* **49**, 320–331.

Waddell, W.J., Lech, J.J., Marlowe, C., Kleinow, K.M. and Friedman, M.A. (1990). *Fund. Appl. Toxicol.* **14**, 84–87.

Woodhead, A.D., Setlow, R.B. and Scully, P.M. (1979). *Cancer Res.* **39**, 2698–2703.

PART 6

Experimental Models

Contents

CHAPTER 36 **Cancer** **591**

CHAPTER 37 **Toxicology** **617**

CHAPTER 38 **Cell and tissue culture** **631**

CHAPTER 36

Cancer

Paul C Baumann
Field Research Station, USGS,
The Ohio State University, Columbus, Ohio, USA

Mark S Okihiro
School of Veterinary Medicine,
University of California at Davis, Davis,
California, USA

Introduction

While mammals have been used in cancer research primarily because of their assumed similarity to humans in anatomy and physiology, this has not been the history of fish as a cancer research model. Cancer research using fish as a model was prompted by finding fish with tumors in the natural environment. Descriptions of wild fish with tumors appeared in the literature as early as 1793 (Bell, 1793), and a population of barbels (*Barbus fluviatilis*) with an elevated prevalence of epidermal **papillomas** on the lips was described from the Mosel River, Germany in 1908 (Keysselitz, 1908). The author found intracellular inclusion bodies within the tumor cells, and suggested a viral etiology. Fifty-five years later Hueper (1963) suggested that chemicals could cause **neoplasia** in wild fish. The next year Dawe *et al.* (1964) published their study of liver neoplasms in white sucker (*Catostomas commersoni*) and brown bullhead from Deep Creek Lake, Maryland, and hypothesized that carcinogenic contaminants were the cause. Thus cancer epizootics in wild fish were caused by the same vectors that operated in mammals.

As research on the mechanisms of carcinogenesis in fish proceeded, the similarities to mammalian systems became more apparent. The major route of excretion of high molecular mass polynuclear aromatic hydrocarbons (PAHs) including benzo(a)pyrene (BaP) is by the bile both in mammals (Kotin *et al.*, 1959; Falk *et al.*, 1962) and in fish (Solbakken and Palmork, 1981; Varanasi and Gmur, 1981a). A number of metabolites determined to be mutagenic (Nagao and Sugimara, 1978) or proximate carcinogens (Kapitulnik *et al.*, 1977) in mammals were also produced by the livers of BaP-exposed English sole (*Parophrys vetulus*) (Varanasi and Gmur, 1981a,b). Subsequent experiments demonstrated the *in vivo* binding of BaP metabolites to hepatic DNA in English sole. Furthermore, the amount bound was similar to the amount reported bound in mouse skin and mammary gland tissue for which BaP is carcinogenic. The sequence of lesion progression described for English sole was essentially the same as that described in the rat model of hepatocarcinogenesis

Copyright © 2000 Academic Press

(Myers *et al.*, 1987). Finally, the cloning and sequencing of the first fish oncogenes from goldfish (*Crassius auratus*) (Nemoto *et al.*, 1986) and rainbow trout (*Onchorhynchus mykiss*) (Van Beneden *et al.*, 1986) spurred the use of fish as models for molecular studies.

Therefore fish have been studied as a model both for the mechanics of cancer initiation and for progression, and for a model of environmental risk caused by carcinogens in aquatic ecosystems. In the following sections we will elaborate on the fish model and its advantages in exploring these areas.

Fish in carcinogenicity testing

Although phylogenetically distinct from mammals, fish have been used for carcinogenesis research for decades and have made major contributions to the field. Some of the more widely used species include: medaka (*Oryzias latipes*), rainbow trout (*Oncorhynchus mykiss*), zebrafish (*Brachydanio rerio*), guppies (*Poecilia reticulata*), platyfish (*Xiphophorus* spp.) and swordtails (*Xiphophorus* spp.). The majority of early studies focused on simple assessment of carcinogenic activity using known carcinogens. Increasingly laboratory research utilizing fish has begun to explore: (i) their use in evaluation of environmental damage to aquatic habitats; (ii) their potential as adjunct test animals for assessment of compounds intended for human use and consumption; and (iii) their application as unique research tools in the expanding field of carcinogenesis.

Perhaps the most direct application of fish to carcinogenesis research is as sentinels or surrogates in the assessment of contaminated aquatic ecosystems. Epizootics of tumors in wild fish are often associated with heavily contaminated habitats. Unfortunately, obtaining definitive proof (e.g. replicating neoplasms in native fish species with **xenobiotic** extracts) is logistically difficult to achieve. With a few exceptions, use of native species is impractical because of the size of fish involved, difficulty in obtaining controls from pristine reference sites, specialized husbandry requirements, prolonged time-frame (slow onset and progression of tumors), and expense involved. One solution is to use small fish species as either sentinels (caged fish studies) or surrogates to assess suspect carcinogens under controlled laboratory conditions (Hawkins *et al.*, 1985; Wolfe *et al.*, 1991). Medaka have already been used to verify the carcinogenic potential of chemical extracts taken from contaminated sites in the Great Lakes (Fabacher *et al.*, 1991).

A second practical application of fish carcinogenesis research is with evaluation of potentially carcinogenic chemicals intended for human use and consumption. With the vast backlog of chemical compounds waiting to be tested, extended time-frame for conventional toxicity assessment and rising costs of rodent bioassays, fish have come under increasing scrutiny for use as adjunct test species. Small freshwater species of aquarium fish are especially adaptable for carcinogenesis research (Hawkins *et al.*, 1985, 1988b; Hinton *et al.*, 1985; Spitsbergen *et al.*, 1997). The database for most domestic fish is extensive, covering the same aspects of husbandry (housing, nutrition, reproduction, etc.) critical to rodent-based cancer studies. Other advantages include: short generation periods, wide availability and acceptance of commercial diets, and tolerance for housing at high density. Small size allows for large numbers to be exposed with small amounts of carcinogen, at the same time increasing statistical power. In contrast to high prevalence of preneoplastic and neoplastic lesions in controls of many rodent species (Ward, 1981), spontaneous lesions in domestic fish are rare (Hawkins *et al.*, 1988b; Masahito *et al.*, 1989a; Bailey *et al.*, 1996). And while latency period for carcinogen-induced neoplasms can extend for a year or longer in rodents (Lipsky *et al.*, 1981), time to tumor in fish is considerably faster (weeks to months) (Hatanaka, 1982; Hinton *et al.*, 1985).

Although fish will never replace mammalian models of carcinogenesis (simply because they lack breasts, lungs, adrenal glands and lymph nodes), they can greatly augment rodent bioassays currently in use. One way of helping to reduce (or at least prioritize) the backlog of chemicals to be tested would be to use fish as a preliminary screen to quickly assess the potential toxicity of large numbers of chemicals, and then to retest the most carcinogenic compounds using more expensive 2-year rodent assays. In this way, the most dangerous chemicals could be quickly identified, and valuable resources not wasted on less toxic compounds. However, fish tumor models need to be closely examined to determine if sufficient similarities exist between fish and mammals to warrant such use. If it can be shown that fish models are analogous to carcinogenesis in rodents, then continued use of fish would be justified.

The question of how similar fish tumor models are to those in mammals has only been partially addressed. Numerous studies have demonstrated that many teleost fish species are sensitive to the same classes of carcinogens which affect mammals (Hendricks et al., 1984; Hinton et al., 1985; Harada et al., 1988; Hawkins et al., 1990; Bailey et al., 1996). Far fewer, however, have sought to examine the critical steps between exposure and tumor formation in fish. Despite the relative paucity of studies, however, it is apparent that the three basic stages of carcinogenesis (initiation, promotion and progression) are essentially the same in fish and mammals.

Initiation is the complex first stage of carcinogenesis requiring uptake of the carcinogen, transport to target organs, bioactivation, DNA adduction and gene mutation. While pharmacokinetics of carcinogen uptake are obviously different in fish, gills are analogous to lungs and uptake following aqueous exposure is dependent on surface area of gill filaments and skin.

Uptake and metabolism

Since the topics of contaminant uptake and metabolism are too vast to cover in their entirety, these subjects will, for the most part, be confined to discussions of **PAHs**, particularly BaP. Early tissue analysis of fish from areas subjected to oil spills demonstrated the presence of PAHs with one to three benzenoid rings, but PAHs with four to six rings, the group containing most of the carcinogens, were seldom found (Neff, 1979). However, when fish were analyzed from areas heavily contaminated by coal tars associated with coking or creosote facilities, the larger four and five ring compounds were present, although at lower concentrations than PAHs with fewer rings (Black et al., 1981; Baumann and Harshbarger, 1985). Laboratory experiments established that a variety of fish could acquire radiolabeled BaP from water (Balk et al., 1984), sediment (Stein et al., 1984; Djomo et al., 1996), and diet (Varanasi et al., 1989a). Using all three routes of exposure radioactivity was highest in the bile, with liver tissue being either second (sediment and diet) or third (water) (Varanasi et al., 1989a).

The bile contained metabolites of BaP, including BaP-7,8-diol and its conjugates (Gmur and Varanasi, 1982). BaP-7,8-diol is the precursor of (+)-anti-BaP-7,8-diol-9,10-epoxide (or (+)-anti-BaPDE) which has been shown to be the most carcinogenic metabolite of BaP (Jerina et al., 1984). Thus the mechanism by which a species metabolizes PAHs is important. A number of excellent reviews exist on the metabolic processes used to transform PAH (Bend et al., 1977; Lech et al., 1982; Tan and Melius, 1986; James, 1987; Stegeman, 1987; Varanasi, 1989; George, 1994; Stegeman and Hahn, 1994). Here we will briefly overview the series of steps which have been conventionally divided into two 'phases': phase I reactions (oxidation) and phase II reactions (conjugation and detoxification) (Stegeman and Hahn, 1994), using BaP as our primary example.

Phase I: Metabolism

Oxidation of PAH is carried out by the heme protein (cytochrome P-450) monooxygenases. These P-450 enzymes catalyze two basic types of reactions in fish: those which synthesize or break down endogenous compounds and those which metabolize xenobiotics (Stegeman and Hahn, 1994). The P-450 group is really a superfamily of proteins with hundreds of members. However, the most important subfamily in this group both in terms of carcinogenesis and in terms of research to date is **CYP1A**. The primary role of CYP1A in metabolizing BaP had been demonstrated by several lines of research. The metabolism of BaP is almost entirely stopped in most fish by treatment with antibodies to CYP1A (Stegeman et al., 1985). On the other hand inducers of CYP1A increase rates of BaP metabolism (Stegeman, 1981). Subcellular fractions of fish liver can activate BaP into mutagenic derivatives which are consistent with those seen in mammals (Balk et al., 1982; Michel et al., 1992). Finally fish can metabolize the (−) BaP-7,8-dihydrodiol to (+)-anti-BaPDE (Zaleski et al., 1989). The dominant pathway of BaP metabolism in fish is: (i) a CYP1A mediated addition of oxygen to form a (+)7,8-epoxide; (ii) an epoxide hydrolase mediated alteration to a (−)7,8-dihydrodiol; and (iii) a CYP1A mediated addition of a second oxygen to form the (+)7,8-diol-9,10-epoxide (Stegeman and Hahn, 1994). Channel catfish (Ictalurus punctatus) may have a lower susceptibility to liver cancer because reduced CYP1A and epoxide hydrase activity results in more phenols and fewer diols produced compared to brown bullhead (Yuan et al., 1997). While BaP metabolism has been the most studied, other PAH compounds may also be metabolized to carcinogenic products by CYP1A.

Two other subfamilies of the P-450 complex have been implicated in activating carcinogens in fish. A fungal derivative, aflatoxin, caused a widespread liver cancer epizootic in hatchery trout during the 1950s (Harshbarger, 1997). Trout activate aflotoxin B1 to the carcinogenic 2,3 epoxide by using CYP2K (Williams and Buhler, 1983). Recent research has demonstrated that medaka are capable of metabolizing trichloroethylene (TRI) (Lipscomb et al., 1997a). Livers samples of medaka transformed TRI to chloral hydrate (CH), trichloroethanol (TCOH) and trichloroacetic acid (TCA) *in vitro* (Lipscomb et al., 1998); TCA has been linked to liver tumors in mice (Daniel et al., 1992). In mammals CYP2E seems to be responsible for the initial oxidation of TRI to CH (Lipscomb et al., 1997b). Using a sensitive ELISA method, Lipscomb et al. (1998) were able to demonstrate a low level of CYP2E in medaka microsomes. Also a CYP2E protein and mRNA have been found in the liver of *Poeciliopsis*, a small live-bearing fish (Kaplan et al., 1991). Tumors in closely related species have been induced by exposure to dimethylnitrosamine and diethylnitrosamine (Khudoley, 1984), both of which are activated by CYP2E in mammals (Camus et al., 1993).

Phase II: Metabolism

The phase II enzymes continue the process of transforming xenobiotics into more water-soluble substances which can be eliminated. There are three major groups of phase II enzymes: the glutathione S-transferases (GSTs), the UDP-glucuronosyltransferases (UDPGTs), and sulfotransferases (STs). These enzymes can catalyze reactions at several stages during PAH metabolism (Foureman, 1989). The PAH epoxide formed by the (CYP1A catalyzed) addition of the first oxygen, and the diol epoxide formed after the addition of the second oxygen can be conjugated to glutathione by GSTs. Thus GSTs are capable of neutralizing these two potential carcinogens. GSTs are ubiquitous and have been demonstrated in all fish species tested (James, 1987; Nimmo, 1987).

If the PAH epoxide rearranges into a phenol, or if it is catalyzed into a diol by epoxide hydrolase, these oxidative metabolites can be conjugated with glycosides (such as UDP-glucuronic acid) by UDPGTs (Foureman, 1989). This is another very important mechanism for the detoxifying and excreting xenobiotic compounds in fish (Clarke et al., 1991). Finally these same PAH phenols and diols can be conjugated with activated 3′phosphoadenosine 5′phosphosulfate (PAPS) to form a sulfate monoester in a reaction catalyzed by STs (Jakoby, 1989). The importance of STs varies greatly even among closely related species. For instance while southern flounder (*paralichthys lethostigma*) conjugated more of the carcinogenic (+)-anti-BaPDE via STs than via UDPGTs (Pritchard and Bend, 1984), sulfates were only minor metabolites for other flatfish including English sole and starry flounder (*Platichthys stellatus*) (Varanasi, 1989). Obviously differences in phase II conjugation reactions may play a role in the varying susceptibility of fish species to chemically induced carcinogenesis (Yuan et al., 1997; Smeets et al., 1999).

Initiation and DNA adducts

The initiating step in the chemical carcinogenesis model is the covalent binding of a carcinogen to DNA (Reichert et al., 1998). Thus the identification of such DNA adducts points toward cause and effect for a compound and neoplasms, and the amount of such adducts may be an index of both carcinogen exposure and neoplasm prevalence. The most common method used to detect adduct formation in fish exposed to environmental carcinogens is ^{32}P-postlabeling (Reichert et al., 1998). Although several other methodologies have also been used with fish (Maccubbin, 1994), ^{32}P-postlabeling is the most sensitive, being able to potentially detect one adduct in 10^9–10^{10} bases (Gupta, 1985; Reddy and Randerath, 1986). This process involves the enzymatic hydrolyzation of DNA to 3′monophosphates, followed by the removal of normal nucleotides to concentrate the DNA adducts (Gupta and Randerath, 1988). These adducts are then labeled by [^{32}P]phosphate and separated by thin layer chromatography or on polyethyleneimine-modified cellulose sheets. Quantification of adducts is accomplished by autoradiography and scintillation counting or by image analysis (Reichert et al., 1992).

The use of the ^{32}P-postlabeling technique was especially attractive with field studies, where a measure of DNA adducts provided an integration of the inherent variations in uptake, metabolism and repair (Maccubbin, 1994). Thus adducts provided an objective measurement of genetic damage (Dunn, 1990). The technique was first used with fish from areas where epizootics had been reported, such as English sole from Puget Sound (Varanasi et al., 1986) and brown bullheads from the Buffalo and Detroit Rivers (Dunn et al., 1987). English sole from PAH contaminated areas were found to have elevated DNA adducts compared to sole from relatively uncontaminated areas (Varanasi et al., 1989b). Similarly, brown bullhead from areas with PAH contaminated sediment were also found to have elevated DNA adducts compared to fish from reference locations (Dunn et al., 1990; Maccubbin et al., 1990). This same methodology has been used in studies of European fish fauna from polluted environments (Poginsky et al., 1990; van der Oost et al., 1994). In contrast, over 28 species (from 22 genera) collected from relatively pristine areas (as documented by sediment analysis), have had either no detectable adducts or minimal adduct levels (Reichert, 1998). Although liver tissue has been used in the vast majority of studies, other tissues may also have detectable levels of adducts. Intestinal tissue from brown bullhead in the Buffalo River was found to have elevated levels of DNA adducts compared to the same tissue taken from fish inhabiting a pristine lake (Maccubbin, 1994).

Field studies have also demonstrated a more refined correlation between sediment concentrations of PAH and DNA adducts. Oyster toadfish (*Opsanus tau*), a benthic, territorial species with a small home range, were collected from a number of sites with various degrees of creosote contamination (Collier et al., 1993). Their adduct concentrations were correlated with sediment PAH values over almost five orders of magnitude ($p < 0.001$). However, red drum, a more widely ranging species collected from Charleston Harbor, SC, did not have adduct levels that correlated directly to the areas in the harbor where they were captured, although levels reflected exposure to carcinogens (Reichert et al., 1998). English sole from Puget Sound demonstrated a relationship between adduct levels and the prevalence of some hepatic lesions (Myers et al., 1995, 1997). In these fish levels of PAH/DNA adducts were a significant risk factor for the occurrence of nuclear pleomorphism, megalocystic hepatosis and preneoplastic foci of cellular alteration, the latter often being an early stage in the development of hepatic neoplasms.

Laboratory experiments on adduct formation were able to ask more specific questions than those asked in field studies, where fish were exposed to complex mixtures of organics that could not be completely characterized (Reichert, 1998). Laboratory studies allowed the specific identification of major adducts (Maccubbin, 1994). A variety of carcinogens have been examined, and considerable information exists on several including aflotoxin B1 (Bailey et al., 1988; Loveland et al., 1988). However BaP adducts have attracted the most research to date. The major BaP/DNA adduct formed *in vivo* by bluegill (Shugart et al., 1987), English sole (Varanasi et al., 1989b) and brown bullhead (Sikka et al., 1990) was *anti*-BPDE/dG, the same adduct formed in experiments with mouse skin (Brooks, 1979; Ashurst et al., 1983). However, some studies with trout have revealed larger amounts of different BaP adducts (Ahokas et al., 1979; Smolarek et al., 1987). Once BaP adducts were formed in fish they tended to be persistent, lasting for 60 days in brown bullhead (Sikka et al., 1990) and 70 days in English sole (James et al., 1991). This persistence of DNA adducts in fish is believed to be caused, at least in part, by slow DNA repair mechanisms compared to mammals (Bailey et al., 1988; Fong et al., 1988).

The amount of carcinogen binding to liver DNA is also influenced by route of exposure and by the metabolic reactions inherent in a given species. DNA adduct formation increased at least 10-fold in English sole given an i.p. injection of BaP as compared to those exposed through their diet (Varanasi et al., 1989a). Research on southern flounder demonstrated binding of BaP to intestinal DNA (James et al., 1991), thereby demonstrating one mechanism by which these routes would differ. Species differences also exist. *In vivo* experiments with English sole and starry flounder demonstrated that the former had two to four times the level of adduct formation as the latter (Varanasi et al., 1986). This conforms to the greater tumor susceptibility of the former, but was in contrast to results obtained from earlier *in vitro* studies (Varanasi and Gmur, 1980; Varanasi et al., 1980). The lower adduct formation by starry flounder in the *in vivo* experiment was also accompanied by an enhanced conjugation of reactive metabolites to glutathione, thus indicating that more efficient phase II conjugation may reduce covalent binding (Stein et al., 1990).

Oncogenes and tumor suppressor genes

Two gene classes playing crucial roles in neoplastic transformation, following DNA adduction, are **oncogenes** and tumor suppressor genes. Although typically associated with tumor formation, both are also intimately involved with normal cellular proliferation and differentiation. In general, oncogenes serve as positive modulators of cellular growth, while tumor suppressor genes have negative inputs, regulating mitotic activity. Activation of proto-oncogenes (normal homologs of transforming oncogenes) involves mutation, DNA rearrangement, or gene amplification, and leads to either overexpression of normal proteins or production of altered proteins associated with signal transduction. In contrast, tumor induction associated with tumor suppressor genes involves inactivation or gene deletion leading to loss of regulatory control.

Approximately 200 oncogenes (subdivided into 15 classes) and close to 30 tumor suppressor genes have already been identified in mammals (Hesketh, 1997). Major oncogene classes include those coding for: growth factors, tyrosine kinases, G-proteins and transcriptional regulators. Of these, only a small fraction have been identified in teleost fish and even fewer definitively linked with tumor induction in fish (Van Beneden, 1993; Van Beneden and Ostrander, 1994). Oncogenes identified in fish tissue include: *ras*, *myc*, *src*, *erb*, *yes*, *fyn*, *abl*, *wnt*, *fgf*, *pea*, *ret*, *fos* and *Xmrk*. Tumor suppressor genes identified in fish include: *p53*, *Rb* and a *CDKN2*-like gene. Both *Xmrk* and the *CDKN2*-like gene are associated with hereditary melanoma in *Xiphophorus* hybrids and will be discussed separately.

A recent explosion in zebrafish research has led to many oncogene discoveries, including identification of: *myc* (Cheng et al., 1997b); *fgf-3* (Keifer et al., 1996a,b); *wnt* (Molven et al., 1991; Dorsky et al., 1998); N-*ras* (Cheng et al., 1997a); *ret1* (Bisgrove et al., 1997; Marcos-Gutierrez et al., 1997); *pea3* (Brown et al., 1998); and *thra1/erbA* (Essner et al., 1997). The push in zebrafish research has focused on cloning and sequencing of proto-oncogenes from embryonic cDNA libraries, and primarily evaluates expression during normal development. While there has been little direct study of zebrafish oncogenes in association with neoplastic transformation, zebrafish are viewed as a potentially valuable fish cancer model (Spitsbergen et al., 1997) and these baseline studies will provide a substantial framework from which to examine the molecular events involved with carcinogenesis in fish.

Several proto-oncogenes have also been identified in normal trout (embryonic and adult) including: *tgf-β* (Hardie et al., 1998); *myc* (Van Beneden et al., 1986; Panno and McKeown, 1995, 1997); *fos* (Matsuoka et al., 1998); and *ras* (Mangold et al., 1991). Tumor suppressor genes identified in trout include: *Rb* (Van Beneden and Ostrander, 1994) and *p53* (Soussi et al., 1990; Caron de Fromentel et al., 1992). There has only been limited assessment of activated oncogenes in trout neoplasms, but increased expression of myc and ras oncoproteins has been detected in two trout cell lines derived from tumors (Carter et al., 1996). In addition, activated Ki-*ras* oncogenes have been identified in trout liver, stomach and swimbladder tumors induced by AFB1 (Chang et al., 1991), DMBA (Fong et al., 1993), DEN (Hendricks et al., 1994), dehydroepiandrosterone (Orner et al., 1995), MNNG (Bailey et al., 1996) and dibenzo[a]pyrene (Bailey et al., 1996).

Ras proteins are membrane-bound GTPases involved with control of cellular growth, proliferation and differentiation. The *ras* gene family includes three genes (Ha-*ras*, Ki-*ras* and N-*ras*) and are the most commonly encountered oncogenes in human tumors (Anderson et al., 1992; Lowry, 1993). In addition to trout tumors, mutated Ki-*ras* has also been detected in DEN-induced medaka liver tumors (Torten et al., 1996) and in liver tumors from wild Atlantic tomcod (*Microgadus tomcod*) from the Hudson River (Wirgin et al., 1989) and winter flounder (*Pseudopleuronectes americanus*) from Boston Harbor (McMahon et al., 1988, 1990).

Despite these findings, detection of activated *ras* oncogenes is neither pathognomonic for liver tumors in fish nor in mammals. Even in studies where *ras* oncogenes are found, mutated *ras* cannot be demonstrated in 100% of tumors and in a recent survey of wild English sole (*Pleuronectes vetulus*) from Puget Sound, researchers were unable to demonstrate Ki-*ras* mutations in liver tumors (Peck-Miller et al., 1998). Studies with transgenic mice have revealed that expression of activated *ras* alone is insufficient for neoplastic transformation and that co-expression of v-Ha-*ras* and *myc* results in a synergistic increase in tumor induction (Sinn et al., 1987; Mangues et al., 1992).

There is a strong likelihood that activation multiple oncogenes and/or tumor suppressor genes are necessary for tumor induction, with as many as 12 different mutations being required for genesis of a single cancer. There are currently 10 oncogenes (*MYB*, *MYC*, *CCND1*, *CCND3*, *EGFR*, *HER2*, *HRAS*, *HSTF1*, *INT2* and *YES1*) and three tumor suppressor genes (*p53*, *RB1* and *BRCA1*) associated with human breast cancer (Hesketh, 1997). Increased expression of multiple oncogenes (v-H-*ras*, *myc* and *src*) has been detected in Northern blots of medaka tumor cell lines derived from melanomas and hepatocellular adenomas (Muraiso et al., 1987). In addition, immunohistochemical analyses of DEN-induced hepatic neoplasms from rivulus (*Rivulus ocellatus marmoratus*) has revealed increased expression of four oncogenes (*ras*, *myc*, *fos* and *egfr*) and one tumor suppressor gene, *p53* (Goodwin and Grizzle, 1994).

Detection of altered *p53* expression in rivulus tumors is not surprising in that *p53* is the most frequently mutated tumor suppressor gene associated with human cancer (Greenblatt et al., 1994). However, other researchers assessing DEN-induced hepatic neoplasms from medaka (personal communication, C. Davis, University of California at Davis), MNNG-induced non-hepatic neoplasms in medaka (Krause et al., 1997), and AFB1-induced liver tumors in trout (Bailey et al., 1996) have thus far been unable to detect either *p53* mutations or altered gene expression.

Another tumor suppressor gene which has at least been partially characterized in fish is *Rb*, the gene definitively linked with hereditary retinoblastoma in humans. Western blot analysis has revealed widespread tissue distribution of the *Rb* gene product (p105Rb) in rainbow trout, English sole, coelacanth and medaka (Fekete et al., 1993; Van Beneden and Ostrander, 1994). And although loss or defect in gene expression has not been demonstrated in any fish species, medaka could potentially serve as an animal model for the study of retinoblastoma. Medaka do develop retinoblastoma following exposure to MNNG (Ostrander et al., 1992) and is the only vertebrate species in which the tumor is chemically inducible.

Obviously, however, there is great need for further evaluation of the role of Rb with medaka retinoblastoma, as well as for a much broader assessment of oncogene activation with fish carcinogenesis overall. The discovery rate of new oncogenes and tumor suppressor genes, in fish, has accelerated greatly over the past few years. For current information regarding teleost oncogene research, interested parties are referred to three websites: the Zebrafish Homepage (zfish.uoregon.edu); the Medaka Homepage (biol1.bio.nagoya-u.ac.jp:8000); and the National Center for Biotechnology Information Genbank (ncbi.nlm.nih.gov/Genbank).

Promotion

Although there is some evidence that, under certain circumstances, initiated cells can progress directly into malignant neoplasms (Sell, 1993; Sell and Pierce, 1994), the general consensus is that for the majority of organ systems, tumors must first pass through an intermediate stage of development called promotion (Moore and Kitagawa, 1986; Farber and Sarma, 1987). After DNA lesions – induced by adducts – are 'fixed' in the genome following a wave of mitotic activity (Cayama et al., 1978), transformed cells are believed to undergo clonal expansion to form preneoplastic foci. Foci are not overt neoplasms, but clearly have some selective growth advantage over surrounding normal tissue. Histologically, they are characterized by altered phenotypes and some loss of normal organization, but with minimal **dysplasia**, little propensity towards invasion and no metastasis. Biochemically, foci have been shown to demonstrate altered protein synthesis (often involving fetal proteins) and enzyme activity (Bannasch et al., 1985; Beer and Pitot, 1987; Tatematsu et al., 1988).

In fish, preneoplastic lesions have been well documented in skin (pigment cell hyperplasia preceding chromatophoromas) and liver (foci of cellular alteration preceding hepatocellular adenomas), and undoubtedly develop in other teleost organ systems (Okihiro et al., 1993; Teh and Hinton, 1993; Bailey et al., 1996). While hepatic foci of cellular alteration (FCA) have been extensively studied in a variety of marine and freshwater fish species, and are readily inducible via exposure to known carcinogens, the precise mechanism underlying transition from foci to tumors is still unclear. Much of the difficulty with research in liver carcinogenesis stems from the large number and variety of preneoplastic lesions. Similar to mammalian models, following carcinogen exposure, fish develop dozens (perhaps hundreds or thousands in larger species) of FCA for each hepatic neoplasm. Obviously, many foci must regress and/or be destroyed, leaving only a select few to be promoted to benign tumor status. To complicate matters further, hepatic foci (in fish and mammals) can be classified into 4 to 11 separate categories based on just tinctorial staining properties with standard HE histologic preparations (Boorman et al., 1997; Okihiro, personal observations). And while hepatic foci in fish have been assessed for alterations in enzyme expression and immunohistochemically, with antibodies specific for PCNA, intermediate filaments and a variety of oncoproteins, definitive answers with respect to which subclass (or classes) of foci are promoted are still lacking (Teh and Hinton, 1993; Bunton, 1994, 1995; Van Beneden and Ostrander, 1994; Bailey et al., 1996).

Experiments designed to chemically modulate (positive and negative) hepatic, gastrointestinal and renal tumor production have been extensively conducted in trout (Bailey *et al.*, 1996). However, because these studies focus on final tumor production (number, size and prevalence), they provide little insight with respect to the specific promotional aspects of preneoplastic development. One of the few studies to address neoplastic promotion in fish was conducted in medaka (Cooke and Hinton, 1999). Working under the hypothesis that DEN-induced **basophilic** FCA are preferentially promoted in female medaka because of endogenous estrogen (E2), they morphometrically demonstrated a similar promotional effect using exogenous E2 (in the diet) and sequential sampling following an initial subcarcinogenic dose of **DEN** (aqueous exposure). Studies are currently underway, in the same laboratory, to determine if differences exist with regard to E2 receptor levels and expression of E2-mediated proteins (choriogenin and vitellogenin) in medaka FCA (Corrine Davis, personal communication).

Although the selective growth advantage of basophilic foci in medaka liver appears to be hormone-mediated, the ability of specific subclasses of preneoplastic lesions to out-compete their neighbors is probably governed by genetics, and involves oncogenes and/or tumor suppressor genes. Just as the initial impetus for neoplastic transformation is specific mutations, it is also likely that additional mutations (whether spontaneous – because of an unstable genotype – or induced following additional carcinogen exposure) are necessary for the jump from preneoplastic lesion to benign tumor. These additional mutations confer additional growth and survival advantages to already transformed cells, allowing subpopulations within preneoplastic clones to overgrow and replace lesser cells. Mammalian researchers, using rodent models, have already found evidence of tumor in focus lesions which support this hypothesis (Solt *et al.*, 1977; Farber and Sarma, 1987; Scherer, 1987; Tatematsu *et al.*, 1988). Although in-depth promotional studies with fish models are notably lacking, the good news is that almost all the techniques and assays currently focused on overt tumors can also be used to assess fate of preneoplastic lesions.

Progression

Neoplastic progression is the process whereby slow growing benign tumors are transformed into rapidly growing malignant neoplasms that are invasive and metastasize. Uncontrolled growth is predicated on both a rapid rate of cell division and having an adequate blood supply. Tumor-induced neovascularization (angiogenesis) not only supplies proliferating cells with oxygen and nutrients, but provides new avenues of escape. Angiogenesis is dependent on a host of promotional and inhibitory factors, many of which are linked to oncogenes or tumor suppressor genes. Oncogenic H-*ras* and Ki-*ras*, for example, markedly increase expression of vascular endothelial growth factor (VEGF) in transformed cells (Rak *et al.*, 1995). Invasion and subsequent metastasis to distant sites are strongly correlated with changes in cell surface proteins involved with cell adhesion (e.g. cadherins, integrins, CD44 and the immunoglobulin superfamily) and with production of proteolytic enzymes (e.g. matrix metalloproteinases, heparanase and cystein and aspartic proteinases), allowing cancer cells to degrade basement membrane and intercellular connective tissue (Hesketh, 1997).

Although malignant neoplasms are commonly identified in fish, diagnoses have largely been based on histologic criteria (cellular atypia, anaplasia, pleomorphism) with little or no assessment of long-term biologic behavior. Determination of malignant potential (especially gauging metastatic capability of various tumor types) is time consuming, expensive and, for the majority of fish studies, superfluous to the primary goal of tumor induction. Increasingly, however, as fish cancer models are refined to cover broader aspects of cancer biology, greater emphasis needs to be placed on critical evaluation of malignant progression.

One of the few studies to systematically assess neoplastic progression in fish was recently conducted using four groups of medaka: two larval and two adult (Okihiro and Hinton, 1999a). Larvae were exposed to 350–500 ppm DEN for 48 h, while adults were exposed to 50 ppm DEN for 5 weeks in aqueous bath. Fish were subsequently maintained as long as possible (until fish were moribund or had died) to determine malignant potential of resultant neoplasms. A total of 423 medaka with 106 hepatic neoplasms were examined. Benign hepatocellular adenomas were only observed in DEN-exposed larvae and were slow growing (average age = 297 days) and non-metastatic. In contrast, adult-exposed fish had higher prevalences of hepatocellular carcinomas (100% of hepatocellular tumors in adult-exposed medaka were malignant versus 51.5% in larval-exposed medaka) and mixed hepato-biliary carcinomas (24.3% of total tumors were mixed in adult-exposed versus 3% in larval-exposed medaka). In addition, a unique hepatocellular lesion termed 'nodular proliferation' was only observed in adult-exposed medaka. The lesion was characterized by small size (50–300 μm), complete loss of normal tubular architecture and variable **megalocytosis**. Nodular proliferation was distinct from preneoplastic foci of cellular alteration (FCA) and may represent microcarcinomas. There was stepwise increase in mean tumor size (diameter in microns) with age (days post-exposure), from nodular proliferation (174 μM, 17 days) to hepatocellular carcinoma (1856 μM, 62 days) and mixed carcinomas (3209 μM, 93 days) in adult-exposed medaka. Metastasis was observed with 19 neoplasms and tumors with the highest metastatic potential were hepatocellular and mixed carcinomas. The most common form of metastasis was **transcelomic** (detachment of tumor emboli in the abdominal cavity), followed by direct invasion and distant metastasis, presumably via the vascular route.

Differences in tumor prevalence and rate of progression between medaka exposure groups are believed to be the result of differing target cells and duration of DEN exposure, rather than age of fish at the time of exposure. In larval medaka with brief (48 h) DEN exposure, neoplasms are thought to be the result of dedifferentiation of initiated hepatocytes, with slow progression of **FCA** to benign and then malignant tumors. In contrast, with adult medaka and prolonged (5 week) DEN exposure, neoplasms are believed to result from initiation of committed stem cells (capable of differentiating into either biliary epithelial cells or hepatocytes) and formation of microcarcinomas ('nodular proliferation'), before progressing to larger hepatocellular and then mixed carcinomas. The current hypothesis is that different stages of neoplastic growth are dependent on number and possibly type of DNA mutations. Although Ki-*ras* oncogene mutations have been identified in DEN-induced hepatic neoplasms from both trout (Hendricks *et al.*, 1994) and medaka (Torten *et al.*, 1996), the goal now is to determine if specific mutations are associated with the progression of FCA to adenomas, adenomas to carcinomas and hepatocellular to mixed carcinomas.

The only fish model where the molecular events associated with progression have been investigated is hereditary melanoma in *Xiphophorus* hybrids. Progression of melanoma (similar to that of liver neoplasia) occurs in stages, with initial formation of hyperplastic lesions, transition to benign melanomas and then final development of invasive malignant melanomas which metastasize and are fatal (Vielkind and Vielkind, 1982). Histologic, ultrastructural and cytochemical criteria have been established for all stages of tumor formation. Malignant melanomas are characterized by small spindle cells with scant, incompletely pigmented melanosomes and strong tyrosinase activity, while benign lesions are composed of large, highly dendritic chromatophores with well-pigmented melanosomes and moderate tyrosinase activity. The *Xmrk* oncogene, which is not expressed in normal skin, is demonstrable at a low level in benign lesions and is highly expressed in malignant melanomas (Mäueler *et al.*, 1993). And while there is no difference in the amount of STAT5 protein in benign versus malignant melanoma, the level of STAT5-tyrosine phosphorylation (mediated by Xmrk kinase activity) is much higher in malignant tumors (Wellbrock *et al.*, 1998b). Activated STAT5 is translocated to the nucleus and induces expression of target genes *cis*, *osm* and *pim*-1. The precise function of these target genes with respect to malignancy, however, is still unknown.

Medaka and trout fish tumor models

Although a diverse array of fish have been used for carcinogenesis research (Hawkins *et al.*, 1985; Bogovskii, 1995), the two best studied and most widely used species are medaka (*Oryzias latipes*), a small warmwater fish native to Japan and Asia, and rainbow trout (*Oncorhynchus mykiss*), a coldwater salmonid fish species native to North America. Adult medaka have an upper size limit of approximately 1 g, mature in 2 to 3 months, are sexually dimorphic, continuous spawners and tolerant of a wide range of water temperatures and conditions. In sharp contrast, adult trout can reach upwards of 10 kg, mature in 2 to 3 years, are sexually dimorphic only when spawning, annual spawners and generally intolerant of changes in temperature and water quality. Despite these differences, both have proven to be exceptional subjects for carcinogenesis research.

The utility of medaka for carcinogenesis research was initially demonstrated by Ishikawa *et al.* (1975), and since then medaka have repeatedly shown themselves to be an ideal test species (Masahito *et al.*,

1996). Small size and high tolerance to marginal water quality allow medaka to be housed in high density, increasing statistical power at the same time lowering maintenance costs and decreasing the amount of carcinogen needed to conduct studies. Small size also decreases costs associated with histological processing, allowing complete survey of major organs on a single slide and, with serial sections, morphometric analysis (Hinton et al., 1984). Early maturation and sexual dimorphism allow for manipulation of gender to assess sex-linked and hormonal influences (Nakazawa et al., 1985; Teh and Hinton, 1998), and continuous spawning ensures year-round availability.

Medaka are sensitive to a diverse array of carcinogens including: diethylnitrosamine (DEN) (Ishikawa and Takayama, 1979); methylazoxymethanol acetate (MAMA) (Aoki and Matsudaira, 1977; Hawkins et al., 1986); aflatoxin (Hatanaka et al., 1982); N-methyl-N'nitro-N-nitrosoguanidine (MNNG) (Bunton and Wolfe, 1996); polyaromatic hydrocarbons (PAHs) (Hawkins et al., 1990); and ethylene dibromide (EDB) (Hawkins et al., 1998). Although liver is the primary target organ for many carcinogens (Aoki and Matsudaira, 1977; Hatanaka et al., 1982), neoplasms have also been induced in eye (Hawkins et al., 1986), gonad (Hawkins et al., 1996), skeletal muscle (Hawkins et al., 1988a), gill (Brittelli et al., 1985), swimbladder (Bunton and Wolfe, 1996), pancreas (Hawkins et al., 1991), kidney (Hawkins et al., 1985) and skin (Hyodo-Taguchi and Matsudaira, 1984).

Another key advantage with medaka is a short time to tumor. Latency periods following brief carcinogen exposure are of the order of months. With longer exposure protocols, tumor induction can take as little as 6 weeks (Ishikawa and Takayama, 1979; Hatanaka et al., 1982). And unlike many rodent models, spontaneous neoplasms are rare in medaka (Hawkins et al., 1988b; Masahito et al., 1989a).

The same qualities which make medaka an ideal test animal for rapid carcinogen screening also make it useful for basic research. Medaka embryos have been used in transgenic studies to assess the role of oncogenes in hereditary melanoma of *Xiphophorus* hybrids (Winkler et al., 1994). Carcinogen-exposed medaka have been used to evaluate DNA repair mechanisms and unscheduled DNA synthesis (Ishikawa et al., 1984; Nakatsuru et al., 1987). Enzyme histochemistry has been employed to assess preneoplastic foci in liver (Nakazawa et al., 1985; Teh and Hinton, 1993) and intermediate filament-specific antibodies used to characterize normal and neoplastic tissue (Bunton, 1994, 1995). Tumor cell lines have been established from carcinogen-induced neoplasms (Etoh et al., 1985; Muraiso et al., 1987) and inbred strains developed to study tumor pathogenesis with transplantion experiments (Hyodo-Taguchi and Matsudaira, 1984; Hyodo-Taguchi et al., 1988). Clearly though, there are disadvantages to using a small fish species. Medaka tissue samples must frequently be pooled for biochemical and molecular analyses. Blood samples from medaka are measured in microliters, making assessment of carcinogen uptake and metabolism difficult or impossible. And small size makes surgical manipulations problematic at best.

In contrast, large blood samples are easily obtained from trout and, with limitations, can be taken multiple times. Surgical manipulations, such as hepatectomy, bile duct ligation and organ biopsy can be performed relatively easily in trout and ID-tags or fin clips used to follow individual fish (Ostrander et al., 1992; Okihiro and Hinton, 1999b). Non-invasive laparoscopy could potentially be used to follow growth or regression of individual lesions subsequent to use of modulatory agents. Macroscopically visible tumors are of tremendous benefit in studies seeking to assess biochemical data from single tumors or when attempting to develop tumor cell lines. Larger sample sizes also allow multiple assays to be conducted on single organs or tumors, enabling researchers to correlate histologic or ultrastructural assessment with biochemical and molecular data.

Rainbow trout were initially recognized as potential test animals following accidental induction of liver tumors with aflatoxin-contaminated feeds (Ashley and Halver, 1961; Ashley et al., 1965), and over the past four decades 28 different compounds have been demonstrated as carcinogenic in trout (Bailey et al., 1996). Although trout exposure facilities are still restricted by availability of space and chilled water, many logistical problems associated with a larger fish species have been overcome with use of embryos or sac fry. In addition, microinjection protocols have been refined such that computer-assisted microsyringes can precisely deliver microliter volumes to embryonic yolk sacs, allowing carcinogen dosages to be tapered to the nanogram level (Dashwood et al., 1994). Confirmed trout carcinogens include MAMA, MNNG, dehydroepiandrosterone (DHEA), six aflatoxins, three nitrosamines and a variety of PAHs. The primary target organ for almost all assayed carcinogens is liver, but neoplasms have also been induced in kidney (nephroblastoma), stomach (papillary adenomas) and swimbladder (adenomas).

Variations in exposure protocols (altering carcinogen class, dosage and routes of exposure) combined with use of different stages of development (embryonic, larval and adult) have resulted in an extensive database with regard to target organs, tumor prevalence and latency. Many of the differences in carcinogen susceptibility center around liver organogenesis and maturation, and researchers have helped define the role of trout cytochrome P-450s in bioactivation of aflatoxin, PAHs and DEN. In addition, DNA adduction and repair, as well as mutations in Ki-*ras* oncogenes, have also been assessed following carcinogen exposure in trout (Chang *et al.*, 1991; Fong *et al.*, 1993; Hendricks *et al.*, 1994; Orner *et al.*, 1995).

Data derived from baseline studies have provided a solid foundation from which to examine tumor modulation. Positive (increase in tumor prevalence, number, size, or shorter latency) and negative modulation have both been documented with a wide range of chemicals (Bailey *et al.*, 1987; Bailey and Hendricks, 1988). Chemicals promoting AFB1, MNNG, or DMBA-mediated carcinogenesis include: β-naphthoflavone (BNF), indole-3-carbinol (I3C), β-estradiol, perfluorooctanoic acid, hydrogen peroxide and DHEA (Bailey *et al.*, 1989; Nunez *et al.*, 1989; Dashwood *et al.*, 1991; Kelly *et al.*, 1992; Orner *et al.*, 1996a, 1998).

Interestingly, three promotional agents (BNF, I3C and DHEA) have also been found to be tumor inhibitors depending on: (i) what initiating carcinogen was used; (ii) how the modulator was administered; and (iii) which organ system was examined (Nixon *et al.*, 1984; Goeger *et al.*, 1988; Dashwood *et al.*, 1994; Orner *et al.*, 1996b). The anticarcinogenic effects of BNF and I3C are mediated through induction or inhibition of cytochrome P-450s (Nixon *et al.*, 1984; Goeger *et al.*, 1988). In contrast, chlorophyllin inhibition of AFB1-induced hepatocarcinogenesis is based on non-covalent complex formation with AFB1, effectively reducing the number of DNA adducts formed (Breinholt *et al.*, 1995a,b).

In summary, both medaka and rainbow trout are excellent cancer models. Both have low rates of spontaneous neoplasms, are sensitive to mammalian carcinogens and have short latency periods. The small size and short lifespan of medaka (2–5 years) make it ideal as a screening tool (to quickly assess large numbers of potential carcinogens) and for lifetime carcinogen bioassays. Fast maturation and sexual dimorphism in medaka allow for assessment of gender related influences and the greater diversity of tumor types induced in medaka potentially make it a valuable tool in the study of neoplasms which are simply not manifested in trout. On the other hand, the trout model clearly holds an advantage with respect to: assessment of carcinogen uptake and metabolism; correlation of data from individual lesions; and studies assessing the impact of tumor modulators. There simply is no one best fish model for carcinogenesis research, and studies involving medaka and trout, as well as other fish species, should be used to complement one another.

Field studies of liver cancer epizootics

Field studies of tumor epizootics present an opportunity to gauge the effects of exposure to complex mixtures of xenobiotics, compare susceptibilities of related species, and research population level effects of carcinogens. These epizootics have usually involved liver, skin, or chromatophore lesions. Since, as opposed to skin cancers, liver cancers in fish are believed to be primarily carcinogen induced (John Harshbarger, Director, Registry of Tumors in Lower Animals, George Washington University, personal communication), we will only review field studies of liver neoplasms and chromatophoromas in this section. There have been epizootics of liver neoplasia reported involving at least nine different species of fish at 21 different locations in North American waters alone (Baumann, 1998). Three of the more intensive case studies: one marine, one freshwater, and one estuarine, will provide an overview of fish as a monitor of natural systems.

English sole in Puget Sound

Probably the most comprehensive series of studies of environmentally induced cancer in vertebrates was conducted on Puget Sound by personnel of the Northwest Fisheries Science Center, NMFS. In 1975 English sole (*Pleuronectes vetulus*) from the urban Duwamish Waterway had a significantly elevated prevalence (12%) of liver neoplasms (McCain *et al.*, 1977). Sediments in that waterway contained high levels of polycyclic aromatic hydrocarbons (PAHs) as well as PCBs and heavy metals. Research on these liver neoplasms and their cause expanded over the next decade and a half to include over 45 sampling stations in Puget Sound with a wide range of toxicant levels and a diversity of toxicant profiles (Myers *et al.*, 1990). Although several benthic species were examined, English sole proved to be the most susceptible to tumor development and was therefore the main focus of most studies.

Locations where English sole had a significantly elevated neoplasm prevalence were either contaminated urban areas (Elliot Bay and Commencement Bay) or sites contaminated by creosote (Malins *et al.*, 1984). At Mukelteo English sole had a 7.5% incidence of hepatic neoplasms and over a 16% incidence of foci of hepatocellular alteration, potential preneoplastic lesions (Malins *et al.*, 1985a). Elevated levels of PAHs were found both in the sediment and in the stomach contents of English sole. Sole from Eagle Harbor had an even higher liver neoplasm prevalence (27%) (Malins *et al.*, 1985b). Sediments from the harbor contained particularly high concentrations of aromatic hydrocarbons characteristic of creosote including the carcinogens benzo(a)pyrene (BaP) and benz(a)anthracene (BaA).

A statistical analysis of potential etiological factors affecting the wild population was conducted by Myers *et al.* (1990). Results from six different studies which had sampled a total of 61 stations were used to correlate the occurrence of neoplasms with PAH concentrations. The adjusted significance level from Fisher's combined probability test was 0.0001. A logistic regression model further compared the PAHs, PCBs, chlorinated butadienes (CBDs) and fish size (age) in their influence on neoplasm incidence. Fish size (age) had the greatest effect (15% of the variability, $p < 0.00001$). PAHs were the major etiologic agent, accounting for 12% of the variation ($p < 0.0001$). The only other positive correlation (PCBs) explained only 1% of the variability (Myers *et al.*, 1990). An epizootiologic study also investigated the odds of ratios of significant risk factors for hepatic lesions (Rhodes *et al.*, 1987). The Duwamish Waterway had an odds ratio of over 8 for hepatic neoplasms as compared to the reference station of Port Madison in Puget Sound.

The presence of biliary metabolites has been used to confirm exposure to PAHs for wild fish. Fluorescent aromatics in bile samples taken from English sole showed significant positive correlations ($p < 0.002$) with the prevalence of idiopathic liver lesions (Krahn *et al.*, 1984, 1986). This correlation works well for site comparisons, but cannot be used to predict neoplasms in individual fish, since the bile metabolites are indicators of recent exposure, while neoplasms reflect past exposure (Myers *et al.*, 1990).

Brown bullhead in the Black River

Fish populations make excellent indicators of carcinogenic risk for a given ecosystem through time. Benthic fish such as bullhead, which are resident rather than migratory, make good synthesis models for effects of bioavailable carcinogens in water and sediment. A series of studies beginning in 1980 focused on liver neoplasms in brown bullhead (*Ameiurus nebulosus*) living in the Black River, Ohio. In the early 1980s the Black River received effluent from a coking facility associated with a US Steel Plant (Baumann *et al.*, 1982). Livers of brown bullhead sampled in 1982 ($N = 125$) were examined histopathologically. The population had a liver neoplasm prevalence of 56% for age 3 fish and 62% for age 4 fish (Baumann *et al.*, 1990). Sediment samples taken near the coke plant outfall in 1980 were very high in PAH, with carcinogens such as benzofluoranthenes, BaA, and BaP being present in levels of 43 to 75 ppm. Similarly whole fish from the Black River in 1980 had phenanthrene at ppm levels, while BaA ranged from 3 to 33 ppb and BaP from 0.7 to 18 ppb.

Two major changes affecting carcinogen exposure have occurred on the Black River since 1982. The first was the closure of the coking facility in 1983. Four years later in 1987 surficial (top 5 cm) sediment PAH levels had declined to 0.4% of the 1980 concentrations (Baumann and Harshbarger, 1995). Similarly neoplasms had declined to 16% for age 3 fish and 41% for age 4 fish. Frank carcinomas also declined by 74% over this period.

Secondly in the early 1980s a Consent Decree between the US EPA and US Steel included a provision to dredge the area with the most highly contaminated sediments. This dredging, using an open clam shell dredge, finally took place in 1990. Fish sampled in 1992 and 1993 revealed that the prevalence of liver neoplasms in mature fish was again elevated to the levels seen in the early 1980s (59% in 1992 and 64% in 1993) (Baumann and Harshbarger, 1998). However, in 1994 liver neoplasms became less frequent. This decline was primarily due to a much lower neoplasm frequency in age 3 fish. Three-year-old fish were the youngest age group used in the tumor surveys, since neoplastic lesions are seldom found in fish of age 2 or less (Baumann *et al.*, 1982). Neoplasm frequencies in fish of age 3 or older were 46% in 1992, 61% in 1993, and 0% in 1994 (Baumann and Harshbarger, 1998). The 1994 sample was the first which included fish (age 3) that were not present during the dredging in 1990 (having hatched in 1991). Limited sampling in 1995 and 1996 indicated that neoplasm frequencies remained low.

The Black River provided a natural experiment, with the hypothesized cause of the liver neoplasms (carcinogenic PAH) first being reduced by the coke plant closure, then briefly spiked by the remedial dredging. PAH concentrations in 1992 did show an approximately fourfold increase over 1987 levels, although they did not approach the levels recorded in the early 1980s (Baumann and Harshbarger, 1998). Having these two events spaced 7 years apart allowed time for the pathological response in the fish to be clear, even given the slow burial of contaminated sediment by new sediment, and the latent period between exposure/induction and tumor growth common for chemically induced neoplasms.

Mummichog in the Elizabeth River

Mummichog (*Fundulus heteroclitus*), a common estuarine species along the Atlantic coast, make an excellent vertebrate monitor because of their exceedingly small home range of only 30–40 m (Lotrich, 1975). Mummichog captured in the Elizabeth River, Virginia from a site adjacent to an active wood treatment facility had a 35% prevalence of hepatic neoplasms (N = 60). Those (N = 30) taken from a reference site directly across the Elizabeth River (approximately 600 m) had no hepatic neoplasms (Vogelbein *et al.*, 1990). PAH levels near the wood treatment plant were over 2000 mg kg^{-1} dry wt, approaching two orders of magnitude higher than PAH concentrations at the reference location (Vogelbein *et al.*, 1990).

In 1991 a second study again compared neoplasm prevalence and sediment chemistry between the creosote contaminated site and a nearby reference site (Vogelbein *et al.*, 1997). Those from the contaminated site (N = 60) had over a 50% prevalence of hepatic neoplasms, while those from the reference site (N + 60) had none. In addition the mummichog from the polluted location also had over a 10% incidence of exocrine pancreatic neoplasms and over a 35% incidence of vascular neoplasms, lesions which did not occur in fish from the reference site (Vogelbein *et al.*, 1997).

This dramatic difference in tumor prevalence for two closely located populations demonstrates how a fish species with a small range can help determine causal relationships in field situations.

Exposure experiments based on field samples

A natural extension of a field study that demonstrates a high tumor prevalence in a wild population is a laboratory study exposing fish to sediment or sediment extract. The first such experiments used sediment from the Black River, Ohio (Baumann *et al.*, 1982) and the Buffalo River, New York (Black, 1983). Sediment extracts were characterized by HPLC and used to treat brown bullhead (Black *et al.*, 1985). Buffalo River extract was used in two experiments, one where animals were painted on the head between the upper lip and the occiput, and another where fish were fed a diet containing sediment extract dissolved in corn oil. Fish painted with the extract had a greater than 35% prevalence of papillomas after two years, while those painted with solvent did not develop neoplasms. Although only small numbers of bullhead were used in the feeding study, one of the five bullheads sacrificed at 4 months had developed hepatocellular carcinoma, and three of six fish sacrificed at 7 months had also developed neoplasms in the liver (Black *et al.*, 1985). Laboratory studies undertaken in conjunction with the field studies on in Puget Sound exposed English sole to reference site sediment with added [^{14}C]naphthalene (NPH) and [^{3}H]BaP (dissolved in 1% crude oil), sediment from the contaminated Duwamish waterway, and virtually clean reference site sediment (Varanasi *et al.*, 1987). Sole were shown to readily take up BaP from sediment and to extensively metabolize the compound. Hepatic DNA was altered after 2 to 4 weeks following a single exposure to BaP. BaP was metabolized into essentially a single adduct identified by the authors as (+) *anti*-7,8-diol-9,10-epoxy-7,8,9,10-tetrahydroBaP-dG.

Sediments from Hamilton Harbor, another location with a history of external and liver neoplasms in wild fish (Cairns, 1983; Smith *et al.*, 1986), were also used for mutagenicity and carcinogenicity tests. Hamilton harbor sediments were low in PCBs and other organochlorines, but high in PAHs, with BaP concentrations at about 4.8 ppm (Metcalfe *et al.*, 1988). In the Ames assay the Hamilton Harbor extract was mutagenic using the TA100 *Salmonella* strain both with and without the addition of S-9. A low

PAH reference site did not show any mutagenic activity. The extracts were also used to microinject rainbow trout sac fry. After one year trout dosed at higher levels with Hamilton Harbor extract developed carcinomas in the 8 to 9% prevalence range (Metcalfe et al., 1988). Sac fry dosed with reference harbor sediment failed to develop any neoplasms.

Chromatophoromas in wild fish

Chromatophoromas are pigment cell tumors arising from dermal chromatophores in the skin of fish, amphibians and reptiles (Bagnara, 1966, 1986; Schliwa, 1986; Masahito et al., 1989b). Although analogous to mammalian melanoma, pigment cell neoplasms in lower vertebrates are broadly classified as **chromatophoromas** because they can potentially arise from more than a single pigment cell type. The four chromatophore classes commonly found in fish are: melanophores with black or brown pigment (melanin); iridophores with colorless pigments (purines); erythrophores with red pigments; and xanthophores with yellow pigments. Xanthophores and erythrophores are closely related and both contain carotenoids, pteridines and flavins.

Chromatophoromas are common fish tumors and large epizootics, involving thousands of fish, have occurred worldwide. Epizootics in marine waters include: croakers (*Nibea mitsukurii*) off Japan (Kimura et al., 1984); deepwater redfish (*Sebastes mentella*) in the North Atlantic (Bogovski and Bakai, 1989); butterflyfish (*Chaetodon multicinctus* and *C. miliaris*) from the Hawaiian islands (Okihiro, 1988); Pacific rockfish (five *Sebastes* spp.) off California (Okihiro et al., 1993); and deepwater sablefish (*Anoplopoma fimbria*) off the Farallon Islands (unpublished data, Okihiro). Freshwater epizootics have involved freshwater drum (*Aplodinotus grunniens*) from the Great Lakes (Okihiro and Baumann, unpublished data), domestic goldfish (*Carassius auratus*) (Etoh et al., 1983), and possibly gizzard shad (*Dorosoma cepedianum*) from three lakes in Oklahoma, Texas, USA (Jacobs and Ostrander, 1995; Ostrander et al., 1995; Geter et al., 1998). Gizzard shad lesions are described as pigmented subcutaneous spindle cell neoplasms, and histologic and ultrastructural features are consistent with poorly differentiated (or **amelanotic**) melanophoromas. However, researchers studying this phenomenon did not feel there were enough data to definitively rule out neoplasms originating from nerve sheath, fibroblasts, or some other poorly differentiated cell type.

Subclassification of chromatophoromas is primarily based on identification of specific pigment organelles (Okihiro, 1988; Okihiro et al., 1993). Melanophoromas, the most common tumor type encountered, are diagnosed at the light or electron microscopic (EM) level via detection of electron-dense melanosomes. Iridophoromas are characterized at the light level by the presence of olive-green granular pigment, which is birefringent with polarized light.

Ultrastructurally, in unstained sections, iridophores have stacked arrays of reflecting platelets. Routine EM staining with uranyl acetate and lead citrate, however, results in dissolution of reflecting platelets and empty latice-works of open membranes. Xanthophore and erythrophore lesions are difficult to diagnose as pigments characteristic of these cell types are undetectable with routine histologic processing and staining. Ultrastructurally, xanthophores and erythrophores can be identified by detection of lipid (presumptive carotenoid pigment) and/or **pterinosomes** (specific organelles associated with pteridine pigments). However, because these cell types share many of the same pigments, their tumors cannot be differentiated ultrastructurally. Subclassification of xanthophore and erythrophore tumors therefore is best made on the basis of gross coloration.

Hyperplastic lesions are likely precursors to chromatophoromas in some fish species, and progression appears to occur stepwise from mild to severe hyperplasia, dysplasia and eventual neoplasia (Bogovski and Bakai, 1989; Okihiro et al., 1993). In other epizootics, neoplasms appear to arise without early onset of preneoplastic hyperpigmentation. While the majority of chromatophoromas are benign and restricted to the dermis, some of the largest are malignant as evidenced by increasing **anaplasia**, invasion, and occasional metastasis to liver or gill (Okihiro et al., 1993). Anaplastic chromatophoromas are quite variable histologically and, with severe pigment loss, must be differentiated from other cutaneous spindle cell tumors including fibromas, neurofibromas and schwannomas (Escourolle and Poirier, 1978).

Although etiology for the majority of chromatophoroma epizootics is unknown, epidemiologic surveys from many studies are at least suggestive of possible exposure to anthropogenic carcinogens. Potential sources for xenobiotic exposure include Kraft pulp mill effluent in Japan (Kimura et al., 1989), agricultural pesticides in Hawaii (Okihiro, 1988) and a variety of radioactive and chemical compounds in California and the Great Lakes (Okihiro et al., 1993). Unfortunately, while polyaromatic hydrocarbons (PAHs) have been proven to induce melanoma in mammals (Knox et al., 1988), follow-up studies to definitively link pigmented lesions with xenobiotic exposure have been lacking in most fish studies. One notable exception has been with researchers studying tumor epizootics off Japan. Researchers were able to induce chromatophore hyperplasia and neoplasia in laboratory-reared croakers using 7,12-dimethylbenz-[a]anthracene and N-methyl-N'nitrosoguanidine (Kimura et al., 1984). The same researchers were also able to induce melanophore hyperplasia in 70–100% of marine catfish (*Plotosus anguillaris*) and chromatophoroma in one croaker using effluent from a pulp mill epidemiologically linked with identical lesions in wild fish (Kinae et al., 1990). The Japanese study is one of the few examples where the hypothesis that chromatophoromas in wild fish are caused by exposure to carcinogenic compounds has been tested under laboratory conditions. Although it appears likely that chemical carcinogens are responsible for pigment cell neoplasms in Japanese croakers, a number of other potential etiologies (oncogenic viruses, genetic predisposition and ultraviolet radiation) must be fully evaluated before xenobiotic compounds can be definitively implicated as causative agents in other chromatophoromas epizootics.

Damselfish neurofibromatosis

One of the most extensively studied epizootics of neoplasms in wild fish is damselfish neurofibromatosis (**DNF**). DNF affects bicolor damselfish (*Pomacentrus partitus*) native to coral reefs off southern Florida (USA), and is characterized by multiple nerve sheath tumors (neurofibromas and neurofibrosarcomas) and, to a lesser extent, cutaneous chromatophoromas (Schmale et al., 1986, 1996). Although Schwann cells are an integral component of DNF and some early tumors were diagnosed as schwannomas (Schmale et al., 1983), DNF is distinct from schwannomas in that lesions are multifocal and composed of mixtures of Schwann cells, nerves, fibroblasts, collagen fibers and pigment cells. Specific lesion classification is based on the degree of malignancy (neurofibrosarcomas exhibiting greater pleomorphism, disorganization and invasiveness) and pigmentation (chromatophoromas having a preponderance of pigment cells – usually melanophores). The disease has been proposed as an animal model of human neurofibromatosis type 1 (von Recklinghausen's disease) (Schmale et al., 1983, 1986). Similarities include tumor types involved, multifocal nature and presence of eosinophilic granular leukocytes (putative piscine mast cells) (Vicha and Schmale, 1994).

Schwann cells from normal damselfish are antigenically (S100 positive; fibronectin and glial fibrillary acidic protein negative) and morphologically similar to mammalian Schwann cells (Schmale et al., 1994). DMF Schwann cells are distinguished by increased reactivity to S100 antibodies, variable morphology, and ability to proliferate in cell culture without mitogens. Six Schwann cell lines have been derived from DNF fish.

It appears likely that DNF has a viral etiology. Initial epidemiologic surveys revealed a distribution pattern (clusters of cases) consistent with an infectious etiology (Schmale, 1991). DNF is transmissible via crude tumor homogenates (Schmale and Hensley, 1988) and cell-free filtrates from homogenized tumors or cell lines derived from tumors (Schmale, 1995). In addition, type C retroviral particles (90–110 nm) have been observed budding from cultured DNF cells and exhibit reverse transcriptase activity (Schmale et al., 1996).

Melanomas in *Xiphophorus* hybrids

Freshwater fish belonging to the genus *Xiphophorus* have been used in genetic and carcinogenesis research for over 70 years (Vielkind and Vieldkind, 1982; Vielkind and Dippel, 1984; Wittbrodt et al., 1989). Native to Central America, these small viviparous fish are commonly known as platyfish and swordtails. It has long been known that interspecific hybrids produce heritable melanoma and several cancer models using *Xiphophorus* fish have been established. The most extensively studied of these is the Gordon-Kosswig melanoma model. In this model, female platyfish (*X. maculatus*) homozygous for the pigment allele spotted dorsal (Sd) are crossed with male swordtails (*X. helleri*). The Sd allele on the X chromosome encodes for macromelanophore (a large polyploid, melanin-bearing pigment cell) proliferation in the dorsal fin of platyfish. Backcrossing F_1 hybrids to swordtails results in melanoma formation in a specific subset of fish and conforms to simple Mendelian inheritance involving two gene loci: a sex-linked tumor locus (*Tu*) and an autosomal recessive tumor suppressor locus (*Diff*), originally described as a regulatory locus responsible for macromelanophore differentiation (Vielkind, 1976).

Extensive scrutiny of these gene loci has led to the identification of Tu as a specific oncogene, the *Xiphophorus* melanoma receptor kinase (*Xmrk*) gene. *Xmrk* is a sex-linked tyrosine kinase receptor gene closely related to the human epidermal growth factor (EGF) receptor gene (Morizot et al., 1998). The two versions of *Xmrk* are the melanoma inducing oncogene, ONC-*Xmrk*, and a proto-oncogene, INV-*Xmrk*. The evolutionary origin of the *Xmrk* oncogene is associated with non-homologous recombination whereby INV-*Xmrk* was translocated downstream of an anonymous donor gene locus, D, which has only been partially characterized (Förnzler et al., 1996). Translocation resulted in *Xmrk* obtaining a novel promoter (from D) and led to *Xmrk* gene duplication. The novel promoter differs markedly from that in the proto-oncogene and is active only in hybrids, leading to marked overexpression of ONC-*Xmrk* in both tumors and melanoma cell lines (Schartl et al., 1994; Baudler et al., 1997). IVN-*Xmrk*, in contrast, is expressed at very low levels.

Sequence comparisons have revealed that ONC-Xmrk protein is distinct from the proto-oncogene receptor because of several amino acid exchanges (Wellbrock et al., 1997; Dimitrijevic et al., 1998). These 'activating mutations' result in strong ligand-independent tyrosine phosphorylation. The Xmrk kinase is the most abundant phosphotyrosine protein in melanomas and studies with melanoma-derived cell lines have demonstrated marked receptor autophosphorylation (Malitschek et al., 1994).

The oncogenic potential of ONC-*Xmrk* has been confirmed with transgenic studies involving microinjection of *Xmrk* genes into medaka (*Oryzias latipes*) embryos. **Ectopic** expression of ONC-*Xmrk* results in transgenic embryos with high incidences of embryonic tumors, while INV-Xmrk transgenic embryos are unaffected (Winkler et al., 1994; Dimitrijevic et al., 1998). Additional support for the importance of ONC-*Xmrk* in neoplastic transformation has come from studies using an *Xmrk* specific tyrphostin (AG555) to inhibit kinase activity. In fish melanoma cell lines, AG555 inhibition of Xmrk autophosphorylation correlates strongly with blockage of DNA synthesis and cessation of tumor cell growth (Wellbrock et al., 1998a).

AG555 has also aided the elucidation of Xmrk-induced signal transduction. One of the proteins strongly tyrosine phophorylated in the melanoma cell line PSM is the signal transducer and activator of transcription STAT5. Using murine pro B-cells and Xmrk kinase in a chimeric receptor, tyrosine phosphorylation of STAT5 is followed by nuclear translocation, binding to target genes (*cis*, *osm* and *pim*-1) and gene expression (Wellbrock et al., 1998b). In PSM and melanoma cells, STAT5 has also been shown to localize to nuclei and induce *pim*-1 expression and STAT5 DNA localization can be blocked by AG555-mediated inhibition of Xmrk kinase activity (Wellbrock et al., 1998b).

In contrast to the extensive research into elucidation of the *Tu* locus/*Xmrk* gene, relatively little is known about Diff, the other major determinant of *Xiphophorus* hybrid melanoma. While *Diff* has long been known as an autosomal recessive locus associated with tumor suppression, it has never been fully investigated. Recently, however, linkage group V (on which *Diff* is located) was carefully re-mapped and 20 different loci identified (Kazianis et al., 1998). Testing of loci for association with degree of pigment pattern and melanoma susceptibility revealed strong linkage with *CDKN2X* (a *CDKN2* gene family member) and this gene has been proposed as the classically defined *Diff* tumor suppressor locus. Additional support for this hypothesis comes from studies involving UVB-inducible melanoma in spotted side (Sp) *Xiphophorus* hybrids (Nairn et al., 1996). Testing for joint segregation of genetic markers revealed linkage of a *CDKN2*-like DNA polymorphism with UVB-induced melanoma, and this *CDKN2*-like gene was also proposed as a candidate for the *Diff* tumor suppressor locus. Significantly, *CDKN2A* is a tumor suppressor gene associated with human melanoma.

Although the *CDKN2*-like gene and *Xmrk* are clearly the major determinants in hereditary melanoma of *Xiphophorus* hybrids, several other oncogenes also appear to play significant roles with respect to progression. *Xiphophorus* melanomas have elevated levels of $pp60^{c-src}$ related kinase activity (Barnekow et al., 1982; Schartl et al., 1982), and both *src* expression and src-related kinase activity are strongly correlated with malignancy in a variety of hybrids (Schartl *et al.*, 1985; Mäueler *et al.*, 1993). The *SRC* family consists of eight oncogenes, and Northern blot analysis has revealed that, in addition to *src*, *fyn* and *yes* are also expressed in *Xiphophorus* melanomas and the melanoma cell line PSM (Hannig *et al.*, 1991). Fyn-kinase activity is highly elevated in melanoma cells, and affinity binding assays have revealed that Xmrk-kinase has a higher affinity for the SH2 domain of *fyn*, compared to that of *src* and *yes* proteins (Wellbrock *et al.*, 1995). In addition, stimulation of Xmrk-kinase activity in melanoma cells results in a concurrent increased *fyn* activity. There is also a possibility that *ras* may play a role, at least with secondary events of melanoma progression, but that expression of proto-oncogenes *erb-B*, *c-myc*, *sis* and *fms* does not appear to be related to tumor formation or progression (Mäueler *et al.*, 1993).

Viral tumors

Typically, fish tumors are first linked to a virus either by experimental transmission of the neoplasm or by identifying a virus in tumor cells using an electron microscope. For instance papillomas in white suckers have been transmitted by injection of filtrates (Premdas and Metcalfe, 1996), thereby arguing for a viral etiology, even though no virus particles have been identified. On the other hand, electron microscope images are often reported first, sometimes simply as 'virus-like particles'. Viruses from four families have been specifically identified in fish lesions: herpesvirus, adenovirus, papovavirus and retrovirus (Anders and Yoshimizu, 1994). Members of these same four viral families are also involved in the etiology of tumors in humans (zur Hausen, 1980). Two of these families are involved in the majority of tumors in fish with viral etiologies, the herpesvirus and the retrovirus.

Herpesvirus

Herpesvirus is by far the most common virus associated with tumors in fish. Twelve herpesviruses have been detected in benign epidermal lesions, ranging from hyperplasia to papilloma as reviewed by Anders and Yoshimizu (1994). Both freshwater and saltwater species from nine different genera have had tumors attributed to herpes. Two of these have been proven to be oncogenic through isolation in cell culture and the fulfillment of River's postulates (Anders and Yoshimizu, 1994).

Oncorhynchus masou virus (OMV) was initially isolated from masu salmon in Japan (Kimura *et al.*, 1981). The virus was independently isolated from immature masu salmon with jaw papillomas (Sano *et al.*, 1983). OMV has been used experimentally to induce epidermal papillomas in three additional *Oncorhynchus* species (Yoshimizu *et al.*, 1987), and was shown to be pathogenic to a fourth (Anders and Yoshimizu, 1994). Tumor tissues of coho salmon (*O. kisutch*) cultured both in fresh water and in ocean water were also used to successfully isolate OMV (Kimura and Yoshimizu 1991). Infected fish often develop epidermal papillomas around the mouth (Yoshimizu *et al.*, 1987) or elsewhere on the body surface.

The other virus with proven oncogenicity is cyprinid herpesvirus (**CHV**), which causes the papillomas seasonally found on common carp and referred to as 'carp pox'. This virus had been reported as early as 1896 (Hofer, 1896). The virus was first described in electron micrographs of common carp tumors (Schubert, 1964, 1966). Isolation in cell culture was not successful until the virus was obtained from papillomas of Japanese fancy carp (Sano *et al.*, 1985a,b). The isolate was used to infect both adult carp (fancy and mirror) and carp fry (fancy and common). Fry suffered a greater mortality rate than adults, and of the surviving fry, 83% developed epidermal papillomas (Sano *et al.*, 1985a,b). The virus multiplies and produces tumors in the temperature range of 10–25°C (Morita and Sano, 1990). At temperatures above 20°C lymphocytes cause sloughing of the lesions. However, dormant virus remains in ganglion cells so lesions often reappear when temperatures cool down again (Sano *et al.*, 1993).

Retrovirus

Retroviruses have been associated with both benign tumors and cancers in fish; Anders and Yoshimizu (1994) reviewed 10 different retroviruses or retrovirus-like particles which have been described from skin tumors. Species with such skin tumors include white sucker (*Catostomus commersoni*) (Sonstegard, 1973, 1977), walleye (Walker, 1969; Yamamoto *et al.*, 1976), and smelt (Anders, 1989). Both lymphoma in northern pike (Nigrelli, 1952) and leukemia in chinook salmon (Harshbarger, 1982) have also been linked to a retrovirus. As opposed to benign tumors, induction of cancers in fish has only been achieved using retroviruses of the subfamily Oncovirinae (Anders and Yoshimizu, 1994).

Two of three benign skin lesions occurring on walleye are caused by retroviruses. Discrete epidermal hyperplasia, first reported by Walker (1969), consists mainly of cuboidal-shaped cells. The margins of the lesion are raised and distinct (Bowser et al., 1998). Cells from such epidermal hyperplasias on walleye were found to be associated with surface-budded, retrovirus-like particles occurring in the expanded intercellular spaces (Yamamoto et al., 1985). Further research identified two related retroviruses, designated walleye epidermal hyperplastic virus types 1 (WEHV1) and 2 (WEHV2) (Martineau et al., 1991a; LaPierre et al., 1998). Recently young-of-the-year walleye were injected with cell-free filtrates from walleye epidermal lesions known to contain WEHV1 and WEHV2 (Bowser et al., 1998). Fish injected with a control filtrate from normal skin did not develop lesions, while 97% of those inoculated with the viral filtrates developed discrete epidermal hyperplasia. Polymerase chain reaction assays (PCR) identified WEHV2 in 100% of the lesions and WEHV1 in 69% of the lesions (Bowser et al., 1998). The authors stated that these data indicated that WEHV2 could induce the tumors independently of WEHV1.

Walleye dermal sarcoma is a benign lesion described by Walker in 1947, who also reported the presence of virions in the sarcomas (Walker, 1961). Electron microscopy allowed Yamamoto et al. (1976) to describe the sarcoma virus as 135 nm, and more regularly spherical with a better defined core, envelope, and peplomer layer compared to the discrete epidermal hyperplasia virus. Transmissions of the disease using injections of fingerling walleye with cell-free filtrate were reported by Martineau et al. (1990) and Bowser et al. (1990). Reverse transcriptase activity was also found in tissue extract of walleye dermal sarcomas by Martineau et al. (1991b). Molecular characterization indicated that the retrovirus may be of the subfamily Spumavirinae (Martineau et al., 1991a).

Lymphomas

Lymphoid tumors have been reported from several different species of fish and retrovirus-like particles were noted in some of these cases (Hoffmann et al., 1988; Harada et al., 1990). However, the best known of the lymphomas attributed to retrovirus are those of the pike family (genus *Esox*). Epizootics have been reported in northern pike (*Esox lucius*) from Scotland (Haddow and Blake, 1933); Ireland (Mulcahy, 1963); brackish waters of the Baltic Sea coast in Sweden (Ljungberg and Lange, 1968; Ljungberg, 1976), Finland (Thompson, 1982) and Estonia (Bogovski, 1988); and in both northern pike and muskellunge (*Esox masquinongi*) from North America (Brown et al., 1973; Sonstegard, 1975, 1976). This wide distribution of a naturally occurring epizootic in a species which is both easily captured in the wild and easily bred in captivity demonstrates again the value of fish as a model.

Serial transmission of lymphoma was first achieved by Mulcahy and O'Leary (1970) using a cell-free tumor homogenate. Sonstegard (1976) noted that transmission trials involving filtrate from wild specimens were only successful in spring; all trials using filtrate from specimens collected in late summer failed. Ljungberg (1976) achieved both serial transmission through four repetitions and contact transmission. Viruses were first documented in electron micrographs of skin sarcomas from Swedish pike (Winqvist et al., 1973). C-type virus particles were also found in tumors from North American pike (Papas et al., 1976), and the authors further demonstrated that reverse transcriptase was associated with the virus. The reverse transcriptase was also found to be similar (except for optimum temperature) to the reverse transcriptase of viruses found in avian and mammalian neoplasms (Pappas et al., 1977; Sonstegard and Papas, 1977).

Attempts to classify the lymphomas have become increasingly more sophisticated. Thompson (1982), using light and electron microscopy, classified the tumors as lymphoblastic or stem cell lymphoma based on cell morphology. The presence of the intermediate filament protein vimentin (Thompson et al., 1987a), groups of lysosomes, and abundant cytoplasmic lipid droplets led to a diagnosis of true histiocytic lymphoma (Thompson et al., 1987b; Thompson and Miettinen, 1988). Bogovski (1988) described three histological types of lymphoma in Estonian pike using standard staining techniques. Subsequent immunohistochemistry using rabbit anti-human polyclonal antibodies against lysozyme (LYS), alpha-1-antichymotrypsin (ACT) and S-100 protein led to a diagnoses of histiocytic lymphoma, histiocytoid-lymphocytic lymphoma, and lymphocytic lymphoma (Bogovski et al., 1994).

Environmental effects on viral lesions

The fact that fish are, in general, easier to capture in large numbers in the wild than mammals, makes them a better model to study the interaction of environmental variables and carcinogenic vectors. One of the most consistent features of viral tumor diseases in fish is that they are seasonal. Anders and Yoshimizu (1994) list at least eight different species in which virally derived skin tumors are more prevalent during the spring spawning season. They also point out that some salmonid viral tumors are more common during **smoltification**, another time of altered endocrine activity and, presumably, altered immune system function.

Some viral lesions have been more prevalent in populations from polluted areas, such as white sucker papillomas (Smith *et al.*, 1989; Premdas *et al.*, 1995), and lymphosarcoma in northern pike from America (Brown *et al.*, 1973) and from Estonia (Paakspuu, 1987). Thus persistent tumor viruses may be activated at times of immune system dysfunction or inhibition caused by a variety of potential agents, including carcinogenic or immune-suppressing compounds (Anders and Yoshimizu, 1994). Other stress factors may also affect the immune system; Thompson (1989) pointed out that high salinity in areas of pike epizootics around the Aland Islands might have increased susceptibility to the disease. Etiologies of such tumors would therefore be multifactoral, with the induction and progression of a virally induced lesion being influenced by environmental factors (Premdas and Metcalfe, 1996).

Summary: advantages of fish cancer models

Within this chapter, we have pointed out numerous advantages of fish carcinogenesis models, both in the field and laboratory. Major points are significant and worth emphasizing. In the field, teleost fish are probably most beneficial in their role as indicators of xenobiotic contamination of aquatic habitats, with complex mixtures of carcinogens and/or promotional agents. Tumor epizootics not only indicate threats to wild fish populations, but help alert us to the potential for human carcinogen exposure, whether via the food chain or through direct contact. In addition, epizootics identify those resident fish species (i.e. flatfish in Puget Sound, butterflyfish in Hawaii, bullheads in Great Lakes tributaries) which are most sensitive, yet still present in large enough numbers to allow monitoring as barometers of ecosystem health.

Laboratory studies, primarily using domestic freshwater fish species, have served to complement field research. In addition to helping confirm etiology of some tumor epizootics (by serving as surrogates or sentinels), laboratory studies utilizing rainbow trout and small aquarium fish (medaka, zebrafish, guppies) have greatly expanded our existing knowledge of carcinogens, promoters, and anti-carcinogens. And although fish will never replace rodent cancer models (simply because of the lack of complete organ homology), in many ways fish are ideal test subjects with high sensitivity to a broad range of carcinogens, extremely low prevalence of background tumors, and short tumor latency. Additional advantages to domestic fish include: (i) established husbandry requirements and ease of maintenance; (ii) availability of numerous inbred and outcrossed strains; (iii) tolerance for stocking at high density, increasing statistical power; and (iv) significantly lower costs to perform carcinogen bioassays, when compared to rodents.

The benefits of using fish in carcinogenesis research, however, are not just limited to use as simple chemical screening tools. Many fish models offer unique insights into etiology, pathogenesis and progression of cancer. The *Xmrk* gene responsible for hereditary melanoma in *Xiphophorus* hybrids, for example, was the first dominant oncogene discovered that is transmitted through the germ line. Damselfish neurofibromatosis, a naturally occurring disease of bicolor damselfish, is one of the few animal models for human neurofibromatosis type 1. Medaka are the only animal species where retinoblastomas (the most common intraocular tumor of children) are inducible with chemical carcinogens. And rainbow trout, with their predisposition towards mixed liver tumors, are an ideal host in which to evaluate the oval/bipolar stem cell theory of hepatocarcinogenesis.

In summary, fish cancer models not only serve as ecological indicators of environmental contamination and health, but could serve as adjunct test animals (helping reduce the vast backlog of potentially carcinogenic chemicals destined for human use or consumption), and greatly augment mammalian models seeking to understand how normal cells are transformed into malignant neoplasms.

References

Ahokas, J.T., Saarni, H., Nebert, D.W. and Pelkonen, O. (1979). *Chem. Biol. Interact.* **25**, 103.

Anders, K. (1989). In *Viruses of Lower Vertebrates* (eds W. Ahne and E. Kurstak), pp. 184–197. Springer-Verlag, Berlin.

Anders, K. and Yoshimizu, M. (1994). *Dis. Aquat. Org.* **19**, 215–232.

Anderson, M.W., Reynolds, S.H., You, M. and Maronpot, R.M. (1992). *Environ. Hlth. Perspect.* **98**, 13–24.

Aoki, K. and Matsudaira, H. (1977). *J. Natl. Cancer Inst.* **59**, 1747–1749.

Ashley, L.M. and Halver, J.E. (1961). *Fed. Proc.* **20**, 290.

Ashley, L.M., Halver, J.E., Gardner, W.K. and Wogan, G.N. (1965). *Fed. Proc.* **24**, 627.

Ashurst, S.W., Cohen, G.M., Nesnow, S., DiGiovanni, J. and Slaga, T.J. (1983). *Cancer Res.* **43**, 1024.

Bagnara, J.T. (1966). *Int. Rev. Cytol.* **20**, 173–205.

Bagnara, J.T. (1986). In *Biology of the Integument. 2. Vertebrates* (eds J. Bereiter-Hahn, A.G. Matoltsy and K.S. Richards), pp. 136–149. Springer-Verlag, Berlin.

Bailey, G.S. and Hendricks, J. (1988). *Aquat. Toxicol.* **11**, 69–75.

Bailey, G.S., Selivonchick, D. and Hendricks, J. (1987). *Environ. Hlth. Perspect.* **71**, 147–153.

Bailey, G.S., Williams, D.E., Wilcox, J.S., Loveland, P.M., Coulombe, R.A. and Hendricks, J.D. (1988). *Carcinogenesis* **9**, 1919.

Bailey, G.S., Goeger, D.E. and Hendricks, J.D. (1989). In *Metabolism of Polynuclear Aromatic Hydrocarbons in the Aquatic Environment* (ed. U. Varanasi), pp. 253–268. CRC Press, Boca Raton, Florida.

Bailey, G.S., Williams, D.E. and Hendricks, J.D. (1996). *Environ. Hlth. Perspect.* **104**, 5–21.

Balk, L., DePierre, J.W., Sundvall, A. and Rannug, U. (1982). *Chem. Biol. Interact.* **4**, 1.

Balk, L., Meijer, J., DePierre, J.W. and Appelgren, L.E. (1984). *Toxicol. Appl. Pharmacol.* **74**, 430.

Bannasch, P., Zerban, H. and Hacker, H.J. (1985). In *Monographs on Pathology of Laboratory Animals, Digestive System* (eds T.C. Jones, U. Mohr and R.D. Hunt), pp. 10–30. Springer-Verlag, Berlin, Heidelberg, New York, Tokyo.

Barnekow, A., Schartl, M., Anders, F. and Bauer, H. (1982). *Cancer Res.* **42**, 2429–2433.

Baudler, M., Duschl, J., Winkler, C., Schartl, M. and Altschmied, J. (1997). *J. Biol. Chem.* **272**(1), 131–137.

Baumann, P.C. (1998). *Mutat. Res.* **411**, 227–233.

Baumann, P.C. and Harshbarger, J.C. (1985). *Mar. Environ. Res.* **17**, 324–327.

Baumann, P.C. and Harshbarger, J.C. (1995). *Environ. Hlth. Perspect.* **2**, 168–170.

Baumann, P.C. and Harshbarger, J.C. (1998). *Environ. Monit. Asses.* **53**, 213–223.

Baumann, P.C., Smith, W.D. and Ribick, M. (1982). In *Polynuclear Aromatic Hydrocarbons: Sixth International Symposium on Physical and Biological Chemistry* (eds M.W. Cooke, A.J. Dennis and G.L. Fisher), pp. 93–102. Battelle Press, Columbus, Ohio.

Baumann, P.C., Harshbarger, J.C. and Hartman, K.J. (1990). *Sci. Total Environ.* **94**, 71–87.

Beer, D.G. and Pitot, H.C. (1987). In *Mouse Liver Tumors* (eds P.L. Chambers, D. Henschler and F. Oesch). *Arch. Toxicol.* Suppl. 10, 68–80, Springer-Verlag, Berlin.

Bell, W. (1793). *Philos. Trans.* 7–9. Cited by B. Lucke and H.G. Schlumberger (1948).

Bend, J.R., James, M.O. and Dansette, P.M. (1977). *Ann. NY Acad. Sci.* **298**, 505.

Bisgrove, B.W., Raible, D.W., Walter, V., Eisen, J.S. and Grunwald, D.J. (1997). *J. Neurobiol.* **33**(6), 749–768.

Black, J.J. (1983). In *Polynuclear Aromatic Hydrocarbons: Formation, Metabolism, and Measurement* (eds M.W. Cooke and A.J. Dennis), pp. 99–111. Battelle Press, Columbus, Ohio.

Black, J.J., Hart, T.F., Jr, and Evans, E.E. (1981). In *Chemical Analysis and Biological Fate: Polynuclear Aromatic Hydrocarbons* (eds M. Cooke and A.J. Dennis), pp. 343–355. Battelle Press, Columbus, Ohio.

Black, J.J., Fox, H., Black, P. and Bock, F. (1985). In *Water Chlorination Chemistry: Environmental Impact and Health Effects* (eds R.L. Bolley, R.J. Bull, W.P. Davis, S. Katz, M.H. Roberts Jr. and V.A. Jacobs), pp. 415–427. Lewis Publishers, Chelsea, Michigan.

Bogovski, S.P. (1988). *Aquat. Toxicol.* **11**, 421.

Bogovski, S.P. (1995). *Laboratornye Zhivotnye* **5**(1), 5–15.

Bogovski, S.P. and Bakai, Y.I. (1989). *J. Fish Dis.* **12**, 1–13.

Bogovski, S.P., Rossi, L., Bocchini, V., Lastraioli, S., Aiello, C. and Santi, L. (1994). *J. Fish Dis.* **17**, 557–566.

Boorman, G.A., Botts, S., Bunton, T.S., Fournie, J.W., Harshbarger, J.C., Hawkins, W.E., Hinton, D.E., Jokinen, M.P., Okihiro, M.S. and Wolfe, M.J. (1997). *Toxicol. Pathol.* **25**(2), 202–210.

Bowser, P.R., Martineau, D. and Wooster, G.A. (1990). *J. Aquat. Anim. Hlth.* **2**, 157–161.

Bowser, P.R., Earnest-koons, K.A., Wooster, G.A., LaPierre, L.A., Holzschu, D.L. and Casey, J.W. (1998). *J. Aquat. Anim. Hlth.* **10**, 282–286.

Breinholt, V., Hendricks, J., Pereira, C., Arbogast, D. and Bailey, G. (1995a). *Cancer Res.* **55**(1), 57–62.

Breinholt, V., Schimerlik, M., Dashwood, R. and Bailey, G. (1995b). *Chem. Res. Toxicol.* **8**(4), 506–514.

Brittelli, M.R., Chen, H.H.C. and Muska, C.F. (1985). *Cancer Res.* **45**(7), 3209–3214.

Brookes, P. (1979). *Cancer Lett.* **6**, 285.

Brown, E.R., Hazdra, J.J., Keith, L., Greenspan, I., Kwapinski, J.B. and Beamer, P. (1973). *Cancer Res.* **33**, 189–198.

Brown, L.A., Amores, A., Schilling, T.F., Jowett, T., Baert, J.L., de Launoit, Y. and Sharrocks, A.D. (1998). *Oncogene* **17**(1), 93–104.

Bunton, T.E. (1994). *Exp. Toxicol. Pathol.* **46**(4–5), 389–396.

Bunton, T.E. (1995). *Carcinogenesis* **16**(5), 1059–1063.

Bunton, T.E. and Wolfe, M.J. (1996). *Toxicol. Pathol.* **24**(3), 323–330.

Cairns, V.W. (1983). Abstract, 113th Annual Meeting Amer. Fish. Soc., Milwaukee, Wisconsin.

Camus, A.M., Geneste, O., Honkakoski, P., Bereziat, J.C., Henderson, C.J., Wolf, C.R., Bartsch, H. and Lang, M.A. (1993). *Mol. Carcinogen.* **7**, 268–275.

Caron de Fromentel, C., Pakdel, F., Chapus, A., Baney, C., May, P. and Soussi, T. (1992). *Gene* **112**, 241–245.

Carter, C.A., Ellington, W.W. and Van Beneden, R.J. (1996). *Toxicol. Pathol.* **24**(3), 339–345.

Cayama, E., Tsuda, H., Sarma, D.S.R. and Farber, E. (1978). *Nature* **275**, 60–61.

Chang, Y.-J., Mathews, C., Mangold, K., Marien, K., Hendricks, J. and Bailey, G. (1991). *Mol. Carcinogen.* **4**, 112–119.

Cheng, R., Bradford, S., Barnes, D., Williams, D., Hendricks, J. and Bailey, G. (1997a). *Mol. Mar. Biol. Biotechnol.* **6**(1), 40–47.

Cheng, R., Ford, B.L., O'Neal, P.E., Mathews, C.Z., Bradford, C.S., Thongtan, T., Barnes, D.W., Hendricks, J.D. and Bailey, G.S. (1997b). *Mol. Mar. Biol. Biotechnol.* **6**(2), 88–97.

Clarke, D.J., George, S.G. and Burchell, B. (1991). *Aquat. Toxicol.* **20**, 35.

Collier, T.K., Stein, J.E., Goksoyr, A., Myers, M.S., Gooch, J.W., Huggett, R.J. and Varanasi, U. (1993). *Environ. Sci.* **2**, 161–177.

Cooke, J.B. and Hinton, D.E. (1999). *Aquat. Toxicol.* **45**, 127–145.

Daniel, F.B., DeAngelo, A.B., Stober, J.A., Olson, G.R. and Page, N.P. (1992). *Fundam. Appl. Toxicol.* **19**, 159–168.

Dashwood, R.H., Fong, A.T., Williams, D.E., Hendricks, J.D. and Bailey, G.S. (1991). *Cancer Res.* **51**, 2362–2365.

Dashwood, R.H., Fong, A.T., Arbogast, D.N., Bjeldanes, L.F., Hendricks, J.D. and Bailey, G.S. (1994). *Cancer Res.* **54**, 3617–3619.

Dawe, C.J., Stanton, M.V. and Schwartz, F.J. (1964). *Cancer Res.* **24**, 1194.

Dimitrijevic, N., Winkler, C., Wellbrock, C., Gomez, A., Duschl, J., Altschmied, J. and Schartl, M. (1998). *Oncogene* **16**(13), 1681–1690.

Djomo, J.E., Garrigues, P. and Narbonne, J.F. (1996). *Environ. Toxicol. Chem.* **15**(7), 1177–1181.

Donnarumma, L., De Angelis, G., Gramenzi, F. and Vittozzi, L. (1988). *Ecotoxicol. Environ. Saf.* **16**, 180.

Dorsky, R.I., Moon, R.T. and Raible, D.W. (1998). *Nature* **396**(6709), 370–373.

Dunn, B.P. (1990). In *In Situ Evaluation of Biological Hazards* (ed. S.S. Sandhu), 177 pp. Plenum Press, New York.

Dunn, B.P., Black, J.J. and Maccubbin, A. (1987). *Cancer Res.* **478**, 6543–6548.

Dunn, B.P., Fitzsimons, J., Stalling, D., Maccubbin, A.E. and Black, J.J. (1990). *Proc. Am. Assoc. Cancer Res.* **31**, 96.

Escourolle, R. and Poirier, J. (1978). In *Manual of Basic Neuropathology*, pp. 18–58. W.B. Saunders Co., Philadelphia.

Essner, J.J., Breuer, J.J., Essner, R.D., Fahrendrug, S.C. and Hackett, P.B., Jr. (1997). *Differentiation* **62**(3), 107–117.

Etoh, H., Hyoda-Taguchi, Y., Aoki, K., Murata, M. and Matsudiara, H. (1983). *J. Natl. Cancer Inst.* **70**, 523–528.

Etoh, H., Watai, Y., Hyodo-Taguchi, Y., Suyama, I. and Matsudaira, H. (1985). *Zool. Sci. (Tokyo)* **2**(6), 895.

Fabacher, D.L., Besser, J.M., Schmitt, C.J., Harshbarger, J.C., Peterman, P.H. and Lebo, J.A. (1991). *Arch. of Environ. Contam. Toxicol.* **21**(1), 17–34.

Falk, H.L., Kotin, P., Lee, S.S. and Nathan, A. (1962). *J. Natl. Cancer Inst.* **28**, 699–724.

Farber, E. and Sarma, D.S.R. (1987). *Lab. Invest.* **56**, 4–22.

Fekete, A., Shim, J.-K., Blair, J.B., Hawkins, W.E. and Ostrander, G.K. (1993). *Proc. 84th Annual Meeting of the American Association for Cancer Research* **34**, 109.

Fong, A.T., Hendricks, J.D., Dashwood, R.H., Van Winkle, S. and Bailey, G.S. (1988). *Food Chem. Toxicol.* **26**, 699.

Fong, A.T., Dashwood, R.H., Cheng, R., Mathews, C., Ford, B., Hendricks, J.D. and Bailey, G.S. (1993). *Carcinogenesis* **14**(4), 629–635.

Fornzler, D., Altschmied, J., Nanda, I., Kolb, R., Baudler, M., Schmid, M. and Schartl, M. (1996). *Genome Res.* **6**(2), 102–113.

Foureman, G.L. (1989). In *Metabolism of Polycyclic Aromatic Hydrocarbons in the Aquatic Environment* (ed. U. Varanasi), pp. 185–202. CRC Press, Boca Raton, Florida.

George, S.G. (1994). In *Aquatic Toxicology: Molecular, Biochemical, and Cellular Perspectives* (eds D.C. Malins and G.K. Ostrander), pp. 37–85. Lewis Publishers, Boca Raton, Florida.

Geter, D.R., Hawkins, W.E., Means, J.C. and Ostrander, G.K. (1998). *Environ. Toxicol. Chem.* **17**(11), 2282–2287.

Gmur, D.J. and Varanasi, U. (1982). *Carcinogenesis (London)* **3**, 1397.

Goeger, D.E., Shelton, D.W., Hendricks, J.D., Pereira, C. and Bailey, G.S. (1988). *Carcinogenesis* **9**, 1793–1800.

Goodwin, A.E. and Grizzle, J.M. (1994). *Carcinogenesis* **15**(9), 1993–2002.

Greenblatt, M.S., Bennett, W.P., Hollstein, M. and Harris, C.C. (1994). *Cancer Res.* **54**, 4855–4878.

Gupta, R.C. (1985). *Cancer Res.* **45**, 5656–5662.

Gupta, R.C. and Randerath, K. (1988). In *DNA Repair* (eds E.C. Friedberg and P.H. Hanawalt), pp. 399–418. Marcel Dekker, New York.

Haddow, A. and Blake, I. (1933). *J. Pathol. Bacteriol.* **36**, 41–47.

Hannig, G., Ottilie, S. and Schart, M. (1991). *Oncogene* **6**, 361–369.

Harada, T., Hatanaka, J. and Enomoto, M. (1988). *J. Comp. Pathol.* **98**(4), 441–452.

Harada, T., Hatanaka, J., Kubota, S.S. and Enomoto, M. (1990). *J. Fish Dis.* **13**, 169–173.

Hardie, L.J., Laing, D.J., Daniels, G.D., Grabowski, P.S., Cunningham, C. and Secombes, C.J. (1998). *Cytokine* **10**(8), 555–563.

Harshbarger, J.C. (1982). In *Advances in Comparative Leukemia Research 1981* (eds D.S. Yohn and J.R. Blakeslee), pp. 39–46. Elsevier Biomedical, New York.

Harshbarger, J.C. (1997). In *Spontaneous Animal Tumors: A Survey* (eds L. Rossi, R. Richardson and J. Harshbarger), pp. 41–53. Press Point di Abbiategrasso, Milano, Italy.

Hatanaka, J., Doke, N., Harada, T., Aikawa, T. and Enomoto, M. (1982). *Jpn. J. Exp. Med.* **52**, 243–253.

Hawkins, W.E., Overstreet, R.M., Fournie, J.W. and Walker, W.W. (1985). *J. Appl. Toxicol.* **5**(4), 261–264.

Hawkins, W.E., Fournie, J.W., Overstreet, R.M. and Walker, W.W. (1986). *J. Natl. Cancer Inst.* **76**, 453–465.

Hawkins, W.E., Fournie, J.W., Overstreet, R.M. and Walker, W.W. (1988a). *J. Fish Dis.* **11**(3), 259–266.

Hawkins, W.E., Overstreet, R.M. and Walker, W.W. (1988b). *Aquat. Toxicol.* **11**, 113–128.

Hawkins, W.E., Walker, W.W., Overstreet, R.M., Lytle, J.S. and Lytle, T.F. (1990). *Sci. Total Environ.* **94**, 155–167.

Hawkins, W.E., Fournie, J.W., Battalora, M.S.J. and Walker, W.W. (1991). *J. Aquat. Anim. Hlth.* **3**(3), 213–220.

Hawkins, W.E., Fournie, J.W., Ishikawa, T. and Walker, W.W. (1996). *J. Aquat. Anim. Hlth.* **8**(2), 120–129.

Hawkins, W.E., Walker, W.W., James, M.O., Manning, C.S., Barnes, D.H., Heard, C.S. and Overstreet, R.M. (1998). *Mutat. Res.* **399**(2), 221–232.

Hendricks, J.D., Meyers, T.R. and Shelton, D.W. (1984). *Natl. Cancer Inst. Monogr.* **65**, 321–336.

Hendricks, J.D., Cheng, R., Shelton, D.W., Pereira, C.B. and Bailey, G.S. (1994). *Fund. Appl. Toxicol.* **23**(1), 53–62.

Hesketh, R. (1997). In *The Oncogene and Tumour Suppressor Gene FactsBook* pp. 1–549. Academic Press, San Diego.

Hinton, D.E., Lantz, R.C. and Hampton, J.A. (1984). *Natl. Cancer Inst. Monogr.* **65**, 239–249.

Hinton, D.E., Hampton, J.A. and McCluskey, P.A. (1985). In *Water Chlorination Chemistry, Environmental Impact and Health Effects* (eds R.L. Jolley, R.J. Bull, W.P. Davis, S. Katz, M.H. Roberts, Jr and V.A. Jacobs), pp. 439–450. Lewis Publishers, Chelsea.

Hofer, B. (1896). *Allg. Fisch. Ztg.* **1**, 2–3.

Hoffmann, R.W., Fisher-Scherl, T. and Pfeil-Putzien, C. (1988). *J. Fish Dis.* **11**, 267–270.

Hueper, W.E. (1963). *Ann. NY Acad. Sci.* **108**, 963–969.

Hyodo-Taguchi, Y. and Matsudaira, H. (1984). *J. Natl. Cancer Inst.* **73**, 1219–1227.

Hyodo-Taguchi, Y., Sasaki, T. and Yamagami, K. (1988). *Zool. Sci.* **5**(6), 1322.

Ishikawa, T. and Takayama, S. (1979). *J. Toxicol. Environ. Hlth.* **5**(2–3), 537–550.

Ishikawa, T., Shimaamine, T. and Takayama, S. (1975). *J. Natl. Cancer Inst.* **55**, 909–916

Ishikawa, T., Masahito, P., Matsumoto, J. and Takayama, S. (1978). *J. Natl. Cancer Inst.* **61**, 1461–1470.

Ishikawa, T., Masahito, P. and Takayama, S. (1984). *Natl. Cancer Inst. Monogr.* **65**, 34–43.

Jacobs, A.D. and Ostrander, G.K. (1995). *Environ. Toxicol. Chem.* **14**(10), 1781–1787.

Jakoby, W.B. (1989). *The Sulfotransferases*, Francis and Taylor, London.

James, M.O. (1987). *Environ. Hlth. Perspect.* **71**, 97.

James, M.O., Schell, J.D., Boyle, S.M., Altman, A.H. and Cromer, E.A. (1991). *Chem. Biol. Interact.* **79**, 305.

Jerina, D.M., Yagi, H., Thakker, D.R., Sayer, J.M., van Bladeren, P.J., Lehr, R.E., Whalen, D.L., Levin, W., Chang, R.L., Wood, A.W. and Conney, A.H. (1984). In *Foreign Compound Metabolism*, vol 1 (eds J. Caldwell and G. Paulson), 257 pp. Taylor & Francis, London.

Kapitulnik, J., Levin, W., Conney, A.H., Yagi, H. and Jerima, D.M. (1977). *Nature* **266**, 378–380.

Kaplan, L.A.E., Schults, M.E., Schultz, R.J. and Crivello, J.F. (1991). *Carcinogenesis* **12**, 647–652.

Kazianis, S., Gutbrod, H., Nairn, R.S., McEntire, B.B., Della, Colleta, L., Walter, R.B., Borowsky, R.L., Woodhead, A.D., Setlow, R.B., Schartl, M. and Morizot, D.C. (1998). *Genes, Chrom. Cancer* **22**(3), 210–220.

Keifer, P., Mathieu, M., Mason, I. and Dickson, C. (1996a). *Oncogene* **12**(7), 1503–1511.

Keifer, P., Strahle, U. and Dickson, C. (1996b). *Gene* **168**(2), 211–215.

Kelly, J.D., Orner, G.A., Hendricks, J.D. and Williams, D.E. (1992). *Carcinogenesis* **13**, 1639–1642.

Keysselitz, G. (1908). *Arch. Protistenkd.* **11**, 326–333.

Khudoley, V.V. (1984). *Natl. Cancer Inst. Monogr.* **65**, 65–70.

Kimura, T. and Yoshimizu, M. (1991). In *Proc. Second Int. Symp. Viruses of Lower Vertebrates*, 29–31 July 1991 (ed. J.L. Fryer), pp. 183–189. Oregon State University, Corvallis.

Kimura, T., Yoshimizu, M. and Tanaka, M. (1981). *Fish Pathol.* **15**, 149–153.

Kimura, I., Taniguchi, N., Kumai, H., Tomita, I., Kinae, N., Yoshizaki, K., Ito, M. and Ishikawa, T. (1984). In *Correlation of Epizootiological Observations with Experimental Data: Chemical Induction of Chromatophoromas in the Croaker, Nibea mitsukurii* (ed. K.L. Hoover). *Natl. Cancer Inst. Monogr.* **65**, 139–154.

Kimura, I., Kinae, N., Kumai, H., Yamashita, M., Nakamura, G., Ando, M., Ishida, H. and Tomita, I. (1989). *J. Invest. Dermatol.* **92**(5), 248–254.

Kinae, N., Yamashita, M., Tomita, I., Kimura, I., Ishida, H., Kumai, H. and Nakamura, G. (1990). *Sci. Tot. Environ.* **94**, 143–153.

Knox, K., Lill, P.H., Foster, S.A., Saftlas, A.F., Pitcher, H.M., Fabo, E.D., Buncher, R.C., Warshawsky, D. and Ward, W.D. (1988). In *Ultraviolet Radiation and Melanoma* (ed. J.D. Longstreth), pp. 1–13. US EPA, Washington, DC.

Kotin, P., Falk, H.L. and Busser, R. (1959). *J. Natl. Cancer. Inst.* **23**, 541–555.

Krahn, M.M., Meyers, M.S., Burrows, D.G., Malins, D.C. (1984). *Xenobiotica* **14**, 633–646.

Krahn, M.M., Rhodes, L.D., Myers, M.S., Moore, L.K. and MacLeod, W.D., Jr and Malins, D.C. (1986). *Arch. Environ. Contam. Toxicol.* **15**, 61–67.

Krause, M.K., Rhodes, L.D. and Van Beneden, R.J. (1997). *Gene* **189**, 101–106.

LaPierre, L.A., Holzschu, D.L., Wooster, G.A., Bowser, P.R. and Casey, J.W. (1998). *J. Virol.* **72**, 3484–3490.

Lech, J.J., Bodicnik, M.J. and Elcombe, C.R. (1982). In *Aquatic Toxicology*. Raven Press, 107 pp.

Lipscomb, J.C., Garrett, C.M. and Snawder, J.E. (1997a). *Toxicol. Appl. Pharmacol.* **142**, 311–318.

Lipscomb, J.C., Buttler, G.W. and Confer, P.D. (1997b). *Ann. Clin. Lab. Sci.* **27**, 158–163.

Lipscomb, J.C., Confer, P.D., Miller, M.R., Stamm, S.C., Snawden, J.E. and Bandiera, S.M. (1998). *Environ. Toxicol. Chem.* **17**(2), 325–332.

Lipsky, M.M., Hinton, D.E., Klaunig, J.E., Goldblatt, P.J. and Trump, B.F. (1981). *J. Natl. Cancer Inst.* **67**, 377–392.

Ljungberg, O. (1976). *Prog. Exp. Tumor Res.* **20**, 156–165.

Ljungberg, O. and Lange, J. (1968). *Bulletin de l'Office International des Epizooties* **69**, 1007–1022.

Lotrich, V.A. (1975). *Ecology* **56**, 191–198.

Loveland, P.M., Willcox, J.S., Hendricks, J.D. and Bailey, G.S. (1988). *Carcinogenesis* **9**, 441–446.

Lowy, D.R. (1993). *Annu. Rev. Biochem.* **62**, 851–891.

Lucke, B. and Schlumberger, H.G. (1941). *J. Exp. Med.* **74**, 397–408.

Maccubbin, A.E. (1994). In *Aquatic Toxicology: Molecular, Biochemical, and Cellular Perspectives* (eds D.C. Malins and G.K. Ostrander), pp. 267–294. Lewis Publishers, Boca Raton, Florida.

Maccubbin, A.E., Black, J.J. and Dunn, B.P. (1990). *Sci. Total Environ.* **94**, 89–95.

Malins, D.C., McCain, B.B., Brown, D.W., Chan, S., Myers, M.S., Landahl, J.T., Prohaska, P.G., Friedman, A.J., Rhodes, L.D., Burrows, D.G., Gronlund, W.D. and Hodgins, H.O. (1984). *Environ. Sci. Technol.* **18**(9), 705–713.

Malins, D.C., Krahn, M.M., Brown, D.W., Rhodes, L.D., Myers, M.S., McCain, B.B. and Chan, S. (1985a). *J. Natl. Cancer Inst.* **74**(2), 487–494.

Malins, D.C., Krahn, M.M., Myers, M.S., Rhodes, L.D., Brown, D.W., Krone, C.A., McCain, B.B. and Chan, S. (1985b). *Carcinogenesis* **6**(10), 1463–1469.

Malitschek, B., Wittbrodt, J., Fischer, P., Lammers, R., Ullrich, A. and Schartl, M. (1994). *J. Biol. Chem.* **269**(14), 10423–10430.

Mangold, K., Chang, Y., Mathews, C., Marien, K., Hendricks, J., Bailey, G. (1991). *Mol. Carcinogen.* **4**, 97–106.

Mangues, R., Seidman, I., Gordon, J.W. and Pellicer, A. (1992). *Oncogene* **7**(10), 2073–2076.

Marcos-Gutierrez, C.V., Wilson, S.W., Holder, N. and Pachnis, V. (1997). *Oncogene* **14**(8), 879–889.

Martineau, D., Bowser, P.R., Armstrong, G.A. and Wooster, G.A. (1990). *Vet. Pathol.* **27**, 230–234.

Martineau, D., Bowser, P.R., Renshaw, R.R. and Casey, J.W. (1991a). *J. Virol.* **66**(1), 596–599.

Martineau, D., Renshaw, R., Williams, J.R., Casey, J.W. and Bowser, P.R. (1991b). *Dis. Aquat. Org.* **10**, 153–158.

Masahito, P., Aoki, K., Egami, N., Ishikawa, T. and Sugano, H. (1989a). *Jpn. J. Cancer Res.* **80**, 1058–1065.

Masahito, P., Ishikawa, T. and Sugano, H. (1989b). *J. Invest. Dermatol.* **92**(5), 266–270.

Masahito, P., Aoki, K. and Ishikawa, T. (1996). *Fish Biol. J. MEDAKA* **8**, 15–19.

Matsuoka, I., Fuyuki, K., Shoji, T. and Kurihara, K. (1998). *Biochim. Biophys. Acta* **1395**(2), 220–227.

Maueler, W., Schartl, A. and Schartl, M. (1993). *Int. J. Cancer* **55**, 288–296.

McCain, B.B., Pierce, K.V., Wellings, S.R. and Miller, B.S. (1977). *Bull. Environ. Contam. Toxicol.* **18**, 1–2.

McMahon, G., Huber, L.J., Stegeman, J.J. and Wogan, G.N. (1988). *Mar. Environ. Res.* **24**, 345–350.

McMahon, G., Huber, L.J., Moore, M.J., Stegeman, J.J. and Wogan, G.N. (1990). *Proc. Natl. Acad. Sci. USA* **87**(2), 841–845.

Metcalfe, C.D., Cairns, V.W. and Fitzsimons, J.D. (1988). *Can. J. Fish Aquat. Sci.* **45**, 2161–2167.

Michel, X.R., Cassand, P.M., Ribera, D.G. and Narbonne, J.F. (1992). *Comp. Biochem. Physiol.* **103C**, 43.

Molven, A., Njolstad, P.R. and Fjose, A. (1991). *Eur. Mol. Biol. Organ. J.* **10**(4), 799–808.

Moore, M.A. and Kitagawa, T. (1986) *Int. Rev. Cyt.* **101**, 125–174.

Morita, N. and Sano, R. (1990). *J. Fish Dis.* **13**(6), 505–511.

Morizot, D.C., McEntire, B.B., Della Coletta, L., Kazianis, S., Schartl, M. and Nairn, R.S. (1998). *Mol. Carcinogen.* **22**(3), 150–157.

Mulcahy, M.F. (1963). *Proc. Roy. Irish Acad.* 63 Section B, 103–127.

Mulcahy, M.F. and O'Leary, A. (1970). *Experientia* **26**, 891.

Muraiso, C., Nemoto, N., Asami, K., Etoh, H. and Matsudaira, H. (1987). In *Invertebrate and Fish Tissue Culture* (eds Y. Kuroda, E. Kurstak and K. Maramorosch), pp. 211–214. Springer-Verlag, Berlin, Heidelberg, New York, London, Paris, Tokyo.

Myers, M.S., Rhodes, L.D. and McCain, B.B. (1987). *J. Natl. Cancer Inst.* **78**(2), 333–363.

Myers, M.S., Landahl, J.T., Krahn, M.M., Johnson, L.L. and McCain, B.B. (1990). *Sci. Total Environ.* **94**, 33–50.

Myers, M.S., Anulacion, B.F., French, B., Hom, T. and Collier, T.K. (1995). 16th Annual Meeting of the Society of Environmental Toxicology and Chemistry. Vancouver, British Columbia, Canada.

Myers, M.S., Johnson, L.J., Hom, T., Collier, T.K., Stein, J.E. and Varanasi, U. (1997). *Mar. Environ. Res.* **45**, 47–67.

Nagao, M. and Sugimara, T. (1978). In *Polycyclic Hydrocarbons and Cancer*, vol. 2 (eds H.V. Gelboin and P.O.P. Ts'U), pp. 99–121. Academic Press, New York.

Nairn, R.S., Kazianis, S., McEntire, B.B., Della Colleta, L., Walter, R.B. and Morizot, D.C. (1996). *Proc. Natl. Acad. USA* **93**(23), 13042–13047.

Nakatsuru, Y., Nemoto, N., Nakagawa, K., Masahito, P. and Ishikawa, T. (1987). *Carcinogenesis* **8**(8), 1123–1127.

Nakazawa, T., Hamaguchi, S. and Kyono-Hamaguchi, Y. (1985). *J. Natl. Cancer Inst.* **75**(3), 567–573.

Neff, J.M. (1979). *Polycyclic Aromatic Hydrocarbons in the Aquatic Environment*. Applied Science Publishers, London, England.

Nemoto, N., Kodama, K., Tazawa, A., Prince Masahito and Ishikawa, T. (1986). *Differentiation* **32**, 17–24.

Nigrelli, R.F. (1952). *Ann. N. Y. Acad. Sci.* **54**, 1076–1092.

Nimmo, I.A. (1987). *Fish Physiol. Biochem.* **3**, 163–171.

Nixon, J.E., Hendricks, J.D., Pawlowski, N.E., Pereira, C.B., Sinnhuber, R.O. and Bailey, G.S. (1984). *Carcinogenesis* **5**, 615–619.

Nunez, O., Hendricks, J.D., Arbogast, D.N., Fong, A.T., Lee, B.C. and Bailey, G.S. (1989). *Aquat. Toxicol.* **15**, 289–302.

Okihiro, M.S. (1988). *Vet. Pathol.* **25**, 422–431.

Okihiro, M.S. and Hinton, D.E. (1999a). *Carcinogenesis* **20**(6), 933–940.

Okihiro, M.S. and Hinton, D.E. (1999b). *Toxicol. Pathol.* (in press).

Okihiro, M.S., Whipple, J.A., Groff, J.M. and Hinton, D.E. (1993). *Cancer Res.* **53**, 1761–1769.

Orner, G.A., Mathews, C., Hendricks, J.D., Carpenter, H.M., Bailey, G.S. and Williams, D.E. (1995). *Carcinogenesis* **16**(12), 2893–2898.

Orner, G.A., Donohoe, R.M., Hendricks, J.D., Curtis, L.R. and Williams, D.E. (1996a). *Fund. Appl. Toxicol.* **34**(1), 132–140.

Orner, G.A., Hendricks, J.D., Arbogast, D. and Williams, D.E. (1996b). *Toxicol. Appl. Pharmacol.* **141**(2), 548–554.

Orner, G.A., Hendricks, J.D., Arbogast, D. and Williams, D.E. (1998). *Toxicol. Appl. Pharmacol.* **19**(1), 161–167.

Ostrander, G.K., Shim, J.-K., Hawkins, W.E. and Walker, W.W. (1992). *Proceedings of the 83rd Annual Meeting of the American Association for Cancer Research*, vol. 33, p. 109.

Ostrander, G.K., Hawkins, W.E., Kuehn, R.L., Jacobs, A.D., Berlin, K.D. and Pigg, J. (1995). *Carcinogenesis* **16**(7), 1529–1535.

Paakspuu, V. (1987). In *Loodusevaatlusi (Nature Observations)* (ed. E. Magi), pp. 20–30. Valgus, Tallinn (in Estonian).

Panno, J.P. and McKeown, B.A. (1995). *Biochim. Biophys. Acta* **1264**(1), 7–11.

Panno, J.P. and McKeown, B.A. (1997). *Mol. Cell. Endocrinol.* **134**(2), 81–90.

Papas, T.S., Dahlberg, J.E. and Sonstegard, R.A. (1976). *Nature* **261**, 506–508.

Papas, T.S., Pry, T.W., Shafer, M.P. and Sonsgard, R.A. (1977). *Cancer Res.* **37**, 3214–3217.

Peck-Miller, K.A., Meyers, M., Collier, T.K. and Stein, J.E. (1998). *Mol. Carcinogen.* **23**, 207–216.

Poginsky, B., Blomeke, B., Hewer, A., Phillips, D.H., Karbe, L. and Marquardt, H. (1990). *Proc. Am. Assoc. Cancer Res.* **31**, 96.

Premdas, P.D. and Metcalfe, C.D. (1996). *Can J. Fish. Aquat. Sci.* **53**, 1018–1029.

Premdas, P.D., Metcalfe, T.L., Bailey, M.E. and Metcalfe, C.D. (1995). *J. Great Lakes Res.* **21**, 207–218.

Pritchard, J., Bend, J. (1984). *Drug Metab. Rev.* **15**, 655–667.

Rak, J., Mitsuhashi, Y., Bako, L., Filmus, J., Shirasawa, S. and Sasazuki, T. (1995). *Cancer Res.* **55**(20), 4575–4580.

Reddy, M.V. and Randerath, K. (1986). *Carcinogenesis* **7**, 1543–1551.

Reichert, W.L., Stein, J.E., French, B., Goodwin, P. and Varanasi, U. (1992). *Carcinogenesis* **13**, 1475–1479.

Reichert, W.L., Myers, M.S., Peck-Miller, K., French, B., Anulacion, B.F., Collier, T.K., Stein, J.E. and Varanasi, U. (1998). *Mutat. Res.* **411**, 215–225.

Rhodes, L.D., Myers, M.S., Gronlund, W.D. and McCain, B.B. (1987). *J. Fish Biol.* **32**, 395–407.

Sano, T., Fukuda, H., Okamoto, N. and Kaneko, F. (1983). *Bull. Jap. Soc. Scient. Fish* **49**, 1159–1163.

Sano, T., Fukuda, H. and Furukawa, M. (1985a). *Fish Pathol.* **20**, 381–388.

Sano, T., Fukuda, H., Furukawa, M., Hosoya, H. and Moriya, Y. (1985b). In *Fish and Shellfish Pathology* (ed. A.E. Ellis), pp. 307–311. Academic Press, London.

Sano, N., Moriwake, M. and Sano, T. (1993). *Fish Pathol.* **28**(4), 171–175.

Schartl, M., Barkenow, A., Bauer, H. and Anders, F. (1982). *Cancer Res.* **42**, 4222–4227.

Schartl, M., Schmidt, C-R, Anders, A. and Barnekow, A. (1985). *Int. J. Cancer* **36**, 199–207.

Schartl, A., Dimitrijevic, N. and Schartl, M. (1994). *Pigment Cell Res.* **7**(6), 428–432.

Scherer, E. (1987). In *Mouse Liver Tumors* (eds P.L. Chambers, D. Henschler and F. Oesch). *Arch. Toxicol.* Suppl. 10, 81–94, Springer-Verlag, Berlin.

Schliwa, M. (1986). In *Biology of the Integument. 2. Vertebrates* (eds J. Bereiter-Hahn, A.G. Matoltsy and K.S. Richards), pp. 65–77. Springer-Verlag, Berlin.

Schmale, M.C. (1991). *Mar. Biol.* **109**, 203–212.

Schmale, M.C. (1995). *Dis. Aquat. Organ.* **23**, 201–212.

Schmale, M.C. and Hensley, G.T. (1988). *Cancer Res.* **48**(13), 3828–3833.

Schmale, M.C., Hensley, G. and Udey, L.R. (1983). *Am. J. Pathol.* **112**(2), 238–241.

Schmale, M.C., Hensley, G.T. and Udey, L.R. (1986). *Ann. NY Acad. Sci.* **486**, 386–402.

Schmale, M.C., Gill, K.A., Cacal, S.M. and Baribeau, S.D. (1994). *J. Neurocytol.* **23**(11), 668–681.

Schmale, M.C., Aman, M.R. and Gill, K.A. (1996). *J. Gen. Virol.* **77**(6), 1181–1187.

Schubert, G. (1964). *Z. Naturforsch* **19b**, 675–682.

Schubert, G. (1966). *Bull. Off. Int. Epiz.* **65**, 1011–1022.

Sell, S. (1993). *Environ. Health Perspect.* **101**, 15–26.

Sell, S. and Pierce, G.B. (1994). *Lab. Inves.* **70**, 6–22.

Shugart, L., McCarthy, J., Jimenez, B. and Daniels, J. (1987). *Aquat. Toxicol.* **9**, 319.

Sikka, H.C., Rutkowski, J.P., Kandaswami, C., Kumar, S., Earley, K. and Gupta, R.C. (1990). *Cancer Lett.* **49**, 81.

Sinn, E., Muller, W., Pattengale, P., Tepler, I., Wallace, R. and Leder, P. (1987). *Cell* **49**(4), 465–475.

Smeets, T.M.W., Voormolen, A., Tillitt, D.E., Everaarts, J.M., Seinen, W. and Van Den Beug, M. (1999). *Environ. Toxicol. Chem.* **18**(3), 474–480.

Smith, I.R., Ferguson, H.W., Rokosh, D.A. and Hayes, M.A. (1986). Abstract, 7th Annual Meeting Society of Environ. Tox. and Chem., Alexandria, VA.

Smith, I.R., Baker, K.W., Hayes, M.A. and Ferguson, H.W. (1989). *Dis. Aquat. Org.* **6**, 17–26.

Smolarek, T.A., Morgan, S.L., Moynihan, C.G., Lee, H., Harvey, R.G. and Baird, W.B. (1987). *Carcinogenesis* **8**, 1501.

Solbakken, J.E. and Palmork, K.H. (1981). *Comp. Biochem. Physiol.* **70C**, 21–26.

Solt, D.B., Medline, A. and Farber, E. (1977). *Am. J. Pathol.* **88**, 595–618.

Sonstegard, R.A. (1973). In *Viruses in the Environment and their Potential Hazard* (eds M.S. Mahdy and B.J. Dutka), pp. 119–129. Burlington, Ontario.

Sonstegard, R. (1975). In *The Pathology of Fishes* (eds W.E. Ribelin and G. Migaki), pp. 907–924. University of Wisconsin Press. Madison, Wisconsin.

Sonstegard, R.A. (1976). *Prog. Exp. Tumor Res.* **20**, 141–155.

Sonstegard, R.A. (1977). *Ann. NY Acad. Sci.* **298**, 261–269.

Sonstegard, R.A. and Papas, T.S. (1977). Proceedings of the 68th Annual Meeting of the American Association for Cancer Research, Denver, Colorado, 18–21 May.

Soussi, T., de Fromentel, C. and May, P. (1990). *Oncogene* **5**, 945–952.

Spitsbergen, J., Tsai, H., Reddy, A. and Hendricks, J. (1997). Proceedings of the American Association for Cancer Research Annual Meeting, vol. 38, p. 354.

Stegeman, J.J. (1981). In *Polycyclic Hydrocarbons and Cancer* (eds H.V. Gelboin and P.O.P. Ts'o). Academic Press, New York.

Stegeman, J.J. (1987). In *Pollutant Studies in Marine Animals* (eds C.S. Giam and L.E. Ray), p. 65, CRC Press, Boca Raton, Florida.

Stegeman, J.J. and Hahn, M.E. (1994). In *Aquatic Toxicology: Molecular, Biochemical, and Cellular Perspectives* (eds D.C. Malins and G.K. Ostrander), pp. 87–206. Lewis Publishers, Boca Raton, Florida.

Stegeman, J.J., Woodin, B.R., Park, S.S., Kloepper-Sams, P.J. and Gelboin, H.V. (1985). *Mar. Environ. Res.* **17**, 83.

Stein, J.E., Hom, T. and Varanasi, U. (1984). *Mar. Environ. Res.* **13**, 97.

Stein, J.E., Reichert, W.L., Nishimoto, M. and Varanasi, U. (1990). *Sci. Total Environ.* **94**, 51.

Tan, B. and Melius, P. (1986). *Comp. Biochem. Physiol.* **83C**, 217.

Tatematsu, M., Mera, Y., Inoue, T., Satoh, K. and Ito, K.S.N. (1998). *Carcinogenesis* **9**, 215–220.

Teh, S.J. and Hinton, D.E. (1993). *Aquat. Toxicol.* **24**, 163–182.

Teh, S.J. and Hinton, D.E. (1998). *Aquat. Toxicol.* **41**(1–2) 141–159.

Thompson, J.S. (1982). *J. Fish Dis.* **5**, 1–11.

Thompson, J.S. (1989). Ph.D. Thesis. Department of Limnology, University of Helsinki.

Thompson, J.S. and Miettinen, M. (1988). *J. Fish Dis.* **11**, 47–55.

Thompson, J.S., Kostiala, A.A.I. and Miettinen, M. (1987a). *J. Fish Biol.* **31** (Suppl. A), 167–173.

Thompson, J.S., Virtanen, I. and Lehto, V.P. (1987b). *J. Comp. Pathol.* **97**, 257–266.

Torten, M., Liu, Z., Okihiro, M.S., Teh, S.J. and Hinton, D.E. (1996). *Mar. Environ. Res.* **42**, 93–98.

Van, Beneden, R.J. (1993). In *Biochemistry and Molecular Biology of Fishes*, vol. 2 (eds P.W. Hochachka and T.P. Mommsen), pp. 113–136. Elsevier.

Van Beneden, R.J., Watson, D.K., Chen, T.T., Lautenberger, J.A. and Papas, T.S. (1986). *Proc. Natl. Acad. Sci., USA* **83**(11), 3698–3702.

Van Beneden, R.J. and Ostrander, G.K. (1994). In *Aquatic Toxicology: Molecular, Biochemical, and Cellular Perspectives* (eds D.C. Malins and G.K. Ostrander), pp. 295–326, Lewis Publishers, Boca Raton, Florida.

van der Oost, R., van Schooten, F.J., Ariese, F. and Heida, H. (1994). *Environ. Toxicol. Chem.* **13**, 859–870.

Varanasi, U. (1989). *Metabolism of Polyaromatic Hydrocarbons in the Aquatic Environment*. CRC Press, Boca Raton, Florida.

Varanasi, U. and Gmur, D.J. (1980). *Biochem. Pharmacol.* **29**, 753.

Varanasi, U. and Gmur, D.J. (1981a). In *Chemical Analysis and Biological Fate: Polynuclear Aromatic Hydrocarbons* (eds M. Cooke and A.J. Dennis), pp. 367–376. Battelle Press, Columbus, Ohio.

Varanasi, U. and Gmur, D.J. (1981b). *Aquatic Toxicol.* **1**, 49–67.

Varanasi, U., Gmur, D.J. and Krahn, M.M. (1980). In *Polynuclear Aromatic Hydrocarbons: Chemistry and Biological Effects* (eds A. Bjorseth and A.J. Dennis), p. 455. Battelle Press, Columbus, Ohio.

Varanasi, U., Nishimoto, M., Reichert, W.L. and Le Eberhart, B. -T. (1986). *Cancer Res.* **46**, 3817.

Varanasi, U., Stein, J.E., Nishimoto, M., Reichert, W.L. and Collier, T.K. (1987). *Environ. Hlth. Perspect.* **71**, 155–170.

Varanasi, U., Stein, J.E. and Nishimoto, M. (1989a). In *Metabolism of Polycyclic Aromatic Hydrocarbons in the Aquatic Environment* (ed. U. Varanasi), pp. 93–150. CRC Press, Boca Raton, Florida.

Varanasi, U., Reichert, W.L. and Stein, J.E. (1989b). *Cancer Res.* **49**, 1171.

Vicha, D.L. and Schmale, M.C. (1994). *Anticancer Res.* **14**(3), 947–952.

Vielkind, U. (1976). *J. Exp. Zool.* **196**, 197–204.

Vielkind, J.R. and Dippel, E. (1984). *Can. J. Genet. Cytol.* **26**, 607–614.

Vielkind, J.R. and Vielkind, U. (1982). *Can. J. Genet. Cytol.* **24**, 133–149.

Vogelbein, W.K., Fournie, J.W., Van Veld, P.A. and Huggett, R.J. (1990). *Cancer Res.* **50**, 5978–5986.

Vogelbein, W.K., Zwerner, D.E., Unger, M.A., Smith, C.L. and Fournie, J.W. (1997). In *Spontaneous Animal Tumors: a Survey. Proceedings of the First World Conference on Spontaneous Animal Tumors* (eds L. Rossi, R. Richaardson and J.C. Harshbarger), pp. 55–63. Press Point de Abbiategrasso, Milan, Italy.

Walker, R. (1947). *Anat. Rec.* **99**, 559–560.

Walker, R. (1961). *Am. Zool.* **1**, 395–396. Abstract.

Walker, R. (1969). *Natl. Cancer Inst. Monogr.* **31**, 195–207.

Ward, J.M. (1981). *Virchows Arch.* **390**, 339–345.

Wellbrock, C., Fischer, P. and Schartl, M. (1998a). *Int. J. Cancer* **76**(3), 437–442.

Wellbrock, C., Geissinger, E., Gomez, A., Fischer, P., Friedrich, K. and Schartl, M. (1998b). *Oncogene* **16**(23), 3047–3056.

Wellbrock, C., Lammers, R., Ullrich, A. and Schartl, M. (1995). *Oncogene* **10**(11), 2135–2143.

Wellbrock, C., Gomez, A. and Schartl, M. (1997). *Pigment Cell Res.* **10**(1–2), 34–40.

Williams, D.E. and Buhler, D.R. (1983). *Cancer Res.* **43**, 4752.

Winkler, C., Wittbrodt, J., Lammers, R., Ullrich, A. and Schartl, M. (1994). *Oncogene* **9**(6): 1517–1525.

Winqvist, G., Ljungberg, O. and Hellstroem, B. (1968). *Bull. Off. Int. Epiz.* **69**, 1023–1031.

Winqvist, G., Ljungberg, O. and Ivarsson, B. (1973). In *Unifying Concepts of Leukemia* (eds R.M. Dutcher and L. Chieco-Bianchi), pp. 26–30. Karger, Basel.

Wirgin, I., Currie, D. and Garte, S.J. (1989). *Carcinogenesis* **10**, 2311–2315.

Wittbrodt, J., Adam, D., Malitschek, B., Maueler, W., Raulf, F., Telling, A., Robertson, S.M. and Schartl, M. (1989). *Nature* **341**, 415–421.

Wolf, K. (1988). In *Fish Viruses and Fish Viral Diseases*, pp. 51–68. Ithaca and London, Cornell University Press.

Wolfe, M.J., Gardner, H.S. and Van der Schalie, W.H. (1991). *Can. Tech. Rep. of Fish. Aquat. Sci.* **1774**(1–2), 587.

Yamamoto, T., MacDonald, R.E., Gillespie, D.C. and Kelly, R.K. (1976). *J. Fish. Res. Bd. Can.* **33**, 2408–2419.

Yamamoto, T., Kelly, R.K. and Nielsen, O. (1985). *J. Fish Dis.* **8**, 425–436.

Yoshimizu, M., Tanaka, M. and Kimura, T. (1987). *Fish Pathol.* **22**, 7–10.

Yuan, Z-X., Kumar, S. and Sikka, H.C. (1997). *Environ. Toxicol. Chem.* **16**(4), 835–836.

Zaleski, J., Steward, A.R. and Sikka, H. (1989). *Mar. Environ. Res.* **28**, 153.

zur Hausen, H. (1980). *Adv. Cancer Res.* **33**, 77–107.

CHAPTER 37

Toxicology

Chris D Metcalfe
Environmental and Resource Studies,
Trent University, Peterborough, Ontario, Canada

Introduction

The discipline of toxicology involves studying the nature and mechanisms of toxic lesions, and evaluating in a quantitative manner the spectrum of biological changes produced by exposure to chemicals. In this chapter, we will mainly concern ourselves with the methods and techniques used to quantitatively evaluate toxicological responses in fish. It is important to realize that every chemical can be toxic to fish under certain exposure conditions. For every chemical there should be an exposure condition (i.e. dose or concentration) that is 'safe' and an exposure condition that is 'toxic' to fish (Sprague, 1971). As illustrated in Table 37.1, the range of concentrations or doses that are toxic to fish may span several orders of magnitude. Therefore, toxicology is primarily concerned with establishing relationships between 'dose (concentration) and effect' for single test organisms or 'dose (concentration) and response' for a population of test organisms. It is also important to determine toxic 'thresholds'; that is, concentrations or doses above which toxicity occurs and below which, it does not.

Until relatively recently, toxicological studies with fish focused almost exclusively on very toxic substances which produce 'acutely lethal' responses; that is, mortalities in fish exposed to chemicals for only short periods. Recently, we have become concerned with substances that may produce 'sublethal' responses in fish after 'chronic' exposure. Much of the terminology associated with toxicology is relatively specialized, so it is useful to start this chapter with a glossary of terms commonly used in fish toxicology.

TABLE 37.1: Approximate toxic concentrations of chemicals in water for early life stages of fish

Chemical	Toxic concentration ($\mu g\,L^{-1}$)
Nitriloacetic acid (NTA)	100 000
Fluoride	10 000
Zinc	1 000
Cadmium	100
DDT	10
Methyl mercury	1
PCB congener 126	0.1
2,3,7,8-tetrachlorodibenzo-*p*-dioxin	0.01

Source: Estimates of toxicity were prepared from data presented by McKim (1985) and Harris *et al.* (1994a,b) and from unpublished data by the author on the toxicity of DDT to Japanese medaka.

Glossary of terms used in fish toxicity studies

- *Dose* Describes quantitatively the amount of chemical to which the fish has been exposed. Experimentally, exposure is usually done by dosing the food (per os), force-feeding solid material (gavage) or liquid through a tube (intubation), or by injecting the chemical into an egg (*in ovo* injection) or into the fish intraperitoneally (i.p.), intramuscularily (i.m.) or sub-cutaneously (s.c.). Dose is usually calculated as the weight of chemical (ng, μg, mg) administered per gram or kilogram of body weight. $1\,\mu g\,kg^{-1} = 1\,ng\,g^{-1} = 1\,ppb$; $1\,mg\,kg^{-1} = 1\,\mu g\,g^{-1} = 1\,ppm$.
- *Concentration* Fish are often exposed to chemicals dissolved in water, and the exposure conditions are described according to the weight of the chemical dissolved in a volume of water. $1\,\mu g\,L^{-1} = 1\,ng\,mL^{-1} = 1\,ppb$; $1\,mg\,L^{-1} = 1\,\mu g\,mL^{-1} = 1\,ppm$.
- *Toxicity* The capacity of a chemical to cause injury to a living organism. Lethality is a specific term refering to the capacity of a chemical to cause death to a living organism. Toxicity or lethality cannot be defined for fish without providing information on:
  - The species and life stage of the fish.
  - The dose or concentration of the chemical to which the fish were exposed.
  - The distribution of exposure in time (i.e. single, repeated or continuous dose).
  - The type and severity of injury (i.e. lethal response, sublethal response).
  - The time needed to produce toxicity (i.e. acute, chronic).
- ED_{50} (EC_{50}) Toxicity is often assessed quantitatively as a 'dose–response' or 'concentration–response'. The ED_{50} and EC_{50} are the median effective dose and concentration, respectively; that is, the dose or concentration of a chemical at which 50% of the test population demonstrates a toxic effect. A specific response involving death of the organism is the median lethal concentration or dose (i.e. LC_{50} or LD_{50}). The duration of exposure must be specified (e.g. 96-h LC_{50}).
- *LOED (LOEC)* The 'lowest-observed-effect' dose or concentration is the lowest dose or concentration of a chemical that causes injury to a fish.
- *NOED (NOEC)* The 'no-observed-effect' dose or concentration is the highest dose or concentration of a chemical that does not cause injury to a fish.
- *MATC* The 'maximum acceptable toxicant concentration' is the hypothetical toxic threshold lying in a range bounded by the LOEC and the NOEC.

Factors affecting toxicity to fish

There are several chemical, physical and biological factors that influence the toxicity of chemicals to fish, including the properties of the chemical in water, the water quality conditions, the route of exposure, and the species and life stage of the fish being tested. Therefore, there are several parameters that must be taken into account or controlled when designing fish toxicity tests if we are to eliminate or reduce variability in the toxic responses due to these factors (Table 37.2). In this chapter, we will examine the chemical, physiological and biological bases of why these factors influence toxicity.

TABLE 37.2: Parameters that must be controlled or standardized in toxicity tests with fish

Abiotic parameters	Biotic parameters
Water temperature	Fish species and source
Photoperiod	Life stage
Dissolved oxygen concentration	Sex
pH	Size and condition
Alkalinity	Reproductive status
Hardness	Health
Salinity	Acclimation period
Dissolved organic carbon	Nutrition

Solubility in water

The solubility of a chemical in water governs the extent to which the chemical mixes with the aqueous phase to form a homogeneous solution. Since fish live in water, the extent to which fish are exposed to a chemical is dependent on aqueous solubility. The solubility of non-ionic chemicals, such as organic compounds and elemental forms of toxic metals (e.g. $Hg°$), is influenced by the polarity of the compound. The intermolecular forces that occur between non-ionic molecules are dipole–dipole interactions for polar compounds, and van der Waals interactions for non-polar molecules; the former being much stronger than the latter (Connell, 1997). When a chemical is added to water, the solute will not dissolve in water to any appreciable extent unless it is sufficiently polar to overcome the dipole–dipole interactions of adjacent water molecules. Thus, polar solutes such as methanol readily dissolve in water.

Non-polar solutes such as carbon tetrachloride do not dissolve readily in water, but tend to partition into other environmental media, such as lipids and sediments. This is of interest in fish toxicology because biotic membranes are composed primarily of lipid. Thus, the tendency for a non-ionic chemical to partition between water and fish is governed by the polarity of the compound (Gobas and Russell, 1992). Thus, non-polar chemicals are termed, 'hydrophobic' or 'lipophilic', whereas polar chemicals are 'hydrophilic' or 'lipophobic'.

A partition coefficient which is used as an indicator of lipophilicity is the 'octanol–water partition coefficient' (K_{ow}). This coefficient describes the distribution of a chemical between water and a non-polar substance, octanol. Chemicals that tend to dissolve in the octanol rather than water have high K_{ow} and are classified as hydrophobic. The importance of hydrophobicity in determining the toxicity of organic compounds in fish is illustrated in studies where the lethal toxicity of organic compounds to fish has been shown to increase with $\log K_{ow}$ (Yoshioka et al., 1986) or decrease with water solubility (Neely, 1984). In fact, theoretical relationships have been developed which predict toxicity based upon estimates of how readily chemicals bioconcentrate in fish and reach 'critical body residues' (Barron et al., 1997).

The solubility of ionic chemicals, which include most salts of toxic metals and some ionic organic compounds, is usually much higher than that of non-ionic compounds. Ions tend to dissociate freely in aqueous solution, as polar water molecules can overcome the forces between ions. Ions are not actually 'free' within an aqueous matrix, but are associated with water molecules as an 'aqua ion complex'. In many cases, the toxic form of ionic chemicals is the 'free ion'; that is, the cation (Me^+), anion (Me^-), or oxyanion (MeO_3^-). Therefore, the amount of free ion in solution is an important factor in the toxicity of the element (Campbell, 1995). The degree of dissociation into the free ionic species is pH dependent. For instance, under acidic conditions, free cations (e.g. Zn^{2+}, Al^{3+}) should predominate. Therefore, the process of 'speciation' of ionic chemicals in water is an important factor governing toxicity (Wang, 1987).

Water quality conditions

As mentioned above, ions may be dissolved in water in non-toxic forms. For instance, ions may form complexes with inorganic and organic 'ligands'. Inorganic ligands for cations in fresh water include carbonate (CO_3^{2-})*, sulfate (SO_4^{2-}), and fluoride (F^-) ions, and Cl^- is an important ligand in saline water (Morel and Hering, 1993). Complexes between cations and inorganic ligands tend to be fairly 'labile' or reversible, depending on the concentration of the ion and the ligand, and the pH. However, complexes with organic ligands, such as humic acids, tend to be relatively non-labile. 'Alkalinity', which is primarily the concentration of carbonate ions in solution, is an important measure of the cation binding capacity of fresh water. Alkalinity and pH are important variables influencing the toxicity of metal ions to fish (Lauren and McDonald, 1985).

Transformations of chemicals dissolved in water can occur by hydrolysis, photolysis and oxidation (Connell, 1997). Photolytic transformation of some photoactive compounds, such as polynuclear aromatic hydrocarbons (PAHs) can increase toxicity to fish (Oris and Geisy, 1987). However, in general, hydrolysis is the most important transformation process affecting toxicity. Hydrolysis occurs with both organic and inorganic chemicals, but hydrolysis of metal salts is a very important process governing toxicity to fish:

$$Me^{n+} + H_2O \longrightarrow Me(OH)^{(n-1)+} + H^+$$

e.g.

$$Zn^{2+} + H_2O \longrightarrow Zn(OH)^+ + H^+$$

If a metal dissolves in distilled water, without any ligands being present, the metal will still form hydroxy complexes through hydrolysis. Metals differ in their stability constants as hydroxy complexes, so some metals form stable hydroxy complexes at relatively low pH, while others require higher pH. In general, trivalent cations such as Al^{3+} form significant levels of hydroxy complexes (e.g. $AlOH^{2+}$, $Al[OH_2]^+$) at the pH of natural water (Campbell, 1995). These hydroxy complexes are relatively non-toxic relative to the free cation, so hydrolysis, which is pH dependent, is an important factor influencing toxicity.

Water 'hardness', which is essentially the concentration of calcium and magnesium cations in water, affects fish toxicity for physiological reasons. Calcium may competitively bind to proteins on cell membranes. This binding may reduce the transport of toxic cations (e.g. copper, cadmium) across membranes such as the gill; thus reducing toxicity (Campbell and Stokes, 1985). Figure 37.1 illustrates the ameliorative effect of calcium ion concentration on the toxicity of aluminum to early life stages of rainbow trout (*Oncorhynchus mykiss*). Since high calcium concentrations usually co-occur with high carbonate concentrations in natural waters, it is often difficult to separate the effects of hardness and alkalinity on toxicity to fish (Welsh *et al.*, 1996).

Overall, if we accept that the toxicity of ionic chemicals is usually dependent upon the concentrations of the free ion in solution, then various factors that affect speciation of ions will affect toxicity; including pH, alkalinity, hardness, and concentrations of organic ligands (Playle *et al.*, 1992, 1993). Figure 37.2 illustrates the speciation processes that affect the toxicity of ionic chemicals to fish in water.

The toxicity of non-ionizable chemicals, such as organic compounds, is affected to a lesser extent by water quality conditions such as pH, alkalinity and hardness. However, dissolved and particulate organic material in water can alter the toxicity of organic compounds by acting as ligands for hydrophobic substances.

Biological interactions

A chemical can be toxic to a fish in two possible ways. It may affect tissues on the surface of the organism (e.g. gill epithelium) or the chemical may enter the organism and cause toxicity. The epithelial and endothelial integument of fish is usually thickened and relatively impermeable to chemicals, except in the gill tissues, which are specialized for gas exchange, and in the gastrointestinal tract. Thus, branchial or gastrointestinal uptake routes are the most efficient mechanisms for uptake of toxic chemicals into fish. 'Bioaccumulation' of chemicals represents the uptake and retention of chemical from the environment into fish via any pathway (e.g. food, water), whereas, 'bioconcentration' represents uptake and retention of a chemical directly from water into fish.

A toxic chemical must pass through cell membrane barriers to reach 'target' organs or tissues. Therefore, uptake of chemicals by fish can be influenced by both the lipophilicity and molecular size of the chemical (Protic and Sabljic, 1989). The membranes consist of a bimolecular layer of phospholipids,

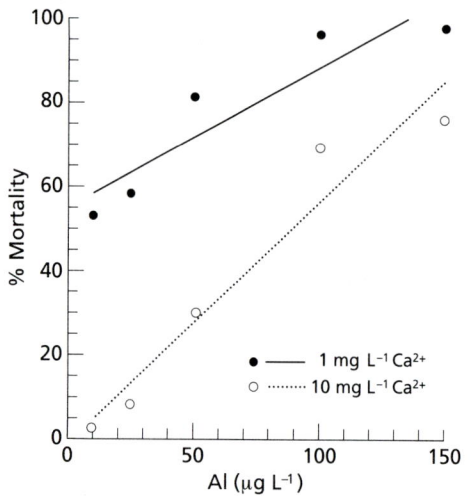

Figure 37.1 Percent mortalities of eggs and fry of rainbow trout exposed to four concentrations of aluminum (25, 50, 100, 150 µg L^{-1}) under conditions of 1,4 and 10 mg L^{-1} of calcium (unpublished data by the author).

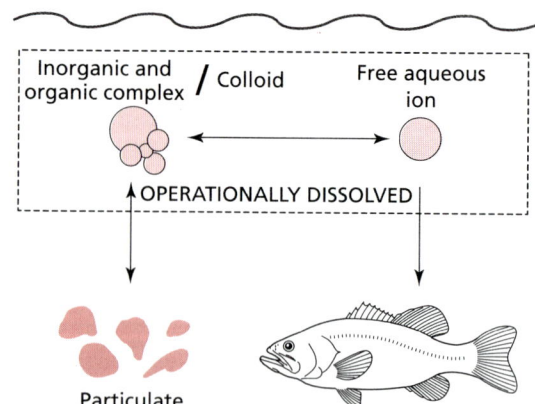

Figure 37.2 Speciation processes that influence the toxicity of chemical ions in tests with fish, including binding to suspended particulates, binding to dissolved organic and inorganic ligands, formation of colloids and formation of toxic free ions.

with integral and peripheral proteins. Because lipids are relatively non-polar, lipophilic chemicals will pass readily through them. Generally, the more lipophilic the chemical, the more readily it passes through the lipid-rich cell membranes into the organism (Gobas and Russell, 1992), but some large 'superhydrophobic' molecules cannot pass readily through membranes because of their molecular size (Bruggeman et al., 1984). The condition of the fish, and especially the fat content, can often influence the toxicity of hydrophobic chemicals since these compounds will tend to partition into fats, keeping them away from the target tissues and organs.

Inorganic contaminants in the elemental form (e.g. $Hg°$) can diffuse through lipid membranes, but inorganic cations and anions must pass through the cell membrane by routes that do not involve the lipid component of the membrane. There are several models proposed for this process, but the most likely mechanism involves binding of ions with amino acid or protein ligands within the membrane matrix, and passive diffusion of the complex through the membrane, or 'carrier mediated diffusion' (Astrue, 1989).

For some chemicals, the rate of uptake is strongly influenced by the physiology of the fish. Gill ventilation rates and dietary intake are governed by the metabolic rates of fish. There is considerable variation in the metabolism of fish; from fast-swimming **pelagic** predators to slow-swimming **benthivores**, so toxicity thresholds may vary considerably, depending on the fish species tested. In **poikilothermic organisms** such as fish, metabolism changes with the water temperature, so temperature may be an important factor influencing toxicity (Hodson and Blunt, 1986). Similarly, dissolved oxygen concentrations may influence gill ventilatory rates. Early life stages of fish tend to have higher rates of metabolism than later life stages (McKim, 1985).

Freshwater fish and saltwater fish have very different physiological mechanisms for regulating levels of essential ions such as Na^+, K^+, Cl^- (Evans, 1980). As illustrated in Figure 37.3, freshwater fish are constantly pumping ions into the body and excreting water out. Therefore, some metal cations and anions can be taken into the body along with essential ions (Na^+, Cl^-). In salt water, aquatic organisms are constantly pumping ions out into the surrounding water, and imbibing large quantities of water. There may be transport of chemicals into the body along with the water in saltwater fish. **Anadromous** fish (e.g. salmon) and **catadromous** fish (e.g. eels) go through changes in ionoregulatory physiology when they move from fresh water to salt water and vice versa. There can be large differences in the toxicity of chemicals to these fish species, depending upon the salinity of the water and the degree of acclimation of the test species (McDonald et al., 1989). Similar physiological adaptations must be taken into account in toxicity tests with **euryhaline** fish, which can tolerate a wide range of salinities.

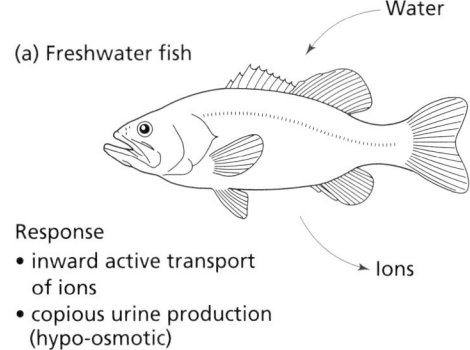

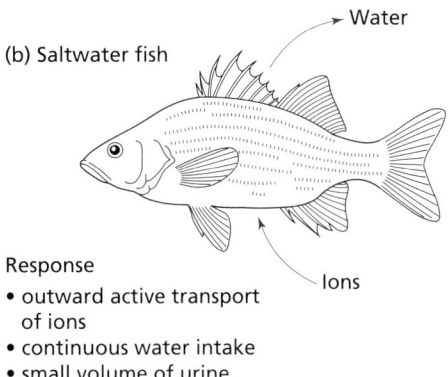

Figure 37.3 Flux of essential ions and water and mechanisms for ionoregulation and osmoregulation in freshwater and saltwater fish.

Fish possess metabolic pathways capable of transforming chemicals. Oxidation of chemicals through the activity of cytochrome P-450 monooxygenases is an important biotransformation pathway in fish (Stegeman and Hahn, 1994), and binding of chemicals to proteins or other large biomolecules (i.e. conjugation) is another transformation process (Di Giulio et al., 1995). In terms of affecting toxicity, biotransformation can result in: (i) inactivation of the chemical; (ii) activation of the chemical to form reactive metabolites; or (iii) maintenance of activity. Any factors that influence the rates of metabolic transformations of chemicals, such as water temperature, the sex and reproductive state of the fish, nutrition and the acclimation period, will often influence toxicity.

Toxicity testing with fish

Throughout this volume, there are detailed descriptions of the anatomy of fish and some of the pathological conditions that can occur as a result of exposure to chemicals. A comprehensive review of all of the possible 'endpoints' of toxicity that can be used with fish is beyond the scope of this chapter. However, as illustrated in Table 37.4, quantitative estimates of chemical toxicity to fish vary according to the toxic endpoint and species tested. In general, studies with adult fish using lethality as a toxic endpoint and acute (short-term) exposures are the least sensitive tests, but they are the most commonly used methods for assessing toxicity because they are relatively rapid, simple and inexpensive.

Toxicological studies are generally divided into four categories, depending on the duration of exposure:

1. Acute toxicity studies which involve either a single administration of the chemical to an organism, or continuous exposure over a very short period (e.g. 24–96 h).
2. Short-term toxicity studies which involve continuous, repeated or 'pulsed' exposure to a chemical over a period of about 10% of the life span.
3. Chronic toxicity studies which involve continuous or repeated exposures over a period of the entire life span of the test organism. If exposures occur over only a major portion of the life span, some researchers may classify these studies as long-term toxicity studies.

The life stage of the fish is a major determinant of toxicity. In general, the most sensitive period for chemically induced toxicity in fish is the embryolarval or early juvenile stages. This is illustrated by the data on the median lethal concentration in Japanese medaka exposed to aqueous solutions of nonylphenol (Table 37.3), where the LC_{50} for embryolarval stages and adults exposed to nonylphenol were $460\,\mu g\,L^{-1}$ (Gray and Metcalfe, 1997) and $1400\,\mu g\,L^{-1}$ (Yoshimura, 1986), respectively. Acute exposure of early life stages (ELS) of fish to a chemical generally provides a good estimate of the MATC obtained in chronic toxicity tests with later life stages (McKim et al., 1985). The high rate of metabolism of ELS may partially explain this sensitivity, but it is largely due to the fact that eggs, larvae and fry are going through many developmental events over a short space of time – several involving biochemical or hormonal events that depend on critical timing.

Fish species differ widely in their sensitivity to the toxic effects of chemicals, as illustrated in Table 37.3. In fact, there may be greater variability in toxicity associated with the test species than associated with differences in test conditions (Sprague, 1985). Some of this variability can be attributed to differences in metabolic rates and physiology of the fish, or the presence of **alloenzyme** genotypes that confer sensitivity or tolerance (Gillespie and Guttman, 1989). In some cases, whole fish taxa are more tolerant than other taxa. For instance, Spear and Pierce (1979) observed that centrarchids (i.e. Sunfish) are about 15 times more tolerant of copper that salmonids or minnows. It is important to note that previous exposure to chemicals may increase resistance to toxicity in fish (Duncan and Klaverkamp, 1982).

TABLE 37.3: Median lethal concentrations (LC_{50}) for nonylphenol determined in toxicity tests with various species of fish

Test species	LC_{50} ($\mu g\,L^{-1}$)
Winter flounder (*Pseudopleuronectes americanus*)	17[a]
Atlantic salmon (*Salmo salar*) – juvenile	130[b]
Sheepshead minnow (*Cyprinodon variegatus*)	210[a]
Fathead minnow (*Pimephales promelas*)	300[a]
Japanese medaka (*Oryzias latipes*) – embryolarvae	460[c]
Japanese medaka (*Oryzias latipes*) – adult	1,400[d]
Cod (*Gadus morhua*)	3,000[a]

[a]Brooke (1993); [b]Talmage (1994); [c]Gray and Metcalfe (1997); [d]Yoshimura (1986).

TABLE 37.4: Nominal threshold concentrations for lethality and various sublethal endpoints of toxicity in fish species exposed to aqueous solutions of nonylphenol

Species	Endpoint	Toxic concentration ($\mu g\,L^{-1}$)
Japanese medaka (adult)	Lethality	1400 (LC_{50})[a]
Rainbow trout	Lethality	194 (LC_{50})[b]
Japanese medaka (males)	Induction of intersex in gonad	50 (LOEC)[c]
Rainbow trout (males)	Inhibited testicular development	54 (LOEC)[d]
	Induction of vitellogenin (VtG)	20 (LOEC)[d]
	Induction of VtG mRNA precursor	14 (EC50)[e]
Rainbow trout	Inhibition of growth	10 (LOEC)[b]

[a]Yoshimura (1986); [b]Brooke (1993); [c]Gray and Metcalfe (1997); [d]Jobling *et al.* (1996); [e]Lech *et al.* (1996).

4. Life cycle toxicity studies which involve continuous or repeated exposures over at least two generations of the test organism – usually 'embryo to embryo'.

Of course, chronic toxicity studies are very expensive and time consuming in comparison to acute toxicity studies. In order to reduce costs, chronic studies are typically done with few doses or concentrations. These generally include a control and at least a low dose and a high dose. Acute toxicity tests usually involve a large number of doses in order to determine more precisely the thresholds for a toxic response.

Lethality testing

In lethality testing, the median lethal concentration or dose (LC_{50} or LD_{50}) is the usual method of reporting results, and a time period must be specified for the parameter (e.g. 96-h LC_{50}). This parameter is determined by observing whether there is increasing lethality in organisms exposed to a range of test doses or concentrations (i.e. a 'dose–response' relationship). Toxicant concentrations usually span a range which includes a concentration causing zero mortality, and a concentration causing complete mortality. Good lethality estimates are possible when at least one concentration or dose (and preferably two) kill some, but not all of the test organisms in the maximum observation period (i.e. 'partial mortalities').

When lethality is monitored, a plot of the frequency of mortalities in the test population generally forms an S-shaped curve (Figure 37.4). The central portion of the curve may be sufficiently straight to estimate a median response (e.g. LC_{50}), but if the curve can be 'straightened', a better estimate can be made of

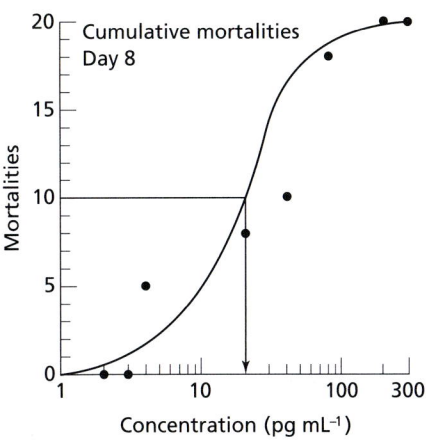

Figure 37.4 Cumulative mortalities observed over 17 days between fertilization of eggs and swim-up (start of exogenous feeding) in embryos and larvae of Japanese medaka (n = 20) in a control treatment and a treatment with aqueous solutions of 2,3,7,8-TCDD (30 pg mL^{-1}). (Data from Metcalfe *et al.*, 1997.)

the 50% response point. A common method of curve straightening is the 'probit' method, where probit units are assigned to deviations around the mean (Table 37.5). There are other statistical methods for analyzing lethality data, including moving average interpolation, non-parametric methods (e.g. trimmed Spearman–Karber method) and binomial confidence intervals (Ellerseik and La Point, 1995).

The basic data for lethality tests are tabulations of percent cumulative mortalities at each concentration, as observed at different times throughout the assay. These data can be used to calculate lethal concentrations. The concentration or dose at which 50% of the population dies at time *t* can be determined by using a computer program that uses a probit transformation model, or by plotting data on probit paper. In order to illustrate the methods used in acute lethality testing, consider the data in Table 37.6. For each test concentration, there are several observation times. In the example, probit transformed percent cumulative mortality data are used to calculate LC_{50}s and 95% confidence limits around these estimates for each observation time over 1 to 18 days.

The LC_{50}s are then plotted against observation time to determine a 'lethality line', as shown in Figure 37.5. Note that the lowest concentration of the chemical (i.e. 4 pg mL^{-1}) caused less than 50% mortality (i.e. 8 of 20 fish) at the maximum observation time of 18 days. This point is plotted against the maximum observation time as ●. A lethality line allows the

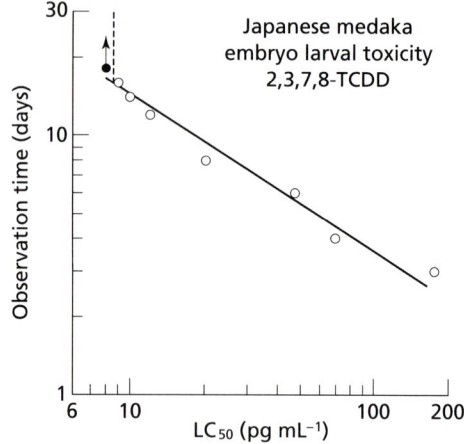

Figure 37.5 A lethality line plotted as LC_{50} vs. observation time from the data presented in Table 37.6 on the lethality of 2,3,7,8-TCDD to embryolarvae of Japanese medaka. The lowest concentration causing less than 50% mortality at the maximum observation time is plotted as a filled circle.

TABLE 37.5: Probit units used to transform cumulative toxicity data for estimates of median toxic responses

% Response	Deviations	Probit unit
0.1	−3	2
2.3	−2	3
15.9	−1	4
50.0	0	5
84.1	1	6
97.7	2	7
99.9	3	8

TABLE 37.6: Cumulative mortality data for the lethality of 2,3,7,8-TCDD (0–200 pg mL^{-1}) to embryolarvae of Japanese medaka (n = 20) over 18 days between fertilization of eggs and swim-up. The LC_{50}s plus upper and lower 95% confidence limits (in brackets) were calculated for each observation period by probit analysis

	Conc. (pg mL^{-1})						
	0	4	20	40	80	200	
	Cumulative mortalities (n = 20 per treatment)						LC_{50} (pg mL^{-1})
Observation Time (days)							
1	0	0	0	0	2	4	>200
2	0	0	2	4	5	6	>200
3	0	0	2	4	6	12	176 (155–204)
4	0	2	3	4	10	18	69 (45–119)
6	0	3	4	5	12	20	47 (31–75)
8	0	5	8	10	18	20	20 (31–75)
Hatch							
12	0	8	9	13	20	20	12 (6–19)
14	0	8	9	16	20	20	10 (5–16)
16	0	8	9	20	20	20	9 (4–13)
18	0	8	10	20	20	20	8 (4–12)
Swim-up							

Source: Original data are from the study published by Metcalfe *et al*. (1997).

researcher to extract the maximum amount of information from the lethality test. Instead of estimating an LC_{50} from a single observation time (which may have anomalous data), all of the observation times can be utilized. From a single lethality line, we can get estimates of the LC_{50}s at 4 days, 10 days, 15 days, etc. We can also estimate a 'lethal threshold', or 'incipient LC_{50}' as the highest concentration at which less than 50% of the animals die. This is calculated as the mean of the highest concentration showing < 50% mortality (i.e. 4 pg mL^{-1}) and the lowest LC_{50} in the bioassay (i.e. 8 pg mL^{-1}). The lethal threshold is sometimes used as an estimate of the MATC, although it is preferable to estimate this latter parameter from chronic toxicity data.

Sublethal toxicity testing

It is beyond the scope of this chapter to review in detail all of the toxicological models for evaluating sublethal effects in fish. The rapid developments in this area of fish toxicology, particularly for evaluating molecular, biochemical, immunological and histological responses to chemical exposure, have been documented over the past 15 years in several reviews (Thomas, 1990; Hinton and Lauren, 1990; Stegeman and Hahn, 1993; Arkoosh et al., 1994; Bunton, 1996; Di Giulio et al., 1995; Metcalfe, 1998). A summary of the most commonly used methods for evaluating sublethal responses of fish to chemical exposure is provided in Table 37.7, along with a limited number of references that illustrate the methods used in the sublethal tests.

The experimental models vary in the duration of exposures and the complexity and time required to evaluate the endpoints. For instance, responses such as induction of hepatic ethoxyresorufin-deethylase (EROD) or inhibition of brain acetylcholinesterase (AchE) can be evaluated after acute exposures of fish for 24–72 h, and the biochemical methods required to evaluate these responses are relatively rapid and simple. However, chronic exposures over several months may be required to evaluate endpoints such as carcinogenicity or alterations to reproductive potential in fish. In addition, some sublethal endpoints are indicators of exposure to specific classes of chemicals, such as the induction of DNA adducts by aromatic hydrocarbons or the induction of vitellogenin (VtG) synthesis in the livers of male fish by estrogenic substances. Other sublethal responses can occur in fish exposed to many chemical classes, as well as environmental stressors (e.g. disease, nutrition, temperature, reproductive status). Obviously, it is important to control for all environmental stressors that may influence sublethal responses in these experimental models.

Recently, there has been a great deal of interest in endocrine disrupting (modulating) substances, particularly those that alter sexual development and reproductive success in fish and wildlife (Hose and Guillette, 1995). There are a variety of sublethal endpoints that have been used as indicators of endocrine disruption in fish, including: (i) altered serum steroid levels; (ii) synthesis of egg-yolk protein (vitellogenin) in male fish; (iii) alterations to the development of gonadal tissues; and (iv) effects on reproductive success. However, it is fair to say that the process of identifying the most appropriate *in vivo* models and fish species for assessing the endocrine modulating potential of chemicals is ongoing (Ankley, 1998). For some classes of endocrine disrupting substances, such as those that exhibit anti-androgenic activity, including p,p'-DDE and the fungicide, vinclozolin (Monosson et al., 1997), there are few *in vivo* models currently available for determining endocrine modulating effects on fish.

Bioassays

A 'bioassay' is a test in which the quantity of material is determined by the response of a living organism to it. For instance, we may evaluate the concentration of toxic constituents in an industrial effluent by the lethal or sublethal toxic response of organisms to it. The term 'bioassay' should not be confused with toxicity test. Bioassays with fish and other aquatic organisms are commonly used for regulatory purposes to ensure that toxic substances in industrial effluents and wastewaters are not being discharged in excessive quantities. There are standard methods required by regulatory agencies for these bioassay procedures (e.g. ASTM, 1991).

Bioassays are also useful for 'Toxicity Identification and Evaluation' (TIE), which is a process of systematic treatment of environmental samples (e.g. pH adjustment, filtration, subfractionation) followed by bioassays for toxicity to identify the chemicals in the sample that are responsible for lethal or sublethal effects (US, EPA, 1991). TIE approaches with fish bioassays have been used to identify carcinogens in sediments (Balch et al., 1995), embryotoxic compounds in extracts from Lake Ontario trout (Harris et al., 1994a) and the EROD-inducing compounds in a lampricide formulation (Hewitt et al., 1997).

TABLE 37.7: Sublethal toxicity endpoints with fish species

Endpoint	Reference
(I) Enzyme/protein induction	
(a) CYP4501A monooxygenases:	
EROD	Hodson et al. (1997)
AHH	Janz and Metcalfe (1991)
mRNA	Williams et al. (1997)
(b) Glutathione:	Nishimoto et al. (1995)
(c) Metallothionein:	Hamilton and Mehrle (1986)
(II) Enzyme inhibition	
(a) AchE:	Richmonds and Dutta (1992)
(b) Phosphatases:	Sharma (1990)
(III) Enzyme release	
Lysosomal:	Dixon et al. (1985)
Hepatic:	Dixon et al. (1987)
(IV) Disturbance in metabolism	
Adenine nucleotides	Vetter et al. (1996)
Bilirubin	Mattsoff and Oikari (1987)
(V) Ionoregulation	McDonald et al. (1989)
(VI) Endocrine disruption	
(a) Retinoid:	Palace and Brown (1994)
(b) Thyroid:	Ruby et al. (1993)
(c) Catecholamine:	Black and Nilsson (1990)
(d) Sex steroid:	MacLatchy and Van Der Kraak (1995); Miranda et al. (1992)
(e) Vitellogenin:	Jobling et al. (1996)
(VII) Reproduction	Arcand-Hoy and Benson (1998); Donaldson (1990)
(VIII) Gonad development	Gray and Metcalfe (1997); Gimeno et al. (1997)
(IX) Oxidative damage	Di Giulio et al. (1993); Scarano et al (1994)
(X) Growth	Benton et al. (1994)
(XI) Behavior	
(a) Courtship:	Schroder and Peters (1988)
(b) Avoidance:	Morgan et al. (1991)
(c) Swimming:	Little and Finger (1990)
(d) Ventilatory:	Diamond et al. (1990)
(XII) Immunomodulation	Anderson and Zeeman (1995)
	Zeeman and Brindley (1993)
	Zelikoff et al. (1991)
	Arkoosh et al. (1994)
(XIII) Carcinogenesis	Bunton (1996); Metcalfe (1989)

Table 37.7: (Continued)	
Endpoint	**Reference**
(XIV) Genotoxicity	
(a) DNA adducts:	Stein et al. (1994)
(b) DNA damage:	Pandrangi and Petras (1995)
(c) Clastogenic:	Al-Sabti and Metcalfe (1995)
(XV) Histopathology	
	Hinton and Lauren (1990)
	Metcalfe (1998)

Toxic equivalency factors (TEFs)

Many of the toxic responses observed in vertebrates exposed to planar halogenated aromatic hydrocarbons (HAHs) such as 2,3,7,8-TCDD are mediated through binding to a specific cellular protein called the *Ah* (aryl hydrocarbon) receptor. One group of proteins synthesized as a result of this molecular response is the hepatic monooxygenases belonging to the cytochrome P4501AI (CYP1AI) subgroup (e.g. EROD, AHH). The degree of induction of CYP4501AI monooxygenases in fish has been used as a biochemical indicator of the toxic potency of PCB **congeners** (Janz and Metcalfe, 1991; Harris et al., 1994b). In order to standardize estimates of the toxic potency of planar HAHs, Safe et al. (1990) proposed toxic equivalency factors (TEFs) based upon the potencies of various HAHs relative to a reference compound, 2,3,7,8-TCDD for several *Ah*-mediated responses in mammalian models.

As outlined above, much of the data describing the structure–activity relationships for the toxicity of planar HAHs has been determined with mammalian *in vivo* and *in vitro* models. Since the toxicity of coplanar PCBs and other planar HAHs may vary between vertebrate taxa, there have been several studies focused on determining TEFs for fish. The most complete set of TEF data was developed from embryolarval lethality tests with salmonid species (Walker and Peterson, 1991; Zabel et al., 1995). The TEF data summarized in Table 37.8 were calculated by dividing the LD_{50} for embryolarval mortalities obtained by *in ovo* injection of a planar HAH congener (nmol) by the LD_{50} (nmol) for 2,3,7,8-TCDD. These TEFs indicate that several dioxin and dibenzofuran congeners and non-*ortho* (coplanar) PCB congeners are extremely embryotoxic to fish. The PCB congeners with chlorine substitution in *ortho* positions all have very low toxic potencies in fish relative to the non-*ortho* PCB congeners. It is likely that the 'TEF approach' for total planar HAHs will be increasingly used to regulate discharges or to legislate remedial action for dioxin, dibenzofuran and PCB contaminants in the environment.

Overview

Although teleost *in vitro* models with cell lines, primary cell cultures and tissue explants are being used increasingly, there are several reasons why *in vivo* models will remain in use for toxicity testing of aquatic contaminants. As described earlier in this chapter, the uptake, metabolism and clearance of chemicals in fish are complex processes that cannot be reproduced easily with *in vitro* models. Chemical toxicity to fish is often affected by external factors, such as photoperiod, temperature, salinity, reproductive status, disease and exposure to other external stressors, and these factors cannot be replicated in *in vitro* tests. In addition, biochemical, physiological and endocrine systems in fish are often complex and are controlled by positive and negative feedback processes that involve several different tissues and organs. Once again, these conditions are difficult to replicate in *in vitro* models.

In vitro toxicity tests are very useful to initially screen chemicals for toxic potential or to evaluate the mechanisms of toxicity. However, it is safe to assume that researchers and regulators will continue to require *in vivo* toxicity testing with fish for definitive evidence of the toxicity of chemicals to aquatic organisms. Indeed, regulatory agencies in some countries require *in vivo* toxicity testing with 'native' fish species before toxicity data are accepted. Techniques for *in vivo* toxicity testing with fish continue to be

TABLE 37.8: Toxic equivalency factors (TEFs) for selected polychlorinated dibenzo-*p*-dioxins (PCDDs), dibenzofurans (PCDFs) and PCB congeners as determined from embryolarval mortality data for salmonids

Congener	TEF	Congener	TEF
PCDDs:		Non-ortho PCBs:	
2,3,7,8-TCDD	1.0	Congener 126	0.004 9
1,2,3,7,8-PeCDD	0.659	Congener 81	0.000 62
1,2,3,4,7,8-HxCDD	0.263	Congener 77	0.000 18
1,2,3,6,7,8-HxCDD	0.020	Congener 169	0.000 04
1,2,3,4,6,7,8-HpCDD	0.0015		
PCDFs:		Ortho PCBs:	
2,3,4,7,8-PeCDF	0.339	Congener 28	< 0.000 009
1,2,3,4,7,8-HxCDF	0.240	Congener 118	< 0.000 003
1,2,3,7,8-PeCDF	0.032	Congener 105	< 0.000 002
2,3,7,8-TCDF	0.030	Congener 156	< 0.000 001

Source: Zabel *et al.*, 1995.

developed and refined – especially for sublethal toxicity tests using biochemical indicators. However, we need much more basic research on the biochemical and physiological bases of toxicity in fish in order to evaluate the significance of these types of toxicity data.

References

Al-Sabti, K. and Metcalfe, C.D. (1995). *Mutat. Res.* **343**, 121–135.

Anderson, D.P. and Zeeman, M.G. (1995). In *Fundamentals of Aquatic Toxicology*, 2nd edn (ed. G.M. Rand), pp. 371–404. Taylor and Francis, London.

Ankley, G. *et al*. (1998). *Environ. Toxicol. Chem.* **16**, 68–87.

Arcand-Hoy, J.D. and Benson, W.H. (1998). *Environ. Toxicol. Chem.* **16**, 49–57.

Arkoosh, M.R., Stein, J.E. and Casillas, E. (1994). In *Modulators of Fish Immune Responses: Models for Environmental Toxicology, Biomarkers, and Immunostimulators*, vol. 1 (eds J.S. Stolen and T.C. Fletcher), pp. 33–48, SOS Publ., Fair Haven, New Jersey.

ASTM (1991). In *Annual Book of ASTM Standards, Volume 11.04, Pesticides, Resource Recovery, Hazardous Substances and Oil Spill Response, Waste Disposal and Biological Effects*, pp. 378–397. American Society for Testing and Materials, Philadelphia, Pennsylvania.

Astrue, M. (1989). In *Aquatic Ecotoxicology: Fundamental Concepts and Methodologies*, vol. I (eds A. Boudou and F. Ribeyre), pp. 97–106. CRC Press, Boca Raton, Florida.

Balch, G.C., Metcalfe, C.D. and Huestis, S.Y. (1995). *Environ. Toxicol. Chem.* **14**, 79–91.

Barron, M.G., Anderson, M.J., Lipton, J. and Dixon, D.G. (1997). *Environ. Res.* **6**, 47–62.

Benton, M.J., Nimrod, A.C. and Benson, W.H. (1994). *Ecotoxicol. Environ. Safety* **29**, 1–12.

Block, M. and Nilsson, G.E. (1990). *Aquat. Toxicol.* **16**, 1–8.

Brooke, L.T. (1993). Report to the US EPA for work assignment No. 02 of contract No. 68-C1-0034. Lake Superior Research Institute, University of Wisconsin-Superior, Bunton, T.E. (1996) Superior, Wisconsin 30pp.

Bruggeman, W.A., Opperhuizen, A., Wijbenga, A. and Hutzinger, O. (1984). *Toxicol. Environ. Chem.* **7**, 173–189.

Bunton, T.E. (1996). *Toxicol. Pathol.* **24**, 603–618.

Campbell, P.G.C. (1995). In *Metal Speciation and Bioavailability in Aquatic Systems* (eds A. Tessier and D.R. Turner), pp. 45–102, John Wiley, New York.

Campbell, P.G.C. and Stokes, P.M. (1985). *Can. J. Fish. Aquat. Sci.* **42**, 2034–2049.

Connell, D.W. (1997). *Basic Concepts of Environmental Chemistry*. Lewis Publ., Boca Raton, Florida. 506pp.

Diamond, J.M., Parson, M.J. and Gruber, D. (1990). *Environ. Toxicol. Chem.* **9**, 3–11.

Di Giulio, R.T., Habig, C. and Callagher, E.P. (1993). *Aquat. Toxicol.* **26**, 1–22.

Di Giulio, R.T., Benson, W.H., Sanders, B.M. and Van Veld, P.A. (1995). In *Fudamentals of Aquatic Toxicology*, 2nd edn (ed. G.M. Rand), pp. 523–560. Taylor and Francis, London.

Dixon, D.G., Hill, C.E.A., Hodson, P.V., Kempe, E. and Kaiser, K.L.E. (1985). *Environ. Toxicol. Chem.* **4**, 789–796.

Dixon, D.G., Hodson, P.V. and Kaiser, K.L.E. (1987). *Environ. Toxicol. Chem.* **6**, 789–796.

Donaldson, E.M. (1990). *Am. Fish. Soc. Symp.* **8**, 109–122.

Duncan, D.A. and Klaverkamp, J.F. (1982). *J. Fish. Aquat. Sci. Can.* **40**, 128–138.

Ellersieck, M.R. and La Point, T.W. (1995). In *Fundamentals of Aquatic Toxicology*, 2nd edn (ed. G.M. Rand), pp. 307–344. Taylor and Francis, London.

Evans, D.H. (1980). In *Environmental Physiology of Fishes* (ed. M.A. Ali), pp. 93–122. Plenum Press, New York.

Gillespie, R.B. and Guttman, S.I. (1989). *Environ. Toxicol. Chem.* **8**, 305–317.

Gimeno, S., Komen, H., Venderbosch, P.W.M. and Bower, T. (1997). *Environ. Sci. Technol.* **31**, 2884–2890.

Gobas, F.A.P.C. and Russell, R.W. (1991). *Comp. Biochem. Physiol.* **100C**, 17–20.

Gray, M.A. and Metcalfe, C.D. (1997). *Environ. Toxicol. Chem.* **16**, 1082–1086.

Hamilton, S.J. and Merhle, P.M. (1986). *Trans. Am. Fish. Soc.* **115**, 595–609.

Harris, G.E., Metcalfe, T.L., Metcalfe, C.D. and Huestis, S.Y. (1994a). *Environ. Toxicol. Chem.* **13**, 1405–1414.

Harris, G.E., Kiparissis, Y. and Metcalfe, C.D. (1994b). *Environ. Toxicol. Chem.* **13**, 1393–1404.

Hewitt, L.M., Munkittrick, K.R., Scott, I.M., Carey, J.H., Solomon, K.R. and Servos, M.R. (1997). *Environ. Toxicol. Chem.* **15**, 894–905.

Hinton, D.E. and Lauren, D.J. (1990). *Am. Fish. Soc. Symp.* **8**, 51–66.

Hodson, P.V. and Blunt, B.R. (1986). *J. Fish Biol.* **29**, 37–46.

Hodson, P.V., Maj, M.K., Efler, S., Burnison, B.K., Van Heiningen, A.R.P., Girard, R. and Carey, J.H. (1997). *Environ. Toxicol. Chem.* **16**, 908–916.

Hose, J.E. and Guillette, L.J. (1995). *Environ. Hlth. Perspect.* (Suppl. 4)**103**, 87–91.

Janz, D.M. and Metcalfe, C.D. (1991). *Environ. Toxicol. Chem.* **10**, 231–239.

Jobling, S., Sheahan, D., Osborne, J.A., Matthiessen, P. and Sumpter, J.P. (1996). *Environ. Toxicol. Chem.* **15**, 194–202.

Lauren, D.J. and McDonald, D.G. (1986). *Can. J. Fish. Aquat. Sci.* **43**, 1488–1496.

Lech, J.J., Lewis, S.K. and Ren, L. (1996). *Fund. Appl. Toxicol.* **30**, 229–232.

Little, E.E. and Finger, S.E. (1990). *Environ. Toxicol. Chem.* **9**, 13–19.

MacLatchy, D.L. and Van Der Kraak, G.J. (1995). *Toxicol. Appl. Pharmacol.* **134**, 305–312.

Mattsoff, L. and Oikari, A. (1987). *Comp. Biochem. Physiol.* **88C**, 263–268.

McDonald, D.G., Reader, J.P. and Dalziel, T.R.K. (1989). In *Acid Toxicity and Aquatic Animals* (eds R. Morris, E.W. Taylor, D.J.A. Brown and J.A. Brown), pp. 221–242. Cambridge University Press, Cambridge, UK.

McKim, J.M. (1985). In *Fundamentals of Aquatic Toxicology* (eds G.M. Rand and S.R. Petrocelli), pp. 58–94. Hemisphere Publ., Washington, DC.

Metcalfe, C.D. (1989). *CRC Crit. Rev. Aquat. Sci.* **1**, 111.

Metcalfe, C.D. (1998). In *Fish Diseases III: Noninfectious Disorders* (eds J.F. Leatherland and P.T.K. Woo). CAB International, New York.

Metcalfe, C.D., Metcalfe, T.L., Cormier, J.A., Huestis, S.Y. and Niimi, A.J. (1997). *Environ. Toxicol. Chem.* **16**, 1749–1754.

Miranda, C.L., Henderson, M.C., Wang, J-L., Chang, H-S., Hendricks, J.D. and Buhler, D.R. (1992). *Comp. Biochem. Physiol.* **103C**, 153–157.

Monosson, E., Kelce, W.R., Mac, M. and Gray, L.E. (1997). In *Chemically Induced Alterations in Functional Development and Reproduction of Fishes* (eds R.M. Rolland, M. Gilbertson and R.E. Peterson), pp. 53–58. SETAC Press, Pensacola, Florida.

Morel, F.M.M. and Hering, J.G. (1993). *Principles and Application of Aquatic Chemistry*, Wiley and Sons, NY.

Morgan, J.D., Vigers, G.A., Farrell, A.P., Janz, D.M. and Manville, J.F. (1991). *Environ. Toxicol. Chem.* **10**, 73–79.

Neely, W.B. (1984). *Chemosphere* **13**, 813–819.

Nishimoto, M., Le Eberhart, B-T., Sanborn, H.R., Krone, C., Varanasi, U. and Stein, J.E. (1995). *Environ. Toxicol. Chem.* **14**, 461–469.

Oris, J.T. and Geisy, J.P. (1987). *Chemosphere* **16**, 1395–1404.

Palace, V.P. and Brown, S.B. (1994). *Environ. Toxicol. Chem.* **13**, 473–476.

Pandrangi, R. and Petras, M. (1995). *Environ. Mol. Mutagen.* **26**, 345–356.

Playle, R.C., Gensemer, R.W. and Dixon, D.G. (1992). *Environ. Toxicol. Chem.* **11**, 381–391.

Playle, R.C., Dixon, D.G. and Burnison, K. (1993). *Can. J. Fish. Aquat. Sci.* **50**, 1–11.

Protic, M. and Sabljic, A. (1989). *Aquat. Toxicol.* **14**, 47–64.

Richmonds, C.R. and Dutta, H.M. (1992). *Bull. Environ. Contam. Toxicol.* **49**, 431–435.

Ruby, S.M., Idler, D.R. and Peng, So Y. (1993). *Aquat. Toxicol.* **26**, 91–102.

Safe, S. (1990). *CRC Crit. Rev. Toxicol.* **21**, 51–88.

Scarano, L.J., Calabrese, E.J., Kostecki, P.T., Baldwin, L.A. and Leonard, D.A. (1994). *Ecotoxicol. Environ. Safety* **29**, 13–19.

Schroder, J.H. and Peters, K. (1988). *Bull. Environ. Contam. Toxicol.* **40**, 396–404.

Sharma, R.M. (1990). *Bull. Environ. Contam. Toxicol.* **44**, 443–448.

Sleiderink, H.M., Beyer, J., Scholtens, E., Goksoyr, Nieuwenhuize, J., Van Liere, J., Everaarts, J.M. and Boon, J.P. (1995). *Aquat. Toxicol.* **32**, 189–209.

Spear, P.A. and Pierce, R.C. (1979). Copper in the Aquatic Environment: Chemistry, Distribution and Toxicology. National Research Council of Canada Environmental Secretariat, NRRC No. 16454.

Sprague, J.B. (1971). *Water Res.* **5**, 245–266.

Sprague, J.B. (1985). In *Fundamentals of Aquatic Toxicology* (eds G.M. Rand and S.R. Petrocelli) pp. 124–163. Hemisphere Publ., Washington, DC.

Stegeman, J.J. and Hahn, M.E. (1993). In *Aquatic Toxicology: Molecular, Biochemical and Cellular Perspectives* (eds D.C. Malins and G.K. Ostrander, G.K.), pp.87–206. Lewis Publ., Boca Raton, Florida.

Stein, J.E. and Reichert, W.L. (1994). *Environ. Hlth. Perspect.* **102** (Suppl. 12), 19–23.

Talmage, S.S. (1994). *Environmental and Human Safety of Major Surfactants: Alcohol Ethoxylates and Alkylphenol Ethoxylates.* Lewis Publ., Boca Raton, FL, 374pp.

Thomas, P. (1990). *Am. Fish. Soc. Symp.* **8**, 9–28.

USEPA (1991). Technical Report 05-91, Report EPA/600/6-91/005, US Environmental Protection Agency, National Affluent Toxicity Assessment Centre, Duluth, Minnesota.

Vetter, R.D., Wang, H-M. and Hodson, R.E. (1986). *Trans. Am. Fish. Soc.* **115**, 47–51.

Walker, M.K. and Peterson, R.E. (1991). *Aquat. Toxicol.* **21**, 219–238.

Wang, W. (1987). *Environ. Int.* **13**, 437–457.

Welsh, P.G., Parrott, J.L., Dixon, D.G., Hodson, P.V., Spry, D.J. and Mierle, G. (1996). *Can. J. Fish. Aquat. Sci.* **53**, 1263–1271.

Williams, J., Courtenay, S.C. and Wirgin, I.I. (1997). *Environ. Toxicol. Chem.* **16**, 241–244.

Yoshimura, K. (1986). *J. Am. Oil Chem. Soc.* **63**, 1590–1596.

Yoshioka, Y., Mizuno, T., Ose, Y. and Sato, T. (1986). *Chemosphere* **15**, 195–203.

Zabel, E.W., Cook, P.M. and Peterson, R.E. (1995). *Aquat. Toxicol.* **31**, 315–328.

Zeeman, N.G. and Brindley, W.A. (1993). In *Immunologic Considerations in Toxicology*, vol. II (ed. R.P. Sharma), pp. 1–47. CRC Press, Boca Raton, Florida.

Zelikoff, J.T., Enane, N.A., Bowser, D., Squibb, K.S. and Frenkel, K. (1991). *Fund. Appl. Toxicol.* **16**, 576–589.

CHAPTER 38

Cell and Tissue Culture

Rosemarie C Ganassin, Kristin Schirmer
and Niels C Bols
Department of Biology, University of Waterloo,
Waterloo, Ontario, Canada

Introduction

Cell cultures can be prepared from different tissues and organs of many fish species (Bols and Lee, 1991; Fryer and Lannan, 1994). As with other animals, two types of cultures can be distinguished: primary cultures and cell lines. Primary culture refers to the *in vitro* maintenance of cells taken directly from fish, while cell lines are cultures that arise from the subculturing or passaging of primary cultures (Schaeffer, 1990). For mammals, cell lines are described as being either continuous, which means that the cells are immortal and can be grown indefinitely, or finite, which means that the cells age or senesce after a certain number of passages. This distinction is rarely made with fish cell lines and most are assumed to be immortal.

The uses of fish cell cultures are many, varied, and growing, but they can be grouped into six general categories. One general use is in the study of infectious diseases affecting fish. Originally, piscine cell cultures were developed in the 1950s and 1960s to identify and study fish viruses, and cell lines continue to have valuable roles in diagnosing fish viral diseases (Wolf and Ahne, 1982). More recently, cell cultures have been used to study fish diseases caused by microorganisms, such as bacteria (Garces *et al.*, 1991) and protozoa (Wongtavatchai *et al.*, 1994). Studies into the biochemistry and cell biology of fish are a second type of general application. This includes studies on fatty acid metabolism (Tocher *et al.*, 1995), the heat-shock response (Bols *et al.*, 1992) and radiation biology of fish cells (Mitani *et al.*, 1996). A third category of application is in fish molecular biology and genetics. Some examples of this are the use of cell cultures to study the strength of gene promoters (Moav *et al.*, 1992), the generation of cell hybrids for gene mapping (Chevrette *et al.*, 1997), and the creation of transgenic fish (Collodi *et al.*, 1992; Hong *et al.*, 1998). Investigation of cell differentiation is a fourth use of fish cell cultures. The differentiation of embryonic stem cells (Hong *et al.*, 1996), neural crest cells (Matsumoto *et al.*, 1989), myosatellite cells (Koumans *et al.*, 1993), blood cells (Ganassin and Bols, 1996), oligodendrocytes (Jeserich and Stratmann, 1992), and spermatocytes (Saiki *et al.*, 1997) have all been studied in culture. A fifth application is as a tool in basic research disciplines that until recently have used mainly whole fish. For this purpose, primary cultures in particular have been important, and these are being used in fish endocrinology (Montero *et al.*, 1996), in neurobiology (Schwalb *et al.*, 1995) and in physiological studies of the gill

Copyright © 2000 Academic Press

(Wood and Part, 1997) and swimbladder (Pelster and Pott, 1996). As well, both primary cultures and cell lines have been valuable in fish immunology (Neumann et al., 1995; Clem et al., 1996). A sixth general use is in the closely related fields of aquatic toxicology (Babich and Borefreund, 1991) and ecotoxicology (Bols et al., 1997). This area has perhaps shown the largest growth in the last 15 years (Malins and Ostrander, 1994). Closely linked with many of these goals is an aquaculture purpose, as the knowledge gained through the use of fish cell cultures can be applied in the future to the aquaculture industry (Bols, 1991) or in some cases provide a product for the industry. An example of the latter is the production of viruses for use in vaccines (Leong et al., 1997).

Although the culturing, development and monitoring requirements will vary with the goals, cell type and species under investigation, much can be learned from the protocols for developing specific types of fish cell cultures and for monitoring general and specific functions in fish cell cultures. In this chapter, methods for developing primary long-term hemopoietic cultures from fish are described. As well, step-by-step descriptions are given for methods of measuring general and specific cell parameters useful in both basic research and aquatic toxicology. These are cell viability and the induction of 7-ethoxyresorufin-o-deethylase (EROD) activity in fish cell cultures. The species used is rainbow trout (*Oncorhynchus mykiss*) because it is the most commonly used species in fish research (Wolf and Rumsey, 1985). Although this position is being challenged by other species, such as zebrafish, rainbow trout is still the most studied representative of coldwater species. The expectation is that the techniques described here can be used with only slight modification in basic and toxicological research programs with other teleosts.

A model of rainbow trout hemopoiesis – long-term hemopoietic culture

Long-term bone marrow cultures (Dexter and Testa, 1976) have been valuable research tools as a model of hemopoiesis in mammals. These cultures consist of two components: an attached, multilayer stroma and liquid growth medium. Successful *in vitro* hemopoiesis depends upon both cell-to-cell contacts between the developing cells and the cells of the stromal layer, and diffusible factors supplied both by the stromal cell layer and by high concentrations of serum, usually fetal bovine serum (FBS), supplementing the medium. Stem cells in these cultures differentiate to produce a non-adherent cell population consisting of blood cells in varying stages of maturity.

In teleost fish, hemopoiesis does not occur in the bone marrow, but instead occurs in the pronephros and spleen. Long-term hemopoietic cultures (LTHC) from the pronephros and spleen have been reported by several authors (Diago et al., 1993; Siegl et al., 1993; Ganassin and Bols, 1996; Chilmonczyk et al., 1997).

Preparation of stock solutions

All solutions used for the preparation and maintenance of primary cultures and cell lines must be sterile. If autoclaving is damaging to a solution, it must be filter sterilized, using a 0.22 μm filtration device. These are available in syringe tip sizes, and larger bottle top sizes, from manufacturers including Gelman and Millipore. Chemicals used should ideally be tissue culture grade, which are less likely to contain trace contaminants harmful to cells. We recommend the use of tissue culture grade water (Canadian Life Technologies, Burlington, ON, Canada), if there is any question about the quality of your local water.

Phosphate buffered saline (PBS)

PBS for general use in fish cell culture can be prepared with calcium and magnesium (PBS) or without (PBS−) from the same stock solutions.

1. To prepare PBS stock solutions, mix together the following: *Solution A*: NaCl (6.0 g), KCl (1.5 g), Na_2HPO_4 (8.625 g), KH_2PO_4 (1.5 g). Add water to a volume of 2000 mL. Dispense in 400 mL aliquots. *Solution B*: $CaCl_2 \cdot 2H_2O$ (0.75 g). Add water up to a volume of 375 mL. Dispense in 75 mL aliquots. *Solution C*: $MgCl_2 \cdot 6H_2O$ (0.75 g). Add water up to a volume of 375 mL. Dispense in 75 mL aliquots. Autoclave solutions A, B and C, and store at room temperature.

2. To prepare 1500 mL of PBS, add sterile water to solution A to bring volume to 1200 mL. Bring solution B to 150 mL, and solution C to 150 mL. Mix well.

3. To prepare 1500 mL of PBS−, bring the volume of solution A to 1500 mL with sterile water.

Primary culture initiation solution (PBS, with gentamicin and fungazone)

1. To prepare 100 mL, mix together 100 mL PBS, 100 μL gentamicin (Sigma G1397) (final concentration 50 μg mL^{-1}) and 4 mL Fungizone (Gibco 15295-017) (final concentration 100 μg mL^{-1}).

Collagenase A (500 mL)

The type of collagenase used is critical to the successful establishment of actively hemopoietic cultures (Ganassin and Bols, 1996). The PBS used must contain calcium and magnesium, as collagenase requires a magnesium co-factor for activity. We have had greatest success with collagenase A (Boehringer Mannheim, Dorval, Que).

1. To prepare a 1 mg mL^{-1} stock solution, add the contents of a 500 mg bottle to 500 mL of PBS.
2. To sterilize, prefilter the solution through a glass fiber filter, then filter with a 0.22 μm filter. The solution is viscous and could require an additional prefiltering step, using a larger pore filter, prior to the 0.22 μm filter.
3. Dispense in single-use aliquots, label and store at −20°C, where it is stable for 1 year.
4. Thaw the enzyme just before use. Collagenase should be stored below 0°C and should not be subjected to repeated freeze–thaw cycles.

Growth medium (100 mL)

A number of different basal media can be used to culture fish cells, but Leibovitz's L-15 (Leibovitz, 1963, 1977) offers the advantage of not requiring a CO$_2$ incubator. For hemopoietic cultures, the medium must be supplemented with a high concentration of fetal bovine serum. Only selected batches will support hemopoiesis. We have had the best results with FBS (catalog number 200-6140) from Canadian Life Technologies (Burlington, ON).

1. To prepare 100 mL of growth medium, mix 70 mL L-15 with L-glutamine (Gibco 11415-023), 1.4 mL penicillin/streptomycin (Gibco 15140-015) and 30 mL fetal bovine serum.

Initiation of spleen or kidney cultures

Long-term hemopoietic cultures can be initiated either from the pronephros (head or anterior kidney) or from the spleen of the rainbow trout (Ganassin *et al.*, 1996). In general, kidney cultures are more easily established and produce more progeny cells.

1. Sedate rainbow trout with 1 : 10 000 MS-222 or clove oil (Anderson *et al.*, 1997).
2. Remove blood by caudal puncture or by bleeding from the tail. Bleed fish as thoroughly as possible, to reduce the red blood cell volume of the kidney and spleen. Red blood cells will interfere with cell attachment.
3. For spleen cultures, swab the outside of the fish with 70% ethanol, make a ventral incision, and aseptically remove the spleen. Transfer the spleen into a 60 mm Petri dish containing 5 mL of initiation solution. Go to step 7.
4. To remove the head kidney, swab the outside of the fish with 70% ethanol. Make an incision from just ventral to the anus to past the operculum. Make a second cut straight down just dorsal to the anus, and a second incision parallel to the backbone, then cut the body wall to meet the first incision. This frees the intestine and all internal organs.
5. Lift all of the internal organs upwards toward the head of the fish, exposing the kidney. Cut through the retroperitoneum overlying the kidney, and move it aside.
6. Scoop the head kidney tissue into a 60 mm Petri dish containing 5 mL of initiation solution.
7. Mince the tissue into small pieces, approximately 1 mm^3 in diameter, using a sterilized scalpel and small scissors.
8. Add 5 mL of collagenase solution to the dish, to give a final concentration of 0.5 mg mL^{-1} collagenase, cover and place at 10°C for 10–12 h. The time may vary depending upon the batch of collagenase. The collagenase digestion step is not absolutely necessary for kidney cultures, but is essential for those from spleen. Kidney tissue can also be dissociated by pushing the kidney tissue through a 40 mesh screen directly into 5 mL of culture medium.
9. Remove from incubator, and pipette into a 15 mL centrifuge tube. At this point, you should have some single cells, but mostly small fragments of tissue. As a guideline, you should be able to draw

the suspension into a 10 mL serological pipette, but it is very important not to create a single cell suspension.
10. Centrifuge for 5 min at 300g to pellet cells and tissue fragments.
11. Discard supernatant, and add 6 mL of L-15 containing 30% FBS. Pipette up and down several times to dissociate the tissue, and distribute evenly into three 12.5 cm^2, two 25 cm^2, or one 75 cm^2 culture flask(s). 12.5 cm^2 flasks are available only from Falcon (Becton Dickinson Labware, Franklin Lakes, NJ, USA), but other sizes from other manufacturers are acceptable. We have not observed any difference for this application with the use of flasks specially treated for primary culture. Bring medium volume to 10 mL for 75 cm^2 flasks, and to 5 mL for smaller sizes.
12. Incubate culture flasks at 18–22°C for 2 weeks without disturbing.
13. Remove all non-adherent material, and rinse surface gently with culture medium. Add fresh L-15 with 30% FBS. This removes any cells that have not adhered to the surface.
14. Other authors have used a supplement of up to 5% rainbow trout, serum (RTS) for long-term hemopoietic cultures. Low (0.5–5%) RTS concentrations are effective culture supplements, but the toxicity of higher RTS concentrations to cultures of rainbow trout cells has often been described (Collodi and Barnes, 1990). If used, RTS should be heated to 46°C for 30 min to inactivate complement. Inactivation at 56°C, as used for mammalian sera, may compromise the growth-supporting ability of RTS (Yano, 1995).
15. Head kidney cultures will produce progeny cells more rapidly than spleen cultures, which can take several months to become established and productive. The production of progeny cells by kidney cultures is dispersed over the entire stromal layer (Figure 38.1A). For spleen cultures, monitor the flasks regularly, by observing with a phase contrast microscope, for the appearance of hemopoietic foci, which are areas of small, round, phase bright cells associated with the stromal layer (Figure 38.1B). As the foci increase in size and the cells become more mature, they will be released into the culture medium.
16. Harvest non-adherent progeny cells by removing half of the culture medium, and centrifuging to collect the cells. Replace with fresh medium.
17. During the period of active hemopoiesis, the stromal layer cells do not proliferate. As a culture declines, fewer progeny cells are produced. The decline of a spleen culture is often signaled by the appearance of islands of cells that proliferate rapidly and have an epithelial morphology. The decline of head kidney cultures is evident when the cells forming the stromal layer begin to detach from the culture surface.

Even if all reagents are ideal, hemopoietic cultures will not result from every fish from which cultures are initiated. The reasons for this are unclear, but it is also true of mouse long-term bone marrow culture, where there is much variation both within and between strains of mice (Spooncer et al., 1993).

Examination and quantification of cellular products

The number of cells produced with different culture supplements is one parameter that can be used to evaluate hemopoietic cultures. Cells can be counted with a hemocytometer, or using a Coulter Counter or other particle counter. Instructions for the use of a hemocytometer and a particle counter can be found in Freshney (1994), or in the series edited by Griffiths et al. (1994).

The types of cells produced by hemopoietic cultures can be identified in cytocentrifuge preparations by staining with a blood stain such as Wright–Giemsa, or by leukocyte enzyme cytochemistry. For example, rainbow trout neutrophils show myeloperoxidase activity while monocytes and macrophages are non-specific esterase positive.

Preparation of solutions

Wright–Giemsa staining solution (WG)

WG solution should be prepared fresh daily. To prepare 50 mL, add 1 mL Wright stain (Sigma WS16), and 3 mL Giemsa stain (Sigma GS500), to 46 mL distilled water.

Myeloperoxidase fixative solution (MF)

MF solution should be prepared fresh daily. To prepare 50 mL, mix 5 mL of 37% formaldehyde with 45 mL of 95% ethanol.

Daimobenzidine (DAB) staining solution

Prepare solution using Fast DAB tablets (Sigma D-4293) according to package directions.

Non-specific esterase activity assay (NSE) fixative and staining solutions

1. Prepare fixative by mixing 25 mL of citrate solution, 65 mL acetone and 8 mL of 37% formaldehyde.
2. Place in a glass bottle, cap tightly, and store in refrigerator.
3. Warm to room temperature prior to use.
4. Prepare staining solution by dissolving 10 mg α-naphthyl acetate in 250 µL of acetone, diluting to 20 mL with PBS, and adding 20 mg of Fast Blue (Sigma FBS-25).

Staining of cellular progeny

1. Remove non-adherent cells from a culture flask and place 300 mL of cell suspension directly into each chamber of a cytocentrifuge (Shandon Instruments). Centrifuge at 500 rpm for 10 min to deposit the cells on to microscope slides. Air dry. If a cytocentrifuge is not available, cells can be deposited to the slide as if preparing a blood smear.
2. To stain with Wright–Giemsa stain to distinguish general morphology, fix the slides for 5 min at room temperature in absolute methanol, air dry, and cover with WG staining solution for 20 min. Rinse with distilled water and examine immediately, or mount a coverslip with Permount (Fisher Scientific).
3. To stain for the leukocyte enzyme myeloperoxidase, which identifies neutrophils, fix for 30 s at room temperature in fixative solution. Wash in gently running tap water and air dry in the dark for 10 min. Incubate washed, fixed slides in prewarmed DAB solution for 30 min in the dark in a 37°C waterbath. Rinse slides in tap water, allow to air dry and counterstain with Acid Hematoxylin Solution for 10 min. Brown cytoplasmic granules indicate myeloperoxidase. Positive cells are neutrophils (Figure 38.1C).
4. To stain for non-specific esterase activity, fix freshly prepared smears or cytocentrifuged slides by immersing for 30 s in fixative. Drop the staining mixture on to the fixed slides, and incubate for 15 min at room temperature. Counterstain with Methylene Blue for 5 min. Rinse stained slides three times with deionized water, and mount a coverslip with glycerol/PBS (1:1). Positively stained cells are macrophages.

Flow cytometry and immunocytochemistry

These techniques, which can definitively identify leukocyte types, require antibodies that recognize fish-specific antigens, and these are relatively scarce. As monoclonal antibodies to fish leukocyte surface markers become more readily available, identification of specific cell types and stages will be greatly simplified. One useful antibody that is currently available is 1:14, developed by DeLuca et al. (1983), a monoclonal antibody to rainbow trout immunoglobulin M which is found primarily on the surface of B lymphocytes.

Use of 1:14 to stain the B lymphocyte population

These instructions, while specifically written for staining of rainbow trout lymphocytes with the monoclonal antibody 1.14, are broadly applicable to the staining of any cell surface antigen with any antibody. Stained cells can be examined by microscope, or by flow cytometer.

1. Remove the cell-containing supernatant from an actively hemopoietic kidney or spleen cell culture. Preparations of leukocytes freshly isolated from rainbow trout blood, spleen or kidney can be stained using the same method for comparison.
2. Wash cells three times in tissue culture medium by centrifugation (150g for 10 min at 5°C).
3. Resuspend cells in 10 mL of medium, and count the number of cells per milliliter using a hemocytometer.
4. Pipette three aliquots of 10^7 leukocytes of each cell suspension into microcentrifuge tubes and centrifuge each to obtain a pellet (150g for 10 min at 5°C).
5. Add 0.1 mL of the monoclonal antibody 1.14, diluted 1:100 in L-15 medium, to each sample.
6. Incubate for 30 min, on ice to prevent endocytosis.
7. Wash twice by centrifugation (150g for 10 min at 5°C), to remove unbound antibody.
8. Add 1 mL of a secondary antibody, generated against mouse immunoglobulin and fluorochrome conjugated (for example, a goat anti-mouse FITC conjugate) diluted 1:100 in L-15 medium, to each sample.

9. Incubate for 30 min, on ice to prevent endocytosis.
10. Wash twice by centrifugation (150g for 10 min at 5°C), to remove unbound antibody.
11. For microscopic examination, resuspend after the final centrifugation by adding 1 drop of glycerol/PBS, 1:1, to the dry cell pellet, and mix thoroughly.
12. Put one small drop of cell suspension on a slide, add a coverslip and ring with nail polish.
13. Examine under an incident light UV microscope, and identify B lymphocytes, which will be visible as green rings.
14. To calculate the percentage B lymphocytes, count the number of fluorescently stained lymphocytes in a total of 200 cells, and multiply this number by 0.5.
15. For flow cytometric examination, resuspend the cell pellet after step 10 in 1 mL of PBS or L-15.

Other identifying characteristics

Rainbow trout macrophages can be identified by the presence of the scavenger receptor for lipoproteins, which is indicated by their ability to take up 1,1'-dioctadecyl-3,3,3',3'-tetramethyl-indocarbocyanine perchlorate (DiI-Ac-LdL) (Molecular Probes, Eugene, OR).

The 'scavenger' receptor was so named because it is found on cells of the macrophage/monocyte lineage. The 'scavenger' cells and these cells accumulate large amounts of DiI-Ac-LdL, resulting in bright focal fluorescence (Goldstein et al., 1979). Endothelial cells (Voyta et al., 1984) also accumulate this compound, but this results in a more diffuse, less intense staining. DiI-Ac-LdL is an unstable compound, and must be stored refrigerated and used within 4–6 weeks.

1. Allow cells to adhere to chamber slides (Falcon), or stain in suspension in a microcentrifuge tube in growth medium.
2. Add $10\,\mu g\,mL^{-1}$ DiI-Ac-LdL at room temperature in normal media for 4 h.
3. For slide flasks, remove medium and wash the cells once with probe-free media for 10 min, rinse with PBS and fix with 10% buffered formalin for 5 min. Mount a coverslip with 10% PBS in glycerol and view with a standard rhodamine filter. For cells stained in suspension, deposit them to a slide using the cytocentrifuge.
4. Uptake of DiI-Ac-LdL is evident as bright red focal fluorescence.

Functional assays

Cells can also be identified and studied by their functional characteristics. For example, macrophages exhibit phagocytosis and respiratory burst activity, while neutrophils exhibit differential adhesion to extracellular matrix proteins (Ganassin et al., 1996). Instructions for observing phagocytosis are given here.

Phagocytosis assay

1. The particles for phagocytosis are carboxylated fluorescent latex beads $1.0\,\mu m$ in diameter (Polysciences Inc., Warrington, PA).
2. Prepare a suspension of beads by adding $2\,\mu L$ of the commercial bead preparation (2.5% solids latex) to 5 mL of growth medium.
3. Replace the regular growth medium in cultures to be studied with this preparation.
4. Rinse the culture four times with PBS to remove non-ingested beads, after various periods of incubation. Macrophages will take up significant numbers of beads within 15 min of addition, but cultures can be left for longer periods.
5. Observe the cells using an inverted fluorescent microscope with phase optics and by fluorescence (UV filter), to reveal the ingested beads (Figure 38.1D).

Discussion

The hemopoietic culture system may be used to study the interactions of the stromal cell layer with developing blood cells, and also provides a source of blood cells for other experimental purposes. An active hemopoietic culture will continuously produce blood cells for a period of up to six months. The influence of various additions to the basic culture medium, for example, toxicants, growth factors and cytokines, will provide information about the factors that perturb blood cell formation. Passaging or subcultivating (see below) of hemopoietic cultures can give rise to cell lines. RTS11 (Figure 38.1E), a monocyte–macrophage-like cell line (Ganassin and Bols, 1998), and RTS34st (Figure 38.1F), a spleen stromal cell line (Ganassin and Bols, 1999), are examples of cell lines that arose from primary cultures.

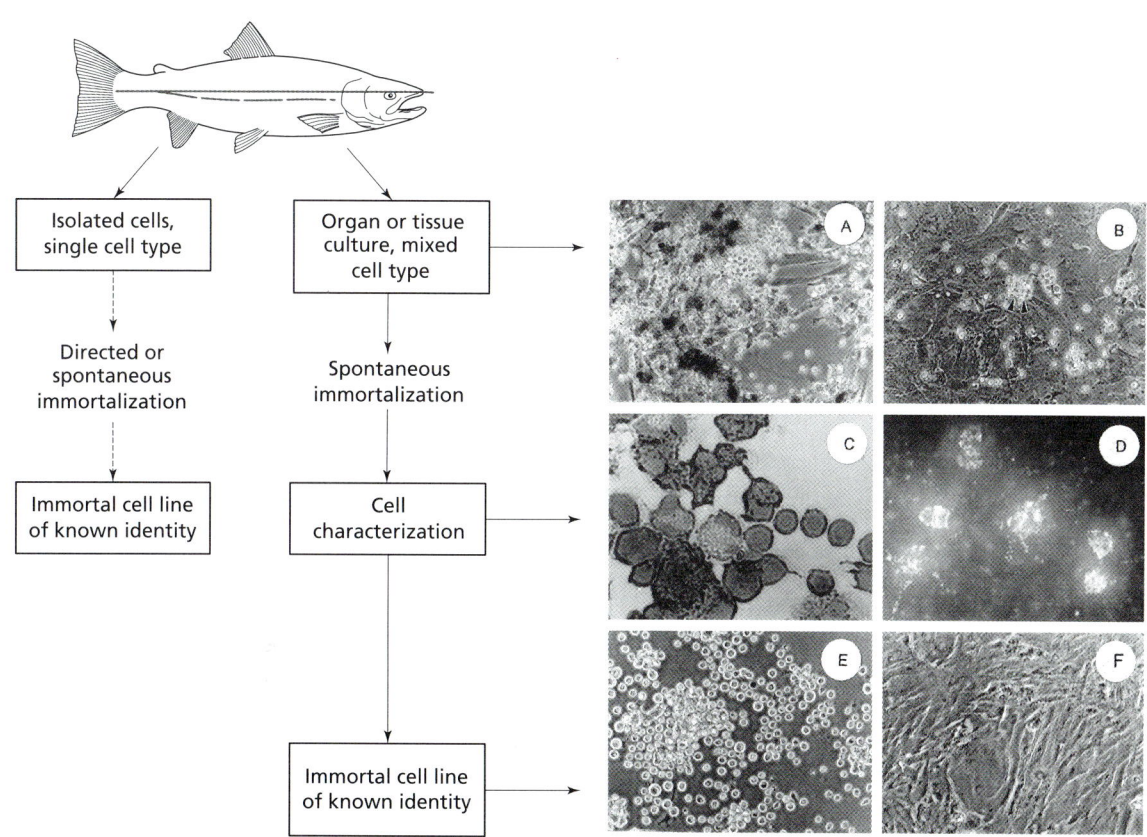

Figure 38.1 The progression from primary hemopoietic culture to cell lines. Cultures of multiple cell types from head kidney (A) or spleen (B) will produce progeny cells for periods of several months. In kidney cultures, the progeny cells develop scattered over the surface, while in spleen cultures, they develop in discrete hemopoietic foci (arrowheads). The progeny cells can be identified by their cytochemistry (C), or by their functional characteristics (D). The example shown in (C) is a myeloperoxidase stained preparation of head kidney progeny cells. Cells exhibiting dark granules are neutrophils. In (D), an example of phagocytosis of fluorescent latex beads by progeny cells from a kidney culture is shown. The cells that have taken up numerous beads can be identified as macrophages. From the original mixed cultures, cell lines consisting of one or several cell types often arise. Shown here is RTS11 (E), a monocyte/macrophage cell line (Ganassin and Bols, 1998), and RTS34st (F), a stromal cell line (Ganassin and Bols, 1999), both of which arose from long-term hemopoietic cultures of spleen. Cell lines very rarely arise from cultures started from a single isolated cell type.

Primary cultures may be initiated from other organs using a similar procedure (Ganassin and Bols, 1997). Changes in the basal medium and supplement, dissociation method, time and temperature of incubation, and culture medium used may be necessary for other applications. These primary cultures of whole organs are viable for a period of up to several months, and may spontaneously give rise to cell lines. These cell lines are usually epithelial-like or, less commonly, fibroblast-like in morphology, but require further characterization to identify the specific cell type they represent.

Care of fish cell lines

Several useful resources are available for people culturing fish cells. Instructions for set-up and maintenance of a tissue culture laboratory, aseptic technique, and routine care of cell lines are available in *A Manual of Cell and Tissue Culture* (Freshney, 1994). A series of volumes edited by Griffiths *et al.*, entitled *Cell and Tissue Culture: Laboratory Procedures*, is a good general reference, and includes a chapter specifically concerned with culture of cells from poikilotherms (Ganassin and Bols, 1997). Finally, the paper by Lannan (1994) provides excellent, specific instructions for the care and maintenance of fish cells in culture. As a starting point, instructions are given here for subculture of a continuous fish cell line with an adherent pattern of growth. For other circumstances, refer to the listed references.

Subculturing or passaging fish cells

1. **Note:** All components used must be sterile (filter sterilized or autoclaved), and manipulations must be carried out in a laminar flow hood. Instructions are for passaging a 75 cm^2 flask of cells. Reduce volumes accordingly for smaller flasks.

2. Maintain a separate stock of culture medium for each cell line used, to prevent cross-contamination between cell lines. In addition, to prevent the possibility of microbial contamination completely destroying a cell line, never manipulate the entire stock of a particular cell line on the same day.

3. Remove fresh medium, Versene solution (Canadian Life Technologies) and trypsin solution from refrigerator, and warm to room temperature before proceeding. Turn on the laminar flow hood, and wipe all surfaces with 70% ethanol solution.

4. Examine the flask to be passaged under the phase contrast microscope. Note the general health of the cells, and check for contamination before proceeding. Cells to be passaged should be healthy, contamination free and confluent (i.e. covering the bottom of the flask completely).

5. Under the laminar flow hood, aspirate off the old medium, and add 1.5 mL of Versene to the flask. Swirl it around gently, leave for a minute, and aspirate it off. Repeat.

6. Add 1.5 mL of trypsin solution to the flask, leave for 15–30 s, and aspirate it off. Replace the cap, and observe under the phase contrast microscope. The cells will begin to detach from the culture surface. The rate of detachment will depend on the particular cell line, and also on the age of the culture. Older cultures that have not been passaged for a long period of time are sometimes difficult to detach. Ideally, the cells will detach individually, and form a single cell suspension. The cells should not be left in trypsin too long, as it will begin to damage them. After 10 min, if the cells have not detached, use a sterile cell scraper to scrape them from the culture surface.

7. Add 10 mL of serum containing medium to the flask. (Trypsin inhibitors in the serum stop the action of the trypsin.) Pipette the medium up and down, directing the stream towards the bottom of the flask, to make sure that all cells are dislodged and resuspended in the medium.

8. Transfer the cell suspension to a sterile 15 mL centrifuge tube, and centrifuge in the table-top centrifuge at 200g for 5 min.

9. Aspirate the supernatant from the centrifuge tube, being careful not to aspirate the cell pellet. Leave a small amount of supernatant (0.25 mL) over the cell pellet.

10. Flick the centrifuge tube with your finger to resuspend the cells in the small volume of medium. Add 10 mL of fresh medium to the centrifuge tube, and transfer 5 mL to each of two culture flasks. Add 5 mL of medium to each. Rapidly growing cell lines can be split into more than two new flasks, e.g. 1 : 4.

11. Label each flask with the name of the cell line, the passage number and the date.

Examine flasks under the phase contrast microscope. Note whether cells have detached as single cells or in clumps, and whether the cell suspension has been evenly divided between the two flasks. Return the flasks to the incubator.

Cell viability assays

Assays of cell viability measure plasma membrane integrity and/or metabolic integrity (Shaw, 1994). They have two broad applications in fish cell culturing. One is as a general monitoring tool to evaluate the success of a particular culturing protocol. An example of this would be the use of trypan blue to monitor the preparation of fish hepatocytes (Cheng *et al.*, 1993; Flouriot *et al.*, 1993; Ostrander *et al.*, 1995). The other general use is in the discipline of *in vitro* toxicology. For example, we have been using cell viability assays to evaluate the cytotoxic and photocytotoxic potential of the US Environmental Protection Agency's (EPA) 16 priority polycyclic aromatic hydrocarbons (PAHs) to fish cells (Schirmer *et al.*, 1998a,b).

Although numerous assays of cell viability have been developed, those that use fluorometric indicator dyes are perhaps best. Firstly, more and more dyes are becoming available commercially to evaluate different cellular parameters, which ultimately are measures of the integrity of metabolism and the plasma membrane (Haugland, 1996). Secondly, the development of fluorometric multiwell plate readers has made the use of fluorometric dyes easy and rapid. The microwells conserve material resources by reducing the number of cells needed and increasing the number of replicates. The plate readers have the potential for high interlab reproducibility and can be coupled to computers to rapidly and easily manage data. Several different companies manufacture fluorometric multi-well plate readers.

We have developed cell viability assays with two fluorometric dyes: alamar Blue and 5-carboxyfluorescein diacetate acetoxymethyl ester (CFDAAM) (Schirmer et al., 1997). The assays were developed with the rainbow trout gill cell line, RTgill-W1 (Bols et al., 1994a). A gill cell line was used because the fish gill epithelium has been identified as one of the major target sites of photoinduced PAH toxicity (Oris and Giesy, 1985; Weinstein et al., 1997), but the assays can be used generally with fish cell lines. Here we describe the details of these assays, including adding the two dyes together, and illustrate the value of multiple assays by testing 9,10-phenanthrenequinone (PHEQ), which is the primary product from photomodifying phenanthrene with UV radiation (McConkey et al., 1997).

Cell culture

Cell viability assays are commonly performed in 48-well microplates with confluent cell cultures.

1. Grow cells to confluent monolayers in tissue culture plates in L-15 medium without phenol red (#21083-027 from Canadian Life Technologies, Burlington, ON) with 10% FBS.
2. For rainbow trout cell lines, add 5×10^4 cells in 500 μL of growth medium to 40 wells of a 48-well tissue culture plate. The remaining eight wells receive just growth medium and will be the blanks. If 48-well plates are not compatible with the available multiwell plate reader, the overall protocol can still be used but the volumes will have to be adjusted.
3. Allow the cells to grow for 3–4 days in the dark at 18–22°C to form a confluent cell monolayer. Growing of cells over the 3–4 day period allows for a more consistent cell density and better adherence of cells than newly initiated, dense cultures. Subsequently, results will be less variable once the test compound is added to the cultures (see below).

Exposure to test compound(s)

Conditions for exposing cell cultures to potential toxicants will vary depending on the nature of the compound(s) under study. For PAHs, details on how to dissolve them can be found in Schirmer et al. (1997, 1998a,b). Here we specifically describe the exposure medium and the addition of PAHs to this medium. The exposure medium is a novel one and might be generally useful for compounds that exert their toxicity through the generation of reactive oxygen species (ROS). However, the use of this medium requires that the addition of compounds in dimethylsulfoxide (DMSO) be done carefully as documented below.

1. The medium for exposing cells to PAHs is a modification of the basal medium, L-15. This modified medium, which is designated L-15/ex, contains the salts, galactose and pyruvate as in L-15 but no vitamins or amino acids. Salts, galactose and pyruvate can be prepared in cell culture grade water and mixed together as recommended by Leibovitz (1963, 1977) for the preparation of complete L-15. The advantage of L-15/ex is that it is suitable for both studies on the cytotoxicity of PAHs as well as on the photocytotoxicity of PAHs (Schirmer et al., 1997). For photocytotoxicity, L-15/ex eliminates the generation of toxic products that can arise from UV modification of components, such as vitamins and amino acids, in conventional growth media (Stoien and Wang, 1974). For cytotoxicity and for photocytotoxicity, L-15/ex eliminates the protection afforded to cells by media components, such as the amino acids tryptophan and histidine, against reactive oxygen species (ROS), which appear to be part of the cytotoxic mechanism.
2. Remove growth medium from plates by inverting over a catch basin. Drain plates further for a few seconds on a paper towel.
3. Rinse each culture well with 500 μL of L-15/ex, then add 500 μL of L-15/ex to each well.
4. Expose cells to a PAH by adding the PAH in 2.5 μL of DMSO to each well. This is done with a Nichiryo Model 800 positive displacement digital micro pipette (Nichiryo Co. Ltd, Tokyo, Japan) in a vertical laminar flow hood such as the Sterilgard Class II Type A/B3 (The Baker Company, Sanford, ME, USA). Proceed as described in step 5 on page 645 with two additional precautions. Because cells in L-15/ex are particularly sensitive to DMSO, tilt the plates before adding the PAH to increase the volume of medium above the cells as a protective layer. In addition, reduce the light level in the flow hood to avoid irradiation of cells in the presence of the PAH. Dose six wells (five with confluent cell monolayers and one without cells to serve as a blank) for each of seven concentrations of PAH and the DMSO control. Wrap plates in

parafilm to prevent evaporation and in aluminum to fully protect them from light.

5. Place plates in an incubator at 18–22°C for 2 h. This incubation time is appropriate for studies on the direct cytotoxicity of PAHs because it allows for the distribution of the PAH between the three compartments of a culture well: the cells, the medium and the tissue culture surface (Schirmer et al., 1997). As well, 2 h is likely too short an exposure period for the cells to metabolize the PAH to a toxic metabolite(s). Therefore, any cytotoxicity is attributed directly to the intact PAH.

6. Following the 2 h of exposure to the PAH, cell viability can be measured using the alamar Blue and CFDA-AM fluorescent indicator dyes.

Alamar Blue

Alamar Blue™ (Alamar Biosciences Inc., Sacramento, CA, USA) is reduced by cellular dehydrogenases, specifically targeting the mitochondrial electron transport chain (Alamar Biosciences Inc, 1997). Alamar Blue is water soluble in both its oxidized (blue, non-fluorescent) and its reduced (red, fluorescent) form, and can diffuse freely along the concentration gradient (Goegan et al., 1995). No extraction procedures are necessary and cells remain viable whether exposed to alamar Blue continuously or repeatedly (Ahmed et al., 1994; Alamar Biosciences Inc., 1997). Alamar Blue can be measured spectrophotometrically as the difference in absorbance at 570 nm (absorbance maximum of the reduced form) and 600 nm (absorbance maximum of the oxidized form), or fluorometrically at respective excitation and emission wavelengths of 560 and 590 nm. Pagé et al. (1993) found that the fluorescence assay is 10 times more sensitive than the colorimetric assay and is therefore preferable. The preparation of alamar Blue for cell viability measurements on fish cell cultures is outlined below.

1. Alamar Blue can be purchased from Immunocorp. Science Inc. (Montreal, PQ, Canada) as a ready to use solution in 25 mL and 100 mL quantities. When stored in the dark at 2–8°C and kept aseptically, alamar Blue can be used for at least one year.

2. Prepare a working solution of alamar Blue in L-15/ex (see step 1 on page 639). This will minimize changes of cell metabolism due to dye exposure. Using L-15/ex for the working solution also avoids potential interference between alamar Blue and extracellular proteins, such as has been observed with FBS (Goegan et al., 1995).

3. Although the manufacturer recommends 10% (v/v) alamar Blue in the working solution, 5% (v/v) is sufficient for treating fish cell cultures for up to 3 h. This might be due to the lower metabolic rate of fish cells as compared to mammalian cells.

4. Prepare enough of the 5% v/v alamar Blue working solution in L-15/ex so that the cell monolayer can be fully covered with the dye solution. For wells of a 48-well plate, 100 to 150 µL per well is sufficient.

5. Incubate the plates in the dark at 18–22°C for 30 min. Longer incubation times are also possible.

CFDA-AM

CFDA-AM is an esterase substrate that is converted by the non-specific esterases of living cells from a non-polar, non-fluorescent dye into a polar, fluorescent dye (Haugland, 1996). The substrate diffuses into cells rapidly whereas product diffuses out of cells slowly. The preparation of CFDA-AM for cell viability measurements on fish cell cultures is outlined below.

1. CFDA-AM can be purchased in crystalline form from Molecular Probes (Eugene, OR, USA). Add anhydrous DMSO (#27,685-5, Aldrich, Milwaukee, WI, USA) directly to the CFDA-AM vial to give a 4 mM CFDA-AM stock solution. Store desiccated at −20°C to avoid ester hydrolysis due to moisture.

2. Prepare a working solution of 4 µM CFDA-AM by diluting the CFDA-AM stock solution 1 : 1000 in L-15/ex (see step 1 on page 639). CFDA-AM should be added to cells in serum- and amino acid-free medium (Haugland, 1996).

3. Add 100 to 150 µL CFDA-AM working solution to each well of a 48-well plate and incubate the plates in the dark at 18–22°C. After 30 to 120 min terminate the assay in one of two ways: either step 4 or 5.

4. Aspirate the dye solution, rinse wells and add 100–150 µL L-15/ex. Read fluorescence as described in Fluorometric readings below. When fluorescent readings have been compared before and after removing the dye, little difference has been observed for culture wells that had been exposed to the dye for 30 min. However, for culture wells that had been exposed to the dye for 120 min, the fluorescence readings were reduced by 70% after removal of the CFDA-AM solution. These results

indicate that the fluorescent product carboxyfluorescein is retained entirely within the cells during a 30 min incubation but slowly diffuses out during the course of a longer incubation. Therefore, when the dye solution is removed prior to reading the culture wells, the fluorescence value is a measure of the ability of cells to both produce and retain carboxyfluorescein.

5. Without the removal of the CFDA-AM solution, the microplate is placed directly into a fluorometric reader (see below). When done this way, the fluorescence value is a measure of the ability of cells to produce carboxyfluorescein (i.e. total esterase activity) because both intra- and extracellular compartments of carboxyfluorescein are detected. This variation of the protocol is advantageous if the same culture well is going to be assayed again at a later date (see pages 641–642), because it reduces the number of rinsing steps. Each aspiration and rinsing risks mechanical dislodgement of some cells from the culture well. As a result, we favor reading the wells in the presence of the CFDA-AM solution.

When the two methods (step 4 vs. step 5) were compared in an experiment that was assayed with CFDA-AM for 120 min, the percent cell viability was the same despite approximately a 70% reduction in fluorescent readings in the step 4 protocol. The results are the same because both methods are assessing plasma membrane integrity. In the case of step 5, an intact plasma membrane is needed to maintain a cytoplasmic milieu to support esterase activity. In the case of step 4, an intact plasma membrane is needed to both support esterase activity and retain the fluorescent product.

Combining alamar Blue and CFDA-AM

We have found that because the fluorescent products of alamar Blue and CFDA-AM can be detected at different emission wavelengths, both dyes can be added together to perform the two assays in a single step (Schirmer et al., 1997). The advantage of this protocol is the saving of material and time as fluorescent readings are taken on the same culture wells. As a test of this approach, we prepared standard curves of fluorescent units vs. cell number for each dye individually and together. In all cases fluorescent readings increased linearly with the number of cells (Figure 38.2). The slopes of these lines did not differ significantly whether or not the dyes were used singly or together and r-squared values were similar in all cases. However, a slight decrease in slope was noted for the CFDA-AM standard curve (Figure 38.2). This could be due to alamar Blue absorbing radiation emitted by CFDA-AM. These results led to the two dyes being used together in many experiments (Schirmer et al., 1998a,b) using the following protocol.

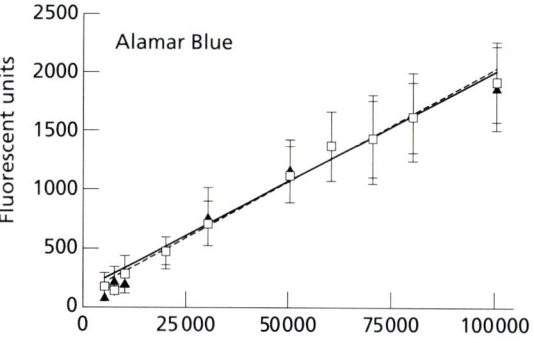

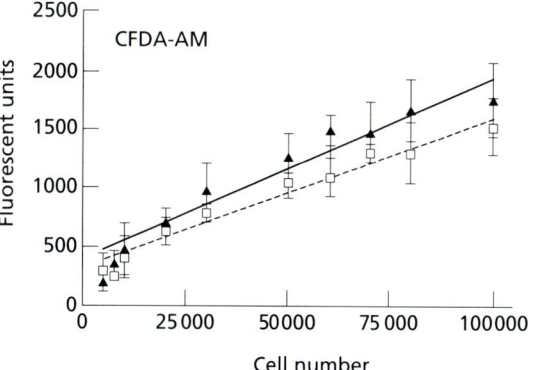

Figure 38.2 Relationship between number of viable cells and fluorescent units from assays with alamar Blue and CFDA-AM indicator dyes used singly or together. A range of 5000 to 100 000 cells per well of 48-well plates was plated and cells allowed to attach for 24 h. Wells were rinsed free of culture medium and exposed to either alamar Blue (top graph), CFDA-AM (bottom graph), or a mixture of both dyes (both graphs). After 3 h, both dyes were analyzed for in the single dye preparations (▲) as well as in the mixture (□). Five representative experiments were used to calculate mean and standard deviation of fluorescent units for each cell density. Regression analysis of the resulting fluorescent units versus cell number standard curves revealed r-squared values ranging from 0.81 to 0.86. No significant differences were observed for the slopes of the standard curves for each dye individually versus in the mixture of dyes ($\alpha = 0.05$).

1. Prepare the dye working solutions from stock solutions as above while cells are being incubated with the PAH.

2. Following the 2 h PAH exposure, remove microwell plates from the incubator and empty them by inversion as described in step 2 on page 639. The medium should be disposed of in a manner recommended by institutional health and safety guidelines.
3. Each culture well receives 150 µL of the alamar Blue and CFDA-AM prepared together in L-15/ex. The two are combined by adding 8 µL of the CFDA-AM stock solution and 0.4 mL of alamar Blue to 7.6 mL L-15/ex for each plate to yield a mixture of 5% (v/v) alamar Blue and 4 µM CFDA-AM.
4. Incubate the plates in the dark at 18–22°C for 30 min. Longer incubation times are also possible (see steps 3–5 on pages 640–641).
5. Obtain fluorescent readings as described below.
6. For measurements on the same plates at a later time point, aspirate the dye solutions and add 500 µL per well of complete L-15 medium with 10% FBS.
7. Twenty-four hours later, or whenever desirable, repeat the fluorometric assays again on the same plates.

Fluorometric readings

The fluorometric plate readers that we routinely use for the cell viability assays and for the EROD assay (see page 646) are the CytoFluor 2350 and CytoFluor 4000 by PerSeptive Biosystems (Framingham, MA, USA). Both scanners can be programmed to read the same plate repeatedly at similar or different filter settings. Using this option, the alamar Blue and the CFDA-AM indicator dyes can both be read on the same plates within minutes. As well, the plates can be read either with or without a lid. If repeated measurements are to be taken on the same plate, it is preferable to leave the lid on during measurement to ensure sterility. Readings are taken at excitation and emission wavelengths of 530 and 595 nm respectively for alamar Blue, and 485 and 530 nm for CFDA-AM. For the 2350, the alamar Blue assay is read at sensitivity 2 (longer exposure times) to 3 (shorter exposure times) and the CFDA-AM assay is read at sensitivity 3. For the 4000, both dyes are read at gain 50 for a 30 min exposure period.

Expressing results as percent of control

1. Import the CytoFluor files with the raw numbers of fluorescent units into a spreadsheet, such as Excel.
2. Calculate the mean fluorescent units and standard deviation for the five replicate cultures for each of the seven PAH concentrations and the DMSO control.
3. From these mean values of fluorescent units, subtract the blank values of fluorescent units in wells without cells. If fluorescent units of the blanks are the same for the DMSO control and the seven PAH concentrations without cells, subtract the mean blank value. If the different PAH concentrations yield different blank values, subtract the corresponding single blank value. Although we have not yet observed interference between the PAHs and the indicator dyes, it could arise from phenomena such as the autofluorescence of PAHs. At least two agents, methanol and hydrogen peroxide, have been found to increase the blank values in the CFDA-AM assay.
4. Express the average fluorescent units and standard deviations for each PAH concentration as a percent of the control value for DMSO alone.
5. Import the data, expressed as percent of control, and the PAH concentrations into a graphics program, such as SigmaPlot (Jandel Scientific, San Rafael, CA, USA).
6. Plot the percent of control as the ordinate, and the concentrations on a logarithmic scale as the abscissa (Figure 38.3).

Analysis of dose–response curves

PAH cytotoxicity dose–response curves take a sigmoid form (Figure 38.3). As for the dose–response curves of EROD activity (see page 648), the EC_{50} value is a useful parameter to compare the cytotoxic potency of PAHs. Firstly, EC_{50} values can be used to compare the effect of a PAH on cellular endpoints, as measured by the different fluorescent indicator dyes. Secondly, EC_{50} values can be used to compare the effect of different PAHs for either the same or different cellular endpoints. As described below (see page 648), EC_{50} values are calculated from an estimated best-fit sigmoid curve, using non-linear regression analysis. However, because cytotoxicity data are expressed on a 0–100% basis, the four parameter logistic equation simplifies to:

$$f(x) = \frac{100}{1 + e^{b(x-c)}}$$

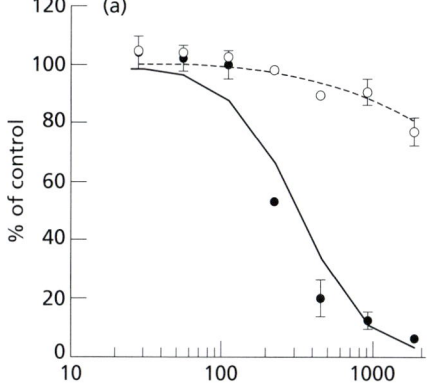

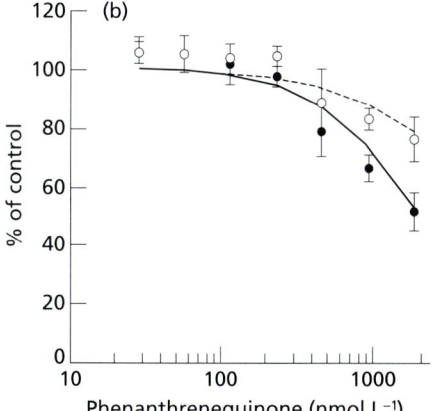

Figure 38.3 Cell viability in RTgill-W1 cultures immediately and 24 h after being exposed to increasing concentrations of 9,10-phenanthrenequinone (PHEQ). Confluent cultures were exposed to PHEQ in L-15/ex and kept in the dark for 2 h. Immediately afterwards (A) and 24 h later (B), effects on cells were assayed with a mixture of alamar Blue (●) and CFDA-AM (○). The results were expressed as a percentage of the readings in control cultures and analyzed as described in the text. One representative experiment is shown. Each data point is the mean of four culture wells with the vertical lines indicating the standard deviation.

where $f(x)$ is the % cell viability, x is the PAH concentration, b is the slope coefficient, and c is the EC_{50}.

For the example in Figure 38.3, EC_{50} values are obtained for the cytotoxicity of PHEQ as measured by the alamar Blue assay. These EC_{50} values are 320 nmol L^{-1} of PHEQ for the 2 h immediate cytotoxicity and 1982 nmol L^{-1} of PHEQ for the cytotoxicity 24 h later.

Interpretation of results

With either fluorescent dye, a reduction in fluorescent unit readings in treated cultures relative to the readings in control cultures indicates cytotoxicity. The use of multiple dyes has the potential of revealing the mechanism(s) behind the cytotoxicity. Thus understanding the cellular function being monitored with each indicator dye is critical to the interpretation of mechanisms.

The CFDA-AM assay appears to monitor impairment to the plasma membrane. However, behind this general statement are potentially two very different modes of action. When the assay is carried out as described here, a decrease in fluorescent readings directly measures a decline in cellular esterase activity. One possible cause for this diminution in activity is the loss of plasma membrane integrity. Subsequently, this would lead to the loss of the cytoplasmic milieu necessary to maintain maximal esterase activity. Support for this comes from the observation that freeze–thaw disrupted cells show little esterase activity (Persidsky and Baillie, 1977). In addition, we have observed that disrupting cell membranes in monolayer cultures with digitonin profoundly diminishes esterase activity as measured with CFDA-AM (unpublished observation). Therefore, the measurement of impaired plasma membrane integrity is the interpretation that we currently favor for the CFDA-AM assay. However, an alternative cause for the decline in esterase activity could be impaired uptake of the substrate, CFDA-AM, across the plasma membrane. This has some merit for the PAHs because they partition into lipophilic structures, and in doing so, could impair the entry of other molecules. Finally, detrimental changes in the cytoplasmic milieu caused by events other than changes in the integrity of the plasma membrane could cause a decline in esterase activity. One possible set of events would be impairment to the membrane integrity of certain organelles, such as lysosomes and endosomes. These events could lead specifically or non-specifically to a reduction in esterase activity by changing enzyme stability or catalytic activity through the inappropriate release into the cytosol of proteases, protons or ions. Support for this idea comes from the observation that some PAHs preferentially accumulate in lysosomes (Allison and Mallucci, 1964; Kocan et al., 1983). Another possible set of events would be the generation of products from the PAHs such as metabolites or lipid peroxides that either specifically or non-specifically reduce esterase activity. Clearly, all these possibilities should be kept in mind as this assay is applied to different classes of toxicants. The potential complexity in interpreting the results of this single assay points to the importance of performing multiple cytotoxicity assays as a route to understanding cytotoxic mechanisms.

Alamar Blue appears to indicate an impairment of mitochondrial function. This is because alamar Blue is a substrate of oxidoreductases, such as NADH–ubiquinone reductase (Alamar Bioscience Inc., 1997). However, although a decrease in fluorescent readings can be linked to impaired electron transport, by itself this decrease in fluorescent units cannot be used to distinguish several pathways that might lead to it. One possible cause is a direct action on electron transport. This could involve the inhibition of one or several oxidoreductases, or the passage of electrons to compounds that structurally mimic components of the electron transport chain. Another possible cause is a general disruption of the mitochondrial membrane, which also would disrupt electron transport. Thus, although interpreting the results of the alamar Blue assay appears to be less complex than for the CFDA-AM, multiple cytotoxicity assays are still important.

In our experience with PAHs, the alamar Blue and CFDA-AM assays give broadly similar results. This is true for the direct cytotoxicity and for the photocytotoxicity of PAHs. For some PAHs, alamar Blue is a little more sensitive than CFDA-AM, but the results are too close to conclude that these PAHs specifically act on mitochondria. Likely, PAHs almost simultaneously target all cellular membranes, including the mitochondrial membrane (Schirmer et al., 1998a,b). However, in a few situations alamar Blue is significantly more sensitive than CFDA-AM. An example of this occurs with PHEQ (Figure 38.3). After 2 h of exposure, the cytotoxicity of PHEQ was found to be at least 10-fold higher with alamar Blue than with CFDA-AM. In this case, the quinone appears to impair the mitochondrial electron transport chain well before it apparently alters plasma membrane integrity. One explanation for this is that PHEQ is reduced by NADH–ubiquinone oxidoreductase and undergoes redox cycling (Di Giulio et al., 1989), thereby diminishing the flow of electrons through the respiratory chain. Such specific actions would be missed if only one assay had been performed. On the other hand, 24 h after the first measurements, little difference was seen between alamar Blue and CFDA-AM when the assays were performed again on the same cultures (Figure 38.3). Because the PHEQ-containing medium had been removed prior to the 2 h assays, this result indicates that the effect of PHEQ on the mitochondrial electron transport chain was not permanent. Such specific behavior would have been missed if the assays were performed only once.

In the future, combining three or more fluorescent indicator dyes, each measuring a different cellular function, and taking repeated measurements on the same cultures could allow the rapid determination of cytotoxic mechanisms.

Induction of 7-ethoxyresorufin (EROD) activity

Induction of 7-ethoxyresorufin-o-deethylase (EROD) activity in piscine cell cultures can be used to study the actions of dioxin-like compounds in fish (Bols et al., 1997). Dioxin-like compounds encompass members of several chemical classes, including dioxins, furans, polychlorinated biphenyls (PCBs) and polycyclic aromatic hydrocarbons (PAHs). In most vertebrates, the toxicity of these compounds is mediated by the aryl-hydrocarbon receptor (AhR) signal-transduction pathway (Okey et al., 1994). One cellular event also mediated by the AhR is the induction of CYP1A (P4501A), which is a member of the hemoprotein superfamily, the cytochromes P-450. In fish, EROD has been shown to be a catalytic measure of cytochrome CYP1A1 (P4501A1) (Stegeman, 1989), and recently, also of CYP1A2 (P4501A2) (Gooneratne et al., 1997). In rainbow trout, these two proteins, which have been renamed CYP1A3 and CYP1A1 respectively (Nelson et al., 1996), appear to be so closely related that they are difficult to distinguish catalytically or physically (Gooneratne et al., 1997). Despite this complexity, the induction of EROD activity in fish cells by a compound is a simple and easy measure of the relative ability of a compound to activate the AhR signal-transduction pathway. This has been studied in primary hepatocyte cultures (Miller et al., 1993a,b; Cornell et al., 1995) and in cell lines (Clemons et al., 1994, 1996, 1997). Procedures for studying this in fish cell lines are presented here.

Stock solutions of inducers

Commonly, the standard inducing compound is 2,3,7,8-tetrachlorodibenzo-p-dioxin (TCDD) because this is the most toxic of the AhR-active compounds (Okey et al., 1994). PAHs constitute another group of EROD inducers and references to the preparation of PAH stock solutions have been given above. Cambridge Isotopes Laboratory (Andover, MA) supplies TCDD either dissolved in nonane solution (ED-901) or as a crystalline solid (ED-901-C). Consult your institutional health and safety office on the handling of TCDD. An advantage of using the nonane solution is that the TCDD can be transferred to different carrier solvents without having to weigh out TCDD. We compared dosing cells with TCDD in isooctane and in DMSO and found the results better with DMSO (Clemons et al., 1994). However, this issue is worthy of more investigation as differences in the potency of some compounds could be related to the carrier solvent (Tillitt et al., 1991; Clemons et al., 1996). Transfer of TCDD (ED-901) to DMSO is done by adding isooctane, evaporating the solvent in a stream of nitrogen gas, and reconstituting in DMSO in a fume hood using appropriate safety precautions.

Induction conditions

1. We most commonly use the rainbow trout liver cell line, RTL-W1 (Lee et al., 1993), but comparable EROD induction has been observed with other salmonid cell lines, including the rainbow trout gonadal cell line (RTG-2), available from the American Type Culture Collection (ATCC). However, EROD activity is not induced in all salmonid cell lines. One example of this is the Chinook Salmon Embryo cell line (CHSE-214) (Lee et al., 1993).
2. For the induction period, grow the cells in Leibovitz's L-15 without phenol red (21083-027) from Canadian Life Technologies (Burlington, ON, Canada), supplemented with 5% or 10% FBS, at 18 to 22°C.
3. Add 3×10^4 cells of the chosen cell line in 500 μL of growth medium to each well of a 48-well tissue culture plate. A number of companies, including Falcon and Costar (Corning Inc, Acton, MA, USA) manufacture 48-well plates. It is critical to make sure the plates are compatible with the multi-well plate reader. If your plate reader handles only 96-well plates, the overall protocol can still be used but the volumes will have to be reduced. The edge of the plate should be wrapped with parafilm to prevent evaporation.
4. Allow the cells to grow for 2 to 4 days to form a confluent monolayer of cells over each microwell, or measure induction in newly initiated or less dense cultures. However, dose–response curves are more similar from experiment to experiment if they are obtained from cultures of a similar cell density. A confluent monolayer provides a cell density that can be obtained most consistently.
5. Add the potential inducing agent in 2.5 μL of DMSO to each well. This dosing of microwells is done in a vertical laminar flow hood (see step 4 on page 639) with the Nichiryo micro pipette. Place the pipette tip near the center of the well, and dispel the fluid as close to the surface of the medium as possible, to allow the surface tension to help disperse the DMSO solution rapidly and evenly through the well. Failing to do this near the surface can result in a blob of DMSO falling directly on to the monolayer and causing the immediate death of all or part of the monolayer. Dose at least three wells for each concentration and at least three wells with only DMSO (control). For TCDD and salmonid cells, the dose typically ranges from 1 to 388 pmol L^{-1}. For a PAH, the dose typically ranges from 10 to 2000 nmol L^{-1}. Rewrap the edge of the plate in parafilm.
6. Allow induction to proceed for 24 to 72 h. A dose–response curve can be seen as early as 6 h after exposure to a potent inducer like TCDD but the response is not strong (Clemons et al., 1997). If a compound is an inducer, the strongest responses are consistently seen after exposure periods of between 24 and 72 h. For longer exposure periods, the strength of the response varies with the class of the inducing agent.
7. Terminate the induction period by removing the growth medium. If only a few plates are being handled at one time, aspirate off the medium from each well. If many plates are being handled, empty all the wells of a plate rapidly by inverting the plate over a foil lined catch basin. Dispose of the growth medium in a manner recommended by institutional health and safety guidelines.
8. Rinse each culture well with 100 μL of phosphate buffered saline (PBS, pH 7.5) prior to the assay of EROD activity. Culture wells can also be rinsed with other solutions, such as basal medium, before the EROD assay.

EROD assay

The assay of EROD activity in live cells as outlined here has arisen from a few modifications to methods published for the cells of other vertebrates. Assaying EROD activity in live cells was reported first in 1992. The fluorescent product, resorufin, was measured with a fluorescence spectrophotometer for human cells (Hammond and Strobel, 1992) and with a fluorescence microplate reader for rodent cells (Donato et al., 1992). Subsequently, a microplate reader was used to monitor EROD activity in intact cells of the topminnow cell line (PLHC-1) (Hahn et al., 1996). In these assays, cells appeared to generate sufficient endogenous NADPH to support the deethylation of exogenous 7-ethoxyresorufin. This is also true of salmonid cell lines.

For the salmonid cell lines, the reaction mixture is 0.825 µmol L^{-1} ethoxyresorufin (ER) in Dulbecco's Modified Eagle's Medium without phenol red (DMEM). Both ER (E-3763) and DMEM (5-5921) are available from Sigma, St Louis, MO, USA. The stock ER solution is made up in methanol to a concentration of 40 mg mL^{-1} and is stored in a dark bottle at room temperature.

1. Initiate the reaction by adding 250 µL of reaction mixture to each well with cells.
2. Incubate the microwell plate at room temperature on a shaker set at 100 rpm.
3. Allow the reaction to proceed for 15 min. In maximally induced cultures, the reaction is linear for at least 20 min; in submaximally induced or control cultures, the reaction is linear for up to 60 min (Figure 38.4).
4. Measure fluorescent units in each well with a fluorometric microplate reader.
5. To measure protein in each well go to step 3 on page 647.

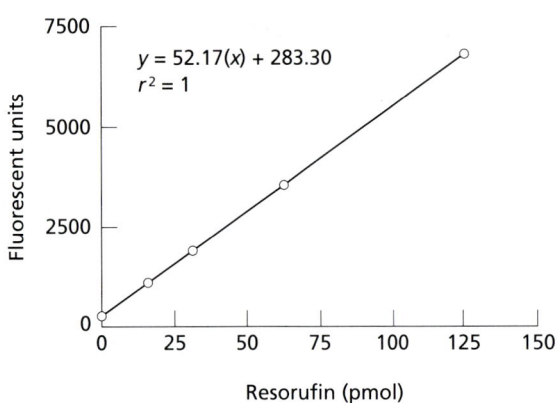

Figure 38.4 Resorufin concentration versus fluorescent units. A series of resorufin concentrations were prepared in 48-microwell plates as described in the text and a CytoFluor 2350 fluorometric plate reader used to measure fluorescent units. Each point represents the mean of three culture wells. Standard deviations are smaller than the symbols.

Fluorometric readings

The EROD assay is monitored with either the CytoFluor 2350 or CytoFluor 4000 fluorometric microplate reader. Both models can use plates in a 96-, 48-, 24-, 12- or 6-well format. We routinely use the 48-well format for the EROD assay, but the other formats are possible. For either machine, the assay is read at an excitation wavelength of 530 and emission of 595 nm. For the 2350, the sensitivity used is 4, while with the 4000, gain is set at 64. Turn on the machines for approximately 30 min prior to use. Remove lids before placing the plates into the machine. From entry into the reader to the export of the data to a personal computer in Excel files, the reading of a single plate takes approximately 1.5 min in the 2350 and 2.5 min in the 4000. The data appear as fluorescent units (FUs).

Calculation of resorufin produced

To convert FU readings to picomoles of product, a standard curve of resorufin amounts versus FUs is needed. Construct the curve as follows.

1. Prepare 10 mL of 1000 mmol L^{-1} resorufin (Sigma, R-3257) in methanol. Vortex vigorously for 10 min in order to completely dissolve the resorufin. Use the solution on the day of preparation and vortex immediately before the next step.
2. Mix 8 µL of this resorufin solution with 1992 µL of the reaction mixture (see above) (final concentration 4000 nmol L^{-1}).
3. The 4000 nmol L^{-1} resorufin solution is diluted serially six times in reaction mixture to yield seven solutions. After each dilution, the solution is vortexed vigorously. The four most dilute solutions are used and range from a high of 500 nmol L^{-1} to a low of 62.5 nmol L^{-1} resorufin.
4. For each of these four solutions, 250 µL is added to each of three wells in a 48-well microplate. This gives a range of 125 to 15.625 pmol of resorufin per well. In addition three wells receive 250 µL of reaction mixture without resorufin. All wells have confluent cell monolayers. The monolayers are established as described above except that they usually are not exposed to an inducer.
5. Scan the plate in the CytoFluor under the conditions described above. Calculate the mean FUs per well for each resorufin concentration.

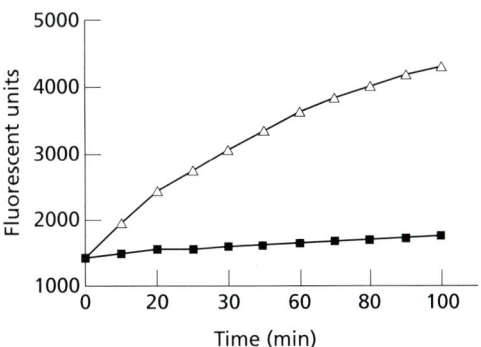

Figure 38.5 Fluorescent units versus time of EROD assay. Cultures of the rainbow trout liver cell line RTL-W1 were initiated in a 48-microwell plate with 80 000 cells per well and exposed to either 97.6 pmol L^{-1} 2,3,7,8-tetrachlorodibenzo-p-dioxin (TCDD, △) in dimethylsulfoxide (DMSO) or DMSO alone (control, ■). After a 24 h exposure, the culture medium was removed and replaced with 7-ethoxyresorufin in DMEM, which is the 7-ethoxyresorufin-o-deethylase (EROD) reaction mixture. Resorufin formation was followed by recording the fluorescent units (FU) in the same wells repetitively with time after beginning the reaction. A CytoFluor 2350 fluorometric plate reader was used to record FU as described in the text.

6. Plot mean FUs versus picomoles of resorufin per well (Figure 38.5). Use a linear regression to determine the slope and intercept, and use the following equation to calculate the amount of product, resorufin, formed in the EROD assay:

$$\text{Resorufin (pmol)} = \big(\text{FUs (EROD assay)} - y \text{ intercept} \times (\text{resorufin standard curve})\big) / \text{Slope (resorufin standard curve)}$$

Calculation of specific activity

Conventionally, enzymatic activity is expressed relative to the amount of protein. In many cases the results can just as well be expressed as resorufin formed per culture well or per number of cells. A confluent monolayer of rainbow trout cells in a 48-well plate after an induction period of 24 to 48 h usually contains about 8×10^4 cells. Whether the results are expressed per cell or per milligram of protein, the shape of the dose–response curve and the concentration eliciting 50% of the maximal response (EC$_{50}$) are the same. One potential advantage of expressing the results per milligram of protein is that it corrects for any loss of protein in cultures due to toxicity of the inducing agent. However, in our experience toxicity is rarely, if ever, observed at the doses required to elicit an EROD response. Cytotoxicity of the inducing treatment can be checked by the assays mentioned above. In addition, expressing the results per milligram of protein could be helpful in situations where the induction period is very long and the inducer might have had a cytostatic effect. If the results are to be expressed per milligram of protein, a microwell fluorometric assay for protein can be used to calculate the amount of protein per well (Lorenzen and Kennedy, 1993). This procedure uses fluorescamine and is outlined below.

1. Fluorescamine (20,165-0) is purchased from Aldrich Chemical Co. (Milwaukee, WI, USA) and a solution is prepared fresh daily at a concentration of 0.3 mg mL^{-1} in acetone. One 48-well plate requires 15 mL of fluorescamine solution.
2. Prepare a standard curve (fluorescent units versus concentration) of bovine serum albumin (BSA) in 48-well plates as described by Lorenzen and Kennedy (1993). Analyze the data by a linear regression and record the slope and intercept for use in step 8.
3. Measure protein in an experimental plate from step 4 on page 646 by removing the reaction mixture and rinsing the well. Rinse by adding and immediately removing 500 µL of PBS (pH 7.6) per well.
4. Add 250 µL of deionized water per well and let the plate sit at room temperature for 15 min before putting the plate into a −80°C freezer. Remove the plate as early as 30 min later and as long as months later and thaw the contents at room temperature to disrupt the cells.
5. After cell disruption, add 500 µL of phosphate buffer (7.49 g K$_2$HPO$_4$ and 0.925 g KH$_2$PO$_4$ in 1 litre of H$_2$O, pH 8.0) per well. Shake the plate for approximately 1 min on a Rotator V orbital shaker (American Dade, Miami, FL, USA) at 100 rpm and then add 250 µL of fluorescamine solution per well and continue shaking.
6. After 5 min of shaking, place the plate without the lid into the CytoFluor and read. The filter combination is excitation 360 nm and emission 460 nm. For the CytoFluor 2350 the sensitivity is set at 5, while for the CytoFluor 4000, gain is set at 55.
7. Both the fluorescamine solution and the plate are hazardous waste. The solution is placed into a jar for toxic organic solvents and the plate is treated as solid hazardous waste.

8. Use the slope and intercept from the BSA standard curve in step 2 to convert the fluorescent units from step 6 to micrograms of protein in the following equation:

$$\text{protein (μg) per well} = (\text{FU per well} - y \text{ intercept from BSA standard curve}) / \text{Slope from BSA standard curve}$$

9. Divide the protein content by 10^3 to express the value as mg/well, and use this value together with the amount of resorufin produced from step 6 on page 647 to calculate the specific EROD activity in the following equation:

$$\text{EROD activity} = \frac{\text{Resorufin (pmol) produced}}{\text{Protein (mg)} \times \text{Reaction time (min)}}$$

Expressing specific activity per milligram rather than per microgram of protein makes the results more comparable to the older literature where EROD induction was studied in much larger Petri dish cultures and milligrams of protein were present in a culture.

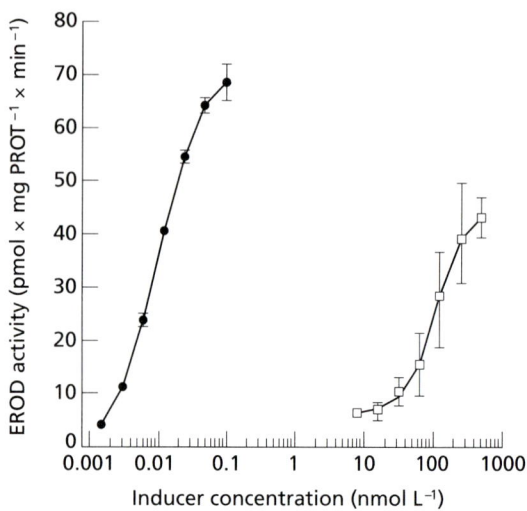

Figure 38.6 Induction of EROD activity versus concentration of TCDD or DMBA. Cultures of the rainbow trout liver cell line RTL-W1 in 48-microwell plates were exposed to 2,3,7,8-tetrachlorodibenzo-p-dioxin (TCDD, ●) or 7,12-dimethylbenz[a]anthracene (DMBA, □). After an induction period of 48 h, EROD activity was measured as described in the text. EC_{50} for TCDD was 10.12 pmol L^{-1}; EC_{50} for DMBA, 112.58 nmol L^{-1}.

For the example in Figure 38.6, the EC_{50}s for EROD induction in RTL-W1 by TCDD and DMBA are 10.12 pmol L^{-1} and 112.58 nmol L^{-1} respectively.

Analysis of dose–response curves

EROD dose–response curves take the form of a normal sigmoid (Figure 38.6). A useful parameter from these dose–response curves is the concentration eliciting 50% of the maximal response (EC_{50}). The EC_{50} is used to compare the potency of different inducers and the dioxin responsiveness of different cell lines. Several computer programs could be used to estimate a sigmoid curve that best fits the experimental data and from which the EC_{50} is calculated. We usually use the SigmaPlot non-linear curve fitting module (Jandel Scientific, San Rafael, CA, USA). The function is a four parameter logistic equation:

$$f(x) = \frac{a}{1 + e^{b(x-c)}} + d$$

where $f(x)$ is the EROD response, x is the inducer concentration, a is the range from the EROD minimum to the EROD maximum, b is the slope coefficient, c is the EC_{50}, and d is the minimum EROD response.

Calculation of induction potency

Expressing the inducing ability of a compound relative to TCDD has been referred to variously as the toxic equivalency factor (TEF), TCDD equivalency factor (TEF), induction potency and relative potency (REP). Usually the TEF is calculated by dividing the EC_{50} for TCDD by the EC_{50} for the compound of interest. From the EC_{50}s for TCDD and DMBA in the dose–response curves of Figure 38.6, the TEF for DMBA is 0.000 09. TEFs can be applied in the toxic equivalency concept to calculate the toxicity of complex mixtures of dioxin-like compounds, such as might be found in environmental samples (Eadon et al., 1986; Safe, 1994; Bols et al., 1997). When TEFs for some common dioxin-like compounds were derived from EROD induction in rainbow trout and rat liver cell lines, the potency of a few were substantially different in trout cells than in rat cells (Bols et al., 1997). An interesting future study would be to derive TEFs for these compounds in cell cultures from other fish species to see if the induction potencies of these compounds are generally similar in fish and different from mammals.

General discussion

Future cytotechnology research should improve the overall utility of fish cell cultures. Cytotechnology is the science of culturing cells from multicellular animals independent of the whole organism and is most developed for a few mammals and some insects. Progress with the cells of other animals can provide some direction for improving the culturing of piscine cells but features unique to fish can be anticipated. Certainly, improvements are still needed in generally developing, maintaining and monitoring fish cell cultures.

For the routine development of cell lines, techniques for directly immortalizing fish cells are needed. Up to this point, most fish cell lines have arisen through spontaneous immortalization (Clem et al., 1996; Ganassin and Bols, 1997), but success this way might vary considerably with cell type and species. Many directed immortalization procedures have been tried with mammalian cells and could be explored with piscine cells (Bols and Lee, 1994). However, of these, only oncogene transfer has been reported with fish cells (Tamai et al., 1993). Perhaps, the most interesting method to be explored in the future is the introduction into cells of telomerase (Bodnar et al., 1998). This approach will likely require more knowledge about senescence, telomeres and telomerase in fish cells. In addition, as this approach relies on transfection of the telomerase gene, the development of improved techniques for transfecting fish cells for this and other purposes also would be useful.

Improvements in growth media will likely improve the usefulness of fish cell cultures. Currently, the basal media that are employed have been optimized only for mammalian cells, but several observations suggest that the requirements for growth and for specific functions of fish cells might be slightly different. For example, salmonid cell lines can be grown in the complete absence of glutamine (Bols et al., 1994b), and goldfish pituitary cells are responsive to gonadotropin-releasing hormone in the basal medium 199 but not in L-15 (Chang and Jobin, 1994). The serum supplement, which is most commonly some type of mammalian serum, complicates the interpretation of some experiments by altering the composition of polyunsaturated fatty acids in cell membranes (Tocher et al., 1995). Eventually, serum should be replaced by specific growth factors to give completely defined media for piscine cells.

Progress in monitoring fish cell cultures for general cell parameters is advancing steadily but progress is slow for cell-specific parameters. For general parameters, fluorescent indicator dyes are continually being developed to assay a variety of cellular functions, such as the cell viability assays discussed in this chapter. The results of these assays can be measured rapidly and inexpensively in microwell cultures with a multiwell fluorometric plate reader, improving the utility of fish cell cultures in many disciplines but especially in toxicology. The monitoring of specific cell types and their functions in culture has been more difficult because of the paucity of antibodies to fish proteins. In some cases, commercially available antibodies to mammalian polypeptides have been used successfully on fish material. For example, antibodies to mammalian cytokeratins have been used to reveal the histological distribution of cytokeratins in rainbow trout (Markl and Franke, 1988). However, in this case, cytokeratin expression is different in fish than in mammals and cannot easily be used to identify epithelial cell types in teleosts (Hermann et al., 1996; Groff et al., 1997). As more fish genes are cloned and antibodies are raised to their protein products, gene expression by specific cell types will be more easily studied in fish cell cultures.

References

Ahmed, S.A., Gogal, R.M., Jr. and Walsh, J.E. (1994). *J. Immunol. Meth.* **170**, 211–224.

Alamar Biosciences Inc. (1997). User Instructions.

Allison, A.C. and Mallucci, L. (1964). *Nature* **203**, 1024–1027.

Anderson, W.G., McKinley, R.S. and Colavecchia, M. (1997). *N. Am. J. Fish Manage.* **17**, 301–307.

Babich, H. and Borenfreund, E. (1991). *Toxicol. in Vitro* **5**, 91–100.

Bodnar, A.G., Ouellette, M., Frolkis, M., Holt, S.E., Chiu, C.P., Morin, G.B., Harley, C.B., Shay, J.W., Lichtsteiner, S. and Wright, W.E. (1998). *Science* **279**, 349–352.

Bols, N.C. (1991). *Biotech. Adv.* **9**, 31–49.

Bols, N.C. and Lee, L.E.J. (1991). *Cytotechnology* **6**, 163–187.

Bols, N.C. and Lee, L.E.J. (1994). In *Biochemistry and Molecular Biology of Fishes*, vol. 3 (eds

Bols, N.C., Mosser, D.D. and Steels, G.B. (1992). *Comp. Biochem. Physiol.* **103A**, 1–14.

Bols, N.C., Barlian, A., Chirino-Trejo, S.J., Caldwell, S.J., Goegan, P. and Lee, L.E.J. (1994a). *J. Fish Dis.* **17**, 601–611.

Bols, N.C., Ganassin, R.C., Tom, D.J. and Lee, L.E.J. (1994b). *Cytotechnology* **16**, 159–166.

Bols, N.C., Whyte, J.J., Clemons, J.H., Tom, D.J., van den Heuvel, M. and Dixon, G.D. (1997). In *Ecotoxicology: Responses, Biomarkers and Risk Assessment* (eds J.T. Zelikoff, J. Shepers and J. Lynch), pp. 329–350. SOS Publications, Fair Haven, New Jersey.

Chang, J.P. and Jobin, R.M. (1994). In *Biochemistry and Molecular Biology of Fishes*, vol. 3 (eds P.W. Hochachka and T.P. Momemsen) vol. 3, pp. 205–213. Elsevier, Amsterdam.

Chevrette, M., Joly, L., Tellis, P. and Ekker, P. (1997). *Biochem. Cell Biol.* **75**, 641–649.

Cheng, L.-L., Bowser, P.R. and Spitspergen, J.M. (1993). *J. Aquat. Anim. Hlth.* **5**, 119–126.

Chilmonczyk, S., Voccia, I., Tarazona, J.V. and Monge, D. (1997). In *Ecotoxicology: Responses, Biomarkers and Risk Assessment* (eds J.T. Zelikoff, J. Shepers, and J. Lynch), pp. 171–184. SOS Publications, Fair Haven, New Jersey.

Clem, L.W., Bly, J.E., Wilson, M., Chincar, V.G., Stuge, T., Barker, K., Luft, C., Rycyzyn, M., Hogan, R.J., van Lopik, T. and Miller, N.W. (1996). *Vet. Immunol. Immunopathol.* **54**, 137–144.

Clemons, J.H., van den Heuvel, M.R., Stegeman, J.J., Dixon, D.G. and Bols, N.C. (1994). *Can. J. Fish. Aquat. Sci.* **51**, 1577–1584.

Clemons, J.H., Lee, L.E.J., Myers, C.R., Dixon, D.G. and Bols, N.C. (1996). *Can. J. Fish. Aquat. Sci.* **53**, 1177–1185.

Clemons, J.H., Dixon, D.G. and Bols, N.C. (1997). *Chemosphere* **34**, 1105–1119.

Collodi, P. and Barnes, D.W. (1990). *Proc. Natl. Acad. Sci. USA* **87**, 3498–3502.

Collodi, P., Kamei, Y., Sharpes, A., Weber, D. and Barnes, D. (1992). *Mol. Mar. Biol. Biotechnol.* **1**, 257–265.

Cornell, N.W., Hahn, M.E. and Martin, H.A. (1995). *Biol. Bull.* **189**, 227–228.

DeLuca, D., Wilson, M. and Warr, G.W. (1983). *Eur. J. Immunol.* **13**, 546–551.

Dexter, T.M. and Testa, N.G. (1976). In *Methods in Cell Biology*, vol. XIV (ed D. M. Prescott), pp. 387–405. Academic Press, New York.

Diago, M.L., Lopez-Fierro, M.P., Razquin, B. and Villena, A. (1993). *Exp. Hematol.* **21**, 1277–1287.

Di Giulio, R.T., Washburn, P.C., Wenning, R.J., Winston, G.W. and Jewell, C.S. (1989). *Environ. Toxicol. Chem.* **8**, 1103–1123.

Donato, M.T., Castell, J.V. and Gomez-Lechon, M.J. (1992). *J. Tiss. Cult. Meth.* **14**, 153–158.

Eadon, G., Kaminsky, L., Silkworth, J., Aldous, K., Hilker, D., O'Keefe, P., Smith, R., Gierthy, J., Hawley, J., Kim, N. and DeCaprio, A. (1986). *Environ. Hlth. Perspect.* **70**, 221–227.

Flouriot, G., Vaillant, C., Salbert, G., Pelissero, C., Guiraud, J.M. and Valotaire, Y. (1993). *J. Cell Sci.* **105**, 407–416.

Freshney, R.I. (1994). *Culture of Animal Cells, A Manual of Basic Technique*, 3rd edn. Wiley-Liss, New York.

Fryer, J.L. and Lannan, C.N. (1994). *J. Tiss. Cult. Meth.* **16**, 87–94.

Ganassin, R.C. and Bols, N.C. (1996). *Fish Shell. Immunol.* **6**, 17–34.

Ganassin, R.C. and Bols, N.C. (1997). In *Cell and Tissue Culture: Laboratory Procedures* (eds J.B. Griffiths, A. Doyle and D.G. Newell), pp. 23:A1.1–23A:1.9, John Wiley, Chichester, UK.

Ganassin, R.C. and Bols, N.C. (1998). *Fish Shell. Immunol.* **8**, 457–476.

Ganassin, R.C. and Bols, N.C. (1999). *In Vitro Cell. Dev. Biol.* **35**, 80–86.

Ganassin, R.C., Brubacher, J.L. and Bols, N.C. (1996). In *Modulators of Immune Responses; The Evolutionary Trail* (eds J.S. Stolen, T.C. Fletcher, C.J. Bayne, C.J. Secombes, J.T. Zelikoff, L.E. Twerdok and D.P. Anderson), pp 267–279. SOS Publications, Fair Haven, New Jersey.

Garces, L.H., Larenas, J.J., Smith, P.A., Sanddino, S., Lannan, C.N. and Fryer, J.L. (1991). *Dis. Aquat. Organ.* **11**, 93–97.

Goegan, P., Johnson, G. and Vincent, R. (1995). *Toxicol. in Vitro* **9**, 257–266.

Goldstein, J.L., Ho, Y.K., Basu, S.K. and Brown, M.S. (1979). *Proc. Natl. Acad. Sci. USA* **76**, 335–337.

Gooneratne, R., Miranda, C.L., Henderson, M.C. and Buhler, D.R. (1997). *Xenobiotica* **27**, 175–187.

Griffiths, J.B., Doyle, A. and Newell, D.J. (eds) (1994). *Cell and Tissue Culture: Laboratory Procedures*. John Wiley, Chichester, UK.

Groff, J.M., Naydan, D.K., Higgins, R.J. and Zinkl, J.G. (1997). *Cell Tissue Res.* **287**, 375–384.

Hahn, M.E., Woodward, B.L., Stegeman, J.J. and Kennedy, S.W. (1996). *Environ. Toxicol. Chem.* **15**, 582–591.

Hammond, D.K. and Strobel, H.W. (1992). *Toxicol. in Vitro* **7**, 41–46.

Haugland, R.P. (1996). In *Handbook of Fluorescent Probes and Research Chemicals*, pp. 365–398, 549. Molecular Probes, Inc., Eugene, Oregon.

Hermann, H., Munick, M.D., Brette, M., Fouquet, B. and Markl, J. (1996). *J. Cell Sci.* **109**, 569–578.

Hong, Y., Winkler, C. and Schartl, M. (1996). *Mech. Dev.* **60**, 33–44.

Hong, Y., Winkler, C. and Schartl, M. (1998). *Proc. Natl. Acad. Sci. USA* **95**, 3679–3684.

Jeserich, G. and Stratmann, A. (1992). *Brain Res. Dev. Brain Res.* **67**, 27–35.

Kocan, R.M., Chi, E.Y., Eriksen, N., Benditt, E.P. and Landolt, M.L. (1983). *Environ. Mutagen.* **5**, 643–656.

Koumans, J.T.M., Akster, H.A., Booms, R.G.H. and Osse, J.W.M. (1993). *Differentiation* **53**, 1–6.

Lannan, C.N. (1994). *J. Tiss. Cult. Meth.* **16**, 95–98.

Lee, L.E.J., Clemons, J.H., Bechtel, D.G., Caldwell, S.J., Han, K.-B., Pasitschniak-Arts, M., Mosser, D.D. and Bols, N.C. (1993). *Cell Biol. Toxicol.* **9**, 279–294.

Leibovitz, A. (1963). *Am. J. Hyg.* **78**, 173–180.

Leibovitz, A. (1977). *TCA Manual* **3**, 557–559.

Leong, J.C., Anderson, E., Bootland, L.M., Chiou, P.W., Johnson, M., Kim, C., Mourich, D. and Trobridge, G. (1997). *Dev. Biol. Stand.* **90**, 267–277.

Lorenzen, A. and Kennedy, S.A. (1993). *Anal. Biochem.* **214**, 346–348.

Malins, D.C. and Ostrander, G.K. (1994). *Aquatic Toxicology: Molecular, Biochemical and Cellular Perspectives.* CRC Press, Boca Raton, Florida.

Markl, J. and Franke, W.W. (1988). *Differentiation* **39**, 97–122.

Matsumoto, J., Wada, K. and Akiyama, T. (1989). *J. Invest. Dermatol.* **92**, 2555–2560.

McConkey, B.J., Duxbury, C.L., Dixon, D.G. and Greenberg, B.M. (1997). *Environ. Toxicol. Chem.* **16**, 892–899.

Miller, M.R., Saito, N., Blair, J.B. and Hinton, D.E. (1993a). *Exp. Mol. Pathol.* **58**, 127–138.

Miller, M.R., Wentz, E., Blair, J.B., Pack, D. and Hinton, D.E. (1993b). *Exp. Mol. Pathol.* **58**, 114–126.

Mitani, H., Uchida, N. and Shima, A. (1996). *Photochem. Photobiol.* **64**, 943–948.

Moav, B., Liu, Z., Groll, Y. and Hackett, P.B. (1992). *Mol. Mar. Biol. Biotechnol.* **1**, 338–345.

Montero, M., Le Belle, N., Vidal, B. and Dufour, S. (1996). *Gen. Comp. Endocrinol.* **104**, 103–115.

Nelson, D.R., Koymans, L., Kamataki, T., Stegeman, J.J., Feyereisen, R., Waxman, D.J., Waterman, M.R., Gotoh, O., Coon, M.J., Estabrook, R.W., Gunsalus, I.C. and Nebert, D.W. (1996). *Pharmacogenetics* **6**, 1–42.

Neumann, N.F., Fagan, D. and Belosevic, M. (1995). *Dev. Comp. Immunol.* **19**, 473–482.

Okey, A.B., Riddick, D.S. and Harper, A. (1994). *Trends Pharmacol. Sci.* **15**, 226–232.

Oris, J.T. and Giesy, J.P. (1985). *Aquat. Toxicol.* **6**, 133–146.

Ostrander, G.K., Blair, J.B., Stark, B.A., Marley, G.M., Bales, W.D., Veltri, R.W., Hinton, D.E., Okihiro, M., Ortego, L.S. and Hawkins, W.E. (1995). *In Vitro Cell. Dev. Biol. Anim.* **31**, 367–378.

Pagé, B., Pagé, M. and Noël, C. (1993). *Int. J. Oncol.* **3**, 473–476.

Pelster, B. and Pott, L. (1996). *Am. J. Physiol.* **270**, R578–R584.

Persidsky, M.D. and Baillie, G.S. (1977). *Cryobiology* **14**, 322–331.

Safe, S.H. (1994). *Crit. Rev. Toxicol.* **24**, 87–149.

Saiki, A., Tamura, M., Matsumoto, M., Katowgi, J., Watanabe, A. and Onitake, K. (1997). *Dev. Growth Differ.* **39**, 337–344.

Schaeffer, W.I. (1990). *In Vitro Cell. Dev. Biol.* **26**, 97–101.

Schirmer, K., Chan, A.G.J., Greenberg, B.M., Dixon, D.G. and Bols, N.C. (1997). *Toxicol. in Vitro* **11**, 107–119.

Schirmer, K., Chan, A.G.J., Greenberg, B.M., Dixon, D.G. and Bols, N.C. (1998a). *Toxicology* **127**, 129–141.

Schirmer, K., Dixon, D.G., Greenberg, B.M. and Bols, N.C. (1998b). *Toxicology* **127**, 143–155.

Schwalb, J.M., Boulis, N.M., Gu, M.F., Winickoff, J., Jackson, P.S., Irwin, N. and Bernowitz, L.I. (1995). *J. Neurosci.* **15**, 5514–5525.

Shaw, A.J. (1994). *ATLA* **22**, 124–126.

Siegl, E., Albrecht, S. and Ludtke, B. (1993). *Comp. Haematol Int.* **3**, 168–173.

Spooncer, E., Eliason, J. and Dexter, T.M. (1993). In *Haemopoiesis: A Practical Approach.* (eds N.G. Testa and G. Molineux.), pp. 55–73. IRL Press, Oxford.

Stegeman, J.J. (1989). *Xenobiotica* **19**, 1093–1110.

Stoien, J.D. and Wang, R.J. (1974). *Proc. Natl. Acad. Sci. USA* **71**, 3961–3965.

Tamai, T., Shirahata, S., Sato, N., Kimura, S., Nonaka, M. and Murakami, H. (1993). *Cytotechnology* **11**, 121–131.

Tillitt, D.E., Giesy, J.P. and Ankley, G.T. (1991). *Environ. Sci. Technol.* **25**, 87–92.

Tocher, D.R., Dick, J.R. and Sargent, J.R. (1995). *Prostaglandins Leukot. Essent. Fatty Acids* **53**, 365–375.

Voyta, J.C., Via, D.P., Butterfield, C.E. and Zetter, B.R. (1984). *J. Cell Biol.* **99**, 2034–2040.

Weinstein, J.E., Oris, J.T. and Taylor, D.H. (1997). *Aquat. Toxicol.* **39**, 1–22.

Wolf, K. and Ahne, W. (1982). *Adv. Cell Culture* **2**, 305–328.

Wolf, K. and Rumsey, G. (1985). *J. Appl. Ichthyol.* **1**, 131–138.

Wongtavatchai, J., Conrad, P.A. and Hedrick, R.P. (1994). *J. Tiss. Cult. Meth.* **16**, 125–132.

Wood, C. and Part, P. (1997). *J. Exp. Biol.* **200**, 1047–1059.

Yano, T. (1995). *Fish Pathol.* **30**, 151–158.

Glossary

Terms defined in the glossary are emboldened in the main text.

α and β cells: Various examples (e.g. types of chloride cells, cells found on Islets of Langherhans in the pancreas and involved in glucose/glycogen regulation)

β-LPH: β-Lipotropin; lacks hormonal activity in fish, but produces β-MSH and β-endorphin upon further proteolytic processing

17α,20β-DP: 17α,20β-Dihydroxy-4-pregnen-3-one or MIH

20α-HSD: 20α-Hydroxysteroid dehydrogenase; cyprinid equivalent to 20β-HSD

20β-HSD: 20β-Hydroxysteroid dehydrogenase; enzyme which produces MIH from 17α-progesterone

accessory annuli: Rings or encircling structures

ACE: Angiotensin-converting enzyme; converts Ang I into Ang II

achondral: Dermal or membrane bones having no topographical relationship with cartilage

ACTH: Adrenocorticotrophic hormone

activin: TGF beta superfamily growth and differentiation factor associated with somatic and germ cell differentiation

adenylyl cyclase: Also called adenylate cyclase; enzyme which catalyzes the formation of the second messenger, cAMP, from ATP

agnathan: Jawless fish (e.g. lamprey eels and hagfishes)

alevins: Larvae stage of salmonid fishes

alloenzyme: Regulatory enzymes whose catalytic activity is modulated by binding of a metabolite at a site other than the catalytic site

allograph: Graft of tissue between individuals of the same species but different genotype

allometric: Measurement of the relationship between two varying dimensions, i.e., shape and size

allometric analysis: Process commonly used in systematic analysis

amacrine: Having no long processes

amacrine cells: Retinal neurons without axons

amelanotic: Without melanin

ametropia: Discrepancy between size and optical resolution

anabantid: Fishes (e.g. climbing perches and gouramies) belonging to the suborder Anabantoidei that are distinguished by a labyrinth organ

anadromous: Fishes that spend most of their lives in marine waters and return to freshwater to spawn (e.g. salmon)

analgesia: Absence of pain sensitivity

anaplasia: Dedifferentiation with loss of normal cellular relationships

ANF: Atrial natriuretic factor; mammalian hormone released from the atria

Ang I/Ang II: Angiotensin I/angiotensin II; active peptide hormone products of angiotensinogen

ankyloses: Fusion of bones or joints

anlage: Primordium

anlagen: Segmented region of kidney, lacking in caudal region of kidney of fishes
ANP: Atrial natriuretic peptide; analogous to mammalian ANF that is released from fish atria
aponeurosis: Flat, ribbon-like tendinous expansion
arterious: Pertaining to the arteries, arterial
asepsis: Condition without possibility of infection
AT: Angiotensin receptor; there are two well-characterized mammalian subtypes, AT_1 and AT_2
atheriniform: Order of relatively advanced fishes characterized by two dorsal fins (first with true spine)
ATP: Adenosine $5'$-triphosphate
atresia: Degenerative process by which ovarian follicles lose their integrity and are eliminated prior to ovulation
atretogenic factors: Hormones and other cues involved in the regulation and control of apoptotic cell death in ovarian follicles of higher vertebrates
atrial systole: Contraction of the artria propelling blood into the ventricles
atriopeptin: Synonymous with ANF or ANP
autotrophs: Self nourishing microorganisms that can build their organic constituents from carbon dioxide and inorganic salts
axenic: Refers to pure cultures of microorganisms
azurophilic: Microorganisms or cells that stain well with blue aniline dyes
bacterins: A bacterial vaccine
barbels: Extension of skin that serves as accessory feeding structure and contains sensory organs
basolateral membrane: Pertaining to a membrane lining the base and lateral surfaces of said structure
basophilic: Term applied to tissues, cells, or microorganisms that stain readily with basic dyes
bathypelagic: Deep water zone that is not penetrated by sunlight (ca. 1000–3000 m)
benthic: Referring to a region of a body of water at or near the bottom
benthivores: Organisms that feed exclusively on bottom dwelling organisms in aquatic ecosystems
benthivory: Life-history trait of feeding on bottom dwelling organisms in aquatic ecosystems
betadine: Providone-iodine; a disinfectant
bioaccumulated: Typically refers to accumulation of toxicant(s) in tissues over time
biogenic: Refers to compounds that alter growth rates of organisms
biomass: Refers to the entire assemblage of living organisms in a particular location
biotransformed: Compounds that have been modified by various enzyme systems *in vivo*
Brockmann bodies: Found exocrine pancreatic tissues in fishes and involved in various secretions including glucagon-like peptides
BTUs: British Thermal Units
C-reactive protein: (CRP) recognizes and binds to molecular groups found on a variety of bacteria and fungi
C_{18}: Family of steroid hormones that have 18 carbons and includes the estrogens
C_{19}: Family of steroid hormones that have 19 carbons and includes the androgens
C_{21}: Family of steroid hormones that have 21 carbons and includes the progestins and cortisol
cAMP: Cyclic $3',5'$-adenosine monophosphate or cyclic AMP
capacitance: Property of a circuit element that allows for the storage of a charge; blood vessels (i.e. veins) serving as reservoirs
carcinogenesis: The process of development of cancer
carotenoid vesicles: Type of chromatosome
catadromous: Fishes that spend much of their lives in a freshwater environment and return to sea to spawn (e.g. some eels)
caveolae: A small pit or depression
CCK: Cholecystokinin
cDNA: DNA coding for a gene that has been obtained from reverse transcription of mRNA and therefore lacks introns; commonly used to produce large quantities of the 'cloned' gene product
ceca: Pleural of cecum
cephalic: Pertaining to the head
ceratioid: Refers to type of anglerfish (Lophiiformes) in which males are obligate parasites on females
cGMP: Cyclic $3',5'$-guanosine monophosphate or cyclic GMP

cGnRH: The forms of GnRH originally isolated from chicken, although certain fish species such as salmonids also express these forms
CGRP: Calcitonin gene-related peptide which is produced by processing of the calcitonin prohormone
chemosensory cell: Perceptive to chemical substances
chemotactic cells: Chemical signal that moves cells toward or away in a concentration gradient
choroid gland: See choroid reet
choroid reet: Horseshoe-shaped body around the optic nerve that serves to maintain high interocular partial pressure of oxygen
chromatic aberration: Occurs when light of different wavelengths focuses at different distances from a lens along its main axis
chromatophores: Pigmented epithelial cells
chromatophoromas: Neoplasms of the pigmented epithelial cells
CHV: Cyprinid herpesvirus
chyme: Homogenous material produced by action of gastric acid on food
circuli: Calcified concentric ridges found on elasmoid scales
cisternae: Reservoirs such as for lymph or spinal fluid
CLIP: Corticotropin-like intermediate lobe protein; inactive peptide produced by further processing of the POMC-derived ACTH
club cell: Large epidermal cell; alarm cells
clupeoid: Refers to order that contains herring and herring-like fishes
CNP: Type C natriuretic peptide; released from brain in fish
COD: Chemical oxygen demand
collothalamic: Region of thalamus that receives inputs relayed through the roof of the midbrain
commissural vessel: Lying midway between the pseudobranch and choroid in fishes, this vessel connects the two ophthalmic arteries
commissures: Linking of parts of the nervous or lateral-line systems
conductance: Capacity or ability to convey
congeners: Closely related (i.e. members of the same genus)
conus arteriosus: Anterosuperior portion of the heart
copeptin: A peptide of unknown function produced by processing of the vasotocin and isotocin prohormones
Coriolis circulation: Refers to the circulation of water in circular aquaria about a central standpipe
COS: An immortal mammalian cell line derived from monkey tissue used frequently as a recombinant protein expression system
cosmoid scale: Scale found in extinct lobe-finned fishes with a layer of cosmine beneath a thin outer layer of vitrodentine
CRF: Corticotropin-releasing factor; a family of peptide hormones that includes CRF1, CRF2 and urotensin I
CRH: Corticotropin-releasing hormone which is a mammalian equivalent to CRF
cryopreservation: Process of freezing tissue or cells for later analysis or use
cryptorchidism: Developmental defect characterized by failure of testes to descend into scrotum
ctenoid: Scale type of advanced bony fishes characterized by minute spines on outer edge
cupula: Cup shaped structure of lateral line that surrounds sensory filaments of receptor cell; structure of ear
cyclostomes: Living agnathan (jawless) fishes
CYP1A: Cytochrome P450 isozyme involved in biotransformation of xenobiotics
cytophaga: Ingestion of cells by phagocytes
D1: Type I outer ring deiodinase; produces T_3 from T_4
D2: Type II outer ring deiodinase; produces T_3 from T_4
D3: Type III outer ring deiodinase; produces rT_3 from T_4
DAG: Diacylglycerol; second messenger which stimulates activity of PKC in the presence of calcium and phosphatidylserine
decussations: Crossing over (e.g. optic chiasma formed by optic nerves)

demersal eggs: Fish eggs that are heavier than water and tend to sink to the substrate
DEN: Diethylnitrosamine, a common carcinogen
depulsator: Windkessel vessel, bulbus arteriosus in fish serves as depulsator
detritivore: Organism that feeds exclusively or nearly exclusively on detritus
dichromatism: Existing or presenting two different colors, members of a species may present as different color morphs
dioecious (gonochoristic) anatomy: Sexually distinct; male and female genitalis occur in separate individuals
distal tubule: Region kidney tubule between loop of Henle and collecting tubule involved in resorption of water for urine formation
DNF: Damselfish neurofibromatosis
DOC: Dissolved organic carbon
dynein: Large multi-subunit protein complex involved in movement of subcellular materials toward the center of the cell
dysplasia: Alteration in development, alteration in size, shape and organization of cells
E_2: Estradiol
ectopic: Located away from the normal position (e.g. ectopic pregnancy)
ectotherm: Poikilothermic; "cold-blooded"
ectothermic: Regulates body temperature by external environment
elasmodine: Multiple layers of collagen fibers found in ganoid scales
elasmoid scales: Platelike cycloid scales found in lungfishes
endocrine: Cells whose function is to secrete a hormone into the blood or lymph
endomysium: Fibrous tissue separating parallel muscle fibers
enterochromaffin cells: Neuroendocrine intestinal cells which stain with chromium salts
enterocytes: Intestinal epithelium
enzootic: Present in an animal community at low incidence at all times
epimysium: Forms the thick epicardial and thinner endocardial layer in all hearts
epiphysis: Articular end of a long bone
epitheliomas: Benign tumors of the epidermis
epizootic: A disease of high morbidity only occasionally present in an animal community
epizootics: Plural of epizootic
erythrosomes: Type of chromatosome containing pteridine pigment
ethological: Pertaining to animal behavior
eurydendroid cells: Widely branching cells
euryhaline: Adapts to a range of salinity
euthanized: To humanely destroy
eutherian: True placental mammals
exophthalmic: An abnormal protrusion of the eye
extant: Species with living representatives, as opposed to extinct
fascicles: Bundles or clusters
fasciculi: Pleural of *fasciculus*
FCA: Focus of cellular alteration
foam fractionation: Refers to removal of dissolved organic compounds in recirculating water systems
foam fractionators: Apparatus for distilling foam fraction from recirculating water system
FSH: Follicle-stimulating hormone (piscine equivalent is referred to as GTH-I)
funiculi: Plural of *funiculus*, general terms for multiple cord-like structures or parts
furunculosis: Fish disease caused by bacterial infection (i.e. *Aeoromonas*)
fusiform: Body form of fish; streamlined configuration with elliptical to round cross section characteristic of open water fishes
G-protein: Guanine nucleotide binding protein
G_i: G protein named for its inhibitory action on adenylyl cyclase, thereby decreasing cAMP production

G$_s$: G protein named for its ability to stimulate activity of adenylyl cyclase and increase cAMP production
ganoid scale: Found in primitive extant fishes; outer layer of ganoin with cosmine-like dentine layer beneath
GAP: GnRH-associated peptide; produced from processing of GnRH preprohormone
geniculate: Bent, like a knee
GH: Growth hormone
GHRH: Growth hormone-releasing hormone; also known as growth hormone-releasing factor (GRF) or somatocrinin
gill rakers: Accessory strictures on gill arches that aid in filtering food from water; size, number and shape vary with species
GIP: Gastric inhibitory peptide
gizzard: Strong muscular stomach found in some fishes (e.g. milkfish, some herrings, and gizzard shad)
GLP: Glucagon-like peptide; produced from proteolytic processing of the glucagon preprohormone and may possess independent hormone activity from glucagon
glycocalyx: Polysaccharide and glycoprotein matrix that surrounds many cells
GnRH: Gonadotropin-releasing hormone
gonochoristic: Differentiation of gonads with normal development of the reproductive organs appropriate to the sex
GR: Glucocorticoid receptor; only type of corticoid receptor found in fish; mediates metabolic effects of glucocorticoids in fish
granular cell: Cell characterized under light microscopy as possessing many granules or grains
ground fault interrupters (GFI): Electrical devices that detect rapid draw in electricity and shut down the system immediately, essential safety feature when working with electrical devices around aquaria
GTH: Gonadotropin (LH and FSH are the major two mammalian subtypes of this hormone)
GTH-I: Piscine equivalent to FSH
GTH-II: Piscine equivalent to LH
GTH-RI: Gonadotropin receptor type I
GTH-RII: Gonadotropin receptor type II
guanylyl cyclase: Also called guanylate cyclase; enzyme which catalyzes the formation of the second messenger, cGMP, from GTP
gynogenetic: Development of egg following stimulation of sperm in the absence of participation of the sperm nucleus
habenular: Pertaining to the habenula, frenum or rein-like structures such as found in the cochlea
halastasis: Halastasis-mechanism for bone resorption, involves removal of the unstable phase of calcioum phosphate without lysis of the matrix
heterochrony: Pertaining to temporal differences, different stages of development or ages
heterotropic: Failure of visual axes to remain parallel
holocentrid: Member of the family Holocentridae, squirrel fishes
HOX genes: Code for homeotic proteins that are sequence-specific transcription factors which are part of a developmental regulatory system that provides cells with specific positional identities on the anterior–posterior axis
HRE: Hormone-responsive element; sequence of DNA immediately upstream from a gene to which hormone/receptor complexes bind to promote/suppress transcription of the gene
HSP: Heat shock protein, e.g. hsp56, hsp70 and hsp90
hyaline: Glassy and/or nearly transparent
hyaloine: A non-osseous, non-collagenous, hypermineralized tissue that forms the superficial layer of the scutes (bony dermal plates) of armored catfishes.
hypaxial: Ventral to the long axis of the body
hypocalcin: Another name for stanniocalcin
hypophysectomy: Surgical removal of the hypophysis or pituitary gland
idiopathic: Disease or condition to occur spontaneously
IGF: Insulin-like growth factor; fish possess two types, IGF-I and IGF-II

interplexiform cells: A layer of retinal cell processes, including dendrites of the ganglion cells
intromittant organ: Permanent feature of external anatomy of some male fishes for transfer of sperm (e.g. modified anal fin ray of the male guppy or urogenital papilla of the male channel catfish)
ionocytes: Ion transporting cells (e.g. chloride cells)
IP$_3$: The second messenger, inositol-1,4,5-trisphosphate, which stimulates release of calcium from intracellular stores
iridophores: Non-dendritic (usually) type of chromatophore found in the dermis
JAK: Janus kinase which is involved in the growth hormone and prolactin signaling cascades
kinesin: Pertaining to movement; causes movement
kinocilia: Motile, protoplasmic filament on the free surface of a cell
kymograph: Instrument which records variations or undulations
lacunae: Small pit or hollow cavity
laparotomy: To open the abdomen
lemniscal: Pertaining to a band of fibers in the central nervous system
lemnothalamic: Rostral band of fibers traversing the thalamus
Leydig cells: Testicular interstitial cells which produce testosterone
LH: Luteinizing hormone (piscine equivalent is referred to as GTH-II)
ligated: Tied off
macula/e: A spot distinguishable from the surrounding tissue
masculinization: To convert to male characteristics
Matthiessen's ratio (Matthiessen, 1882): Is the focal length divided by the lens radius. This is constant for any given species
MCH: Melanin-concentrating hormone
megalocytosis: Enlarged cells
meninx: Meninges; three layers of tissue surrounding the brain
Mgrp: The tilapia equivalent to trout NEV and mammalian NEI
MHC: Major histocompatibility complex
microencapsulation: Encasement of feeds to prevent leaching
microridge: Small elevations (e.g. in the skin of scaleless fish)
MIH: Maturation-inducing hormone or 17α,20β-dihydroxy-4-pregnen-3-one or 17α,20β-DP; named for its ability to induce oocyte maturation
monoclinic crystal: The vertical axis is inclined to one lateral axis but at right angles to the other
monongean: Common fish parasites, typically found attached to skin or gills
MPF: Maturation-promoting factor or factor that promotes germinal vesicle breakdown and oocyte maturation; composed of cyclin B and cdc2-kinase in fish
MR: Mineralocorticoid receptor; mediates osmoregulatory effect of aldosterone in mammals, but has not been found in fish
mRNA: Messenger ribonucleic acid
MS-222: See tricaine methane sulfonate
MSH: Melanocyte-stimulating hormone; fish express two isoforms, α-MSH and β-MSH
mucous cells: Cells capable of secreting mucous
myocomma: A myotome or muscle segment in fishes
myographic: Pertaining to myograph: A device for recording the effects of muscular contraction
myoids: Contractile portion of the inner member of rods (eye), resembling or like a muscle
myology: The scientific study of muscles
nauplii: Early larval stage of crustaceans
necrotroph: Organisms that feed on dead or decaying organisms
NEI: Neuropeptide-E-I is the mammalian equivalent to trout NEV; both are peptides derived from the MCH prohormone
neoplasia: Formation of a neoplasm, multiplication of cells under conditions that would cease growth of normal cells
neurocrine: Referring to endocrine influence on or by the nerves; pertaining to neurosecretion

neuroepithelial cells: Epithelium made up of specialized cells that serve as sensory cell receptors
neurofibromatosis: Multiple pedunculate soft tumors (neurofibromas) distributed over the entire body and associated with areas of pigmentation
neurohemal organ: Organized capillary network that projects from the hypothalamus
neuromasts: Mechanoreceptors of the acousticolateralis system
neurophysin: A cysteine-rich peptide produced by processing of the vasotocin and isotocin prohormones
NEV: A trout peptide derived from the MCH prohormone that may be co-released with MCH; possibly related to the mammalian NEI
nitrification: To treat or combine with nitrogen
nitrifiers: Nitrobacteria capable of oxidizing compounds to nitric acid, nitrous acid or any nitrate to nitrite
NLT: Nucleus lateralis tuberis; a region in fish brain involved in endocrine function
NPY: Neuropeptide Y
NRL: Nucleus reticularis lateralis; a region in fish brain involved in endocrine function
odontodes: Placoid scales in chondrichthyans, dermal denticles on the body surface of some osteichthyans
oligovillous cells: Chemosensory epidermal cells found in lamprey and elasmobranchs
oncogenes: Genes who protein products, when over-expressed, lead to a malignant phenotype
opercular: Lid or covering, refers to the operculum which is the structure covering the gills of fishes
opisthonephric: Adult type of kidneys found in most fishes
opsin: A protein found in the rods and cones of the retina that combines with 11-*cis*-retinal to form visual pigments
opsonization: The rendering of microorganisms and other cells subject to phagocytosis
optic tectum: Region of the roof of the midbrain in fishes that primarily receives visual input
osteoblasts: A cell that arises from fibroblasts and is associated with the production of bone
osteocytes: Osteoblasts which as they mature, produce bone
osteon: The unit structure of compact bone
otolith: A bony mass in the inner ear
otolithic: Referring to otolith
ovocoel: A cavity within the ovum
ovotestis: Gonad with both ovarian and testicular tissue
ovoviviparity: Bearing live young which hatch from eggs within the maternal organism (e.g. sharks)
oxyntopeptic: Acid secreting
ozonation: Increasing levels of ozone, step in water purification
P-glycoprotein: Integral membrane protein function as an energy-dependent efflux pump for certain lipophilic aromatic compounds
PACAP: Pituitary adenylate cyclase-activating polypeptide; produced by processing of the GH prohormone
PAHs: Polycyclic aromatic hydrocarbons
papillomas: Frond-like benign tumors consisting of an epithelial covering and a connective tissue stalk
parachondral: Refers to the relationship of bones surrounding cartilage, such bones add strength to the cartilaginous skeleton
paracrine: Function in which a hormone is synthesized and released from cells and binds to receptors in nearby cells to influence their function
parasagittal: A section to the side of the median plane of the body
parasympathetic: Arm of the autonomic nervous system
parietal: Pertaining to the walls of a cavity e.g. the parietal pleura
parthenogenesis: Modified form of sexual reproduction with the development of the gamete without fertilization
pelagic: Inhabiting open waters
percid: Referring to the perches
percomorph: Pertaining to an order of fishes with liver oil rich in vitamins A and D
pericaryon: The cell body as distinguished from the nucleus and the processes; applied particularly to neurons

perikarya: Area around the nucleus (e.g. cytoplasm)
perimysial: Fibrous tissue surrounding a fascicle of skeletal muscle fibers
perineurium: Protective sheath surrounding blood vessels and nerve bundles
photic information: Information transmitted in the form of light (i.e. vision)
photophores: Specialized organs found in fish that are capable of producing light
physoclists: Fishes containing a physoclistus swimbladder (no connection between swimbladder and gut)
phystostomes: Fishes containing a physostomous swimbladder (connection between swimbladder and gut)
pillar cells: Spool-shaped cells with a cylindrical body, delimit the lamellar vascular compartment
PIP_2: Phosphatidyl inositol-4,5-bisphosphate; a phospholipid found in plasma membranes of cells
piriform: Pear shaped
PKA: CAMP-dependent protein kinase or protein kinase A
PKC: Protein kinase C or calcium/lipid-dependent protein kinase
PKG: CGMP-dependent protein kinase
placodes: Plate like structure; thickening of ectoderm of early embryo that gives rise to sensory organ
placoid scales: Tiny bony scales found on sharks
planktivory: Feeding on plankton
PLC: Phospholipase C; enzyme which catalyzes the breakdown of PIP_2 to produce the second messengers, IP_3 and DAG
poikilothermic: "Cold-blooded"; inability regulate body temperature independent of surrounding environment
poikilotherms: "Cold-blooded" organisms
polarization: To move toward opposite poles; to cause to concentrate about two conflicting positions/locations
POMC: Pro-opiomelanocortin; this fish prohormone contains sequences for ACTH, α-MSH, β-MSH, β-LPH and β-endorphin once proteolytically processed
porphyropsin: A purple pigment in the rods of some freshwater fishes
PPSS: Preprosomatostatin; fish possess two different genes that encode for PPSS-I and PPSS-II
proctodeum: Invagination of the ectoderm of the embryo that gives rise to the anus
prohormone convertase: Family of proteolytic enzymes whose actions convert inactive peptide preprohormones into active peptide hormones
prolarvae: Larval fish that still retains the yolk sac
pterinosomes: Specific organelles associated with pteridine pigments
pterygiophores: Chondral bony blade-like elements that support dorsal and anal rayed fins, also serve as attachment sites for the erector and depressor muscles that control fin movement
PUFA: Polyunsaturated fatty acid
PVR1: A subtype of PACAP receptors
pyloric caecae: Blind pouches located at the distal portion of the stomach
rami: Plural of ramus; multiple branches (e.g. blood vessels, nerves)
raphe: A general term for the line of union of the halves of two symmetrical parts
renin: Enzyme that converts angiotensinogen into Ang I
retia mirabilia: "Wonderful network", countercurrent vascular networks
retinoblastoma: A malignant tumor of the eye that arises from the retinoblasts
retinotopic map: Virtual image of visual space transmitted to the brain
retinotopy: Pertaining to visual pathways
rheotaxis: The orientation of an organism into a stream of liquid, with its long axis parallel to the direction of flow
riverine: Pertaining to rivers
rodlet cell: Controversial type of cell found in fish tissues that has been named as a parasite (*Rhabdospora thelohani*) which may be of host origin
rT_3: Reverse T_3 or $3,3',5'$-triiodothyronine; inactive metabolite of T_4
rugae: Elevations or ridges in an organ (e.g. intestinal rugae)
sacciform cells: Type of unicellular gland in fish skin distinct from goblet cells
saprotroph: Organism that feeds on dead and decaying flesh

Schreckstoff system: Found in ostariophysans, functioning of club cells to produce alarm substance (phermones)
sclerad: Pertaining to the sclera which is the fibrous outer covering of the eyeball
scleroblasts: Scale forming cells
scoliosis: Lateral curvature of the spine
scutes: Bony "scale-like" structures lacking ganoine or dentine found in a few primative groups of fish (e.g. sturgeons)
sentience: To be able to sense, feel
serosa: A serous lining surface
Sertoli cells: Structural and nutritive supportive cells for the germinal epithelium of the seminiferous tubules of the testis
sesamoid bones: Small bones embedded in a tendon or joint capsule
sexual dimorphism: Having sexual characteristics of both sexes; physical or behavioral differences associated with sex
sGnRH: Form of GnRH originally isolated from salmon, but species other than salmon also express this form
sideroblastic anemia: Anemia with red blood cells containing cytoplasmic iron granules
silastic elastomer: Synthetic rubber made of polymeric silicone
sinoatrial ostium: Opening between the sinus venosus and the atrium
sinus venosus: Venous receptacle attached to the posterior wall of the atrium
skein cells: Modified club cells found in lamprey eels
smoltification: Smolts are young salmon (4–8 inches) that migrate to sea
SMRT: Silencing mediator of retinoic acid and thyroid hormone receptor; a protein that acts as a co-repressor of thyroid hormone-responsive genes
somata: Pertaining to the body
species: A taxonomic category subordinate to genus
spermiation: Freeing of mature spermatozoa from the Sertoli cells
spherical aberration: Occurs when light enters eye at different distance from the optic axis and is focused at different distances from the lens
spinal transection: Transverse sectioning of the spinal cord
spiral valve: Modification of the intestine to increase the surface area for digestion/resorption
SRIF: Somatostatin release-inhibiting factor
SST: Somatostatin; SST-14 and SST-22 are the most important variants in fish
stenohaline: Organisms that can only live within a narrow range of water salinity
stomodeum: An invagination of the ectoderm of an embryo that will give rise to the mouth
sulci: Plural of sulcus: groves, trenches or furrows (e.g. depressions on the surface of the brain)
supersaturated gases: Refers to water being more highly concentrated than normally possible with a particular gas (e.g. oxygen) under given conditions of temperature and pressure
sympathetic: Pertaining to the sympathetic nervous system, autonomic nervous system
symphysis: Growing together; site formed by joining of two bones, a type of cartilage joint
synapse: The region of junction between processes of two adjacent neurons
T: Testosterone
T_3: $3,5,3'$-triiodothyronine; active metabolite of T_4
T_4: Thyroxine or $3,5,3',5'$-tetraiodothyronine
tectum: A roof-like structure (e.g. optic tectum)
teleocalcin: An older name for stanniocalcin
theront: Small infective stage of ciliate protozoan *Ichthyophthiriasis*
tiPRL: Prolactin isoforms isolated from tilapia; $tiPRL_1$ and $tiPRL_2$ isoforms have been characterized
TMS: See tricaine methane sulfonate
tomite: Free swimming stage of *Ichthyophthiriasis* life cycle
tonofilaments: Slender thread-like organelles in the cytoplsm of cells of the basal layer of the epidemis
topological: Refering to the anatomy of a specific area of the body (e.g. brain)
toxicopathic: Capable of inducing a disease by a posion

trabecular: Pertaining to supporting or anchoring strand of connective tissue
trabeculated: Having a trabecular structure, multiple foci of connective tissue support
transcoelomic: Within the body cavity (e.g. detachment of tumor emboli in the abdominal cavity)
transcription factor: Protein/chemical whose binding to the 5′-flanking region of a gene's promoter region affects the transcription rate of that gene
transponders: Devices that can be inserted under fish skin or in abdominal cavity for field tracking
transporters: Molecules responsible for movement of materials in biological systems
TRF: Thyrotropin-releasing factor
TRH: Thyrotropin-releasing hormone (same as TRF)
tricaine methane sulfonate: Commonly used fish anesthetic
trocar: A canula with a sharp-pointed end for piercing the wall of a cavity
trophic: Pertaining to nutrition, often word termination denoting relation to nutrition (e.g. heterotrophic)
trophonts: Feeding stage of certain ciliated protozoa
TSH: Thyroid-stimulating hormone
tubulated scales: Modified fish scales that provide surface outlets for the lateral line sensory canal
urophysis: Caudal neurosecretory neurons in teleosts
urotensins: Cyclic neuropeptides first isolated from the caudal neurosecretory system of teleosts
V_{1a} receptor: Vasopressin-1a receptor subtype; similar to the fish vasotocin receptor
vagosympathetic trunk: Vagus and sympathetic nerve roots
varicosities: Varicose; unnaturally and permanently distended
vasa protein: Essential for the assembly a specializied cytoplasm (pole plasm) found in the posterior portion of the egg and early embryo
VIP: Vasoactive intestinal peptide
vitellogenic: Stimulating the production of egg yolk
vitread: Pertaining to the vitreous body of the eye
viviparity: Giving birth to live young
VNP: Ventricular natriuretic peptide; released from fish ventricles
vomeronasal: Pertaining to the flat bone forming the inferior and posterior portion of the nasal septum (vomer), and the nasal bone
white cells: Leukocytes; the cells of the blood excluding the red blood cells
xanthosome: Yellow pigment granule
xenobiotic: Chemical foreign to the biological body

Index

abdominal vein 169
abducens nerve 140
Acanthaluteres vittiger, spinous scale 105
acetic acid 572
achondral bones 309
acousticolateral system 347
acrodont dentition 175
actinopterygian fish
  brain structure 130
  sound pressure reception 257
Actinopterygii 1
Actinopterygii, renal tubule transition 403
activin 198
adductor arcus palatini (AAP) 121
adductor mandibulae complex 120
adductor operculi (AO) 121
adductores (Ad) 124
adenohypophysial hormones 192
adenyl cyclase 113
adenylyl cyclase 433
Adinia xenica 45
adrenaline, in stress response 504
adrenocorticotrophic hormone (ACTH) 194, 418, 422, 423
adrenocorticotropin *see* adrenocorticotrophic hormone (ACTH)
Aeromonas salmonicida 79, 86
  phagocytosis by 442
AFB1 601
aflatoxin 594, 600
AG555 606
agnathans, brain structure 130
Alamar Blue 640, 641, 643
alar plate, elaboration 351
alar plate, variations 147
alarm pheromone 283
  components 290
albino species 6
aldehydes, safety data 576
alimentary canal 173–4
alimentary canal
  necropsy resection 544

necropsy separation 544
tissue layer organization 379
alkalinity 619
alloenzyme genotypes 622
allografts 220, 443
allometric analysis 132
alpha cells 362
alpha-melanocyte-stimulating hormone (MSH) 422–3
amacrine cells 233, 460
Amazon molly 6
Ameiurus nebulosus, environmentally induced cancer studies 602
American Fisheries Certified Fish Health Pathologist 90
American Veterinary Medical Association 510
Ames assay 603
ametropia 230
amino acid
  requirements 71
  sequence of
    GnRH hormones from fish species 419
    vasopressin and isotocin 424
ampullary receptor cells 347
anabantids, sound-pressure reception 257
anadromous species
  ionocytes in 286
  ionoregulation 621
anaplasia 604
anesthesia
  drugs for 507–9
  induction 507
  inhalation 506
  overdose 507
  practical aspects 506–7
  pre-handling applications 507
  regulations 509
  sentience 509–10
  for surgery 506
  survival 506
ANF 432
angiogenesis 598

angiotensins
  gene expression and regulation 431–2
  hormone effectors in target tissues 432
  molecular structure 431
Anguilla anguilla, urinary system 185
Anguilla australis, skin layers 5, 97
anlagen 184, 248
ANP 432
  prohormone 432–3
anteroventral arteries 164
antibiotics 91
  resistance tests 91
anticoagulants 513–15
antifreeze proteins 96–7
aorta 165
  ventral and dorsal 373
apoptosis 194
  hormonal regulation 198
appetite
  loss in stress response 504
  post-surgical 565
aqua ion complex 619
aquaculture 65
  swimbladder hyperinflation syndrome 564
aquarium fish 68
Aristotle 2
armored catfish, scutes 105–6
arterial system 163
arteries 372–4
  intrarenal, vessel wall architecture 385–6
arterioles 374
  renal vascular system 385–6
aryl-hydrocarbon receptor (AhR) 644
asepsis 558
astaxanthin 75
Atlantic croaker
  gap junctions 497
  ovarian follicle 495
Atlantic salmon 66
ATP 417

atresia 198–9
  follicular 496
atretogenic factors 198
atrial systole 162
atriopeptin 432
auditory nerve 237
auditory pathways 349
auditory system 251
autonomic nervous system 130
autoradiography 579
  basic considerations 579
  components 580
  computer-assisted image analysis 586
  electron microscope (EM) 581
  exposure arrangements 580
  fish freezing 582
  light microscopy (LM) 581
  macroscopic 581
    X-ray films 582
  processing steps 580
  quantification 586
  quantitative analysis 586
axons 334
  conduction 334
azurophilic cytoplasmic granules 442

β cells 362
B lymphocytes 441
  stain technique 635
bacteria
  identification tests 82
  in laboratory stock 61
bacterial diseases, systemic 80–1
bacterins 220
barbels 105
basal cells 473
basal plate, motor-related 350
basement membrane 291
  functions 292
basolateral membrane 362
bathypelagic spp. 246
benthic species
  epidermis 271
  sacciform cells 281
benthivore
  feeding 248
  metabolism 621
benz(a)anthracene (BaA), in sediment 602
benzo(a)pyrene (BaP) 591
  acquisition from water 593
  adducts 595
  English sole sedimentary uptake 603
  fish uptake 593
  in sediment 602
betadine 525
bile
  BaP metabolites in 593
  necropsy harvest 545

bioaccumulation of contaminated prey 533
bioassays 625
biogenic compounds 5
bipolar cells 233, 457
Black River, Ohio, sediment experiments 603
Blennius pavo, dorsal segmental vessels 170
blood
  chemistry, in stress response 504
  collection 513
    cardiac puncture 516
    serial sampling techniques 516
    single sample 515
    ventral aorta 516
  samples for research 600
blood vessels 441, 372
body musculature 125
bone
  acellular 313
  architecture 312
  cellular 313
  cytological structure and function 312
  matrix deposition 313
  molecular structure and function 312
  resorbing cells 313
  structure and function 309
  tissue mineralization 314
  vascular canal 312
Bouin's fixative 572–3
  whole small fish 575
Bowman's capsule
  parietal cells 395
  parietal layer 389
Brachydanio rerio 6, 43–4
  carcinogenicity testing 592
brain
  atlas 43
  development, relative and differential 147
  divisions 335
  elaborated 146
  evolution 146
  laminar 146
  necropsy 548, 554–5
  organization 146
  structure 131
  structure, variations 130
  weight 131
brain-pituitary axis, functional structure 490
brain/body ratios 131
brainstem 141
  electrosensory area 348
  mechanosensory area 348
  octavolateral area 348
branchial arches 139
branchial arteries 155, 163

branchial muscle plates 120
branchial nerves 156
branchial veins 156
breeding, selective 90
brine shrimp
  hatching container 53
  as live food 68
  stocks 53–4
Brockmann bodies 206
brown bullhead
  in Black River, environmentally induced cancer studies 602
  sediment experiments 598
Buffalo River, New York, sediment experiments 603
bulbus arteriosus 372–3

C18 425
C19 425
C21 430
C-reactive protein 446
calcitonin 205, 434
calcium
  metabolism, hormones 314
  muscle activation 327
  water concentrations 620
calcium-regulating hormones 204
Cambridge Isotopes Laboratory 645
cAMP 417
Canadian Council on Animal Care 509, 558
cancer
  chemical cause 591
  malignant, histologic criteria 598
  progression 598
    assessment in fish 598–9
  transcelomic metastasis 599
cannula
  chemical agent injection 534
  designs 516
  insertion 517, 518, 536
  intraperitoneal 536
  intraperitoneal, insertion position 536
  preparation 535
cannulation
  caudal vein 519
  dorsal aorta 517–8
  isolated vessels 519
  ventral aorta 518
capacitance 162
capillaries 374
Carassius auratus 6, 66
Carassius auratus gibelio, renal portal system 187
carbohydrate requirements 72
carbon dioxide 508–9
  excretion 358
carcass composition modification 72

carcinogenesis
    assays, fish eggs 537
    initiation 593–4
    mechanisms 591
    promotion 597–8
        dysplasia 597
    stages in fish and mammals 593
carcinogenicity
    chemicals evaluation 592
    control difficulties 592
    testing 6, 592
cardiocytes 371
    contractile proteins 371
carotenoid vesicles 294
carotenoids 75
carp 66
    ascorbic acid synthesis 74
    dietary requirements 74
    male sperm cell development 498
    muscle anatomy 321
    ovarian follicles, growth and atresia 493–4
    testis, cross-section 498
carp pox 607
cartilage
    calcification 310
    cranial 308
    hyaline–cell (HCC) 310
    nutrition 309
    resorption 309
    structural diversity 308
    tissues, distribution and function 309
catadromous species
    ionocytes in 286
    ionoregulation 621
catecholamines 203
    actions on gill function 203
    and stress response 203
catfish
    amino acid sequences 419
    computer–assisted image analysis 586
    cryosection of intestinal mucosal fold 584, 586
    dietary requirements 74
    enteric septicemia 81
    heart electron micrograph 163
catheterization, restrained and free-swimming fish 520–1
catheters
    external 521
    post-bladder 521
    urinary bladder 519
caudal fin muscles 127
caudal peduncle, severance 516
caudal propulsion 110
caudal puncture 515
caudal puncture, chemical agent injection 534
caveolae 371

cDNA 421
cell cultures 631
    cellular progeny staining 635
    exposure to test compounds 639
    functional assays 636
    growth media 633, 649
    hemopoietic cell culture to cell lines 637
    monitoring 649
    product examination and quantification 634
    uses 631
cell lines 631
    care 637
    development 649
    laboratory experiments 447–8
Cell and Tissue Culture: laboratory procedures 637
cells
    subculturing or passaging 638
    surface protein changes 598
    viability assays 638
        cell culture 639
        fluorometric readings 642
        interpretation of results 643
        as percent of control 642
cement lines 313
central nervous system 130
central venous system 167
Cephalaspidomorphi 1
cephalic musculature 119
ceratioid spp. 246
cerebellar corpus 135
cerebellum 135, 344
CFDA-AM 641, 643
cGMP 433
CGRP 434–5
channel catfish 66
characteristics 3–5
cheek muscles 120
chemicals
    acute exposure of early life stages (ELS) 622
    bioaccumulation 620
    bioconcentration 620
    biotransformation 533
    capsule administration 534
    carrier mediated diffusion 621
    contaminants and environment 4–5
    contaminated prey consumption 533
    dissolved in water 530
    environmental, endocrine function modulation 209
    exact dosage 534
    fish testing 592
    gill
        extraction efficiency calculation 531
        uptake 530

    intramuscular injection 536
    intraperitoneal injection 535
    intravascular injection 534
    ionic
        solubility 619
        speciation processes and toxicity 620
    laboratory exposure 530
    membrane transport 620
    metabolic transformation 621
    oral administration 532
    pellet implants 538
        release rate 538
    radiolabeled, in food 533
    routes of administration 529
    routine feeding with impregnated food 532
    solubility in water 619
    systemic adsorption 533
    topical exposure 540
    toxic concentrations in water 617
    transformations by hydrolysis 619
    trophic transfer 533
    uptake 620
        physiology 621
        in vehicle for injection 535
    water-borne exposures 529
chemo-therapeutants 90
chemoreception 246, 471
    processes 474
chloride cells 361
chlorinated butadienes (CBDs), neoplasm occurrence 602
cholecystokinin (CCK) 418
Chondrichthyes 1
    brain structure 130
chondrocytes 308
chondroid bone 314
Chordata 1
choroid gland 164
choroid reet 230
chromatic aberration 228
chromatophores 287, 293
    types 293
chromatophoromas 604
chyme 380–1
ciliated non-sensory cells 473
circulation
    peripheral 372
    secondary 169, 375
        characteristics 375
        systemic 376
circulatory system
    overview 162, 370
cisternae 371
clove oil compounds 508
club cells
    morphology 282
    secretions 291
    secretory vacuole 283

clupeoids 243
cold water disease 80, 83
cold water disease, skin erosion 83
coldwater species 67
collagen 291, 312
    cartilage-type 308
collagenase A, preparation 633
collecting duct
    cytological organization 408
    ontogenetic considerations 408
collecting tubule-collecting duct (CT/CD) system 408
    functional considerations 411
color, and stress 504
columnar epithelial cells 21
Columnaris bacteria 83
Columnaris infections 80, 82–5
commissures 334
commisural vessel 164
communication 96
comparative studies 5–6
complement (C) 442
conductance 162
confinement stressors, immune reactions 444
congeners 627
conger eels, male, sound production 259
connective tissue 372
conus arteriosus 163
Coommassie Blue 86
copepods 69
copeptin 424
corpus cerebelli 344
corticosteroid receptors 445
corticosteroids 202
corticotropin-like intermediate lobe protein (CLIP) 422
corticotropin-releasing hormone (CRH) 192
    gene expression and regulation 431
    molecular structure 430
    in target tissues 431
cortisol 202
    hormone effectors in target tissues 430
    in immune reactions 445
    molecular structure and regulation 430
    in stress response 504
    and stress response 202
Corydoras aeneus, scutes 105–6
cosmoid scale 99
costiasis 80, 84
cranial nerves 237, 342
    components 137
    embryological studies 137
    gustatory components 141
    motor component classification 139
    nuclei 345
    retinal axons 140
    sensory component classification 138
    visual 139
cranial vault 111
cryopreservation 525
cryosection 584, 586
ctenoid scales 102
culture, fish 51–60
cultures
    cell and tissue 631
    stock solutions preparation 632–3
cupula 239, 253
cuticle 278
    components 290
cyclostomes 393
CYP1A 593, 644
cyprinid herpesvirus (CHV) 607
Cyprinodon variegatus 6, 42, 45
Cyprinus carpio 66
Cyprinus carpio
    male sperm cell development 498
    ovarian follicles 493–4
    testis, cross-section 498
cytochrome P-450 593, 601
CytoFluor 642
cytokines 441
cytophaga 82

Dactylgyroidea 80, 84
daimobenzidine (DAB) staining solution 635
damselfish 7
    neurofibromatosis (DNF) 605
Danio rerio 43, 66
    olfactory epithelium 471
database, domestic fish 592
Davidson's fixative 573–4
decussations 334
deiodinase I, II, III (D1, D2, D3) 427
demersal eggs 262
dendrites 333
dental tissues 314
dentine 315
depulsator 372
dermal endothelium 292
dermal skeleton 307
    non-osseous tissues 315
dermis 97–105, 291
detritivore species 67
DHEA 601
diacylglycerol (DAG) 417
diamond killifish 45
dichromatism 263
diencephalon 131, 133, 339, 340
diet 65–75
    history 66–4
    larvae and fry live food 68–9
    manufactured feeds 69–70
    post-surgical 565
diethylnitrosamine (DEN) 600–1
    exposure time 599
    medaka susceptibility 44, 600
    neoplastic promotion 598
Dietrich's fixative 573
digestive system 173–7
    accessory organs 383–6
    anatomy 379–7
    functions 173
    regions and components 173–7
$17\alpha 20\beta$-dihydroxy-4-pregnen-3-one ($17\alpha, 20\beta$–DP) 425
dilator operculi (DO) 121
dioecious (gonochoristic) anatomy 1
dioxin, medaka embryo assessment 43
Dipnoi
    collecting tubule-collecting duct (CT/CD) system 409
    intercalated cells 410
diseases
    prevention 90
    systemic bacteremial 85–8
    treatment 90–1
    summary 80–1
diuresis, after stress 521
diversity 1
DMBA 601
DNA
    adducts 594
    detection 594
    laboratory studies on formation 595
    quantification 594
    carcinogen binding 594
    hepatic alteration 603
dogfish, amino acid sequences 418
dogfish shark, fin tissue 98
dosage
    by contaminated prey 533
    gavage with impregnated food 534
    gelatin capsules 534
    injection techniques 534
    for microinjection 537
    result variations 534
dose-response curve analysis 642–3
drug assessment model 592
Dulbecco's Modified Eagle's Medium (DMEM) 646
dyes for cell viability assay 638, 640–1
dynein 295

ear 480–8
    afferent fiber connections 485
    angular and linear acceleration effects 480
    efferent system 486
    evolution 486
    hair cell
        as mechanosensory transducer 481
        organization 481
        polarity patterns 484–5
    macula 251, 482, 484
    morphological changes 251–2

papilla neglecta 251
saccular recess 254
semicircular canals 251, 482
    function 252
    length and width 253
    organization 482–3
sensory systems 251
statolithic organs 484
types
    actinopterygian 251
    chondrichthyan 251
    jawless vertebrate 251
    sarcopterygian 251
utricular recess 253
ecosystems, contaminated, assessment 592
ectopic expression 606
ectothermic species
    glomerular aspects 395
    surgery 557
ED50 618
Edinger-Westphal nucleus 140
EDTA 515
    preparation 516
Edwardsiella ictaluri 81, 85–9
eel, urinary system 185
eggs
    biology 262
    buoyant 72
    demersal 262
    fertilized, collection 525
    fungal infection treatment 91
    hatch time 531
    hatching chamber 530
    medaka 42–4
    microinjection 537
    mummichog 3
    rainbow trout, mortality with aluminum and calcium 619
    stripping 525
    topical toxicant exposure 540
eicosanoids 110
Elasmobranchiomorphii, proximal tubule, segment II 402
elasmobranchs, heating-related specializations 258
elasmodine 312, 315, 297
elasmoid scales 99, 100, 297
    development 299
electric eel 66
electric organ discharges (EODs) 348
electricity
    ground fault interrupters (GFI) 560
    for surgical suite 560
electron microscopy, fixation 574
electronarcosis 509
Electrophorus electricus 66
electroreceptor organs 142
electrosensory fishes 135
electrosensory lateral line lobes (ELLL) 344, 345

embryo, transverse section 139
embryology 2
endochondral bones 311
endocrine modulating chemicals (EMCs) 209, 625
    effects 211
    mechanisms 209
endocrine modulating chemicals (EMCs)-receptor complex 211
endocrine receptor signal transduction 416
endocrine stimuli 374
endocrine structures and regulation 265
endocrine system 189, 415–35
endorphin 422
endoskeleton 307
endothelium 372
energy
    dietary 65
    immediate nutritional needs 70
English sole 5
English sole
    Puget Sound
        environmentally induced cancer studies 601
        laboratory studies 603
enteric nervous system 383
enteric red mouth 80, 85
enteric septicemia of catfish 85–6
enterochromaffin cells 383
enterocytes 381
environmental change 4
environmental chemicals
    estrogenic 211
    exposure, field necropsy 543
environmental effects on lesions 609
environmental factors, sex differentiation and determination 493
environmental risk model 592
environmental stressors 5
environmental stressors, endocrine function modulation 209
environmental toxicants, reproduction effect studies 199
enzootic infection 543
enzyme-linked immunosorbent assay (ELISA) 190
epaxialis (Epx) 126
epidermal cells, solitary 286
epidermal growth factor (EGF) 197
epidermis 272
epidermis
    cell types 272
    club cells 274
    epithelial cells 274
    goblet cell 273
    granular cell 273
    keratinized epidermal plaque 274
    Merkel cell 274

pavement cells 276
sacciform cells 273–4
secretions 288
sensory cells 274
thickening 277
epimere 139
epimysium 372
epiphysis, in lampreys 134
epithalamus 339
epithelial cells 272
    ion exchange 277
    microridges 276
    mucigenic 273, 279
    nucleus 275
    phagocytic activity 277
    size and shape 275
    types 273
    ultrastructure 275
    wound repair mechanism 276
epithelium 358
epizootic infection 543
erythrocytes 441
erythrophores 294, 604
erythrosomes 294
esophagus 175
    microscopic anatomy 380
Esox lucius, kidney location 162
17β-estradiol (E2) 425
ethanol 571
    safety data 577
7-ethoxyresorufin (EROD)
    activity induction 644
    assay 645
    calculation of
        induction potency 648
        production 646
        specific activity 647
    dose-response curves analysis 648
    fluorometric readings 646
    induction conditions 645
ethylene dibromide (EDB) 600
eurydendroid cells 344
euryhaline fish, physiological adaptations 621
euryhaline teleosts, glomerular size 392
euthanasia 79, 509
eutherian mammals 255
evagination 145
eversion 145
excretion 96
experimental use 5
exposure chambers 49, 530
    field 531
        cage design 531
    in-situ, for larval fish 532
    in-vitro perfusion, research trials 538
    physiological 531
    specialized 530
exposure laboratories 49
extant fishes, brain evolution 130, 132

extracellular matrix (ECM) 308
eye
   bisection 555
   resection 555
   size and light collection 227

facial lobes 135
facial nerves, motor components 142
facilities, small fish models 47–51
fascicles 334
fathead minnow 6, 66
   freshwater bioassay studies 44–5
   hermaphroditic gonads 493
   ovarian differentiation 491
   testicular differentiation 492
fatty acids 72
   requirements 72
feces
   catheters 524–5
   collection 522
      gravity-settled 522
   manual stripping 524
feeding behavior
   and diencephalon 339
   visually-guided 341
feeds
   flake 68
   practical and commercial 70
   purified 70
   semipurified 70
field exposure *see* exposure chambers, field
field samples, exposure experiments 603
filtration slit 393–4
fin rays 300
Fish Net website 47
Fish in Research 2
fixatives
   concentration 575
   criteria 571, 575
   electron microscopy 574
   mixtures 572
   simple coagulant 571
   simple non-coagulant 572
   types 571
flatfish
   kidney necropsy 553
   visceral mass removal 552
Flavobacterium branchiophilum 83
Flavobacterium columnarae 80
Flexibacter maritimus 83
Flexibacter psychrophilus 80, 83
flow cytometry 635
flow-through exposure systems 530
fluorescamine 647
fluorometric indicator dyes 638
foci of cellular alteration (FCA) 597
   basophilic 598
   DEN exposure time 599
   preneoplastic 598

forebrain *see* diencephalon; telencephalon
formaldehyde 571–2, 574
   safety data 576
formalin 571–2
   whole small fish fixation 575
freshwater fish
   glomerular filtration barrier 396
   ionoregulation and osmoregulation 621
fright pheromones 96
FSH 418
Fugu rubripes 7
Fundamentals of Aquatic Toxicology 2
Fundulus grandis 5–6
Fundulus heteroclitus 45, 366
Fundulus heteroclitus, hepatic neoplasms 603
fungal infections 80, 85
funiculi 334
furunculosis 80, 86, 220

G-proteins 417, 474
gall bladder
   necropsy 553
   necropsy resection 545
Gambusia affinis 6, 46–7
gametes, collection 525
ganglion cells 233, 454–5, 460
ganoid scales 99, 297
ganoid scales, development 297
ganoine 37–9
gap junctions (GJ) 496
Gasterosteus aculea 66
gastric inhibitory peptide (GIP) 429
gastro-entero-pancreatic hormones 205
gastrointestinal tract
   endocrine cells 206
   hormones 206
gavage 534
genes
   Diff 606
   expression and regulation 416
   Xmrk 606
geniculate ganglion sensory neurons 482
genome model 7
germinal vesicle 496
gill 219
   bacterial infections 83
   blood vessels 364
   chemicals uptake 529
   color 79
   damage treatment 91
   dorsal arch musculature 123
   embryonic 152
   epithelial cells, electron micrograph 361
   exposure area 529
   functions 152
   infections 79–82
   interlamellar circulation 376

   interlamellar vessels 365
   necropsy 550
   nutrient circulation 375
   nutrient vessels 365
   parasites 2, 79
   parasitic infestations 83–5
   secondary circulation 156, 375
   ventral arch musculature 124
gill arch 153
   arrangement 152
   innervation 156
   musculature 154
   nutrient supply 156
   orientation 152
   vasculature 155
gill filament 153, 156, 358
   arteries 155
   arterio-arterial pathway 157
   arteriovenous pathway 158
   blood vessels 156
   capillaries 366
   cartilage and muscle 156
   function 153
   interlamellar system 158
   nerves 159
   respiratory lamellae 156
   vasculature 157
   veins 158
gill rakers 82, 153, 156
gill slits 153
gizzard 381
glomerular basement membrane (GMB) 393, 395
glomerulo-tubular contact 411
glomerulus 389
   capillary, Prussian carp 394
   environmental changes 397
   freshwater adaptation to saltwater 396
   lamprey type 390–1
   marine elasmobranch type 391
   marine teleost type 392
   myxine type 390, 396
   Prussian carp 394
   skate 392
   structural-functional correlation 395
   types 390
glossopharyngeal nerves, motor components 142
glucagon 206
   gene expression and regulation 429
   hormone effectors in target tissues 429
   molecular structure 429
glucagon-like peptide (GLP) 429
glucocorticoid (GR) 430
glutaraldehyde
   for electron microscopy 574–5
   safety data 576
glycocalyx 382
goblet cells 277, 473
   environmental changes 279
   holocrine glands 280

male and female fish 279
   secretions 280
   stressors 279
   types 278
goldfish 6, 66
gonadal hormones 194, 425
gonadotropin (GTH) 192
   gene expression and regulation 420
   hormone effectors in target tissues 420
   molecular structure 419
   receptor subtypes 420
   seasonal profiles 193
gonadotropin-I 420
gonadotropin-II 420
gonadotropin-releasing hormone, salmon type (sGnRH) 419
gonadotropin-releasing hormone (GnRH)
   effectors in target tissues 419
   gene expression and regulation 419
   molecular structure 418
   neurons 473, 490
   pellet implants 538
gonadotropin-RI 420
gonads 490
   hermaphroditic 493
   necropsy 547, 552
   primordial 491
gonochoristic species 490
Gordon-Kosswig melanoma model 605
granular cells 284, 363
gravel station 59–60
growout methodology 60–1
growout systems 50
growth hormone (GH) 193, 416–17
   effectors in target tissues 418
   gene expression and regulation 418
   molecular structure 418
growth hormone-releasing hormone (GHRH)
   effectors in target tissues 417
   gene expression and regulation 417
   molecular structure 417
growth media, improvement 649
guanylyl cyclase 433
Guelph system for fecal collection 523
Guide to the Anatomy of Fishes: the Musculoskeletal System 109
Gulf Coast Research Laboratory (GCRL) 47
Gulf killifish 45
guppy 6, 66
guppy
   see also king cobra guppy
   breeding 56–7
   carcinogenicity testing 592
   husbandry 56
   as live-bearer 46
gustation 475
gustatory cells 11–22

gustatory nerves 27–6
gustatory organ 248
gustatory system 347
   related lobes 345
gut-associated lymphoid tissue (GALT) 445
gynogenetic spp. 261
Gyrodactyloidea 80, 84

habenular commissures 477
habenular nuclei 134, 339
habitat and stress 505
hagfish, slime glands, cell types 291
hair cells
   sensory 463
      distribution and morphology 464
      morphological variation 464
      physiology 464
halastasis 314
halogenated aromatic hydrocarbons (HAHs) 627
Hamilton Harbour, sediments and fish neoplasms 603
handling
   conditioning to 504
   immune reactions 444
Haplochromis burtoni 231
Haversian bone 312
head
   kinematic 110
   lateral line system, morphology 238
   neuromast receptors 239
   venous drainage 166
hearing 251
   in water 255
heart 162, 370
   innervation 372
   necropsy 546, 551
   parasympathetic ganglion 372
   sympathetic nerves 372
hemopoiesis, rainbow trout model 632
heparin for blood collection 513–14
hepatoerythropoietic porphyria model 7
hermaphrodites 261
hermaphroditism 490, 493
herpesvirus tumors 607
herrings, swim bladder ear connections 257
Herteropneustes fossilis, heart electron micrograph 163
heterochrony 240
heterotropic thyroid tissue 200
high performance liquid chromatography (HPLC) 190
hindbrain 135
   reticular formation 344
histological information, World Wide Web access to 577
historical use 2–3
History of Biology 2

Hitra disease 88
holding facilities
   post-operative 565
   water chemistry 565
holding systems 50–1
'hole-in-the-head' syndrome 86
holocentrid fish, sound pressure reception 258
horizontal cells 233, 458
hormones 560
   action 211
   biosynthesis 210
   clearance 210
   effectors 416
   receptors, signaling schemes 416
   transport 210
HOX genes 468
HREs 427
HSPs 427
humic acids 619
husbandry 17, 50, 56
hyaline 310
hyaloine 300
hydrolysis 619
hydroxyapatite 314
20β-hydroxysteroid dehydrogenase (20β-HSD) 425
hyohyoides abductors (HhAb) 122
hyohyoides adductors (HhAd) 122
hyohyoides inferioris (Hhl) 122
hyoid arch 139
hyoid bars 112–13
hyoid muscle plate 120
hypaxial muscle
   necropsy incision 544
   resection 550
hypaxialis (Hpx) 126
hyperplasia 604
hypocalcin 314, 435
hypodermis 292
hypophysectomy 193
hypothalamo-hypophysial axis 191
hypothalamo-pituitary-interrenal (HPI) axis 445
hypothalamus 133, 339
hypoxia 559

Ichthyobodo necator 80, 84
ichthyophthiriasis 83
Ichthyophthirius multifiliis 80–3
Ictalurus punctatus 66
idiopathic disease 543
immune reactions 444
immune system 219, 441
   intestinal tract in 221
   kidney in 220
   liver in 446
   spleen in 446–8
   thymus in 220
immunocytochemistry 190, 635
immunoglobulins 441

immunohistochemical techniques, fixation for 576
implants 538
indole-3-carbinol (I3C) 601
inducers, stock solutions 644
infectious pancreatic necrosis virus (IPNV) 81, 89
inflammatory cells 559
injection techniques, dosage 534
innervation, vagosympathetic trunk 372
insecticide toxicity 6
insulin 206
    gene expression and regulation 428
    hormone effectors in target tissues 428
    molecular structure 428
insulin-like growth factor (IGF) 418
    gene expression and regulation 428
    hormone effectors in target tissues 428
    molecular structure 428
insulin-like growth factor (IGF-1) 197–8
integumentary system *see* skin
inter-renal hormones 200
inter-renal tissue, diagrams 205
interbranchial septum 154
interferons 445
interleukin IL-1 445
intermandibularis (IM) 122
interplexiform cells 458
interstitium 186
intestine 221, 176
    accessory organs 177
    anatomy 381
    defense mechanisms 383
    dietary epithelial damage 383
    diverticula 176
    endocrine cells 383
    enzymes 382
    functional anatomy 177
    in immune system 221
    infections 222
    length and configuration 176
    ligation 522
    location 221
    macromolecular absorption 382
    necropsy 545
    osmoregulation 383
    paracrine cells 383
    protein absorption 382
    spiral valve 177
intraepithelial lymphocytes 383
intromittent organ 262
intrusive cells 287
ion transport 358
ionocytes 285, 358
    morphological changes 286
ions, regulation 621
IP3 417
iridophores 294–5
iridophoromas 604

isotocin 191–2
    gene expression and regulation 424
    hormone effectors in target tissues 424
    molecular structure 423
isotopes 579

janus kinase (JAK) 418, 434
Japanese medaka 42–3
    see also medaka
    median lethal concentration 622
    mortality data 624
jawless species 272
    olfactory system 133

Karnovsky's fixative 574–5
keratinization 276
kidney 220
kidney
    anterior 221
    arterial system 185
    bacterial disease 81, 87
    cell culture initiation 633
    comparative anatomy 181–2
    excretory system 182
    filtration barrier 396
    immune functions 221
    location 220
    necropsy 548
    ongoing nephrogenesis 395
    opisthonephric 391
    shapes 183
    types 186
    vascular system 184
killifish 6, 66
killifish, *see also* diamond killifish; Gulf killifish
kinesin 295
king cobra guppy 42
    see also guppy
kymograph 48
L-carnitine 74
laboratory
    acclimation 5
    rearing 5
    stock health 61
laboratory exposure to water-borne chemicals 530
Lactococcus garviae 87
lagena 253–5
lamellar bone matrix 312
Lampetra tridentata, skin layers 96
lamprey I and III, amino acid sequences 419
laparotomy
    surgical instruments 561
    technique 563
larvae, feeding 68
lateral line
    head, morphology 238
    mechanosensory 237
    neuroanatomy 240

    specializations 147
lateral line bones, development 466
lateral line electrosensory receptors 142
lateral line nerves 141, 242
lateral line system 96
    mechanosensory, functions 242
lateralis superficialis (LtS) 127
LC50 618
    various species 622
Lebistes reticulatus 6, 46
lemnothalamic structure 347
lens
    accommodation 230
    quality 228
    refractive index 229
lethality line 624
lethality testing 623
    basic data 623
    Japanese medaka mortality data 624
    statistical methods 622
leukocyte signaling molecules 445
leukotriene 442
levamisole 223
levator arcus palatini (LAP) 121
levator operculi (LO) 121
levator posterior (LP) 123
levatores externi (Le) 123
levatores interni (Li) 123
Leydig cells 197
LH 418
light
    absorption 453
    collection 227
    focusing 228
    and stress 491
lighting, post-surgical 565
Lillie's fixative 572–3
    whole small fish 575
lipid requirements 72
β-lipotropin (β-LPH) 27–8
liver 222
    cancer
        epizootics and field studies 601
        nodular proliferation 598
    in digestive system 384
        cells 384
    DNA, carcinogen binding 595
    foci of cellular alteration (FCA) 597
    immune function 222
    in immune system 446
    neoplasms 5
    preneoplastic lesions 597
    tissue fixation 575
    vitellogenin 494
long-term hemopoietic culture (LTHC) 632–3
luminescent organs 287
lymphatic vessels 186

lymphatics 169, 186, 375, 441
lymphocytes 441
lymphoma
    classification 608
    and retrovirus 608
    serial transmission 608

macrophages 444
macrophages
    identification 636
    intestinal 446
macula 253
    common 253
magnetorecptor cells 473
major histocompatibility complex (MHC) 442
mammal substitutes 6–7
mandibular arch 139
mandibular muscle plate 120
mangrove rivulus 46
Manual of Cell and Tissue Culture 637
Manufacturers' Material Safety Data Sheets (MSDS) 576
marine fish, glomerular filtration barrier 396
marine fish food 68
masculinization 47
MATC 625
mating call generation 327
mating interactions, sound production 258
Matthiessen's ratio 228–9
maturation-inducing hormone (MIH) 196, 425, 496
maturation-promoting factor (MPF) 197, 427
mechanosensory lateral line 463
    canal morphogenesis 467
    development mechanisms 468
    system development 465
medaka 6, 66
medaka
    see also Japanese medaka
    abnormalities 60–1
    for basic research 600
    breeding 14
    cancer progression assessment 598
    carcinogen bioassay 601
    carcinogenicity testing 592
    dose-response studies 43–4
    as egg-layer 42–3
    eggs 43–4
    foci of cellular alteration (FCA) 598
    fry 43
    fry
        culture 52
        diet 52
        husbandry 51–4
        inbred strains 6
        oncogene research websites 597
        rapid tumor induction 600
    retinoblastoma after MNNG exposure 597
    trichlorethylene (TRI) metabolism 594
    as tumor model 599
    tumor promotion 598
    whole small, fixation 575
medial octavolateralis nucleus (MON) 241
median fin muscles 127
medulla 200
    transection 549
medullary cavity 311
megalocytosis 599
melanin 6
melanin-concentrating hormone (MCH)
    effectors in target tissues 423
    gene expression and regulation 423
    molecular structure 423
melanoma
    hereditary 609
    in *Xiphophorus* hybrids 605
melanophores 294
melanophoromas 604
melanotropin (MSH) 194
melatonin 209
Menidia beryllina 46
Menidia menidia 45
Menidia peninsulae 45–6
meninges 143
Merkel cells 287
mesencephalon 342
metabolism
    phase I 593
    phase II 594
        enzymes 594
methylazoxymethanol acetate (MAMA) 600
Mgrp 423
microencapsulated diets 70
microinjection
    apparatus 537
    technique 538
Micropogonias undulatus
    gap junctions 497
    ovarian follicle 495
microridge 360
microscope
    electron (EM)
        autoradiographic approaches 585
        emulsions for 585
        loop method for slides 585
        tissue preparation 585
    light (LM)
        autoradiographic approaches 583
        conventional processing 583
        cryosection-based techniques 584
midbrain 134
    reticular formation 342
    roof structures 342
milt
    collection by stripping 526
    collection by syringe and catheter 526
    cryopreservation methods 526
    sample preservation 526
mineralocorticoid (MR) 430
minerals 75
mini-osmotic pumps 539–40
minimum convex polygon 131
MNNG 600–1
models
    cancer, fish advantages 609
    for carcinogenesis 592
    experimental toxicity 625
    fish as 3
    fish, advantages 592
    fish tumors 592
    Gordon-Kosswig melanoma 605
    medaka tumor 599
    muscle function 319
    rainbow trout hemopoiesis, long-term hemopoietic culture (LTHC) 632
    rainbow trout tumor 599–600
    small fish 41–61
        species 42
    spleen section immunization 222
    toxicological evaluation 625
molecular techniques, fixation for 576
monoclinic crystal 295
mormyrids
    sound production 258–9
    swim bladder ear connections 256
morphological color changes 295
mortality and loss of mucus and scales 505
mosquitofish 6
motor innervation 130
motor specialization 351
motor system 346, 350
    organization 335
motor-related feedback pathways 350
mouth 174
mRNA 415
MS-222 508–9, 520, 524
MSH 422–3
mucosal associated lymphoid tissue (MALT) 383
mucous cells 358, 363, 473
mucus, components 290
mucus sampling 220
mudminnow 6
mummichog 345
    in Elizabeth River, hepatic neoplasms 603
muscle energetics 329
muscle fiber
    activation 327
    force production 328
    mechanical properties 321
    metabolic properties 321

muscle fiber (*Continued*)
   relaxation 327
   twitch tension 328
   type use 322
   types 320
   V/Vmax 326
muscle function 120
   fish advantages 320
   model 319
muscle plates 120
muscle structure 322
   cellular level 322
   gross level 322
   ultrastructural/molecular level 323
muscle velocity, V/Vmax 325
muscular system 119–27
   design 323
Mustelus canis, fin tissue 97
myeloperoxidase fixative solution (MF) 634
myocardial fibers 370
myofilament overlap 324
myofilaments, overlap 325
myographic studies 373
myoids 232
myology 119
Myxini 1

Nagoya University Medakafish Group 43
β-naphthoflavone (BNF) 601
National Cancer Institute Monograph, Hoover 1984 6
National Center for Biotechnology Information Genbank 597
National Research Council, reports on nutritional requirements 74
natriuretic peptides 203
   gene expression and regulation 433
   hormone effectors in target tissues 433
   type C (CNP) 432
nauplii 68
necropsy
   field 543
   pleuronectiform
     external examination 550
     internal examination
   entry 550
   incision 544
   visceral mass examination 553
   visceral mass resection 553
   salmoniform
     buccal cavity examination 544
     external examination 544
     external examination
   head region 544
   ventral 544
     gill examination 544
     internal examination
   entry 544
   organs 544

technique 543
necrotroph 85
nematodes, culture 54
neoplasia, *see* cancer
nephron 386–7
   arrangement 390
   microdissected 388–9
   nomenclature 417
   phylogeny 388
   types 387
nervous system 130, 332
netting, conditioning to 504
neural crest 137, 139
neural tube afferents 140
neurocranium 110–11
neurocrine stimuli 374
neuroepithelial cells 358, 363
neurofibromatosis model 7
neurohemal organ 191
neurohypophysial hormones 191
neuromast 286, 463
   canal, morphology 465
   hair cells in 466
   location 466
   receptor organs 463
     anatomy 463
   superficial 240
   superficial 465
   transformation to juvenile 466
neuromast receptors, head, morphology and distribution 239
neurons 333
   bipolar 130, 137, 143
   cell bodies 130, 334
   classification 334
   definition 333
   dendritic arborizations 334
   fusiform 334
   gustatory input 135
   piriform 334
neuropeptide Y (NPY) 418
neuropeptide-E-I (NEI) 423
neurophysin 424
neutrophilic granulocyte 442
NEV 423
NLT 423
non-specific esterase activity assay (NSE) fixative 635
Northwest Fisheries Science Center 601
nose 245
NRL 423
nutrition 65

obliqui dorsales (ObD) 124
obliqui ventrales (ObV) 125
obliquus posterior (ObP) 124
octanol-water partition coefficient (Kow) 619
oculomotor nerves 140
odontoblasts 313
odontodes 314

odor molecules 474
olfaction 474
olfactory bulb 132, 476
   functional organization 477
olfactory epithelium 19
olfactory lamella 19
olfactory nerve 140, 475
   primary, extrabulbar projection 478
olfactory neurons 19
olfactory organ, development 249
olfactory receptor cells 19, 24
   regeneration 472
olfactory rosette 248
olfactory sac 246
olfactory system 346
olfactory tract 477
   functional separation 477
oligovillous cells 286
oncogenes 595, 606
   classes 596
   cloning from goldfish 592
   in fish tissue 596
Oncorhynchus masou virus (OMV) 607
Oncorhynchus mykiss 6
   carcinogenicity testing 592
   eggs and fry mortality with aluminum and calcium 619
   olfactory receptor cells 472
   skin layers 192
   taste bud anatomy and distribution 249
   as tumor model 599, 601
Oncorhynchus rhodurus, gonadotrops 490
Oncorhynchus spp. 66
ontogeny 182
oocyte
   maturation 195
   maturation competence (OMC) 496
   meiosis 491
oogenesis
   hormonal control 195
   stages 195
   two-cell type model 195
oogonia 491
opecular-branchiostegal series 112
operating equipment 560
opercular cavity 120
opisthonephric ducts 548
opisthonephros 184
Opsanus tau, muscle experiments 327–9
opsin 453
   DNA sequence 454
opsonization 442
optic chiasm 140
optic nerve 140
   transection 549
optic tectum 135, 144, 342–3, 350, 460
   growth 460
optic tract 347

oral cavity 174
oral jaws 113
Oreochromis spp. 66
organic compounds 65
ornamental fish 67
orobranchial cavity 120
Oryzanias javanicus 43
Oryzias latipes 6, 42–3, 66
osmium tetroxide
   for electron microscopy 574, 576
   safety data 576–7
osmoregulation 621
   hormones 430
osmotic pumps
   design 539
   implantable 539
ostariophysans
   club cells 283
      alarm pheromone 290
osteoblasts 313
osteocytes 309, 313
osteoid 309
osteon 312
otolithic organs 251
   sensitivity to linear acceleration 255
otoliths 252, 480
   at necropsy 554
   effects on movement 254
ovarian ducts 265
ovarian hormones 194
   plasma levels 197
ovary 493
   differentiation 491
   follicles, growth and atresia 494
   necropsy 552
ovocoel 264, 492
ovotestis 494, 496
ovoviviparity 194
oxidative stress response models 7
oxygen, and stress 505
oxygen transport 358
oxyntopeptic cells 381
oyster toadfish, creosote contamination 595

P-glycoprotein 404
^{32}P-postlabeling 594
PACAP
   gene expression 417
   hormone effectors in target tissues 418
   molecular structure 418
pacemaker cells 350
Pacific lamprey, skin layers 98
Pacific salmon 66
   vitamin needs 74
pain response 509
pallium 131–2
   development 144
pancreas 178, 206
   cells 383

digestive secretions 384
functions 383
hormonal production 383
papilloma 591
parachondral bones 311
paracrine stimuli 374
parasites
   gills 42
   infestations 80
   removal 91
paraventricular organ 339
Parophrys vetulus 5
parthogenesis 43
pavement cells 360
pectoral fin muscles 127
pectoral girdle 116
pelagic fish
   metabolism 621
   scaleless 99
pellet implants
   cellulose cholesterol 538
   chemicals formulation 539
   injection site 539
   insertion 539
pelvic fin muscles 127
pelvic girdle 116
peptide hormones, structure 190
Perca flavescens
   female abdominal organs 265
   skin 97
perikaryon 395, 458
perimysial sheaths 372
peripheral nervous system 130
peripheral olfactory organ 246
peripheral spinal nerves 142
pH 619
phagocytosis 442, 444
   assay 636
pharyngeal jaw apparatus 114
pharyngeal muscles
   dorsal view 124
   ventral view 125
pharyngeal musculature 122
pharyngeal transport phases 123
pharyngoclavicularis externus (PcE) 125
pharyngoclavicularis internus (PcI) 125
pharyngohyoideus (PhH) 125
pharynx, anatomy 380
phenotype, sexual, manipulation 489
phosphate buffered saline (PBS), preparation 632
phospholipase C (PLC) 417
phospholipids 72
phosphorus requirements 75
photic information 455
photolytic transformation 619
photophores 96, 227, 287
photoprotective substance 6

photoreceptor cells 453, 457
   rod and cone comparison 454
   types 453
photoreceptors 231
phototransduction 452
physiological chamber, divided 521–2
physiological chambers, divided, for fecal collection 524
picric acid 572
pigment cell neoplasms *see* chromatophoromas
pigmented epithelial cell 457
pigmented epithelium 231
pike
   kidney location 162
   postsurgery social requirements 559
pillar cells 365
Pimephales promelas 6, 44–5, 66
   hermaphroditic gonads 493
   ovarian differentiation 491
   testicular differentiation 492
pineal gland 339
PIP2 417
Piscirickettsia salmonis 81, 88
pituitary gland 377
PKA 417, 419
PKC 417
PKG 433
placodes 140, 237, 465
placoid scale 296
placoid scales 99–100
planktivory feeding 248
Platichthys stellatus 5
platyfish 66
   carcinogenicity testing 592
   hybrids 6
   melanoma 605
Pleuronectes americanus, kidney location 162
Pleuronectes vetulus, environmentally induced cancer studies 601
pleuronectiform fish type, necropsy 549
Poecilia formosa 6
Poecilia latipinna 47
Poecilia reticulata 42, 46, 66
Poecilia reticulata, carcinogenicity testing 592
Poeciliopsis 6
poikilotherms 396, 574
   dietary energy requirements 70
   metabolism 621
polarization 226
polychlorinated biphenyls, neoplasm occurrence 602
polycyclic aromatic hydrocarbons (PAHs) 600
   metabolism 594
   neoplasm occurrence 602, 603
   oxidation 593

polycyclic aromatic hydrocarbons
    (PAHs) (Continued)
  sediment concentrations and DNA
    adducts 595
  uptake 593
polycyclic aromatic hydrocarbons
    (PAHs)/DNA adducts as risk
    factor 595
polyethylene (PE) tubing
  specifications 517
  for urinary catheters 519
polymerase chain reaction (PCR) 90
  assays 608
  fixation for 576
polynuclear aromatic hydrocarbons
    (PAHs) 591
Polypteridae
  collecting tubule-collecting duct
    (CT/CD) system 409
  intercalated cells 410
polyunsaturated fatty acids (PUFA) 72
Pomacentrus partitus 7
porphyropsin 454
posterior tuberculum 340
posterodorsal arteries 164
potassium permanganate 91
preglomerular nuclear complex 340
preglomerular sphincters 386
preoptic area 339, 340
preprosomatostatin (PPSS) 429
primary culture initiation solution (PBS
    with gentamicin and fungazone)
    633
primary stress response 202
pro-opiomelanocortin (POMC)
  gene expression and regulation 422
  molecular structure 422
  proteolytic processing scheme 422
proctodeum 379
profundus nerve 140
prolactin 193, 314, 434
proliferating cell nuclear antigen
    (PCNA) 576
prostaglandin 442
proteins
  requirements 71
  transporters 381
proteolytic enzymes 598
proto-oncogenes 596
protractor hyoidei (PrH) 121
Prussian carp, renal portal system
    187
pterinosomes 294, 604
pterygiophores 126
pufferfish genome 7
Puget Sound, environmentally induced
    cancer studies 601
pulp mill effluent, white sucker
    apoptosis 199
Purkinje cells 344
PVR1 418

pyloric ceca 177
pyloric ceca, in necropsy 544

quarantine 5

radiation effects 5–6
radio transmitters
  surgical implantation 564
    complications 564
radioimmunoassay (RIA) 190
radioisotopes 579
  autoradiographic procedures 580
  background 581
  considerations 580
  efficiency 581
  resolution 580
radiolabeled compounds for
    cryosectioned tissue 584
rainbow trout 6, 66
  carcinogen bioassay 601
  carcinogenicity testing 592
  eggs and fry mortality with aluminum
    and calcium 619
  gut-associated lymphoid tissue
    (GALT) 445
  hemopoiesis model, long term
    hemopoietic culture \(LTHC)
    632
  hormone and growth factor studies
    199
  macrophage identification 636
  olfactory bulb 477
  olfactory receptor cells 472
  splenic macrophages 447
  splenic tissue 447
  thymus sections 443
  as tumor model 599–600
  urinary catheters 520
  vitamin needs 74
Raja erinacea, arterial system 186
ras proteins 596
ray 66
razorback suckers 6
rectal fluid collection 525
recti ventrales (RcV) 125
red sea bream 68
renal corpuscles
  cytological organization 393
  dogfish 392
  endothelial cells 393
  mesangial cells 393
  podocytes 393–4
renal portal system 185
renal tissue
  organization 388
  zonation 8–9
renal tubule
  distal tubule 405
    basolateral interdigitation 405
    basolateral invaginations 405
    early (EDT) 405–7

    late (LDT) 407
    intercalated cells 410
    intermediate segment 404
    microdissected 388–9
    neck segment 398
    proximal tubule 397, 400
    proximal tubule
      functional considerations 403
      segment I 399
      segment II 401
      sexual dimorphism 403
renal vascular system 385
Renibacterium 81
Renibacterium salmoninarum 87
renin-angiotensin system 203
reproductive state evaluation 564
reproductive systems 261, 489–99
resorufin 645
respiration, cutaneous 96
respiratory burst 444
respiratory gases, gill diffusion 530
respiratory lamellae 156, 358, 365
respiratory lamellae
  anatomy 156
  arterioles 157
  shape 157
respiratory system 152, 358
retia mirabilia 377
reticular formation 350
retina 230, 452
  axons 346
  cell phenotypes 231
  cell types and connections 457
  development 460
  ethological context 457
  growth by stretching 460
  growth in teleosts 459
  information flow 454
  input 342
  macroscopic structure 231
  projections 346–7
  structure 455
  variability 234
retinal 453–4
retinoblastoma model 7
retinomotor movements 231
  evolution 231
retinopic map 343
retinotopy 460
retractor dorsalis (RD) 124
retrovirus tumors 607
Rhabdoviruses 81, 89
rheotaxis 243
rhombencephalon 344
Rivulus marmoratus 646
rodlet cell 288, 359, 364
  appearance 289
  possible functions 288
rotifers, as live food 68
rT3 427
rugae 380

sacciform cells 281
   different names 281
   function 281
   light microscopy 281
   secretions 290
sacrotroph 85
sailfin molly 337
Salmo salar 66
salmon
   amino acid sequences 419
   culture development 67
   postsurgery social requirements 559
salmoniform fish type, necropsy 544
saltwater fish, ionoregulation and osmoregulation 621
Saprolegnia 80
Saprolegnia parasitica 85
Sarcopterygii 1
scales 97, 296
   accessory annuli 103
   circuli 101
   cteni 102
   development 296
   functions 95
   growth 299
   imbricated 103–4
   layers and components 272
   modified 105
   patterns 103–4
   regenerated pattern 300
   regenerating, microscopic structure 298
   sulci 101–2
   in surgery 560
   tubulated 105
scavenger receptor 636
Schesny rings 58–9
Schreckstoff system 283
sclerad 232
sclerad retinal surface 456
scleroblasts 300
scutes 98, 105–6
scutes
   development 300
   layers and components 296
   location 300
   tissues 315
secretin 178
sediment dredging and neoplasm reduction 602
segmental arteries 165
sensory cells
   epithelial 286
   solitary 287
sensory epithelium, distribution patterns 19
sensory inputs 133
sensory strip 464
sensory system 346
   organization 335
serosa 380

Sertoli cells 197, 498
sesamoid bones 309
sex differentiation, gonadal 491
sex steroid hormones 265
   biosynthesis 426
   effectors in target tissues 425
   expression and regulation 425
   molecular structure 425
   production 492
sGnRH 418
Sharpey's fibers 312
sheepshead minnow 42
   bioassay studies 45
   breeding 58–60
   husbandry 57–8
shelter, post-surgical 565
short-fin eel, skin layers 99
sideroblastic anemia model 7
signal transduction, molecular basis 474
silastic 538
silversides 45–6
sinoatrial ostium 162
sinus venosus 162
skate, arterial system 186
skeleton
   arch support 153
   craniofacial 308
   remodeling 312
   system 109–10, 307
   tissues 307
skin 95–107, 271
   body temperature 96–7
   colours 295
      changes 295
   damage treatment 91
   electrosensory receptors 96
   in excretion 96
   extensions 105–6
   functions 95–7
   fungal infections 85
   gas exchange across 96
   general structure 97–106
   infections 79
   lateral line 219
   and locomotion 96
   parasitic infestations 83
   preneoplastic lesions 597
   protective patterns 95–6
   taste buds 96
   tumors and retrovirus 607
   water and mineral balance 96
skull 110–11
   ethmoid region 110
   infraorbital bridge 110–11
   lateral line canals 238
   supraorbital bridge 110
smoltification 86, 194, 609
   hormonal regulation 200
SMRT 427
sodium cacodylate 574–6

safety data 576
solitary chemoreceptor cells (SCCs) 248, 474
somata 476
somatic efferent (SE) fibers 139
somatic indices, field necropsy 543
somatic nervous system 130
somatolactin 190, 194, 433
somatosensory feedback pathways 350
somatostatin (SST) 209
   gene expression and regulation 429
   hormone effectors in target tissues 429
   molecular structure 429
somatotropin release-inhibiting factor (SRIF) 418, 429
somitomeres 139
sonic muscles 258
sound
   direct and indirect 255
   extraction in near field 255
   generation 351
   production 258
   selective frequencies 255
sound-pressure receivers 255
sound-pressure reception 255, 258
spawning
   artificial, techniques 48
   in lab aquaria 525
speciation 619
species in research 66
spectral sensitivity 227
sperm 496
spermatocyst 498
spermatogonia 498
spermiation 197, 498
spherical aberration 228
sphincter oesophagi (SpO) 124
spinal cord 142, 345
spinal cord, organization 346
spinal transection 522
spiral valve 381
spleen 222
   cell culture initiation 633
   ellipsoids 447
   in immune system 446–8
   pulp tissue 447
stanniocalcin 190, 204, 435
starry flounder 5
STAT5 599, 606
statolithic organs, function 253
steelhead trout, skin layers 98
sternohyoideus (StH) 122
steroid hormones 190
steroidogenesis 195, 197
stickleback 66
stomach
   anatomy 380
   necropsy 545
   shapes 175
stomodeum 379

stratum compactum 292
stratum spongiosum 292
streptococcal bacteria 81
Streptococcus spp. 81, 87
stress
   and phagocytosis 445
   post-surgical reduction 565
   response 504
      mitigation 504
stress factors
   and apoptosis 199
   control 201
   and viral lesions 609
stress indices, field necropsy 543
sturgeon, ascorbic acid synthesis 74
subpallium 131
suction feeding, muscle activity 120
sulfotransferases (STs) 594
supplementary feeds 67
surgery
   behavioral needs after 559
   complications 560, 564
   convalescence management 565
   elective and emergency 563
   equipment 561–2
      drapes 561–2
      steel staples 562
      sterile surgical packs 561
      sutures and needles 561
   field 564
   incision healing 559
      temperature role 559–60
   incisions 563
   preparation 562
   procedures 562
   protocols 558
   scale removal 562
   scalpels 563
   site disinfection 562
   techniques 557, 562
   water flow requirements 558
surgical glues 562
surgical suite, safety 560
surgical tables 560
   flatfish 561
   foam sponge 561
   plastic PVC pipe 561
   water flow-through system 561
suspensoria 111–12
sutures, tension 564
swimbladder
   connection to ear 255, 257
   hyperinflation 564
   muscles 258, 328
   necropsy 548
   in sound production 258
swimming
   design parameters 324
   movements 324
   pictures 323
swordtail 66

carcinogenicity testing 592
   hybrids 6
   melanoma 605
syringe
   for chemical agent injection 535
   injection locations 535
systemic arteries 164

T lymphocytes 441
tagging 564
taste buds 137, 249, 364
TCDD 644, 648
tectum 341
teeth 174, 314
   attachment 315
   enamel 315
tegmentum, midbain 135
telencephalon 131, 336
   area dorsalis 337
   area ventralis 338
   development 144
   development, group differences 145
   olfactory projections 336–7
   transverse sections 336–7
   visual pathways 347
teleocalcin 435
teleosts 1
   aglomerular 403
   apoptotic DNA fragmentation 198
   calcium levels 314
   carcinogen sensitivity 593
   collecting tubule-collecting duct (CT/CD) system 409
   complement (C) system 442
   diencephalon 339
   endocrine hormones 189–212
   epidermal cell types 274
   epithelial cells 272
   female reproductive physiology 489
   gill innervation 156
   gills 154
   gonad structure and brain-pituitary axis 489
   gonadotropins 192, 490
      immunocytochemical localization 490
   granular cells 283
   gustatory region 135
   head canal systems 239
   head musculature
      lateral view 120
      ventral view 122
   hormone-producing cells 490
   inter-renal tissue anatomy 201
   ionocyte functions 286
   kidney 444
   kidney, shapes 184
   lateral line canals, patterns 242
   muscular system 120
   myomeres 126

neural organization 475
neuroanatomy 335
olfaction 474
olfactory bulbs 132
olfactory organs 246
olfactory receptor cells 472
oncogene research websites 597
ovaries 492–3
pituitary gland 490
reproductive system 489
retinal growth 459
retinal structure 452
   and habitat 456
sexual anatomy
   external 262
   female structures 264
   internal 263
   male structures 263
skeletal system 109–18
spleen 446
telencephalon 337
testis 264
thyroid anatomy 200
ventral thalamus 339
telomerase 649
temperature, and stress 504
terminal nerve 139
test materials, evaluation 639
testicular differentiation, fathead minnow 492
testicular hormones 197
   plasma levels 198
testing systems, small fish models 49–50
testis 496
testosterone (T) 425
thalamotelencephalic visual projections 347
thalamus 339
   dorsal
      collothalamic part 339
      lemnothalamic part 339
theronts 84
thread cells 284
thymocytes 443
thymus 220, 442
thyroid hormones 199, 427
   biological actions 200
thyroid-stimulating hormone (TSH) 192, 420
   stimulation 200
thyrotropin, molecular structure 421
thyrotropin-releasing hormone (TRH) 192
   effectors in target tissues 421
   gene expression and regulation 421
   molecular structure 421
thyroxine (T4) 200
   expression and regulation 427
   hormone effectors in target tissues 427
   molecular structure 427

Tilapia spp. 66
tiPRL 434
tissue, freezing 584
tissue culture 631
tissue fixation 569
   chemistry 570–1
   immunohistochemical techniques 576
   liver tissue 575
   molecular techniques 576
   procedures 570
   protocols 570
   purpose and principles 572
   safety 576
   samples 570
   whole small fish 575
tissue synthesis 71
tomites 84
tonofilaments 275
topminnows 47
Torpedo spp. 66
torus longitudinalis 344
torus semicircularis 135, 348
   laminar organization 349
toxic equivalency factors (TEFs) 627, 648
   salmonid embryolarval mortality as toxic response 627
Toxicity Identification and Evaluation (TIE) 625
toxicity studies 617
   biological interactions 620
   categories 622
   chemical lethal and sublethal concentrations 623
   control parameters 618
   factors affecting 80
   fish life stage 622
   glossary of terms 618
   ionic chemicals 620
      speciation processes 620
   non-ionizable chemicals 620
   overview 627
   quantitative estimates of chemicals 622
   sublethal endpoints 625
   'sublethal' responses 617
   sublethal testing 625
toxicology 617
toxicopathic disease 543
trabecular bone 311–12
trabeculated myocardium 163
transcription factor 416
transponders 562
transversi dorsales (TrD) 124
transversi ventrales (TrV) 125
TRF *see* thyrotropin-releasing hormone
tricaine methane sulfonate (TMS) 509, 559
Trichodina 80, 84
trigeminal nerve 141
   visceral arch muscles 141
trigeminal sensory nucleus 141
triiodothyronine 200
3,5,3'-triiodothyronine (T3)
   expression and regulation 427
   hormone effectors in target tissues 427
   molecular structure 427
trinitrophenol 572
trocar 517
trochlear nerve 140
trophonts 84
troponin 327
trout
   free-swimming, urinary catheter 520
   head, central venous system 167
   juvenile brain section 266
   segmental vessels 166
   systemic arteries 165
   systemic veins 167
   tumor production modulation 598
trunk
   lateral line canals
      electron micrograph 241
      norphology 240
   venous drainage 168
tuberous receptor cells 347
TUF column system for fecal collection 523–4
tumor necrosis factor 445
tumor suppressor genes 595
turbot 68

ultraviolet-B (UVB) radiation 5
Umbra limi 6
Umbra spp 6
University of Oregon Stock Center 47
urinary bladder 411
urinary tract 181, 385
   collecting duct system 386–7
urine
   collection 519
      continuous 521
      normally vented 521
   production 404
urophysis 190, 204
urotensins 190, 204
   gene expression and regulation 431
   hormone effectors in target tissues 431
   molecular structure 430
US Environmental Protection Agency (EPA) 638
utricular macula 254–5

V1a receptor 425
vaccination preparations
   anal 221
   oral 221
vaccines 90
   preventive immersion 85
vagal lobes 135
vagal lobes, laminar organization 345
vagal nerves, motor components 142
valvula 135, 344–5
vasa protein 491
vascular endothelial growth factor (VEGF) 598
vasotocin 3–4
   gene expression and regulation 424
   hormone effectors in target tissues 424
   molecular structure 423
veins 375
   cutaneous 169
   intrarenal 386
venom cells 284
venoms, components 291
venous hepatic portal system 168
venous renal portal system 168
venous system 167
ventricular system 143
vertebral column 114–15
Vertebrata 1
vestibulocochlear nerve 141
Vibrio anguillarum 81, 88
Vibrio salmonicida 88
vibriosis 81, 88
VIP 418
viral hemorrhagic septicemia virus (VHSV) 89
viral infections 81, 88
viral infections, diagnosis 88
viral tumors 607
visceral arches 139
   muscles, embryonic origin 139
visceral arteries 166
visceral efferent (VE) fibers 139
viscerosensory nucleus 142
vision 225, 451
   neural remodeling 462
   underwater 225
visual implant tags 564
visual system 346
vitamin C, requirements 74
vitamin D3 205
vitamin E 72
vitamin test diet 66
vitamins 73
   needs 71
vitelline envelope 494
vitellogenesis 195
vitellogenic phase 494
vitellogenin, as biochemical marker 211
vitread retinal surface 457
viviparity 194
VNP 432
vomeronasal organ 246

walleye epidermal hyperplastic virus types 608
warmwater species 67
water
   brackish, benefits 505
   for culture 47–8
   'hardness' 620
   hearing in 255
   in holding facilities 565
   quality 48
   quality conditions 619
   small fish models 47–8
   surgical suite safety 560
   temperature, and sex determination 493
   toxic concentrations of chemicals 617
Weberian ossicles 256
   organization 254
western mosquitofish, as live-bearer 46–7
white spot disease 80

white sucker, reproductive damage from mill effluent 199
winter flounder, kidney location 162
wound repair mechanism 276
Wright–Giemsa staining solution (WG) 634

X-ray films for macroscopic autoradiography 582
xanthophores 294, 604
xanthosomes 294
xenobiotics 584, 592
xenobiotism 594
xenoestrogens 493
Xiphophorus
   hereditary melanoma, molecular events 599
   hybrids, melanoma 605
   as malignant melanoma model 7
Xiphophorus spp 6, 661
   carcinogenicity testing 592
Xmrk oncogene 599
Xyrauchen texanus 6

yellow perch
   female abdominal organs 265
   skin, vertical section 97
yellowtail 68
Yersinia ruckeri 80, 85
yersiniosis 85
yolk vesicle 494

zebrafish 6, 66
   carcinogenicity testing 592
   as egg-layer 43
   mutant model 7
   olfactory bulb 477
   olfactory epithelium 19
   research 596
Zebrafish Book 43
Zebrafish Science Monitor 47
zinc fingers 425
zona radiata 494
zooplankton enrichment 69
Zosterisessor ophiocephalus, dorsal segmental vessels 45, 170